Cyanophyta, Bacteria, Algae, and Fungi

Liverworts

Mosses

Microphyllophytes

Arthrophytes

Ferns

Cycadophytes

Ginkgophytes

Coniferophytes

Gnetophytes

Angiosperms

MORPHOLOGY OF PLANTS AND FUNGI

MORPHOLOGY OF PLANTS AND FUNGI
Fifth Edition

HAROLD C. BOLD
C. L. Lundell Professor Emeritus of Systematic Botany

CONSTANTINE J. ALEXOPOULOS
Late, Professor of Botany

THEODORE DELEVORYAS
Professor of Botany

The University of Texas at Austin

HARPER & ROW, PUBLISHERS, New York
Cambridge, Philadelphia, San Francisco, Washington,
London, Mexico City, São Paulo, Singapore, Sydney

1817

Cover photographs (clockwise, beginning top right): *Schizophyllum commune* (courtesy of Ruth McVaugh Allen); *Physarum roseum; Sphagnum lescurii,* pseudopodium and sporophyte; *Rhodymonia palmata; Pinus taeda;* microstrobili.

MORPHOLOGY OF PLANTS AND FUNGI,
Fifth Edition

Copyright © 1987 by Harper & Row, Publishers, Inc.

Library of Congress Cataloging-in-Publication Data

Bold, Harold Charles, 1909–
 Morphology of plants and fungi.

 Bibliography: p.
 Includes index.
 1. Botany—Morphology. 2. Plants—Reproduction.
I. Alexopoulos, Constantine John, 1907–1986.
II. Delevoryas, Theodore, 1929– . III. Title.
QK641.B596 1987   581.4   86-14294
ISBN 0-06-040839-1

Harper International Edition
ISBN 0-06-350197-X

86 87 88 89 9 8 7 6 5 4 3 2 1

Acknowledgments begin on page 884.

Sponsoring Editor: Claudia M. Wilson
Project Editor: Holly D. Gordon
Text Design Adaptation: Leon Bolognese
Cover Design: Wanda Lubelska Design
Text Art: J & R Services, Inc.
Production: Kewal K. Sharma
Compositor: Black Dot, Inc.
Printer and Binder: R. R. Donnelley & Sons Company

CONTENTS

PREFACE

Among the justifications for preparing a new edition of a textbook are improvements in the organization, lucidity, and clarity of the text and of the illustrations; and inclusion of material and references thereto that have appeared in the literature since the preceding edition was prepared. The authors have attempted to meet these and other goals in preparing the present account of the structure and reproduction of plants and fungi. The comments and responses of students have also been taken into account.

No drastic changes have been made in the format and organization of the book. The text is not encyclopædic or comprehensive. It has been designed for an upper-division undergraduate course or a first graduate course of a year's duration. Such a course is, of course, divisible into semesters or quarters. The book emphasizes the organization and reproduction of representatives of the plant kingdom and of the Kingdom Monera (bacteria and blue-green algae) and the Kingdom Myceteae (Fungi). Although critics have sometimes recommended deletion of most of the information on the representative genera and even its relegation to an appendix, it seemed that this would result in a textbook of plant morphology without plants. Furthermore, it would provide only shadow and little substance, or stones instead of bread. The discussions of representative genera, according-

ly, have been retained in the text. As the first of the undersigned wrote in the preface for the first edition:

One who attempts the task of summarizing the structure and reproduction of representatives of the entire plant kingdom within the covers of a single volume is faced constantly with the problems of scope and degree of coverage. He must satisfy certain minima, but at the same time avoid overwhelming the student with detail. He also is liable to criticism from his colleagues for what may seem to them drastically abbreviated presentations. The phycologist and mycologist, for example, will deplore the omission of certain organisms which they deem important, and the bryologist and phanerogamic botanist will condemn the treatment of algae and fungi as too exhaustive. The writer can only protest that the present text is designed to serve as an introduction on the basis of which more specialized treatments of the several groups may become more intelligible.

It is anticipated that undergraduates and graduates majoring in biology, especially in plant science, having covered the material in this volume, may later elect specialized courses dealing with one or another of the several groups of plants—such courses as phycology, mycology, bryology, and pteridology.

Great effort has been devoted to including in the text relevant citations and information from research papers published in the field since the last edition went to press. The Bibliography, accordingly, has been expanded gradually in successive editions.

Critics, including friendly ones, of the book during its first four editions have often expressed regret that the discussions of plant evolution and phylogeny were somewhat brief and noncommittal. In response, these have been somewhat expanded, but not inordinately so, lest the reader conclude *that we really know with certainty the actual course of evolution and the interrelationships of the higher taxa or categories (orders, classes, and divisions) of plants.* Furthermore, the classification and suggestions presented in this book regarding phylogenetic relationships need not be accepted as authoritative. In this connection, the undersigned offer several quoted statements from eminent students of plant morphology, which ought to be the occasion of reflection. One (Kaplan, 1977) writes: "There has been too much emphasis on attempts to answer largely unanswerable questions, such as whether a particular trait is 'primitive' or 'advanced' within some established phylogenetic scheme, rather than going back to first principles and evaluating structural relationships on a more objective, evidential basis." Another (Wagner, 1977) states [in reference to *Psilotum* (p. 57)]: "Since the nature of the spore-bearing appendage is disputed . . . I shall conceive of the structure of the appendage *as is* rather than what it *might be* to avoid having speculation creep into classification." And (p. 55): "*Concurrence of the scientific public—majority opinion—*may be a dangerous influence in restraining authors from new ideas and teachers from accepting them." And further (p. 55): "*History and tradition,* for example, have always been influential in how we classify plants, and it often seems difficult for us to break away from time-honored custom." And finally (p. 56): "Taxonomy must be based upon hard facts rather than speculations."

In the preparation of this fifth edition the authors acknowledge constructive suggestions and aid from: Professors Vernon Ahmadjian, Lewis Anderson, Barbara Crandall-Stotler, Margaret Dietert, Linda E. Graham, James Kimbrough, John La Claire II, Lynn Margulis, Karl Mattox, Phillip Miles, Ulrich Näf, Knut Norstog, Tommy L. Phillips, Rudolph Schuster, Richard C. Starr, Kenneth Stewart, Raymond Stotler, Thomas N. Taylor, Indra K. Vasil, Dean Whittier, and Michael J. Wynne.

The authors are grateful to Mrs. Frances Denny and John Allensworth for assistance in the preparation of the manuscript and index.

Finally, we acknowledge financial assistance in preparing the manuscript from funds provided by Professor C. L. Lundell in connection with the Professorship established by him and held by the first of the undersigned.

Harold C. Bold
Constantine J. Alexopoulos
Theodore Delevoryas

ABOUT THE AUTHORS

HAROLD C. BOLD

Harold C. Bold, C. L. Lundell Professor Emeritus of Systematic Botany, was born June 16, 1909, in New York City, attended elementary and high school there, received degrees from Columbia University (B.A., 1929; Ph. D., 1933) and the University of Vermont (M.S., 1931). He has taught at the University of Vermont, Barnard College, Columbia University, Vanderbilt University, and The University of Texas at Austin. At The University of Texas he served as Chairman of the Department of Botany, Chairman of the Division of Biological Sciences, Secretary of the General Faculty, and Chairman of the Graduate Assembly. Dr. Bold also taught during the summers at the University of Tennessee, Fordham University, Chesapeake Biological Laboratory, and the Marine Biological Laboratory at Woods Hole, Massachusetts. He served as Secretary, Vice-President, and President of the Botanical Society of America; Editor-in-Chief of the American Journal of Botany; and President of the Phycological Society of America. He holds memberships in Phi Beta Kappa, Sigma Xi, Phi Kappa Phi, Omicron Delta Kappa, Delta Chi, the National Academy of Sciences, and the American Academy of Arts and Sciences.

Dr. Bold has authored or coauthored five books, three of which have had worldwide publication. His *Plant Kingdom* has been translated into seven foreign languages. He has also been very active in research and has earned an international reputation as a distinguished phycologist, having pioneered in the laboratory culture of the green algae. For this and other researches he was awarded the Botanical Society of America Certificate of Merit and has recently been honored by the University of Vermont as one of its distinguished graduates.

Dr. Bold has also excelled as a teacher and was awarded the Standard Oil Company of Indiana award for teaching excellence.

CONSTANTINE J. ALEXOPOULOS[1]

Constantine J. Alexopoulos, late Professor of Botany at The University of Texas at Austin, was born in Chicago in 1907. He graduated with honors from the University of Illinois in 1927 and, studying on a University Fellowship in Botany, was awarded the M.S. and Ph.D. degrees from the same institution in 1928 and 1932. He married Juliet Dowdy, M.M., A.A.G.O., in 1939.

After receiving the doctorate, Dr. Alexopoulos taught at the University of Illinois, Kent State University, Michigan State University,

---

[1]Deceased, May 15, 1986.

and the University of Iowa, where he served as Head of the Department of Botany for six years before going to the University of Texas at Austin as Professor of Botany in 1962. He served as Fulbright Senior Research Fellow at the National University of Athens, Greece, in 1954–1955. He also taught at the Mountain Lake Biological Station of the University of Virginia and at the University of Washington in Seattle.

Dr. Alexopoulos authored or coauthored four books. Among these, his *Introductory Mycology*, now in its third edition, has been translated into five languages and is considered the standard textbook in mycology throughout the free world. Also, the *Biology of the Myxomycetes*, which he coauthored with W. D. Gray, and *The Mycomycetes*, a world monograph, which he coauthored with G. W. Martin, are the standard references for these organisms. Dr. Alexopoulos also authored or coauthored more than 80 technical articles on fungi and contributed articles on fungi to four encyclopædias, including the *Encylopædia Britannica*.

Through his research and publications on slime molds and fungi, he earned a worldwide reputation as a mycologist; in recognition of this status the Botanical Society of America awarded him its Certificate of Merit and elected him President. He also served as Secretary-Treasurer, Vice-President and President of the Mycological Society of America and was the first President of the International Mycological Association from 1971 to 1977. He was a Life Member of the Mycological Society of America, a Fellow of the American Academy of Arts and Sciences, and a Corresponding Member of the Academy of Athens.

THEODORE DELEVORYAS

Theodore Delevoryas was born in Chicopee Falls, Massachusetts, on July 22, 1929. He received a B.S. degree from the University of Massachusetts (summa cum laude) and an M.S. (1951) and Ph.D. degree (1954) from the University of Illinois. During 1954–1955 he was a Rockefeller Fellow in the Natural Sciences of the National Research Council at the University of Michigan. He taught at Michigan State University, the University of Illinois, and Yale University before joining the faculty of The University of Texas at Austin in 1972. He served as Chairman of the Department of Botany from 1974 to 1980. He has been Chairman of the Division of Biological Sciences since 1982. He was awarded a Guggenheim Foundation Fellowship for study in England during 1964–1965. Other honors include election as first alumnus member of Phi Beta Kappa at the University of Masachusetts, election as Fellow of the Linnean Society, election as Fellow of the Palæobotanical Society (India), and an Award of Merit from the Botanical Society of America. He is currently Editor-in-Chief of the American Journal of Botany.

He is the author of two other books, *Morphology and Evolution of Fossil Plants* and *Plant Diversification*. Professor Delevoryas served as Treasurer, Vice-President, and President of the Botanical Society of America and was President of the International Organization of Palæobotany from 1978 to 1981.

MORPHOLOGY OF PLANTS AND FUNGI

Chapter 1
THE DIVISIONS OF
PLANT SCIENCE

If each of us were to attempt to summarize the treatment of plant science in the introductory botany or biology course in which he or she had been enrolled, we would probably agree that we had devoted the greater part of it to studying the structure and functioning of what is often referred to as the "higher plant"[1] or "flowering plant." In some cases, this may have been supplemented by consideration of such topics, among others, as cellular organization, biochemistry and metabolism, transpiration, translocation, tropisms, and growth-regulating mechanisms. Some portion of the course might also have included a brief and cursory "survey" of other representatives of the plant kingdom from among the so-called lower plants, and possibly, in addition, consideration of such important topics as organic evolution and inheritance. In such a course, as a matter of fact, we would have made brief excursions into several fields of plant science, namely, **morphology, physiology, genetics, ecology,** and others.

These brief excursions may have given us some insight into the nature and significance of these disciplines, but further discussion may afford us perspective as we begin a somewhat more intensive study of one of them, morphology. At the outset we must realize that these various divisions of plant science are not separate or mutually exclusive and that most of the fundamental advances in our knowledge have been attained by correlation and synthesis of the contributions of the several fields. Furthermore, no one of them is intelligible without reference to the others. The physiologist, for example, would be hard pressed to achieve real understanding of transpiration if he did not understand the structure of the leaf. Conversely, the leaf structure described by the morphologist lacks significance until one considers the functions of that organ, the genetic factors involved in its differentiation, and the variations a leaf may undergo under different environmental conditions.

The classification of plant science into major and minor subdivisions depends in some measure on the biological vocation and interests of the classifier. Probably no one classifica-

---

[1]The terms "lower" and "higher" to describe plants, regrettably still often used, should probably be replaced by others, such as "simple" and "more complex" or "primitive" and "evolutionarily advanced" or "reduced."

tion could be devised that would be acceptable to all botanists. However, it is possible to distinguish between the various fields of technological, economic, humanistic, or applied botany, on the one hand, and those of basic botany, the so-called pure science, on the other. Among the former may be listed such major divisions as **agriculture, horticulture, floriculture, plant breeding, forestry,** and **plant pathology.** The scope of each of these is well known or may be ascertained from the dictionary, encyclopedia, and other sources. In a sense these divisions constitute the anthropocentric aspects of plant science—plants in their relation to man. Pure, or basic, botany is the study of plant life without exclusive interest in and reference to its relations to man, although some of them are alluded to in this book. It includes such major divisions as **taxonomy, morphology, physiology, ecology, phytogeography,** and **genetics.** In reality, it is often impossible to distinguish absolutely between pure and applied botany. The results of the researches in pure science achieved today often become the basis of applied science tomorrow, and thus are relevant.

Taxonomy, currently amplified and enriched and more often called systematic botany, is probably the oldest division of plant science, inasmuch as primitive man, in his survival efforts, must have learned early to distinguish between edible or other useful plants and poisonous or otherwise noxious species. Taxonomy deals with the identification, naming, and classification of the diverse types of plants that populate the earth, with their patterns of distribution and putative phylogenetic relationships. Its goal, in the eyes of most modern taxonomists, is the achievement of a natural or phylogenetic classification, which implies that the groupings indicate actual relationship by descent from common precursors. Until relatively recently, classification of plants was based largely on characteristics of structure and form—that is, on morphological features. Currently, however, in systematic studies of all plant (and animal)

groups, a great range of additional data is being appraised, including chemical, genetic, physiological, cytological, and ecological attributes. Consideration of this broad spectrum of characteristics often modifies earlier taxonomic treatments based largely on morphological criteria.

Physiology is concerned with the fundamental nature of life itself, with vital activities, and with the mechanisms by which the living plant maintains itself. More and more, the physiological aspects of plants and animals are being investigated and explained as chemical and physical phenomena. While these aspects of biological science are making impressive progress, they do not provide, as some of their more zealous proponents imply, the sole and final solution to all biological problems. The latter, of course, will always require study at different levels—organismal and cellular as well as molecular—with different tools and different techniques. In this connection, the words of the eminent biochemist Sir Frederick Gowland Hopkins (1931) are especially relevant:

It is only necessary for the biochemist to remember that his data gain their full significance only when he can relate them with the activities of the organism as a whole. He should be bold in experiment but cautious in his claims. His may not be the last word in the description of life, but without his help the last word will never be said.

Modern biologists studying phenomena at the organismal level, however, are becoming increasingly impressed by the truth of the last sentence of the preceding quotation.

Ecology deals with the interrelation of the plant and its environment. **Phytogeography** is the study of plant distribution on the earth. The mechanisms and laws governing the transmission of structural and functional attributes of individual plants to their offspring form the subject matter of **genetics,** and recently the structure and code of the deoxyribonucleic acid (DNA) molecule and its relationship to enzymes and development have been added.

It is relatively easy for a morphologist to summarize or to define the scope of the other divisions of plant science but more difficult for him to describe the nature of **plant morphology**. As understood by many botanists, modern morphology represents a study of the development, form, and structure of plants, and, by implication, an attempt to interpret these on the basis of similarity of plan and origin. Morphology is comparative. Both the reproductive and vegetative (somatic) organs are studied, as are the reproductive processes, normal life cycles, and deviations therefrom. Morphology embraces various levels of study of structure, organization, and development, such as ultrastructure, cytology, histology, and anatomy, between which the boundaries are daily becoming less clear. **Ultrastructure** and **cytology** are both studies of cellular organization, the former using electron microscopy and the latter, light microscopy. The study of aggregations of cells in groups or tissues and in organs constitutes the fields of **histology** and **anatomy**; these are often considered jointly in plant science. These categories represent subdivisions of morphology. All of them have contributed richly to our understanding of the organization and reproduction of plants. The morphologist, in his preoccupation with studies of plant life cycles, reproduction, and organization at different levels, appraises all these data for clues regarding the origin, development in time (evolution), and possible relationships (phylogeny) among plant groups. Furthermore, he has come to realize that the real significance of plants of the present must be sought, in part, in plants of the past that have been preserved as fossils. **Paleobotany** as a segment of paleontology, therefore, is an important adjunct to, if an actual component of, plant morphology.

In addition to the areas just cited as constituting plant morphology, a somewhat different concept of its scope has been developed by some botanists. To them, morphology is dynamic and experimental, and they emphasize the development of the organism and the interplay of factors involved in its morphogenesis. In this area, especially, the morphologist has begun to use the methods of the biophysicist and biochemist. Although originally and still, of necessity, descriptive, morphology is becoming increasingly experimental in subject matter and method.

On the basis of the data derived from several divisions and subdivisions of plant science, the morphologist often augments the methods of direct observation, perception, and experiment and becomes a speculative philosopher. At this state, he constructs logical hypotheses regarding relationships among plants and their component structures and attempts to construct a **phylogeny**, or history of the origin and development, of extant plants in light of the past. *The student must critically distinguish such speculations and hypotheses from verified conclusions based on direct observation and experiment; nevertheless, speculation and hypothesis enhance the progress of science and are essential aspects of its method.*

The several divisions of plant science discussed in earlier paragraphs are based in each case on one or another aspect of plant life. In contrast, there have also developed a number of divisions of plant science that reflect investigators' interests in one or another plant group. Thus, we have the areas of plant science known as **phycology, mycology, bryology**, and **pteridology** among others, and the corresponding vocations of **phycologist, mycologist, bryologist,** and **pteridologist** for specialists in algae, fungi, bryophytes, and ferns.

One should not conclude from the foregoing discussion that modern botany is composed of categories or subdivisions that are sharply delimited and segregated like the cubicles in a warehouse. The cooperative investigations and methods of various specialists in plant science, together with those from other natural sciences, especially chemistry, physics, and geology, have resulted in the great advances of our knowledge of plant life. The student of biological science, no matter of what phase, is (or should be) first of all a biologist with a broad comprehension of living organisms; secondarily, he may become a specialist in one or more of the several subdivisions of plant (botany) or animal (zoology) science.

THE ORIGIN AND DEVELOPMENT OF LIVING THINGS

Man has always been interested in the question of the origin of all living organisms, although historically and for obvious reasons he has been most interested in the origin of the human species. Inasmuch as the method of origin of the first living matter on this earth is no longer subject to observational verification, no explanation of the origin of life based exclusively on observation is possible. As a result, there have emerged a number of hypotheses and speculations regarding both the origin and course of development of life on the earth.

However, experimental procedures have produced molecules of biological significance (without the mediation of living organisms) in relatively simple systems and from such simple substances as water, methane, hydrogen, and ammonia; the experimental conditions simulated those of the earth's supposed primitive atmosphere. These investigations indicate clearly that experimentation has an important role in our understanding of the origin of life and the conditions under which it began. In the experiments referred to above, and by further study, it has been possible to synthesize amino acids and other essential molecules of living organisms, even the important nucleotides and the energy-rich compound adenosine triphosphate (ATP). Still more significant are the polymerization and combination of the simpler units into increasingly complex macromolecules that constitute living matter, such as proteins, which

have been achieved by the chemist. (See Bernal and Synge, 1972, and Cairns-Smith, 1982, for a succinct discussion of the origin of life.)

Hypotheses that have been offered in solution to the question of the origin and development of life may be grouped into two categories: creationism and evolution. To many, these categories seem to be in violent contradiction and mutually exclusive, whereas others do not find them so. Creationist hypotheses postulate that living matter arose first by an act of intervention on the part of a force extraneous to the natural universe itself—a supernatural force. Some of them imply, furthermore, that the present species of animals and plants in all their diversity were called into existence in their present form at approximately the same time and that they have persisted in essentially that form until the present. By further implication, it would seem to be impossible that new types of living things are appearing now or that they may appear in the future, or that natural relationships exist between species and other taxa. These last points of view represent extremes in creationist theories.

Evolution, on the contrary, emphasizes the changes that have modified species of living organisms, and everything else on earth, over long periods of time. The population of living things on the earth at a given instant is considered by the evolutionist to represent the more or less modified descendants of organisms that existed earlier. These processes of change and modification of extant species are continuing in the world today, and species are not fixed and immutable but rather are plastic and changing, according to evolutionary theory.

The evidence for evolution is incontrovertible. One can observe changes, or mutations, in living organisms occurring spontaneously in nature, and they can be evoked experimentally in the laboratory by means of chemical and physical agents. Mutations are transmissible from one generation to another. Their combination, sub-

4

sequent segregation, and recombination in sexually reproducing populations result in variability among individuals. Mutations, their segregation, recombination, and natural selection by the environment, repeated and continuing through the approximately 3 billion years during which life has evidently existed on earth, are considered to be the explanation of the current diversity of living things.

While there can no longer be doubt that living species are changing constantly and that mutations are being transmitted and selected for in successive generations, it is quite a different matter to trace in detail and with assurance the *course* of the resulting diversification of species and especially of the larger groups (higher categories) such as genera, families, orders, classes, and divisions or phyla—categories that, *by extrapolation*, are postulated to have developed by the same mechanisms of change demonstrated at the individual and specific level. Although the evidences for such extrapolation are not as compelling as one might wish, no other satisfactory alternative scientific hypothesis to explain the diversity of living things has as yet been suggested.

Schemes, diagrams, evolutionary "trees," and statements regarding the *course* of evolution and relationship among the higher categories are of necessity largely speculative and have not been emphasized in this book. Such phylogenetic syntheses often vary with the individual biologist who proposes them, and they are subject to continuing modification as new evidences become available.

What are the evidences on which evolutionary relationship is postulated? The most trustworthy are certainly the fossil record and the comparative morphology of both extinct and extant organisms. Comparative morphology includes, especially, comparative studies of the ontogeny or development of individual living plants. In addition, present, as compared with earlier, geographical distribution of plants is helpful in providing clues to relationship. Although classical phylogenetic pathways of the nineteenth and early twentieth centuries were postulated largely on morphological criteria, increasing attention is now being devoted to comparative physiology, chemistry, and serology of both living and extinct organisms. Differences of opinion in interpretation of available evidence are reflected in differences in phylogenetic systems of classification, a topic to which we shall return in the next section.

In summary, then, species are demonstrably changing, nonstatic, and heterogeneous, and they differ increasingly from successively earlier precursors. (For a more extensive, general discussion of evolution, see Grant, 1977 and 1985.)

One difficulty experienced by many students when they first become aware of the hypothesis of organic evolution as an explanation for the diversity of living organisms is the stated or implied mechanistic philosophy of most evolutionists. Mechanistic evolutionists state or imply that the changing manifestations of life have occurred without supernatural cause or intervention. They hold a similar view of the origin of life. To them, all life and living things do not differ fundamentally from inorganic matter and phenomena except in details of physical and chemical structure and activity. These physical and chemical attributes, they are convinced, ultimately will be completely understood as science progresses. Physical and chemical phenomena, they urge, will adequately explain life and all its manifestations. Mechanistic biologists neither require nor postulate final or supernatural causes. The immediate has become final for them.

In any event, we are confronted with a great diversity of living organisms that populate the earth, notwithstanding our speculations regarding their origin and history. It is the purpose of the present volume to survey this diversity in the plant world and to attempt to reduce the apparent chaos to some semblance of order.

5

THE CLASSIFICATION OF ORGANISMS

In the nineteenth and early twentieth centuries all living organisms (viruses not included) were classified either as plants or animals. However, a third category, Protista, was suggested for flagellate organisms, which seemingly intergraded between the other two of plants and animals. With the advent of the electron microscope, a firm, compelling, and somewhat different set of criteria emerged, which demanded classification of all living organisms into two separate groups, the prokaryotes and eukaryotes.[2] To the former belong those organisms that lack membrane-bounded nuclei, mitochondria, and Golgi apparatus, while all these are present in the eukaryotes. Margulis (1979) states, "That the profound differences between prokaryotes and eukaryotes should provide the foundation for any modern classification is taken as axiomatic," and with this many biologists apparently agree. Whittaker and Margulis (1978) and Margulis (1979) have also summarized various classifications of living organisms that vary in the number of kingdoms proposed from 2 (plant and animal) to 13. In this volume, although it is entitled *Morphology of Plants and Fungi*, members of three kingdoms are discussed—Monera, Phyta (Plantae), and Myceteae (Fungi).[3] The discontinuity between prokaryotes and eukaryotes is considered of sufficiently fundamental importance to warrant the classification of living organisms into two "superkingdoms," the Prokaryonta and the Eukaryonta, as Edwards (1976) and Whittaker and Margulis (1978) have proposed.[4] Under the former are included the bacteria and blue-green algae in the Kingdom Monera, while the Eukaryonta are here represented by the Kingdoms Phyta (Plantae) and Myceteae (Fungi). The gross classification used in this book may be characterized in general terms as follows:

Superkingdom I. Prokaryonta

Organisms lacking in their cells: membrane-bounded respiratory organelles (mitochondria); ribosome-associated endoplasmic reticulum; Golgi apparatus; DNA fibrils of 24 Å unit diameter; ribosomes circa 18 nm with a mass of 2.8 megadaltons. Genetic recombination unidirectional or virus-mediated.

Kingdom A. Monera (Bacteria and Blue-Green Algae) with the characteristics of superkingdom I.

Superkingdom II. Eukaryonta

Organisms having in their cells: membrane-bounded DNA (nuclei); mitochondria; ribosome-associated endoplasmic reticulum; Golgi apparatus; DNA unit fibrils of 100 Å diameter and cytoplasmic ribosomes circa 20–22 nm with a mass of 4 megadaltons. Genetic recombination in most involving karyogamy and meiosis.

Kingdom A. Phyta (Plantae)[5]

Having chlorophyll *a* and accessory chlorophylls in membrane-bounded plastids.[6]

---

[2]Except for the absence of mitochondria in certain protozoa, red blood cells, and anaerobic yeasts.

[3]Whittaker and Margulis (1978) designated these kingdoms "Plantae" and "Fungi," respectively. They recognized two kingdoms not included (as such) in this book: the Protoctista and Animalia. The former, as proposed by Whittaker and Margulis (1978), is a rather heterogeneous assemblage. The authors prefer to include the algae, eukaryotic organisms containing chlorophyll *a* and accessory pigments, in the Kingdom Phyta (Plantae) which are also eukaryotic and similarly pigmented.

[4]Edwards (1976) uses the terms "Procaryota" and "Eucaryota," while Whittaker and Margulis (1978) use "Prokaryota" and "Eukaryota."

[5]The International Code of Botanical Nomenclature is available in a publication edited by Stafleu et al. (1978).

[6]In all, additional, accessory chlorophylls. Exceptional genera of Phyta—for example, a few algae (such as *Polytoma*) and *Monotropa*, Indian pipe—have seemingly secondarily lost their chlorophyll (although they have retained their plastids).

Kingdom B. Myceteae (Fungi)[7]

Without chlorophyll but with mostly chitinous, sometimes cellulosic, cell walls, or with both cell walls lacking in the Gymnomycota and a few other species.

With reference to the discontinuity between Prokaryonta and Eukaryonta, it has been hypothesized that the chloroplasts and mitochondria of eukaryotic cells represent symbionts, formerly free-living prokaryotic cells that were incorporated long ago and that have persisted within other cells. Evidence cited in support of this hypothesis is the occurrence of DNA in both chloroplasts and mitochondria and their capacity to metabolize when they are freed from the cells that contain them. In this connection, the number and diversity of symbiotic relationships between algae and other organisms (see p.183) seem significant. For more extensive discussion of this endosymbiotic hypothesis (of the origin of eukaryotic cells) see, for example, Margulis (1970, 1975, 1980); Raven (1970); Whitton et al. (1971); Schnepf and Brown (1971); and Gibbs (1981).

An alternate hypothesis, the "direct filiation" hypothesis, proposes that eukaryotic cells arose by internal compartmentalization of originally prokaryotic cells (see for example, Bogorad, 1975; Cavalier-Smith, 1975).

With reference to the preceding discussion, the reader may wonder why a book on morphology of plants contains chapters on bacteria, blue-green algae, and fungi. The explanation relates to history and tradition. These three groups of organisms were originally studied by plant scientists. The study of bacteria soon became the separate province of bacteriology (more recently, microbiology), but they are treated briefly in this volume because of the biological principles they illustrate and because of the characteristics they have in common with blue-green algae. The latter are included because most of the literature about them and their taxonomy is still of interest to botanists. Finally, some botanists are of the opinion that the fungi have strong relationships to the algae, and in view of the supposed relationships, they would consider them to be members of the plant kingdom.

DISCUSSION QUESTIONS

1. How might one define or explain the term "science"?
2. What aspects of science distinguish it from other fields of knowledge?
3. Are all phenomena subject to analysis by the methods of science?
4. Is a distinction between pure botany and applied botany always possible?
5. Do you have reasons for believing that the study of one is more valuable than the study of the other?
6. Can you cite examples indicating that researches in pure botany have led directly to important applications?
7. How may one distinguish among the several divisions of plant science? Are they mutually exclusive? Explain.
8. What is meant by "creationism"? "evolutionism"? "mechanism"? "vitalism"?
9. Do you think that the various taxonomic categories, or taxa, such as species, genus, and so on, exist in nature? Explain.
10. Why are "two-kingdom" classifications of organisms no longer appropriate?
11. In what respects do Prokaryonta differ from Eukaryonta?
12. On the basis of your present knowledge, how would one distinguish the Myceteae (Fungi) from the Phyta (Plantae)?

---

[7]In this book the slime molds, Gymnomycota, are included in the fungi.

Chapter 2
SUPERKINGDOM PROKARYONTA, KINGDOM MONERA: BACTERIA AND CYANOPHYTA[1]

Plants, all of which contain chlorophyll *a* (except for a few with unpigmented plastids, e.g., *Polytoma Monotropa*), are primarily photosynthetic, although a number have requirements for vitamins. The bacteria (except the photosynthetic bacteria), slime molds (p.688), and fungi (Chapters 28–35), by contrast, lack chlorophyll *a*. They are, accordingly, dependent for their nutrition on other organisms and obtain their energy from organic compounds; this type of nutrition is called **heterotrophism**. In contrast, relatively self-sufficient organisms, which obtain their energy from sources other than organic compounds and their carbon from CO_2, are classified nutritionally as **autotrophic** (Gr. *autos*, self, + Gr. *trophe*, nourishment). Autotrophic organisms differ in the source of energy that drives their syntheses. A number of autotrophic bacteria, like green plants, are **photoautotrophic**, that is, **photosynthetic**, and use light energy through the agency of their bacterial chlorophyll. These organisms are green, brown-red, or purple, and can be cultivated in an entirely inorganic medium in the absence of oxygen, provided the cultures are illuminated. They contain a magnesium porphyrin compound that is similar to, but not identical with, chlorophyll. Their photosynthesis differs from that of chlorophyllous organisms in two respects: Oxygen is not liberated during the process, and the hydrogen donor is not water but usually a sulfur compound. The photosynthesis of the green and purple bacteria involves a photochemical oxidation of hydrogen sulfide into sulfur or sulfuric acid.

A small number of autotrophic bacteria are **chemosynthetic**, in that (lacking chlorophyll) they use chemical energy freed by them in the oxidation of inorganic compounds such as ammonia, nitrite, and sulfur in synthesizing their protoplasm. Among these may be mentioned *Nitrosomonas*, which oxidizes ammonium salts to nitrites; *Nitrobacter*, which oxidizes nitrites to nitrates; and *Thiobacillus thiooxidans*, which oxidizes sulfur under acid conditions.

Although microorganisms may be shown to lead a purely autotrophic existence in the laboratory test tube, since they may be made to multiply in the absence of organic substances,

---

[1] See Lewin (1984) and Bold and Wynne (1985) for a discussion of an additional, relatively newly discovered, division of Prokaryonta, the Prochlorophyta.

it is doubtful that all lead an exclusively autotrophic existence in nature, because their environment usually contains many soluble organic substances, some of which probably are used by the organisms. In connection with photoautotropism, one further point is noteworthy. It cannot be inferred with certainty merely from the presence of chlorophyll that an organism is photoautotrophic. For example, a number of chlorophyllous algae, such as *Euglena* (p. 107), a species of *Chlorococcum* (p. 64), and *Eremosphaera* (p. 68), require vitamins that they are unable to synthesize, while one species of *Ochromonas* (p. 139) requires both vitamins and organic carbon sources for maximum growth.

The vast majority of bacteria are heterotrophic (as are animals and fungi). The energy they use in building their protoplasm is derived not from light or from the oxidation of inorganic compounds but from breaking down complex organic substances produced by other organisms. Heterotrophic species that are associated with living organisms are known as **parasites** (Gr. *parasitos*, eating at another's table); those that utilize either nonliving organisms or the products of living organisms are known as **saprophytes** (Gr. *sapros*, rotten, + Gr. *phyton*, plant) and are said to be **saprobic** in nutrition. In the account of the fungi (Chapters 28–35), a number of examples of both parasitic and saprobic organisms will be described. Among the bacteria, some of the species present in human and animal bodies are parasitic; saprobic species are more widely distributed. Certain organisms other than bacteria that ingest or engulf particles of organic matter are said to be **holozoic** (Gr. *holos*, whole, + Gr. *zoion*, animal) or **phagotrophic** (Gr. *phagein*, to eat, + Gr. *trophe*, nourishment). This form of nutrition, of course, is characteristic of most animals and of certain primitive flagellates.

Microorganisms frequently serve as the point of departure for discussions concerning the ultimate origin of life and the nature of primitive life on earth. In such speculations, knowledge of the types of nutrition is an important prerequisite. Some have postulated that chemoautotrophic organisms represent the most primitive living organisms, inasmuch as they could have existed in darkness and in the purely inorganic environment of the cooling earth's crust. According to this view, as the atmosphere cleared sufficiently for the penetration of light rays, photosynthetic organisms, also requiring purely inorganic substances, would have been able to exist. In the final stage, it is postulated that heterotrophic organisms arose as degenerate forms that secondarily lost the ability to chemosynthesize or photosynthesize and grew dependent on other organisms or their products.

These conjectures have been questioned by those who believe that the evolution of nutritional and energy relations has proceeded in exactly the opposite direction. Their basic assumption is that "organic" substances were present on the earth before the appearance of living organisms. Therefore, they argue, the most primitive organisms would be those that could use organic substances to build protoplasm, much as certain of our heterotrophic bacteria and fungi do at present. Furthermore, such organisms would require less complex enzyme systems than do chemosystematic and photosynthetic organisms, which may start their synthetic chain with substances as simple as water and carbon dioxide. Evolution, beginning with the primitive heterotrophic organisms, proceeded in the direction of increasing capacity for effecting complex biosynthesis from decreasingly complex environmental substances. It culminated in the type of nutrition exhibited by chemosynthetic bacteria, which are able to obtain the carbon for cellular biosynthesis from CO_2, using chemical energy from their oxidation of inorganic compounds. The final step, according to this hypothesis, was the appearance of photosynthetic organisms, which developed a capacity for a similar synthesis, using the energy of light.

Thus, the course of evolution of nutrition and metabolism, like that of many morphological attributes, has been interpreted by different scholars as having proceeded in opposite directions. However, the metabolism of microorganisms remains of fundamental importance in all discussions regarding the origin of life.

INTRODUCTION TO PROKARYONTA

As stated in Chapter 1, a great discontinuity exists, with respect to cellular organization, among living organisms, some (relatively) few being prokaryotic and most eukaryotic. This discontinuity has been recognized in the assignment of the former to the Superkingdom Prokaryonta and the latter to the Superkingdom Eukaryonta in several modern classifications of living organisms.

The Prokaryonta and its single Kingdom Monera comprise two major groups: Bacteria and Blue-Green Algae (Cyanophyta). The latter are considered by many to be a group of chlorophyllose, photosynthetic bacteria and, accordingly, are called "Cyanobacteria" (Kondratyeva, 1982). In the present chapter, both groups are discussed, but for the reasons cited on p. 19, the blue-green algae are classified in a division of their own, the Cyanophyta. The Prokaryonta are discussed first in this book because the fossil record indicates that they are more ancient than the Eukaryonta (Schopf, 1970).

DIVISION[2] BACTERIA

General features. Bacteria, along with blue-green prokaryotes, were probably the first organisms present on earth (Schopf, 1970). Fig-

---

[2]The category "division" is in accordance with the International Code of Botanical Nomenclature (Stafleu et al., 1978).

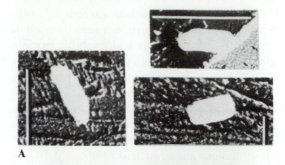
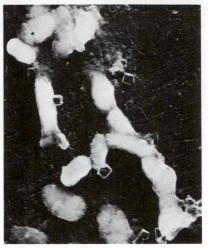

FIGURE 2-1
A. *Eobacterium isolatum.* Bacterial cells about 0.6 μ long from South African chert 3.1 billion years old. **B.** Fossil bacteria from the Gunflint Iron Formation, about 2 billion years old. ×12,000. (**A.** *After Barghoorn and Schopf.* **B.** *After Schopf et al.*)

ure 2-1A shows *Eobacterium isolatum* in black chert from the eastern Transvaal, approximately 3.1 billion years old. Figure 2-1B illustrates fossil bacteria present in Precambrian chert from southern Ontario, about 2 billion years old. These fossil bacteria seem to be morphologically similar to their living descendants.

In a volume treating the morphology of plants, the discussion of the bacteria is necessarily brief, for several reasons. First, none of the several groups of plants, other than

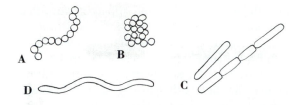

FIGURE 2-2
Some representative genera of bacteria. **A.** *Streptococcus.* **B.** *Staphylococcus.* **C.** *Bacillus.* **D.** *Spirillum.* ×2000.

blue-green algae, approaches the bacteria in apparent simplicity of structure, although there is evidence of a higher degree of organization than hitherto has been suspected. Second, the relative simplicity of their morphology has stimulated and necessitated study of the physiology of bacteria, with the result that their classification is based in large measure on their extremely diverse physiological activities. The study of bacteria, originally initiated by botanists, later came to be recognized as a separate field of biology, namely, **bacteriology**, now a major part of microbiology. Nevertheless, in a treatment of representatives of the plant kingdom and their possible relationships, the bacteria must be reckoned with, particularly in discussing such fundamental concepts as the nature of primitive life on earth. Furthermore, there are certain cytological and biochemical aspects of bacteria that seem to parallel those of the Cyanophyta.

These are considered by some authorities to be sufficiently significant to warrant classification of the two groups in a single division. Approximately 190 genera and 1700 species of bacteria have been described. It is debatable, however, whether bacterial species correspond precisely in scope to those of plants.

Although bacteria were known to Leeuwenhoek as early as 1683, detailed knowledge of their structure and nutrition goes back only to the last decades of the nineteenth century. Bacteria sometimes are considered to be unicellular fungi that reproduce only by simple fission. A number of genera, however, consist of multicellular chains or filaments, or sporangiate aggregations suggestive of those of certain slime molds, to be discussed in Chapter 29. In general, bacteria are the smallest living organisms visible with the ordinary light microscope; they frequently have dimensions that range between 0.5 and 2.0 μm in width and 1.0 and 8.0 μm in length. As to cell form, many bacteria fall into three groups (Figs. 2-2, 2-3): those with spherical cells, the **cocci** (Gr. *kokkos*, berry); those

FIGURE 2-3
Photomicrographs of three common forms of bacterial cell. **A.** Bacillus. **B.** Coccus. **C.** Spirillum. ×2000. (*After J. Novak.*)

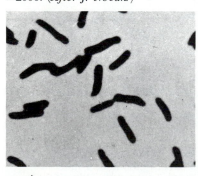

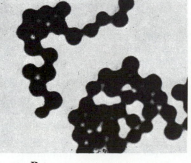

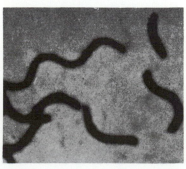

A   B   C

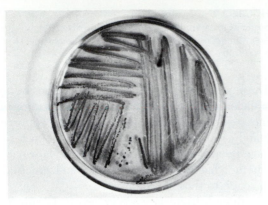

FIGURE 2-4
Serratia marcescens, growing on a streaked plate of nutrient agar. ×½.

in cellular form, it became apparent early that additional criteria must be employed to recognize bacterial genera and their species; these criteria are largely physiological. Many bacteria grow readily on a variety of organic culture media solidified with agar (Fig. 2-4), a colloidal derivative of seaweeds. The growth habit varies considerably among different species.

The individual bacterial cell is delimited from its environment by a wall and contains cytoplasmic and nuclear material (Fig. 2-5). The existence of a wall has been demonstrated by plasmolysis, by microdissection, lysis of the cells, and electron microscopy. The cell wall is rigid and confers shape to the cell. It contains a peptidoglycan that is composed of N-acetylmuramic acid, N-acetylglusamine, L-alanine, D-alanine, D-glutamic acid, and lysine or diaminopimelic acid. These are arranged as repeating structural units cross-linked to form the three-dimensional cell wall. This cell wall is surrounded, in turn, by a layer of slimy material of variable thickness, which may be present as a recognizable **capsule** (Fig. 2-6). The peptidogly-

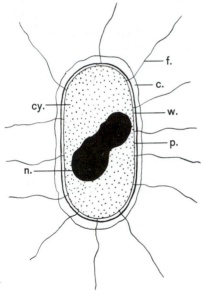

FIGURE 2-5
Cellular organization in bacteria, schematized. c., capsule; cy., cytoplasm; f., flagellum; n., nuclear material; p., plasma membrane; w., wall. (*Modified from Clifton.*)

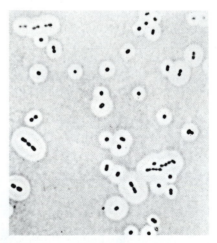

FIGURE 2-6
Bacterium anitratum: Phase-contrast photomicrograph of cells mounted in India ink showing capsules. (*Courtesy of Dr. Elliot Juni.*)

whose cells are short rods or cylinders, the **bacilli** (L. *bacillus*, little stick); and those with their cells curved and twisted, the **spirilla** (Gr. *speira*, coil). Because many bacteria are similar

can layer of the wall serves to prevent the cell from bursting from the high internal osmotic pressure generated by the high metabolic rate and need for concentrating materials to sustain this rate. Some bacteria, the *Mycoplasm* group, have no cell walls and grow more slowly.

Bacterial cytoplasm contains small particles, **ribosomes**, which are a part of the cell machinery involved in protein synthesis. In addition, storage products are often present in the form of inclusions and granules. These include polyhydroxybutyric acid bodies, glycogen, fat, sulfur, and polyphosphate granules. The cytoplasm is bounded externally by a cytoplasmic membrane.

The question of the organization of the nuclear material was long one of the most controversial aspects of the morphology of the bacterial cell. Inasmuch as bacteria maintain certain specific, inheritable attributes through countless series of cell generations, our knowledge of other organisms would indicate that the bacteria possess some physical mechanism of inheritance —in other words, a nuclear mechanism. As a result of the study of a number of bacteria by special microchemical methods for the detection of deoxyribonucleic acid, which seems to be present universally in the nuclei of plants and animals, it was demonstrated that this substance is also present in bacterial cells (Figs. 2-5, 2-7). In those bacteria thoroughly studied, the DNA has been found in a single linear structure ("chromosome"), usually a closed loop. It may take the form of a somewhat dumbbell-shaped mass lying parallel to the transverse axis of the cell (Fig. 2-5), or it may be more dispersed. The mass, a chromosomelike body, divides longitudinally prior to cell division; it thus has one of the attributes of chromosomes of eukaryotic organisms. The nuclear material appears as a region of extremely delicate threads of DNA lacking a bounding membrane. There is apparently a certain precocity in the division of the chromosomal element of some bacteria. In these species, recently divided chromosomes may become separated by partition of the cytoplasm while actual secretion of the dividing septum may be delayed. Thus, a single cell, during active growth, may have its protoplast divided into a number of segments, in some of which the chromosomal element is dividing in preparation for an ensuing cytokinesis, while formation of the transverse walls may lag behind karyokinesis and cytokinesis. Actual separation of daughter protoplasts occurs only after they have secreted these walls. It should be noted that bacterial cells, like those of Cyanophyta (p. 19), lack mitochondria, Golgi bodies, and endoplasmic reticulum, which characterize the cells (Fig. 4-3E) of other living organisms.

Some bacteria are nonmotile, but others are actively motile by means of flagella; their presence can be confirmed only by special illumination or methods of staining and electron microscopy. A single flagellum or group of flagella may be present at one or both poles of the cell, or the cells may be covered uniformly with flagella (Fig. 2-8). Various arrangements have been described.

Many of the rod-shaped bacteria have the capacity of forming **endospores** (Fig. 2-9). There is some evidence that these are produced in response to depletion of the nutrients in the surrounding medium. Each spore contains nuclear material and cytoplasm and at maturity is enclosed by complex coating structures. Spores are extremely resistant to such unfavorable environmental conditions as high temperatures and desiccation. Upon germination (Fig. 2-9D), each spore produces a single vegetative cell. Spore formation in bacteria, therefore, usually is not a means of increasing the number of individuals but rather a mechanism for survival in adverse environmental conditions.

SEXUALITY

Mechanisms for gene interchange, or sexual phenomena, were unknown in bacteria until several decades ago. As compared with sexual phenomena in other organisms, those of bacteria are subtle and, for the most part, not obvious cytologically. Furthermore, all the mechanisms for genetic interchange that occur in bacteria differ from those in other organisms in that complete union of the two nuclei does not occur; hence, zygotes are not produced, but only DNA-modified (genetically) cells known as **merozygotes** (partial zygotes).

The sexual process of *Escherichia coli* most closely approaches that in nonbacterial organ-

isms. It involves the temporary (up to 2 hours) union of two cells by a tube that arises from a donor ("male") cell, which attaches to a recipient ("female") cell (Fig. 2-10) by a **pilus**. There are, in addition to sexually functioning (in conjugation) pili, ordinary pili, which result in cellular cohesion; these are shorter than flagella. Depending on the duration of the union, varying amounts of donor DNA are injected into the recipient cell, where they may replace allelomorphic genes present therein. Thus, the genetic constitution of the donor cell becomes

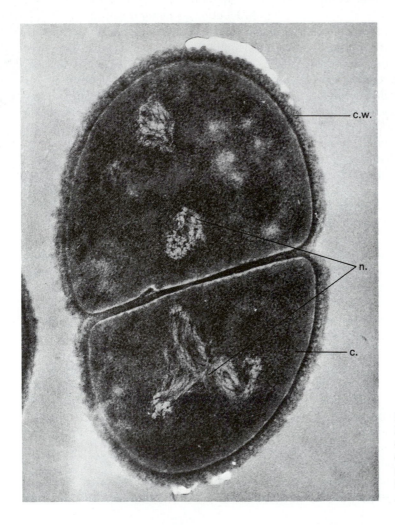

FIGURE 2-7
Micrococcus radiodurans,
electron micrograph. c.,
cytoplasm; c.w., cell wall; n.,
nuclear material. ×65,000.
(*Courtesy of Dr. R. G. E.
Murray.*)

14

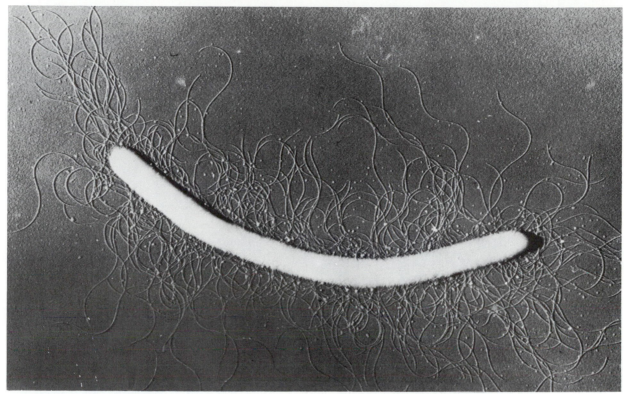

FIGURE 2-8
Proteus vulgaris. Electron micrograph shadowed with chromium; note flagella. (*S.A.B. photograph LS-258, after Houwink and van Iterson.*)

modified by incorporation of new genes, and its progeny are accordingly changed.

Two other types of transfer of genetic materials occur in bacteria: transformation and transduction. In **transformation**, free DNA of a given species is taken up by competent recipient bacteria of the same species with modification of their genetic constitution. Thus, streptomycin resistance can be incorporated into some individuals of a nonstreptomycin-resistant population by growing the latter in a culture medium containing streptomycin-resistant DNA.

In **transduction**, viruses enter a bacterial cell, and DNA from their preceding host is carried and may replace in part that of the new host, which accordingly becomes modified. Also transferred by conjugation or by certain bacterio-phages are the genetic elements (as DNA loops) called **plasmids**. These plasmids can replicate in the cytoplasm of the bacterial cell independently of the cell chromosome, and their genes then confer new characteristics on the bacterial host —for example, multiple antibiotic resistance, production of certain toxins, and new biochemical capabilities. By means of genetic and biochemical manipulations, microbiologists can incorporate foreign genes into a plasmid and then insert the latter into bacterium, where the foreign genes may be expressed ("genetic engineering"). For example, the rat gene specifying insulin has been incorporated into an *E. coli* plasmid, so that the bacterium then can produce quantities of this animal hormone.

In summary, it should again be emphasized

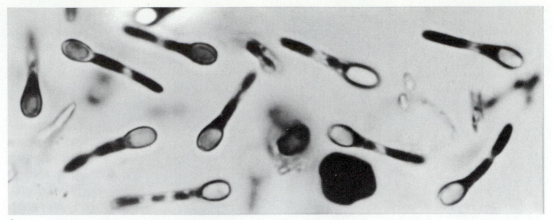

A

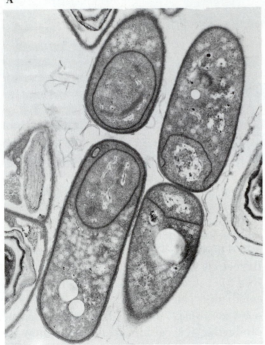

B

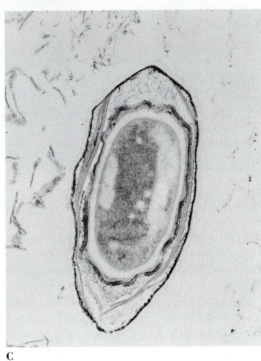

C

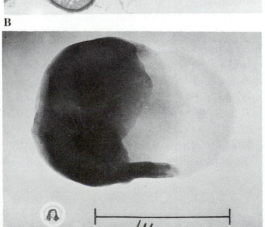

D

FIGURE 2-9
A. *Clostridium pectinovorum* mounted in Lugols'
iodine, showing spores. The dark (stained) material
is starchlike granulose. **B, C.** *Bacillus megaterium.*
B. Spore formation; note residual cytoplasm. **C.**
Cell containing mature spore. **D.** *Bacillus mycoides.*
Spore germination. **A.** ×3160. **B.** ×10,520. **C.**
×21,720. (**A.** *Courtesy of Professor C. Robinow.* **B,**
C. *Courtesy of Dr. H. Stuart Pankratz.* **D.** *S. A. B.*
photograph 203, after Knaysi, Baker, and Hillier.)

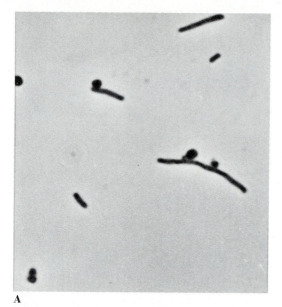

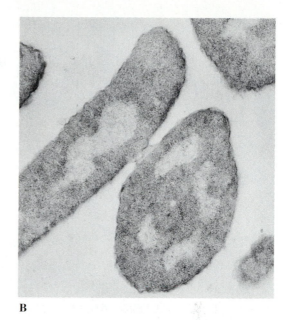

A

B

FIGURE 2-10

Escherichia coli K 12. **A.** Conjugating cells of two strains that differ in cell form, phase-contrast photomicrograph. **B.** Electron micrograph of a thin section of a mating pair that has come into direct contact through retraction of the F. pilus. **A.** ×2500. **B.** ca. ×25,000. (*Courtesy of Prof. Lucien Caro.*)

that conjugation, transformation, and transduction in bacteria produce only incomplete zygotes, or merozygotes, in which a few genes are substituted for their alleles. In contrast to this, sexuality involving nuclear union, as it does in plants and animals, results in the association of two complete sets of parental genes in the zygote.

THE ACTINOMYCETES

The organisms described in the preceding paragraphs are representative of the "true" bacteria, usually classified together in a single order distinct from several orders of "higher" bacteria, which are morphologically more complex. One group of the latter, commonly called the actinomycetes, also prokaryotic, is of special interest in several respects. In the first place, the organisms are sometimes incorrectly classified with the fungi because they are filamentous in organization (Fig. 2-11A), like those of many fungi. Furthermore, they produce chains of minute, dustlike spores, **conidia** (Fig. 2-11B), which are analogous to those of many fungi. However, the cytological organization of actinomycetes is similar to that of bacteria rather than to that of the fungi. The filaments of actinomycetes, which rarely exceed 1 μm in diameter, form radiating colonies in agar cultures. They are distributed widely in soil, and many strains have been isolated into pure cultures. Three of them, among others—*Streptomyces griseus, S. aureofaciens*, and *S. rimosus*—yield, respectively, the antibiotics streptomycin, aureomycin, and terramycin. More than 80 other antibiotic substances have been obtained from species of the genus *Streptomyces*. There is every reason to believe that actinomycetes play an important role in the biology of soil.

17

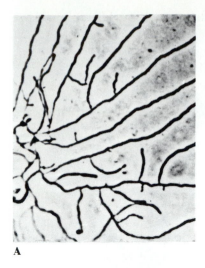

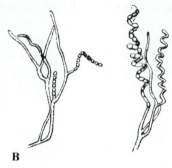

B

FIGURE 2-11
Actinomycetes. **A.** *Streptomyces coelicolor*, mycelium. ×690. **B.** Spore formation. (**A.** *After Hopwood.*)

A

ACTIVITIES OF BACTERIA AND CONCLUSION

Although relatively simple morphologically, the prokaryotic bacteria are exceedingly diverse physiologically and biochemically. Because of this diversity, they have become ubiquitous and of tremendous biological importance especially in their relation to human beings. Their role in the degradation of complex organic materials is of primary importance in maintaining a supply in water and soil of nutriments for other organisms. This same proclivity, of course, is responsible for their destructive role in food spoilage. There is a great diversity of biochemical attack and degradation of virtually all naturally occurring organic compounds, and many synthetic organic compounds, but bacteria are not effective against some detergents and halogenated aromatic hydrocarbons (insecticides). Bacteria of one or another type are able to grow in great extremes of the environment—in heat of 100°C and cold of −20°C, as in saline ponds in the Antarctic. They occur in environments with pHs from 1.0 to 11.5, and some species *require* saturated brine for their growth. Many species require an anaerobic environment, and some

are very easily injured by traces of free oxygen. The pathenogenicity of certain bacteria for plants and animals, including human beings, results in a great variety of diseases. On the other hand, the role of certain bacteria, both free-living and when associated with legumes, in enriching soils in nitrogen fixation is of tremendous benefit. Antibiotics such as streptomycin and subtilin, among others, are of bacterial origin. The ability of some bacteria (Fig. 2-12) to

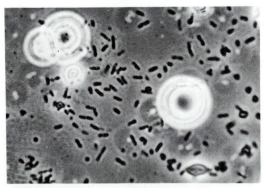

FIGURE 2-12
Oil-degrading bacteria among oil droplets. ×2500. (*Courtesy of Dr. Carl Oppenheimer, after Hunnicutt.*)

degrade the major types of oil will no doubt prove to be an important factor in control of pollution by oils. The great physiological and biochemical diversity of bacteria constitutes much of the subject matter of bacteriology and microbiology.

DIVISION CYANOPHYTA

The Cyanophyta (Gr. *kyanos*, blue + *phyta*, plants) with the single class Cyanophyceae (Gr. *kyanos* + Gr. *phykos*, seaweed, hence alga), are commonly known as blue-green algae. Some authors prefer the class name Cyanophyceae (Gr. *kyanos* + Gr. *phykos*, alga or seaweed), whereas others use Schizophyceae (Gr. *schizo*, cleave, + Gr. *phykos*).

Although traditionally interpreted as algae and classified as such, the evidence from electron microscopy and biochemistry strongly indicates an affinity between blue-green "algae" and bacteria. This evidence will be presented below. However, the authors are as yet unwilling to go as far as Stanier et al. (1971), who stated:

These organisms are not algae; their taxonomic association with eucaryotic groups is an anachronism, formally equivalent to classifying the bacteria as a constituent group of the fungi or the protozoa. In view of their cellular structure, blue-green algae can now be recognized as a major group of bacteria, distinguished from other photosynthetic bacteria by the nature of their pigment system and by their performance of aerobic photosynthesis.

The authors' reluctance resides in the absence from bacteria of chlorophyll *a*, universally present in green plants and in Cyanophyta. Those bacteria that are photosynthetic have a magnesium porphyrin compound that is similar to, but not identical with, chlorophyll *a*. In addition, many blue-green "algae" are more highly differentiated morphologically than most bacteria in having such specialized cells as heterocysts and akinetes (p.28). Finally,

the photosynthetic bacteria differ from blue-green algae in that oxygen is not evolved in the photosynthesis of the bacteria, although this may be induced to occur in laboratory cultures of certain blue-green algae. However, because of the striking similarities between the blue-green "algae" and bacteria, the two groups are herein considered to represent two related evolutionary lines (divisions or phyla), the Cyanophyta (blue-green "algae") and Schizonta (bacteria) in the Kingdom Monera.

Cyanophyta are ubiquitous, occurring in aerial, terrestrial, and aquatic habitats. Approximately 150 genera with 1500 species have been named, although many of these are suspect as having been inadequately described (Stanier et al., 1971). The current tendency is toward reducing the number of genera and species by amalgamation of taxa, many of which probably represent environmentally induced variants (Drouet and Daily, 1956, 1957; Drouet, 1968, 1973, 1978, 1981). These organisms frequently form extensive strata on moist, shaded, bare soil, and may appear as gelatinous incrustations on moist rocks, other inanimate objects, and plants. Aquatic species inhabit both marine and fresh waters; they are either attached to submerged objects or free-floating. Cyanophyta often represent a major component of the plankton. As such, they are important, along with other planktonic algae and minute animals, as the direct or ultimate source of food for more complex aquatic animals. In axenic cultures of three blue-green algae, vitamin B_{12} and panthothenic acid are secreted into the culture medium during active growth of the algae (Shah and Vandya, 1977). This would have a significant ecological impact in natural habitats.

Organisms such as *Microcystis* and *Anabaena* occur frequently as dominant organisms in water blooms (Fig. 2-13). A few species are endophytes, living within cavities in other plants. Species of *Nostoc*, for example, occur within the plant bodies of such liverworts as *Blasia* and *Anthoceros* (see p.244), and species

19

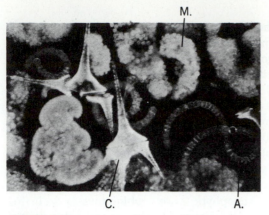

FIGURE 2-13
Sample of a water bloom from a Wisconsin lake;
the dominant organisms present are species of the
cyanochloronts *Anabaena*, A, and
Microcystis, M, and the dinoflagellate
Ceratium, C, ×1600. (*Courtesy of Dr. G. P.
Fitzgerald.*)

of *Anabaena* grow within the water fern *Azolla*
(see p.449) and in the roots of cycads (Peters,
1978). Species of some genera, such as *Calothrix*
and *Chamaesiphon*, live as epiphytes on algae
or other aquatic plants. Members of the Cyanophyta may serve as the components of lichens;
these are discussed in Chapter 36. A number
grow in waters of hot springs, where they
deposit rocklike strata composed of carbonates.
Some of these are present in water that attains a
temperature of 73–74°C (Castenholz, 1969).

Color alone is insufficient to distinguish
Cyanophyta from the algae, for their color,
especially *en masse*, may be blue-green, black,
dark purple, brown, or even red. This range in
color is occasioned by the presence of varying
proportions of several pigments. These are chlorophyll *a* and carotenoid pigments, as well as
three other pigments—the phycobilins,
phycocyanin (Gr. *phykos*, seaweed, + Gr.
kyanos, blue), **allophycocyanin,** and
phycoerythrin (Gr. *phykos* + Gr. *erythros*,
red). These pigments are biliproteins, and each

consists of an open-chain tetrapyrrole attached
to a protein. Species of Cyanophyta that are
reddish contain a large proportion of phycoerythrin, while dark-colored species contain a
preponderance of phycocyanin. Both the
phycoerythrin and phycocyanin are water-
soluble. They readily diffuse out of plants killed
by boiling, leaving the unmasked chlorophyll.
Various carotenoid pigments also are present.
The relation of this pigment complex to the
course of photosynthesis has been investigated
in Cyanophyta grown in pure culture.
Phycoerythrin and phycocyanin absorb light
rays of wavelengths not absorbed by chlorophyll
a. The light energy absorbed by phycoerythrin
is transferred to phycocyanin and, in turn, to
chlorophyll *a*, the primary agent of photosynthesis.

A number of species are photoautotrophic
(Gr. *autos*, self, + Gr. *trophe*, food); they are
able to grow and reproduce in culture media
containing only inorganic compounds, using
light energy. A few, such as strains of *Nostoc
commune* and *N. muscorum*, can grow slowly in
darkness, using certain sugars as a carbon and
energy source. Heterotrophic growth occurs in
darkness, in the presence of sugars, of a strain of
Nostoc isolated from cycad roots (Hoare et al.,
1971). It has been demonstrated (Van Baalen et
al., 1971) that some species that cannot grow
photoautotrophically in dim light (10 ft-c) can
grow heterotrophically under such conditions
with glucose as the carbon source. Khoja and
Whitton (1971) have reported that 17 strains of
Cyanophyta studied by them can grow heterotrophically in darkness, albeit slowly, in a medium containing 0.01 M sucrose. The accessory
pigment phycocyanin (perhaps also allophycocyanin and phycoerythrin) has been identified especially as a light absorber for one of the
two photoreactions occurring in photosynthesis.

A number of Cyanophyta having heterocysts
(see p. 28 and Figs. 2-25, 2-26) have the capacity
to grow in the absence of combined nitrogen
and are able to use atmospheric nitrogen in

their metabolism. Numerous heterocysts are induced to form when the nitrogen level in the culture medium falls. Such utilization of gaseous nitrogen and its reduction within the organism to a combined state are known as **nitrogen fixation** (see Stewart, 1980, and Bothe, 1982, for summaries). The formation of nitrogenase (the primary enzyme involved in nitrogen fixation) is correlated with the presence of heterocysts in Cyanophyta (Neilson et al. 1971). Nitrogen fixation also characterizes the metabolism of certain bacteria (see p. 18). As is the case with the latter, both free-living and endophytic (Millbank, 1974) nitrogen-fixing Cyanophyta are known (see pp. 31 and 35). There is a strong indication that heterocystous species are active in fixing nitrogen in hot springs (Stewart, 1970). They also may grow on mosses (Scheirer and Brasell, 1984).

About 50 heterocystous Cyanophyta have been demonstrated to be nitrogen fixers. Species of *Nostoc*, *Anabaena*, and *Calothrix* are especially efficient in fixing nitrogen. There is considerable evidence (Stewart et al., 1969; and Thomas, 1970) that the heterocysts are the site of nitrogen fixation. The addition of ammonium nitrogen to culture media inhibits the formation of heterocysts and, concomitantly, nitrogen fixation. Nitrogen fixation is also inhibited by the presence of oxygen (Stewart and Pearson, 1970), which is seemingly not abundant in heterocysts; the latter do not contain phycocyanin, which is necessary for the liberation of oxygen in the photosynthesis of blue-green algae (Thomas, 1970). Furthermore, the enzyme nitrogenase has been proved to be present in heterocysts but not in vegetative cells under aerobic conditions. Vegetative cells of filamentous blue-green algae can produce nitrogenase under anaerobic conditions (Smith and Evans, 1970). It seems probable, therefore, that under aerobic conditions, nitrogen fixation occurs largely in heterocysts.

However, it has also been demonstrated (Wyatt and Silvey, 1969; Stewart and Lex, 1970)

that two nonheterocystous species, *Gloeocapsa* sp. and *Plectonema boryanum*, have nitrogenase activity; that is, they can fix atmospheric nitrogen. It is clear, too, that this is inhibited in the presence of oxygen in the case of *Plectonema*. Furthermore, nitrogen fixation occurs in two strains of *Gloeocapsa* under aerobic conditions (Rippka et al., 1971).

For more comprehensive discussions of the blue-green algae refer to Wolk (1973), Carr and Whitton (1973, 1982), Fogg et al. (1973), Stewart (1974, 1977), Bold and Wynne (1985), and Stanier and Cohen-Bazire (1977).

CELLULAR ORGANIZATION

The **protoplast** is the unit of protoplasm within the walls of a single cell. The protoplast of Cyanophytan cells is surrounded by a wall external to which a layer of slimy material of varying thickness and consistency is often present, the so-called **sheath** (Figs. 2-15 through 2-17, 2-23, 2-25A). Electron microscopy has revealed that the latter is composed of fibrillar material embedded within an amorphous matrix. The presence or absence of a sheath may be demonstrated readily with light microscopy by immersing the organisms in diluted India ink or by staining with methylene blue. Masses of cells with sheaths are slimy to the touch, and this characteristic suggested one class name, Myxophyceae.

Between the sheath, if one occurs, and the **plasma membrane** (plasmalemma) of the protoplast, a wall of varying complexity is present. In most cases it is composed of at least three layers (Allen, 1968; Lang, 1968): (1) an outer ("double track") layer, (2) an electron-dense middle layer, and (3) an inner, electron-dense, mucopolymer layer. It is the innermost layer that gives rigidity to the cell. Analysis of the cell wall, which is composed of murein (peptidoglycans or mucopeptides), indicates the presence of as many as eight amino acids and, in

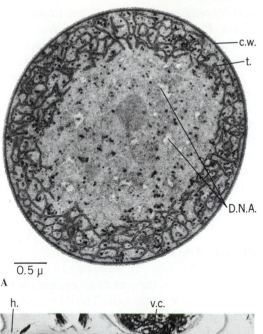

A

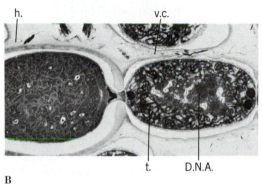

B

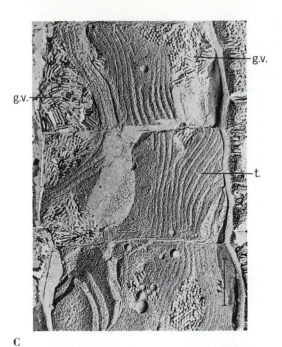

C

FIGURE 2-14
Electron micrographs of Cyanophytan cells. **A.**
Nostoc sp., endophytic in the roots of *Macrozamia*
sp.: transection of a cell. **B.** *Anabaena* sp.,
longisection of a vegetative cell and heterocyst. **C.**
Oscillatoria rubescens, longisection of three cells of
a trichome. c.w., cell wall; D.N.A.,
deoxyribonucleic acid; g.v., gas vacuole; h.,
heterocyst; t., thylakoid; v.c., vegetative cell. (**A.**
After Hoare et al. **B.** *Courtesy of Prof. Howard J.*
Arnott. **C.** *Courtesy of Balzers Aktiengesellschaft.*)

addition, muramic and α,ε-diaminopimelic acid
and glucosamine. It is of phylogenetic interest
that the cell walls of both blue-green algae and
bacteria (at least 50% in Gram-positive ones,
less in Gram-negative ones) contain these same
substances. Furthermore, the mucopolymers of
both Cyanophyta and bacteria are digested by
lysozyme (muramidase), a bacterial enzyme,
which degrades the innermost layer of
Cyanophytan cell walls (Holm-Hansen et al.,
1965; Allsopp, 1969).

The organization of the protoplast of
Cyanophyta differs markedly from that of algae
and, in fact, from that of all other plants and
animals except bacteria. The nuclear material
(DNA) is not delimited from the cytoplasm by a
nuclear membrane and often is dispersed as a
network through it (Fig. 2-14A). Furthermore,
well-defined nucleoli are absent. This lack of
membrane-bounded nuclei, and with it typical
mitosis, has suggested the group name
Prokaryota[3] for these blue-green organisms and
bacteria. A second noteworthy feature of these

[3]From the Greek *karyon*, kernel or nut, referring to the
nucleus.

Cyanophytan cells is that the flattened, chlorophyll-containing sacs, the **thylakoids**, are not segregated from the cytoplasm by a membrane (Fig. 2-14) as in other chlorophyllous plants, so that true chloroplasts are absent. The simple thylakoids here correspond to the saclike structures of photosynthetic bacteria. Although they often occur largely in the periphery of the cell, the pigment-containing thylakoids may be present in the central region as well. The accessory pigments are present on the outer surface of the thylakoids, as they are in red algae (Gantt and Conti, 1969); they occur as small bodies called **phycobilisomes**. These are arranged in parallel rows on the surface of the thylakoids. Large, central aqueous vacuoles, so characteristic of most algal cells, also are absent from those of Cyanophyta, as are endoplasmic reticulum, mitochondria, and Golgi apparatus, all absent also from bacterial cells but present in most eukaryotic cells.

The excess photosynthate is stored as minute granules, sometimes called **Cyanophycean starch**, or alpha granules. These are apparently glycogen (Chao and Bowen, 1971) in *Nostoc muscorum* or glucopyranosides with $\alpha - 1 : 4$ and some $-1 : 6$ linkages.

Various other granules are present in the cells. Most conspicuous of these, even with light microscopy, are the refractive Cyanophycin granules, which contain arginine and aspartic acid peptides. These are often present near the transverse walls of certain filamentous species. Polyglucoside and polyphosphate granules also may be present. Finally, a number of planktonic species contain peripheral **gas vacuoles**, which appear as reddish refractive areas. These disappear if the cells are subjected to pressure.

There have been two suggestions regarding the function of gas vacuoles: (1) they are responsible for the buoyancy of the cells in which they occur (Walsby, 1972, 1978) and (2) they reduce the intensity of light reaching the photosynthet-

ic thylakoids (Van Baalen and Brown, 1969). Evidence for this has been presented by Waaland et al. (1971).

ILLUSTRATIVE REPRESENTATIVES

Three types of organization occur in the Cyanophyta: unicellular, colonial, and filamentous. In unicellular species the organism is a single cell, either free-living or attached (Figs. 2-15 through 2-18). As a result of more or less temporary coherence of several generations of recently divided cells, incipient colonial forms may arise. Permanently colonial species are those in which a number of cells grow together within a common sheath that is augmented by the secretions of the individual cells (Fig 2-19A). The cells of filamentous species are joined in unbranched (Figs. 2-21, 2-22A, 2-23), branched (Fig. 2-34), or falsely branched (p.32) (Fig. 2-29) chains. Unbranched filaments arise as a result of restriction of cell division to only one direction. The unicellular and colonial types usually are considered to be primitive and the filamentous ones as derived from unicellular precursors.

Reproduction in blue-green algae is by cell division in the unicellular species and by fragmentation in the colonial and filamentous ones. Some blue-green algae produce minute spores (e.g., **exospores**, Fig. 2-18C). Specialized cells, the **heterocysts** (p.28) and **akinetes** (p.29), under certain conditions and in certain species, also serve as agents of reproduction.

Sexual reproduction by union of gametes to form zygotes has not been observed in blue-green algae. Some indication that genetic recombination occurs has been reported by Bazin (1968) and Orkwizewski and Kaney (1974). Ladha and Kumar (1978) include a discussion of recombination in their review of the genetics of blue-green algae.

Chroococcus, *Gloeocapsa*, and *Chamaesiphon*

Chroococcus[4] (Gr. *chros*, color, + Gr. *kokkos*, berry), *Gloeocapsa*[5] (Gr. *gloia*, glue, + L. *capsa*, a box or case), and *Chamaesiphon*[6] (Gr. *chamai*, on the ground, hence sessile, + Gr. *siphon*, a tube) will be described as representatives of the more simple unicellular types. *Chroococcus* (Figs. 2-15, 2-16) frequently occurs sparingly intermingled with other algae in the sludge at the bottoms of quiet bodies of water; *Gloeocapsa* (Fig. 2-17), along with *Chroococcus*, is frequently encountered on moist rocks and walls and on flowerpots in greenhouses.

It is difficult, sometimes, to find single, isolated cells of *Chroococcus* and *Gloeocapsa* in a particular sample, because of the abundance of cell division and the tendency of the daughter cells to cohere at its conclusion. Cell division in both these genera is accomplished by the centripetal growth of a surface furrow, which ultimately divides the cell (Fig. 2-15).

New walls are synthesized around the daughter protoplasts within the persistent wall of the mother cell, which becomes distended as the division products increase in size. Nuclear material apparently is distributed to the juvenile cells at cytokinesis. Thus, as in most unicellular organisms, cell division effects reproduction or multiplication of the individual. Subsequent growth of the division products results in their achieving the size characteristic of the species.

The sheaths of *Gloeocapsa* may be colored in the living cells and usually are thicker and more prominent than in *Chroococcus*; the incipient colonies of *Gloeocapsa* generally are more complex than those of *Chroococcus*, being composed of more individual cells. Careful micro-

---

[4]Included in *Anacystis*, by Drouet and Daily (1956).

[5]Included in *Anacystis*, by Drouet and Daily (1956).

[6]Included in *Entophysalis*, by Drouet and Daily (1956).

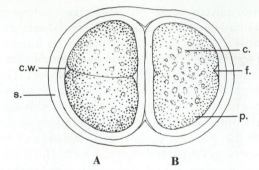

FIGURE 2-15
Chroococcus turgidus. Dividing cell. **A.** Surface view. **B.** Median optical section. c., colorless central region; c.w., cell wall; f., incipient cleavage furrow; p., pigmented cytoplasm; s., sheath. ×1700.

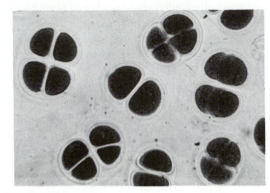

FIGURE 2-16
Chroococcus turgidus. Living cells. ×1000.

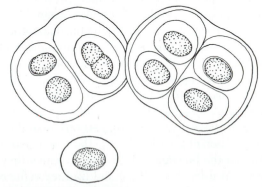

FIGURE 2-17
Gloeocapsa sp. ×1700.

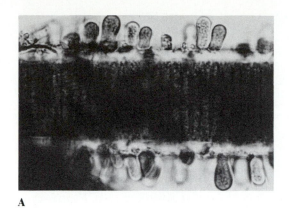

A

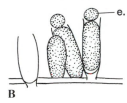

B

FIGURE 2-18
Chamaesiphon sp. **A.** Cells epiphytic on a filamentous blue-green alga, *Porphyrosiphon* sp. **B.** Liberation of exospores. **A, B.** ×1500.

scopic study of such aggregations of *Gloeocapsa* cells usually reveals that each cell has an individual sheath (which may be lamellated) at the conclusion of cell division. The sheaths of the parent cells of *Gloeocapsa* and *Chroococcus* stretch and may persist (Figs. 2-15 through 2-17).

Chamaesiphon (Fig. 2-18) is epiphytic on other algae and aquatic flowering plants and is frequently present on such plants in aquaria. The cells of *Chamaesiphon* are attached to their hosts by holdfasts. The cells are enlarged distally from a tapering base. Division of one cell into two equal daughter cells does not occur in *Chamaesiphon*. Instead, a series of small walled cells is delimited from the distal portion of the individual (Fig. 2-18B, C), and these are gradually discharged through a terminal opening in the cell wall. These small cells are called **exospores** and presumably float to suitable substrata, where they germinate into new individuals.

COLONIAL ORGANISMS

Microcystis[7] and *Merismopedia*

Microcystis (Gr. *micros*, small, + Gr. *kystis*, bladder, hence, cell) and *Merismopedia* (Gr. *merismos*, division, + Gr. *pedion*, a plain) illustrate the colonial type of plant body among Cyanophyta. In both these genera the cells are surrounded by a common envelope. In *Microcystis* the densely cellular colonies vary in shape from spherical to irregular (Figs. 2-13, 2-19), whereas in *Merismopedia* the colony is a flat-

[7]Included in *Anacystis*, by Drouet and Daily (1956).

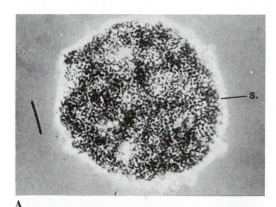

A

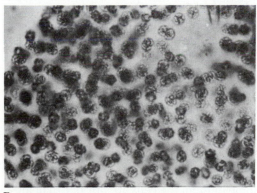

B

FIGURE 2-19
Microcystis aeruginosa. **A.** Single living colony mounted in India ink. **B.** Portion of the same more highly magnified. **A.** ×125. **B.** ×1600.

25

tened or slightly curved plate (Fig. 2-20). *Microcystis aeruginosa* frequently is a component of **water blooms** (Fig. 2-13). The individual cells are minute and spherical and usually contain refractive pseudovacuoles, which are filled with gas, a common attribute of planktonic organisms.

The ellipsoidal cells of *Merismopedia*[8] are arranged in flat colonies in which the individual cells occur in rows (Fig. 2-20). This regularity of arrangement arises from the limitation of cell division to two directions. Numerous dividing cells often are visible within the colony.

In these colonial genera, in contrast to strictly unicellular organisms, cell division results in increase in colony size rather than in multiplication of the individual. The latter is accomplished by fragmentation of larger colonies, the fragments continuing to increase in size by cell division.

FILAMENTOUS ORGANISMS

Oscillatoria, Lyngbya, Anabaena, Nostoc, Scytonema, Rivularia, Gloeotrichia, Calothrix, and *Hapalosiphon*

The coherence of cells after completion of cell division, mostly in one direction, results in the production of another type of plant body, the **trichome** (Fig. 2-21). In the literature of the Cyanophyta the term "trichome" is limited to the chain of cells; the term **"filament"** refers to both the trichome and its enclosing sheath. Where cell division is entirely restricted to a single direction, an unbranched trichome results, as in *Oscillatoria* (L. *oscillare*, to swing) and *Lyngbya*[9] (in honor of Lyngbye, a Danish phycologist). *Oscillatoria* (Figs. 2-21, 2-22), which occurs floating in aquatic habitats or on damp soil, is a genus containing many species. In some of them the cells are broader than long,

---

[8]Included in *Agmenellum*, by Drouet and Daily (1956).

[9]Included in *Oscillatoria*, by Drouet (1968).

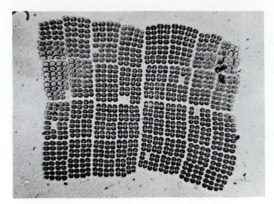

FIGURE 2-20
Merismopedia sp., living colony. ×250.

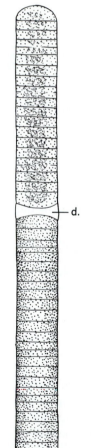

d.

FIGURE 2-21
Oscillatoria limosa. Single trichome; lower portion in surface view, upper in median longisection. d., dead cell. ×700.

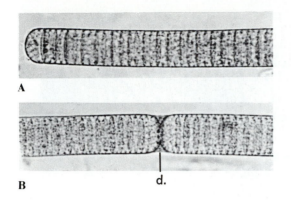

A

B

d.

C

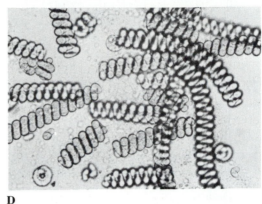

D

FIGURE 2-22
Oscillatoria sp. **A, B, C.** Three stages in hormogonium formation. d., dead cell. ×600. **D.** *Arthrospira* sp. ×500.

whereas in others the reverse is true. In *Oscillatoria* and *Lyngbya* there is no differentiation among the component cells of a trichome, except that the apical cell may differ in shape from the other vegetative cells, and its wall thickens. Sheaths usually are not demonstrable around the trichomes of *Oscillatoria*. When they are observed in aqueous media, a number of the trichomes frequently exhibit an oscillating motion, as well as rotation and forward and backward movement along their long axes. The mechanism of these movements is not understood completely. In *Oscillatoria princeps*, the filaments undergo gliding movement (Halfen and Castenholz, 1971), which is defined as active movement of an organism lacking both organs of locomotion (cilia or flagella), and amoeboid motion when in contact with a solid substrate. In *O. princeps* the sheathed filaments move at a maximum rate of 11.1 μm/sec. An intermediate wall layer has continuous fibrils on its surface in a helical pattern. Halfen and Castenholz propose that "gliding is produced by unidirectional waves of bending in the fibrils which act against the sheath or substrate, thus displacing the trichome." It has been suggested

that the movement of filamentous Cyanophyta is occasioned by the secretion of polysaccharides through pores in their cell walls. Electron microscopy has revealed the occurrence of such pores, but theoretical considerations of the rate of movement (forward and backward), which is 5–6 μm/sec in a genus related to *Oscillatoria*, indicate that impossibly large amounts of colloids would have to be secreted to account for the movement. In *Arthrospira* (Gr. *arthron*, joint, + Gr. *speira*, helix) the trichomes are helical (Fig. 2-22D).

Lyngbya (Figs. 2-23, 2-24), which occurs in both fresh and salt water, differs from *Oscillatoria* in that the trichomes are surrounded by rather firm, clearly visible sheaths. In both *Oscillatoria* and *Lyngbya* cell division is generalized, all the cells of the trichomes being capable of division. Cell division here, as in colonial genera, results in increase in the size of the individual. Multiplication of the filaments takes place by a type of fragmentation called **hormogonium formation** (Gr. *hormos*, chain, + Gr. *gonos*, reproductive structure). In this process, because of the death of one or more cells in the trichome (Figs. 2-22, 2-24), the chains of

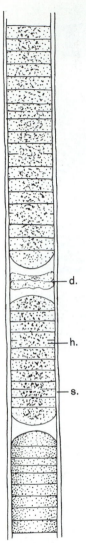

FIGURE 2-23
Lyngbya sp. Segment of a filament; the lowermost portion in surface view, the remainder in median optical section. d., dead cells; h., hormogonium; s., sheath. ×700.

cells break up into multicellular fragments, the **hormogonia**. These are usually motile and capable of forming new trichomes. In *Oscillatoria* and *Lyngbya* the rupture is evoked by the death of a cell (Figs. 2-22B, 2-24). It has been demonstrated that the rupture of the dead cell, the

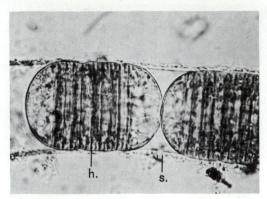

FIGURE 2-24
Lyngbya sp. Living filament with hormogonium. h., hormogonium; s., sheath. ×1400.

necridium, takes place by the tearing of pores that occur opposite each other in the innermost layer of the wall at the transverse wall septa (Lamont, 1964).

It is of interest that extracts of *Lyngbya* (and several other blue-green algae) are active against a strain of lymphocytic leukemia in mice (Mynderse et al., 1977).

Anabaena (Gr. *anabainein*, to arise) and *Nostoc* (name used by Paracelsus), although unbranched like *Oscillatoria* and *Lyngbya*, have several features not present in the latter. *Anabaena* (Figs. 2-25, 2-26), a genus that contains both planktonic species and some that form coatings on other aquatic vegetation, is widespread in bodies of fresh and salt water. *Nostoc*, "starjelly," or "witches' butter" (Figs. 2-27, 2-28), includes a number of both aquatic and terrestrial or rock-inhabiting species in which the tortuous, threadlike filaments are grouped together in matrices of macroscopically recognizable form. The plant mass may be spherical, ovoidal, or sheetlike. Cell division is generalized also in *Anabaena* and *Nostoc*, but two manifestations of differentiation occur, namely, the heterocyst and the akinete.

Heterocysts (Figs. 2-25, 2-27C, 2-28) (Gr. *heteros*, different, + Gr. *kystis*, cell) arise from transformation of vegetative cells (Haselkorn,

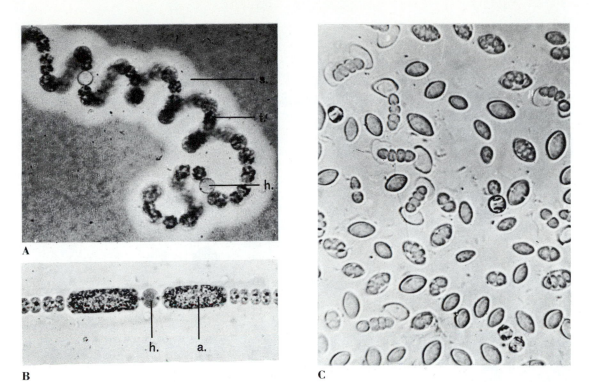

FIGURE 2-25
Anabaena flos-aquae. **A.** Single filament mounted in India ink. **B.** *Anabaena* sp. **C.** *A. doliolum*
Bharadwaja, germinating akinetes. a., akinete; h., heterocyst; s., sheath; t., trichome. **A, B.** ×540. **C.** ×500.
(**C.** *Courtesy of Dr. R. N. Singh.*)

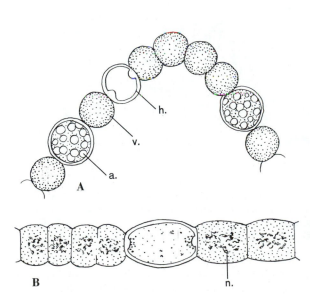

1978). The term **"vegetative cell"** is used by botanists in much the same sense that "somatic" is used by zoologists—that is, as the antonym to "reproductive," or "germ," cells. This may involve cell enlargement, synthesis of a multilayered wall, decrease in granular cellular inclusions, and reorganization of the thylakoids into concentric or reticulate pattern. The heterocyst wall thickens uniformly except in the regions of its contacts with adjacent cells. Here the wall is somewhat modified to form the **polar nodules**. Lang and Fay (1971) and Fay and Lang (1971) have reported on the ultrastructure of intact and

FIGURE 2-26
Anabaena circinalis. **A.** Segment of living filament. **B.** *Anabaena* sp. Segment of acetocarmine-stained filament. a., akinete; h., heterocyst; n., nuclear material; v., vegetative cell. ×850.

29

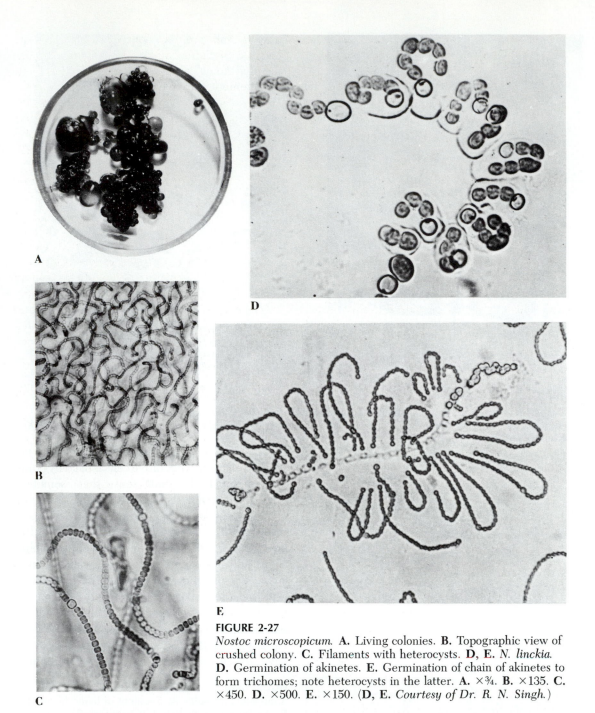

FIGURE 2-27

Nostoc microscopicum. **A.** Living colonies. **B.** Topographic view of crushed colony. **C.** Filaments with heterocysts. **D, E.** *N. linckia.* **D.** Germination of akinetes. **E.** Germination of chain of akinetes to form trichomes; note heterocysts in the latter. **A.** ×¾. **B.** ×135. **C.** ×450. **D.** ×500. **E.** ×150. (**D, E.** *Courtesy of Dr. R. N. Singh.*)

FIGURE 2-28
Nostoc sp. Single trichome, with heterocysts and vegetative cells, two of the latter dividing. ×850.

isolated heterocysts. In addition to the original vegetative cell wall, the heterocyst in *Anabaena cylindrica* develops three additional, peripheral surface layers. Furthermore, these authors demonstrated clearly that heterocysts are not closed systems but are in communication by means of microplasmodesmata with adjacent vegetative cells. The function of the heterocyst remained an enigma until quite recently, when evidence was educed that heterocysts may be the agents of nitrogen fixation (p.21) and of transfer of combined nitrogen to the vegetative cells intervening between heterocysts. Furthermore, there is some evidence that they may evoke the transformation of vegetative cells adjacent to them into akinetes (Wolk, 1966, 1967). The presence of ammonium nitrogen inhibits the formation of heterocysts and of nitrogen fixation, while it stimulates the germination of the heterocysts into new filaments (Wolk, 1965; Singh and Tiwari, 1970) (Fig. 2-32D).

Light is required for the formation of heterocysts, which develop 27 to 33 hours after the cultures have been illuminated (Bradley and Carr, 1977).

The vegetative cells of *Nostoc* and *Anabaena* also may become transformed into **akinetes** (Gr. *akinesia*, absence of motility). Formation of akinetes in *Nostoc* begins in vegetative cells remote from heterocysts and proceeds toward them (Ahluwalin and Kumar, 1982). In this case also, cell enlargement may occur while the walls

thicken and proteinaceous cyanophycin granules accumulate. Unlike those of heterocysts, the thylakoids in the akinete retain the same arrangement they have in the vegetative cells. Akinetes are highly resistant to environmental adversities and have germinated after 87 years of air-dry storage. Upon germination, akinetes give rise to filaments (Figs. 2-27D, E, 2-32B, C).

The development of the mature, ensheathed colony in certain species of *Nostoc* is controlled by light, which affects the motile trichomes (hormogonia). In *N. commune*, white fluorescent light prevents the development of motile trichomes, while it evokes their development in *N. muscorum* (Lazaroff, 1966; Lazaroff and Vishniac, 1961, 1964; Lazaroff and Scheff, 1962; Robinson and Miller, 1970). Green light and red light are efficacious in motile trichome formation. The motile trichomes excrete a substance into the culture medium that causes motility in nonmotile cultures.

Scytonema (Gr. *skytos*, hide or skin, + Gr. *nema*, thread), a genus widely distributed on moist rocks and soil, where it forms dark, blackish, felty coatings, is of interest in two respects. Its sheaths are thick, firm, sometimes lamellated, and yellow-brown in older parts of the plant body (Fig. 2-29). Furthermore, there is a strong tendency for cell division to be restricted to cells near the apices of the filaments. However, heterocyst formation and renewal of cell divisions in intercalary vegetative cells between heterocysts often result in modifications of the trichome. Inasmuch as the heterocysts are firmly attached to the sheath, pressure generated by intercalary cell division results in the rupture of the sheath and the emergence of pairs of trichomes (Fig. 2-29). In this way, although cell division is limited to one direction, as in all the preceding filamentous genera, **false branching** takes place.

Rivularia (L. *rivulus*, a small brook) (Fig. 2-30) and *Gloeotrichia* (Gr. *gloios*, gelatinous, + Gr. *thrix*, hair) (Figs. 2-31, 2-32)

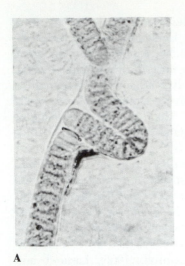

A

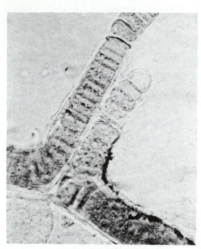

B

FIGURE 2-29
Scytonema myochrous. **A.** Origin of false branches.
B. False branches emergent. A, B. ×750.

FIGURE 2-30
Rivularia sp. Several
trichomes. ×350.

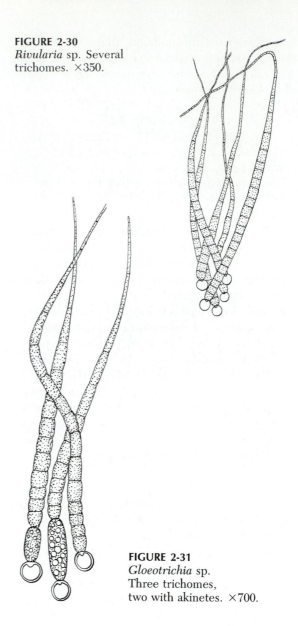

FIGURE 2-31
Gloeotrichia sp.
Three trichomes,
two with akinetes. ×700.

also have falsely branched filaments, but
they differ from *Scytonema* in that their filaments are united in spherical attached
colonies and they taper from base to apex. The
apices are composed of long, almost colorless,
hairlike cells. The basal vegetative cell of each
filament becomes transformed into a heterocyst

(Figs. 2-30, 2-32). In *Gloeotrichia*, one or more
enlarged akinetes usually are developed from
the vegetative cells in the vicinity of the heterocyst; akinetes are lacking in *Rivularia*. The
akinetes germinate into new trichomes under
certain conditions (Fig. 2-32B, C). In both of
these genera the sheaths of the individual fila-

ments are partially confluent, thus contributing to the common matrix; however, remnants of the individual sheaths are usually apparent at the base of the plants.

The related genus *Calothrix* (Gr. *kalos*, beautiful, + Gr. *thrix*) (Fig. 2-33) is frequently encountered as an epiphyte on aquatic plants, or grows abundantly on rocks and pilings exposed to seawater and spray, and on soil.

True branching, resulting from the division of certain cells of a trichome in a direction different from that of the majority, characterizes the genus *Hapalosiphon* (Gr. *hapalos*, gentle, + Gr. *siphon*, tube) (Fig. 2-34). Species of *Hapalosiphon* often form extensive coatings on aquatic vegetation.

Summary and classification

In summary, it should be emphasized that the Cyanophyta represent a fairly large but somewhat anomalous group. They are unique among all organisms in being at the same time prokaryotic and in containing chlorophyll *a*, and in evolving oxygen during photosynthesis. Their cellular organization is more nearly like bacterial cells than that of any group of algae. Distinctive attributes, as compared with algae, include their prokaryotic cellular organization—characterized by the absence from the cells of definitely delimited plastids, membrane-bounded nuclei, large aqueous vacuoles, mitochondria, Golgi apparatus, and endoplasmic

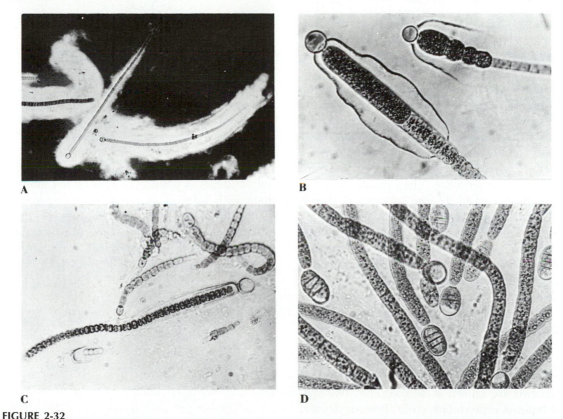

A

B

C

D

FIGURE 2-32
Gloeotrichia natans. Filaments mounted in India ink; note sheaths and basal heterocysts. **B–D.** *G. ghosei* sp. **B, C.** Stages of germination of basal akinete *in situ.* **D.** Germination of heterocysts *in situ.* **A.** ×175. **B.** ×650. **C.** ×550. **D.** ×630. (**B–D.** *After Singh and Tiwari.*)

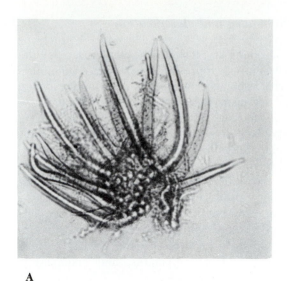

A

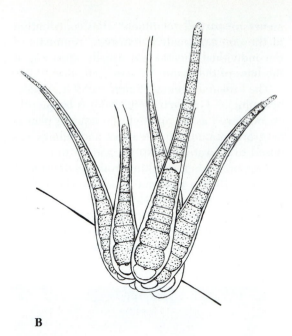

B

FIGURE 2-33
Calothrix fasciculata. **A.** Epiphytic on *Nemalion*. **B.** Epiphytic on *Enteromorpa* sp. **A.** ×250. **B.** ×375.

reticulum—and the presence of phycocyanin and phycoerythrin, in addition to chlorophyll *a* and carotenoid pigments. Heterocysts do not occur in plants other than Cyanophyta. Their stored photosynthate, called Cyanophycean starch, a glycogenous substance, is different from that of other algae. The capacity of heterocystous species to fix free nitrogen is shared only by certain bacteria.

The cell walls of Cyanophyta apparently lack cellulose. Those that have been carefully analyzed contain eight amino acids in their peptides and, in addition, α,ε-diaminopimelic acid and glucosamine. Their combination of pigments also is characteristic.

Reproduction in Cyanophyta is accomplished by cell division in unicellular genera and by various types of fragmentation in colonial and filamentous ones. In the latter, the filamentous fragments are known as hormogonia. Two special types of reproductive cells, heterocysts and akinetes, are developed by many of the filamentous genera. With few exceptions, cell division in multicellular types is generalized, not local-

ized. *Scytonema, Calothrix, Rivularia,* and *Gloeotrichia* differ from other filamentous genera in this respect. Growth in *Scytonema* is largely apical, whereas in *Calothrix, Rivularia,* and *Gloeotrichia* it is basal.

Although many Cyanophyta grow on damp soil, students of plant evolution do not look on them with interest as possible progenitors of higher forms of plant life because of their prokaryotic cellular organization, in which they differ markedly from algae as well as from other plants. They have no clear kinship with other

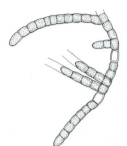

FIGURE 2-34
Hapalosiphon sp. Segment of a branching filament. ×205.

34

living organisms, unless it is with the bacteria, a suggestion first made more than 100 years ago.

Among eukaryotic plants, the chloroplasts contain α-linolenic and related polyunsaturated fatty acids. These occur also in the cells of certain *Gloeocapsa*-like organisms. This furnishes support for the hypothesis that the chloroplasts of eukaryotes may have originated from modified endosymbiotic, blue-green algae (Kenyon and Stanier, 1970; Margulis, 1970, 1980; Raven, 1970).

The classification of the Cyanophyta is in a state of flux. On the basis of studies largely of herbarium material, Drouet (1981) consolidated genera, families and orders, some of which are cited below. In contrast, using microbiological methods of study, the late Professor Stanier and his associates have been reclassifying organisms available in culture (Stanier et al., 1981). The traditional taxonomic categories presented below may ultimately be modified. The orders and families, of which representative genera have been chosen for this chapter, may be classified as follows:

Division: Cyanophyta
  Class 1. Cyanophyceae
    Order 1. Chroococcales
      Family 1. Chroococcaceae
        Genera: *Chroococcus, Gloeocapsa,*
            Microcystis, Merismopedia
    Order 2. Chamaesiphonales
      Family 1. Chamaesiphonaceae
        Genus: *Chamaesiphon*
    Order 3. Oscillatoriales
      Suborder 1. Oscillatorineae
        Family 1. Oscillatoriaceae
          Genera: *Oscillatoria, Lyngbya*
      Suborder 2. Nostochineae
        Family 1. Nostocaceae
          Genera: *Anabaena, Nostoc*
        Family 2. Scytonemataceae
          Genus: *Scytonema*
        Family 3. Stigonemataceae
          Genus: *Hapalosiphon*

Family 4. Rivulariaceae
    Genera: *Rivularia, Gloeotrichia,*
        Calothrix

The orders are readily distinguishable from one another. The Chamaesiphonales alone produce endospores and exospores. The remaining unicellular and colonial species are members of the Chroococcales; the filamentous genera belong to the Oscillatoriales. Two series are distinguishable in the order Oscillatoriales, depending on the absence (suborder Oscillatorineae) or presence (suborder Nostochineae) of heterocysts. The several families of the Nostochineae are recognizable by the following unique attributes: The Nostocaceae are unbranched, with trichomes of uniform diameter; false branching occurs in the Scytonemataceae, which also have trichomes of uniform diameter. True branching characterizes the Stigonemataceae, while the trichomes of the Rivulariaceae, which may be falsely branched, are tapered.

DISCUSSION QUESTIONS

1. What characteristics are shared by bacteria and Cyanophyta? How do they differ?
2. Define and explain the terms autotrophic, heterotrophic, chemoautotrophic, photoautotrophic, chemosynthetic, photosynthetic, parasitic, saprophytic, phagotrophic.
3. Discuss genetic interchange in bacteria.
4. Are bacteria the smallest living organisms? Explain.
5. Can most bacteria be identified specifically by microscopic examination alone? Explain.
6. How would you prove that bacterial cells have walls?
7. What types of nutrition occur in bacteria?
8. What type of nutrition do you consider to be the most primitive? Give reasons for your answer.

9. Obligate parasites are those that cannot be cultivated except in a living host. Can you suggest an explanation for this?

10. In what respects are bacterial spores different from those of endospores of Cyanophyta?

11. List some harmful and some beneficial activities of bacteria.

12. What characteristics distinguish Cyanophyta from other algae?

13. For what reasons is the cell structure of Cyanophyta said to be prokaryotic?

14. What possible functions can you see for the copious sheaths of certain Cyanophyta?

15. Explain how cell division is related to the form of the plant body of Cyanophyta, mentioning illustrative genera.

16. Define or explain the following terms: heterocyst, spore, akinete, vegetative cell, hormogonium, somatic, germ cell.

17. Do you consider colonies such as *Microcystis* and *Merismopedia* to be multicellular organisms? Give reasons for your answer.

18. Where would you look for Cyanophyta in nature?

19. What evidence associates heterocysts with nitrogen fixation?

20. What is a taxonomic "key"?

21. Construct a dichotomous key to all the genera of Cyanophyta cited in this chapter.

22. What characteristics do Cyanophyta share with bacteria?

23. Does genetic interchange occur in blue-green algae? Explain.

Chapter 3
INTRODUCTION TO THE
SUPERKINGDOM EUKARYONTA
AND KINGDOM PHYTA

EUKARYONTA

General features. The Eukaryonta, which comprise by far most living organisms, include the Kingdoms Protoctista, Phyta (Plantae), Myceteae (Fungi), and Animalia. The members of this superkingdom differ from those of the Superkingdom Prokaryonta, having in their cells membrane-bounded, histone-coated DNA; mitochondria;[1] ribosome-associated endoplasmic reticulum; Golgi apparatus; and cytoplasmic ribosomes circa 20 to 22 nm, with a mass of 4 megadaltons. Of the component kingdoms just cited, only the Phyta are discussed in this chapter, and these with respect to their classification.

THE CLASSIFICATION OF PLANTS

The organisms discussed in the preceding chapter are all members of the Superkingdom Prokaryonta. It will be recalled from Chapter 1 that most living organisms are members of the Superkingdom Eukaryonta, their cells characterized by having such membrane-bounded structures as nuclei, chloroplasts (Kingdom Plantae), mitochondria, Golgi apparatus, and also endoplasmic reticulum. This chapter and Chapters 4 to 27 discuss the classification and diversity of members of the Kingdom Phyta, both extinct and extant, while Chapters 28 to 35 do so for the Kingdom Myceteae (Fungi).

When one considers that more than 350,000 species of plants have been described, one sees that it is impossible to become familiar with all of them. As a beginning, therefore, one is driven to the expedient of selecting representatives that illustrate fundamental characteristics of larger groups of organisms. The more diversified the group of plants under consideration, the more representatives it will be necessary to study. However, in attempting to include a survey of the morphology of representatives from the entire plant kingdom in one volume,

---

[1]Except certain protozoa, red blood cells, and anaerobic yeasts.

the present authors have obviously had to select some types and exclude others. It is hoped that familiarity with the chosen representatives of the diverse plant groups will form a sound foundation and perhaps kindle the reader's interest in a more extensive study of one or several of them.

In classifying living organisms, it is possible to set up the categories in several different ways, depending on the purpose of the classifier. In the first place, **artificial systems** may be devised, the primary purpose of which is ease and convenience of grouping and segregation, other considerations being minor. The classification of vascular plants into trees, shrubs, and herbs, and into annuals, biennials, and perennials exemplifies an artificial system. **Phylogenetic systems** of classification, on the other hand, endeavor to arrange organisms so as to indicate real relationship and affinity based on evolutionary development. The closeness of the supposed relationships is implied by the proximity of the taxa to one another in the system. The system itself, of course, is (or should be) based on data available from paleontology, comparative morphology, genetics, and all other possible lines of evidence.

Classification of plants is subjective in large measure. It is small wonder, therefore, that few students of plant science have reached unified conclusions regarding the classification of members of the plant kingdom. Furthermore, classifications are constantly altered by the discovery of new facts; they are fluid and dynamic.

In spite of the subjective nature of classification, anarchy is no more desirable in the grouping of plants than it is elsewhere. If his or her efforts are to be recognized, the classifier is bound to make the taxonomic categories, called taxa, conform to the legislation of the International Botanical Congress (Stafleu et al., 1978), which has stated that "every individual plant belongs to a species, every species to a genus, every genus to a family, every family to an order, every order to a class and every class to a

division." The highest (i.e., most inclusive) category, the "division," corresponds to the category "Phylum" in the Kingdom Animalia. A number of botanists, the authors[2] included, have urged that "Phylum" replace "Division" also in the classification of plants (Bold et al., 1978), but this was not approved.

In addition to the *prescription* of the International Botanical rules regarding the hierarchy of taxonomic categories, there are recommendations regarding the endings of the names of the higher taxa (divisions and classes, among others), which have been followed in the present volume. The classification and arrangement of the plant types to be described here deviate to some degree from those in many other current texts. The table on the inside back cover summarizes the present and certain other systems of classification in comparative fashion.

While the relative merits of various systems of classification (see Dodson, 1971; Leedale, 1974; Edwards, 1976; Margulis, 1980) might be discussed at this point, profitable consideration of this subject presupposes considerable knowledge of the plants to be classified, knowledge that becomes available only after study of the plants themselves. For this reason, the system of classification presented here will receive minimum comment at this point, more extensive discussion being deferred to the final chapter (Chapter 37). In anticipation of this, it should be stated that the very existence of various systems of classification is eloquent evidence of divergence in interpreting data on relationship.

The system of classification here presented will serve also as a sort of table of contents, for the illustrative genera will be discussed in an order that approximates that in which the divisions and their component classes are listed in Table 3-1.

To those familiar with older systems of classification (see inside back cover), in which the

---

[2]H. C. B. and C. J. A.

TABLE 3-1

Classification of plants through the level of class, as discussed in this book

Kingdom Phyta (Plantae)
  Division 1. Chlorophyta (green algae)[a]
    Class 1. Chlorophyceae
  Division 2. Charophyta (stoneworts)
    Class 1. Charophyceae
  Division 3. Euglenophyta (euglenids)
    Class 1. Euglenophyceae
  Division 4. Phaeophyta (brown algae)
    Class 1. Phaeophyceae
  Division 5. Chrysophyta (golden algae)
    Class 1. Xanthophyceae (yellow-green algae)
    Class 2. Chrysophyceae (golden-brown algae)
    Class 3. Bacillariophyceae (diatoms)
  Division 6. Pyrrhophyta
    Class 1. Dinophyceae (dinoflagellates)
  Division 7. Rhodophyta (red algae)
    Class 1. Rhodophyceae
  Division 8. Hepatophyta (liverworts)
    Class 1. Hepatopsida
  Division 9. Anthocerotophyta (hornworts)
    Class 1. Anthocerotopsida
  Division 10. Bryophyta (mosses)
    Class 1. Sphagnopsida (peat mosses)
    Class 2. Andreaeopsida
    Class 3. Bryopsida ("true" mosses)
  Division 11. Rhyniophyta (Rhyniophytes)
    Class 1. Rhyniopsida
  Division 12. Zosterophyllophyta (Zosterophyllophytes)
    Class 1. Zosterophyllopsida
  Division 13. Trimerophytophyta (Trimerophytophytes)
    Class 1. Trimerophytopsida
  Division 14. Microphyllophyta[b] (lycopods)
    Class 1. Aglossopsida
    Class 2. Glossopsida
  Division 15. Arthrophyta[c] (arthrophytes)
    Class 1. Arthropsida
  Division 16. Pteridophyta[d] (Ferns)
    Class 1. Pteridopsida
  Division 17. Psilotophyta[e] (whisk ferns)
    Class 1. Psilotopsida
  Division 18. Progymnospermophyta (progymnosperms)
    Class 1. Progymnospermophyta
  Division 19. Pteridospermophyta (seed ferns)
    Class 1. Pteridospermopsida
  Division 20. Cycadophyta (cycads)
    Class 1. Cycadopsida
  Division 21. Cycadeoidophyta (cycadeoids)
    Class 1. Cycadeoidopsida
  Division 22. Ginkgophyta (*Ginkgo*)
    Class 1. Ginkgopsida
  Division 23. Coniferophyta (conifers)
    Class 1. Coniferopsida
  Division 24. Gnetophyta (gnetophytes)
    Class 1. Gnetopsida
  Division 25. Anthophyta (flowering plants)
    Class 1. Anthophyta

[a]Blue-green algae are discussed in Chapter 2. The first six divisions are all (with the possible exception of Charophyta) at the algal level of organization.

[b]Some authors use "Lycopodophyta" or "Lycophyta."

[c]Some authors use "Sphenophyta."

[d]Some authors use "Filicophyta."

[e]Some authors use "Psilophyta."

plant kingdom was divided into four divisions—Thallophyta (including the algae and fungi), Bryophyta (liverworts and mosses), Pteridophyta (ferns and their "allies"), and Spermatophyta (seed plants)—the classification summarized in Table 3-1 will seem longer and more complicated. At first glance there is merit in this complaint, for the 4 divisions have been replaced by 25.[3]

The student may well wonder on what criteria the divisions of the plant kingdom are defined. To some extent, the criteria vary with the classifier, but in general most botanists seem to

agree that the division is the largest phylogenetic taxon in which should be grouped organisms that seem to be related because they share common basic characteristics. This definition itself probably will not satisfy all botanists.

When one examines the old division Thallophyta (inside back cover) in the light of this concept of the division, serious difficulties become apparent. For example, the division Thallophyta, usually defined as a group of organisms lacking stems and leaves, nonetheless included such brown algae as the kelps, in some of which leaflike, stemlike, and rootlike organs are developed. It includes as well simpler unicellular, colonial, and filamentous algae and

---

[3]Not including bacteria, blue-green algae, and fungi.

39

fungi. Furthermore, grouping the algae and fungi together in the same division indicated that they are closely related groups. Modern study of these plants has not provided strong support for this view. In addition, there is good evidence that the several groups of algae, in the past included in a formal category (class) under the name Algae (inside back cover), are themselves diverse and fundamentally different from one another physiologically and biochemically.

The division Bryophyta, which in other systems (inside back cover) includes liverworts, hornworts, and mosses, here is conceived in a more restricted sense. For reasons to be discussed in Chapter 12, the writers are of the opinion that the liverworts, hornworts, and mosses are not such close "allies" as was implied by earlier views of the scope of the division Bryophyta. They have, therefore, segregated the liverworts in the division Hepatophyta and the hornworts in the division Anthocerotophyta and restricted the division Bryophyta to the mosses alone.

It will be recalled also that the division Pteridophyta (inside back cover) formerly included the ferns and their "allies." That these supposed alliances are as nebulous and untrustworthy as certain political groupings seems clear from a comparison of the morphology of the plants themselves. Divisions 11 to 17 have been proposed, therefore, in place of the old division Pteridophyta.

Probably no two botanists would agree on the disposition of the higher categories of plants that formerly constituted the division Spermatophyta (inside back cover), the seed plants. In some current textbooks they are included in the same division with the ferns (inside back cover, middle column). For reasons that can be presented profitably only after the student has become familiar with the groups involved, the present authors have assigned the plants in the old division Spermatophyta to the seven divisions listed as 18 to 25. A more detailed presentation of the evidence on which the present

system of classification is based is included in the discussion of the several groups and in Chapter 37.

The system of classification to be followed in this text is admittedly polyphyletic in that the divisions Thallophyta, Bryophyta, Pteridophyta, and Spermatophyta, and even certain classes (Algae and Fungi) of other systems (inside back cover), have been broken down into smaller units, which themselves have been elevated to divisional rank. Polyphyletic classifications are those in which several seemingly independent evolutionary lines are recognized, in contrast to monophyletic systems, in which the several divergent lines are considered to have had a common origin. To the writers, classifications at present are of necessity tentative and speculative; this relatively polyphyletic classification seems to them truly conservative in not classifying within the same division organisms whose fossil records represent unconnected lines of development, insofar as that record is currently known.

The resolution of monophyletic versus polyphyletic origin of living things is, in the last analysis, related to the origin of life itself. If, in fact, life arose only once in time and substance, as some biologists have concluded, then the present diversity of living things is unequivocally monophyletic in origin. Strong evidences of the unity of living things support the hypothesis of monophyletic origin. These evidences include organizational manifestations at several levels, such as general cellular organization, mitochondria, Golgi bodies, plastids, vacuoles, endoplasmic reticulum, other membranes, and flagellar structure; and chromosomal, nuclear, and cellular organization and replication among Eukaryotes. Furthermore, basic biochemical patterns in metabolism and the widespread occurrence of such compounds as starch, cellulose, and chlorophyll *a* in a vast group of plants are evidences of common origin. On the other hand, those who speculate that life may have originated more than once would interpret the unifying phenomena listed above as evidences of parallel development. Although speculations regarding such ultimate questions as the origin of life would have been considered well beyond experiment at one time, advances in biochemical techniques, such as syntheses of DNA from its components in cell-free systems, indicate that resolution of the origin of life is more susceptible of solution than we used to think.

Finally, before leaving the subject of classification for the present, the writers cannot emphasize too strongly the futility of attempting to memorize the system of classification (Table 3-1) at this point. The several divisions, together with their component taxa, will be learned more readily in connection with the discussion of their morphology, in each case.

DISCUSSION QUESTIONS

1. Should generic names of plants always be capitalized?

2. Do the names of plants have meaning? Where can one find their meaning?

3. What syllables usually end the name of the plant family? the order? Are these endings subject to change by individual botanists?

4. Where can one find a printed copy of the International Rules of Botanical Nomenclature?

5. Distinguish between the terms "monophyletic" and "polyphyletic."

6. Inasmuch as the formal category "Algae" has been abandoned, may one still speak of algae? Explain.

7. What is the meaning of "phyceae" in the algal classes and divisions in Table 3-1?

8. Who has the "right" to propose modifications in plant classification?

Chapter 4
INTRODUCTION TO THE ALGAE; DIVISION CHLOROPHYTA

DEFINITION OF ALGAE[1]

As noted in the preceding chapter, the term "Algae" has been abandoned as a formal taxon or category in modern classifications of the plant kingdom (inside back cover). However, the word is still useful in grouping informally the series of algal divisions, which, in spite of marked diversity, have certain characteristics in common that distinguish them from other chlorophyllous plants. It is necessary to cite several technical aspects of the reproduction of algae if we are to delimit them reliably from the other chlorophyllous plants (Fig. 4-1). Algae differ from the latter in that (1) the organisms themselves (as in some unicellular algae) may function directly as sex cells, or **gametes**, and unite in pairs to form **zygotes** (Fig. 4-1A); (2) the gametes may be produced within specialized, unicellular **gametangia** (Fig. 4-1B); (3) the gametes may develop in multicellular gametangia, every cell of which is fertile (that is, every cell produces a gamete) (Fig. 4-1C); (4) the spores develop either within unicellular containers called **sporangia** or in multicellular ones in which every cell is fertile.

What of those organisms that are algal in organization but either lack sexual reproduction or have a different method of genetic interchange from those referred to in defining algae? The large group of blue-green algae (Chapter 2) belongs to this category, and because of their cellular organization (p. 21), they are by some biologists (e.g., Stanier et al., 1971) considered not to be algae but rather bacteria. However, there remain many other eukaryotic algae, mostly green algae, in which sexuality has not been demonstrated or in which it does not occur. Inasmuch as sexuality may be present in some species and absent in other species of a single genus (e.g., *Chlamydomonas*, p. 51), those that lack it are classified as algae on the basis of their similar morphology. Algal reproduction will be discussed more fully in the chapters that follow.

---

[1]For more comprehensive accounts of the algae and their biology, consult Smith (1950), Stewart (1974), Lewin (1976), and Bold and Wynne (1985).

A

B

C

D

E

F

HABITAT

Algae are largely aquatic in habitat, occurring in waters of a wide range of salinity, in fresh, brackish, and marine waters, and in brines so concentrated that their solutes are crystallizing. Some algae (certain unicellular types and *Entromorpha*, for example) can tolerate diversified salinities, while others (desmids, *Spirogyra*, and *Oedogonium*, for example) are restricted to fresh waters.

An increasingly large number of algae (especially green algae and diatoms) have been found to be regular inhabitants of the soil surface and considerable depths below it (Bold, 1970; Metting, 1981), while many others occur on moist pebbles and rocks, tree bark, woodwork, crushed-stone roofs, and even on rocks that are subject to prolonged desiccation. Investigation of atmospheric dusts has revealed that these carry a rich algal flora, derived largely, of course, from soil. There is some indication that algae (and other microorganisms) may multiply in clouds (Parker, 1970*b*).

A number of algae, the so-called **cryoflora**, live on and within long-persistent patches of snow (Wharton and Vinyard, 1983). Several species live within and among the cells of other plants and animals; special instances of these are described later in this and in other chapters, as are other algae of more bizarre habitats.

Aquatic algae may be attached to rocks, wood, or other aquatic vegetation or may be free-floating. Attached algae in marine habi-

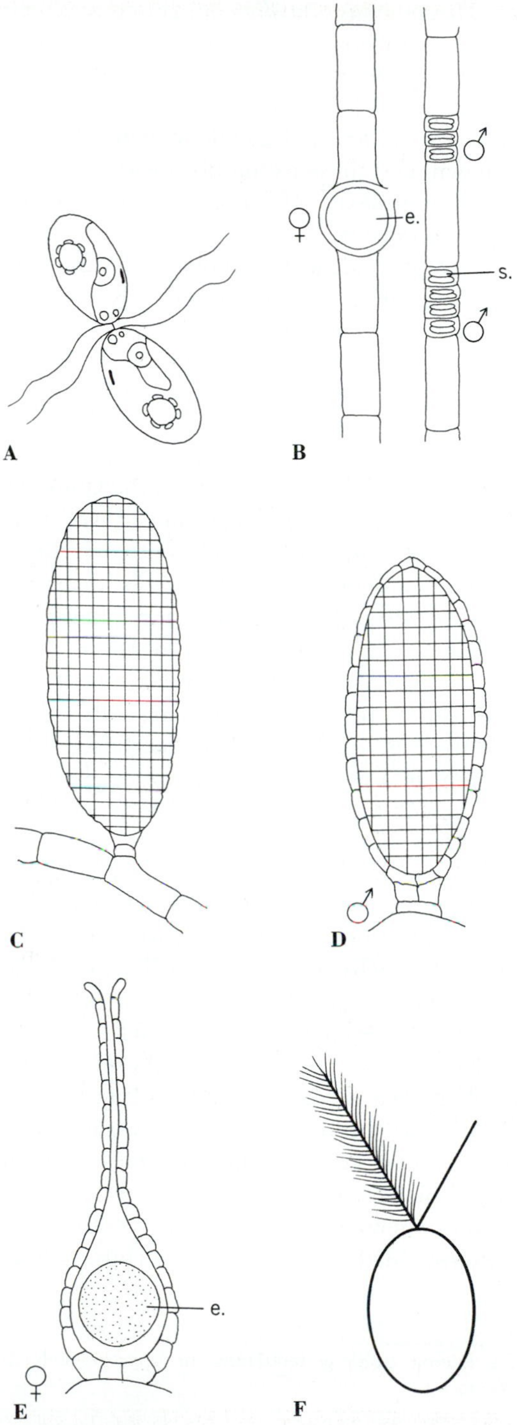

FIGURE 4-1
Sexual reproduction in algae and nonalgae, diagrammatic. **A.** Uniting gametes (individual organisms) of a unicellular motile alga. **B.** Alga with unicellular male (♂) and female (♀) sex organs. **C.** Multicellular sex organ of an alga, every cell of which is fertile. **D,** Male and, **E.** female sex organs of nonalgae. **F.** Whiplash and tinsel flagella, diagrammatic. Note peripheral sterile cells; in D and E. e., egg; s., sperm.

tats often exhibit orderly zonation when their substrates are exposed at low tides. A number of genera, such as *Porphyra*, *Enteromorpha*, *Fucus*, and *Ascophyllum*, grow in the intertidal zone; they are thus subject to, and able to withstand, periods of desiccation. In contrast, contrast, other attached algae, the kelps and *Polysiphonia*, for example, are usually ublittoral.

A vast array of unicellular, colonial, and delicate filamentous algae occurs permanently suspended in water, where they may be associated with bacteria, fungi, protozoa, and other minute animals to form a community known as the **plankton** (Gr. *planetis*, wanderer). Planktonic algae, under unusually favorable conditions, may multiply rapidly and become strikingly abundant as **water blooms** (Fig. 2-13), of which "red tides" are an example. A variety of algae may be present in blooms, or only a single organism may predominate. Certain blue-green algae (Chapter 2) and dinoflagellates, when concentrated in blooms, are toxic (Gorham, 1964; Steidinger, 1973). Planktonic algae are of tremendous importance as the basis of the food chain for larger animals in aquatic environments. Thus, it has become routine practice to add commercial fertilizer to tanks and ponds stocked with fish; this enhances the growth of algae, which, in turn, augments the basic food supply for the smaller animals on which the fish feed. Other aspects of the economic importance of algae are discussed in Chapter 10.

ORGANIZATION OF THE PLANT BODY

As a group, algae are paradoxical in that they include both minute organisms and probably the largest chlorophyllose plants. Thus, certain species of *Chlorella* approach the larger bacteria in size (2–5 μm),[2] while the smallest known alga, *Micromonas pusilla*, a green alga, is $1 \times 1.5\,\mu m$. By contrast, certain species of kelp may attain a length of 65 m, a length equal to that of larger forest trees.

Between 19,000 and 25,000 species of algae are known, and these exemplify a wide range in size and complexity of form. Several types of organization occur.

The simplest algae morphologically are unicellular (e.g., Figs. 4-2, 4-17), the cells cohering in groups only temporarily after cell division, after which they separate. In another series the cells undergo divisions to form incipient tissue complexes (Fig. 4-13D). A different level of organization is the colonial type (e.g., Figs. 4-8, 4-9), which probably arose through the failure of cells to separate at the conclusion of cell division. Colonies may be undifferentiated (Figs. 4-8, 4-20), or they may contain more than one kind of cell (Fig. 4-9), exemplifying differentiation, specialization, and division of labor.

Occurrence of cell division predominantly in one direction results in chains of cells called **filaments** (Fig. 4-25). Division of certain cells of a filament in a new direction produces branching (Fig. 4-30). In a number of algae that are filamentous in their juvenile stages, abundant cell divisions in two or more planes result in a leaflike or membranous structure one or several layers of cells thick (Figs. 4-40, 4-41A), the cells forming **parenchyma** tissue. Finally, in a large number of green (and a few yellow-green) algae, the plant body is composed of a vesicle, or tube, with few, if any, septations. In these, a large central vacuole is surrounded by a thin layer of multinucleate protoplasm. These vesicles and tubes (Figs. 4-42, 4-44) may be measured in terms of inches and feet, respectively, and are known as **coenocytes** (Gr. *coenos*, common, + Gr. *kystos*, bladder; hence, cell). Thus, among

---

[2]A micron (μm) μ 0.001 mm, or approximately 1/25,000 inch.

algae, five major types of body form have developed: unicellular, colonial, filamentous, membranous, and tubular or coenocytic.

CLASSIFICATION OF ALGAE

The system of classification of the algae varies with the classifier. Current systems of classifications recognize between four and nine divisions. This variation is occasioned, in part, by the niche assigned to the blue-green algae. Although historically and traditionally the blue-green algae have been considered a series of the algae and thus long classified with them, as was done in the first three editions of this book, the overwhelming evidence from electron microscopy and biochemistry now indicates that their affinities are with the bacteria (Stanier et al., 1971; Stanier and Cohen-Bazire, 1977). Accordingly, the prokaryotic organisms (blue-green "algae" and bacteria) have been classified together in the Superkingdom Prokaryonta and described in Chapter 2.

Lewin (1975, 1977, 1984) discovered and described a remarkable organism, *Prochloron didemni*. This alga is prokaryotic like blue-green algae (p. 21) but, like the green algae, has both chlorophylls *a* and *b* and lacks the phycobilin pigments (p. 20). For this unique organism he proposed a new division of algae, the Prochlorophyta. Lewin's proposal, however, has met with criticism (Antia, 1977).

Another example of divergent opinion occasioning differences in number of algal divisions is the treatment accorded the stoneworts (p.100). In this text they are classified in a division of their own (Charophyta), but in many others they are included as a special class (Charophyceae) of the green algae.

The divisions of algae have been organized with respect to differences in pigmentation, nature of the stored food reserves, chemical nature of the cell wall, and presence or absence of flagella, and when the latter are present, with respect to their number, length, and site of insertion. Two basic types of flagella occur in algal (and fungal) cells, the whiplash and the tinsel types. In the latter the flagellar sheath bears one or more rows of appendages (Fig. 4-1F). Table 4-1 summarizes the algal divisions, as recognized in this text, and their characteristics. It should be emphasized that the characteristics ascribed to the several algal divisions have in many cases been extrapolated from results of studies of relatively few genera. Furthermore, new data are continually being made available regarding the chemistry of algal cell walls, algal pigments, and algal reserve products. Accordingly, as these new data become available, Table 4-1 will require modification. Ragan and Chapman (1978) discussed the evolutionary significance of some of these characteristics.

Of the algal divisions and classes cited in Table 4-1, the major ones are Chlorophyta, Phaeophyta, and Rhodophyta and the Bacillariophyceae of the Chrysophyta. The Chlorophyta will be discussed in the remainder of this chapter, while the others form the subject matter of Chapters 5 to 9. Chapter 10 recapitulates the entire discussion of algae, their role in nature, and their economic significance.

DIVISION CHLOROPHYTA

General features. The division Chlorophyta (Gr. *chloros*, green, + Gr. *phyton*, plant), as presented here, includes a single class of algae, the Chlorophyceae (Gr. *chloros* + Gr. *phykos*, seaweed). The Chlorophyta include all plants usually designated as "green algae" with the exception of the stoneworts—*Chara, Nitella,* and related genera—which in this text are grouped in a separate division, the Charophyta (see Chapter 5).

The green algae are widespread, occurring in both fresh and marine waters, on moist wood

TABLE 4-1
Summary of some algal divisions and their more noteworthy attributes

Division	Common name	Pigments and plastid organization	Stored food	Cell wall[a]	Flagellar number and insertion[b]	Habitat[c]
Chlorophyta	Green algae	Chlorophyll a, b; α-, β-, and γ-carotenes + several xanthophylls; 2–6 thylakoids/band[d]	Starch, resembling that of land plants, i.e., α-1 : 4-glucopyranosides = amylose and glucopyranosides with α-1 : 4 and some 1 : 6 linkages	Cellulose in many = β-1 : 4-glucopyranoside, hydroxyproline glycosides; or wall absent; siliceous and/or calcified in some[e]	1, 2–8, many, equal, apical	f.w., b.w., s.w., t., a.
Charophyta	Stoneworts	Chlorophyll a, b; α-, β-, and γ-carotenes + several xanthophylls; thylakoids variably associated	Starch, resembling that of land plants (amylose and amylopectin)	Cellulose (= β-1 : 4-glucopyranoside)	2, equal, subapical	f.w., b.w.
Euglenophyta	Euglenids	Chlorophyll a, b; β-carotene + several xanthophylls; 2–6 thylakoids/band, sometimes many	Paramylon (= β-1 : 3-glucopyranoside) and fats	Absent	1–3 apical, subapical	f.w., b.w., s.w., a.
Phaeophyta	Brown algae	Chlorophyll a, c; β-carotene + several xanthophylls including fucoxanthin; 3 thylakoids/band	Mannitol, laminaran (= β-1 : 3-glucopyranoside, predominantly)	Cellulose, alginic acid, and sulfated mucopolysaccharides	2, unequal,[h] lateral	f.w. (rare) b.w., s.w.
Chrysophyta	Golden algae (including diatoms)	Chlorophyll a, c (e in some); β-carotenes + several xanthophylls; 3 thylakoids/band	Oil chrysolaminaran (β-1 : 3-glucopyranoside, predominantly)	Cellulose silicon, mucilaginous substances, and some chitin	1–2, unequal or equal, apical	f.w., b.w., s.w., t., a.

TABLE 4-1
(*Continued*)

Division	Common name	Pigments and plastid organization	Stored food	Cell wall[a]	Flagellar number and insertion[b]	Habitat[c]
Pyrrhophyta	Dinoflagellates, in part	Chlorophyll *a, c;* β-carotene + several xanthophylls; 2–3 thylakoids/ band	Starch (probably like land plants), fats and oils	Cellulose or absent; mucilaginous substances	2, one trailing, one girdling	f.w., b.w., s.w., a.
Rhodophyta	Red algae	Chlorophyll *a, d,* (in some); *C,* and *R*-phycocyanin, allophycocyanin; *C-* and β-phycoerythrin; *α-* + β-carotene + several xanthophylls; 1 thylakoid/ band	Floridean starch[f] (glycogenlike) (glucopyranosides with α-1 : 4 and some 1 : 6 linkages)	Cellulose,[g] xylans, pectin, calcified in some	Absent	f.w. (some), b.w., s.w. (most)

[a]In terms of cell wall chemistry, the vegetative cells have received most attention. Spores, akinetes, dormant zygotes, and other resting stages have not been studied, but it is clear that their walls may contain other substances, e.g., waxes and other nonsaponifiable polymers and phenolic substances. (See also Parker, 1970*b;* Roberts et al., 1982.)

[b]In motile cells, when these are produced.

[c]f.w. = fresh water; b.w. = brackish water; s.w. = salt water; t. = terrestrial; a. = airborne.

[d]Based on Gibbs, 1970.

[e]Others are wall-less or have xylans, mannans, other glucans, some silica, or even protein. Also, nearly all skeletal polysaccharides (cellulose, xylans, mannans) are accompanied by one or more mucilaginous substances (e.g., arabinogalactans, sulfated mucopolysaccharides, etc.).

[f]Stains wine red with iodine.

[g]Except in *Porphyra,* which has mannans and xylans (personal communication, Dr. B. C. Parker; see, however, Gretz et al., 1984).

[h]Except for the uniflagellate sperms of the Dictyotales.

and rocks, and on the surface of and within soil. Many green algae have been recovered from airborne dusts. More than 450 genera and 7500 species of Chlorophyta have been described. They are well represented in the plankton and may occur in bodies of water ranging in size from small temporary pools to oceans. A number of species are epiphytic on other algae, on aquatic flowering plants, and on animals, and still others (*Chlorella,* for example) are endophytic in the cells of certain protozoa, coelenterates, and sponges. Marine species are frequently attached to rocks, pilings, or larger algae, or grow on the sandy bottoms of quiet estuaries, often on shells. Planktonic species occasionally form water blooms.

The Chlorophyta are usually grass-green during their vegetative stages, except for a few species with tannin-containing, purple vacuolar pigments or orange carotenoid pigments, which mask their green color. The cells contain both chlorophylls *a* and *b* as well as α- and β-carotenes and certain xanthophylls (Table 4-1). The predominance of chlorophyll pigments accounts for the typical green color of members of the division. The pigments of the Chlorophyta are restricted to cytoplasmic organelles called **chloroplasts**, which are delimited by double membranes (Fig. 4-2). Each chloroplast encloses a number of flattened, saclike **thylakoids** (Fig. 4-3) associated in groups of two to six. The chloroplasts exhibit a great range of form, varying from large, urnlike structures (Figs. 4-2, 4-6) to planar or twisted ribbons (Fig. 4-35), to minute, lens-shaped bodies (Fig. 4-18). As noted earlier, a number of investigators have suggested that chloroplasts originated as symbiotic blue-green algae within originally colorless, eukaryotic cells (see, e.g., Whitton, Carr, and Craig, 1971; Schnepf and Brown, 1971; Margulis, 1970, 1980; Gibbs, 1981; Whatley and Whately, 1981, 1982, 1984). **Pyrenoids** (Gr. *pyren*, fruit stone) are present in the chloroplasts of most Chlorophyta. They may well be centers of formation of the enzyme amylose synthetase, which combines glucose molecules into starch. Holdsworth (1971) has isolated the pyrenoids of *Eremosphaera* and analyzed their protein. Golgi apparatus, mitochondria, and endoplasmic reticulum are present in the cells of Chlorophyta (Figs. 4-2, 4-3).

Of all the algae, the Chlorophyta (and Charophyta, p. 46) are the most like land plants in that their excess carbon compounds are stored within the plastids as starch. The nuclei of Chlorophyta are similar in organization to those organisms other than Cyanophyta and bacteria in having two-layered, perforate nuclear membranes (Figs. 4-2, 4-3) and one or more nucleoli in addition to DNA. Mitotic nuclear division has been observed in many species. In some, the nuclear membrane remains essentially intact during mitosis, and paired centrioles are present at the apices of the spindles. The nuclei (Figs. 4-2, 4-3, 4-35C, 4-37D) are embedded in colorless cytoplasm and are always centripetal to the chloroplasts.

Cell walls (Roberts et el., 1982), although absent in a few flagellate green algae, are present in a majority of Chlorophyta. Many have somewhat rigid cell walls containing cellulose, and external to this an amorphous, mucilaginous layer. Cellulose is lacking from the tubular (siphonous) green algae (p.91). Small amounts of silicon may be present (*Pediastrum*).

Large central vacuoles occur in the cells of many Chlorophyta, and small **contractile vacuoles**, which rid the cells of excess water, are present universally in the motile cells of all freshwater species.

The Chlorophyta exhibit a wide range and complexity of structure and reproduction. The details will be discussed in connection with the several illustrative genera. The latter exemplify five types of organization: (1) motile unicellular and colonial organisms, (2) nonmotile unicellular and colonial organisms, (3) filamentous organisms, (4) membranous organisms, and

FIGURE 4-2
Chlamydomonas reinhardtii. **A.** Photograph of cell in motion prepared with interference microscope. **B.** Electron micrograph of median longitudinal section of cell. **C.** Transection of flagellum; note two central fibrils surrounded by nine (double) ones. c.l., chloroplast lamella; c.w., cell wall; cy., cytoplasm; f., flagellum; n., nucleus; nu., nucleolus; p.pa., plasma papilla; py., pyrenoid; s., starch grain; w.pa., wall papilla. **A.** ×1900. **B.** ×23,000. **C.** ×100,000. (*Courtesy of Dr. David L. Ringo.*)

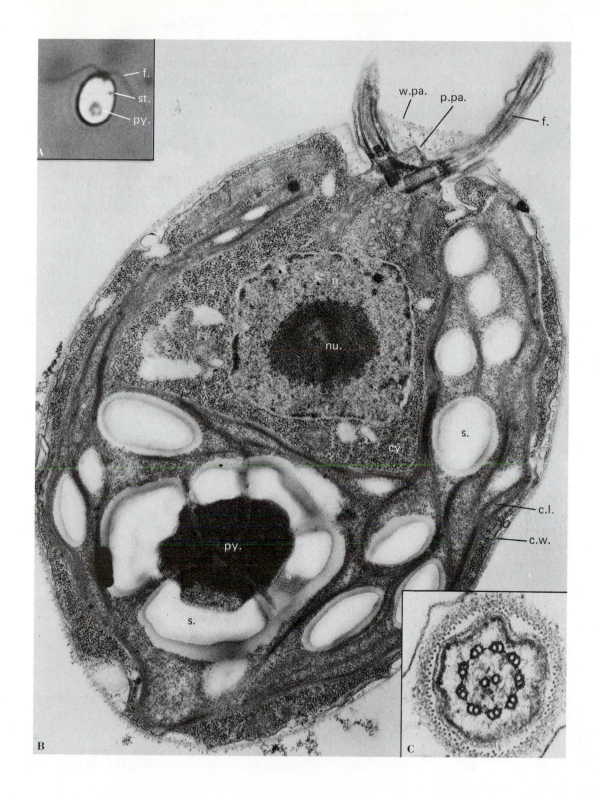

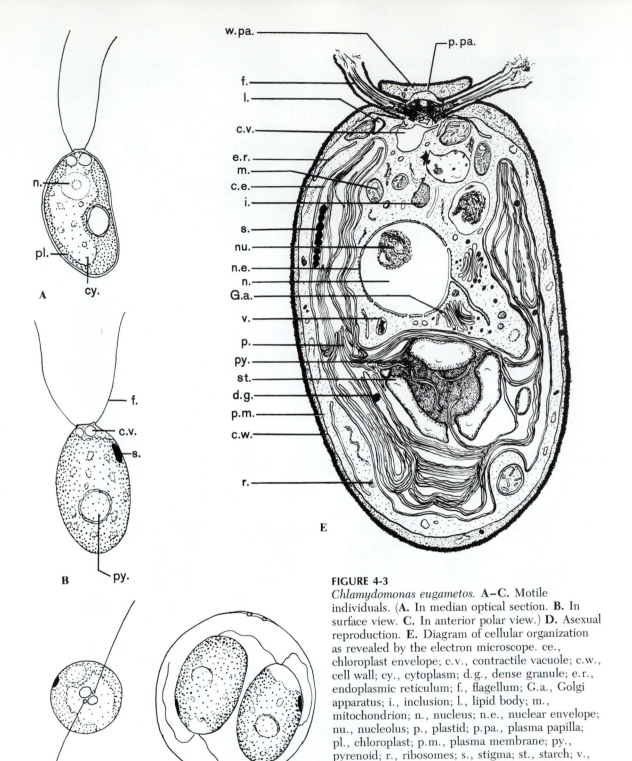

FIGURE 4-3

Chlamydomonas eugametos. **A–C.** Motile individuals. (**A.** In median optical section. **B.** In surface view. **C.** In anterior polar view.) **D.** Asexual reproduction. **E.** Diagram of cellular organization as revealed by the electron microscope. ce., chloroplast envelope; c.v., contractile vacuole; c.w., cell wall; cy., cytoplasm; d.g., dense granule; e.r., endoplasmic reticulum; f., flagellum; G.a., Golgi apparatus; i., inclusion; l., lipid body; m., mitochondrion; n., nucleus; n.e., nuclear envelope; nu., nucleolus; p., plastid; p.pa., plasma papilla; pl., chloroplast; p.m., plasma membrane; py., pyrenoid; r., ribosomes; s., stigma; st., starch; v., vesicle; w.pa., wall papilla. **A–D.** ×2250. **E.** ×8000. (**E.** *Courtesy of Dr. Patricia L. Walne.*)

(5) coenocytic and tubular organisms. The classification of the illustrative-type organisms is presented on p. 97.

I. MOTILE UNICELLULAR AND COLONIAL ORGANISMS

A number of unicellular and colonial green algae are motile throughout their existence by means of flagella. Flagella[3] are protoplasmic extensions of the cell. In the Chlorophyta, the flagella always consist of a sheath surrounding a complex of fibrils. Electron microscopy has revealed a remarkable uniformity of flagellar organization throughout the plant and animal kingdoms in that the flagellum, in most cases, has been shown to consist of two central fibrils surrounded by nine double ones,[4] all surrounded by a sheath (Fig. 4-2C) that is an extension of the cell's plasma membrane.

In having flagella, motile algae resemble certain Protozoa, with which they are sometimes classified. The occurrence of flagellate reproductive cells in the life cycles of many nonmotile algae suggests that the latter may have evolved from motile ancestral precursors. In this connection, two genera, *Chlamydomonas* and *Carteria*, are sometimes thought to be especially significant.

A. UNICELLULAR ORGANISMS

Chlamydomonas

Chlamydomonas (Gr. *chlamys*, mantle, + Gr. *monas*, single organism) is widespread in aquatic habitats and soil and has been recovered from airborne dusts. The structure and reproduction of this organism will be described in considerable detail for a number of reasons: (1)

---

[3]Singular, flagellum (L. *flagellum*, a small whip).

[4]Except bacterial flagella.

It is readily available in pure cultures for laboratory study; (2) many of its attributes are shared by other genera of Chlorophyta; (3) the environment can be manipulated so as to evoke asexual or sexual reproduction as one wills; (4) it is currently the experimental organism in a number of biochemical and genetical investigations.

The cells of most species of *Chlamydomonas* (Figs. 4-2 through 4-5) do not exceed 25 μm in length. The organisms are surrounded by a wall through which two flagella protrude anteriorly. Motility is effected by the lashing movements of these organelles. Each cell contains a single massive chloroplast, which may be urn-, cup-band-, or H-shaped or stellate. The chloroplast may contain one or more pyrenoids, as well as a red pigment body often called the **eyespot**, or **stigma** (Gr. *stigma*, mark or brand). With high magnification, one can frequently observe that there is an area of clear cytoplasm subtended by the concave stigma. It has been suggested that this functions as a primitive lens, and experiments with related organisms containing stigmata indicate that the stigma is indeed a site of light perception. It has been demonstrated that cells of *Chlamydomonas* with stigmata react with greater rapidity to the stimulus of light than those that lack them. The single nucleus lies in the colorless cytoplasm and often is obscured by the chloroplast in living cells, but it may be demonstrated by staining. Two or more **contractile vacuoles** are present near the anterior pole of each cell (Fig. 4-3). There is good evidence that they play a role in the elimination of excess fluids from the cells. Electron microscopy (Figs. 4-2, 4-3) reveals at increased magnification the complexity of the green algal cell as compared with that of prokaryotic cells.

Multiplication of the organism is accomplished by cell division involving nuclear and cytoplasmic division. The process is illustrated in part in Fig. 4-3D. Two or more cellular progeny may arise within a single parent cell by repeated bipartition. The flagella of the parent cell degenerate at the beginning of division. The young cells emerge, after becoming motile within the parent cell, by the enzymatic degradation of the parent cell wall (Schlösser, 1981). The liberated individuals gradually grow to the size characteristic of the species and then divide again.

Under a combination of suitable environmental and protoplasmic conditions not yet completely understood, certain populations undergo sexual reproduction. This process is made manifest in many species by the rapid aggregation of a number of individuals in groups, **clumps,** or **clusters** (Fig. 4-5B, C). In *C. moewusii*, the latter arise as the flagella of compatible individuals in crowded cultures become attached at their tips to those of one or more other cells. The individuals in a clump become paired (Figs. 4-4A, 4-5D) because of a chemical attraction between the flagella. Such pairs are later held together by a delicate apical protoplasmic thread that connects them at the bases

of their flagella (Figs. 4-4A, 4-5D, G); the latter are not entangled, except briefly, after the cells have paired. This connecting strand arises from the union of the plasma papillae (Fig. 4-3) of the paired gametes. As soon as the cells become united by the connecting strand of protoplasm, the paired flagella become free. In *Chlamydomonas moewusii* and in *C. eugametos*, the flagella of only one member of the gametic pair are motile and thus propel the pair in one direction. The cell walls of each member of the pair are ultimately dissolved at the anterior poles, and the protoplasts emerge, gradually uniting to form a single unit (Figs. 4-4B, C, 4-5F, H); the four flagella gradually shorten and are abscised. Soon after union of the naked protoplasts, a new wall is formed around the fusion product. The discarded individual cell walls persist for some time in the vicinity of the fusion cell. Stained preparations of the uniting cells reveal that the two nuclei, which are thus

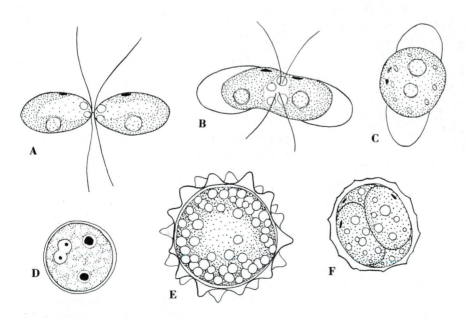

FIGURE 4-4
Sexual reproduction in *Chlamydomonas*. **A.** Pair of isogametes recently emergent from a clump. **B,** **C.** Stages in plasmogamy. **D.** Karyogamy. **E.** Dormant zygote. **F.** Zygote germination. (All except **D,** *Chlamydomonas eugametos.* **D,** *C. chlamydogama.*) ×1275.

brought together in one protoplast, unite to form one large nucleus (Fig. 4-4D). The chloroplasts also unite subsequently.

Electron and phase-contrast microscopy have provided additional data regarding sexual reproduction in *Chlamydomonas*; in *C. reinhardtii*, a fertilization tube connects the gametes (Friedmann et el., 1968), while none is present in *C. moewusii* (Brown et al., 1968; Mesland, 1976; and Blank et al., 1978).

The process just described represents the **sexual reproduction** of *Chlamydomonas*. It is characterized by the union of two cells, the union of their nuclei, and the association within the fusion nucleus of the chromosome complements, hence, genomes, of the two uniting cells. Each of the latter that undergoes union is called a **gamete** (Gr. *gamos*, marriage), or sex cell. The product of the sexual union is a **zygote** (Gr. *zygon*, yoke). In *Chlamydomonas moewusii* and many other algae, the zygote develops a thick wall, which may be variously ornamented (Figs. 4-4E, 4-5I), and undergoes a period of dormancy.

It is obvious that the zygote, a product of the union of two cells and their nuclei, will contain in its nucleus two sets of parental chromosomes and the genes they carry. Whenever this association of two sets of parental chromosomes occurs, there follows inexorably, sooner or later in the life cycle, a phase in which the chromosomes and their genes are redistributed and segregated in different nuclei. The type of nuclear division that accomplishes this is known as **meiosis** (Gr. *meiosis*, diminution). Meiosis involves not only a quantitative reduction of the chromosome number by one-half but a qualitative segregation of genes as well. That meiosis has taken place may often be tentatively inferred from the occurrence of two rapidly successive nuclear divisions and cytokineses in which four daughter cells are produced from a single cell.[5] Cytological and genetical studies of

Chlamydomonas (Maguire, 1976; Triemer and Brown, 1977) have demonstrated that meiosis occurs at the time the dormant zygote germinates. Meiosis in this case, therefore, is said to be **zygotic** with reference to its site of occurrence in the life cycle. The germinating zygote of *Chlamydomonas* gives rise to four[6] motile cells (Figs. 4-4F, 4-5I), which are liberated by rupture and/or dissolution of the zygote wall. These cells are **haploid** as to chromosome constitution, since they have received only a single basic set of chromosomes in meiosis. The zygote itself, having both sets of parental chromosomes before its meiosis, is said to be **diploid**. Accordingly, in the life cycle of *Chlamydomonas*, only the zygote is diploid; all other cells are haploid.

The phenomenon of sexual reproduction is relatively uniform in its essentials at the cellular level in most living organisms. It seems imperative, therefore, to emphasize certain of its significant features. Cytological investigation has demonstrated that the process usually involves four components: (1) the union of two cells (gametes) to form a zygote, a process known as **plasmogamy**; (2) the union of their nuclei, or the process of **karyogamy**; (3) the association of parental chromosomes (and genes) within the zygote nucleus; and (4) their segregation in **meiosis**.

Primitive organisms such as *Chlamydomonas* have been receiving intensive study with respect to their sexual reproduction, inasmuch as the process in unicellular organisms is not obscured or complicated by secondary morphological features. Although the origin of sexual reproduction remains unknown, it is clear that the process is not as indispensable to the maintenance of the species as it is in land plants and in animals, because most unicellular organisms also multiply indefinitely by asexual means.

Figures 4-4A and 4-5F illustrate pairs of

---

[5]As, for example, in spermatogenesis in animals and in sporogenesis in many plants.

[6]In some species of algae, mitosis after meiosis may result in the formation of more than four cells, as in the fungal ascus (p. 733).

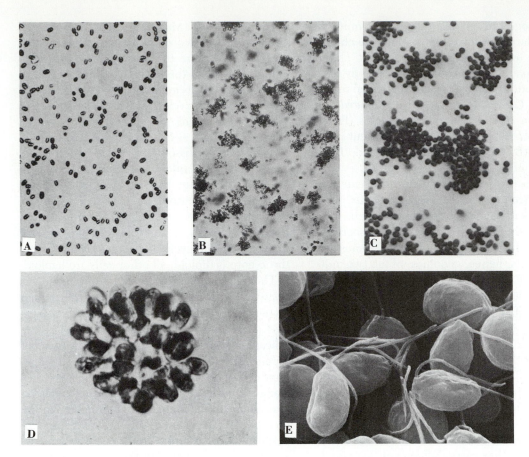

FIGURE 4-5
Micrographs of sexual reproduction in *Chlamydomonas*. **A.** Motile individuals of one mating type. **B–D.** Clump or group formation when two mating types are mixed. **E.** Details of clump; flagellar agglutination 15 sec. after mixing. **F.** Paired gametes (arrow 1) and plasmogamy (arrow 2). **G.** Electron micrograph of apices of paired gametes. **H.** Immature zygote. **I.** Dormant zygotes (left) and zygote germination. **J.** *C. pseudogigantea*, oogamy; note flagellated sperm (arrow) attached to nonflagellated egg. c.s., connecting strand; f., flagellum. (**A–F, H, I,** *C. eugametos*; **G,** *C. moewusii*.) **A, B.** ×125. **C, D.** ×200. **E.** ×3000. **F.** ×500. **G.** ×37,000. **H, I.** ×1200. **J.** ×1700. (**C, E.** *After Mesland;* **G.** *After Lewin and Meinhart;* **J.** *Courtesy Dr. Byron Van Dover.*)

uniting gametes of *C. eugametos*. In this and other species of sexually reproducing algae, the members of the pair are similar in size[7] and in

other morphological attributes and are, therefore, said to be **isogamous** (Gr. *isos*, equal, + Gr. *gamos*, marriage). That isogamy is more apparent than real is indicated by the following: In clonal cultures of isogamous species of *Chlamydomonas*, sexual reproduction may not occur. A **clonal culture** is one in which all the individuals of the population are genetically

[7]In isogamous species, sporadic differences in size of the uniting gametes are occasioned by the union of cells of different age.

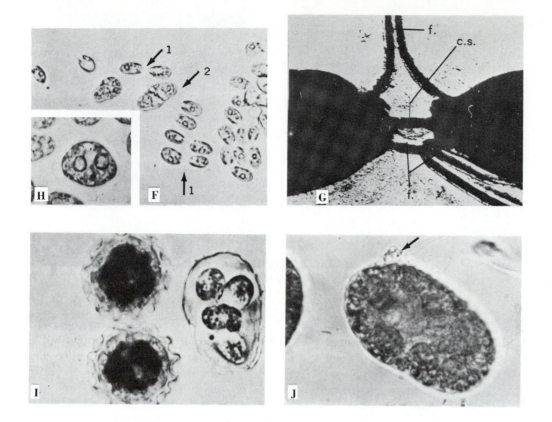

homogeneous, all being derived from a single cell. When some clonal cultures are mixed with other appropriate clonal cultures of compatible mating type, sexual reproduction occurs. Such clones are **heterothallic**, that is, self-incompatible: although they are similar in appearance, the gametes are, in fact, different. Extracts of one of such compatible clonal cultures or of their flagella will cause agglutination or clump formation of the opposite mating type. This is a crude but incontrovertible manifestation of the fact that the isogametes differ chemically. It is evidence, furthermore, that gametic attraction resides, at least in part, in the complementary chemical nature of the flagella (Wiese, 1984). In *C. eugametos* and in *C. moewusii,* as stated earlier, the flagella of only one pair of gametes beat after their union; accordingly, the gametic pair moves in one direction; this also is a manifestation of difference between the gametes.

In other species of *Chlamydomonas* and in other algae (Fig. 4-43B, C) the uniting gametes are always different in size. This results, in part, from a difference in the number of nuclear divisions and cytokineses in the gamete-producing cells. Such gametes are known as **anisogametes** (Gr. *anisos,* unequal, + Gr.

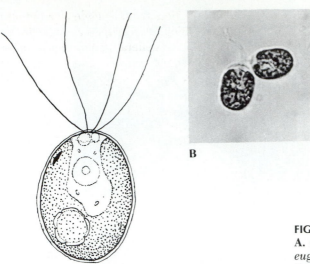

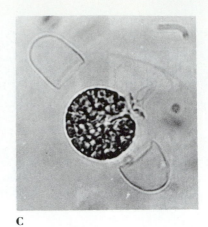

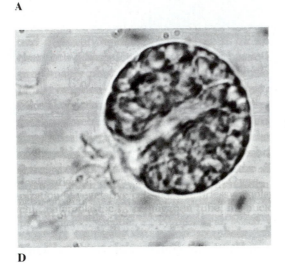

A

B

C

D

FIGURE 4-6
A. *Carteria* sp., motile vegetative cell. **B–D.** *C.
eugametos* var. *contaminans.* **B.** Uniting isogametes.
C. Zygote with gametic walls. **D.** Zygote; note two
sets of four flagella. **A.** ×2300. **B.** ×1200. **C.**
×1700. **D.** ×3300. (**B–D.** *Courtesy of Dr. Byron
Van Dover.*)

other large and nonmotile; this condition is
called **oogamy** (Gr. *oön*, egg, + Gr. *gamos*),
and the smaller gamete is designated the
sperm, or **antherozoid**, and the larger one, the
egg. Within the single genus *Chlamydomonas*,
therefore, there occur among the numerous
species isogamy, anisogamy, and oogamy.

Reference was made above to the fact that in
clonal cultures of certain *Chlamydomonas* species, sexual reproduction would not occur unless the clonal culture were mixed with one of a
compatible mating type. Such species are composed of individuals of two different types, often
designated + and −. In other species of
Chlamydomonas, and in some other algae, sexual reproduction does take place within single
clonal cultures. This indicates that compatible
mating types are present in a single clonal
population; the clone, accordingly, is **homothallic**.

In summary, the gametes may be *morphologically* similar, as in isogamy, or differentiated, as

gamos); their union is called **anisogamy**. Probably on the basis of analogy with sexual reproduction in animals, the smaller of the gametes in
anisogamy is referred to as male and the larger
as female.

Finally, in other species of *Chlamydomonas*, and in other algae (Fig. 4-5J, 4-11H),
the differences in the pairing gametes are more
pronounced, one being small and motile and the

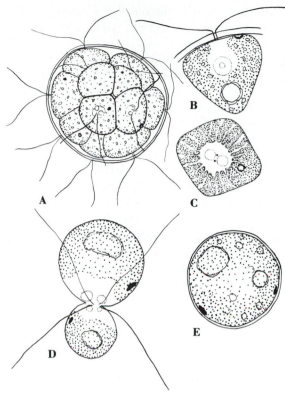

FIGURE 4-7
Pandorina sp. **A.** Mature colony, surface view. **B.** Vegetative cell, median optical longisection. **C.** Vegetative cell, anterior polar view (flagella omitted). **D.** Gamete union (isogamy but gametes illustrated of different age). **E.** Zygote. **A.** ×700. **B–E.** ×1700.

in anisogamy and oogamy. In the algae, when compatible gametes occur in a single clone, such a clone is said to be homothallic. When they occur in separate clones, the condition is called **heterothallism**. When both male and female gametes occur on the same individual, the latter is said to be **monoecious** (or hermaphroditic); if they are on different male and female individuals, the organism is said to be **dioecious**.

The foregoing account of sexuality in *Chlamydomonas* is somewhat protracted and detailed, not only because it is designed to present

specific information about a single alga but also because the biological principles and terminology involved are of general application to other groups of plants and animals.

A number of factors affect the sexual process in *Chlamydomonas* (and other algae). Thus, in several species (e.g., *C. eugametos*), light, adequate CO_2, temperatures below 25°, and depleted nitrogen[8] content of the medium enhance the sexual process. However, in at least one species (Sager and Granick, 1954), gametic union will occur in darkness if the cells are nitrogen-starved.

Carteria

The genus *Carteria* (Fig. 4-6) differs from *Chlamydomonas* in certain ultrastructural details in the wall, chloroplast, pyrenoid, and stigma and in that its cells are quadriflagellate. It is less frequently encountered in nature but occurs in both soil and water. The structure and asexual and sexual reproduction of *Carteria* are fundamentally similar to those of *Chlamydomonas*. Sexual reproduction in *Carteria* is illustrated in Fig. 4-6B–D.

B. COLONIAL ORGANISMS[9]

The origin of the colonial type of algal organization probably resides in the tendency for recently divided cells of unicellular organisms to remain associated after cell division. Examples of this are abundantly evident in populations of *Chlamydomonas* and *Carteria*, among others.

Pandorina (mythology: reproduction like the opening of Pandora's box) and *Volvox* (L. *volvere*, to roll) (Figs. 4-7A, 4-9) are two widely distributed genera that illustrate the motile colonial type of plant body. Both are sometimes

---

[8]Decrease in the concentration of the laboratory culture medium, which usually greatly exceeds that available in natural habitats.

[9]See Coleman (1979), Kochert (1982), and Starr (1984).

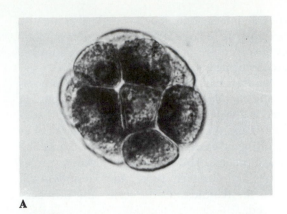

A

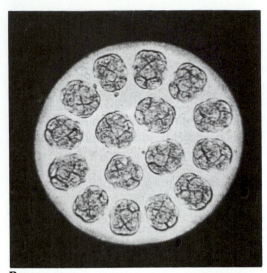

B

FIGURE 4-8
Pandorina morum. **A.** Living colony (flagella not visible). **B.** Daughter-colony formation (India ink preparation). **A.** ×700. **B.** ×375.

present in water blooms. The individual cells, which are included in a common matrix composed of polysaccharides (Crayton, 1982), show many morphological features reminiscent of *Chlamydomonas,* such as massive chloroplasts, stigmata, and contractile vacuoles. Multiplication in all these genera is effected by repeated division of cells of the parent colony into miniature daughter colonies, which are liberated

ultimately by dissolution of the matrix of the parent colony (Figs. 4-8B, 4-11B). The young colonies increase in cell size, but not in cell number, until the dimensions characteristic of the species have been attained. Such a colony in which the cell number is fixed is called a **coenobium**.

Pandorina

The mature colonies of *Pandorina* (Figs. 4-7A, 4-8A) usually consist of 16 cells arranged in an almost solid, ovoidal colony. Each cell is flattened at its anterior pole and narrowed posteriorly (Fig. 4-7B, C). The chloroplast is massive and contains a prominent stigma and basal pyrenoid. In anterior view (Fig. 4-7C), two alternately pulsating contractile vacuoles are visible in the opening of the plastid at the base of the two flagella. The single nucleus is in the colorless central cytoplasm. Although all the cells of the colony are similar in size, a definite polarity is present, as evidenced by the fact that the stigmata of the more anterior cells are larger than those in the posterior part of the colony.

After attaining the maximum size characteristic of the species, the colonies sink to the bottom of the pond or culture vessel and initiate autocolony formation (Fig. 4-8B). An **autocolony** is a miniature of a parental colony. In autocolony formation, each of the parental cells undergoes repeated nuclear and cytoplasmic division until miniature 16-celled colonies are produced. The minute cells of the autocolonies then develop flagella, and the colonies begin to move slowly within the matrix of the parent colony until liberated by its dissolution. Under certain conditions, colonies of *Pandorina morum* exhibit isogamous sexual reproduction (Fig. 4-7D, E).[10] It has been demonstrated that both homothallic and heterothallic clones of *P. morum* occur in nature. Meiosis is zygotic.

Pandorina morum produces a toxin to which

---

[10]See footnote 7.

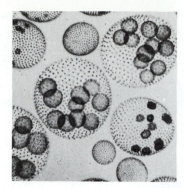

FIGURE 4-9
Volvox aureus. Colonies containing daughter colonies of various ages. ×40.

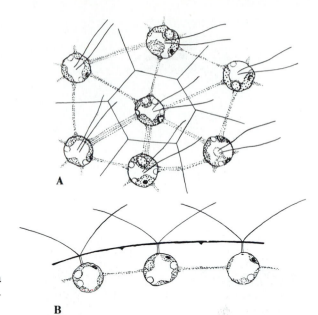

FIGURE 4-10
Volvox aureus. Vegetative cells. **A.** In surface view. **B.** In vertical section. ×700.

some other freshwater algae, aerobic bacteria and a few flowering plants are sensitive (Patterson and Harris, 1983).

Volvox

Volvox is perhaps the most spectacular of the motile colonial Chlorophyceae, since its slightly ovoidal colonies may contain thousands of cells arranged at the periphery of a matrix (Figs. 4-9, 4-10). The plant is readily visible to the unaided eye and has been known for several hundred years. In a number of species of *Volvox*, the protoplasts of the individual cells are connected by delicate protoplasmic extensions (Figs. 4-10, 4-11A). In ontogeny, the young colonies turn themselves inside out (Fig. 4-11C). In *V. aureus*, the cells are entirely similar to one another during the early stages. However, a dimorphism soon becomes apparent in that certain cells enlarge and become slightly depressed beneath the surface (Fig. 4-11F). As the colonies move, it becomes evident that these larger cells lie in the posterior hemisphere. As in *Pandorina*, the stigmata of the posterior vegetative cells are smaller than those in the anterior hemisphere. The enlarged cells, the **gonidia**, alone are capable of dividing into daughter colonies (Figs. 4-9, 4-11B); the remaining cells are purely vegetative and disintegrate when the

adult colony liberates its daughter colonies by enzymatic action (Jaenicke and Waffenschmidt, 1982).

Sexual reproduction in *Volvox* is oogamous. In some species—for example, *V. rousseletii*—the individual colonies are dioecious; that is, they produce either sperms or eggs. In other species (*V. globator*) the individual colonies produce both sperm and eggs; that is, they are monoecious.

In species such as *V. rousseletii*, special cells enlarge and give rise to gametes as the colonies become sexually mature (McCracken and Starr, 1970). The sperms are borne in compressed spherical packets, each of which may contain as many as 512 male cells. As in *V. aureus* (Fig. 4-11E), the intact sperm packets are liberated from the male parent colony and swim to female colonies. There the sperm packets dissolve a hole in the female colony (Fig. 4-11G). The many biflagellate sperm cells dissociate and

59

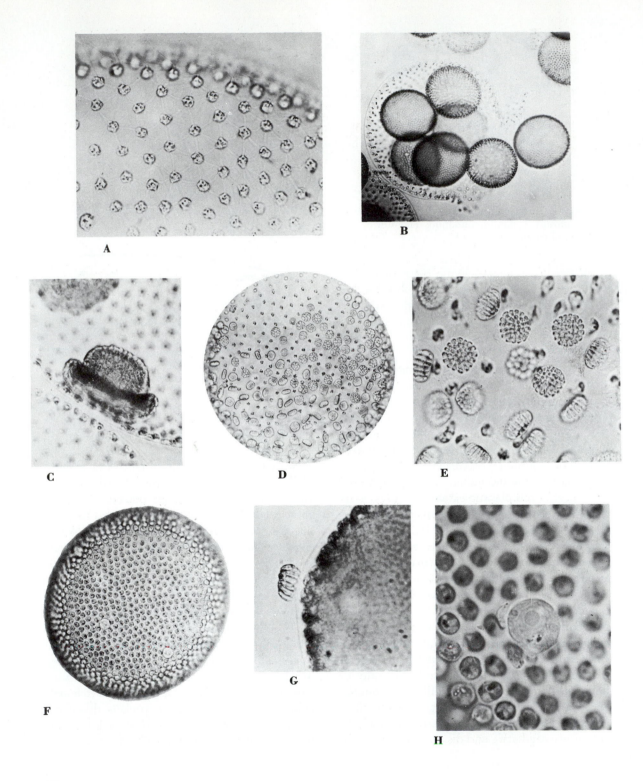

A

B

C

D

E

F

G

H

60

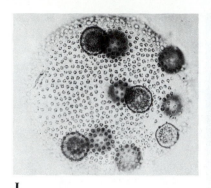
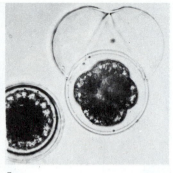
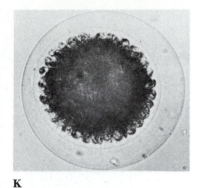

I   J   K

FIGURE 4-11

Volvox aureus. Photomicrographs of living colonies. **A.** Surface view of colony; note protoplasmic connections. **B.** Autocolony release. **C.** Autocolony undergoing eversion. **D.** Male colony with sperm platelets. **E.** Sperm platelets (colonies) more highly magnified; note flagella. **F.** Colony with gonidia (daughter-colony initials) which may function as eggs. **G.** Sperm colony approaching female colony. **H.** Fertilization; note sperm adpressed to egg. **I.** Female colony with mature zygotes. **J, K.** Stages in zygote germination. **A.** ×500. **B.** ×100. **C.** ×200. **D.** ×250. **E.** ×700. **F.** ×250. **G.** ×800. **H.** ×900 **I.** ×250. **J, K.** ×500. (*Courtesy of Dr. William A. Darden.*)

penetrate to the vicinity of the eggs so that **fertilization** (oogamous gametic union) takes place within the female colony. The zygotes develop thick walls and undergo a period of dormancy, during which they are set free by disintegration of the female colony in which they are formed (Fig. 4-11I).

Darden (1966, 1970), Kochert (1968, 1975), McCracken and Starr (1970), Starr (1969, 1971), and Starr and Jaenicke (1974) investigated the control of sexual reproduction in various species of *Volvox.* In a population of *V. aureus,* the gonidia of young asexual colonies may function as eggs and may be fertilized (Fig. 4-11H) by sperm to form smooth-walled zygotes (Fig. 4-11I, J) (Darden, 1966). In *V. aureus* filtrates from cultures with male colonies induce the formation of male colonies in other populations in which asexual colonies would otherwise have been formed. In *V. carteri,* an inducing sub-

stance from male populations evokes the formation of male colonies in the male strain and female colonies in the female strain (Kochert, 1968; Starr, 1969). The active agent from the male populations is glycoprotein, and in the strain studied by Starr and Jaenicke (1974) it is 100% effective at concentrations less than 3×10^{-15}M.[11]

Other colonial organisms

In addition to *Pandorina* and *Volvox,* there are other types of motile coenobia that vary in form and arrangement of the cells. Thus, in *Gonium* (Fig. 4-12A) the organism is a slightly curved, platelike colony. In *Eudorina* (Fig. 4-12B), *Pleodorina* (Fig. 4-12D), and *Platy-*

---

[11]For further research on sexuality in *Volvox,* see McCracken and Starr (1970), Kochert (1975), and Starr (1971).

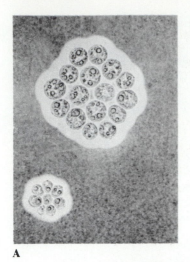

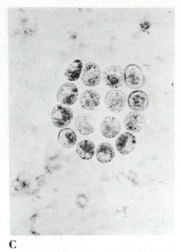

A

B

C

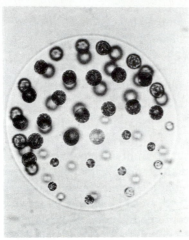

D

FIGURE 4-12
Miscellaneous coenobic, motile green algae. **A.**
Gonium multicoccum. **B.** *Eudorina elegans.* **C.**
Platydorina caudata. **D.** *Pleodorina californica.* **A.**
×475. **B, C, D.** ×600. (**A.** *Courtesy of Dr. Richard
C. Starr.*)

II. NONMOTILE UNICELLULAR AND COLONIAL ORGANISMS

A. UNICELLULAR ORGANISMS

dorina (Fig. 4-12C) the cells are arranged in alternating superficial rings within the colonial matrix, but in *Platydorina* the spherical colony becomes flattened during ontogeny. In *Pleodorina californica* the cells in the anterior half of the coenobium are smaller than those in the posterior half, and only the latter are usually capable of functioning as gonidia. In sexual reproduction, *Gonium* is isogamous, while *Platydorina, Eudorina,* and *Pleodorina* are anisogamous. All apparently undergo zygotic meiosis.

Of the vast assemblage of commonly encountered unicellular, nonmotile green algae, two series may be distinguished on the basis of whether or not they produce flagellated cells in their life cycles. The first series produces flagellated cells called **zoospores**. In the second series, zoospores are not produced, but, instead, asexual reproduction is accomplished by autospores (p.67) or autocolonies (p.70). In this second series, zoospores are absent and flagellated gametes are produced by a very few species (see p.70).

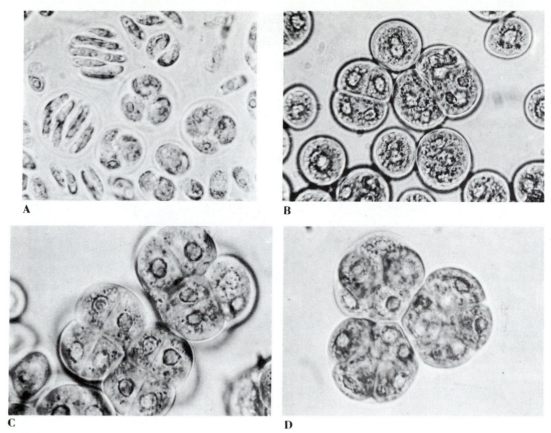

FIGURE 4-13
Tetracystis sp. **A.** Stages in zoosporogenesis. **B.** Vegetative cells in tetrad formation. **C.** Mature tetrads. **D.** Tetrad complexes. ×560.

1. ZOOSPORE PRODUCERS

Tetracystis

The soil-inhabiting alga *Tetracystis* is significant in its combination of seemingly primitive and advanced[12] attributes. *Tetracystis* produces

biflagellate zoospores (Fig. 4-13A); the structure of the latter corresponds in all respects to that of *Chlamydomonas*. After motility, the zoospores lose their flagella, and the cells enlarge and sooner or later become partitioned so that two or four nonmotile daughter cells are formed (Fig. 4-13B, C). This cell partitioning is similar to that in multicellular plants in that the parental cell walls persist after the daughter cell walls have formed. The component cells of the tetrads enlarge and may again divide into tetrads, and this process may continue, thus giving rise to tissuelike complexes of cells (parenchyma) (Fig. 4-13D). We see in this unicellular green alga, accordingly, the potentiality of forming paren-

---

[12]In evolutionary biology, characteristics present in ancient organisms, or those clearly little modified from those of such organisms, are said to be "primitive." Modifications of primitive attributes are designated as "advanced," "specialized," or "derived." Secondary simplification of an attribute is referred to as "reduction," and the simplified attribute as "reduced."

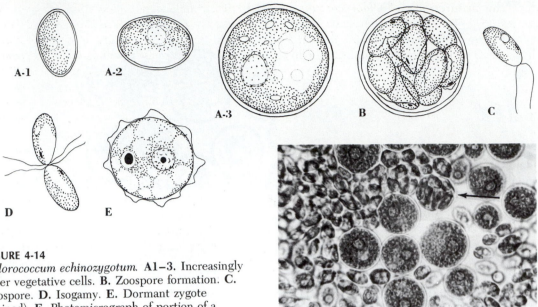

FIGURE 4-14
Chlorococcum echinozygotum. **A1–3.** Increasingly older vegetative cells. **B.** Zoospore formation. **C.** Zoospore. **D.** Isogamy. **E.** Dormant zygote (stained). **F.** Photomicrograph of portion of a population; note vegetative cells of various ages and zoosporangium (arrow). **A–E.** ×1500. **F.** ×500.

chymatous[13] aggregations of cells such as occur in membranous green algae (p.91) and in the nonalgal groups of chlorophyllous plants. Incipient tissue formation was a highly significant evolutionary modification. These cellular complexes of *Tetracystis* may dissociate into smaller cell clusters or even to the unicellular level by degradation and/or splitting of old cell walls. The individual cells of the complex may form zoospores (Fig. 4-13A) either before or after dissociation.

In some species of *Tetracystis*, the zoospores may function as isogamous gametes that unite to form zygotes. This occurs in many algae. Whether we consider such zoospores to be facultative gametes or to be gametes that may develop parthenogenetically[14] if they fail to

unite with another gamete seems immaterial. Sexuality in many algae such as *Tetracystis* seems to be incipient or sporadic rather than obligate.

Chlorococcum

Chlorococcum (Gr. *chloros*, green, + Gr. *kokkos*, berry) (Fig. 4-14) also is an inhabitant of fresh water and soils (both undisturbed and cultivated). Thirty-six species are known and available in culture.[15]

Chlorococcum is difficult to distinguish from other nonmotile, spherical unicellular Chlorophyta, unless one cultivates it in unialgal cul-

---

[13]"Parenchymatous," adjective of "parenchyma," plant tissue composed of thin-walled living cells.

[14]Parthenogenesis is the development of a gamete, without uniting with another gamete, into a new individual.

---

[15]Cultures are populations of microorganisms maintained in the laboratory. Unialgal cultures contain only one species of alga, although bacteria, fungi, protozoa, or all three may be present. Axenic or pure cultures contain populations of only one species of organism. Clonal cultures of a given organism are those in which the population has risen from one individual.

tures. The mature cells of *Chlorococcum* are spherical (Fig. 4-14), unless they have become polyhedral by mutual compression. Each has a cell wall and a protoplast containing a hollow, spherical chloroplast, usually with one aperture. One or more pyrenoids are embedded in the chloroplast, opposite the aperture.

Division of one cell into two or four nonmotile daughter cells, as in *Tetracystis* (Fig. 4-13), is absent in *Chlorococcum*. Instead, the vegetative cells divide and form a number of biflagellate zoospores (Fig. 4-14B), which, again, are almost identical with cells of *Chlamydomonas*. A nonmotile cell that produces zoospores is called a **zoosporangium** (Fig. 4-14B, F). After a period of motility, the duration of which is affected by such environmental factors as light intensity, temperature, and composition and concentration of the culture medium, the zoospores aggregate in the most brightly illuminated portion of the culture vessel, lose their flagella, and grow into new vegetative cells (Fig. 4-14, A-1 through A-3). Movements by organisms to or away from stimuli are **taxes**; the movement of the zoospores of *Chlorococcum* in the present instance exemplifies **positive phototaxis**. When the zoospores settle, the stigma disappears, but in some species the contractile vacuoles persist. Under certain conditions, potential zoospores are not freed from the parental cells but develop within directly into young, nonmotile vegetative cells, or **aplanospores**. The latter rupture the parental wall as they grow and thus are freed.

In *Chlorococcum*, *Tetracystis*, and other algae, production of zoospores by nonmotile vegetative cells is interpreted as a reversion to a primitively motile condition. This might be cited as an example of the biogenetic law that states that ontogeny recapitulates phylogeny.

In some species of *Chlorococcum*—*C. echinozygotum*, for example—the zoospores may function as isogametes and unite to form spiny-walled dormant zygotes (Fig. 4-14D, E). These give rise upon germination to four zoospores, which, upon liberation, develop into vegetative cells. Meiosis is thought to be zygotic. The capacity of the motile cells of *Chlamydomonas*, *Tetracystis*, and *Chlorococcum* to function either sexually or asexually suggests a primitive grade of development, perhaps incipient sexuality, in contrast to those algae in which gametes and zoospores differ morphologically.

Protosiphon

Protosiphon botryoides (Figs. 4-15, 4-16), a widespread soil alga, is included here because of its interesting morphology and life cycle and the marked modifications thereof evoked by environmental stimuli. *Protosiphon* is a terrestrial alga of cultivated soils (where it often grows intermingled with *Botrydium*, p.135). After rain, it forms dark-green patches, which become orange-red as the soil dries. The mature plants are saclike, with a single basal, rhizoidal protuberance that penetrates the substrate (Fig. 4-15A). The upper, bulblike portion contains alveolar cytoplasm with a diffuse chloroplast containing many pyrenoids; the sacs are multinucleate (coenocytic) and, when adequate nitrogen is available, may exceed 1 mm in length.

When it rains, or when agar cultures are submerged, the sacs form numerous biflagellate zoospores (Fig. 4-15B, C, 4-16-2), which, here again, may function as gametes. These zoospores, as in *Tetracystis* and *Chlorococcum*, may be transformed into aplanospores if the moisture level falls.

On the other hand, sacs on drying soil or agar become subdivided into a small number of large, multinucleate cells called **coenocysts** (Figs. 4-13, 4-15D, 4-16). The walls of these thicken as the moisture level falls, and the chlorophyll is replaced or masked by an orange-red carotenoid pigment, the chemical nature of which has not been determined. The coenocysts (and thick-walled zygotes) retain their viability during long periods of desiccation. When moisture again becomes available, they may become

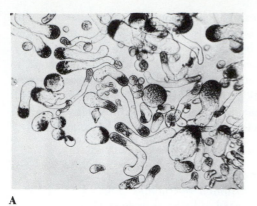

A

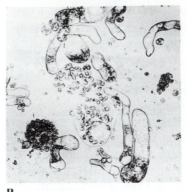

B

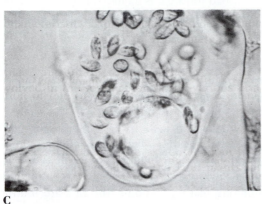

C

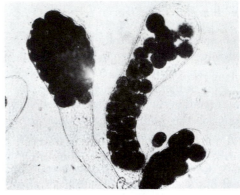

D

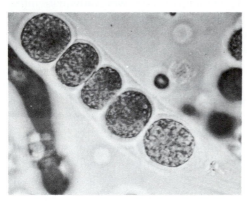

E

FIGURE 4-15
Protosiphon botryoides. **A.** Various stages of sac development. **B, C.** Zoospore formation. **D, E.** Coenocyst formation. **A, B.** ×25. **C.** ×150. **D.** ×100. **E.** ×150.

green and develop directly into new sacs, or the coenocysts may form zoospores or gametes. The complicated life cycle, as it is determined by availability of moisture, is summarized in Fig. 4-16.

2. AZOOSPORIC ORGANISMS

Chlorella

Chlorella (Gr. *chloros*, green, + L. *ella*, diminutive) (Fig. 4-17) is widespread in fresh and salt water and also in soil. It appears often with surprising rapidity in laboratory vessels in which distilled water or inorganic salt solutions are stored. Like *Chlorococcum*, *Chlorella* is most successfully studied in unialgal cultures. *Chlorella* was the first alga to be isolated and

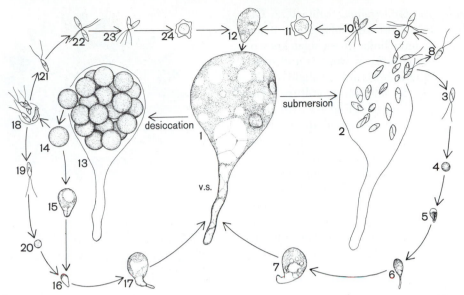

FIGURE 4-16
The life cycle of *Protosiphon botryoides* 1. Vegetative cell or sac. 2. Zoospore/gamete formation and release. 3–7. Stages in asexual development of sacs from zoospores. 8–12. Sexual cycle including gamete union (9, 10). 11, 12. Zygote and its germination. 13. Coenocyst formation. 14–17. Direct development of coenocysts into vegetative sac. 18–20. Asexual cycle of zoospores formed by coenocysts. 18–24. Sexual cycle of gametes from coenocysts. v.s., vegetative sac.

reproductive cells, which have no capacity for motility, are known as **autospores**, because they resemble, in miniature, the mother cells that produce them. Sexual reproduction has not been observed in *Chlorella*. A number of species of *Chlorella* have been grown in pure culture and have provided the material for experimental studies of photosynthesis.

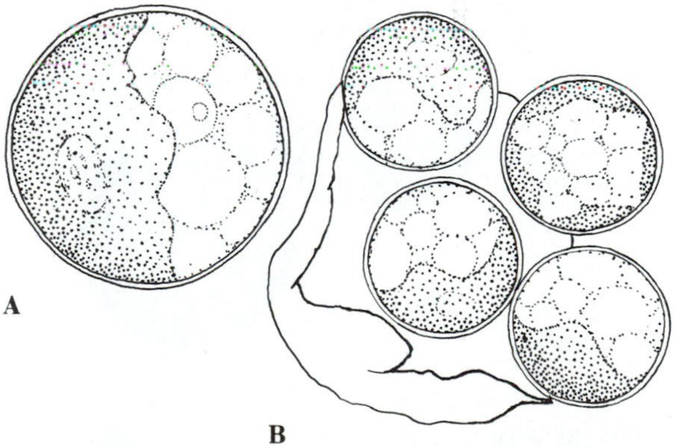

grown in axenic culture; this was accomplished in 1890 by the Dutch microbiologist Beijerinck. The cells of most species of *Chlorella* are minute green spheres (Fig. 4-17) in which the details of cell structure are seen best under high magnification. The protoplast is composed of a cuplike chloroplast, which may or may not contain a pyrenoid, and of colorless central cytoplasm in which the minute nucleus is embedded. A series of bipartitions may occur, forming four or eight protoplasts endogenously. These develop delicate cell walls, and after they have begun to enlarge, they are liberated by rupture of the mother cell wall (Fig. 4-17B). Such asexual

FIGURE 4-17
Chlorella sp. **A.** Vegetative cell. **B.** Autospore liberation. **A, B.** ×1850.

Eremosphaera (Gr. *eremos*, solitary, + Gr. *sphaira*, ball) (Fig. 4-18) is one of the largest and most spectacular unicellular green algae known. It occurs on the bottoms of swamps and quiet ponds in which the water is at least slightly acid. It grows readily in laboratory culture. The individual cells of *E. viridis* (Fig. 4-18) are large enough to be visible to the unaided eye. Each contains many small, pyrenoid-bearing chloroplasts and a prominent, central nucleus suspended by threads of streaming colorless cytoplasm. Reproduction is by the division of the parent cell into two, four, or eight (rarely) nonmotile daughter cells (autospores) or by union of biflagellate, almost colorless sperms and autosporelike cells that function as eggs.

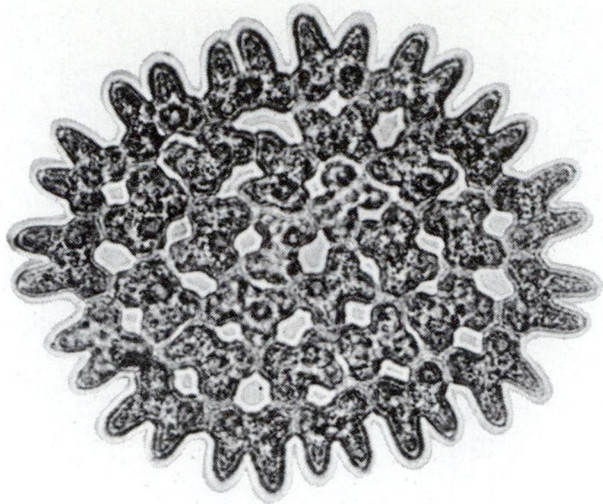

FIGURE 4-19
Pediastrum duplex. ×300.

B. COLONIAL TYPES

The nonmotile colonial Chlorophyta parallel the nonmotile unicellular forms in that two series of genera are known. In one, represented here by *Pediastrum* and *Hydrodictyon*, flagellate reproductive cells (zoospores and/or gametes) occur, while in the other, illustrated here by *Scenedesmus* and *Coelastrum*, motile cells are *usually* absent.

1. ZOOSPORE PRODUCERS

Pediastrum and *Hydrodictyon*

The coenobic colonies of *Pediastrum* (Gr. *pedion*, plane, + Gr. *astron*, star) (Fig. 4-19) grow on the bottoms of quiet pools and lakes as well as in their plankton and may readily be grown in laboratory culture. The coenobia of *Pediastrum* are flat plates. *Pediastrum* reproduces by zoospores, which form daughter colonies within a vesicle released from the parent cell. It also undergoes isogamous sexual reproduction.

Hydrodictyon reticulatum (L. *hydro*, water, + Gr. *dictyon*, net) (Fig. 4-20), commonly known as the "water net," often appears in great abundance in pools, lakes, and quiet streams. The mature colonies are composed of large cylindrical cells joined together in polygonal configurations, the whole colony being cylindrical. The young cells (Fig. 4-20A, B) are uninucleate and delicate green. They ultimately

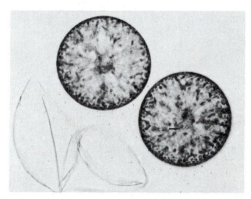

FIGURE 4-18
Eremosphaera viridis. Two recently released young vegetative cells (autospores). ×2000. (*Courtesy of Dr. Richard L. Smith.*)

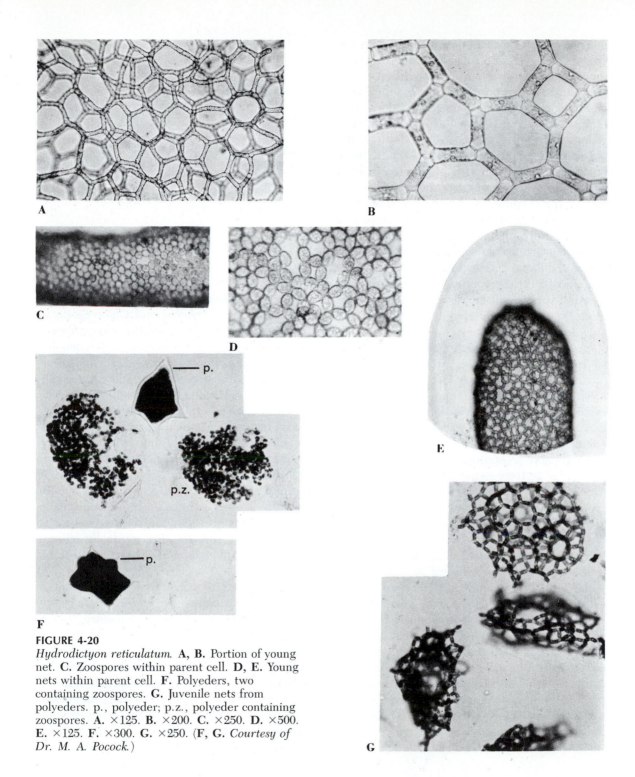

FIGURE 4-20

Hydrodictyon reticulatum. **A, B.** Portion of young
net. **C.** Zoospores within parent cell. **D, E.** Young
nets within parent cell. **F.** Polyeders, two
containing zoospores. **G.** Juvenile nets from
polyeders. p., polyeder; p.z., polyeder containing
zoospores. **A.** ×125. **B.** ×200. **C.** ×250. **D.** ×500.
E. ×125. **F.** ×300. **G.** ×250. (**F, G.** *Courtesy of
Dr. M. A. Pocock.*)

69

enlarge many times and develop numerous nuclei and large central vacuoles; the latter force the cytoplasm into a peripheral position.

In asexual reproduction, the mature cells cleave into smaller and smaller portions until uninucleate segments result (Fig. 4-20C). Marchant and Pickett-Heaps (1970, 1971, 1972) and Hawkins and Leedale (1971). Each of these functions as a zoospore (Fig. 4-21A). As motility abates, the zoospores are arranged in groups of four to nine, typically six, within the cylindrical parent cell (Fig. 4-20D, E), which serves as a mold for the young net of the next generation. After the flagella have disappeared, the cells begin to enlarge and assume a cylindrical form (Fig. 4-20A, B). By continuous increase in cell size and rupture of the parent wall, nets more than 75 cm in length may develop under uncrowded conditions.

Sexual reproduction is isogamous. Unlike the zoospores, which they resemble morphologically, the gametes are liberated from the parent cells. Meiosis is zygotic, and the germinating zygotes develop four zoospores, which grow into nonmotile, polyhedral cells known as **polyeders** (Fig. 4-21B). These enlarge, undergo cleavage and zoosporogenesis, and liberate a number of actively swimming zoospores within a gelatinous vesicle (Fig. 4-20F). These zoospores arrange themselves as a hollow sphere (which may be flattened), or as a flat plate, lose their flagella, and grow into cylindrical cells typical of the adult plant. This juvenile colony (Fig. 4-20G) is not cylindrical as in the adult, although its component cells are cylindrical. Of great interest is the fact that its marginal cells may each bear two *Pediastrum*-like protuberances (Fig. 4-21C). These phenomena often are interpreted as evidence of a common ancestry for *Hydrodictyon* and *Pediastrum*. In this connection, it is of interest that two other species of *Hydrodictyon* have adult colonies that are flattened rather than cylindrical, probably because they lack asexual reproduction.

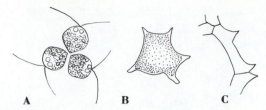

FIGURE 4-21
Hydrodictyon reticulatum. **A.** Motile zoospores. **B.** Polyeder. **C.** Marginal cell of juvenile net; note *Pediastrum*-like protuberances. **A, C.** ×700. **B.** ×300.

2. AZOOSPORIC ORGANISMS

Scenedesmus

With respect to zoospores, *Scenedesmus* stands in relation to *Pediastrum* and *Hydrodictyon* as *Eremosphaera* and *Chlorella* do to *Chlorococcum*. As in *Chlorella*, zoospores are not produced in *Scenedesmus*. This coenobic alga is ubiquitous, occurring abundantly in almost every freshwater habitat and occasionally in soil. It consists of a colony comprised of four or more elongate cells united laterally (Fig. 4-22). In some species the terminal cells have spinelike processes. The uninucleate cells have a parietal chloroplast containing a single pyrenoid. Reproduction, which in many species is entirely asexual, is by the formation of **autocolonies** within each cell of the adult (Fig. 4-22B). These are liberated by the rupture of the parent cell wall and then gradually achieve the size and ornamentation characteristic of the species.

It has been demonstrated that isogamous sexual reproduction occurs under certain conditions in *S. obliquus* (Trainor and Burg, 1965). Clonal cultures are heterothallic, and the two compatible mating types must be mixed before clump formation and gamete union occur. The zygotes enlarge and later germinate to form 40 or more cells. When these are taken into clonal culture, the clones are unisexual; this is genetic evidence of zygotic meiosis.

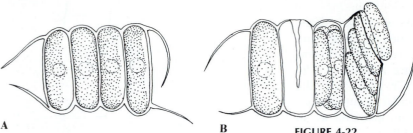

A   B

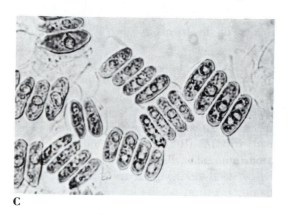

C

filamentous organisms among the green algae;
division of certain cells in the filament in a plane
at 90° or less, with reference to the prevailing
direction, initiates branching.

The genera to be discussed in this section
may be grouped into two categories according
to the criterion of whether or not they produce
flagellate cells or zoospores.

In light of these observations, it has been
suggested that *S. obliquus* and other species
that produce motile cells should be classified
with coenobic algae such as *Hydrodictyon* and
Pediastrum.

A. ZOOSPORE PRODUCERS

The illustrative genera in this group include
both unbranched (*Ulothrix* and *Oedogonium*)
and branched (*Stigeoclonium* and *Cladophora*)
plants.

Coelastrum

The coenobia of *Coelastrum* (Gr. *kollos*,
hollow, + Gr. *astron,* star) are hollow spheres
of up to 128 cells, the latter contiguous or united
by protuberances of the wall (Fig. 4-23). The
sole method of reproduction is by autocolony
formation (Starr, 1984), in which the individual
cells divide into miniature colonies, which are
liberated by the breaking of the parent cell wall.

III. FILAMENTOUS ORGANISMS

Restriction of cell division to one direction
and coherence of the daughter cells result in

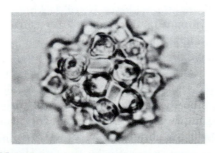

FIGURE 4-23
Coelastrum sp. ×500.

Ulothrix

The unbranched filaments of *Ulothrix* (Figs. 4-24, 4-25A, B) grow attached to stones and other submerged objects in cold-water streams and lakes; several species are marine. In *Ulothrix* (Gr. *oulos*, woolly, + Gr. *thrix*, hair), the cells of each filament are similar to one another except that the basal cell is modified as an attaching structure, the **holdfast** (Figs. 4-24A, 4-25A, F). The cells contain partial or complete band-shaped chloroplasts (Fig. 4-24B) with more than one pyrenoid and are uninucleate.

After a period of vegetative growth by cell division and elongation, asexual reproduction by zoospores may occur (Figs. 4-24C, 4-25E). Zoospores may be produced singly or in multiples of two from each vegetative cell. The liberated zoospores exhibit the usual attributes of motile cells, such as stigmata, contractile vacuoles, and four flagella (Figs. 4-24C, 4-26A), in which respect they resemble *Carteria* (Fig. 4-6). After a motile period, the zoospores settle on submerged objects with their flagellate poles foremost (Figs. 4-25F, 4-26A, B), lose their flagella, and attach themselves. Elongation and division of the original zoospore produce a vegetative filament (Figs. 4-25, 4-26C, D).

Isogamous sexual reproduction (Fig. 4-26F) also occurs in *Ulothrix*. The zygote (Fig. 4-26G) may undergo a prolonged period of vegetative development and germinate ultimately into four or more zoospores or aplanospores. The site of meiosis is presumed to be zygotic, but cytological and genetic evidence is still lacking.

Oedogonium

Oedogonium (Gr. *oedos*, swelling, + Gr. *gonos*, reproductive structure) (Figs. 4-27 through 4-29) differs from *Ulothrix* in that its zoospores possess a crown of about 120 flagella (Fig. 4-28C). Furthermore, although the quadriflagellate and biflagellate cells of nonmotile algae suggest, respectively, the unicellular motile genera, *Carteria* and *Chlamydomonas*, no multiflagellate unicellular genus corresponding

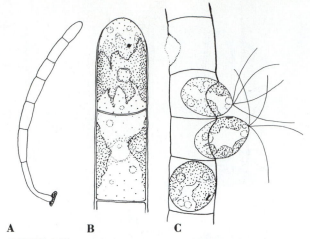

FIGURE 4-24
Ulothrix fimbriata. **A.** Young plant attached to particle of debris by a holdfast. **B.** Cellular organization at apex of filament; apical cell, three-dimensional, second cell in optical section. **C.** Zoospore formation. **A.** ×215. **B.** ×700. **C.** ×770.

to the motile cells of *Oedogonium* is known. Because of its multiflagellate motile cells, *Oedogonium* and its relatives are sometimes grouped in a distinct class of green algae. However, both biflagellate and multiflagellate motile cells may be produced in the life cycle of another green alga (*Derbesia*); there thus seems to be no reason to create a special class. *Oedogonium* grows frequently as an epiphyte on other algae and aquatic angiosperms (Fig. 4-17A). It may also be attached to stones or free-floating. Growth is intercalary and localized in certain cells, on which annular scars indicate the number of cell divisions that have occurred (Fig. 4-27D) (Hill and Machlis, 1968; Pickett-Heaps and Fowke, 1969). The cells contain segmented, netlike chloroplasts with pyrenoids.

Asexual reproduction is effected by the formation and liberation of single zoospores from vegetative cells (Figs. 4-27E, F, G, 4-28C). Pickett-Heaps (1971, 1972) has described zoosporogenesis on the basis of electron microscopy.

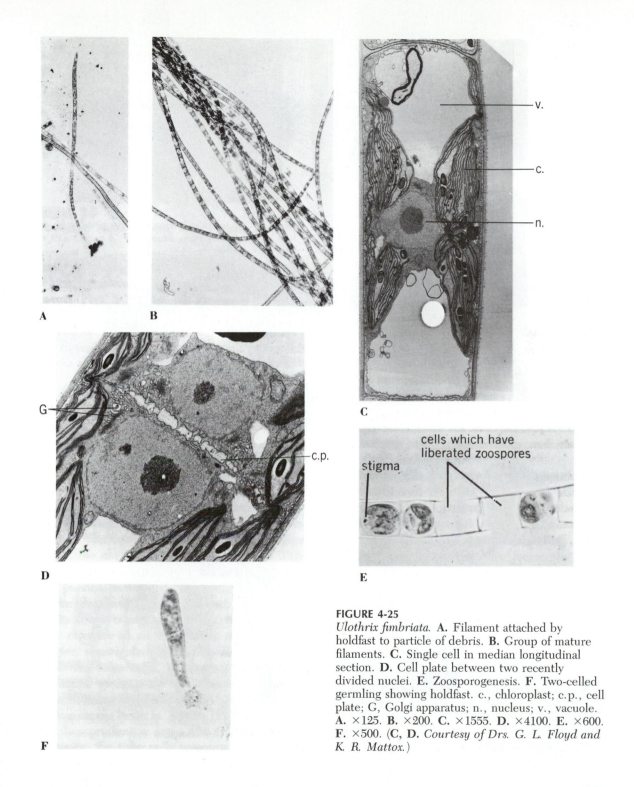

A

B

C

v.

c.

n.

G

c.p.

D

E

stigma

cells which have
liberated zoospores

F

FIGURE 4-25
Ulothrix fimbriata. **A.** Filament attached by
holdfast to particle of debris. **B.** Group of mature
filaments. **C.** Single cell in median longitudinal
section. **D.** Cell plate between two recently
divided nuclei. **E.** Zoosporogenesis. **F.** Two-celled
germling showing holdfast. c., chloroplast; c.p., cell
plate; G, Golgi apparatus; n., nucleus; v., vacuole.
A. ×125. **B.** ×200. **C.** ×1555. **D.** ×4100. **E.** ×600.
F. ×500. (**C, D.** *Courtesy of Drs. G. L. Floyd and
K. R. Mattox.*)

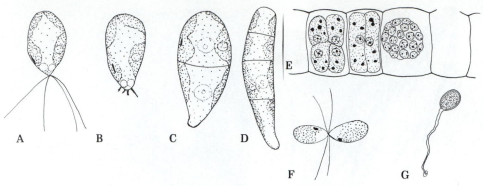

FIGURE 4-26
Ulothrix. **A–D.** Development of zoospores of *U. fimbriata* into a young plant. **E, F.** *U. zonata.* **E.** Gametogenesis. **F.** Isogamy. **G.** Zygote. **A–C.** ×700. **D.** ×400. **E, F.** ×315. **G.** ×250.

FIGURE 4-27
Oedogonium sp. **A.** Habit of growth on water lily petiole. **B.** Two vegetative filaments with recently quiescent zoospore attached to one. **C.** Vegetative cell; note segmented chloroplast and nucleus (arrow). **D.** Vegetative cell with apical caps. **E.** Zoosporogenesis, early stage. **F, G.** Release of zoospore. **H.** Zoospore germination; note lobed holdfast. **A.** ×⅓. **B.** ×150. **C, D.** ×1200. **E–G.** ×1000. **H.** ×1200.

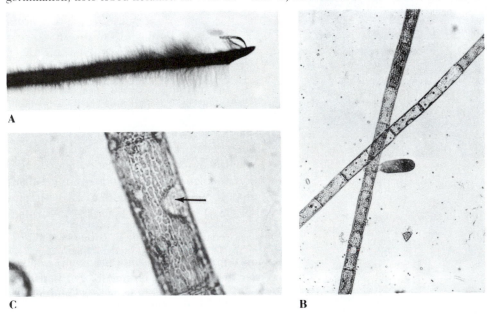

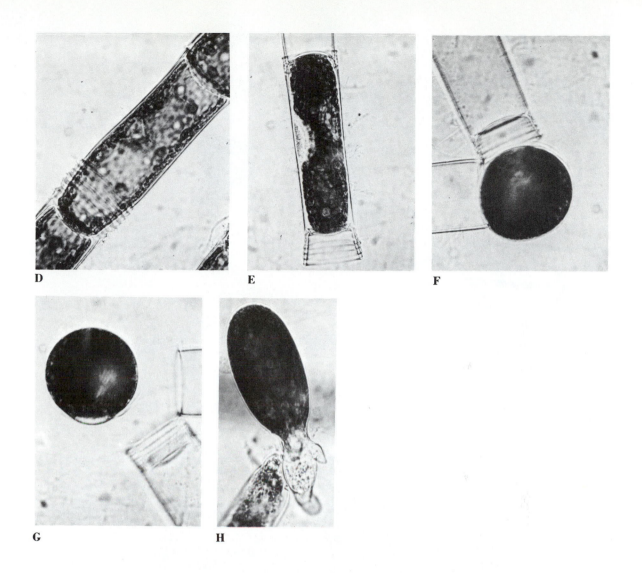

D   E   F

G   H

Sexual reproduction in *Oedogonium* is ooga-mous. The egg is produced in an enlarged gametangium, the **oogonium**, which opens by a pore or fissure just before fertilization (Figs. 4-28B, D, 4-29C, F, G). The sperms arise in pairs in short, boxlike cells, the **antheridia** (Figs. 4-28B, 4-29B, D, E). The sperms also are multiflagellate (Fig. 4-29F). After fertilization (Fig. 4-29F), the zygote develops a wall com-posed of two or more layers, often becomes reddish, and enters a period of dormancy (Fig. 4-29G). It is liberated by the disintegration of the oogonial wall. The germination of the zy-gote[16] into four zoospores is preceded by meio-sis.

---

[16]The zygote formed in oogamous reproduction was sometimes called an oospore.

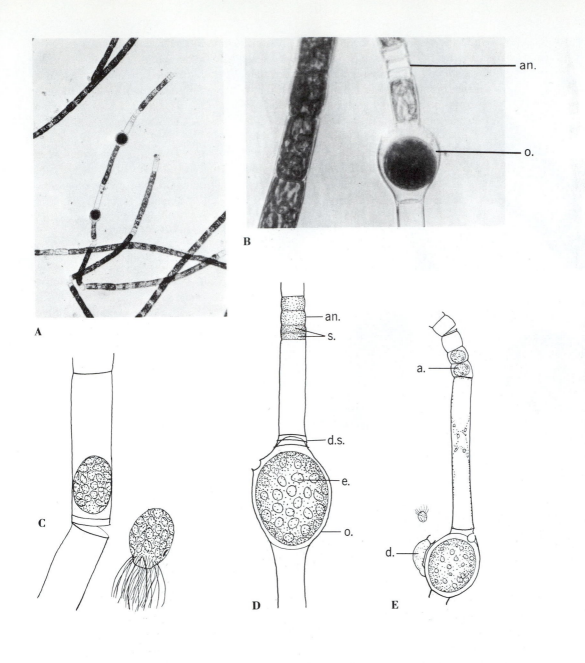

FIGURE 4-28

Oedogonium intermedium. **A, B.** Sexually mature filaments. **C.** *Oe. cardiacum*, zoospore formation and single, free-swimming zoospore. **D.** *Oe. foveolatum*, a bisexual species; **E.** *Oedogonium* sp., a gynandrosporous form. a., androspore; an., antheridium; d., dwarf males; d.s., division scars; e., egg; o., oogonium; s., sperm. **A.** ×125. **B.** ×625. **C–E.** ×315.

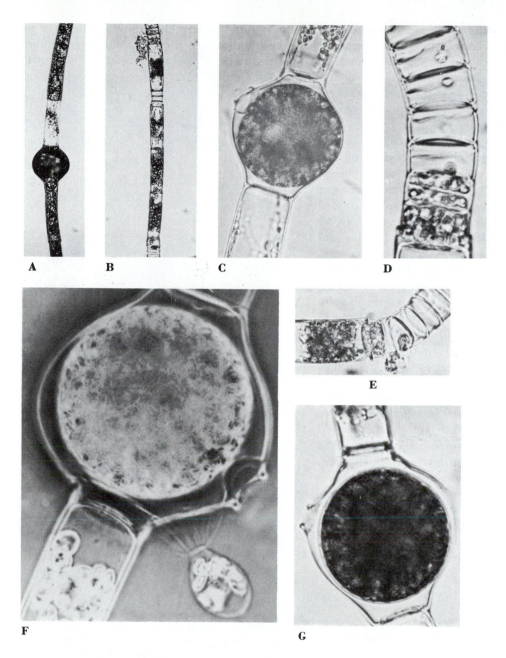

FIGURE 4-29

Oedogonium cardiacum. **A.** Female filament with immature oogonium. **B.** Male filament with three series of antheridia. **C.** Oogonium with egg ready for fertilization; note oogonial pore. **D.** Antheridia, two with sperm, the rest after sperm discharge; note sperm in antheridium. **E.** Liberation of sperm. **F.** Prefertilization, sperm approaching oogonium. **G.** Young zygote (oospore). **A, B.** ×150. **C.** ×700. **D.** ×1500. **E.** ×700. **F.** ×150. **G.** ×1000. (**F.** *Courtesy of Dr. L. R. Hoffman.*)

Considerable variation occurs among the numerous species of *Oedogonium* with respect to the location of the sex organs.

Both antheridia and oogonia may occur on the same filament (Fig. 4-28A, B), in which case, of course, the species is monoecious. In species with unisexual filaments, a plant that develops from a single zoospore produces either antheridia or oogonia (Fig. 4-29A, B), never both. The male filaments may be slightly narrower than the female but are almost the same diameter and length, that is, **macrandrous**.

It has been demonstrated that in species with unisexual filaments, the mature oogonia produce a substance that chemically attracts free-swimming sperm (Hoffman, 1960); this substance is an example of an **erotactin** (Machlis, 1972), a substance that attracts sperm.

In a number of **nannandrous** *Oedogonium* species, the male filaments are epiphytic, dwarf filaments; they develop from special **andro-spores** that attach themselves to the cell, which divides to form the oogonium and its supporting cell (Fig. 4-28E). The flagellate androspores are attracted to this site by a substance, also an erotactin, secreted by the oogonial mother cell (Rawitscher-Kunkel and Machlis, 1962). The androspores may arise in the female filament (**gynandrospory**) as in a special, androspore-forming filament (**idioandrospory**). The direction of growth of the dwarf males, in turn, is determined by a hormone from the oogonium and its supporting cell only after the dwarf males probably produce a chemical that evokes division of the oogonial mother cell.

Stigeoclonium

Occasional cell division in a second direction produces a branching filamentous plant body in *Stigeoclonium* and *Cladophora*.

Stigeoclonium (L. *stigens*, sharp, + L. *clonium*, branch) is a plant widely distributed in lakes and streams, where it grows attached to stones and vegetation. In most species the plant consists of two portions (Fig. 4-30A): a pros-

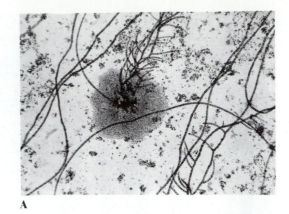

A

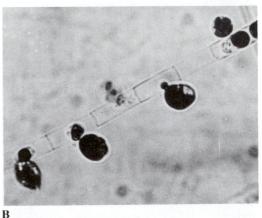

B

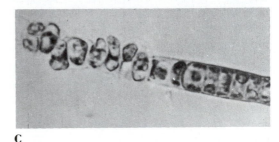

C

FIGURE 4-30
Stigeoclonium farctum. **A.** Cultured plant showing heterotrichy; note prostrate basal and erect systems. **B, C.** Zoospore release. **A.** ×75. **B, C.** ×800. (*Courtesy of Dr. Elenor R. Cox.*)

trate, and probably perennial, system of irregularly branched filaments or a disc is attached to the substratum; from one or another of these

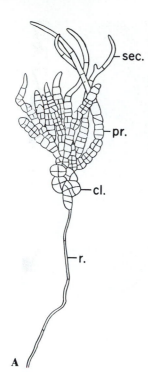

B

FIGURE 4-31

Fritschiella tuberosa. **A.** Small, but mature, plant. **B.** Plants growing on agar. cl, prostrate system; pr., primary erect branch; r., rhizoid; sec., secondary erect branch. **A.** ×80, **B.** ×3. (**A.** *After Singh;* **B.** *Courtesy of Dr. E. William Ruf, Jr.*)

A

prostrate phases, elongate branching filaments grow out into the water. These are attenuated and may end in hairlike branches. Because they have two branch systems, the plants are said to be **heterotrichous.** The cells of *Stigeclonium* are uninucleate and contain single chloroplasts with one or several pyrenoids.

In asexual reproduction, quadriflagellate zoospores are liberated singly from the cells of the plant body (Fig. 4-30B, C). After a period of activity, these become attached to the substratum by the formerly flagellate pole, secrete a wall, and develop into new plant bodies.

Union of biflagellate gametes has been described for some species; in others the gametes are reported to be quadriflagellate. The life cycle of *Stigeoclonium* has not been satisfactorily elucidated. Apparently, meiosis is zygotic in most species. As in *Ulothrix,* the zygote germinates to form four zoospores, which ultimately grow into new plants.

Fritschiella[17]

A rather remarkable alga, *Fritschiella tuberosa,* somewhat like *Stigeoclonium* and originally described from several locations in India, occurs on moist soil in several places in the United States. The branching, filamentous plants consist of subterranean rhizoids, a prostrate system at, or just beneath, the soil surface, and erect branches of two orders (Fig. 4-31); they are thus more differentiated than other green algae. The life cycle has been reported to be the D, h + d type (Chart 4-1) but this requires confirmation; quadriflagellate zoospores are produced on some plants and biflagellate gametes on others. The zygotes develop without dormancy directly into new plants. The subterranean rhizoids, prostrate system, and erect branches suggest the organization of land plants, as does the parenchymatous construction, but the reproduction is clearly algal.

---

[17]Named in honor of the late distinguished English phycologist F. E. Fritsch.

79

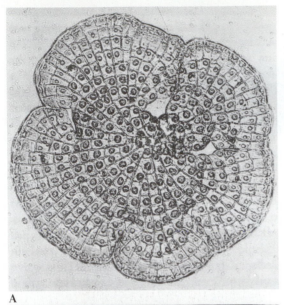

A

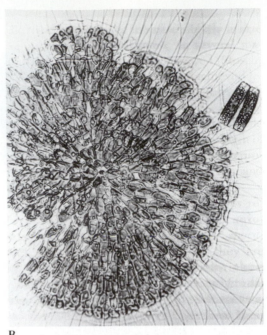

B

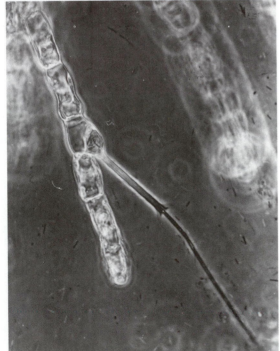

C

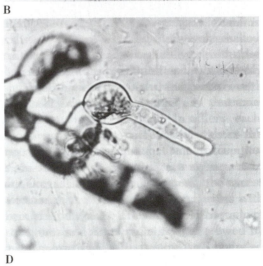

D

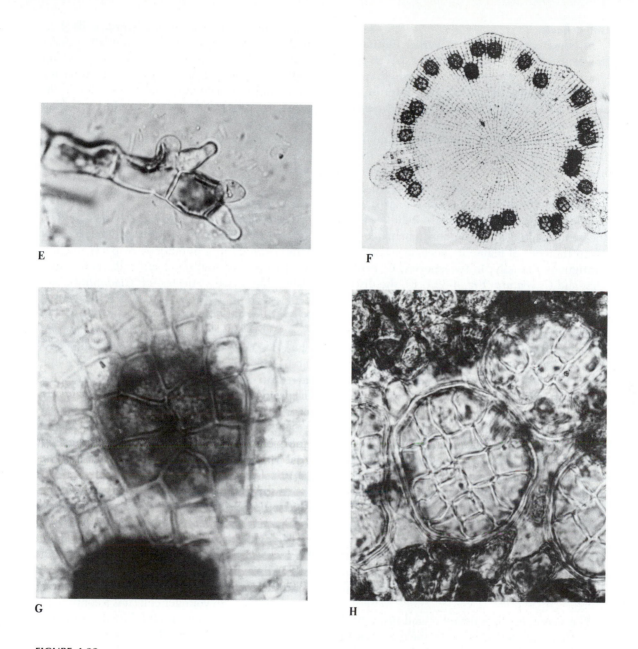

FIGURE 4-32

Coleochaete. **A.** *C. scutata*, vegetative plant in culture. **B.** *C. soluta* showing hair cells. **C–E.** *C. pulvinata.* **C.** Sheathed hair. **D.** Oogonium with trichogyne. **E.** Portion of a plant with antheridia. **F.***C. scutata*, plant with marginal zygotes. **G.** *C. orbicularis*, zygote covered by a layer of vegetative cells. **H.** *C. soluta*, zygote germination; zoospores (meiospores) have been liberated, leaving empty chambers. **A.** ×360. **B.** ×100. **C.** ×4000. **D.** ×1300. **E.** ×64. **F.** ×64. **G.** ×600. **H.** ×320. (**A–E, G, H.** *Courtesy of Dr. L. E. Graham.* **F.** *After Bold and Wynne.*)

Coleochaete

Although not included in earlier editions of this book, the genus *Coleochaete* (Gr. *koleon*, sheath + Gr. *chaetos*, hair), is here discussed not only because of its inherent interest, but also because of its possible phylogenetic implications (see p.309).

The various species, which are epiphytic on freshwater aquatic plants, vary from freely branching filaments to richly branched ones with the branches so closely adpressed as to form a pseudoparenchyma. (In true parenchyma, adjacent cells have arisen by cell division of the same parent cell; pseudoparenchyma arise when the cells of filaments become secondarily contiguous.) Finally, in two species, *C. orbicularis* and *C. scutata* (Fig. 4-32A, F), the discoid plants are truly parenchymatous (Graham, 1982) with nuclear and cell division occurring in the outermost cells of the discs. These radial divisions are like those of land plants in that they involve a cell plate and phragmoplast (Marchant and Pickett-Heaps, 1973). Some of the cells bear long, sheathed hairs (Fig. 4-32C).

Asexual reproduction is accomplished by biflagellate zoospores whose ultrastructural organization is complex (Sluiman, 1983).

Sexual reproduction is oogamous. The antheridia (Fig. 4-32E) are small cells each of which produces a single biflagellate sperm; in the parenchymatous species *multicellular* antheridia (consisting of 4 spermatogenous cells and 2 vegetative cells) are present. The sperm of *Coleochaete* are in a number of respects organized like those of land plants (Graham and McBride, 1979; Graham, 1980; Duckett et al., 1982).

The young oogonia develop, and presumably have their eggs fertilized, near the meristematic periphery of the plant body in discoid species like *C. orbicularis*. In some species there is a longer or shorter receptive protuberance, the **trichogyne**, on each oogonium (Fig. 4-32D). The cells surrounding the zygote undergo nuclear and cell divisions after fertilization and encase the zygote in a jacket of sterile cells (Fig. 4-32F, G). Where the walls of the latter make contact with the wall of the zygote, they develop digitate ingrowths, an apparent adaptation for increasing plasmalemma-wall surface for facilitating transfer of nutrients to the enlarging zygote (Graham and Wilcox, 1983). Meiosis is zygotic, and 8–32 biflagellated zoospores are released by the germinating zygotes (Fig 4-32H).

Cladophora

Cladophora (Gr. *klados*, branch, + Gr. *phoros*, bearer) (Figs. 4-33, 4-34) differs from *Stigeoclonium* in a number of respects, including larger size, multinucleate cells, and, especially, life cycle. Species of *Cladophora* are widespread in both fresh and marine waters, where they may be free-floating or attached to rocks or vegetation. The plants often are anchored to the substratum by rhizoidal branches. The latter are perennial and persist through adverse conditions. Growth of the branching filaments is localized near the apices of the filaments, in contrast with the generalized growth of *Ulothrix* and the intercalary growth of *Oedogonium*. In many species the branches arise as eversions from the upper portions of the lateral walls of relatively young cells (Fig. 4-33A, arrow). When they have achieved a certain length, they are delimited from the parent cell by an annular ingrowth of the wall.

The cylindrical cells of *Cladophora* are much larger than those of *Stigeoclonium*, and their cell walls are thicker and stratified (Fig. 4-34A). The structure of the chloroplast varies with the age of the cell. In younger cells it is a continuous network, but in older ones it is largely peripheral and composed of irregular segments, in some of which pyrenoids are embedded. Segments of the chloroplast may extend toward the center of the cell. Frothy cytoplasm, with numerous nuclei (Fig. 4-34A) suspended in its meshes, fills the center of the cell. Mitosis and cytokinesis are entirely independent processes in *Clado-*

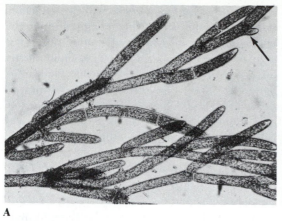

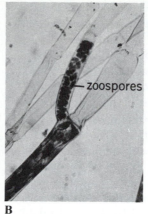

zoospores

FIGURE 4-33
Cladophora. **A.** Vegetative
filaments; note incipient
branch at arrow. **B.**
Zoospore formation. ×125.

A   B

phora, in contrast with their rather close relationship in most plants and animals with uninucleate cells.

Asexual reproduction is accomplished by uninucleate, quadriflagellate zoospores (Figs. 4-33B, 4-34B–D). These arise by the cleavage of the protoplasts of terminal and near-terminal cells into uninucleate segments. Each segment develops four flagella, and the mature zoospores

are liberated through a pore in the zoosporangial wall. After a period of motility, the zoospores grow into new plants. The young germlings are uninucleate, but the **coenocytic** (multinucleate) condition is soon initiated (Fig. 4-34D) by continuation of mitosis without ensuing cytokinesis. *Cladophora* plants also produce biflagellate isogametes (Fig. 4-34E) in sexual reproduction. These also are formed in the terminal and

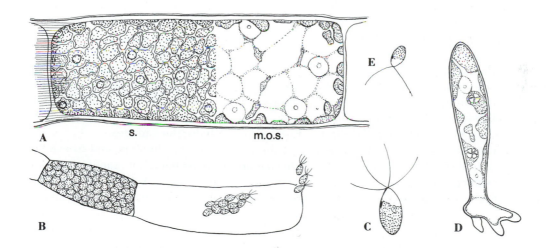

A   s.   m.o.s.   E

B   C   D

FIGURE 4-34
Cladophora. **A.** Cellular organization; surface view at s., median optical section at m.o.s. **B.**
Zoospore formation and liberation in a marine species. **C.** Zoospore. **D.** Germling from zoospore. **E.**
Gamete. **A.** ×770. **B.** ×315. **C.** ×770. **D.** ×550. **E.** ×770.

near-terminal cells. The zygotes germinate without a period of dormancy and grow directly into new plants.

Cytological studies of plants in nature and in culture indicate that a somewhat complicated cycle occurs in *Cladophora*.[18] In *C. suhriana* and other marine species, it has been shown that two types of plant occur in nature. These are morphologically indistinguishable but differ in chromosome complement and the nature of their reproductive cells. One type of plant is diploid and produces only zoospores from cells in which meiosis precedes the cleavage into zoospores. The latter, accordingly, are haploid and develop into haploid plants, morphologically similar to the diploid ones. However, the haploid plants, which are probably heterothallic, produce only biflagellate gametes at maturity. The gametes unite in pairs to form zygotes, which develop without meiosis into diploid filaments. Meiosis in *Cladophora*, therefore, is **sporic**—rather than zygotic as in all the other green algae discussed so far.

Two final points regarding the life cycle of *Cladophora* are relevant. (1) Gametes that fail to unite with other gametes under suitable environmental conditions develop asexually into new haploid, gametophytic plants of the same type as those that produced the gametes. This is an example of **parthenogenesis**, development of a gamete into a new organism without sexual union. (2) Although *Chlamydomonas* and *Carteria* are classified as being different largely on the criterion of having two and four flagella, respectively, both quadriflagellate and biflagellate motile cells occur in *Cladophora*. This is an example of the subjective nature of many schemes of biological classification.

Organisms like *Cladophora* with two distinct, free-living plants in the life cycle are **dibiontic**.[19] Other genera, in which only one free-living organism occurs in the life cycle, are **monobiontic**. Dibiontic life cycles are but one manifestation of the larger phenomenon of **alternation of generations** (see Chart 4-1). They are a specialized example of the latter in which the alternates are free-living individuals. In contrast, alternation of generations of another type, in which the alternates are physically connected, occurs in red algae, liverworts, mosses, and in some vascular plants.

Monobiontic life cycles may be thought of as exhibiting only alternation of nuclear (haploid and diploid) phases, one alternate, either the zygote or the gametes, consisting merely of a single cell and not of a free-living plant. This condition occurs in *Chlamydomonas* and *Ulothrix*, among others (type M, h, Chart 4-1). In these genera the entire life cycle consists of haploid individuals, with the exception of the diploid zygotes. It should be noted that the terms "haploid" and "diploid" are used with reference to chromosome constitution. **Haploid** organisms contain a single basic complement of chromosomes in their nuclei; **diploid** individuals have nuclei with two such sets.

Quite the reverse type of monobiontic life cycle (type M, d, Chart 4-1) is present in coenocytic green algae such as *Codium* (p. 94) and *Caulerpa* (p. 93). In these the diploid plant body is dominant, and the haploid phase is represented only by the gametes.

Dibiontic life cycles often are said to exhibit **alternation of morphological generations** (as well as cytological), since in this type both the diploid and haploid phases occur as morphologi-

---

[18]It was reported long ago that one freshwater species of *Cladophora* was a diploid organism with gametic meiosis, but this requires confirmation, because another freshwater species has a dibiontic life cycle (Shyam, 1980).

[19]Literally, two living individuals. The terms *dibiontic* and *monobiontic* are used in this edition (instead of *diplobiontic* and *haplobiontic*) to preclude the readers' possible confusion with diploid and haploid.

CHART 4-1
Three basic types of life cycle in Algae[a]

Type M, h: Monobiontic, organism haploid

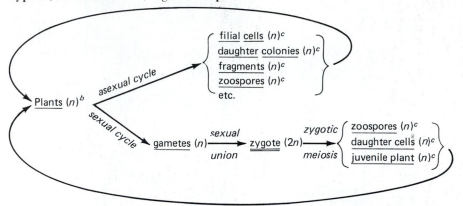

Type M, d: Monobiontic, organism diploid

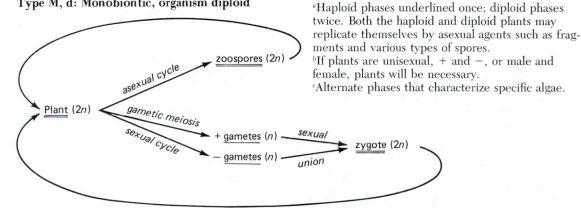

[a]Haploid phases underlined once; diploid phases twice. Both the haploid and diploid plants may replicate themselves by asexual agents such as fragments and various types of spores.
[b]If plants are unisexual, + and −, or male and female, plants will be necessary.
[c]Alternate phases that characterize specific algae.

Type D, h + d: Dibiontic, one alternate haploid, the other diploid

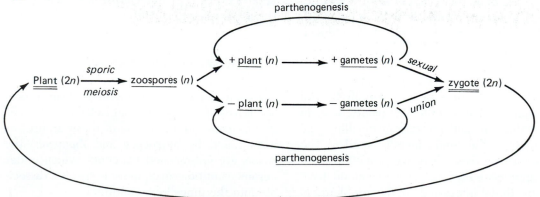

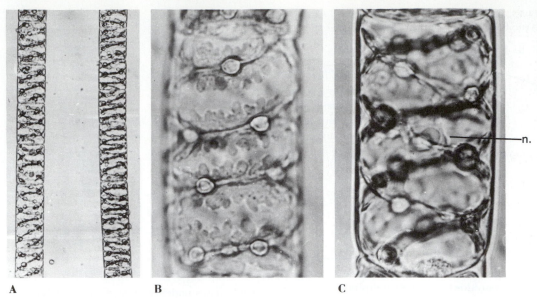

A   **B**   **C**

FIGURE 4-35
Spirogyra. **A.** Vegetative filaments. **B.** Cell in
surface view. **C.** The same cell in optical section.
n., nucleus. **A.** ×450. **B, C.** ×1750.

cally recognizable plants. The alternates may be
equal in stature and morphologically similar, as
in *Cladophora*. In this case alternation is said to
be **isomorphic**. When the alternating phases are
dissimilar morphologically (as in *Bryopsis halymeniae* [p.92], in *Laminaria*, a brown alga
[p.115], and in the fern, for example), the
alternation is called **heteromorphic**.

Alternation of generations itself is a sequence
in which a diploid phase gives rise by meiosis to
haploid spores, which initiate a haploid,
gamete-producing phase. The zygote produced
in sexual reproduction of the latter reinitiates
the diploid phase.

One can compare the two alternating generations of *Cladophora* with the sporophytic and
gametophytic phases of the land plants, since in
origin, function, and position in the life cycle
they seem to be fundamentally similar. The
significance of these facts has sometimes not
been appreciated as widely as it should have
been by those interested in the problems of

evolution, of alternation of generations and phylogeny of plants. While an alternation of generations like that described as type D, h + d
(Chart 4-1) has long been known in the land
plants, some students of these groups, in specu-
lating about their origin, have not considered
the significance of the occurrence of a similar
type of life cycle in the Chlorophyta. This
problem will be discussed again in our consideration of the land plants. For further discussion of
this topic refer to p. 309.

B. AZOOSPORIC ORGANISMS

Finally, among the filamentous algae, representatives of two groups that lack flagellate cells
will be considered. In one of these groups, as
exemplified by *Spirogyra* and *Zygnema*, the
plants are unbranched filaments, which under
certain conditions may, in some species, dissociate into the unicellular condition.

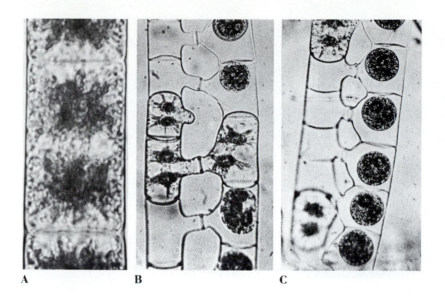

FIGURE 4-36
Zygnema. **A.** Vegetative cells. **B, C.** Stages in conjugation. **A.** ×1250. **B, C.** ×500.

Spirogyra and *Zygnema*

Spirogyra (Gr. *speira*, a coil, + Gr. *gyros*, curved) and *Zygnema* (Gr. *zygon*, yoke, + Gr. *nema*, thread), but especially *Spirogyra* (Fig. 4-35), usually are familiar to everyone who has studied biology. These genera often form floating, bright-green, frothy, and/or slimy masses in small bodies of water in the spring of the year and are frequently referred to as "pond scums" by laymen. The unbranched filaments, generally unattached, grow by generalized cell division and cell elongation. Masses of the plants are slimy to the touch, because the filaments are surrounded by pectic sheaths, demonstrable by India ink and methylene blue dye. Secretion of pectin by the filaments is involved in the movement of some species (Yeh and Gibor, 1970).

The cell structure of *Spirogyra* is familiar to many, at least superficially, because of the spiral arrangement of the ribbonlike chloroplast or chloroplasts (Figs. 4-35, 4-37D). However, this familiarity and ability to recognize the plant readily often result in a failure to appreciate the many details of cellular organization clearly observable by those determined to see them. A careful study of the cells of *Spirogyra* in the living condition, as revealed by an oil-immersion objective, is not only a good test of one's power of perception but also affords an opportunity for observation of detail in three-dimensional relations, which can be transferred, with profit to the observer, to the study of all cells. Species containing one or a few chloroplasts in each cell are especially favorable for study of the cellular organization, which is illustrated in Figs. 4-35 and 4-37D. The living cells of *Spirogyra* are excellent for observation of protoplasmic streaming, or **cyclosis**. *Zygnema* (Fig. 4-36A) differs from *Spirogyra* in having two stellate chloroplasts in each of its cells.

Aside from fragmentation of the filaments, no method of asexual reproduction occurs in *Spirogyra* and *Zygnema*. In *Spirogyra*, after a period of vegetative development, the filaments tend to become apposed. Adjacent cells of contiguous filaments produce papillate protuberances, which meet and elongate, thus forcing the filaments apart (Fig. 4-37B). Ultimately, the terminal walls of the contiguous papillae are dissolved, and one of the protoplasts of the pair of connected cells, both of which have lost much of the cell sap from their large vacuoles, initiates

87

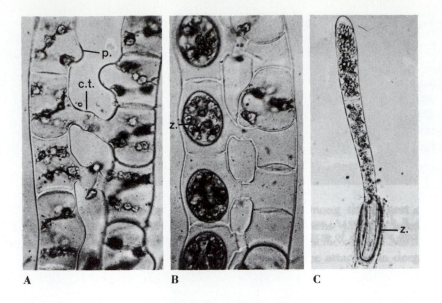

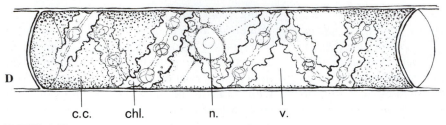

FIGURE 4-37

Spirogyra. **A, B.** Early and late stages in conjugation. **C.** Zygote germination. **D.** Organization of vegetative cell. c.c., colorless cytoplasm; chl., chloroplast; c.t., conjugation tube; n., nucleus; p., papilla; v., vacuole; z., zygote. **A, B.** ×500. **C.** ×250. **D.** ×1000.

movement through the tubular connection. Contractile vacuoles, usually present only in flagellate cells, appear in the protoplasts during dehydration and play an important role in that process. The two protoplasts and their nuclei unite, and the resultant zygote develops a thick wall and enters a period of dormancy (Fig. 4-37B). There is evidence that the chloroplast of the migrant cell subsequently disintegrates. Sexual reproduction of this type is interpreted as morphological isogamy with physiological an-

isogamy, the migrant protoplast being considered a male gamete. In defined inorganic medium in laboratory culture, pH was seemingly the controlling factor in evoking sexual reproduction (Grote, 1977).

At the conclusion of dormancy, the zygote, which has previously been liberated from the cell wall of the vegetative cell, germinates into a new filament (Fig. 4-37C). Meiosis precedes germination, but only one filament emerges from each zygote, because three of the four products of meiosis disintegrate before germination. In one species, meiosis occurs soon after karyogamy in the zygote and before its dormancy is initiated (Godward, 1961); Tatuno and Liyama (1971) have also reported zygotic meiosis and chromosome numbers of $n = 2$ in three species, $n = 4$ in two species, and $n = 24$ in one species.

In some species of *Spirogyra*, papillae from adjacent cells of the same filament establish contact, and the protoplasts of alternate cells function as male gametes with respect to the next cell of the filament so that zygotes occur in alternate cells. This is known as **terminal conjugation**, in contrast with the previously described **scalariform**, or ladderlike, pattern. With respect to the life cycle, *Spirogyra* belongs to the M, h type (Chart 4-1).

Desmids: Closterium, Cosmarium, and Micrasterias

Although they are represented by both unicellular and filamentous genera, the desmids are similar to *Spirogyra* and its relatives because of their structure and sexual reproduction. They differ markedly, however, in their cell division. The name **desmid** (Gr. *desmos*, bond) is ascribable to the fact that the cells of a majority of these plants are organized as two semicells that are mirror images of each other; the connecting region is known as the **isthmus**. As in *Spirogyra*, flagellate motile cells are absent. *Micrasterias* (Gr. *micros*, little, + Gr. *asterias*, star) (Fig. 4-38A), *Closterium* (Gr. *kloster*, spindle) (Fig. 4-38B–E), and *Cosmarium* (Gr. *kosmos*, an ornament) (Fig. 4-39) are widely distributed genera representative of the unicellular desmids, although *Micrasterias* cells sometimes are connected in chains. The two semicells of *Micrasterias* (Fig. 4-38A) are separated

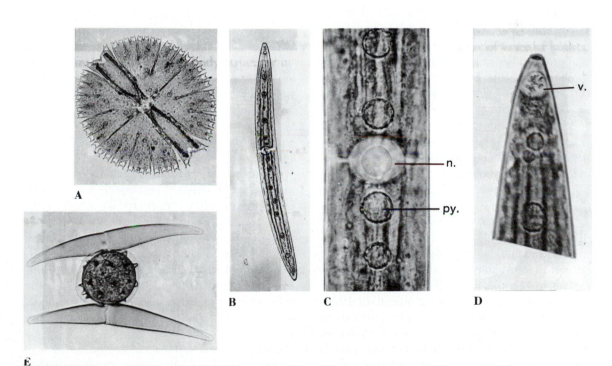

FIGURE 4-38
Desmids. **A.** *Micrasterias thomasiana.* **B–D.** *Closterium* sp. **B.** Vegetative cell. **C.** Enlarged view of isthmus and adjacent area. **D.** Cell apex. **E.** *Closterium calosporum.* var. *matus*, sexual reproduction; note dormant zygote between two empty cells. n., nucleus; py., pyrenoid; v., vacuole. **A, B.** ×250. **C, D.** ×750. **E.** ×175. (**E.** *After Cook.*)

89

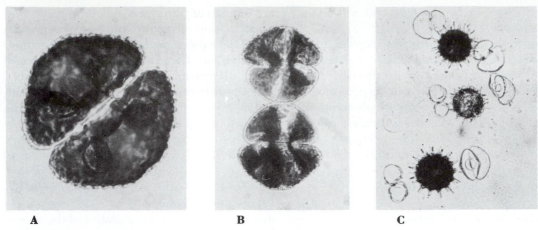

FIGURE 4-39
Cosmarium **A.** Vegetative cell. **B.** Late stage in cell division. **C.** Sexual reproduction; note dormant zygotes between pairs of empty cells. **A.** ×880. **B.** ×440. **C.** ×220.

by a deep incision, or **sinus**. The nucleus lies in the isthmus. The cell wall of a desmid is composed of several layers, the outermost of which is a rather diffluent pectin. It has been shown in some species that localized secretion of pectin through pores in the wall layers results in movement of the cells (Yeh and Gibor, 1970).

Asexual reproduction of unicellular genera is by cell division preceded by mitosis. In *Micrasterias* and *Cosmarium*, the two products of cytokinesis, the semicells, each containing a nucleus, regenerate new matching semicells (Fig. 4-39B).

Cosmarium and *Closterium* are ubiquitous desmids that have been receiving intensive genetic and cytological analysis. Both homothallic and heterothallic strains are known in *Closterium* and *Cosmarium*. When abundant sexually compatible individuals are mixed in laboratory cultures provided with adequate carbon dioxide, the cells pair within about 48 hours. This is accomplished through slow movements affected by secretion of pectin at one pole of the cell. The cells lie in a common mass of pectin after they have paired. This is followed by opening of the cell walls at the isthmus and liberation of the protoplasts, which unite to

form a spiny zygote (Fig. 4-39C). After a period of dormancy, the zygote germinates, producing two daughter cells. Meiosis occurs during germination, but two of the four products of meiosis disintegrate. A similar mode of gametic union and zygote germination occurs in *Closterium* (Fig. 4-38E).

IV. MEMBRANOUS ORGANISMS
Ulva

The plant body of a membranous, parenchymatous, green alga such as *Ulva* (L. *Ulva*, marsh plant) (Fig. 4-40A), since its life cycle is dibiontic, may be initiated either by a haploid zoospore or by a zygote. In either case the initial cell develops at first as an unbranched *Ulothrix*-like filament, from which longitudinal divisions later form in a flat blade. In addition, all the cells undergo one division parallel to the blade surface; thus, the plant body becomes two-layered (Fig. 4-41A).

Ulva, the sea lettuce, is a familiar alga of marine and brackish waters. It is a widely distributed alga that grows attached to rocks, woodwork, and larger marine algae in quiet estuaries. In *U. lactuca* the plant body is blade-like, often lobed and undulate; in some varieties

90

A

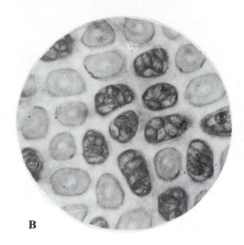

B

FIGURE 4-40
Ulva lactuca. **A.** Living plant. **B.** Zoosporogenesis; note emergence pores in empty zoosporangia. **A.** ×¹⁄₁₂. **B.** ×250.

it may exceed 1m in length. Each plant is anchored to the substratum by a small multicellular holdfast composed of cells with rhizoidal protuberances. The cell walls of *Ulva* (Figs. 4-40B, 4-41B) are rather thick, a probable correlative of the fact that the plants can withstand some desiccation when exposed at low tide. Each cell contains a single laminate chloroplast with one or more pyrenoids (Fig. 4-41B). The cells of the blade are uninucleate, but those of the holdfast may have several nuclei in their rhizoidal processes. Growth of the plant is generalized.

Ulva reproduces by zoospores (Figs. 4-40B, 4-41C, D) and anisogametes (Fig. 4-41G–J). If one gathers *Ulva* plants of sufficient maturity,[20] permits them to dry slightly, and then immerses them individually in dishes of seawater under strong unilateral illumination, such dishes soon become green with liberated motile cells. The latter manifest strong positive phototaxis. Careful microscopic examination of the motile cells of the several dishes reveals that three types are produced, each by different plants. Some of them liberate large zoospores (Figs. 4-40B, 4-41C, D), which are quadriflagellate; a second group of plants shed biflagellate gametes of two distinct sizes (Fig. 4-41G–J). The small male gametes arise from plants other than those that produce the female ones, so that *U. lactuca* is dioecious. Thus, three types of plants, one zoosporic and two gametic, constitute a population of *Ulva lactuca.*

The zoospores grow directly into new plants (Fig. 4-41F). These plants are the male and female gametophytes, which liberate the anisogametes at maturity. The zygotes develop into diploid, zoospore-producing plants (Fig. 4-41 L–M). Meiosis occurs in the first two nuclear divisions in the cells that produce the zoospores, which accordingly are haploid; half of the zoospores of a given zoosporangium develop into male and the other half into female plants; the gametophytes, of course, are haploid. Gametes that fail to unite may grow parthenogenetically into new plants. The life cycle of *Ulva* is entirely similar to that of *Cladophora*, being dibiontic and isomorphic; it clearly belongs to the D, h + d type (Chart 4-1).

V. COENOCYTIC AND TUBULAR ORGANISMS

In addition to the unicellular, colonial, filamentous, and membranous types of plant bod-

---

[20]These are readily recognizable by their nongreen margins, an indication that reproductive cells have been liberated.

91

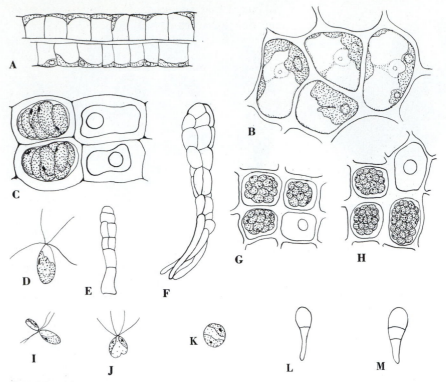

FIGURE 4-41

Ulva lactuca. **A.** Transection of plant. **B.** Cellular organization. **C.** Zoosporogenesis; note liberation pores for zoospores. **D.** Zoospore. **E, F.** Germling from zoospores. **G.** Female gametangia and gametes. **H.** Male gametangia and gametes. **I–K.** Stages in anisogamous gamete union. **L, M.** Germlings from zygote. **A, E, F, L, M.** ×315. The remainder ×770.

ies of Chlorophyta already described, brief mention must be made of one additional type, the coenocytic, tubular, or siphonous. These green algae are marine, with but a single exception. They may be simple, bilaterally symmetrical, tubular, pinnately branching plants such as *Bryopsis* (Fig. 4-41) or radially symmetrical as in *Acetabularia* (Fig. 4-45A). In all these plants the unit of structure is a coenocytic tubular "cell," the multinucleate protoplasm of which is peripherally disposed around a large central vacuole. Transverse septa usually occur only at sites of injury or when reproductive organs are delimited from the vegetative branches.

Bryopsis

Bryopsis (Gr. *bryon*, a moss, + Gr. *opsis*, appearance) (Fig. 4-42) grows attached to rocks in shallow marine waters. Growth is apical. At maturity, certain of the branches of *Bryopsis* become segregated from the main axis by the formation of septa and become transformed into gametangia (Fig. 4-43). The plants are dioecious and produce either male or female gametes.[21]

---

[21]*Bryopsis hypnoides* produces both male and female gametes on the same individual, and in the same gametangium, according to Burr and West (1970).

FIGURE 4-42
Bryopsis plumosa. Apex of living plant. ×60.

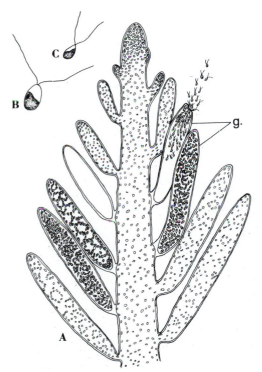

FIGURE 4-43
Bryopsis corticulans. **A.** Shoot with gametangia (g.).
B, C. Anisogamous gametes. **A.** ×30. **B, C.** ×550.
(*Modified from G. M. Smith.*)

Gametes that do not unite to form zygotes do not develop further.

Two kinds of life cycle have been found to occur in *B. plumosa* (Rietema, 1969, 1970, 1971), depending on where the plants are growing. In northern European waters (Holland) the zygotes develop directly into new diploid plants; in this case the life cycle is probably of the M, d type. On the coast of France, by contrast, the zygotes develop into branched tubular structures 4–6 mm long, which produce lateral zoosporangia. The latter form zoospores with a crown of flagella, as in *Oedogonium* (p. 72), and these grow into *Bryopsis* plants. This type of life cycle is dibiontic and heteromorphic; that is, the alternating plants differ in morphology. Both types of life cycle occur in the same populations of *B. plumosa* from Naples. The postzygotic tubular phase suggests the branching siphonous alga *Derbesia*. In the case of *B. halymeniae*, the sporophytic alternate in life cycle is *D. neglecta* (Hustede, 1964). The site of meiosis in these species of *Bryopsis* is not known with certainty; it was reported long ago to take place during gametogenesis, but this requires substantiation. (For a further discussion of life cycles in *Bryopsis*, see Bold and Wynne, 1985.)

Caulerpa

In *Caulerpa* (Gr. *kaulos*, a stem, + Gr. *herpo*, to creep) (Fig. 4-44), also monobiontic and diploid, the coenocytic plant bodies are composed of large-diameter, sometimes flattened tubes, the vacuoles of which are traversed by supporting ingrowths of the wall. These tubes may colonize extensive areas and simulate the vascular plants with their stemlike, rootlike, and leaflike branches.

As the plants mature sexually, the protoplasm of the "leaf" or "stem" cleaves into biflagellate gametes. These are released through extrusion papillae on the plant surface and are clearly anisogamous. The plants may be monoecious or

A

B

FIGURE 4-44
Caulerpa taxifolia. **A.** Habit of growth. **B.** Slightly
enlarged view of a branch. **A.** ×½. **B.** ×3.
(*Courtesy of Drs. C. J. Alexopoulos and L.
Almodovar.*)

dioecious, depending on the species (Goldstein
and Morrall, 1970). The zygotes develop directly into the branching plant (Price, 1972).

Codium

In *Codium* (Gr. *codion*, a fleece) (Fig. 4-45A),
the branched, ropelike plants are composed of a
longitudinal axis of small-diameter tubes that
bear vesicular branches (utricles) all over their
surface (Fig. 4-45B). The chloroplasts lack pyrenoids. At the bases of these, gametangia are
delimited by walls as in *Bryopsis.*

Plants of *C. fragile* are dioecious (Borden and
Stein, 1969), and the gametes are anisogamous.

A

B

FIGURE 4-45
Codium. **A.** *Codium fragile* growing
on a scallop. **B.** Surface view of a
branch of *C. repens.* **A.** ×½. **B.**
×25. (**A.** *After Ramus.*)

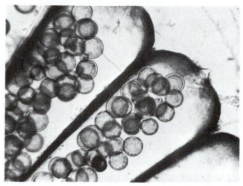

A

FIGURE 4-46
Acetabularia crenulata. **A.** Group of plants. **B, C.**
Enlarged view of cap. **C.** Almost mature cysts. **A.**
×1. **B, C.** ×100. (**A.** *Courtesy of the New York Botanical Garden.*)

B

C

The female gametangia are dark green, and the male gametangia are bright yellow. The zygotes develop into small plants with utricles, but a new generation of sexually mature plants with utricles has not yet been grown in culture, so the life cycle is incompletely known. *Codium* can rapidly colonize new areas from which it had been absent (i.e., the New England coast) and is highly destructive of oysters, clams, and scallops (Ramus, 1971).

Acetabularia

Finally, among the marine, coenocytic algae, mention must be made of *Acetabularia* (Fig. 4-46) (L. *acetabulum,* vinegar cup), the "mermaid's wine goblet," or "mermaid's parasol," a calcified organism widely distributed in subtropical and tropical waters. The radially symmetrical plants arise from zygotes that become attached to calcareous substrates and differentiate slowly (Fig. 4-47A–D) into a rhizoidal por-

tion, an erect axis, and, ultimately, a disclike cap. Throughout the major portion of this development, the zygote nucleus, although it enlarges, remains undivided in the rhizoidal portion. As the cap and its branching appendages mature (Fig. 4-47A–D), the zygote nucleus is reported to divide meiotically and then mitotically into numerous secondary nuclei, which become distributed, presumably by cytoplasmic streaming, throughout the plant. Microspectrophotometric analysis indicated that all the nuclei except the primary zygote nucleus are haploid; accordingly, meiosis seemingly occurs as the zygote nucleus forms secondary nuclei. Koop (1975*a,b*) investigated the life cycle of *A. mediterranea.*

In the chambers of the cap, each nucleus becomes a center about which cytokineses delimit a number of **cysts** (Figs. 4-46C, 4-47D).

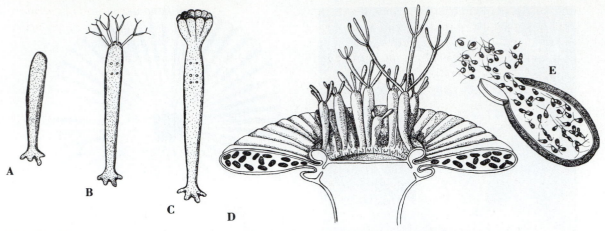

FIGURE 4-47

Acetabularia. **A–C.** Successive stages in maturation of the plant, diagrammatic. **D.** Organization of mature cap; note cysts. **E.** Liberation of gametes by germinating cysts. **D.** ×8. (*Modified from G. M. Smith.*)

These become multinucleate later in development. Subsequent disintegration of the old plants liberates the cysts, in which numerous biflagellate gametes arise (Fig. 4-47E) after cytokinesis. The zygotes formed by the union of the gametes from different cysts again initiate a new generation of plants.

Acetabularia, of which there are a number of species, has been intensively investigated with respect to morphogenesis and metabolism. Graft hybrids between different species and nuclear-transplant studies have been made successfully and have provided basic information regarding the role of nuclei and cytoplasm in morphogenesis and in inheritance.

The existence of large and complex bodies composed of coenocytic tubes from which transverse septa are lacking, except during reproduction, has led to speculation regarding the relation of such plants to other types of plant body. To some, the extensive growths represent single giant multinucleate cells. According to others, they are to be interpreted as acellular plant bodies, the individual nuclei and their surrounding cytoplasm being considered as repre-

senting cellular units not delimited by cell walls. More extensive accounts of *Acetabularia* have been published by Brachet (1965), Gibor (1966), and Puiseux-Dao (1970).

Summary and classification

One may wonder at the relative length of the discussion of the division *Chlorophyta* just concluded, as compared with other algal groups, but if one reviews the number and significance of the biological phenomena that they exhibit, the length of the discussion is justified. The *Chlorophyta* are distinguished from other algae by their pigments, by their apically flagellate cells, and by their storage of starch.

The Chlorophyta comprise a series of genera with a wide range of body form, including motile unicellular and colonial organisms, non-motile unicellular and colonial types, branched and unbranched filaments, and membranous and tubular organisms. The component cells may be uninucleate or multinucleate.

The various genera may undergo asexual reproduction by cell division (unicellular forms), by fragmentation (colonial and filamentous types), or by the production of such special reproductive cells as zoospores, aplanospores, and autospores. Colonial genera reproduce by daughter-colony formation.

Many genera of Chlorophyta exhibit sexual reproduction that involves the union of two cells, their nuclei, the association of their chromosomes and genes, and meiosis. Variations of sexuality, such as isogamy, anisogamy, and oogamy, may be observed in the several illustrative genera. In organisms such as *Chlamydomonas* and *Chlorococcum*, in which gametes are morphologically indistinguishable from asexual cells, it is possible that sexuality is primitive and incipient. The gametes of both sexes may occur on one individual or be present in its asexual descendants, a clone, or they may be segregated in different individuals or clones. The Chlorophyta also illustrate a variety of reproductive cycles, as follows: type M, h: organisms that are monobiontic and haploid, with zygotic meiosis; type M, d: organisms that are monobiontic and diploid, with gametic meiosis; type D, h + d: organisms that are dibiontic, with sporic meiosis (Chart 4-1).

Alternation in the life cycle in the first two types (M, h and M, d) is often called cytological, in contrast with that of D, h + d, which is morphological (as well as cytological) in the sense that two plants are involved.

The genera of Chlorophyta discussed in the present chapter may be classified as follows:

Division Chlorophyta
  Class 1. Chlorophyceae
    Order 1. Volvocales
      Family 1. Chlamydomonadaceae
        Genera: *Chlamydomonas* (M, h),[22]
            Carteria (M, h)
      Family 2. Volvocaceae

        Genera: *Pandorina, Gonium, Eudorina, Platydorina, Pleodorina, Volvox* (all M, h)
    Order 2. Chlorosarcinales
      Family 1. Chlorosarcinaceae
        Genus: *Tetracystis* (M, h)
    Order 3. Chlorococcales
      Family 1. Chlorococcaceae
        Genus: *Chlorococcum* (M, h)
      Family 2. Hydrodictyaceae
        Genera: *Pediastrum* (M, h), *Hydrodictyon* (M, h)
      Family 3. Chlorellaceae
        Genera: *Chlorella; Eremosphaera* (M, h)
      Family 4. Scenedesmaceae
        Genera: *Scenedesmus* (M, h), *Coelastrum*
    Order 4. Ulotrichales
      Family 1. Ulotrichaceae
        Genus: *Ulothrix*
      Family 2. Chaetophoraceae
        Genera: *Stigeoclonium* (M, h [?]), *Fritschiella* (D, h + d)(?)
      Family 3. Coleochaetaceae
        Genus: *Coleochaete* (M, h)
    Order 5. Ulvales
      Family 1. Ulvaceae
        Genus: *Ulva* (D, h + d)
    Order 6. Cladophorales
      Family 1. Cladophoraceae
        Genus: *Cladophora* (D, h + d)
    Order 7. Caulerpales
      Family 1. Bryopsidaceae
        Genus: *Bryopsis* (D, h + d and M, d)[23]
      Family 2. Caulerpaceae
        Genus: *Caulerpa* (M, d)
      Family 3. Codiaceae
        Genus: *Codium* (M, d)
    Order 8. Dasycladales
      Family 1. Dasycladaceae
        Genus: *Acetabularia* (M, h)
    Order 9. Oedogoniales

---

[22]The letters in parentheses after certain genera indicate the type of life cycle confirmed or postulated; where these letters are absent, sexual reproduction is unknown.

---

[23]Depending on the species.

97

Family 1. Oedogoniaceae
    Genus: *Oedogonium* (M, h)
Order 10. Zygnematales
  Family 1. Zygnemataceae
    Genera: *Spirogyra, Zygnema* (M, h)
  Family 2. Desmidiaceae
    Genera: *Micrasterias, Closterium,*
       Cosmarium (M, h)

The distinguishing characteristics, on the basis of which the orders, families, and their component genera are delimited, are discussed in several of the publications listed in the Bibliography (e.g., Bold and Wynne, 1985). Such characteristics as presence or absence of motility in the vegetative or reproductive phases, structure of the plant body, nature of the chloroplast, and other morphological aspects are involved in the segregation of the various taxa.

In concluding this account of the Chlorophyta, it should be emphasized that their pigmentation, cellulosic walls (in many), and their storage of excess photosynthate as starch seem to link them (physiologically) to the land plants more closely than do the attributes of any other group of algae except the Charophyta (Chapter 5). For this reason, most speculations regarding the origin of the more complex groups of plants always involve consideration of the morphology, physiology, and biochemistry of the division Chlorophyta. In this connection, it may be noted that within the past decade and a half, a great volume of ultrastructural, and some biochemical, data have been amassed, on the basis of which the class Chlorophyceae of the division Chlorophyta has been more narrowly circumscribed. Some of its members were transferred to two other classes, Ulvophyceae and Charophyceae. The latter are considered to be the ancestral group from which the liverworts, hornworts, mosses, vascular cryptogams and seed plants evolved (Stewart and Mattox, 1975, 1978; Mattox and Stewart, 1984; and Syrett and Al Houty, 1984).

DISCUSSION QUESTIONS

1. Does habitat distinguish algae from other chlorophyllous plants? Explain your answer.
2. What characteristics distinguish algae from other chlorophyllous plants?
3. It has been demonstrated that some algae retain their chlorophyll and multiply in darkness beneath the soil surface. What possible explanations are there for these observations?
4. What types of plant body occur among algae?
5. Define the terms "plankton" and "water bloom."
6. In what respects are plankton algae of biological and economic importance?
7. What organisms were formerly included in the group Thallophyta? (See Chapter 3 and inside back cover.)
8. What is meant by the following terms: growth, reproduction, vegetative reproduction, asexual reproduction, sexual reproduction, cytokinesis?
9. What types of growth can you distinguish on the basis of location in the plant body?
10. Cite the characteristics that distinguish the Chlorophyta from prokaryotic organisms.
11. On what basis is motility considered to be a primitive attribute?
12. Explain the meaning and use of the following terms: sexual and asexual reproduction; isogamy, anisogamy, and oogamy; zygote and oospore; zoospore and zoosporangium; monobiontic, dibiontic, haploid, diploid, meiosis, mitosis, and cytokinesis; plasmodesma; stigma, pyrenoid, contractile vacuole; flagellum; protoplast; coenocytic; coenobium; gamete and gametangium; antheridium, oogonium, sperm, egg; isomorphic and heteromorphic alternation of generations.

13. What evidence can you cite that indicates that sexuality in genera such as *Chlamydomonas* may be incipient? How would you investigate this experimentally?

14. Can you make a statement regarding the relative advantages of isogamy and oogamy? Explain.

15. Why is the stigma sometimes called the "red eyespot"? Consult your instructor about the experiments of Engelmann, Hartshorne, Mast, Wolken, Cobb, Diehn, and Batra and Tollin.[24]

16. What significance do you attach to the fact that many nonmotile green algae produce motile reproductive cells with two or four flagella?

17. What result would be realized, in your opinion, if the motile zoospores of *Hydrodictyon* were to be released from the parent cell? Verify by releasing some, if material is available.

18. Give examples of cellular differentiation of "division of labor" in the Chlorophyta.

19. Describe cell division in *Micrasterias* or *Cosmarium*. Plan an experiment to obtain data on its frequency.

20. In scalariform conjugation in *Spirogyra*, conjugation tubes are often established between more than two filaments. What is the disposition of the zygotes when this occurs? What are the implications?

21. What genetic effect is produced by the disintegration of three of the four nuclei arising by meiosis in *Spirogyra*? Can you cite examples of a similar phenomenon elsewhere in the plant or animal kingdom?

22. Construct a dichotomous key to the genera of Chlorophyta discussed in this chapter.

23. Define the terms unialgal culture, axenic culture, pure culture, and clonal culture.

24. Define or explain the terms life cycle or life history, alternation of generations, isomorphic and heteromorphic alternation, cytological and morphological alternation, alternate.

25. Do you consider *Bryopsis*, *Codium*, *Caulerpa*, and *Acetabularia* to be unicellular, multicellular, or acellular? Explain.

26. Place all the algae with sexual reproduction described in this chapter in types M, h; M, d; or D, h + d with respect to their life cycles.

[24]T. W. Engelmann, "Über Licht- und Farbenperception niederster Organismen," *Arch. f. d. ges. Physiol.* 29:387–400 (1882); J. N. Hartshorne, "The Function of the Eyespot in *Chlamydomonas*, *New Phytol.* 52:292–297 (1953); S. O. Mast, "Structure and Function of the Eye-spot in Unicellular and Colonial Organisms," *Arch. f. Protistenk.* 60:197–220 (1928); J. J. Wolken, "Photoreceptors: Comparative Studies," in *Comparative Biochemistry of Photoreactive Systems*, "Symposia on Comparative Biology 1." Academic, New York (1960); H. D. Cobb, "An *In Vivo* Absorption Spectrum of the Eyespot of *Euglena mesnili*, *Texas J. Sci.* 15:231–235 (1963); D. Diehn, "Action Spectra of the Phototactic Responses in *Euglena*." *Biochem. Biophys. Acta* 177: 136–143 (1969); P. P. Batra and G. Tollin, "Phototaxis in *Euglena*. I. Isolation of the Eye-spot Granules and Identification of the Eye-spot Pigments." *Biochem. Biophys. Acta* 79:371–387 (1964).

Chapter 5
DIVISION CHAROPHYTA

The **stoneworts** and **brittleworts**, represented here by the genera *Chara* (Latin name) and *Nitella* (L. *nitella*, splendor), are sometimes classified in the division Chlorophyta as a class coordinate with the Chlorophyceae. This reflects the point of view that their morphological deviations from the Chlorophyceae are of insufficient magnitude to warrant their removal to a separate division. In this text, however, for reasons to be enumerated below, the stoneworts are considered to represent a group of divisional rank coordinate with the Chlorophyta. The division Charophyta contains the single class Charophyceae, order Charales, and family Characeae.

Chara (Figs. 5-1, 5-2) and *Nitella* grow in the muddy or sandy bottoms of clear lakes and ponds, or in limestone streams and quarry basins. In the latter habitats, certain species have the capacity of precipitating calcium carbonate from the water and covering themselves with calcareous surface layers. This last attribute has suggested the names stoneworts and brittleworts. Calcareous casts of the oogonial and vegetative branches of stoneworts have been preserved abundantly as fossils (see Fig. 10-6).

Unlike most freshwater algae, the Charophyta are plants with macroscopically distinctive features. The markedly whorled branching (Figs. 5-1, 5-2), the organization of the plant body into regular nodes and internodes, and its regular pattern of ontogeny from the single apical cell (Figs. 5-2, 5-4) are all features that suggest the Arthrophyta (Chapter 15) and/or *Ephedra* (Chapter 24) among the vascular plants.

The plant consists of a branching axis on which arise whorls or smaller branches of limited growth, often called "leaves." The lower portions of the axes are anchored to the substratum by branching filaments, the **rhizoids**. The rhizoids serve as organs of vegetative propagation, giving rise to erect green shoots. Branches arise at the nodes among the leaf bases.

Median longitudinal sections through the apex of the axis (Figs. 5-3, 5-4) (Pickett Heaps, 1967) reveal the very regular manner of development that occurs in the stoneworts. All the cells have their origin from the descendants of a prominent, dome-shaped apical cell that cuts off derivatives in a transverse direction, parallel to its basal wall (Figs. 5-3, 5-4). Each of these segments divides again transversely into a nodal and an internodal cell. The internodal cells elongate tremendously and may remain uncovered, as in *Nitella*; or, as in most species of

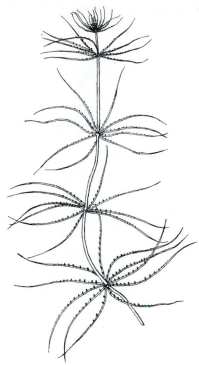

FIGURE 5-1
Chara sejuncta. Portion of axis. ×1.

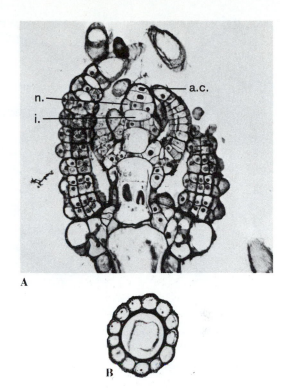

A

B

FIGURE 5-3
Chara sejuncta. **A.** Median longitudinal section of apex. **B.** Transection of internode with cortications. a.c., apical cell; i., internodal cell; n., nodal cells. **A.** ×135. **B.** ×90.

FIGURE 5-2
Chara contraria. Bacteria-free culture; note vegetative propagation by rhizoids. ×⅓. (*Courtesy of Dr. Eugene Shen.*)

Chara, the internodal cells become clothed with corticating cells that arise from the node above and below a given internode (Figs. 5-3, 5-4).

Nuclear division in the internodal cells is amitotic, and more than 2000 nuclei may occur within a single cell (Shen, 1967*b*). The protoplasm of the internodal cells streams rapidly in a direction parallel to the long axis of the cell (Kamitsubo, 1980); the minute peripheral chloroplasts are embedded in stationary cytoplasm. A prominent vacuole occupies the central portion of the elongate internodal cells, which are multinucleate. The nodal initials divide in such fashion as to form two central cells surrounded

101

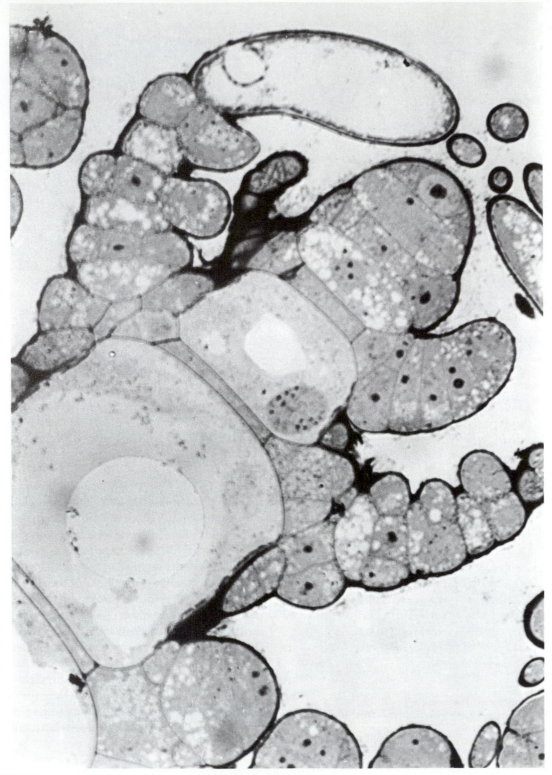

FIGURE 5-4
Chara sp. Median longisection of the apex; note apical cells; nodal and internodal derivatives; origin of branches ("leaves"), also by apical growth. ×4300. (*After Pickett-Heaps.*)

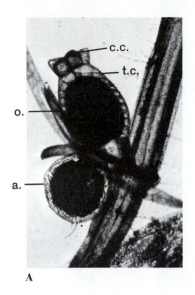

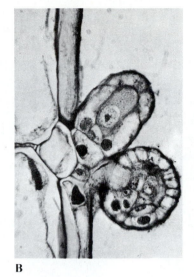

FIGURE 5-5
Chara sp. **A.** Segment with oogonium and antheridium. **B.** Section of early stages in development of oogonium and antheridium (below). a., antheridium; c.c., crown cells; o., oogonium; t.c., tube cells. **A.** ×60. **B.** ×125.

A   B

by one or more rings of cells. The outermost of these are the precursors of the whorled lateral branches of "leaves." The latter develop nodes, internodes, and cortications (Fig. 5-3A) like those of the main axes in *Chara*; in *Nitella* the leaves are uncorticated.

Reproduction in the Charophyta is strictly oogamous, and the gametes are produced in specialized complex structures usually called antheridia and oogonia, but sometimes designated the **globule** and **nucule**, respectively.

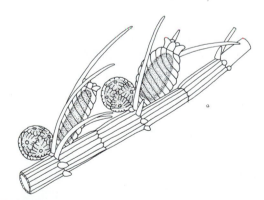

FIGURE 5-6
Chara contraria. Portion of axis with sex organs. ×50. (*Courtesy of Dr. Takashi Sawa.*)

Some species are dioecious, while others are monoecious. The reproductive structures are borne on the leaves (Figs. 5-5, 5-6). A mature plant furnishes a rather complete series in the ontogeny of the sex organs, if one examines leaves of successively older nodes. The younger sex organs are green, but as development proceeds, the antheridia become orange-red and the oogonia rather blackish brown (after fertilization) in many species.

The mature male reproductive organ (Figs. 5-6, 5-7, 5-8A, B) consists of chains of colorless cells, each of which produces a single sperm, surrounded by several types of sterile accessory cells; the whole structure is stalked. Its surface is composed of eight large, epidermislike **shield cells**, which are orange-red at maturity and contain incomplete, anticlinal[1] septa (Figs. 5-5B, 5-7A). To the inner tangential surface of each of these is attached a prismatic cell, the **manubrium** (L. *manus*, hand), to which, in turn, are attached one or more isodiametric cells, the **primary capitulum** (L. *caput*, head) (Fig. 5-7A). The primary capitula are all contiguous at the center and give rise to secondary

[1]Perpendicular to the surface.

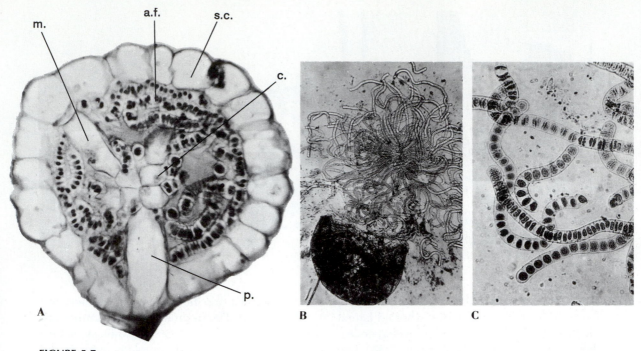

FIGURE 5-7
Chara sp. Organization of the antheridium. **A.** Longitudinal section. **B.** Crushed antheridium showing mass of antheridial filaments. **C.** Antheridial filaments in nuclear division. **A.** ×250. **B.** ×100. **C.** ×200.

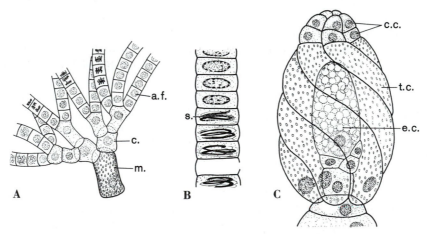

FIGURE 5-8
Nitella opaca. **A.** Manubrium with capitula and antheridial filaments. **B.** Segment of antheridial filament enlarged; note empty cell, which has released its sperm. **C.** Living, immature oogonium. a.f., antheridial filament; c., capitulum; c.c., crown cells; e.c., egg cell; m., manubrium; s., sperm; t.c., tube cell. **A.** ×500. **B.** ×1800. **C.** ×400. (*Courtesy of Dr. Takashi Sawa.*)

104

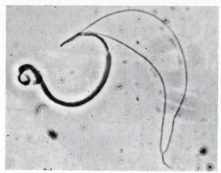

FIGURE 5-9
Nitella sp. Mature sperm. ×5400. (*After Turner.*)

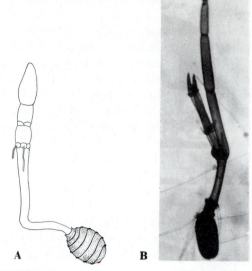

FIGURE 5-10
Nitella sp. Germination of the oospore (zygote). **B.** *Chara* sp. Oospore germination. **A.** ×10. **B.** ×12. (**A.** *Courtesy of Prof. John Dodd.*)

and, in some cases, tertiary capitular cells, which generate the colorless **antheridial filaments** (Figs. 5-7, 5-8A, B). These are composed of boxlike cells coiled up within the cavities formed in the male organ by the enlargement of the developing shield cells. A single biflagellate sperm emerges through a pore in the wall of each antheridial cell at maturity. The sperms are liberated by the partial separation of contiguous shield cells.

The sperms of Charophyta (Fig. 5-9) are quite complex (Turner, 1968; Pickett-Heaps 1968*a, b*; Moestrup, 1970*a*). They consist of a "head," an intermediate portion, and a "tail." The anterior "head" consists of the flagella, their basal bodies, and a row of mitochondria; the slightly unequal flagella emerge at the base of the "head." The intermediate region consists largely of the elongate, condensed nucleus. The posterior "tail" contains plastids, mitochondria, and possibly residual cytoplasm.

The female reproductive organ of the Charophyta is less complicated (Figs. 5-5, 5-8C). It, too, consists of a fertile cell, the **oogonium** proper, surrounded by spirally elongate sterile cells, the **tube cells**. The apices of these are delimited to form the five cells (*Chara*) (Fig. 5-5A) of the **corona** or **crown**; there are two tiers or ten crown cells in *Nitella* (Fig. 5-8C). The female reproductive organ also is pedicellate. At maturity, the tube cells separate from one an-other immediately under the corona, thus providing pathways for the entrance of sperms. The single large egg is uninucleate and contains abundant starch grains.

After fertilization, the zygote develops a thickened wall, and the oogonium is abscised from the leaf. The inner walls of the tube cells also thicken and persist as spiral markings on the dormant zygote (oospore) surface (Fig. 5-10A).

After a period of dormancy, which is probably followed by meiosis (Shen, 1967*a*), the zygote germinates (Fig. 5-10B) into a juvenile plantlet, all the nuclei of which are the descendants of one of the products of meiosis, as in *Spirogyra*.

Comparison of the morphology of *Chara*, *Nitella*, and other genera of Charophyta with that of the Chlorophyta provides few points of similarity. The complexity of the plant body in the Charophyta is unparalleled among the Chlorophyta, except, perhaps, in certain marine, siphonous genera. Furthermore, such fea-

105

tures as division into nodes and internodes, cortication of the axes and leaves, and the occurrence of special cellular sheaths around the sexual organs are absent among Chlorophyta.[2] For these reasons, among others, it is thought that the stoneworts represent a distinct phyletic line and that, as such, they should be placed in a division separate from the Chlorophyta. It is argued by some that their sex organs suggest affinity with the Hepatophyta (liverworts, Chapter 11), or Bryophyta (mosses, Chapter 12). This claim is denied by others on the ground that the sex organs of the Charophyta are really unicellular, while those of the Hepatophyta and Bryophyta are multicellular. It is not clear to the writers, however, why the egg protoplast enclosed in a cell wall and surrounded by tube cells in the Charophyta should all together be considered "unicellular," while the egg protoplast of a liverwort or moss, enclosed in its cell wall, surrounded by the venter, and associated with neck canal and neck cells, should be considered "multicellular." Even if they are not homologous, both are apparently multicellular organs.

---

[2]Segregation into nodes and internodes may be observed in members of the Chlorophyta such as *Chaetophora*, *Draparnaldia*, and *Draparnaldiopsis*. Sterile cellular sheaths grow around the oogonia of *Coleochaete* after fertilization.

If the term "antheridium" is applied in the Charophyta to one of the colorless cells of the antheridial filament, because it produces one sperm, by those who wish to reserve the older terms "globule" and "nucule," application of similar reasoning would restrict the use of the term "antheridium" to what is now called a spermatogenous cell or, possibly, an androcyte, in liverworts, mosses, and other plants. Is this not, perhaps, an eloquent example of the statement that "nature mocks at human categories"?

DISCUSSION QUESTIONS

1. Why are the Charophyta known as stoneworts and brittleworts? To what other algae might this name also be applied for a similar reason?
2. Describe the ontogeny of the plant body of *Chara* from the apical cell.
3. Describe the structure of the male and female reproductive organs of *Chara*. How does the female organ of *Nitella* differ from that of *Chara*?
4. How do the sperms reach the egg of *Chara*? What becomes of the zygote, or oospore?
5. Do you consider the Charophyta to be closely related to the Chlorophyta? Give all the reasons for your answer.

Chapter 6
DIVISION EUGLENOPHYTA

In addition to the Chlorophyta, two other groups of organisms, the Euglenophyta and Charophyta, often considered to be algae, have chloroplasts that contain chlorophylls *a* and *b* as well as a largely similar array of carotenes and xanthophylls. The Euglenophyta will be discussed in this chapter.

DIVISION EUGLENOPHYTA

The Euglenophyta contain approximately 11 chlorophyllous genera and 25 colorless ones, seemingly with a number of characteristics in common (Leedale, 1967). More than 800 species have been described, but the number will certainly be revised downward when the organisms have been studied critically in culture.

The green genera contain chlorophylls *a* and *b*, β-carotene, antheraxanthin, neoxanthin, and several other carotenoids and quinones in their plastids (see Table 4-1). *Euglena*, *Trachelomonas*, and *Phacus*, all chlorophyllous, are included in the following discussion, which emphasizes *Euglena*. *Euglena* (Gr. *eu*, good, + Gr. *glene*, eyeball), *Phacus* (Gr. *phakos*, lentil), and *Trachelomonas* (Gr. *trachelos*, neck, + Gr. *monas*, single organism) are widely distributed in freshwater pools, often in such abundance as to form water blooms. A comprehensive summary of the biology of *Euglena* has been published by Buetow (1968). At first glance, one would be inclined to classify *Euglena* and *Phacus* (Figs. 6-1, 6-4) as members of the Chlorophyta, but closer study reveals a number of respects in which they differ. In the first place, the protoplast in these genera is unwalled and is bounded externally by the plasma membrane. Just beneath this, a specialized series of ridged and grooved strips (Fig. 6-2A, B) forms a structure that together with the plasma membrane is called the **periplast.** The latter is largely proteinaceous but contains a small amount (ca. 6–17%) of lipids and/or carbohydrates. The periplast is apparently elastic, and cellulose is entirely absent. The strips of periplast are arranged in a helix in most species (Arnott and Walne, 1967). A row of mucilage-producing bodies, each connected by a canal to a strip of periplast, is present along each of the latter (Fig. 6-2B). The strips may be ornamented with warts or lumps. In a number of species of

gellum bears many very fine hairs, sometimes in a single row; the hairs are usually not visible with ordinary light microscopy. The emergent flagellum has a lateral swelling at about the level of the eyespot. At one side, or slightly posterior to the reservoir, occurs a large contractile vacuole, which discharges into the reservoir usually once every 15–60 sec.

Chloroplasts of varying form (discoid, elongate, platelike, according to the species) are present in *Euglena*, and these, except for the small discoid type, contain pyrenoids. The chloroplasts of members of the *Euglena* group, like those of dinoflagellates (p. 150), differ from those of all other algae in being surrounded by three membranes (Schantz et al., 1981). Mitochondria, Golgi apparatus, and endoplasmic reticulum also are present. The large nucleus is usually central.

Unlike the stigma of the Chlorophyta, that of *Euglena* is not included in a plastid but lies free in the cytoplasm in the vicinity of the **flagellar swelling** (parabasal body) (Figs. 6-1, 6-2). The stigma, or eyespot, is composed of 20–60 orange-red droplets, each one of which, or groups of which, is bounded by a membrane (Arnott and Walne, 1967). The pigments ascribed to the eyespot include echinenone along with at least two other carotenoids and possibly others.

Investigations of and speculations regarding the function of the stigma of *Euglena* and related genera have been numerous and contradictory. It seems clear that the positive movement of euglenoid cells toward light of moderate intensity (positive phototaxis) and their movement away from intense light (negative phototaxis) are connected with the eyespot, directly or indirectly. According to one hypothesis, the eyespot is the primary site of light perception, and this is supported by the fact that the action and absorption spectra of the eyespot pigments (ca. 485 nm) coincide; on the other hand, it has been postulated that the true site of reception of light stimuli is the flagellar swelling and that the

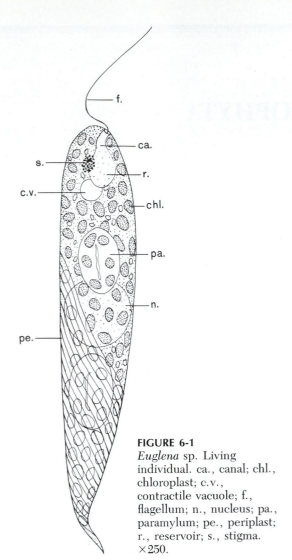

FIGURE 6-1
Euglena sp. Living individual. ca., canal; chl., chloroplast; c.v., contractile vacuole; f., flagellum; n., nucleus; pa., paramylum; pe., periplast; r., reservoir; s., stigma. ×250.

Euglena the periplast is nonrigid, and the organisms undergo marked changes in body form, sometimes called **metaboly**.

The anterior portion of a *Euglena* cell contains a slightly excentric opening, which leads through a **canal** to a **reservoir** (Fig. 6-1). Attached separately to the base of the latter are two flagella, one long and emergent and one much shorter and nonemergent. The long fla-

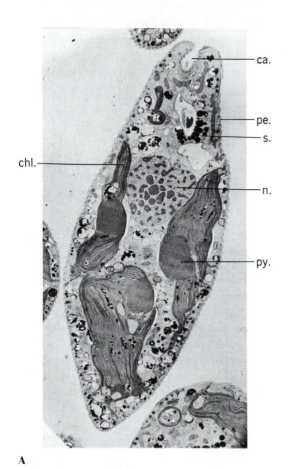

A

eyespot serves merely as a shield or shade for the organelle (Leedale, 1967; Diehn, 1969).

The reserve polysaccharide of euglenoid cells is a β-1 : 3-linked glucan called **paramylon**, which does not stain with iodine. It will be recalled that most Chlorophyta form starch, an α-1 : 4-linked glucan. Unlike the starch of green algae, paramylon forms in the colorless cytoplasm outside the chloroplast, although in some cases it may arise in close association with pyrenoids but outside the chloroplast envelope. In some species the cells may contain two large paramylon grains, one anterior to and one posterior to the nucleus.

Reproduction in *Euglena* (Fig. 6-3) is by cell division, which is preceded by nuclear division (mitosis). The latter differs from that in many other organisms in the persistence and division of the nucleolus and the persistence of the

FIGURE 6-2
Euglena granulata. Electron micrographs. **A.** Longisection. **B.** Highly magnified section of cell surface. ca., canal; chl., chloroplast; m., microtubule; p., plasma membrane; pe., periplast; py., pyrenoid; s., stigma. **A.** ×2500. **B.** ×110,000. (**A.** *Courtesy of Prof. Howard J. Arnott;* **B.** *Courtesy of Profs. H. J. Arnott and P. L. Walne.*)

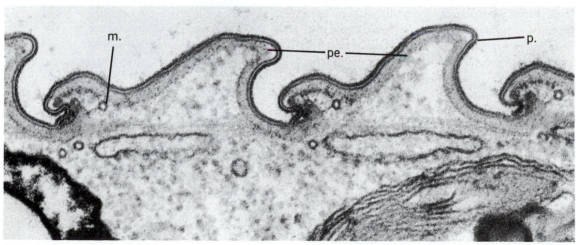

B

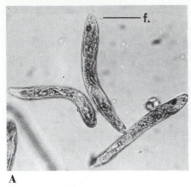
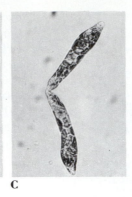

A B C

FIGURE 6-3
Euglena mesnili. Living
specimens. **A.** Mature
individuals. **B, C.** Division
stages. f., flagellum. ×175.

nuclear membrane during mitosis. There is no incontrovertible evidence of sexual reproduction in the Euglenophyta.

With respect to nutrition, *Euglena* and the related green species show some variation, although none has been proved to be completely photoautotrophic or phagotrophic. Photoautotrophic organisms can synthesize their protoplasm *entirely* from inorganic substrates using light energy. Phagotrophy is the animallike ingestion of food particles. Almost all euglenoids *require* vitamins B_1 and B_{12}, and they need reduced nitrogen compounds, not nitrates. A number can use, but seemingly do not require, amino acids and proteins as nitrogen sources. Chlorophyllous euglenids may be facultatively heterotrophic in darkness, using organic compounds such as acetate, organic acids, and alcohol, and their growth rate may be increased by addition of organic compounds to light-green cultures.

In relation to nutrition and also to phylogeny, it is of interest that certain strains of *Euglena* (*E. gracilis*) can be transformed experimentally into colorless organisms by cultivating them at high temperatures (35° C), or by subjecting them to ultraviolet radiation or to antibiotics (streptomycin, etc.). In such a regime the division of the chloroplast is retarded, while cell division continues, so that some of the cellular division products ultimately lack chloroplasts. From such individuals, colorless races may be devel-

oped and maintained, of course, in media with organic carbon sources. These colorless races of *Euglena* correspond to the colorless genera *Khawkinea* and *Astasia*, which are clearly protozoa. Similarly, both *Phacus* (Fig. 6-4), a flattened, rigid euglenoid, and *Trachelomonas* (Fig. 6-5), a euglenid living in a pectic shell, or **lorica**, have colorless counterparts (*Hyalophacus* and *Hyalotrachelomonas*, respectively).

The division Euglenophyta contains the sin-

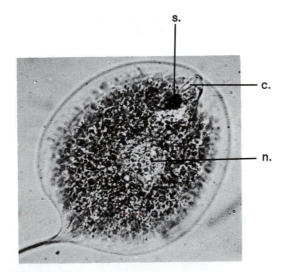

FIGURE 6-4
Phacus pleuronectes. Living individual. c., canal; n., nucleus; s., stigma. (Flagellum not visible.) ×2400.

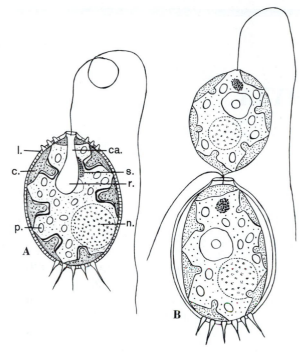

FIGURE 6-5
Trachelomonas armata. **A.** Longisection of a single individual. **B.** Reproduction. c., chloroplast; ca., canal; l., lorica; n., nucleus; p., paramylum; r., reservoir; s., stigma. ×700. (*Courtesy of Dr. Kamala Prasad Singh.*)

gle class Euglenophyceae, which in turn contains six orders (Leedale, 1967), distinguished largely in flagellation. The genera discussed above, except *Khawkinea* and *Astasia*, all belong to the class Euglenophyceae, order Euglenales, and family Euglenaceae.

DISCUSSION QUESTIONS

1. Why are the Euglenophyta placed in a division other than the Chlorophyta?
2. Summarize the plantlike and animallike attributes of *Euglena*.
3. How would you prove whether or not a given species of *Euglena* is autotrophic?
4. What facts concerning the occurrence of *Euglena* in nature might lead one to question its capacity for autotrophic nutrition?
5. How do you interpret the presence of a canal and reservoir in *Euglena*?
6. Why are bacteria-free cultures necessary for the study of physiological problems in algae?
7. How might one determine whether or not meiosis had occurred other than by counting chromosomes or using genetic evidence?

111

Chapter 7
DIVISION PHAEOPHYTA

The division Phaeophyta (Gr. *phaios*, dusky, + Gr. *phyton*, plant), the brown algae, includes a single class, the Phaeophyceae (Gr. *phaios* + Gr. *phykos*). The Phaeophyta are marine in habitat, with the exception of about four genera. They occur in the open ocean as well as in quiet estuaries and may be abundant on the muddy bottoms of salt marshes. Many grow attached to rocks, shells, or coarser algae such as the kelps. Approximately 250 genera and 1500 species of Phaeophyta have been described. In general, brown algae flourish in colder ocean waters and on rocky coasts, where many grow attached in relatively shallow water in the intertidal and sublittoral zones. A number of genera are able to withstand exposure to the atmosphere during low tide, whereas others are sublittoral and continuously submerged. Some of the large genera live in shoal waters, and most thrive in waters with considerable current. Few brown algae occur at great depths. Both annual and perennial genera are known.

The brownish shades of the plant reflect the abundant presence in the plastids of the xanthophyll, **fucoxanthin**, which is dominant over chlorophylls *a* and *c*, the other xanthophylls, and β-carotene (Table 4-1). The plastids are single, few (Fig. 7-2A), or numerous in each cell, and may be elaborate in form. No starch occurs in Phaeophyta; instead, the excess photosynthate accumulates as a carbohydrate, *laminaran* (a mixture of polysaccharides with 1 : 3 and 1 : 6 linkages), as mannitol, or in the form of fat droplets. The nuclei of Phaeophyta are prominent structures. In many genera, centrosomes and astral radiations, as viewed by light microscopy, appear during mitosis, as in many animal cells. The protoplast is bounded by a primary wall and middle lamella composed of a gummy substance, alginic acid. This may represent 10–25% of the dry weight. Alginic acid is a polymer of D-mannuronic and L-guluronic acids. Alginic acid has considerable commercial importance as a stabilizer, emulsifier, and coating for paper (Boney, 1965, 1966; Chapman, 1970).

The motile cells of brown algae are distinctive and differ from those of the Chlorophyta in that they are laterally or subapically biflagellate (Fig. 7-2E). The longer, usually anterior, flagellum is of the tinsel type, and the shorter, posterior one is whiplash (Loiseaux and West, 1970). Although it has been suggested that the Phaeophyta originated from unicellular motile

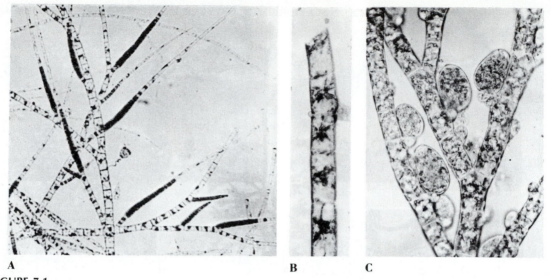

A   B   C

FIGURE 7-1

Ectocarpus siliculosus. **A.** Living plant with plurilocular reproductive organs. **B.** Segment of filament. **C.** Plant with unilocular zoosporangia in various stages of development. **A.** ×135. **B, C.** ×300. (**B.** *Courtesy of Dr. Charles Yarish.*)

organisms with similar lateral or subapical flagellation, no such organisms have yet been discovered.

The simplest type of plant body in the group is the branched filament (Fig. 7-1). Many Phaeophyta have considerable complexity of structure, as manifested in their leaflike, stemlike, and rootlike organs (Fig. 7-11), which exhibit obvious histological differentiation. Certain brown algae, the giant kelps, which may attain a length of 50 m, rival forest trees in stature. For a more extensive discussion of the brown algae, consult Bold and Wynne (1985).

Ectocarpus

Ectocarpus (Gr. *ektos*, outside, + Gr. *karpos*, fruit) (Figs. 7-1 through 7-3) is a relatively simple brown alga commonly growing on stones and shells or epiphytically on larger marine algae. *Ectocarpus* is a branching filamentous plant in which erect filaments arise from an attached prostrate branch system, much as in

the green alga *Stigeoclonium.* Its growth is diffuse or generalized. The mature cells contain band-shaped plastids with pyrenoidlike bodies (Figs. 7-1B, 7-2A); the function of the latter is not known.

The life cycle and reproduction of *Ectocarpus* are fundamentally similar to those in *Ulva* and *Cladophora*—namely, type D, h + d (p. 85)—in that meiosis is sporic during an alternation of isomorphic generations. In the diploid sporophytic plants, the terminal cells of lateral branchlets enlarge, and their protoplasts segment into approximately 32–64 zoospores (Figs. 7-1C, 7-2C, D, 7-3B), each of which becomes pear-shaped and laterally biflagellate. These are discharged through an apical pore, and after a period of motility they begin to develop into new filaments. The organs producing these zoospores are called **unilocular zoosporangia**, inasmuch as the zoospores lie within a single cavity. It has been shown that meiosis occurs during the first two nuclear divisions in

113

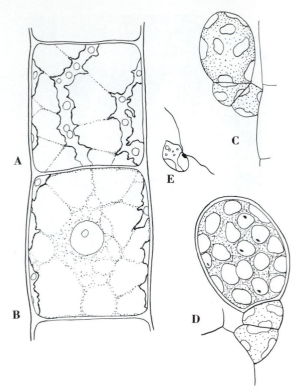

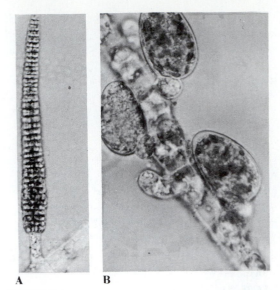

FIGURE 7-3
Ectocarpus sp., living. **A.** Plurilocular organ. **B.** Almost-mature zoosporangia. **A.** ×450. **B.** ×650.

FIGURE 7-2
Ectocarpus siliculosus. **A.** Surface and **B.** Median optical view of vegetative cell. **C, D.** Stages in development of unilocular zoosporangia. **E.** Motile cell from plurilocular organ. ×600.

the unilocular zoosporangium, so that the zoospores produced from these structures are haploid.[1]

These zoospores grow into plants morphologically similar to the zoospore-producing sporophytes, but these produce multicellular gametangia on lateral branches. Every cell of these plurilocular gametangia (Figs. 7-1A, 7-3A) produces a laterally biflagellate gamete.

Although morphologically isogamous, the compatible gametes of *E. siliculosus*, which develop on different plants (heterothallism), are

functionally anisogamous, since some gametes (the female) settle on a substrate and attract other motile gametes (the male). The male gametes become attached to the female by the tip of their long flagellum.

The attractant produced by the female gametes is volatile and is known as **Ectocarpene** (All-*cis*-1[cycloheptadien-2′,5′-y1]-butene-1). Its chemical structure (Fig. 7-4) was elucidated in 1971 by Müller et al. Its action and that of other sex hormones in brown algae are discussed by Jaenicke (1977) and Boland et al. (1980). After plasmogamy has occurred between the female gamete and one male, the female gamete no longer attracts males. The life cycle (type D, h + d) of *Ectocarpus* is summarized in Chart 4-1 on p. 85.

FIGURE 7-4
Ectocarpene, structure.

---

[1]Plurilocular zoosporangia that duplicate the diploid phase may also be produced by the sporophyte.

The life cycle here described occurs in *E. siliculosus*. Few other species have received adequate study. There is some indication that the life cycle of *E. siliculosus* varies with geographical location, being different in Naples and England from that reported above for the coast of Massachusetts. For example, Müller (1966, 1967) reported an exceedingly complex life cycle for *E. siliculosus* on the basis of field studies and laboratory cultures of material from Naples. According to him, in addition to the D, h + d type of life cycle, several additional alternatives occur. These include the occurrence of haploid, diploid, and tetraploid sporophytes, all of which produce unilocular zoosporangia, although meiosis occurs only in those of diploid and tetraploid plants. The haploid sporophytes arise from gametes that develop parthenogenetically. Gametophytes develop only from zoospores released by unilocular zoosporangia. That sporophytes may be morphologically sporophytes and yet haploid, diploid, or tetraploid is evidence (more of which will be cited and discussed later) that chromosome constitution *per se* does not determine whether gametophyte or sporophyte develops.

The Kelps

The kelps are of interest not only because of the complexity of their vegetative structure but also because their life cycle is representative of a type that occurs in few Chlorophyta but is similar in many respects to that of ferns and other vascular plants.

Laminaria

Laminaria (Figs. 7-5, 7-6) (L. *lamina,* blade) occurs attached to rocks that are usually submerged, even at extreme low tide. The plant consists of a branching **holdfast**, a **stipe**, and an expanded **blade.** Growth occurs at the junction of the stipe and is, therefore, **intercalary.** The oldest portion of the blade is the apex. Both blade and stipe are quite complex histologically.

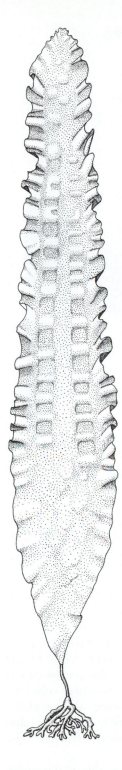

FIGURE 7-5
Laminaria agardhii (young specimen). Note blade, stipe, and attaching organs. × ½.

115

B

A

FIGURE 7-6
A. *Laminaria agardhii* (left) and *L. digitata* (right), freshly collected at Sandwich Beach, Mass. **B.** *L. agardhii*, detail of attachment. **C.** *Pelagophycus porra*, the elk kelp. **D.** *Laminaria groenlandica*, young sieve element (living). **B.** ×1/12. (**C.** *Courtesy of Dr. B. C. Parker.* **D.** *After Schmitz and Srivastava.*)

Only the more superficial cells of both stipe and blade are photosynthetic, other cells having very few plastids (Fig. 7-7). The central part of the blade is composed of elongate, colorless, filamentous cells constituting the medulla. Some of these, the **trumpet hyphae**, have flaring ends and function as sieve elements (Fig. 7-6D) (Schmitz, 1981).

Late in the growing season, during the winter and spring on the eastern coast of North America, certain superficial cells of the blade elongate

and become transformed into unilocular zoosporangia (Fig. 7-8). These occur in extensive groups, or **sori**. Between the zoosporangia occur sterile filaments called **paraphyses**. Paraphyses may occur among sex organs or sporangia. Meiosis is sporic and occurs in the unilocular zoosporangia, as in *Ectocarpus*. Sex chromosomes of the X-Y type have been described in *Laminaria* and other kelps (Evans, 1965), so that sexuality of the gametophytes is under genetic control and determined at meiosis. Each zoosporangium produces 32–64 zoospores, which are liberated and develop asexually into prostrate, *Ectocarpus*-like branching filaments that ultimately produce gametangia (Figs. 7-9, 7-10).

The *Laminaria* plant is diploid and sporophytic. The prostrate, branching filaments that develop from the haploid zoospores are gametophytic, haploid, dioecious, and oogamous. The

C

D

15 µm

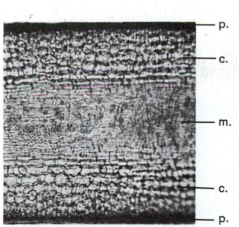

p.

c.

m.

c.

p.

FIGURE 7-7
Laminaria agardhii. Transection of blade. c.,
cortex; m., medulla; p., photosynthetic tissues.
×90.

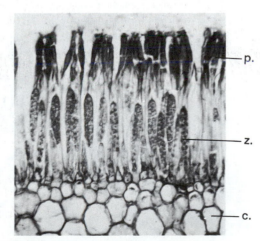

p.

z.

c.

FIGURE 7-8
Laminaria sp. Transection of fertile area. c., cortex;
p., paraphyses; z., zoosporangium. ×300.

117

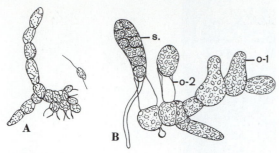

FIGURE 7-9
Laminaria japonica. **A.** Male gametophyte; note cluster of antheridia and sperm. **B.** Female gametophyte. o-1, immature oogonium; o-2, oogonium with extruded egg; s., young sporophyte at mouth of oogonium. **A.** ×450. **B.** ×350. (*Modified from Kanda.*)

male gametophytic filaments are of smaller diameter than the female and often grow intertwined with them. The antheridia (Fig. 7-9A) are produced as lateral cells on the male gametophyte; each antheridium produces a minute,

laterally biflagellate sperm. The oogonia (Figs. 7-9B, 7-10E–G) produce single eggs, which are released from but remain attached to a tubular protuberance of the oogonium, so that fertilization and the development of the embryonic sporophyte occur *in situ* (Fig. 7-9B). Without undergoing a dormant period, the zygote grows into a new sporophyte, which ultimately develops the form typical of the species. The life cycle[2] of *Laminaria* is summarized in Chart 7-1.

In laboratory culture at 10°C, gametophytes of *L. saccharina* produced antheridia before oogonia matured; antheridia matured within 23 days, and oogonia within 28 days, after the cultures had been started from zoospores. A photoperiod of 18 hours daily was most favorable for evoking sexual reproduction (Hsiao and Druehl, 1971).

---

[2]Diploid phases are underscored with double lines, haploid ones with single lines.

FIGURE 7-10
A–H. *Laminaria saccharina.* **A.** Gametophyte with extruded egg. **B.** Male gametophyte. **C, D.** Germination of zoospores into ♂ and ♀ gametophytes. **E–G.** States in extrusion of the egg from the oogonium. **H.** ♂ gametophyte releasing sperm (arrow). **I.** Egg cell surrounded by attracted sperms. **A, B.** ×360. **C–I.** ×571. (**A, B.** *After Hsiao and Druehl.* **C–I.** *Courtesy of Dr. K. Lüning and Dr. H. H. Huenert and the Institut für den Wissenschaftlichen Film, Göttingen, W. Germany.*)

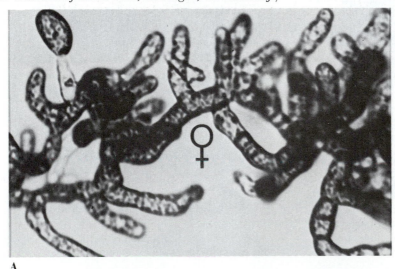

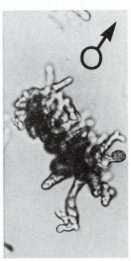

A

B

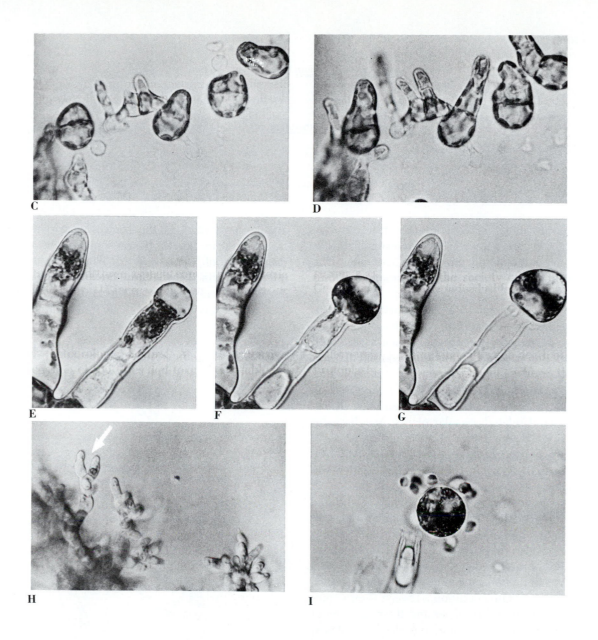

C

D

E

F

G

H

I

The life cycle of *Laminaria* and other kelps is instructive in a number of respects. It is similar to that of *Ulva*, the marine species of *Cladophora* and *Ectocarpus*, in that all have sporic meiosis and alternation of free-living, sporophytic and haploid, gametophytic generations (type D, h + d, Chart 4-1). However, in *Laminaria* and other kelps, the sporophyte and game-

CHART 7-1
Life cycle of *Laminaria*.

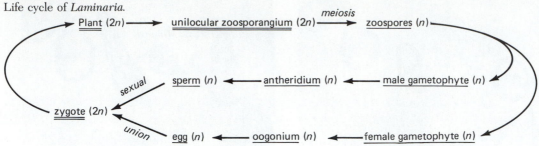

Plant (2*n*) → unilocular zoosporangium (2*n*) —*meiosis*→ zoospores (*n*)

sperm (*n*) ← antheridium (*n*) ← male gametophyte (*n*)

sexual

zygote (2*n*)

union

egg (*n*) ← oogonium (*n*) ← female gametophyte (*n*)

tophyte differ markedly in size, structure, and longevity; alternation here is **heteromorphic,** as in some species of *Bryopsis* (Chart 4-1). The sporophyte of *Laminaria* and other kelps is a large, complex, perennial plant, dominant in the life cycle, whereas the gametophytes are microscopic, few-celled, branching filaments, and relatively ephemeral. It should be noted that both generations are free-living plants and presumably photoautotropic. In balance of the two generations, the life cycle of *Laminaria* and other kelps is practically identical with that of ferns and related vascular plants. As a basis for theoretical discussions of the origin and relation of the alternating generations, it must be borne in mind that various genera of algae in the same aquatic environment illustrate alternation of both similar and dissimilar generations. Among the land plants, the alternating generations always are markedly dissimilar morphologically. The partial retention of the egg, and consequently the zygote, within the oogonium, which occurs in *Laminaria* and other kelps, represents an intermediate condition between their expulsion in many other algae and their permanent retention in *Coleochaete* and the land plants. The significance of these features will be referred to later in our discussion of the land plants.

Other Kelps

Plants of *Macrocystis* (Gr. *makros*, long, + *kystis*, bladder), the giant kelp (Fig. 7-11), are among the largest of the brown algae, specimens 45 m in length being common on the Pacific coast of North America. The large plants are attached in deep water, and the branching stipes bear leafy blades, at the base of each of which is a gas-filled bulb, the **pneumatocyst**

A   B

FIGURE 7-11
Macrocystis integrifolia. **A.** Portion of plant showing holdfasts and branching stipes with air bladders and blades. **B.** Single segment enlarged. **A.** ×1/12. **B.** ×1/4. (*Modified from G. M. Smith.*)

120

FIGURE 7-12
Nereocystis luetkeana. Stipe, pneumatocyst (float)
and blade of freshly collected plant. ×¹⁄₁₆.
(*Courtesy of Prof. Michael J. Wynne.*)

(Fig. 7-11B). *Macrocystis, Nereocystis* (Fig.
7-12), and *Pelagophycus* (Fig. 7-6C), among
other kelps, are of interest in that they have
specialized series of cells, called **sieve tubes**, in
their stipes. These are very similar to the sieve
tubes of the phloem of the vascular plants
(Chapter 13, p. 324). The word "sieve" refers to
the pores present in the terminal walls (Fig.
7-13B) of the adjacent cellular components of
the sieve tube. The sieve tubes here, as in
vascular plants, lack nuclei at maturity.

It has been shown that ^{14}C-labeled products of
photosynthesis move through the stipes of
Macrocystis at rates comparable to movement
of substances in the phloem of vascular plants.

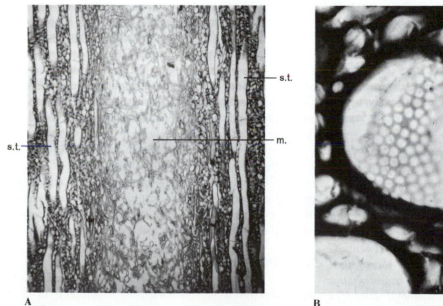

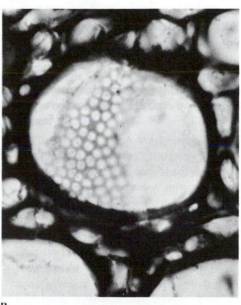

A

B

FIGURE 7-13
Macrocystis pyrifera. **A.** Longisection of stipe showing central medulla and two groups of adjacent
sieve tubes. m., medulla; s.t., sieve tubes. **B.** Transection of same showing portion of sieve plate. **A.**
×75. **B.** ×700. (*Courtesy of Dr. B. C. Parker.*)

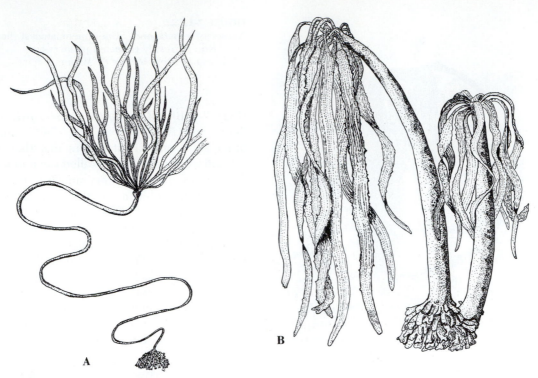

FIGURE 7-14
A. *Nereocystis luetkeana.* **B.** *Postelsia palmaeformis.* **A, B.** ×⅓. (**B.** *Modified from G. M. Smith.*)

Nicholson and Briggs (1972) have demonstrated movement of ¹⁴C-labeled photosynthate through the medulla (containing sieve tubes) of *Nereocystis* at an average rate of 37 cm/hour. There is evidence that this movement occurs in the sieve tubes (Schmitz, 1981).

In *Nereocystis* (Gr. *Nereus*, god of the sea, + Gr. *kystis*, bladder) (Figs. 7-12, 7-14A), a large kelp of the North Pacific, the stipe is usually unbranched and terminates in a large pneumatocyst that bears a profusion of blades.

Postelsia (after A. Postels, German naturalist), the sea palm (Fig. 7-14B) of the Pacific coast of the United States and Canada, grows in the intertidal zone. The stout and fleshy unbranched stipes are anchored to the rocky substrate and may be 60 cm tall. Each bears distally a number of leaflike blades.

In *Macrocystis, Pelagophycus, Nereocystis,* and *Postelsia*, as in *Laminaria* and other kelps, the diploid plants become fertile and by meiosis produce haploid zoospores, which grow into microscopic, oogamous gametophytes; some of the latter bear eggs and others bear sperms. In all the kelps, accordingly, the life cycle is diplobiontic and heteromorphic. A comprehensive account of the kelp beds of California was published by North (1971).

Dictyota and *Padina*

In contrast to the other genera of Phaeophyta discussed in this chapter, which live in cold ocean waters, *Dictyota* (Fig. 7-15A) and *Padina* (Fig. 7-15B) grow in tropical and subtropical latitudes. Both are lithophilic and continuously submerged.

122

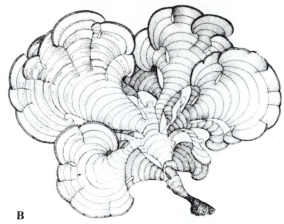

FIGURE 7-15
A. *Dictyota dichotoma*, growth habit. **B.** *Padina vickersiae*, growth habit. **A, B.** ×½. (**B.** *After Taylor.*)

Growth in the blades of *Dictyota* is geometric in precision and may be traced to a prominent, dome-shaped apical cell (Fig. 7-16). The derivatives of this cell divide twice in a plane parallel to the surface of the blade so as to form two cortical layers of small, plastid-rich cells and, between these, a medullary layer of larger cells with fewer plastids (Fig. 7-17B).

The life cycle of *Dictyota* was long ago shown to be D, h + d and isomorphic. The gametophytes are dioecious and oogamous (Fig. 7-17A–D) and the sperms uniflagellate with a tinsel flagellum. Both oogonia and antheridia, the latter like plurilocular gametangia, are borne in groups, or sori, on the surface of the plants. The zygote grows rapidly into the sporophytic phase, which, except for chromosome constitution and reproductive cells, is identical

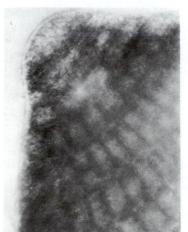

FIGURE 7-16
Dictyota sp. **A.** Surface view of tip of living plant. **B.** Origin of dichotomy by equal division of apical cell, diagrammatic. **A.** ×¼. (**B.** *After Oltmanns.*)

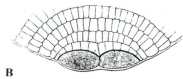

123

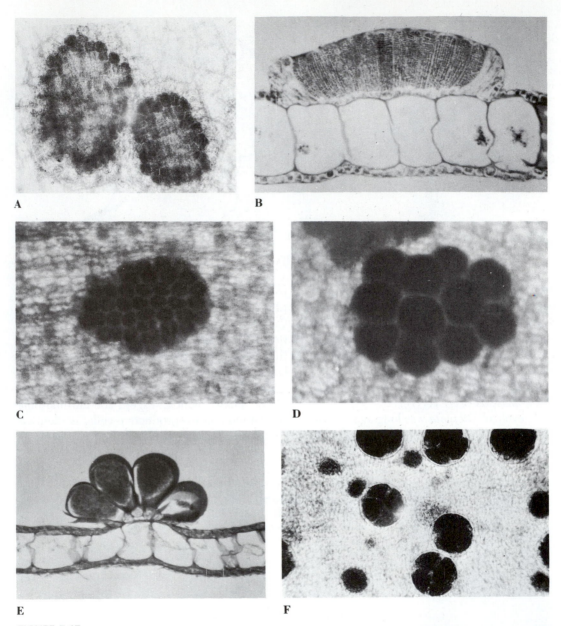

FIGURE 7-17

Dictyota binghamiae (**B–E**) and *D. dichotoma* (**A, F**), reproduction. **A.** Sorus of male gametangia. **B.** Section through a blade and male sorus. **C, D.** Oogonial sori. **E.** Section of a blade within mature tetrasporangial sorus. **F.** Tetrasporangia on thallus surface. **A.** ×100. **B.** ×165. **C.** ×100. **D, E.** ×165. **F.** ×150. (**B–E.** *Courtesy of Drs. M. Foster, M. Neushul, and E. Y. Chi.*)

with the gametophytes. The unilocular sporangia develop on the surfaces of the sporophyte and after meiosis produce four nonflagellate spores (Fig. 7-17E–F), which develop into gametophytes. The life cycle of *Padina*, the "sea fan," is like that of *Dictyota*.

The Rockweeds: *Fucus* and *Sargassum*

The widely distributed genera *Fucus* (L. *fucus*, from Gr. *phykos*, seaweed) (the rockweed) (Figs. 7-18 through 7-20) and *Sargassum* (Sp. *sargazo*, seaweed) (Fig. 7-25) represent still a third type of life cycle that occurs among the

A

B

FIGURE 7-19
Fucus vesiculosus. **A.** Plants attached to rocks, incipiently emergent with falling tide. **B.** Plants exposed at low tide; note receptacles and air bladders. (**B.** *Courtesy of Dr. Arthur Cronquist.*)

Phaeophyta. While four species of *Fucus* have been grown from zygotes in the laboratory, only one, *F. distichus*, attained sexual maturity (McLachlan, Chen, and Edelstein, 1971).

Fucus commonly grows attached to rocks in the intertidal zone (Fig. 7-19), where the plants are exposed at low tide. The plant body, which may attain a length of 2 m in certain species, is leathery, flattened, and dichotomously branched. Growth is initiated by the divisions of several clearly differentiated apical cells, derivatives of which, by subsequent division, enlarge-

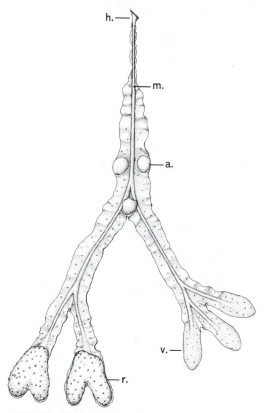

FIGURE 7-18
Fucus vesiculosus. a., air vesicle; h., holdfast; m., midrib; r., receptacle with conceptacles; v., vegetative apex. ×½.

125

ment, and differentiation, build up a rather complex plant body. The plants are attached by multicellular holdfast discs. Prominent **midribs**, **cryptoblasts** (probably sterile conceptacles, the latter described below), and **air bladders** occur in some species (Fig. 7-19A, B).

The plant body of *Fucus* and other rockweeds is quite complex histologically. The derivatives of the apical cells differentiate into an epidermis, a cortex, and a central region of branching filaments and fibers, the last two embedded in a slimy, polysaccharide matrix. The walls of all the cells are thick and fibrillar and are composed of alginic acid, while the amorphous matrix seems to be fucoidin, a sulphated polysaccharide. The latter is produced within the cells, as is the alginic acid, and the fucoidin is secreted through the fibrillar (alginic acid walls) into the intercellular spaces (McCully, 1966).

The production of reproductive cells is localized at the tips of the branches in fertile areas called **receptacles** (Figs. 7-18, 7-19B), which become enlarged and distended because of the internal secretion of large quantities of hydrophilic compounds (Fig. 7-20). The receptacles bear scattered, pustulelike cavities, the **conceptacles**, which communicate with the surrounding water through narrow ostioles, through which tufts of colorless filaments protrude (Fig. 7-20). At maturity, the conceptacles bear eggs and sperms; either these may be in the same conceptacle (monoecism), or those that produce the eggs may be on different plants (dioecism) from those producing sperms, depending on the species. Thus, plants of *F. vesiculosus* are usually dioecious (Figs. 7-20, 7-21); certain other species found along the Atlantic coast of North America, *F. spiralis* and *F. distichus*, are monoecious.

The sperms are laterally biflagellate and produced in groups of 64 from antheridia developed on branching filaments from the wall of the conceptacle (Figs. 7-20A, 7-21A, 7-22). In *Fucus*, each oogonium, also an outgrowth from

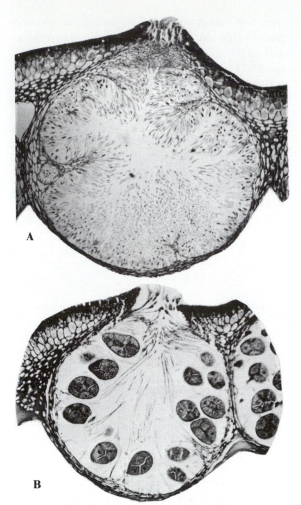

FIGURE 7-20
Fucus vesiculosus. **A.** ♂ (antheridial) conceptacle in median section. **B.** Median section of ♀ (oogonial) conceptacle. **A.** ×60. **B.** ×45.

the conceptacle wall, produces eight eggs (Figs. 7-20B, 7-21B). The conceptacles contain colorless paraphyses, which undergo basal growth. Young oogonia and antheridia are uninucleate (Fig. 7-22A). Meiosis occurs during the first two nuclear divisions in these structures, the plants themselves being diploid. Large amounts of

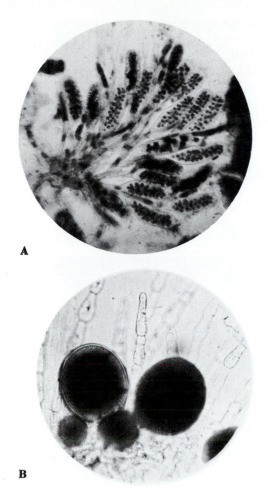

A

B

FIGURE 7-21
Fucus vesiculosus. **A.** Stages in development of
antheridia (acetocarmine preparation). **B.**
Developing oogonia and paraphyses (living). **A.**
×320. **B.** ×135.

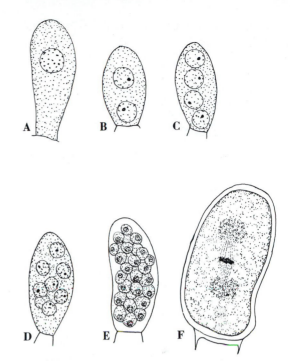

FIGURE 7-22
Fucus vesiculosus. **A–E.** Stages in development of
an antheridium. **F.** Young oogonium, first meiotic
metaphase; note polar asters. ×770.

alginic acid and fucoidin are synthesized, especially within the oogonia, and contribute to the
development of the three-layered oogonial wall
(McCully, 1968).

Liberation of gametes is closely connected
with tidal conditions in some species. At low
tide, when the plants are exposed to the drying
action of the air, shrinkage of the plant body

may be accompanied by extrusion of ripe oogonia and antheridia in slimy masses through the
ostioles to the surface of the plant. The in-
coming tide, in submerging these droplets
containing the sex organs, effects swelling and
dissolution of their walls, so that the individual
gametes are set free in the water. However, in
other species, extrusion of gametes occurs in
continuously submerged plants.

The eggs are large, spherical, and nonmotile
and are penetrated by individual sperms, which
swarm about the eggs in great numbers (Fig.
7-23A). The sperms of *Fucus* are unusual in that
the posterior flagellum is longer than the anterior (Fig. 7-24E). Fertilization in *F. distichus*

127

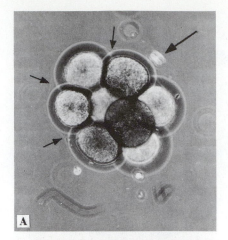

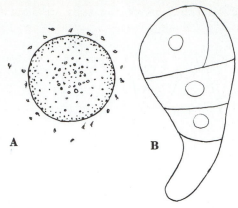

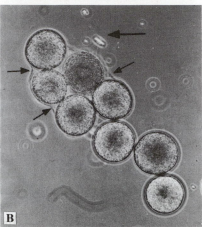

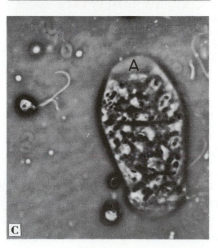

FIGURE 7-23
A. *Fucus serratus.* Egg surrounded by sperm. **B.**
Fucus vesiculosus. Young plant from germinating
zygote. **A.** ×250. **B.** ×300. (**A.** *After Thuret and
Bonnett.* **B.** *After Nienburg.*)

has been studied by Pollock (1970), whose illustrations are produced as Fig. 7-24 and explained
in its caption. The outermost layer of the three-
layered oogonium ruptures so that the eggs are
enclosed by two layers when they leave the
conceptacles. In *F. distichus*, about 20 min
elapse before the sperms emerge from the freed

FIGURE 7-24
Fucus distichus. Discharge of eggs and fertilization.
A. Freshly discharged group of eight eggs still
enclosed by an oogonial membrane (arrows); the
large arrow indicates a pore through which sperm
can enter. **B.** Eggs being released from oogonial
membrane. **C.** Freshly liberated antheridium
containing nonmotile sperm; a motile sperm with
crossed flagella is visible at the left. **D.**
Protuberance on an egg opposite the pore in the
oogonial membrane. **E.** Motile sperm; note greater
length of trailing flagellum. **F.** Unfertilized and **G.**
fertilized eggs in saturated ZnCl solution. **A.**
antheridium; **H, L.** Blebs from egg surface; **M.**
oogonial membrane; **E.** egg. **A, B.** ×424. **C.**
×2900. **D.** ×2280. **E.** ×5000. **F, G.** ×1125. (*After
Pollock.*)

128

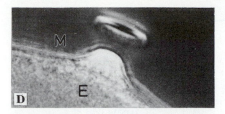

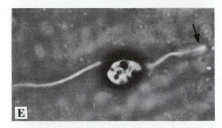

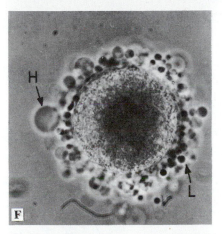

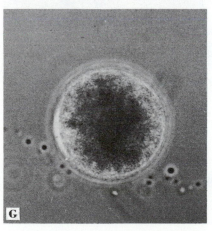

antheridia and before the eggs are liberated from the remaining oogonial wall layers (Fig. 7-24).

The eggs of *Fucus serratus* also produce a volatile sperm attractant designated **Fucoserratene**. Chemically this is 1,3-trans-5-cis-octatriene (Jaenicke, 1977). Nuclear union follows plasmogamy, and the resulting zygote secretes a thin wall and germinates (Fig. 7-23B), without a period of dormancy, into a new *Fucus* plant. The life cycle[3] of *Fucus* is summarized in Chart 7-2.

It seems clear from this summary that the life cycle of *Fucus*, like that of some species of *Bryopsis, Codium, Caulerpa,* and other siphonous green algae, falls into type M, d (p. 85),[4] in which a diploid organism undergoes gametic meiosis and the zygote grows directly into the new plant.

---

[3]Diploid phases are underscored with double lines, haploid ones with single lines.

[4]The four nuclei present in the oogonia and antheridia at the conclusion of the first two nuclear divisions in these organs have been interpreted by some to be homologous with the nuclei of microspores and megaspores (produced in groups of four from their mother cells) in heterosporous land plants (p. 356). If this were true, one would have to look on the "oogonium" and "antheridium," up to the four-nucleate stage, as a "megasporangium" and "microsporangium," respectively. The ensuing nuclear division, therefore, would be analogous to those that occur in microspores and megaspores in their production of gametophytes. In the case of *Fucus*, however, the gametophytic phase would be markedly abbreviated and would approach a condition most similar to that in certain seed plants, in which gametophytes complete development while the spores that produce them are still retained in their sporangia. According to this interpretation, the alternation of generations in *Fucus*, like that in the flowering plants, involves a dominant diploid sporophyte and a much-reduced gametophyte, the latter represented by only a few nuclear and cell generations.

CHART 7-2
Life cycle of *Fucus*.

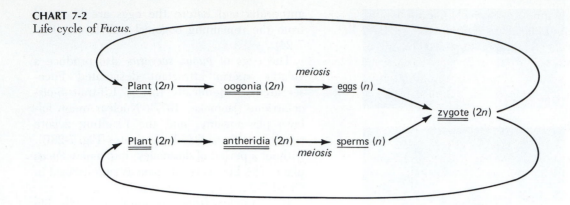

Plant (2*n*) ⟶ oogonia (2*n*) ⟶ *meiosis* ⟶ eggs (*n*) ⟶ zygote (2*n*)

Plant (2*n*) ⟶ antheridia (2*n*) ⟶ *meiosis* ⟶ sperms (*n*) ⟶ zygote (2*n*)

Sargassum, unlike *Fucus*, is, with the exception of one or two species, largely a plant of warm marine waters. *Sargassum* is complex in having leaflike organs, stalked air bladders, and much-modified receptacles (Fig. 7-25). In *Sargassum*, seven of the eight nuclei disintegrate, so that each oogonium gives rise to only a single egg. *Sargassum natans*, the "Gulf weed," is abundant in the Gulf of Mexico, and tremendous quantities of it are often washed onto Gulf beaches after tropical storms.

Summary and classification

The above account of *Ectocarpus*, *Laminaria* (and other kelps), *Dictyota* and *Padina*, *Fucus*, and *Sargassum* summarizes three basic types of life cycles and reproduction that occur among the Phaeophyta. Most of the remaining genera of this large alliance correspond in reproduction to one of the types described above; the generic differences are based largely on vegetative morphology. The simplest among them are filamentous, unicellular and colonial species being unknown. As a group, the Phaeophyta are sharply characterized and distinct from other algae in their pigmentation and photosynthate,

FIGURE 7-25
Sargassum filipendula. Portion of plant. a., air vesicle; r., immature receptacles; v., leaf-like vegetative branch. ×¾.

in their almost exclusively marine habitat, their laterally biflagellate reproductive cells, and the complexity of the plant body in size and internal differentiation that is achieved in certain genera. Their life cycles have counterparts in certain Chlorophyta, on the one hand, and in certain vascular plants, on the other. However, no phaeophycean genus with zygotic meiosis has yet been described. The representative genera of Phaeophyta discussed in this chapter may be classified as follows:

Division Phaeophyta
  Class 1. Phaeophyceae
    Order 1. Ectocarpales
      Family 1. Ectocarpaceae
        Genus: *Ectocarpus*
    Order 2. Laminariales
      Family 1. Laminariaceae
        Genus: *Laminaria*
      Family 2. Lessoniaceae
        Genera: *Macrocystis, Nereocystis,*
           Pelagophycus, Postelsia
    Order 3. Dictyotales
      Family 1. Dictyotaceae
        Genera: *Dictyota, Padina*
    Order 4. Fucales
      Family 1. Fucaceae
        Genera: *Fucus, Sargassum*

DISCUSSION QUESTIONS

1. Describe the life cycle and reproduction of *Ectocarpus, Laminaria, Dictyota,* and *Fucus.*
2. Compare these genera with respect to form of plant body and localization of growth.
3. What possible significance can you attach to the partial retention of the egg, zygote, and young embryo at the mouth of the oogonium of *Laminaria* and other kelps?
4. Do you think that Phaeophyta may have originated from the Chlorophyta? Give reasons for your answer.
5. If the Chlorophyta and Phaeophyta are not closely related in your opinion, how do you interpret the occurrence of the filamentous habit, holdfasts, zoospores, oogamous reproduction, and alternation of generations in both groups?
6. Distinguish between cytological, morphological, isomorphic, and heteromorphic alternation of generations, giving examples from genera of Phaeophyta.
7. How do the unilocular sporangia of *Dictyota* differ from those of other Phaeophyta?

Chapter 8
DIVISIONS CHRYSOPHYTA AND PYRRHOPHYTA

DIVISION CHRYSOPHYTA

The division Chrysophyta includes three classes of algae in the plastids of which carotenes and xanthophylls are prominent in relation to the chlorophylls; in this respect (and others) they resemble the brown algae (Phaeophyta, Chapter 7) (Table 4-1, p. 46). The cells, accordingly, are varying shades of yellow-green and brown. Their excess photosynthate is never stored as starch but in the form of another carbohydrate or oil. The cell walls contain cellulose, may be silicified in some, and may be composed of two articulated portions. The Chrysophyta, as here discussed, may be subdivided as follows:[1]

Class 1. Xanthophyceae, the yellow-green algae
Class 2. Chrysophyceae, the golden-brown algae
Class 3. Bacillariophyceae, the diatoms

A brief account of representatives of these three classes is given in the present chapter. More than 325 genera and 6000 species of Chrysophyta have been described.

CLASS 1. XANTHOPHYCEAE

General features.   The Xanthophyceae (Gr. *xanthos*, yellow, + Gr. *phykos*), with about 80 genera and 450 species, are called the "yellow-green algae" because their color is distinctly that hue, especially if they are compared directly with members of the Chlorophyta. They were classified formerly among the Chlorophyceae until it was recognized that several of their attributes differ markedly from those of that group. They are sometimes assigned to divisional rank as the Xanthophyta.

The yellow-green color results from a combination of pigments, among them chlorophyll α and β-carotene, and several xanthophyll pigments. The pigments are localized in plastids that

---

[1]Six classes have been recognized (Bold and Wynne, 1985).

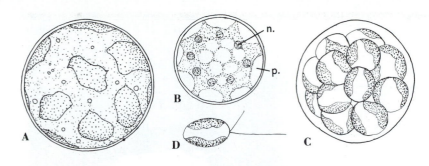

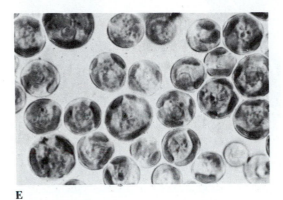

FIGURE 8-1
Botrydiopsis arhiza. **A.** Vegetative cell, surface view. **B.** Stained cell. **C.** Zoospore formation. **D.** Single liberated zoospore. **E.** Photomicrograph of living vegetative cells. n., nucleus; p., plastid. **A–D.** ×1500. **E.** ×800.

are usually lens- or disc-shaped. Droplets of oil and granules of a substance called **leucosin**, or **chrysolaminaran**, related chemically to phaeophycean laminaran, are frequently observable in cells of Xanthophyceae. In many genera—*Tribonema*, for example—the cell wall is not homogeneous but is composed of overlapping segments (Figs. 8-2B, 8-4). This attribute is not demonstrable in all Xanthophyceae, however, except by special chemical treatment. The cell wall is silicified in some genera. The flagella of Xanthophyceae are of unequal length (Figs. 8-1D, 8-3), the longer tinsel and the shorter whiplash in organization, as in the Phaeophyta.

The Xanthophyceae are predominantly freshwater organisms, but they may be aerial (moist rocks and other vegetation) or terrestrial in habitat. A few are marine. A number of species have been isolated into culture from subterranean soil samples. The genera of Xanthophyceae, now grouped in a separate class, for-

merly were distributed among the orders and families of Chlorophyceae, the members of which they resemble in body structure. Removal of these xanthophycean genera into a separate class revealed that they comprise a series of body types largely parallel to those described in the Chlorococcaceae. Space does not permit discussion of a complete array of parallel genera, but four commonly occurring and readily available xanthophycean organisms will be described.

Botrydiopsis

Botrydiopsis (Gr. *botrydion*, in clusters, + Gr. *opsis*, resemblance), the simplest of these, is unicellular. *Botrydiopsis* (Fig. 8-1) occurs on and in soil, from which it may be readily isolated into unialgal culture. The spherical cells are thin-walled and contain an increasingly large number of lenticular plastids as the cells grow older and larger. The cells are multinucleate

133

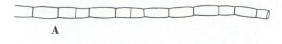

A

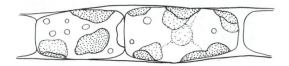

B

FIGURE 8-2
Tribonema sp. **A.** Outline drawing of a segment of a filament. **B.** Two vegetative cells, the one at the left in surface view, the other in median optical section. **A.** ×75. **B.** ×500.

(Fig. 8-1B). As in *Chlorococcum*, each cell undergoes cleavage to form a number of zoospores (Fig. 8-1C, D), the number varying with the size of the cell. The zoospores lack walls and have two flagella of unequal length; the longer of these bears two rows of stiff hairs. After a short period of motility, they become spherical, develop walls, and grow. Sexual reproduction has not been observed in *Botrydiopsis*.

Tribonema

Tribonema (Gr. *tribein*, to rub, + Gr. *nema*, thread) (Figs. 8-2 through 8-4) is representative of the unbranched, filamentous Xanthophyceae. *Tribonema* is cosmopolitan and occurs as floating masses and as overgrowth on submerged sticks and aquatic vegetation during the cooler months of the year. The uniseriate cells are often shaped like slightly inflated cylinders. Each contains a single nucleus and several discrete, discoidal, decidedly yellow-green plastids (Fig. 8-2B). *Tribonema* clearly illustrates the fact that certain Xanthophyceae have walls composed of overlapping halves. When the filaments break apart or dissociate, the wall sections may readily be observed to consist of H-shaped segments (Fig. 8-4), as viewed in

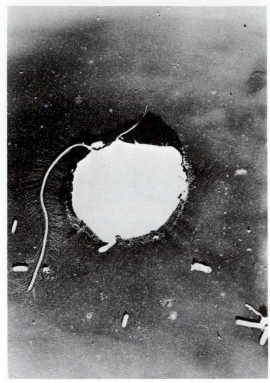

FIGURE 8-3
Tribonema vulgare. Shadowcast zoospore showing two unequal flagella, the longer bearing two rows of hairs. ×3000. (*After Massalski and Leedale.*)

optical section. They actually consist of segments of cylinders joined together by a planar, disclike wall. At the conclusion of cytokinesis in a given vegetative cell, the two daughter protoplasts form such a wall segment within the original wall of the parent cell. As in *Ulothrix* among the Chlorophyceae, *Tribonema* reproduces by forming zoospores, which arise within the vegetative cells; *Tribonema* zoospores (Fig. 8-3) have flagella of unequal length, the longer with two rows of appendages (Massalski and Leedale, 1969). The longer flagellum is forward during motion, and the shorter is directed posteriorly or laterally. The germling produced by a zoospore has a holdfast, but the mature filaments are rarely encountered in an attached condition. Union of isogamous gametes has also been reported in *Tribonema*.

Botrydium

The terrestrial genus *Botrydium* (Gr. *botry-dion*, in clusters), of widespread occurrence on damp soil, often grows in association with *Proto-siphon*, its Chlorophytan counterpart, with which it was long confused. The cells of *Botry-dium* (Figs. 8-5, 8-6) consist of an inflated, epiterranean vesicle and a rhizoidal system; the latter is usually richly branched. Under appro-priate conditions of laboratory culture (Fig. 8-6A) the vesicular portion of the plant is promi-nent, and in nature it may attain a size of 2 mm; it is frequently ornamented with granules of calcium carbonate (Fig. 8-6B). Mature plants of *Botrydium* contain a thin peripheral layer of protoplasm surrounding an extensive central vacuole. The protoplasm is composed of a su-perficial layer of chloroplasts (Fig. 8-5B), and slightly centripetal to these are numerous min-ute nuclei embedded in colorless cytoplasm. The rhizoidal portion of the plant contains few, if any, plastids and is filled with highly vacuolate protoplasm. *Botrydium* reproduces by zoospore and aplanospore formation. The zoospores (Figs. 8-5C, 8-6D) arise by cleavage of the protoplast when the vesicles are submerged in water during rains. The zoospores are faculta-tive, isogamous gametes, which may unite in pairs to form zygotes. Zoospores may also devel-op into new plants without union. Meiosis in *Botrydium* is probably zygotic, but this has not been demonstrated unequivocally.

Vaucheria

Vaucheria (in honor of Vaucher, a Swiss phy-cologist), the "water felt" (Figs. 8-7 through 8-9), is a widely distributed member of the Xanthophyceae; the genus was formerly includ-ed among Chlorophyta. Careful examination of its pigments, of its photosynthetic storage prod-

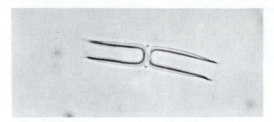

FIGURE 8-4
Tribonema sp. H-shaped wall segment. ×1000.

ucts, and of the flagellation of its motile cells (Fig. 8-7D, E) demonstrated that the affinities of *Vaucheria* are with the Xanthophyceae. Spe-cies of *Vaucheria* may be amphibious, like cer-tain liverworts. Some flourish as dark-green mats in running water or floating or submerged in quiet pools or on moist, undisturbed soil like that in greenhouse flowerpots. The plant body consists of an elongate, sparingly branched tube (Fig. 8-9) from which septations are absent except in the reproductive stages or as a re-

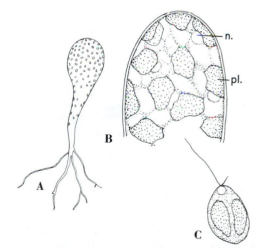

FIGURE 8-5
Botrydium granulatum. **A.** Small vegetative plant. **B.** Apex of young sac. **C.** Zoospore; n., nucleus; pl., plastid. **A.** ×60. **B, C.** ×1500.

135

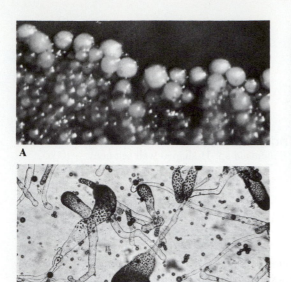

A

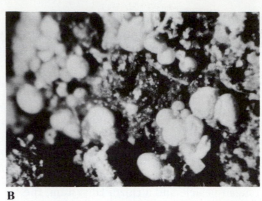

B

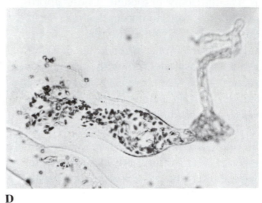

D

C

FIGURE 8-6
Botrydium granulatum. **A.** Portion of an agar
culture. **B.** Calcified plants growing on soil. **C.** Sacs
from an agar culture. **D.** Zoosporogenesis. **A.** ×5.
B. ×7. **C.** ×15. (**C.** *Courtesy of Mrs. Elena Sun.*)

A

B   C

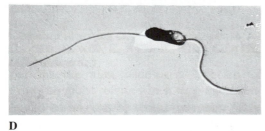

D

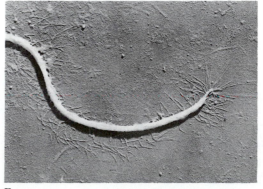

E

FIGURE 8-7
Vaucheria sp. **A.** Apex of plant; the tip in surface
view, the remainder in optical section. **B.**
Germinating zoospore. **C.** Stained sperm of *V.
pachyderma.* **D, E.** *V. sescuplicaria.* **D.** Sperm. **E.**
Anterior sperm flagellum. **A.** ×350. **B.** ×75. **C.**
×1750. **D.** ×3000. **E.** ×10,000. (**C.** *After Koch.* **D,
E.** *After Moestrup.*)

136

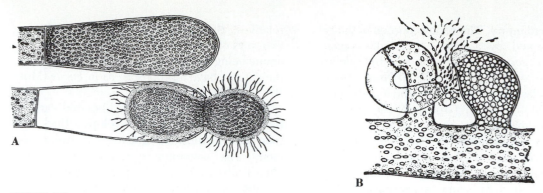

FIGURE 8-8
A. *Vaucheria* sp. Zoospore formation. **B.** *V. sessilis.* Mature antheridium (left) and oogonium. **A.** ×286. **B.** ×185. (*From G. M. Smith after Couch.*)

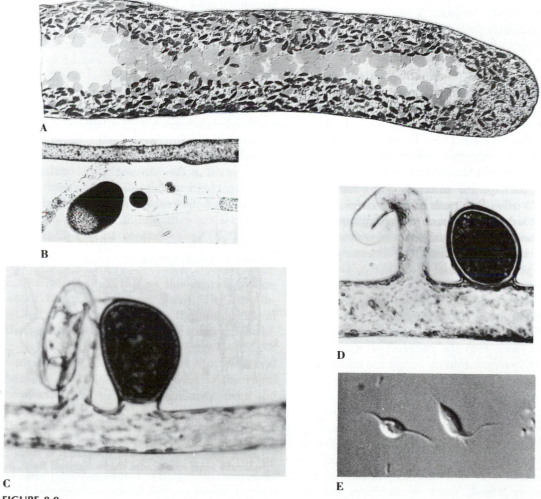

FIGURE 8-9
Vaucheria. **A.** *V. pachyderma*, longisection of vegetative branch. **B.** *V. sessilis*, recently liberated zoospore. **C.** *V. sessilis* var. *sessilis*, oogonium containing zygote and antheridial branch with empty antheridium. **D.** *V. sessilis* var. *clathrata*, oogonium with zygote, ♀ branch with empty antheridium. **E.** *V. sessilis*, sperm; note unequal flagella. **A.** ×612. **B.** ×175. **C.** ×340. **D.** ×210. **E.** ×6250. (*Courtesy of Dr. D. W. Ott.*)

sponse to injury. The central portion of the tube is occupied by a large, continuous vacuole, which is separated from the wall by a delicate peripheral layer of protoplasm. Numerous discoidal plastids, which overlie the minute nuclei, occur in this layer. Growth of the siphonlike tubes is apical.

Asexual reproduction in the coenocytic *Vaucheria* plant is effected by the formation of large zoospores at the tips of the tubes, which are delimited by septa as **zoosporangia** (Fig. 8-8A). The protoplast in the sporangium contracts, and an interchange of position between nuclei and plastids occurs, so that the nuclei are now nearer the surface. A pair of slightly unequal flagella is then generated from the surface of the protoplast in the region of each beak-shaped nucleus (Fig. 8-8A). The large compound zoospore is liberated from the terminal zoosporangium and undergoes rather slow, narrowly circumscribed movements. It soon loses its flagella and germinates, frequently from both poles, to form a new *Vaucheria* siphon.

Sexual reproduction, which is oogamous, is rather striking in *Vaucheria* because of the large size of the sex organs (Figs. 8-8B, 8-9C, D). These may be sessile on the main siphons, or they may occur in groups on special reproductive branches. Both sex organs arise as protuberances, into which the streaming protoplasm carries numerous nuclei and plastids. They become segregated from the subtending branch relatively late in their ontogeny. The oogonium is at first multinucleate, but prior to formation of the delimiting septum, all the nuclei except one migrate back into the subtending siphon. At maturity, each oogonium contains a single uninucleate egg cell. The sperms enter through a pore in the oogonial wall in a special receptive region. The antheridium, which is multinucleate when it is delimited by a septum from its subtending branch, produces a large number of minute, almost colorless, unequally flagellate sperms (Fig. 8-9E). The flagellation of the sperms is lateral, much as in *Fucus* and its

relatives, the short anterior flagellum having two rows of appendages and the longer posterior one being without appendages. Furthermore, the sperms have a proboscis, as in *Fucus* (Moestrup, 1970*b*).

In nature, the sperms are liberated early in the morning. The zygote develops a thick wall soon after fertilization and loses its green pigment. The oogonium containing the zygote often is abscised from the parent branch. Germination into a few filaments after a period of dormancy probably involves zygotic meiosis.

Summary and classification

This brief account of *Botrydiopsis*, *Tribonema*, *Botrydium*, and *Vaucheria* has been presented to provide some insight into the attributes of the yellow-green algae, the Xanthophyceae. In spite of the paucity of genera described, it should be apparent that the Xanthophyceae exhibit parallelisms with the Chlorophyceae insofar as body form is concerned. This parallelism is reflected in the classification of the Xanthophyceae into orders that correspond to those of the Chlorophyceae. The genera described above may be classified as follows:

Division Chrysophyta
    Class 1. Xanthophyceae
        Order 1. Mischococcales
            Family 1. Pleurochloridaceae
                Genus: *Botrydiopsis*
        Order 2. Tribonematales
            Family 1. Tribonemataceae
                Genus: *Tribonema*
        Order 3. Vaucheriales
            Family 1. Botrydiaceae
                Genus: *Botrydium*
            Family 2. Vaucheriaceae
                Genus: *Vaucheria*

The Mischococcales correspond to the Chlorococcales of the Chlorophyceae; similar pairs of orders are the Tribonematales and Ulotrichales and the Botrydiales and Caulerpales. In spite of the parallelisms, the Xanthophyceae are clearly recognizable by their lack of starch, their pigmentation, their unequal flagellation, and the overlapping construction of the cell walls in many genera. There is no impressive evidence of relationship between the Xanthophyceae and the Chlorophyceae, but the former do have characteristics in common with the Chrysophyceae (see below).

On the basis of cytology and ultrastructure, it has been found (Hibberd and Leedale, 1970) that the Xanthophyceae is not a homogeneous class but contains two series. One of these, it is suggested, should be removed as a distinct class, Eustigmatophyceae. In these organisms the stigma occurs outside the plastid, there is a single emergent flagellum on the motile cells, girdle lamellae are absent from the plastids, and an unusual type of stalked pyrenoid is present in the vegetative cells.

CLASS 2. CHRYSOPHYCEAE

General features. The Chrysophyceae (Gr. *chrysos*, gold, + Gr. *phykos*), or golden-brown algae, consisting of about 70 genera and 325 species, are widely distributed in fresh and salt water, but with the exception of a few genera they are rarely encountered in any great number. Many of the species are planktonic and flourish in bodies of cold water or only in the colder months of the year. The golden-brown color is the result of a combination of pigments, including chlorophylls *a* and *c*, β-carotenes, and several xanthophylls (Table 4-1, pp. 46–47), the abundance of β-carotene and of the xanthophylls masks the chlorophyll in most species. The excess photosynthate is stored in the form of oil droplets or as rather large granules of chrysolaminaran. In a great majority of genera the cells contain one or two parietal plastids and are uninucleate. The surface of the protoplast is often ornamented with small siliceous scales, or an internal siliceous skeleton may be present.

The Chrysophyceae, like the Xanthophyceae, represent a series in which types of plant body have developed in a manner parallel to that observed in the Chlorophyceae. In addition, in spite of the fact that the cells contain pigmented plastids, a number of motile genera carry on phagotrophic nutrition. Many Chrysophyceae are capable of forming siliceous cysts, which are often ornamented in various ways. Although a considerable number of genera and species of Chrysophyceae have been found in this country and abroad, with few exceptions they do not seem to be organisms that appear with frequency in collectors' jars. Furthermore, few of them have been grown in culture in the laboratory. For these reasons, only a few, relatively widely distributed genera—*Ochromonas* (Gr. *ochros*, pale yellow, + Gr. *monas*, single organism), *Synura* (Gr. *syn*, together, + Gr. *oura*, tail), and *Dinobryon* (Gr. *dinos*, whirling, + Gr. *bryon*, moss)—will be described.

Ochromonas

Ochromonas (Fig. 8-10A) is a unicellular motile organism that varies from a spherical to a somewhat irregular shape. The naked cells contain one or two pale-yellow plastids, which are curved and parietal. Two very unequal flagella, the longer tinsel and the shorter whiplash,

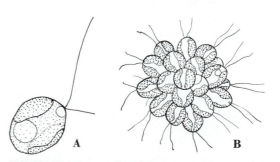

FIGURE 8-10
A. *Ochromonas* sp. Single vegetative cell. **B.** *Synura uvella.* Single colony. **A.** ×1500. **B.** ×700.

emerge from the anterior portion of the protoplast. Each cell contains a single nucleus; contractile vacuoles and a stigma are present in the cells of certain species. Actively photosynthetic cells usually contain a single, large, posterior grain of excess photosynthate, called chrysolaminaran. Reproduction is by cell division. The cells may form siliceous cysts with prominent pores.

Various species of *Ochromonas* differ physiologically. In two species of *Ochromonas*, *O. danica* and *O. malhamensis*, distinct metabolic differences occur, which are directly reflected in the relative amounts of growth. Growth of *O. danica* is fairly rapid in inorganic medium, while *O. malhamensis* will not grow well unless organic carbon sources (sugars, etc.), amino acids, and vitamins are added to the growth medium. The mode of nutrition also varies between the above species, because *O. malhamensis* and *O. danica* (Cole and Wynne, 1974) have the capacity to ingest particulate food materials, like protozoa. *Ochromonas danica* secretes carbohydrates, nucleic acidlike molecules, proteins, and lipids into its environment (Aaronson et al. 1971).

Synura

Synura (Fig. 8-10B) is a motile colonial organism in which the individual cells are stipitate and united into spherical clusters. The cells are ovoid to elongate, and the anterior pole is broader than the stipitate posterior pole. Each is biflagellate and covered with delicate siliceous scales. The protoplast contains two parietal, concave plastids, a single nucleus, and contractile vacuoles. Cell division augments the number of individuals in a colony. Multiplication is accomplished by fragmentation of the colony and continued growth of the fragments.

Dinobryon

The colonies of *Dinobryon* (Fig. 8-11) are branching and composed of a series of urn- or bell-like loricas, usually widely separated from

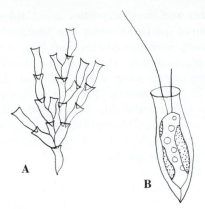

FIGURE 8-11
Dinobryon setularia. **A.** Arborescent colony. **B.** Single lorica with protoplast enclosed. **A.** ×225. **B.** ×625. (**A.** *From G. M. Smith.*)

their contained protoplasts except at the base. Each protoplast has two apical flagella (one of which, a tinsel flagellum, is markedly longer than the other, which is of the whiplash type), usually two plastids, a single nucleus, contractile vacuoles, and a stigma. After longitudinal division, one of the daughter cells moves to the mouth of the lorica, becomes fixed there, and forms a new lorica. Continuation of this process through a number of divisions results in the formation of dendroid colonies. These fragment readily, and the fragments continue to grow as individual colonies. Isogamous union of two vegetative individuals has been reported in *D. borgei*.

Summary and classification

In addition to the genera described above, nonmotile unicellular colonial and filamentous forms of Chrysophyceae are known. One group of unicellular forms has cells that are amoeboid; the production of pseudopodia effects their movement and is involved in the ingestion of solid foods, an animallike attribute. The motile Chrysophyceae are most abundant in numbers of genera and species. However, the occurrence of filamentous types indicates that the group contains plantlike organisms as well as flagellates. The organisms considered in the present account are usually classified as follows:

Division Chrysophyta
  Class 2. Chrysophyceae
    Order 1. Ochromonadales
      Family 1. Ochromonadaceae
        Genera: *Ochromonas, Dinobryon*
      Family 2. Synuraceae
        Genus: *Synura*

CLASS 3. BACILLARIOPHYCEAE

General features. The diatoms, Bacillariophyceae, are at once the best known, the most numerous (in number of genera, species, and individuals, about 200 genera, 5000 species), and, economically, the most important members of the division Chrysophyta. Diatoms are the despair of the amateur and the joy of the professional microscopist because of their structural complexity. Their great beauty and perfection of design rival those of the desmids, but the beauty of diatoms is perhaps more subtle, since it is mostly confined to their coverings or frustules. To appreciate this beauty in full measure, it is usually necessary to dissolve the protoplast with acid and to mount the cells in a highly refractive medium, a process known as "cleaning" the diatoms.

Although some diatoms are bottom dwellers and epiphytes in salt and fresh water, a great number occur in the plankton, where they are of inestimable and basic value in the nutritional cycle of aquatic animals. They have been extant at least since the Jurassic period (inside front cover). The abundance of diatoms in earlier geological periods is attested by the finding of great deposits of their **frustules.** As individual diatoms died in certain bodies of water, they sank to the bottom; there the protoplasts disintegrated, leaving the siliceous frustules. In this way there were built up great deposits (Fig.

FIGURE 8-12
Great deposit of diatomaceous earth being mined at Lompoc, California. (*Courtesy of Johns Mansville Products Corp.*)

8-12), which were exposed in later geological periods and which are now mined as **diatomaceous earth** for use in industrial and technical processes. The economic importance of this substance is tremendous, and its uses are many.[2]

At the present time, living diatoms are ubiquitous and important components of algal vegetation. In bodies of fresh water they seem to be more abundant when the temperature is low. In marine habitats they often cover other algae with a heavy epiphytic growth. In running water they often form a brownish coating on submerged rocks and other vegetation.

---

[2]For a fascinating and informative account of this and other aspects of diatoms, see Paul S. Conger, *Significance of Shell Structures in Diatoms*, Smithsonian Report, 1936, pp. 325–344.

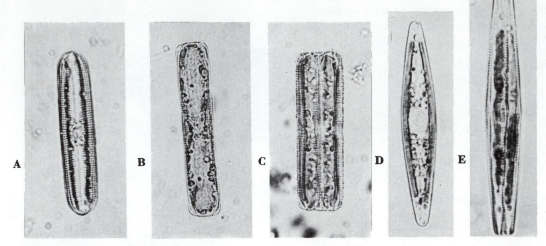

FIGURE 8-13
Living pennate diatoms. **A.** *Pinnularia* sp., valve view. **B.** *Pinnularia* sp., girdle view. **C.** *Pinnularia* sp., recently divided cells. **D.** *Navicula* sp., valve view. **E.** *Navicula* sp., recently divided cells, girdle view. ×500.

Diatoms may be strictly unicellular, colonial, or filamentous. They are divided into two types on the basis of symmetry. In the first, the **pennate diatoms**, exemplified by such genera as *Navicula* (L. *navicula*, small ship) and *Pinnularia* (L. *pinnula*, small feather) (Fig. 8-13), the symmetry is bilateral. The second group, the **centric diatoms**, to which many marine genera belong, is characterized by radial symmetry. *Melosira* (Gr. *melos*, jointed, + Gr. *seira*, rope) (Fig. 8-17), a common genus in both fresh and salt water, and *Coscinodiscus* (Gr. *koskinon*, sieve, + NL. *discus*, disc), *Arachnodiscus*, and *Stictodiscus* (Fig. 8-18B, C), usually marine, illustrate this type.

The taxonomy of diatoms is based almost exclusively on differences in the structure and ornamentation of the frustules, rather than on attributes of the living protoplasts. The wall is impregnated with polymerized, opaline silica ($SiO_2 \cdot nH_2O$), which is assimilated at the surface of the cell from the silicic acid of the environment. In several diatoms (e.g., *Navicula pelliculosa*) the silicon is deposited within a very delicate, flattened surface sac, the **silicolemma** (Reiman et al., 1966), as are the plates of "armored" dinoflagellates (p. 152). The diverse types of marking represent ridges, thin places, or minute pores in the frustules; these are rather constant in arrangement and form the basis for delimitation of species. The transverse lines seen on the valves of many pennate diatoms represent lines of pores, which can be resolved as such with good oil-immersion lenses as well as with the electron microscope.

The frustule in all diatoms is composed of two overlapping portions, the **valves**; one is usually slightly larger than the other, much like the bottom and cover of a box (Figs. 8-13, 8-14). The larger, coverlike portion is called the **epitheca** (Gr. *epi*, over, + Gr. *theke*, case); the smaller is known as the **hypotheca** (Gr. *hypo*, under, + Gr. *theke*). The two valves, instead of overlapping each other directly, frequently are attached to **girdle bands** (Fig. 8-14C); these are composed of two overlapping portions. Supernumerary bands may be intercalated between the valves and girdle-band segments; hence, the

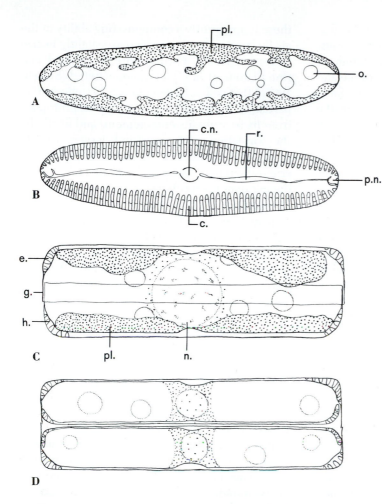

A

B

C

D

FIGURE 8-14
Pinnularia streptoraphe.
Organization of living cells. **A.** Valve
view. **B.** Valve view, protoplast
omitted. **C.** Girdle view. **D.**
Recently divided cells. c., costae;
c.n., central nodule; e., epitheca;
g., girdle band; h., hypotheca; n.,
nucleus; o., oil droplet; pl., plastid;
p.n., polar nodule. ×600.

epitheca and hypotheca may be quite widely separated in certain species. The cells are shaped like elongate boxes in *Navicula* (Fig. 8-13D, E) and *Pinnularia* (Fig. 8-13A, B). When viewed from either above or below (valve view), *Pinnularia* cells usually have parallel sides and rounded polar portions (Figs. 8-13A, B, 8-14). A prominent line, the **raphe** (Gr. *raphe*, seam) traverses each valve (Fig. 8-14B).[3] When observed in lateral aspect (girdle view) (Figs. 8-13B, 8-14C), the same cells appear rectangular. Each valve of *Pinnularia* is marked by two

series of prominent riblike lines, the **costae** (Fig. 8-14B). In *Navicula*, rows of pores, or punctae, extend from the margin toward the central region of the wall, which, as in *Pinnularia*, is traversed by the raphe. The raphe is connected to a **central nodule** and to two **polar nodules** (Fig. 8-14B), all of which open to the external aqueous medium. The raphe is wedge-shaped or ridge-and-valleylike, as viewed in transverse section. It has been reported that the secretion of an adhesive from the raphe and its subsequent hydration account for movement of diatoms with raphes (Drum and Hopkins, 1966; Hopkins and Drum, 1966).

---

[3]A raphe is absent in nonmotile diatoms.

The most conspicuous structures of the diatom protoplasts are the brownish plastids (Figs. 8-13A, 8-14C), which are reported to contain, in addition to chlorophylls *a* and *c*, β-carotene and a number of xanthophylls, including fucoxanthin, which is largely responsible for the color of diatoms; the latter is also present in Phaeophyta. The plastids may be few in number and massive, as in the pennate diatoms, or numerous and discoidal, as in the centric types. The excess photosynthate is stored in the form of oil or chrysolaminaran, which is frequently conspicuous in the living cells. The single nucleus is usually readily observable in the center of the cells in pennate genera. In addition to the nucleus, which lies in a bridge of colorless cytoplasm, the central portion of the cell is occupied by a large central vacuole.

Asexual reproduction in unicellular diatoms is effected by nuclear and cell division. Immediately after cytokinesis, the original frustule contains two protoplasts, each approximately half the volume of the parental one. As the filial protoplasts enlarge, each develops a new valve, using the half-frustule of the parent cell as the epitheca (Figs. 8-13C, 8-14D). It is obvious, therefore, that one of the filial cells thus formed will be slightly smaller than the parent cell, and that if this process were to continue, some of the progeny would become progressively smaller. While there is evidence that this may occur,

there is apparently a compensating ability of the girdle band and valve to increase slightly in size in some species. Furthermore, in several marine diatoms it has been demonstrated (Stosch, 1965*a*) that cell enlargement can be achieved by partial or complete extrusion of the protoplast from its frustule (its enlargement) and finally by the formation of a new and larger frustule around the extruded protoplast. The problem of size reduction in the progeny of asexually reproducing diatoms has been discussed by Rao and Desikachary (1970).

Sexual reproduction in pennate diatoms results in the formation of zygotes called **auxospores** (Gr. *auxo*, increase, + Gr. *spora*, spore). These are so called because they are naked protoplasts that increase markedly in size after their formation. Sexuality is *superficially* similar to that in certain desmids, because the protoplasts of two cells may escape from their walls and unite directly to form a zygote. It differs, however, in that the pennate diatoms are diploid during their vegetative stages and sexual reproduction is immediately preceded by meiosis. In some species—*Navicula halophila*, for example—each member of the pair of vegetative cells forms two gametes, so that two zygotes are formed (Fig. 8-15). In others, each cell forms only one gamete. The zygotes increase rapidly in size and become auxospores that are larger than the parent cells whose

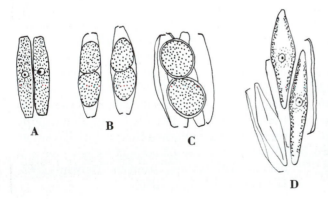

FIGURE 8-15
Navicula halophila. Stages in sexual reproduction. **A.** Pairing of cells. **B.** Formation of two gametes in each cell. **C.** Two auxospores (zygotes) formed. **D.** Elongation of auxospores. **A–C.** ×175. **D.** ×260. (*After Subrahmanyan.*)

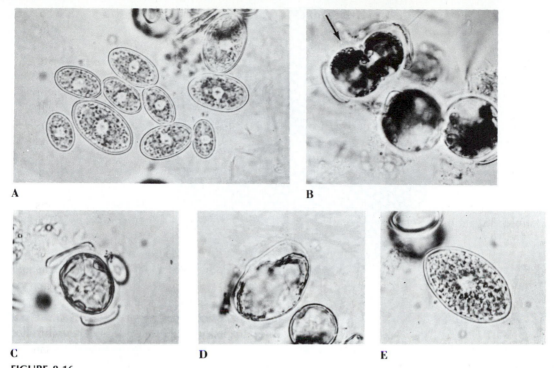

A   B

C   D   E

FIGURE 8-16
Cocconeis placentula. **A.** Group of vegetative cells. **B, C.** Release and union of gametic protoplasts (arrow) (only two of four valves visible in **B.**). **D, E.** Auxospore formation. Note difference in size of vegetative cells in **A** and **E. A–E.** ×600. (*With the assistance of Dr. Dianna Tupa.*)

protoplasts functioned as gametes. In this way, when the auxospore has formed new valves, it again achieves the maximum size characteristic of the species and functions as a vegetative cell (Fig. 8-15D). Granetti (1968) has demonstrated a similar cycle in clonal cultures of *N. minima.*

Sexual reproduction has been observed to occur abundantly when leaves of aquatic plants on which the epiphytic *Cocconeis placentula* (Fig. 8-16A) is growing are transferred to petri dishes containing water from the natural habitat enriched with inorganic culture medium and/or soil extract. The diatoms move from the host in great numbers and attach to the bottom of the petri dish, where they multiply rapidly. Within several days, many of the cells pair, their frustules open, and their protoplasts unite to form zygotes (Fig. 8-16B, C). These enlarge rapidly

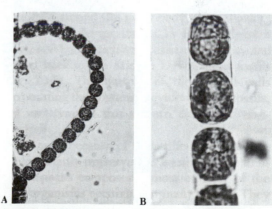

A   B

FIGURE 8-17
Melosira sp. A living marine species, girdle view.
A. ×150. **B.** ×450.

145

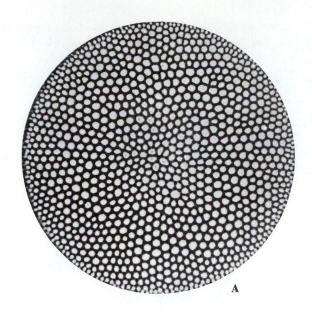

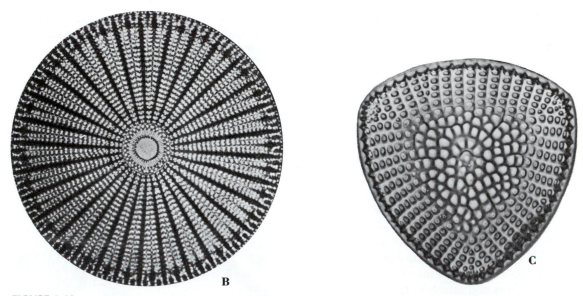

FIGURE 8-18
Frustules of marine, centric diatoms. **A.** *Coscinodiscus radiatus.* **B.** *Arachnodiscus indiscus.* **C.** *Strictodiscus johnsonianus.* **A.** ×700. **B.** ×180. **C.** ×540. (*Courtesy of Dr. Ruth F. Patrick, Jay Sacks, and the Academy of Natural Sciences of Philadelphia.*)

as auxospores (Fig. 8-16D, E) and within 48–72 hours secrete a new frustule and initiate cell division. Parthenogenetic auxospores also are formed.

The centric diatoms are largely marine, but a few also occur in fresh water. Of the latter, the filamentous genus *Melosira* (Fig. 8-17), with marine and freshwater species, is by far the most widely distributed. It is frequently present in the plankton in sufficient abundance to color the water, but benthic species are also common. The elongate cells, when joined together, are observed in girdle view. The cells contain many small, brownish plastids and are uninucleate. The epitheca and hypotheca are joined by connecting bands that seem scarcely to overlap. The valve view of *Melosira* cells, which is circular, becomes apparent when the filaments dissociate into individual cells.

Coscinodiscus (Gr. *koskinon*, sieve, + L. *discus*, disc), *Arachnodiscus* (Gr. *arkys*, net, + *discus*), and *Stictodiscus* (Gr. *stiktos*, spotted, + *discus*) (Fig. 8-18) are representative of numerous marine species of centric diatoms. Their cells are markedly flattened in the manner of extremely shallow petri dishes. Sexual repro-

duction has been reported to be oogamous for a number of centric forms. Our knowledge of sexual reproduction in centric diatoms was greatly augmented by investigations (Stosch and Drebes, 1964; Drebes, 1966) of certain marine species, such as *Stephanopyxis* (Gr. *stephanos*, crown, + M.L. *pyxis*, box) *turris* in laboratory culture. In this organism (Fig. 8-19) the large eggs, which are little-modified vegetative cells, are fertilized by uniflagellate sperms. Both eggs and sperms arise as a result of (gametic) meiosis. Only one of the nuclear products of meiosis survives in the oogonium. After fertilization, the zygotes (auxospores) increase greatly in size and form two new valves. Figure 8-20 and its caption illustrate and explain sexual reproduction in *Stephanopyxis palmeriana*, also a marine species.

As a group, diatoms are perhaps economically the most important of the algae because of their role in the food cycle of aquatic animals and because of the many uses of diatomaceous earth. While diatoms have received considerable taxonomic study, many other aspects require further investigation. The Bacillariophyceae are clearly delimited from other Chrysophyta by

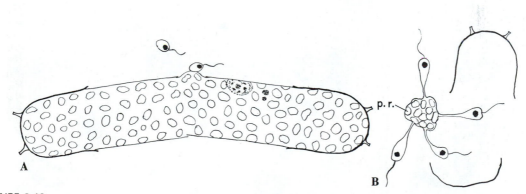

FIGURE 8-19
Sexual reproduction in *Stephanopyxis turris*, a centric diatom. **A.** Protoplast of vegetative cell functioning as an oogonium, sperm making contact. **B.** Liberation of sperm. p.r., plastid remnant. Diagrammatic. **A.** ×1000. (*After von Stosch and Drebes.*)

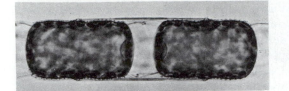

A

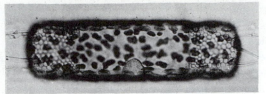

B

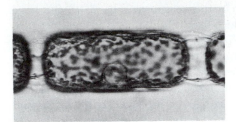

C

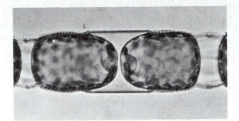

D

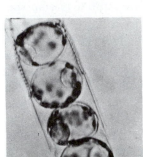

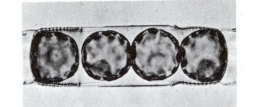

E

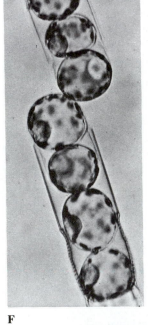

F

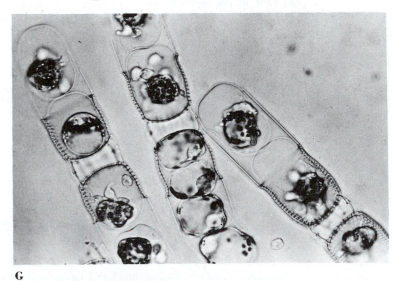

G

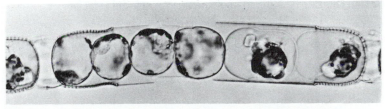

H

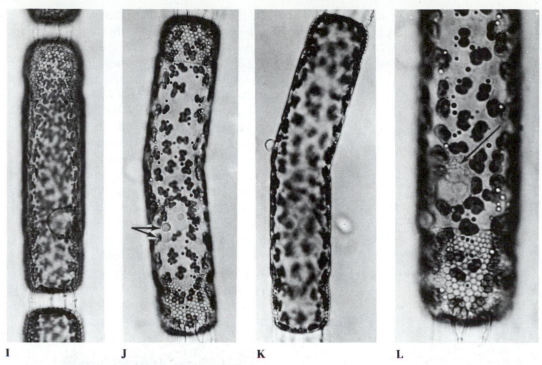

I J K L

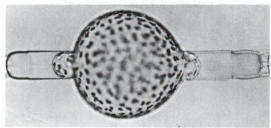

M

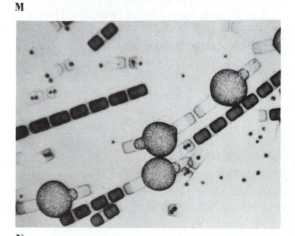

N

FIGURE 8-20

Stephanopyxis palmeriana. **A.** Two vegetative cells with nuclei in an adjacent parietal arrangement. **B.** Cell elongation prior to division. **C–F.** Successive stages in the formation of sperm mother cells (**spermatogonia**), small cells which arise as a result of mitotic divisions; meiosis occurs in the enlarging spermatogonia in **F. G, H.** Completion of meiosis and formation of four colorless sperms by each spermatogonium; note single flagellum visible on one sperm in **H** and protoplasmic remnants containing plastids. **I–J.** Oogenesis. **I.** Oogonium in meiotic prophase; note enlarged nucleus. **J.** Mature egg; note large egg nucleus and two degenerate nuclear products of meiosis (arrows). **K.** Oogonium, egg and sperm at its surface. **L.** Fertilization: karyogamy; union of large egg nucleus with smaller sperm nucleus (arrow). **M.** Enlarged zygote or auxospore. **N.** Four auxospores and remnants of spermatogonial cytoplasm scattered in the field. **A–K.** ×450. **L.** ×750. **M.** ×216. **N.** ×55. (*After Drebes.*)

their specialized wall structure, method of wall formation, and sexual reproduction. For more comprehensive discussions of diatoms see Patrick and Reimer (1966, 1975), Werner (1977), and Bold and Wynne (1985).

Classification of Diatoms

The genera of diatoms discussed in the preceding account may be classified as follows:

Division Chrysophyta
    Class 3. Bacillariophyceae
        Order 1. Centrales
            Family 1. Coscinodiscaceae
                Genera: *Melosira, Coscinodiscus,*
                          Arachnodiscus, Stictodiscus,
                          Stephanopyxis
        Order 2. Pennales
            Family 1. Naviculaceae
                Genera: *Navicula, Pinnularia*
            Family 2. Achnanthaceae
                Genus: *Cocconeis*

Summary of Chrysophyta

In concluding this account of the division Chrysophyta, the question remains regarding the propriety of grouping the three classes Xanthophyceae, Chrysophyceae, and Bacillariophyceae in one division, a practice that implies their relationship. Among the attributes these classes share may be listed the yellowish and/or brownish color of the plastids, which results from predominance of carotenes and xanthophylls, the frequent deposition of siliceous material on the cell surfaces, and the storage of the excess photosynthate as oils. Furthermore, in all genera of one class (Bacillariophyceae) and many of the other two, it has been demonstrated that the cell is covered by two overlapping segments. In some genera of all three classes there has been observed the formation of a unique type of silicified resting cyst, which is not found among other algae.

Furthermore, like the Chrysophyceae, the Xanthophyceae have their thylakoids in groups of three in the plastid with seemingly continuous peripheral "girdling lamellae," and the longer of their unequal flagella with two rows of hairlike appendages. Both Chrysophyceae and Xanthophyceae are like the Bacillariophyceae in that all can store fat or oil. In view of these common attributes, the tentative grouping of these classes into one division, Chrysophyta, seems to be justified. It is perhaps significant that the Phaeophyta have a number of similar features (3-thylakoid groups and girdling lamellae in the plastids, similarity of eyespots, and storage of a β-1 : 3-glucan, laminaran). On the basis of these and other considerations, the Chrysophyceae, Xanthophyceae, Bacillariophyceae, and Phaeophyceae have been grouped in a single division, "Chromophyta" (Bourrelly, 1968).

DIVISION PYRRHOPHYTA

General features. Like the Euglenophyta, the division Pyrrhophyta (Gr. *pyr*, fire, + Gr. *phyton*, plant), contains both chlorophyllous and colorless organisms. The Pyrrhophyta, as represented in this account by the single class Dinophyceae (Gr. *dinein*, to whirl, + Gr. *phykos*), the **dinoflagellates**, are often considered to be members of the Protozoa; they are still so classified by many protozoologists. However, nonmotile unicellular and filamentous genera have been discovered that produce motile stages in which the cells are dinoflagellate in character; this indicates an affinity with plant-like organisms.

The carotenoid pigments in the dinoflagellates are present in sufficient abundance (as in the Bacillariophyceae, Chrysophyceae, and Phaeophyta) to mask the chlorophylls *a* and *c*. In addition to β-carotene, several xanthophylls are present (Table 4-1), including peridinin and dinoxanthin, which are peculiar to the group. Starch accumulates both within the chloroplast and colorless cytoplasm, and oils may also be stored.

Dinoflagellates are abundant in both fresh and marine waters, where they frequently form an important constituent of the plankton (Steidinger and Williams, 1970). Marine water blooms of the genera *Gonyaulax* and *Gymnodinium* in the Gulf of Mexico and elsewhere may form "red tides" and cause widespread destruction of fish. *Gonyaulax catenella* produces a toxin that paralyzes shellfish, while species of *Gymnodinium* produce other types of toxin (Loeblich, 1966). Steidinger (1975) has distinguished three aspects common to all toxic red tides as: (1) increase in population size or initiation; (2) support (i.e., proper salinity, temperature and growth factors); and (3) maintenance and transport of blooms by hydrologic and meteorologic factors. A more extensive discussion of red tides and toxin is available on pp. 492–495 of Bold and Wynne (1985).

The motile genera as well as the motile cells of nonmotile genera are characterized by the arrangement of their flagella (Fig. 8-21). One flagellum is elongate, usually extending posteriorly with reference to the direction of motion. The second flagellum, which emerges from the same point as the first, is ribbonlike and lies in a transverse groove in which it undergoes undulating movement. Both flagella are of the tinsel type. The furrow divides the cell into anterior and posterior portions, which may be unequal in size in some genera. A few dinoflagellates, such as *Gymnodinium* (Gr. *gymnos*, naked, + Gr. *dinein*, to whirl) (Fig. 8-21), lack cell-wall material, but in others the wall is prominent (Fig. 8-22) and composed of regular platelike segments. The plates that form the wall are in vesicles inside the plasma membrane (Dodge and Crawford, 1970) and not outside it, as are the cell walls of most algae. Genera with walls of the latter types are often said to be "armored."

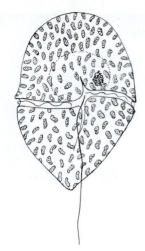

FIGURE 8-21
Gymnodinium sp. Single individual from freshwater plankton. ×770.

The walls probably contain cellulose; the empty walls persist for some time after the death of the protoplasts.

The nuclei of dinoflagellates are highly interesting in a number of respects. Unlike those of eukaryotic organisms, the chromosomes remain condensed and clearly recognizable individually throughout the cell cycle. Furthermore, they contain no RNA or histone, or any other protein component, in which respect they are similar to prokaryotic organisms. The chromosomes of a number of species are clearly fibrillar in organization (Zingmark, 1970a). For these reasons they are sometimes called Mesokaryotes. Dodge (1971a) has reviewed the fine structure of the Pyrrhophyta.

Ceratium (Gr. *keration*, little horn) (Fig. 8-23), *Gymnodinium*, and *Peridinium* (Gr. *peridines*, whirled around) (Figs. 8-21, 8-22) are representatives of the motile type of dinoflagellates of common occurrence in the plankton in bodies of cold water or during the colder seasons of the year. The protoplasts contain discoid plastids, which are yellow-brown to dark brown. The excess photosynthate is stored as starch or

151

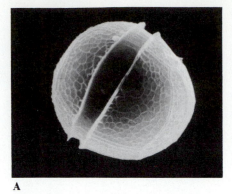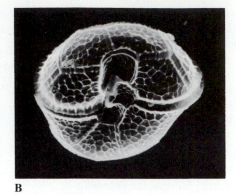

A   B

FIGURE 8-22
Peridinium cinctum var. *ovaplanum,* cell walls in surface view (**A.** in dorsal, **B.** in ventral aspect); note sculptured plates that compose the wall. ×800. (*Courtesy of Drs. Cox and Arnott.*)

oil. Each cell has a single prominent nucleus, and most genera have prominent stigmata as well. Small, noncontractile vacuoles are frequently visible. There is some evidence that certain species of dinoflagellates, although photosynthetic, may undergo phagotrophic nutrition and ingest solid foods, but the exact mechanism of ingestion is not clearly understood.

Asexual reproduction is by cell division, frequently of cells in a motile condition. The nuclear membrane persists during mitosis. In plate-walled genera, each daughter cell receives a portion of the original cell wall.

Sexual reproduction has been observed in several species of dinoflagellates. In the genus *Ceratium* it has been reported that small individuals function as male gametes. In *C. horridum,* a marine species, fusion of male and female gametes, the latter similar in size to normal vegetative cells, has been observed (Stosch, 1964). The zygote in this species remains motile. In *C. cornutum,* a freshwater species, the zygote becomes dormant and enclosed in a cyst. At its germination, meiosis occurs (Stosch, 1965a). Pfiester (1975, 1976) induced sexual reproduction of *Peridinium* in a

nitrogen-deficient medium. In this process small, naked individuals unite in pairs. Here, too, meiosis is zygotic. It should be noted that some dinoflagellates apparently are diploid. In a colorless, luminescent marine dinoflagellate, *Noctiluca miliaris,* uniflagellate isogamous gametes unite to form zygotes (Zingmark, 1970b).

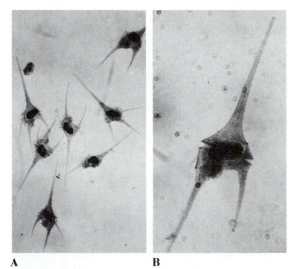

A   B

FIGURE 8-23
Ceratium sp. stained to show nuclei. **A.** ×125. **B.** ×250.

Classification of Pyrrhophyta

The chlorophyllous genera of dinoflagellates included in this brief discussion may be classified as follows:

Division Pyrrhophyta
  Class 1. Dinophyceae
    Order 1. Gymnodiniales
      Family 1. Gymnodiniaceae
        Genus: *Gymnodinium*
    Order 2. Peridiniales
      Family 1. Gonyaulacaceae
        Genus: *Gonyaulax*
      Family 2. Peridiniaceae
        Genus: *Peridinium*
      Family 3. Ceratiaceae
        Genus: *Ceratium*

DISCUSSION QUESTIONS

1. What features distinguish the Xanthophyceae from the Chlorophyta?
2. Cite instances of parallel body structure in these two algal classes.
3. Do you think that Xanthophyceae, like the terrestrial *Botrydium*, may be like the ancestors of the "higher" land plants? Give the reasons for your answer.
4. What rapid laboratory technique can you suggest for distinguishing Xanthophyceae from Chlorophyceae?
5. In what location and under what conditions would you attempt to collect members of the Chrysophyceae?
6. List the respects in which diatoms are of economic importance.
7. Describe asexual reproduction in unicellular diatoms such as *Pinnularia*.
8. How could you prove that diatom frustules are impregnated with silicon?
9. Speculate on what might happen if you cultivated diatoms in a silicon-free medium.
10. Distinguish between "pennate" and "centric" diatoms.
11. Why is the zygote of diatoms called an auxospore?
12. To what genera of Chlorophyceae and Phaeophyceae are the pennate diatoms similar in respect to type of life cycle?
13. On the basis of supplementary reading, discuss the economic importance of both living and fossil diatoms.
14. How does sexual reproduction of centric diatoms differ from that of the pennate type?
15. What significance do you attach to the observation, by a European phycologist, that a brownish, branching filamentous alga produced dinoflagellate-type zoospores?

Chapter 9
DIVISION RHODOPHYTA

The division Rhodophyta (Gr. *rhodon*, rose + Gr. *phyton*, plant) contains a single class, the Rhodophyceae (Gr. *rhodon* + Gr. *phykos*), commonly called the red algae. About 400 genera and 3900 species are known. The Rhodophyceae, like the Phaeophyceae, are predominantly marine organisms; however, several genera, such as *Batrachospermum* (Gr. *batrachos*, frog, + Gr. *sperma*, semen) (Fig. 9-12), are widely distributed in freshwater streams, lakes and springs. Marine Rhodophyceae flourish in both littoral and sublittoral zones. Rhodophyceae are very abundant in tropical seas, where they often grow at great depths in clear waters. A number of red algae precipitate calcium carbonate on their cell surfaces and become calcareous. These are important in reef formation. Many of the marine Rhodophyceae are strikingly beautiful, both in living condition and when mounted on herbarium sheets.

In most genera, chlorophylls *a* and *d* (when present) and the carotenoids are largely concealed by a red pigment, **phycoerythrin**, and sometimes by the blue pigments **phycocyanin** and **allophycocyanin** (Table 4-1, p. 46). These pigments absorb light energy, which they transfer to chlorophyll *a*. There is also evidence that, in addition, the accessory pigments may have a more direct role in the photosynthetic process. They are apparently associated with the surfaces of the plastid lamellae (Fig. 9-2). The numerous genera exhibit a range of color; various shades of red are common, and some plants are almost black. Some species of *Batrachospermum* (Figs. 9-13, 9-14), on the other hand, are markedly blue-green. The pigments are localized in plastids, which may be massive, with a single plastid in each cell (Fig. 9-2), or the plastids may be numerous and discoid (Fig. 9-20A). The excess photosynthate is stored as a complex carbohydrate, called **Floridean starch**, composed of approximately 15 glucose units; grains of this substance stain slightly red with iodine—potassium iodide solution. They usually occur in the colorless cytoplasm at the surface of plastids rather than within them. The accumulated carbohydrate is often associated with the nucleus rather than with the pyrenoid when the latter is present. Floridean starch is similar to the amylopectin fraction of green algal and other green-plant starches but differs in requiring prolonged boiling for gelatination; glycogen may also be present.

The vegetative cells of Rhodophyceae may be uninucleate (as in *Nemalion* and *Batrachos-*

permum) or multinucleate (as in *Polysiphonia, Callithamnion,* and *Griffithsia*). The vacuole of large, multinucleate cells is more prominent than in the uninucleate cells of genera such as *Nemalion.* The cellulose cell wall is often surrounded by a slimy layer. Other components such as xylans are present in the walls of some red algae. In a number of genera, such as *Nemalion,* the filaments are covered, in addition, by copious gelatinous material of rather firm consistency. The hydrocolloids of red algae have been classified as agars, carrageenans,[1] and gelans. Agars are of prime importance in biology as relatively inert agents for solidification of culture media. The walls between two adjoining cells in most of the Rhodophyta contain structures known as **pit connections** (Fig. 9-16B). The latter are really specialized plugs in an intercellular wall, secreted after cytokinesis; this results in incomplete walls. A special function for these structures has not been elucidated.

Although a few unicellular and colonial rhodophycean genera have been described, the vast majority are filamentous (Figs. 9-3, 9-9) or membranous and foliaceous plants (Fig. 9-4). The basic pattern, however, is filamentous, and this can be demonstrated even in many membranous types. The development of the plant body is initiated by the activity of one or more apical cells. The membranous Rhodophyceae are less complex internally than similar types of Phaeophyceae.

Most red algae have dibiontic life cycles, with sporic meiosis occurring in the sporangia of a special alternate, the **tetrasporophyte.** The details of the reproductive process are discussed with the illustrative genera.

The Rhodophyta, with the single class Rhodophyceae, are usually further classified into the subclasses Bangiophycidae and Florideophycidae; the latter contains the majority of red algae.

---

[1]See also p. 184

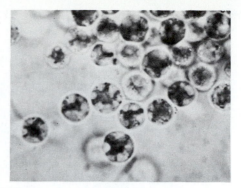

FIGURE 9-1
Porphyridium purpureum. Living cells; note asteroidal plastids. ×1200.

CLASS 1. RHODOPHYCEAE

SUBCLASS 1. BANGIOPHYCIDAE

The Bangiophycidae take their name from the genus *Bangia* and include unicellular (Fig. 9-1), filamentous (Fig. 9-3), and foliose (Fig. 9-4) plants. In some of the last two types, pit connections (see p. 167) are absent between cells, and the female sex organs, if present, are not highly differentiated. Only a few genera are known to have sexual reproduction.

Porphyridium

The unicellular red alga *Porphyridium* (Gr. *porphyra,* purple, + Gr. *idion,* similarity) is an inhabitant of moist soils, on which it forms shiny patches, and also occurs in fresh, brackish, and marine waters. The wall-less, spherical, uninucleate cells of *Porphyridium* (Figs. 9-1, 9-2) contain a stellate plastid with a central pyrenoid. The cells in mass may be blood-red or bluish green, depending on the relative amounts of phycocyanin and phycoerythrin present, and these occur as minute particles, the phycobilosomes, on the surface of the thylakoids (Gantt and Conti, 1965, 1967). The cells have colloidal sheaths, which under some circumstances may have stalklike protuberanc-

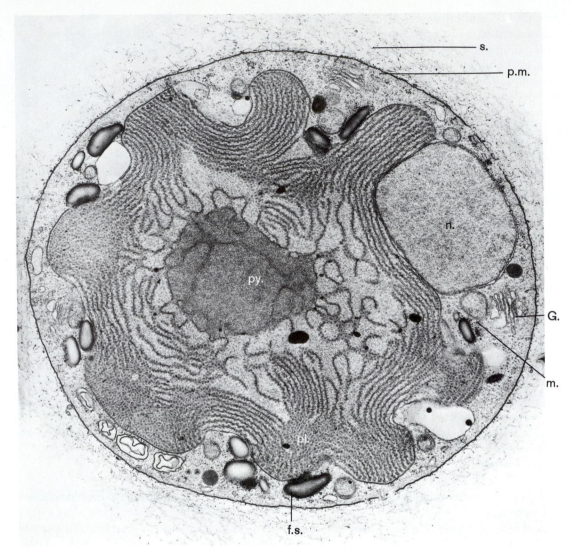

FIGURE 9-2
Porphyridium purpureum. Electron micrograph. f.s., Floridean starch; G., Golgi apparatus; m., mitochondrion; n., nucleus; pl., plastid; p.m., plasma membrane; py., pyrenoid; s., colloidal sheath. ×7000. (*After Gantt and Conti.*)

es. Reproduction is by cell division into two or more cells (Sommerfeld and Nichols, 1970*a*). Certain isolates of *Porphyridium* apparently move by means of unipolar secretion of a slime exudate.

Bangia

Bangia (after N. H. Bang, a Danish botanist) (Fig. 9-3) is an unbranched, firmly sheathed filamentous plant, which may become pluriseriate as the filaments age. Here again, each

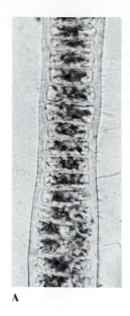

A

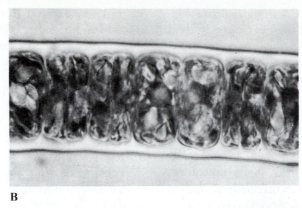

B

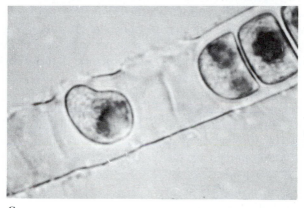

C

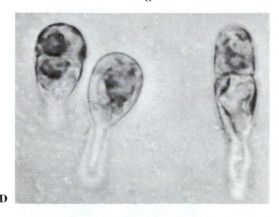

D

FIGURE 9-3
Bangia atropurpurea. **A.** Uniseriate and pluriseriate portions of one strand. **B.** Segment of filament more highly magnified; note asteroidal plastids and pyrenoid (p). **C.** Monospore release. **D.** Monospore germination. **A.** ×550. **B.** ×1960. **C, D.** ×1000. (**B–D.** *After Sommerfeld and Nichols.*)

uninucleate cell contains a massive, stellate plastid with central pyrenoid. *Bangia atropurpurea* grows on rocks and woodwork, is often exposed at low tide, and hence is periodi-cally wet by both fresh and salt water. It is also increasingly common in the Great Lakes (Lin and Blum, 1977).

In asexual reproduction, the protoplasts of

cells in both uniseriate and pluriseriate filaments are liberated from their enclosing walls as **monospores** (Fig. 9-3C), which may be amoeboid. Cells that produce monospores are called **monosporangia.** The monospores germinate and grow into new plants (Fig. 9-3D).

Two phases occur in the life cycle of *B. atropurpurea*, the macroscopic, filamentous phase (uniseriate and pluriseriate) and a microscopic, branching filamentous state most frequently growing on animal shells in nature. The microscopic phase was originally described as a free-living alga, *Conchocelis* (Gr. *conche*, conch, + Gr. *kele*, tumor). The development of each phase in the life cycle is markedly affected by the duration of light during each 24-hour period (Richardson, 1970). If the cultures are illuminated for less than 12 hours, the *Bangia* phase develops; longer periods of illumination evoke the development of the *Conchocelis* phase. Both phases can reproduce by monospores, and the nature of the product of monosporic germination is determined by the duration of illumination. The *Conchocelis* phase can be maintained under a regime of 16 hours of light followed by 8 hours of dark, and its monospores duplicate the *Conchocelis* phase. The *Bangia*-filament stage prevailed and reproduced by monospores under less than 12 hours of light diurnally.

In addition to monospores, the *Bangia* phase forms more than one spore per cell. These are of two types, which differ in size. The smaller, lighter-colored spores arise in groups of more than 16 per parent cell. In *Bangia* they are seemingly functionless and do not germinate. The larger spores, formed 16 or fewer per cell, develop into the *Bangia* or *Conchocelis* phase, depending on the photoperiod. Sexual reproduction has not been demonstrated unequivocably in all species of *Bangia*, and the chromosome numbers in both the *Bangia* phase and *Conchocelis* phase have been reported to be $n = 3$, so that if sexual reproduction does occur,

it must be followed by zygotic meiosis (Sommerfeld and Nichols, 1970*b*); Richardson (1970), however, reported a chromosome number of 10 in both the *Bangia* and *Conchocelis* phases. However, Cole and Conway (1980) reported a sexual process in some populations of *B. atropurpurea*.

Porphyra

Porphyra (Gr. *porphyra*, purple) also is representative of a group of algae that are considered to be primitive Rhodophyceae. The brown-purple or rose-tinted, *Ulva*-like plant bodies of *Porphyra* (Fig. 9-4) grow attached to rocks or to larger marine algae. They are often inhabitants of the intertidal zone. The fronds may become more than a foot in length. They are composed of one or two layers (depending on the species) of cells, which are surrounded by thick, colloidal cell walls (Fig. 9-5). The plants are attached to the substratum by rhizoidal holdfasts.

FIGURE 9-4
Porphyra purpureo-violacea. Living plant. ×⅙.

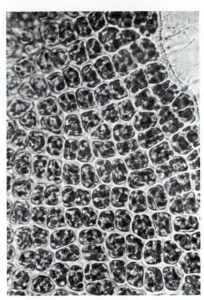

FIGURE 9-5
Porphyra purpureo-violacea. Vegetative cells; note asteroidal plastid. ×350.

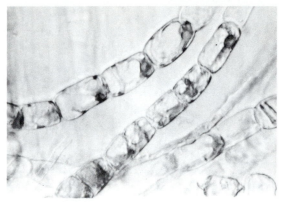

FIGURE 9-6
Conchocelis phase of *Porphyra leucosticta.* ×800. (*Courtesy of Prof. Michael J. Wynne.*)

The uninucleate cells of *Porphyra* contain one or two[2] prominent, stellate chloroplasts, which are central (Fig. 9-5). The cells of *P. leucosticta* have pit connections in the *Conchocelis* phase (Lee and Fultz, 1970) characteristic of so many Floridean (p. 160) Rhodophyceae. Growth in *Porphyra* is generalized; that is, cell division is not restricted to a certain region of the maturing plant body.

A number of careful and intensive investigations in field and laboratory have added a great deal to our knowledge of the life cycle of *Porphyra.* It is known with certainty that in several species spores shed from membranous plants germinate into branching filaments (Fig. 9-6), long ago described, from examples growing on discarded mollusk shells, as a distinct genus, *Conchocelis.* Spores ("Conchospores") of

the latter, in culture, have given rise, in turn, to the membranous *Porphyra* phase. *Conchocelis* grows on rocks as well as on shells.

The reproductive cycle of *Porphyra,* then, is much like that in *Bangia* in that there is an alternate *Conchocelis* phase (Fig. 9-6). The spores from the latter develop into *Porphyra* plants. In some species, it has been shown that the *Porphyra* plants are haploid, while the *Conchocelis* filaments are diploid. Meiosis occurs in the fertile cells of *Conchocelis* (Giraud and Magne, 1968), which give rise to the spores destined to form the *Porphyra* plants (gametophytes). The occurrence of meiosis is evidence that sexual union has preceded it. The *Porphyra* plants produce minute, almost colorless cells, the spermatia[3] (Fig. 9-7) and also groups of 8–16 larger spores, often called carpospores, an implication that they are products of a zygote, although conclusive evidence of sexual union has until recently rarely been presented (Yabu, 1969; Bold and Wynne, 1985). The diploid

---

[2]Depending on the species.

[3]They are sometimes considered to be vestigial male gametes.

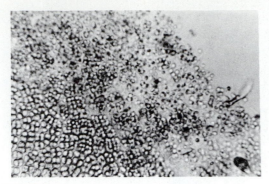

FIGURE 9-7
Porphyra purpureo-violacea. Spermatium formation and discharge. ×300.

carpospores develop into the *Conchocelis* phase. In addition, the *Porphyra* plantlets can arise directly from the *Conchocelis* filaments, and both phases may perennate. Chen et al. (1970) have investigated the life cycle of *P. miniata* in culture. As in *P. purpurea-violacea,* the bladelike plants are haploid, while the *Conchocelis* phase is diploid. Kito et al. (1971) report chromosome numbers of *n* = 3,4 and 5 in three different Atlantic species of *Porphyra.* For one species, *P. gardneri,* a thoroughly convincing report of sexual reproduction has been published (Hawkes, 1978). This report includes micrographs showing spermatia attached to, and contributing nuclei to, carpogonia. It also provides evidence that the *Conchocelis* phase is diploid.

SUBCLASS 2. FLORIDEOPHYCIDAE

The distinctions cited between the Bangiophycidae and Florideophycidae are increasingly less impressive. Most of the latter have apical growth, pit connections, and, except for *Nemalion* and related forms, discoid plastids. Furthermore, the sex organs are highly differentiated.

The greater number of genera and species of red algae are classified in the Florideophycidae, and space permits the discussion of just a few representatives, namely, *Nemalion, Batrachospermum, Griffithsia, Callithamnion,* and *Polysiphonia.* In the Florideophycidae prominent female gametangia, called **carpogonia,** are developed on branches, usually recognizable because of their pale color, which is occasioned by the arrested development of their plastids. In the great majority, the life cycle is a slightly modified D, h + d type, which may be isomorphic or heteromorphic. The "modification" is the occurrence, between the zygote and sporophytic plant, of a group of diploid carpospores on the gametophyte, primarily at the site where the zygote is formed. This intercalated, diploid phase is often called the **carposporophyte.** It has been suggested that intercalation of the diploid carposporophyte is in a way a compensation for the absence of flagellated sperm in red algae. Fewer fertilizations can still effect the development of many individuals by means of the carpospores (Searles, 1980).

Nemalion

Nemalion (Gr. *nema,* thread) differs from *Porphyra* in a number of important respects. In *Nemalion* growth is strictly apical and traceable to several apical cells. Prominent pit connections are present (Fig. 9-10D, F).

Nemalion helminthoides is a marine organism that grows attached to rocks that may be exposed at low tide; the living plants have the appearance and texture of gelatinous wormlike branching strands (Fig. 9-8); they are anchored to rocks by discoidal bases, which may perennate. In median longitudinal sections or in crushed preparations of apices, it is apparent that the plant body is composed of a number of colorless, central, longitudinal filaments, the tips of which elongate through the activity of apical cells. From the central filaments, tufts of lateral photosynthetic filaments arise in dense

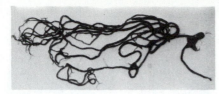

FIGURE 9-8
Nemalion helminthoides. Living plants. × ½.

whorls (Fig. 9-9A). The axial and photosynthetic systems are both embedded in a rather firm colloidal material. The uninucleate photosynthetic cells are beadlike, each with a single starlike chloroplast in which a single pyrenoid is embedded (Figs. 9-9, 9-10). The function of the latter and its relation, if any, to Floridean starch have not yet been ascertained. The apices of the photosynthetic filaments may terminate in hair-like cells.

The reproductive organs of *Nemalion* and those of most Rhodophyta are unlike those of other algae. In some respects, however, they resemble those of certain ascomycetous and basidiomycetous fungi (Chapters 33, 34). *Nemalion* is monoecious. The female reproductive organ, here called the **carpogonium**,[4] is an oogonium with a more or less well-developed protuberance, the **trichogyne** (Gr. *thrix*, hair, + Gr. *gyne*, female). The carpogonia are borne on almost colorless lateral branches, which arise near the center of a tuft of photosynthetic filaments (Figs. 9-9B, C, 9-10B, D). Other branches on the same plant, by successive divisions, produce male sex organs, the **spermatangia** (Figs. 9-9D, 9-10C), which are analogous to the unicellular antheridia of other algae. Each spermatangium produces a single **spermatium**, a male gamete, which is discharged at maturity.

---

[4]The terminology for the male and female reproductive organs in algae and fungi reflects the fact that the names were assigned prior to any considerations of homology and/or analogy.

The spermatia, which are produced and liberated in large numbers, are transported by water currents; they may be slightly amoeboid. When a spermatium makes contact with the trichogyne (Figs. 9-9C, 9-10D), its nucleus divides. The walls then dissolve at the point of contact, and one of the spermatial nuclei enters the trichogyne, migrating to the base of the carpogonium (Fig. 9-10D), where union of a spermatial nucleus with the carpogonial nucleus ensues. Union of the sperm and egg in oogamous reproduction is called **fertilization.** Soon after fertilization, the trichogyne withers. A series of mitoses and cell divisions follow, which result in the production of a tuft of short filaments (Figs. 9-9E, 9-10F–H), the cells of which become naked **carpospores** (Fig. 9-9H, I).

The carpospores of *N. helminthoides*, upon germinating in laboratory culture, give rise to branching filaments, putatively the **tetrasporophyte**, which so far have not developed into characteristic *Nemalion* plants (Fries, 1967). The carpospores of *N. helminthoides* are diploid (Magne, 1961), so that meiosis probably is not zygotic. In a Japanese species, *N. vermiculare* (Umezaki, 1967; Masuda and Umezaki, 1977), it has been demonstrated that the carpospores germinate into uniseriate, microscopic filaments, which produce **tetrasporangia**, each of which produces four **tetraspores** (Fig. 9-11). These are released and germinate to form prostrate, branching filaments from which the typical, multiaxial erect branches of *Nemalion* arise. The tetraspore-bearing phase of red algae is called the **tetrasporophyte**. The tetrasporophyte of *Nemalion* may duplicate itself through the agency of monospores. The life cycle of *Nemalion*, accordingly, is dibiontic and heteromorphic, inasmuch as the large *Nemalion* plants (gametophytes) alternate with minute, filamentous tetrasporophytes. Although meiosis may be presumed to occur in the tetrasporangia, as it does in many other red algae, it remains to be demonstrated in *Nemalion*.

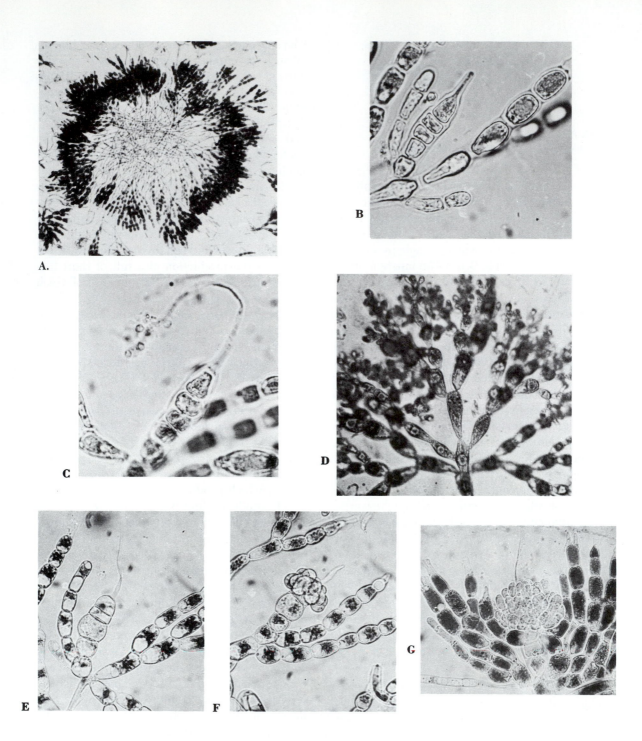

A.

B.

C.

D.

E.

F.

G.

162

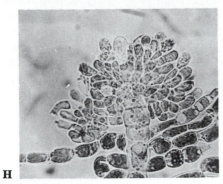
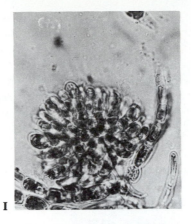

H **I**

FIGURE 9-9
Nemalion helminthoides. **A.** Transection; note axial filaments supporting photosynthetic ones. **B.** Immature carpogonial branch; note distal carpogonium and trichogyne and absence of abundant pigment. **C.** Fertilization; note abundant spermatia on trichogyne. **D.** Spermatangial branches. **E.** Early postfertilization; end of first division of zygote. **F–I.** Stages in development of carpospores. **A.** ×60. **B, C.** ×500. **D.** ×400. **E.** ×500. **F–I.** ×400.

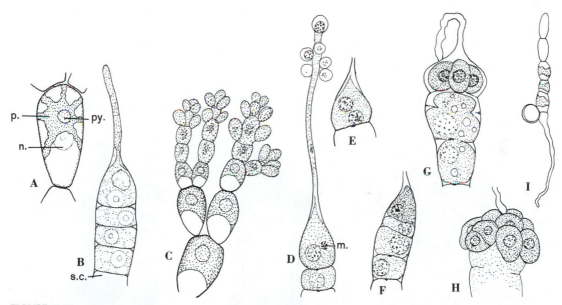

FIGURE 9-10
Nemalion helminthoides. **A.** Organization of a cell of photosynthetic filament. **B.** Mature carpogonial branch; note rudimentary plastid in carpogonium. **C.** Branch with spermatangia. **D.** Fertilization. **E.** End of first division in zygote. **F.** Completion of first cytokinesis of zygote. **G, H.** Early stages in development of carposporangia. **I.** Germinating carpospore. **A–C.** living; the remainder from acetocarmine preparations. m., spermatial nucleus; n., nucleus; p., plastid; py., pyrenoid; s.c., supporting cell. **A–H.** ×770. **I.** ×315.

163

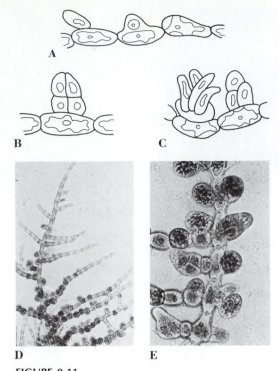

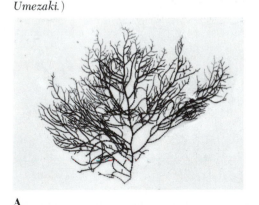

FIGURE 9-11

Nemalion vermiculare. **A.** Portion of filament. **B.** Tetrasporangium. **C.** Tetraspores germinating *in situ.* **D, E.** Photomicrographs of tetrasporophyte. **D.** Plant with young tetrasporangia. **E.** Mature tetrasporangia. **A–C,** ×227. **D.** ×85. **E.** ×434. (**A–C.** *After Umezaki;* **D, E.** *After Masuda and Umezaki.*)

Batrachospermum

Batrachospermum (Fig. 9-12), a freshwater red alga, differs from *Nemalion* in vegetative structure. Instead of the numerous axial filaments and apical cells of the latter, *Batrachospermum* develops from a single apical cell; derivatives of the latter form an axial filament and tufts of photosynthetic filaments that are loosely verticilate and embedded in slime. Plants of *Batrachospermum* may be blue-green or deep wine red, depending on the ratios between phycocyanin and phycoerythrin.

The carpogonial branches of *Batrachospermum* (Fig. 9-13) may occur on the photosynthetic filaments or on a modified branch of limited growth. The spermatia (Fig. 9-14A) are larger than those of *Nemalion.* Huge masses of carpospores (Fig. 9-14B) develop after fertilization. These germinate, developing into prostrate branching filaments that may be crustlike.

From these diploid filaments, the typical haploid thalli of *Batrachospermum* arise. In this process an apical cell of the diploid filaments undergoes meiosis and forms two small "polar bodies."[5] The now haploid apical cell develops

---

[5]So called on the basis of analogy with the maturation of animal eggs.

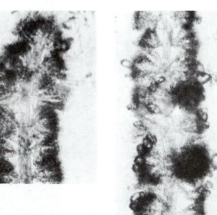

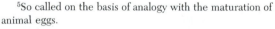

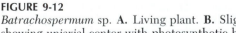

FIGURE 9-12

Batrachospermum sp. **A.** Living plant. **B.** Slightly flattened branch showing uniaxial center with photosynthetic branches. **C.** Slightly flattened branch with two groups of carpospores. **A.** ×0.75. **B, C.** ×50.

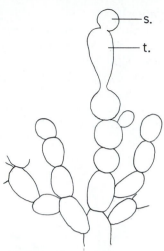

FIGURE 9-13
Batrachospermum sp. Outline drawing of carpogonial branch at fertilization. s., spermatium; t., trichogyne. ×580.

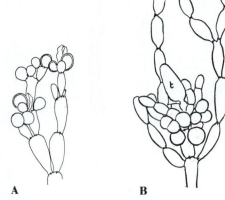

A   B

FIGURE 9-14
Batrachospermum **A.** *B. moniliforme,* branch with spermatangia. **B.** *B. anatinum,* early stages in development of the carpospores. **A.** ×800. **B.** ×600. (*After Bourrelly.*)

into a new haploid axis with whorled photosynthetic laterals (von Stosch and Theil, 1979; Balakrishnan and Chaugule, 1980). The life cycle is, accordingly D, h + d and heteromorphic.

Callithamnion

Alternation of generations in *Bangia, Porphyra,* and *Nemalion* is dibiontic and heteromorphic. By contrast, isomorphic dibiontic life cycles occur in a majority of Florideophycidae, here exemplified by *Callithamnion, Griffithsia,* and *Polysiphonia.*

Callithamnion (Gr. *kallos,* beauty, + Gr. *thamnion,* little shrub) (Fig. 9-15) is widespread in marine waters and often grows epiphytically on other, larger algae. The plants consist of delicate, branching filaments with pronounced apical growth. The cells are multinucleate at maturity and contain numerous lenticular plastids. As populations of *Callithamnion* mature in the field or in laboratory cultures, male and

female sexual plants and morphologically similar tetrasporophytes develop (Edwards, 1969) (Fig. 9-15). The tetraspores (Fig. 9-15D) germinate into equal numbers of male and female gametophytes (Fig. 9-15A, B), which at maturity produce spermatangia and carpogonia on separate individuals. The spermatangia develop near the distal terminal walls of the branch cells, and the carpogonia arise at a corresponding location on the main axes of the female gametophytes; the carpogonia are readily recognizable because of their extremely elongate trichogynes. In cultures containing male and female plants, fertilization, development of the carposporophyte, and release of carpospores require 5 days. The carpospores develop into tetrasporophytes; in the tetrasporangia of the latter, meiosis is inferred to occur on the basis of the segregation and tetraspores into male- or female-producing types. The entire life cycle in laboratory culture requires one month for completion (Edwards, 1969).

165

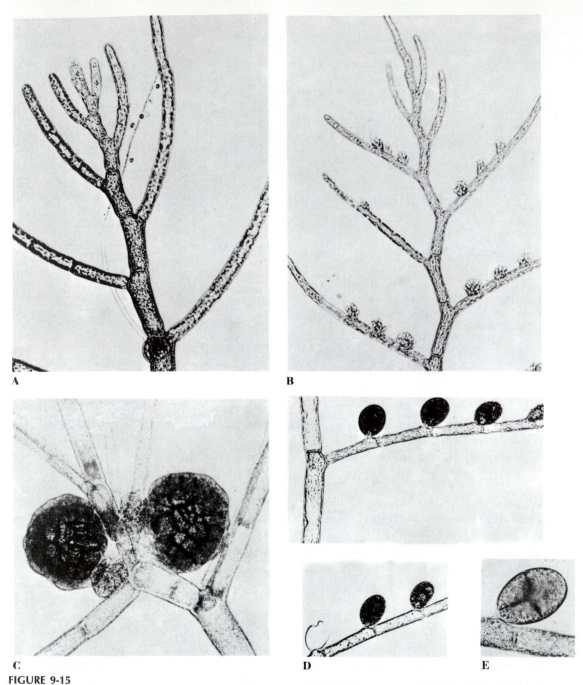

FIGURE 9-15

Callithamnion byssoides, life history in culture. **A.** Female gametophyte; note long trichogyne with spermatia attached. **B.** Male gametophyte with clusters of spermatangia containing spermatia. **C.** Two cystocarps (carposporophytes) on ♀ plants resulting from fertilizations. **D.** Tetrasporophyte: branch with tetrasporangia above, one empty tetrasporangium below. **E.** Tetrasporangium, enlarged. **A.** ×150. **B.** ×170. **C.** ×170. **D.** ×75. **E.** ×170. (*After Edwards.*)

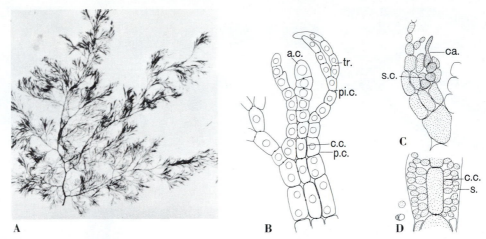

FIGURE 9-16
Polysiphonia. **A.** Habit of growth of a vegetative plant. **B–D.** *P. harveyi.* **B.** Apex, showing ontogeny. **C.** Carpogonial branch (densely stippled). **D.** Median longisection of spermatangial branch. a.c., apical cell; ca., carpogonium; c.c., central or axial cell; p.c., pericentral cell; pi.c., pit connection; s., spermatangium; s.c., supporting cell; tr., trichoblast. **A.** ×¾. **B.** ×388. **C, D.** ×770.

Polysiphonia

Polysiphonia (Gr. *polys*, many, + Gr. *siphon*, tube) (Figs. 9-16, 9-17, 9-18) is also widely distributed in marine waters, where it grows both epiphytically on larger algae and aquatic flowering plants and also on rocks and woodwork. The life cycle of *P. boldii* was elucidated in laboratory cultures from tetraspore to tetraspore (Edwards, 1968).[6]

Polysiphonia (Figs. 9-16, 9-17) is a branching filamentous plant. Growth is strictly apical, and the derivatives of the apical cell segment in a regular pattern (Fig. 9-16A), thus forming the multiseriate axis, which in some species achieves considerable complexity through superficial cortication. Pit connections are clearly visible between contiguous cells of *Polysiphonia*, and delicate, hairlike branches, the **trichoblasts**, may be present (Fig. 9-17A) in some species.

Polysiphonia is dibiontic. The gametophytes are dioecious, spermatia and carpogonia being produced on different individuals. The sex organs arise from derivatives of the apical cell near the tips of the branches (Fig. 9-17). The curved carpogonial branch of *Polysiphonia* (Fig. 9-16C) at fertilization consists of four almost colorless cells, the most distal one of which develops a trichogyne. This branch arises from a pericentral cell known as the **supporting cell.** Certain cells at the base and above the carpogonial branch grow up around it, forming an urn-shaped envelope, the **pericarp** (Gr. *peri*, around, + Gr. *karpon*, fruit) (Fig. 9-17F). The spermatangia and spermatia are borne on lateral branches (Fig. 9-17D, E), the pericentral cells of which produce spermatangial mother cells, which give rise to large numbers of colorless spermatangia that are abscised and function directly as spermatia[7] (Fig. 9-16D). As in all Rhodophyta, the spermatia are borne passively

[6]Referred to as *P. denudata* in this paper; later recognized to be *P. boldii.*

[7]In *Polysiphonia* and *Griffithsia*, unlike *Nemalion*, the entire spermatangium is abscised and functions as spermatium.

167

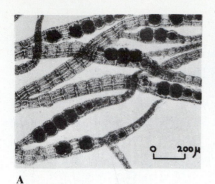

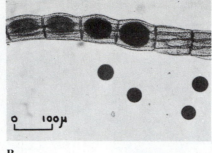

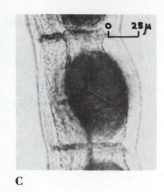

A

B

C

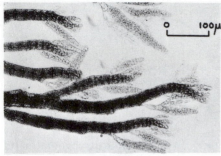

D

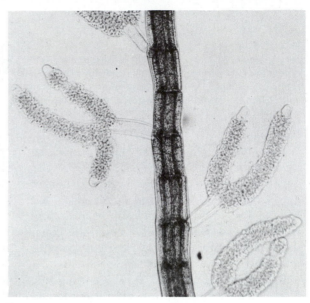

FIGURE 9-17

Polysiphonia boldii. **A.** Tetrasporophyte with developing tetrasporangia. **B.** Portion of tetrasporophyte more highly magnified, showing four tetraspores just liberated from the empty tetrasporangium at the right. **C.** Single tetrasporangium. **D.** Male gametophyte with colorless spermatangia branches. **E.** Portion of ♂ gametophyte, more highly magnified. **F.** Female gametophyte with carpogonial branch; note spermatia attached to long trichogyne. **G.** Cystocarp liberating carpospores. **A.** ×60. **B.** ×100. **C.** ×200. **D.** ×60. **E.** ×100. **F.** ×100. **G.** ×125. (*After Edwards.*)

E

in the trichogyne by water currents. After attachment of spermatia to the trichogyne (Fig. 9-17F), their contents flow into the latter. The spermatial nucleus migrates down through the trichogyne and ultimately unites with the carpogonial nucleus, as in *Nemalion*.

Postfertilization development is rather complicated in detail. It involves extensive cell fusions, degeneration of the carpogonial branch, and migration of the zygote nucleus into an **auxiliary cell**, which arises after fertilization. Mitotic divisions of the zygote nucleus give rise to a number of diploid nuclei, which are present in a large fusion cell, the **placental cell** (Fig. 9-18A). Carpospores, each with a diploid nucleus, are abstricted from this cell and develop in a series (Fig. 9-18A).

The mature carpospores are liberated through the terminal opening of the urnlike pericarp (Fig. 9-17G) and germinate into diploid plants under suitable conditions. The carpo-

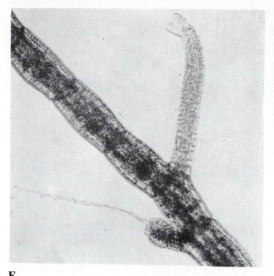

F

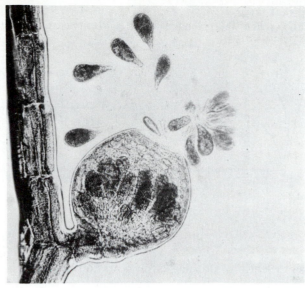

G

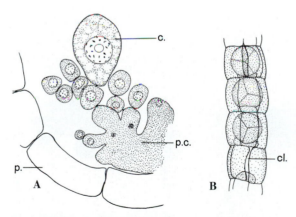

FIGURE 9-18

Polysiphonia harveyi. **A.** Portion of cystocarp with placental cell and developing carposporangia. **B.** Portion of tetrasporic plant. c., carposporangium; cl., cleft through which tetraspores were liberated; p., pericarp cell; p.c., placental cell. **A.** ×500. **B.** ×125.

spores surrounded by the pericarp are called the **cystocarp.** These plants, the tetrasporophytes, are similar in size and general appearance to the male and female gametophytes, but at maturity they produce tetrasporangia (Fig. 9-17A), which arise on short stalk cells. Cytological investigation has demonstrated that the two successive nuclear divisions in the tetrasporangium accomplish meiosis, so the four spores produced are haploid. The tetrasporangial wall breaks open, and the tetraspores are shed through clefts between the pericentral cells (Figs. 9-17B, 9-18B). It has been shown by culture methods that tetraspores develop into gametophytic plants. This rather complicated life cycle of *Polysiphonia* (and of *Callithamnion* and *Griffithsia*) is summarized in Chart 9-2.

Griffithsia

Griffithsia (after Mrs. Amelia Griffiths) is similar in life cycle to *Polysiphonia*, but its vegetative structure is simpler, because the branching filaments are composed of large cells that are not covered by pericentral cells or cortications (Fig. 9-19). *Griffithsia globulifera*,

169

CHART 9-1

Life cycle of *Polysiphonia* (and of *Callithamnion* and *Griffithsia*).

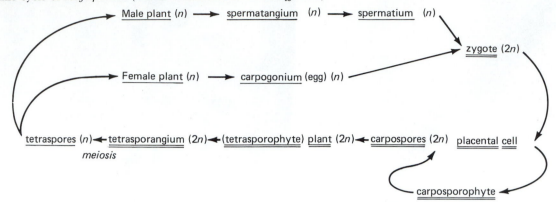

an Atlantic coast species, appears during the summer, when the water temperature is relatively high. The bushy growths are attached to stones and pilings in sublittoral habitats and are rosy pink in color. The multinucleate cells are very large, those near the base of the branches attaining a length as great as 5 mm. Growth is apical, from a multinucleate cell with dense cytoplasm and without a vacuole. The apparently dichotomous branching develops because of the upgrowth of the lateral surface of the cell below an apical cell to form a new growing point. In mature cells the cytoplasm is peripheral; it contains beautiful ribbonlike, segmented plastids, which are arranged in curved mosaics (Fig. 9-20A). The attaching system of *Griffithsia*

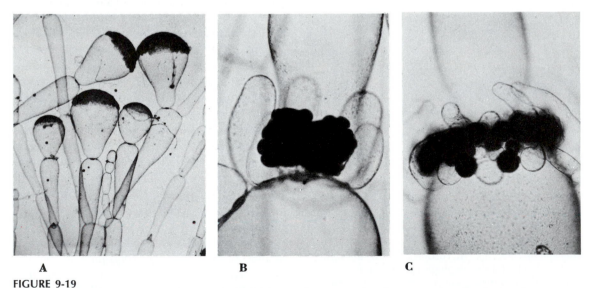

A   **B**   **C**

FIGURE 9-19

Griffithsia globulifera. Living plants. **A.** Male plants with spermatangial caps. **B.** Female plant with carposporophyte; note involucral cells. **C.** Node of tetrasporic plant with tetrasporangia. **A.** ×12.5. **B, C.** ×60.

170

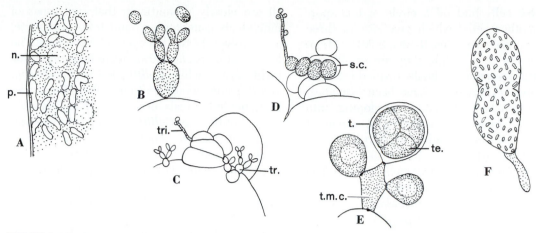

FIGURE 9-20

Griffithsia globulifera. **A.** Portion of living vegetative cell, surface view. **B.** Detail of single spermatangial branch. **C, D.** Details of ♀ reproductive branches and fertilization. **E.** Tetrasporangium mother cell with three tetrasporangia. **F.** Germinating carpospore. n., nucleus; p., plastid; s.c., supporting cell; t., tetrasporangium; te., tetraspore; t.m.c., tetrasporangium mother cell; tr., trichoblast; tri., trichogyne with spermatia attached. **A, B.** ×700. **C–F.** ×300.

is weak and restricted, as evidenced by the frequency with which free-floating plants are encountered. Trichoblasts—colorless, branched, hairlike filaments—are present on the axes.

The life history of *Griffithsia* involves three types of free-living plants, as in *Polysiphonia* and *Callithamnion*—namely, male and female gametophytes and tetrasporophytes, the female plants bearing cystocarps at maturity. The male plants are readily recognizable (Fig. 9-19A) because of their caplike mantles of spermatangial filaments, which produce spermatia (Fig. 9-20B) in enormous numbers and because cell size decreases from apex to base. The carpogonial branches arise from special three-celled, short, lateral branches, which originate on the free distal surface of vegetative cells (Fig. 9-20C, D). The earliest stages in the development of these branches, and of the carpogonial branches they produce, occur near the growing point of each main axis. The supporting cell generates a four-celled carpogonial branch,

which is recurved, as in *Polysiphonia*. After fertilization, the supporting cell gives rise to the auxiliary cell, with which the fertilized carpogonium becomes united. The zygote nucleus passes from the carpogonium into the auxiliary cell, after which the carpogonial branch withers and is abscised. Further development involves extensive cell fusions and the ultimate formation of a placental cell, as well as generation of diploid nuclei within the auxiliary cell and the contiguous regions of the placental cell. Later, the formation of carpospores is initiated. These are borne in groups (Fig. 9-19B). Meanwhile, soon after fertilization, the basal cell of the short lateral axis that bears the carpogonial branch gives rise to a number of curved, elongate, sausage-shaped pericarp cells that partially conceal the developing carpospores (Fig. 9-19B). The latter are shed at maturity and then germinate (Fig. 9-20F).

There is good evidence that the carpospores germinate into plantlets that mature as tetrasporophytes. In these, the distal portions of

171

vegetative cells bud off a circle of tetrasporangial mother cells, which give rise to three tetrasporangia each (Figs. 9-19C, 9-20E). Meiosis occurs in these while they are small, and each tetrasporangium is finally divided into four tetraspores. When these have been liberated, they germinate, ultimately developing into male and female gametophytes. A circle of involucral cells surrounds each fertile, tetrasporangium-bearing node (Fig. 9-19C).

The *Polysiphonia-Callithamnion-Griffithsia* type of life cycle is most similar to that observed among such Chlorophyta as *Cladophora* and in such Phaeophyta as *Ectocarpus* (D, h + d, p. 114). It differs, however, in one important respect: In *Polysiphonia, Callithamnion, Griffithsia,* and *Nemalion* there is intercalated between fertilization and the developing tetrasporophyte a series of cell generations consisting of the carpospores. This phase has been interpreted by some as still a third alternating phase, the **carposporophyte.** According to this point of view, alternation in *Polysiphonia, Callithamnion, Griffithsia,* and *Nemalion* involves the haploid gametophytes, the diploid carposporophyte, and finally the diploid tetrasporophyte. There is no phase similar to the carposporophyte in groups of algae other than Rhodophyta, nor does it occur among the land plants; the moss sporophyte is somewhat analogous in its location. The closest approach to it, perhaps, is found in the postfertilization development in certain ascomycetous fungi (Chapter 33). It should be emphasized that interpolation of the diploid carposporophyte makes possible the development of many tetrasporophytes (rather than only one) as a result of a single act of fertilization and from one zygote.

Summary and classification

With the exception of *Porphyridium*, the genera of Rhodophyta discussed in this chapter all are clearly dibiontic in that two types of individuals (gametophyte and tetrasporophyte) are involved in the life cycle of *Nemalion*. The latter and *Batrachospermum* are heteromorphic in life cycle, while *Callithamnion, Griffithsia,* and *Polysiphonia* all are isomorphic. The gametophytic individuals of *Nemalion* are monoecious, but those of *Callithamnion, Polysiphonia,* and *Griffithsia* are genetically dioecious, which is determined by meiosis in the tetrasporangium. *Bangia* and *Porphyra* also are clearly dibiontic and heteromorphic, but for most species sexual reproduction and meiosis have not been clearly demonstrated.

As a group, the Rhodophyta are sharply segregated from other eukaryotic algae by their pigmentation, storage synthate (Floridean starch), and characteristic reproductive structures, which appear elsewhere only in certain ascomycetous and basidiomycetous fungi (Chapters 33, 34).

The development of a special series of terms for the reproductive organs and cells of the Rhodophyta is unfortunate in some ways, because it occasions confusion in the minds of those approaching the study of the group for the first time. It seems clear that the carpogonium and spermatangium correspond in function to the oogonium and antheridium, respectively, of other oogamous algae. The permanent retention of the zygote on the gametophyte of red algae, however, seen also in *Coleochaete*, marks a deviation rarely seen in other groups of algae, although it is suggested in the kelps. The continued production of carpospores for considerable periods following fertilization possibly is correlated with this retention, since although the postzygotic cells may contain pigments, they are nourished largely as a result of their organic connection with the parent gametophyte. Thus, a single act of fertilization, with retention of the zygote, results in the potential production of many more individuals than in those plants in which the zygotes are promptly separated from the gametophyte. Many other data about red algae have been summarized by Dixon (1973) and Bold and Wynne (1985).

The genera of Rhodophyta discussed in this chapter may be classified as follows:

Division Rhodophyta
  Class 1. Rhodophyceae
    Subclass 1. Bangiophycidae
      Order 1. Porphyridiales
        Family 1. Porphyridiaceae
          Genus: *Porphyridium*
      Order 2. Bangiales
        Family 1. Bangiaceae
          Genera: *Bangia, Porphyra*
    Subclass 2. Florideophycidae
      Order 1. Nemaliales
        Family 1. Heminthocladiaceae
          Genus: *Nemalion*
      Order 2. Batrachospermales
        Family 1. Batrachospermaceae
          Genus: *Batrachospermum*
      Order 3. Ceramiales
        Family 1. Ceramiaceae
          Genera: *Griffithsia, Callithamnion*
        Family 2. Rhodomelaceae
          Genus: *Polysiphonia*

DISCUSSION QUESTIONS

1. How does the life cycle of *Polysiphonia* or *Griffithsia* differ from that of dibiontic Phaeophyta and Chlorophyta?

2. Can you see any advantage to the plant in carpospore production?

3. The sex organs of the Rhodophyta are called spermatangia and carpogonia, although they are similar to the antheridia and oogonia seen in other oogamous algae. Can you suggest a reason for this?

4. If you were to isolate and cultivate separately the four tetraspores from a single tetrasporangium, what result could be expected? Can you cite similar phenomena from other groups of plants or animals?

5. While most Rhodophyta are marine, a few genera such as *Batrachospermum* are common in freshwater streams. What might account for this?

6. Yamanouchi counted 20 chromosomes in nuclear division in the vegetative cells of the female gametophyte of *Polysiphonia flexicaulis*. What would be the chromosome number in the nuclei of the following structures in the plant: spermatium, carpospore, carpogonium, apical cell of the tetrasporophyte, an individual tetrasporangium, the pericarp, the basal cell of the carpogonial branch?

7. Summarize the life cycles of *Bangia* and *Porphyra*. How do they differ from that of other Rhodophyta discussed in this chapter?

Chapter 10
THE ALGAE: RECAPITULATION

Chapters 2–9 have presented a brief account of the several divisions of Algae and a discussion of some representative organisms from each of the groups. Now that the reader has gained some degree of familiarity with these through mastery of the material in these chapters and, more important, by laboratory study of the living plants themselves, certain general topics regarding the algae will be considered in the present chapter.[1]

FOSSIL ALGAE

There is both direct and indirect evidence that algae were present on the earth in Precambrian times (see inside front cover). It is generally believed that algae were responsible (by removing carbon dioxide from water during photosynthesis) for precipitation of large quantities of calcium carbonate, which eventually became limestone. Especially prominent deposits, presumably the result of this activity of the algae, are visible at Glacier National Park. The process of carbonate precipitation is still carried on today by green and red algae in coral reefs. In fact, in many instances the bulk of the reef is a result of algal activity rather than of the corals themselves. Blue-green prokaryotes may precipitate travertine rock in the waters of hot springs, such as those at Yellowstone National Park. In some parts of the world, mounds of layered calcium carbonate, called **stromatolites**, are the result of precipitation of calcium carbonate by blue-green prokaryotes. There are a number of recorded occurrences of ancient stromatolites in Precambrian sediments (Schopf et al., 1971).

The fossil record of ancient algae and other organisms has been summarized by Schopf (1975). A more recent paper (Knoll and Barghoorn, 1977) reports small (average diameter 2.5 μm), flattened spheres that may represent organisms that existed almost 3½ billion years ago (Fig. 10-1). While these structures could have been abiologically produced, a number of them have been found that resemble modern unicells in various stages of binary fission.

---

[1]Because the blue-green prokaryotes (Cyanophyta, Chapter 2) have so long been considered to be algae, some references to them are included in the present chapter.

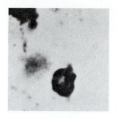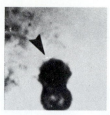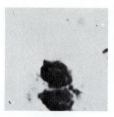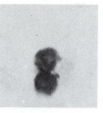

FIGURE 10-1
Stages in presumed cell division of microfossils (algae?) from rocks almost 3.5 billion years old. (*After Knoll and Barghoorn.*)

Other early organisms from the Precambrian have been compared with unicellular, blue-green, prokaryotic organisms, but on the basis of form alone. *Archaeosphaeroides barbertonensis*, from rocks about 3.1 billion years old from South Africa, is represented by spherical cells, and although there is some similarity to members of the Cyanophyta, definite assignment to that group is not possible (Fig. 10-2).

By the middle Precambrian, between 1.6 and 2.0 billion years ago, both unicellular and filamentous organisms resembling blue-green prokaryotes were growing on underwater boulders. Some of the filamentous forms produced what look like heterocysts, while others lacked them. A rich microfossil flora from the late Precambrian (about 0.9 billion years ago) has been described from north-central Australia. Dozens of organisms have been described (Schopf, 1968; Schopf and Blacic, 1971; Oehler, 1976) from these Precambrian cherts, including not only well-preserved unicells and filaments resembling members of the Cyanophyta (Fig. 10-3) but also remains of organisms that certain authors believe to be eukaryotic green algae.

Certain unicellular forms have conspicuous dark bodies distributed one per cell, and it is tempting to compare them with nuclei of eukaryotic organisms. Caution must be exercised for the time being, however, because there may still be other explanations of these bodies not involving nuclear material. Similarly, tetrads of spherical cells from the same late Precambrian deposits have been interpreted as examples of the end of meiotic divisions (Fig. 10-3D), thus signifying the reduction of chromosome number from the diploid state to the haploid. If some phase of the life cycle had indeed been diploid, sexual reproduction must have taken place. Again, a word of caution needs to be introduced, and it is premature to attempt to identify the time of the first appearance of sex on earth. Simply because cells are produced in clusters of four is no proof, in itself, that meiosis was involved.

If some of the earliest forms of living organisms from the Precambrian actually represent remains of the Cyanophyta, then that division has been in existence longer than any other organisms except bacteria. These Precambrian fossils indicate that photosynthetic organisms have been oxygenating the atmosphere for more

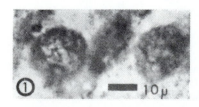

FIGURE 10-2
Archaeosphaeroides barbertonensis; a coccoid, possibly Cyanophytan fossil, several specimens (1–3). (*After Schopf and Barghoorn.*)

175

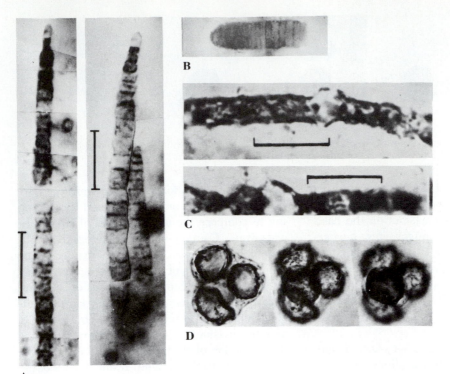

A

FIGURE 10-3
A–C. Late Precambrian Cyanophyta. **A.**
Cephalophytarion grande. **B.** *Palaeolyngbya minor.*
C. *Archaeonema longicellularis,* a heterocystous
species. **D.** *Eotetrahedrion princeps* (individual
cells ca. 10 µm). (**A, B, D.** *After Schopf and
Blacic.* **C.** *After Schopf.*)

than 3 billion years; they would also suggest
that prokaryotic organisms preceded the eukaryotic.

Other forms from the late Precambrian deposits from Australia include organisms that
have certain resemblances to green algae, red
algae, and dinoflagellates as well as fungi. These
latter have only superficial morphological similarities and offer no proof of the existence of
those groups.

The degree of confidence concerning the
presence of various algal groups increases in the
Paleozoic Era. *Courvoisiella ctenomorpha* (Fig.

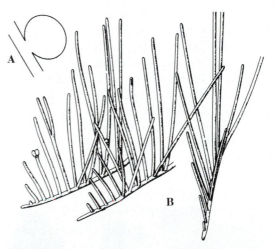

FIGURE 10-4
Courvoisiella ctenomorpha, a fossil siphonous alga
from the Upper Devonian. Gametangium enlarged
at **A. B.** Habit. **A.** ×1200. **B.** ×650.

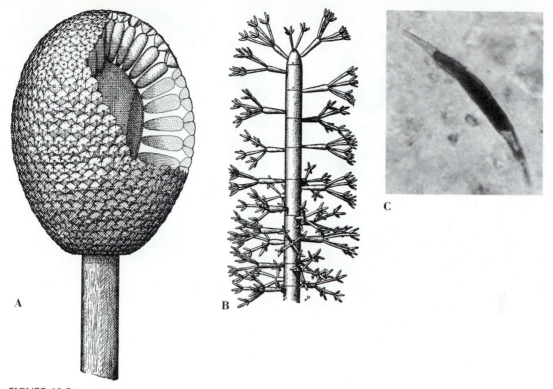

FIGURE 10-5
A, B. Ordovician algae; fossil Chlorophyta. **A.** *Goniolina geometrica.* **B.** *Primicorallina trentonensis.*
C. *Paleoclosterium leptum,* a Middle Devonian desmid. (**A, B.** *After Pia in Hirmer.* **C.** *After Baschnagel.*)

10-4) is an alga, Upper Devonian in age, consisting of a procumbent tubule from which arise two rows of projecting tubules. On the basis of structure and chemical data derived from analysis of the cell wall, it seems quite possible that *Courvoisiella* represents a siphonous member of the Chlorophyta. Because of their characteristic morphological features, certain desmids would be identifiable in the fossil record, and there are reports of that group in rocks as far back as the Devonian period (Fig. 10-5C).

Tubular, coenocytic green algae related to *Codium* and *Acetabularia* are frequently found as fossils because of the persistent nature of the calcium carbonate that precipitated about the plant body (Fig. 10-5A, B). These calcium carbonate incrustations are found as far back as the Ordovician. Members of the Charophyta are also known to cause precipitation of calcium carbonate around the plant body, thus allowing the plants to be preserved as fossils. Zygotes (oospores) are particularly resistant to decay and extend back to the Middle Devonian as fossils (Fig. 10-6). The pattern of the helical cells surrounding the oogonia is evident in the fossils, and variations of this pattern are useful taxonomically.

Diatoms, because of their siliceous frustules, are frequently preserved as fossils, with extensive deposits composed almost entirely of the dead organisms occurring in many places. They are found first in the fossil record in the Jurassic (although it is suspected by many that vagaries of preservation may have been responsible for

177

FIGURE 10-6
Oogonia of the Pennsylvanian fossil charophyte, *Catillochara moreyi*. **A, C.** In lateral aspect. **B.** Median longisection. **D–F.** Anterior polar view. ×60. (*After Peck and Eyer.*)

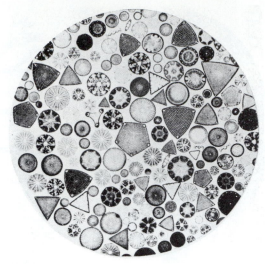

FIGURE 10-7
Fossil marine diatoms from Hungary, grouped for illustration. ×45. (*After Mann.*)

absence of these fossils in older rocks), and persist through the Cenozoic to the present time. Most of the fossil forms are marine Centrales (Fig. 10-7).

One unusual genus of fossil plants is of interest because of its possible relationship with Phaeophyta. The Devonian *Prototaxites* (Fig. 10-8) is found as massive, trunklike fossils that are often misidentified as tree stumps. Sections show that the plant body consisted of elongated tubes of two types. One type is thick-walled and seems to lack cross walls. The more delicate, slender tubes intertwine among the large ones, and have cross walls with septal pores, much like those in filaments of certain algae and fungi. *Prototaxites* is thought by many to be related to brown algae because no other known group of algae has massive axes such as the stipes of some of the kelps. Furthermore, the pseudoparenchymatous nature of the plant body of *Prototaxites* is reminiscent of the structure of some brown algal bodies.

Finally, calcareous red algae, because of the persistence of precipitated calcium carbonate, have a long fossil record from the Devonian to the present.

From this brief discussion of the fossil record of the algae, except for the demonstration that prokaryotic blue-green algalike organisms are the most ancient, no clues are provided as to the pathways and mechanisms of diversification through which the extant algae evolved. It is also clear that most of the major groups of algae were in existence early in the Paleozoic Era. Flügel (1977) has edited a volume summarizing recent advances in the study of fossil algae.

ALGAL PHYLOGENY AND CLASSIFICATION

A phylogenetic classification[2] is one in which the arrangement of the various groups, or taxa—the genera, families, orders, classes, and divisions—signifies degree of kinship. To include all the algae in a single class, Algae, implies that they all had a common ancestry, in spite of their present divergences. This would signify, for example, that the blue-green "algae," which lack plastids, Golgi apparatus, mitochondria, endoplasmic reticulum, and the

---

[2]Sometimes called a natural classification.

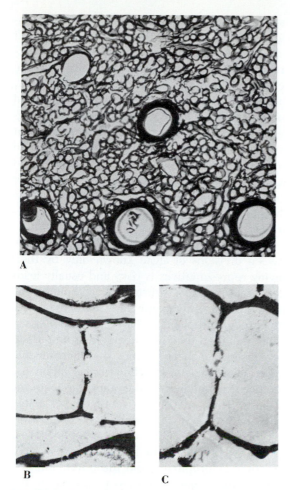

FIGURE 10-8
Prototaxites southworthii. **A.** Transverse section of plant body with large and small tubes. ×275. **B, C.** Longitudinal sections of slender filaments showing septal pores on end walls. ×4500. (*Courtesy of Dr. Rudolf Schmid.*)

sex organs and gametes is of insufficient magnitude to indicate multiplicity of origin. Most modern phycologists are unwilling to accept these implications. The formal class Algae was erected many years ago, before the diversity of the morphological and physiological attributes of the algae had been clarified by the studies of many botanists and biochemists. The various types of pigmentation (Table 4-1) and storage products are now considered to be characteristics of sufficiently fundamental importance to separate the old class Algae into distinct phyletic lines or divisions. The remaining attributes—flagellation, wall structure, habitat, body organization, and structure of the sex organs—have also been used as criteria on the basis of which to delimit taxa of divisional rank or lower. For example, the orders Oedogoniales and Ulotrichales among the Chlorophyceae are segregated on such criteria as cell-wall structure and flagellation. It must be emphasized once again, and it should be clear from the discussion in earlier chapters, that the classification of plants is to some extent subjective. The classification represents a system of presumed relationship carefully elaborated by one or more individuals on the basis of evaluation of the available evidence. The scheme frequently is cemented together by the classifier in a framework of speculation. In a group such as the algae, for which the fossil record has not demonstrated the lines of development clearly, the classifier is bound to rely heavily on the comparative morphology and physiology of the extant genera. It is quite possible, however, that genetic study of physiological attributes of algae may yet furnish evidence that the diversities of pigmentation and photosynthate, currently considered to be of such fundamental importance as to delimit divisions among algae, may represent merely small mutations from an ancestral type. Should this possibility be realized, it might be necessary to replace our currently polyphyletic interpretation of the algae by a more conservative one.

type of nuclear organization characteristic of other algae, nevertheless arose from the same stock. Furthermore, it would imply that the range of variations (Table 4-1, p. 46) described in the preceding chapters with respect to pigmentation, type of reserve photosynthate, flagella number and insertion, cell-wall structure, habitat, body organization, and structure of the

The spectrum of criteria considered in formulating phylogenetic classifications of organisms is constantly broadening; this is especially evident in discussions of the classification of algae. Largely on the basis of morphological characteristics, the organisms described in Chapters 4–9 and their relatives were in the nineteenth and early twentieth centuries considered to be a homogeneous group, accordingly designated as the "class Algae" (within the division Thallophyta) in classifications of this period (see p. 39) and inside back cover). While morphological characteristics, both gross and at the light-microscopic levels, continue to be important considerations in classification, ultrastructural organization and biochemical criteria are currently also affecting classification.

For example, the prokaryotic organization of the Cyanophyta (Chapter 2) clearly distinguishes them from algae, as do the chemistry of their walls, their ability to fix free nitrogen (some species), and their pigmentation. Concomitantly, all but the last of these argue for phylogenetic connection of Cyanophyta and bacteria; the primary photosynthetic pigment of the Cyanophyta is chlorophyll a, like that of algae and unlike bacterial chlorophyll. However, the combination of pigments in Cyanophyta, which is so similar to that in Rhodophyta, has impelled some biologists to postulate a phylogenetic connection with the Rhodophyta in spite of the eukaryotic organization of the latter. In addition to similarity in pigmentation, the occurrence of trehalose in both divisions (and not in others) and their lack of flagella are cited. Although the Rhodophyta have membrane-bounded plastids, their thylakoids are single, not grouped, as they are also not grouped in Cyanophyta. Furthermore, α-carotene is more concentrated than β-carotene in both the blue-green eukaryotes and red algae, and the glycogen of the former suggests the Floridean starch of the latter. Finally, the reproductive cells (spermatia and spores) of Rhodophyta undergo gliding movement, as do many blue-green prokaryotes. In spite of these common attributes, most phycologists do not consider the blue-green and red algae to be closely related because of the occurrence of highly developed sexuality and the eukaryotic organization of the latter. However, as pointed out earlier, eukaryotic algae (and other chlorophyllous plants), it has been suggested, arose very early from the permanent symbiosis of prokaryotic organisms (pigmented and colorless) with eukaryotic cells, and the red algae may have retained many of the attributes of blue-green prokaryotic symbionts while becoming markedly divergent in other respects.

Speculation regarding interrelationships between divisions of algae known only by their living representatives is risky and usually leads to shadow, not substance. Although the Chlorophyta, Euglenophyta, Charophyta, and the green land plants all have chlorophyll a and chlorophyll b, they are divergent in important morphological characteristics, and the Euglenophyta are divergent in cellular organization and in their paramylon from the other starch-storing groups. It is, of course, generally postulated that the Chlorophyta represent the progenitors of the other divisions of chlorophyllous land plants (often called "the higher plants").

While chlorophyll b and the phycobilin pigments are major accessory pigments in the green algae and blue-green prokaryotes, respectively, carotenoid pigments are the major accessory pigments in the Chrysophyta and Phaeophyta. In these groups there is an excess of carotenoid pigments over chlorophyll; β-carotene is the dominant carotene. The xanthophyll fucoxanthin occurs as an important accessory pigment in the Chrysophyceae, Bacillariophyceae, and Phaeophyta. Oil is stored in these groups, as are the carbohydrates chrysolaminaran and laminaran (Phaeophyta), which are polyglucans with β-1:3 and β-1:6 linkages. Furthermore, the occurrence of chlorophyll c, the high carotene–chlorophyll ratio, and the storage of oil in some Pyrrhophyta may be significant in linking the latter with the Chrysophyta and Phaeophyta.

On the basis of these and some additional data from biochemical, comparative physiological, and ultrastructural studies, several groupings of the divisions of algae have been proposed. For example, Bourrelly (1968, 1970, 1972) assigns the blue-green prokaryotes (herein called Cyanophyta) as a class to the division Schizophyta, which also includes the bacteria. In the division Chlorophyta he includes Charophyceae, while grouping the carotenoid-rich Chrysophyceae, Xanthophyceae, Bacillariophyceae, and Phaeophyceae together in the division Chromophyta. He retains the red algae, dinoflagellates, and euglenoids as separate divisions. Fott's (1971) classification is largely similar, except that he maintains a group for flagellate algae, to which he assigns the euglenoids. Inasmuch as the present text is not designed to include an in-depth treatment of the algae and their classification has not been emphasized in earlier chapters, see Bold and Wynne (1985).

PIGMENTATION AND STORAGE PRODUCTS

Review of the type of pigmentation in the several groups of algae reveals that all contain chlorophyll *a*, but that only the Chlorophyta and Euglenophyta (and Charophyta) contain chlorophyll *b* as well. Apparently, the occurrence of a unique combination of pigments in all the members of a given group indicates that the development of the pigmentation preceded evolutionary morphological diversification of the group. Chlorophyll *c* replaces chlorophyll *b* as an accessory pigment in diatoms, dinoflagellates, and Phaeophyta, while chlorophyll *d* is an accessory pigment in some Rhodophyta. Chlorophyll *a*, alone of the chlorophylls, is present in Cyanophyta and Xanthophyceae.

Carotenoid pigments are present throughout the algae and, like the accessory chlorophylls, function in absorbing light energy from regions of the spectrum from which chlorophyll *a* does not absorb. β-carotene is universally present in algae (Table 4-1); α-carotene occurs in Chlorophyta, Euglenophyta, Charophyta, and Rhodophyta, whereas γ-carotene is present in trace amounts in green algae, Euglenoids, and Charophyta. The xanthophylls (oxidized carotenoids) are numerous and widely distributed and also function in transferring light energy to chlorophyll *a*. Fucoxanthin is abundant in Phaeophyta and Chrysophyta (Chrysophyceae and Bacillariophyceae), while diatoxanthin occurs in diatoms and brown algae. Many other types of xanthophylls are present in the divisions of algae.

Phycobilin pigments (phycocyanin, allophycocyanin, and phycoerythrin) occur mostly in Cyanophyta and Rhodophyta. These pigments also function as accessory to chlorophyll *a* in absorbing light energy.

ORGANIZATION OF THE ALGAL PLANT BODY

Comparison of the several groups of algae with reference to the degree of organization of the plant body is instructive. In every division except the Charophyta and Phaeophyta, unicellular genera have been described. Of the algal divisions containing unicellular genera, flagellate motile forms are absent only in the Rhodophyta. Gelatinous and nongelatinous cell aggregates—colonies—occur in several classes of algae. There is evidence that colonial plant bodies represent unicellular types in which the products of cell division have failed to separate. Similarly, the unbranched filament characteristic of a number of genera of Chlorophyta, Xanthophyceae, Bacillariophyceae, and a few Rhodophyta may arise in ontogeny from a single cell in which cytokinesis is restricted to one direction. Initiation of cell division in a second direction by certain cells of unbranched fila-

181

ments results in the branched-filamentous type of algal plant body (*Stigeoclonium, Coleochaete, Cladophora, Ectocarpus, Griffithsia*). In such genera as *Ulva* and some species of *Porphyra*, continuous cell division in two directions perpendicular to each other and one division in a third direction by each of the cells have resulted in two-layered, sheetlike expanses. Such plant bodies may also arise ontogenetically by continuous branching of primarily filamentous axes in certain Rhodophyceae. Of all the algae, the Phaeophyta have the largest and most complex bodies.

Growth in algae may be **generalized** (e.g., *Ulva* and *Spirogyra*) or **localized.** In the latter case, two types of localization are present in the representative genera that have been described. **Apical growth** occurs in such plants as *Cladophora, Chara, Nemalion, Batrachosperum, Polysiphonia, Callithamnion,* and *Griffithsia.* **Intercalary growth** is characteristic of *Oedogonium* and of kelps.

ALGAL REPRODUCTION

The preceding chapters have cited examples of a number of methods of asexual reproduction. Among these may be mentioned cell division, which occurs in a majority of unicellular genera; fragmentation, in colonial and multicellular types; and formation of specialized types of asexual reproductive cells. These may be motile zoospores or nonmotile aplanospores or autospores. All of these share the negative attribute that no union of cells and nuclei is involved in their development into new individuals.

Sexual reproduction—the union of protoplasts and nuclei and the association of chromosomes and meiosis—is absent only in the Euglenophyta of the major groups of algae. In the other groups, sexuality ranges through isogamous, anisogamous, and oogamous types. The Charophyta and Rhodophyta are unique in having only oogamous reproduction. The male gamete is actively motile by means of flagella in every group but the Rhodophyta. Both flagellate (*Chlamydomonas*) and nonflagellate (*Spirogyra*) gametes may display evidence of amoeboid activity during union. In algae, as in animals, certain secondary characteristics may accompany sexuality. Among the motile unicellular genera, the gametes may be morphologically indistinguishable from vegetative cells. However, in other algae, the gametes are markedly differentiated from vegetative cells, and they may be produced frequently in specialized structures, the **gametangia** (Fig. 4-1B). The gametangia in oogamous reproduction are known as antheridia and oogonia. Both sex potentialities may occur on the same individual, or each may be present on a separate individual.

The product of sexual union, the **zygote,** is free-floating in the water in many algae (*Chlamydomonas, Ulothrix,* etc.). In others— *Spirogyra, Oedogonium,* and the *Charophyta,* for example—the zygote is formed within the protective envelope of a parent cell or cells, which, however, ultimately disintegrate, thereby freeing the zygote. In *Laminaria* and other kelps and in all the Rhodophyta, the zygote is retained on the gametophyte after fertilization.

The behavior of the zygote after fertilization varies in different genera of algae, and this is associated with the development of several types of life cycle, shown in Chart 4-1, p.85. In what is regarded as the most primitive and simplest cycle (type M, h [p. 84]), the zygote undergoes meiosis, often after a period of dormancy, so that the product or products of its germination are haploid (*Chlamydomonas, Volvox, Oedogonium,* and *Spirogyra*) and the zygote represents the only diploid cell in the life cycle. At the other extreme is the condition (type M, d, p. 84) in which the zygote, without undergoing dormancy, grows directly into a new individual, which is diploid; meiosis occurs in this case when the diploid individual forms gametes (*Codium, Caulerpa,* and diatoms). Here the gametes, alone of all the cells involved in the life cycle, are haploid. Both of these types of life cycle may be termed **monobiontic**, because only one recognizable plant body type is present in nature. Monobiontic organisms undergo alternation only of cytological states, because the diploid or haploid phase, as the case may be, is represented only by a single cell—the zygote or gamete, respectively.

Intermediate between these two extremes are the various modifications of **dibiontic** life cycles (type D, h + d). In these the zygote, without undergoing meiosis, develops into a diploid individual. Meiosis takes place in the latter in connection with the formation of asexual reproductive cells (spores or zoospore), which, therefore, are haploid. These develop into haploid, sexual, gamete-producing individuals; union of their gametes gives rise to zygotes. In this type of life cycle, two sets of independent individuals occur in nature. These may be morphologically similar (*Ulva, Cladophora, Ectocarpus, Griffithsia*) or divergent (*Laminaria, Porphyra,* and *Nemalion*). Hence, in morphological alternation the cycle may be isomorphic or heteromorphic. The dibiontic Rhodophyta have a life cycle like that summarized in type D, h + d, with the interpolation of diploid spores between the zygote and the diploid plant. The origin and relationship of the alternating generations have been referred to briefly in preceding chapters. This is a question of fundamental importance in any speculation dealing with the development of the so-called higher plants, and it will be referred to again in Chapter 12.

ECONOMIC AND BIOLOGICAL IMPORTANCE OF ALGAE

Although this volume is devoted primarily to morphological data, brief mention must be made of the relation of algae to other organisms, and especially of their economic importance and uses, excellent summaries of which have been prepared for the marine algae by Boney (1965); Levring, et al. (1969); Zajic (1970); Krauss (1978) and Waaland (1981). That the algae have a two-fold basic biological role becomes clear at once if we speculate regarding the outcome if all aquatic algae were to disappear from the earth. It is evident that all aquatic animal life would also soon be eradicated, for, ultimately, it is dependent on algal green pastures, not only as a basis of the food chain, but because of the role of algae in maintaining an adequate level of oxygen in the animals' environment. The widespread presence of algae in soils suggests that their occurrence there is not fortuitous, and, as a corollary, that they must play some as yet undiscovered role in the society of soil organisms.

A number of examples of algal associations with other organisms are known. *Chlorella*-like algae occur within certain species of *Paramecium, Hydra* (Neckelmann and Muscatine, 1983) and certain freshwater sponges, where their role is not entirely clear. Certain algae live in intimate association with corals, where the photosynthesis and oxygen production of the algae are of fundamental importance in the biology of the reef community. It has been reported that larvae of *Convoluta roscoffensis*, a flatworm, do not survive unless they become infected with a *Carteria*-like alga.

Of great interest and biological significance is the endozoic occurrence of unicellular green (Muscatine, 1971) and golden-brown algae ("zooxanthellae") within marine animals. It has been demonstrated (Trench, 1971) that ^{14}C-labeled products of the symbiotic algae diffuse from their cells and that they appear rapidly in the lipids and proteins of the host animals; these extracellular products of the algae range between 20 and 50% of the algal photosynthate. In axenic culture, the extracellular products fall markedly but increase greatly if juices of the host animal are added to the culture medium. Still more remarkable are the symbiotic chloroplasts of certain green algae that live within the

183

cells of certain marine animals (opisthobranchs and slugs) and photosynthesize there (Taylor, 1970; Trench et al., 1972). These chloroplasts are from species of *Caulerpa* and *Codium*, on which the animals feed. Smith (1973*a*) has summarized our knowledge of symbiosis of algae with invertebrates.

It has been shown that when unicellular green algae and zooxanthellae are present in the sea anemones, the green algae do not release significant amounts of their photosynthate. Zooxanthellae, however, have been shown to release up to 60% of their fixed carbon to the coral (Taylor 1969). A comprehensive summary of the subject of nutritive relations between autotrophic and heterotrophic organisms in close associations has been prepared by Smith, Muscatine, and Lewis (1969). A further example of the physiological interactions among organisms is the recent demonstration of the secretion by living cells of DNA, RNA, carbohydrates, lipids and proteins, including enzymes and vitamins by the chrysophyte *Ochromonas danica* and other algae (Aaronson et al., 1971; Aaronson et al., 1977); this report summarizes in tabular form information regarding extracellular secretions by algae. Smith (1973*a*) and Muscatine and Porter (1977) have summarized many aspects of the symbiosis in corals.

The value of algae directly as food (Krauss, 1962), indirectly as vegetable manure, and for a variety of other purposes has been appreciated for centuries in Oriental countries and to a more limited extent in the Western world. Many species of algae are regularly used in the diet of Oriental peoples. For example, a species of *Porphyra* known as "**laver**" has long been cultivated in Japan by sinking nets attached to bamboo poles in shallow estuaries; the *Porphyra* spores settle on the nets and develop into the bladelike plants, which are subsequently harvested. On the coast of France (Brittany), marine brown algae are used as food for cattle and as a soil conditioner. In the latter connection,

marine algae (mostly kelps) mixed with rock particles have made possible the formation of soils by inhabitants of the rocky Aran Islands of Ireland. The toxins produced by blue-green algae were discussed by Moore (1977) and Gorham and Carmichael (1979). In contrast, Ehresmann et al. (1978) have demonstrated that some California marine algae contain antibiotics. The active agent may be debromoaplysiatoxin; this substance has recently been shown to exercise antileukemia activity in mice (Kashiwagi and Norton, 1977).

Various products of alginic acid, derived from the cell walls of coarser Phaeophyta, are used for several purposes in the textile industry, among them waterproofing cloth. Alginic acid has also been used to improve the texture of commercial ice cream.

Colloidal extracts from a number of marine Rhodophyta, but in this country especially *Gelidium robustum*, are the basis for the purified product known as agar-agar or simply **agar**. This substance is of paramount importance as the relatively inert agent of solidification of microbiological culture media, but it has a number of additional commercial and medicinal applications. Another red alga, *Chondrus crispus*, Irish moss, often called "carragheen," is used in certain coastal localities in this country and abroad as the basis for blancmange and other confections. Colloids extracted from *Chondrus* are widely used in food products such as chocolate milk and ice cream. More than 3 million pounds (dry weight) of *Chondrus* were harvested commercially in Nova Scotia in 1960, and Japan alone harvested 285,000 metric tons of marine algae the same year. These examples indicate the importance of algae in the economy of certain populations. (For additional information on the uses of marine algae, see Lobban and Wynne, 1981.)

In addition to the uses of these living algae or products derived from them, algae are important in many other connections. Reference has already been made to the numerous uses of the fossil remains of diatoms known as **diatomaceous earth** (Conger, 1936).

Airborne algae have been implicated as responsible for inhalant allergies in human beings (McElhenney, et al., 1962), while the blue-green prokaryote *Lyngbya majuscula* has been the cause of contact dermatitis of swimmers in Hawaii (Moikeha et al., 1971; Moikeha and Chu, 1971).

Michanek (1975) has surveyed the seaweed resources of the world's oceans, while Krauss (1978) provided data regarding those of the northwest Pacific Coast of the United States. Other examples of the toxicity of algae and their relations to human and animal welfare have been summarized by Gorham (1964) and by Schwimmer and Schwimmer (1964). Bernstein and Safferman (1970) have reported the presence of algae (*Chlorella*, *Chlorococcum*, *Chlamydomonas*, and *Planktosphaeria*, together with several Cyanophyta, etc.) in house dusts from 41 homes in the Cincinnati area. Extracts of some of these algae gave positive skin-test results in some patients being treated for allergies occasioned by house dusts.

Finally, unicellular algae (*Chlorella*, etc.) are currently the experimental organisms in several very important lines of research that have obvious implications for human welfare. The first concerns the mechanism and energy relations of photosynthesis. The second concerns the large-scale cultivation of the organisms in a controlled environment for the purpose of obtaining maximum yields to provide possible supplementary sources of food and possibly fuel. Furthermore, algae are extremely important organisms for the study of such life processes as differentiation, nutrition, sexuality, and so on. More extensive discussions of the topics and organisms alluded to in this chapter and in Chapters 4–9 will be found in Bold and Wynne (1985) and other citations in the Bibliography.

DISCUSSION QUESTIONS

1. What plants were formerly included in the division Thallophyta? What did this practice imply?
2. Define or explain the terms ontogeny, phylogeny, monophyletic, polyphyletic.
3. What evidence can you cite to indicate that the algae are a polyphyletic group?
4. Discuss pigmentation and reserve photosynthates as they occur in algae. Compare with the land plants in these respects.
5. List the divisions and classes of algae and the characteristics that distinguish them.
6. Summarize the types of life cycle that are present in algae and name illustrative genera.
7. Speculate regarding the possible development and relationships of the various types of plant body structure observed in the algae.
8. Discuss sexual reproduction with respect to possible origin, significance and distribution of the sexes.
9. (a) Where may the algal zygote be located during its formation, at maturity and at germination.
   (b) What correlations can you make between its location and further development?
10. Distinguish between morphological and cytological alternation of generations; between isomorphic and heteromorphic alternation; between monobiontic and dibiontic life cycles; between haplonts and diplonts; between haploid and diploid.
11. Discuss the algae with respect to habitat.
12. How would you go about proving that aquatic animals depend on algae (or other submerged plants) for their existence?
13. Cite some examples of the biological importance of algae.

14. In what respects are algae of economic importance?

15. What is meant by a "culture" of algae? What are pure cultures, unialgal cultures, clonal cultures, and axenic cultures?

16. In agricultural areas in many parts of the country it has become the practice to excavate for tanks or ponds in which to raise fish for food. Once the water has filled these artificial ponds, commercial fertilizer is added. Exactly how does this affect the fish?

17. What significant data are contributed by the fossil record of the algae?

18. Because algae and other organisms are classified in separate divisions, is one to infer that there were many separate origins of life? Give evidence for and against such an inference, and explain what is really meant by our segregative system of classification.

Chapter 11
INTRODUCTION TO THE LAND PLANTS: DIVISIONS HEPATOPHYTA AND ANTHOCEROTOPHYTA

INTRODUCTION TO THE LAND PLANTS

Chapters 4–10 included a discussion of the algae, which were formerly considered to be a class (with the class Fungi), division Thallophyta (see inside back cover). It will be recalled that the algae are chlorophyllous and photoautotrophic,[1] whereas the fungi, and a few types of bacteria, are achlorophyllous, heterotrophic organisms. With the rarest exceptions, the plants to be described in the present and succeeding chapters are, like the algae, also photoautotrophic. Although the algae are largely aquatic organisms and include relatively few terrestrial representatives,[2] the remaining groups of photosynthetic plants are primarily land dwellers. However, almost every division provides examples of genera that probably have become secondarily adapted to the water from an originally terrestrial habitat; a few may even be primitively aquatic.

The land plants, other than terrestrial algae and the flowering plants, are often spoken of as "archegoniates" because of their characteristic female gametangium (Fig. 4-1E). It has been suggested that almost as characteristic as the archegonium for these organisms is the organization of the flagellate sperm which contains a "multilayered structure" also present in a few algae (e.g., *Chara*, *Coleochaete*).

In this connection it is of interest that "land plants" are often spoken of as "vascular plants," which in current usage means they have *lignified* water-conducting cells and cells specialized for the conduction of organic compounds *phloem* (p. 324). It should be noted, however, that mosses (and a few liverworts) have conducting cells, hydroids (unlignifed) and leptoids, the latter phloemlike, although they are not usually grouped with vascular plants (Scheirer, 1980).

---

[1] Most of those so far investigated are photoautotrophic.

[2] In spite of the ubiquity of soil algae.

When one compares the organization of the aquatic algae with that of terrestrial plants, it becomes evident that most of the complexities of the latter are manifestations of adaptation to existence in a drying atmosphere. Compared to its postulated algal ancestors, the land plant is a pioneer in a harsh and unfavorable environment. Aquatic plants are immersed continuously in a solution containing the materials used in the synthesis of their protoplasm. On the contrary, large portions of the body surface of terrestrial plants are not in direct contact with water and dissolved nutrients, and, furthermore, may lose a large percentage of the water that has been absorbed. Study of the structure of land plants reveals a variety of both morphological and physiological adaptations to terrestrial life. The degree and efficiency of these adaptations are undoubtedly correlated with the habitats of land plants in relation to moisture and with the very survival of the plants themselves.

In these considerations, it should not be overlooked that many green (and other) algae, both aquatic and edaphic, are well adapted, both morphologically and physiologically, to periodic desiccation. The selective process in evolution, apparently, resulted in the development of land plants, when morphological and physiological characteristics became combined in individuals that would both grow vegetatively and reproduce in a wide range of habitats differing in availability of water.

In addition to the morphological, there must have been modifications of physiological nature that enabled land plants, or at least some of their reproductive agents, to withstand periodic desiccation. Many liverworts and mosses and xerophytic plants in every group clearly have such physiological adaptations. It is known, for example, that when certain mosses (and lichens) dry to about 90% below saturation with water, photosynthesis ceases. However, even after long periods of such desiccation, they can re-sume the full photosynthetic rate within 30 seconds. Furthermore, the production of spores and seeds resistant to desiccation is an important adaptation to life on land.

The liverworts,[3] hornworts, and mosses are classified by many, but not all, botanists in a single division, Bryophyta (*sensu lato*) (Schuster, 1966). This reflects their judgment that these three groups of plants are more closely related to one another, phylogenetically, than any one group is related to any other division of plants. While they are, indeed, similar in life cycle, the plant being in each case a free-living gametophyte on which a partially or completely parasitic sporophyte is borne, the morphological differences among them are, in the authors' judgment, sufficiently significant to warrant the classification of the liverworts, hornworts, and mosses as separate divisions of the plant kingdom. Stotler and Crandall-Stotler (1977) and Crandall-Stotler (1980) used such a classification, which is followed in this book as well. The classification of organisms is subjective and should not be regarded as inviolate and inflexible. Classifications are but approximations that purport to reflect the *course* of evolution of living organisms. With respect to the relationships between the "higher categories" of plants, our *factual knowledge*, as distinct from *speculation*, is woefully inadequate. "It is axiomatic that absolute truth cannot be realized and that individual interpretations of the same biological phenomena can produce different systems of classification and phylogenetic schemes" (Mitler, 1979).

The Divisions Hepatophyta (liverworts) and Anthocerotophyta (hornworts) are discussed in this chapter; the Bryophyta (mosses) are described in Chapter 12. A discussion of alterna-

---

[3]The name "liverwort" was first applied to the genus *Conocephalum* because of a fancied resemblance of its lobes to those of a liver.

tion of generations and the phylogeny of these three groups is deferred to the end of Chapter 12.

DIVISION HEPATOPHYTA[4]

The Hepatophyta (Gr. *hepar*, liver) or liverworts, containing over 300 genera and 6,000–10,000 species (Sharp, 1974*b*), are certainly among the most simple extant land plants. They vary in size from minute leafy forms such as *Cephaloziella* (150–350 μm wide × 2–10 mm long) to large, lobed types such as *Dumortiera* (2–3.5 × 15–25 cm) and *Monoclea* (5 × 20 cm). The earliest fossil liverworts are Devonian (see inside front cover). Several of the genera included in this group are aquatic organisms; the remainder, with few exceptions, are restricted to moist habitats. In some of them, such as *Sphaerocarpos*, *Pellia*, and *Takakia*, the vegetative plant body is scarcely more complex, internally and externally, than that of such algae as *Ulva*.[5]

The division Hepatophyta here is considered to include the single class, Hepatopsida. A chromosome number of $n = 9$ occurs in 75% of the liverworts studied, but species with $n = 4$, 8, and 10, and multiples thereof, are also known.

CLASS 1. HEPATOPSIDA[6]

General features. The gametophyte is the dominant phase in the heteromorphic alternation of the Hepatopsida or liverworts. They are grouped into orders and families on the basis of differences in the structure of their gametophytic and sporophytic phases. If one attempts to list first in the class the most primitive organisms exhibiting the greatest simplicity, one is at a loss to make a decision, for low and high degree of complexity in both sporophyte and gametophyte do not coincide in the same organism. Furthermore, attributes that some morphologists interpret as simple may have been interpreted as either primitive or reduced by others. Thus, on the basis of structure of the sporophyte alone, for example, if simplicity is the criterion, *Ricciocarpus* and *Riccia* deserve first place, but this simplicity is interpreted as secondary by some and considered to represent evolutionary reduction.

The internally differentiated gametophytes of *Ricciocarpus* (Fig. 11-2), *Riccia* (Fig. 11-12), and *Oxymitra* (Fig. 11-17) are more complex than those of genera such as *Sphaerocarpos* (Fig. 11-36), *Pellia* (Fig. 11-42B), and *Pallavicinia* (Fig. 11-47).

Some cells of many Hepatophyta (90% of those investigated in this respect), in contrast to those of the Anthocerotophyta and Bryophyta, contain oil bodies. These are various in form (Fig. 11-54D–E) and may be colorless, gray, brown, or bright blue. They consist of a number of types of ethereal oils, the function of which has not been determined. Oil bodies of this type do not occur elsewhere in the plant kingdom.

The Hepatopsida may be divided into seven orders: Marchantiales, Sphaerocarpales, Monocleales, Metzgeriales, Jungermanniales, Haplomitriales, and Takakiales. Representative genera of the first five orders will be emphasized in the following account, with briefer treatment of the Haplomitriales and Takakiales. For a summary of their organization see Crandall-Stotler (1981).

---

[4]For an in-depth discussion see Schuster (1984 *a, c*) and Crandall-Stotler (1984).

[5]Some would take exception to this statement arguing that *Ulva*, although exhibiting desmoschisis, lacks plasmodesmata.

[6]Stotler and Crandall-Stotler (1977) divide the Hepatophyta into three classes, the Haplomitriopsida, the Jungermanniopsida, and the Marchantiopsida.

Order 1. Marchantiales

The order Marchantiales (95 genera, 2,000 species [Sharp, 1974*b*]) includes thallose or bladelike, flat, dorsiventral organisms, usually with some degree of internal differentiation, in the form of photosynthetic cells, air chambers, and storage tissues. Some of the morphological differentiations in the Marchantiales apparently enhance the vital functions but are not essential for survival, while some are seemingly disadvantageous (Bischler and Jovet-Ast, 1981). In all Marchantiales the sporophyte is compact with no seta or a short one and with the capsule (sporangium) wall one-layered. Representatives of three families of Marchantiales are considered in the following discussion.

Family 1. Ricciaceae
Ricciocarpus

Ricciocarpus natans (Fig. 11-1), the only species of the genus, is amphibious. It grows readily in laboratory cultures (Lorenzen et al., 1981). However, it usually fails to form sex organs when it is growing on soil (Fig. 11-1B).

Vegetative morphology. The individual plants of *Ricciocarpus* (Fig. 11-1) taper posteriorly from a broad, dichotomously lobed anterior growing region. The posterior portion of the plant is continually sloughing off as growth occurs at the tip. Increase in number of individuals occurs when branches become separated or when death and decay extend to the region of a dichotomy. Plants that contain mature sporophytes may attain a length of 2.5 cm. Four or more rows of **ventral scales** are present on the lower surface of each plant (Kronestedt, 1981). The scales, which are a single cell layer thick, are often purple. Unicellular protuberances, or **rhizoids,** emerge from the cells of the lower surface of terrestrial plants and penetrate the substratum.

The apparent midrib in *Ricciocarpus* is in reality a furrow that extends deep into the plant body from the dorsal surface (Fig. 11-2A, C). Observed in transverse section (Fig. 11-2A), this furrow is approximately in the form of an inverted Y. The lower portion of the plant is

A   B   C

FIGURE 11-1
Ricciocarpus natans. **A.** Floating plants. **B.** Plants on soil. **C.** Single plant showing dorsal furrow and ventral scales. **A.** ×0.75. **B.** ×0.37. **C.** ×80. (**A.** *Courtesy of Dr. William C. Steere.* **C.** *After Kronestedt.*)

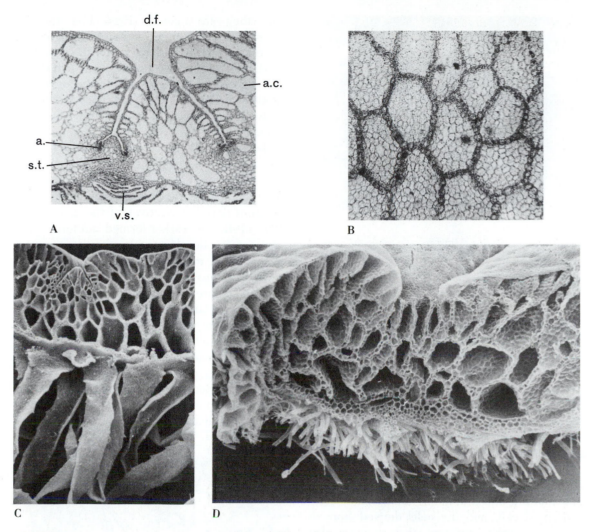

FIGURE 11-2
Ricciocarpus natans. **A.** Transection of a plant in the region of a dichotomy; note two Y-shaped dorsal furrows, a., archegonium; a.c., air chamber; d.f., dorsal furrow; s.t., storage tissue; v.s., ventral scales. **B.** Dorsal surface of plant showing air chambers. **C, D.** Transections. Note ventral scales. **A.** ×30. **B.** ×60. **C, D.** ×40. (**C, D.** *After Kronestedt.*)

composed of rather compact parenchyma cells, which contain few chloroplasts and function as storage cells; these are covered above by several tiers of air chambers (Fig. 11-2). The latter are responsible for the spongy appearance of the plants when they are viewed in dorsal aspect under low magnification (Fig. 11-2B). The walls of the air chambers are built up of photosynthetic parenchyma (chlorenchyma) cells rich in chloroplasts. The cells of the upper surface contain only a few poorly developed plastids and function as an epidermis. That their outer walls are cutinized at least delicately is suggested by the occurrence of apertures, one corresponding

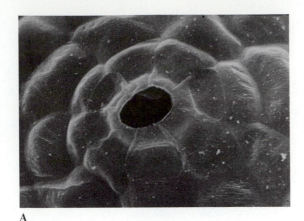

A

B

FIGURE 11-3
Ricciocarpus natans. **A.** Surface view of a pore. **B.** Sectional view of a pore and surrounding cells. **A.** ×1200. **B.** ×410. (**A.** *After Kronestedt.*)

to each air chamber, on the upper surface of the plant body (Fig. 11-3). The epidermal cells surrounding the aperture are slightly modified in structure.

Growth of the plant body of *Ricciocarpus*, and of almost all land plants, is localized at the apex, where one prominent meristematic cell, the apical cell, and the cells derived from it undergo orderly divisions that augment the tissues of the plant. A median sagittal section through the plant demonstrates its ontogeny (Fig. 11-4). Dorsal derivatives of the apical cell form chlorophyllous tissue, while the ventral derivatives give rise to the storage tissue, ventral scales, and rhizoids.

Reproduction: the gametophyte.   Other than vegetative reproduction by fragmentation, increase in number of individuals in *Ricciocarpus* is the ultimate result of sexuality. The reproductive organs are sunken in chambers on the floor and walls of the dorsal furrow by the

time they are mature (Figs. 11-4 through 11-7A). They arise from single cells and at first protrude from the surface of the furrow (Fig. 11-4); they become secondarily sunken in chambers because of the upgrowth of the surrounding tissues. The liverwort plant body is gametophytic. *Ricciocarpus* is monoecious, the male and female sex organs arising on the same plant.

The **antheridia,** or male reproductive organs, appear first in young plants and are followed by the **archegonia**, the female reproductive organs in the land plants. Since both types of sex organs are from dorsal derivatives of the apical cells (Fig. 11-4), the earlier-formed antheridia are posterior in the furrow to the later-formed archegonia in monoecious species. Plants such as *Ricciocarpus*, in which the male sex organs develop before the female, are said to be **protandrous** (Gr. *proteros*, prior, + Gr. *ander*, male); those in which archegonia are produced first are said to be **protogynous** (Gr. *proteros* + Gr. *gyne*, female). Sex organs usually appear only in floating plants of *Ricciocarpus*, but they may develop occasionally in terrestrial individuals. Floating plants that bear only antheridia are young individuals in which archegonia have not yet developed. Plants bearing only archegonia

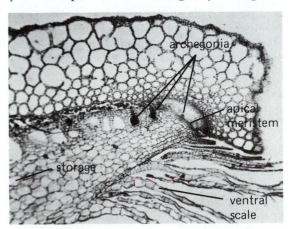

FIGURE 11-4
Ricciocarpus natans. Median longitudinal (sagittal) section of plant apex cut parallel to the dorsal furrow; note air chambers above. ×30.

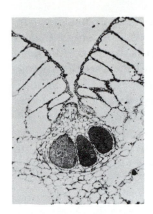

FIGURE 11-5
Ricciocarpus natans.
Transection of plant
with almost-mature
antheridia. ×60.

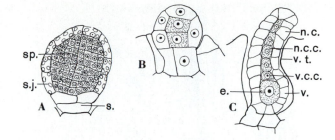

FIGURE 11-6
Ricciocarpus natans. Sex organs in median
longisection. **A.** Immature antheridium. **B, C.**
Stages in the development of the archegonium. e.,
egg cell; n.c., neck cell; n.c.c., neck canal cell; s.,
stalk; s.j., sterile jacket; sp., spermatogenous tissue;
v., venter; v.c.c. ventral canal cell; v.t., vegetative
tissue of plant body. ×546.

are older ones in which the antheridia have
been lost through decay of posterior portions of
the individuals. It should be noted that the
floating plants are usually abundant if not actually contiguous.

The antheridia and archegonia are arranged
in three or more ill-defined rows in the dorsal
furrow (Figs. 11-4, 11-5). They differ from the
gametangia of the algae—with the exception of
the plurilocular gametangia of the Phaeophyta,
the antheridia and oogonia of the Charophyta,
and those of certain fungi—in their multicellular
construction. Furthermore, unlike the plurilocular gametangia of the Phaeophyta, the sex
organs are composed, in part, of sterile cells.[7]
The antheridium (Figs. 11-5, 11-6A) consists of a
short stalk and a single layer of surface cells, the
sterile jacket, which encases the small, cubical
spermatogenous cells in which the **sperms** are
organized. The archegonium (Figs. 11-6B, C,
11-7A) is flasklike, consisting of an axial row of
cells, including the **egg, ventral canal cell,** and
neck canal cells, surrounded by a jacket of six
rows of sterile cells, which form the slender

neck and the **venter**[8]—the latter, the enlarged
basal portion of the archegonium. As the archegonium matures, the neck canal cells and ventral canal cell disintegrate and are extruded as
slime, thus leaving a moist canal (Fig. 11-7A)
through which the sperm swims to the egg,
probably through chemotactic attraction.

Reproduction: the sporophyte. The zygotes of Hepatophyta (and all other archegoniate plants) are retained on the gametophytes
upon which they were formed, as in
Rhodophyta and *Coleochaete.* Shortly after its
formation, the zygote of *Ricciocarpus* (Fig.
11-7A) undergoes a series of nuclear and cell
divisions that produce a spherical mass of tissue
within the venter of the archegonium (Fig.
11-8). As in such algae as *Cladophora, Ulva,
Ectocarpus, Laminaria, Griffithsia,* and *Polysiphonia,* the divisions of the zygote nucleus are
mitotic rather than meiotic, so that diploid

---

[7]Some students of phylogeny, who postulate an algal
ancestry for the land plants, have been impressed by an
apparent simplicity between the sex organs of the latter and
the plurilocular gametangia of Phaeophyta. The current
view that the Phaeophyta and Chlorophyta are parallel,
rather than closely related, groups offers scant encourage-
ment to such speculations.

[8]The ventral canal cell and neck canal cells are considered by some to represent vestigial female gametes that
have become sterile, a view which is fraught with mechanical difficulties unless one postulates that at the time they
were still functional all the female gametes were shed into
the water; otherwise, only the most distal (represented by
the first neck canal cell near the neck orifice) could have
been reached by a sperm.

193

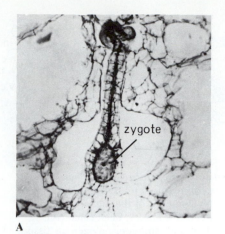

A

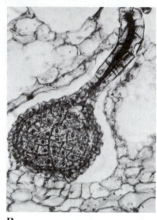

B

FIGURE 11-7
Ricciocarpus natans.
A. Median longitudinal section of archegonium with zygote; note opened apex of archegonium. **B.** Archegonium containing young sporophyte at about the stage shown in Fig. 11-9B. **A.** ×200. **B.** ×250.

tissue is formed. The venter of the archegonium becomes two-layered by periclinal (Gr. *peri*, around, + Gr. *klino*, bend) divisions that occur immediately after fertilization (Fig. 11-8); by continuous anticlinal (Gr. *anti*, against, + Gr. *klino*) divisions, the archegonial venter keeps pace with the growth of the diploid tissues within it (Figs. 11-8, 11-10A). After fertilization, such an enlarging archegonium surrounding a sporophyte is called the **calyptra** (Gr. *kalyptra*, veil). A differentiation of the sporophyte occurs during development as a result of periclinal divisions in the outermost layer of cells, which segregate a peripheral layer, the **amphithecium** (Gr. *amphi*, around, + Gr. *thekion*, case), from

the central mass of tissue, the **endothecium** (Gr. *endon*, within, + Gr. *thekion*) (Fig. 11-8B). When the endothecium has increased to about 400 cells, enlargement of the endothecial cells of the diploid tissue occurs, and the cells separate and become suspended in the liquid within the amphithecial wall. Traces of a green pigment, possibly chlorophyll, in the form of dispersed or aggregated droplets appear in both the amphithecial and endothecial cells during their development.

The endothecial cells now undergo two successive divisions, during which the chromosome number of the diploid tissue ($2n = 16$) is reduced ($n = 8$). Each of the cells now contains four haploid nuclei, arranged tetrahedrally. Cytokinesis follows, and the original cells are transformed into groups of four coherent haploid cells, that is, **spore tetrads** (Figs. 11-9B, 11-10A). Each member of the tetrad later becomes invested with a thick, black wall (Fig. 11-9C).

The inner layer of the archegonial venter disintegrates at about this stage, with the result that the spore tetrads are surrounded by the diploid amphithecium and the outermost layer of the archegonial venter; these usually cohere.

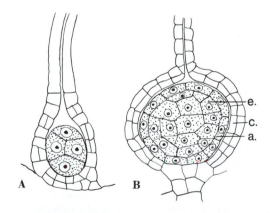

A   B

FIGURE 11-8
Ricciocarpus natans. Development of the sporophyte. **A.** Four-celled stage. **B.** Later stage. a., amphithecium; c., calyptra; s., endothecium. ×250.

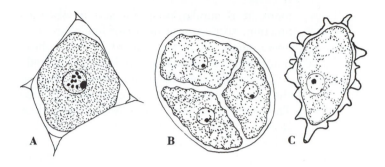

FIGURE 11-9
Ricciocarpus natans. Sporogenesis.
A. Sporocyte in prophase of meiosis.
B. Spore tetrad. **C.** Section of mature
spore. ×500.

Meiosis in *Ricciocarpus* and in all land plants is sporic rather than zygotic or gametic. The spores, accordingly, are **meiospores**. The cells that undergo meiosis and form spore tetrads are usually called **spore mother cells** or **sporocytes** in the land plants. The four products of meiosis

and cell division of each sporocyte become the **spores (meiospores)**.

The spherical mass of diploid tissue within the enlarged archegonial venter is called the **sporophyte**. One unfamiliar with life cycles in the algae, on the one hand, and with those in

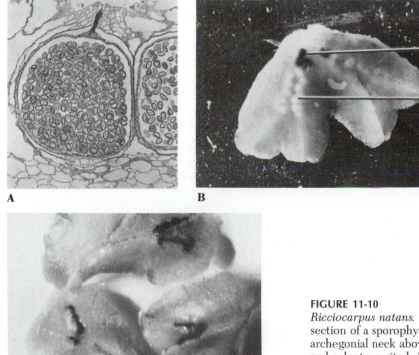

FIGURE 11-10
Ricciocarpus natans. Sporophytes. **A.** Median section of a sporophyte with spore tetrads; note archegonial neck above; remains of amphithecium and calyptra united. **B.** Plant with sporophytes of several ages. **C.** Plant with all mature sporophytes. i.s., immature sporophyte; m.s., mature sporophyte. **A.** ×60. **B.** ×2. **C.** ×3.

195

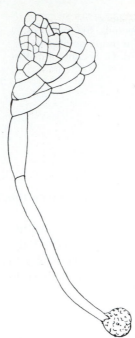

FIGURE 11-11
Ricciocarpus natans. Spore germination and young gametophyte. ×62.

the vascular land plants, on the other, would be at a loss to understand this designation, literally "spore plant," because the almost colorless, spherical mass of tissue would scarcely pass muster as a "plant." However, comparison of its attributes with those of the free-living sporophytes of certain algae and of vascular plants provides evidence that it has the characteristics of diploidy and spore production generally associated with free-living sporophytes. Further-

more, it is similar in origin and in ultimate function, so that on grounds of homology (Gr. *homo*, same, + Gr. *logos*, word, study) application of the designation "sporophyte" is justified.

No special mechanism for dissemination of the spores is present in *Ricciocarpus* and *Riccia*. The older, posterior sporophytes mature first (Fig. 11-10B), as evidenced by their blacker color, which results from the thickening deposited on the spore walls. As the thallus grows apically and decays in the older portions, the spores are liberated and are usually disseminated by water.

The germination of the spores (Fig. 11-11) of *Ricciocarpus* occurs in the southern United States late in the summer or in the autum on damp soil on the margins of ponds as the water level recedes. The young plants remain terrestrial until they become submerged by the rise in the water level that accompanies autumn and winter rains and spring thaws. The germinating spores give rise to plants like the original gametophytes. When submerged by the rising water level, the apices, which are not anchored by rhizoids, become detached and float to the surface, where they initiate large colonies of floating plants, which multiply early in the spring by fragmentation. A typical hepatophytan life cycle is summarized in Chart 11-1.[9]

---

[9]Note that haploid phases are underlined once, diploid phases twice.

CHART 11-1
Life cycle of Hepatophyta.

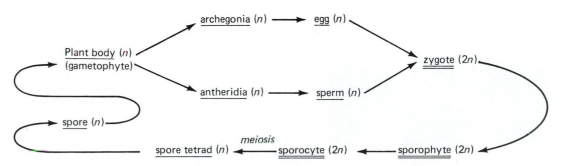

A

B

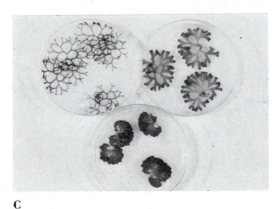

C

FIGURE 11-12
Various species of *Riccia*. **A.** *R. beyrichiana*, on soil. **B.** *R. hirta* on soil. **C.** Three species growing on agarized culture medium in the laboratory: above left, *R. fluitans;* above right, *R. dictyospora;* below, *R. curtisii*. **A.** ×3.5. **B.** ×4. **C.** ×0.25. (**C.** *Courtesy of Dr. Carol Woodfin*).

of the ventral scales is the result of pigmentation in the cell walls.

The plants may be uni- or bisexual, the archegonia and antheridia arising near the apex and often in alternating groups. They become immersed in the thallus as the latter matures, the archegonial necks protruding. The opening (**ostiole**) of the antheridial cavity is slightly ele-

Riccia

Most species of *Riccia* are terrestrial, forming rosettes or mats (Fig. 11-12) on moist soil, but several, such as *R. fluitans* (Fig. 11-13), are amphibious, growing both submerged and on moist soil and, rarely, on decaying wood. The air chambers in *Riccia* may take the form of fissures between vertical columns of cells, or they may be polyhedral, depending on the species (Fig. 11-14). Some species become spongy by decay of the upper cells roofing the air chambers. Pegged[10] and smooth-walled rhizoids and ventral scales (often purplish) occur in many species, and are, indeed, characteristic of the Marchantiales in general. The purple color

---

[10]See rhizoids of *Marchantia*, p. 205.

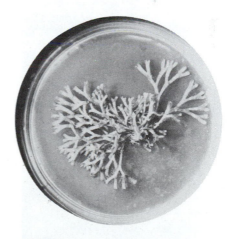

FIGURE 11-13
Riccia fluitans. Culture on agar; note dichotomous branching. ×1.

197

A

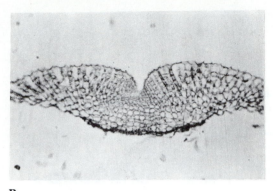

B

FIGURE 11-14
Riccia austinii. **A.** Surface view. **B.** Transection. **A.** ×6. **B.** ×18.

vated or quite elongate and prominent. The sporophytes (Fig. 11-15) are spherical, their endothecium entirely fertile, as in *Riccio-carpus*, and the amphithecium remains monostromatic. The spores are large, black or dark brown, and variously ornamented. (Stein-kamp and Doyle, 1979; Thaithong, 1982). Those in *R. curtisii* remain adherent in tetrads (Fig. 11-16), so that at germination the male and

female thalli arise in groups. The spores are set free as the thalli decay. Woodfin (1972, 1976) analyzed comparatively in axenic culture the factors affecting growth and reproduction in a number of species of *Riccia*.

Family 2. Oxymitraceae
Oxymitra

Oxymitra paleacea (Fig. 11-17) is scattered in distribution but occurs, among other places, in Texas and Oklahoma and in southern Europe,

FIGURE 11-15
Species of *Riccia* with sporophytes. **A.** *R. duplex-like.* **B, C.** *R. dictyospora* (2-month-old cultures). **A.** ×5. **B.** ×2. **C.** ×6. (*Courtesy of Dr. Carol Woodfin.*).

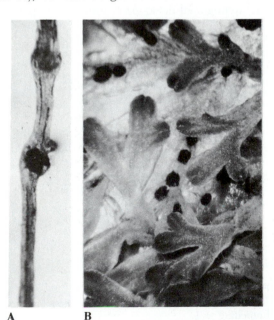

A   B

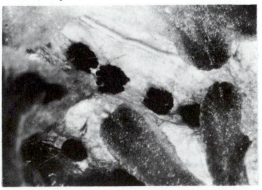

C

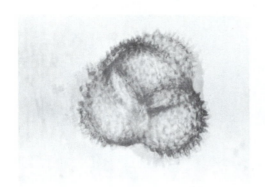

FIGURE 11-16
Riccia curtisii: spores permanently united in tetrads. ×350.

A

B

FIGURE 11-17
Oxymitra paleacea. **A, B.** Plants on soil; note marginal scales and sporophytes in dorsal midline in **B. A.** ×1.3. **B.** ×2.5.

where it grows on intermittently moist soil. It is readily distinguishable from *Riccia* by its conspicuous white marginal scales (Fig. 11-17B), which become still more apparent as the plants dry, and by the hexagonal air chambers, each with an air pore surrounded by six specialized cells. The thalli are deeply grooved dorsally. The sex organs occur in the dorsal groove of the bisexual thalli; the archegonia are surrounded by ovoid **involucres,** which are absent in the Ricciaceae; the involucres grow as the sporophytes increase in size. As in *Riccia* and *Ricciocarpus,* the sporophytes are spherical and the endothecium completely fertile except for a few sterile cells. The amphithecium may be more than two cell layers thick at the base and apex of the sporophytes; the latter liberate the spores when the thallus rots and/or dies.

Family 3. Marchantiaceae
Marchantia

Habitat and vegetative morphology. The manifestations of internal differentiation and specialization already noted in *Riccia,* *Ricciocarpus,* and *Oxymitra* are still more pronounced in *Marchantia* (after N. Marchant, a French botanist), a genus usually terrestrial in habitat. *Marchantia polymorpha* (Fig. 11-18),

FIGURE 11-18
Marchantia polymorpha. Note gemma cups. ×0.33. *(Courtesy of Dr. F. R. Trainor.)*

199

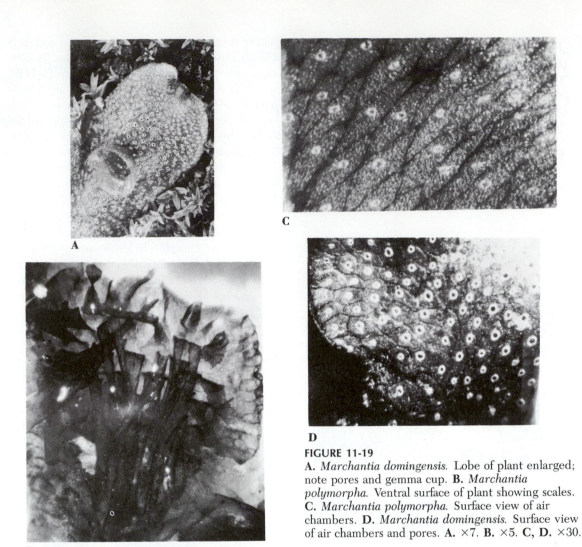

FIGURE 11-19
A. *Marchantia domingensis.* Lobe of plant enlarged;
note pores and gemma cup. **B.** *Marchantia
polymorpha.* Ventral surface of plant showing scales.
C. *Marchantia polymorpha.* Surface view of air
chambers. **D.** *Marchantia domingensis.* Surface view
of air chambers and pores. **A.** ×7. **B.** ×5. **C, D.** ×30.

perhaps the most widespread species, often
grows on moist soil on which the vegetation has
been burned; *M. paleacea* grows on moist limestone rocks, and *M. domingensis* (Fig. 11-19A)
lives on moist clay banks as well as on calcareous
rocks. The last two species are southern in
distribution in the United States, while *M.
polymorpha* is more northern.

   The plant body of *M. polymorpha* is larger
than that of *Ricciocarpus*, *Riccia*, and *Oxymitra*;
under favorable conditions of moisture and nu-

trition, it may exceed 4 inches in length. It also
is dichotomously branched, exhibiting the apical growth and posterior decay common to most
Hepatophyta. The upper surface of the plant is
conspicuously divided into polygonal air chambers, each with a central pore (Figs. 11-19A–D
through Fig. 11-21). The development of the
pores and air chambers in *M. paleacea* has been
studied by Apostolakos et al. (1982). The surface
of the plant is cutinized, as are the cells of the
air pores, the cutinization preventing the en-

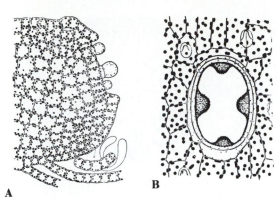

A

B

FIGURE 11-20
Marchantia polymorpha. **A.** Sagittal section of apex. **B.** Surface view of air pore. **A.** ×250. **B.** ×410. (*From a Kny chart.*)

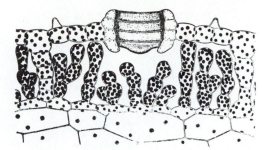

FIGURE 11-21
Marchantia polymorpha. Longisection of one air chamber and pore. ×350. (*From a Kny chart.*)

trance of water (Schonherr and Ziegler, 1975). Caldicott and Eglinton (1976) found cutin acids also in the related Marchantialean genera *Conocephalum* (p. 211) and *Asterella.*

Numerous rhizoids emerge from the ventral surface of the thallus; both smooth-walled and tuberculate ("pegged") (Fig. 11-25F) rhizoids are produced. The smooth-walled rhizoids penetrate the soil, while the pegged type run horizontally along the plant's ventral surface, where, as wicklike strands, they are seemingly held in place by the ventral scales. The latter in *M. polymorpha* (Fig. 11-19B) are arranged in six or eight rows. The tuberculate rhizoids originate from the portions of the ventral surface under the scales or near them. The smooth-walled rhizoids develop near the midportion of the ventral surface. Bundles of tuberculate rhizoids from below the scales converge toward the midrib. It has been shown that the scales and bundles of rhizoids are involved in the rapid conduction of water by capillarity over the ventral surface.

It is apparent from sections (Fig. 11-22) that the upper portion of the plant is composed of a single layer of air chambers surrounded by photosynthetic cells and covered by an epidermis, and that the remainder is made up of

densely arranged parenchymatous cells, which contain a few chloroplasts and probably serve as storage cells. The cells forming the walls and floor of the air chamber, on the other hand, are rich in chloroplasts. Although the photosynthetic portion of *Marchantia* is more restricted in proportion to total thickness than that in *Ricciocarpus* and *Riccia,* the richly branched cactus-like, dwarf, chlorenchyma filaments of *Marchantia* constitute a photosynthetic region of great extent. The specialized cells surrounding the pore of each air chamber are typically arranged in four or five superimposed tiers of

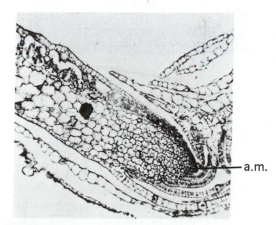

a.m.

FIGURE 11-22
Marchantia paleacea. Median longitudinal (sagittal) section of apex; note apical meristem (a.m.) to the right. ×135.

201

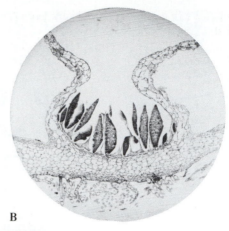

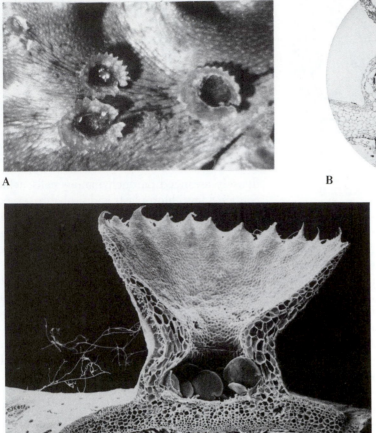

A

B

C

FIGURE 11-23
Marchantia polymorpha. **A.**
Gemma cups (almost all gemmae
shed). **B.** Transection of plant in
region of gemma cup; note
gemmae in sectional view. **C.**
Scanning electron micrograph of
a bisection of a gemma cup; note
scales, rhizoids and air chambers.
A. ×5. **B.** ×20. **C.** ×15. (**C.**
*Courtesy of F. R. Turner and E.
W. Ruf.*)×135.

four in *M. polymorpha* (Figs. 11-20B, 11-21).
The uppermost and lowermost tiers are arranged in circles of smaller diameter; hence, the
whole structure is barrel-shaped. The tiered
cells contain only a few chloroplasts. Cells of the
lowermost tier may protrude into the barrellike
cavity (Figs. 11-20B, 11-21); with loss of water
these lower cells become juxtaposed, while
under saturated conditions they separate to
form a cruciform opening. The pores thus are
analogous to stomata.

Median longitudinal (sagittal) sections cut
perpendicular to the surface of the plant body
demonstrate the method of development of the
various tissues and the origin of the scales (Fig.
11-22). As in *Riccia*, *Ricciocarpus*, and
Oxymitra, the air chambers and their chlorenchymatous boundaries originate from dorsal derivatives of the apical cells, while the storage
region, rhizoids, and ventral scales are from the
ventral derivatives. Fowke and Pickett-Heaps
(1978), on the basis of electron microscopy,

reported that cell division in *Marchantia* is like that of more complex land plants.

Reproduction: the gametophyte. In their development, the young plants of the *Marchantia* are at first entirely vegetative. Later on, vegetative plants produce special propagative structures, the **gemma** cups (Fig. 11-23A), on their dorsal surfaces. Each gemma cup gives rise to a continuously developing series of minute, doubly notched, lens-shaped bodies, the **gemmae** (L. *gemma*, bud), which are attached to the bottom of the cup by short stalks (Fig. 11-23B). When the cups have been flooded with water, the gemmae are rapidly released by the swelling of gelatinous substances secreted by cells that line the base of the cup. They float away from the parent plants, and those that are carried to favorable substrata undergo bipolar growth from the two apical notches, ultimately forming pairs of young plants. It has been observed that splashing raindrops may cause the ejection of gemmae for distances as great as 1 m. **Gemmae** are special agents of the parent plant that can regenerate new plants, a phenomenon widespread in the Hepatophyta and Bryophyta, and similar in principle to the fragmentation of certain algae, on the one hand, and to vegetative propagation of vascular plants, on the other.

Development of *Marchantia polymorpha* (from gemmae) is under photobiological control (Courtoy, 1966a,b; Terni, 1981). Both duration and quality of light are involved. Fluorescent lights for 16 hours daily, at 20–24°C, lacking wavelengths above 7000 Å, support only vegetative growth, while incandescent light (with rays above 7000 Å) stimulates the development of the gametophores, which appear as primordia 30 days after gemmae have been planted. Addition of 1% sucrose to the basal mineral medium hastens the appearance of the gametophore primordia. Paucity of nitrogen may also be involved in evoking sexuality.

The rather generalized distribution of sex organs along the dorsal surface of the plant body observed in *Riccia* and *Ricciocarpus* is absent in *Marchantia*. In the latter, the fertile regions of the plant body are restricted to specialized erect branches often called **gametophores**.

In *M. polymorpha*, the plants are strictly dioecious, the antheridia arising on stalked, disc-headed branches, the **antheridiophores** (Fig. 11-24A), and the archegonia on umbrella-like heads, the **archegoniophores** (Fig. 11-24B), the spokelike processes of which are seven to ten in number. That these rather bizarre structures are to be interpreted as modified branches of the plant body is indicated by the presence of rhizoids and scales on their lower surfaces and of air chambers and pores on their upper ones. Furthermore, the stalks of these sexual branches show dorsiventrality in transection (Fig. 11-25D), and their scales and rhizoids also give evidence of the branchlike structure. Both the archegoniophores and antheridiophores originate as minute, buttonlike excrescences at the apices of certain branches; further elongation of these branches in a horizontal plane is thus terminated. Careful study of the ontogeny of archegoniophores and antheridiophores has revealed that their lobing is an expression of dichotomy repeated in rapid succession.

The antheridia occur in chambers in radiating rows just below the upper surface of the disc of the antheridiophore, the youngest and last-formed near the margins (Fig. 11-25A, C). At maturity, each antheridium is sunken in a chamber, which is connected by a narrow canal to a surface pore. Spermatogenesis has been investigated electron microscopically by Carothers and Kreitner (1967, 1968), Kreitner (1977a, b) and Carothers and Duckett (1978).

Discharge of the biflagellate sperms takes place through these canals and pores. In at least one species of *Marchantia*, there is evidence that the discharge of sperms in the presence of water involves a centripetal distention of the wall cells of the antheridium and of the air chamber, as well as a swelling of the walls of the spermatogenous cells. The pressure generated

FIGURE 11-24
Marchantia polymorpha. Sexual branches. **A.** Plants with antheridiophores. **B.** Plants with archegoniophores. ×1.3. (*Courtesy of Dr. A. R. Grove.*)

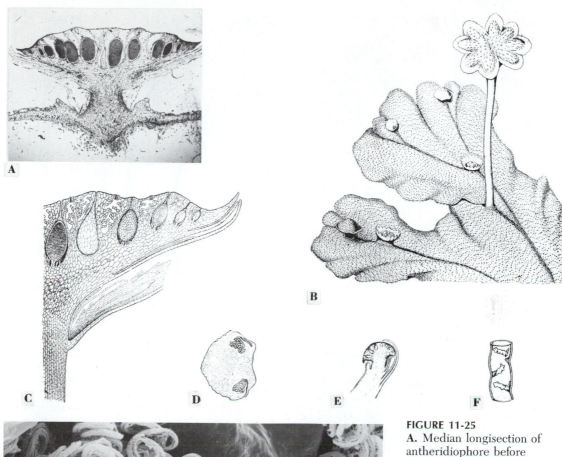

FIGURE 11-25
A. Median longisection of antheridiophore before elongation of its stalk; note younger antheridia at the margins. **B.** Male plant with antheridiophore and gemma cups. **C.** Hemilongisection of antheridiophore; note air chambers above storage tissues and ventral scales. **D.** Transection of antheriodiophore stalk. **E.** Sagittal section of young antheridiophore before elongation. **F.** Segment of a tuberculate rhizoid. **G.** Scanning electron micrograph of sperms. **A.** ×30. **B.** ×2. **C.** ×100. **D.** ×35. **E.** ×1. **F.** ×45. **G.** ×1200. (**A–F.** *From a Kny chart.* **G.** *Courtesy of F. R. Turner and E. W. Ruf.*)

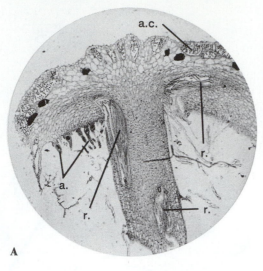

A

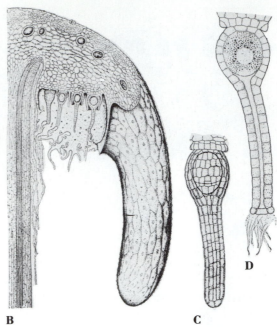

B

C

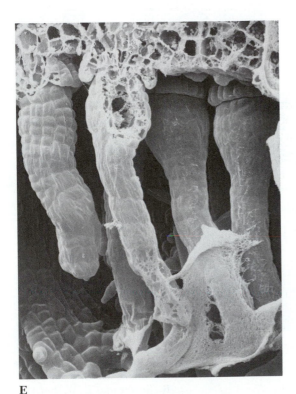

E

D

FIGURE 11-26

Marchantia polymorpha. **A.** Median longisection of an archegoniophore. **B.** Diagrammatic portion of a longisection. **C.** Surface view of immature archegonium. **D.** Median longisection of an archegonium at fertilization. **E.** Scanning electron microscope micrograph of archegonia. a., archegonia; a.c., air chambers; r., rhizoids. **A.**×60. **B.** ×100. **C**, **D.** ×230. **E.** ×400. (*B–D. From a Kny Chart.* **E.** *Courtesy of F. R. Turner and E. W. Ruf.*)

in this manner ruptures the upper cells of the antheridial wall, and the sperms ooze out. The sperms may be observed actively rotating in their mother cells, from which they ultimately are liberated.

The archegonia also occur in rows but are located on the apparent ventral surface of the archegoniophores between the proximal portions of the radiating spokelike rays (Fig. 11-26A, B). The words "apparent ventral surface" are used because in reality the archegonia arise from dorsal derivatives of apical cells, but excessive growth of the dorsal surface results in

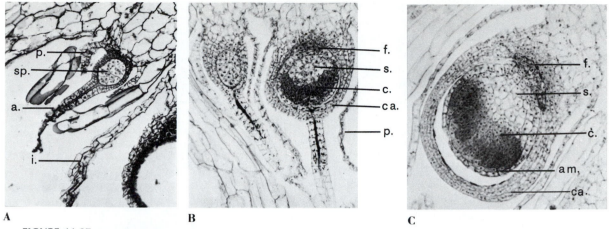

FIGURE 11-27

Marchantia polymorpha. **A, B, C.** Stages in the development of the sporophyte. a., archegonium; am., amphithecium; c., capsule region; ca., calyptra; f., foot; i., involucre; p., pseudoperianth; s., seta region; sp., sporophyte. **A, B.** ×125. **C.** ×175.

an inversion of the portion that bears archegonia. The latter thus become suspended, neck downward, in radiating rows from the undersurface of the archegoniophore (Fig. 11-26). Here again, there are six rows of neck cells. The rows of archegonia, with the last-formed in each group nearest the stalk of the archegoniophore, are separated by fringed **pseudoperianths.**[11] The antheridia and archegonia of *Marchantia* are quite similar to those of *Riccia* and *Ricciocarpus* but somewhat larger. Zinsmeister and Carothers (1974) have described the organization of the archegonia at the ultrastructural level.

Fertilization of the eggs of the first-formed archegonia occurs before the elongation of the stalk of the archegoniophore, and because of the differing degrees of maturity of the antheridia in a single antheridiophore, discharge of the sperms, effected by flooding, probably continues over a considerable period. Fertilization of the later-formed archegonia probably takes place after their elevation above the plant body.

This indicates either that the sperms remain viable for a long period or that they reach the archegonia through splashing or by swimming through the surface films of water on the surface of and in the canals of the stalks of archegoniophores. Raindrops falling on antheridiophores have been observed to splash the sperms for distances up to 60 cm. That the elongation of the archegoniophore stalk is not dependent on fertilization is evidenced by its occurrence even in segregated female plants in which fertilization has not taken place.

Reproduction: the sporophyte.   Although the eggs of many archegonia on an archegoniophore may be fertilized, not all the zygotes develop into mature sporophytes, possibly because of crowding and insufficient nutrients. The more mature sporophytes usually occur near the periphery of each radiating group, corresponding in position to that of the first mature archegonia. Successive transverse and longitudinal divisions of the zygote result in the formation of a spherical mass of diploid tissue, the sporophyte, enclosed in the venter (Fig. 11-27A), as in *Ricciocarpus* and *Riccia*. The

---

[11]True perianths are derived from leaves.

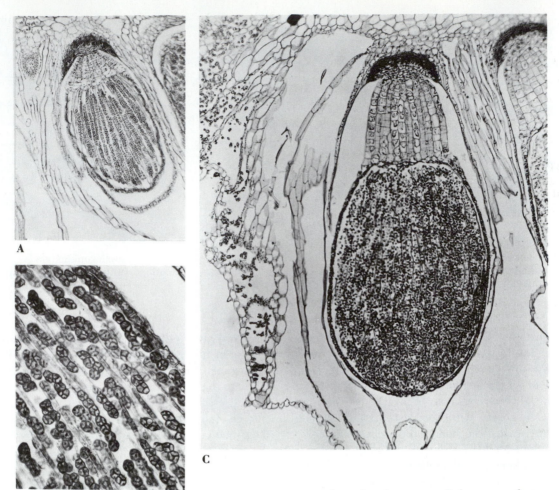

FIGURE 11-28

Marchantia polymorpha. Late stages in sporophyte ontogeny in median longisection. **A.** At the spore-tetrad stage; note alternating strips of tetrads and elaters, enlarged at **B. C.** Maturing sporophyte; note foot, elongating seta, and capsule with mature spores and elaters. **A, C.** ×30. **B.** ×400.

sporophyte undergoes cell divisions, increases in size, and remains as a covering layer, the **calyptra,** as in *Ricciocarpus.* During these stages, a collarlike layer, the **involucre,** grows down from the receptacle enclosing each archegonium (Fig. 11-27, 11-30D) with two flaps of tissue.

Further development of the sporophyte results in differentiation and specialization not present in the sporophytes of *Ricciocarpus* and *Riccia.* Certain cells of the young spherical sporophyte now grow through the base of the venter into the compact storage tissue of the archegoniophore, forming an anchoring and absorptive organ, the **foot.** Enlargement of the opposite pole also occurs, so that the sporophyte is differentiated ultimately into three regions (Figs. 11-27B, C, 11-28A): an enlarged **foot;** an intermediate short cylindrical region, the **seta** or stalk; and a sporogenous region, the **capsule** or sporangium. In the capsule the differentiation of the tissues into amphithecium and endo-

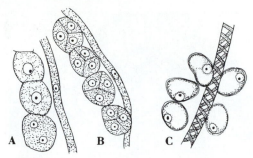

FIGURE 11-29
Marchantia polymorpha. Sporogenesis and elater development. **A.** Sporocytes and elater precursor. **B.** Tetrads with elater precursor. **C.** Mature spores and elater. ×700.

through the seta into the capsule, it is also quite evident that the sporophyte of *Marchantia* (and that of all Hepatophyta) is to some degree photoautotrophic (Thomas et al., 1979). When one compares the independent, photoautotrophic sporophytes of such Chlorophyta as *Cladophora* and *Ulva*, which have no physical connection with the gametophyte, to those of *Ricciocarpus* and *Marchantia*, one may interpret the paucity or absence of chlorophyll, and the photosynthetic activity in *Marchantia*, in one of two ways: either the paucity is primitive and the presence of the chlorophyll is incipient, or these are manifestations of reduction in

thecium (Fig. 11-27C), observed in *Ricciocarpus*, also takes place. At first, all the cells in the endothecial portion of the capsule are similar, but differentiation into rows of spherical cells follows (Fig. 11-28A). The rows of spherical cells function as sporocytes (Fig. 11-29A), and, as in *Ricciocarpus*, undergo two successive nuclear and cell divisions during which meiosis is accomplished (Fig. 11-29B). The individual spores thicken their walls, and the tetrad finally dissociates (Fig. 11-29C). The remaining cells of the endothecium, which have elongated, secrete spirally arranged thickenings on the inner surfaces of their walls, after which their protoplasts disintegrate (Figs. 11-28C, 11-29C, 11-30F). These pointed, elongated cells are called **elaters** (Gr. *elater*, driver) and are sensitive at maturity to slight changes in atmospheric moisture. They seem to have a role in effecting gradual, rather than simultaneous, spore dispersal and in loosening up the spore mass.

Examination of sections of living sporophytes during their development reveals that they contain abundant chloroplasts, often with enclosed starch grains, an indication that photosynthesis occurs actively. While there can be little doubt that organic substances as well as water and dissolved salts are absorbed from the parent gametophyte through the foot and transported

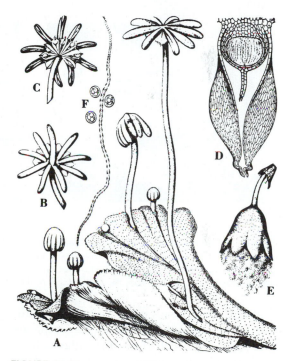

FIGURE 11-30
Marchantia polymorpha. **A.** Female plant with archegoniophores in various stages of elongation. **B, C.** Dorsal and ventral views of archegoniophore apex. **D.** Median longisection of young sporophyte within calyptra and pseudoperianth. **E.** Dehiscent capsule. **F.** Elater and spores. **A.** ×2. **B, C.** ×1.5. **D.** ×25. **E.** ×10. **F.** ×40. (*From a Kny chart.*)

209

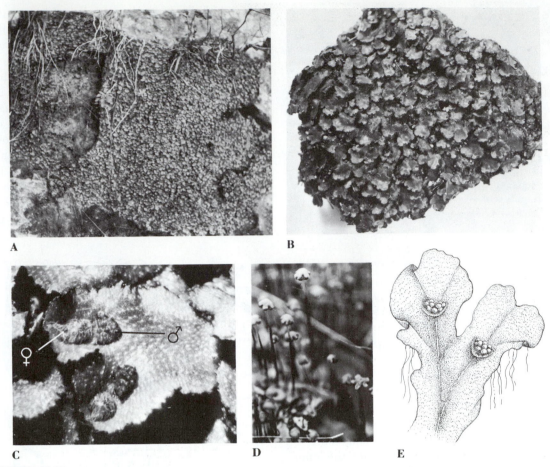

FIGURE 11-31

Reboulia hemisphaerica. **A.** Habit of growth on limestone cliff. **B.** Portion of colony more highly magnified. **C.** Thallus apices with still-sessile archegoniophores, posterior to which are the antheridial receptacles. **D.** Archegoniophores with protuberant sporophytes. **E.** *Lunularia cruciata,* with gemma cups. **A.** ×0.165. **B.** ×0.5. **C.** ×8. **D.** ×½. **E.** ×8.

Ricciocarpus. It seems to the writers that in the light of comparison with the carposporophytes of the algae, especially those of the Rhodophyta, it is probable that permanent retention of the sporophyte of the Hepatophyta (and Bryophyta) on the gametophyte has been accompanied by degeneration of green tissue and adoption of a facultative heterotrophic form of nutrition by the sporophyte. This is cited by some hepaticologists as evidence that these liverworts are reduced, not primitive (see p. 257).

Up to the time sporogenesis has been completed, the sporophytes are surrounded by several protective layers—the calyptra, the involucre, and the two pseudoperianths—which seem to function in preventing premature drying. When the spores have matured, limited elongation of the seta pushes the capsule of the sporophyte out through the calyptra, the pseudoperianths, and the involucre (Fig. 11-30E). As elongation ceases, sporophyte growth being strictly determinate, the capsule dehisces into a

210

A

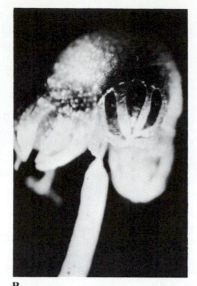

B

FIGURE 11-32
Asterella tenella. **A.**
Elongating archegoniophores
with sporophytes. **B.**
Archegoniophore heads more
highly magnified showing
sporophytes. **A.** ×¾. **B.** ×10.

number of petallike segments from within which the elaters and spores form a protuberant mass. The spores are carried away by air currents; those that fall on favorable substrata germinate and ultimately form new gametophytes. Two of the spores of each tetrad presumably grow into male and two into female gametophytes. Similarly, the gemmae from potentially male plants and those from potentially female plants grow into male and female plants, respectively.

Other Marchantiaceae

A number of genera related to *Marchantia* occur in moist habitats in North America. Among them may be mentioned *Reboulia* and *Lunularia*[12] (Fig. 11-31), *Asterella* (Fig. 11-32), and *Conocephalum* (Fig. 11-33), the last-named being of special interest because cell divisions take place within the spores while they are still enclosed in the capsule. This endosporic germination is similar to the primary germination that occurs in *Pellia* (Fig. 11-46), *Frullania*, and in such vascular plants as some species of *Selagi-*

---

[12]Rare in nature, widespread in greenhouses.

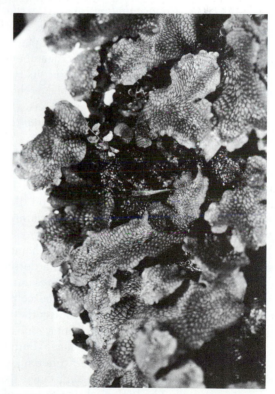

FIGURE 11-33
Conocephalum conicum. ×0.5. *(Courtesy of Prof. Marjorie Behringer.)*

FIGURE 11-34
Dumortiera hirsuta, archegoniophores with
protuberant sporophytes. ×3.

nella (p. 358) and the seed plants. *Cono-cephalum* has a distinctive mushroomlike odor
because of the presence of an oil in certain of its
cells. *Reboulia*, *Conocephalum*, and *Asterella*
are common on calcareous rocks and soil.
Dumortiera, considered to be a reduced plant,
although Marchantiacious, lacks air chambers
(or has mere vestiges of them) and pores; its sex
organs, however, are borne on special archegoniophores and antheridiophores, as are its
sporophytes (Fig. 11-34), just as in *Marchantia*
and *Asterella*.

Marchantia is one of very few genera in the
family in which both the archegoniophores and
antheridiophores are raised above the thalli. In
most, the antheridiophore is sessile on the
gametophyte. Only *Lunularia* and *Marchantia*
produce special gemma cups.

In summary, the *Marchantiales* are characterized by thallose gametophytes with a dorsal
epidermis usually interrupted by pores surrounded by one or more tiers of special cells.
One or more layers of ventilated photosynthetic
tissue are usually distinguishable from a zone of
more compact, storage parenchyma. Ventral
scales and both smooth-walled and tuberculate
rhizoids are present. The archegonia have six
rows of neck cells. The biological significance of
morphological features in the Marchantiales has
been appraised by Bischler and Jovet-Ast (1981).

In Marchantialean sporophytes the seta is
absent (Ricciaceae, Oxymitraceae) or short.[13]
The capsule wall is monostromatic; elaters may
or may not be present, the sporocytes are
unlobed, and capsular dehiscence is irregular.

Order 2. Sphaerocarpales

Members of the very small order Sphaerocarpales (2 genera, 20 species [Sharp, 1974*b*])
exhibit several interesting morphological, cytological (Carothers et al., 1983), and ecological
features.

Family 1. Sphaerocarpaceae
Sphaerocarpos

Habitat and vegetative morphology.
Sphaerocarpos texanus is widely distributed in
the southern United States, where it occurs on
bare soil in fields and on the alluvial deposits of
streams. It is a fall-winter-spring annual and
disappears as the soil dries. It grows readily on
inorganic culture media. Another species, *S.
michelii*, has similar distribution.

The thalli of *Sphaerocarpos* (Gr. *sphaira*,
sphere, + Gr. *karpon*, fruit) are bilaterally
symmetrical (Figs. 11-35, 11-36A), consisting of
a midportion several cells thick and monostromatic, lobed, rather than leaflike, wings. They
are not unlike certain fern gametophytes (Fig.
17-16) in their gross, thallose appearance. Ventral scales and cuticle are absent, and the plants
bear only simple, nontuberculate rhizoids.
There is no obvious internal differentiation.
Growth originates from a transverse row of
apical cells which generate cells dorsally and
ventrally as in the Marchantiales. All cells except the rhizoids are rich in chloroplasts.

Plants of *Sphaerocarpos* often become
branched. The branches are readily detached
and, being regenerative, grow rapidly into new
plants.

---

[13]As compared with that in Metzgeriales and Jungermanniales.

212

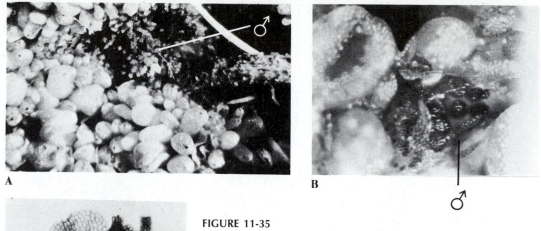

A

B

C

FIGURE 11-35

Sphaerocarpos texanus. **A.** Female plants with ♂ plants, arrow. **B.** Male plants (arrow). **C.** More highly magnified view of ♂ plants; note antheridia within flask-shaped involucres. **A.** ×5. **B.** ×10. **C.** ×30.

Reproduction. *Sphaerocarpos* is strictly dioecious and sexually dimorphic, the male plants being markedly smaller than the female and often purplish, while female plants are green. Both the antheridia and archegonia protrude from the dorsal surface, and both are individually surrounded by involucres (Figs. 11-35A–C, 11-36A). The archegonial involucre is inflated and cylindrical, with an apical ostiole. The involucre surrounding each antheridium (Fig. 11-35B, C) is flask-shaped; it is smaller than the archegonial involucre.

The archegonial necks are composed of six rows of cells as in the Marchantiales. The archegonia (Fig. 11-36B) mature, and fertilization occurs before their involucres have grown as tall as the archegonia. The antheridia (Fig. 10-36D, E) are almost globose and are borne on stalks which are partially embedded in the thallus. Diers (1967) and Zimmerman (1973) have investigated the ultrastructure of the sperms of *S. donnellii.*

Schieder (1968) has reported for *S. donnellii* the production of a sperm attractant by the archegonia; the attractant from one archegonium may be functional for 10 days.

Fertilization stimulates an increase in size and inflation of the archegonial involucres, but they enlarge whether or not fertilization occurs. The zygote undergoes mitotic nuclear division and cytokineses to form a chain of four cells (Fig. 11-37A). During these divisions the venter becomes two-layered. In each of four cells of the young sporophyte, two perpendicular vertical

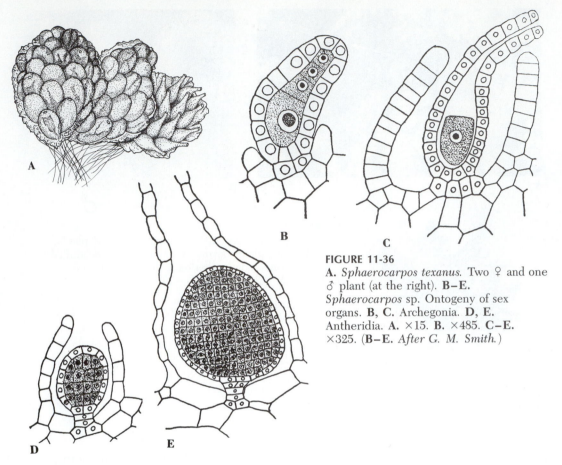

FIGURE 11-36
A. *Sphaerocarpos texanus.* Two ♀ and one ♂ plant (at the right). **B–E.** *Sphaerocarpos* sp. Ontogeny of sex organs. **B, C.** Archegonia. **D, E.** Antheridia. **A.** ×15. **B.** ×485. **C–E.** ×325. (**B–E.** *After G. M. Smith.*)

divisions occur, to make a 16-celled sporophyte. Additional cell divisions in the most basal tier develop a small foot, while from the next upper tier a rudimentary seta is formed. The upper two tiers divide rapidly to form a globose capsular region, which is early differentiated into an endothecium and single-layered amphithecium, the latter being rich in chloroplasts and photosynthetic.

Electron microscopy has revealed that the walls of the superficial cells of the foot and of the contiguous gametophyte cells are thickened in an undulate pattern, thus increasing the surface area for translocation and absorption (Kelley, 1969).

In the endothecium, the cells separate, be-

come spherical, and float in fluid. Some of them function as sporocytes that undergo meiosis, forming tetrads of black spiny-walled spores. Others, which contain chloroplasts, fail to undergo meiosis but are gradually absorbed, thus contributing to the nutrition of the developing spores. Kelley and Doyle (1975) described at the ultrastructural level differentiation of the sporocytes and nutritive cells, prior to meiosis, in the sporophytes of *S. donnellii.* Figure 11-37B shows a median section of an immature sporophyte.

The amphithecium is persistent and indehiscent, and the spores are set free only upon disintegration of the capsule wall; the seta fails to elongate.

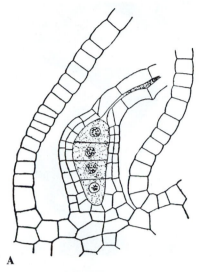

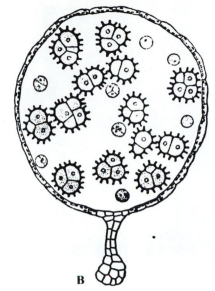

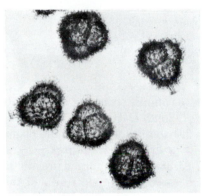

FIGURE 11-37
Sphaerocarpos sp. **A.** Young, four-celled stage in sporophyte development. **B.** Median longisection of sporophyte at the tetrad stage; note degenerating sporocytes. **C.** Tetrads of spores. **A.** ×200. **B.** ×60. **C.** ×125. (**A, B.** *After G. M. Smith.*)

In *Sphaerocarpos texanus* the spores remain coherent in tetrads (Fig. 11-37C). Thus, upon germination, male and female gametophytes are in close proximity, since two spores develop into male and two into female gametophytes. Spores of *S. texanus* present in the soil germinate only after the soil has become thoroughly moist, late in the autumn or during the winter and early spring. The resultant gametophytes rapidly grow into sexually mature plants.

Sphaerocarpos has been the object of numerous genetic and cytological studies, which have revealed that the unisexual gametophytes differ in their chromosome constitution. The nuclei of the female plants contain a special X, or female-determining, chromosome, and those of the male plants contain a Y, or male-determining, chromosome. *Sphaerocarpos* was the first plant in which sex chromosomes were discovered. It is also the first organism in which was shown the actual (as distinct from statistical) segregation of maleness and femaleness. This was done by following the development and maturation of the four progeny from single tetrads. Thus, in meiotic divisions during sporogenesis, the descendants of the parental sex chromosomes,

215

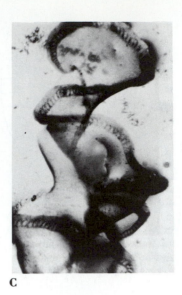

A   B   C

FIGURE 11-38

Riella americana. **A.** Segment of plant, near apex. **B.** Older portion of plant. **C.** Male plant with marginal antheridia. **D.** Axis with archegonia, enlarged at **E. F.** Female plant with four sporophytes in various stages of development. **G.** *Riella affinis,* habit of growth in culture tube. **A, B.** ×3. **C.** ×6. **D.** ×20. **E.** ×125. **F.** ×6. **G.** ×⅔.

brought together at fertilization, are segregated, so that two of the spores of the tetrad are male-producing and two are female-producing. While this type of chromosome mechanism in relation to sex suggests that which is present in such insects as *Drosophila*, the fruit fly, it should be noted that sex in the gametophyte of *Sphaerocarpos* is manifested only after the sex chromosomes have been segregated, whereas in the diploid insect body it is expressed only when they become associated in the same nuclei.

Schieder (1973, 1974, 1976) has produced and investigated nutritional mutants of *S. donnellii.* Of 22,553 plants, thirteen mutants were obtained that required nicotinic acid, choline, or arginine. He also obtained a diploid hybrid by somatic protoplast fusion.

Family 2. Riellaceae
Riella

The single genus *Riella*, with two species, *R.*

americana and *R. affinis*, occurring in the United States (Texas, South Dakota, and California), is localized in distribution. However, the plants grow luxuriantly in laboratory cultures.

Riella (Fig. 11-38) is a submerged aquatic in which the plant body is asymmetrical and the life cycle is completed under water. The plant is composed of an erect, sparingly branched axis from one side of which an *Ulva*-like wing, one cell thick, emerges. A few simple rhizoids at the base of the axis anchor the plant to the substratum. Delicate ventral (one row) and lateral (two rows) scales occur on the axis and along its junction with the blade. Growth is strictly apical.

Vegetative propagation occurs by gemmae (Fig. 11-39C, D) borne among the scales; the gemmae grow into adult gametophytes. Individual plants of *Riella americana* are dioecious, while those of *R. affinis* are bisexual. The antheridia occur in groups along the margin of the blade (Figs. 11-38C, 11-39A). The archego-

216

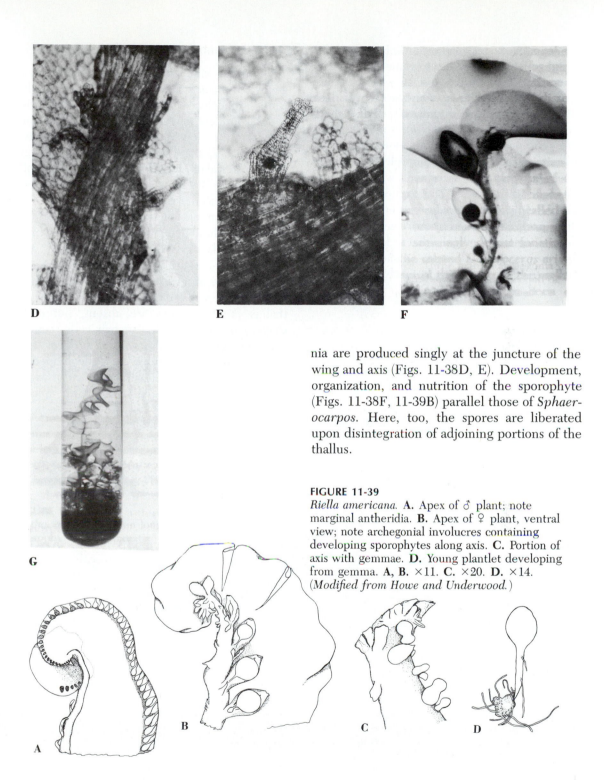

D

E

F

G

nia are produced singly at the juncture of the wing and axis (Figs. 11-38D, E). Development, organization, and nutrition of the sporophyte (Figs. 11-38F, 11-39B) parallel those of *Sphaerocarpos*. Here, too, the spores are liberated upon disintegration of adjoining portions of the thallus.

FIGURE 11-39
Riella americana. **A.** Apex of ♂ plant; note marginal antheridia. **B.** Apex of ♀ plant, ventral view; note archegonial involucres containing developing sporophytes along axis. **C.** Portion of axis with gemmae. **D.** Young plantlet developing from gemma. **A, B.** ×11. **C.** ×20. **D.** ×14. (*Modified from Howe and Underwood.*)

A

B

C

D

In summary, the small order Sphaerocarpales includes plants with delicate gametophytes and smooth rhizoids; pores and air chambers are absent. The archegonia, with six rows of neck cells, are surrounded by inflated, urnlike involucres; the antheridial involucres are flasklike. The sporophytes lack active, elongating setae and elaters and contain chlorophyllous sterile cells, and their spores are shed when the capsule wall disintegrates. This group seems to be related to the Marchantiales in producing sporocytes that are not lobed prior to meiosis; in both orders specialized oil cells that lack chloroplasts may occur scattered in the gametophyte. In both orders the seta is short or absent.

FIGURE 11-40
Monoclea forsteri. **A.** Male thallus showing antheridial receptacles of three different ages. **B.** Female thallus with three archegonial receptacles; the central one shows an immature sporophyte enclosed by the calyptra; the one at the right has an emergent, dehiscent sporophyte. **A, B.** ×0.75. (*After Johnson.*)

Order 3. Monocleales

The order Monocleales includes but a single genus, *Monoclea* (Figs. 11-40, 11-41) in the family Monocleaceae. Two species, *M. gottschei* and *M. forsteri*, have been described, the former from the New World tropics and the latter from New Zealand and Patagonia. Although living material of *Monoclea* may not be available for laboratory study, it is discussed briefly here because of its interesting morphology.

Monoclea is one of the largest liverworts, a thallose, dichotomously branched plant with a Marchantialike growth habit. However, the thallus is histologically homogeneous, and air chambers are absent. Chloroplasts are concentrated in the cells of the upper portion of the thallus. Ventral scales are absent, but club-shaped, mucilage-producing hair cells recurve over the apex from the ventral surface. Two kinds of unicellular rhizoids are developed: a slender type with thickened walls and a wider kind with thin walls and branching apices. Large brown oil bodies are present in some of

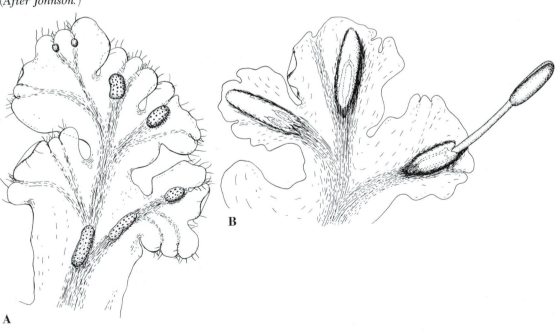

A

B

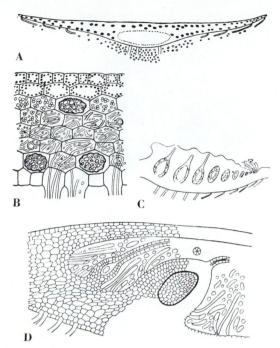

FIGURE 11-41
Monoclea forsteri. **A.** Transection of thallus; black dots represent oil bodies; the dotted line encloses the mycorrhizal region. **B.** Transection of central region of thallus; note oil bodies and mycorrhizal fungi in thallus cells and rhizoids. **C.** Sagittal section, diagrammatic, of male receptacle; note antheridia. **D.** Sagittal section of female receptacle showing two archegonia, one containing a sporophyte. **A.** ×14. **B.** ×80. **C.** ×14. **D.** ×60. *(After Cavers.)*

the cells of the thallus, while fungi may occur within the thallus cells (Fig. 11-41A, B).

Monoclea is dioecious. The antheridia occur in moundlike receptacles (Fig. 11-40A, 11-41C), while the archegonia are produced in elongate troughs covered by an involucrelike flap of the dorsal surface of the thallus (Fig. 11-41D). The archegonia have elongate necks with six rows of neck cells, while the antheridia are similar to those of the Marchantiales.

One to several sporophytes may develop in association with an archegonial receptacle. The sporophyte has a foot, elongate seta, and a cylindrical capsule (Fig. 11-40B). The latter dehisces by one longitudinal fissure. The capsule, whose wall is composed of a single layer of cells, contains elongate elaters. The cells of the capsule wall have thickened fibrils in their walls.

Earlier classified as a member of the Marchantiales, *Monoclea* was later placed in a separate order for several reasons. Although the plants are coarse and thallose, and their sex organs are similar to those of the Marchantiales, they differ in lacking air chambers, ventral scales, and archegoniophores; and the location of the archegonia is unique among the Hepatopsida. Furthermore, the elongate seta suggests the Metzgeriales, but the method of dehiscence of the capsule and its one-layered wall differ from the sporophytic features in that order.

Order 4. Metzgeriales

Formerly included as a suborder in the Jungermanniales, the plants so classified are sometimes, as here, elevated to ordinal rank—as the Metzgeriales, with about 20 genera and 550 species (Sharp, 1974*b*). To this order belong simple thallose genera that lack air chambers, air pores, ventral scales, and pegged rhizoids. The sex organs and sporophyte are sessile on the gametophyte, archegoniophores and antheridiophores being absent; elevation of the capsule is accomplished by extreme elongation of the seta of the sporophyte. These plants are quite distinct from the leafy liverworts, with which they were formerly classified, because they are **anacrogynous**. By this is meant that the apical cell of a fertile branch does not itself become a sex organ (archegonium), so that growth of the fertile branch continues after the sex organs have matured (Fig. 11-45).

The representative genera chosen for discussion here, *Pellia*, *Pallavicinia*, *Fossombronia*, and *Petalophyllum*, are members of two different families.

219

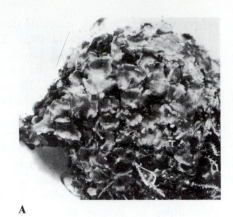
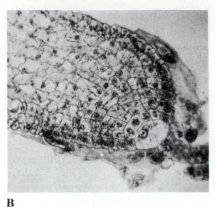

A   B

FIGURE 11-42
Pellia epiphylla. **A.** Mat of living plants from a cranberry bog, Cape Cod. **B.** Sagittal section of apex; note homogeneous tissue, apical cell, and the mucilage glands near the latter. **A.** ×0.5. **B.** ×175.

Family 1. Pelliaceae

Pellia (after Leopoldi Pelli-Fabrone, a Florentine lawyer) is a rather widespread member of the Pelliaceae and is here discussed as representative of the family.

Pellia

Habitat and vegetative morphology. *Pellia* (Fig. 11-42A) is encountered frequently on moist sandstone rocks and on stream banks in shady woods where the soil is alkaline, neutral, or acid in reaction. It may be submerged during high water. The irregularly dichotomously branching plant body of *Pellia* is smaller, smoother in appearance, and usually a brighter and more translucent green than that of *Marchantia polymorpha*. The smooth appearance of the plant is due to the absence of pores and air chambers on the upper surface. Numerous nontuberculate rhizoids arise from the ventral surface of each branch along the thickened central portion. Ventral scales are absent, but mucilage-secreting, glandular hairs occur in the region of the growing points (Fig. 11-42B). The latter has a dome-shaped apical cell at its apex.[14] The

margins of the plants may be slightly lobed and ruffled. As in many other Hepatophyta, anterior growth and branching and posterior decay of the plant body result in vegetative multiplication by fragmentation; gemmae are absent in *Pellia*.

Growth is strictly apical and may be traced to the division of a single, hemidiscoid apical cell with a curved base (Fig. 11-42B). This cell undergoes mitoses and cytokineses; the derived cells add to the tissues of the plant body by subsequent divisions. Transverse sections of a branch reveal the absence of any considerable internal differentiation, as compared with *Marchantia*, although the cells near the center may have thickened bands in their walls. There is a gradual diminution in thickness of the branches from the center toward each margin, so that the latter are monostromatic. Essentially all the cells of the plant body contain chloroplasts, and all contain many small, glistening oil bodies. The superficial layers are not noticeably differentiated as epidermal cells. The *Pellia* plant is scarcely more differentiated than the sea lettuce, *Ulva*, but differs from the latter in having apical growth.

Reproduction. *Pellia epiphylla,* a species widely distributed in the United States, is mo-

---

[14]For a discussion of apical cells in liverworts, hornworts, and certain vascular cryptogam, see Gifford (1983).

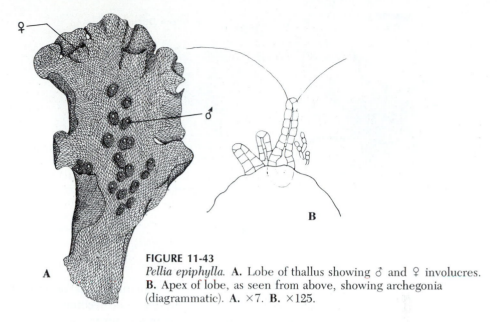

FIGURE 11-43
Pellia epiphylla. **A.** Lobe of thallus showing ♂ and ♀ involucres.
B. Apex of lobe, as seen from above, showing archegonia
(diagrammatic). **A.** ×7. **B.** ×125.

noecious and strongly protandrous. However, the antheridia are slow in maturing and still contain viable sperms when the archegonia have matured. In habitats where the plants have overwintered, the old, dark thalli produce light-green branches early in the spring. After a period of vegetative development, these attain sexual maturity, some of the dorsal derivatives of the apical cell differentiating as antheridia (Fig. 11-43A). These occur scattered on the dorsal surface of the plant in the central portion of each branch. Each globose antheridium is protected by a moundlike layer of cells with a circular pore. Somewhat later in the season, the same plants produce archegonia at their apices (Fig. 11-43A, B). The archegonia arise on a mound of tissue formed from derivatives of the apical cell. The entire group, composed of 15 or more archegonia, is covered by a protuberant involucral flap (Fig. 11-43A). The archegonia of *Pellia* differ from those of the Marchantiales and Sphaerocarpales in that they have short stalks and the necks are composed of only five vertical rows of neck cells.

It has been demonstrated that rapid move-

ment of films of capillary water takes place on both surfaces of the *Pellia* plant. There is little doubt that such films of water, as well as heavy rains and dew, accumulate in the space between the antheridial wall and the antheridial chamber and aid in the dehiscence of the antheridia and liberation of the sperms. The latter are among the largest in the Hepatophyta and Bryophyta, attaining a length of 95 μm (Duckett et al., 1981). They are released by rupture of the antheridial wall and emerge in gelatinous spheres, from which their movements ultimately liberate them. The close proximity of the thallus branches and of the antheridia and archegonia on a single thallus ensure abundant fertilizations. Although the eggs of several of the archegonia on one receptacle may be fertilized, normally only one develops into a mature sporophyte (Fig. 11-44). The other eggs and zygotes abort but remain recognizable during the early stages of development of the functional sporophytes.

A transverse division of the zygote initiates the development of the sporophyte, which is later differentiated into foot, seta, and capsule

221

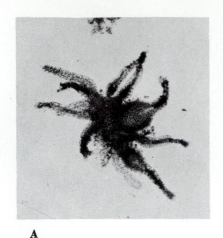
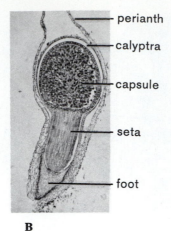

perianth

calyptra

capsule

seta

foot

FIGURE 11-44
Pellia epiphylla.
A. Archegonial complex; the two enlarged archegonia contain sporophytes. **B.** Median longisection of sporophyte. **A.** × 60. **B.** × 12.5.

A   B

regions. Growth of the sporophyte of *P. epiphylla* in the eastern United States progresses as far as the sporocyte stage before midwinter; development is then arrested until the following spring. The dormant sporophytes are covered by the basal portion of the receptacles, by their calyptras, and also by the involucral flap (Fig. 11-44B). As in *Marchantia*, the cells of all the regions of the sporophyte contain abundant starch-filled chloroplasts, which indicates active occurrence of photosynthesis during ontogeny. The capsule wall in the immature sporophyte is amphithecial in origin and consists of several layers of cells, in contrast to the single-layered amphithecium of the Marchantiales and Sphaerocarpales. The interior of the capsule becomes differentiated into deeply four-lobed sporocytes (Suire, 1970) and elongate cells, the latter maturing as elaters. Some of the elaters are oriented with one end, in each case, attached at the base of the capsule, thus forming an **elaterophore**. Following meiosis in the four-lobed sporocytes, the members of the spore tetrads separate and, as in *Conocephalum*, each spore undergoes a limited number of nuclear and cell divisions within the spore wall, so that by the time of their dissemination the spores have already begun development. As the sporo-

phyte matures, the outermost layer of cells of the capsule becomes thickened and brown, except for four vertical rows of cells that remain thin-walled; the latter foreshadow the lines of dehiscence of the capsule. When the spores are multicellular and the capsule mature, the seta lengthens and raises the capsule above the gametophyte (Fig. 11-45). Elongation of the seta is very rapid; it may increase in length within 3 or 4 days from 1 to 80 mm. Setae 110 mm long

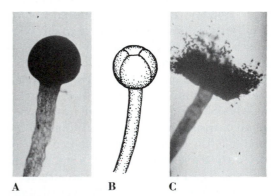

A   B   C

FIGURE 11-45
Pellia epiphylla. **A.** Distal portion of seta and capsule. **B.** Incipiently dehiscent capsule, diagrammatic. **C.** Capsule of **A.** At dehiscene. ×12.5.

222

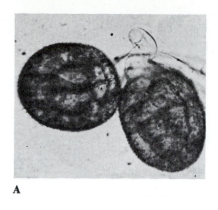

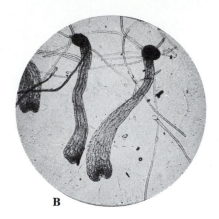

FIGURE 11-46

Pellia epiphylla. **A.** Spore germination, 48 hours after spores had been planted on agar. **B.** Young plants, 45 days later. **A.** ×400. **B.** ×30.

are common on plants in shady ravines. The spherical capsule dries when elongation has been completed; it dehisces violently (Fig. 11-45) when contraction ruptures the four rows of thin-walled cells. The spore mass is loosened by the hygroscopic movements of the elaters, and many of the spores are simultaneously shed.

The spores of *Pellia* initiate nuclear and cell divisions and gametophyte development within the spore wall (Fig. 11-46A) and before the spores are shed from the sporangia. This precocious, endosporal "germination" occurs also in *Conocephalum* (p. 211), *Porella* (p. 230), and many vascular plants (e.g., some species of *Selaginella*, p. 358). Other spores remain in the meshes of the twisted elaters, which are attached at the base of the opened capsule. The spores are principally water-dispersed, being too heavy to be carried by air currents, and those that settle in proper habitats continue their development into new *Pellia* thalli (Fig. 11-46B).

Family 2. Pallaviciniaceae
Pallavicinia

Habitat and vegetative morphology. *Pallavicinia* (after L. Pallavicini, Archbishop of Genoa) (Figs. 11-47 through 11-51) occurs on

moist or boggy, nonalkaline soils. The plants are densely overlapping in mats.

Pallavicinia is similar to *Pellia* in a great many respects. Its elongate, rather sparingly branched, ribbonlike thalli, approximately 4 mm wide, occur on moist, humus-rich soil. The plants can withstand periodic submersion. Branching is largely from the ventral surface of the gametophyte in *P. lyellii*, the only North American species; the branches originate from the ventral side of the thallus near the prominent midrib. Numerous smooth-walled rhizoids develop along the midrib region and penetrate the substratum; growth is apical (Fig. 11-48A). The plant body is clearly differentiated into midrib and lateral wings; the latter are one cell layer thick, except toward the midrib, and somewhat undulating. Growth of each branch is localized in the region of the prominent lens-shaped apical cell, which is slightly sunken in a notch because of the more rapid growth of the wings near the apex. The central portion of the midrib (Fig. 11-49B) is occupied by a strand of elongate pitted-walled cells up to 280 μm long without nuclei or cytoplasm at maturity, which have been shown experimentally to function in the conduction of water. Although these cells are strikingly like xylem tracheids and/or vessels

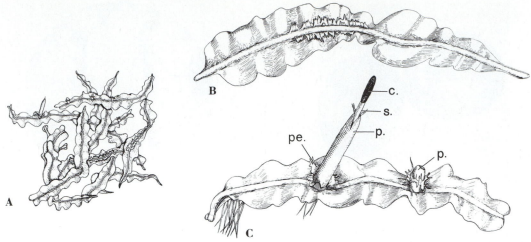

FIGURE 11-47
Pallavicinia lyellii. **A.** Male and ♀ plants in mat. **B.**
Male plant enlarged. **C.** Enlarged view of ♀ plant
with emergent sporophyte. c., capsule;
p., pseudoperianth; pe., perichaetial scales;
s., seta. **A.** ×1. **B, C.** ×3.

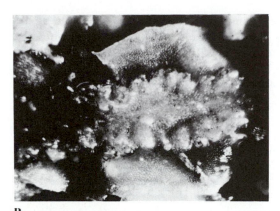

FIGURE 11-48
Pallavicinia lyellii. **A.** Apices. **B.** Male plant
showing antheridia in two rows along the midrib.
C. Female plant with three archegonial receptacles.
A. ×6. **B, C.** ×12.

224

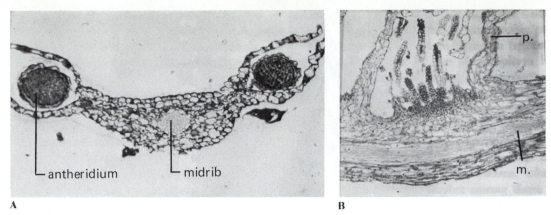

A   B

FIGURE 11-49
Pallavicinia lyellii. **A.** Transection of ♂ plant. **B.** Portion of sagittal section of ♀ plant. m., midrib; p., pseudoperianth. **A.** ×18. **B.** ×27.

(p. 326) in structure and function, their walls, unlike those of tracheids, are not lignified. However, some of the pits lack closing membranes, so that a vessellike organization is approached. Like xylem vessels, the mature cells are not living (Smith, 1966).

The wings of the thallus are composed of photosynthetic parenchyma cells rich in chloroplasts.

Reproduction. *Pallavicinia* is dioecious. The male plants are slightly smaller than the female, with which they are usually intermingled in dense colonies (Figs. 11-47, 11-48B, C). The antheridia are produced on the dorsal surface of the thallus in linear order on both sides of the midrib (Figs. 11-47B, 11-48B). Each antheridium is short-stalked and is partly covered during development by a scalelike **perigonium.** The fringed perigonial scales of several antheridia may become more or less confluent (Figs. 11-47B, 11-48B). As is obvious from the figure, formation of archegonia does not limit elongation of the branch; *Pallavicinia*, therefore, like *Pellia*, is anacrogynous. The archegonia, which are also produced above the midrib, are borne in groups in receptacles surrounded by fused, leaflike bracts called **perichaetial scales** (Fig. 11-47C). The archegonia of a given group, be-

tween 18 and 30 in number, mature at different rates. As in *Pellia*, the archegonia are stalked, with elongate necks composed of five vertical rows of neck cells. As the first archegonia mature, a ringlike layer, the **pseudoperianth**, appears at their base. Immediately after fertilization, the pseudoperianth grows rapidly, forming a prominent cylindrical structure within the perichaetial scales (Fig. 11-47C).

Development of the sporophyte follows much the same pattern as that described for *Pellia*, the first division of the zygote being transverse. Here, too, usually only one sporophyte matures on one receptacle (Fig. 11-47C). The calyptra attains a thickness of four or five cells and increases in size with the developing sporophyte (Fig. 11-50A), which is surrounded by three protective sheaths of gametophytic origin —the calyptra, the pseudoperianth, and the perichaetial scales (at its base). Meiosis of the lobed sporocytes (Fig. 11-51A) reduces the chromosome number from $2n = 16$ to $n = 8$. Following elongation of the seta, the spores are shed by the longitudinal divisions of the capsule wall along predetermined dehiscence lines into four segments (Fig. 11-51B) called **valves**, which may cohere in pairs. In most individuals these valves remain united at the apex so that

225

A

B

FIGURE 11-50
Pallavicinia lyellii. **A.** Immature living sporophyte within calyptra; note aborted archegonia at base. **B.** Spores and elaters. **A.** ×12.5. **B.** ×250.

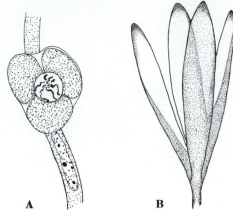

A

B

FIGURE 11-51
Pallavicinia lyellii. **A.** Lobed sporocyte with elater precursor. **B.** Dehiscent capsule. **A.** ×770. **B.** ×25. (**B.** *Courtesy of Dr. H. W. Bischoff.*)

the capsule opens by four longitudinal slits. The spores remain spherical and one-celled to the time of dehiscence of the capsule. They grow into a new generation of ribbonlike *Pallavicinia* plants.

Family 2. Fossombroniaceae
Fossombronia and *Petalophyllum*

Lobing and folding of the histologically simple thallus characterize the genus *Fossombronia* (Fig. 11-52). In *Petalophyllum* (Gr. *petalon*, lamella, + Gr. *phyllon*, leaf) (Fig. 11-53) fusion of the leaflike lobes results in a corrugated aspect. These plants are considered especially significant in phylogenetic speculations regarding the liverworts. Some morphologists see in the lobing and folded thalli of these plants suggestions of the leafy liverworts (Jungermanniales). To some the lobing and folding represent precursors of the leafy habitat, while others consider them as reductions from it. Still others are of the opinion that the Metzgeriales and Jungermanniales are not closely related.

Fossombronia (in honor of B. Fossombroni, an

Italian statesman) (Fig. 11-52A) is a widespread thallose genus in which the various species exhibit degrees of lobing and incision of the wings that suggest an approach to the leafy organization characteristic of the Jungermanniales.

Fossombronia brasiliensis grows readily on agar in laboratory culture and forms sex organs abundantly under 8–12 hours of illumination at 20°C (Bostic, personal communication). The sex organs appear after the cultures have been maintained for one month under these conditions, and sporophytes mature within sixty days. The antheridia and archegonia are interspersed along the axes. The capsules of some species of *Fossombronia* dehisce into four parts, but that of *F. brasiliense* dehisces irregularly.

Petalophyllum ralfsii (Fig. 11-53) occurs on moist calcareous soils and rocks.

In summary, the Metzgeriales are characterized by having dorsiventral thalli, the latter sometimes markedly lobed, relatively undifferentiated, lacking collenchyma or with a strand of elongate conducting cells; all the cells usually contain oil bodies. Ventral scales, tuber-

226

A

B

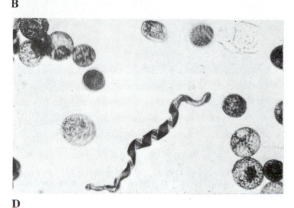

D

C

FIGURE 11-52
Fossombronia brasiliensis. **A.** Plants with sporophytes.
B. Sporophytes with elongating setae. **C.** Dehiscent capsule.
D. Spores and elater. **A.** ×8. **B.** ×16. **C.** ×25. **D.** ×300.

FIGURE 11-53
Petalophyllum ralfsii. **A.** Plants with antheridia,
more highly magnified at **B.** **A.** ×5. **B.** ×9.

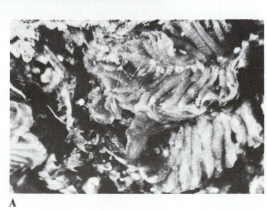

A

B

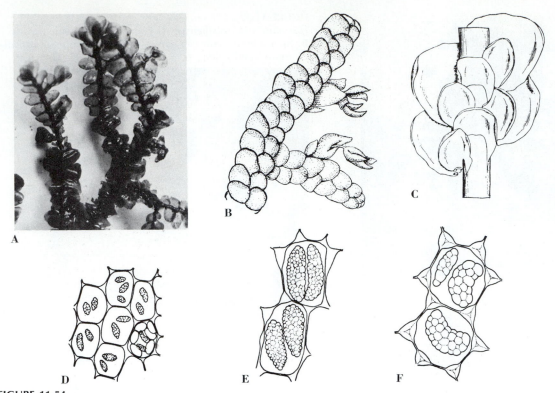

FIGURE 11-54

Porella platyphylloidea. **A.** Portion of plant, dorsal view. **B.** The same with dehiscent sporophytes. **C.** Portion of plant, ventral view. **A.** ×4. **B.** ×4.5. **C.** ×18. **D, E.** Various types of oil bodies in leaf cells. **D, E.** *Frullania kunzei.* **F.** *Cheilolejeunea clausa.* (**C.** *After Evans.* **D, E.** *After Schuster.*)

culate rhizoids, pores, and air chambers are absent. The gametophytes are anacrogynous. The sporophytes have long setae and the capsule walls are more than one layer of cells thick, the capsule usually dehiscing into four portions. The sporocytes and later the spores occur among elongate, spirally thickened elaters. In their lobed sporocytes, numerous oil bodies in the chlorophyllous cells, their elongated setae, and several-layered capsule walls, they resemble the Jungermanniales.

Order 5. Jungermanniales

The order Jungermanniales, the largest among the Hepatophyta (over 180 genera and about 7500 species) as presently constituted, includes the "leafy liverworts." Most of these plants are dorsiventral, composed of axes with two rows of delicate, monostromatic lateral leaves and often a third row of ventral leaves (underleaves, or **amphigastria**), which may differ in size or shape, or both, from the lateral leaves. Exceptionally, the plants are erect and radially symmetrical, the leaves being equal in size (isophyllous) or nearly so; the plants then resemble mosses.

Inasmuch as the leaves and stems lack a cuticle, absorption in many leafy liverworts occurs directly through the leaf and stem cells which, except their growing tips, are often closely appressed to the moist substratum.

228

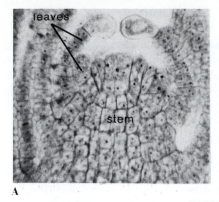

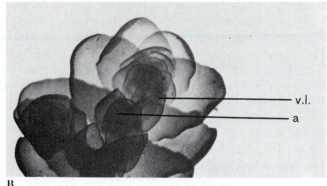

A

B

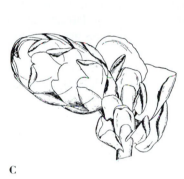

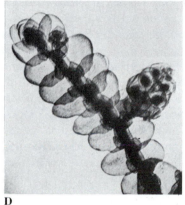

C

D

FIGURE 11-55
Porella platyphylloidea. **A.** Median longisection of the apex; note apical cell. **B.** Ventral view of plant near apex. **C.** Ventral view, living ♂ branch. **D.** Ventral view of stained ♂ plant; note two lateral, antheridium-bearing branches. a., amphigastrium; v.l., ventral lobe. **A.** ×250. **B.** ×12. **C.** ×18. **D.** ×5.

Moreover, the dense colonial habit, coupled with the partial overlapping of the leaves as well as the folded lobing of the leaves (Fig. 11-55B), provides extensive appressed surfaces between which water may be held by capillarity. Finally, the plants grow in densely interwoven mats, which contributes to their water-holding capacity.

Berrie and Eze (1975) have demonstrated that in *Radula flaccida*, a leafy liverwort epiphytic on leaves, P[32] can move from the host plant to the liverwort.

The cells of the leaves of many Jungermanniales contain characteristic oil bodies (Figs. 11-54D, E, 11-60E), which are useful taxonomically.

A few, such as some species of *Frullania* (Fig. 11-63B–D), may be xerophytic. Several, such

as *Porella pinnata*, are periodically submerged aquatics. The vast majority are abundant in regions of heavy rainfall, fog, or high humidity, where they are ubiquitous on tree bark, fallen logs, soil, stones, and, in tropical latitudes, on the twigs and leaves of vascular plants. They may grow even on the bodies of large, long-living weevils in tropical regions.

The young leaves overlap a meristematic stem tip in which a prominent, apical cell gives rise to cellular progeny oriented parallel to three of its surfaces (Fig. 11-55A). Each of these derivative cells undergoes further division to form a several-celled unit of organization called a **merophyte**. Each merophyte normally develops one-third portion of the axis and a leaf (Fig. 11-55A). Branching may be pinnate or more irregular. The branches may originate near the

229

growing tip or in a more mature region of the axis. Their origin may be from the internal or superficial layers of the stem (Crandall, 1969).

The leaves, when the plant is viewed in dorsal aspect, may be oriented in one of two basic patterns if they overlap. In some genera—*Porella*, for example—the lower portion of a given leaf is overlapped by the upper portion of the next older leaf below it (Fig. 11-54B). In others, the upper portion of a given leaf is overlaid by the lower portion of the next young leaf, just above it—that is, near the apex—as are the shingles of a roof. The first arrangement is said to be **incubous** (Fig. 11-54B) and the second **succubous** (Fig. 11-60B, C).

The stems of the many genera vary in internal complexity, in some being virtually undifferentiated, while others have different kinds of cells in a medulla surrounded by a cortex.

In addition to spore formation preceded by sexual reproduction, a number of mechanisms of asexual reproduction occur. These include special deciduous propagative leaves and gemmae of various types. The earliest stages (primordia) of leaves are always bilobed, and in many this lobing persists in the mature organism. There is, however, an exceedingly wide range of leaf morphology among the numerous genera. The leaves are normally monostromatic, lack a costa or midrib, and may contain collenchymatous tissue composed of chlorophyllous cells thickened at their contiguous corners. Oil bodies of characteristic form are present in the leaf cells of many species. The rhizoids are smooth-walled, variable in abundance, and associated with the underleaves or scattered on the ventral surface of the stem. A chromosome number of $n = 9$ characterizes the group, although $n = 8$ (e.g., *Porella*) is frequent.

The morphology of the Jungermanniales is here exemplified by *Porella*, following which other representative genera are discussed.

Family 1. Porellaceae
Porella

Vegetative morphology. *Porella* (Figs. 11-54 through 11-58) and other leafy liverworts superficially resemble certain mosses. Their leaves are unequally bilobed, folded, and only one cell layer thick. The plant body is composed of rather flattened, branching leafy axes, from the undersurfaces of which emerge scattered rhizoids, which may penetrate the substratum. The relative paucity of rhizoids indicates that they play a minor role in water absorption. In many cases their major role seems to be anchorage.

The plant body develops from an apical cell with three cutting faces (two lateral and one ventral), from which derivatives are successively cut off in three directions (Fig. 11-55A). The plants are strongly dorsiventral, and the third row of leaves, called **underleaves** or **amphigastria,** is visible only ventrally (Fig. 11-55B). The underleaves are markedly smaller and different in shape from the laterals, a condition known as **anisophylly.** Not all leafy Hepatophyta have underleaves (e.g., *Scapania nemorosa, Nowellia curvifolia, Plagiochila artica,* etc.). Evidence that some insight into the phylogeny of liverworts may possibly be gained by experimental procedures is provided by the work of the Basiles (summarized, 1983, 1984): low concentrations of hydroxy-L-proline and 2.2^1-dipyridyl evoked the formation of underleaves, normally absent, in five species of leafy liverworts. The symmetry was transformed from dorsiventral to triquetrous in some individuals. This was interpreted as a reversion to a more primitive condition, as postulated by some students of the Hepatophyta (Schuster, 1979).

When rhizoids are present, they rise from the basal portions of the amphigastria. In ventral aspect, plants of *Porella* seem to have five rows of leaves (Fig. 11-54C), but in reality only three are present. The impression that there are five rows is due to the occurrence of ventral lobes on each of the lateral leaves; the small ventral lobes are more or less closely pressed against the larger dorsal portions of the leaves (Fig. 11-54C). Such leaves are said to be **complicate bilobed.** The leaves themselves consist of a single layer of rather uniform chlorenchymatous cells. The stems show scarcely any internal differentiation. *Porella* and many other leafy Hepatopsida are remarkable in their tolerance of prolonged and periodic drying. The osmotic properties and water relations of the component cells have not been fully investigated, but there is every indication that they must differ from those of other plants (such as *Haplomitrium*, p. 240) that seemingly also lack cuticles but that are unable to withstand desiccation.

Reproduction.   All species of *Porella* are strictly dioecious, the male plants being slightly narrower than the female, a condition suggestive of *Pallavicinia* and certain macrandrous species of *Oedogonium* (p. 72). The antheridia are borne singly in the axils of densely overlapping bractlike leaves on projecting, conelike, compact branches (Fig. 11-55C). Each globose antheridium, which is attached to the leaf axil by a rather long stalk two cells in width, bears numerous biflagellate sperms. At maturity, if sufficiently moist conditions prevail, the upper jacket cells of the antheridium separate from one another at its apex, curve back, and liberate the sperms. The archegonia are produced on short, lateral branches (Fig. 11-56) of the female plants. The archegonial branches are scarcely different from young vegetative branches until later in their development, and, therefore, are difficult to recognize. Development is **acrogynous,** so that elongation of archegonial branches is stopped because the last-formed archegonium develops from the apical cell itself. After fertilization, the development of an inflated **perianth** surrounding the group of terminal archegonia adds to the prominence of the archegonial branch. Eight to ten archegonia occur in each group. The archegonium is

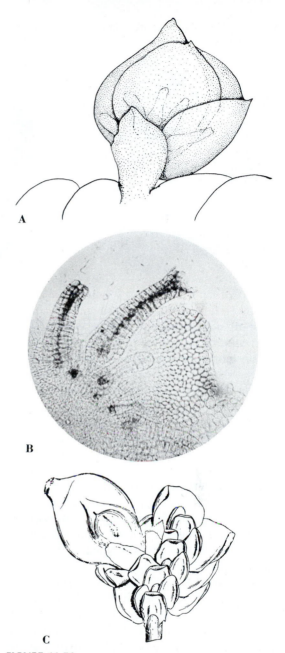

FIGURE 11-56
Porella platyphylloidea. **A.** Female lateral branch with archegonia. **B.** Living archegonia. **C.** Ventral view of archegonial branch with sporophyte within. **A.** ×50. **B.** ×125.

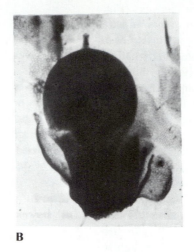

A

B

FIGURE 11-57
Porella platyphylloidea. **A.**
Female plant with immature
sporophyte at f. **B.** Living
sporophyte within calyptra. **A.**
×7. **B.** ×45.

stipitate, with a scarcely swollen venter and five rows of neck cells (Fig. 11-56B).

The abundance of maturing sporophytes, which may be observed in female plants of *Porella* collected early in the spring, indicates that numerous fertilizations have occurred the preceding autumn. Little information is available regarding the mechanism by which the sperms reach maturing archegonia; but the fact that male and female plants often grow intermingled in dense mats, as well as the extensive, continuous capillary spaces that exist between the plants, the substratum, and the overlapping leaf surfaces, suggest that a convenient avenue is available for the dissemination of the motile sperms.

Although the eggs of several archegonia in each branch may be fertilized, only one of the zygotes usually develops into a sporophyte (Figs. 11-57B, 11-58A, B). The others abort but often persist at the base of the calyptra of the fertile archegonium. The cells of the sporophyte contain abundant chloroplasts during the later stages of their development and are undoubtedly photoautotrophic to some degree. The base of the sporophyte is firmly embedded by its foot in

the enlarged stem tip and is surrounded distally by the enlarging calyptra, by the common archegonial envelope or perianth, which is somewhat trihedral in form, and by the leaflike basal female bracts and bracteole of the archegonial branch. The sporophyte is differentiated ultimately into a capsule, a seta, and a rather poorly developed foot (Fig. 11-58A, B). The sporocytes are lobed like those of *Pallavicinia* and other Metzgeriales, and the capsule wall is two to four layers in thickness. After sporogenesis and maturation of the spores and elaters (Fig. 11-58D), the seta cells elongate, thrusting the capsule out from its protective envelopes (Fig. 11-58B, C). Thomas (1975, 1976, 1977b) and Thomas and Doyle (1976) have investigated elongation of the seta in the Jungermannialean *Lophocolea heterophylla.* The seta cells elongate 50-fold in 3 to 4 days; this elongation is associated with increase of wet weight due to water absorption. Total wall carbohydrate increased 1.8 to 2-fold during elongation, while protein content and phospho- and glycolipids also increased; it is suggested that this reflects an increase in membranes.

Dehiscence of the capsule of *Porella* then

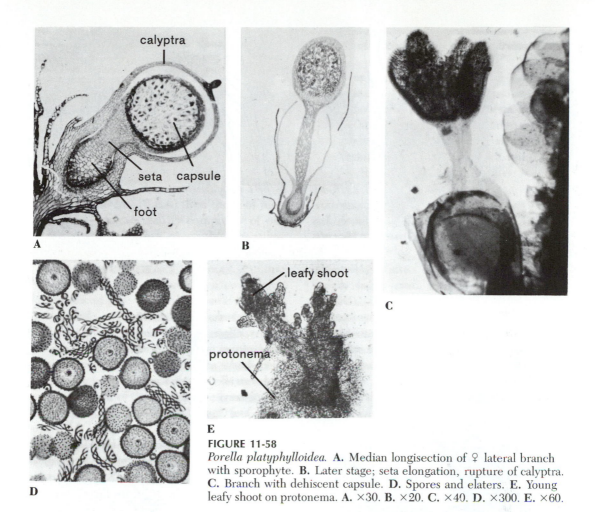

FIGURE 11-58

Porella platyphylloidea. **A.** Median longisection of ♀ lateral branch with sporophyte. **B.** Later stage; seta elongation, rupture of calyptra. **C.** Branch with dehiscent capsule. **D.** Spores and elaters. **E.** Young leafy shoot on protonema. **A.** ×30. **B.** ×20. **C.** ×40. **D.** ×300. **E.** ×60.

takes place along four vertical rows of thin-walled cells, the four dehiscence lines (Fig. 11-58C). Spirally thickened elaters are present among the spores (Fig. 11-58D).

The rather large spores of *Porella*, as in *Pellia* and *Conocephalum*, usually undergo precocious and endogenous divisions before they shed from the capsule. They develop into globose structures, each called a **protonema**; this gives rise to one leafy axis (Fig. 11-58E) after a tetrahedral apical cell has been organized.

More than ten types of protonema, including both filamentous and globose, have been distinguished in the Jungermanniales.

Other Jungermanniales

As stated above, the Jungermanniales comprise more species of liverworts than any of the other orders. In his comprehensive summary of the order, Schuster (1966, 1969, 1974) classified the Jungermanniales in 38 families. These are segregated from one another by characteristics that are largely gametophytic. These include branching pattern and location of branch origin; presence or absence of stolons; leaf characteristics (presence or absence of underleaves, insertion, orientation, form [simple, lobed, complicate-bilobed, water sacs]; cells thickened where the walls meet or not;

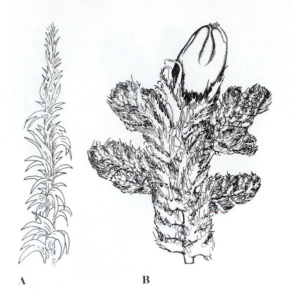

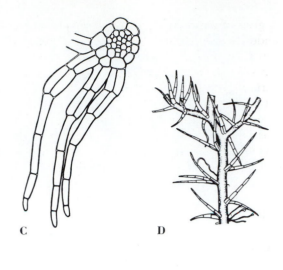

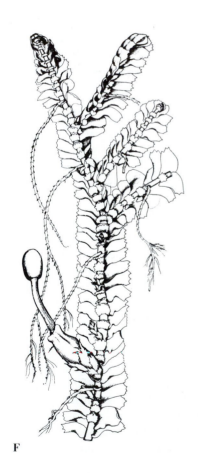

FIGURE 11-59
Various Jungermanniales. **A.** *Herberta adunca.* **B.**
Ptilidium pulcherrimum. **C, D.** *Telaranea*
nematodes. **C.** Transection of stem with "leaf." **D.**
Segment of shoot showing branching. **E, F.**
Bazzania trilobata. **E.** Dorsal view. **F.** Ventral
view. **A.** ×3. **B.** ×11. **C.** ×90. **D.** ×20. **E.** ×2.3.
F. ×3. (**A.** *Modified from Watson.* **B–D, F.** *After*
Schuster. **E.** *After Grout.*)

degree of anisophylly; presence or absence of underleaves); rhizoids (present or absent; at bases of underleaves or scattered); gemmae (present or lacking); location of sex organs; perianth (cylindrical or flattened); and perigynium (present or lacking). Such sporophytic characteristics as nature of the seta (delicate or massive) and elaters (free or attached) are also involved. Mues and Zinsmeister (1978) discuss and review the occurrence of phenolics in Jungermanniales in relation to their taxonomy. In order to expand the reader's concept of the order beyond the genus *Porella*, which has been described in considerable detail in the preceding pages, representatives of some of the other families of Jungermanniales will be considered at this point.

Herberta (Herbertaceae)

Herberta (after George Herbert) (Fig. 11-59A) exemplifies leafy liverworts on which the plants are erect, or nearly so, or pendant. The branches arise from prostrate branches that have reduced leaves. The axes bear three rows of bifid[15] leaves that are, unlike most other Jungermanniales, almost equal in size and therefore only slightly anisophyllous. The symmetry, accordingly, is radial. The leaves of *Herberta* are decidedly mosslike. *Herberta* grows on rocks and bark in acidic habitats. In the United States it occurs in the southern Appalachians and along the Pacific Coast from Washington (through Canada) to Alaska. The leaf cells are collenchymatous, that is, thickened at their corners. The mosslike habit, slight anisophylly, and radial symmetry are especially striking features of *Herberta*.

Ptilidium (Ptilidiaceae)

On *Ptilidium* (Gr. *ptilidion*, a small feather) (Fig. 11-59B), a prostrate plant, the leaves are deeply lobed and have fringed margins. The underleaves are only slightly smaller than the dorsal leaves, so that the anisophylly and dorsiventrality are not pronounced. *Ptilidium pulcherrimum*, which is widely distributed in the northern regions of the world, is dioecious. The archegonia, and hence the sporophytes, occur at the apices of the main branches. The perianth is characterized by three distal folds (Fig. 11-59B).

Telaranea (Lepidoziaceae)

Telaranea (L. *tela*, web, + *aranea*, spider) is a minute and delicate liverwort, the plants being up to 2.5 cm long and 300–460 μm broad. The axes are pinnately branched. The leaves are divided into filiform segments that consist of single rows of cells at their tips. The plants have rhizoids at the bases of the underleaves. *Teleranea nematodes* (Fig. 11-59C, D), grows on wet soil in the southern United States and West Indies. It is monoecious.

Bazzania (Lepidoziaceae)

Bazzania (after M. Bazzani, Italian anatomist), one of the largest leafy liverworts, is widespread in moist, shady woods, where it carpets acidic soil, logs, and stones. Several hundred species have been described. *Bazzania trilobata*, up to 6 mm wide (Fig. 11-59E, F) occurs in the eastern United States and westward to Minnesota, Arkansas, and Kentucky. The 2- to 3-toothed leaves are incubously oriented, and oil bodies occur in some of the leaf cells. The underleaves are prominent (Fig. 11-59F). The dioecious plants frequently produce sex organs, but sporophytes are of infrequent occurrence.

Calypogeia (Calypogeiaceae)

Calypogeia (Gr. *kalos*, beautiful, + *hypogeios*, subterranean) (Fig. 11-60A) is a large (90 species) genus of creeping, prostrate leafy liverworts whose species are worldwide in distribution. The somewhat translucent plants form mats on moist soils rich in humus. The

---

[15]Divided into two equal lobes.

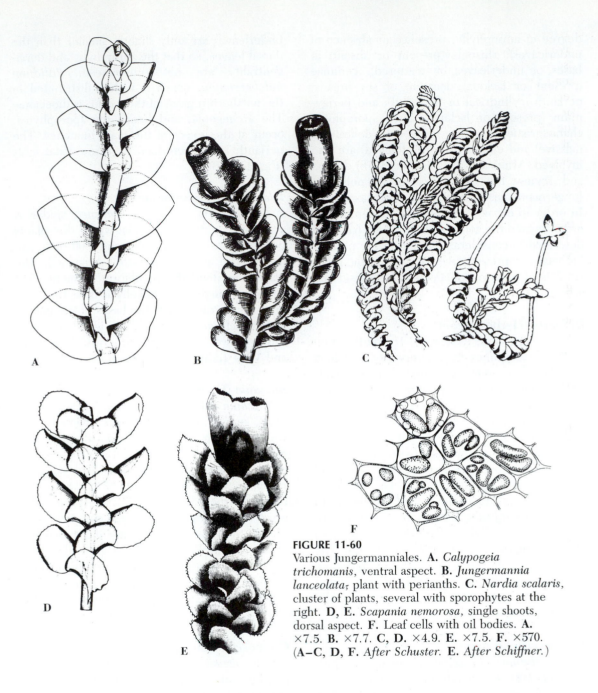

FIGURE 11-60
Various Jungermanniales. **A.** *Calypogeia trichomanis*, ventral aspect. **B.** *Jungermannia lanceolata;* plant with perianths. **C.** *Nardia scalaris*, cluster of plants, several with sporophytes at the right. **D, E.** *Scapania nemorosa*, single shoots, dorsal aspect. **F.** Leaf cells with oil bodies. **A.** ×7.5. **B.** ×7.7. **C, D.** ×4.9. **E.** ×7.5. **F.** ×570. (**A–C, D, F.** *After Schuster.* **E.** *After Schiffner.*)

leaves are incubously oriented. The under-leaves are prominent, often bifid, and at their bases prominent rhizoids arise. *Calypogeia trichomanis* (Fig. 11-60A) is bluish to grayish green, the individual plants up to 3 mm wide. The apices of the living plants are turquoise blue because of the deep blue oil bodies in the leaf cells. *Calypogeia trichomanis* is monoecious. The sporophytes appear in the spring and in New England liberate their spores in May and June.

Jungermannia (**Jungermanniaceae**)

Plants of *Jungermannia* (J. Jungermann, seventeenth-century German botanist) are prostrate and sometimes have brownish secondary pigmentation. The rhizoids are scattered along the stems. The leaves are obviously succubously oriented (Fig. 11-60B) and not lobed. Under-leaves are absent except on the erect, gemma-bearing shoots; on the latter, one- or two-celled gemmae arise in groups on the margins and lower surface of the leaves. These are agents of asexual propagation. North American species are monoecious. The perianths are prominent, cylindrical, constricted distally, and beaked and are quite distinctive. *Jungermannia lanceolata* is widely distributed in the eastern part of North America on damp sandstone rocks and soil.

Nardia (**Jungermanniaceae**)

Nardia (after S. Nardia, an Italian abbot), with seven North American species, is a good example of a genus with succubously oriented leaves (Fig. 11-60C) that are bilobed or unlobed and orbicular, or broader than they are long. The underleaves are rather narrow (lanceolate) and small. Oil bodies are present in the cells. The plants are monoecious or dioecious, depending on the species. The archegonia are terminal on the main shoots. The sporophyte in some species is enclosed by a complex calyptra (**perigynium**) derived not only from the archegonial wall but also by elongation of the recep-tacular tissue at the base of the archegonia. *Nardia scalaris* (Fig. 11-60C) grows on wet noncalcareous rocks in eastern North America.

Scapania (**Scapaniaceae**)

Scapania (Gr. *skapanion*, a spake or a hoe, in reference to the often flattened perianth) (Fig. 11-60D), like *Porella* and *Frullania*, has complicate-bilobed leaves, but it is distinctive in that the upper lobe of the leaf is smaller than the ventral one. Underleaves are lacking. *Scapania* has 27–28 species in North America. The species are common on moist rocks and acidic soils, rarely on logs or stumps or even on the bark of trees. The plants are dioecious, and their leaves contain oil bodies (Fig. 11-60E). *Scapania nemorosa* is widely distributed and abundant on moist rocks and cliffs with water seepage.

Cephalozia (**Cephaloziaceae**)

The genus *Cephalozia* (Gr. *cephale*, head, + Gr. *ozis*, branch, in reference to the short female branches) has 30–40 species in North America. The minute plants are mostly less than 1500 μm wide. The leaf cells have thin or slightly thickened walls and lack oil bodies. The leaves are bifid. *Cephalozia* is monoecious or dioecious, depending on the species, two of which, *C. bicuspidata* and *C. connivens*, are illustrated in Fig. 11-61A, B. These plants are widespread on rotting logs.

Nowellia (**Cephaloziaceae**)

Nowellia (after John Nowell, an English botanist) is *Cephalozia*-like but the basal portions of the leaves are dilated and strongly concave (Fig. 11-61C, D); they seem to function as water sacs. The rhizoids are scattered along the stems. Asexual reproduction is by unicellular gemmae borne at the apices of the branches. *Nowellia curvifolia* (Fig. 11-61C, D), although widely distributed, is limited to decaying logs. The plants are dioecious, the male more slender than the female as in some species of *Oedogo-*

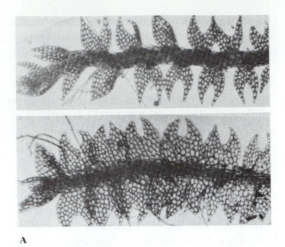

A

B

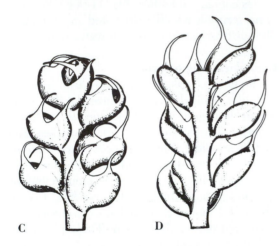

C   D

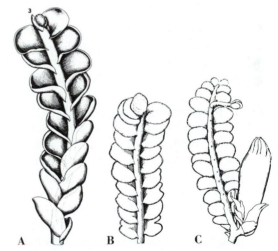

A   B   C

nium. Schertler (1977) summarized the morphology and development of *N. curvifolia* in axenic culture.

Odontoschisma (Adelanthaceae)

Plants of *Odontoschisma* (Gr. *odons*, tooth, + Gr. *schisma*, a split, in reference to the mouth of the perianth in most species) (Fig. 11-62) are prostrate with ascending tips. The leaves are rounded and entire and their cells contain prominent oil bodies, the latter filling the cell lumen. Bilobed underleaves are present, as are gemmae on the margins and surfaces of the

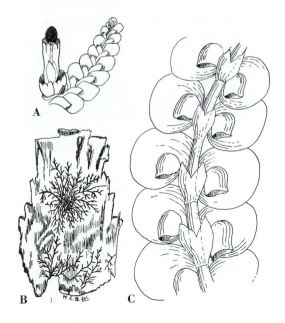

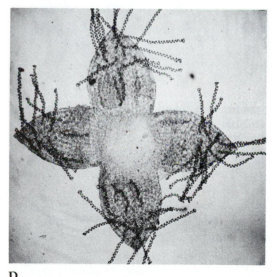

D

FIGURE 11-63

Various Jungermanniales. **A.** *Radula complanata*, ventral view; calyptra within perianth, involucre at base of the latter. **B–D.** *Frullania*. **B.** *F. eboracensis* on the bark of birch. **C.** *F. dilatata*. Note saclike lobules. **D.** *Frullania* sp., frontal view of dehisced capsule; note four valves with attached elaters. **A.** ×7.5. **B.** ×1. **C.** ×12. **D.** ×24. (**A, B.** *After Grout.* **C.** *Modified from Watson.*)

leaves in some species. The leaves are succubous in orientation. The genus is worldwide in distribution on moist logs and peat. *Odontoschisma denudatum* is characterized by its erect, gemmiferous shoots.

Radula (**Personniellaceae**)

Plants of *Radula* (L. *radula*, a scraper) (Fig. 11-63A) are prostrate and/or pendulous and once or twice pinnate. The genus contains several hundred species. The leaves are incubously oriented, the rhizoids occur in tufts at the bases of the underleaves. The leaves are complicately bilobed and appressed to the stem. The antheridia are terminal on main shoots or lateral branches. The organism reproduces by discoid gemmae.

Frullania (**Frullaniaceae**)

Frullania (after Leonardo Frullani of Florence, Italy) is a very large genus with about 700 described species of much branched plants. The plants are green, reddish-brown, or black and live on tree bark (Fig. 11-63B) and/or stone often in xeric habitats; the complicate bilobed leaves are striking because their ventral lobes (lobules) are more or less saclike (Fig. 11-63C) in many species. The antheridia are borne on lateral branches, while the archegonia may be terminal on the main axes or on lateral branches. The capsules dehisce into four valves, some elaters remaining attached at their tips (Fig. 11-63D).

Order 6. Haplomitriales

Gametophytes of the Haplomitriales (2–3 genera, 9 species) are erect or ascending with unlobed leaves that do not pass through a bilobed phase in ontogeny. The leaves are **isophyllous,** or in those that are anisophyllous, the dorsal leaves are smaller. The leaves lack collenchyma, and the plants are without rhizoids.

239

A

FIGURE 11-64
Haplomitrium. **A, B.** *H. mnioides.* **A.** Growing *in situ* in Japan; note emergent sporophytes. **B.** Left to right, ♂ plant, ♀ plant, ♀ plant with sporophyte. **C, D.** *Haplomitrium hookeri.* **C.** Female plant with sporophytes. **D.** Male plants. **A, B.** ×3. **C.** ×4.8. **D.** ×10. (**A, B.** *Courtesy of Drs. Zennoske Iwatsuki and A. J. Sharp.* **C. D.** *After Schuster.*)

Family 1. Haplomitriaceae
Haplomitrium

Haplomitrium[16] (Gr. *haplos*, one, + Gr. *metra*, headband or girdle) (Fig. 11-64) is an

---

[16]According to Schuster (1966), species of *Calobryum* and *Haplomitrium* form an intergrading series, so that he considers the two genera synonymous and discusses them as *Haplomitrium*. Within the single genus *Haplomitrium*, there is a gradual transition from acrogyny (typical of Jungermanniales) to anacrogyny (typical of Metzgeriales).

important morphological type with certain mosslike characteristics. Schuster (1966), Carothers and Duckett (1979), and Crandall-Stotler (1984) consider it to be related to the Metzgeriales. *Haplomitrium mnioides*, native to Japan, is an erect plant, the branches of which are about 3–5 cm tall. These develop from a subterranean, branched, rhizomelike axis that lacks rhizoids, as do the aerial stems. The latter bear their leaves in three rows. The leaves of one row, considered dorsal, may be slightly smaller than those of the other two. In a study of living and herbarium specimens of four species of *Haplomitrium* (= *Calobryum*), Grubbs (1970) reported that these plants probably demonstrate modification of their leafless, underground axes ("rhizomes"), which leads to the presence of "roots" in *H. intermedium*. The "rhizomes" and "roots" may be covered by

240

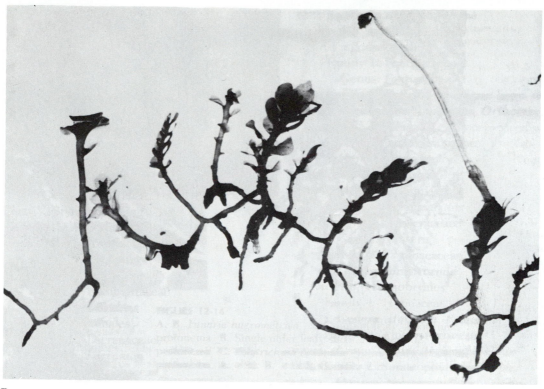

B

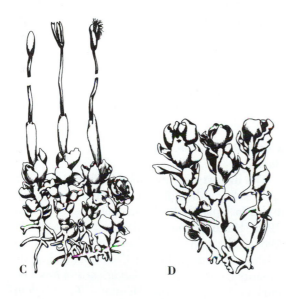

C   **D**

mucilaginous sheaths. The "roots" lack root caps
when growing in moist air. The central cells of
the axes are larger than the peripheral ones.

In *H. gibbsiae* the end walls of the elongate
conducting cells, transverse or oblique, are
interrupted by numerous pores that effect continuity (Burr, Butterfield, and Hébant, 1974).

Plants of *Haplomitrium* are dioecious. The
male plants terminate in a rosette of prominent
leaves (Fig. 11-64D) surrounding a convex-to-
flattened receptable on which the stalked, ovoid
antheridia are borne. The apices of the female
shoots also may end in juxtaposed leaves forming a rosette surrounding a flat receptable on
which six to many archegonia are borne. The
archegonial necks are unusual in being very
long and twisted like those of mosses, but unlike

241

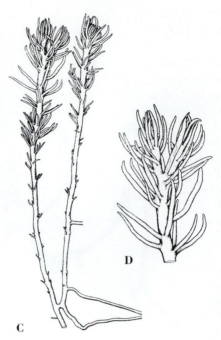

FIGURE 11-65
Takakia lepidozioides. **A, B.** Living plants. **C.** Habit of growth. **D.** Apex more highly magnified. **A, B.** ×75. **C.** ×130. (**C.** *After Schuster.*)

A

B

C

D

either mosses or other liverworts, they are composed of four vertical rows of neck cells.

A single sporophyte (Fig. 11-64)—occasionally two or even three—develops from the group of archegonia. The capsule is elongate-cylindrical, and its wall is a single layer thick except at the apex. After elongation of the massive, greenish seta, dehiscence by four, less often two, valves liberates the spores and elaters. There is no perianth as such, but the rather thick covering of the sporophyte is a shoot-calyptra (composed of the stem and archegonial venter) with many (often) deformed, unfertilized archegonia over its surface. The sporocytes are lobed as in the Metzgeriales and Jungermanniales.

The absence of rhizoids, erect shoots with leaves often almost equal in size, the ovoid antheridium, and unusual archegonia distinguish *Haplomitrium* from other Hepatophyta.

Unlike the Jungermanniales, the leaves are unlobed throughout their development.

Order 7. Takakiales

The single genus *Takakia*, with two species, is the only member of its order and family. It is sometimes included in the order Haplomitriales.

Family 1. Takakiaceae
Takakia

The liverwort genus *Takakia* (in honor of Dr. K. Takaki), although first collected at high altitudes in Japan in 1951, was not described until 1958 (Hattori, Sharp, Mizutani, and Iwatsuki, 1968). The genus has also been found in the offshore islands of British Columbia and in the mountains of Borneo. *Takakia ceratophylla* occurs in Nepal and the Aleutians. *Takakia lepidozioides* is an extremely simple plant (Fig.

242

11-65), which superficially might be confused with the leafy liverworts (Jungermanniaceae). It has sometimes been included in its own order Takakiales. The plant consists of erect, delicate branching axes, up to 1 cm tall, which lack rhizoids, although an associated fungus is thought to function in absorption. Like *Haplomitrium*, *Takakia* plants have prostrate and/or underground axes ("rhizomes"), which may bear branching "roots" (Grubbs, 1970). The smaller cylindrical branches, usually in groups of two or three, have been called phyllids (Fig. 11-65D); they are almost certainly homologous with leaves of *Haplomitrium*. Archegonia occur singly; they have six rows of neck cells. Antheridia and sporophytes have not yet been found.

Takakia is currently thought by some to be an extremely primitive plant, perhaps similar to the ancestors of modern Hepatophyta and Bryophyta. On the other hand, it may be a reduced, aberrant leafy liverwort. The occurrence of non-living conducting cells with pores in their terminal walls in the central strand of *Takakia* (Hébant, 1975), however, argues against the primitive condition of these plants. The chromosome number of *Takakia* is $n = 4$, unique among Hepatophyta.

Takakia shares with *Haplomitrium* the lack of collenchyma and rhizoids and is similar in its isophylly, radial organization, and in the mode of branching; both genera develop peculiar leafless, rootlike branches.

DIVISION ANTHOCEROTOPHYTA

Stotler and Crandall-Stotler (1977) have, in the writers' opinion, justifiably elevated the former class Anthocerotopsida to divisional rank, as Anthocerotophyta, as distinct from the liverworts, division Hepatophyta. The writers follow their example in this text. For a more detailed account of the division, consult Renzaglia (1978), Crandall-Stotler (1981), and Schuster (1984).

CLASS 1. ANTHOCEROTOPSIDA

The Anthocerotopsida, or hornworts, are given ordinal rank and included in the Hapatopsida by some authorities. However, while in a number of respects suggestive of certain thallose Hepatophyta, the members of the Anthocerotopsida differ in numerous important characters, and their segregation into a separate division appears to be warranted. The features in which they differ from members of the Hepatophyta are both gametophytic and sporophytic. Haploid chromosome complements of $n = 5$ are characteristic of the group.

Order 1. Anthocerotales

The Anthocerotales include four or five genera, of which three are fairly widely distributed in the United States. These are *Anthoceros*, *Phaeoceros*,[17] and *Notothylas; Dendroceros* and *Megaceros* are tropical. The last two genera have spirally thickened elaters and green spores, in contrast to *Anthoceros* and *Phaeoceros*. The gametophytes of *Dendroceros* have a prominent midrib, which is lacking in those of *Megaceros*.

Family 1. Anthocerotaceae
Anthoceros and *Phaeoceros*

Habitat and vegetative morphology. *Anthoceros* (Gr. *anthos*, flower, + Gr. *keras*, horn) and *Phaeoceros* (Gr. *phaios*, dusky, + Gr. *keras*) (Fig. 11-66) are widely distributed over the world. Some species of *Phaeoceros*, like *Sphaerocarpos*, grow as winter annuals in the

---

[17]Species of *Anthoceros* and *Phaeoceros* were formerly included in *Anthoceros*. The primary difference is the occurrence of mucilage and air cavities, often tubelike, other than ventral ones, within the gametophytes of *Anthoceros*, and their absence in *Phaeoceros*. The antheridial jackets are clearly composed of four tiers of elongate cells in *Anthoceros* and of many smaller cells in *Phaeoceros*. The spores of *Anthoceros* (*sensu stricto*) are dark brown or black, while those of *Phaeoceros* are a translucent yellow.

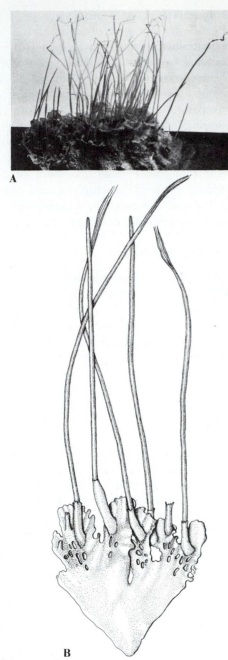

FIGURE 11-66
Phaeoceros laevis. **A.** Group of plants with sporophytes. **B.** Lobes with sporophytes more highly magnified; note antheridial craters. **A.** ×0.66. **B.** ×2.

southern United States, often developing on moist roadside cuts, eroded areas, and in unplowed fields; they occur later in the season northward. Southward, the gametophytes appear on the soil in the late autumn and produce sex organs during April and May, after which they disappear. Other species are perennial, occurring in permanently moist areas.

In uncrowded conditions in the field, the thallus of *P. laevis* develops an orbicular form. The dark-green plants have a rather dull, greasy appearance and are somewhat fleshy and brittle in texture. In the vegetative condition their texture and color suggest such thallose genera as *Pellia* and related forms; however, they can be distinguished readily by features to be described below.

The orbicular shape of the thallus reflects frequently repeated dichotomies during the early phases of growth; somewhat laciniate lobes may project from the margins of the plants. The gametophytes are anchored to the soil by nontuberculate rhizoids that arise from the ventral surface. The lower surfaces of the plants are interrupted by minute fissures (stomata) that communicate with chambers frequently containing colonies of the Cyanophytan *Nostoc* (Fig. 11-67), which fixes free nitrogen. Experimentation has revealed that some species can grow in media lacking combined nitrogen when *Nostoc* is present in the gametophytes or merely in the cultures. Control plants without *Nostoc* became chlorotic (Ridgway, 1967*a*). When within the hornwort, *Nostoc* produces more heterocysts than when it is free-living. The host cells become ultrastructurally modified (Duckett et al., 1977; Honneger, 1980). Both free-living strains of *Nostoc* from nature and those isolated from lichens (p. 801) and cycads (p. 529) were able to establish symbiotic associations with *Anthoceros* in experimental cultures (Enderlin and Meeks, 1983).

Rodgers and Stewart (1977) and Stewart and Rodgers (1977) have investigated in depth the relationship between the liverwort *Blasia*[18] *pus-*

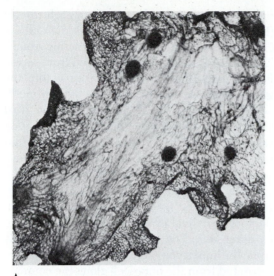

A

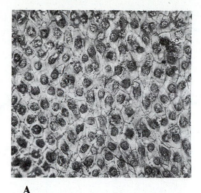

A

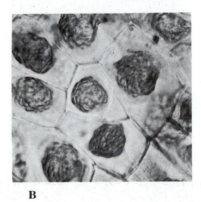

B

FIGURE 11-68
Phaeoceros sp. **A, B.** Plant surface at increasingly higher magnification; note the single, massive chloroplast in each cell. **A.** ×75. **B.** ×500.

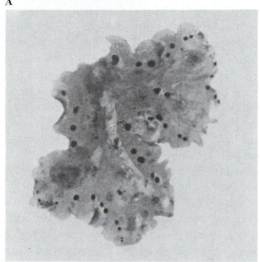

B

FIGURE 11-67
Nostoc colonies in hornworts and liverworts. **A.** In *Anthoceros punctatus.* **B.** In *Blasia pusilla.* **A.** ×13.5. **B.** ×5.4. *(After Stewart and Rogers.)*

illa and *Anthoceros punctatus* (Fig. 11-67) with various strains of *Nostoc* isolated from these plants and from others (e.g., cycads, p. 529).

---

[18]Metzgeriales; *Blasia* is not discussed in this text, but see Fig. 11-67.

They found that these organisms, when they contained *Nostoc sphaericum*, were active in fixing $^{15}N_2$ and that this was transferred from the gametophyte to the sporophyte. If the gametophytes were grown in darkness but the sporophytes illuminated, photosynthate was transferred from the latter to the gametophytes.

With the exception of the ventral chambers, the gametophytes of *Phaeoceros* lack marked internal differentiation (Fig. 11-70). The component cells are chlorenchymatous and differ from those of the Hepatophyta in containing single massive chloroplasts (Fig. 11-68) in which are embedded a number of proteinaceous segments that have been designated pyrenoid bodies be-

245

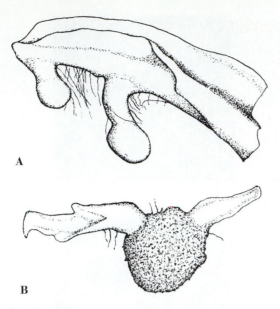

FIGURE 11-69
Anthoceros phymatodes, tubers. **A.** Lateral view of plant showing development of a tuber. **B.** Older tuber sprouting. **A, B.** ×18. (*After Howe.*)

cause of their position and suggested role in condensing glucose units into starch. The internal cells of the thallus may contain two chloroplasts or one bilobed one. No oil bodies or oil cells are ever present, in contrast to the Hepatophyta, in which 90% of those investigated by Schuster (1966) contained oil bodies. Vegetative reproduction occurs by the separation of marginal lobes from the parent thallus and, in some species, by the formation of tuberlike bodies which penetrate the soil and later regenerate new gametophytes (Fig. 11-69).

Reproduction: the gametophyte.   Careful studies of laboratory cultures of a number of species of *Anthoceros* and *Phaeoceros* have revealed that some are dioecious, whereas others (*P. laevis*) bear antheridia and archegonia on the same gametophyte. The sex organs arise in rows from dorsal derivatives of the marginal apical cells with two cutting faces from which derivative cells are formed. *Phaeoceros laevis* is usually markedly protandrous, so that its antheridia

mostly lie back from the apices on each rosette (Fig. 11-70A, B); however, antheridia and archegonia may be borne in alternating zones. Although the sex organs are derived from superficial dorsal cells, they are sunken within the gametophyte at maturity because of their method of development. It has been shown that antheridium formation in six species of *Anthoceros* and *Phaeoceros* is evoked by photoperiods of 4–12 hours; none developed when the plants were illuminated for 18 hours (Ridgway, 1967*b*).

The antheridium arises from a superficial cell recently derived from the apical cell. This undergoes periclinal division to form an inner and an outer cell. The antheridium arises entirely from the innermost cell, which early becomes separated from the outer cell along the common periclinal wall. This space enlarges to form the cavity in which the antheridia lie. The outer cell divides anticlinally and periclinally to form the roof of the chamber as the antheridia enlarge. From one to more than 25 antheridia may develop in a single cavity (Fig. 11-70). Cavities containing mature antheridia are recognizable to the unaided eye as small, orange-yellow pustules. The color results from the transformation of the chloroplasts in the antheridial jacket cells into chromoplasts. When the antheridia are mature, the superficial cells of the antheridial cavity break down, permitting water to come in contact with the antheridia. These dehisce at their apices and liberate large numbers of extremely minute biflagellate sperms. Duckett (1974*b*), Moser et al. (1977), Carothers et al. (1977), and Duckett et al. (1980) have investigated spermatogenesis in *Phaeoceros laevis* at the ultrastructural level. Information on the details of spermatogenesis in liverworts, hornworts, and mosses is available in the papers by Carothers (1975) and Carothers and Duckett (1978).

The archegonia are also completely sunken within the dorsal surface (Fig. 11-71A); they are hard to recognize because of their sunken location. They differ from those of liverworts and mosses in not having a free, discernible jacket

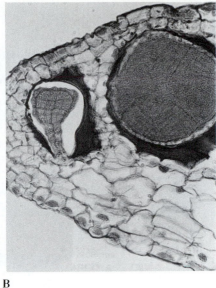

FIGURE 11-70
Phaeoceros laevis. **A.** Living antheridia, surface view. **B.** Section of plant with two stages in antheridial development. **A.** ×3. **B.** ×60. (**B.** *After Crandall-Stotler.*)

A   B

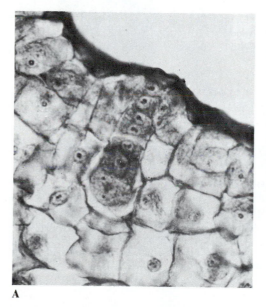

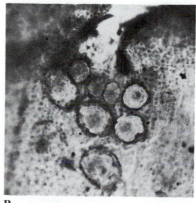

A

B

FIGURE 11-71
Phaeoceros sp. **A.** Embedded, immature archegonium in medium longisection. **B.** Surface view of thallus with archegonia; the circles represent transparent slimy globules over each archegonium. **A.** ×600. **B.** ×45.

layer. The apex of the archegonial neck is surmounted by four cover cells; the neck itself is composed of six vertical rows of neck cells. The latter are difficult to recognize, because they are so closely associated with the surrounding cells of the gametophyte, but they are evident in a surface view of the thallus after the cover cells have disintegrated. The axial row of cells of the archegonium is composed of the egg, ventral canal cell, and four or five neck canal cells

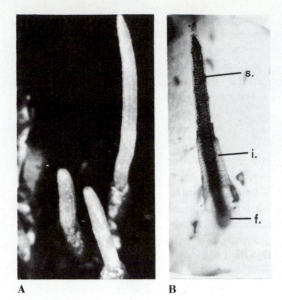

A   **B**

FIGURE 11-72
Phaeoceros laevis. **A.** Living sporophytes
magnified. **B.** Cleared sporophyte showing foot
embedded in gametophyte. f., foot; i., involucre;
s., sporophyte. ×12.5.

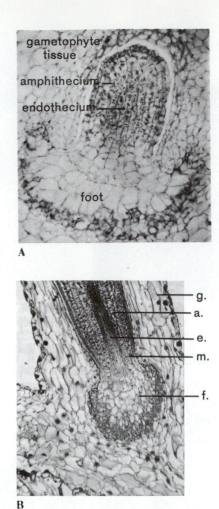

A

B

FIGURE 11-73
Phaeoceros laevis. **A.** Median longisection of young
sporophyte before emergence from the
gametophyte. **B.** Median longisection of basal
region of an older sporophyte. a., amphithecium;
e., endothecium; f., foot; g., gametophyte; m.,
meristematic zone. **A.** ×125. **B.** × 90.

(Fig. 11-71A). As the archegonium matures, the
tip of its neck becomes covered with a mound of
mucilage as the disintegrating ventral canal and
neck canal cells are extruded, after the cover
cells have separated (Fig. 11-71B).

Reproduction: the sporophyte.   After fertilization, numerous zygotes scattered over the
plant complete their development into sporophytes. The latter are elongate, green, cylindrical, needlelike structures, which project from
the upper surface of the thallus (Figs. 11-66,
11-72). In its ontogeny, the zygote undergoes a
longitudinal division[19] followed by two additional divisions that form eight cells, in two tiers,
within the archegonial venter. The lowermost
tier of the octant, by further multiplication,
gives rise to the sterile bulbous foot of the

sporophyte (Fig. 11-73). The upper tier undergoes a series of transverse divisions to form a
columnar structure, which becomes differentiated, by periclinal divisions, into an outer,
amphithecial layer and an inner, endothecial
zone. The amphithecial layer, by further pericli-

---

[19]The division is also reported to be transverse (Schuster, 1966).

248

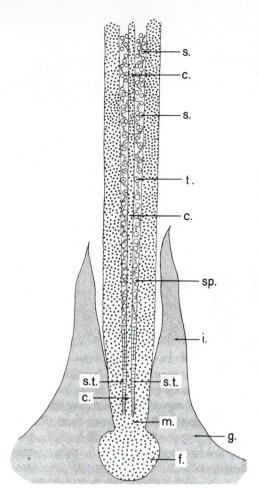

FIGURE 11-74

Anthoceros sp. Median longisection of sporophyte, diagrammatic; c., columella; f., foot; g., gametophyte; i., involucre; m., meristematic zone; s., spore; sp., sporocyte; s.t., sporogenous tract; t., spore tetrad.

nal divisions, ultimately becomes about four to six cells deep (Fig. 11-73).

In *Phaeoceros* and *Anthoceros*, very early in the development of the sporophyte the region that corresponds to the liverwort seta becomes actively meristematic and functions as an intercalary meristem, much like that in *Laminaria*, so that the sporophyte increases in length by growth near its base. As a result, the cylindrical sporophyte emerges from the gametophyte,

pushing through a sheathlike portion of the latter (Fig. 11-72). The elongation of the sporophyte by intercalary growth may continue for several months. Sporogenesis and spore dissemination are continuous and progressive in *Anthoceros* and *Phaeoceros*, not simultaneous as they are in the sporophytes of all Hepatopsida.

Considerable histological differentiation is present in the more mature regions of the sporophyte (Figs. 11-74, 11-75A). The entire endothecium remains sterile and is called the **columella**. It is composed of elongate, thin-walled cells that seemingly do not function in conduction. In the related *Dendroceros crispus* partial, double-spiral thickening of the walls of the outermost columella cells has been observed (Proskauer, 1960).

The surface layer of the amphithecium is cutinized; it serves as an epidermis and develops **stomata,** the guard cells of which are thickened along the stomatal aperture (Fig. 11-75C). The stomata seemingly remain open until late in development (Paton and Pearce, 1957). The wall layers of the sporophyte are also photosynthetic and play an important role in its nutrition. The innermost layers of the amphithecium develop into sporogenous tissue, many cells of which, in older regions of the developing sporophyte, separate, become spherical, and function as sporocytes (Figs. 11-76, 11-77), each giving rise to a tetrad of spores as a result of the meiotic process. It is worthy of note that one of the earliest accounts of the origins of cells by division, published in 1820, was based on a study of living sporocytes of *Anthoceros*. Certain groups of potentially sporogenous cells remain sterile but divide and may elongate (Fig. 11-75B). Because of their multicellular condition, when the component cells cohere, they are sometimes referred to as **pseudoelaters.** As the spores near the apical portion of the sporophyte mature, the surrounding tissues lose their chlorophyll and become dry and brown. During ontogeny, the walls of the epidermal cells of the more mature portions of the sporophyte thicken, the common

249

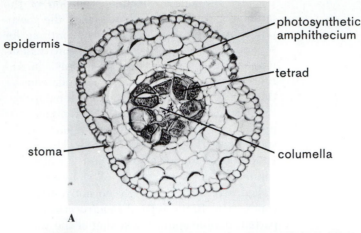

epidermis

photosynthetic amphithecium

tetrad

stoma

columella

A

FIGURE 11-75
Anthoceros sp. **A.** Transection of sporophyte at tetrad level. **B.** Columella, spores and pseudoelaters. **C.** Epidermis and stoma of sporophyte. e., pseudoelater. **A, C.** ×250. **B.** ×300.

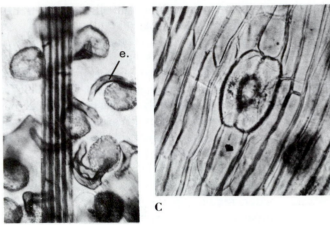

e.

C

B

walls between the vertical rows of cells in one or two shallow grooves on the surface of the sporophyte remaining thin-walled. Dehiscence of the sporophyte valves begins near the apex of the sporophyte and extends toward the base as development proceeds. The two valves and sometimes the pseudoelaters exhibit twisting, hygroscopic movements, which have a role in the dissemination of the spores. The valve apices may remain adherent at the apex and twist about each other, or they may separate (Fig. 11-66). The columella persists and appears as a dark thread between the open valves. In the mature regions of the sporophytes, spores and pseudoelaters may be brown-black (*Antho-*

ceros), or the spores may be yellow (*Phaeoceros*).

Thomas et al. (1978) have investigated the nutritional relationships between sporophyte and gametophyte in *Phaeoceros laevis*. Photosynthesis in the sporophyte is at about twice the rate of that of gametophytes, but the carbon fixed by the sporophyte is sufficient for maintenance but not for sustained growth; organic nutriments from the gametophyte seem to be required.

It has also been demonstrated that branched cells of the base of the sporophyte are associated with specialized transfer cells of the gametophyte, circumstantial evidence that transport of

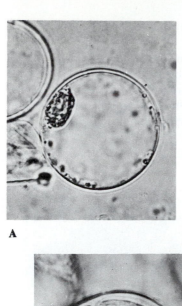

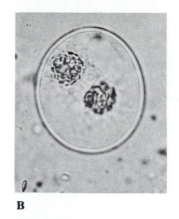

B

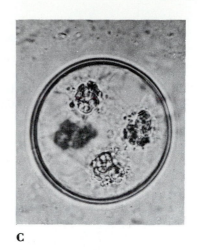

A

B

C

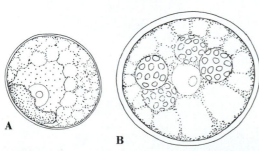

D

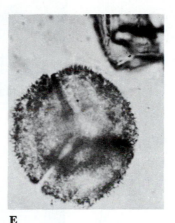

E

FIGURE 11-76

Anthoceros sp. Sporogenesis (photomicrographs of living material). **A.** Young sporocyte with single chloroplast which masks the nucleus. **B.** End of first division of chloroplast; sporocyte nucleus between plastids. **C.** Sporocyte with four plastids and central nucleus. **D.** Spore tetrad within sporocyte. **E.** Mature spore tetrad; sporocyte wall has disappeared; spore walls have thickened. ×770.

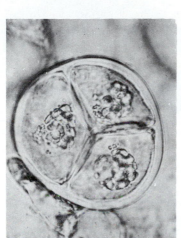

A

B

FIGURE 11-77

Anthoceros sp. Sporocytes. **A.** With single laminate chloroplast and nucleus. **B.** After chloroplast division. **A.** ×770. **B.** ×1000.

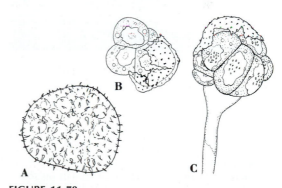

B

A

C

FIGURE 11-78

Phaeoceros laevis. **A.** Mature spore, surface view. **B, C.** Stages in spore germination. **A.** ×770. **B, C.** ×315.

251

substances occurs from gametophyte to sporophyte (Gambardella et al., 1981).

Stages in spore germination and the development of young gametophytes are illustrated in Fig. 11-78. It has been reported that sex chromosomes are present in certain species of Anthocerotopsida.

Family 2. Notothylaceae
Notothylas

The single genus *Notothylas* (Gr. *nota*, posterior parts or back, + Gr. *thylakion*, bag), clearly a hornwort, differs in a number of respects from *Anthoceros* and *Phaeoceros*. The orbicular gametophytes (Fig. 11-79) occur on

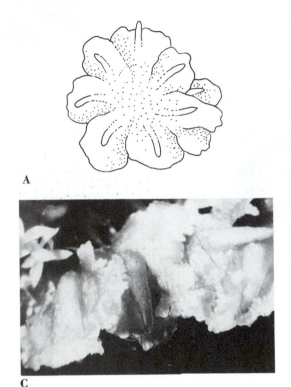

A

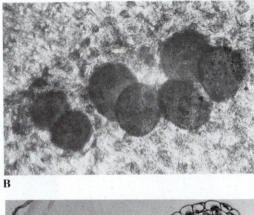

B

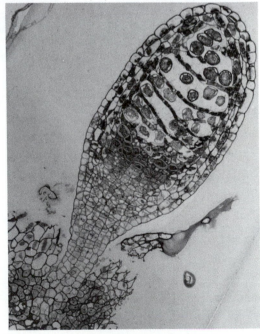

D

FIGURE 11-79
Notothylas orbicularis. **A.** Habit of growth on soil.
B. Antheridia. **C.** Margin of plant with maturing sporophytes. **D.** Longisection of a meiotic sporophyte. **A.** ×¾. **B.** ×55. **C.** ×4. **D.** ×100. (**D.** *Courtesy of Roy C. Brown.*)

C

252

bare, moist soil. In development and organization, including single plastids in the cells and internal cavities in some species, *Notothylas* resembles *Phaeoceros* and *Anthoceros*, as it does with respect to its archegonia and antheridia.

The sporophyte (Fig. 11-79) is less elongate than in *Phaeoceros* and *Anthoceros* and contains a central columella surrounded by sporogenous tissue several layers of cells thick, more extensive transversely than either the columella or sterile amphithecium. Pseudoelaters are present among the sporocytes and spores; the latter are in alternating tiers. The intercalary meristematic zone between the foot and capsule regions of the sporophyte is active for just a short time, so that the sporophytes are short-cylindrical and project only a little from the gametophyte. Guard cells and stomata are lacking, and the foot of the sporophyte is small. At maturity, the sporophyte dehisces into two valves.

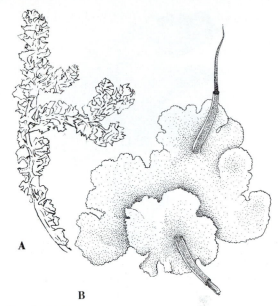

FIGURE 11-80
A. *Dendroceros* sp. B. *Megaceros tjibodensis*, thalli with sporophytes. A. ×3. B. ×2. (*After Campbell.*)

Other Anthocerotales

The tropical genera *Dendroceros* and *Megaceros* are somewhat intermediate in the characteristics of the sporophytes between *Anthoceros*, *Phaeoceros*, and *Notothylas*. This is true of the secondary growth of the sporophyte through the division of the intercalary meristem, which persists longer than in *Notothylas* but often not as long as in *Anthoceros* and *Phaeoceros*. The gametophyte of *Dendroceros* (Fig. 11-80A) has a prominent, ventrally projecting midrib with monostromatic wings, which are lobed and folded as in *Fossombronia*. The antheridia occur singly in rows along the midrib. In *Megaceros* (Fig. 11-80B) the cells of some species contain more than one chloroplast, but these are larger than the chloroplasts of the Hepatophyta. The chloroplasts of *Megaceros* lack pyrenoid bodies (Burr, 1970). The gametophyte of *Megaceros* may achieve dimensions of 5 × 1 cm, and the sporophytes may be 9 cm long.

Stomata are lacking from the epidermis of the sporophytes of *Dendroceros* and *Megaceros*, as in *Notothylas*, but both the former have spirally thickened elaters. The spores of *Dendroceros*, like those of *Pellia* and *Conocephalum*, contain chloroplasts and germinate endosporically.

FOSSIL HEPATOPHYTA

The fossil record of liverworts, like that of other nonvascular plants, is relatively sparse and fragmentary. A recent summary lists 9 species of liverworts from the Paleozoic; about 40 from the Mesozoic; yet only about 22 from the Tertiary (inside front cover). The liverworts and mosses are separate lines as far back as the fossil record is available. Until relatively recently, Hepatophyta were not known earlier than the Carboniferous. The form genus *Hepaticites*

253

FIGURE 11-81
Hepaticites kidstonii. A fossil liverwort from the Carboniferous. ×30. (*After Walton.*)

FIGURE 11-82
Pallaviciniites devonicus. Fragment of a Devonian thallose liverwort. ×7.5. (*From Hueber.*)

(Fig. 11-81) includes thallose, leafy, and intermediate types. Five species have been described from the Carboniferous period. However, an earlier form is now known. *Pallaviciniites devonicus* (Schuster, 1966) (Fig. 11-82), first called *Hepaticites devonicus* (Hueber, 1961), is a thallose liverwort from the Devonian of New

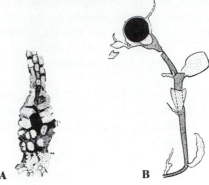

FIGURE 11-83
Naiadita lanceolata, a fossil liverwort. **A.** Archegonium. **B.** Thallus with emergent sporophyte. (*After Harris.*)

York State. It closely resembles the extant *Pallavicinia.*

Fossil Ricciaceae are known from the older Mesozoic and Marchantiaceae from the lower Jurassic of Australia and the Lower Cretaceous of Patagonia. Metzgerialean fossils are present in the Middle and Upper Carboniferous (Boureau, 1967).

A leafy liverwort, *Naiadita* (Fig. 11-83), of seemingly Sphaerocarpalean affinity, with axial archegonia, gemma cups, and globose sporophytes, has been described from the Upper Triassic (inside front cover). Genera from more recent strata, for example, *Jungermannites* and *Marchantites,* are increasingly like extant liverworts.

Fossil liverworts shed no light on what liverwort precursors were like. They provide evidence that the group is an ancient one and that diversification into thallose and leafy types had already occurred by the Carboniferous. The fossil history of the Anthocerotopsida is unknown.

Finally, mention should be made of *Sporogonites exuberans* (Fig. 11-84) from the Lower Devonian of Belgium. This fossil, as reconstructed by Andrews (1960), suggests a plant having attained a level of organization similar to

254

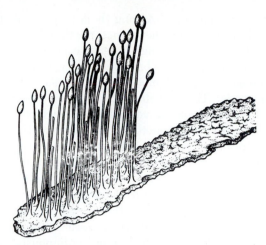

FIGURE 11-84
Sporogonites exuberans. Restoration of part of a plant; note sporophytes arising from thallus. (*After Andrews.*)

that of thallose liverworts with what may be numerous attached, erect sporophytes.

Summary and classification

In summarizing this account of the representatives of the Hepatophyta and Anthocerotophyta discussed in this chapter, attention will be devoted first to the question of their classification, now that the basic facts of their morphology have been presented. As noted in the introductory section, the liverworts, hornworts and mosses were formerly classified together, almost universally, in a single division, Bryophyta. The reasons for placing these groups in separate divisions, Hepatophyta, Anthocerotophyta and Bryophta, as is done in this text, will be enumerated at the conclusion of the following chapter. A comprehensive survey of the complete classification of these organisms presupposes familiarity with a wider range of genera than has here been described. However, the criteria used to delimit the higher categories among the liverworts and hornworts should be sufficiently familiar at this stage to permit some fruitful discussion.

No single scheme of classification of the Hepatophyta and Anthocerotophyta has met with universal approval. Some authorities consider that the liverworts and the hornworts are sufficiently different morphologically to warrant their separation into separate divisions of coordinate rank, as has been done in this text. Others reduce the Anthocerotophyta to ordinal rank and include them in the liverwort class. The following classification includes the representative genera discussed in this chapter and is arranged in the same order as the discussion.

Division 1. Hepatophyta
Class 1. Hepatopsida (Liverworts)
  Order 1. Marchantiales
    Family 1. Ricciaceae
      Genera: *Ricciocarpus, Riccia*
    Family 2. Oxymitraceae
      Genus: *Oxymitra*
    Family 3. Marchantiaceae
      Genera: *Marchantia, Dumortiera*
    Family 4. Conocephalaceae
      Genus: *Conocephalum*
    Family 5. Rebouliaceae
      Genera: *Reboulia, Asterella*
    Family 6. Lunulariaceae
      Genus: *Lunularia*
  Order 2. Sphaerocarpales
    Family 1. Sphaerocarpaceae
      Genus: *Sphaerocarpos*
    Family 2. Riellaceae
      Genus: *Riella*
  Order 3. Monocleales
    Family 1. Monocleaceae
      Genus: *Monoclea*
  Order 4. Metzgeriales
    Family 1. Pelliaceae
      Genus: *Pellia*
    Family 2. Pallaviciniaceae
      Genus: *Pallavicinia*
    Family 3. Fossombroniaceae
      Genera: *Fossombronia, Petalophyllum*
  Order 5. Jungermanniales
    Family 1. Herbertaceae
      Genus: *Herberta*

Family 2. Ptilidiaceae
  Genus: *Ptilidium*
Family 3. Lepidoziaceae
  Genera: *Telaranea, Bazzania*
Family 4. Geocalycaceae
  Genus: *Lophocolea*
Family 5. Calypogeiaceae
  Genus: *Calypogeia*
Family 6. Jungermanniaceae
  Genera: *Jungermannia, Nardia*
Family 7. Scapaniaceae
  Genus: *Scapania*
Family 8. Cephaloziaceae
  Genera: *Cephalozia, Cephaloziella, Nowellia*
Family 9. Adelanthaceae
  Genus: *Odontoschisma*
Family 10. Personniellaceae
  Genus: *Radula*
Family 11. Porellaceae
  Genus: *Porella*
Family 12. Frullaniaceae
  Genus: *Frullania*
Order 6. Haplomitriales
  Family 1. Haplomitriaceae
    Genus: *Haplomitrium*
Order 7. Takakiales
  Family 1. Takakiaceae
    Genus: *Takakia*
Division 2. Anthocerotophyta.
  Class 1. Anthocerotopsida
    Order 1. Anthocerotales
      Family 1. Anthocerotaceae
        Genera: *Anthoceros, Phaeoceros, Megaceros, Dendroceros*
      Family 2. Notothylaceae
        Genus: *Notothylas*

In discussing a classification such as this, a question at once arises regarding the criteria that were considered in grouping or segregating the several categories. Whether they are considered in relation to ordinal, class, or divisional rank, the characteristics that distinguish the liverworts and horned liverworts are fairly striking and numerous. Among them may be cited such features of the gametophytic phase of *Anthoceros* as its single[20] massive chloroplasts (containing pyrenoid bodies) in each cell and the endogenous development of antheridia and archegonia. The gametophytes of some species have stomata on the ventral surface, which are absent from the gametophytes of liverworts. As contrasted with those of Hepatophyta, the sporophytes of the Anthocerotophyta, as exemplified by *Anthoceros* and *Phaeoceros*, are elongate cylindrical structures with marked internal complexity and long-continued development. The latter is effected by a basal intercalary meristematic zone which adds continuously to the sporophytic tissues; thus, spore production and dissemination are gradual processes. These include a cutinized epidermis with functional stomata, a zone of photosynthetic parenchyma cells, a fertile sporogenous layer, and a central sterile columella composed of elongate cells that arise from the endothecium. Furthermore, unlike those of the Hepatophyta, the sporophytes of *Anthoceros* and *Phaeoceros* have a sterile endothecium and are dehiscent into two valves. These characteristics, among others, are interpreted as manifestations of fundamental dissimilarity from the members of the Hepatophyta.

Within the class Hepatopsida, the characteristics that distinguish the seven orders are also both gametophytic and sporophytic. The gametophytes of the Marchantiales, unlike those of the other orders, are always thallose, with some degree of internal differentiation into photosynthetic and storage regions. The presence of two kinds of rhizoids (tuberculate and nontuberculate) and ventral scales are also consistent and characteristic features of this order. A specially differentiated epidermis with air pores may be present (Marchantiales), as well as ventral scales. Both tuberculate and nontuberculate rhizoids occur in the Marchantiales, while only

---

[20]Some species of two related genera may have up to eight.

nontuberculate rhizoids are present in the members of the other orders that also lack ventral scales. The Haplomitriales lack rhizoids, and their gametophytes are erect, radially symmetrical, and more or less isophyllous. The leafless rhizomelike branches are nearly unique among the Hepatophyta and Bryophyta.

The sporophytes of the Marchantiales consist solely of ephemeral, spherical, indehiscent capsules in which sporogenesis takes place (Ricciaceae, Oxymitraceae); or they may be differentiated into foot, seta, and capsule, as are the sporophytes of all other liverworts. The capsules are indehiscent or irregularly dehiscent,[21] and their walls are monostromatic. As noted earlier, the Monocleales are quite distinctive.

On the other hand, the thallose gametophytes of the Sphaerocarpales, Monocleales, and Metzgeriales show little or no internal differentiation into photosynthetic and storage regions; they lack air chambers, but a conducting strand is present in some (e.g., *Pallavicinia*). The archegonia of Metzgeriales and Jungermanniales have five rows of neck cells, in contrast to the six rows in those of Marchantiales, Sphaerocarpales, and Monocleales. The archegonia of Haplomitriaceae have only four rows of neck cells; the antheridial stalks also are in four rows. In the Sphaerocarpales, Metzgeriales, and Monocleales, the plant bodies are typically thallose or ribbonlike, with smooth or undulate margins. In the Jungermanniales, the plants are normally leafy, with leaves in two lateral rows; frequently, a third, ventral row is present. In a few of these (*Herberta*), the leaves are isophyllous or almost so. In the Metzgeriales and Jungermanniales, spore dissemination is usually preceded by splitting of the capsule into four valves. Elaters, which aid in spore dissemination, are universally present in these orders. Furthermore, the capsule wall is always more than one cell layer thick. In the Monocleales the capsule

has one fissure and the wall is one layer thick.

The arrangement of classes and orders presented above is based on the opinion (speculation) that evolution in the Hepatophyta has been in the direction of increasing complexity in both sporophyte and gametophyte. This view is not necessarily correct, nor is it accepted by all contemporary hepaticologists (e.g., Schuster, 1979, 1983, 1984). That such an arrangement is not always feasible is demonstrated by the fact that the Ricciaceae and Oxymitraceae, which have the simplest sporophytes among those in the land plants, are listed before the Sphaerocarpales and Metzgeriales, which have gametophytes simpler than those of the Ricciaceae and Oxymitraceae. The sporophytes of the Anthocerotophyta are, apparently, the most complex. It will be clear from what follows that the evolutionary sequence has been read in the opposite direction by some morphologists. Further discussion of these topics is deferred to the end of Chapter 12.

DISCUSSION QUESTIONS

1. With reference to the structure of the gametophytic phase, which genera of the Hepatophyta do you consider to be most simple? Most complex? Give the reasons for your answers.
2. With reference to the sporophytic phase, which of the Hepatophyta do you consider to be most simple? Most complex? Give the reasons for your answers.
3. What phenomenon among Chlorophyta can you cite as similar to gemma formation?
4. To what type of algal life cycle is that of the Hepatophyta most similar? How does it differ?
5. What characteristics distinguish sporophytes and gametophytes? Do these alternate phases occur among algae? Explain.
6. Can the terms "monobiontic" and "dibion-

---

[21]Into four segments in *Lunularia*.

257

tic" be applied appropriately to the Hepatophyta? Explain.

7. What factors seem to be involved in effecting the distribution of liverworts?

8. Can you suggest a mechanism, and explain its operation, for the occurrence of dioecism in *Sphaerocarpos*? In *Marchantia*? How, then can you explain monoecism in genera such as *Ricciocarpus*?

9. Can you suggest any biological advantages, with respect to survival of the organism, that may accrue from the fact that in Hepatophyta and other land plants meiosis is delayed until sporogenesis?

10. What significance do you attach to the occurrence of chlorophyll in the sporophytes of the Hepatophyta and Anthocerotophyta? What is the source of the inorganic salts, carbon dioxide, and oxygen used by the sporophyte?

11. In what respects do rhizoids differ from roots and rhizomes?

12. How do the sex organs of Hepatophyta and Anthocerotophyta differ from those of algae? Are there exceptions?

13. According to some morphologists, the sporophytes of Hepatophyta afford evidence that there has occurred a progressive sterilization of potentially sporogenous tissue to form vegetative or somatic tissues. Cite evidence in support of this statement.

14. In the same connection, it has been postulated that all vegetative tissue of the sporophyte has arisen by sterilization of sporogenous tissue. What evidence do the algae provide in this connection?

15. Define or explain sporophyte, gametophyte, homology, apical growth, complicate-bilobed, amphigastrium, foot, elater, sporocyte, tetrad, calyptra, protandrous, periclinal, anticlinal, incubous, succubous.

16. In your opinion, to the gametophytes of which of the Hepatophyta is that of *Anthoceros* most similar? How does it differ?

17. On what grounds is one justified in segregating *Anthoceros* from liverworts?

18. What differences are present in the sporophyte of *Anthoceros* as compared with those of Hepatophyta?

19. How could one distinguish a vegetative gametophyte of *Anthoceros* from that of *Pellia*?

20. What is the origin of the pseudoelaters of *Anthoceros*?

21. How would you plan an experiment to clarify the role of endophytic *Nostoc* that occurs in *Anthoceros*?

22. Each young sporocyte of *Anthoceros* contains one chloroplast. Each young spore also contains one. Explain. Why do the cells of the sporophyte not contain two chloroplasts, since they are diploid?

23. What significance do you attribute to the occurrence of stomata on the *Anthoceros* sporophyte?

24. Describe an experiment you could design to ascertain the degree of autotrophism of the *Anthoceros* sporophyte.

25. Describe an experiment that might prove whether or not the columella functions in conduction.

26. How could you prove whether or not the gametophyte of *Anthoceros* is photoautotrophic?

27. Give the classification of the Hepatophyta as presented in this chapter; cite distinguishing attributes and illustrative genera.

28. Of what possible phylogenetic significance are the Haplomitriales?

Chapter 12
DIVISION BRYOPHYTA[1]

As noted at the beginning of the preceding chapter, the division Bryophyta, as conceived by many taxonomists, includes the liverworts, the hornworts, and the mosses. In the present text, on the other hand, the liverworts and hornworts have been classified in separate divisions, the Hepatophyta and Anthocerotophyta; hence, the division Bryophyta, as here constituted, has narrower limits, since it includes only the mosses. According to one estimate, some 800 genera (Crosby and Magill, 1978) and 10,000–12,000 species (Anderson, personal communication) are included in the Bryophyta. It is generally agreed that there are three basically different morphological types among mosses. This occasions the division of the group into three classes as follows:

Class 1. Sphagnopsida, the peat mosses
Class 2. Andreaeopsida
Class 3. Bryopsida, the "true" mosses

For a summary of Eastern North American mosses consult Crum and Anderson (1981); the contributors to Smith (1982) discussed ecological aspects of bryophytes.

CLASS 1. SPHAGNOPSIDA

Habitat and vegetative morphology. The class Sphagnopsida includes only a single order, family, and genus—*Sphagnum* (Gr. *sphagnos*, kind of moss)—but many species. The spongy, pale-green mats and mounds of *Sphagnum* (Fig. 12-1) are familiar to all who have frequented the out-of-doors, especially in those regions where the soil is not markedly alkaline. *Sphagnum* typically is an inhabitant of pools, bogs, and swamps and often occurs abundantly around the shores of ponds and lakes. Its rapid growth under such conditions and its great water-holding capacity frequently combine to "fill in" completely fairly large bodies of water. At one stage of this process, in which the plants form a dense surface mat over the

---

[1]For more discussion of some topics see certain chapters in Schuster (1979, 1983, 1984).

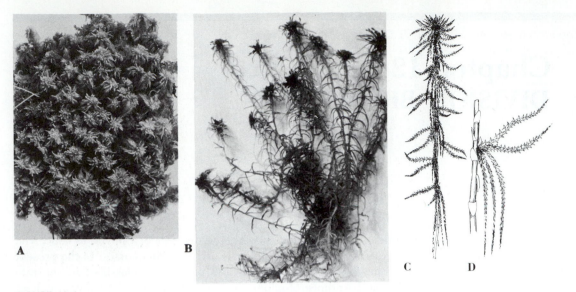

FIGURE 12-1
A, B. *Sphagnum* sp. **A.** Mat of plants *in situ.* **B.** Plants separated to show branching. **C, D.** *S. squarrosum.* **C.** Habit of one axis. **D.** Portion of main stem showing branches **A.** ×0.33. **B.** ×0.5. **C.** ×0.2. **D.** ×0.6. (**C, D.** *After Cavers*).

water below, so-called quaking bogs are formed. In certain parts of the world—Ireland, for example—the plant is gathered, dried, compressed, and used as fuel. Its antiseptic and highly absorptive qualities have been responsibe for its use in a number of circumstances. Among these may be cited its use as dressing for wounds, especially during wars, as packing material about the roots of living plants in transit, and as colloidal material for increasing the water-holding capacity of soils as well as acidifying them; it has also been used as wicks in lamps. Various species of *Sphagnum* have been shown, when tested, to hold as much as 16 to 26 times as much water as their dry weight. For example, a bale of dried *Sphagnum* weighing 20 lbs weighed 500 after it had been soaked. *Sphagnum* is important in northern latitudes in controlling and impeding drainage. Crum (1976) discussed this and many other uses of *Sphagnum*.

The bogs inhabited by *Sphagnum* are extremely acidic, which discourages growth of the bacteria and fungi that cause decay. Hence, the dead *Sphagnum* and other plants accumulate in thick deposits called *peat*. Examination of cores made from such deposits reveals through the pollen grains present at different depths the composition of the flora in a given location for thousands of years in the past; this indirectly provides information on changes in climate.

Peat is used as a soil conditioner because of its water-holding capacity and acidity and, ultimately, because it provides nutrients upon decay. Organisms that die in *Sphagnum* bogs may be preserved without decay for long periods. Thus, clothed bodies about 2000 years old have been recovered from peat bogs in western Denmark.

The individual plants of *Sphagnum* are closely matted together, but careful dissection reveals several kinds of branching (Fig. 12-1B–D). The

260

main axes are of unlimited growth and sometimes dichotomously branched. Each plant terminates in a dense tuft of apical branches, the **capitulum**. In addition, the stem bears other branches of two kinds. The ascendant branches are more or less horizontal and project outward from the main axes; the other branches are pendulous and usually twisted about the axes (Fig. 12-1B–D). The densely intertwined condition of the individual plants, the wicklike action of their pendulous branches, the overlapping leaves, and finally, the special cellular modifications of the latter all increase the water-holding capacity of the *Sphagnum* plant. Mature plants lack rhizoids, and all absorption takes place through the leaf and stem surfaces.

The development of the individual plant may be traced to a single apical cell at the tip of the stem (Fig. 12-2A). This cell is triangular in transverse section, and three rows of derivative cells are regularly produced. By further cell divisions, these give rise to the young leaves and tissues of the stem. The three-ranked origin of the leaves becomes obscured in older parts of the branches.

The stem of the main axes is composed of a central region surrounded by a cortex of one to five layers of hyaline cells (Fig. 12-2B); the cortex of the branches also is similar but usually has only one layer of hyaline cells. The hyaline cells are similar to those of the leaves (see below). The cortical cells are primarily water-storage cells. It is doubtful whether the cells of the central strand function efficiently in conduction. Instead, fluids are conducted by the wicklike branches and the numerous capillary surfaces in the densely interwoven plants.

The **leaf primordium** (L. *primordium*, beginning or origin) develops into a mature leaf through the activity of an apical cell, the derivatives of which arise as a result of cell division in two directions.

The leaves on the main axes are attached at some distance from each other and in many

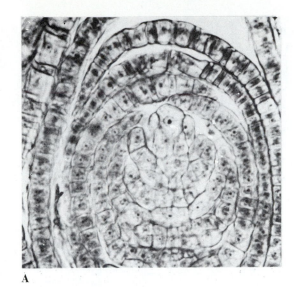

A

B

FIGURE 12-2
Sphagnum sp. **A.** Median longisection of stem apex; note apical cell and stages in leaf development. **B.** Transection of stem; note four peripheral layers of hyaline water-storage cells. **A.** ×330. **B.** ×120. (**B.** *After Cavers.*)

species differ morphologically from those of the branches, which are closely overlapping. The leaves on the branches that bear sex organs may differ in size from the vegetative leaves.

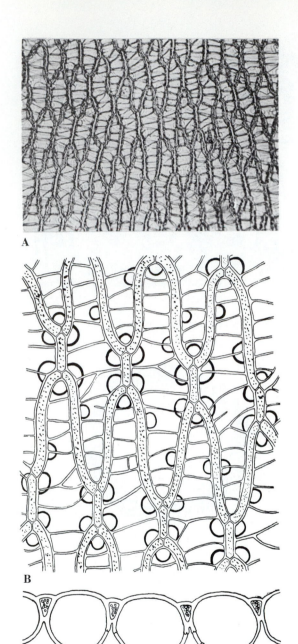

FIGURE 12-4
Sphagnum imbricatum. Partial dissection of ♂ branch showing antheridia. ×25. (*Courtesy of Professor Lewis Anderson.*)

FIGURE 12-3
A. *Sphagnum* sp. Leaf cell dimorphism. General view of surface. **B, C.** *S. acutifolium.*
A. Dimorphic leaf cells, enlarged; note hyaline water-storage cells surrounded by photosynthetic cells. **C.** Transverse section of leaf. **A.** ×137.
B, C. ×600. (**B, C.** *After Cavers.*)

The cells of young leaves are at first uniform in size and shape, but as development proceeds, cell division occurs in such a pattern that the mature leaf is composed of large, barrel-shaped, hyaline cells between which there are smaller, photosynthetic cells (Fig. 12-3). The colorless cells are nonliving at maturity, are in many species thickened with annular-helical[2] ridges (spiral fibrils), and frequently are perforated by circular or elliptical pores. These pores may be thickened around their periphery. Portions of the walls of the hyaline cells may be eroded or resorbed, in part or more extensively. They store large quantities of water. The smaller size of the photosynthetic cells and the abundance of the colorless water-storage cells account for the pale-green color of the mature plants.

Reproduction: the gametophyte. The leafy *Sphagnum* plant is sometimes called the **leafy gametophore** (Gr. *gamos*, marriage, + Gr. *phora*, bearer), since it bears the sex organs when mature. Some species are monoecious and others dioecious. The antheridia occur in short lateral branches (Fig. 12-4) near the apex of a main axis and are reddish or light purple.

---
[2]Usually incompletely helical.

The leaves of the branch are closely overlapping and suggest the antheridial branches of the leafy liverworts. The antheridia are borne on the stem in the axils of the leaves, which are said to be **perigonial** (Fig. 12-4, 12-5A). The antheridium is rather long-stalked, as in *Porella*, but differs from the antheridium of the Hepatopsida in that its development involves the divisions of an apical cell. The sperms of *Sphagnum* are biflagellate.

The archegonia (Fig. 12-5B–D) are also borne on very short lateral branches at the

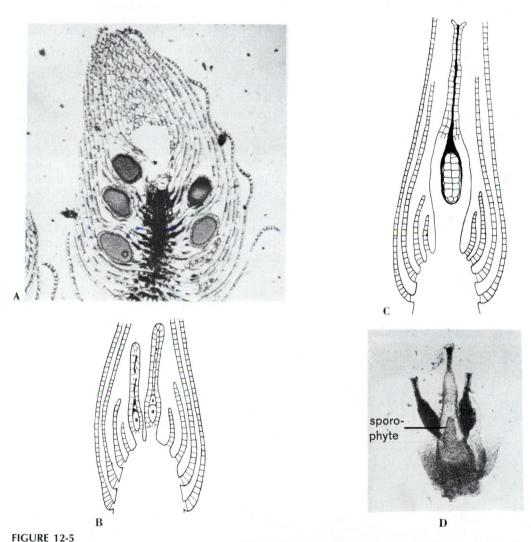

FIGURE 12-5
A. *Sphagnum palustre.* Longisection of antheridial branch; note stalked antheridia in leaf axils. **B, C.** *S. compactum.* **B.** Longisection of apex ♀ branch with two archegonia. **C.** Longisection of apex of ♀ branch; archegonium (calyptra) containing young sporophyte. **D.** *S. palustre.* Dissection of apex of living fertile ♀ branch; note calyptra with embryo in the center and two infertile archegonia. **A.** ×30. **B, C.** ×30. **D.** ×30. (**B, C.** *After Cavers.*)

263

center of the capitulum. The leaves which surround the archegonia are said to be **perichaetial**. The apical cells of these branches, as in acrogynous liverworts, ultimately give rise to an archegonium, so that increase in length of the branch ceases, and ontogeny of this (terminal) archegonium involves an apical cell. The two lateral archegonia do not have apical cells in their ontogeny. Mature archegonia are massive and stalked. Their necks are elongate and curved and are composed of five or six rows of neck cells. The eight or nine neck canal cells become disorganized as the archegonium matures.

Reproduction: the sporophyte. Fertilization has been infrequently observed in *Sphagnum* but seems to occur in the late autumn and winter in the eastern portion of the United States. At Highlands, North Carolina, for example, at an altitude of over 1300 m, the young sporophytes of *S. palustre* are already in the sporocyte stage early in May, but dehiscence of the capsule and dissemination of the spores do not occur until late June or early July. *Sphagnum squarrosum* and other species have mature sporophytes in July in northern Michigan, while those of *S. lescurii* mature in March

in central Texas. The zygote of only one of the archegonia usually develops into a sporophyte; the other two archegonia may persist for some time in association with the fertile one (Fig. 12-5C, D).

The first division of the zygote is transverse, and further divisions in the same direction result in the formation of a short chain of diploid cells. Approximately the upper half of this chain continues nuclear and cell division to form the capsule region of the sporophyte; an extremely short seta and haustorial foot develop from the lower cells. In the early stages of development, the foot and seta regions of the sporophyte exceed the precursor of the fertile region in size. The cells of the sporophyte contain actively photosynthetic chloroplasts throughout their development. The sporophyte remains covered by the calyptra and leaves of the gametophore until just before spore dissemination. An endothecium and amphithecium are differentiated in the upper portion of the sporophyte early in its development. The sporogenous tissue, except for that above the columella, arises from the innermost layer of the amphithecium and becomes four layers deep; it occupies a domelike position within the capsule (Fig. 12-6). The

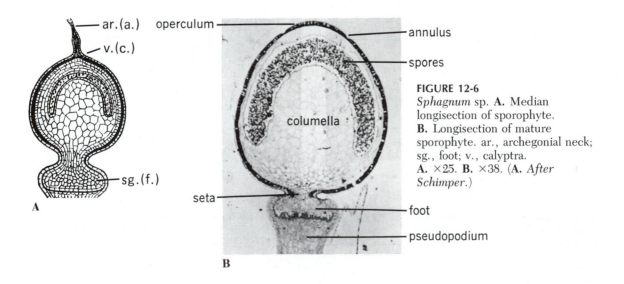

FIGURE 12-6
Sphagnum sp. **A.** Median longisection of sporophyte. **B.** Longisection of mature sporophyte. ar., archegonial neck; sg., foot; v., calyptra. **A.** ×25. **B.** ×38. (**A.** *After Schimper.*)

A   B

C

FIGURE 12-7
A–C. *Sphagnum squarrosum*. Sporophytes. **A, B.** Before dehiscence. **C.** After spore dispersal; note annulus and operculum in **B. A, B.** ×3.5. **C.** ×4.

central sterile tissue, which arises from the endothecium, is called the **columella**, as in *Anthoceros*. The sterile cells of the amphithecium are covered by an epidermis in which pairs of apparently nonfunctional guard cells develop. The basal portion of the sporophyte consists of a short seta region and an enlarged foot (Fig. 12-6).

All the potentially sporogenous cells undergo sporogenesis and form tetrads of spores. Meiosis is accomplished during these divisions, so that the spores are haploid. No elaters or other sterile cells, such as those observed in many Hepatophyta, occur in Bryophyta.

As the spores mature, they secrete a brown, sculptured wall (Brown et al., 1982), and the sterile tissues within the capsule become dehydrated. Meanwhile, the walls of the cells composing the outer part of the capsule thicken and also become brown. A circular layer of cells, the **annulus** (Figs. 12-6, 12-7B), near the apex of the capsule remains thin-walled and is torn when the upper portion of the capsule, the **operculum**, and spores are explosively shed (Fig. 12-7C). Just before this, the stem of the lateral branch that bears the sporophyte elongates rapidly (Fig. 12-7A–C), thus raising the entire

sporophyte above the gametophore. This elongated gametophytic stalk is called the **pseudopodium**. Its function is similar to that of the seta of the Hepatophyta and Bryophyta other than *Andreaea* and *Sphagnum*. The explosive discharge of the spores is audible, and they may be ejected for distances as great as 15 cm. This is brought about by expansion of air within the capsule, under pressure up to 2 atm (Ingold, 1974), and by rupture of the annulus.

The spores of *Sphagnum* can germinate immediately after being shed from the capsule. If they do so in crowded cultures, they form algalike filaments, which ultimately develop flattened, spatulate apices one layer of cells thick. A spore that is well separated from others forms a minute thallose structure, the **protonema** (Gr. *protos*, first, + Gr. *neme*, thread) (Fig. 12-8), very early in its development. The posterior marginal cells of this structure produce multicellular rhizoids that anchor it to the substratum. Each protonema ultimately gives rise to a single leafy shoot, or **gametophore** (Fig. 12-8), which develops from one basal, marginal protonematal cell. The first few leaves of the young gametophore lack the cellular dimorphism characteristic of *Sphagnum* leaves; the

265

A   B

FIGURE 12-8
Sphagnum lescurii. Products
of spore germination. **A.**
Young thallose protonema,
after 21 days. **B.** Protonema
with single young leafy shoot
after 48 days. ×100.

latter attribute appears gradually, beginning with the fourth or fifth leaf. The multicellular rhizoidal branches of the protonema frequently give rise to secondary protonemata from their apical cells.

Summary

In summarizing the morphology and life cycle of the peat moss, *Sphagnum,* a number of noteworthy features may be emphasized. Although the mature plants (leafy gametophores) lack rhizoids and specialized conducting tissue, they attain a stature that exceeds that of most Hepatophyta and rivals that of the largest mosses, which have well-developed rhizoids and supporting and conducting tissues. This probably is effected by their growth in dense mats in boggy soil or water, where the individual plants are able to furnish each other with mutual mechanical support. The numerous adaptations for storage and conduction of water are significant. Among them may be cited the matted growth of the leafy gametophores, the overlapping of the leaves and branches, the presence and wicklike action of the pendulous branches, and the occurrence of special water-storage cells in the leaves and cortex of the stems.

The stalked antheridia suggest those of leafy liverworts; the long-necked stalked archegonia are more massive than those of other Bryophyta. The activity of an apical cell in the formation of the antheridia and one[3] of the archegonia is characteristic of the Bryophyta and absent in the Hepatophyta, except in acrogynous species such as *Porella.*

The structure and dehiscence of the sporophytic capsule are strikingly different from those of the Hepatophyta and Anthocerotophyta. It should be noted that relatively less sporogenous tissue develops in the capsule of *Sphagnum* than in that of the Hepatophyta, and that sporogenesis is simultaneous, unlike that in *Anthoceros* and *Phaeoceros.* As in the latter, however, the central region of the capsule is occupied by a sterile columella which involves the entire endothecium, the sporogenous tissue of *Sphagnum* also arising from the innermost layer of the amphithecium, except for the portion over the columella. The entire sporophyte is raised by elongation of the gametophytic pseudopodium, and the spores are explosively discharged. Eddy (1978) stated that there is no evidence to link *Sphagnum* with other bryophytes.

---

[3]In *Sphagnum.*

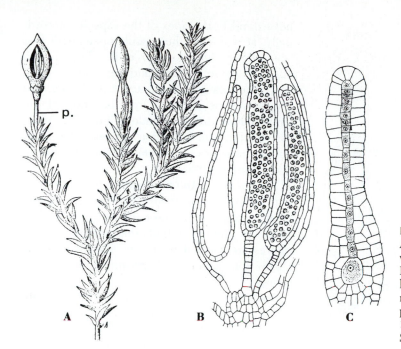

FIGURE 12-9
Andreaea rupestris. **A.** Plant
with maturing sporophytes.
B. Antheridia. **C.** Median
longisection of an almost-
mature archegonium. p.,
pseudopodium. **A.** ×3. **B.**
×180. **C.** ×350. (*After G. M.
Smith.*)

CLASS 2. ANDREAEOPSIDA

Andreaea

This class of mosses contains only a single order and family and only two genera, of which one (*Neuroloma*) is monotypic. The other, *Andreaea* (after J. G. R. Andreae) (Fig. 12-9), contains more than 100 species, of which some are known as "granite mosses." These plants are abundant in cold regions on soil but do occur in the temperate zone at higher altitudes on siliceous rocks.

The plants are blackish green, sympodially branched, and anchored to the substrate by rhizoids. Growth is apical, and the crowded leaves, which have thickened cell walls and may lack midribs, are borne in three rows. The leaf cells are rich in oils. The stems lack a central strand and their cells are all thick-walled.

The various species are monoecious or dioecious, but in the former case, as in *Funaria* (Fig. 12-18), the archegonia and antheridia (Fig. 12-9B, C) are borne at the tips of separate branches. The digitate antheridia are long-stalked and the archegonia massive. **Paraphyses** are present among both the antheridia and archegonia.

An apical cell functions in development of the sporophyte. An absorbing foot that penetrates into the gametophore is also formed.

The sporogenous tissue arises from the outermost layer of the endothecium, while the remainder of the structure forms a sterile, elongate, central columella (Figs. 12-10A, 12-11C); the sporogenous tissue overarches the columella. No air chambers are present between the sporogenous tissue and capsule wall as they are in most Bryopsida. The amphithecium is four or more layers thick and forms the capsule wall. Until late in development, the tissues are rich in chloroplasts. When sporogenesis has been completed, the surface cells of the capsule develop thick walls, except for four vertical strips of cells, the **dehiscence lines**, which, however, do

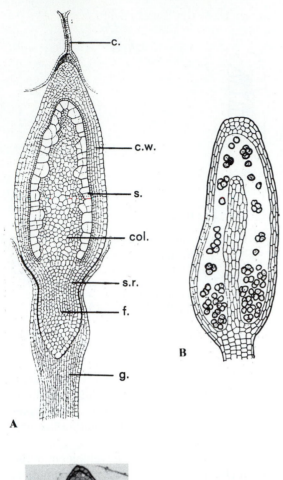

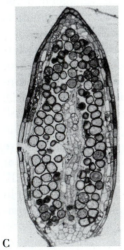

not extend to the apex of the capsule. Accordingly, at maturity, the capsule opens with four (sometimes eight to ten) clefts (Fig. 12-9A, 12-10D). An operculum and peristome are absent. The seta is rudimentary in *Andreaea*, and the capsule is elevated by elongation of the gametophore axis as pseudopodium. The process of spore wall formation has been described on the basis of electron microscopy by Brown and Lemmon (1984).

The spores of *Andreaea* develop into branching, straplike or platelike protonemata somewhat suggestive of those of *Sphagnum* (Fig. 12-11). A number of buds that initiate gametophores are formed on these. Consult Nishida (1978) for patterns of germination of moss spores.

Andreaea has attributes suggestive of both the Sphagnopsida and Bryopsida. Among the former may be cited the flattened protonema, the elevation of the capsule on a pseudopodium, the absence of a peristome, the long-stalked antheridia, and the overarching of the columella by the sporogenous tissues. Characteristics that suggest the Bryopsida are the gross morphology and the origin of the sporogenous tract from the endothecium. The partial dehiscence of the capsule by four clefts is unique among the Bryophyta and unknown in the Sphagnopsida and Bryopsida.

FIGURE 12-10
A. *Andreaea* sp. Median longisection of immature sporophyte. **B.** *A. rupestris.* Median longisection of maturing capsule. **C, D.** *A. rothii.* **C.** Longisection of capsule. **D.** Mature, dehisced capsule. c., calyptra; c.w., capsule wall; col., columella; f., foot; g., gametophore; p., pseudopodium; s., sporogenous tract; s.r., seta region. **A.** ×100. **B.** ×105. **C.** ×125. **D.** ×10. (**A.** *After Kühn, from Bower.* **B.** *After G. M. Smith.* **C.** *After Brown and Lemmon.* **D.** *After Flowers.*)

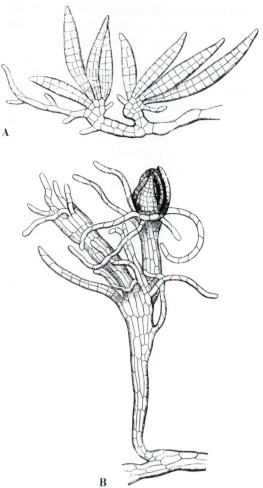

FIGURE 12-11
Andreaea sp. **A.** *A. blytii.* Leaflike branches on prostrate protonema. **B.** *A. petrophila.* Erect branches on protonema with one leafy shoot. (**A.** *After Berggren, from Ruhland.* **B.** *After Kühn, from Ruhland.*)

CLASS 3. BRYOPSIDA

Habitat.   The Bryopsida, the so-called true mosses, include the largest number of genera and species among the Bryophyta. In spite of their abundance, they display a remarkable

uniformity in structure and life cycle. While a number of representative genera (*Funaria, Polytrichum, Weissia, Mnium,* and *Physcomitrium*) have been selected to illustrate certain aspects of the Bryopsida, *Funaria* and *Polytrichum* will be emphasized in the following account. *Polytrichum* and other members of its family have been summarized by G. L. Smith (1971).

Although *Sphagnum* is relatively limited in distribution, being confined to markedly acidic, aquatic, or boggy habitats, representatives of the Bryopsida may be collected from xeric, mesic, and hydric environments. The great majority, however, live under moderately moist conditions rather than in extremely wet habitats. *Fissidens, Amblystegium,* and *Fontinalis* are often submerged in small streams. *Mnium* and certain species of *Bryum* are mosses of very moist substrata but are not submerged. *Orthotrichum* and *Grimmia,* on the other hand, are examples of the numerous xerophytic mosses. Terrestrial species grow on various substrata such as rock, tree bark, wood, and moist soil. Members of the group are often pioneers on freshly exposed, bare soil surfaces, where they rapidly carpet the substratum with their filamentous, branching protonemata. In this connection they are no doubt of considerable importance in preventing incipient erosion. Most mosses are perennial, forming increasingly dense mats each year on a given substratum. A few are annuals, frequently developing in the fall or winter in milder climates.

Several mosses grow in unusual habitats; for example species of *Splachnum* (p. 299) grow on dung, while *Schistostega* (p. 300) occurs in caves. Certain mosses, such as *Grimmia,* are pioneers on bare rocks, which they can colonize by vegetative propagation of fragments. *Physcomitrium* and *Funaria* (Fig. 12-29, 12-38) are among the first pioneers to recolonize burned-over soil. Mosses are used in boreal regions for "chinking" log structures (Leurs, 1981).

Xerophytic mosses are extremely resistant to desiccation and have been revived after dry storage for periods between 8 months and 19 years (Crum, 1976, p. 138; and Richardson, 1981). Mosses of drier habitats have proportionately thick cell walls as compared with the area of their protoplasts (Proctor, 1979).

ORGANIZATION AND DEVELOPMENT

Although a number of genera of mosses are cited in this section, *Funaria* (Order Funariales) and *Polytrichum* (Order Polytrichales) will be emphasized because of their widespread occurrence, and, often, ready availability, and because *Funaria* may readily be grown and maintained in culture in the laboratory (Dietert, 1977).

As in the genus *Sphagnum*, the gametophyte of the Bryopsida is represented in the life cycle by two phases, the protonema and the leafy gametophore. In most genera the protonema is a uniseriate, branching filament (Figs. 12-12, 12-13), in contrast to the spatulate protonema of *Sphagnum*. The protonema, in most mosses, ultimately produces a number of leafy shoots, which are often known as **leafy gametophores**

because they produce sex organs at maturity. In a few mosses, such as *Buxbaumia* and a few species of *Pogonatum*, the protonematal stage is long-lived and persistent and the main site of photosynthesis; the leafy gametophoric phase in such genera is reduced and consists of as few as one leaf borne on the protonema (*Buxbaumia*, male plants) (Fig. 12-45B). The zygote develops into a sporophyte borne on the female gametophores (leafy stem). The complexity of the sporophyte in most genera in the Bryopsida is unparalleled among Hepatophyta, Anthocerotophyta, and other Bryophyta.

Because of the large number of available genera, which furnish suitable material for the study of the moss life cycle, a number of them will be referred to in the following account. Moss spores shed from the capsule of a mature sporophyte germinate promptly if they are carried to a suitable environment. In a few mosses, the spores have already undergone endogenous divisions before they are shed (Schofield, 1981). Paolillo and Jagels (1969) and Paolillo and Payne (1970) investigated photosynthesis, respiration, and CO_2 exchange in germinating spores of *Polytrichum juniperinum* and *P. ohiense*. The first authors found that respiration increased rapidly and reached a maximum in 10 minutes. Photosynthesis increased to a maximum at

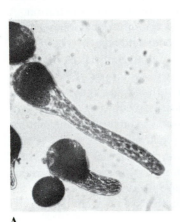

A

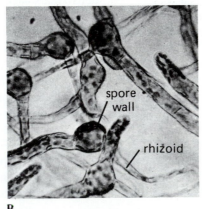

B

spore wall

rhizoid

FIGURE 12-12
Funaria hygrometrica. Spore germination. **A.** 48 hours after spores were planted. **B.** At 96 hours. **A.** ×80. **B.** ×60.

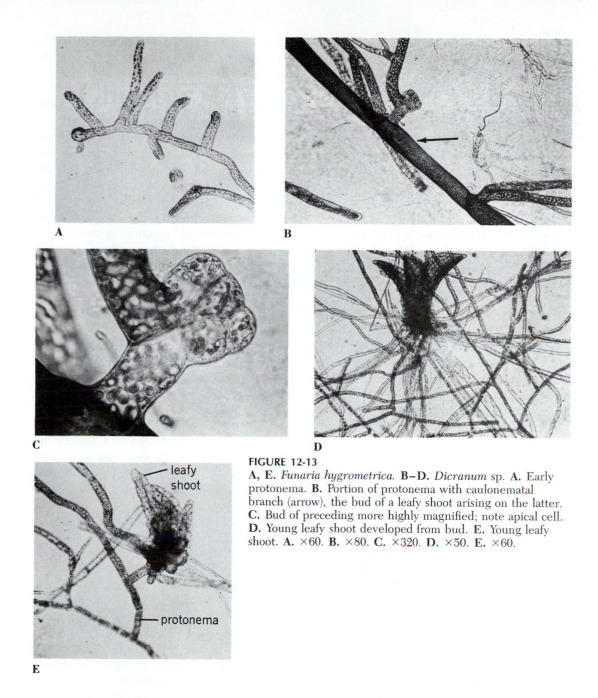

A

B

C

D

FIGURE 12-13

A, E. *Funaria hygrometrica.* **B–D.** *Dicranum* sp. **A.** Early protonema. **B.** Portion of protonema with caulonematal branch (arrow), the bud of a leafy shoot arising on the latter. **C.** Bud of preceding more highly magnified; note apical cell. **D.** Young leafy shoot developed from bud. **E.** Young leafy shoot. **A.** ×60. **B.** ×80. **C.** ×320. **D.** ×50. **E.** ×60.

leafy
shoot

protonema

E

about 80 minutes in continuous light, after the spores had been hydrated.

Dietert (1977) reported that spores from freshly matured sporophytes of *Funaria* (L. *funarius*, pertaining to a cord or rope)[4] germinated within 24 hours. Spores from older capsules require 72 hours to a week for germination. Their germination and subsequent development may be followed readily in laboratory cultures in which the spores have been sown on agar containing inorganic salts. Kass and Paolillo (1974*a,b*) and Paolillo and Kass (1977) reported that the number of chloroplasts in germinating spores and in the antheridial jacket is directly correlated with cell size.

In *Funaria hygrometrica* each spore is covered with a brown outer wall, which ruptures (Fig. 12-12A) as a result of the swelling of the spore protoplast and inner spore wall. The spores of many strains of *Funaria* will not germinate unless they are illuminated. However, they will germinate in darkness if provided with glucose, but under such conditions the protonema is almost colorless. Electron microscopy and cytochemical tests reveal that in germination, even before the spore wall cracks, active metabolic processes are in progress (Monroe, 1968). In at least one moss, a species of *Drummondia sinensis*, the spores germinate before they are shed from the capsule (Nishida, 1978).

At germination, the spore protoplast may protrude at both poles of the spore. The germ tube enlarges rapidly and soon gives rise to a branching filamentous system (Figs. 12-12B, 12-13) as a result of repeated nuclear and cell division.

Growth of the protonema is apical, that is, localized in the terminal cells of the branching filaments. Its cells are rich in lens-shaped chloroplasts. The protonema may be readily distinguished from terrestrial green algae by the

discrete, lenslike plastids, together with the oblique position of the end wall of some of its component cells. The protonema from a single spore of *F. hygrometrica* has been shown to cover an area 16 inches in diameter within several months; it had been transferred to fresh, agarized culture medium several times during that period.

Some branches of the protonema are superficial, whereas others penetrate the substratum, often developing brown walls. Transitional types between subterranean and surface-growing protonematal branches are common.

The development and physiology of protonema have been intensively studied in *Funaria* by a number of investigators (Schmiedel and Schnepf, 1979*a, b*). Among the latter, Bopp (1963, 1976) has shown that there are two phases in protonema development, the **chloronema** stage and the later **caulonema**; these occur in other mosses (Fig. 12-13B). In the former, which develops first from the spore, the cell walls are colorless, the chloroplasts are lenticular, and the transverse walls are for the most part perpendicular to the longitudinal walls. The caulonema develops from the chloronema, has brownish walls (the transverse ones oblique), contains spindle-shaped plastids, and grows on, or just below, the surface of the substrate.

After a period of growth and vegetative activity, certain cells of the caulonema undergo nuclear and cell divisions in which a tetrahedral apical cell is differentiated (Fig. 12-13C); this gives rise to three series of derivative cells. These apical cells and their derivatives which may occur on both surface and subterranean branches of the protonema, soon are organized as minute buds (Fig. 12-13B, C). This occurs about 14 days after spore germination in *Funaria* under laboratory conditions.

With the continued activity of the apical cell, each bud may form a young leafy gametophore (Figs. 12-13D, 12-14). In laboratory cultures at 22°C, on inorganic salt-agar medium illuminat-

---

[4]The name refers to the twisted dry setae. *Funaria* is also called "Cinderella" because of its occurrence in ashes and burned-over soil.

FIGURE 12-14
A, B. *Funaria hygrometrica.* **A.** Group of leafy shoots on protonema. **B.** Single older leafy shoot attached to protonema. **C.** *Polytrichum commune:* young leafy shoots on protonema. **A.** ×20. **B.** ×12.5. **C.** ×6.

ed at 150 ft-c intensity, leafy (shoot) gametophore formation is usually initiated within 30 days after spore germination.

In *Physcomitrium turbinatum*, it has been demonstrated (Nebel and Naylor, 1968*a,b*) that the initiation of buds is not directly dependent on the extent of protonema development. Instead, the total dosage of light is the limiting factor in bud formation, no matter how intermittent the light. It has been suggested that a morphogenetic substance that evokes bud formation is made in correlation with total light, and that this substance is stored up during periods of darkness. Addition of sugar to the medium speeded up the formation of buds. Formation of buds was initiated on transplanted protonemata within 19 days with a light intensity of 30 ft-c; but at a light intensity of 715 ft-c, buds started to form in 5 days.

Bopp (1963) and Bopp et al. (1977) have found that the chloronema and caulonema of *Funaria* both produce a diffusible substance, called factors F and H, respectively. Factor F, from the chloronema, is thermolabile and stimulates growth and development of the caulonema. Factor H is thermostable and stimulates production of buds on the caulonema, at the same time inhibiting its growth. Bud formation depends on a proper balance of factors F and H. Both factors F and H diffuse into the culture medium, can be removed from it, and show their effects on other protonemata. Bud formation can also be evoked by application of gibberellic acid, bacterial extracts, and a cytokinin (benzyladeninebenzl-7-^{14}C), the effect of the last demonstrated by Brandes and Kende (1968) and Nehlsen (1979).

Spiess, Lippincott, and Lippincott (1971) found that the addition of a virulent strain of *Agrobacterium tumefaciens* to cultures of moss protonema (*Pylaisiella seluryii*) markedly hastened the production of buds and the total number produced, as compared with axenic controls (Spiess et al., 1971). Later it was shown that 35 strains of *Agrobacterium* and 5 of *Rhizobium* and the virulent strains of the former and all species of the latter induced bud formation on the protonema. Some required contact with the latter, but in others the inducer was diffusible (Spiess et al., 1977, 1983).

The protonema may be filamentous at the outset and later give rise to a thallose structure from which the leafy shoots arise, as in *Tetraphis pellucida* (Schneider and Sharp, 1962).

A protonema arising from a single spore can in most mosses produce a large number of leafy gametophores[5] rather than one, as is produced by *Sphagnum*; this accounts for the densely colonial growth of moss plants. The protonematal branches function as absorbing organs for the very young gametophores. As the latter increase in age and stature, slender rhizoids, which are much like subterranean protonematal branches, arise from the bases of the stems (Fig. 12-14B). Moss gametophores usually grow in cushions, turfs or mats; a few are dendroid.

Continued development of the gametophore is effected by the activity of the apical cell with three cutting faces and the derivatives therefrom. In a majority of mosses, the leaves in young plants are arranged in three rows, the result of the order in which the derivatives are cut off from the apical cell of the stem. In older plants, the three-ranked leaf arrangement is

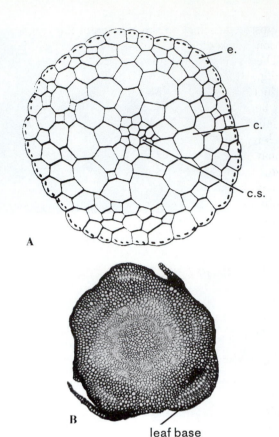

FIGURE 12-15
A. Transection of stem of *Funaria* sp.
B. Transection of stem of *Polytrichum* sp. c., cortex; c.s., central strand; e., epidermis.
A. ×60. **B.** ×130. (**A.** *After Parihar.*)

usually disturbed. In a few mosses—*Fissidens* (Fig. 12-36) for example—the occurrence of leaves in two ranks is correlated with the presence of an apical cell from which only two rows of derivatives are produced.

Stems of many mosses (Fig. 12-15) are differentiated into three regions: a superficial epidermal layer, a thick cortex, and a central strand. Branches in mosses arise below leaves, not in their axils (Schofield, 1981).

Hébant (1977, 1979) summarized our knowledge of the structure and function of the conducting tissues of the stems of liverworts and

---

[5]For a more comprehensive discussion of the moss gametophore see Schofield and Hébant (1984).

mosses. Much of our knowledge in this connection is based on his own incisive investigations. The stems of some mosses (and a few liverworts, e.g., *Pallavicinia*, p. 223) have a central conducting strand. By the latter is meant a region where nonliving, empty, water-conducting cells, called **hydroids**, are present. Sieve-element-like, food-conducting cells, **leptoids**, may also be present (as in *Polytrichum, Dawsonia*, and related mosses). A conducting strand may be present both in the axes of the gametophytes (gametophore) and in the sporophyte (in the seta).

Leptoids are elongate cells with inclined terminal walls and thickened lateral walls. Plasmodesmata pass through the terminal walls of leptoids and may be surrounded by pores as they thicken. The nucleus of the leptoid ultimately disappears. In *Polytrichum* (p. 276), the slanting, terminal walls of hydroids are markedly hydrolyzed, which probably enhances their permeability (Hébant and Johnson, 1976). The lateral walls are thickened with polyphenolic polymers (rather than with lignin as in vascular plants), which renders them less permeable (Hébant, 1977) at maturity. Hydroids, like the tracheids of vascular plants, are nonliving cells. Stevenson (1974) considers the thickened leptoids of *Atrichum undulatum* entirely similar to sieve elements of vascular plants.

Scheirer and Goldklang (1977) presented evidence that the hydroids are pathways of conduction of water and solutes, although external conduction occurs in nature in many mosses (and liverworts). Evidence for conduction of organic substances by the leptoids and associated parenchyma cells is not as clear-cut, and further investigations of translocation in bryophytes are greatly to be desired. For a more complete discussion of the organization and functioning of the conducting system of bryophytes, refer to Hébant (1977, 1979) and to Schofield and Hébant (1984).

From the central conducting cells of the stem, branches of conducting cells connect with the leaves, but the connection of the traces to the conducting tissues of the stem is not extensive (Eschrich and Steiner, 1967, 1968). However, in a related, large moss, *Dawsonia intermedia*, when the distal portions of the leafy gametophores were exposed to $^{14}CO_2$, no evidence for the transport of labeled substances was obtained, although slower movement, by diffusion, is not precluded (Thrower, 1964).

Eschrich and Steiner (1967), however, reported that C^{14}-labeled $NaHCO_3$ placed on a leaf of *Polytrichum commune* was transported upward through the leptoids at the rate of 32 cm/hour. There is evidence that sugars move through leptoids (Reinhart and Thomas, 1981).

With respect to the water translocation in some mosses, experimental evidence indicates there is little internal movement of water through the tissues of the stem, so that the thin-walled cells of the central strand are not efficient conducting cells. On the contrary, available evidence indicates that most water moves over and is absorbed by the external surfaces of these moss plants. The prime source of water used by mosses is rain, which is held with great avidity by the densely matted plants in small capillary spaces between leaves and stem and within rolled leaves. Experimental analyses make it doubtful that absorption of water vapor is adequate to keep mosses turgid under mesophytic conditions (Klepper, 1963).

Development of the leaves in the Bryopsida may be traced to the activity of an apical cell. The young leaf is composed of only a single layer of cells, but in most genera the central region becomes thicker, forming a midrib. This consists of layers of elongate cells, some of which are thickened and function in support. *Funaria* is typical of the mosses in which leaf structure is relatively simple (Fig. 12-16A). However, it has been shown that the leaf consists of five kinds of cells: upper and lower epidermis, hydroids, leptoids and sterids (supporting cells) (Wiencke and Schulz, 1983). *Polytrichum* and *Atrichum* leaves exhibit considerable complexity (Fig.

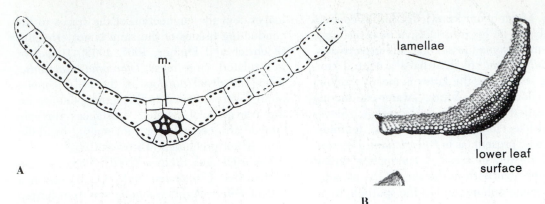

A

lamellae

lower leaf
surface

B

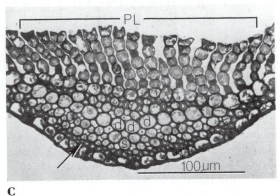

C

FIGURE 12-16
A. Transection of young leaf of *Funaria* sp. **B, C.**
Transections of leaf of *Polytrichum.* m., midrib. **A.**
×250. **B.** ×125. **C.** ×450. (**A.** *After Parihar.* **C.**
After Scheirer.)

12-16B, C) (Scheirer, 1983). In these, the many-layered midrib is considerably expanded and sclerotic. From its upper surface arise a number of parallel lamellae of thin-walled photosynthetic cells, which are separated from each other by narrow fissures. These are capillary spaces in which water is stored after rains. The photosynthetic cells are protected from desiccation by inrolling of the sclerotic leaf margins.

The mature leafy plants of most Bryopsida are anchored to the substratum by systems of multicellular rhizoids with oblique terminal walls. As noted above, the rhizoidal system of the young plant is composed largely of protonematal branches. Secondary rhizoids may arise from the superficial cells of the stems and leaf bases. In some genera (*Polytrichum*) they become

twisted into ropelike masses. Older rhizoids are usually brown-walled and contain few chloroplasts. It has been demonstrated in many mosses that such environmental stimuli as wounding may evoke the production of protonematal branches from almost any portion of the plant body, including the stem, leaves, and even parts of the sporophyte. The rhizoids may give rise to secondary protonemata and ultimately to young gametophores.

Reproduction: the leafy gametophyte. Vegetative reproduction—that is, propagation of the individual from fragments of leaves, stems, and segments of protonema—is widespread in mosses. Miller and Ambrose (1976) found as many as 4000 viable propagules of mosses (mostly leafy shoot tips) and liverworts present per cubic centimeter of granular snow in the Canadian Arctic. The regenerative potentialities of mosses are phenomenal, so that fragments of various organs, including paraphyses and the sterile portions of the sex organs themselves (see below), give rise to protonemata and, ultimately, to leafy gametophores when conditions are favorable. In addition, some species produce special agents of propagation, **gemmae**, which replicate the gametophytic phase.

Bryophyta and Hepatophyta have remarkable powers of regeneration. Small fragments (even single cells, in some cases) of both liverworts and mosses can regenerate to form new plants. It was first demonstrated in 1876 that fragments of sporophytic setae could regenerate diploid protonemata, an example of apospory that favors the homologous theory of alternation of generations (p. 309). From such protonemata, diploid gametophores matured and gave rise to tetraploid sporophytes.

With respect to regeneration, not only leaves, stems, rhizoids, and setae but also paraphyses, archegonial neck and venter cells, and antheridial stalk and jacket cells, even single cells, develop protonemata in culture, and these ultimately produce leafy shoots. However, all attempts to achieve regeneration from the egg and spermatogenous tissue have so far met with failure.

It is of interest in this connection that in mosses such as *Funaria*, in which the bisexual individuals produce archegonia and antheridia on different branches (Fig. 12-17), regeneration of either the male or female branches gives rise to bisexual individuals. However, in another genus, *Splachnum*, with bisexual individuals, Bauer (1963) has found cases in which a stable male race can be developed by regeneration from a male branch.

The leafy plants of most mosses have been observed to produce sex organs at maturity (Ramsay and Berrie, 1982). In many mosses— *Funaria* and *Polytrichum*, for example—the leafy shoots are erect, unbranched or sparingly branched, and the sex organs are borne at the apices of the leafy shoots (Fig. 12-17). Such mosses are said to be **acrocarpous**; this results, of course, in the sporophytes' occurring at the tips of the leafy shoots. Many other mosses have

prostrate, richly branched leafy gametophores (Fig. 12-44), and in these the sex organs develop at the tips of the lateral branches; such mosses are designated **pleurocarpous**. In some pleurocarpous mosses the leaves are dimorphic, those on the branches differing somewhat from those on the main axes. In some the underleaves are smaller than the others and suggest the underleaves of certain Jungermannialean liverworts.

Wyatt (1977) investigated the proximity of male and female patches of the dioecious moss *Atrichum angustatum*. He reported that extensive colonies or patches of exclusively male or female plants resulted from vegetative propagation of clones, rather than from a special pattern of spore dissemination. Numbers of male and female shoots at one site in nature were approximately equal. Wyatt determined that the gametes (sperm) were dispersed to a maximum distance of 11.00 cm.

Polytrichum and *Atrichum* are dioecious, and this is manifested by the dimorphism of the male and female plants (Fig. 12-17A, B). Plants of *Funaria*, although monoecious, produce antheridia and archegonia in separate branches of the same plant (Figs. 12-17D, E; 12-18A, B). *Funaria* is protandrous, and the male branch at first overshadows the female, which is a lateral branch of it (Figs. 12-17D, E, 12-18A, B). In bacteria-free laboratory cultures of *Funaria*, initiation of sex organs in experimental laboratory cultures occurred at temperatures of 10°C or below, when the cultures were illuminated between 6 and 20 hours daily (Monroe, 1965).

Newton (1972) reported that an endogenous rhythm of sexual maturation persisted in greenhouse-grown plants of *Mnium*. This rhythm could be modified, to some extent, by varying temperature and photoperiod. Dietert (1980) concluded that temperature, rather than day length, was a critical factor affecting production of gametangia in *Funaria*, the gametangia developing in laboratory cultures at temperatures between 5 and 15°C. She reported, however, that populations from a greenhouse

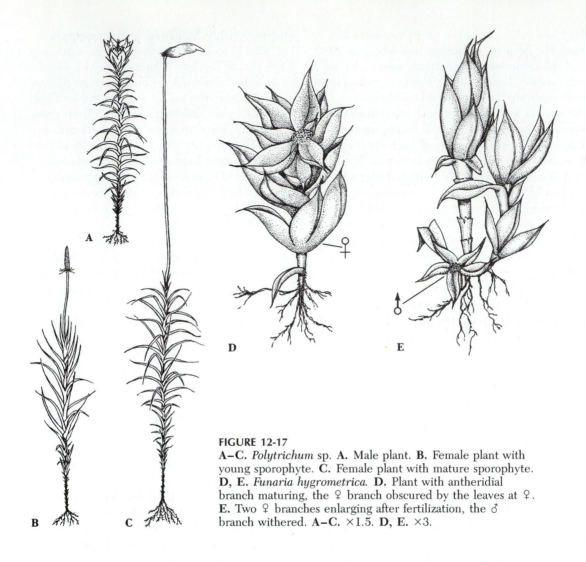

FIGURE 12-17
A–C. *Polytrichum* sp. **A.** Male plant. **B.** Female plant with
young sporophyte. **C.** Female plant with mature sporophyte.
D, E. *Funaria hygrometrica.* **D.** Plant with antheridial
branch maturing, the ♀ branch obscured by the leaves at ♀.
E. Two ♀ branches enlarging after fertilization, the ♂
branch withered. **A–C.** ×1.5. **D, E.** ×3.

formed gametangia at 20°C. The leafy gametophores required from 6 to 16 weeks after spore
germination to reach sexual maturity; the variation correlated with the source of the spores and
ambient temperature. It has been reported that
Polytrichum, on the other hand, while also not
strongly affected by photoperiod, forms gametangia best at 21°C. For a review of factors
affecting gametangial formation see Chopra and
Bhatla (1981, 1983).

In certain species of *Mnium*, antheridia and
archegonia occur in the same group at the stem
apex. Sexual, as distinct from vegetative, apices
in mosses may often be recognized by the
occurrence of somewhat modified leaves about
the sex organs. The apices of male individuals or
branches (Fig. 12-17A, D) appear to be cuplike
because of their closely packed leaves, which
may be red or purple. Certain identification of
branches bearing archegonia is more difficult; it
is best made by periodic study of apparently
vegetative apices of a given species at a time

278

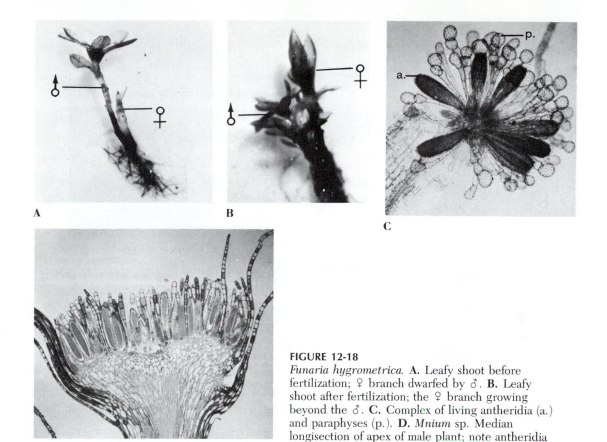

FIGURE 12-18
Funaria hygrometrica. **A.** Leafy shoot before fertilization; ♀ branch dwarfed by ♂. **B.** Leafy shoot after fertilization; the ♀ branch growing beyond the ♂. **C.** Complex of living antheridia (a.) and paraphyses (p.). **D.** *Mnium* sp. Median longisection of apex of male plant; note antheridia and paraphyses. **A.** ×6. **B.** ×12. **C.** ×60. **D.** ×30.

when some individuals of the same species bear antheridia. Periods when sex organs are present differ in various moss species, in different seasons, and in different latitudes. After they have produced sex organs and sporophytes, new vegetative shoots may proliferate through the old sexual apices, or new branches may arise below the apices that have borne sporophytes.

In some mosses, as in certain species of the alga *Oedogonium*, there is a marked dimorphism of the male and female plants, the male being much smaller and, in some cases, borne on the leaves of the female plant or just below the leaves of the female plant (Fig. 12-19) that surround the archegonia. The dwarf males die soon after their sperms have been shed. Vitt

(1968) summarized the various patterns of location of sex organs in mosses. In some species the dwarf males are reported to develop from the leaves or rhizoids of the female plants. There is evidence in other species, however, that the dwarf males grow from spores. Thus, in *Macromitrium blumei* and *M. braunii* (Fig. 12-19), the spores are of two sizes; the larger develop into female plants and the smaller into male plants located on the leaves of the female. A similar situation is suggested in the location of the male plants in *Trismegistia brauniana*. The difference in spore size in *M. blumei* and their behavior at germination suggest the heterospory of certain vascular plants (p. 356), but Vitt (1968) correctly pointed out that the spores of differing size

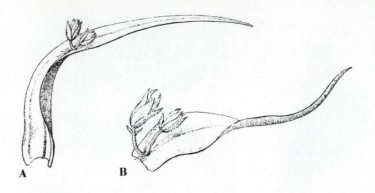

C

D

FIGURE 12-19
Dwarf ♂ plants in mosses. **A.** *Macromitrium braunii.*
B. *M. javanicaum,* dwarf ♂ plants borne on leaves.
C. *Holomitrium* sp., dwarf ♂ in hairy filaments (tomentum)
at base of ♀ branch. **D.** Dwarf ♂ in similar location in
Dicranum spurium. **C.** ×3. **D.** ×2. (**A, B.** *After Brotherus.*
C. *Courtesy of Dr. Dale Vitt.* **D.** *Courtesy of Dr. Lewis
Anderson.*)

occur within the same sporangium (capsule) in
mosses, rather than in separate ones as they do
in true heterospory (Fig. 15-28) (see also Ramsay, 1979). Furthermore, in some species of
Macromitrium with dwarf males the spores are
uniform in size. Accordingly, he designates the
production of different-sized spores in mosses as
anisospory, in contrast to heterospory. Loveland (1956) made an intensive investigation of
the dwarf males in six species of *Dicranum*,
which he found could be identified on the
differing characteristics of the dwarf males. He
found that the dwarf males grew only on the
female plants. Spores sown in culture failed to
produce sexually mature plants, but those planted on female plants produced fertile dwarf
males. Dwarf males, epiphytic near the tips of
the female gametophyte, occur in *Dicranum
spurium.* Dr. Lewis Anderson has shown (personal communication) that this species has the
XY system of sex determination. Figure 12-20

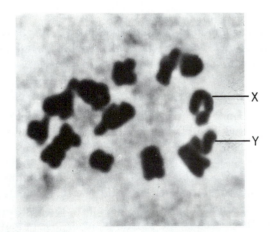

FIGURE 12-20
Dicranum spurium: paired meiotic chromosomes
from a sporocyte; note sex chromosomes (labeled X
and Y). ×4600. (*Courtesy of Dr. Lewis Anderson.*)

280

a'.

a.

B

FIGURE 12-21
A. *Funaria hygrometrica.* Three mature archegonia just after fertilization. **B.** *Mnium* sp. Longisection of apex of ♀ plant showing archegonia (a., a'.) and paraphyses. **A.** ×120. **B.** ×60.

A

illustrates a preparation of the meiotic chromosomes in which the X and Y chromosomes are labeled. X and Y sex chromosomes occur in certain species of *Macromitrium*, which is also anisosporous. Ramsay (1966, 1979) summarized sex determination in certain liverworts and mosses and cited references to their sexual dimorphism.

Development of both the antheridia and archegonia in the Bryopsida involves the activity of apical cells. Sterile hairlike or bulbous filaments and modified leaves, all called **paraphyses**, occur among the sex organs of the Bryopsida (Fig. 12-18C, D). It has been suggested that the paraphyses function in preventing drying of the sex organs by increasing the surface on which capillary water may be held. The antheridia (Fig. 12-18C, D) and archegonia (Fig. 12-21) are massive, visible to the unaided eye in many cases, and always considerably larger than those of the Hepatophyta and Anthocerotophyta. Both types of sex organs are

stalked. The archegonia have extremely long, often twisted, necks composed of six vertical rows of neck cells that enclose a correspondingly long series of neck canal cells. These and the ventral canal cell disintegrate when the archegonium is mature, thus providing an unobstructed passageway to the egg. The latter secretes mucilage into the venter cavity (Lal et al., 1982).

In many mosses the chloroplasts of the antheridial jacket cells are transformed into chromoplasts when the antheridia are mature; hence, ripe antheridia may be recognized by their orange-red color. The spermatogenous cells within the jacket layer divide repeatedly, forming a columnar mass of rather minute cubical cells. At the conclusion of these divisions, the protoplast of each minute cell becomes organized as a biflagellate sperm.

Paolillo (1977*b*) reported, as did earlier investigators, that the release of sperm from the antheridia of *Funaria* depends on the swelling of the sperm mass following swelling of the walls of the cells at the apex of the antheridial jacket. Paolillo (1975, 1977*a,b,c*), Paolillo and Cukierski (1976), and Hausmann and Paolillo (1977,

281

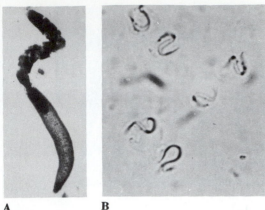

A **B**

FIGURE 12-22
Funaria sp. **A.** Antheridium shedding sperm.
B. Living sperm. **A.** ×60. **B.** ×800.

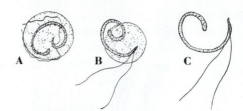

FIGURE 12-23
Funaria hygrometrica. Final stage in development
of motile sperm. **A.** Sperm with vesicle; note
partially free flagella. **B.** Flagella entirely free. **C.**
Free-swimming sperm. ×1200.

1978*a,b*) clarified our knowledge of the organization of the antheridium and of the mechanism
of dehiscence and liberation of sperm in the
moss genera *Polytrichum*, *Atrichum*, and
Mnium, which differ from those in *Funaria*. As
the antheridia mature, fluid accumulates at the
base of the antheridium between the proximal
portion of the spermatogenous tissue and the
sterile jacket. When the cells of the latter at the
tip open, the antheridial jacket contracts; this
forces out much of the sperm mass because the
fluid at the base acts as a "hydraulic ram." As
this fluid absorbs more water, the remainder of
the sperms are forced out into the surrounding
medium. The sperms at the surface of the mass
become motile and are shed from the cells that
produced them, in a hyaline, vacuolate vesicle
(Figs. 12-22B, 12-23). The flagella, which are
attached to the spirally coiled nucleus, project
through, and are attached partially to, the surface of the plasma membrane on which they
undergo undulating movements, which cause
the vesicle to turn in the water. Ultimately the
vesicle disappears, and the sperms swim rapidly
and freely in the water (Fig. 12-23C).

The sperms of mosses are rather complex in
their ontogeny and at maturity (Paolillo, 1967;

Paolillo et al., 1968; Lal and Bell, 1975). An
apical body of mitochondrial origin is joined to
an elongate, dense, coiled nucleus (Fig. 12-24).
The cytoplasm, which is sloughed off at maturity, contains the plastids and mitochondria
that were not involved in forming the apical
body.

The way in which sperms reach the archegonial necks of bisexual moss plants can be readily
understood. Probably the moisture of a heavy
dew filling the capillary spaces between the
apical leaves and sex organs is sufficient to effect
antheridial dehiscence and suffices for the
sperms to swim to the necks of the proximate

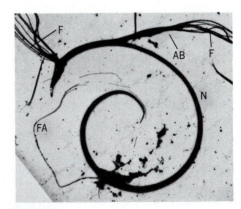

FIGURE 12-24
Polytrichum juniperinum: electron micrograph of a
mature sperm. AB, apical body; F, flagellum; FA,
filamentous appendage; N, nucleus. ×5500. (*After
Paolillo et al.*)

archegonia. In the case of species in which the colonies of male and female gametophores frequently are separated by considerable distances, it is more difficult to explain the apparent frequency of fertilization as evidenced by the production of abundant sporophytes on the female plants. It is well known, however, that splashing raindrops probably account for the distribution of the sperms. In view of the paucity of cytological investigations of mosses, the possibility of parthenogenetic (Gr. *parthenos*, virgin, + Gr. *genesis*, origin) development of the egg in some cases cannot be excluded but has not been demonstrated. The ecological significance of the moss gametophyte was discussed by Schofield (1981).

Reproduction: the sporophyte. The ontogeny of the sporophyte from the zygote has been investigated more thoroughly in *Funaria* than in any other moss, and the following account emphasizes development in that genus. After fertilization, the male branch withers, while the female branch enlarges and overgrows it (Fig. 12-17D, E; Fig. 12-18A, B). The first

division of the zygote is transverse. An apical cell is differentiated early by divisions in the upper hemisphere. Subsequent development of the young sporophyte is traceable to the activity of the apical cell and its derivatives. The apical growth of the sporophyte of the Bryopsida is a departure from that in the Hepatophyta and from the transient activity of the apical cell in the young sporophyte of *Sphagnum*.

Serial transverse sections, beginning at the apex of the cylindrical sporophyte, reveal that the derivatives of the apical cell divide so as to form an endothecium that is at first composed of four quadrately arranged cells surrounded by eight primary amphithecial cells (Fig. 12-27A). By continued periclinal and anticlinal divisions, the amphithecium and endothecium increase in thickness and circumference.

The apical development of the sporophyte of *Funaria* results in the formation of a spindlelike (biconical) structure, the embryonic sporophyte, within the old archegonium (Fig. 12-25B, C). The base of the sporophyte grows through the archegonial stalk into the stem

 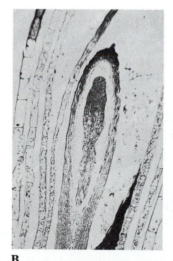 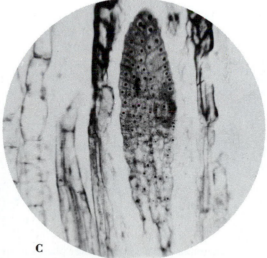

A   **B**   **C**

FIGURE 12-25
Funaria sp. Early development of the sporophyte. **A.** Early, biapical sporophyte. **B.** Older embryonic sporophyte within enlarging calyptra. **C.** Enlarged view of sporophyte of **B**; note apical cell above. **A.** ×600. **B, C.** ×125. (**A.** *After Campbell.*)

tissue of the gametophore, where it functions as a foot that anchors and absorbs. The apical cell of the sporophyte ceases to function when the latter is only a few millimeters long.

Meanwhile, the cells near the junction of the archegonial venter and its stalk undergo rapid divisions and form an inflated **calyptra** (Fig. 12-26D–H), so that the young sporophyte is enclosed for a time; but the rapid and inexorable

enlargement of the needlelike, cylindrical sporophyte soon causes the calyptra to rupture at its base. In *Funaria* this occurs when the needlelike sporophyte is about 6 mm long. In this manner the archegonial neck and upper portion of the calyptra are raised above the leaves of the gametophore. In *Polytrichum* (Fig. 12-33C) the major portion of the calyptra is composed of thick-walled protonemalike branches, which

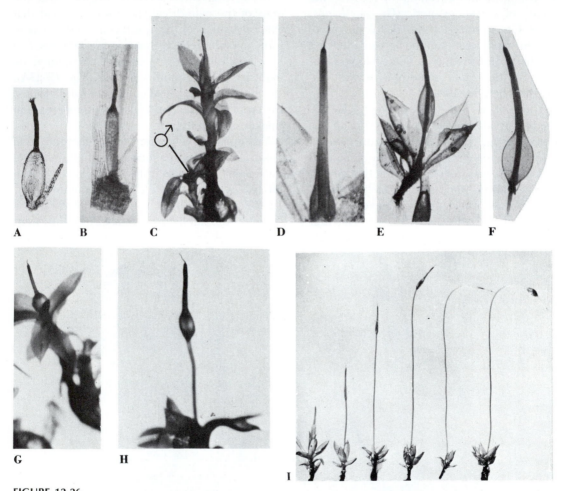

FIGURE 12-26

Funaria sp. Photographs of successively later phases in sporophyte development. **A, B.** Early stages; note aborted archegonium in **A. C.** Female branch with sporophyte (within calyptra) at apex; note branch at left. **D.** Apex of **C**, dissected and enlarged. **E.** Rupture of inflated calyptra by sporophyte elongation. **F.** Similar stage, showing foot. **G, H.** Stages in seta elongation; note elevation of calyptra. **I.** Stages in sporophyte development. **A, B.** ×30. **C.** ×3. **D.** ×8. **E.** ×6. **F.** ×8. **G, H.** ×3. **I.** ×1.8. (**I.** *After French and Paolillo.*)

284

arise from the venter soon after fertilization and expand during development of the sporophyte. There is also present an inner layer, which is closely adpressed to the apex of the sporophyte. Confinement of the sporophyte by this inner layer of the calyptra during development and later splitting of the calyptra are important factors in determining the form of the capsule (Paolillo, 1968). French and Paolillo (1976a,b) reported that removal of the calyptra from the young (needle-stage) sporophyte results in marked thickening of the seta as compared with that in the intact sporophyte. Such thickened setae underwent indeterminate growth in culture provided they were excised before they had developed capsules at their tips. The same investigators found (1975b) that the guard cells on the capsule are normally oriented with their long axes parallel to the long axis of the capsule and that removal of the calyptra during capsule development disturbs this arrangement. They reported (1975d) that the calyptra must be present for normal capsular development, and that its role seems to be physical restraint. French and Paolillo (1975c) also reported that decapitated elongating sporophytes stopped elongating, but that elongation could be restored by supplying growth regulators (benzyl adenine or indoleacetic acid) exogenously. Finally, in their studies of the development of the sporophyte (1970b), they concluded that a supply of carbohydrates limited the dry weight of capsules under growing conditions in a greenhouse.

French and Paolillo (1975a,b,c,d; 1976a,b) have carefully investigated the ontogeny of the sporophyte of *Funaria* soon after it becomes visible above the leaves of the female leafy gametophore. The elongation of the seta results from the growth of the derivatives of an intercalary meristem below the apical region of the sporophyte. At first the sporophyte is entirely erect, but later in development the seta bends and tilts the enlarging apical region, which becomes the upper part of the capsule, into a horizontal or recurved position (Fig. 12-26I). The apical region and seta are early differentiated into an endothecium and amphithecium (Fig. 12-27A).

The seta of mosses may be complex histologically (Hébant, 1977). For example, Favali and Bassi (1974) reported that the seta of *Polytrichum commune* is composed of an outer cortex of thick-walled cells and an inner cortex of thin-walled cells, both filled with starch. The central core of the seta contains hydroids and leptoids.

Weincke and Schulz (1975) studied the organization and function of the foot (**haustorium**) in the development of the sporophyte of *Funaria*. According to them, the basal part (0.5 mm) of the foot is embedded in the central strand of the leafy gametophore, and its main function is water absorption. The middle portion of the foot functions mostly in absorption of nutrients, while the function of the uppermost portion is mechanical support. The ultrastructure of the haustorial transfer cells is specialized (Lal and Chavham, 1981; Caussin et al., 1983). Specially oriented plasmodesmata seemingly provide a pathway for the movement of nutrients to the leptoids of the seta.

In the maturing and mature capsule, three regions become distinguishable (Fig. 12-30B): (1) An apical, domelike region, which gradually is differentiated into a cover or **operculum, peristome,** and **annulus** (see below); (2) a more extensive region consisting of a sterile central mass of tissue, the **columella,** surrounded by sporogenous tissue, both of these derivatives of the endothecium; and surrounded by nutritive tissue (**tapetum**), a space or lacuna transversed by filamentous trabeculae, and the capsular wall; and (3) the basal sterile region of the capsule, slightly enlarged, the **apophysis,** in the epidermis of which occur stomata. The first two regions are derived from the apical region of the young sporophyte, while the apophysis is derived in part from the apical region and in part from derivatives of the intercalary meristem.

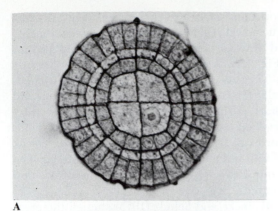

A

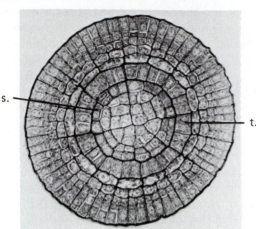

s.

t.

B

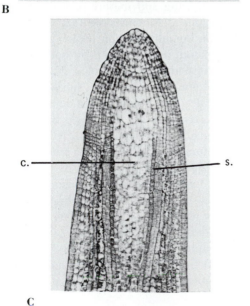

c.

s.

C

FIGURE 12-27
Funaria, development of the sporophyte.
A. Transection near middle of apical region showing four central endothecial cells, the remainder amphithecial derivatives (from 8–12-mm class sporophyte). **B.** Transection of apical region of 20–28-mm class sporophyte. **C.** Longisection of capsular region of sporophyte on the first day of expansion. c., columella; s., sporogenous tissue; t., trabecular layer. **A.** ×370. **B.** ×370. **C.** ×165. (*After French and Paolillo.*)

The mature sporophytes of *Funaria hygrometrica* average 34 mm + or − 7 mm in length.

Dietert (1977) found that development of the sporophyte was affected by temperature: Maximum elongation of the sporophyte required from 6 to 16 weeks after fertilization, while 6 to 10 additional weeks were required for maturation of the capsules.

Although the moss sporophyte is chlorophyllose throughout development, there can be little question that elaborated foods as well as water and inorganic salts from the gametophore are transferred to the young sporophyte through the foot (Hancock and Brassard, 1974).

Precise and incisive investigations of the morphology and physiology of the developing sporophyte of *Funaria* (Browning and Gunning, 1979 *a*, *b*, *c*) have revealed that there are specialized transfer cells in both the base of the developing sporophyte and the surrounding gametophytic cells. There is evidence that these cells with labyrinthine wall amplification facilitate the transport of solutes between sporophyte and gametophyte. Labeled products of gametophytic photosynthesis are transported through the foot and seta of the sporophyte.

It was demonstrated (Freeland, 1957) that the gametophytes and sporophytes of five mosses contain the same pigments: chlorophylls *a* and *b*, carotene, lutein, violaxanthin, and zeaxanthin. In quantitative studies of the relative rates of photosynthesis and respiration and of their chlorophylls *a* and *b*, Rastorfer (1962) found the ratio of chlorophyll *a* and *b* to be approximately 2.5:1 in both sporophyte and gametophyte, but the total chlorophyll content of the gametophyte was greater than that of the sporophyte. The ratio of photosynthesis to respiration in the sporophyte did not exceed 1.6:1, while in the gametophyte it varied between 2.8 and 6.3:1. It was concluded that the sporophyte was not totally self-sufficient (insofar as photosynthate is involved) but its nutriment was augmented by the gametophyte.

Paolillo and Bazzaz (1968) have measured the rate of photosynthesis, as compared with respiration, in the sporophytes of *Polytrichum juniperinum* and *Funaria hygrometrica*. Only in the latter did the rate of photosynthesis exceed that of respiration. Krupa (1969), however, reported that the sporophyte of *Funaria* depends on the gametophyte for some carbohydrates.

The sporophyte of most Bryopsida exceeds in stature and complexity that of any Hepatophyta or other Bryophyta. In duration of development, it is surpassed only by the sporophytes of *Anthoceros* and *Phaeoceros*. The sporophytes of *Funaria* (Figs. 12-28, 12-30) may exceed 2 inches in length, and those of species of *Polytrichum* (Fig. 12-31B) may attain a length of 6 inches.

The sporophyte remains a needlelike cylindrical structure until the distal, apical elongation has ceased (Figs. 12-26I, 12-31A). At that time the distal portion of the sporophyte becomes much enlarged and differentiated into the **capsule** (Fig. 12-29). Krisko and Paolillo (1972) investigated the process of capsular expansion in *Polytrichum* and found that light was a necessary factor. The major portion of the

A

B

FIGURE 12-29
Funaria hygrometrica. **A.** Photograph of living plants with sporophytes. **B.** Capsules and calyptras. **A.** ×0.75. **B.** ×4.

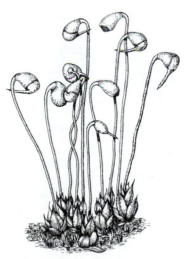

FIGURE 12-28
Funaria hygrometrica with maturing sporophytes; habit sketch. ×1.

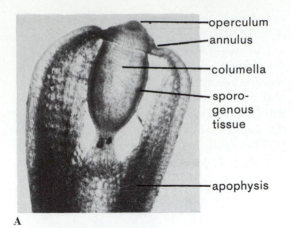

A

operculum
annulus
columella
sporo-
genous
tissue

apophysis

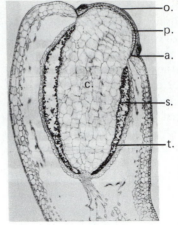

B

o.
p.
a.

c.

s.

t.

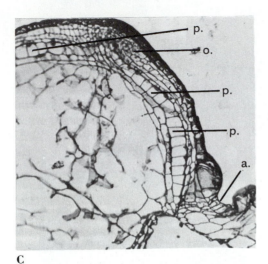

C

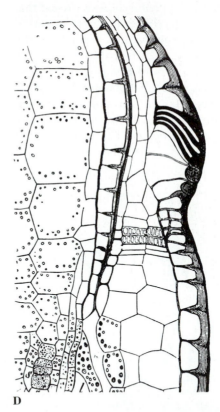

D

FIGURE 12-30

Funaria sp. Morphology of the capsule.
A. Freehand bisection of still-green capsule.
B. Microtomed section of maturing capsule.
C, D. Sections of apices of maturing capsules, showing development of the peristome.
E. Transection of capsule at the level of the operculum; note wall thickenings on 16 developing peristome teeth and segments. **F, G.** *F. hygrometrica.* Portion of peristome from mature capsule. **G.** Three-dimensional view of teeth and segments. a., annulus, c., columella, o., operculum, p., peristome, s., sporogenous tissue, t., tapetum. **A.** ×12. **B.** ×18.5. **C.** ×125. **D.** ×150. **E.** ×12. **F.** ×125. **G.** ×275. (**B.** *After French and Paolillo.* **D.** *After Sachs.* **E.** *After Schulz and Schmidt.* **G.** *After Lantzius-Benninga.*)

288

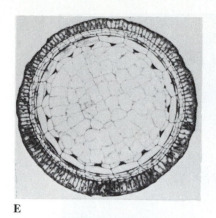

E

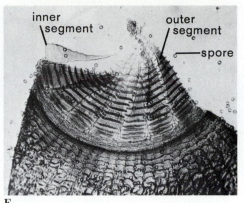

inner segment   outer segment   spore

F

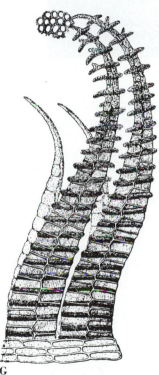

G

sporophyte below the capsule functions as a **seta**; the short **foot** is embedded in the gametophore. The central cells of the seta are thin-walled and probably function in conduction. As the sporophyte of *Funaria* matures, profound

changes occur in the capsular region; it becomes a rather complicated, highly differentiated structure, the organization of which is illustrated in Fig. 12-30. In *Funaria* and certain other mosses, the basal portion of the capsule remains sterile, enlarges somewhat, and is actively photosynthetic; this region is known as the **apophysis**. Its epidermis includes guard cells and stomata. (The development of the latter in *Funaria* has been investigated in detail by Sack and Paolillo, 1983 *a, b, c*). The stomata of most mosses are always open during development and apparently close only upon extreme desiccation of the capsule. Neither carbon dioxide concentration nor light intensity noticeably affects the condition of the guard cells and stomata. In some moss capsules stomata are absent (Paton and Pearce, 1957).

The upper portion of the capsule contains both sterile and fertile cells. The latter (Fig. 12-30B), ultimately two layers (four layers in some mosses [Mueller, 1974]), in extent, arise from the outermost layers of the endothecium and are arranged in the form of a hollow urn, with the distal portion wider than the proximal. The cells within the region of sporogenous tissue in the central portion of the capsule form a columella, which represents the old endothecium, except the sporogenous tissue. The cells

A

B

FIGURE 12-31
Polytrichum sp. **A.** Female plants with young, just-emergent sporophytes. **B.** Female plants with maturing sporophytes. **A.** ×0.75. **B.** ×0.5.

external to the sporogenous tissue remain sterile and form photosynthetic tissue and the capsular wall (Fig. 12-30A, B). The columella and surrounding sporogenous tissue in *Funaria* and many other mosses is separated from the capsular wall by a cylindrical air chamber (Fig. 12-30A). Relatively late in development, the sporogenous cells function as sporocytes, each undergoing meiosis to form a tetrad of spores (Fig. 12-32).

It was demonstrated (Kumra and Chopra, 1980) that in laboratory cultures of *Funaria*, the capsules may produce protonema (apospory) or additional capsules; the latter has been considered as **apogamy**, the development of a sporophyte without the union of gametes, but an alternate view would consider it as vegetative propagation of the sporophyte.

In *Polytrichum* the sporogenous tissue belt is cylindrical in form. Paolillo (1964, 1969) has followed changes in chloroplast form and presumed activity during sporogenesis in *Polytrichum*. The young spores contain little or no chlorophyll, but as the spore matures the plastid regreens and divides so that 2 to 4 chloroplasts

are present in each spore at shedding. Mueller (1974) reported that the spores of *Fissidens limbatus* had three wall layers, centrifugally, the intine, the exine, and the perine. The first two are products of the spore protoplast, but the perine is deposited, apparently from sources external to the spores. The latter, as in *Polytrichum*, contain several chloroplasts.

In *Funaria*, walls of the fourth to sixth layer of cells from the surface layer of the operculum become differentially thickened and at maturity

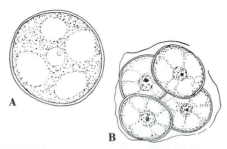

FIGURE 12-32
Funaria hygrometrica. **A.** Sporocyte. **B.** Spore tetrad. ×700.

dry, forming a ring of toothlike segments, the **peristome**[6] (Gr. *peri*, around, + Gr. *stoma*, mouth) (Figs. 12-30C–E) (Schulz and Schmidt, 1974). This thickening also extends for a short distance centripetally from the tangential walls along the horizontal walls. As a matter of fact, since the thickening of the tangential (periclinal) walls occurs also on the cell layers adjacent to the fifth (the fourth and sixth), three cell layers are actually involved in peristome formation in *Funaria*. The vertical, radial (anticlinal) walls of the fifth layer remain unthickened. As these cells dry, they split along the thin radial walls, thus freeing the outer tangential walls from the inner tangential walls. In *F. hygrometrica* the peristome is double, consisting of 16 **teeth** and 16 inner, more delicate **segments** (Figs. 12-30F, G, 12-33A). The peristome teeth and segments thus are acellular and consist only of the inner and outer common tangential walls of the fourth layer of cells and the contiguous walls of the fifth and sixth layers. The peristome is attached to a ring of thick-walled cells that form the rim of the capsule. In mosses with a single peristome (e.g., *Ceratodon* [Evans and Hooker, 1913]), thickening occurs only in the common periclinal walls of the fourth and fifth layers of cells. Thus, upon drying, the peristome is single.

Meanwhile, the thin-walled cells within the capsule dry, and it ultimately contains the powdery mass of cellular debris intermingled with spores. The peristome teeth are hygroscopic, responding to slight changes in humidity by expansion and resultant curving. As they dry, they become somewhat arched and lift the operculum from the capsule apex. The teeth remain arched and separated from one another (Fig. 12-33A) during periods of low humidity, but in dampness or rain they expand longitudinally and laterally and thus cover the mouth of the capsule. Because of such mechanisms, spore

dissemination in the Bryopsida is almost always a gradual process.

As in *Sphagnum*, elaters are absent in the Bryopsida. The apical portion of the capsule is entirely sterile and undergoes considerable differentiation. Its outer layers thicken and are shed ultimately as a caplike **operculum**; this is loosened by the drying and collapse of thin-walled cells below the rimlike annulus at its base. In those mosses that have an annulus, it is hygroscopic, and its movements result in the shedding of the operculum.

Garner and Paolillo (1972*a,b*) reported that 65 to 70 days are required from the time the sporophyte of *Funaria* becomes just visible at the apex of the female shoot until the capsule matures. This was under greenhouse conditions. Elongation of the sporophyte occurred in 2 to 5 weeks, while expansion of the capsule required 9 days. The stomata of the capsule opened and began to function (open and close) on the 4th day and continued actively from the 5th through the 10th day of capsular expansion. Although the guard cells are completely separated from one another by a longitudinal wall when they first form, this wall later disappears at both ends, so that the guard cells become continuous. Maier (1967, 1973*a,b,c*) has carefully analyzed the process of dehiscence in the capsules of *Funaria hygrometrica*.

The variation in structure and mechanism of the peristome of mosses is a fascinating subject that can be studied readily in the field at low magnification, with the aid of only a hand lens. In some genera, such as *Atrichum* (Fig. 12-33B) and *Polytrichum* (Fig. 12-33D–G) and their relatives, the short peristome teeth are cellular instead of being composed only of portions of cell walls, as in *Funaria*. Furthermore, in these plants the teeth are short and are attached to a membranous layer, the **epiphragm**, which covers the mouth of the capsule. A number of other widely distributed mosses have double, acellular peristomes like those of *Funaria*. Among

---

[6]For a more extended discussion of the peristome consult Edwards (1984).

A

B

C

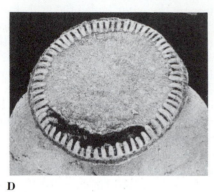

D

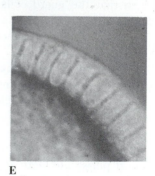

E

FIGURE 12-33

Capsule morphology. **A.** *Funaria hygrometrica*. Frontal view of (single) peristome; note spores. **B.** *Pogonatum dentatum*. Apex of capsule showing peristome and epiphragm. **C.** *Polytrichum* sp. Organization of the capsule. **D.** *Polytrichum* sp. Frontal view of peristome and epiphragm. **E.** Same as **D**, enlarged. **F–H.** *Polytrichum juniperinum*. **F.** Dorsal view of three peristome teeth. **G.** Transection showing five peristome teeth. **H.** Longisection of peristomatal region of capsule. **I.** Capsule apex of *Tortula norvegica*; twisted, hairlike peristome. a., apophysis; ann., annulus; c., calyptra; epi., epiphragm; o., operculum; per., peristome. **A.** × 47. **B.** × 30. **C.** × 4. **D.** ×16. **E.** × 45. **F.** ×90. (**A, B, D.** *Courtesy of Dr. Dale Vitt.* **F–I.** *After Flowers.*)

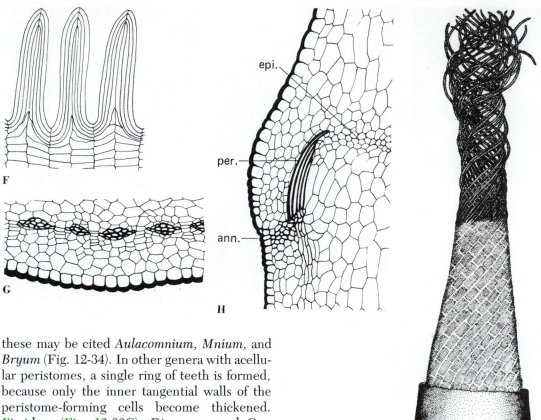

F

G

H

epi.

per.

ann.

I

these may be cited *Aulacomnium, Mnium,* and *Bryum* (Fig. 12-34). In other genera with acellular peristomes, a single ring of teeth is formed, because only the inner tangential walls of the peristome-forming cells become thickened. *Fissidens* (Fig. 12-36C), *Dicranum,* and *Ceratodon* are mosses with single peristomes. Mueller (1973), who investigated the ontogeny of the peristome in *Fissidens,* reported that the teeth walls are composed of cellulose and hemicellulose. Such features as seta length and peristome organization have been modified in evolution in accordance with the amount of moisture in the environment (Vitt, 1981).

Paolillo and Kass (1973) found that permeability of spores of *Polytrichum* increases as the spores mature after meiosis; immediately post-meiotic spores failed to germinate but did so to some degree when supplied with sucrose.

OTHER BRYOPSIDA

Funaria and *Polytrichum,* emphasized in the preceding account of the development and re-

production of Bryopsida, are representative of only two of 14 or 15[7] orders in which the genera of Bryopsida are usually classified (Brotherus, 1924–1925). Relatively few genera from these orders have been studied developmentally and morphologically, most emphasis having been taxonomic. The orders have been delimited on the basis of both gametophytic (largely with respect to the leafy gametophores) and sporophytic characteristics. Among the former are: (1) the persistence of the gametophores, whether

---

[7]Fifteen if the Archidiales are separated from the Dicranales, which their leafy gametophores resemble (Snider, 1975a).

A

B

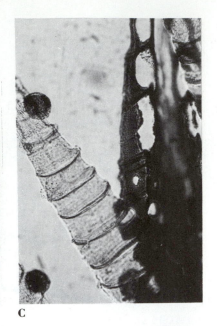

C

D

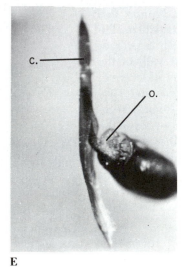

E

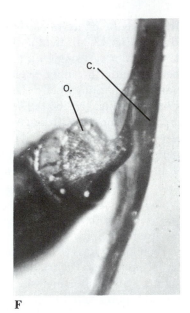

F

G

294

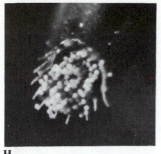

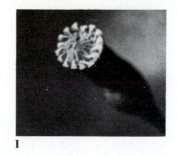

H

I

J

FIGURE 12-34

A–C. *Mnium cuspidatum*: peristome structure. **A.** Frontal view of capsule showing double peristome. **B.** Portion of peristome with spores. **C.** Portion of preceding more highly magnified; note thick-walled outer tooth, more delicate inner peristomatal segment, and spores. **D–H.** *Weissia controversa*. **D.** Several sporophytes with calyptras and opercula intact. **E.** Shedding of operculum and teeth. **F.** Preceding, more highly magnified. **G.** Opened capsule. **H.** Another capsule, showing spores. **I, J.** *Sematophyllum adnatum*. **I.** Capsule with double peristome, its calyptra and operculum in **J.** having been shed as a unit. **A.** ×10. **B.** ×80. **C.** ×180. **D.** ×3. **E.** ×5. **F.** ×12. **G.** ×7. **H.** ×15. **I, J.** ×5.

annual, biennial, or perennial; (2) the size and persistence of the gametophores, whether well developed or rudimentary and whether persistent or ephemeral; (3) arrangement and structure of the leaves, their shape, histology, margins and presence or absence of the midrib; and (4) in a few cases, such protonematal characteristics as luminescence and occurrence of thallose branches. Sporophytic characteristics used in delimiting the orders of Bryopsida, include, among others: (1) position of the sporophyte, whether acrocarpous[8] or pleurocarpous[9]; (2) degree of development of the seta and whether it is smooth or papillose; (3) shape of the capsule and its position (i.e., whether erect or pendulous and cylindrical or dorsiventral); (4) dehiscence or indehiscence of the capsule; (5) presence or absence of an operculum; and (6) the peristome, whether present, rudimentary, or absent; single or double; and cellular or composed only of cell walls.

Discussion and illustration of representatives of all of the orders of Bryopsida are beyond the scope of this book, but are available in those of Brotherus (1924–1925), Grout (1928–1941), Watson (1964), Flowers (1974), Crum (1976), and Crum and Anderson (1981), among others. However, a few additional genera will be described and illustrated at this point in order to amplify the reader's concept of the Bryophyta.

Archidium (Archidiales)

Species of *Archidium* (Gr. *archidion*, primitive) (Fig. 12-35) are minute (3 to 20 mm long), acrocarpous inhabitants of bare soil. They may be ephemeral or perennial. Their sporophytes differ from those of all other mosses in lacking a columella and a differentiated layer of sporoge-

---

[8]In acrocarpous mosses, the sporophytes are at the apices of the main, erect shoots.

[9]In pleurocarpous mosses, the sporophytes are at the apices of short, lateral branches.

A

B

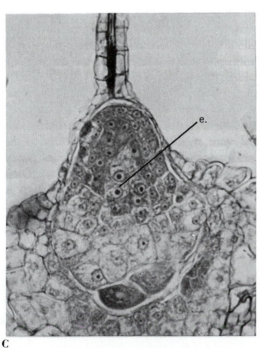

C

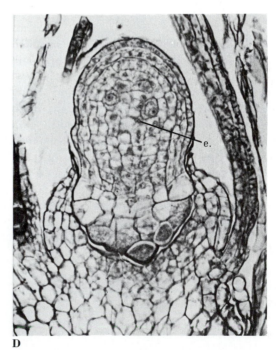

D

FIGURE 12-35
Archidium, habit of growth. **A.** *A. dinteri.* **B–D.** *A. alternifolium.* **C.** Longisection of young sporophyte; note early endothecium and three layers of amphithecium. **D.** Longisection of young sporophyte; note beginning of differentiation of endothecium into presporocytes. e., endothecium. **A.** ×18.7. **B.** ×14.3. **C, D.** ×360. (*After Snider.*)

nous tissue. From 1 to 44 cells of the endothecium function as sporocytes, which give rise to spores larger (50 to 310 μm) than those of most mosses. The capsule wall is composed of one layer at maturity, and seta, operculum, annulus, stomata, and peristomes are lacking. The spores are released by irregular rupture of the thin capsular wall. *Archidium* has been considered to be a reduced plant, in which case it has been classified in the Dicranales, but Snider (1975*a,b*) considered it to be primitive and

assigned it to the order Archidiales. Twenty-six species of *Archidium* have been recognized.

Fissidens (Fissidentales)

Species of *Fissidens*[10] (L. *fissus*, split, + L. *dens*, tooth) (Fig. 12-36) occur in quite moist niches. Their apical cells give rise to cells in two directions so that the leaves are arranged in two alternating rows. The leaves are quite distinctive in having a dorsal, winglike lobe (Fig. 12-36A), and the leaves of most species have midribs. The sporophytes are lateral or terminal, depending on the species. The capsules are obovate to shortly cylindrical, erect, or inclined with a conic operculum (Fig. 12-36B). The distinctive peristome is single, composed of 16 deeply bifid, often bright-reddish teeth (Fig. 12-36C).

---

[10]Generic name refers to bifid peristome teeth.

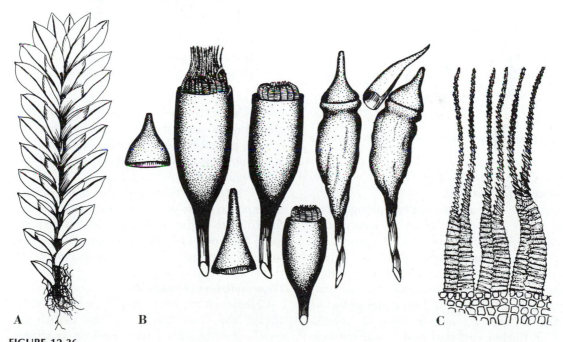

FIGURE 12-36
Fissidens limbatus. **A.** Leafy shoot. Note specialized leaves. **B.** Capsule morphology. **C.** Portion of peristome; note bifid teeth. **A.** ×6. **B.** ×8. **C.** ×60. (*After Flowers.*)

FIGURE 12-37
Dicranum scoparium. Habit ×0.4. (*After Flowers.*)

FIGURE 12-38
Physcomitrium pyriforme. Moist plant at the left; dried plant at the right. ×4. (*After Flowers.*)

Dicranum (Dicranales)

Dicranum (Gr. *dicranon,* a two-pronged fork) (Fig. 12-37) is a large genus of acrocarpous, sparingly branched mosses with narrow, secund,[11] sickle-shaped leaves that have narrow midribs. The various species grow on soil, humus-covered rocks, tree bark, and decaying logs. The cells at the leaf bases, **alar cells,** are square in outline, mostly two- to several-layered, thick-walled and brown near the mar-

gins. The male plants (as in the algal genus *Oedogonium*) are either similar in length to the female, but more slender, or dwarfed, budlike, and epiphytic on densely hairy stems of the female gametophores.

The setae are erect and the more or less asymmetric capsules may be inclined. Each of the 16 teeth of the single peristome is divided, for ½ to ⅔ of its length, into narrow segments; the teeth have bands of short, vertical markings.

Physcomitrium (Funariales)

Colonies of *Physcomitrium*[12] *pyriforme* (Fig. 12-38), the "urn moss," or "top moss," are widely distributed on bare soils on roadsides

---

[11]Turned to one side; sometimes called "broom moss" because of the secund, windswept leaves.

[12]Gr. *physce,* the large intestine + Gr. *mitrion,* a cap; refers to the inflated calyptra of the young sporophyte.

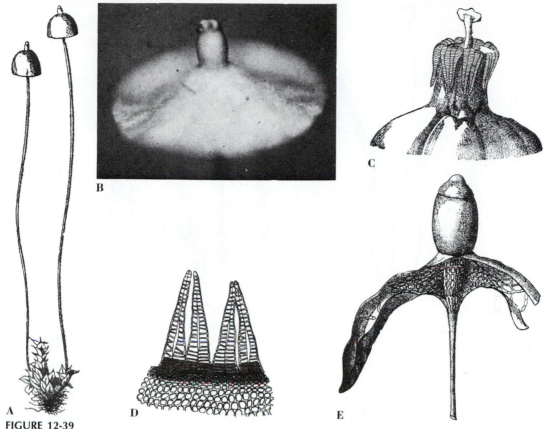

FIGURE 12-39

Splachnum luteum. **A.** Fruiting plants.
B. Photograph of living capsule. **C.** Capsule,
mouth, enlarged. **D.** Peristome teeth.
E. Capsule with dissected apophysis. **A.** ×1.
B. ×2.5. **C–E.** Magnification not stated.
(**A, C–E.** *After Brotherus, from Bryol. Europ.*
B. *Courtesy of Dr. W. C. Steere.*)

and old fields. The small, *Funaria*-like gametophores are light green or yellowish and acrocarpous. As in *Funaria*, the antheridia and archegonia are borne in separate branches on the same
leafy gametophore. The capsules are globose-
pyriform and erect on the setae, have an operculum, but lack a peristome. The calyptra is
divided at its base into several segments. Apogamy in one species of *Physcomitrium* has been
reported. *Physcomitrium* has been reported as
containing more than 70 species.

Splachnum (**Funariales**)

Splachnum[13] (Fig. 12-39) is a small genus with
approximately 8 species, which usually grow on
dung, bones, or other organic material. The
plants are light green or yellow-green and tuft-
ed and, according to the species, are monoecious or dioecious. The most striking features of
Splachnum are sporophytic. Unlike those of
other mosses, the setae continue to elongate
after the spores have matured. The base of the
capsule, the **hypophysis,** becomes somewhat or
greatly expanded, depending on the species,
and may be skirt- or umbrellalike, this having
suggested such vernacular names as "umbrella

[13] The generic name was used by Dioscorides for some
cryptogam, presumably the lichen *Lobaria pulminata*,
which has a "lunglike" appearance similar to that of the
apophysis of *Splachnum* when wrinkled and dry.

299

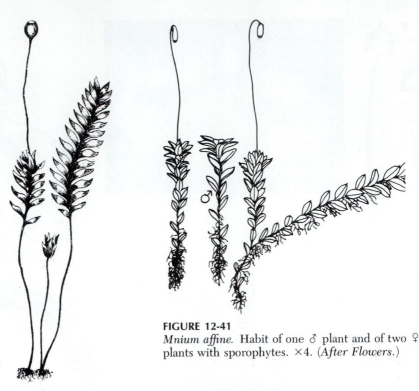

FIGURE 12-40
Schistostega osmundacea.
Left, ♀ plant with
sporophyte. Right,
♂ plant magnified. (*After
Braithwaite.*)

FIGURE 12-41
Mnium affine. Habit of one ♂ plant and of two ♀
plants with sporophytes. ×4. (*After Flowers.*)

moss" and "petticoat moss." The expanded hypophysis has stomata and in some species contains spongy photosynthetic tissue reminiscent of the spongy mesophyll of leaves. The peristome is single with 16 teeth. It has been suggested that the bright color, expanded form, and musty odor of some *Splachnums*, together with their habitat in dung, all are adaptations for dispersal of the spores by insects.

Schistostega (Schistostegales)

Schistostega[14] (Fig. 12-40) has such common names as "luminous moss," "cave moss," and "goblin gold." These vernacular names were inspired by the luminous nature of the persistent protonema. The luminous characteristic is

based on the reflection of light from the globose cells in branches of the protonema. The moss occurs on sandstone rocks, in caves, and in other fairly dark locations such as under buildings. The leafy gametophores are small, 4 to 7 mm long, and lack leaves below. The sterile gametophores have two-ranked leaves, confluent at their bases, while dioecious fertile ones are smaller and their leaves many-ranked. The capsules are small (ca. 0.5 mm high) and borne on short setae. No peristome is developed.

Mnium (Eubryales)

The genus *Mnium* (Gr. *mnion*, moss) (Fig. 12-41) contains approximately 80 species and is worldwide in distribution, its species occurring in swamps, on soil, on rocks, and on tree bark. The leafy gametophores are erect or in some species, especially the sterile axes, are prostrate and spreading (as in *M. cuspidatum*). The plants vary in coloration, being various shades of

---

[14]Gr. *schisto*, split, + Gr, *stegos*, a lid, which would imply that the operculum splits radially into toothlike sections, but this is erroneous!

300

green, yellowish brown or reddish. The leaves have more or less isodiametric cells, but the marginal cells are differentiated and more elongate; the various species may be monoecious or dioecious. The capsules may be almost erect or pendent, their stomata sunken and limited to the basal part of the capsule. The peristome is double, both circles of teeth being of the same length. The opercula have long beaks.

Bryum (Eubryales)

Bryum (Gr. *bryo*, moss) (Fig. 12-42) is a large genus of widely distributed species. The stems are erect, often branched and angular in transection. The leaves have midribs and may or may not have a border of thick-walled linear cells; the rhizoids are often reddish. In some species branching filaments occur in the axiles of the leaves and apparently serve, when detached, as agents of asexual propagation. The various species are monoecious or dioecious. The capsules are pear-shaped, horizontal, or pendulous. The yellow peristome is double, the inner segments arising from a high basal collar. The calyptra is hoodlike.

FIGURE 12-43
Fontinalis antipyretica. **A.** Habit of growth.
B. Capsule mouth showing peristome. **A.** ×0.2.
(**A.** *After Flowers.* **B.** *After Bryol. Europ.*)

FIGURE 12-42
Bryum capillare. Habit of plants with sporophytes.
×1.

Fontinalis (Isobryales)

Fontinalis (L. *fontinalis*, pertaining to a spring or fountain) (Fig. 12-43A) is an aquatic moss more than 50 species of which occur submerged and attached in running water. The leafy gametophores are elongate (up to 15 cm) with the leaves three-ranked. The leaves lack midribs. Most species are dioecious, the gametangia being borne on the main or secondary branches, the antheridial branches short, the

301

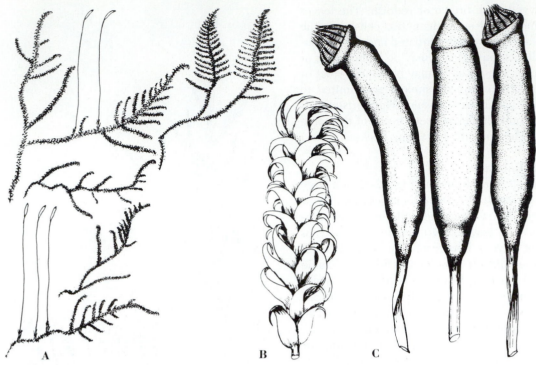

FIGURE 12-44
Hypnum revolutum. **A.** Habit of growth. **B.** Portion of shoot, magnified. **C.** Capsules. **A.** ×0.36.
B. ×3.6. **C.** ×7.2. (*After Flowers.*)

archegonial elongated and pointed. The sporophytes are almost hidden or barely emergent from the perichaetial leaves. The capsule is ovoid to cylindrical with rudimentary stomata. The peristome is double, both cycles of 16 teeth being equal in length. The inner segments and **cilia**[15] are united into a cone-shaped lattice or trellis (Fig. 12-43B). The calyptra is quite short.

Species of *Fontinalis* are known as "water mosses" or "brook mosses." The presence of a well-developed peristome in the submerged aquatic *Fontinalis* has suggested that it has become secondarily aquatic and a derivative from a terrestrial ancestor.

---

[15]Cilia are delicate, threadlike structures among the segments of the inner peristome.

Hypnum (**Hypnobryales**)

The genus *Hypnum* (Gr. *hypnos*, sleep; applied in antiquity to certain mosses and lichens used in medicine) (Fig. 12-44) contains more than 60 species of coarse or delicate, shiny, sod- or mat-forming pleurocarpous mosses. They are green, yellow-green, brownish green, or golden brown and prostrate, some branches ascendant only when crowded. Rhizoids may be absent. The leaves are mostly flattened and sickle-shaped, their midribs short and double, or lacking. Most species are dioecious but a few are monoecious. The capsules are inclined to horizontal, elongate to cylindrical, more or less bent and asymmetric, seldom straight, but mostly smooth. The operculum is conical, sometimes with a short beak. Stomata are present at the base of the capsule. The peristome is double,

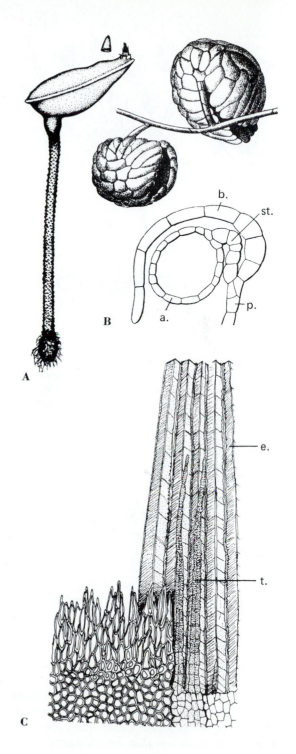

FIGURE 12-45
Buxbaumia aphylla. **A.** Sporophyte, lateral view. **B.** Protonema with two ♂ plants, the one below in section. **C.** Peristome. Note teeth and pleated inner peristome. a., antheridium; b., leaf; e., inner peristome; t., peristome tooth. **A.** ×10. **C.** ×50. (**A, C.** *After Flowers.* **B.** *After Goebel.*)

the keeled segments united in a basal membrane. Filiform structures, the cilia, in twos and threes, are present between the segments.

Buxbaumia (Buxbaumiales)

Approximately five species of *Buxbaumia* (after Johann Christian Buxbaum, a German physician-botanist who discovered it in 1712 at the mouth of the Volga River but did not name it) have been described. *Buxbaumia aphylla* (Fig. 12-45), a pioneer on disturbed soil, logs, and stumps in the northern United States, Canada, Greenland, Europe, and Asia, is an annual moss. Its unusual appearance inspired such common names as "bug-on-a-stick," "hump-backed elves," and "elf-cap moss." The minute leafy gametophores are dioecious and are borne on a more or less persistent protonema. The leaves lack midribs. The male plants are very small and have a single shieldlike leaf, which encloses a simple antheridium raised on a curved stalk. The female plants bear archegonia at their apices. The sporophytic setae are between 4 and 11 mm long and bear single asymmetric capsules at approximately 30–45° angles to the setae. The peristome is double, the outer consisting of 16 short teeth and the inner consisting of a folded membrane (there are 16 folds) opening at the apex. Mueller (1972) reported the presence of prominent plasmodesmata in the transverse walls of the protonema.

Tetraphis (Tetraphidales)

Tetraphis (Gr. *tetra*, four)[16] is a small genus with approximately four species; *T. pellucida* (Fig. 12-46) is a widely distributed species that

---

[16]Referring to the peristome teeth.

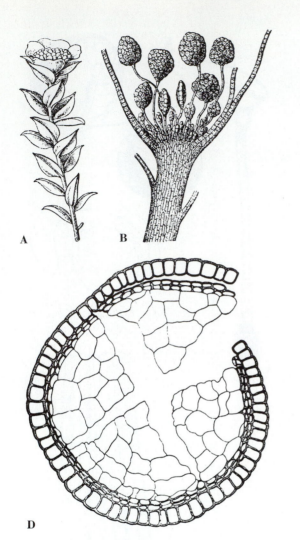

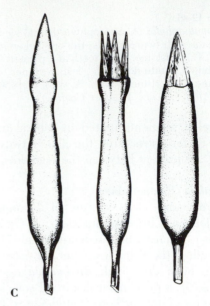

FIGURE 12-46
Tetraphis pellucida. **A.** Habit of a gemmiferous branch. **B.** Longisection of a gemmiferous branch. **C.** Capsules. **D.** Transection of apex of capsule; note sections of four teeth. **A.** ×20. **B.** ×50. **C.** ×10. **D.** ×75. (**A.** *After Ruhland.* **B.** *After Sachs.* **C.** *After Flowers.* **D.** *After Lantzius-Benninga, from Ruhland.*)

grows typically on humus and rotten logs. The leafy gametophores are small, from 8 to 15 mm high and acrocarpous. Some of the gametophores have the apical leaves arranged to form a splash cup within which stalked gemmae are borne (Fig. 12-46A, B). Presumably the gemmae are dispersed by rain drops like the sperms of *Polytrichum* and gemmae of *Marchantia* (p. 202). *Tetraphis* is monoecious, the antheridia occurring in branches that arise from the bases

of sterile leafy gametophores or on separate, branched male gametophores.

The sporophytes of *Tetraphis* are distinctive in several respects, including the lack of stomata on the capsule and the massive peristome composed of four large cellular teeth, which separate as the conical operculum is shed (Fig. 12-46C). In having a peristome of cellular teeth, *Tetraphis* is similar to *Polytrichum* (p. 293), but the teeth are more massive, not composed of concentric cells, and an epiphragm is lacking in *Tetraphis*.

Classification[17]

The genera of mosses discussed in this chapter are classified as follows:

Division Bryophyta
  Class 1. Sphagnopsida
    Order 1. Sphagnales
      Family 1. Sphagnaceae
        Genus: *Sphagnum*
  Class 2. Andreaeopsida
    Order 1. Andreaeales
      Family 1. Andreaeaceae
        Genera: *Andreaea, Neuroloma*
  Class 3. Bryopsida
    Order 1. Archidiales
      Family 1. Archidiaceae
        Genus: *Archidium*
    Order 2. Fissidentales
      Family 1. Fissidentaceae
        Genus: *Fissidens*
    Order 3. Dicranales
      Family 1. Dicranaceae
        Genus: *Dicranum*
      Family 2. Ditrichaceae
        Genus: *Ceratodon*
    Order 3. Pottiales
      Family 1. Pottiaceae
        Genus: *Weissia*
    Order 4. Funariales
      Family 1. Funariaceae
        Genera: *Funaria, Physcomitrium*
      Family 2. Splachnaceae
        Genus: *Splachnum*
    Order 5. Schistostegales
      Family 1. Schistostegaceae
        Genus: *Schistostega*
    Order 6. Eubryales
      Family 1. Mniaceae
        Genus: *Mnium*
      Family 2. Bryaceae
        Genus: *Bryum*

      Family 3. Aulacomniaceae
        Genus: *Aulacomnium*
    Order 7. Isobryales
      Family 1. Fontinalaceae
        Genus: *Fontinalis*
      Family 2. Orthotricaceae
        Genera: *Macromitrium, Orthotrichum*
      Family 3. Leucodontaceae
        Genus: *Leucodon*
    Order 8. Grimmiales
      Family 1. Grimmiaceae
        Genus: *Grimmia*
      Family 4. Neckeraceae
        Genus: *Neckera*
      Family 5. Cryphaeaceae
        Genus: *Forrstroemia*
    Order 8. Hypnobryales
      Family 1. Hypnaceae
        Genera: *Hypnum, Pylaisiella*
      Family 2. Amblystegiaceae
        Genus: *Amblystegium*
      Family 2. Sematophyllaceae
        Genera: *Sematophyllum, Trismegistia*
    Order 9. Buxbaumiales
      Family 1. Buxbaumiaceae
        Genus: *Buxbaumia*
    Order 10. Tetraphidales
      Family 1. Tetraphidaceae
        Genus: *Tetraphis*
    Order 11. Polytrichales
      Family 1. Polytrichaeceae
        Genera: *Polytrichum, Atrichum, Pogonatum*
    Order 12. Dawsoniales
      Family 1. Dawsoniaceae
        Genus: *Dawsonia*

---

[17]For current summaries of the classification of Bryopsida consult Miller (1974), Crum and Anderson (1981), and Vitt (1984).

Summary

A number of features in the morphology and reproduction of the Bryopsida are worthy of note. The gametophyte includes two separate phases, the protonema, with few exceptions a branching filament (in contrast to the spatulate protonema of *Sphagnum*), and a leafy gametophore that is produced from buds on the protonema. The gametophores develop sex organs at maturity. The sex organs are large and stalked and develop through the activity of apical cells. The zygote gives rise to the embryonic sporophyte, development of which is apical. The mature sporophyte is composed of foot, seta, and capsule, the last more complex than that of the Sphagnopsida and Hepatophyta; the capsule is elevated by gradual elongation of the seta. An apical cell and intercalary meristem are involved in its development, in contrast to that of the sporophytes of the Hepatophyta. Except for the foot cells, the sporophyte is actively photosynthetic from the earliest stages of development, and particularly so in the apophysis of the capsule, where stomata and guard cells are present in many genera. The sporogenous tissue is restricted in amount and arises as a double-layered, hollow cylinder from the outermost cells of the endothecium; unlike that of the Sphagnopsida, it does not overarch the columella. A complicated mechanism, the peristome, related to spore dissemination, is organized at the mouth of the capsules in most Bryophyta; this becomes operative after the operculum has been shed. In mosses that lack a peristome, it is thought to have been lost through reduction. The sporophyte in the Bryopsida exceeds that of other Bryophyta in stature and complexity. Among the Bryophyta, *Sphagnum* is clearly the most important economically. However, the Bryopsida also have both economic and biological significance. They have important roles in preventing erosion, in holding moisture, in contributing to humus, and in forming soils from rock. They are used by horticulturalists as decoration and as "liners'" for hanging baskets. In northern Europe they have been used for filling spaces in chimneys and walls and even as bedding and/or stuffing in mattresses. Mats and colonies provide a habitat for small animals, and the immature capsules serve as food for insects, birds, and mice. Although mosses are often thought to detract from the beauty of lawns, in Japan, lawns composed entirely of moss are cultivated in some places. Finally, the pleasing esthetic effects of Bryopsida on many landscapes and habitats should not be overlooked.

FOSSIL BRYOPHYTA

While fossil Hepatophyta existed as early as the Upper Devonian (inside front cover), the occurrence of fossilized mosses before the Carboniferous has not been demonstrated unequivocally. Reports of mosses from the Ordovician cannot be regarded seriously. The genus *Muscites*, established to include fossils that are obviously bryophytes but whose affinities are uncertain, includes two species from the upper part of the Carboniferous. One of these has leaves resembling those of the extant *Rhizogonium* or *Polytrichum*. The order Protosphagnales of the class Sphagnopsida includes fossil mosses from the Permian of the Soviet Union that have some resemblance to members of the Sphagnales. The oldest, *Vercutannularia* (Fig. 12-47), is early Permian, and has helically arranged leaves that appear to be in rosettes at the tips of axes (Neuburg, 1960). Cells of the leaves are of two types, reminiscent of the chlorophyllous photosynthetic cells and water-storage cells in leaves of *Sphagnum*. The Upper Permian *Junjagia* and *Protosphagnum* have similar leaves, with those of the latter showing a main vein and small laterals. The genus *Sphagnum*, itself, does not appear until the Mesozoic Era. There are a number of other Permian genera of mosses from the Soviet Union, including the Lower Permian *Intia*, which is known from leaves on fragments of axes. There is some similarity to *Mnium* or *Bryum*.

Most fossil remains of the Bryophyta are only vegetative fragments. The discovery of a moss

A

A

B

FIGURE 12-47
Vorcutannularia plicata. **A.** Tips of axes.
B. Clearing of portion of leaf showing cells of two types. **A.** ×1.5. **B.** ×150. (*From Neuberg.*)

B

FIGURE 12-48
A. *Palaeohypnum arnoldianum,* a Miocene moss.
B. *Hypnites arkansana,* a Lower Eocene moss.
A. ×1.25. **B.** ×5. (**A.** *After Steere and Arnold.*
B. *After Wittlake.*)

capsule from the Permian of India is of interest; some of its features correspond to those of the Sphagnales, Andreaeales, and Bryales.

Mesozoic bryophytes are relatively uncommon in the fossil record, but by Tertiary times, many of the modern families were in existence. Some Tertiary fossil mosses are illustrated in Fig. 12-48.

The fossil record provides no clues regarding the possible relationships among the three classes (Sphagnopsida, Andreaeopsida, and Bryopsida) of the division Bryophyta, because the known fossilized remains of Bryophyta are largely like modern genera. It is interesting to note the early appearance of members related to the Sphagnopsida, but the importance of this fact in regard to the interrelationships of this class with other groups is still unclear. As is true of algae and fungi, the fossil record indicates that extant mosses are little modified from their ancient fossilized relatives.

ALTERNATION OF GENERATIONS AND PHYLOGENY OF LIVERWORTS, HORNWORTS, AND MOSSES

Although several types of life cycles, including those involving alternations of generations, were described in the chapters on algae, a more general discussion of alternation has been deferred to this point purposely. This is because of its occurrence also, albeit in somewhat different patterns, in the liverworts and hornworts, and, in fact, in all the remaining groups of land plants as well as in some fungi.[18]

It is generally assumed that asexual reproduction preceded sexual reproduction and that the latter, involving genetic interchange, was a later development. If this is true, the earliest organisms were, of course, haploid. Diploid phases and organisms are conjectured to have evolved when some zygotes, instead of undergoing meiosis immediately, delayed that process (return to haploidy and segregation) until several to numerous intervening diploid mitoses had occurred, then producing meiospores. Whether initiation of diploidy occurred only once among early plants, so that all plants with diploid phases have a common origin, or more than once, is unknown.

However, if it occurred more than once, it *could* have done so either in free-living zygotes

(e. g., *Chlamydomonas*) or in zygotes retained in the parental sexual plant (gametophyte) as in *Coleochaete* (and most Rhodophyta). Thus, the diploid phases (sporophytes) could be free-living, as in *Cladophora* and *Ulva*, like the gametophyte, or borne upon the latter and nourished wholly or in part by it, as in Rhodophyta, Hepatophyta, Anthocerotophyta and, in fact, even if only transitorily, in all the remaining groups of land plants. It may be noted, in anticipation, at this point that in all the seed plants the gametophyte phases are borne upon and nourished by the sporophyte!

Until the 1920's and 1930's, it had generally been assumed that the green algae (Chlorophyta, the most probable ultimate precursors of *all* green land plants)[19] lacked organisms with alternation of two morphological phases, entities, or "generations," in their reproductive cycles. Accordingly, it had been suggested that alternation of generations in green plants evolved as plants became terrestrial. Botanists speculated that in the course of evolution of the green algal precursors into primitive land plants, the zygote became retained permanently on the parent organism (gametophyte), as in *Coleochaete* (p. 82). It was suggested additionally that, as a result of delay in meiosis along with intervening mitoses and cytokineses, a diploid tissue (homologous with the sporophyte) arose *within* the gametophyte. The cells of this ultimately underwent meiosis and produced haploid spores (**meiospores**) which, in turn, developed into the haploid phase. Increasing morphological complexity and size of this primitive sporophyte and its ultimate independent existence (as in vascular cryptogams and seed plants) were viewed as successive stages in the evolutionary development of the sporophyte. These suggestions regarding the origin and evolution of alternation of generations have been designated the "antithetic" and/or "interpola-

---

[18]See Williams (1981) for a discussion of ecological implications of alternation.

[19]See following page.

tion" theory, in emphasis that the sporophyte was something "new and different," *not* phylogenetically a modified gametophyte.

Several phenomena discovered before and after the 1930's seemed to support, albeit in somewhat modified form, an alternate concept of the nature of alternation of generations, namely, the homologous or transformation theory. According to this, sporophytes (of land plants) are not new and fundamentally different phases of the life cycle, but merely transformed or modified gametophytes with respect to their morphology. This view was seemingly supported by the discovery that not all green algae are monobiontic and haploid, but that a number of species have dibiontic cycles, both isomorphic and heteromorphic, with morphologically similar or dissimilar gametophytes and sporophytes, respectively. A second source of support for the homologous theory was the discovery of apospory (development of gametophytes directly from sporophytes [not from their spores]) and apogamy (p. 290) which indicated to some botanists that gametophytes and sporophytes are fundamentally similar, inasmuch as the successive phases develop from each other without intervening meiosis and/or gametic union. For example, excised setae of the sporophytes of a liverwort produced diploid gametophytes, an example of apospory (Matzke and Raudzens, 1968). By contrast, sporophytes can develop from the protonema, gametophores and leaves of several mosses, examples of apogamy (Lal, 1961; Bauer, 1966; Chapra and Rashid, 1967; Rashid and Chapra, 1969; and Menon and Lal, 1974-1977).

However, recent evidence from the ultrastructural studies has provided possible additional support for the antithetic theory and suggests the origin of the sporophyte in the green algae, one of them extant. In *Coleochaete* (p. 82), it will be recalled, the zygote, unlike that of other green algae, is permanently retained on the gametophyte. This seemingly stimulates the neighboring gametophytic cells to di-

vide and form a sterile, protective covering around it (Fig. 4-32). Futhermore, these cells, ultrastructurally, provide evidence that they are involved in providing nutrition for the zygote and its progeny (which, however, are chlorophyllous), as do the gametophytic cells ("transfer cells") surrounding certain hepatophytan sporophytes and the bases of the sporophytes of mosses (p. 250). A mutation delaying the site and time of meiosis in *Coleochaete* or a *Coleochaete*-like green alga would have resulted essentially in the simple type of sporophyte seen in such liverworts as *Riccia, Ricciocarpus* and *Oxymitra*.[20] Finally, it has been demonstrated that in *Coleochaete*, cytokinesis involves a phragmoplast (Fig. 36-10), such as is present in all land plants, and that the ultrastructural features of the sperm resemble those of land plants. These considerations suggest that the antithetic theory cannot be dismissed as having less support than the homologous theory (Graham, 1984). Wahl (1965) presented an interesting summary of alternation of generations.

The Hepatophyta, Anthocerotophyta and Bryophyta share the same pattern of life cycle in which the photosynthetic (with few possible exceptions, e.g., *Riccia, Ricciocarpos,* and *Oxymitra*) sporophyte is *permanently* borne upon the dominant gametophyte. What was the origin of this pattern? Of course, no unequivocal answer can be given to this question, but several hypotheses have been offered in answer to it. One early answer (Hypothesis I) suggested that in some organism's green algal ancestor, the zygote remained attached to the parent gametophyte and that a rudimentary sporophyte evolved as a result of diploid mitoses sooner or later succeeded by meiotic divisions to form spores. Thus, the primary function of the sporo-

---

[20]The cells of *Coleochaete* have single, large chloroplasts while those of most land plants, except Anthocerotophyta, have numerous lenticular plastids. In *Selaginella*, however, the plastid number in the leaf cells varies from one to several.

phyte was considered to be spore (meiospore) production (Bower, 1908). According to this viewpoint, increasing morphological (and physiological, e.g., photosynthesis) complexity of the sporophyte (and its independence at maturity in vascular cryptogams and seed plants) was interpreted as advancement in evolutionary development. Thus, the more primitive liverworts were considered to be those with simple spherical sporophytes of the *Ricciocarpus-Riccia-Oxymitra* type; those with more highly differentiated sporophytes (with foot, seta and capsule regions, elaters and/or nurse cells) were considered to be evolutionarily derived from more simple forms.

In opposition to this it is often argued that these putative primitive Marchantialean liverworts arose later in the fossil record (Permian to Tertiary) than leafy (Jungermannialean) forms (Devonian). Such an argument assumes, of course, that the fossil record is complete and that representatives of *all* types of plants have been preserved.

A second hypothesis (II), is a somewhat modified version of the first, summarized above. According to it, the precursors of Hepatophyta were branching, parenchymatous, cylindrical green algae with isomorphic alternation of generations. These supposedly gave rise to erect gametophytes with attached, similar sporophytes. From such radially symmetrical precursors, dorsiventral leafy and later thallose Hepatophyta are inferred to have evolved. It has been suggested further that the precursors could also have given rise to the mosses. The evolutionary appearance of the earliest liverwort and moss precursors is to be contemporaneous with that of earlier vascular plants.

Accordingly it was postulated (Schuster, 1979, 1984c; Clark and Duckett, 1979) that the primitive liverworts (like many mosses) were erect, radially symmetrical and leafless (as in *Calobryum-Haplomitrium* stolons). One evolutionary pattern in liverworts involved reduction phenomena in which the gametophytes became flattened (some internally differentiated) and the sporophytes retained on and within the gametophytes until mature enough for spore dissemination. This is in contrast to the mosses in most of which (except, e.g., *Sphagnum*) the sporophyte emerges from the gametophyte long before sporogenesis has been completed.

With reference to the Anthocerotophyta, there is no evidence of their origin available from the fossil record, and their cellular organization and that of their sperms, and their unique sporophytes, among other characteristics, mark them as derived from a precursor other than that (those) which gave rise to Hepatophyta and Bryophyta.

Speculation regarding the origin of mosses (as distinct from liverworts) suggests algal progenitors to some. This is based on the branching filamentous form of the protonema in most Bryopsida and on the necessity of water to effect fertilization. It should be noted that no extant branching, filamentous, freshwater green algae with discoid chloroplasts and oblique terminal cell walls are known. Similarities in the ultrastructure of the sperm suggests to some (Duckett and Carothers, 1979, 1984) relations between liverworts and mosses but the differences between these groups (Table 12-1) argue in favor of parallelism, in the writers' opinion.

Alternatively (Hypothesis III) in view of the occurrence of strands of conducting cells in certain Hepatophyta and Bryophyta (Scheirer, Schofield and Hébant), it has been suggested that they represent reduced vascular plants (those with xylem and phloem, Chapter 13) (Miller, 1979).

Because of the differences among the Hepatophyta, Anthocerotophyta and Bryophyta (Table 12-1) and the divergent conclusions regarding their origin and phylogenetic relationship, why then were and are they sometimes classified together in a single division, Bryophyta? After all, such grouping implies *at the least* morphological parallelism or, to some, real relationship and common origin. In the

TABLE 12-1

Comparison of some characteristics of Hepatophyta, Anthocerotophyta and Bryophyta[a]

Hepatophyta	Anthocerotophyta	Bryophyta
1. Growth of protonema, if present, apical and intercalary; walls not oblique.	1. No protonema.	1. Growth of protonema only apical in most part of protonema at least with oblique cross walls.
2. Especially differentiated oil bodies in the cells of most.	2. No oil bodies.	2. Differentiated oil bodies lacking.
3. Rhizoids absent or unicellular.	3. Rhizoids unicellular.	3. Rhizoids multicellular filaments.
4. Leaves, if present, bilobed, at least in early ontogeny, ranked.	4. No leaves.	4. Leaves usually not bilobed, even in early ontogeny, spiral.
5. Many lenticular plastids per cell.	5. Single, massive plastid per cell, or up to 8 large ones.	5. Many lenticular plastids per cell.
6. Both leafy and thallose genera and intermediates; the leafy most abundant.	6. Thallose.	6. All leafy.
7. Protonema from one spore mostly producing a single adult gametophore.	7. No protonema.	7. Protonema from one spore producing more than one leafy plant in most.
8. Cells of protonema with thin, colorless walls.	8. No protonema.	8. Some protonematal cells with thickened brown walls (caulonema).
9. Sex organs free, emergent.	9. Sex organs sunken in thallus.	9. Sex organs free, emergent.
10. Sporophyte enclosed within calyptra and perianth or pseudoperianth during development, maturing within them: sporogenesis simultaneous.	10. Sporophyte not so enclosed; sporophyte continuously undergoing ontogeny and sporogenesis from an intercalary meristem.	10. Sporophyte in most emerging early from calyptra, maturing after its rupture, sporogenesis simultaneous.
11. Seta of sporophyte, if present, elongating only at maturity.	11. No seta.	11. Seta present in sporophyte, elongating during ontogeny of most Bryopsida.
12. Elaters present in capsule of most.	12. Pseudoelaters present.	12. Elaters absent.
13. All sporophytes lacking operculum, peristome; annulus absent from capsule.	13. All sporophytes lacking operculum peristome; annulus absent from capsule.	13. Most Bryopsida with operculum, peristome and annulus; but a number lacking one, two or three; peristome lacking in some and in Sphagnopsida and Andreaeopsida.
14. Chromosome numbers: $n = 9$ in 75%.	14. $n = 5–6$.	14. $n = 6, 7, 10, 11, 12, 20,$ and 26.

[a]See also Crandall-Stotler (1980).

writers' opinion there is only one major feature uniquely characteristic of liverworts, hornworts and mosses: the pattern of the life cycle.

Mosses have certain cytological features in common, including some small "m" chromosomes and a heteromorphic bivalent chromosome. These cytological attributes, it is reported, do not occur elsewhere in the plant kingdom.

While such characteristics as biflagellate sperms, multicellular sex organs, sporic meiosis, aerial spore dissemination, and the terrestrial habitat may seem significant as important common attributes, one or more of them also occur in the algae or in the vascular plants. In the final analysis, it is largely the life cycle and the relative balance between the alternating generations on which the inclusive concept of the division Bryophyta is based. While common habitat is sometimes proposed as an indication of relationship of liverworts, hornworts, and mosses, it should be noted that algae, fungi, lichens, certain ferns, and seed plants share the same ecological niches.

In light of the diversities among liverworts, hornworts, and mosses summarized in Table 12-1 and of the failure of the fossil record to provide evidence of their relationships, it is questionable whether their common characteristics—life cycle and cytological organization—provide sufficiently compelling evidence for inferring phylogenetic unity.

The great German morphologist Goebel wrote many years ago, "Between Hepaticae (liverworts) and Musci (mosses) there are no transition forms; as there are none between Bryophyta and Pteridophyta, and as there never were such transitions their absence is not caused by their having died out." In the writers' opinion, no evidence has been educed in the interim to contradict these views.[21] While they are not as certain as Goebel professed to be

regarding the lack of possible phylogenetic relation, following Watson's injunction (see below), the writers prefer to withhold their judgment. For these reasons, the liverworts, hornworts, and mosses have been assigned to separate divisions in this volume.

In the concluding chapter of his book *The Structure and Life of Bryophytes,* Watson (1964) stated:

When, however, we seek enlightenment from contemporary or recent assessments of bryophyte interrelationships, we must surely turn away in disappointment. . . . Anyone reading the varied viewpoints currently put forward . . . would be justified in referring to this as the Age of Speculation, where interpretative morphology is concerned. Perhaps the only honest conclusion to draw is that we do not know how these organisms are interrelated. Nor do we know from what earlier organisms the remote ancestors of present-day mosses and liverworts came. We do not know, because there is insufficient reliable evidence at hand. . . . Meantime, the contemporary botanist is wise to withhold judgment. . . . [And further:] Can it be that morphologists, spurred on by the possession of a definite viewpoint, have not always been sufficiently clear as to the true nature of evidence?

In the last analysis, whether one classifies liverworts, hornworts, and mosses in one or several divisions (phyla) may be a reflection of personal opinion regarding what constitutes a division!

DISCUSSION QUESTIONS

1. What characteristics distinguish the Bryophyta from the Hepatophyta and Anthocerotophyta?
2. What characteristics do the Hepatophyta, Anthocerotophyta, and Bryophyta share in common?
3. Which of the Bryophyta, in your opinion, has the most highly developed gametophyte with reference to size and/or tissue differentiation? Which has the most highly developed sporophyte?

---

[21]See also W. C. Steere (1958, 1969), Fulford (1964, 1965), and R. S. Chopra (1968).

4. On what grounds can you support separation of the Sphagnopsida from the Bryopsida?

5. What evidence can you cite in support of an algal origin for Hepatophyta, Anthocerotophyta, and Bryophyta? What alternate hypothesis is there, and on what evidence is it based?

6. Can you suggest an explanation for the scarcity of liverwort and moss fossils?

7. Describe the modifications related to water absorption in *Sphagnum*.

8. How does *Polytrichum* withstand drought?

9. Of what theoretical significance is the observed fact that wounded moss setae produce protonemata? What mechanism is involved?

10. Can you suggest a procedure for obtaining diploid moss gametophytes? Triploid moss sporophytes?

11. Review your knowledge of the structure and nutritional arrangements in the sporophytes of algae, Hepatophyta, Anthocerotophyta, and Bryophyta. Then state whether or not you are of the opinion that the evidence supports the interpolation (antithetic) theory of alternation of generations, especially its doctrine of progressive sterilization. Give the reasons for your answer.

12. Can you suggest an experimental approach for obtaining evidence that might support the homologous, or transformation, theory of alternation of generations?

13. With the aid of labeled diagrams, describe the structure and reproduction of *Sphagnum* and one of the Bryopsida.

14. Observe mosses in the field and examine various species for the presence of sex organs and sporophytes.

15. How do the Andreaeopsida differ from the Sphagnopsida? From Bryopsida?

16. What seems to be the primary pathway of water absorption and movement in Bryopsida?

17. *Funaria*, although bisexual, produces antheridia and archegonia on separate branches of the same plant. If leaves, stem, archegonia, antheridia, and paraphyses were to regenerate, would the resulting plants be, in each case, male, female, or both? Discuss.

Chapter 13
INTRODUCTION TO VASCULAR PLANTS

It is generally agreed that life arose in an aquatic environment and then flourished during more than 3 billion years. Terrestrial life, both plant and animal, is relatively recent in geologic time (inside front cover).

In the plant kingdom, vascular plants (plants having the conducting tissues, xylem and phloem) evolved, colonized, and have come to dominate terrestrial habitats, so much so, in fact, that they are sometimes spoken of as "the land plants," a designation that quite overlooks many algae, fungi, liverworts, and mosses that share the same habitat.

The fossil record indicates that the oldest undisputed vascular plants originated in late Silurian times, and that by the Devonian period (inside front cover) considerable diversification of these plants had already occurred, so that members of at least seven great lines of vascular plants were represented in Devonian floras. Origin of vascular plants may have been **monophyletic** (that is, originating only one time from one group of ancestors) or **polyphyletic** (having had more than one origin in the past). Several evolutionary lines present in Devonian floras are represented by members of the divisions Rhyniophyta, Zosterophyllophyta, Trimerophytophyta, Microphyllophyta, Arthrophyta, Pteridophyta, Progymnospermophyta, and Pteridospermophyta; these will be discussed in chapters to follow. Before discussing these groups in detail, some features of the classification and organization of vascular plants will be reviewed in this chapter.

CLASSIFICATION OF VASCULAR PLANTS

Consideration of the modern period of plant classification may begin appropriately with the classification of Linnaeus, published in 1753 in his *Species Plantarum* and in 1767 in his *Systema Naturae*. In Linnaeus' summary, the vascular plants are dominant, since all but one of the 24 major categories of classification included vascular plants only. The twenty-fourth category, the Cryptogamia (Gr. *kryptos*, hidden, + Gr. *gamos*, marriage), Linnaeus characterized as plants having "flowers" scarcely visible to the naked eye, their sexual reproduction concealed. In contrast, the remaining 23 divisions, often called "Phanerogams" (Gr. *phaneros*, apparent, + *gamos*, marriage), have flowers plainly visible and sexual reproduction (as manifested by stamens and pistils, according to Linnaeus) obvious or apparent. In the period since Linnaeus, his group Cryptogamia (which included ferns and fernlike plants, mosses, liverworts, algae, and fungi) has been dismembered into a number of separate categories (inside back cover). Furthermore, it has become clear that their sexual reproduction is not concealed or obscure.

The terms "Cryptogamia" used by Linnaeus and "Phanerogamae" of earlier plant taxonomists[1] are no longer used as formal taxa in classification, nor do we use them in their original literal etymological sense. Cryptogamic plants do not have "hidden" gametangia nor, conversely, are those of phanerogams exposed or obvious. In fact, quite the opposite is the case, because stamens and pistils are spore-producing organs (p. 630). The term **cryptogam** in current usage designates a plant lacking seeds. All **phanerogams** are seed-bearing and vascular; however, both nonvascular (algae, liverworts, and mosses) and vascular (Psilotophyta, Microphyllophyta, Arthrophyta, and Pteridophyta) cryptogams are known among the extant flora.

At present, there is a diversity of opinion regarding classification of vascular plants. According to one viewpoint, they are monophyletic, and xylem and phloem—postulated to have evolved just once—are the manifestation of their common origin. Proponents of this doctrine, accordingly, group all vascular plants in a single division, Tracheophyta (inside back cover).

Another group of botanists, impressed by (1) the evidence of the fossil record, which clearly indicates that at least six or seven distinct lines of vascular plants had become established by the Middle Devonian (inside front cover), and (2) the fundamental diversities among the extant and extinct members of these lines (and later

series), refuse to accept the unidivisional Tracheophyta concept. Whether or not one accepts the monophyletic theory or the polyphyletic theory, it must be recognized that organization into distinct lines happened early and that the various lines of vascular plants are distinct enough to warrant assigning them to several divisions. Examination of the table on the inside back cover reveals that, for the most part, the divisions of vascular plants correspond to subdivisions often included within the single division Tracheophyta.

ORGANIZATION OF THE VASCULAR PLANT BODY

Most vascular plants are composed of well-differentiated organs, both **vegetative** and **reproductive**. The former comprise the stem, root, and leaf; the latter include spore-bearing organs, seeds, and fruits.

THE STEM

ONTOGENY OF THE PRIMARY STEM

An individual vascular plant may originate either sexually, from a zygote, or asexually, by growth of a fragment of the plant body. The fragment may be a totipotent single cell,[2] a group of cells (such as a callus or tissue culture), or stem, leaf, or root cutting.

In the first alternative, the zygote develops initially into a seemingly undifferentiated[3] mass of cells, which subsequently differentiates both

---

[1]The formal taxon "Phanerogamae" was established by Eichler in 1883.

[2]Professor F. C. Steward first accomplished this experimentally for a vascular plant, using carrot phloem tissue culture. See F. C. Steward et al., "Growth and Development of Cultured Plant Cells," *Science* 143:20–27 (1964).

[3]Undifferentiated in the sense that all the cells are meristematic. In some taxa, certain cells of the young embryo are differentiated in that they have very fixed patterns of division that anticipate the later organization of distinct meristematic apices.

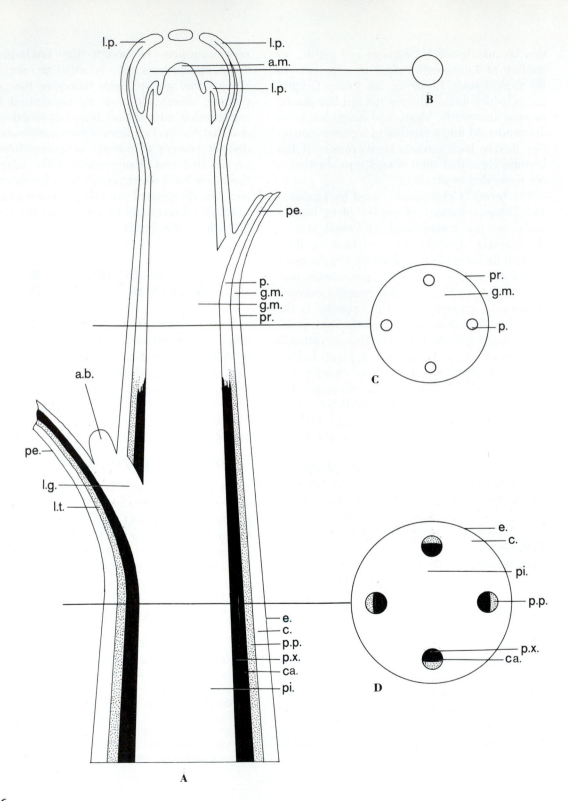

316

FIGURE 13-1
Stem ontogeny. **A.** Diagrammatic median longisection of a shoot apex. **B–D.** Transections of same at levels indicated by lines. a.b., branch primordium; a.m., apical meristem; c., cortex; ca., vascular cambium; e., epidermis; g.m., ground meristem; l.g., leaf gap; l.p., leaf primordium; l.t., leaf trace; p., procambium; pe., petiole; pi., pith; p.p., primary phloem; pr., protoderm; p.x., primary xylem.

externally into organs (leaves, stems, roots) and internally with respect to component tissues. These diverse tissues arise from localized groups of embryonic or **meristematic cells**, which are replicating themselves frequently. Such meristematic cells occur at the stem and root tips and in the young leaves.

In a sense, the stem and root apices of vascular plants recapitulate continually the embryogenic development from the zygote, since these apical groups of undifferentiated cells continue to reproduce themselves throughout the life of the plant, while some of their progeny differentiate. It is most convenient to study development or ontogeny of the individual by means of serial transverse and longitudinal sections of the plant body.

The axis or stem is apparently the fundamental organ in many vascular plants.[4] A majority of vascular plants, however, have axes with both leaves and roots. Leafy axes are often called shoots, as distinct from roots.

Stems may be both subterranean and aerial. Horizontal stems, which are often, but not always, fleshy, are often known as **rhizomes**. Aerial stems are photosynthetic, at least early in their development.

Figure 13-1 represents diagrammatically a median longitudinal section and increasingly

older transections of a developing shoot; Fig. 13-2 is a photomicrograph of a median longitudinal section of such a stem. The apex is usually covered by overlapping leaves to form a **terminal bud.** Leaves in an orderly sequence arise by localized cell division at various points on the stem surface. Each immature leaf is called a **leaf primordium**, and in its axil a small group of meristematic cells represents a **branch primordium,** the precursor of an axillary bud, which may develop into a branch. Origin of branches in this manner is typical of seed plants. The convex stem apex is composed of actively dividing cells accordingly said to be meristematic or embryonic. This region of the stem is designated the **apical meristem,** or, sometimes, the promeristem. In a number of vascular cryptogams a large, prominent **apical cell** may be present (Fig. 19-5A). Apical cells may be tetrahedral and pyramidal, giving rise to cells on three faces (Fig. 13-3), or three-sided or lenticular, with two cutting faces. In many Anthophyta (flowering plants) the apex is differentiated into a superficial layer, the **tunica** (in which most divisions are anticlinal), surrounding an internal **corpus;** the cell form and division patterns differ in these (Fig. 13-4).

At varying distances (in different species) from the apical meristem (Figs. 13-1, 13-2), the cells, although still meristematic, elongate or enlarge, thereby changing form. Thus, there are organized at a level below the apical meristem the three **primary meristems,** namely, the **protoderm,** the **procambium** (sometimes called the provascular tissue), and the **ground meristem** (Figs. 13-1C, 13-2). The procambium, depending on the species, may have the form of a solid central strand, a ring of strands, or a hollow cylinder. From these three regions, as cells mature, there differentiate the **primary mature tissues** of the stem.

The protoderm becomes differentiated into the **epidermis** (Fig. 13-1D). This may produce various types of appendages such as glandular and nonglandular hairs and scales. At maturity

---

[4]Although this has been challenged in the case of certain ferns (p. 433), a number of primitive vascular plants, both living and extinct, consist solely of leafless and rootless axes (Fig. 19-1).

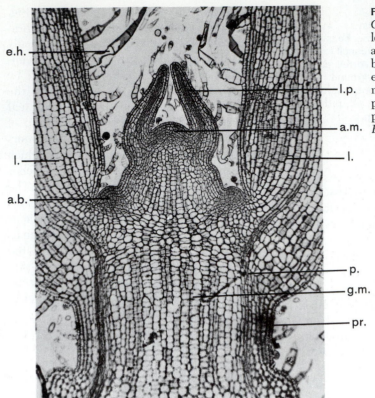

FIGURE 13-2
Coleus blumei. Median longisection of stem apex. a.m., apical meristem; b.p., branch primordium; e.h., epidermal hair; g.m., ground meristem; l., leaf; l.p., leaf primordium; p., procambium; pr., protoderm. ×60. (*After Bold.*)

e.h.

l.p.

a.m.

l.

l.

a.b.

p.

g.m.

pr.

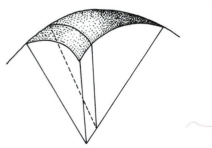

FIGURE 13-3
Tetrahedral type of apical cell from three faces of which derivative cells arise in stems (from all four sides in certain roots). Note that a cell has been cut off from one of the flat faces of the apical cell. (*Modified from Schüepp.*)

the living epidermal cells have their walls impregnated with the fatty substance **cutin,** and a common waxy layer, the **cuticle,** may cover the outer epidermal walls. The epidermis is interrupted periodically by intercellular fissures surrounded by pairs of **guard cells** (Figs. 13-5, 13-19B). The opening and guard cells together form the **stoma** (pl., **stomata**) (Gr. *stoma, mouth*).[5]

Van Cotthem (1970*a,b*) and Dilcher (1974) have summarized the organization, terminology, and classification of stomata. The stomatal complex includes the stoma (pore and guard cells) and those surrounding cells that partici-

_____
[5]Sometimes the term "stoma" is used to refer only to the fissure itself.

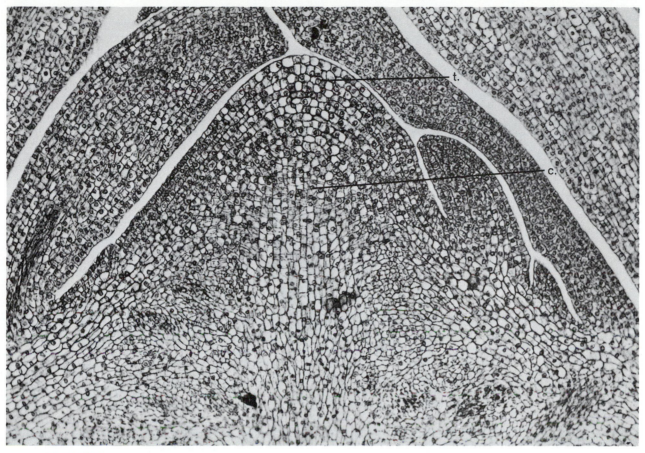

FIGURE 13-4
Yucca whipplei. Median longisection stem apex. c., corpus; t., tunica. ×92.5. *(Courtesy of Dr. H. J. Arnott.)*

pate in the mechanism of opening and closing the pore. In some plants (e.g., *Equisetum*, Fig. 15-5C) these become modified as **subsidiary cells.**

The ground meristem matures into the regions called the **cortex** and **pith.** The outermost cells of the cortex usually contain chloroplasts and, accordingly, are photosynthetic. The inner layers of the cortex often function in storage. The cortex is composed mostly of thin-walled living cells of a type referred to as **parenchyma tissue.** Certain zones of the cortex may be composed of thick-walled cells, which provide support and rigidity to the stem. The cell walls of the innermost layer of the cortex in certain cryptogams and in the rhizomes of other vascular plants may become specially thickened or otherwise modified to form an **endodermis** (Fig. 20-6B).

The pith, when present, occurs in the center of the stem. It is usually composed of relatively thin-walled parenchyma which functions in storage.

The procambium or provascular tissue ulti-

319

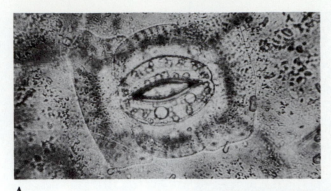

A

FIGURE 13-5
Stomata. **A.** Surface view of stoma from stem epidermis of *Zebrina pendula;* note pore, guard cells, and accessory cells. **B.** Electron micrograph of transection of stoma of *Lemna minor.* **A.** ×250. **B.** ×4500. (*Courtesy of Dr. H. J. Arnott.*)

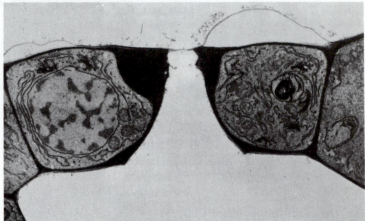

B

mately matures into vascular[6] tissue, the **primary xylem** and **primary phloem,** the former lignified and functioning in the conduction of water and inorganic salts, and the latter cellulosic and functioning in the conduction of sugars and other complex compounds.

STELES

The xylem and phloem, the vascular tissue of the region of primary permanent tissues, together form the **stele** (L. *stela*, rod or column). Provascular tissue assumes different patterns in the various kinds of plants, and when vascular tissue differentiates, different stelar configurations result. These different types will be sum-

marized at this point because they are often useful attributes in discussions of classification and relationship of vascular plants.

The following types of stele (Fig. 13-6) occur: (A) the **protostele** (including the haplostele, the actinostele, and the plectostele), (B) the **siphonostele** (both amphiphloic and ectophloic), including the **dictyostele,** (C) the **eustele,** and (D) the **atactostele.**

Protostele

When the provascular tissue is in the form of a solid cylinder, the xylem and phloem that mature from it will also be in the shape of a solid cylinder, usually with the phloem surrounding the xylem. The type of stele formed in this fashion is called a **protostele.** Protosteles are

---

[6]And other closely associated tissues such as the pith in certain roots.

320

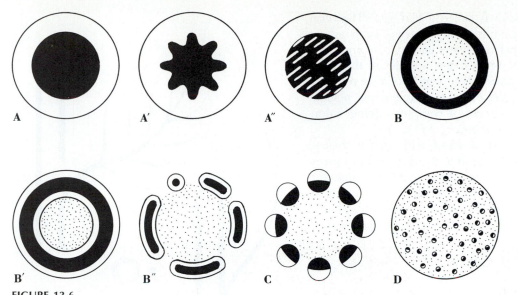

FIGURE 13-6
Types of stele, diagrammatic transections. **A–A″**. Protosteles. **A**. Haplostele. **A′**. Actinostele. **A″**.
Plectostele. **B**. Ectophloic siphonostele. **B′**. Amphiphloic siphonostele or solenostele. **B″**.
Dictyostele. **C**. Eustele. **D**. Atactostele. Xylem, black; phloem, white; pith, stippled.

considered the most primitive, occurring in the earliest vascular plants, and among extant vascular plants considered to be primitive. If the xylem of the protostele is round in cross section, it is called a **haplostele** (Fig. 13-6A). An **actinostele** (Fig. 13-6A′) is a type of protostele in which the xylem cylinder has ridges and furrows, with the phloem concentrated within the furrows. A **plectostele** is a protostele in which phloem is interspersed in masses among xylem cells (Fig. 13-6A″). Among extant plants, protosteles occur in many Microphyllophyta, in some members of the Pteridophyta, and in the Psilotophyta, as well as in the juvenile stems of other groups. Almost all roots are also protostelic.

Siphonostele

In a **siphonostele**, the xylem and phloem form a cylinder around a central pith. In **ectophloic siphonosteles** (Fig. 13-6B) the phloem is restricted to the outer surface of the xylem. An

amphiphloic siphonostele is one in which phloem is on the outside and on the inside of the xylem (Fig. 13-6B′). An amphiphloic siphonostele is sometimes called a **solenostele**. Solenosteles are widely distributed, occurring frequently in ferns.

The separation of **leaf traces** (branches of xylem and phloem to leaves) often influences the configuration of the stele. In some plants the traces are single and delicate, composed of only a few xylem and phloem cells, and their separation from the stele occasions little or no "disturbance" in the vascular tissue of the stem (Fig. 13-7A, C). This is true of the leaf traces of the Microphyllophyta, for example.

In the Pteridophyta, a significantly large branch or several branches of vascular tissue pass out as leaf traces (Fig. 13-1, 13-8B, D). Although we speak of traces as "passing out," they, of course, do not move but differentiate in position from precursor strands of procambium. The apparent interruption of the cylinder of vascular tissue is called a **leaf gap**. The term "gap" may be misleading: there is a gap only in the sense that the xylem and phloem are interrupted, but parenchyma cells fill in this area. Thus, in a given cross section of a stem with a solenostele, the cylinder of xylem and phloem in the stele may not be complete, being "interrupted" by a leaf gap.

In siphonostelic stems with closely spaced leaves, leaf gaps tend to overlap, and in a given cross section of the stem, more than one gap may be seen. The several gaps make it appear as if the siphonostele is dissected. Such a siphonostele is called a **dictyostele**[7] and characterizes many ferns.

Eustele

Where the xylem and phloem occur in discrete collateral (phloem toward the outside of the xylem) or bicollateral (phloem toward the outside *and* the inside of the xylem) strands or bundles, the arrangement is called a **eustele** (Fig. 13-6C). The separations or gaps between adjacent bundles in the cylinder are not the result of emergent leaf traces as they are in a dictyostele. A eustele occurs in the stems of *Equisetum* (Fig. 15-5A) and in most gymnosperms and dicotyledonous flowering plants. Certain of the strands pass into leaves as traces.

Atactostele

An **atactostele** appears to be composed of many (thousands in some plants) vascular strands throughout the living tissue of a stem (Fig. 13-6D). According to some authors (e.g., Beck, Schmid and Rothwell, 1983) an atactostele may actually be regarded as a modified

---

[7]The original definition of a dictyostele restricted the usage to netlike steles in which the separated strands of the dissected solenostele had, in transection, a concentric arrangement (phloem surrounding xylem) of vascular tissue resulting from the dissection of a solenostele. The term is sometimes used less restrictively to include also those gap-dissected siphonosteles with collateral xylem and phloem, derived presumably from ectophloic siphonosteles.

FIGURE 13-7
A, B. Segments of longitudinal bisections of stems. **A.** Stem with microphyllous leaf. **B.** Stem with megaphyllous leaf. **C, D.** Transections at levels of **A** and **B** indicated by dotted lines. Steles, traces, and veins indicated in heavy black; note leaf gaps on **B** and **D** and their absence in **A** and **C**.

eustele, with some of the strands following courses toward the center of the stem before bending outward and some functioning as leaf traces. Atactosteles occur primarily in stems of monocotyledonous flowering plants.

The experimental investigations of Wetmore and Rier (1963) on the causal relationship in the differentiation of vascular tissues have given us some insight into the factors involved. Using undifferentiated, parenchymatous callus in tissue culture, Wetmore and his associates have demonstrated that grafting (insertion) of buds of the same species will induce the differentiation of nodules of vascular tissue in the callus which would not otherwise develop. This indicates

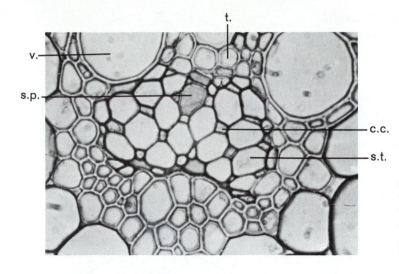

FIGURE 13-8
Zea mays; corn. Transection of portion of phloem region of vascular bundle. c.c., companion cell; s.p., sieve plate; s.t., sieve tube; t., tracheid; v., vessel. ×300.

clearly that vascular differentiation is under chemical control. These investigators have shown further that localized application of auxin and sugar will result in localized induction of vascular tissues. The concentration of sugar (glucose or sucrose) appears to be critical. At 1.5 to 2% sucrose, xylem alone develops. If the concentration is increased to 3 to 3.5%, the vascular nodules develop xylem toward the center of the callus and phloem toward the periphery, and a cambium may develop and join the ring of vascular nodules. Concentration of 4 to 4.5% sucrose stimulates production of a preponderance of phloem. Application to the callus of the auxin-sugar mixture through capillary pipettes resulted in the formation of a cylinder of xylem and phloem with cambium (see below) between; these simulated a siphonostele. These investigations are of significance in having demonstrated a causal role of a growth hormone (auxin) and sugars, the latter in varying concentrations, in the differentiation of vascular tissues.

What is the evolutionary relationship among these various types of stele? It has been hypothesized that siphonosteles developed by progressive replacement of xylem elements by parenchyma cells in the center of protosteles. This hypothesis is supported by the occasion of persistence of xylem elements (tracheids) in pith. This process may be designated **medullation,** a reference to the pith as "medulla." The dissection of siphonosteles in plants with short internodes by emergent leaf traces and associated gaps in many ferns suggested to some botanists that eusteles and atactosteles represent phylogenetic end results of the process of dissection. An alternate scheme (Namboodiri and Beck, 1968) proposes that eusteles, especially those of gymnosperms, arose by fragmentation of lobed prosteles into discrete vascular strands.

NATURE OF VASCULAR TISSUE

At any given level, the primary phloem usually differentiates from cells of the procambium that are situated toward the outer part of the stem or root axis. The first phloem to mature, the **protophloem,** does so before elongation of the stem has ceased. Thus, much of the protophloem may be crushed and obliterated. The **metaphloem** matures from procambium after elongation of the stem has stopped.

In the seedless vascular plants and in gymnospermous seed plants, the conducting agents of the phloem are **sieve cells.** In the contiguous (lateral and/or terminal) walls of these, specialized **sieve areas** occur. These are places on the wall with fine pores through which delicate strands of protoplasm effect continuity of the cells. In almost all flowering plants (angiosperms), the conducting elements of the phloem, **sieve tube members,** are joined in a series, the **sieve tube;** the individual members are separated terminally by **sieve plates.** The latter may consist of single sieve areas (simple sieve plates) or of several sieve areas (complex sieve plates) (Fig. 13-8). In either case, the pores are usually larger than those in the sieve areas of nonangiosperms; they are lined with callose and transversed by relatively thick strands of protoplasm. On the lateral walls of the sieve tube members are nonfunctional sieve areas resembling those of nonangiosperms. For a succinct and more comprehensive review of phloem, see Evert (1984).

The sieve cells and sieve tube members are usually enucleate at maturity and, in the case of sieve tubes, are flanked by smaller, nucleate **companion cells** (Fig. 13-8). In addition, parenchyma cells (nonspecialized living cells, which here function in storage) and lignified fibers (supporting cells) may be present in phloem.

With respect to the pattern or direction of differentiation of primary xylem from procambium, four relationships are common (Fig. 13-9). Their recognition depends on a distinction between two kinds of primary xylem, protoxylem and metaxylem. **Protoxylem** is the first-formed primary xylem. Its elements are usually smaller in diameter, having differentiated before elongation of the organ has ceased. Lignin is deposited within the cellulose wall in the form of rings

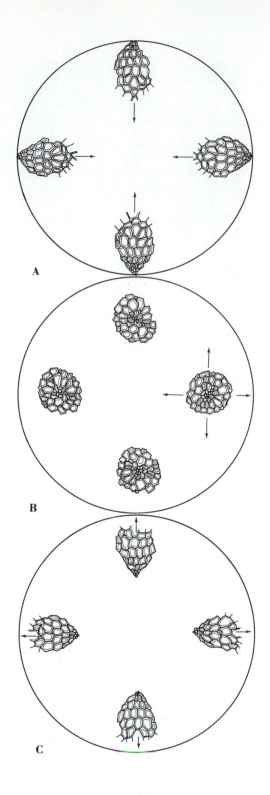

FIGURE 13-9
Diagrammatic illustrations indicating (by arrows) some common patterns of primary xylem differentiation from procambium. **A.** Exarch. **B.** Mesarch. **C.** Endarch. Protoxylem cells small, metaxylem elements large, in each case. Synchronous type, consisting of metaxylem only, not illustrated.

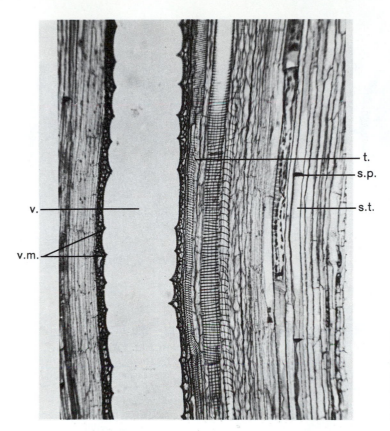

FIGURE 13-10
Cucurbita sp. Longisection of vascular tissue. s.p., sieve plate; s.t., sieve tube; t., tracheid, v., xylem vessel; v.m., limits of one vessel member. ×300.

(annual) or helices. In contrast, the **metaxylem** differentiates after the protoxylem, its elements are larger in diameter, and they mature only after elongation of the organ has ceased. There is considerably more secondary wall material in the form of horizontal bars of lignin on each wall, connected at the corners (scalariform). In the most specialized metaxylem, the entire wall is covered with lignin. In places the secondary wall is separated from the primary wall in domed structures, each with a perforation at the top of the dome. These structures are called **circular bordered pits.** However, there may be transitions between protoxylem and metaxylem. The four common patterns of primary xylem differentiation are **exarch,**[8] **mesarch, en-**

---

[8]"Arch" here means first; hence, first point of origin.

darch, and **synchronous.** Three of these are illustrated in Fig. 13-9 and explained in its legend. An endarch protostele is frequently referred to as being **centrarch.**

The number of points at which protoxylem differentiates varies with the organ and the species. An axis with only one group of protoxylem is said to be **monarch;** those with two are **diarch,** with three, **triarch,** and so on.

The primary xylem may contain the water-conducting elements tracheids and/or vessels (Figs. 13-10, 13-11), parenchyma cells, and fibers. One or more of these elements may be absent.

Tracheids are elongate, with tapering, interlocking walls, nonliving at maturity, their secondary walls occurring in various patterns of cellulose and lignin deposition. These patterns of

325

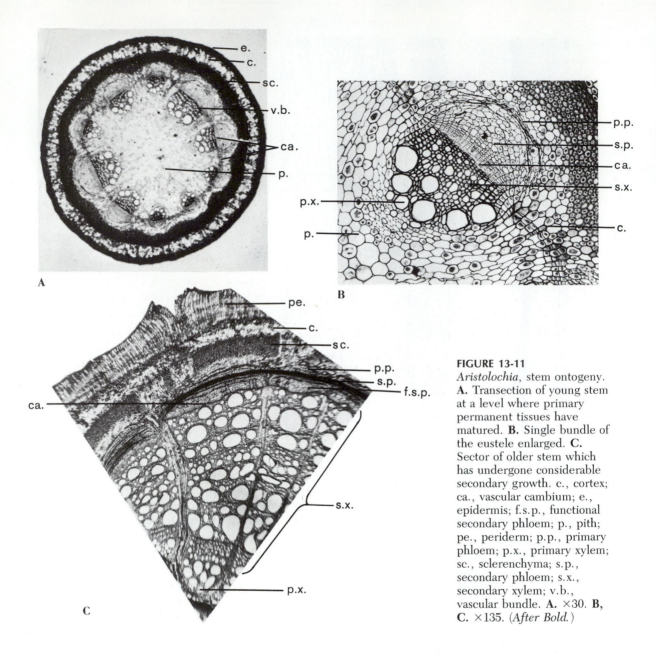

FIGURE 13-11
Aristolochia, stem ontogeny.
A. Transection of young stem at a level where primary permanent tissues have matured. **B.** Single bundle of the eustele enlarged. **C.** Sector of older stem which has undergone considerable secondary growth. c., cortex; ca., vascular cambium; e., epidermis; f.s.p., functional secondary phloem; p., pith; pe., periderm; p.p., primary phloem; p.x., primary xylem; sc., sclerenchyma; s.p., secondary phloem; s.x., secondary xylem; v.b., vascular bundle. **A.** ×30. **B, C.** ×135. (*After Bold.*)

lignification, in order of increasing amounts of lignin deposited, are annular, helical, scalariform, and pitted (elongate to circular, often transitional) (Fig. 13-10). Intermediate patterns occur.

Vessels are composed of series of tracheidlike members, the **vessel members,** in which part of or all of the end walls have been dissolved, leaving perforations (Fig. 13-10); these are thought to facilitate movement of liquids. Vessel members or elements differentiate from a series of procambium cells, while each tracheid arises from only a single one. There is strong evidence that vessels have arisen phylogenetically from tracheids by gradual dissolution of part of the sloping, contiguous terminal walls of adjacent tracheids. Primitive vessels are composed of elongate tracheidlike segments, angular in transection, with oblique perforation plates. The segments of the most specialized vessels are short, oval to circular in transection, have single large terminal perforations, and are quite unlike tracheids. Vessels have arisen both from tracheids with circular and from those with elongate pits, as well as from elements with secondary thickenings characteristic of protoxylem and early metaxylem. The changes in cellular organization during xylem differentiation have been summarized by Torrey, Fosket, and Hepler (1971).

Structural support of the primary stem is effected by the amount of xylem and the degree of its lignification as well as by the presence of supporting cells within the vascular tissue or elsewhere. Fibers are elongate, heavily lignified cells, often with interlocking ends, which strengthen stems. Fibers and other lignified cells in the cortex (collectively termed **sclerenchyma**) also may contribute to support of the primary stem.

SECONDARY GROWTH

In most vascular cryptogams and in many flowering plants the axis is entirely primary in ontogeny—that is, the component tissues differentiate entirely from the promeristem and primary meristems. In a few vascular cryptogams (e.g., *Isoetes*, p. 363, *Botrychium*, p. 392) and in many seed plants, however, the primary tissues (Fig. 13-11A, B) are augmented by the formation and differentiation of additional cells throughout the life of the individual. This is called **secondary growth** because the added cells are derived from the cambium and cork cambium, which are secondary meristems.

The vascular **cambium** is a zone of meristematic cells between the primary xylem and phloem. At least that part of the cambium that lies between the primary xylem and phloem originates from procambium. In axes with siphonosteles, the entire cambium would have such an origin. In eusteles and other dissected steles, the parenchyma cells between the vascular strands become meristematic at the inception of cambial activity within the strand so that a complete cylinder of cambium arises (Fig. 13-11). A preponderance of cells formed by tangential divisions of the cambium differentiate into the **secondary xylem,** which lies between the cambium and the primary xylem (Fig. 13-11C). Fewer of the outer derivatives of the cambium differentiate into **secondary phloem;** this lies between the vascular cambium and primary phloem (Fig. 13-11C). Accordingly, the vascular tissues increase radially in thickness as long as secondary growth continues.

Where this is of long duration, stresses arise that rupture the epidermis and cortical tissue. This or other factors stimulate the dedifferentiation of cells in the outer cortex into meristematic cells. This peripheral zone of meristematic cells, the **cork cambium**, or **phellogen**, is either a continuous cylinder or in the form of localized strips. The outer derivatives of the phellogen are **cork cells**, or **phellem**, the impervious walls of which are impregnated with suberin. The internal derivatives are parenchyma cells, which may augment the cortex of the stem. As the corky layers replace the epidermis, there are developed **lenticels** that are loosely organized derivatives of the cork cambium (Fig. 13-12). They apparently replace the stomata in function. Long-continued cambial activity results in woody stems as in trees and shrubs.

THE ROOT

The stems of all vascular plants except the Psilotophyta (Chapter 19) and *Salvinia* and a few unusual ferns (Chapters 17, 18) bear roots. Roots differ externally from stems in not bearing

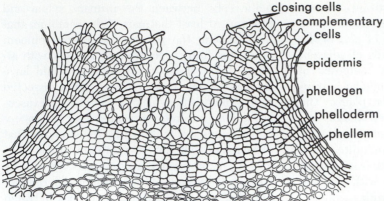

closing cells
complementary cells
epidermis
phellogen
phelloderm
phellem

FIGURE 13-12
Transection of a lenticel of *Prunus avium* stem. (*After Devaux, from Eames and McDaniels.*)

leaves and in lacking nodes and internodes. (There are some primitive plants with leafless stems, however.) Some of the epidermal cells of most roots, furthermore, produce absorptive protuberances called **root hairs.** The apical meristem of the root is covered by a mantle of cells, the **root cap** (Fig. 13-13).

Branch roots, in contrast to appendages of the stem, originate endogenously by proliferation of cells of the **pericycle** (Fig. 13-14), a zone surrounding the vascular tissues of the root stele and just within the endodermis; in some species they arise from the endodermis. They grow through the cortex and epidermis into the soil.

The primary vascular system in many roots is organized with a solid, central strand of xylem, which is exarch (Fig. 13-15). Roots may be monarch to polyarch. Between the protoxylem points in roots occur groups of primary phloem cells, so that primary xylem and phloem are radially alternate in roots (Fig. 13-15B) rather than collateral or bicollateral as they are in

FIGURE 13-13
Zea mays, corn. Median longisection of root tip. a.m., apical meristem; g.m., ground meristem; p. protoderm; pr., procambium; v.p., xylem vessel precursors; r.c., root cap. ×40. (*Courtesy of Dr. W. G. Whaley.*)

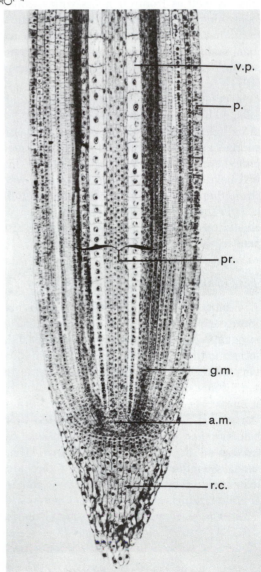

v.p.

p.

pr.

g.m.

a.m.

r.c.

328

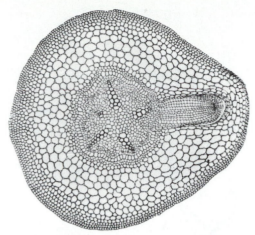

FIGURE 13-14
Phaseolus sp., bean. Transection of root showing origin of branch roots. ×65. (*After Gibbs.*)

stems. The roots of some plants have a central pith.[9]

_____

[9]This is true of the roots of *Ophioglossum* (p. 389) and the Marattiales (p. 395) as well as of certain flowering plants. Monocotyledonous plants with pith-containing roots also have smaller roots that lack pith.

In all woody plants and in many herbaceous types, cambial activity in the root also results in secondary growth and increase in diameter and woodiness of that organ. Here the outer tissues (cortex and epidermis) may be sloughed off during secondary growth and replaced by corky layers, formed from a phellogen that originates in the pericycle.

THE LEAF

Leaves arise ontogenetically as meristematic emergences or primordia at the surface of young stems (Figs. 13-1, 13-2). Several layers near the stem surface may undergo active division to initiate the leaf. The primordium at first consists largely of axis, but later marginal meristems, from which the bulk of the leaf blade develops (Fig. 13-16), are organized on opposite sides of the axis. In some leaves—those of many ferns, for example—a prominent apical cell functions in leaf ontogeny.

FIGURE 13-15
Ranunculus sp. **A.** Transection of root. **B.** Stelar region of preceding. c., cortex; e., epidermis; en., endodermis; p., primary phloem; pe., pericycle; s., stele; x., primary xylem. **A.** ×25. **B.** ×200.

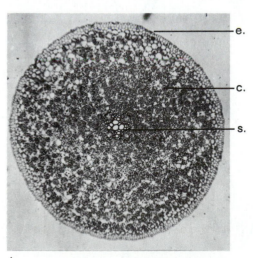

A

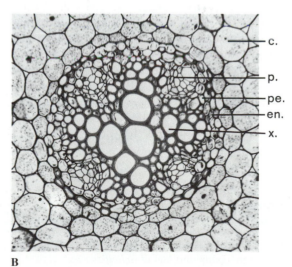

B

329

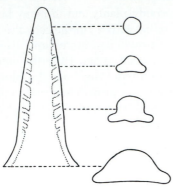

Tobacco: leaf ontogeny. Diagrams of leaf
primordium and transections at four levels from
apex to base; note midrib and marginal meristem in
three lower transections. (*After Avery.*)

At maturity, leaves vary widely in both external and internal organization (Fig. 13-17). They
may be **simple leaves,** with undivided blades, or
variously **compound,** with blades divided into
leaflets. They may be stalked (**petiolate**) or
sessile (lacking petioles) in their attachment to
the axis.

The vascular tissue of the leaf, the veins, may
be unbranched (Microphyllophyta, *Equisetum*)
or variously branched. Open **dichotomous** venation (Fig. 13-18A) occurs in many ferns and
Ginkgo. In many monocotyledonous flowering
plants, the venation is **striate** (Fig. 13-18B), the
main veins being elongate and connected at
intervals by delicate, transverse, ladderlike
veins. In certain ferns and in many dicotyledonous flowering plants, venation is **reticulate,**
the veins being abundantly branched and anastomosing (Fig. 13-18C). Intermediate types of
venation also occur.

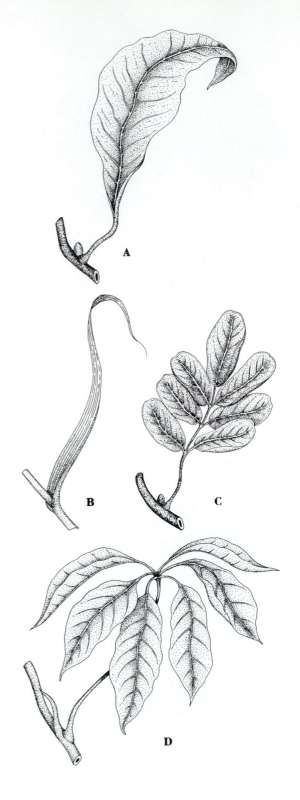

FIGURE 13-17
Leaves; gross morphology. **A.** Simple petiolate leaf
of croton. **B.** Sessile, simple leaf of a grass. **C.**
Pinnately compound leaf of a legume, *Sophora.* **D.**
Palmately compound leaf of *Schefflera* sp.

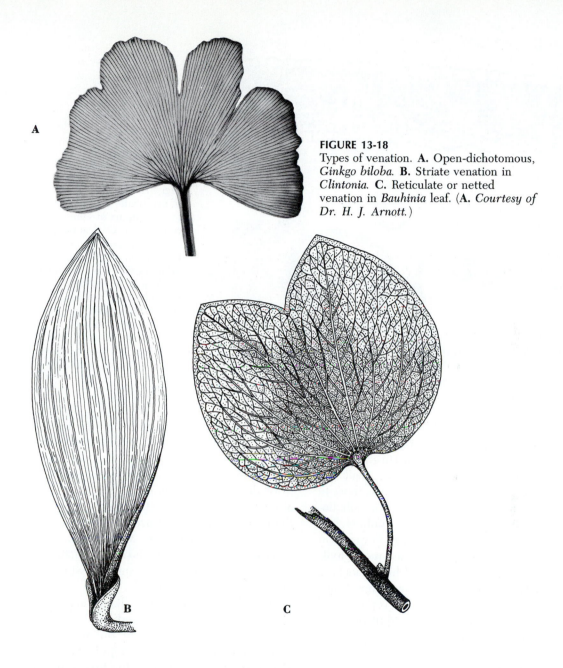

FIGURE 13-18
Types of venation. **A.** Open-dichotomous, *Ginkgo biloba.* **B.** Striate venation in *Clintonia.* **C.** Reticulate or netted venation in *Bauhinia* leaf. (**A.** *Courtesy of Dr. H. J. Arnott.*)

Histologically, most leaves are relatively simple, being covered by a more or less heavily cutinized epidermis (with stomata most frequently on the lower surface, although they may occur in the upper surface, or both epidermal layers); veins and photosynthetic cells, the **mesophyll,** lie within. In some species the latter is differentiated into palisade and spongy zones (Fig. 13-19A, C).

Comparative morphology distinguishes be-

331

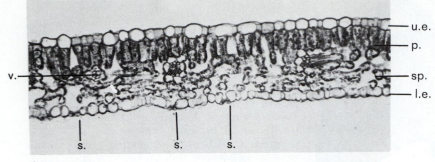

A

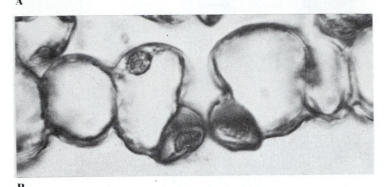

B

FIGURE 13-19
Ligustrum sp. **A.** Transection of leaf blade. **B.** Enlarged view of stoma of same transection. **C.** Scanning electron micrograph of a transection of a mulberry leaf. (*Morus alba*). l.e., lower epidermis; p., palisade mesophyll; s., stoma; sp., spongy mesophyll; u.e., upper epidermis; v., vein in transection. **A.** ×30. **B.** ×550. **C.** ×135. (**A, B.** *After Bold.* **C.** *Courtesy of Prof. H. J. Arnott.*)

tween two categories of leaves in vascular plants, **microphylls** (in the Microphyllophyta) and **megaphylls**, which occur in all other groups.[10] **Microphylls** are usually of small size (although there are exceptions, such as those of *Isoetes*, p. 363, and certain fossil Microphyllophyta, p. 496) and have single, unbranched veins, traces to which leave no gap in the stem stele (Figs. 13-9A, C). By contrast **megaphylls** are usually large and have richly branching veins, and their leaf traces leave gaps (Figs. 13-9B, D) in the stem stele.[11] Microphylls and megaphylls may be of quite different phylogenetic origin (see pp. 472–473).

---

[10]Leaves of Arthrophyta were apparently derived from minor branch systems.

[11]This last attribute is not always valid: Leaf gaps are absent in megaphyllous plants with protosteles and eusteles, although grooves (corresponding to gaps) may be present in the steles on protosteles of megaphyllous plants.

The scalelike emergences on the stems of certain Psilotophyta (Figs. 19-1 to 19-3), which lack veins, may well represent precursors of microphyllous leaves. In some species, traces leave the stem stele but end in the cortex short of the leafy scale base. Microphylls differ from these only in that the trace continues into the leaf to form its single vein. Microphylls then, if this view is correct, would be simple enations or emergences from the stem surface, which ultimately became vascularized.

Megaphylls, on the other hand, are usually considered to represent modified stems that originated in evolution from the more distal branches of axes. In this development (pp. 474, 475), it has been suggested, the first step was a change from dichotomous to monopodial branching by dominance and **overtopping** of one dichotomy. The second modification was restriction of the remaining dichotomies to one plane, **planation.** Finally, there occurred addi-

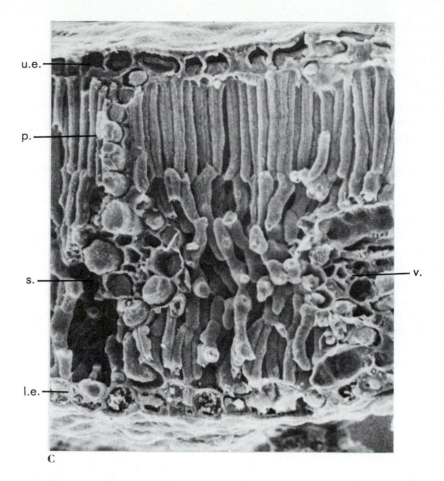

u.e.

p.

s.

l.e.

v.

C

tion of tissue between the axes (**webbing**) to form the leaf blade. These did not all necessarily occur in the same plant or in that order. The presence of open dichotomous venation in the leaves of many ferns and the prolonged apical growth in development of fern leaves are cited as evidence for this interpretation. Although it was stated above that the supposed phylogenetic origin of the microphyll was quite different from that of megaphylls, there is an alternative interpretation, namely, their origin by reduction (Fig. 20-5). The phenomena of overtopping, planation, webbing, and reduction are discussed further in Chapter 20.

The preceding paragraphs have presented a brief summary of some fundamental concepts regarding the organization of the vegetative body of the vascular plants. The organization and functioning of the reproductive structures of the vascular plants are discussed with the representative types where these are described in the remaining chapters of this text.

More detailed and comprehensive discussion of the material briefly referred to in this chapter will be found in the texts by Esau (1965), Cutter (1969, 1971), and Fahn (1982).

DISCUSSION QUESTIONS

1. Define vascular tissue and state the functions of its components.

333

2. What is usually meant by the term "land plants"? What divisions of the plant kingdom have terrestrial representatives?

3. What divisions of vascular plants had evolved by the Devonian?

4. What did Cryptogamia mean to Linnaeus?

5. What are cryptogams in current usage?

6. On what characteristics is the division Tracheophyta based?

7. How does vascular tissue originate in ontogeny?

8. Define or explain: rhizome, bud, axillary bud, leaf primordium, meristematic, promeristem, apical cell, primary meristems, tunica, corpus, protoderm, procambium, ground meristem, epidermis, cuticle.

9. Define or explain: primary xylem and phloem, secondary xylem and phloem; protophloem and metaphloem; protoxylem and metaxylem; sieve cell, sieve tube, companion cell, tracheid, vessel; exarch, endarch, and mesarch xylem development; monarch, diarch, and triarch protosteles; node, internode, leaf trace, leaf gap.

10. Distinguish between microphylls and megaphylls. What hypotheses are there in explanation of the origin of these leaf types?

11. What is the supposed phylogenetic relationship between tracheids and vessels?

12. Define and make three-dimensional diagrams of a protostele, haplostele, plectostele, actinostele, siphonostele, solenostele, dictyostele, eustele, and atactostele.

13. Define or explain: cambium, phellogen, cork cambium, collateral bundle, venation.

14. In what respects, internal and external, do roots and stems differ?

Chapter 14
VASCULAR CRYPTOGAMS I:
DIVISION MICROPHYLLOPHYTA

This chapter and Chapters 15–19 will discuss divisions of cryptogams with living representatives. Members of the division Microphyllophyta (Gr. *mikros*, small, + Gr. *phyllon*, leaf, + Gr. *phyton*, plant), a group accorded only subdivisional rank under the name Lycopsida in many schemes of classification (inside back cover), are readily distinguishable in having vascularized leaves with single veins,[1] having roots, and by the intimate association of their sporangia with fertile leaves known as **sporophylls** (Gr. *spora*, spore, + Gr. *phyllon*, leaf). Some of these plants are commonly known as club and spike mosses,[2] because their small size and their mosslike leaves, closely arranged on the stems, suggest mosses; furthermore, the aggregation of the sporophylls of certain species into terminal groups (Figs. 14-1B, 14-2) has suggested the terms "club" and "spike."

Whether they are given class or divisional rank, two series are usually distinguished in the classification of these plants (p. 39). In one series, exemplified in this chapter by the extant *Selaginella* (Figs. 14-17 through 14-35) and *Isoetes* (Fig. 14-36) and by the extinct *Selaginellites* and the Lepidodendrales (p.496), each leaf produces a small, basal protuberance, the **ligule** (L. *ligula*, little tongue) (Fig. 14-24C). These genera are grouped in the class Glossopsida (Gr. *glossa*, tongue, + Gr. *opsis*, appearance of), sometimes called the Ligulatae; in the other series, the Aglossopsida, the ligule is absent. The extant *Lycopodium* and *Phylloglossum* (Fig. 14-1) and the extinct *Protolepidodendron* (p. 494) belong in the latter category.

CLASS 1. AGLOSSOPSIDA

The *Aglossopsida* include the single order Lycopodiales, family Lycopodiaceae with extant species.

---

[1]See, however, p.351.

[2]They were, in fact, classified with the mosses by Linnaeus.

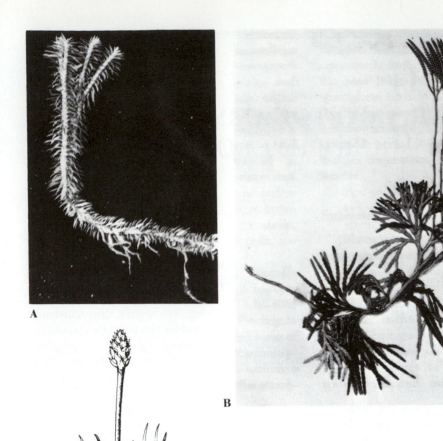

FIGURE 14-1
A. *Lycopodium lucidulum.* Portion of plant; note aerial branches, rhizome, and roots. **B.** *Lycopodium digitatum;* note strobili on special branches. **C.** *Phylloglossum drummondii.* **A, B** ×0.5. **C.** ×4. (**C.** *After Bower, from Engler and Prantl.*)

Lycopodium[3]

Introduction and vegetative morphology.

Lycopodium (Gr. *lykos*, wolf, + Gr. *pous*, foot) (Figs. 14-1A, B, 14-2) and *Phylloglossum* (Gr. *phyllos*, leaf, + Gr. *glossum*, tongue) (Fig. 14-1C), the only living genera included in this class, comprise 200 (Sporne, 1970) to 400 (Bierhorst, 1971*a*) species. *Phylloglossum*, of which only a single species is known, occurs only in Australasia. Species of *Lycopodium* are widely distributed and are familiarly known as "ground pines," "trailing evergreens," and "club mosses." Most species are perennials, living on the forest floor in temperate climates, but some tropical species are pendulous epiphytes. The plants are rather firm herbs with dichotomously (Fig. 14-1A) or pseudomonopodially (Fig. 14-1B) branched stems. In the latter case the

[3]The critical reading of this section by Professor D. P. Whittier is greatly appreciated by the authors.

336

A

B

branch system consists of a main axis that supports minor branches. The latter represent members of dichotomies that failed to develop to as great an extent as their counterparts.

The genus *Lycopodium* is a large one and it has not infrequently been suggested that it be divided into several genera. Wilce (1972) has discussed this but recommends retention of the diverse species within the single genus *Lycopodium* in three subgenera—*Urostachya, Lepidotis,* and *Lycopodium.* Bruce (1976*b*) has tabulated the sporophytic and gametophytic characteristics of these three subgenera.

The leaves are small (Figs. 14-1, 14-2, 14-3,

A

B

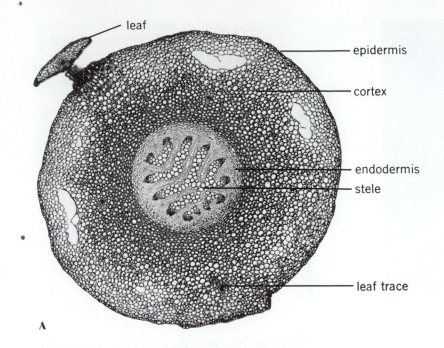

leaf

epidermis

cortex

endodermis

stele

leaf trace

A

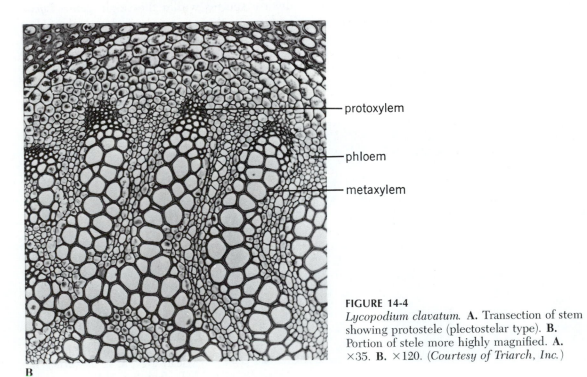

protoxylem

phloem

metaxylem

B

FIGURE 14-4
Lycopodium clavatum. **A.** Transection of stem
showing protostele (plectostelar type). **B.**
Portion of stele more highly magnified. **A.**
×35. **B.** ×120. (*Courtesy of Triarch, Inc.*)

14-6), sessile, and spiral or whorled in arrangement. Chu (1974) described the anatomy of the leaves of 31 species. The stomata may be adaxial, abaxial, or both, depending on the species. In *L. digitatum,* the leaves are much reduced and almost scalelike (Fig. 14-3B). All species have branching rhizomes from which aerial branches develop. The roots are delicate and dichotomously branched and are scattered along the underground portions of the stem (Fig. 14-1A). The roots of some species of *Lycopodium* arise deep within the stems, not superficially like those of most vascular plants. They grow through the cortex and emerge into the soil at points somewhat removed from the level of their origin. Stevenson (1976) described the phyllotaxis, stelar morphology, and gemmae of *L. lucidulum.*

The pattern of stem and leaf development was investigated by Freeberg and Wetmore (1967), who emphasized that it is difficult to analyze. Development of both the stem and root of *Lycopodium* may be traced to an apical group of meristematic cells, no one of which is specially differentiated. A central procambium strand develops some distance from the promeristem in both stem and root and ultimately gives rise to an exarch[4] protostele in both (Figs. 14-4, 14-5). The stem stele is bounded by several layers of pericycle cells and sharply delimited by a well-developed endodermis. The epidermis contains stomata and guard cells. The arrangement of xylem and phloem in the stem varies in different species, in accordance with the degree of ridging or dissection of the central xylem mass. It also varies within a species in accordance with stem diameter. In species in which the xylem seems to consist of discrete masses in transverse section (Fig. 14-4), serial transverse sections reveal that the apparently discrete units are lobes that join each other at different levels of the stem. In species with dissected xylem masses, phloem cells develop

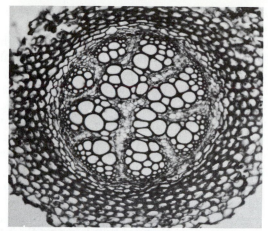

FIGURE 14-5
Lycopodium digitatum.
Transection of root. ×60.
(Courtesy of Mrs. Elizabeth Pixley.)

between them, forming a **plectostele.** In species with continuous xylem, the phloem is present external to the xylem. Functional phloem is, however, always separated from the xylem by a layer of parenchyma cells. The conducting elements of the xylem are all tracheids. The first formed are lignified in annular fashion; these are followed later in development by tracheids with circular bordered pits, and last, by those with elongate (scalariform) bordered pits. The sieve cells of the phloem have tapered ends.

The sieve cells may have thickened walls early in development, but in those that do, the wall is eroded, seemingly by the protoplast, during maturation (Warmbrodt and Evert, 1974).[5] Sieve areas with plasmodesmata and pores occur in both the terminal and lateral walls of the sieve cells in *L. lucidulum.* The ultrastructural organization of the phloem of *L. clavatum* has been described by Hébant et al. (1978).

Two types of mucilaginous canals are pres-

---

[4]Sometimes weakly mesarch.

---

[5]For a discussion of phloem in "lower" vascular plants consult Warmbrodt (1980) and Evert (1984).

ent in some species (subgenus *Lepidotis*) or only one in the subgenus *Lycopodium* (Bruce, 1976a). They are **veinal,** that is, associated with the leaf vein, or **basal,** the latter a cylindrical network in the cortex of the strobilar axis.

The root stele varies with the size of the root in a given species and also with the species. The exarch stele may be crescentic with phloem in the concavity of the xylem crescent, as in *L. lucidulum,* or a plectostele, as in *L. digitatum* (Fig. 14-5); other configurations occur.

The tip of the root is covered by a root cap. The root hairs are anomalous in that they develop in pairs from the epidermal cells. Roots of *Lycopodium* usually branch dichotomously; the branches originate through the reorganization of the apical meristematic cells into two groups.

The leaves of *Lycopodium* arise by localized growth of groups of superficial cells near the stem apex. As they grow, each develops a central procambium strand, which finally differentiates into a vascular strand of tracheids surrounded by scattered sieve elements and parenchyma. The chlorenchyma of the leaf is rather uniform in structure, small intercellular spaces being present. Depending on the species, the stomata occur either on the epidermis of both leaf surfaces or only on the lower epidermis. The vascular supply of the leaf is connected to the stem stele by a leaf trace (Fig. 14-4A), which is connected with the protoxylem of the stem stele. Leaves with a single unbranched vein, the traces of which leave no parenchymatous gap in the stele above its point of departure, are said to be **microphyllous.** It should be emphasized at this point that small size is merely a secondary attribute of some microphylls; their single, unbranched veins and the absence of leaf gaps near the point of departure of their traces from the stem stele are their distinguishing features.

Reproduction: the sporophyte. Species of *Lycopodium,* such as *L. lucidulum* (Fig. 14-1A), in which the sporophylls are not localized in compact aggregations, in many instances pro-

FIGURE 14-6
Lycopodium lucidulum. b., propagative gemma or bulbil; s., axillary sporangia (their sporophylls removed). ×4.

duce special mechanisms for vegetative propagation, namely, **gemmae** or **bulbils** (Fig. 14-6). These consist of a proximal, enlarged base and a distal, short axis with several pairs of leaves. The distal portion is abscised and may develop into a young sporophyte under favorable conditions.

The sporophyte of *Lycopodium* at maturity produces spores in rather massive, kidney-shaped sporangia which are often borne on short

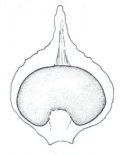

FIGURE 14-7
Lycopodium digitatum. Adaxial view of sporophyll. ×8.

A   B

C

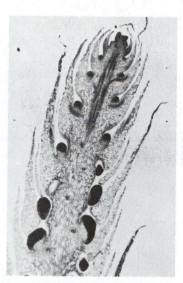

FIGURE 14-9
Lycopodium sp. Median longisection of immature
strobilus; note developing sporangia and
sporophylls. ×30.

stalks either on the leaf base or in the axil of the
leaf and stem. In some species, *L. lucidulum,*
for example, the fertile, sporangium-bearing
leaves, the **sporophylls,** are entirely similar to
sterile leaves and occur in zones among them
(Fig. 14-6); such species are considered to be
primitive. In species such as *L. digitatum,* on
the other hand, localization of the sporophylls
(Fig. 14-7) into a terminal conelike structure,
the **strobilus** (Figs. 14-1B, 14-2, 14-8), is accompanied by their modification into nonphotosyn-
thetic, scalelike structures that are reduced in
size. This condition is interpreted as the most
advanced in the genus. A **strobilus** is a stem
having short internodes and fertile appendages
(Fig. 14-9). In the Microphyllophyta these are
fertile leaves or **sporophylls.** Some morphologists regard the *lucidulum* type as a strobilus
constituting an entire plant. According to this
view, the sterile leaves represent secondarily
sterilized sporophylls.

The strobilus develops from an apical meristem, and its vascular structure is similar to that
of the vegetative axis. Each sporophyll is supplied from the stele by a single trace, as are the
vegetative leaves of the plant.

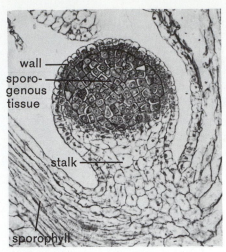

FIGURE 14-10
Lycopodium obscurum. Longisection of immature sporangium. ×250.

Labels on figure: wall, sporogenous tissue, stalk, sporophyll

The individual sporangium (Fig. 14-10) arises by periclinal division of a group of surface (protodermal) cells near the base of an embryonic leaf. In further development it may become shifted to the stem, depending on the species. The outer cells resulting from original periclinal cleavage, as a result of continuing anticlinal and periclinal divisions, form a several-layered sporangial wall and tapetum, while the inner divide to form a mass of sporogenous tissue. The tapetum completely encloses the latter in young sporangia but disintegrates, serving a nutritive function during sporogenesis. The sporangia in some species are stalked.

As the sporangium develops, the sporogenous cells become isolated and spherical and ultimately produce a tetrad of spores each. These finally separate and secrete a wall with ornamentation, which varies with the species. The spore wall consists of two layers: an inner **intine** and the outer **exine.** Pettit (1976*a*) reported that the developing exine of *L. gnidioides* has channels through which substances move into the spore from the locular fluid. The spores are

yellow at maturity. Dehiscence of the sporangium occurs along a line of cells running across the upper surface of the reniform sporangium and hence transverse to the long axis of the sporophyll. In species with compact strobili, this is preceded by slight elongation of the internodes and by drying and spreading of the sporophylls, which are facilitated by the shrinkage of a cavity in the lower side of each sporophyll.

Reproduction: The Gametophyte. Of the approximately 400 species of *Lycopodium,* the gametophytes of only 28 species are reasonably well known. Most of these were described from specimens collected in nature, but the gametophytes of a few have been grown from spores in laboratory culture (Freeburg and Wetmore, 1957; Whittier, 1977, 1981). Most textbook descriptions of the gametophyte draw heavily on the publications of Bruchmann (1898). The latter classified the gametophytes as belonging to five types, but more recent investigations (Bruce, 1979*a,* 1979*b*) seem to indicate there are only four.

Some species (e.g., *L. digitatum*) have subterranean gametophytes which are disc- or carrot-shaped (Fig. 14-11, 14-12), while those of others are epiterranean. In nature, except on rare occasions, they are white, but may develop chlorophyll, at least in laboratory cultures, when exposed to light. Both types of gametophytes contain mycorrhizal fungi which apparently enter initially through the unicellular rhizoids. The fungi may be present in special zones especially within the fleshy base of the gametophytes (Fig. 14-12E).

Of the superficially green type, the gametophytes of *L. carolinianum* are, perhaps, best known (Bruce, 1979*b*). They may reach 2.7 mm in length. Each consists of a tapering basal portion and a more robust portion (Fig. 14-12F) both derived from a *lateral* meristem. The base is radial and has unicellular rhizoids. The main portion of the gametophyte is dorsiventral with dorsal photosynthetic lobes. Both portions of

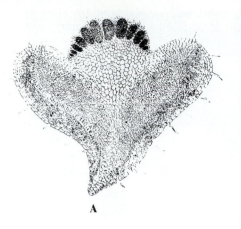

A

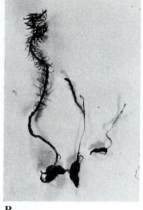

B

the gametophyte contain fungi. The sex organs originate from superficial cells near the meristem and their gradual maturation coincides with that of the lobes. The sperms are biflagellate.

Of species with subterranean gametophytes, those of *L. digitatum*[6] (Fig. 14-12A) and *L. obscurum* (Fig. 14-11B) are best known. These were described by Bruce (1979a) and Whittier (1977, 1981). The latter's accounts are based on axenic laboratory cultures, and the former's on material from nature (Bruce and Beitel, 1979).

The gametophytes of *L. digitatum* are carrot-like, lack lobes and are fleshy throughout. Each consists of a tapering base, a meristematic ring and a gametangial cap. They range from 8 to 25 mm (1 cm, Whittier, 1981) long and 2.5 to 13 mm in width. The gametophytes, although parenchymatous, are differentiated internally into several zones. Sex organs in the gametangial cap are illustrated in Fig. 14-12B, D, G, and H and in Fig. 14-13. The gametophytes of *L. digitatum* from nature and in axenic culture are similar morphologically even though the latter lack fungi. Spore germination occurs only in darkness in laboratory cultures. Gametophytes re-

moved from dark-grown cultures to light become green (Whittier, 1981a).

The gametophytes of *L. obscurum,* both in nature and axenic culture, are compressed-carrot-shaped or disclike (Fig. 14-11A.) In this species, too, spore germination occurs only in darkness. The sex organs develop on the upper surface. The gametophytes of *L. obscurum* also became green when illuminated.

Bruce (1979b) has compared the types of gametophyte that occur in *Lycopodium.* He found the gametophytes of *L. cernuum* to be characteristic of the subgenus *Lepidotis;* those of *L. obscurum* and *L. digitatum* (*L. complanatum*) to be characteristic of the subgenus *Lycopodium;* and those of *L. selago* and *L. phlegmaria* to be characteristic of subgenus *Urostachya.*

Both antheridia and archegonia occur in the crowns or dorsal surfaces (*L. carolinianum*) of the monoecious gametophytes (Fig. 14-12E, G, H). The development of both male and female sex organs may be traced to single superficial cells. In each case, these undergo periclinal division; by further division the inner of the two cells gives rise to the major portion of the antheridium and archegonium. The sex organs, therefore, are embedded as in the Anthocerotophyta.

The antheridia are massive (Fig. 14-12B, G) and produce large numbers of biflagellate

<hr>

[6]Equals *L. flabelliforme* or *L. complanatum* var. *flabelliforme*

343

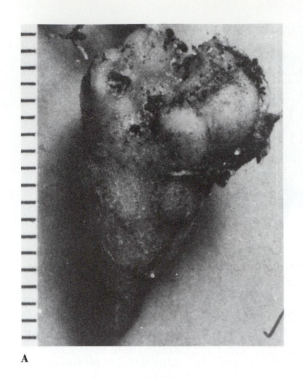

A

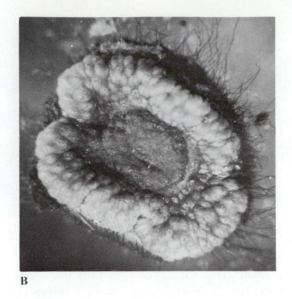

B

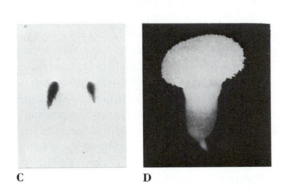

C D

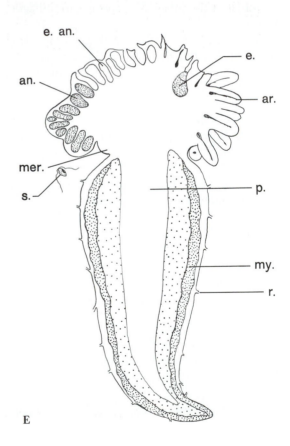

E

F

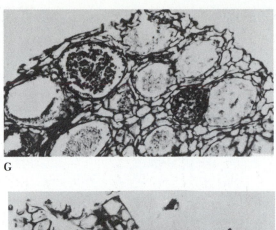

G

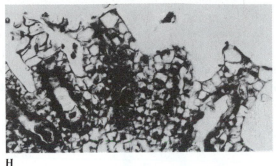

H

FIGURE 14-12
Gametophytes of *Lycopodium*. **A.** *L. digitatum*. **B.**
L. obscurum: apex of gametophyte with many
antheridia. **C, D.** *L. flabelliforme*. **C.** Young
gametophytes (6-month-old cultures) from axenic
culture. **D.** Mushroom-shaped gametophytes; the
expanded top bears numerous archegonia, the
necks of which are just visible (18 months in
culture). **E.** *L. complanatum*. Median longisection
of gametophyte (diagrammatic). **F.** *L. carolinianum*,
from nature. **G, H.** *L. digitatum*. **G.** Section
showing antheridia. **H.** Section showing archegonia.
an., antheridium; ar., archegonium; e., embryo; e.
an., empty antheridium; mer., meristematic collar;
my., mycorrhizal zone; p., central pithlike region;
r., rhizoid; s., sperm. **A.** Scale in mm. **B.** ×30.
C. ×1. **D.** ×0.3. **E., F.** ×25. **G., H.,** ×150. (**A, F,
G, H.** *After Bruce.* **B.** *Courtesy of Dr. D. W.
Bierhorst.* **C, D.** *Courtesy of Dr. Dean Whittier.*
E. *After Bruchmann.*)

sperms, the latter ultimately liberated through
the ruptured wall cells. Electron microscopy
has provided details of spermatogenesis (Robbins and Carothers, 1975, 1978; Haas, 1975).

The archegonial necks (Fig. 14-12H) vary in
length and degree of emergence among the
several species. The necks are composed of four
or five rows of neck cells.

Reproduction: embryogeny. Most of our
knowledge of embryogeny in *Lycopodium* is
based on the investigations of Bruchmann
(1898). More recently Bruce (1979*b*) has reported on some aspects of the embryogeny of *L.
carolinianum*.

The occurrence of *Lycopodium* gametophytes, each with several attached sporophytes
in various stages of development, indicates not
only that the gametophytes are long-lived and
active in the nutrition of the embryonic sporo-

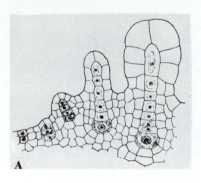

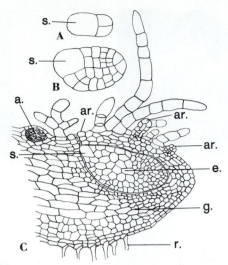

FIGURE 14-14
Lycopodium clavatum. **A–C.** Successively older stages of embryogeny. a., antheridium; ar., archegonium; e., embryo; g., gametophyte; r., rhizoid; s., suspensor. **A, B.** ×120. **C.** ×60. (*After Bruchmann.*)

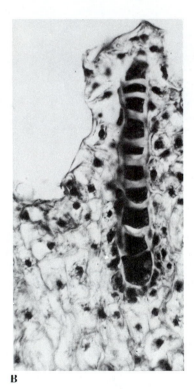

FIGURE 14-13
A. *Lycopodium selago,* successively later stages in the development of archegonia. **B.** *L. flabelliforme.* Longisection of archegonium of gametophyte grown in axenic culture. **A.** ×250. **B.** ×300. (**A.** *After Bruchmann.* **B.** *Courtesy of Prof. Dean Whittier.*)

phyte, but also that the sex organs may function over long periods. The zygote divides by a transverse wall into an outer, **suspensor cell,** and an inner, embryo-forming cell (Fig. 14-14). The latter develops the embryo itself; the former may divide once or twice. The embryo-forming cell gives rise to a massive, sometimes lobed foot, stem, and leaf (Figs. 14-14, 14-15). The root develops adventitiously at the base of the first leaf. The primary axis grows out of the gametophyte and up into the light. Axes and leaves of sporophytes borne on subterranean gametophytes remain colorless until they emerge above the substratum. The gametophyte may persist for a long time attached to the sporophyte (Figs. 14-11, 14-16C, D), but it ultimately disintegrates as the latter becomes established. The first leaves in some species differ from those on the mature plant in their scalelike habit as well as in the absence of vascular tissue and chlorophyll. The sporo-

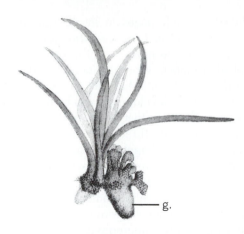

g.

FIGURE 14-15
Lycopodium cernuum: gametophyte (g.) with young sporophyte attached. ×12. (*After Treub.*)

FIGURE 14-16
Lycopodium cernuum. Development of gametophytes and young sporophytes in culture. **A.** Gametophyte on agar; note club-shaped tips and rhizoids. **B.** Gametophyte with several embryos (e.) with first leaves or prophylls. **C.** Young sporophytes arising from gametophytes 2 months after spores were sown. **D.** Plants with immature strobili 28 months after spore germination. g., gametophyte; r., roots or rhizophore. **A.** ×2.3. **B.** ×4. **C, D.** ×0.75. (*Courtesy of Dr. Ralph H. Wetmore.*)

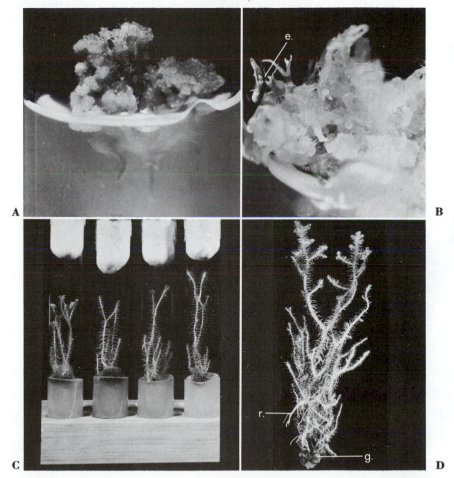

phytes of *Lycopodium* may develop apogamously in laboratory cultures; whether or not this occurs also in nature has not been ascertained.

Summary

The axis of the sporophyte in many *Lycopodium* species is fundamentally protostelic. Its branching may be dichotomous or pseudomonopodial. Although the mature sporophyte has microphyllous leaves and roots, organs that are absent in *Psilotum*, there is evidence that these organs are secondary additions to the axis of *Lycopodium*. This is suggested by the absence of vascular tissue in the early leaves of the embryonic plant and the lack of a specially differentiated root-forming region in the embryo.

The several species show variation in the distribution of sporogenous tissue, apparently progressing from a scattered zonate condition in *L. lucidulum* through stages in localization at the stem apex found in *L. innundatum* and *L. complanatum*. For the most part, the massive, short-stalked, reniform sporangia occur on the adaxial surface of the leaf base at its junction with the stem. Their development can be traced to a group of surface cells that divide periclinally to form a row of cells three tiers deep. The sporangial wall and the tapetum arise from the uppermost layer; the sporogenous tissue may be traced to the eusporangiate condition characteristic of a majority of vascular plants.

The spores of most species require long periods to develop mature gametophytes in nature. *Lycopodium cernuum*, *L. carolinianum*, and *L. innundatum* are exceptional in this respect. The gametophytes in nature are of several kinds, either photosynthetic and partially epiterranean (*L. cernuum*), or completely devoid of chlorophyll and subterranean (*L. digitatum* and *L. obscurum*). In the latter case, they may be

strobilus

A

B

FIGURE 14-17
A. *Selaginella uncinata*. **B.** *Selaginella kraussiana*. ×0.66.

B

FIGURE 14-18
Selaginella pallescens. **A.** Habit of
growth. **B.** Single "frond" with
strobili. **A.** ×0.25. **B.** ×0.33.

A

either tuberous or branching cylindrical structures. Most are infected with mycorrhizal fungus and are saprophytic. The sex organs are
massive and partially embedded; the sperms are
biflagellate. Each gametophyte may produce
several sporophytes at intervals as long as a year
apart. The gametophytes persist during the slow
development of the embryonic sporophyte.

CLASS 2. GLOSSOPSIDA

The Glossopsida include the order
Selaginellales, family Selaginellaceae, and Isoetales, family Isoetaceae with living members.

Selaginella

**Introduction and vegetative morpholo-
gy.** *Selaginella* (dim. of L. *selago*, name of
plant resembling the savin, a juniper, + L. *ella*,
diminutive) (Figs. 14-17 through 14-35) is one of
three extant genera of the class Glossopsida, all
the members of which have ligulate leaves (Fig.
14-24C). Although it has such *Lycopodium*-like
attributes as small, microphyllous leaves, the

herbaceous habit, and strobili composed of
sporophylls bearing a single sporangium on the
adaxial surface, *Selaginella* also has features that
distinguish it from *Lycopodium*. The most significant of these is the production of two kinds of
spores, a condition known as **heterospory,**
which is associated with profound changes in
the morphology and physiology of the gametophyte generation. Furthermore, both the vegetative leaves and the sporophylls of *Selaginella*
have small, tonguelike ligules.

The genus *Selaginella*, sometimes called the
spike moss, is a large one, including approximately 700 species, which are developed most
abundantly in tropical regions with heavy rainfall. In the United States, *S. apoda*, an inhabitant of moist soils, is widely distributed, as is the
xerophytic *S. rupestris*, an inhabitant of exposed rocks. The so-called resurrection plant
(Fig. 14-19), often sold as a novelty, is the
xerophytic *S. lepidophylla*, which is native to
the southwestern United States. A number of
tropical species are cultivated in conservatories
and in Wardian cases because of the beauty of
their branches and foliage; *S. kraussiana* and *S.*

349

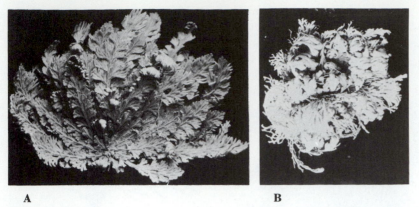

A   **B**

FIGURE 14-19
Selaginella lepidophylla. The "resurrection plant." **A.** Moist. **B.** The same plant, air dried. ×0.33.

uncinata (Fig. 14-17) are encountered frequently as a ground cover in greenhouses.[7]

Most species of *Selaginella* exhibit abundant branching, often in a single plane (Fig. 14-18). The branches are arranged dichotomously or in pseudomonopodial or sympodial fashion, depending on the species. Branching and planation in some species result in the production of frondlike growths (Fig. 14-18B), which may arise from a common center, simulating a fern. Other species are climbers, still others prostrate and creeping (Fig. 14-17).

Development of the stem may be traced either to a single apical cell and its derivatives or to a group of apical meristematic cells, depending on the species. In some species the apical cell is later replaced by a group of meristematic initials. *Selaginella kraussiana* is typical of those with a single apical cell. Some distance back from the apical meristem, the central region of the stem differentiates as procambium, from which the vascular tissues arise. In older regions of the axis (Fig. 14-21) in species such as *S. caulescens*, the central portion is separated from

the cortex by a cylindrical cavity. The cortex and central tissues are connected by elongate endodermal cells and in some species by one to four cortical cells called **trabeculae** (L. *trabecula*, little beam). Casparian thickenings are apparent on the walls of these endodermal cells.

The stele may be circular or ribbon-shaped in transection, depending on the species. The stele in *S. kraussiana* is a dual structure for part of the distance between successive branches. This duality results from a forking of the stele in precocious preparation for departure of the trace to a more distal branch (Fig. 14-22). In other species there may be one or several steles, the polystelic condition. Each stele in *S. kraussiana* is surrounded by a single layer of pericycle cells, immediately within which the phloem is located. According to Burr and Evert (1973), the size of the sieve pores in the lateral and terminal walls of the sieve elements falls in the same size range, so they may be called "sieve cells." The pores occur singly, not in groups, in the terminal and lateral walls. Hébant et al. (1980) reported on the phloem of *S. wildenovii*.

The central portion of each stele contains the xylem, which is exarch and monarch. The xylem of the stem stele is mesarch in some species—in *S. martensii, S. uncinata, S. emmeliana,* and *S.*

---

[7]A magnificent collection of *Selaginella* (and ferns) is maintained in the conservatories of Garfield Park in Chicago.

350

caulescens, among others. The cortex is composed of thin-walled, photosynthetic parenchyma cells, bounded externally by an epidermis without stomata. It should be noted that in some species of *Selaginella* (*S. rupestris*, for example), series of procambium cells differentiate into vessels rather than into tracheids. **Vessels,** it will be recalled, are composed of cell segments in which the common terminal walls have become perforated; they are multicellular in origin, and lacking in all but a few genera of vascular cryptogams. The vessels of *Selaginella* have single terminal perforations. The protoxylem tracheids are lignified in a helical pattern, while those in the later-formed metaxylem have scalariform pits. Some species (e.g., *S. apoda* and *S. ludoviciana*) lack vessels in their xylem (Buck and Lucansky, 1976).

In many species of *Selaginella*, prominent, leafless axes originate near the points of branch origin (Fig. 14-17) and grow toward the substratum. Until recently, these were interpreted formerly as specialized, root-bearing stem branches and termed **rhizophores.** Intensive studies (Webster and Steeves, 1963, 1964, 1967; Karrfalt, 1981; Grenville and Peterson, 1981) of several species, including *S. kraussiana*, indicate that these are adventitious roots that fork repeatedly and dichotomously near their apices.

In *S. kraussiana* (Grenville and Peterson, 1981) the aerial roots have well-developed root caps, with cutinized epidermal cells, before they enter the soil. Their apices bifurcate within the original root cap which ultimately disintegrates and each branch root becomes provided with its own new root cap. As the roots enter the soil, they organize root caps and later produce root hairs. Wochok and Sussex (1976) reported that cultured roots apices of *S. wildenovii* could develop as leafy shoots. Webster and Jagels (1977) have investigated the morphology of the roots as they develop in moist chambers. Growth of the main root and its smaller branches may be traced to a prominent apical cell in some species (e.g., *S. apoda* and *S. ludoviciana*)

according to Buck and Lucansky (1976). In these species the noncutinized epidermis is shed as the root ages, and the next cellular layer centripetally functions as a thick-walled epidermis. The cortex, endodermis, and a two- to four-layered pericycle surrounded the exarch protostele.

The ligulate leaves of *Selaginella* may be arranged either spirally, a primitive attribute, as in *S. rupestris*, *S. riddellii*, and *S. sheldonii* (Fig. 14-20), or spirally and compressed in four rows, as in *S. kraussiana* and *S. apoda* (Fig. 14-23A). In species such as the latter, which are dorsiventral, the two dorsal rows of leaves are smaller than the two rows with ventral or lateral insertion. This condition is known as **anisophylly.**

Dengler (1983*a,b*) investigated the origin, development and histology of the anisophyllous leaves of *S. martensii*. The primordia differ in size. The smaller, dorsal leaves are more precocious than the ventral in histological maturation. Other species of *Selaginella*—*S. riddellii*, for example—are **isophyllous** (Fig. 14-20), like most species of *Lycopodium*. Such species are radially symmetrical.

The leaves are sessile on the stem and alternately inserted, and each is traversed longitudinally by a single unbranched vein that is connected to the stele by a leaf trace (Fig. 14-21). The leaves, therefore, are microphyllous. Recently, however, *S. adunca* A. Br. ex Hieron. and *S. Schaffneri* Hieron. have been reported to have branching leaf veins (Wagner et al., 1982).[8] Mention has already been made of the basal ligule on the adaxial surface of each leaf. Kollenbach and Geier (1970), Horner, Beltz, Jagels, Boudreau (1975), Sigee (1974, 1975, 1976), and Jagels and Garner (1979) have investigated the organization and possible function of the ligule. The ligule consists of an enlarged basal region, a neck, and a tip and is covered

---

[8]The bearing of this on the phylogeny of leaves is discussed on p. 472.

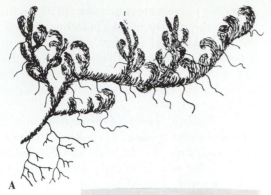

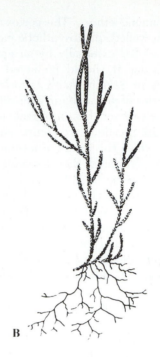

C

D

E

FIGURE 14-20
Isophyllous Selaginellas. **A.** *S. sheldonii.* **B.** *S. riddellii.* **C.** Branch of *S. riddellii.* **D.** Strobilus of *S. sheldonii.* **E.** Strobilus of *S. riddellii.* **A, B.** ×⅓. **C.** ×3.5. **D, E.** ×4. (**A, B.** *After Correll.*)

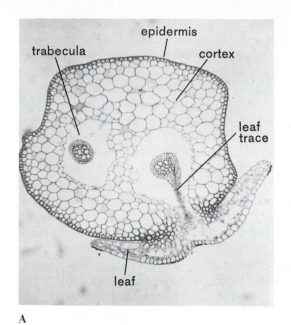

A

B

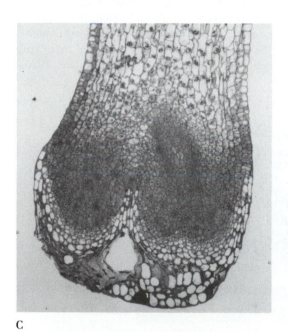

C

FIGURE 14-21
A. *Selaginella caulescens.* Transection of a stem and leaf base. **B, C.** *S. kraussiana.* Longitudinal sections of roots. **B.** Root tip showing several layers of root cap cells. **C.** Longisection of a root just before becoming subterranean; common root cap degenerating, new caps developing. **A.** ×25. **B.** ×200. **C.** ×300. (**B, C.** *After Grenville and Peterson.*)

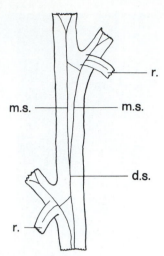

FIGURE 14-22
Selaginella kraussiana. Diagrammatic segment of a main axis, its branches, and roots; note precocious forking of stele in anticipation of branching. d.s., single diarch stele; m.s., two monarch steles; r., root. (*Modified from Webster and Steeves.*)

over its entire surface by a cuticle. The cells of the ligule have few chloroplasts and become inactive early in development. The function of the ligule has not been clearly elucidated. It may be a vestigial structure.

Transverse sections of the leaf (Fig. 14-23B) reveal lower and upper epidermal cells containing chloroplasts, and between them the mesophyll composed of photosynthetic parenchyma cells with intercellular spaces. The leaf cells of different species vary in number of chloroplasts from one to several. In some species—*S. apoda* and *S. uncinata* (Fig. 14-23B), for example—the cells of the adaxial leaf epidermis each contain a single, cuplike chloroplast, while the other cells have several (ca. four) lenslike ones (Jagels, 1970). The plastids are usually larger than those of other vascular plants. Stomata are present on the abaxial surface of the leaf and are localized near the midrib (Fig. 14-23B) in some species (*S. kraussiana* and *S. ludoviciana*), but in others they occur in both surfaces of the dorsal leaves. In *S. apoda* stomata are scattered over the abaxial leaf surface (Buck and Lucansky, 1976).

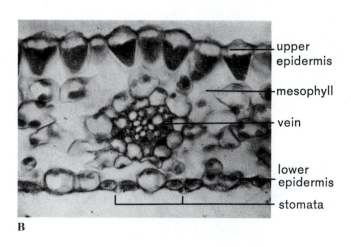

FIGURE 14-23
A. *Selaginella uncinata,* an anisophyllous species. Leaf arrangement and dimorphism. **B.** *Selaginella caulescens.* Transection of a leaf. **A.** ×6. **B.** ×450.

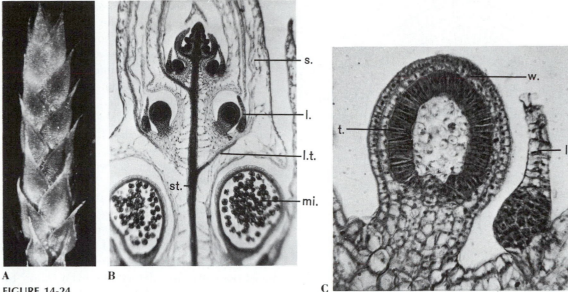

A   **B**

FIGURE 14-24
Selaginella pallescens. **A.** Strobilus, enlarged. **B.**
Median longisection of apex of immature strobilus.
C. Median longisection of immature
megasporangium and ligule. l., ligule; l.t., leaf
trace; mi., microsporocytes; s., sporophyll; st.,
stele; t., tapetum; w., sporangial wall. **A.** ×10. **B.**
×125. **C.** ×250.

The sporogenous tissue is segregated into
individual cells, which function as sporocytes as
development proceeds. It subsequently becomes apparent that there are two types of
sporangia. In some, a small percentage of sporo-

Reproduction: the sporophyte. All
species of *Selaginella* produce their eusporangiate sporangia in strobili (Figs. 14-18, 14-20,
14-24). The sporophylls are scarcely different
from vegetative leaves, and in some species
they are arranged so loosely as to render the
strobili inconspicuous. The sporophylls, like the
vegetative leaves, are ligulate (Fig. 14-24B, C).
Each bears a single stalked sporangium adaxially
near its base (Figs. 14-24, 14-25, 14-27). Growth
of the strobilus is apical (Fig. 14-24B); hence,
median sections of a young strobilus show various stages in the eusporangiate development of
the sporangia. The sporangium wall is two-
layered and separated from the sporogenous
tissue by a tapetum (Fig. 14-24C). The latter
develops from the innermost layer of the sporangial wall.

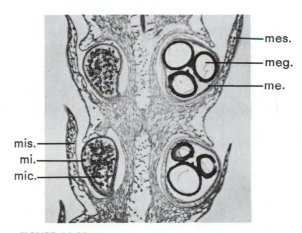

FIGURE 14-25
Selaginella sp. Segment of a longitudinal section of
a strobilus. me., megasporangium; meg.,
megaspore; mes., megasporophyll; mi., microsporangium; mic., microspores; mis., microsporophyll. ×125.

355

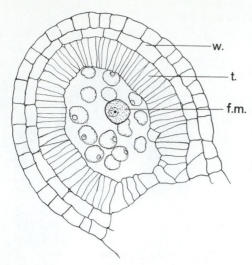

FIGURE 14-26
Selaginella sp. Median longisection of young
megasporangium. f.m., functional megasporocyte
(the remainder, degenerating); t., tapetum; w.,
sporangial wall. ×400.

cytes may degenerate, the remainder undergoing meiosis and cytokinesis to produce many
tetrads of spores (Figs. 14-24B, 14-25). While
still in tetrads and before their exines have
formed, the microspores have been shown to
take in substances from the material that surrounds them (Pettitt, 1974). In other sporangia
(Figs. 14-25, 14-26), usually all except one of the
sporocytes degenerate. The survivor undergoes
meiosis and cytokinesis, producing a single
spore tetrad, the members of which gradually
enlarge, apparently by appropriating the materials made available by the degeneration of the
other sporocytes. They become filled with lipids
and protein granules (Robert, 1971*a*). Ultimately, the four spores in these sporangia grow large
enough to cause bulging of the sporangial wall
(Figs. 14-25, 14-27B). This account of the ontogeny of the two types of sporangia indicates that
they are fundamentally similar during the early
stages of development. However, by the sporocyte stage divergence in development begins.
The greatly enlarged spores are called **mega-**

spores (sometimes, macrospores), and sporangia
in which they develop and the sporophylls that
subtend the sporangia are known as **megasporangia** and **megasporophylls,** respectively. The
smaller spores are called **microspores,** their
sporangia are **microsporangia,** and their sporophylls are **microsporophylls.** The sporocytes
that give rise to megaspores are called **megasporocytes,** while those that form microspores
are known as **microsporocytes.** This dimorphic
condition of the spores is known as **heterospory**
(Fig. 14-28).

Selaginella pilifera is a species in which the
sporangia and sporophylls are decussately arranged. There are consistently two vertical rows
of megasporangia and two of microsporangia.
Three differences between microsporangia and
megasporangia are detectable even prior to
meiosis (Horner and Beltz, 1970): The microsporangia are smaller, *all* their sporocytes develop callose walls (in contrast to one callose-
walled sporocyte in the megasporangium), and
all the microsporocytes stain heavily for RNA,
while only the functional megasporocyte does
so.

The microsporangia may occur in the apical

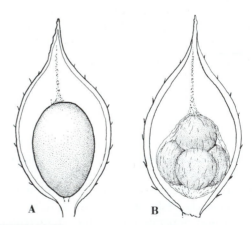

FIGURE 14-27
Selaginella pallescens. **A.** Microsporophyll and
microsporangium. **B.** Megasporophyll and
megasporangium. Both in adaxial view. ×18.

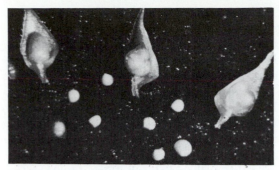

FIGURE 14-28
Selaginella pallescens. Adaxial views of microsporophyll and microsporangium (left) and two megasporangia with megaspores (right); note eight megaspores (and their variant size) and the numerous microspores. ×8.

portion of the strobilus and the megasporangia below, or there may be two vertical rows of microsporophylls and two of megasporophylls; less regular arrangements may occur (Horner and Arnott, 1963).

Pettitt (1977*a*) reported that, frequently, the megasporangia mature megaspores of different sizes. The smaller are abortive and do not develop gametophytes. The difference in fertility of the members of these megaspore tetrads, he reported, was foreshadowed in the differential distribution of the cytoplasm in the megasporocyte that produced them.

Of the living vascular plants discussed so far, *Selaginella* is the first to exhibit heterospory. While the ontogenetic cause of heterospory in *Selaginella* obviously is degeneration of a majority of sporocytes in certain sporangia and increase in size of the survivors, the factors that evoke this condition are obscure. That the number of degenerating megasporocytes is not absolutely fixed is indicated by the presence of as many as 42 and as few as one megaspore in megasporangia of the individuals.

As the microspores and megaspores mature, their walls thicken. Those of the microspores are red in certain species and those of the

megaspores are cream-colored. The tapetum of the megasporangium seems to play a role in the thickening of the megaspore walls. Both microspores and megaspores have prominent **triradiate ridges,** which mark the lines of cytokinesis of the spores within the spore mother cell walls. At dehiscence the spores are ejected through a vertical cleft in a sporangial wall. Ingold (1974) described dehiscence of the sporangia in *Selaginella* and ejection of the spores (microspores to a distance of 1 cm or more, megaspores 10 to 20 times farther because of their size).

Reproduction: the gametophyte. The development of strictly dioecious gametophytes is the usual result[9] of heterospory. Microspores develop into male gametophytes, and megaspores into female gametophytes. Unlike spore germination in *Psilotum* and *Lycopodium*, that of *Selaginella* often is precocious,[10] so that at the time of their ejection the spores are in various stages of gametophyte development. These intrasporangial stages are sometimes said to represent **primary germination.** Under certain conditions, the gametophytes may reach maturity, as manifested by their production of sex organs, by the time of their dissemination from the sporangia. In two extreme cases, which, however, require confirmation, fertilization and embryo development have been reported to occur while the megaspores and their contained gametophytes are still within the megasporangia (Lyon, 1901, 1904).

Ruf (1975) reported that it is probable that Lyon actually saw young lichen thalli growing out of the apices of *S. rupestris.* Apomictic sporelings were observed by Geiger (1934) growing out of undisturbed strobili of *S. anocardia;* he stated that Goebel had also observed this. However, Ruf (personal communication)

---

[9]Except in the fern *Platyzoma*, p. 411.

[10]The spores of *Pellia, Conocephalum,* and *Porella* among the Hepatophyta and certain mosses also exhibit a degree of precocity in their endogenous divisions (Chapter 11).

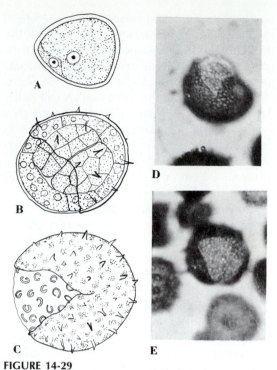

FIGURE 14-29

Selaginella pallescens. Development of microspore into ♂ gametophyte. **A.** Section of microspore; note small prothallial cell. **B.** Partially exposed ♂ gametophyte; note jacket cells and two antheridia. **C.** Microspore with mature ♂ gametophyte ready to liberate sperm. **D, E.** Living microspores with walls open revealing ♂ gametophytes. **A–C.** ×770. **D, E.** ×350.

found sexually initiated sporelings emerging from megaspores in undisturbed strobili of *S. intacta* (Fig. 14-35B). In this species the single large megasporangium is surrounded by two sterile sporophylls, in addition to the megasporophyll; these form an interesting cup-shaped structure into which the microspores can sift to effect fertilization. He inferred that the sporophytes resulted from fertilization because he observed microspores within the megasporangium and also because megaspores, isolated by him in culture, formed gametophytes but no embryos.

Further comparative study of a number of species of *Selaginella* is desirable to establish the factors that effect the liberation of the spores in relation to the degree of maturity of their enclosed gametophytes.

The mature microspore is uninucleate at first. Its development of the male gametophyte is initiated by an internal mitosis and cytokinesis that result in the formation of a small, peripheral **prothallial cell,** or **sterile cell,** and a large cell, the **antheridial cell** (Fig. 14-29A). The prothallial cell has been interpreted as the sole remnant of the vegetative tissue of a free-living gametophyte, although it undergoes no further divisions. The basis for this interpretation is by no means convincing. It has also been suggested that the prothallial cell represents a rhizoid precursor. The antheridial cell, by anticlinal and periclinal divisions (Fig. 14-29B), forms a single-layered jacket enclosing 128 or 256 spermatogenous cells, each of which gives rise to a single biflagellate sperm. The microspore of *S. kraussiana* is shed from the microsporangium before the antheridium is fully formed. The latter matures in the microspores that fall into environments that favor further development. The prothallial cell and wall cells of the antheridium ultimately disintegrate, and the sperms are liberated by rupture of the microspore wall (Fig. 14-29C–E). Chlorophyll is absent from the male gametophyte. Robert (1974, 1977) has investigated spermatogenesis in *S. kraussiana* with the electron microscope. He found that in the mature sperm one flagellum is apical and one median.

The megaspores also begin their development into female gametophytes while still grouped together in the tetrad and before they have attained their maximum size. The young megaspore contains a large central vacuole surrounded by a thin peripheral layer of cytoplasm. The megaspore wall at maturity is composed of two layers, a thick, outer cream-colored **exospore** and a more delicate inner layer, the **endospore** (Sievers and Buchen, 1970). A **me-**

sospore may also be present in some species (Robert, 1971*a*,*b*).

The single nucleus undergoes mitosis that is not followed by cytokinesis. This process continues, and the cytoplasm, which gradually increases in amount, becomes multinucleate (Fig. 14-30A, B). The occurrence of successive mitoses without ensuing cytokineses is known as **free nuclear division.** With continued increase in number of nuclei and amount of cytoplasm, the nuclei lying in the portion of the megaspore near the triradiate ridge are gradually separated by cell walls (Fig. 14-30B, C). This process may continue until much of the megaspore lumen is filled with cellular tissue. The apical tissue of the gametophyte may be separated incompletely from the deeper tissue by a diaphragm (Fig. 14-30B, D) (Robert, 1971*a*,*b*). This may not occur, however, until after fertilization.

The megaspore is finally ruptured in the region of the triradiate ridge by the protrusion of the developing female gametophyte (Figs. 14-30C, 14-31, 14-32). It is in this region that the several archegonia develop. Gametophytes in megaspores that have been shed have been reported to develop chloroplasts and rhizoids if they come in contact with soil in the presence of light. Although rhizoids develop, in the writers' experience the gametophytes are always achlorophyllous. It is probable, moreover, that the gametophyte derives the bulk of its nutriment from the metabolites freed by degeneration of the supernumerary sporocytes, by the activity of the tapetum of the megasporangium, and from the food stored within the megaspores.

A number of superficial cells of the exposed portion of the female gametophyte develop into archegonia (Figs. 14-30E, 14-31B, 14-32, 14-33A). These are largely embedded, except for their short necks, which are two tiers high and composed of four rows of neck cells. If carbohydrates are available and if fertilization is inhibited, the female gametophyte continues to grow and produce more archegonia and rhizoids.

Union of the sperm and egg may occur either after the spores containing the gametophytes have been shed from the strobilus or, in two species, according to an old, unverified report, while the megaspores and their female gametophytes are still within the megasporangia. If, indeed, this does occur (see p.358), such a transfer of microspores and/or sperms would suggest pollination in seed plants.

Reproduction: embryogeny. As in *Lycopodium*, the first division of the zygote in *Selaginella* gives rise to a **suspensor** initially near the neck of the archegonium; the lower cell and its derivatives form the embryo proper. In some species the suspensor remains relatively inactive as it does in *Lycopodium*, whereas in others it undergoes cell division with subsequent elongation, so that the developing embryos are thrust into the nutrient-filled vegetative tissue of the female gametophyte (Fig. 14-33A, B). The portion of the embryo opposite the suspensor becomes organized as a foot, and the remainder develops into an axis consisting largely of primary root (**radicle**) and stem bearing two embryonic leaves (Figs. 14-33C, 14-34). As development continues, the embryo, except for the foot, emerges from the female gametophyte and megaspore, and the young plant is soon established independently. This embryonic sporophyte, attached to the female gametophyte within the megaspore, looks strikingly like a minute seedling at this stage (Fig. 14-35).

Summary

Although *Selaginella* is similar to *Lycopodium* in a number of respects, it differs in having ligulate leaves, frequently polystelic stems, and vessels in its xylem, and especially in its heterospory. As a result of the latter, the gametophytes are strictly dioecious. The ontogeny of the spores of *Selaginella* is instructive in providing a clue as to a possible origin of heterospory. In view of the great similarity in development of both microsporangia and megasporangia

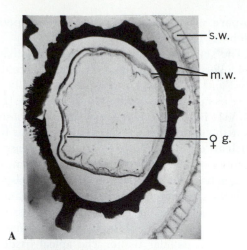

A

FIGURE 14-30
Selaginella sp. **A.** Section of megaspore with free nuclear ♀ gametophyte. **B.** *S. kraussiana.* Section of megaspore showing early cellular ♀ gametophyte. **C.** *Selaginella* sp. Section of megaspore at triradiate ridge; note partially cellular ♀ gametophyte. **D.** Section of megaspore (outer spore wall removed) showing mature cellular ♀ gametophyte with archegonia above and stored nutriments. **E.** Portion of a ♀ gametophyte more highly magnified; note protuberant archegonial necks and egg cells. ar, archegonia; d, diaphragm; e, exospore; ♀g, ♀ gametophyte; Lo, stored lipids; mw, megaspore wall; sw, megasporangial wall; tr, early tissue of ♀g. **A.** ×135. **B.** ×100. **C.** ×250. **D.** ×135. **E.** ×300. (**A, B.** *After Robert.* **D, E.** *From a preparation by Triarch, Inc.*)

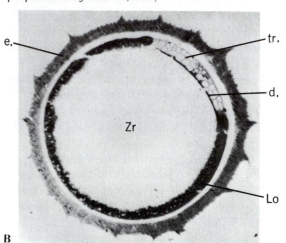

B

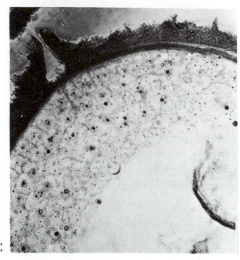

C

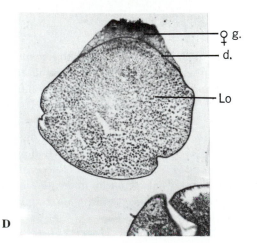

D

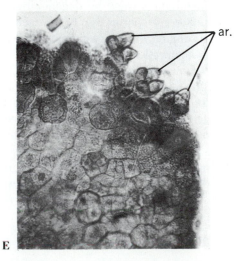

E

A

archegonium

megaspore wall

B

FIGURE 14-31
Selaginella pallescens.
Living megaspores with
♀ gametophytes. **A.**
Female gametophyte,
immature; note
rhizoids. **B.** Mature ♀
gametophyte. **A.** ×35.
B. ×60.

through the sporocyte stage, along with the variation in megaspore number in certain individuals and species, it seems probable that heterospory here is occasioned in part by differences that are initiated prior to sporogenesis. Four significant differences become apparent in mega- and microsporangia even before meiosis occurs: (1) The megasporangia are slightly larger than the microsporangia; (2) cytoplasmic RNA disappears from all but one of the megasporo-

cytes; (3) abortion of most megasporocytes occurs in every megasporangium; and (4) callose walls form only around the sporocytes (mega- and micro-) with the potentiality of undergoing meiosis. The ultimate causes of heterospory are not clear.

The precocious germination, and, in reported instances, of maturation, of the spores into gametophytes before they are shed from the sporangia is a noteworthy departure from the

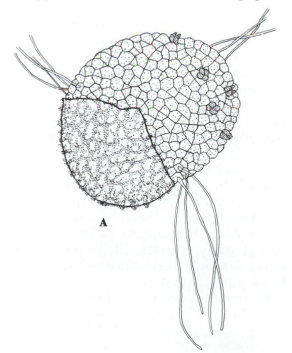

A

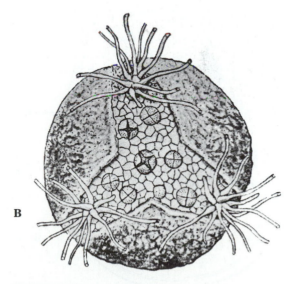

B

FIGURE 14-32
Selaginella pallescens. **A.** Megaspore with mature
♀ gametophyte. **B.** The same stage in *S. martensii.*
A. ×35. **B.** ×175. (**B.** *After Bruchmann.*)

361

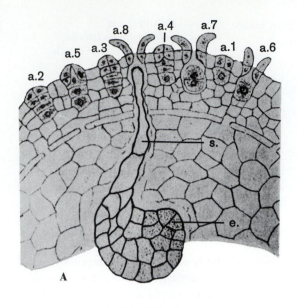

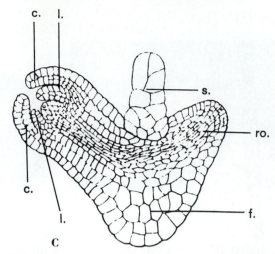

C

FIGURE 14-33

Selaginella. Embryogeny. **A.** Early stages in *S. poulteri.* **B.** *S. martensii.* Median section of megaspore and ♀ gametophyte with embryos. **C.** Median longisection of embryo of *S. martensii.* a.1–a.8, successively more mature archegonia. c., cotyledon or first embryonic leaf; e., successful embryo; e.′, abortive embryo; f., foot; ♀g., ♀ gametophyte; l., ligule; m., megaspore wall; r., rhizoids; ro., primary root or radicle; s., suspensor. **A.** ×225. **B.** ×150. **C.** ×125. (*After Bruchmann.*)

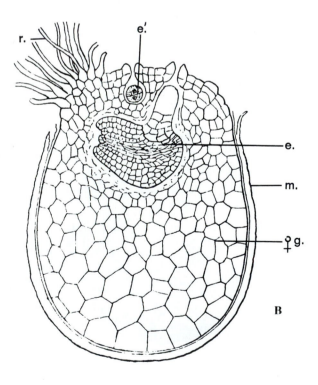

B

reproductive cycle of *Lycopodium* and *Psilotum.* Spore dimorphism in *Selaginella* is correlated with gametophyte dimorphism. Both male and female gametophytes are much reduced in size, duration of existence, and complexity of structure as compared with free-living gametophytes, the male gametophyte especially so. It should be noted that spores of a homosporous plant (e.g., *Polytrichum, Sphaerocarpos,* and *Marchantia*) may develop into dioecious gametophytes. In such homosporous plants, meiosis results in two members of every spore tetrad growing into male and the other two into female gametophytes. In *Selaginella* and other heterosporous plants, however, all the products of meiosis of a single sporocyte form either male or female gametophytes.

The major source of food for the female

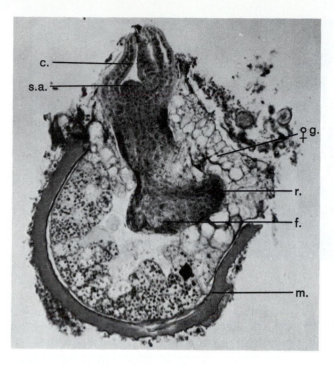

FIGURE 14-34
Selaginella sp. Section of megaspore, ♀ gametophyte and embryo sporophyte. s.a., stem apex or promeristem; other labels same as in Fig. 14-33. (*After Steiner, Sussman, and Wagner.*)

gametophyte, as well as for the developing embryo, is the material stored in the megaspore during its long period of enlargement. The food is sporophytic in origin. While the reported transfer of sperms, or of microspores containing male gametophytes, to the opened megasporangia containing megaspores with female gametophytes (p. 358) would suggest pollination in the seed plants, as the occurrence of fertilization and embryo development within the megasporangium likewise do, there are significant differences. Furthermore, the frequency of occurrence of these phenomena in *Selaginella* requires elucidation.

Isoetes

Isoetes is a member of the order Isoetales, family Isoetaceae.

Introduction and vegetative morphology. Although morphologists differ in their opinion regarding the relation of *Isoetes* to other plants, there is considerable evidence that its affinities may be with the Microphyllophyta, in which division it is included in the present text, although some botanists do not consider that *Isoetes* belongs there (Greguss, 1968). A related South American genus, *Stylites*, with two species, from a lake in the high Andes of Peru, was described in 1954 and has been studied intensively by Rau and Falk (1959) and by Karrfalt and Hunter (1980). *Stylites* is sometimes considered not to be distinct from *Isoetes* (Kubitzki and Borchert, 1964).

Isoetes (L. houseleek; Gr. *isos*, equal, + Gr. *etos*, year = evergreen plant) is a genus containing 40 (Bierhorst, 1971*a*) to 70 (Sporne, 1970) species familiarly known as "quillworts" because of their narrow, elongate leaves, the bases of which are rather spoonlike. Most species are either partially submerged aquatics (*I. engelmannii*) or amphibious; a few, such as *I. butleri*, are terrestrial. *Isoetes butleri* is perennial, but active growth occurs only during the early spring and late autumn rains. *Isoetes*

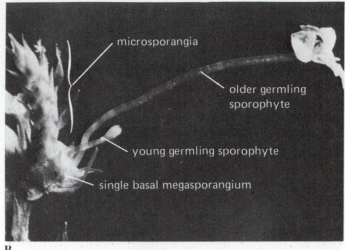

microsporangia

older germling
sporophyte

young germling sporophyte

single basal megasporangium

A   B

FIGURE 14-35

A. *Selaginella pallescens.* Germling sporophyte
attached to ♀ gametophyte (within megaspore). **B.**
S. intacta. Strobilus on living parent plant; note
two juvenile sporophytes emerging from basal
megasporangium. **A.** ×5. **B.** ×3. (*Courtesy Dr. E.
William Ruf, Jr.*)

lithophila (Fig. 14-36) is an inhabitant of shallow
granitic pools and is subject periodically to
desiccation; *I. butleri* occurs on moist soils. The
basic chromosome numbers in the genus has
been reported as x = 11 (Kott and Britton,
1980).

A comprehensive summary of the anatomy
and development of *Isoetes* has been prepared
by Paolillo (1963). The quill-like leaves are
attached in spiral fashion to a subterranean
cormlike structure (Fig. 14-36B). Their spiral
arrangement is readily apparent in transverse
sections through the overlapping leaf bases.
Although the leaves in some species, as in *I.
engelmannii*, attain a length as great as 65 cm,
they are considered to be microphyllous, inasmuch as they have single, unbranched veins and
the traces leave no gap in the vascular tissues of
the stem. The leaves are ligulate (Sharma and
Singh [1984]). It is evident, in transverse section, that each leaf contains four longitudinally

placed lacunae, or air chambers; the vein is
located in the solid tissue in the center of these.
The tissue external to the air chambers is photosynthetic parenchyma. Leaves of terrestrial species have stomata in their epidermis. Kruatrachue and Evert (1977a) have investigated the
sieve elements in the leaves of *I. muricata*.
According to them, the mature sieve element
contains an elongate, degenerate nucleus, and
the lateral walls are traversed by plasmodesmata, the latter or sieve pores being
present in the terminal walls.

The cormlike structure (Fig. 14-37) on which
the leaves are borne is difficult to interpret
morphologically. Its upper portion is considered
to be a much shortened, fleshy, vertical stem
with a broad and sunken apex. The nodes are so
close together that the internodes are practically
obliterated (Fig. 14-37). Elongation of the upper
portion of the axis is very slow, most of the
derivatives of the apical meristem cells becoming involved in the formation of leaves and the
portions of the stems immediately subtending
the leaves. The vascular tissue is arranged as a
central protostele (Fig. 14-37), the xylem of
which consists of a large number of parenchyma cells and relatively few tracheids, a

364

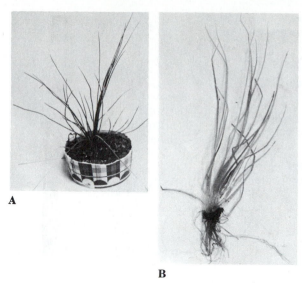

A

B

FIGURE 14-36
Isoetes lithophila. **A.** In soil. **B.** Free of soil, showing "corm" and roots. **A.** ×0.25. **B.** ×0.5.

characteristic of aquatic plants in general. The xylem is surrounded by phloem. Outside the phloem there is a meristematic layer, the exact nature of which is somewhat in dispute. It functions as a **cambium** in that its divisions add to the tissues of the stem (Kruatrachue and Evert, 1977*b*). Apparently, the cambium derivatives develop as sieve elements, occasionally as tracheids, or as parenchyma cells adjacent to the phloem. The secondary sieve elements are similar at maturity to the sieve elements of the leaf except that pores occur in all their walls. Most of the cells produced by the cambium augment the cortex, the outer portion of which, with the older leaf bases, is sloughed off each growing season. No endodermis is present. The surface of the corm is covered by the remains of the leaves of previous seasons. The surface layers of the corm become suberized.

The lower portion of the corm is a bilobed or trilobed organ. It may branch more frequently in some species (Karrfalt and Eggert, 1977) and in *I. tegitiformans* (Rury, 1978) it may be elon-

gate and rhizomatous. Its structure and homologies have received various interpretations. It often is referred to as the **rhizophore**, inasmuch as the delicate roots are borne in orderly series only on this region of the plant. The rhizophore end of the plant develops from its own meristem, which is sunken in a groove (Fig. 14-37B). This is scarcely distinguishable from the cambium. The youngest roots, therefore, occur near the deepest portion of the groove, and the older ones arise from the sides of the rhizophore lobes. The roots are endogenous in origin; each is connected to the central vascular tissue of the rhizophore by a trace (Fig. 14-37).

Karrfalt and Eggert (1977) and Karrfalt (1977) have investigated the comparative morphology and development of *I. nuttallii* and *I. tuckermanii*. The corms of young plants are usually two-lobed at their bases but may become three- and four-lobed as new furrows appear. The bases may branch dichotomously.

The delicate roots are protected by root caps, beneath which a group of apical cells is present; these add to both the root cap and the root itself. The roots branch dichotomously as a result of the organization of two groups of apical initials below the root caps.

Karrfalt (1977) reported that the developing roots of *Isoetes* displace substrate from under the corm and that tension on the roots pulls the corms down into the substrate. Mature roots contain delicate protosteles that are excentric in position (Fig. 14-37C) because of the disintegration of the inner cortical cells on one side of the stele to form a lacuna. The stele is bounded by a well-differentiated endodermal layer. Krautrachue and Evert (1978) have investigated the ontogeny of the sieve elements in the roots of *I. muricata*. These lack nuclei at maturity. The cortex is surrounded by an epidermis. This unusual type of root, with its excentric stele and air chamber, is very similar in organization to the roots of fossil Lepidodendrales, also borne on rhizophores (Fig. 20-36).

365

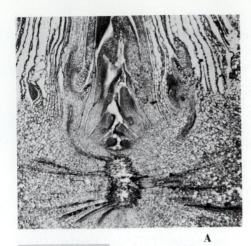

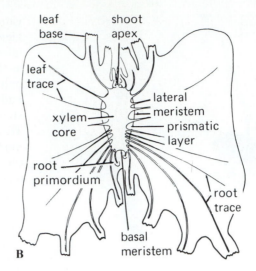

FIGURE 14-37

Isoetes butleri. **A.** Almost median longisection of "corm." **B.** Diagrammatic
median longisection; compare with section in **A. C.** Transection of a root; note
large central lacuna and lateral stele. **A.** ×25. **C.** ×60. (**B.** *After Paolillo.*)

Reproduction: the sporophyte. Every
leaf of *Isoetes* is potentially a sporophyll, and the
plants are heterosporous. The first-formed
leaves of any season, the outermost, are frequently sterile, however. The next older leaves
mature as megasporophylls (Fig. 14-38A) and
are followed by microsporophylls (Fig. 14-38B)
within. The last-formed leaves of the season
frequently bear abortive sporangia. Microsporophylls and megasporophylls are indistinguishable at first. In each case the single sporangium
arises from superficial cells near the adaxial
surface of the spoon-shaped leaf base; the cells
undergo a series of periclinal divisions. These
periclinal divisions give rise externally to a
several-layered wall and internally to sporogenous tissue. Development of the sporangium is
eusporangiate. The sporangium is massive and
larger than those in *Lycopodium* and *Selaginella;* it may attain a length of up to 7 mm. A small
ligule arises just above the apex of the sporangi-

um, and other superficial cells in that region
grow down to form an indusiumlike covering,
the **velum** (L. *velum,* veil) (Figs. 14-38, 14-39A).
The sporangia are incompletely chambered by
strands of sterile tissue, the **trabeculae,** which
extend from the walls partially across the sporangial lumen (Fig. 14-39). The innermost layer
of the sporangial wall differentiates as a two-
layered tapetum. The trabeculae and the portion of the tapetum that covers them are derived
from potentially sporogenous tissue.

As in *Selaginella,* development of both microsporangia and megasporangia is largely similar
through the sporocyte stage. Practically all the
microsporocytes undergo meiosis and form tetrads of microspores; hence, tremendous numbers of spores, estimated to be between 150,000
and 1 million, develop in each microsporangium. Certain of the megasporocytes enlarge, but
only a small number divide and form tetrads;
the remainder disintegrate. Pettitt (1976*b*) has

366

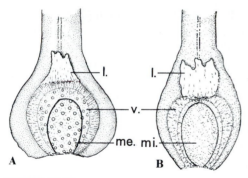

FIGURE 14-38
Isoetes sp. Adaxial views of **A.** megasporophyll, and **B.** microsporophyll. l., ligule; me., megasporangium; mi., microsporangium; v., velum. ×3.

investigated the multiplication of plastids during megasporogenesis in *I. engelmannii*. Megasporangia of the several species produce between 50 and 300 megaspores. Both microspores and megaspores have walls with ornamentation, which varies from species to species; the microspores are often covered with long spines, while megaspores are ridged and grooved. There is no special mechanism of sporangial dehiscence, at least in aquatic spe-

cies. The spores are liberated as the sporophylls and sporangia walls disintegrate at the end of the growing season. The sporophylls may be abscised and float to the surface. The spiny microspores are often found attached by their spines to the megaspores, an important adaptation in dioecious gametophytes (Bierhorst, 1971*a*).

Reproduction: the gametophyte. Unlike *Selaginella*, the spores of *Isoetes* do not begin their development into gametophytes until they have been set free from their sporangia. The spores of *I. lithophila* and *I. engelmannii* (Sam, 1982) require light and variable temperatures (to near freezing) for germination. The male gametophyte, as in *Selaginella*, is entirely enclosed within the microspore wall (Fig. 14-40A). It arises by internal divisions of the microspore protoplast to form a single prothallial cell (sterile cell) and a single antheridial cell. The latter develops into an antheridium consisting of a single-layered wall enclosing four spermatogenous cells. Each of these gives rise to a single multiflagellate sperm at maturity.

Development of the female gametophyte of

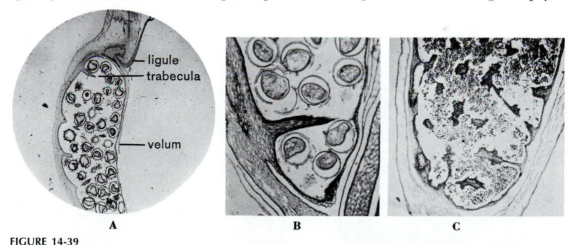

FIGURE 14-39
Isoetes sp. **A.** Sagittal section of leaf base with megasporangium. **B.** Enlarged sectional view of megasporangium; note megaspores and trabecula. **C.** Segmented portion of microsporangium; note trabeculae and microspores. **A.** ×12.5. **B, C.** ×60.

Isoetes, as of *Selaginella,* involves a series of free-nuclear divisions of the megaspore nucleus and its descendants. Cell-wall formation occurs first in the region of the triradiate ridge and gradually extends through the remainder of the female gametophyte. It may still be incomplete in the basal portion of the gametophyte for some time after fertilization. The megaspore wall cracks open, exposing the cellular apex of the developing gametophyte, which develops unicellular rhizoids. Certain superficial cells undergo division to form archegonia, which are largely embedded (Fig. 14-40B, C). Their necks, composed of three or four tiers of four cells, are longer than those of *Selaginella.*

Reproduction: embryogeny. Usually only one zygote in each female gametophyte develops into a sporophyte (Figs. 14-40B, 14-41). Nuclear and cell division by the zygote and its derivatives produce a spherical mass of tissue, which later differentiates into the embryonic regions of the sporophyte. These include a rather massive foot in contact with the starch-filled cells of the female gametophyte, an embryonic root, and a leaf. The stem develops secondarily in the region between the leaf and the root (Fig. 14-40B). Although the young sporophyte soon becomes established as an independent plant, it remains attached to the female gametophyte and megaspore for a considerable period (Fig. 14-41). It should be noted that no suspensor is present in the embryo of *Isoetes.*

Summary

Although the leaves of *Isoetes* are markedly larger than those of *Selaginella* and *Lycopodium,* they are microphyllous, as indicated by their single unbranched veins and the absence of gaps in the stele. All organs of the plant show evidences in their anatomical structure of adaptation to an aquatic habitat. Among these are air lacunae and the paucity of lignified xylem tissue. The leaves of *Isoetes* are ligulate and spirally inserted on a short, underground,

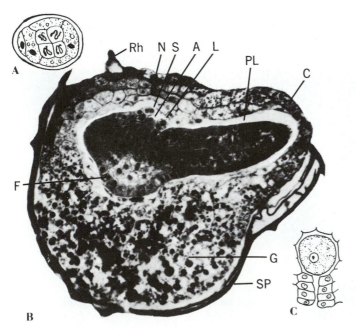

FIGURE 14-40
A. *Isoetes lacustris.* Section of microspore containing mature ♂ gametophyte; note prothallial cell, jacket cells, and four spermatogenous cells. **B.** *Isoetes lithophila.* Section of a megaspore containing a ♀ gametophyte and embryonic sporophytes. A, stem apex; C, calyptra; F, foot; G, ♀ gametophyte; L, ligule; N, archegonial neck; Rh, rhizoids; S, stem apex. **C.** Single archegonium in median longisection. **A.** ×300. **B.** ×250. **C.** ×450. (**A.** *After Liebig.* **B, C.** *After LaMotte.*)

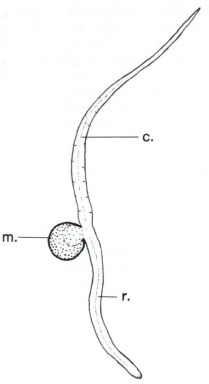

FIGURE 14-41
Isoetes sp. Young sporophyte attached to ♀ gametophyte within megaspore. c., cotyledon; m., megaspore; r., root. ×5.

cormlike structure. Each leaf is potentially a sporophyll. Although growth in length of the corm is very limited, the presence of a cambiumlike layer contributes to its increase in girth. The cortex of the corm is composed of starch-filled cells, an indication that the perennial corm is primarily a storage organ. The stem is protostelic with few tracheids, most of the xylem cells maturing as parenchyma. The lower portion of the corm is called a rhizophore. This is a lobed structure on which the delicate roots are borne in orderly fashion. Both stem and root develop from a group of apical meristematic cells rather than one. Root branching is dichotomous. The roots are characterized by excentric steles and a central air chamber.

The sporangia are massive and heteromorphic. A variable number of megaspores mature in the megasporangium. The number is always larger than in *Selaginella,* and unlike the latter, the spores do not initiate development into gametophytes until they have been shed from their sporangia. Evidences of further reduction of the gametophytes, as compared with those of *Selaginella,* are present in *Isoetes.* Among these, the reduction of spermatogenous cells to four and the failure of the female gametophyte to project from the megaspore wall are noteworthy. The gametophytes of *Isoetes* are colorless and saprophytic, depending on reserves stored within the spores. Unlike both *Selaginella* and *Lycopodium,* the sperms of *Isoetes* are multiflagellate. The development of the embryo differs from these genera in the complete absence of a suspensor and in the long-delayed development of the stem. In spite of the differences just cited, there are many evidences of similarity in morphology between *Isoetes* and *Selaginella* and *Lycopodium.* The roots and rhizophore suggest the fossil Lepidodendrales. Some morphologists, however, on the basis of the multiflagellate sperms and anatomical considerations, are of the opinion that *Isoetes* is more closely related to certain ferns (Greguss, 1968). This view seems to ignore such a fundamental attribute as the microphylly of *Isoetes* as contrasted with the macrophylly of ferns.

DISCUSSION QUESTIONS

1. What characteristics distinguish the Microphyllophyta from the Bryophyta?
2. Define or explain strobilus, leaf trace, microphyll, suspensor, sporophyll.
3. What is meant by eusporangiate sporangium development?
4. Suggest the composition of a culture media suitable for the cultivation of various types of *Lycopodium* gametophytes.

5. What significance may be attached to the absence of vascular tissue from the first-formed leaves on the young sporophytes of *Lycopodium*?

6. On what basis is dichotomous branching considered to be a primitive characteristic in vascular plants?

7. Can you suggest an adaptive advantage of the strobiloid arrangement of sporophylls in certain *Lycopodium* species as compared with the scattered, zonate arrangement as in *L. lucidulum*?

8. Although the stems of *S. kraussiana* are distelic in the mature plants, the embryonic and juvenile stems are monostelic. Of what significance is this?

9. How does the xylem of some species of *Selaginella* differ from that of *Lycopodium*?

10. How do the leaves of some species of *Selaginella* differ from those of *Lycopodium*?

11. What is the origin of the thick megaspore wall in *Selaginella*?

12. What light does ontogeny shed on the possible origin of heterospory? Explain.

13. Why are the gametophytes of *Selaginella* said to be "reduced"?

14. Distinguish between the terms primitive, advanced, specialized, generalized, simple, and reduced, as they are used in comparative morphology.

15. How do *Selaginella* megaspores with attached embryonic sporophytes differ from dicotyledonous seedlings?

16. Why are both *Selaginella* and *Lycopodium* considered to be microphyllous?

17. With the aid of labeled diagrams, illustrate the reproductive cycle in *Selaginella*.

18. Define or explain heterospory, homospory, prothallial cell, free-nuclear division.

19. What characteristics do *Selaginella* and *Lycopodium* have in common?

20. List the characteristics shared by *Isoetes* and other Microphyllophyta.

21. What anatomical evidences of aquatic habitat are present in *Isoetes*?

22. In what respects does the development of gametophytes in *Isoetes* differ from that in *Selaginella*?

23. On what basis are the leaves of *Isoetes* said to be microphyllous?

24. How do the sporangia of *Isoetes* differ from those of other Microphyllophyta?

25. Describe the structure of the corm of *Isoetes*.

26. Why are the roots and rhizophore of *Isoetes* of interest in relation to the fossil record?

Chapter 15
VASCULAR CRYPTOGAMS II: DIVISION ARTHROPHYTA

The division Arthrophyta (Gr. *arthros*, jointed, + Gr. *phyton*, plant) is also a group of vascular cryptogams known from Devonian (inside front cover) strata. Arthrophytan plants were abundant during the Paleozoic, but only a single genus, *Equisetum*, has survived in our present flora.

CLASS 1. ARTHROPSIDA

Order 1. Equisetales

Family 1. Equisetaceae
Equisetum

Equisetum (L. *equus*, horse, + L. *saeta*, bristle) (Figs. 15-1, 15-2), with 10 to 25 species, the only living member of the family Equisetaceae and order Equisetales, is widely distributed; the various species are familiarly known as "horsetails" (the branching species) and "pipes" and "scouring rushes" (the unbranched species). Some, such as the common *E. arvense*, grow both on moist and in somewhat xeric habitats, whereas others, such as *E. sylvaticum*, frequently flourish only in marshy situations. Temperate-zone species are relatively small in stature, rarely exceeding 4 ft in height. The tropical *E. giganteum* may exceed 5 m in height and have stems up to 2.4 cm in diameter. This large species and a related hybrid are used medicinally in Costa Rica as the basis of an infusion that is said to cure kidney ailments (Hauke, 1967a). Some species of *Equisetum* have been reported to be poisonous to livestock (Kingsbury, 1964).

The stem is the dominant organ of the plant body in the genus *Equisetum;* the minute leaves, although photosynthetic for a short period after their formation, soon become dry and scalelike (Fig. 15-3). In *E. arvense* and other species, the plant consists of a subterranean, deep-growing rhizome and an erect aerial stem (Fig. 15-1C). Rhizome systems of *E. arvense* have been observed growing horizontally 6 to 7 ft below the surface of the soil and covering areas of many hundreds of square feet. The rhizomes of some species produce tuberous storage structures at their nodes.

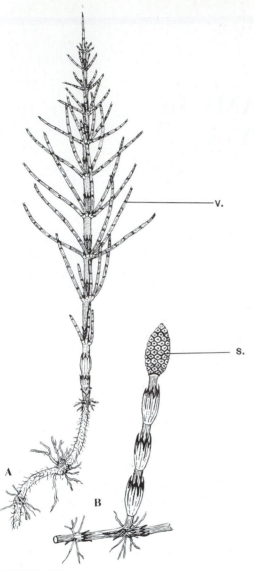

FIGURE 15-1
A–C. *Equisetum arvense.* Segments of rhizome with vegetative (v.) (in **C,** left and right) and fertile shoots with strobili (s.) (s. in center in **C.**). **D–E.** *E. hyemale.* Vegetative shoot with terminal strobilus, the latter enlarged and dehiscent in **E,** ×0.75 (**C.** *From a model, courtesy of the Field Museum of Natural History.*)

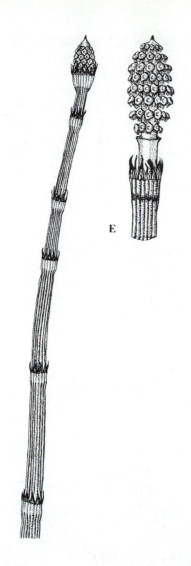

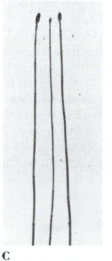

The aerial stem may be richly branched, as in *E. arvense*, or branching may be rare, as in *E. hyemale* (Fig. 15-2A) and *E. kansanum* (Fig. 15-2C). The fact that unbranched species (e.g., *E. hyemale*), under such stimulation as injury to the apex, develop branches at the nodes is often cited as evidence that branching is a primitive attribute in the genus, the branch primordia

FIGURE 15-2
Growth habit of several species of *Equisetum*. **A.** *E. hyemale*, vegetative colony. **B.** *E. hyemale*, shoots with terminal strobili. **C.** *E. kansanum*, shoots with terminal strobili. **D.** *E. kansanum*, single strobilus. **A.** ×1/12. **B, C.** ×1/6. **D.** ×3.

FIGURE 15-3
Equisetum hyemale. Node and portions of adjacent internode, enlarged; note leaves and ridges. ×3.

being present at the nodes. After the strobili of main axes have shed their spores, they often become detached, whereupon branching is initiated at some of the lower nodes of the aerial stems. These branches arise from dormant buds which are present from the time of early development of the stem. Associated with each of the buds is a root meristem, from which all the roots of adult plants arise at the lower, subterranean nodes.

Both the aerial stems and rhizomes have well-defined nodes and internodes (Figs. 15-1, 15-2). The surface of the stem is ribbed or ridged, the ribs of successive internodes being arranged in an alternating pattern. The bases of the leaves are fused into a deep collar, which ensheaths the node and gives the appearance of a scalloped collar. Both rhizomes and aerial stems bear the scalelike leaves, each of which has an unbranched vascular bundle. A few rudimentary stomata are present near the tips of the adaxial surface of the leaves of *E. arvense.* Stomata are present in two or three rows on both sides of the projecting midrib on the adaxial leaf surface. The central region of each leaf is photosynthetic when the leaves first

appear. The individual leaf tips may be united into groups of varying number (Fig. 15-3).

The relation between the branches and leaves in *Equisetum* differs from that in all other vascular plants. In those of the latter that branch at the nodes, stem branches originate from the axils of the leaves, and thus opposite them, whereas in *Equisetum* they emerge from the region of the node between the leaves, and thus alternatively with them. As the branches elongate, they pierce the nodal leaf sheath.

Development of the stem originates in a single pyramidal apical cell (Hébant, Hébant-Mauri, and Barthonet, 1978), which divides regularly in three directions (Fig. 15-4) (Gifford and Kurth, 1983). The derivatives of the apical cell divide in an anticlinal direction soon after they are delimited. One of the cells thus formed in each case contributes by further division to the internodal, and the other to the nodal, portion of the axis. In the development of the stem, certain cells at each node remain meristematic; these constitute the intercalary meri-

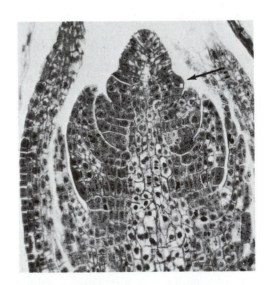

FIGURE 15-4
Equisetum hyemale. Median longisection of vegetative apex; note apical cell and origin of whorled laterals. ×125.

374

stem, which adds to the length of the internodes. The leaves originate as superficial, ringlike outgrowths of the nodal cells (Fig. 15-4, arrow). A short distance back from the apical region, tissue differentiation is initiated in the axis. The surface cells form an epidermis with silicified cells. The distribution of the siliceous material has been investigated by Kaufman et al. (1971) and by Dayanandan and Kaufman (1973). It is deposited mostly as knots and rosettes on the inner walls of the epidermal cells. Certain of the latter undergo two successive divisions to form guard cells and their more superficial accessary cells. The latter are thickened with siliceous ribs (Fig. 15-5C). In *E. arvense* the stomata are superficial, but in *E. hyemale* they are sunken. Stomata are present most abundantly on the slopes of the ridges of the fluted stem surface. Page (1972) has demonstrated with scanning electron microscopy that the stomata are useful criteria in establishing possible phylogenetic relationships among species. Associated with the stoma and guard cells are a pair of subsidiary cells (Fig. 15-5C), which have radiating bands composed in part of siliceous material (Hauke, 1957). The effect of silicon on growth of *Equisetum* in culture has been investigated by Hoffman and Hillson (1979).

Figure 15-5A shows a transverse section of the erect vegetative stem of *E. arvense* taken

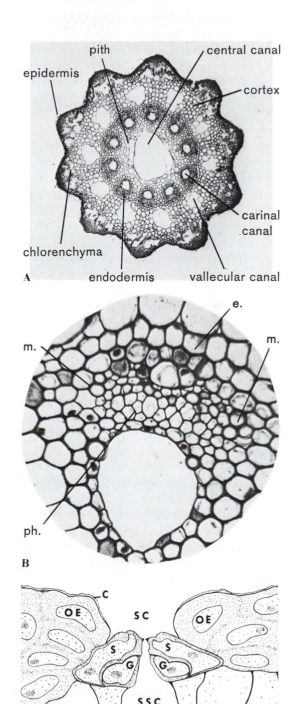

FIGURE 15-5
Equisetum arvense. **A.** Transection of vegetative stem. **B.** Single vascular strand, enlarged. **C.** *E. hyemale:* Transection of epidermis and stomatal apparatus. C, cuticle; e., endodermis; G, guard cell; m., metaxylem; OE, ordinary epidermal cell; ph., phloem; S, subsidiary cell; SC, stomatal crypt; SSC, substomatal chamber. **A.** ×30. **B.** ×250. **C.** ×600. (**C.** *After Kaufman et al.*)

from an internode where differentiation has been completed. The central region of the stem is hollow at maturity and is surrounded by the remains of the parenchymatous pith. This cavity is called the **central canal**. Outside the pith there is a ring of circular canals, the position of which is directly internal to the surface ridges of the stem. These, therefore, are known as **carinal** (L. *carina*, keel) **canals**, and they mark the position of discrete strands of zylem and phloem. Experiments with dyes indicate that these canals have a role in conduction (Bierhost, 1971a). The first-formed annular and helical protoxlyem cells arise near the inner limit of each procambium strand (Fig. 15-5B). This xylem associated with the carinal canal is sometimes called the carinal xylem. With the formation of the carinal canals, after differentiation of the protoxylem, the position of the protoxylem elements is disturbed. The phloem, which lies directly outside each carinal canal, is bordered laterally by two groups of xylem cells; these differentiate from cells of the intercalary meristem (Bierhorst, 1971a). Vessels, with not more than two component elements, occur in the near-nodal xylem; their perforations connect with the carinal canal, which is a water-conducting channel.

A second type of vessel with reticulate perforations occurs in the lateral xylem of the internodal vascular strands of the rhizomes of five species (Bierhorst, 1958). Dute and Evert (1978) reported that sieve elements of the mid-internodal regions are organized as sieve-tube members. The end walls of the sieve elements in the internodal phloem are oblique, and the sieve areas are transversed by plasmodesmata (Dute and Evert, 1977a,b; 1978).

The pericycle is represented by a single layer of cells just within a rather prominent endodermis. In some species of *Equisetum*, each vascular strand is completely surrounded by the endodermis. In others, there are both outer and inner common, continuing, endodermal layers; and in still others, as in the aerial stems of *E. arvense*, only a continuous outer endodermal layer is present (Fig. 15-5A, B).

The cortex (Fig. 15-5A) consists internally of parenchyma cells interrupted by large vallecular or cortical canals, which have a position corresponding with that of the depressions of the stem surface. Groups of other cortical cells contain abundant chloroplasts. The walls of the cortical cells beneath the surface ridges are markedly thickened sclerenchyma and contribute to the support and rigidity of the stem. The epidermal cells are heavily silicified and provided with stomata; the complex organization of their guard cells, overlaid by subsidiary cells with radiating siliceous thickenings, has already been described. The guard cells and subsidiary cells may be at, or sunken below, the surface of the plant (Hauke, 1957). Chen and Lewin (1969) have demonstrated that silicon is required in the normal growth of *Equisetum*.

Although the xylem and phloem are present as discrete strands (eustele) in the internodes of *Equisetum*, serial sections through the nodes reveal that they join there to form a short siphonostele. A diaphragm of tissue at the nodes interrupts the internodal canal. The vascular strands above and below the nodal region are joined to the siphonostele in alternating fashion. A ring of small protoxylem leaf traces leaves the stele at each node; branch traces originate between the leaf traces. The absence of leaf gaps indicates that the leaves are probably to be interpreted as microphyllous. The internodal stele is known as a eustele; here the parenchymatous gaps between the internodal bundles are neither leaf nor branch gaps. It is worthy of note that the internodes of the axis of the embryonic sporophyte at first contain a protostele; the eustele differentiates in later development.

Roots of mature plants are nodal (Fig. 15-1A–C), endogenous, and adventitious in origin and arise at the bases of lateral branches or their bud primordia. The root grows as a result of the activity of a single apical cell (Gifford and Kurth, 1982) and its derivatives and is protostelic and exarch. Specially differentiated cubical cells of the root epidermis give rise to root hairs. The occasional branch roots develop endogenously from the pericycle.

Dute and Evert (1977*a*) have described sieve-element ontogeny in the roots of *E. hyemale* var. *affine*. The roots are characterized by a central metaxylem tracheid and three protoxylem poles. Some vessels occur in the roots. The mature sieve element is enucleate, its protoplast bounded by a plasmalemma, and contains plastids, mitochondria, endoplasmic reticulum, and refractive spherules.

Reproduction: the sporophyte. It is well known that various species of *Equisetum*, such as *E. arvense* and *E. hyemale*, may propagate new individuals from segments of their rhizomes. It has also been reported (Wagner and Hammitt, 1970) that floating segments of the aerial axes of *E. hyemale* in aquatic habitats can regenerate new individuals, which may be dispersed by water as propagules.

Equisetum is homosporous (Duckett, 1970*a*) but in some species, variation in spore size and color has been observed. The sporogenous tissue in all species of *Equisetum* is localized in a strobilus, but the relation of the strobilus to the vegetative branches varies. In some species—*E. hyemale* and *E. kansanum*, for example—the strobili develop at the tips of vegetative axes (Fig. 15-2B, C). In *E. arvense* and *E. telmateia*, however, the strobilus is usually borne on a nonchlorophyllous fertile branch that develops from the rhizome (Fig. 15-1B, C). The rhizome produces green vegetative branches later, as the strobilate branches wither away. *Equisetum sylvaticum* is intermediate between these two extremes in that the unbranched axis that bears the strobilus lacks chlorophyll at first. It becomes green and branched after the spores have been discharged.

In *E. arvense*, the strobilus-bearing branch is formed in the autumn preceding the spring in which it is to appear above the soil. As in the vegetative stem, growth of the strobilus is apical. In some species its axis continues to grow slightly beyond the apex of the strobilus. The axis of the strobilus produces a series of surface enlargements, each of which grows into a spore-producing appendage called a **sporangiophore.** Hauke (1970), on the basis of unusual material of *E. littorale* in which there were strobili with whorls of appendages transitional between leaf sheaths and typical sporangiophores, concluded that the appendages of the strobilus have dual potentialities. Accordingly, he considered the fertile appendages to be sporophylls. The term "sporangiophore" is used by those who are impressed with evidence from certain fossil Arthrophyta in which both branchlike sporangiophores and leaflike whorls occur in the strobilus. Page (1973) discussed hypotheses regarding the organization of the strobilus and its homologies with the vegetative axes.

Mature sporangiophores typically are hexagonal in surface view (Figs. 15-6B, D, 15-8) because of mutual pressure. In young strobili the sporangiophores are moundlike. Between five and ten equidistant single cells on their surface function in initiating sporangia. As they function, expansion of the central tissue of the sporangiophores pushes the sporangial initials and their cellular progeny over the margin to the adaxial surface of the sporangiophore. The walls of immature sporangia are several cells thick but become unistratose at maturity by disintegration of the inner layers. In their origin from single cells, the sporangia of *Equisetum* differ from the sporangia of *Lycopodium*, *Selaginella*, and *Isoetes*. Each sporangiophore is served by a trace that branches distally (Fig. 15-7A) and dichotomizes so that one trace reaches the base of each sporangium. The mature sporangia are elongate and fingerlike (Fig. 15-8).

FIGURE 15-6

Strobili of *Equisetum*. **A–D**, *E. arvense*. **A.** Young (right) and older (left) strobilate branches. **B.** A single, as yet indehiscent, strobilus at greater magnification. **C.** Incipiently dehiscent strobilus. **D.** Magnified view of sporangia and sporangiophores of the strobilus in **C**. **E.** *E. hyemale*, strobili in two stages of maturation. **A.** ×⅓. **B, C.** ×1. **D.** ×3. **E.** ×1.

378

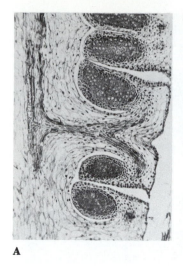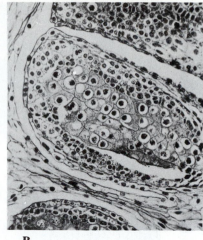

FIGURE 15-7
A, B. *Equisetum hyemale.*
Longisections of
sporangiophores and
immature sporangia; note
traces to sporangia. **A.** ×60.
B. ×125.

A

B

The sporogenous tissue of the young sporangium (Fig. 15-7B) is surrounded by a **tapetum** and a wall several cells in thickness. As development progresses, the walls of the tapetum disintegrate. There is thus formed a **tapetal plasmodium,** which contributes to the nutrition of the sporogenous tissue and formation of the outermost layer of the spore wall. A number of the sporocytes abort, but those that remain undergo meiotic division, each producing a tetrad of spores. The haploid chromosome number is approximately $n = 108$ in *E. arvense* and *E. hyemale.*

The outermost layer of sporangial wall cells thickens spirally and the cells beneath disintegrate. Dehiscence of the sporangium is longitudinal along a vertical line in that portion of the sporangial wall adjacent to the sporangiophore stalk. The wall structure of the mature spores is complex, the outermost wall layer consisting of four spirally arranged portions (Fig. 15-10A). These separate at the time of sporangial dehiscence, so that each spore bears four somewhat spoonlike appendages (Fig. 15-9) sometimes called **elaters.** The latter are hygroscopic and quickly affected by slight changes in humidity.

When the spores are ready for dissemination, the internodes of the strobilus elongate slightly, thus separating the sporangiophores. The stalk of each now increases in length on its lower side, so that the sporangia are approximately perpendicular to the soil surface when the spores have been shed.

Reproduction: the gametophyte. The green spores of *Equisetum* germinate rapidly (Fig. 15-10B) after their dissemination, provided that they are carried to suitable substrates. Lloyd and Klekowski (1970) determined that

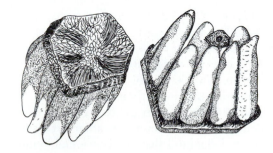

FIGURE 15-8
Equisetum arvense. Enlarged views of sporangiophore and sporangia. ×10.

A

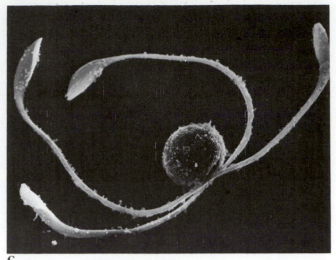

C

B

counts for the formation of extensive colonies.
Germinating spores and mature gametophytes
of several species have been found on moist soil
in nature. Mesler and Lu (1977) found gametophytes of *E. hyemale* 3 cm in diameter in nature
in northern California. Naturally occurring gametophytes of *E. arvense* varied in size from
that of a pinhead to 8 mm in diameter. The
spores of *E. arvense* and *E. hyemale,* among
others, germinate rapidly in laboratory cultures
on suitable media and on moist *Sphagnum* (Fig.
15-10B, C). The developing gametophytes are
extremely sensitive to such unfavorable environmental conditions as crowding, and their
form is modified accordingly. However, well-
isolated spores form disclike or cushionlike
green gametophytes several millimeters in diameter (Fig. 15-11A, B). They are anchored to
the substrate by numerous unicellular rhizoids
that arise from the lower side of the cushion.
The superficial cells of the gametophyte develop
lamellar lobes of photosynthetic cells, which
densely cover the mound- or cushionlike basal

green spores, such as those of *Equisetum, Osmunda* (p. 423), and certain other ferns, germinate in less than 3 days, the mean being 1.46
days. They remain viable, however, for a year or
less (mean 48 days). By contrast, nongreen
spores require 4 to 210 days (mean 9.5) for
germination but retain their viability for a mean
period of 1045 days. In the case of *E. arvense,*
which frequently is found in rather xeric situations, it is doubtful that any considerable number of spores produce mature gametophytes in
such habitats. Vegetative multiplication by
means of the rhizome fragments probably ac-

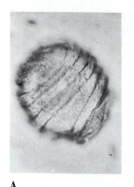
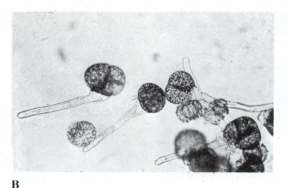
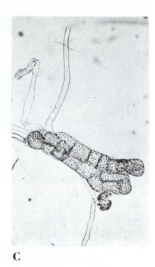

A   **B**

FIGURE 15-10
Equisetum hyemale. **A.** Living spore with spiral "elaters." **B.** Spore germination. **C.** Young gametophyte. **A, B.** ×400. **C.** ×125.

C

portion. This basal portion has a marginal meristem. In the writers' cultures the gametophytes of *E. arvense* sometimes produced antheridia, archegonia, and emergent sporophytes within 21 days after the spores had been planted. Considerable dimorphism of the gametophytes was apparent at this time (Fig. 15-12A).

Hauke (1967*a*, 1968, 1969, 1971) has investigated the gametophytes of *Equisetum* in the field and in laboratory cultures. He reported that the gametophytes of *E. arvense* and *E. bogotense* (of Central and South America) are strictly male or female and somewhat dimorphic. Hauke also reported that light intensity had an effect on the numbers of male gametophytes produced in cultures of *E. arvense* (greater number of males with increasing light intensity); intensity had no effect on the sex ratios in *E. hyemale*, but light quality did (higher number of males in red light), whereas the gametophytes of *E. fluviatile* were unaffected by either light quality or intensity.

Duckett (1970*a,b*) made incisive and comprehensive investigations of sexuality and spore size in five branched species of *Equisetum* (subgenus *Equisetum*). There is no morphological evidence of heterospory. Under laboratory conditions, the antheridia may be produced on

gametophytes only 25 days old (*E. arvense*), but most developed between 40 and 60 days after the spores had been sown.

The male gametophytes were covered with antheridium-bearing lobes and had few sterile lamellae (Fig. 15-11B–E). A majority of the spores developed this type of male gametophyte, which continued to produce antheridia for 300 days in laboratory culture.

By contrast, fewer of the spores grew into gametophytes with sterile lamellae, the archegonia being embedded in the moundlike gametophyte. The archegonial necks protruded between the bases of the sterile photosynthetic lamellae (Fig. 15-11F). Such gametophytes produced archegonia for between 35 and 80 days. Ultimately, however, they developed meristematic antheridial lobes, which continued to produce antheridia, while archegonial development ceased (see also Srinivasan and Kaufman, 1978). All the initially female gametophytes ultimately produced antheridia. Duckett (1970*a,b*) found that the gametophytes of nine species of *Equisetum* are of two kinds—namely, entirely male gametophytes and monoecious ones which are **protogynous** (i.e., they produce archegonia before the antheridia). A similar distribution of sex organs occurs in the fern

381

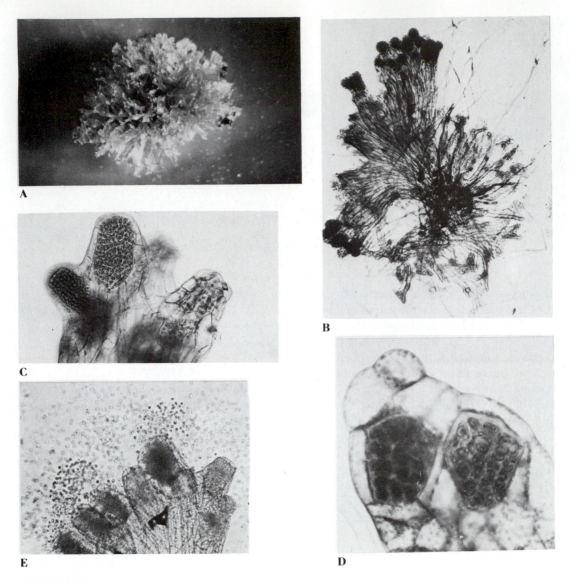

FIGURE 15-11

A–F. *Equisetum hyemale.* Gametophytes and sex organs. **A.** Entire gametophyte; note photosynthetic lobes arising from cushion. **B.** Ventral view of gametophyte showing rhizoids and antheridia. **C.** Two young antheridia and an almost-empty one at the right. **D.** Antheridium (right) with almost mature sperm. **E.** Liberation of sperm. **F.** Overmature archegonium. **G, H.** *E. palustre.* Two stages in the maturation of the archegonium. Note divergent neck cells in **H. A.** ×25. **B.** ×60. **C.** ×120. **D.** ×250. **E.** ×100. **F.** ×175. **G, H.** ×170. (**G, H.** *After Duckett.*)

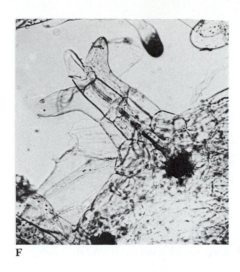

F

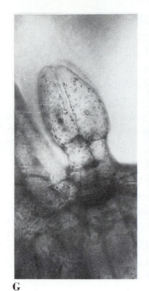

G

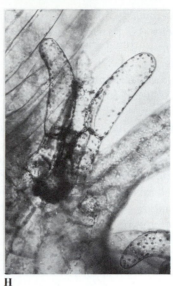

H

A

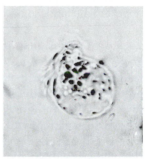

B

FIGURE 15-12
Equisetum arvense: **A.** Cultured gametophytes on agar, approximately 6 weeks old. The three smaller had antheridia and the large one archegonia. **B.** Living sperm. **C.** Fixed sperm. **A.** ×24. **B.** ×300. **C.** ×600.

C

383

Platyzoma (p. 411) and in a number of other species of ferns. It has been demonstrated (Sporne, 1964; Duckett, 1970*b*) that (homozygous) sporophytes can develop as a result of self-fertilization of the bisexual gametophytes.

Environmental factors have a marked effect on the ratio of male to bisexual gametophytes produced. At 32°C, the spores of five species produced only male gametophytes, but at 15°C over 50% of the gametophytes of three of the species were bisexual. High light intensities, which favor vegetative development, evoke the formation of bisexual gametophytes, while high temperatures seem to influence the development of male gametophytes (Duckett, 1970*b*). In another report (Hauke, 1971) it is stated that not only light intensity but also light quality may affect the expression of sexuality in *E. arvense* and *E. hyemale*, while *E. fluviatile* seems not to be markedly affected by these. Hauke (1968), however, has reported that the gametophytes of *E. bogotense* of Central and South America are strictly male or female and somewhat dimorphic.

In 1972 Duckett reported that in the subgenus *Hippochaete* of *Equisetum*, also, the gametophytes were initially male or female and that the males produced only antheridia, but the females finally produced antheridia as well as archegonia. The male gametophytes were smaller and less long-lived than the female. Duckett suggested that Hauke's (1968) report of dioecious gametophytes in *E. bogotense* may have been based on cultures grown for too short a time. The discrepancy between Hauke's and Duckett's investigations has yet to be resolved.

Each antheridium originates from a single superficial cell, which undergoes division into an outer cell and an inner cell. The outer forms the wall of the upper portion of the antheridium; the inner, by successive divisions, forms the spermatogenous tissue (Fig. 15-11C, D). The antheridium produces a large number of multiflagellate sperms, which are liberated explosively when the mature antheridia are moistened (Fig. 15-11E). Duckett (1973, 1975) has

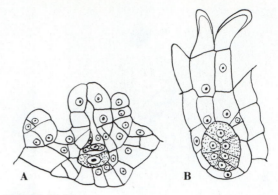

FIGURE 15-13
A. *Equisetum arvense.* Immature archegonia. B. *Equisetum hyemale.* Longisection of archegonium containing young sporophyte. **A, B.** ×150. (*After Jeffrey.*)

investigated spermatogenesis at the ultrastructural level.

According to Bilderback et al. (1973), as the antheridia are immersed in water, their two cover cells separate and the sperms emerge in the passageway thus created. The sperms are released within the spermatid cells, which have 6 to 12 processes. The spermatid cells split open and the sperms escape. They may swim at the rate of 300 µm/sec, or 15 times their own length! The sperms of *Equisetum* are relatively large (Fig. 15-12B, C). Duckett and Bell (1977) studied the mature sperms electron microscopically. The archegonia always occur near the bases of the photosynthetic lobes on the prostrate portion of the gametophyte. The archegonia (Figs. 15-11E, 15-13A) also arise from superficial cells. At maturity, their short necks, consisting of four vertical rows of neck cells, are protuberant and divergent distally; the venter is buried in the thallus. The eggs of several archegonia of a single gametophyte may be fertilized and may develop embryonic sporophytes (Figs. 15-13B, 15-14, 15-15). For additional investigations of the reproduction of *Equisetum* in nature and in the laboratory, consult Duckett (1979 *a, b*) and Duckett and Duckett (1980).

Reproduction: embryogeny. Growth of the sporophyte is initiated by transverse divi-

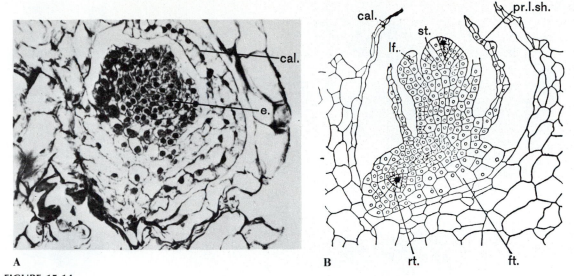

A
B　　　　**rt.**　　　　　**ft.**

FIGURE 15-14
Equisetum arvense. **A.** Almost-median section of embryonic sporophyte. **B.** Median section of older embryo. cal., calyptra; e., embryo; ft., foot; lf., leaf; pr. l. sh., primary leaf sheath; rt., root; st., stem. ×150 (**B.** *After Smith.*)

sion of the zygote, followed by divisions to form a quadrant. The two upper (outer) cells are often smaller than the two lower (inner) cells and develop the first leaf sheath and stem in *E. arvense*. The lower cells develop the foot and the root, no suspensor being present (Fig. 15-14). In some species the primary axis ceases development after it has formed a limited number of nodes and internodes. In such forms, secondary axes are successively formed, one of which ultimately gives rise to the mature axis. The embryonic root grows through the gametophyte into the soil (Fig. 15-15A), thus establishing the independence of the young sporophyte. Each gametophyte may produce many sporophytes, 27 having been reported on one gametophyte of *E. hyemale* (Fig. 15-15D) (Mesler and Lu, 1977). The gametophytes persist for some time in laboratory culture after sporophyte development has been initiated (Fig. 15-15B, C). In the authors' cultures of *E. arvense*, young emergent sporophytes 0.4 cm long became visible 30 days after the spores had been sown.

Summary

Equisetum, the only extant member of the Arthrophyta, differs from other seedless vascular plants, except certain *Lycopodium* species, in the whorled arrangement of its leaves and branches. Furthermore, its alternating arrangement of leaves and branches is anomalous among vascular plants. The leaves are reduced to scalelike appendages; each, however, receives a vascular trace and is provided with stomata. The burden of photosynthesis is borne by the axis. Except for the primary root, the roots are adventitious and arise at the bases of the lateral branches of their primordia. Both roots and stems develop as a result of the activity of apical cells and their derivatives. The root is protostelic, and the stem is eustelic at the internodes and siphonostelic at the nodes. The parenchymatous gaps between the vascular bundles in the internodal steles are not related to either leaf or branch traces.

The sporogenous tissues of *Equisetum* always are localized in strobili; the relation of the latter

385

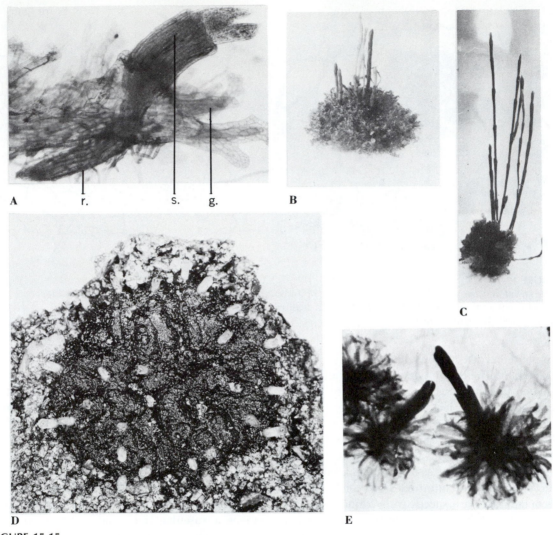

FIGURE 15-15

A–C. *Equisetum hyemale.* **A.** Sporophyte just emerging from gametophyte. **B, C.** Gametophytes with young sporophytes. **D.** *E. hyemale* var. *affine.* Gametophyte on soil from nature; it has 27 emergent sporophytes. **E.** *E. arvense.* Two living gametophytes with emergent sporophytes in culture. **A.** ×60. **B, C.** ×6. **D.** ×1.7. **E.** ×19. (**D.** *After Mesler and Lu.*)

to the vegetative branches is variable in the several species. The sporangium-bearing appendages are called sporangiophores; each has from five to ten sporangia. The sporangiophores differ from sporophylls, which are foliar, in being whorled branches of the strobilus axis. The homosporous spores are anomalous in their possession of four hydroscopic appendages called elaters.

The gametophytes of *Equisetum* are mound-like, chlorophyllous structures with numerous erect lamellae. The archegonia arise at the meristematic margin of the cushion between the bases of the lobes. The gametophytes are

386

either strictly male or protogynous females, the latter ultimately producing antheridia as archegonial formation abates. The antheridia develop at the apices of the lobes. The sperms are multiflagellate. Suspensors are absent from the developing embryos, a number of which may be borne on a single gametophyte.

DISCUSSION QUESTIONS

1. What characteristics distinguish *Equisetum* from other vascular cryptogams?
2. Why are the parenchymatous regions between the internodal vascular bundles in *Equisetum* not considered to be leaf or branch gaps?
3. Define or explain carinal canal, vallecular canal, central canal.
4. What does paucity of xylem in a stem, leaf, or root usually indicate?
5. How does the arrangement of branches with reference to the leaves mark *Equisetum* as unique?
6. Summarize the type of nutrition of the gametophyte in *Lycopodium*, *Selaginella*, *Psilotum*, *Isoetes*, and *Equisetum*.
7. Describe the process of spore dissemination from the strobilus of *Equisetum*.
8. Can you give other examples, among the vascular plants, in which the stems are the chief photosynthetic organs?
9. What evidence can you cite to support the claim that the branched condition is primitive in *Equisetum?*
10. What is the function of the carinal canals?
11. How does the nodal stele differ from the internodal in *Equisetum?*
12. Discuss the distribution of antheridia and archegonia in the gametophytes of *Equisetum*. What significance do you attach to this?

Chapter 16
VASCULAR CRYPTOGAMS III: DIVISION PTERIDOPHYTA I

Because of their number and diversity,[1] this group of vascular cryptogams, the Pteridophyta, are discussed in this and the following two chapters. The Pteridophyta, broadly speaking the ferns, include approximately 10,000 species. Probably precursors of the Pteridophyta flourished during the Middle Devonian (inside front cover). Consideration of these and of the remainder of the fossil record will be deferred to Chapter 20.

Of the divisions of vascular cryptogams with extant members—Psilotophyta, Microphyllophyta, Arthrophyta, and Pteridophyta—only the members of the last are clearly megaphyllous (p. 332), although it is possible that the seemingly microphyllous leaves of *Equisetum* have been reduced from megaphyllous precursors. The leaves in a number of genera are the largest and most complex in the plant kingdom. In many ferns the leaves appear to be the dominant organs of the sporophyte, the stems being smaller and less prominent. All have branching veins, and the traces of all mature fern leaves are associated with gaps in the stem steles, unless the stele is a protostele, in which case a groove is sometimes visible. They share the attribute of megaphylly with the phanerogams or seed plants. Those who classify all vascular plants in the single division Tracheophyta group ferns and seed plants together in the class Pteropsida (inside back cover).

Survey of the Pteridophyta indicates that the group may be divided into two, albeit unequally, with respect to the type of sporangium produced. The first series has a type of sporangium similar in development and organization to that of the Psilotophyta (p. 457), Microphyllophyta, and, probably, the seed plants, that is, **eusporangiate** (see also p. 462). These sporangia are relatively large, sessile or with massive stalks, and contain a large number of spores enclosed by a several-layered sporangial wall, at least during early sporogenesis. In the second series, comprising many times the number of species, the sporangial origin may be traced to single cells that undergo more or less periclinal divisions. After this division the outer cell, by its further divisions and those of its cellular products, forms the bulk of the protuberant sporangium and its delicate sporangial stalk, which is a few

---

[1]See the magnificent volumes of Tryon and Tryon (1982) for ferns and related plants in tropical America.

cells, or even one cell, thick. Here the number of spores per sporangium is usually definite, a multiple of 2 (varying between 16 and 512 in homosporous species). Such sporangia are said to be **leptosporangiate** (see also p. 409). The vast majority of familiar cultivated and wild ferns are leptosporangiate, or show slight modifications thereof. Some botanists (e.g., Fagerlind, 1961; Bierhorst, 1971a) are of the opinion that it is not always possible to distinguish absolutely between some varieties of eusporangiate and leptosporangiate sporangia.

Although the classification of ferns is controversial and in a state of flux, disagreement is most often expressed at the familial level (Jermy et al., 1973). There is rather unified agreement that living ferns may be grouped in five orders—namely, the Ophioglossales, Marattiales, Filicales (sometimes designated Polypodiales), Marsileales, and Salviniales—and representatives of these orders are included in this book. The writers follow Bierhorst (1971a) in grouping the *extant* ferns in three classes, as follows:

Class 1. Ophioglossopsida (order Ophioglossales)
Class 2. Marattiopsida (Order Marattiales)
Class 3. Filicopsida (Orders Filicales, Marsileales, Salviniales)

The Filicales, a very large order of fundamentally leptosporangiate and homosporous[2] ferns, are discussed in Chapter 18 and the leptosporangiate, heterosporous Marsileales and Salviniales in Chapter 18.

CLASS 1. OPHIOGLOSSOPSIDA

The Ophioglossopsida contain but a single order and family, the Ophioglossales and Ophioglossaceae.

---

[2]See, however, *Platyzoma*, p. 411.

Ophioglossum and *Botrychium*

Introduction and vegetative morphology. The genera *Ophioglossum* (Gr. *ophis*, serpent, + Gr. *glossa*, tongue), the adder's-tongue fern (Figs. 16-1, 16-2), and *Botrychium* (Gr. *botrychos*, grape), the grape fern (Fig. 16-3), with between 20 and 30 species each, are rather widely distributed in temperate North America. The former occurs in old fields, meadows, and cedar glades, and the latter is frequently an inhabitant of the forest floor, where it thrives in partial shade. *Ophioglossum engelmannii* is often quite abundant after spring and fall rains in partial shade, near and under cedar trees (*Juniperus virginiana*).

The leaves (Warmbrodt and Evert, 1979) in both genera arise from a rather short, fleshy subterranean stem, which bears fleshy, adventitious roots. The roots of some species of *Ophioglossum* produce adventitious buds, which may

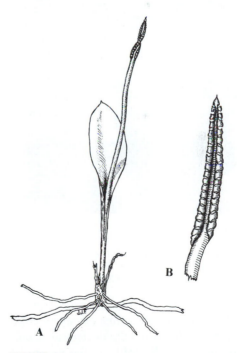

FIGURE 16-1
Ophioglossum engelmannii. **A.** Single fertile plant. **B.** Dehiscent fertile spike. **A.** ×0.5. **B.** ×1.

389

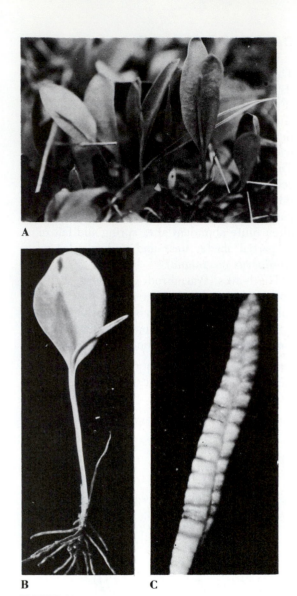

A

B   C

FIGURE 16-2
Ophioglossum engelmannii. **A.** Group of plants
growing near Bastrop, Texas. **B.** Single fertile plant.
C. Fertile spike. **A.** ×⅕. **B.** ×0.5. **C.** ×3.

FIGURE 16-3
Botrychium virginianum. Single plant with fertile
spike. ×0.15. (*From a negative of Dr.
C. J. Chamberlain, courtesy of the Chicago Field
Museum of Natural History.*)

develop new plantlets, a phenomenon that results in the formation of rather extensive colonies. Both *Botrychium* and *Ophioglossum* usually elevate only a single leaf from the perennial
stems each growing season (Figs. 16-1 through
16-3). The leaves are annual in activity. Serial

transverse sections of the axis of *O. engelmannii*
reveal that the leaves arise in spiral order in five
vertical rows, although usually only one leaf is
present at a time. In *B. virginianum* (Fig. 16-3),
the leaf blade is large and dissected, but in other
species of the genus the leaves are smaller and
simpler. The leaf of *O. engelmannii* is simple
and entire. It has been suggested that simple
leaves are reduced rather than primitive in
these ferns. The morphology and histochemistry of the stomata were investigated by Yadaw
and Bhardwaja (1980).

Inasmuch as the leaf is the dominant and most conspicuous organ of these plants, its structure will be described first. The young leaves are rolled up along their longitudinal axes and protected by the remains of the ensheathing stipular sheath of the preceding leaf as they emerge from the soil. The leaflets of *Botrychium* are characterized by open dichotomous venation; the leaf blade in *Ophioglossum* is traversed by a reticulate system of veins, which are mostly united at the leaf margins. The leaves appear as primordia near the slow-growing stem tip several years before they are raised above the ground, and their development is extremely slow. The vascular tissue of the leaf is connected to that in the stem through the petiole. In *O. engelmannii* the double leaf trace branches into four or more strands soon after it enters the petiole. The single trace in *Botrychium* divides into two as it enters the petiole.

Evert (1976) investigated the sieve elements of the petioles and rachises of *B. virginianum*. They are of the **nacreous** types—that is, early in ontogeny the walls thicken with a pearly-looking substance, but the thickening gradually disappears as the sieve elements mature and lose their nuclei. Pores occur in both the lateral and terminal walls of mature sieve elements.

The leaf blade in *Ophioglossum* is covered by epidermal cells above and below (Fig. 16-4); the central portion is composed of photosynthetic parenchyma cells, which are not differentiated into palisade and spongy layers. Stomata occur abundantly on both surfaces of the leaf. The guard cells, alone among the epidermal cells, contain chloroplasts.

The underground stems of both *Ophioglossum* and *Botrychium* are erect, slow-growing, and fleshy. They are covered with the remains of previous seasons' leaves at their summits, and with rather closely arranged, fleshy roots below (Figs. 16-1A, 16-2B), one of the latter associated with each old leaf base. Development of the stem in both genera is localized in the division of a single apical cell and its derivatives. The

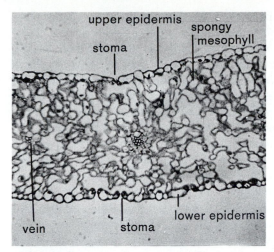

FIGURE 16-4
Ophioglossum engelmannii. Transection of a leaf. ×65.

latter differentiate in older portions of the stem into stelar, cortical, and epidermal regions. The vascular tissue in the primary (embryonic) stems of some species is protostelic, but as the stem grows older, the vascular tissue formed later is arranged as an ectophloic siphonostele. The latter is much dissected (Fig. 16-5) into discrete

FIGURE 16-5
Ophioglossum engelmannii. Transection of stem; note emergent roots and dissection of the ectophloic siphonostele by leaf gaps. ×12.5.

strands in *Ophioglossum* because of the close proximity to each other of the departure of leaf traces, which, in these megaphyllous plants, leave parenchymatous gaps in the stele above the point of their departure. The shortness of the internodes and the overlapping of leaf gaps result in the dissection of the stele, which has endarch maturation of the xylem. The phloem is external to the xylem. Endodermis and pericycle are absent in mature stems of *O. engelmannii*. Webb (1981) investigated stem anatomy and phyllotaxy in *O. petiolatum* and *O. crotalophoroides*. In the stems of *Botrychium*, a cambium adds secondary vascular tissues to the primary ones. The cortex is parenchymatous and starch-filled. The stem surface is suberized in older portions by the formation of a periderm layer. *Botrychium* is the only extant fern genus with secondary vascular tissue and a periderm. Evert (1976) investigated the sieve elements of the petioles and rachises of *Botrychium virginianum*.

The roots in both genera also develop through the activity of single apical cells. They arise endogenously in the rhizome near the phloem in association with the leaves and below them. In *Ophioglossum* the root (Fig. 16-6) may be monarch, diarch, or tetrarch; it is most often tetrarch in *B. virginianum*. The phloem alternates with the xylem, and both are surrounded

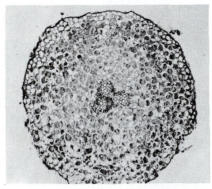

FIGURE 16-6
Ophioglossum engelmannii. Transection of a root; note triarch stele and thick cortex. ×30.

by one or more layers of pericycle cells. The stele is delimited from the cortex by a well-differentiated endodermis. The cortex is extensive and serves as a storage region. The epidermis is devoid of root hairs; in older regions of the root, the surface is suberized. Endophytic fungi usually are present in these genera. Petersen and Brisson (1977) studied the structure of the root cap of *O. petiolatum*.

Reproduction: the sporophyte. Both *Ophioglossum* and *Botrychium* are at once distinguishable from other Pteridophyta by the arrangement of their sporogenous tissue, which is localized in a branched or unbranched **fertile spike** (Figs. 16-1 through 16-3). The latter emerges at the junction of the leaf blade and petiole. Anatomical evidence has been interpreted as indicating that this structure represents a pair of fertile, lateral pinnae (leaflets). Decker-Eisel and Hagemann (1978) discussed its origin. The fertile axis is unbranched in *O. engelmannii* (Figs. 16-1, 16-2) but compound in *B. virginianum* (Fig. 16-3). In the former, it has two longitudinal rows of massive sporangia (Figs. 16-1B, 16-2C, 16-7A). Their walls are composed of several layers of cells (Fig. 16-7B). A branch of vascular tissue runs to the base of each sporangium. There is one- or two-layered tapetumlike zone between the central sporogenous tissue and the wall. As the sporangium develops, the walls of the tapetal cells disintegrate and give rise to a tapetal plasmodium. A few of the sporogenous cells also disintegrate, but the remainder undergo meiosis and give rise to tetrads of spores. The haploid number of chromosomes in *O. vulgatum* is approximately 258. The cells of the sporophyte, therefore, contain more than 500 chromosomes; a tropical species, *O. reticulatum*, with 1260 chromosomes, has the largest number yet observed in a naturally occurring vascular plant. The sporangia dehisce at maturity along a line perpendicular to the long axis of the fertile spike, the line of dehiscence predetermined by the formation of several rows of thin-walled cells. Each sporangi-

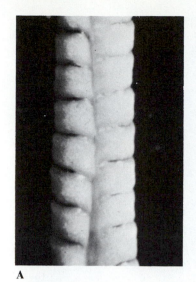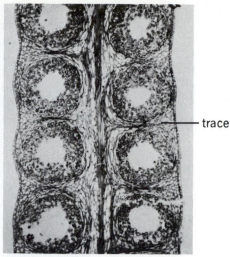

FIGURE 16-7
Ophioglossum engelmanii.
Fertile spike. **A.** At
dehiscence. **B.** Median
longisection of a segment of a
fertile spike. **A.** ×8. **B.** ×25.

— trace

A   B

um may contain as many as 15,000 spores in
some species.

Reproduction: the gametophyte. Development of the gametophyte from the spores has
until recently not been followed completely in
laboratory cultures; hence, our knowledge of it
is based largely on specimens collected in the
field. In both *Ophioglossum* and *Botrychium*
the gametophyte is fleshy, subterranean, and
nongreen and always is infected with the hyphae of an endophytic fungus. In axenic cultures
the fungi are, of course, absent (Gifford and
Brandon, 1978; Whittier, 1981b, 1983). The
gametophytes, except for their apices and sex
organs, are covered with a granular coating that
may have antibiotic properties (Whittier and
Peterson, 1984). *Ophioglossum* gametophytes
are cylindrical (Figs. 16-8, 16-9), up to 5 cm in
length, and branched in some individuals. The
diameter of the largest is about 0.65 cm.

Whittier (1972, 1978, 1981, 1983), succeeded
in growing gametophytes of *B. dissectum* and of
O. engelmannii to sexual maturity in axenic
culture. The spores require darkness for germination. In 11-month-old cultures tracheids developed in the gametophyte, and that of *Botrychium* formed haploid apogamous sporophytes

(Whittier, 1976). Gifford and Brandon (1978)
reported that gametophytes of *B. multifidum*
grown in axenic culture developed only antheridia on a dorsal ridge.

Mesler (1975, 1976) described the gametophytes of *O. palmatum*, a southern Florida
epiphytic species, and those of *O. crotalophoroides* (Fig. 16-10). The gametophytes of the
former are cylindrical and branched, while
those of the latter are globose to hemispherical.
Mesler reported that the spores of *O. crotalophoroides* germinate immediately after their
release in the spring; fertilization occurs about
one year later, and gametophytes with young
sporophytes with their first leaf develop 30
months after spore release. A fungus with nonseptate hyphae was present in the monoecious
gametophytes (Fig. 16-10F) most of which were
2 to 20 mm beneath the soil surface. A few
which were at the surface because of erosion
were pale green.

The nutrition of the gametophyte here, as in
some species of *Lycopodium*, apparently is saprophytic. However, small amounts of chlorophyll may develop in exposed portions of the
gametophyte, the growth of which is apical. The
gametophyte of *O. vulgatum* in the Ryukyu

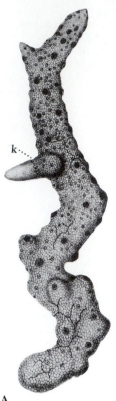

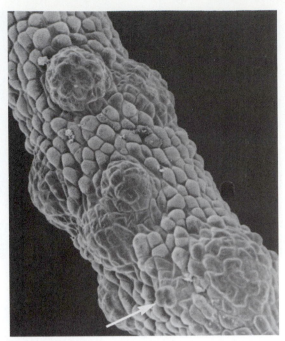

B

C

A

FIGURE 16-8
Gametophytes of *Ophioglossum*. **A.** *O. vulgatum.*
Cylindrical gametophyte with antheridia (larger,
darker) and archegonia; young embryo at k. **B, C.**
O. engelmannii. **B.** Portion of gametophyte with
antheridia and an archegonium (arrow). **C.**
Archegonium; note neck cells. **A.** ×20. **B.** ×130.
C. ×680 (**A.** *After Bruchmann.* **B, C.** *Courtesy of
Dr. D. P. Whittier.*)

Islands contains mycorrhizal fungi (Nozu, 1961).
The authors have searched in vain for gametophytes of *O. engelmannii* in nature but never
have found any, in or near colonies of that
species, in spite of the enormous numbers of
spores shed from their fertile spikes.

Ophioglossum is monoecious, the antheridia
and archegonia occurring together in various
stages of development (Fig. 16-8). Both kinds of
sex organs arise from cells on the surface of the

394

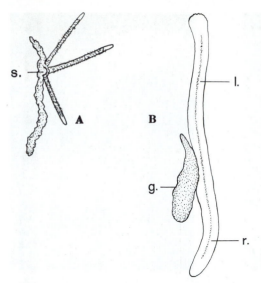

FIGURE 16-9
A. *Ophioglossum vulgatum.* Gametophyte showing roots and minute stem (s.) of young sporophyte. **B.** *O. moluccanum.* Embryonic sporophyte attached to gametophyte. g., gametophyte; l., primary leaf or cotyledon; r., primary root or radicle. **A.** ×2. **B.** ×15. (**B.** *Modified from Smith.*)

gametophyte, but at maturity they are largely sunken within it, as in *Anthoceros.* The antheridia are massive and contain large numbers of multiflagellate sperms (Fig. 16-10E). Only the tips of the archegonial necks (with four rows of neck cells) protrude above the gametophytic surface. Fertilization has been observed to occur in the summer months in the northern hemisphere.

Reproduction: embryogeny. Development of the zygote into an embryonic sporophyte progresses slowly in various species of *Ophioglossum.* A small spherical mass of tissue is formed within the archegonial venter. This becomes differentiated into foot, root, and leaf primordium. The apical initial of the stem appears later, and this perhaps foreshadows the inconspicuous state of that organ in the mature sporophyte. Several years are required in most species for the production of a leaf with a fertile

spike. Figures 16-9 and 16-10B, H illustrate gametophytes with an attached juvenile sporophyte.

CLASS 2. MARATTIOPSIDA

The Marattiopsida contains only a single order, Marattiales, and family, Marattiaceae.

Marattia and *Angiopteris*
Introduction and vegetative morphology. The members of the Marattiales are exclusively tropical. Two of the seven genera, *Marattia* (Fig. 16-11), with about 60 species, and *Angiopteris* (Fig. 16-12), with about 100 species, are here chosen to illustrate the order. *Marattia* and *Angiopteris* are sometimes available in university and other conservatories and as specimens in herbaria.

Species of these genera have the largest and most complex fronds in the family (Stidd, 1974); the fronds are several times pinnately compound. The young leaves are **circinately vernate,** that is, curled in the bud. *Angiopteris* is native to Hawaii, the Philippines, and the western Pacific area, while *Marattia* is more widely distributed in the tropics. The fronds of a New Zealand species are 6 to 9 m long and up to 4.5 m wide. The ultimate leaf divisions (pinnules) have dichotomous venation. Mucilage chambers occur throughout the rather fleshy plants. Warmbrodt and Evert (1979) investigated the leaves of *Angiopteris* and *Marattia* electron microscopically.

The stems are fleshy, tuberous, and erect or trunklike and covered by the persistent paired stipules at each leaf base, from among which the fleshy roots emerge. The roots contain polyarch actinosteles with as many as 12 protoxylem groups (*M. alata*), many more than occur in most ferns.

The young stems develop from single apical cells and in their mature regions contain a single cylinder of vascular bundles. In older stems,

395

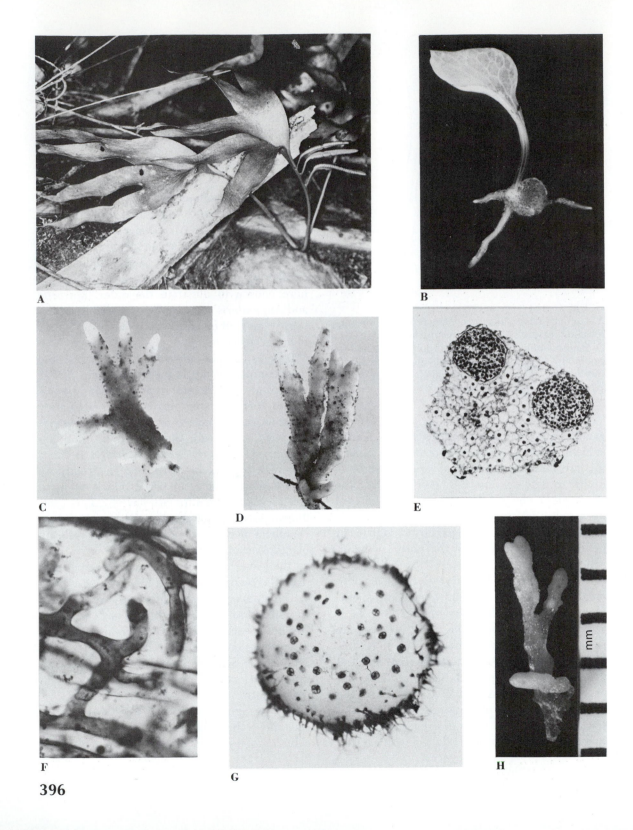

FIGURE 16-10

Ophioglossum. All *O. palmatum* except **B** and **G,** which are *O. crotalophoroides.* **A.** Plant growing on cabbage palmetto. **B.** Gametophyte with young sporophyte. **C, D.** Gametophytes. **E.** Transection of gametophyte with two antheridia. **F.** Mycorrhizal fungus in cells of gametophyte. **G.** Upper surface of mature gametophyte; note sex organs. **H.** Gametophyte with emergent root of young sporophyte (near base). **A.** ×0.3. **B.** ×4.5. **C.** ×8. **D.** ×8. **E.** ×65. **F.** ×560. **G.** ×22. **H.** mm. scale at right. (*All after Mesler.*)

A   B

FIGURE 16-11
A. *Marattia alata.* **B.** *Marattia fraxinea.* Growing in Botanic Garden, Sydney, Australia. **A.** ×0.2. **B.** ×¹⁄₂₄. (**A.** *After G. M. Smith.* **B.** *From a negative of Dr. C. J. Chamberlain, courtesy of the Chicago Field Museum of Natural History.*)

FIGURE 16-12
Angiopteris evecta. Growing in Missouri Botanic Garden. (*From a negative of Dr. C. J. Chamberlain, courtesy of the Chicago Field Museum of Natural History.*)

however, the vascular tissue is present as a polycyclic dictyostele with two or more concentric vascular cylinders. Some of the strands are leaf traces.

Reproduction: the sporophyte.　The eusporangiate sporangia of *Angiopteris* are massive (Fig. 16-13A) and closely appressed laterally. They are borne near the leaf margins along the veinlets in compressed rings of 10 to 20. Approximately 512 sporocytes are differentiated within each sporangium. The latter is thick-walled and provided with a tapetum. The sporangia of *Angiopteris,* although crowded together, retain their individuality and dehisce into two valves by vertical splitting (Fig. 16-13A, B). The sporangia of *Marattia* are united into a compound unit called a **synangium** (Fig. 16-13C). The latter opens as a unit, exposing the vertically split, contiguous sporangia. Sometimes the bivalvular structure is called a **sporocarp.**

Reproduction: the gametophyte.　The gametophytes are massive and liverwortlike (suggestive of *Pellia*), up to several centimeters long (Fig. 16-14). A mycorrhizal fungus is usually present within the gametophytic cells. The gametophytes are long-lived and many cells thick, except at the margins. Antheridia, which produce multiflagellate sperms, may develop on both surfaces of the protandrous gametophytes; the archegonia occur only on the ventral surface. Both types of sex organs are immersed in

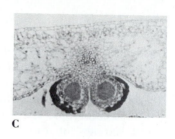

C

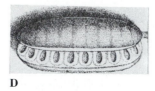

D

FIGURE 16-13

A, B, C. *Angiopteris* sp. Dehiscent sporangia. **C.** Transection of leaf with sporangia. **D.** *Marattia fraxinea.* Dehiscent synangium. **A.** ×5. **B.** ×25. **C.** ×45. **D.** ×5. (**D.** *After Bower, from Bitter.*)

the thallus as in the Ophioglossales and *Anthoceros.*

Reproduction: embryogeny.   Usually only one zygote on each gametophyte develops into a juvenile sporophyte. The Marattiales differ in their embryogeny from the leptosporangiate ferns (p. 418) in that the primary leaf emerges from the dorsal surface of the gametophytes, rather than from the ventral.

Summary

The eusporangiate Pteridophyta include two groups of living ferns exemplified in the preceding account by *Ophioglossum* and *Botrychium,* on the one hand, and by *Marattia* and *Angiopteris,* on the other. The first two genera bear their massive, eusporangiate sporangia on fertile spikes, whereas in the latter they occur separately, though contiguously, or in synangia, on the lower surface near the margins of vegetative leaflets. In both cases, development and maturation of all the sporangia are synchronous, and each sporangium produces large numbers of spores.

The gametophytes of *Ophioglossum* and *Botrychium* are fleshy, subterranean, achlorophyllous, saprophytic structures. That of the former is cylindrical or hemispherical and radially organized, with sex organs well distributed over the surface. In *B. virginianum* the gametophyte has some degree of dorsiventrality, with the sex organs limited to the upper surface. The gametophytes of *Marattia* and *Angiopteris,* on the other hand, are epiterranean, chlorophyllous, and *Pellia*-like, although they also contain a mycorrhizal fungus. The archegonia are usually confined to the lower surface; the antheridia may occur on both surfaces.

These genera have megaphyllous leaves, which are the dominant organs of the sporo-

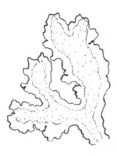

FIGURE 16-14
Marattia sambricina.
Gametophyte 27 months
after spore germination.
×4. (*After Stokey.*)

399

phyte. This is attested by the profound disturbance of the stem stele by leaf gaps and the relatively slow development of the stem itself. The latter is especially apparent in the development of the embryonic sporophyte, in which leaf development far outstrips that of the stem. The first embryonic leaf emerges through the dorsal surface of the gametophyte in the Marattiales.

DISCUSSION QUESTIONS

1. Distinguish between megaphylls and microphylls.
2. How does the venation of *Ophioglossum* differ from that of *Botrychium*? Which is considered to be more primitive, and on what basis?
3. Are microphylls and megaphylls homologous, in your opinion? Explain.
4. Can you suggest an explanation for the relatively small size of the stems in *Botrychium*, *Ophioglossum*, and the Marattiales?
5. Does the approximately simultaneous maturation of the spores in these genera occur in vascular plants previously studied? Can you suggest a biological disadvantage of this habit?
6. Can you suggest similarities between the gametophyte of *Ophioglossum* and that of other seedless vascular plants? Explain.
7. How does spore production in *Marattia* and *Angiopteris* differ from that in *Ophioglossum* and *Botrychium*?
8. In what respects are the gametophytes of these genera different?
9. Where would you look for *Botrychium* and *Ophioglossum* in nature?

Chapter 17
VASCULAR CRYPTOGAMS IV: DIVISION PTERIDOPHYTA II

CLASS 3. FILICOPSIDA

As stated at the beginning of Chapter 16, the division Pteridophyta contains both eusporangiate and leptosporangiate ferns. The former were described in the preceding chapter, and the latter will be discussed in this chapter and in Chapter 18. In leptosporangiate sporangium development (p. 409, Fig. 17-12), it will be recalled, the sporangium arises from a single cell, derivatives of which protrude from the fertile receptacle. Leptosporangiate sporangia are small, produce a definite number of spores (frequently 48 to 64, but up to 512), and usually have a wall one layer of cells in thickness. Other attributes, in addition to sporangial characters, serve to distinguish the eusporangiate from the leptosporangiate ferns. These will be cited in the following account.

In number of genera, species, and individuals, the leptosporangiate ferns far outnumber the eusporangiate. Some 300 genera and 9000 species have been described. These are most abundant and diverse in the tropics, but they are also well represented in the temperate zone.

Two species, *Onoclea sensibilis* and *Pteridium aquilinum*, have been reported to be poisonous to livestock (Kingsbury, 1964). Some ferns are edible by human beings, including the young leaves or "fiddleheads" (see p. 432) of *Osmunda* and those of *Athyrium esculentum*, which reach a length of a decimeter. The latter are cooked and eaten in Malaysia, as is a species of *Ceratopteris* (Copeland, 1942). In habit, they vary from small, delicate, filmy plants to large tropical tree ferns with upright stems and enormous leaves; several are climbers. The leptosporangiate ferns include a number of families, of which many species are inhabitants of the tropics, and, therefore, not readily available for study except in collections in herbaria and occasionally in conservatories. For this reason, the treatment of the group in this text will be restricted to a brief discussion of representatives that are readily available; specialized treatises listed in the Bibliography give a more extensive treatment of the ferns. The books of Bower (1923, 1926, 1928), Bierhorst (1971a), and Foster and Gifford (1974) provide more comprehensive discussions of ferns than does the present volume.

The classification of leptosporangiate ferns with respect to orders and families has been in a state of flux during the past 25 years. The most conservative classification grouped them in a single order, Filicales, which included 19 families. At the other extreme is their classification in 18 orders and 38 families! Most recently, Bierhorst (1971a, p. 292) has recognized three orders (Filicales, Marsileales, and Salviniales) and 27 families. Various schemes of phylogenetic relationships of ferns have been proposed (Jermy, Crabbe, and Thomas, 1973). In the cited reference, however, Holttum writes: "As regards the great majority of ferns, their probable phylogeny can be inferred only from their classification; there is no other evidence available."

Members of the order Filicales will be discussed in this chapter and those of the Marsileales and Salviniales in Chapter 18.

Order 1. Filicales

With the exception of the five genera of "water ferns" classified in the Marsileales and Salviniales, the majority of leptosporangiate ferns are often included in the very large heterogeneous order Filicales, which contains numerous families of homosporous genera and one genus seemingly with incipient heterospory (p. 411). A number of these are known only through tropical representatives from remote places; material of these ferns is not readily available. The present discussion, accordingly, will emphasize certain widely distributed and generally available genera of the family Polypodiaceae and will be followed by briefer treatment of representatives of some of the other families.

Family 1. Polypodiaceae

Introduction and vegetative morphology.
A majority of the most familiar, naturally occurring, and widespread ferns of the temperate zone and tropics, as well as many cultivated varieties, are members of a vast assemblage of polypodiaceous ferns, so named from the genus *Polypodium* (Gr. *polys*, many, + Gr. *pous*, foot). Among these are *Thelypteris* (Gr. *thelys*, female, + Gr. *pteris*), *Adiantum* (Gr. *adiantos*, maidenhair), *Polystichum* (Gr. *polys*, many, + Gr. *stichos*, row), *Woodsia* (after Joseph Wood, an English botanist), *Pteris* (Gr. *pteris*, fern), *Pteridium* (Gr. *pteris*), and *Nephrolepis* (Gr. *nephros*, kidney, + Gr. *lepis*, scale). Most of these genera are plants of moist, mesic woodlands. *Adiantum capillus-veneris* (Figs. 17-1, 17-2A), the Venus maidenhair, however, is an example of a rather hydrophytic species and is widely distributed on wet limestone cliffs; *A. pedatum* (Fig. 17-2B) grows in moist, shady woods. *Pellaea atropurpurea*, the cliff brake, and *Polypodium polypodioides*, the "resurrection fern," on the other hand, are xerophytes. All temperate-zone members of this alliance are perennial, and a few (*Polystichum*, *Cyrtomium*) are evergreen; a number of ferns have mycorrhizal associations (see Cooper, 1977, for the literature).

FIGURE 17-1
Thelypteris normalis and *Adiantum capillus-veneris* growing at Hamiltons' Pool, Hays County, Texas. ×¹⁄₁₅. (*Courtesy of Dr. Dianna Tupa.*)

A

B

FIGURE 17-2
A. *Adiantum capillus-veneris.* **B.** *A. pedatum.*
A. ×0.33. **B.** ×0.1.

Adiantum capillus-veneris (Fig. 17-2A) and *Thelypteris dentata* (Fig. 17-3) will be emphasized as illustrative types in the following account, because they are widespread and readily

FIGURE 17-3
Thelypteris dentata, a species of shield fern. c., circinately vernate leaf. ×0.05.

cultivated in greenhouses or out-of-doors in sufficiently mild climates. In both of them the stem is subterranean, although in the former it may be exposed. The bases of the preceding season's leaves persist indefinitely and form a sort of jagged armor about the stems. The stem in *T. dentata* is ascending and relatively slow-growing, but in *Adiantum* it is a horizontal, rapidly elongating rhizome. Branching of the rhizome is abundant in *Adiantum* and results in colonization of the area around the original plant. The stems of both genera are clothed with a dense mass of adventitious roots, which emerge between the leaf bases.

Growth of the stem in each case may be traced to the activity of a single apical cell and its derivatives. In investigations of stem apices of many ferns, Bierhorst (1977) and Hébant, Hébant-Mauri, and Barthonet (1978) found few whose development did not originate in an apical cell. This apical cell occurs at the stem tip, often concealed by a dense growth of superficial, multicellular, scalelike organs known as **paleae.** The apical cell of the stem of leptosporangiate ferns gives rise to three ranks of derivative cells. Although the vascular tissues

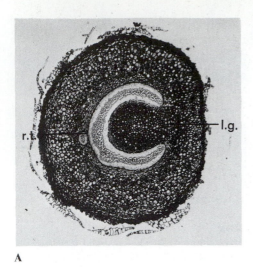

A

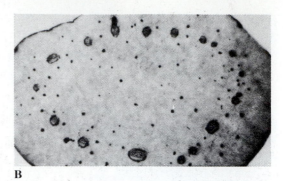

B

FIGURE 17-4
A. *Adiantum* sp. Transection of rhizome; note amphiphloic siphonostele (solenostele).
B. *Thelypteris* sp. Transection of rhizome; note dictyostele. l.g., leaf gap; r.t., root trace. ×20.

in the embryonic stems of all leptosporangiate ferns are reported to be protostelic, there is a gradual transition in older stems to a siphonostelic, solenostelic, or dictyostelic condition, depending on the species. In *Adiantum* (Fig. 17-4A), the vascular tissue is completely surrounded by a well-differentiated endodermis that separates it from the parenchymatous pith and cortex; in *Thelypteris*, the endodermis surrounds each segment of the dictyostele (Fig. 17-4B). Portions of both of these may become sclerotic. *Adiantum* is solenostelic, while *Thelypteris* is dictyostelic. The xylem is mesarch in development and composed chiefly of elongate tracheids with tapered ends with abundant, transversely elongate (scalariform), bordered pits. In a few genera (*Pteridium*), the closing membranes of the pits of the sloping terminal tracheid walls are dissolved at maturity, resulting in direct continuity between adjoining elements. Thus, primitive vessels, previously noted in *Selaginella* and *Equisetum* and characteristic of the xylem of many flowering plants, are present in several ferns. The sieve elements are elongate, lack companion cells,

and have numerous sieve areas in their vertical walls. The sieve elements in some ferns have walls thicker than those of the associated parenchyma cells. These thickenings may be transitory (Evert and Eichorn, 1976).

All fern roots, with the exception of the embryonic radicle, are adventitious (Fig. 17-3) and arise endogenously from the stem. They develop between the leaf bases. Development of the root originates in a single apical cell; in this case, however, the latter, in addition to the usual series, cuts off derivatives in a direction perpendicular to the long axis of the root, thus adding cells to the root cap. The cells of the central procambium differentiate into an exarch protostele, the arrangement characteristic of the roots of most vascular plants. The roots of *Thelypteris dentata* are diarch. The stele is surrounded by a narrow pericycle, a prominent endodermis, cortex, and epidermis, the latter with root hairs. The inner portion of the cortex is often sclerotic. Branch roots originate from endodermal cells opposite the protoxylem groups. Secondary growth is absent in both stems and roots of leptosporangiate ferns.

404

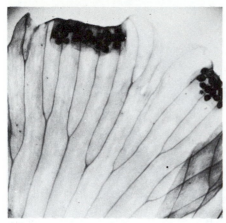

FIGURE 17-5
Adiantum capillus-veneris. Cleared pinnule; note dichotomous venation and sori with false indusia. ×5.

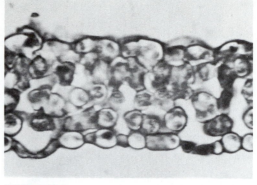

FIGURE 17-6
Thelypteris dentata. Transection of leaf; note upper and lower epidermis, the latter with a stoma and mesophyll. ×400.

As in the eusporangiate ferns, the leaves of *Thelypteris* and *Adiantum* and many other species are the dominant organs of the sporophyte. The large compound leaves, **circinately vernate** in the bud (Figs. 17-3, 17-38), may be highly elaborate when mature. Complex, finely divided fern leaves are considered to represent the primitive condition. Simple leaves in ferns are probably reduced. The primary divisions (leaflets) of a compound fern leaf are called **pinnae.** The ultimate divisions are called **pinnules.** The continuation of the petioles among the leaflets is the **rachis.** The fern leaf is often referred to as a **frond.**

The petiole of each leaf is connected to the stem stele by one or more traces; these run throughout the rachis, giving rise to branches that traverse the pinnae and pinnules. In *Adiantum* the venation is dichotomous (Fig. 17-5), while in *T. dentata* it is pinnate. Reticulate venation is thought to have been derived from the dichotomous pattern by development of branches between the dichotomies. Internally, the leaf blade consists of an upper and lower epidermis, the latter with abundant stomata,

enclosing a relatively undifferentiated mesophyll (Fig. 17-6). Van Cotthem (1970*a,b*; 1971) has made a comprehensive survey (510 species of 240 genera!) of the stomata of ferns and recognizes nine different types. The mesophyll of *Adiantum* is limited in extent; the abundant chloroplasts in the epidermal cells in this genus undoubtedly play a major role in photosynthesis. The veins are collateral in arrangement of xylem and phloem. The leaves arise close to the growing point of the stem and develop at first through the activity of an apical cell with three cutting faces. A new group of leaves is produced each season. The simple leaves of the walking fern, *Asplenium rhizophyllum*, function in propagation (Fig. 17-7).

Reproduction: the sporophyte. After a series of entirely vegetative leaves has been produced from the stem of a young plant, all the leaves that develop subsequently in *Thelypteris* and *Adiantum*, and in most other genera, are usually both vegetative and fertile. In both genera, groups of sporangia known as **sori** (Gr. *soros*, heap) develop on the undersurface of ordinary leaves (Fig. 17-8). Some segregation of

FIGURE 17-7
Asplenium rhizophyllum. The "walking fern"; note simple leaves developing new plantlets at their apices (arrows). ×0.25. (*Courtesy of Dr. Ilda McVeigh.*)

vegetative and reproductive functions is apparent in the distribution of sporogenous tissue in several other genera of polypodiaceous ferns. In the Christmas fern, *Polystichum acrostichoides* (Fig. 17-9), for example, only the distal pinnae of each frond are fertile. *Onoclea sensibilis,* the sensitive fern, and *Pteretis pensylvanicum,* the ostrich fern, superficially suggest *Osmunda cinnamonea,* the cinnamon fern (p. 420), in that they produce sporangia on markedly modified fronds, the remaining leaves being purely vegetative.

In *Adiantum* (Fig. 17-10) and *Pteris* (Fig. 17-11), the sori are almost marginal and are covered during development by a marginal flap of tissue known as a **false indusium.** (L. *indusium,* a woman's undergarment). The sporangia are actually borne on this false indusium in *Adiantum.* In *Thelypteris dentata* (Fig. 17-8)

the sori lie over the veins of the pinnules, and each is covered by a shieldlike flap of tissue, the **true indusium,** a single layer of cells in thickness and of epidermal origin. The sporangia in both genera arise from the lower leaf surface (Fig. 17-8E), and the region of their origin is known as the **receptacle.** The order of development in both cases is said to be **mixed,** since sporangial development follows no definite sequence with relation to position on the receptacle, both immature and mature sporangia being present in contrast to the simultaneous maturation of the sporangia, which is called **simple,** some ferns, e.g., in *Osmunda.*

The process of development of a leptosporangiate sporangium is illustrated in Fig. 17-12. Each sporangium arises from a superficial cell of the receptacle that projects above the surface and undergoes an oblique division. The inner

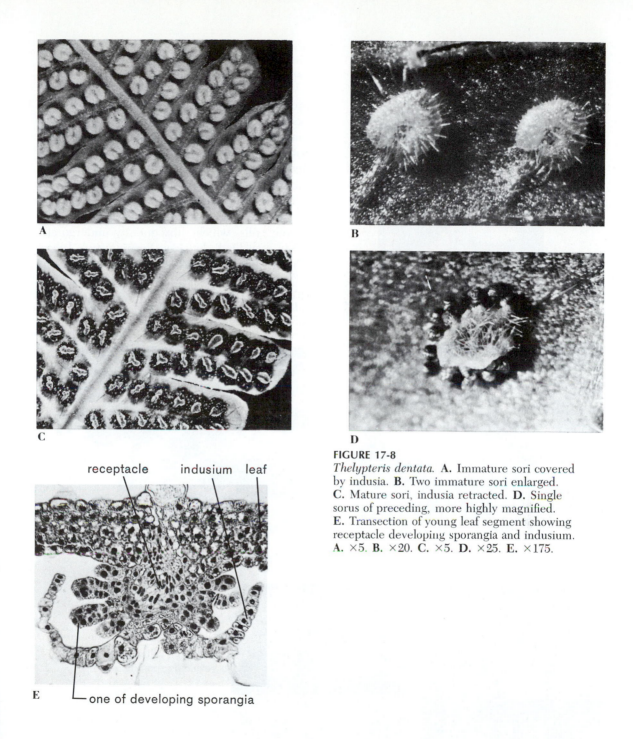

receptacle   indusium   leaf

one of developing sporangia

FIGURE 17-8
Thelypteris dentata. **A.** Immature sori covered
by indusia. **B.** Two immature sori enlarged.
C. Mature sori, indusia retracted. **D.** Single
sorus of preceding, more highly magnified.
E. Transection of young leaf segment showing
receptacle developing sporangia and indusium.
A. ×5. **B.** ×20. **C.** ×5. **D.** ×25. **E.** ×175.

FIGURE 17-9
Polystichum acrostichoides. Portion of frond showing distal, fertile pinnae. ×½.

product of the division either may not divide again or it divides to form two cells, of which the innermost takes no further part in sporangial ontogeny; the outer of these two and the most distal cell undergo intercalary divisions, forming the sporangium, which consists of a stalk and an enlarged distal portion (Fig. 17-12G). A central tetrahedral cell gives rise by periclinal divisions to four cells, which subsequently undergo another periclinal division, forming two layers (Fig. 17-12F, G). The outer usually serves as an

FIGURE 17-10
Adiantum capillus-veneris. Fertile pinnules. **A.** Sori covered by false indusium. **B.** Mature sori, indusium rolled back. ×15.

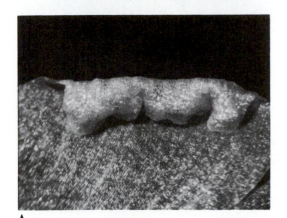

A

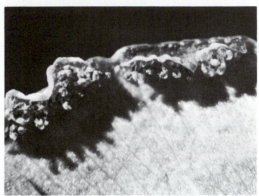

B

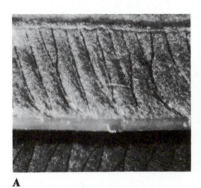

A

B

FIGURE 17-11
Pteris vittata. **A.** Margin of pinna with false indusium covering immature sporangia. **B.** Margin of pinna with indusium rolled back revealing sporangia. ×4.

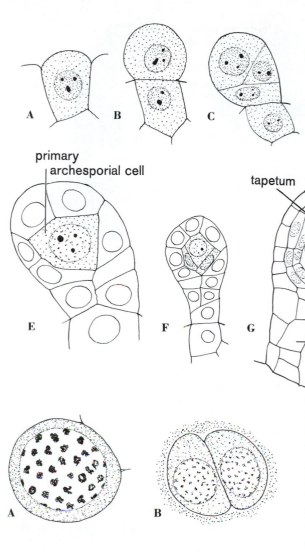

FIGURE 17-12
Thelypteris dentata.
A–G. Successive stages in leptosporangiate sporangium development. **A–E.** ×770.
F. ×410. **G.** ×190.

primary archesporial cell

wall cell (annulus)

tapetum

sporocyte

stalk

FIGURE 17-13
Thelypteris dentata. Sporogenesis. **A.** Sporocyte in prophase of meiosis. **B.** End of meiosis I. **C.** Spore tetrad. **D.** Immature spore. ×1500.

inner layer of the sporangial wall, while the inner layer functions as a **tapetum.** The central cell (**primary archesporial cell**), enclosed by the tapetum, now divides repeatedly and forms the sporocytes (Figs. 17-12G, 17-13A), which vary in number from 12 to 16, the latter the most typical number. Each of these separates and undergoes meiosis, forming a tetrad of spores that ultimately separate and thicken their walls (Fig. 17-13B–D). The chromosome number of the Venus maidenhair, *A. capillus-veneris,* has been determined to be $n = 30$, a rather low number among the leptosporangiate ferns. The tapetal cells elongate radially during sporogenesis and finally become disorganized, presumably functioning in the nutrition of the developing spores.

During sporogenesis, the cells of the sporangial jacket divide by anticlinal division and the products increase in size, as do the cells of the stalk. At maturity, the sporangium is a slightly

409

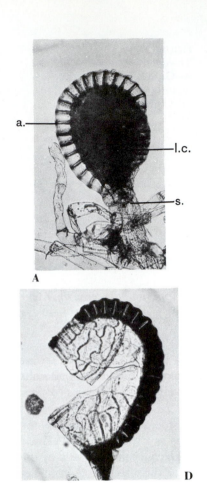

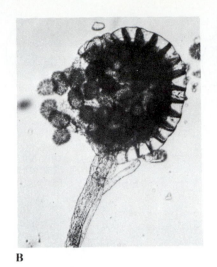

A

B   C

D

FIGURE 17-14
Thelypteris dentata. Sporangial structure and dehiscence.
A. Mature sporangium. **B.** Contraction of annulus and rupture of
lip cells. **C.** Further contraction of annulus and opening of
sporangium. **D.** Reelongation of annulus; sporangium empty.
a., annulus; l.c., lip cell; s., stalk. ×135.

flattened spheroid or ellipsoid (Fig. 17-14A).
The wall cells on the flattened faces are large
and have undulate walls. A vertical row of
specially differentiated cells, the **annulus,** forms
an incomplete ring about the sporangium (Fig.
17-14). The radial and inner tangential walls of
these cells are markedly thickened. Between
the last cell of the annulus and the sporangial
stalk there are a number (often four) of thin-
walled cells, the **lip cells** or **stomium** (Fig.
17-14A, D).

As the spores mature, they become invested
with a yellow, rarely black, wall, and the color of
the entire sorus changes from a delicate whitish-
green to dark brown, rarely black.

The walls of the spores of ferns (and most

other land plants) are differentiated into two
layers, an inner (the endospore, or **intine**) and
an outer (the exospore, or **exine**). The outer
layer is variously ornamented, as revealed by
electron microscopy. In a few ferns a third
layer, the **perispore,** surrounds the exospore.
When the spores on a majority of the sporangia
are mature, the indusium in *Thelypteris* contracts (Fig. 17-8C, D), exposing them to the air.
In *Adiantum* (Fig. 17-10) and *Pteris* (Fig. 17-11)
the revolute leaf edge unfolds slightly when the
spores are mature.

Spore dissemination in both genera is accomplished by dehiscence of the sporangia, which
depends on loss of water. The annulus and lip
cells are directly concerned in the process. The
loss of water through the thin outer tangential
walls of the cells of the annulus shortens its
length so that the sporangial wall is ruptured
transversely in the region of the lip cells (Fig.
17-14B); the outer wall cells of the sporangium
are also ruptured. As the annulus continues to
shorten (Fig. 17-14C), its outer tangential cell

410

walls become increasingly concave, because they are adherent to the water within the annulus cells, the amount of which is lessening through evaporation. Ultimately, a point is reached at which the tensile resistance of the water within the annulus cells is no longer sufficient to prevent the separation of the outer tangential walls from its surface; at this point the water vaporizes, reducing the tension on the outer tangential cell walls of the annulus, which snaps back or catapults to its original position (Fig. 17-14D), thus ejecting the spores (Ingold, 1974).

Although the number of spores produced by a single sporangium in most leptosporangiate ferns is small, frequently 48–64, as compared with those of eusporangiate genera (1500–2000 in *Botrychium*, for example), the large number of sporangia in a sorus, along with the enormous number of sori on the pinnules of a single mature plant, results in the production of a surprisingly large number of spores by a single individual. It has been estimated that one mature plant of *Thelypteris* produces 50 million spores each season. Farrar (1976) found that the ferns in a certain area in central Iowa matured and released most of their spores in a two-week period. This was especially true of *Botrychium* (p. 390) and *Osmunda* (p. 420), which mature all their spores simultaneously. Other species (e.g., of *Adiantum* and *Polypodium*), with nonsimultaneous (mixed) maturation of sporangia, shed their spores during the late summer and autumn.

Unique among the Filicales is an Australian fern, *Platyzoma microphyllum*, which produces spores of two size classes with a volume ratio of 1:7.5; in addition, spores of intermediate size are formed. Thirty-two spores are borne in some smaller sporangia and 16 in other larger sporangia. These spores grow at first into dioecious gametophytes, the smaller into filamentous male gametophytes lacking rhizoids (Fig. 17-15A), and the larger into spatulate ones bearing archegonia (Fig. 17-15B) and, later,

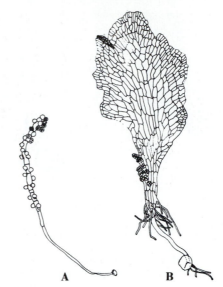

FIGURE 17-15
Platyzoma microphyllum: 4-week-old gametophytes. **A.** Male gametophyte. **B.** Female gametophyte. Note differences in size of spores at the bases of the gametophytes. ×25. (*After Tryon.*)

antheridia (Tryon, 1964; Duckett and Paug, 1984). This distribution of sex organs is reminiscent of that in *Equisetum* (p. 381). Of additional interest is the report that the smaller spores may germinate within the sporangia.

Reproduction[1]: the gametophyte. In spite of the enormous number of spores, relatively few complete their development into mature gametophytes in nature, because so many are borne by air currents to places unfavorable for germination. Gametophytes are sometimes present on moist soil or in rock crevices in the vicinity of fern colonies. As a matter of fact, fern gametophytes seem not to be abundant or conspicuous in the wild, at least in the temperate zone (Wetmore, De Maggio, and Morel, 1963). Germination of fern spores and their development into gametophytes were described first in 1827 by the German, Kaulfuss.

---

[1]Consult De Maggio (1977) for a review of the cytological aspects of reproduction in ferns.

411

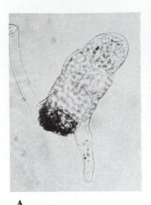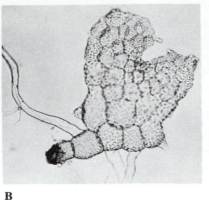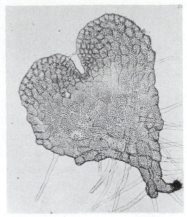

A   **B**

FIGURE 17-16   **C**

Thelypteris dentata. Spore germination and gametophyte
development. **A.** Early germination; note chlorophyllous branch
and rhizoids. **B.** Early gametophyte, 25 days after spores were
planted. **C.** Almost-mature gametophyte (45 days). **D.** Mature
gametophytes on the surface of an agar culture. **A.** ×325. **B.**
×250. **C.** ×45. **D.** ×3.5.

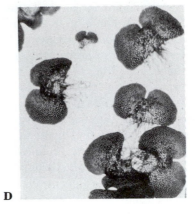

D

Evidence for allelopathic activity (Rice, 1984)
in ferns is increasing. The activity may be that
between sporophytes and sporophytes, gametophytes and sporophytes, sporophytes and gametophytes, and gametophytes and gametophytes.

Davidonis and Ruddat (1973, 1974) reported
that gametophytes of *Thelypteris normalis* do
not grow under the sporophytic plants because
of substances, Thelypterin A and B, diffusing
from the roots. Thelypterin A is an indole
derivative that inhibits cell division, and hence
cell numbers, in the gametophytes of *T.
normalis* and also in the ferns, *Pteris*, and
Phlebodium. It does not inhibit young sporophytes of *T. normalis.* In this connection
Petersen and Fairbrothers (1977, 1980) reported
allelopathy between gametophytes of *Osmunda*
species and *Dryopteris* species as did Munther
and Fairbrothers (1980). The presence of the
former decreases the growth and cell size of the
latter by about 50%. Star and Weber (1978)
found that the powdery, flavonoid exudate on
the leaves of the fern *Pityrogramma calemolanos* inhibits spore germination and normal

development of the gametophytes of the same
species.

Most fern spores require light for their germination, especially red light (500 to 700 nm)
(Mohr, 1963). The processes of spore germination and gametophyte development may be
followed readily in laboratory cultures of spores
of *Thelypteris, Adiantum,* and other leptosporangiate ferns planted on agar on moist soil,
or on the surface of moist flowerpots or other
crockery. Some stages in their development are
illustrated in Fig. 17-16. In crowded cultures
the developing gametophytes may remain irregularly filamentous, but when they are well
spaced they typically become cordate.

412

Spore germination frequently results in the formation of a small, protonemalike chain of bulbous cells rich in chloroplasts (Fig. 17-16A). One or two rhizoids usually develop early from the basal cell, and more rhizoids develop later from the ventral cells of the gametophyte. Dyer and Cran (1976) have demonstrated that the walls of the rhizoids differ from those of other gametophytic cells. The walls of the latter have an oriented, fibrillar organization, while those of the former have a random network of fibrils within which are electron-dense deposits.

The form of the gametophyte and abundance of rhizoids are affected by quality and intensity of light. The spores of most ferns will not germinate in darkness. *Pteridium aquilinum* is exceptional in this respect. Higher intensities of white and blue light evoke the development of the flat, thallose form. They also increase the number of rhizoids. In darkness or in low light intensities, especially of red light, the gametophyte remains filamentous and a few rhizoids form (Fig. 17-17).

Under favorable conditions, oblique divisions in the terminal cell of the young filamentous gametophyte result in the production of an apical cell that cuts off segments in two direc-

tions. The derivatives continue to divide, forming a spatulate plant body (Fig. 17-16B–D) one layer of cells in thickness, except in the central region near the apical notch. Mohr (1963) has reported that blue light (below 500 nm) is required for the initially filamentous stage to develop the spatulate phase of the gametophyte. This type of development is just one of a number of patterns that have been recorded for ferns. The variations have been summarized by Nayar and Kaur (1971) and Atkinson (1973).

It has been demonstrated (Miller and Miller, 1961) that fern gametophytes (*Onoclea*) will grow in red light, provided sucrose is present in the culture medium, although they will not grow in its absence. Accordingly, these gametophytes are facultatively heterotrophic in red light (but not in darkness).

The **prothallia** or prothalli, as the gametophytes are often called, are usually monoecious but usually somewhat protandrous. This is especially noticeable in crowded cultures, in which minute filamentous gametophytes tend to develop antheridia precociously. In uncrowded cultures, the sex organs appear in laboratory cultures on agar in about 35 to 40 days in *T. dentata*. Both types are normally borne on the

FIGURE 17-17
Development of the spores of *Woodwardia radicans* after 3 weeks in: **A.** Darkness. **B.** Light. Note ungerminated spores in **A. A.** ×120. **B.** ×60.

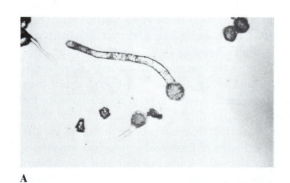

A

B

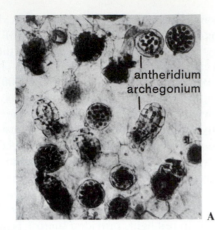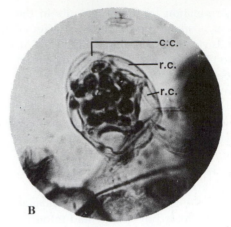

FIGURE 17-18

Thelypteris dentata. **A.** Ventral view of gametophyte near apical notch, showing sex organs. **B.** Antheridium enlarged. c.c., cover cell; r.c., ring cells. **A.** ×135. **B.** ×500.

ventral surface, although deviations may appear in the moist atmosphere of petri dish cultures.

In general, the antheridia of many gametophytes are produced before the archegonia (protandry) and develop nearer the posterior portion of the gametophyte than the archegonia, which develop on the thickened, central cushion nearer the apex (Figs. 17-16D, 17-18A). Like the sporangia, the antheridia are protuberant, whereas only the necks of the archegonia project from the thallus surface (Fig. 17-18A). Both antheridia and archegonia originate from single superficial cells of the gametophyte. The antheridial initial protrudes to form a hemispherical cell that undergoes transverse division. The outer product of division develops into the antheridium, as shown in Fig. 17-18B. The antheridial wall is composed of three cells, two of them ringlike and the uppermost a **cover cell,** or cap cell. Liquid water is necessary for antheridial dehiscence, which is effected by the swelling of the component cells (both wall and sperm) and rupture of the cap cell. The extruded multiflagellate sperms (Fig. 17-19B, C) rapidly become motile and subsequently slough off their spheres of attached cytoplasm. Duckett (1975) and Schrandholf and Richter (1978) de-

scribed the development of the antheridium and organization of the sperms of ferns at the ultrastructural level.

Investigations by Döpp (1962); Näf (1962, 1968, 1969); Näf, Sullivan, and Cummins (1969); Voeller and Weinberg (1969); and Voeller (1971) have uncovered interesting data regarding the control of formation of fern antheridia.

Näf, Nakanishi, and Endo (1975) have summarized our knowledge of this. When spores of the bracken fern, *Pteridium aquilinum,* germinate, some quickly give rise to two-dimensional and cordate gametophytes. These produce a substance, **antheridiogen,** which diffuses through the agar medium (a distance of 25 cm has been recorded) and stimulates other younger gametophytes to develop antheridia (Fig. 17-20). The gametophytes become insensitive to the antheridiogen as they become cordate, and, thus, those that secrete antheridiogen are themselves not induced to form antheridia, but they do form archegonia. Thus, a mechanism for early cross-fertilization is present. Later, those gametophytes that have been induced by the antheridiogen to form antheridia develop archegonia, so that self-fertilization also may occur. Species of 20 genera in four different families of

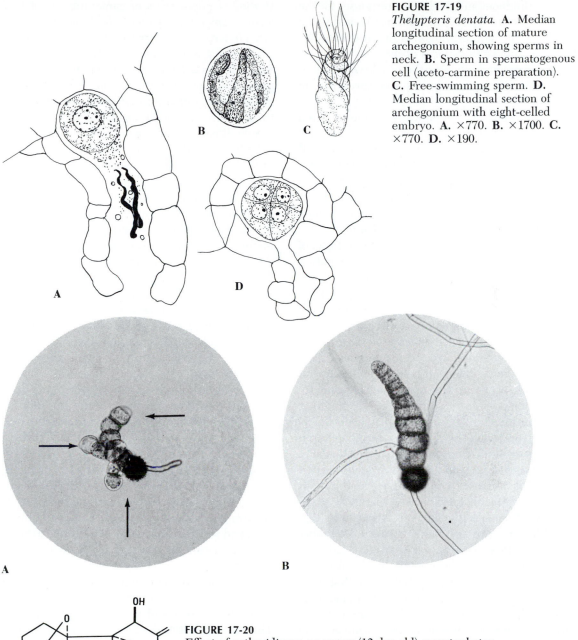

FIGURE 17-19
Thelypteris dentata. **A.** Median longitudinal section of mature archegonium, showing sperms in neck. **B.** Sperm in spermatogenous cell (aceto-carmine preparation). **C.** Free-swimming sperm. **D.** Median longitudinal section of archegonium with eight-celled embryo. **A.** ×770. **B.** ×1700. **C.** ×770. **D.** ×190.

FIGURE 17-20
Effect of antheridiogen on young (12-day-old) gametophytes of *Anemia phyllitides.* **A.** Filamentous gametophyte treated with antheridiogen; note three antheridia (arrows). **B.** Untreated control lacking antheridia. **C.** Chemical structure of antheridiogen. **A, B.** ×125. (**A, B.** *After Näf.* **C.** *After Nakanishi et al.*)

415

ferns produce antheridia in response to the antheridiogen from *P. aquilinum*. An additional effect of this antheridiogen is that it evokes the germination of the spores of these same genera in darkness and obviates the usual requirement for light; *Pteridium* spores can germinate in darkness.

Antheridiogens, chemically different from those of *Pteridium* (Näf, 1968), have been isolated from two other ferns, *Anemia phyllitidis* and *Lygodium japonicum* (Näf, 1959, 1960). Both these antheridiogens are inactive in *Onoclea sensibilis*, the species used to assay for the antheridiogen of *Pteridium aquilinum*. *Anemia phyllitidis*, and *Lygodium japonicum* are distinct entities on the basis both of the criteria of cross-testing and chromatographic separation (Näf, 1968). These antheridiogens are similar in their effect on gametophytic development to gibberellins GA$_4$ and GA$_7$. In addition to stimulating the development of antheridia in some fern species, these substances also can break the dormancy of their spores in darkness. The antheridiogens differ not only chemically but also with respect to the species whose gametophytes and spores they influence. The structure of antheridiogen from *Anemia* has been determined by Nakanishi, Endo, Näf, and Johnson (1971), the first chemical characterization of a fern antheridiogen and the fifth sexually related substance to be so characterized in the plant kingdom (Machlis, 1972). The chemical structure of *Anemia* antheridiogen is shown in Fig. 17-20C, and this substance is related to gibberellin, which also has antheridium-inducing properties. The antheridiogen has been recovered in pure form (Endo, Nakanishi, Näf, McKeon, and Walker, 1972). It is active in inducing antheridial formation at a concentration as low as 10 µg/liter and in stimulating germination of spores in darkness at a concentration of 0.3 µg/liter. That antheridia develop without the addition to cultures of exogenous antheridiogen was reported by Rubin and Paolillo (1981).

Tryon and Vitale (1977), on the basis of their studies of populations of gametophytes of two ferns, *Asplenium pimpernellifolium* and *Lygodium heterodoxum*, reported that an antheridiogen system is seemingly operative in nature. Small (0.3 to 2.2 mm) gametophytes in colonies produced many antheridia, while widely spaced gametophytes of the same size (at least 15 cm from the nearest neighbor) had few.

Schedlbauer and Klekowski (1972) and Schedlbauer (1976) have investigated antheridiogen activity and its results in the fern *Ceratopteris thalictroides*. They found that in multispore cultures bisexual gametophytes develop first; they are the source of the antheridiogen that induces later-developing gametophytes to mature as males. These investigators suggest that spore size is an important factor in this connection, the large spores germinating first into hermaphroditic gametophytes.

The archegonia arise from surface cells close to the growing point of the gametophyte (Fig. 17-18). Each initial undergoes periclinal division into a superficial and a hypogenous cell, the latter dividing only once. The outermost cell undergoes two successive perpendicular anticlinal divisions, forming initials of the four rows of neck cells that arise by division of these initials. The hypogenous cell divides periclinally to form the neck canal nuclei, but cytokinesis is often suppressed. The central cell divides, forming the egg and ventral canal cells. On immersion in water, the distal tiers of neck cells are ruptured, apparently by the swelling of the disintegrated neck and ventral canal cells, which are extruded and leave a canallike passageway to the egg (Fig. 17-19A).

It was demonstrated long ago that the movement of sperms toward archegonia is not fortuitous but the result of a chemotactic response. The chemical stimulant secreted from the archegonium can be replaced by malic acid in laboratory experiments.

Duckett and Bell (1972a,b) and Bell and Duckett (1976) investigated fertilization in the fern *Pteridium aquilinum* with the electron microscope. According to them, the sperms lose their plastid during their passage through the archegonial neck and only one sperm enters the egg cytoplasm.

The moist habitat of the gametophytes, along with the ventral position of the sex organs, enhance the opportunities for fertilization (Fig. 17-19A). Further growth of the gametophyte ceases after its archegonia have been fertilized. Nayar and Kaur (1971) have summarized the organization of the gametophytes in homosporous ferns.

Fern gametophytes are seemingly not obvious in nature, but Farrar and Gooth (1975), on the basis of their investigation of reproduction of ferns in an Iowa canyon, found that numbers of gametophytes in the wild differed in different species. Thus, in plots within 5 meters of sporophytes of *Cystopteris bulbifera*, *Adiantum pedatum*, and *Woodsia obtusa*, between 500 and 3000 gametophytes were present in September, but the numbers diminished because of erosion, and gametophytes were present in only protected pockets by November. Seventy-five gametophytes of *Cystopteris bulbifera* were examined carefully, of which 67 were sexually mature. Of these, 46 bore antheridia and archegonia and 21 only antheridia, so that both self- and cross-fertilization could occur.

Evans and Bazzone (1977) have reported that gametophytes may develop and survive at acid pH's between 5.8 and 2.2 but fertilization, as evidenced by numbers of sporophytes, is reduced below pH 4.2. Thus, acidic precipitation, such as may occur in New England, would limit fertilizations in nature. Although fern gametophytes are usually transitory, there are interesting exceptions. Thus, those of four species (*Gramnitis nimbata*, *Hymenophyllum tunbridgense*, *Vittaria* sp., and *Trichomanes* sp.) grow in nature at Highlands, North Carolina, where they reproduce by gemmae. The corresponding sporophytes are absent or rare (Farrar, 1967).

Farrar (1974) and Emigh and Farrar (1977)

demonstrated that the gametophytes of a number of species of *Vittaria* multiply by filamentous gemmae. Furthermore, the gemmae may be induced to form antheridia precociously if grown in proximity to gametophytes with antheridia. Tests with other species of *Vittaria* indicated that a genus-specific antheridiogen is in operation. Gastony (1977) found specimens of the "Appalachian gametophyte" growing in Crawford County, Indiana. He suggested, on the basis of its chromosome constitution, that it is conspecific or derived from *V. lineata*, but this will require further confirmation. McAlpin and Farrar (1978) have reported *Trichomanes* gametophytes growing in western Massachusetts.

Reproduction: embryogeny. Although it seems probable that more than one egg on each gametophyte is fertilized, only one zygote normally completes its development into an embryonic sporophyte. The zygote secretes a cell membrane soon after fertilization, and after an interval undergoes mitosis and cytokinesis in a direction parallel to the neck of the archegonium. Cell divisions at right angles to the direction of the first division and to each other result in the formation of the quadrant stage (Fig. 17-19D). Subsequent development of the cells of this quadrant indicates that already at this stage differentiation has occurred in preparation for development of the organs of the older embryo. In some species there is doubt that the organs of the juvenile sporophyte are blocked out as early as the quadrant stage. The origin and development of these organs are depicted in Fig. 17-21 and explained in its labeling.

The archegonial venter and sterile gametophytic tissue keep pace for a short time with the developing embryo, but soon the root and primary leaf burst forth (Fig. 17-21B), the former penetrating the substratum and the latter emerging, usually through the apical notch (Fig. 17-22). The gametophyte persists in some cases until several leaves have been formed, but it

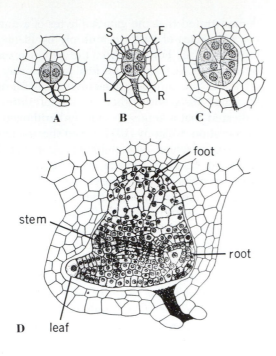

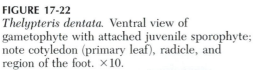

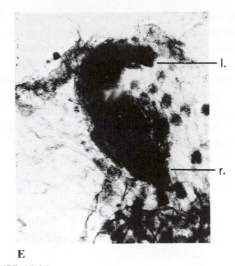

E

FIGURE 17-21

A–D. Fern embryogeny, successively later stages.
E. *Thelypteris dentata.* Ventral view of central
cushion of gametophyte showing emergent
sporophyte. F, foot; L, primary leaf; R, root; S, stem.
A–D. ×160. **E.** ×125.
(**A–D.** *After Haupt.*)

FIGURE 17-22

Thelypteris dentata. Ventral view of
gametophyte with attached juvenile sporophyte;
note cotyledon (primary leaf), radicle, and
region of the foot. ×10.

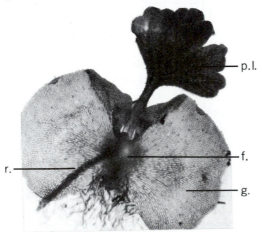

ultimately turns brown, and the embryonic
sporophyte initiates an independent existence.

The early embryonic leaves are simpler in
form than the mature leaves of the parent
species, but leaves are gradually produced that
are characteristic of the mature plant. Other
than the primary root, fern roots are adventitious. Once established, the young sporophyte
embarks on a perennial existence in most
leptosporangiate ferns.

The time required for spores (planted in
inorganic salt agar) of ten species of ferns to
grow into mature gametophytes and to produce
young sporophytes varies from 53 to 113 days.
After this, an additional 136 to 498 days, depending on the species, are required for the
sporophyte to reach maturity (as evidenced by
spore production) (Klekowski, 1967). *Ceratopteris thalictroides,* a floating aquatic fern,
completed a cycle from spore to spore in 3.5
months, the most rapid completion of a com-

plete fern life cycle on record (Klekowski, 1970a); in this fern the gametophytes became sexually mature 15 days after the spores had been sown. De Maggio (1968) was able to induce sporogenesis and meiosis in juvenile, axenic plants of *Todea barbara* growing in an agarized culture medium for 24 months without transfer; control plants which had been transplanted periodically failed to form spores. Analyses of the spore-bearing plantlets revealed that they were senescent and contained less chlorophyll and cellular nitrogen. From this, De Maggio concluded that sporogenesis was evoked by depletion of nutriments. However, it is difficult to apply this conclusion to plants growing in nature or in pots in the greenhouse, especially the latter, where it has been observed repeatedly that new leaves produced on *Thelypteris dentata* develop abundant spores soon after the plants have been transplanted into fresh, newly fertilized soil.

Summary

A number of significant and incisive investigations of the reproductive process and genetics of ferns were reported by Klekowski (1967, 1968, 1969a,b, 1970a,b,c, 1971) and Klekowski and Lloyd (1968). This was summarized to some degree by Lloyd (1974). Klekowski reported that a number of species have gametophytes that undergo self-fertilization, and their zygotes grow into vigorous, completely homozygous sporophytes. One adaptation favoring self-fertilization is the curvature of the archegonial necks toward the antheridia of the monoecious gametophytes. Self-fertilized gametophytes always give rise to homozygous (genetically homogeneous) sporophytes.

In most homosporous ferns the gametophytes are initially antheridial (protandrous) and later bisexual, but a few (e.g., some species of *Woodwardia*) are protogynous. In *Pteridium aquilinum*, the bisexual gametophytes are self-incompatible (see, however, Klekowski, 1972), so that the intergametophytic ("cross") fertilization is obligatory. Of 18 ferns investigated (Klekowski, 1969b), 83.3% exhibited intergametophytic mating.

It was also demonstrated by Klekowski (1970b) that self-fertilizations in gametophytes may not produce as many sporophytes in the population as cross-fertilizations. This is probably a result of the bringing together in self-fertilization of lethal genes which result in sporophyte abnormality or abortion during development. Finally, populations of some ferns (*Osmunda regalis*) are adapted to intergametophytic, rather than to intragametophytic, sexual reproduction (Klekowski, 1970C).

The genera *Thelypteris* and *Adiantum* described above have been chosen to serve as representative of the more advanced leptosporangiate ferns, the Polypodiaceae. Their distinctive characteristics are their method of sporangium development from a single cell; the mixed sori; the size, the annulus position, and the transverse dehiscence of the sporangium; and the small number of spores produced in each. The gametophytes of these genera, as compared with the photosynthetic gametophytes of the eusporangiate Marattiales, differ in their more delicate structure, less massive form, absence of a midrib, and protuberant and smaller antheridia. Furthermore, they are more rapid in their development and more ephemeral in their existence. Their doriventrality is marked. The precocity with which the organs of the embryo are organized in the early divisions of the zygote is also unlike the embryogeny of the eusporangiate ferns and *Osmunda*, in which differentiation appears later.

The independence of chromosome constitution (n or $2n$) of morphological expression of the sporophytic versus the gametophytic phase has previously been emphasized (p. 309). Thus, De Maggio et al. (1971) had in culture $1n$, $2n$, and $4n$ gametophytes of the fern *Todea barbara*, a member of the Osmundaceae. Recent analysis of these indicates that cell and nuclear volume

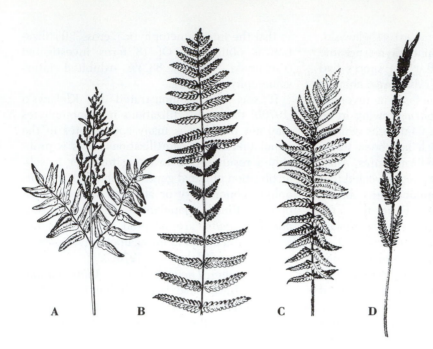

FIGURE 17-23
A. *Osmunda regalis.*
The royal or flowering fern.
B. *Osmunda claytoniana.*
The interrupted fern.
C, D. *Osmunda cinnamomea.* The cinnamon fern;
sterile frond at **C**;
fertile frond at **D**.
A. ×0.3. **B–D.** ×0.12.
(After Gleason.)

increase with increase in chromosome number but such biochemical processes as respiration and photosynthesis are not altered markedly.

Other Filicales

The family Polypodiaceae, as represented by *Adiantum* and *Thelypteris* described above, is the largest family of the Filicales in number of genera and species. Most of the attributes described for *Adiantum* and *Thelypteris* are in marked contrast to those of the eusporangiate ferns. Some of the other families of Filicales to be described briefly below are in some respects intermediate in several of their attributes between the eusporangiate and leptosporangiate ferns.

Family 2. Osmundaceae
Osmunda

Vegetative morphology. In most classifications of the Filicales, the Osmundaceae, including *Osmunda* (*Osmunder*, Saxon equivalent

to the god Thor) and *Todea* (after Henry Tode, German botanist), are placed at the beginning of the series in recognition of certain attributes that are interpreted as being intermediate between the eusporangiate and leptosporangiate Pteridophyta.

The genus *Osmunda* is cosmopolitan. Three species are distributed widely in the United States: *O. regalis*, the royal or flowering fern; *O. cinnamomea*, the cinnamon fern; and *O. claytoniana*, the interrupted fern (Fig. 17-23).

These perennial ferns are found usually in rather hydric habitats, in which the magnificent ascending fronds attain a large size, often reaching 2 m in length. In the northeastern United States, the several species of *Osmunda* are planted as ornamentals in semishaded locations. The striking leaves of *Osmunda* are either once or twice pinnately compound, depending on the species. The ultrastructure of the leaves has been investigated by Warmbrodt and Evert (1979). Each growing season, the underground stem produces a group of circinately coiled fronds, which unfold, photosynthesize, and pro-

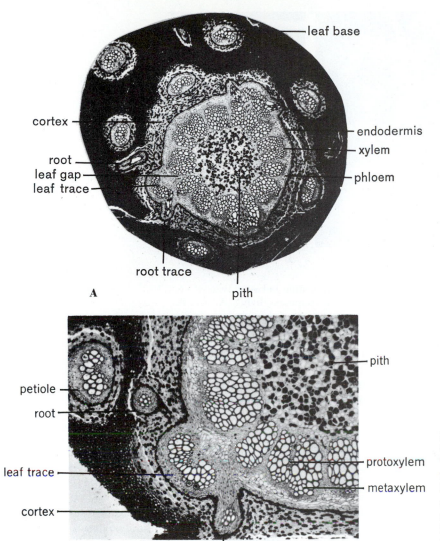

leaf base

cortex

root
leaf gap
leaf trace

endodermis
xylem

phloem

root trace

A

pith

petiole

root

pith

leaf trace

protoxylem

cortex

metaxylem

B

root trace

FIGURE 17-24
Osmunda sp. **A.**
Transection of rhizome;
note associated roots and
leaf bases. **B.** Portion of
stele enlarged; note
mesarch xylem and stages
in departure of leaf and
root traces. **A.** ×18. **B.**
×40.

duce spores during the fifth year after they have
been initiated as leaf primordia in the stem
(Steeves, 1963). The leaves are annual and die at
the end of the season; but the long-persistent
leaf bases surround the stem and, together with
the wiry roots, form prominent mounds. Mature pinnae are rather leathery in texture. Their
venation is open-dichotomous, as in *Botrychium*.

The dense covering of the wiry roots and
persistent leaf bases, noted above, effectively
protect the vertical underground stem of
Osmunda. The roots, leaf bases, and stem are so
intimately associated that sections of all are
usually available for study in a single preparation (Fig. 17-24). The vascular tissue of the
mature *Osmunda* stem, as viewed in transverse
section, is composed of a circle of strands sepa-

421

FIGURE 17-25
Osmunda regalis. **A.** Tip of fertile frond; note transition between vegetative and sporangium-bearing pinnules. **B.** Dehiscent sporangia, living. **A.** ×1. **B.** ×10.

rated by narrow parenchymatous leaf gaps. Leaf traces that have occasioned the gaps may be observed in section at various points in the cortex, through which they pass into the leaf bases. Three-dimensional and longitudinal views of the stele indicate that it is a siphonostele dissected by leaf gaps, as in *Ophioglossum.* The xylem strands are mesarch (Fig. 17-24B), with the protoxylem surrounded by metaxylem, the latter composed largely of scalariform-pitted tracheids. A prominent endodermis delimits the stele from the cortical region of the stem. Within it there are several layers of parenchyma cells, the pericycle, the phloem cells, and the mesarch xylem. The inner cortex is composed of parenchyma and darkly stained sclerenchyma tissue. The spirally arranged leaves are connected to the stele by C-shaped traces. Near each leaf base, two roots, with their vascular supply derived from the leaf trace, originate endogenously (Fig. 17-24) and grow between the leaf bases into the soil.

The roots contain exarch protosteles that are diarch or triarch. The development of both root

and stem is localized in single apical cells. In certain individual roots, however, as many as four apical cells may be present, an attribute suggestive of ferns of the *Marattia* group.

Reproduction: the sporophyte. Further striking modifications of leaf structure are associated with sporangium production in *Osmunda.* The location of the sporangia varies in the several species. In all cases the sporangia do not occur in sori but are in marginal zones without indusia. In *O. regalis,* only the most distal pinnae and pinnules are fertile (Figs. 17-23A, 17-25). In *O. claytoniana* (Fig. 17-23B), on the other hand, certain intermediate pairs or pinnae bear sporangia, whereas the proximal and distal ones are sterile. Finally, in *O. cinnamomea* (Fig. 17-23C), the sporangia are borne only on special, fertile, spore-bearing leaves, the vegetative tissues and functions of which are greatly restricted. Intermediate stages in these dimorphic phenomena are readily observable on examination of several individuals (Fig. 17-25A).

The sporangia themselves originate superfi-

422

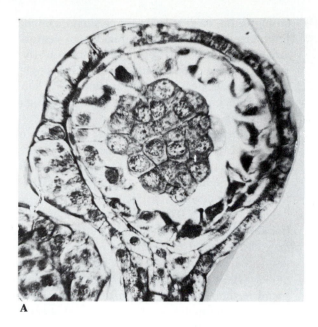

B

FIGURE 17-26
Osmunda regalis. **A.** Section of
sporangium at sporocyte stage; note
two-layered wall and bulging, unilateral
annulus (left), tapetum, sporocytes, and
sporangial stalk. **B.** Wall of mature
sporangium showing unilateral annulus.
A. ×135. **B.** ×60. (**B.** *After Andrews.*)

A

cially on the pinnae surface. Although prominent single initial cells may be present in each
sporangial precursor, adjacent cells also contribute to the formation of the sporangium, which
becomes protuberant and rather massive in
comparison with that in other leptosporangiate
ferns. Furthermore, the sporangial stalk is also
thicker (Fig. 17-26A) than in other leptosporangiate ferns, and the wall is two-layered. A
central sporogenous cell in each sporangium
undergoes repeated divisions until approximately 128 sporogenous cells are produced. The
outermost layer of these functions somewhat as
a tapetal plasmodium during sporogenesis, and,
in addition, a true tapetum is present. The large
and prominent chromosomes are relatively few
in number ($n = 22$). Each sporangium produces
between 256 and 512 spores. These are green at
maturity and are shed by dehiscence of the
sporangium along a vertical line running between a unilateral group of thick-walled annulus
cells and the stalk of the sporangium (Fig.
17-26B).

Reproduction: the gametophyte. As is
the case with the chlorophyllous spores of *Equi-*

setum, those of *Osmunda* do not remain viable
very long. They germinate very readily, however, if transferred to agar, soil, or other environments with adequate moisture (Fig. 17-27).

The young gametophyte at first consists of a
short chain of cells with a basal rhizoid, but an
apical cell is organized soon by successive
oblique division of the terminal cell. This gives
rise to a spatulate gametophyte up to several
centimeters long, at the tip of which several
apical cells may function. Their derivatives
build up a relatively massive green gametophyte (Fig. 17-28) suggestive of those *Marattia*
and *Angiopteris* (Figs. 16-13, 16-14). Prominent

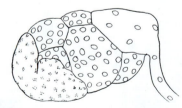

FIGURE 17-27
Osmunda regalis. Spore germination, 2 weeks
after spores were planted on agar. ×315.

423

FIGURE 17-28
Osmunda regalis. Mature gametophyte growing on agar. ×2.

ventrally projecting midribs are present in the gametophytes of *Osmunda*, which are liverwortlike and long-lived and may be perennial. Vegetative propagation of the gametophytes by separation of marginal lobes has been observed.

There is some evidence that the gametophytes are protandrous; well-separated individuals are always monoecious. The antheridia (Figs. 17-29A, 17-31A) are borne on the underside of the gametophyte near the margins, and the archegonia (Fig. 17-29B) are on or near the midrib. The protuberant antheridia have a single layer of jacket cells, one of which, near the apex, is thrown off as the antheridium dehisces. Approximately 100 spermatogenous cells fill the

antheridium; each gives rise to a single, coiled, multiflagellate sperm. The archegonial venters are sunken in the superficial tissue on the ventral surface of the gametophytes. The single binucleate neck canal cell is surrounded by four rows of neck cells about six to eight tiers high (Fig. 17-30).

Reproduction: embryogeny. The zygote undergoes two successive vertical divisions parallel in direction to the axis of the archegonium but perpendicular to each other. The next divisions are transverse, so that an octant stage develops (Fig. 17-31B). The outermost four cells (nearest the neck) gradually develop the stem, leaf, and primary root of the embryo, and the four basal cells form a foot. Growth of the embryo is relatively slow in *Osmunda* as compared with other leptosporangiate ferns. The archegonium functions as a calyptra during early embryogeny, but it is ultimately ruptured by the developing embryo, which protrudes from the gametophyte and becomes established as an independent plant (Fig. 17-31C).

The primary embryonic leaf and those that follow it during the juvenile phases of development are much simpler than those of the adult plant (Fig. 17-31C). The stem of the young sporophyte is protostelic at first, and the dis-

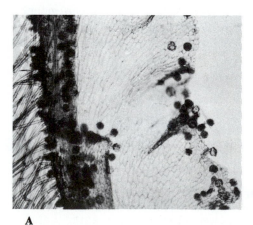

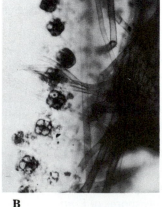

A   B

FIGURE 17-29
Osmunda regalis. **A.** Ventral view of portion of a gametophyte; note antheridia at margins and archegonia (enlarged in **B**) along midrib. **A.** ×45. **B.** ×120.

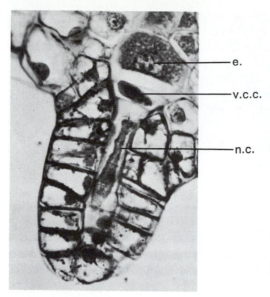

FIGURE 17-30
Osmunda sp. Median longisection of an archegonium. e., egg; n.c., neck canal cell; v.c.c., ventral canal cell. ×500.

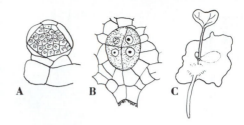

FIGURE 17-31
A. *Osmunda cinnamomea.* Mature antheridium. **B.** *Osmunda claytoniana.* Section of eight-celled embryo in archegonium. **C.** *Osmunda claytoniana.* Gametophyte with young sporophyte emerging. **A.** ×215. **B.** ×175. **C.** ×4. (*From Campbell.*)

sected siphonstelic condition does not arise until considerable development has occurred.

Summary

Osmunda and the Osmundaceae differ from Polypodiaceous ferns such as *Adiantum* and *Thelypteris* in a number of respects. The sporangia are not in sori, and all mature their spores simultaneously as in eusporangiate ferns. Furthermore, the sporangia, which have two wall layers, are not typically leptosporangiate in development because several cells, in addition to the initial cell, function in its development. In addition, as many as 512 spores may mature in each sporangium. The spores are thin-walled and green, like those of *Equisetum*. The sporangium opens by a vertical cleft, and the annulus consists of a terminal and pseudolateral group of thick-walled cells.

The gametophytes are longer-lived and more massive than those of leptosporangiate ferns.

The antheridia produce more sperms and the rhizoids contain chloroplasts. The archegonial necks are straight instead of curved, as in the Polypodiaceae.

The embryo of *Osmunda* differentiates later in development than the quadrant or octant stage. Finally, the primary leaf does not grow up through the apical notch, as in the Polypodiaceae, but laterally around the margin of the gametophyte.

Family 3. Schizaeaceae

Of the genera of this family, *Lygodium* (Gr. *lygodes*, like a willow) and *Anemia* (Gr. *aneimon*, unclad, referring to the naked sporangia) (Fig. 17-32), are perhaps the best known and most readily available for study. Although a single species of *Lygodium*, the climbing fern, *L. palmatum*, is native to the United States, *L. japonicum* (Fig. 17-32B, D) is widely cultivated in conservatories and in many gardens in the southern United States and in some places (Georgia and Texas) has escaped from cultivation to become a wild fern. The petioles and rachises of the compound leaves of *Lygodium* are of almost indeterminate growth (Mueller, 1982*a,b,* 1983*a,b*), in some species forming leaves almost 30 m long, and may grow 3.4–6.5 cm/day for several weeks. The subterranean

425

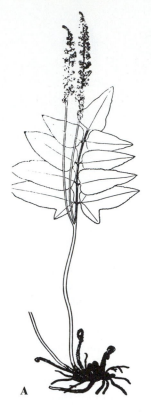

B

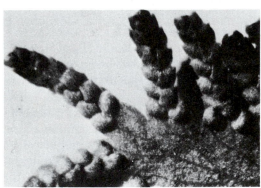

D

A

C

FIGURE 17-32
A. *Anemia mexicana.* **B–D.** *Lygodium japonicum,* climbing fern. **B.** Habit of growth. **C.** Detail of sterile (below) and fertile (above) leaflets. **D.** Detail of fertile pinnules. **A.** ×0.16. **B.** ×0.03. **C.** ×0.5. **D.** ×20. (**A.** *After Correll.* **C.** *Courtesy of Dr. C. A. Brown, after Brown and Correll.*)

426

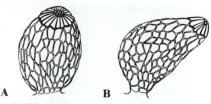

FIGURE 17-33
A. *Anemia mexicana*, sporangium. **B.** *Lygodium palmatum*, sporangium. (*After Andrews.*)

stems are short. *Anemia mexicana* (Fig. 17-32A) is native to the extreme southern United States (southern Florida and Texas); its sporangia occur on much-reduced leaf segments.

In these ferns the sporangia are borne on the leaf margins and either lack an indusium or have a false one. The sporangia produce up to 216 spores; an apical group of thick-walled cells (Fig. 17-33) serves as an annulus. Sporangial dehiscence is vertical. All the sporangia mature simultaneously (simple type), as in the eusporangiate ferns and Osmundaceae. One species of *Schizaea* (Gr. *schizein*, to split) (Fig. 17-34) is native to the United States, occurring in the Pine Barrens of New Jersey.

The gametophytes of *Schizaea*-like ferns are of great interest and of two types. Some, like those of *S. pusilla*, are chlorophyllous and filamentous, much like moss protonemata. Species with such photosynthetic gametophytes have been retained in the genus *Schizaea*. However, a number of ferns, formerly included in *Schizaea*, have been reassigned (Bierhorst, 1971a) to the genus *Actinostachys* (Fig. 17-35A–C), because their gametophytes are subterranean and fleshy, cylindrical structures much like those of *Psilotum* (Fig. 19-11) (Bierhorst, 1971a). Furthermore, subterranean gametophytes of *A. melanesica* contain endophytic fungi. The occurrence of such radially symmetrical gametophytes in the Pteridophyta is interpreted as a primitive attribute which may indicate common ancestry with other vascular cryptogams with similar gametophytes.

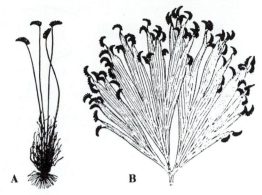

FIGURE 17-34
Schizaea. **A.** *S. pusilla*. **B.** Leaf of *S. elegans*. ×0.3. (*After Smith.*)

Family 4. Stromatopteridaceae

The family Stromatopteridaceae has only a single member, *Stromatopteris moniliformis*, native to New Caledonia; it has been studied intensively by Bierhorst (1968, 1969, 1971a). This plant (Fig. 17-35D) is characterized by lack of clear distinction between leaf petioles and various irregularly branched underground axes and roots. Like the gametophytes of *Psilotum*, *Tmesipteris*, *Ophioglossum*, and *Actinostachys*, those of *Stromatopteris* are subterranean and cylindrical and lack chlorophyll (Fig. 17-35G). Fungi are present in both the subterranean sporophytic axes and gametophytes. As will be noted later, Bierhorst has emphasized a number of similarities between *Psilotum*, *Tmesipteris*, and *Stromatopteris* and considers all these to be primitive Pteridophyta.

Family 5. Hymenophyllaceae

The Hymenophyllaceae, the filmy ferns, are largely tropical and include (among others) two genera of delicate-leaved ferns, *Hymenophyllum* (Gr. *hymen*, membrane, + Gr. *phylon*, leaf) (Fig. 17-36A) and *Trichomanes* (Gr. for waterwort) (Fig. 17-36B), both of which are represented in the flora of the United States. *Trichomanes* occurs in extremely moist habitats such as on rocks bathed in spray from waterfalls.

427

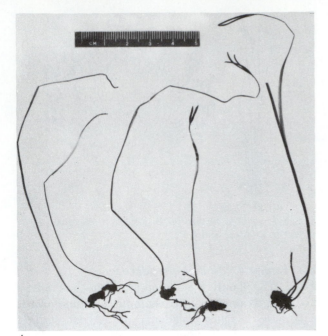

A

B

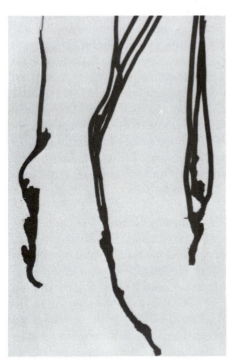

E

F

428

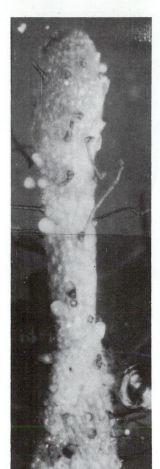

C

D

G

FIGURE 17-35
A. *Actinostachys oligostachys.* Five sporophytes attached to their gametophytes. **B, C.** *Actinostachys melanesica:* two views of the gametophyte; note protuberant antheridia. **D–G.** *Stromatopteris moniliformis.* **D.** Leaf from a young plant and tip of leaf from an older plant. **E.** The axes on which they are borne. **F.** Abaxial view of pinnae bearing sporangia. **G.** Gametophyte. **B.** ×35. **C.** ×18. **D.** ×0.7. **E.** ×0.9. **F.** ×0.7. **G.** ×30. (*After Bierhorst.*)

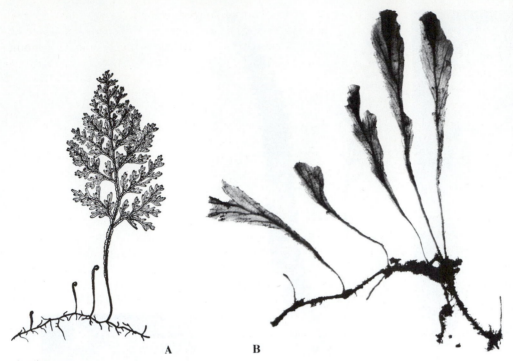

FIGURE 17-36
A. *Hymenophyllum australe.* B. *Trichomanes petersii.* A. ×0.25. B. ×3.
(A. *After Smith.* B. *Courtesy of Dr. C. A. Brown, after Brown and Correll.*)

As the familial name implies, the leaves are delicate, often only a single layer of cells thick except at the midveins. The sporangia are borne at the margin of the leaves on elongate receptacles covered during development by cuplike or two-lipped indusia. Development of the sporangia on the receptacle is orderly, the oldest being at the apex and the younger sporangia arising on the flanks of the receptacle, with the youngest at the base. This represents **gradate** organization of the sorus, in contrast to the simple (all spores maturing simultaneously) sori of the Ophioglossales. Up to 512 spores may be produced in each sporangium. The annulus is obliquely vertical, and the sporangium opens transversely. The spores may undergo the early stages of germination before they are shed from the sporangia.

The green gametophytes are branching and straplike (*Hymenophyllum*) or branching filaments (*Trichomanes*) with erect, fleshy, gemmiferous branches. In *Trichomanes holopterium* the gametophyte is more abundant than the sporophyte and self-reproducing (Farrar and Wagner, 1968).

Family 6. Gleicheniaceae

The Gleicheniaceae are ferns of the tropical and south temperate parts of the earth and comprise more than 100 species, by some classified as four genera, but by others included in the single genus *Gleichenia*. This fern has prostrate, dichotomously branching rhizomes with long internodes. The rhizomes contain haplostelic protosteles with mesarch xylem. The leaves, which are twice or more times pinnately

430

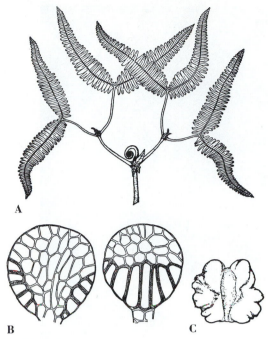

FIGURE 17-37
Gleichenia. **A, B.** *G. pectinata.* **A.** Segment of
leaf. **B.** Sporangium, ventral (left) and dorsal
views. **C.** *G. dichotoma.* Dorsal view of
gametophyte. **A.** ×0.25. **B.** ×90. **C.** ×2.5.
(*After G. M. Smith.*)

compound (Fig. 17-37A), grow for long periods
and are vinelike; they may attain a length of 50
m. Their leaflets have open-dichotomous venation.

The sori, which are borne in two ranks near
the midrib of the pinnule, have few (4 to 8)
sporangia and are not covered with an indusium. An annulus of elongate, thick-walled cells
develops transversely and slightly obliquely the
rather globose sporangium (Fig. 17-37B). This is
correlated with vertical dehiscence of the latter.
Bierhorst (1971*b*) has investigated soral and
pinnule ontogeny in *G. linearis.*

Spore germination, which requires several
weeks, results in a cordate gametophyte (Fig.
17-37C) with a thickened midregion. The antheridia produce large numbers of sperm (up to
500).

The prolonged growth of the leaves, the
protostelic rhizome, and the large number of
spores (up to 1024) in the sporangia, among
other features, suggest that *Gleichenia* is primitive.

Family 7. Cyatheaceae

To this family belong the tree ferns (as well as
some species with massive prostrate stems).
Tree ferns (Fig. 17-38) grow natively in tropical
rain forests, where large specimens may exceed
20 m in height. They are widely cultivated in
conservatories and out-of-doors in mild, moist
climates like that of the Pacific coast.[2] *Cyathea*
(Gr. *kyathos,* cup), *Alsophila* (Gr. *alsos,* grove,
+ Gr. *philein,* to live), and *Cibotium* (Gr.
kibotion, little box) are often present in conservatory collections. The trunks (stems) are rarely
branched and are covered by a dense mat of
adventitious roots, which arise in association
with the leaf bases. Roots at the distal portion of
the trunk never reach the soil. Creeping stems
arise on the trunks some distance above the soil,
run downward and then outward, and initiate
new erect stems. The lower portion of the stem,
which is smaller in diameter than the upper, is
supported by a dense, buttresslike covering of
persistent roots. In spite of their massive size,
tree ferns lack a cambium and secondary
growth. Anatomically, the young stems are simple and protostelic, but as the stem continues to
grow, the stele increases in diameter, becoming
dictyostelic and, finally, polystelic and quite
complex.

The apex of the trunk supports a crown of
leaves, which are circinately vernate when
young. The leaves may approach 4 m in length
and are pinnately compound.

The sporangia are borne in gradate sori. In
the Dicksoniaceae these are marginal and
subtended by cuplike indusia, while in the

---

[2]A group of tree ferns is growing in San Francisco's
Golden Gate Park.

431

FIGURE 17-38
Cyathea sp. Tree ferns. Insert, an enlarged view
of tree fern fiddlehead. (*From a negative of Dr.*
C. J. Chamberlain, courtesy of the Chicago
Field Museum of Natural History.)

Cyatheaceae they are abaxial and may or may
not have indusia. The annulus is obliquely
vertical, and the transverse sporangial dehiscence is initiated in the region of special thin-
walled lip cells or stomium, as in the
Polypodiaceae.

The gametophytes are heart-shaped and organized much like those of the Polypodiaceae.
The archegonia have long, slightly curved necks
like those of *Osmunda*.

Summary of Filicales

The Filicales include all the homosporous
leptosporangiate ferns. They are widely distributed in both the tropics and temperate zones
and are almost exclusively perennial plants. In
colder climates, with few exceptions (*Cyrtomium* [Fig. 17-39] and *Polystichum* [Fig. 17-9], for
example), the leaves die at the end of the
growing season and are replaced by a new set of
leaves the following spring.

The stems of Filicales may be prominent and
trunklike (Cyatheaceae) or rhizomes, or underground, vertical stems. They may be dichotomously or irregularly branched. Their ontogeny

432

may be traced in every case to an apical cell. This may be tetrahedral, with derivatives arising on three sides, or with derivatives arising only on two sides (Bierhorst, 1977). The stele in juvenile sporophytes is usually a protostele. As the stem increases in size, a siphonostele, dictyostele, or complex polystele may develop. Secondary growth is absent.

The roots of Filicales are adventitious. All develop from the activity of prominent (tetrahedral) apical cells and have exarch protosteles.

Filicalean leaves are formed early in ontogeny at the stem apex and have circinate vernation. Compound leaves are considered to be more primitive than simple leaves, which are interpreted as reductions. Similarly, open-dichotomous, as opposed to reticulate, venation is considered to represent the primitive condition. The leaves and stems often have epidermal hairs and/or scales. In a few (e.g., *Lygodium*) the leaves are of indeterminate growth. The epidermal cells of many fern leaves contain chloroplasts; their mesophyll may or may not be differentiated into palisade and spongy regions.

The sporangia occur on the margins or abaxial surfaces of ordinary vegetative leaves or somewhat modified leaves or segments thereof. Marginal sori are considered to be more primitive than superficial ones because they correspond in position, in megaphyllous leaves, to the terminal sporangia of branches. They arise from receptacles in discrete groups or sori, or they may be in continuous zones or rows. They may or may not be covered by indusia during development. The sporangia in certain families (e.g., Osmundaceae and Schizaeaceae) are in some aspects suggestive of eusporangiate sporangia. The sori may be simple, gradate, or mixed in order of development. Dehiscence is either vertical or oblique. This is correlated with the position of thick-walled annulus.

The gametophytes of Filicales show considerable range of organization. There are differences in spore germination and early development (bipolar, tripolar, and unipolar germination and time of initiation of spatulate stage). The gametophytes may be filamentous (*Schizaea*), ribbonlike (*Hymenophyllum*), cylindrical, subterranean, and achlorophyllose (*Actinostachys*, *Stromatopteris*), or cordate, the latter with entire or incised margins, and they may bear unicellular and/or multicellular branched or unbranched hairs. The gametophytes may be massive and long-lived (Osmundaceae) or rather delicate and ephemeral (Schizaeaceae, Hymenophyllaceae, and Polypodiaceae). The rhizoids may be brown-walled or colorless. The gametophytes of almost all Filicales are monoecious. The antheridia may have many wall cells (Osmundaceae) or only three (Polypodiaceae) or four. The archegonia may have long, straight necks (Osmundaceae) or shorter, curved ones (Polypodiaceae). The taxonomy of Filicales is based largely on sporophytic attributes, but gametophytic characteristics are being increasingly studied. Many aspects of the reproductive cycle in ferns have been summarized succinctly by Miller (1968).

Finally, in the embryogeny of the Filicales, differentiation into the organs of the young sporophyte may occur as early as the quadrant or octant stage, or it may be delayed.

THE FERN LIFE CYCLE

The life cycle of *Adiantum* and *Thelypteris* and of most other Filicales is fundamentally similar to that of other vascular cryptogams in that it involves a regular alternation of a diploid, spore-producing generation and a haploid, gamete-producing one. This type of life cycle has been so greatly emphasized in biological teaching that we often gain the impression that it is obligate and inexorable. Furthermore, we are so prone to contrast the two alternates in the life cycle that they sometimes seem like different organisms rather than merely successive morphological expressions of the same one. This tendency to contrast has, of course, been enhanced by the antithetic theory of alternation of generations discussed previously in this text (pp. 308, 309).

The purpose of the following paragraphs is to present a somewhat different interpretation of life cycle phenomena, with the ferns as a point of departure. Bell (1959) and Voeller (1971) have reviewed a number of aspects of the life cycle in ferns.

Deviations from the life cycle described for *Adiantum* and *Thelypteris* occur in nature, or they may be induced. These deviations have been grouped together under the phenomena, apogamy and apospory, alluded to earlier (p. 309). **Apogamy** denotes a deviation from the life cycle in which the transition from the gametophytic phase to the sporophyte is accomplished without sexual union of gametes and their nuclei. By **apospory,** on the other hand, is meant the transition from sporophyte to gametophyte in a manner other than through the medium of spores and occurrence of meiosis.

Apospory is known to occur among both Hepatophyta and Bryophyta, as well as in the vascular cryptogams and seed plants. Injured sporophytes of *Anthoceros* develop gametophytic thalli, and injured moss setae and capsules may develop protonemata. In certain fern varieties, prothallia may develop at the leaf margins or from transformed sporangia, or they may be induced to develop by injuring juvenile leaves. Examples of apogamy are best known from certain varieties of ferns. In most of these, the gametophytes do not produce archegonia, although antheridia may be developed abundantly. The thickened central portion of the gametophyte develops directly into a sporophyte without the occurrence of sexual union.

The phenomena of apogamy and apospory have an important bearing on such fundamental questions as the nature and relation of the alternating generations and the chromosome

A

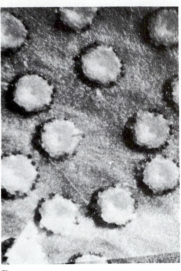

B

FIGURE 17-39
Cyrtomium falcatum, the holly fern, an apogamous species. **A.** Habit of growth. **B.** Sori and indusia. **A.** ×⅛. **B.** ×5.

cycle. Are apogamously produced sporophytes haploid or diploid? If the former is the case, what is the nature of the chromosome behavior during sporogenesis? In the latter case, are the prothallia diploid or haploid, and if diploid, what is the origin of the diploid condition? In spite of the investigations of numerous cytologists, the chromosome cycle in apogamous and aposporous ferns has not been clarified until relatively recently. It has been shown for the holly fern, *Cyrtomium falcatum* (Fig. 17-39), for example, that the chromosome number of both sporophytes and gametophytes is the same, both being diploid. The gametophytes give rise to sporophytes apogamously. However, only the spores from certain types of sporangia in a sorus develop into gametophytes. Spores from other types of sporangia usually abort. In the sporangia that produce fertile spores, the first three divisions of the primary archesporial cell are normal; hence, eight cells are produced. When these undergo nuclear division in preparation for the formation of 16 sporocytes, nuclear division is arrested at prophase or metaphase, with the result that in all eight sporogenous cells the nuclei are precociously reorganized. These contain double the chromosome complement of the original cells. These eight cells (second eight-celled stage) now function directly as sporocytes, undergoing meiosis with regular chromosome pairing and production of 32 spores, each of which, however, contains a nucleus with the chromosome number characteristic of the sporophyte. This temporary doubling, in the abortive fourth nuclear division, makes normal meiosis possible and also explains the similarity of chromosome number in gametophyte and sporophyte. A number of other ferns have been shown to have a similar chromosome cycle.

Another type of apogamous fern life cycle has been described in which the basic (presumably diploid) chromosome number is preserved by mitosis throughout both generations; meiosis is absent. In *Polypodium dispersum*, both sporophyte and gametophyte are diploid, the sporophytes arising from gametophytes lacking sex organs. No change of chromosomes occurs at spore production or at the time the sporophyte is initiated by the gametophyte (Evans, 1964).

Thus, at least three types of life cycle occur among leptosporangiate ferns: (1) the widespread cycle that involves regular alternation of diploid sporophytes and haploid gametophytes, as in *Thelypteris* and *Adiantum* (pp. 412, 419); (2) the *Cyrtomium falcatum* type, in which both

sporophyte and gametophyte are diploid and meiosis occurs; and (3) the *Polypodium dispersum* type, in which both sporophyte and gametophyte are diploid but meiosis is absent. The last two life cycles involve obligate apogamy.

Apogamy may be induced at will in other, normally sexual species. This has been accomplished by withholding water from the surface of gametophyte cultures, thus preventing swimming and functioning of the sperms. It has been demonstrated by Whittier and Steeves (1962) and by Whittier (1964) that when gametophytes of 14 species of ferns (including *Pteridium aquilinum*, *Osmunda cinnamomea*, *Dryopteris* [*Thelypteris*] *marginalis*, and *Adiantum pedatum*) are grown in an inorganic medium supplemented with about 2.5% sugar (glucose, fructose, sucrose, or maltose), they produce sporophytes apogamously. The role of the sugar is not one of effecting a change in osmotic pressure but probably acting as a respiratory substrate. Control cultures grown in an inorganic medium without sugar did not show apogamy.

Elmore and Whittier (1975*a,b* and earlier) found that certain fern gametophytes produce ethylene in culture. Investigating nine gametophytic strains of *Pteridium aquilinum*, the bracken fern, they found that all nine produced ethylene in culture and that, when the cultures were sealed, eight of the nine produced apogamous buds. The same strains responded similarly to exogenous ethylene. Apogamy in this case, according to Elmore and Whittier, depends on capacity of the plants to respond to ethylene.

A further factor effecting apogamy has been shown to be high light intensity and quality (Whittier and Pratt, 1971). It has been suggested that this operates by increasing the rate of photosynthesis. Adding sugars to an inorganic medium may thus replace the factor of intense illumination.

Apospory, which occurs less frequently in nature than apogamy, also may be induced in many ferns (Mehra and Sulklyan, 1969). This is most readily accomplished by removing the leaves from young sporophytes and placing them on agar medium. After an interval, leaf cells in contact with the agar give rise to filaments, which ultimately grow into typical cordate gametophytes.

Hirsch (1976) investigated regeneration of juvenile leaves (from callus) of the fern *Microgramma vaciniifolia*. She found three types of structures developed from regenerative divisions of the epidermal cells of the leaf: apogamous gametophytes, sporophytic plantlets, and intermediates.

Chlorophyllous gametophytes have been induced to develop on the excised, colorless roots in root cultures of the bracken fern (*Pteridium aquilinum*) (Munroe and Sussex, 1969). These gametophytes were diploid and produced sex organs.

The apogamous production of sporophytes and the aposporous development of gametophytes, such that gametophytes and sporophytes have the same chromosomal constitution, are considered by some strong evidence in support of the homologous, or transformation, theory of alternation of generations. The facts that sporophytes can directly transform some of their tissue into gametophytes (aspospory) and that the same chromosome complement may be present in both phases without modifying their fundamental morphology seem to indicate that the two alternating generations are fundamentally similar. In evidence of this may be cited the work of Bristow (1962), who reported that the presence of less than 0.1 g/liter of sucrose, glucose, or fructose caused diploid callus (derived from leaves of *Pteris cretica*) to develop gametophytes, which in a greater concentration of these sugars formed sporophytes. Furthermore, Morel (1963) found that juvenile leaves of *Adiantum pedatum* regenerate gametophytes, while adult leaves regenerate sporophytes in the same culture medium. This indicates an internal change in the leaf tissues as they mature.

These phenomena, then, serve to focus our attention once again on the nature of alternation of generations and on the causal relationships involved. It was suggested early in this century, on the basis of the isomorphic alternation known for certain algae, that the explanation for the profound morphological differences between the sporophyte and gametophyte in land plants was environmental. In the algae, it was argued, both generations develop as free-living plants from reproductive cells set free in the same environment; thus, the morphology of these algal sporophytes and gametophytes was similar. (Heteromorphic alternation in algae such as the kelps, certain red algae, and some species of *Bryopsis* was unknown at the time.) In contrast, it was suggested that while the spore of land plants germinated freely into the gametophyte in nature, the zygote always underwent the early phases of its development into a sporophyte within the confines of a pressure-exerting archegonium or calyptra. This difference of environment during development of sporophyte and gametophyte of land plants was suggested as the cause of their divergence in morphology. A number of investigators have studied the morphogenetic factors involved in fern life cycle, including, among others, Ward and Wetmore, 1954; Ward, 1963; De Maggio, 1963; De Maggio et al., 1963; and White, 1971a,b.

In efforts to test this hypothesis, spores have been injected into the thickened midregion of fern gametophytes; although gametophytes ultimately arose from the injection site, it has not been possible to trace their origin with certainty. On the other hand, surgical removal of the archegonial neck, with less pressure and constraint on the zygote, resulted in the latter's producing a mass of tissue in which differentiation into organs of the sporophyte was somewhat delayed but finally did occur. However, zygotes removed completely from archegonia before they had initiated cell division developed into two-dimensional prothalloid masses, which did not become organized as sporophytes. There is evidence then, that the retention of the zygote within the calyptra and gametophyte has an important morphogenetic role.

It was long ago suggested that the chromosomal differences that normally obtain (the gametophyte being haploid and the sporophyte diploid) determine the difference in morphological expression, and this often is strongly emphasized. Several lines of evidence are available, however, in opposition to this suggestion. The occurrence in nature of gametophytes and sporophytes with the same chromosome complement, as in the case of *Cyrtomium falcatum* and *Polypodium dispersum* already discussed, is one negating factor. A second is the induction in apospory of regeneration of diploid fern leaves into diploid gametophytes. Finally, a polyploid series of haploid, diploid, and tetraploid gametophytes and diploid, tetraploid, and octoploid sporophytes of the same species has been obtained by experimental induction of apospory (followed by fertilizations) in the laboratory, a striking demonstration of the similar potentialities of gametophytes and sporophytes. In these investigations, gametophytes were still gametophytes whether haploid, diploid, or tetraploid, and diploid, tetraploid, and octoploid sporophytes themselves were little modified in morphology. These last results suggest that chromosomal or differences in quantity of DNA of themselves do not explain the differences between sporophyte and gametophyte.

It has been suggested (Bell, 1959) that the cytoplasm of the egg cell itself may be the site of morphogenetic determination. In the later discussion of data from electron microscopy, Bell (1970, 1975) summarized the profound reorganizations that occur during the ontogeny of both the egg (oogenesis) and the spore (sporogenesis) in the fern life cycle. He ascribes to this reorganization the physical basis for differences between gametophyte and sporophyte and discounts other hypotheses that seek to explain them. The most significant causal agent so far

discovered may be the enclosure of the zygote and the young embryo by the archegonial jacket or calyptra. A similar force seems to be involved in the regeneration of juvenile plants from undifferentiated callus of angiosperms (p. 315).

Differences in sporophyte and gametophyte may be occasioned by the "lessened availability of some cell substrate" in the haploid gametophyte that does not permit the development of the more complex sporophyte (De Maggio, 1963). Two phenomena argue against this hypothesis: the greater complexity of the gametophyte, as compared with the sporophyte, in certain liverworts and mosses, and the typical morphology of polyploid gametophytes.

Finally, the increase in apogamous production of sporophytes by addition of sugar to the culture medium has suggested that an adequate nutritional level is a factor in the development of the sporophyte, at least in vascular plants.

In addition to the evidences of similarity of the alternates provided by apogamy and apospory, one additional point is relevant. It has been demonstrated that fern gametophytes may be stimulated to form vascular tissue (tracheids) by treatments with sucrose and an auxin (naphthalene acetic acid). Furthermore, by planting them in an erect position in inorganic agar medium with a 1.0% supplement of sucrose, the gametophytes become cylindrical and develop tracheids (De Maggio et al., 1963). They thus approach in organization the axes of sporophytes and, at the same time, suggest the cylindrical organization of the gametophytes of *Psilotum* and *Ophioglossum*.

Accordingly, although the phenomenon of alternation of generations has been described and investigated for more than a century, we still lack precise data regarding the causal factors involved. It seems increasingly clear, however, on the basis of the evidence at hand, that sporophytes and gametophytes in a given life cycle may not be so fundamentally different from each other as sometimes has been suggested or implied. While the phylogenetic aspects of alternation were emphasized before 1960, an increasing number of experimental morphogenetic investigations have helped since then to shed light on this important phenomenon. White (1971*a,b*) has reviewed a number of experimental studies of the fern sporophyte.

DISCUSSION QUESTIONS

1. How do typical eusporangiate and leptosporangiate sporangium development and sporangia differ? Are the differences always absolute?
2. Define or explain sorus, receptacle, indusium, false indusium, annulus, lip cells, circinate vernation.
3. Define apogamy and apospory. What is the phylogenetic significance of these phenomena?
4. How would you go about inducing apogamy in *Adiantum* and *Thelypteris?*
5. Define and give an example of simple and gradate sori.
6. Why are the sori of *Thelypteris* and *Adiantum* said to be "mixed"?
7. How do the gametophytes of *Thelypteris* and *Adiantum* differ from those of *Ophioglossum* and *Botrychium?* From those of *Osmunda?* From those of *Angiopteris* and *Marattia?* From those of *Schizaea?* From those of *Actinostachys?*
8. Why is the fern gametophyte sometimes called a "prothallus" or "prothallium"?
9. Where would you look for fern gametophytes in nature?
10. How would you go about propagating ferns from spores?
11. What genera of ferns grow in the vicinity of your home or campus?
12. What is meant by the "octant" stage in the embryogeny of leptosporangiate ferns?
13. To what group of nonvascular cryptogams does the fern gametophyte show some similarity? How does it differ?
14. Comment on the significance of the location of sporangia in various species of *Osmunda.*
15. What features of *Osmunda* suggest eusporangiate ferns?
16. How do the gametophytes of *Osmunda* differ from those of *Thelypteris* and *Adiantum?*
17. Could the trunks of tree ferns be used as lumber? Explain.
18. What characteristics are considered primitive in fern sporophytes?
19. What variations occur in the gametophyte generations of ferns?
20. What significance do you attribute to the occurrence of tracheids in fern gametophytes? Under what conditions do tracheids develop? Cite the gametophyte of another vascular cryptogam that may contain xylem.
21. How would you induce apospory in a fern?
22. Comment regarding the causal factors involved in the dimorphism of the alternating generations in ferns.

Chapter 18
VASCULAR CRYPTOGAMS V:
DIVISION PTERIDOPHYTA III

It should be reiterated at this point that the division Pteridophyta contains three classes—Ophioglossopsida, Marattiopsida, and Filicopsida—of which the third was discussed, in part, in the preceding chapter. (The homosporous Filicopsida,[1] the Filicales, were treated in the preceding chapter.) The present chapter includes a discussion of the heterosporous orders, Marsileales and Salviniales, sometimes called the water ferns.

Order 2. Marsileales[2]

The order Marsileales contains the single family Marsileaceae.

Marsilea, Regnellidium, and *Pilularia*

Introduction and vegetative morphology. The genera *Marsilea* (after Count F. L. Marsigli, an Italian naturalist), *Regnellidium* (after A. F. Regnell), and *Pilularia* (L. *pilula,* little ball[3]) constitute the single family Marsileaceae. All are amphibious, growing in shallow ponds and ditches, and are also able to survive when the water level falls, as long as the soil is moist; one species of *Marsilea,* however, is xerophytic. *Marsilea*[4] (Fig. 18-1), with about 70 species, widely distributed throughout the world, occurs in the southern and Pacific regions of the United States; it is emphasized in the following account. *Regnellidium,* with a single species (Figs. 18-2, 18-3A), is native to the Amazon region of Brazil; *Pilularia* (with six species) (Figs. 18-3B, 18-4), the pillwort, occurs in Europe, Africa, and Australia; in the United States

---

[1]Except *Platyzoma.*

[2]The Filicales are the first order of the Filicopsida.

[3]In reference to the sporocarp, the distribution of *P. americana* in North America was described by Dennis and Webb (1981).

[4]For a summary of *Marsilea,* see K. M. Gupta, "Marsilea," *Botanical Monograph No. 2,* Council of Scientific and Industrial Research, New Delhi (1962).

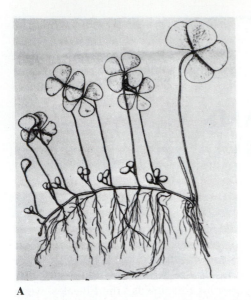

A

B

FIGURE 18-1
A. *Marsilea quadrifolia.* **B.** *Marsilea vestita.* Habit of fertile branch with vegetative leaves and sporocarps. **A.** ×⅔. **B.** ×⅐. (**A.** *After Eames.*)

A

B

it grows natively in Texas, Oregon, California, Kansas, Arkansas, and Georgia. Bonnet (1955) described the organization and reproduction of *P. globulifera* and *P. minuta.*

Marsilea at first glance is decidedly unfernlike in appearance (Fig. 18-1); its compound leaves frequently suggest four-leaf clovers to the uninitiated. It grows readily under greenhouse conditions and is fairly hardy in colder climates, where it perennates by means of its stems, which are embedded in soil or mud.

The plant body consists of an elongate, branching, stolonlike stem (Fig. 18-1B), which grows either on the surface of the mud or slightly below it. The leaves are in two rows, alternately inserted on opposite sides of the stem, from which they arise in circinate fashion. The internodes are very long in aquatic habits. In mature plants, three kinds of leaves may develop: (1) floating leaves, (2) submerged

FIGURE 18-2
Regnellidium diphyllum. **A.** Living plants in pond. **B.** Group more highly magnified. ×0.33.

FIGURE 18-3
A. *Regnellidium diphyllum.* **B.** *Pilularia* sp. Both with sporocarps. ×0.5. (*After Eames.*)

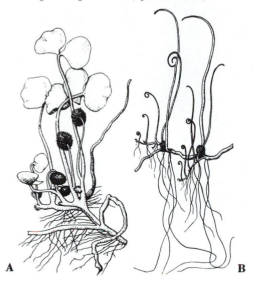

leaves, and (3) aerial leaves (White, 1971*a*). The first two, of course, occur in aquatic habitats and the last when the plant is terrestrial.

In submerged plants the petioles are relatively flaccid and have air chambers, and the leaflets float on the surface of the water; stomata are restricted to the upper surface of the leaflets. The petioles of aerial leaves, however, lack air chambers and are sufficiently rigid to support the leaflets in an erect position; here sunken stomata occur on both the dorsal and ventral leaf surfaces. The four leaflets of each leaf do not actually arise at one locus from the petiole; two are slightly higher than the others and are inserted in alternate fashion, the leaflet veins dichotomously branched but laterally and marginally united to some extent to form a loose reticulum. A transection of the leaf (Fig. 18-5) reveals the presence of an upper and lower epidermis, both with slightly sunken stomata. The mesophyll is differentiated into palisade and spongy areas. Each petiole is traversed by a V-shaped vascular trace that leaves a gap above the place of its departure from the stem stele. The submerged leaves have smaller blades than either of the other two types and lack a cuticle and differentiation into palisade and spongy mesophyll (Gaudet, 1964).

Growth of leaves, stem, and root (Kurth, 1981), originates in apical cells and their derivatives. The vascular tissue of the mature stem (Fig. 18-6A) is arranged in the form of an amphiphloic siphonostele. Both inner and outer

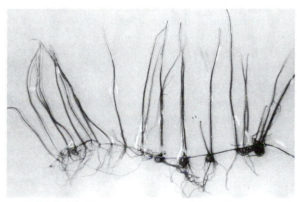

FIGURE 18-4
Pilularia americana. ×½.

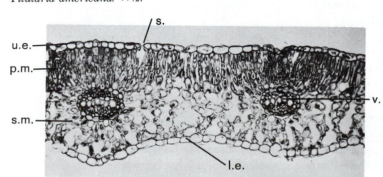

FIGURE 18-5
Marsilea sp. Transection of leaf. l.e., lower epidermis; p.m., palisade mesophyll; s., stoma; s.m., spongy mesophyll; u.e., upper epidermis; v., vein. ×90.

441

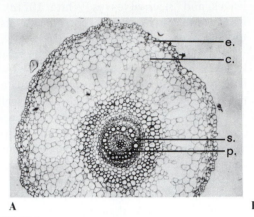

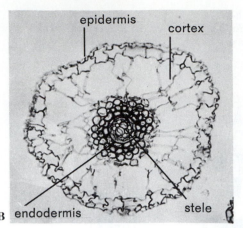

FIGURE 18-6

Marsilea sp. **A.** Transection of stem. **B.** Transection of root. c., cortex with air chambers; e., epidermis; p., pith; s., stele. **A.** ×60. **B.** ×125.

phloem are covered by a single layer of pericycle cells, which, in turn, are covered by a single endodermal layer. In some species the xylem is exarch. The pith may be parenchymatous or sclerotic, usually the latter, in stems growing in nonsubmerged soil. The continuity of the cortex is interrupted by large air chambers, and it is outwardly bounded by an epidermis. Stem branches always arise at the bases of leaves. Stolon ultrastructure was investigated by Miller and Duckett (1979).

The adventitious roots emerge at the nodes and contain monarch or diarch protosteles with exarch arrangement (Fig. 18-6B). Here, too, the stele is surrounded by a single layer of pericycle and an endodermis. The inner cortex is sclerotic; the outer contains air chambers. Vessels are present in *Marsilea* and *Regnellidium* but not in *Pilularia* (Bhardwaja and Takker, 1979).

Reproduction: the sporophyte.   The sporangia of *Marsilea* are borne in specialized structures known as **sporocarps,** which occur on short lateral branches of the petioles (Fig. 18-1). They appear after a long period of vegetative development, usually on terrestrial individuals. Bilderback (1978*a,b*) has reported on the details of their development. The sporocarps at first are relatively soft and green, but they become hard,

brown, and nutlike at maturity. They may be borne singly or in clusters, depending on the species. Anatomical and ontogenetic evidence indicates that the sporocarp probably represents a fertile pinna that has become folded with the margins united, thus enclosing the fertile (abaxial) surface of the leaflet. Each lateral half of the sporocarp bears a row of elongate sori (Fig. 18-7)

FIGURE 18-7

Marsilea sp. **A.** Diagram indicating sectional plane. **B.** Section of sporocarp. i., indusium; me., megasporangium; mi., microsporangium; r., gelatinous ring; re., receptacle; v.b., vascular bundle; w., wall of sporocarp. (*Adapted from Eames.*)

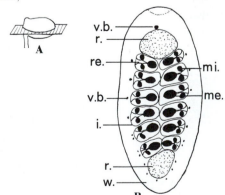

442

A

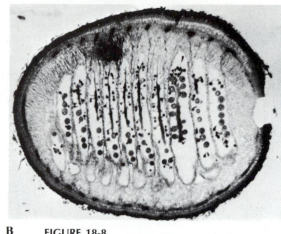

B

FIGURE 18-8
Marsilea sp. **A.** Internal view of germinated sporocarp; note dichotomous venation. **B.** Sagittal section of developing sporocarp; note receptacles, bearing megasporangia and smaller microsporangia, enclosed by indusia. **C.** Portions of three receptacles, enlarged. **A.** ×15. **B.** ×8. **C.** ×30.

C

with ridgelike receptacles. The sori of one half alternate with those of the other, but all are close together and there is overlapping. Each sorus is covered with a delicate indusium; the indusia of adjacent sori are partially fused together, so that each receptacle and its sporangia lie in a cavity (Fig. 18-7).

Sporangial initials develop first at the apex of each ridgelike receptacle; subsequently, additional initials on the flanks initiate sporangium development in **gradate** order (Fig. 18-8B, C).

The sporangia develop according to the leptosporangiate method, and all produce between 32 and 64 spores. In the sporangia at the apex of the receptacle, however, all the spores except one degenerate, their contents mingling with those of the tapetal cells, which have formed a plasmodium late in sporogenesis. The surviving spore, probably as a result of absorbing large quantities of nutriment, increases to many times its original size, becoming somewhat elipsoidal. A single megaspore, therefore, matures in each of the sporangia at the apex of the linear receptacle (Figs. 18-7B, 18-8C). Three abortive megaspores are sometimes visible near the apex of the functional megaspores (Pettitt, 1970). The mature megaspore has a rounded protuberance or papilla at one end on which the triradiate ridges remain visible. The wall in this region is delicate, but over the remainder of the spore protoplast it is extremely thick. The megaspores are sufficiently large to be readily visible to the unaided eye. The single nucleus of the megaspore lies in dense cytoplasm and in the region of the protuberance. The bulk of the spore cavity is filled with starch grains; only a small amount of cytoplasm is present.

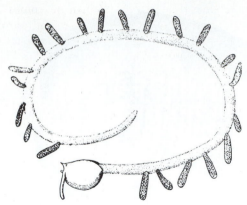

FIGURE 18-9
Masilea sp. Germinated sporocarp; note
sorophore (ring) with sori. ×1.5. (*After Haupt.*)

All the spores in the sporangia on the flanks of
the receptacle mature; they are many times
smaller than the megaspores. Each of these
microspores has a single central nucleus and
rather dense cytoplasm containing starch
grains. Hence, *Marsilea* is heterosporous, as are
Selaginella and *Isoetes* among extant microphyllous plants.

As the spores mature, changes occur in the
sterile tissues of the sporocarp. The outer layers
become stony, and their cell walls thicken markedly. The vacuoles and cytoplasm of a zone of
cells within the sporocarp become filled with
polysaccharide and make up the structure later
to expand as the sorophore (Bilderback,
1978*a,b*). Both microsporangia and megasporangia lack a highly differentiated annulus, an
indication of the fact that the spores are not
discharged vigorously and explosively from the
sporangia.

In nature, the sporocarps persist in the water
and soil after the vegetative portions of the
plants that produced them have disappeared. It
is probable that bacterial action plays a role in
rotting the sporocarp wall and in thus effecting
spore dissemination. This process may be hastened in the laboratory by cutting away a small
portion of the stony wall and immersing the
sporocarps in water. After such treatment,
water is imbibed by the hydrophilic colloids in
the sporocarp. Within a short time, the attend-
ant swelling forces the two halves of the sporocarp apart and a vermiform gelatinous structure,
the **sorophore,** emerges, bearing the sori (Fig.
18-9). The great expansion of the sorophore is
occasioned by the inhibition of water by the
polysaccharides within the sorophore cells,
which results in extreme stretching of their
walls (Bilderback, 1978*b*). Several hours after
their emergence, the sporangial walls become
gelatinous, with the result that large numbers of
free microspores and megaspores are shed into
the common soral cavity still enclosed by the
indusium. The latter disintegrates ultimately,
and the microspores and megaspores are liberated into the water (Fig. 18-10).

Sporocarps are well known for retention of
their viability. Bloom (1955) demonstrated
opening the sporocarp with germination of the
megaspores of sporocarps of *M. quadrifolia* 32
years old; formation of gametophytes and development of embryos followed.

The mature megaspore is surrounded closely
by the megasporangial wall. The megaspore
itself has two wall layers, an inner **endospore**
and outer **exospore**. The latter is composed of

FIGURE 18-10
Marsilea. Recently shed megaspores and
microspores. ×12.

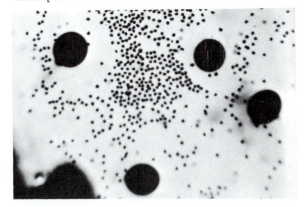

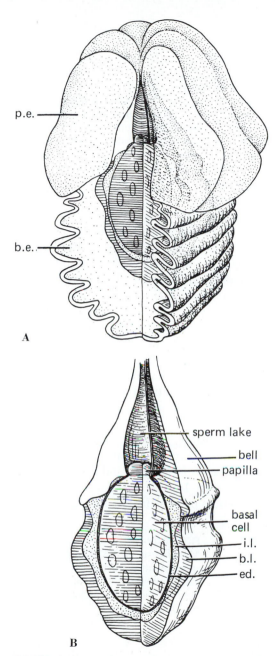

p.e.

b.e.

A

sperm lake

bell

papilla

basal cell

i.l.

b.l.

ed.

B

FIGURE 18-11

Marsilea. Organization of the hydrated megaspore. **A.** Papillar envelope (P.E.) and basal envelope (B.E.) present. **B.** Megaspore with basal and papillar envelopes removed. **A.** ×54. **B.** ×75. (*After Machlis and Rawitscher-Kunkel.*)

five layers (Fig. 18-11). When it comes in contact with water, the sporangial wall unrolls rapidly, and the megaspore wall expands into a complex structure (Fig. 18-11), which has been described by Machlis and Rawitscher-Kunkel (1967a).

Soon after the megasporangial wall is cast off, two extensive gelatinous envelopes appear, the basal envelope (folded horizontally) and the papillar envelope (folded longitudinally) (Fig. 18-14). Within the papilla envelope and over the papilla, a bell-shaped gelatinous structure develops; in its center is an elongate fluid region, the **sperm lake** (Fig. 18-11B). Although sperms may be observed throughout the gelatinous envelopes at the papillar region, they are concentrated in the sperm lake in which they swim actively.

Reproduction: The Gametophytes. Unlike those of *Selaginella* but like those of *Isoetes*, the spores of *Marsilea* do not germinate until they have been shed from the sporocarp. Development into gametophytes is then very rapid.

Hepler (1976), Myles and Hepler (1977), and Myles et al. (1978) have recently investigated incisively the development of the male gametophyte and spermatogenesis in *Marsilea vestita.* Formation of the male gametophyte by the microspore is initiated by a nuclear and cell division to form unequal cells within the microspore wall; the smaller cell is the prothallial cell. As in *Isoetes* and *Selaginella*, this cell fails to divide again and is often interpreted as the sole remnant of vegetative tissue in the male gametophyte. The large cell divides in two, and each of the resulting cells, by further division (Fig. 18-12A, B), forms an antheridium that is covered with jacket cells, which surround the 16 spermatogenous cells. Late in the divisions of the spermatogenous cells, deeply staining organelles, the **blepharoplasts,** arise *de novo* near the nuclei and ultimately function (as in the cycad, *Zamia*, p. 538) in forming the procentrioles and basal bodies of the 100 to 150 flagella of

445

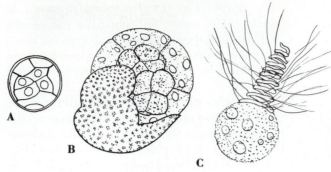

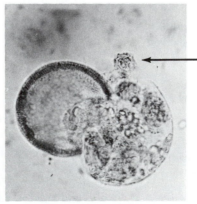

FIGURE 18-12
Marsilea vestita. Development of the ♂ gametophyte. **A.** Section of the microspore with ♂ gametophyte; note prothallial cell and two rudimentary antheridia. **B.** Microspore with emergent ♂ gametophyte. **C.** Living sperm. **A.** ×260. **B.** ×430. **C.** ×2266. (**A.** *Adapted from Sharp.*)

the mature sperms. The microspore wall now ruptures, and the mature antheridia protrude (Figs. 18-12B, 18-13A). The prothallial cells and jacket cells disintegrate, and the large sperms, which have previously developed singly in each spermatogenous cell, become actively motile and swim away (Figs. 18-12C, 18-13B). The sperms are tightly coiled and have their flagella at the apical nuclear portion (Figs. 18-12C, 18-14E). Myles (1975) reported that the motile sperms lose their ball of cytoplasm before they reach the eggs. The development of the male gametophyte and liberation of the sperms may take place in as short a time as from 11 to 15 hours, depending on the species.

In the development of the female gametophyte by the megaspore, a limited number of nuclear and cell divisions occur within the hemispherical protuberance, resulting in the formation of a small amount of vegetative tissue bearing a single apical archegonium (Fig. 18-14A, C). Enlargement of these cells ruptures the megaspore wall, which is delicate in the region of the protuberance. The megaspore that bears the mature female gametophyte becomes surrounded by a gelatinous matrix, which is usually thickest in the region of the female gametophyte (Fig. 18-14). The liberated sperms (Fig. 18-14D), apparently attracted chemotactically, swarm into the matrix in the vicinity of the archegonium. One of them makes its way to the egg cell through the sperm lake, and fertilization occurs.

Reproduction: embryogeny. Development of the embryo is initiated several hours after fertilization by nuclear and cell division of the zygote (Fig. 18-15A). Cytokinesis is usually in a direction parallel to the long axis of the

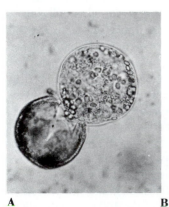

FIGURE 18-13
Marsilea sp. **A, B.** Microspores with maturing ♂ gametophytes, the one at **B** liberating a sperm (arrow). ×175.

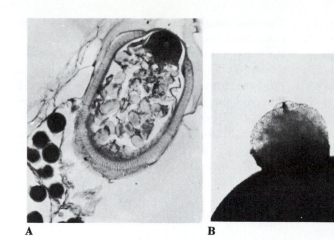

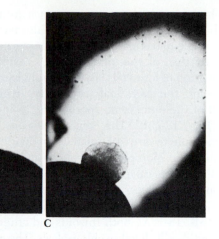

A

B

C

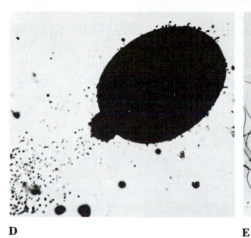

D

E

FIGURE 18-14

Marsilea vestita. **A.** Section of megaspore and immature ♀ gametophyte (dark mass above). **B.** Mature ♀ gametophyte. **C.** The same; India ink preparation showing gelatinous material at apex of megaspore. **D.** Megaspore with protuberant ♀ gametophyte surrounded by sperms. **E.** Single sperm, stained; note coiled nucleus and numerous flagella. **A.** ×65. **B.** ×165. **C.** ×75. **D.** ×60. **E.** ×2500. (**E.** *After Mizukani and Gall.*)

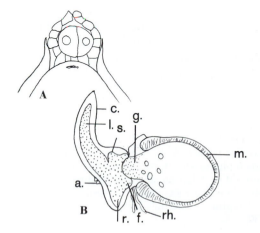

FIGURE 18-15

A. *Marsilea vestita.* Median longitudinal section of ♀ gametophyte; archegonium with early embryo. **B.** *Marsilea* sp. Median longitudinal section of megaspore, ♀ gametophyte, and embryo. a., archegonium; c., calyptralike sheath; g., ♀ gametophyte; l., primary leaf or cotyledon; m., megaspore wall; r., root; rh., rhizoid; s., stem. **A.** ×280. **B.** ×165. (**A.** *From Campbell.* **B.** *From Sachs.*)

447

archegonium and megaspore, which are typically horizontal. A second division, also parallel to the archegonial axis but in a plane perpendicular to that of the first division, results in a formation of four cells. By further division these develop the foot, leaf (cotyledon), stem, and root of the embryonic sporophyte, as indicated in Fig. 18-15B. The vegetative cells of the gametophyte are stimulated to divide as an embryonic sporophyte develops, and they form a sheathing calyptra (Figs. 18-15B, 18-16A) around the latter. The surface cells of this gametophytic tissue develop rhizoids (Figs. 18-15B, 18-16A); both rhizoids and other gametophytic cells develop chloroplasts. The calyptra is relatively persistent, but disappears about the same time that the embryonic root penetrates the substratum. Growth of the embryonic sporophyte is rapid (Fig. 18-16A, B). One series of germling sporophytes of *M. vestita* planted in April produced mature plants bearing sporocarps by the following October.

The genera of Marsileales form an interesting evolutionary series with respect to their leaves.

Marsilea usually has two pairs of pinnae on its compound leaves, but occasionally it may have three pairs. In *Regnellidium* (Figs. 18-2, 18-3A) a deeply bilobed leaf occurs. Finally, in *Pilularia* (Figs. 18-3B, 18-4) no leaf blade develops, and the petiole and rachis are the photosynthetic organs. The organization of the sporocarp, maturation of the gametophytes, and the embryogeny of *Regnellidium* and *Pilularia*, in general, resemble corresponding phases in *Marsilea*.

Summary of Marsileales

Although decidedly not fernlike in appearance, the vegetative structure and sporangial development of *Marsilea* are very similar to those of other leptosporangiate ferns. The cloverlike leaves represent sterile pinnae with a dichotomous venation that is closed at the margins and to some extent laterally. Vernation of the leaves is circinate. The stems are amphiphloic siphonosteles, interrupted by leaf gaps, which usually do not overlap because of the great length of the internodes. The leaves arise alternately in two rows from the stolonlike stems. Both roots and stems afford anatomical evidence of aquatic habitat in the form of air chambers; the vascular tissue, however, is unusually abundant for a hydrophytic plant. The roots contain vessels. *Marsilea* is representative of the leptosporangiate ferns in which heterospory has been developed. The sporangia are borne in sclerotic, modified, fused pinnae which constitute the sporocarp. With it have appeared such concomitant attributes as dioecious game-

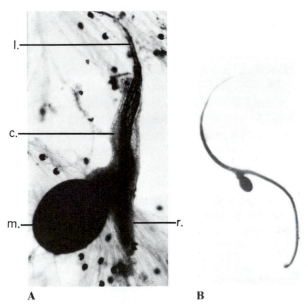

A   B

FIGURE 18-16
Marsilea vestita. **A, B.** Megaspores with ♀ gametophytes after fertilization; the one to the right in A has an advanced embryo, cotyledon, and root, which are visible.
c., calyptra; l., leaf; m., megaspore; r., root.
×30.

448

A

B

C

FIGURE 18-17
Salvinia natans. **A, B.** Floating plants. **C.** Dorsal surface of leaf more highly magnified. Note hairs. **A.** ×0.75. **B.** ×5. **C.** ×18.

tophytes, reduction in vegetative functions and tissue, and duration of the gametophytes. Gametogenesis and embryogeny are extremely rapid in *Marsilea*.

Order 3. Salviniales

The order Salviniales contains two genera, *Salvinia* (after Antonio M. Salvini) and *Azolla* (Gr. *azo*, parched, killed by drought), both small, floating aquatics. Most species of *Salvinia* are native to Africa, but one species occurs in the United States. *Salvinia* is widely distributed in greenhouses and aquaria. Two species of *Azolla* occur in the United States, *A. caroliniana* in the eastern states and *A. filiculoides* in the western states. Ponds and swamps are often colored red by "blooms" of *Azolla* in intense light (Gifford and Polito, 1981*a*).

Salvinia (Fig. 18-17A) is rootless, the apparent roots being morphologically filiform leaf segments. The leaves are in whorls of three, two of which are floating and the third submerged. It is the latter that is divided into rootlike lobes. The floating leaves are concave and covered with hairs and waxy papillae (Fig. 18-17B). The development and organization of both the floating and submerged leaves of *Salvinia* were investigated by Croxdale (1978, 1979, 1981). The slender floating stems of *Salvinia* have a small pith (which is encircled by tracheids) and are, accordingly, siphonostelic. Gaudet (1973) succeeded in growing *S. minima* in axenic culture.

Azolla (Fig. 18-18) consists of branching stems densely clothed with alternate bilobed leaves that suggest those of certain Jungermanniaceae (p. 236). Each leaf is composed of a chlorophyllous dorsal lobe and a submerged achlorophyllous ventral lobe. The dorsal lobes contain cavities in which occur filaments of the blue-green alga *Anabaena* (Fig. 18-18D); there is good evidence that the alga fixes nitrogen (consult Tuug and Shew [1976] and Peters et al. [1978, 1981] for literature).

449

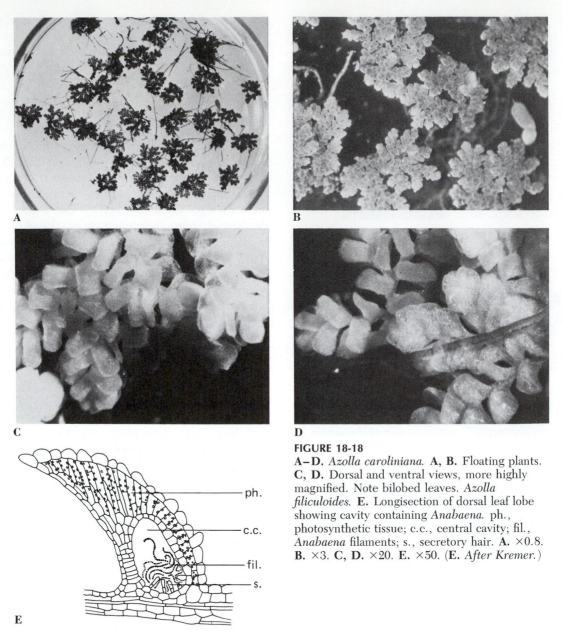

FIGURE 18-18
A–D. *Azolla caroliniana.* **A, B.** Floating plants.
C, D. Dorsal and ventral views, more highly
magnified. Note bilobed leaves. *Azolla
filiculoides.* **E.** Longisection of dorsal leaf lobe
showing cavity containing *Anabaena.* ph.,
photosynthetic tissue; c.c., central cavity; fil.,
Anabaena filaments; s., secretory hair. **A.** ×0.8.
B. ×3. **C, D.** ×20. **E.** ×50. (**E.** *After Kremer.*)

Azolla growing in strong sunlight often becomes orange-red. Holst (1977) has identified
the pigment of *A. mexicana* as an anthocyanin,
luteolinidin-5-glycoside. This pigment is apparently always present in the plants but increases
with the increase of incident light. Pieterse et

al. (1977) reported that anthocyanin production
is influenced by temperature and water composition, not by photoperiod.

True roots, which arise endogenously at intervals along the stem, occur in *Azolla.* The apical
cells are mitotically active in the roots and stems

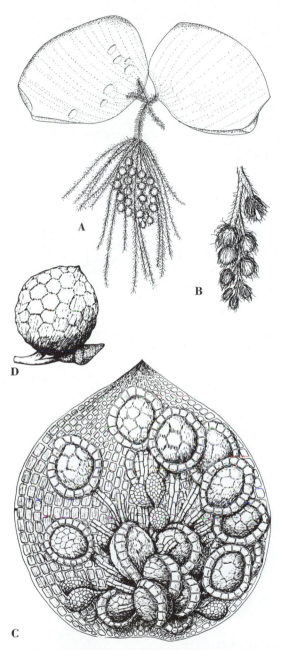

FIGURE 18-19
Salvinia natans. Fertile specimen. **A.** Segment of plant showing submerged sporocarps on leaf segments. **B.** Portion of a sporocarp-bearing branch more highly magnified. **C, D.** *Azolla microphylla,* sporocarps. **A.** Opened sporocarp showing microsporangia and indusium. **D.** Slightly magnified view of a microsporocarp (larger, spherical) and megasporocarp. (**A, B.** *Modified from Bierhorst.* **C, D.** *After Martius from Eames.*)

tively because of their continuing apical growth, branching, and death of the older regions of the plants. Detachment of branches and their subsequent independent growth also increase the population.

Salvinia with 13 species and *Azolla* with 5 species, like the genera of Marsileales, are heterosporous and produce their spores in sporocarps. The latter differ markedly, however, from those of the Marsileales in that they represent single sori enclosed by modified indusium. The sporocarps of *Salvinia* are borne at the tips of the short branches of the submerged, branching rootlike leaves (Fig. 18-19A), while those of *Azolla* are formed in groups of two or four (depending on the species) in the axil of the dorsal lobe of a basal leaf of a branch. The plants are monoecious, but individual sporocarps usually contain only micro- or megasporangia. The reproductive cycle of *Azolla filiculoides* and *Salvinia auriculata* has been elucidated most recently by Bonnet (1955, 1957a). In the former, development of the essentially leptosporangiate sporangia is initiated by a single cell at the apex of a moundlike receptacle (Fig. 18-20A). This first sporangium is potentially a megasporangium, while other sporangia initiated laterally on the receptacle are potentially microsporangia. If the megasporangium matures in a given sporocarp, the microsporangia do not develop; if it fails to mature, the microsporangia do so and the sorus becomes a microsporocarp.

(Gifford and Polito, 1981b, and Nitayankura et al., 1983). The roots bear root hairs as long as they exist. Both stem and root have exarch protosteles.

Both *Salvinia* and *Azolla* reproduce vegeta-

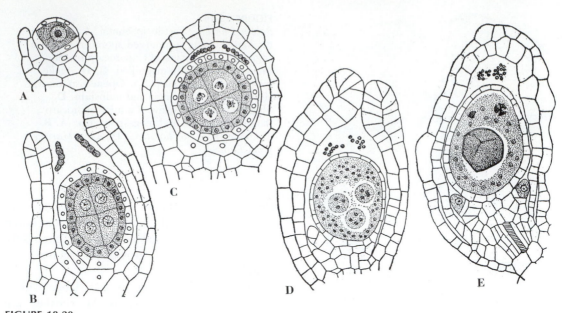

FIGURE 18-20
Azolla filiculoides. Stages in development of the megasporangium and indusium; note included *Anabaena* cells. **A–C.** ×600. **D–E.** ×325. (*After Smith.*)

Eight sporogenous cells in the megasporangium function as megasporocytes, so that 32 potential megaspores are formed. Only one of the potential megaspores survives. It enlarges, filling much of the megasporangium, which enlarges to fill the megasporocarp (Fig. 18-20E). In both mega- and microsporangia the tapetal cells disintegrate to form a tapetal plasmodium. This becomes vacuolate at first, later hardened and alveolar (Fig. 18-20D), is quadripartite in the megasporangium, and sometimes is composed of more lobes in the microsporangium. This hardened alveolar material constitutes the **massulae** (Fig. 18-21A, B). That portion of it immediately surrounding the megaspore is known as the **perispore.** Neither the microsporangial nor megasporangial walls are modified in relation to dehiscence. Lucas and Duckett (1980) investigated cytologically the sporocarps of *Azolla filiculoides.*

The mature sporocarps sink to the bottom of the body of water in which they are growing,

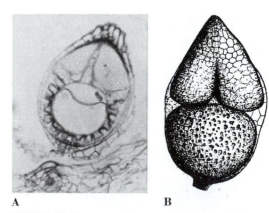

A   **B**

FIGURE 18-21
Azolla caroliniana. **A.** Section of megasporangium with single megaspore; note 3 (of 4) lobes of massulae above.
B. Megasporocarp (wall transparent) showing basal megaspore and apical massulae. (**B.** *From Eames, after Bernard.*)

452

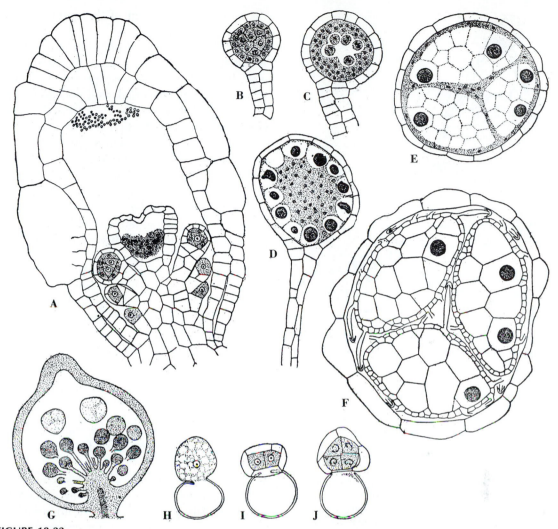

FIGURE 18-22

Azolla filiculoides. **A.** Longisection of immature microsporocarp. **B–F.** Stages in development of microsporangium and microspores, the latter arranged at the periphery of the microsporangium in D–F; note tripartite (in section) massulae. **G.** Longisection of maturing microsporocarp. **H–J.** Development of male gametophyte. **A–F.** ×325. **G.** ×60. **H–J.** ×650. (*After Smith.*)

and the sporocarp and sporangial walls ultimately release the spores there. The megaspores remain attached to their massulae (Fig. 18-23A, C), while the microspores are embedded in the periphery of their massulae. The latter in some species have protuberances known as **glochidia**,

which apparently can become attached to a megasporic massulae or perispore.

Unlike those of many species of *Selaginella*, but like those of *Isoetes* and the Marsileales, the heterospores of *Azolla* and *Salvinia* do not germinate until they are liberated from their spo-

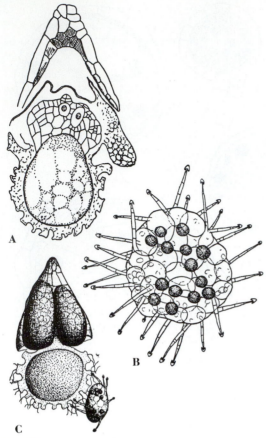

FIGURE 18-23

Azolla filiculoides. **A.** Megaspore with ♀
gametophyte. **B.** Massulae with microspores and
glochidia. **C.** Megaspore attached to massulae
and with a group of microspores and massulae
attached by glochidia. **A.** ×53. (**A.** *From Haupt
after Campbell.* **B, C.** *From Eames after
Bernard.*)

rangia, and development of the gametophytes
takes about one week. As in other heterosporous vascular cryptogams, the gametophytes
(Figs. 18-22H–J, 18-23A) have little vegetative
tissue, that of the female being photosynthetic,
as in *Marsilea.* The male gametophyte of *Azolla*
is largely intrasporal, consisting at maturity of a
prothallial (sterile) cell and one antheridium
with eight sperms. One or more archegonia
develop in the female gametophyte, which is

the site of development of the embryo sporophyte from the zygote.

Reproductive material of *Salvinia* and *Azolla*
is not frequently encountered. It could probably be evoked in laboratory culture where
temperature, photoperiod, and nutrients could
be controlled. It should be clear from the above
brief account that although it is fundamentally
similar to that in other heterosporous vascular
cryptogams, especially the Marsileales, the reproductive process in the Salviniales has become highly specialized.

Salvinia and *Azolla,* although similar in
growth habit and aquatic habitat, differ in a
number of respects as indicated comparatively
below:

Salvinia	*Azolla*
1. Simple, whorled leaves	1. Leaves overlapping, each with floating and submersed lobe
2. Roots lacking, one leaf at each node rootlike	2. Roots with root hairs and root caps present
3. Siphonostelic stem	3. Protostelic stem
4. Sporocarps at branch tips of submersed, rootlike leaves	4. Sporocarps borne on the apex of lobe
5. Sporocarps *either* micro- or mega-sporangiate	5. Sporocarps containing potentially both micro- and megasporangia
6. Several megasporangia present in megasporocarp	6. Only one megasporangium present in a megasporocarp
7. Megaspores germinating at the water surface	7. Megaspores germinating when submersed
8. Female gametophyte strongly photosynthetic	8. Female gametophyte barely green
9. Male gametophyte with two antheridial groups	9. Male gametophyte with one antheridium

PHYLOGENY OF PTERIDOPHYTA

Chapters 16 and 17 and this chapter have presented an account of representative living ferns. The latter are grouped in three classes, the Ophioglossopsida, the Marattiopsida, and the Filicopsida; the first two include a single order each (Ophioglossales and Marattiales) and the last, three orders (Filicales, Marsileales, and Salviniales). The Ophioglossopsida and Marattiopsida are eusporangiate, while the Filicopsida are leptosporangiate, although sporangial development is somewhat modified in some (e.g., *Osmunda*) of the Filicopsida. The most sound evidence for the evolutionary history of a group is the fossil record. Fossils of the Ophioglossopsida are unknown, but members of the Marattiopsida are known from Pennsylvanian[5] times. One family (Osmundaceae) of the Filicopsida is represented beginning with the Permian, but most are Mesozoic or Cenozoic in first appearance. It is important to realize that members of the Polypodiaceae—today the most dominant, widely distributed leptosporangiate ferns —did not appear until the Jurassic, and they seem thus to be as recently evolved as are the angiosperms. For this reason, they are often spoken of as "modern ferns."

The fossil record of ferns and their supposed precursors gives some indication of characteristics that may be considered to be primitive. These include elaborate, compound (branched) leaves, presumably modified from branch systems; erect, epiterranean stems; massive sporangia, lacking annuli, with several-layered walls and indefinitely large numbers of tetrahedral spores; the sporangia borne marginally on the leaves and at the ends of veins. Because these features characterized many ancient ferns, deviations from them have come to be considered to be further evolved or "advanced." Such characteristics are simple leaves, prostrate,

often dorsiventral, rhizomatous stems, small sporangia with annuli, one- to two-layered walls and fewer (32–64) spores, the sporangia borne on the abaxial surface of leaves instead of marginally. A number of additional characteristics have been inferred to be either primitive or advanced. Among the latter are thin-walled, green, rapidly germinating spores; direct formation of a platelike gametophyte (rather than a filament); small, thin, delicate, ephemeral gametophytes lacking a midrib; small antheridia with 24–156 sperms (in contrast to several hundred), with 4 or fewer wall cells; and archegonia with relatively long necks. Many other characteristics have been cited as illustrative of relative primitiveness or advancement. Thus, Wagner (1974) presented, at a symposium on phylogeny of ferns and their classification, a table of over 70 evolutionary "trends" or characteristics, sporophytic and gametophytic, vegetative and reproductive, which he considered valid. He (1974) and others have found the form and ornamentation of fern spores to be important taxonomically and indicative of phylogeny.

Since "nature mocks at human categories," once again we find, as anticipated, that the putative "primitive" and/or "advanced" characteristics are not neatly correlated in one or another genus, family, or order. Although the primitive nature of some characteristics may be reasonably inferred from the fossil record of ferns, the validity of their criteria of advancement and primitiveness has been questioned. Even at the symposium cited above, Holttum, a leading investigator of the taxonomy of ferns, stated: "As regards the great majority of ferns, their probable phylogeny can be inferred only from their classification; there is little other evidence available." At a later point in the same symposium another specialist (Wagner) in pteridology said: "There is simply no 'royal road' to classification, and the way it is done is every man's special prerogative." In an earlier symposium (1964) the same investigator had remarked: "Such patterns as we can construct of

---

[5]See inside front cover.

the origin and evolution of the ferns are based almost exclusively on data from the living members of the class." In conclusion, when one is considering phylogeny or the probable course of evolution among the higher categories (more-embracing taxa), Holttum's statement might well be a sobering consideration. More extensive discussion of the evolution and phologeny of ferns is available in the following references cited in the Bibliography: Delevoryas (1964), Jermy, Crabbe, and Thomas (1973), Bierhorst (1974), Lloyd (1974), Mickel (1974), Stidd (1974), Wagner (1974), and White (1974).

DISCUSSION QUESTIONS

1. How does venation in *Marsilea* differ from that in *Adiantum*? Which do you consider more primitive and for what reason?
2. Describe the structure of the sporocarp of *Marsilea*.
3. What evidence indicates that the sporocarp of *Marsilea* is a folded pinna?
4. What is meant by the terms simple, gradate, and mixed as applied to sporangium development? Illustrate with examples from the genera described above.
5. How does the ontogeny of the functional megaspores differ in *Selaginella* and *Marsilea*?

6. A student argued that inasmuch as heterospory is obviously an advanced condition derived from homospory, *Selaginella*, *Isoetes*, and *Marsilea* should be classified together, as distinct from other extant vascular cryptogams. Give reasons against supporting such a classification.
7. Would you consider the male gametophyte of *Selaginella* or that of *Marsilea* to be more primitive? Give the reasons for your answer.
8. Sporocarps of *Marsilea*, stored in alcohol for 20 years, have germinated when placed in water. How do you explain this?
9. A layman, observing the sporocarps of *Marsilea* for the first time, referred to them as "seeds." How do they differ from seeds?
10. How do the sporangia of *Lygodium* differ from those of *Thelypteris* and *Adiantum*?
11. Name several conservatories and cities of the United States where there are living specimens of tree ferns.
12. Cite examples of hydrophytic, mesophytic, and xerophytic Pteridophyta.
13. Describe the organization of the sporocarps of *Salvinia* and *Azolla*.
14. Describe reproduction in *Salvinia* and *Azolla*.

Chapter 19
VASCULAR CRYPTOGAMS VI:
DIVISION PSILOTOPHYTA

In the earlier editions of this book, and in a number of others, there were included in the division Psilotophyta[1] two genera of living plants (*Psilotum* and *Tmesipteris*) and a rather heterogeneous assemblage of extinct genera, including *Psilophyton, Rhynia,* and *Astero-xylon,* mostly Devonian (inside front cover) plants. Later information obtained by paleobotanists (Banks, 1968) about the latter has resulted in their reclassification into other groups (see Chapter 20, p. 482). Furthermore, it has recently been proposed that the two living genera (*Psilotum* and *Tmesipteris*) have affinities with the ferns (Pteridophyta), as discussed on p. 467. However, for the present, in this text, they are retained in the division Psilotophyta as its sole representatives.

Of the vascular plants, the members of the Psilotophyta have the least complex organization.[2] The plants consist of sparsely, or profusely, dichotomously branched axes arising from subterranean, rootless rhizomes (Fig. 19-1). Their synangia (united sporangia) are borne on short lateral branches and hence are cauline. Their gametophytes are cylindrical, subterranean, and saprophytic, and their sperms are multiflagellate. A single order, Psilotales, and family, Psilotaceae, including two genera, *Psilotum* and *Tmesipteris,* constitute the division as herein conceived.

CLASS 1. PSILOTOPSIDA

Order 1. Psilotales

Family 1. Psilotaceae
Psilotum
Psilotum (Gr. *psilos,* bare) (Figs. 19-1, 19-2), the whisk fern, occurs in Arizona, Florida, Louisiana, and Texas, either epiphytically or in humus-rich soil and also grows in tropical

---

[1]Then called Psilophyta.

[2]Except for certain secondarily (presumably) reduced aquatics.

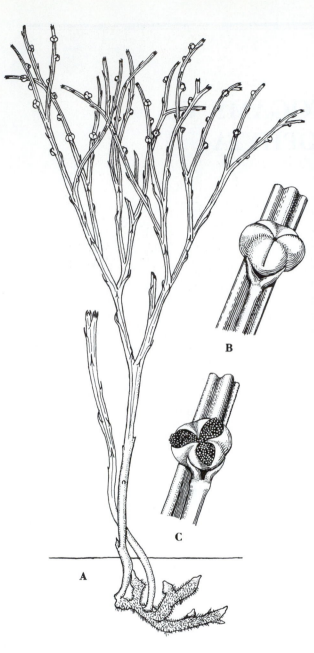

FIGURE 19-1
Psilotum nudum. **A.** Rhizome and aerial branches of sporophyte. **B, C.** Details of synangia and bracts; the synangium in **C**, dehiscent. **A.** ×0.75 **B, C.** ×9.

Asia. This is *P. nudum*, which is widespread in tropical and subtropical regions. Of the two other species, *P. complanatum* (Fig. 19-2B) occurs in Hawaii and is erect, abundant, and widely distributed, and *P. flaccidum* is pendulous. *Tmesipteris* (Gr. *tmesis*, act of cutting, + Gr. *pteris*, fern) grows natively only in Australia, New Zealand, and on other southern Pacific islands. Because of its more ready availability

A

B

FIGURE 19-2
A. *Psilotum nudum.* **B.** *Psilotum complanatum.* ×0.125.

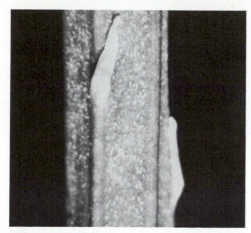

FIGURE 19-3
Psilotum nudum. Close-up of aerial stem showing ribs, stomata, and bracts. ×10.

FIGURE 19-4
A cultivated, naked strain of *Psilotum nudum* called *Bunryu-zan* in Japanese. ×1.7. (*After Rouffa.*)

and ease of culture in the greenhouse, *Psilotum* will be emphasized in the following account.

Vegetative morphology. The conspicuous plant body of *Psilotum*, which is the sporophyte, as is the case in all vascular plants, consists of dichotomously branched aerial axes, 30 cm or more in height, and of subterranean rhizomes which have epidermal cells bearing rhizoids (Fig. 19-1A); the latter may be one to three cells long. The aerial stems are ridged, often pentagonal in cross sections of the lower portions of the axis and triangular above. They bear small, scalelike appendages (Figs. 19-1A, 19-3), **prophylls,** which lack vascular tissue, although vascular traces from the stele emerge toward the bases of some of the scale leaves in both species of *Psilotum*. Roots are absent.

Psilotum used to be prized as a cultivated plant in Japan, and about 100 varieties were known. Rouffa (1971) has described a variety, called *Bunryu-zan* in Japanese, which differs from the usual *P. nudum* in lacking scales or prophylls and in producing multiple, terminal synangia at the tips of some elongate aerial axes (Fig. 19-4).

Both rhizome and aerial branches develop by apical growth that may be traced to the activity of a single tetrahedral apical cell and its derivatives, all of which divide actively. A short distance posterior to the apical meristem, differentiation becomes apparent (Fig. 19-5) in the meristematic tissues, in which three groups of cells are recognizable. In the outermost single layer, the cells are prismatic, radially elongate, and constitute the **protoderm.** The central mass of elongate cells is the **procambium;** the cells between the procambium and protoderm are the **ground meristem.** In these three **primary meristems** (Fig. 19-5C) the component cells continue to divide.

The organization of the aerial axes and rhizomes of *Psilotum* upon completion of differentiation is somewhat variable, according to the level between apex and base and the diameter of the axes. The erect stems may be almost

459

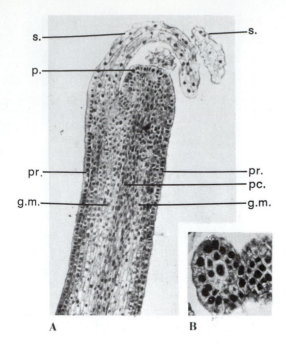

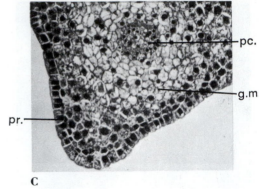

FIGURE 19-5
Psilotum nudum. Ontogeny of the aerial axis. **A.**
Median longisection of branch. **B.** Transection of
promeristem; note apical cell. **C.** Portion of a
transection at the level of the primary meristems.
g.m., ground meristem; p., promeristem (note two
apical cells which initiate dichotomy); pc.,
procambium; pr., protoderm; s., scale or bract. **A.**
×60. **B, C.** ×150.

circular in transection as they emerge from the
soil, but toward the lowermost (first) dichotomy
they are often pentagonal and between the
more distal dichotomies, triangular.

The stele also changes at different levels and
in different individuals. Thus, it may be triarch
or tetrarch at soil level; of much greater diameter and pentarch (Fig. 19-6) to octarch approaching the first dichotomy; and tetrarch,
triarch, or diarch in the most distal dichotomies.
Furthermore, the stele is mesarch at the base of
the aerial shoot, and this condition may extend
beyond the first dichotomy. More distal portions of the axes are exarch.

The stele of *Psilotum* (Fig. 19-6B) has been
interpreted as a protostele (actinostele) by some
and as a siphonostele by others. This is because
the stem center is composed of lignified, sclerenchymatous cells that may belong either to
the xylem (protostelic interpretation) or represent a sclerotic pith (siphonostelic interpretation). It should be noted that the stele always
develops from a solid core of procambium.

The mature phloem cells are devoid of nuclei
and have oblique or transverse terminal walls.
The latter are organized as a large sieve area
(Perry and Evert, 1975). Single, prominent
pores occur in the lateral walls common to the
sieve elements and parenchyma cells. Sieve
cells, with sieve areas in their oblique terminal
walls, are present. Their nuclei disintegrate as
the cells mature. A prominent endodermis surrounds the stele.

The cortex is massive as compared with the
stele (Fig. 19-6A) and consists of a parenchymatous storage region adjacent to the endodermis.
This is surrounded successively by a zone of
sclerenchyma and by several layers of photosynthetic parenchyma cells. The latter represent
the major photosynthetic region of the plant,
and the increase of the internal surface of the
cells by lobing suggests the chlorenchyma in the
leaves of certain conifers. The cortical cells of
rhizomes frequently contain fungi (Fig. 19-7).
The fungi regularly associated internally and
externally with roots and rhizomes are called

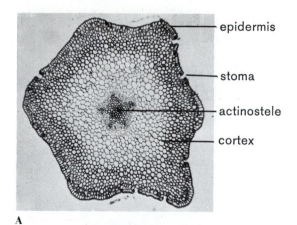

epidermis

stoma

actinostele

cortex

A

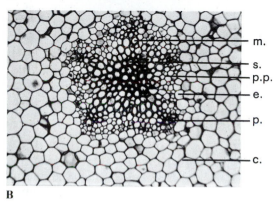

m.

s.

p.p.

e.

p.

c.

B

FIGURE 19-6
Psilotum nudum. **A.** Transection of stem at the level
of the mature primary tissues. **B.** Stele of
preceding, enlarged. c., cortex; e., endodermis; m.,
metaxylem; p., protoxylem; p.p., primary phloem;
s., sclerenchyma. **A.** ×30. **B.** ×90.

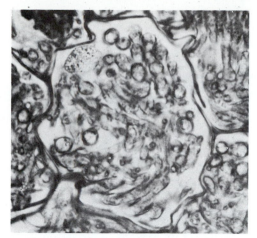

FIGURE 19-7
Psilotum nudum. Cell of cortex of rhizome filled
with mycorrhizal fungus. ×770.

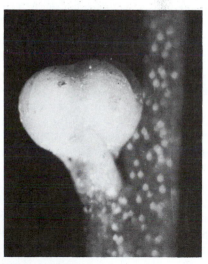

FIGURE 19-8
Psilotum nudum. Mature and dehiscent synangium;
note abundant stomata on axis. ×15.

mycorrhizal fungi and the association is termed
a **mycorrhiza** (Went and Stark, 1968; Harley,
1969). The epidermis is heavily cutinized and
interrupted here and there, between the ridges
of the stem, by stomata and guard cells (Fig.
19-6A), the latter slightly sunken, as in many
xerophytes. A small substomatal chamber is
present beneath each stoma.

The scalelike appendages (Figs. 19-3, 19-5A)
of the stem lack vascular tissue and are made up
entirely of photosynthetic tissues covered by an
epidermis. As noted previously, vascular traces

may extend from the stem stele to the bases of
some of the prophylls.

Reproduction: the sporophyte. Spores
are produced in complex, trilobed structures
called **synangia,** which are borne on the enlarged apices of short lateral branches (Figs.
19-1, 19-8). The synangia are, therefore, **cau-**

461

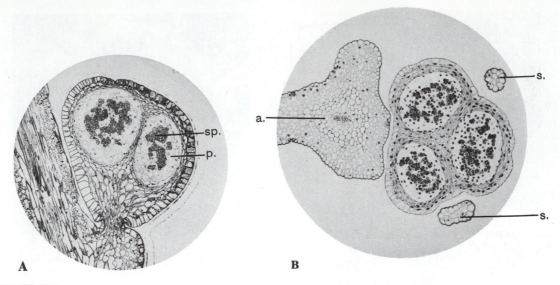

FIGURE 19-9

Psilotum nudum. **A.** Longisection and **B.** Transection of synangium containing sporocytes. a., axis; p., plasmodium; s., scale or bract; sp., sporogenous tissue. ×30.

line. It has been suggested that the synangium represents three phylogenetically united sporangia. Two basal emergences curve about the synangium and its stalk (Fig. 19-1B, C). The latter is transversed centrally by a trace, which is connected with the stele of the main axis. The trace extends into the base of the synangial partition and divides into three branches.

As noted earlier, the synangia usually arise on branches on limited growth, the earliest stages being at the growing apices of the aerial stems. Rouffa (1971, 1978) has reported that in an appendageless Japanese variety (Fig. 19-4), as well as in certain appendaged (with prophylls) races of *P. nudum*, the synangia are terminal on the main axes, sometimes exclusively so.

In development, three superficial groups of cells just below the apex of the immature branches divide **periclinally** (parallel to the surface) into two layers. The outer, by subsequent **anticlinal** and periclinal divisions, forms the several-layered sporangial wall, while the inner cells divide repeatedly to form the sporogenous tissue and **tapetum** (Fig. 19-9A). The latter is several layers of cells thick, but their walls disintegrate during sporogenesis to form a **tape-**

tal plasmodium, a multinucleate mass of nutritive protoplasm.

Later the cells of the sporogenous tissue function as sporocytes. The latter undergo the meiotic process (Fig. 19-10), forming tetrads of spores that ultimately separate. The mature spores (Fig. 19-10G) are colorless and kidney-shaped and have transparent walls. During sporogenesis, the synangial wall undergoes thickening by additional deposition of wall material. However, three vertical layers remain thin-walled and, upon drying, serve as the sites of dehiscence (Fig. 19-1B, C).

The pattern of synangial development just described, which characterizes the development of the sporangia in a number of other vascular cryptogams, is said to be **eusporangiate.** In such a process, as noted, the synangium (or sporangium) arises from superficial cells that divide periclinally. By further divisions, the outer cellular products of the division form a several-layered, sterile enclosure or wall around the sporogenous tissue; the latter develops by further division of the inner (subsuperficial) cell layer.

The dichotomy into wall and tapetal develop-

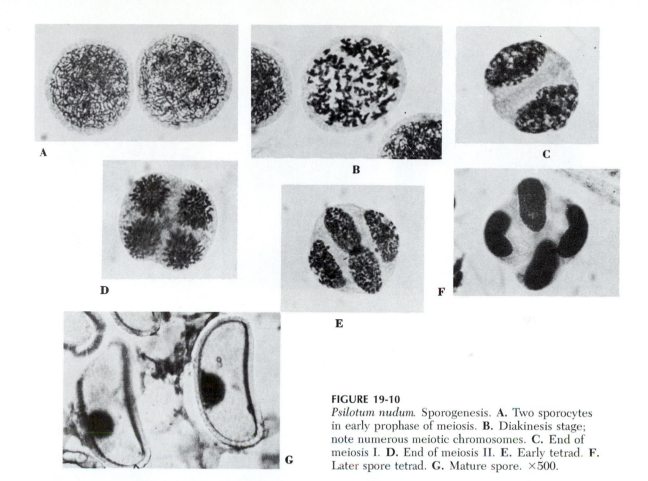

FIGURE 19-10

Psilotum nudum. Sporogenesis. **A.** Two sporocytes in early prophase of meiosis. **B.** Diakinesis stage; note numerous meiotic chromosomes. **C.** End of meiosis I. **D.** End of meiosis II. **E.** Early tetrad. **F.** Later spore tetrad. **G.** Mature spore. ×500.

ment from the outer (more superficial) layer and of sporogenous tissue from the inner, is not absolute in all cases. Some sporogenous cells may arise from cells derived from descendants of the superficial layer and, in plants with a two-layered tapetum, the outer layer may differentiate from the sporangial wall and the inner from the sporogenous tissue.

Reproduction: the gametophyte. The spores of *Psilotum* are slow to germinate. Whittier (1973b, 1975; 1980) and Whittier and Peterson (1980) succeeded in growing the gametophytes to sexual maturity in axenic culture in defined media. Germination occurred only in darkness, although less than 0.1% germination occurred.

The gametophytes (Fig. 19-11) found in nature are cylindrical, rarely more than 2 mm in diameter, are sometimes forked, and are covered with numerous rhizoids. They lack chlorophyll and are saprophytic, possibly because of their association with an endophytic fungus (Fig. 19-11C) (Peterson et al. 1981). They have little internal differentiation. Their cylindrical form, dichotomous branching, and subterranean habitat render them difficult to distinguish from young rhizomes without microscopic examination. The occasional presence of tracheids, phloem (Hébant, 1976), and endodermis in the center of gametophytes, along with their cylindrical form and rhizoids, have been interpreted as an incipient or reduced stele and as

463

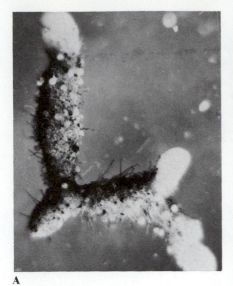

A

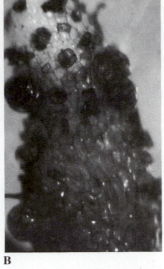

B

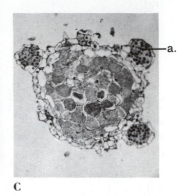

a.

C

FIGURE 19-11
Psilotum nudum. Gametophytes.
A. Forked specimen with rhizoids
and prominent antheridia. **B.**
Apex of gametophyte with
archegonia. **C.** Transection
showing mycorrhizal fungi and
antheridium (a.) **D.** Section of
antheridium. **A.** ×12. **B.** ×42.
C. ×110. **D.** ×450. (*Courtesy of
Dr. D. W. Bierhorst.*)

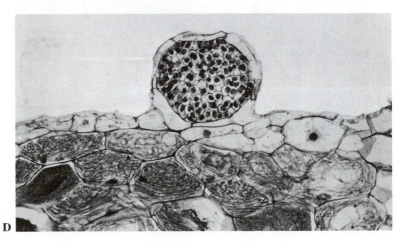

D

support for the homologous theory of alternation of generations, according to which sporophytes and gametophytes are merely different manifestations of a single ancestral plant body. It should be noted that the gametophytes in which vascular tissues have been observed have been shown to be diploid; apparently most greenhouse specimens of *Psilotum* are tetraploid.

The sex organs develop from surface cells of the monoecious gametophytes (Fig. 19-11). The antheridia are hemispherical and slightly protu-

berant, the single layer of jacket cells enclosing a small number of coiled, multiflagellate sperms (Fig. 19-11D). The archegonia (Fig. 19-11B) are partially sunken with the gametophyte and have necks that are much shorter than those in the Hepatophyta and Bryophyta. The necks are composed of only four rows of neck cells; their maturation involves opening of the neck and extrusion of the neck canal cells (Whittier and Peterson, 1980).

Reproduction: embryogeny. Following fertilization, the zygote undergoes transverse

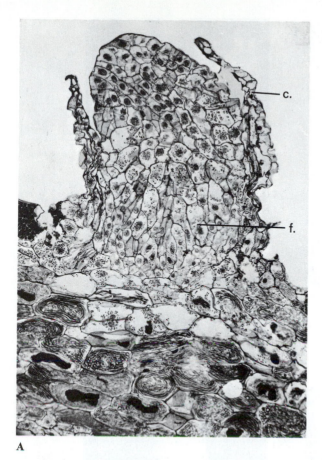

A

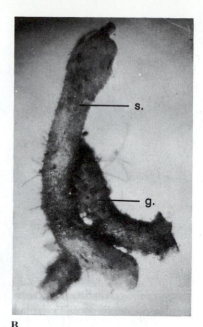

B

FIGURE 19-12
Psilotum nudum. Development of the embryonic sporophyte. **A.** Longisection of barely emergent sporophyte. **B.** Gross appearance of young sporophyte (s.) attached to forked gametophyte (g.); c., calyptra; f., foot. **A.** ×60. **B.** ×12. (*Courtesy of Dr. D. W. Bierhorst.*)

division. The outer product of the division gives rise to the embryonic stem; the derivatives of the inner one are organized as an enlarged foot. The primary stem is a branched rhizome, which develops rhizoids and becomes infected with a fungus as it emerges from the gametophyte (Fig. 19-12). Some of the branch tips become negatively geotropic and produce aerial axes when they are exposed at the soil surface. By this time the embryonic stem usually has separated from the foot, which remains within the gametophyte.

Tmesipteris

Tmesipteris (Fig. 19-13) is a pendulous epiphyte on the trunks of tropical tree ferns, or it may be terrestrial and erect. Both flattened and radially symmetrical species are known. The plant consists of branching, rhizoid-bearing rhizomes, which anchor it, and aerial branches, which may be unbranched or undergo one or several dichotomies, depending on the species. From base to apex, the scalelike, nonvascular appendages are succeeded by small flattened leaves, each of which is transversed by a single, central unbranched vein connected to the stele by a trace. The fact that the stem apex itself is modified as a leaf has suggested that the "leaves" of *Tmesipteris* may represent flattened branchlets.

The synangia of *Tmesipteris* also are subtended by a pair of bracts and are bilobed (Fig. 19-13B). The gametophytes are similar to those of *Psilotum.*

465

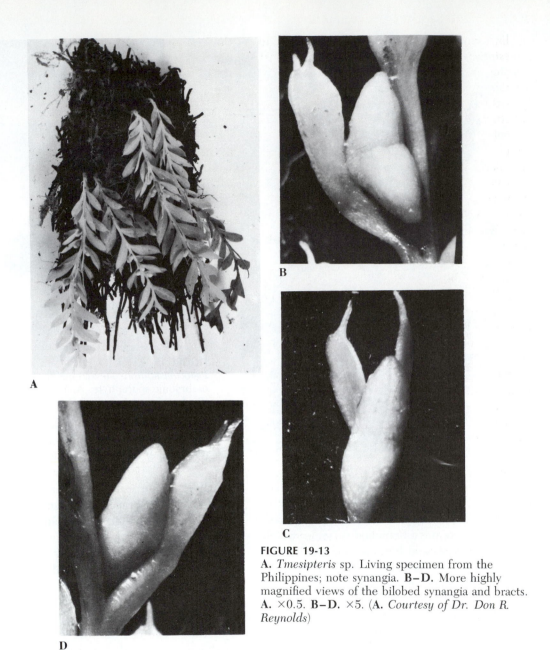

FIGURE 19-13
A. *Tmesipteris* sp. Living specimen from the Philippines; note synangia. **B–D.** More highly magnified views of the bilobed synangia and bracts. **A.** ×0.5. **B–D.** ×5. (**A.** *Courtesy of Dr. Don R. Reynolds*)

Discussion and summary

The more important features of *Psilotum* described above illustrate its anomalous position in comparison with other vascular cryptogams.

In its lack of roots and lack of vascularized leaves it is paralleled only by certain Devonian (see inside front cover) fossil genera such as *Rhynia* (see Chapter 20). An additional characteristic it shares with them is the lateral position of the

466

cauline synangia. In this respect *Psilotum* and *Tmesipteris* are unlike other living vascular cryptogams, with the possible exception of *Equisetum*.[3] The developing spores obtain metabolites through the disintegration of some of the tapetum tissue prior to sporogenesis. The branched cylindrical gametophytes, similar in some respects to the sporophyte, are unusual in the plant kingdom.[4] It has been suggested that it is possible that *Psilotum* represents a living remnant of members of the Devonian flora now extinct, and that it has survived to the present with few modifications from its progenitors. The fossil record sheds no light on this speculation. Those varieties that produce their synangia at the apices of main branches (Rouffa, 1978) suggest *Cooksonia* (p. 486) and, perhaps, *Rhynia* (p. 482), although the latter has sporangia, not synangia.

An alternative, quite different, interpretation of *Psilotum* and *Tmesipteris* was proposed by Bierhorst (1969, 1971, 1977 [in White et al.]) who has studied these plants in field and laboratory for many years. According to him, the proper affinities of *Psilotum* and *Tmesipteris* are with the ferns, and he interprets the shoots of both these plants as fern fronds (leaves). Those of *Tmesipteris* he considers to have laminate pinnae (leaflets), while those of *Psilotum* are composed of nonplanate axes with minute pinnae (the prophylls and scales of this text), which in *Psilotum* and *Tmesipteris* show various degrees of vascularization. Bierhorst, in arriving at this interpretation of *Psilotum* and *Tmesipteris*, was strongly influenced by his investigation of species of two genera of ferns, *Stromatopteris* (p. 429) and *Actinostachys* (p. 429), which, like

the Psilotophyta, have axial-cylindrical, subterranean, mycorrhizal, and saprophytic gametophytes. Furthermore, in *Stromatopteris* roots are absent in young plants and apparently not significantly functional in older ones, and the fronds (leaves) are not clearly differentiated as distinct from, and appendages of, the stems, a primitive condition. Bierhorst (1973, 1977) explained his conclusions comprehensively, but some botanists remain unconvinced (see White et al., 1977).

For example, Wagner (1977) cited more than ten characteristics in which he considers that *Psilotum* and *Tmesipteris* differ from *Stromatopteris*, and, by implication, from other ferns. These include differences in sporangial organization, position, spore number, and vernation of the so-called fronds of the psilotophytes and those of ferns. The authors of this book, like Wagner, are as yet not convinced of the close affinity of Psilotophytes with any other group of vascular cryptogams, and for the present they continue to group them in a separate division, Psilotophyta. For a more complete discussion of this topic, the reader is referred to the collection of papers by White et al. (1977).

Finally, with reference to the phylogenetic relationships of *Psilotum* and *Tmesipteris*, the reports of Cooper-Driver (1977) and of Wallace and Markham (1978*a,b,c*) may be cited. These investigators found that several species of *Psilotum* and *Tmesipteris* synthesized amentoflavones and related biflavonyls, which are absent from putatively primitive ferns. This is further evidence of lack of close relationship.

---

[3]The probable evolutionary derivation of megaphylls from stems (p. 473) weakens the distinction between cauline and foliar sporangia, except in the case of microphyllous plants.

[4]See, however, those of *Ophioglossum* (p. 394), and those of certain species of *Actinostachys* (p. 429), *Stromatopteris* (p. 429), and certain species of *Lycopodium* (p. 344).

DISCUSSION QUESTIONS

1. Define eusporangiate sporangium development.
2. The gametophytes of *Tmesipteris* has never been grown *to maturity*, from spores, in laboratory culture. How would you attempt to accomplish this?

3. What significance has been attached to the occasional occurrence of xylem, phloem, and endodermis in *Psilotum* gametophytes?

4. In what respects is *Psilotum* unusual among vascular plants?

5. Distinguish among rhizoids, rhizomes, and roots.

6. What structural adaptations related to photosynthesis occur in the stems of *Psilotum*?

7. What type of nutrition probably occurs in the *Psilotum* gametophyte? How could you prove it?

8. What does the fossil record indicate regarding the comparative age of algae and vascular plants?

9. What are mycorrhizal fungi and in what parts of the *Psilotum* plant do they occur?

10. Cite evidence for and against interpreting the *Psilotum* stele as a protostele.

11. What kinds of tissues are present in the cortex of the aerial stems of *Psilotum*? State the function of each.

12. What significance might one attach to the sporadic occurrence in *Psilotum* of vascular traces which end near the base of the prophylls?

13. Where does *Psilotum* grow natively in the United States?

Chapter 20
VASCULAR CRYPTOGAMS:
RECAPITULATION
AND FOSSIL RECORD

RECAPITULATION

Chapters 14–19 have presented a discussion of the structure and reproduction of living representatives of the divisions Microphyllophyta, Arthrophyta, Pteridophyta, and Psilotophyta. The present chapter presents a brief comparative summary of the more important features of these organisms, the **vascular cryptogams,** and discusses their phylogenetic implications.

THE LIFE CYCLE

In earlier schemes of classification (other than the one used in this book; see inside back cover), extant seedless vascular plants were grouped together in a single division, Pteridophyta. Although the plants so classified have been regrouped as four separate phyletic lines (divisions) in the present volume, it should be emphasized that they do have in common the same balance between alternating generations or phases in their life cycles. In almost all vascular cryptogams the sporophyte is the dominant and persistent (usually perennial) phase in the life cycle; the gametophyte is clearly simpler and more ephemeral (see, however, p. 417). In all cases, however, the latter is separated ultimately from the parent sporophyte, unlike the female gametophyte in the seed plants, in which it is permanently retained. In the Hepatophyta, Anthocerotophyta, and Bryophyta, on the other hand, the gametophyte is dominant. That the life cycle is not entirely inflexible and obligate is evidenced by numerous examples of apogamy and some of apospory. **Apogamy,** the development of a sporophyte from a gametophyte in which no fertilization takes place, occurs regularly in certain species of polypodiaceous ferns, and there is evidence of its presence in species of *Marsilea* and *Selaginella* as well. Furthermore, apogamy has been induced in normally sexual genera. **Apospory,** the development of a gametophyte directly from sporophytic tissue, and not from a spore, is less widespread. It, too, occurs in nature and may be induced. These phenomena have focused attention on the question of the origin and relation of the two alternating phases and have resulted in speculations regarding the type of life cycle in the first land plants or their immediate ancestors.

There seems to be rather general agreement that primitive plants were at least potentially sexual, gametophytic, and haploid. The initiation of the sexual process would obviously result in the existence of a diploid cell, the zygote. In organisms with zygotic meiosis, the diploid zygote is only a transitory phase, which is obliterated in the meiotic process. It seems clear that the inception of sexuality and the first occurrence of meiosis must have been closely related in time, if not simultaneous, in view of their complementary functions and invariable coexistence. An extensive diploid sporophyte generation appears to have been a secondary adjunct to the life cycle with sexuality. Its origin is postulated to have been occasioned by a delay in meiosis in the zygote and by the interpolation of a more or less extensive series of mitoses and cytokineses between fertilization and meiosis, with the result that diploid cells other than the zygote arose for the first time. Meiosis then occurred in all these diploid cells, however numerous. This, of course, represents the type of life cycle present in all organisms with sporic meiosis. Finally, gradual extension of the duration and importance of the diploid sporophytic phase, concomitant with the suppression of the haploid phase, is thought to have resulted in the type of life cycle in which gametic meiosis occurs. It should be emphasized that the preceding statements are mere conjectures, none of which is verifiable experimentally. The credence with which they frequently are received and repeated might well be tempered by this consideration and others, among them the following: The great majority of animals are diploid organisms with gametic meiosis. It is scarcely conceivable that their present life cycles have evolved from genera with zygotic and sporic meiosis, especially in the absence of any examples of the latter among living animals. Furthermore, that diploidy of zygotes, resulting

from failure of meiosis, might of itself have been insufficient to explain the origin of a sporophyte, especially of the type in heteromorphic alternation, is suggested by polyploidy and the phenomena of apogamy and apospory noted above.

In addition to these speculations concerning the origin of the alternate phases, there have been many conjectures regarding the fundamental relationship between them. Speculations about the relation of the sporophyte and gametophyte to each other often are summarized as two theories—namely, the antithetic and homologous theories, referred to briefly in Chapter 12. According to the **antithetic theory,** the gametophyte and sporophyte are essentially different manifestations of a single organism. The sporophyte is looked upon not as a modified or changed gametophyte but rather as a phase *sui generis*, which has been interpolated between successive gametophytic phases through delay in meiosis, as suggested above. Furthermore, the sporophyte is thought by some proponents of this theory to have been entirely a reproductive phase originally, in which asexual spores in large numbers were produced as a result of delayed meiosis. Assumption by sporophytes of such vegetative functions as photosynthesis and translocation, among others, is viewed as secondary. These are interpreted as acquired functions taken up by formerly sporogenous tissues. It should be emphasized that this view of the sporophyte was originally proposed by a student of land plants long before the complexity of algal life cycles was appreciated fully. The antithetic theory of the nature of alternation of generations, perhaps in somewhat modified form, is still in evidence at present in many morphological discussions. It is sometimes called the **interpolation theory.** Recent investigations of certain green algae, the probable precursors of land plants, seemingly are contributing strong evidence for the interpolation theory. There are features in such an alga as *Coleochaete* (p. 82) that suggest not only a possible origin for the land plant sporophyte,

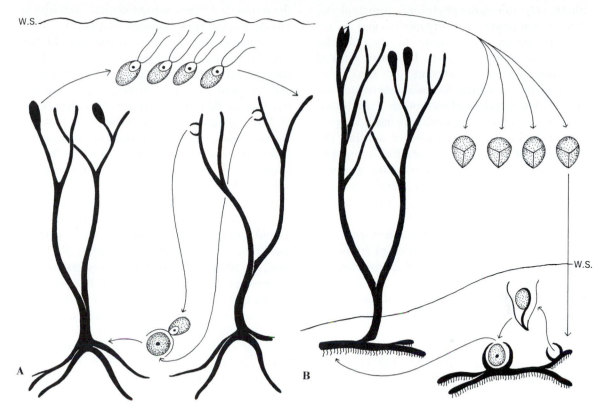

FIGURE 20-1

A. Hypothetical *aquatic* algal ancestor of the vascular plants. The two alternates are isomorphic; the one at the left is the zoospore-producing sporophyte, while the one at the right is the gametophyte, in this case oogamous. **B.** Primitive vascular plant derived from **A.** The sporophyte now has aerial branches and is producing airborne spores (instead of zoospores); these develop into minute, *aquatic*, oogamous gametophytes; alternation here is hetermorphic. w.s., surface of the water. (*After Stewart, from Zimmerman.*)

but also striking similarities in their motile cells (zoospores and gametes) (Graham, 1985).

In opposition to the views summarized above, it has been emphasized by some morphologists that alternating gametophytes and sporophytes are fundamentally similar. The sporophyte is looked upon as a somewhat modified or transformed gametophyte, modified in accordance with its function of spore production. According to this **homologous** or **transformation theory** of the nature of alternation of generations, vegetative and reproductive functions coexist in primitive sporophytes, although the balance between

them may be variable. Certainly algal sporophytes are photosynthetic. Conceding that the alternates among the land plants are, indeed, heteromorphic, possibly because of the stimulus of a terrestrial environment, proponents of this theory also point to the algae (Fig. 20-1), in which numerous cases of both isomorphic and heteromorphic alternation of generations are known in a uniform aquatic habitat. It has been suggested that one important factor effecting heteromorphism is the permanent retention of the sporophyte within or on the gametophyte, as in the diploid carposporophyte of the Rhodophyta and the sporophytes of Hepatophyta, Anthocerotophyta, and Bryophyta, and its retention during embryogeny in the vascular cryptogams and seed plants. However, marked heteromorphism is present in the kelps (*Laminaria*), in which the two generations are entirely independent of each other. This discussion of alternation of generations has been presented not with the purpose of settling the problem, but rather to indicate the complexity of the considerations involved. The apparent independence of alternation of generations from chromosome number has already been emphasized. It seems quite possible that different life cycle patterns may have arisen as a result of different circumstances and different stimuli in different organisms, and that a comprehensive explanation may be unattainable. Wahl (1965) has presented a comprehensive discussion of the phenomenon of alternation of generations.

MORPHOLOGY OF THE VEGETATIVE ORGANS OF THE SPOROPHYTE

Comparison of the several divisions in which the vascular cryptogams have been classified indicates that the stem is certainly the dominant organ of the sporophyte, except in the Pteridophyta, in which it may be equaled or surpassed in stature and anatomical complexity by the megaphyllous leaves. In the living Microphyllophyta, Arthrophyta, and Psilotophyta, examples of erect and ascending, vinelike, rhizomatous, and stolonlike stems occur. With the exception of the tree ferns, the stems of the Pteridophyta are mostly subterranean, vertical or horizontal, or at the surface of the ground. Dichotomous branching of the stems, by the equal division of single apical cells or groups of meristematic cells, is considered primitive as compared with monopodial and sympodial types. *Equisetum,* with its whorls of branches and leaves, the latter alternating in origin with

the former (i.e., they are not actually axillary), is unique among living vascular plants.

A considerable range in disposition of the vascular tissues is apparent in the members of the several divisions. Many primitive genera and the juvenile stages of most others usually have stems that contain protosteles. Examples of increasingly complex types of steles— siphonosteles, both amphiphloic and ectophloic, dictyosteles, eusteles, and atactosteles— have been described. Among the vascular cryptogams, leaf gaps are present only in the Pteridophyta. Thickening of the stem by addition of secondary tissues through the activity of a cambium is rare among living vascular cryptogams, and where it occurs (*Isoetes, Botrychium*), it does not result in any considerable increase in girth. The tracheid is the principal conducting element of the xylem of all these forms. In the stems of *Selaginella, Equisetum,* and *Pteridium,* and in the roots of *Marsilea,* however, perforation of walls of conducting cells has resulted in the formation of vessels.

It is clear that the structures called "leaves" are not necessarily morphologically equivalent in all vascular plants. It will be recalled that in contrast with the microphylls of such plants as *Lycopodium* and *Selaginella,* which are relatively small, determinate in growth, and traversed by unbranched veins that leave no leaf gaps in the steles (Fig. 13-7 A, C), megaphylls are typically larger, their development is more extended, if not indeterminate (*Lycopodium, Gleichenia*), their vascular system is richly branched, and their traces profoundly modify the stele from which they depart, leaving parenchymatous gaps[1] (Fig. 13-7 B, D). Furthermore, the vascular tissue is usually organized in the stalks (petiole, rachis) of megaphylls in a pattern different from that in the stems on which they are borne. Because both these structures are

---

[1]Megaphylls do not leave gaps in protosteles, although grooves may be present.

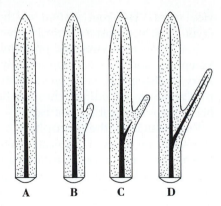

FIGURE 20-2
A–D. Evolutionary origin of a microphyll, according to the enation theory; the emergence at **C** corresponds to that occurring sometimes in *Psilotum;* that at **D** corresponds to leaves of *Lycopodium and Selaginella.* (*After Stewart.*)

called "leaves," are we, therefore, to consider them homologous? Morphologists have given different answers to this question. They have long speculated about the origin and evolution of the plant body of vascular plants. According to the telome theory (Stewart, 1964; Zimmerman, 1952, 1965), five fundamental processes occurred in the evolution of the plant body. These are illustrated in Figure 20-3 and summarized in its caption. Fusion or union of parts of axes is postulated to have occurred both in leaflike (Fig. 20-2C) and stemlike (Fig. 20-2F) portions of axes. The order of the illustrations does not necessarily indicate the order in which these phenomena occurred. Some of them, such as fusion, recurvation, and reduction, also are postulated to have affected the reproductive systems of plants (see Figs. 20-2, 20-3).

Organization of leaves has been considered

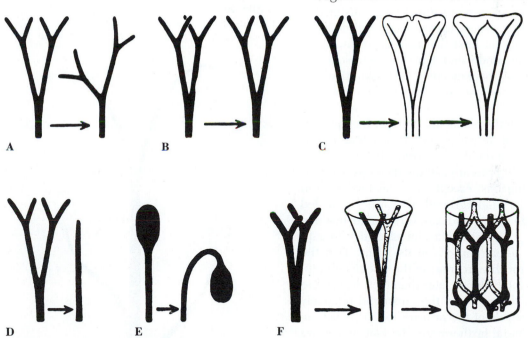

FIGURE 20-3
Five elementary evolutionary processes in the development of the body of vascular plants.
A. Overtopping. **B.** Planation. **C.** Coalescence and webbing in the leaf. **D.** Reduction.
E. Recurvation. **F.** Coalescence in the axis. (*After Zimmerman.*)

by morphologists as affected by these five phenomena. It is argued by some that microphylls are to be interpreted as localized, superficial enations of stems, which ultimately become vascularized (Fig. 20-2). Possible examples of this occur in *Psilotum* (see p. 439). Megaphylls, by contrast, are to be interpreted as branches of stems that have undergone planation by limitation of branching to one plane and webbing by extension of parenchymatous tissues between the branching axes (Fig. 20-3C). The evidence for this interpretation of the megaphyll is based on its extensive, much-branched, vascular supply; on the fact that its trace, like the branch trace of microphyllous plants, effects a break in the stele where it originates; and finally, on its relatively large size and indeterminate development, all supported by the study of the fossil record (see p. 514) and of such primitive living ferns as *Stromatopteris*, in which the distinction between stems and leaves is not well defined. Accordingly, at least the main branching veins even of an undivided fern leaf represent the direction of the branching of a primitive axis which has become webbed between the vein branches. These views regarding leaves are in accord with the telome theory, which emphasizes the shoot or axis as the fundamental unit in the organization of plants (Fig. 20-4).

Roots are present in all the divisions of vascular cryptogams except the Psilotophyta,[2] in which the underground rhizomes, provided with rhizoidal appendages, carry on the function of absorption with sufficient efficiency to support the aerial portions of the plant. That the root is probably an organ of the vascular cryptogams that developed later than their leaves and stems is suggested by the minor and relatively transitory role of the primary root or radicle in many genera; furthermore, the root sometimes arises in embryogeny later than the stem, never

vice versa. The root system of the vascular cryptogams is almost exclusively adventitious in origin. Roots are exarch and protostelic without exception, and they develop entirely from primary meristems. It has been suggested that the root arose as a modified rhizome branch that became covered with a root cap.

Reproduction: the sporophyte. Considerable variation exists in the location and nature of the sporogenous tissues in extant vascular cryptogams. Examples of little or no segregation of vegetative and sporogenous tissues are available in such instances as *Lycopodium lucidulum* and many Polypodiaceous and Marattiaceous ferns. Even in *Psilotum* and *Tmesipteris*, however, sporogenous tissue is restricted to special, determinate lateral branches. The dimorphism

FIGURE 20-4
Telomes and mesomes. f.t., fertile telome; m., mesome; v.t., vegetative or sterile telome.

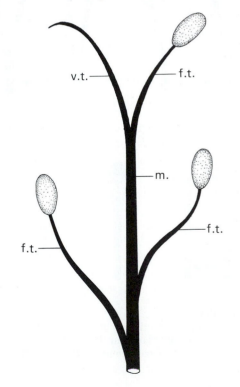

[2]And in mature plants of *Salvinia*, certain Hymenophyllaceae, and *Stromatoperis*.

observable in such plants as *Lycopodium complanatum*, species of *Osmunda*, the fertile spike of the *Ophioglossum-Botrychium* type, and the strobili of the Microphyllophyta and *Equisetum* is a manifestation of segregation of vegetative and reproductive functions at the organ level. Furthermore, examples of both cauline (*Psilotum*) and foliar (Microphyllophyta, Polypodiaceous ferns) sporangia are present.

With reference to their sporangia and method of ontogeny, two types, albeit with intermediates, may be distinguished among the vascular cryptogams. These have usually been designated eusporangiate and leptosporangiate, but Bierhorst (1971*a*, pp. 266–267) argues that the terms ought to be abandoned, because the distinguishing characteristics are not reliable in all cases. This apparent distinction still serves a useful purpose, however.

There is good evidence that heterospory developed independently in the Microphyllophyta, the Arthrophyta (fossil forms), and the Pteridophyta. Spore ontogeny in the heterosporous genera indicates that the *immediate* cause of heterospory lies in differences in spore number and nutrition and that these are expressed ultimately by spore size, but certainly more subtle genetic mechanisms are operating.

Reproduction: gametophyte and embryo. Considerable morphological and physiological variation is apparent among the gametophytes of the vascular cryptogams. Cylindrical form, radial symmetry, and saprophytism characterize the gametophytes of *Psilotum*, of certain species of *Lycopodium*, *Ophioglossum*, *Stromatopteris*, and *Schizaea*. Dorsiventral and photosynthetic gametophytes are present in *Equisetum* and in Marattiaceous and homosporous leptosporangiate ferns. The presence of heterospory is associated with marked reduction in the morphological complexity and duration of the gametophyte, as is evident in *Selaginella*, *Isoetes*, and *Marsilea*. Furthermore, unisexual gametophytes inevitably accompany heterospory. The nutrition in these gametophytes is based largely on metabolites stored by the parent sporophyte in the spores. Although the gametophytes of *Psilotum* and species of *Lycopodium*, *Ophioglossum*, and *Botrychium* are relatively long-lived as compared with those of other genera, the gametophyte generation, in general, is of lesser duration and complexity than are the persistent, perennial sporophytes of the vascular cryptogams.

The archegonia of the female gametophytes produced in heterosporous ferns are in general smaller and less complex than those in homosporous types. Both biflagellate (*Lycopodium*, *Selaginella*) and multiflagellate sperms (*Isoetes*, *Equisetum*, *Pteridophyta*) occur among vascular cryptogams.

The embryogeny of the vascular cryptogams exemplifies both slow (*Lycopodium*, *Psilotum*, *Ophioglossum*) and relatively rapidly growing (*Selaginella*, *Equisetum*, Polypodiaceous ferns) types. Nutrition of the embryo in heterosporous forms is based on metabolites of the parent sporophyte stored in the megaspores.

PHYLOGENETIC CONSIDERATIONS

The preceding paragraphs and Chapters 13–19 have reviewed the ontogeny of the vegetative organs of the vascular plant—its stem, root, and leaf—and the reproductive features of representatives of the several groups of vascular cryptogams. At this point it is appropriate to turn to a consideration of the phylogenetic origin of vascular plants and that of their component organs.

The fossil record indicates clearly that life was aquatic in origin, and that it was largely restricted to an aquatic habitat for about 2.5 billion years. At the end of this period, colonization of the land began. The emigrants were with great probability algae or modified algal descendants. There have been many suggestions regarding the course of evolution of the land flora from its algal progenitors. These have largely been incorporated in a synthesis known as the telome theory by the eminent morphologist Walter Zimmerman (1952, 1965) and discussed by Stewart (1964). The telome theory attempts to interpret extant plants in terms of their origin from extinct precursors. The major components of this concept will now be summarized.

According to the telome concept, the vascular plants evolved from algal precursors (Fig. 20-3). The ultimate branchlets of the dichotomies are called **telomes,** and the connecting axes between dichotomies are called **mesomes.** Telomes may be sterile or fertile (Fig. 20-4).

The Rhyniophyta (p. 482) are, perhaps, closest to the hypothetical ancestor. In these extinct plants, as exemplified by *Rhynia* (Fig. 20-12), there are no leaves or roots, the plant body consisting of a three-dimensional system of mesomes and telomes. The vascular tissue is in the form of a simple protostele. With the land habit, there developed cuticle and stomata on the aerial mesomes and telomes, and rhizoids in the subterranean ones. The gametophytes of primitive land plants generally were not preserved in the fossil record, perhaps because of their delicate nature, or in their algal precursor.

The primitive type of sporophyte just described, which was present also in the Devonian Zosterophyllophyta and Trimerophytophyta, evolved into such diverse plant bodies as those of Microphyllophyta, Arthrophyta and Pteridophyta. The Microphyllophytan, Arthrophytan, and Pteridophytan types of plant body[3] are postulated by the telome concept to have involved the fundamental processes illustrated in Fig. 20-2, among them the following: (1) overtopping, (2) planation, (3) webbing, (4) recurvation, and (5) reduction. These processes occurred independently and in differing order and number in the evolution of vascular plants.

Thus, in the evolution of megaphylls from three-dimensional branches, overtopping and

---

[3]Also those of gymnosperms and angiosperms (Chapters 21–27.)

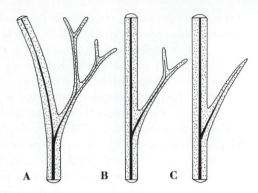

FIGURE 20-5
A–C. Hypothetical origin of a microphyll by reduction; compare with Fig. 20-2. (*After Stewart, from Bower and Zimmerman.*)

planation probably occurred, and in that order. On the other hand, reduction, according to Zimmerman, may account for the origin of microphylls (Fig. 20-5).

Some of these same processes operated in the evolution of the reproductive organs of vascular plants according to the telome concept. The cauline sporangium of the Psilotophyta and the axillary or foliar sporangium of the Microphyllophyta, according to this concept, arose as a result of reduction (Figs. 20-6, 20-7). The arthrophytan sporangiophore, it is proposed, also arose through these modifying forces (Fig. 20-8). In the Pteridophyta, planation and webbing may well have developed the type of sporangial arrangement (Fig. 20-9) seen in *Cladoxylon scoparium* (Fig. 20-61).

The telome concept is a unifying synthesis that seeks to interpret the diversities of organization of extant plants as modifications from a primitive stock of Devonian plants. With respect to its postulates regarding the origin of microphylls by reduction, it is in marked contradiction to the enation theory (p. 474) and the evidence on which the latter is based.

Perhaps, if microphylls and megaphylls are fundamentally similar in organization, differing only in number of component telomes, then we ought not to contrast them as strongly as we do.

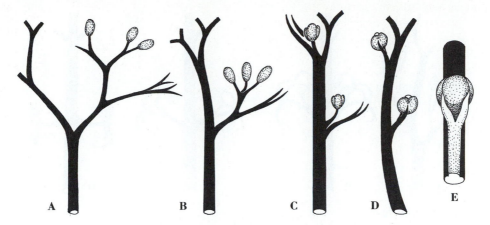

FIGURE 20-6
A–E. Proposed evolutionary steps leading to the development of the *Psilotum* type of synangium. (*After Stewart, modified from Emberger.*)

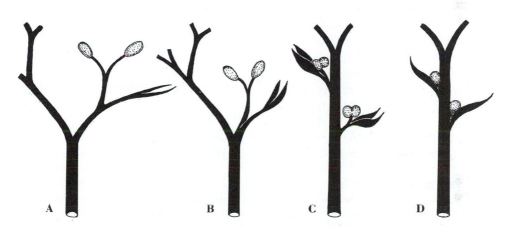

FIGURE 20-7
A–D. Proposed evolutionary steps in modification of a primitive axis (composed of fertile and sterile telomes) into the *Tmesipteris* (**C**) and *Lycopodium* (**D**) type of sporangial position. (*After Stewart, from Zimmerman.*)

The same argument has been applied to sporangial position, which, in the last analysis, is really always at the tip of a telome. Similarly, the telome theory disagrees with the interpretation of the evolution of stelar types accepted by most morphologists. In spite of these difficulties, the telome concept certainly is and has been a stimulating and thought-provoking synthesis that has played and continues to play an important role in plant morphology.[4]

In concluding this brief summary of the extant vascular cryptogams, it seems clear that

_____

[4]For more comprehensive treatments of the telome theory, see Zimmerman (1952, 1965), Wilson (1953), and Stewart (1964).

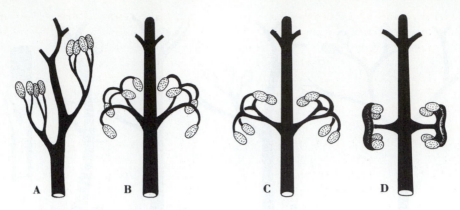

FIGURE 20-8

A–D. Proposed evolutionary steps in development of a primitive axis (**A**), composed of fertile and sterile telomes, into several types of arthrophytan sporangiophores. (*After Stewart, from Zimmerman.*)

comparison of representatives of the divisions Microphyllophyta, Arthrophyta, Pteridophyta, and Psilotophyta offers little evidence for close relationship of these plants. The fossil record indicates that these four phyletic series had been established by the Devonian (inside front cover). Psilotophyta-like organisms may perhaps have been the precursors from which these four lines developed, but, as noted earlier, 400 million years separates these Devonian plants from the extant Psilotophyta. Their diversities

seem clearly to overshadow in significance the combination of attributes they share: vascular tissue, lack of seeds, and similarity in balance of the alternating generations.

THE FOSSIL RECORD

Although comparative morphology of living plants is an indispensable tool in attempts to determine morphological relationships and pos-

FIGURE 20-9

A–C. Proposed course of evolutionary development of a pteridophytan pinnule with marginal sporangia from a primitive axis composed of fertile and sterile telomes. (*After Stewart, from Zimmerman.*)

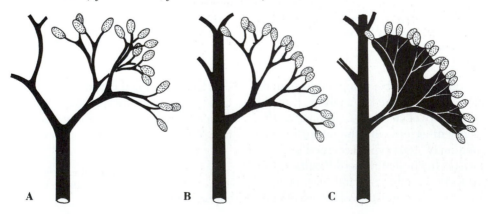

sible evolutionary pathways, the only real evidence concerning the course of change in plant form through time is preserved in the fossil record. This record is, at best, extremely fragmentary; nevertheless we have learned a great deal about the history of the plant kingdom through a study of plant remains preserved in rocks.

A person who studies living plants primarily may be disappointed with the nature of the fossil record because so much is missing. Furthermore, it is seldom that an entire plant is preserved. The conditions under which plants are fossilized frequently involve factors that tend to break the plant into pieces, which are then preserved. Thus, leaves may be separated from stems that are nowhere near reproductive parts that may have been borne on the plants. Also, only those plants that are near appropriate conditions that are conducive to preservation are represented in the fossil record. Undoubtedly countless plants that grew some distance from ideal preservational situations have not entered into the fossil picture.

Plants are preserved in a variety of ways in the rocks. Probably the most frequently occurring type of fossil preservation is one that results from plant parts that are carried into basins (e.g., lagoons, abandoned stream channels in a delta) into which suspended sedimentary particles are also being transported. Accumulation of considerable sediment (e.g., during storms and floods) eventually produced sufficient pressure to squeeze the plant parts until they were nothing but thin films of carbon and other chemicals that are degradation products of the original plant material. This type of fossil is called a **compression** and is valuable in showing external features. At times, however, the cuticle that covers aerial parts of most land plants may resist distortion or decay and may survive the fossilization process intact. This cuticle faithfully reproduces the surface features of a plant organ —such as a leaf, for example—and when removed and mounted for microscopic examination, the cuticle may show quite dramatically the epidermal features of the fossilized plant part (Fig. 20-10).

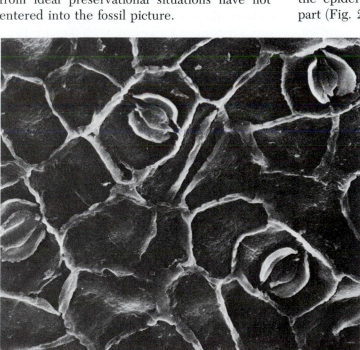

FIGURE 20-10
Cuticle of lower epidermis of Eocene dicotyledonous angiosperm leaf. The view is from the lower surface. ×250. (*Courtesy of Dr. C. P. Daghlian.*)

At times, plant parts that were compressed and preserved as fossils undergo further modification in that the organic material left behind does not hold together and is lost. There is left, then, a negative imprint on the surface of the rock matrix next to where the plant part was. Even when organic material is present, splitting open the rock often leaves one surface devoid of organic remains with an imprint of the flattened plant part. Such a fossil is called an **impression.** Like a compression, an impression shows only surface features. In some instances, the rock matrix is composed of such fine grains (clay, for example) that epidermal features are preserved faithfully as imprints in the rock, and these features may be transferred to a sheet of latex or similar material and then examined with a scanning electron microscope. Even though there is nothing left of the actual plant part, many details of the epidermis may be observed.

When the plant part carried into the basin of accumulating sediment happens to be relatively three-dimensional, in many instances the surrounding sediment is compressed and hardened before the plant fragment was compressed. Subsequently there may have been a decay of the plant part, with a hollow negative resulting. This negative three-dimensional structure in the rock is called a **mold.** Subsequently the mold may have been filled with other sediment that conformed to the contours of the mold and produced a three-dimensional, positive replica of the original plant part. This kind of fossil is referred to as a **cast.** Again, only surface details are evident, and there generally is nothing of the original plant part preserved. (Occasionally a thin film of carbon may surround the cast.)

A very interesting type of process of fossilization permits the observation of internal structure in plant parts. Such a fossil was formed when a plant part settled in a basin of water with a high concentration of dissolved mineral material or, when covered with sediment, was still surrounded by dissolved minerals. After a certain period of exposure to these conditions, the plant part became infiltrated with the liquid with the high charge of mineral material. Subsequently the mineral salts precipitated (no one yet fully understands why) and the plant part became entombed in a solid mass of solidified mineral matter. Most often cellular contents are not preserved, with the mineral having infiltrated into cell lumens and intercellular spaces, leaving the cell wall structure more or less intact. In a few instances, however, nuclei and cytoplasm survive the preservational process. This permineralized fossil is called a **petrifaction.**

Minerals involved in petrifaction include silicon dioxide, calcium, magnesium, or iron carbonate, ferric sulfide, and occasionally others. The most common method of investigation of plant fragments preserved in this manner involves making a thin section of the rock matrix and its contained plant material. This involves cutting a piece of the petrified material with a special saw that incorporates diamond particles in the blade. The piece is then polished on one surface, which is then cemented to a glass slide with a transparent adhesive. With the slide then firmly clamped in the saw, the specimen is cut once again as thin as is reasonably possible. Then while the petrified material is still cemented to the slide, it is ground thinner and thinner with appropriate abrasives until it eventually becomes transparent enough to permit transmission of light. Addition of mounting medium and a cover glass then makes it possible to examine the thin section with a light microscope. This technique is essentially that of geologists who study thin sections of various kinds of rock.

When the embedding matrix is a carbonate (usually calcium carbonate), a simple but ingenious technique may be employed to prepare sections. The cut surface of the petrified material is polished and then etched in dilute hydrochloric acid. This process removes some of the surface carbonate but leaves the cell wall material unaffected. After a short period of etching,

FIGURE 20-11
Cellulose acetate peel preparation of a
fossil fern stem (*Psaronius*). ×1.25.

the cell walls project above the surface of the
rock. The acid is then washed off with water and
the surface allowed to dry. The surface is then
flooded with acetone and a thin sheet of clear
cellulose acetate is laid on the surface, with care
taken to avoid air bubbles. Acetone partially
dissolves the lower surface of the acetate sheet,
and it now envelops the cell walls projecting
from the surface. Because acetone is quite volatile, it evaporates in a short time, once again
rendering the lower surface of the cellulose
acetate into solid form. The cell walls now are
partially embedded in the sheet of cellulose
acetate. After it is completely dry, the sheet
may be pulled away gently, and it then carries
with it a very thin section of the petrified
material embedded in it (Fig. 20-11). This type
of preparation is called a "**peel**," and it permits
examination of the sectioned plant material with
a microscope. If the petrifying mineral is something other than calcium carbonate, other kinds
of acids may be employed for the etching process.

Occasionally, in the past, plant parts may
have been carried into basins of accumulating
sediment and buried by extremely fine-grained
material. Air was excluded and conditions were
unsatisfactory for growth of decaying organisms.
As a result, plant parts may have survived for
millions of years in a relatively unaltered state.
It is possible to section wood or other plant
fragments preserved in such a manner with
equipment used for sectioning tissues of
present-day plants. Plant parts preserved in
peat bogs are also relatively unaltered. Pieces
may be fragmentary or slightly compressed, but
internal tissues are intact and may be sectioned
for microscopic examination.

The above are simply the most frequent types
of preservation of fossil plants. Less common
kinds are sometimes encountered, as well as
combinations of the types cited above.

Despite the vagaries of preservation of plants
of the past and the incompleteness of the fossil
record, a considerable body of knowledge concerning the earliest vascular plants and their
subsequent evolution has been gathered by
paleobotanists. The first indisputable vascular
plants were on land at the end of the Silurian
period (some 405–410 million years ago; see
inside front cover). There are some reports
(e.g., Gray and Boucot, 1971; Mortimer and

481

Chaloner, 1972; Gray, Laufeld, and Boucot, 1974) of the possible earlier existence of vascular plants in the form of resistant spores or peculiar tubelike structures that resemble xylem cells. However, it is necessary to demonstrate more than cutinized spores to prove that they belonged to a vascular plant. Such a resistant spore simply signifies that it was probably exposed to the air during some time in the life cycle of the plant. Furthermore, the tubelike structures that have been compared to tracheids are quite different and are not unquestionably tissues of vascular plants. Admittedly, the first vascular plants may have settled on land before the end of the Silurian, but convincing evidence for that occurrence is still absent.

After the first appearance of vascular plants, there was a rapid diversification, and during the following period, the Devonian (about 345–405 million years ago; see inside front cover), all the major groups of vascular cryptogams were present in the earth's flora. This section will present a brief summary[5] of representative vascular cryptogams. Our knowledge of these ancient plants has been greatly augmented during the past three decades. Many of the early vascular cryptogams belonged to divisions of the plant kingdom that are now extinct. Members of the divisions Rhyniophyta, Zosterophyllophyta, Trimerophytophyta, Microphyllophyta, Arthrophyta, and Pteridophyta will be discussed in the following presentation. Members of all these divisions were present during the Devonian period, while one genus (*Cooksonia*) of the Rhyniophyta, discussed below, occurred in the Silurian.

---

[5]For more comprehensive treatments of fossil plants, consult works of Andrews (1961), Arnold (1968), Banks (1970*b*), Banks et al. (1970), Bierhorst (1971*a*), Boureau (1964–1975), Gensel and Andrews (1984), Stewart (1983), and Taylor (1981), among others cited in the Bibliography.

DIVISION RHYNIOPHYTA

Members of the Rhyniophyta are among the simplest known vascular plants. Their plant bodies consisted of simple, slender, leafless, dichotomizing, photosynthetic axes. Roots were absent, with the anchoring and absorbing functions assumed by the basal portion of the stem system. Sporangia were borne at the tips of some of the ultimate branches. All known members of the Rhyniophyta were homosporous. Rhyniophytes were typically protostelic, with

FIGURE 20-12
Rhynia gwynne-vaughanii. Reconstruction; note vegetative and fertile axes, the latter with terminal sporangia. (*After D. S. Edwards.*)

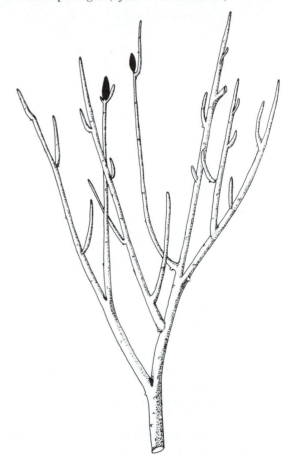

the smallest xylem elements in the center of the stele (centrarch); secondary vascular tissue was absent.

The best preserved members of the Rhyniophyta originate from a chert deposit near the town of Rhynie, in Aberdeenshire, Scotland. This chert, considered to be Lower Devonian in age, represents fossilized remains of a bog that seems to have been preserved with the plants in their growing positions. One genus, *Rhynia* (named for the town of Rhynie), is extremely well known and serves as a typical example of the general habit of plants of the division Rhyniophyta (Fig. 20-12).

Two species of *Rhynia* are known from the Rhynie chert beds: *R. gwynne-vaughanii* and *R. major. Rhynia gwynne-vaughanii* is smaller and more slender, reaching a height of 17 cm, while the larger, more robust *R. major* was about three times as tall. Each plant had a horizontal branching stem from which originated erect, leafless, dichotomously branched aerial axes. *Rhynia gwynne-vaughanii* also bore short, apparently deciduous, lateral branches (Fig. 20-12). These rootless plants had tufts of slender, unicellular rhizoids (Fig. 20-15), each of which was actually an extension of a single epidermal cell. It is assumed that the photosynthetic process took place in the aerial stem system and that these stems were green. There is no evidence of this from the fossil, of course, but the fact that well-preserved stomata occurred in the epidermis provides evidence along those lines.

Internal structure is preserved with amazing fidelity. The vascular system of a *Rhynia* axis is exceedingly simple, consisting principally, in *R. gwynne-vaughanii*, of a slender central strand of just a few tracheids in thickness (Fig. 20-13A). These tracheids are characterized by annular

FIGURE 20-13
Rhynia sp. **A.** Transverse sections of axes of *R. gwynne-vaughanii*. Note small xylem strand. **B.** Transverse section of the centrarch stele of *R. major*. **A.** ×24. **B.** ×150.

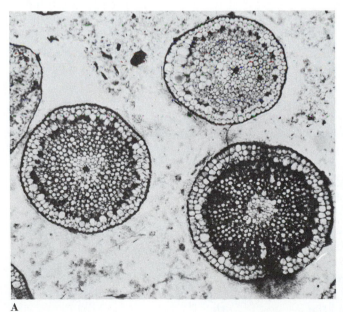

A

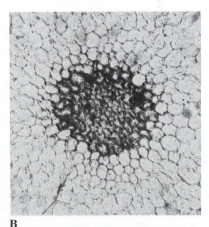

B

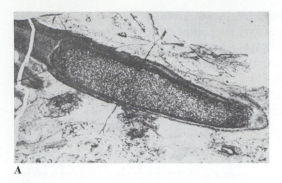

A

FIGURE 20-14
Rhynia. **A.** Longitudinal section of terminal
sporangium of *R. major* filled with spores.
B. *Rhynia* sp. Spore tetrads. **A.** ×4.5. **B.** ×250.
(**A.** *After Kidston and Lang.*)

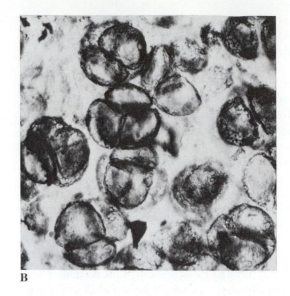

B

secondary thickenings. In the larger *R. major*,
the centrarch central strand of the xylem is
evident (Fig. 20-13B). Satterthwait and Schopf
(1972) described what they considered to be a
phloem zone surrounding the xylem core, but
their conclusions have not found acceptance
among all paleobotanists.

At the ends of some of the axes were borne
sporangia that were actually swollen tips of the
stem system (Fig. 20-14A). The cortex of the
vegetative axis continued into the thick wall of
the sporangium. That these plants were sporophytes is attested to by the occurrence of spores
within the sporangia that, in some instances,
were fossilized in the tetrad stage (Fig. 20-14B).
Hence they were obviously products of meiosis.
Spores were produced in tetrahedral tetrads,
and each shows a prominent trilete mark on the
proximal face, with the flat faces in the angles of
the trilete structure representing planes of contact with adjacent spores in the tetrad.

While sporangium-bearing axes of *Rhynia* are
undoubtedly sporophytes, there is no general
agreement about the nature of the gametophyte
phase. It would be of considerable interest to
know the structure of the gametophyte of these
early vascular plants in order to understand the

nature of the plants that were ancestral to the
first land plants. Pant (1962) and Lemoigne
(1970) feel that certain of the vascularized axes
of *Rhynia* that had been interpreted to be
rhizomes of the sporophyte may actually be
vascularized gametophytes. Swellings on the
surface of the axes with what appears to be a
canal leading to the interior of the axis have
been called archegonia (Fig. 20-15). If they
were, indeed, archegonia, the necks would have
been much more massive than any others

FIGURE 20-15
Rhynia gwynne-vaughanii. Schematic
representation of the "gametophyte."
an., antheridium; ar., archegonium; ex., stele;
me., endodermis; rh., rhizoids; st., stomata.
(*After Lemoigne.*)

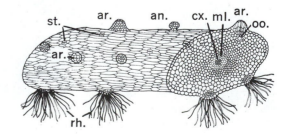

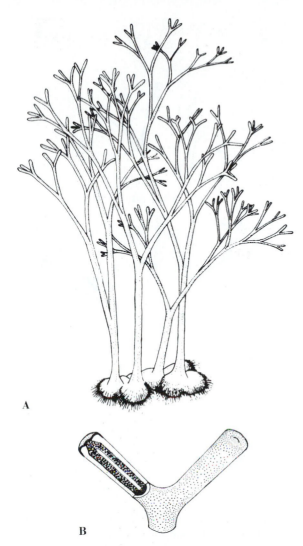

FIGURE 20-16
Horneophyton lignieri. **A.** Habit of plant.
B. Details of forked sporangium. (*After Eggert.*)

known among vascular plants. There is not universal agreement that these structures were archegonia, and Bierhorst (1971*a*) suggests that these "archegonia" may have been hydathodes or some other kind of secretory structures. Similarly, spherical bumps on the surface of certain axes were described as antheridia by

Lemoigne, but not all paleobotanists are prepared to accept that interpretation. If these axes are actually gametophytes, they would have much in common with those of *Psilotum* and *Tmesipteris* as well as those of *Actinostachys* (p. 429).

A second genus, *Horneophyton* (named for a Dr. Horne), resembles *Rhynia* to some extent, but there are some significant differences. In *Horneophyton*, the basal part of the plant consisted of a number of swollen, bulblike structures from which the upright, dichotomizing, aerial shoots arose (Fig. 20-16A). Hairlike rhizoids attached to the bottoms of the swollen portions served as absorbing structures. Terminating some of the aerial shoots were elongated, sometimes branched sporangia that were truncated at the tips (Eggert, 1974) (Fig. 20-16B). Stomata were observed in the sporangium wall (El-Saadawy and Lacey, 1979). A slender columella extended through the middle of each sporangium. The trilete spores were of one type.

Cooksonia (named for Isabel C. Cookson, an Australian paleobotanist) is known from less well-preserved material than that in which the Rhynie chert plants occur, but enough is known to place it in the Rhyniophyta. It is the oldest genus of *bona fide* vascular plants, with two known species from Upper Silurian and Lower Devonian strata. Although the material is fragmentary, the preserved pieces of *C. caledonica* (Edwards, 1970) may be as long as 6.5 cm. These slender, dichotomizing axes were naked, and, like *Rhynia*, bore terminal sporangia that were not so elongated as those in the latter genus (Fig. 20-17A). *Cooksonia hemisphaerica* is illustrated by Ananiev and Stepanov (1969) as having a principal axis with lesser lateral, dichotomously branched fertile branches (Fig. 20-17B).

An example that plants do not always "fit" into categories erected by human beings is the Lower Devonian *Renalia*. It resembles *Cookso-*

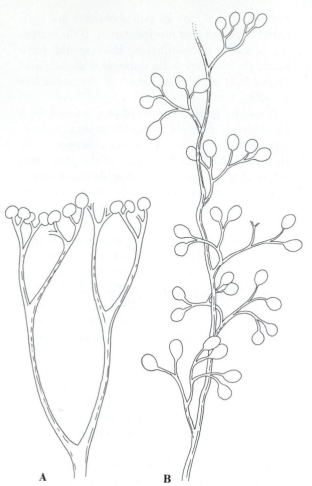

FIGURE 20-18
Renalia hueberi. Habit of portion of plant. (*After Gensel.*)

FIGURE 20-17
A. *Cooksonia caledonica.* **B.** *Cooksonia hemisphaerica.* **A, B.** ×1. (**A.** *After Edwards.* **B.** *After Ananiev and Stepanov.*)

nia in having a dichotomously branched plant body and swollen terminal sporangia (Fig. 20-18). The mode of dehiscence of the sporangium, however, more closely resembles that in the division Zosterophyllophyta (see below) in that splitting occurs along the distal margin, with the line of dehiscence bordered by a row of thick-walled cells. *Renalia* was homosporous.

For more than half a century such genera as

Cooksonia, Rhynia, and *Horneophyton* were considered to be similar, morphologically, to the living genus *Psilotum.* All these genera appear to have axes without vascularized leaves and roots, have rhizoid-bearing rhizomes (actually not demonstrated in *Cooksonia*), and have sporangia terminating branches. These similarities have caused many botanists to consider *Psilotum* as a persistently primitive type of plant that is, in essence, a living rhyniophyte. More likely, however, is the adoption by *Psilotum* of a habit that superficially approximates that of the rhyniophytes.

In summary, it may be stated that members

of the Rhyniophyta exhibit an extremely simple construction with a plant body consisting only of naked, slender, dichotomizing axes. Appendages in the form of leaves and roots were absent. Sporangia were merely swollen tips of branch axes, were thick-walled, and contained spores of only one type. Internal structure, too, is simple, with only a slender solid vascular bundle traversing the axis throughout its length. Secondary thickenings of the xylem elements were annular, the type found on first-formed protoxylem elements in more specialized kinds of vascular plants.

Cooksonia, Rhynia, and *Horneophyton* are classified as follows:

Division Rhyniophyta
  Order 1. Rhyniales
    Family 1. Rhyniaceae
      Genera: *Rhynia, Horneophyton*
    Family 2. Cooksoniaceae
      Genus: *Cooksonia*

DIVISION ZOSTEROPHYLLOPHYTA

Before the work of Banks (1968), in which he reclassified the earliest vascular plants, members of the Zosterophyllophyta were included in the same group of plants as that containing the Rhyniophyta. Contemporaneous in the Lower and Middle Devonian with the Rhyniophyta, the Zosterophyllophyta differed from the former in that their sporangia were borne along the sides of the dichotomously branching stem system rather than terminally. These sporangia, often reniform, opened transversely. Another difference between the two divisions is that in the zosterophyllophytes the xylem of the protostele is exarch. Otherwise there is much similarity, with the body habit of both divisions involving a leafless and rootless stem system that was photosynthetic and had dichotomous or near-dichotomous branching. Some of the members had spinelike projections on the surface of the

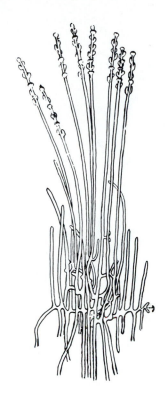

FIGURE 20-19
Zosterophyllum myretonianum. ×¾. (After Walton.)

aerial stem systems, but they were not vascularized, nor were sporangia associated with them in any consistent pattern.

Zosterophyllum (from *Zostera,* eelgrass, a marine angiosperm, + Gr. *phyllon,* leaf) from which the name of the division is derived, was a Lower Devonian genus that had what appeared to have been a prostrate stem system from which arose aerial axes. The basal part demonstrates an unusual kind of branching pattern (Fig. 20-19) in that from one axis a lateral branches off and from that lateral there are frequently two branch axes, one growing upward and one downward so that there is the appearance of "H-branches." It is conceivable that this so-called "H-branching" may be a preservational phenomenon, however.

487

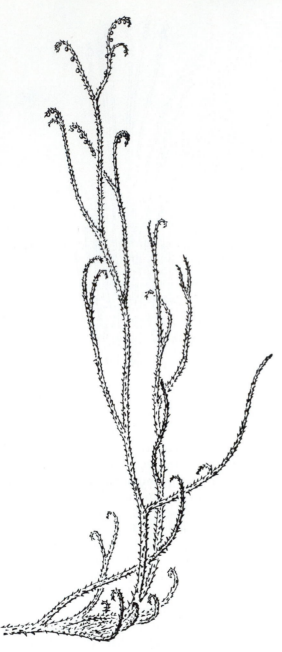

FIGURE 20-20
Sawdonia ornata. ×¾. *(After Andrews and Kasper.)*

Reniform sporangia were aggregated at the tips of some of the aerial shoots and were borne along the surfaces of these axes (Fig. 20-19). These plants were homosporous, and one species is known to have had a cuticle and stomata, thus indicating a terrestrial habit.

Sawdonia ornata is an example of a Devonian zosterophyllophyte that had spiny projections on the stem surface (Fig. 20-20). The dichotomously branched stem bore rounded sporangia along the surface, interspersed among the spines. An exarch protostele extends through the stem.

Kaulangiophyton, described by Gensel, Kasper, and Andrews (1969), occurs in Lower Devonian rocks in northern Maine. It had a robust, prostrate stem system that was supposed to have exhibited "H-branching," from which originated erect axes, 5–9 mm wide. These erect branches bore stout spines with no discernible phyllotaxy and large, ovoid sporangia that seem to have been concentrated in fertile zones (Fig. 20-21). Except for the fact that sporangia had no relationship to leaves, the plant must have resembled quite closely certain creeping species of *Lycopodium.*

One more zosterophyllophyte genus, *Crenaticaulis,* may be mentioned here because of certain peculiarities in its structure. These pseudomonopodially and dichotomously branching plants bore toothlike projections, but only on opposite sides of the axes. Swollen tubercles occurred at the region of branching of the axes, and these tubercles show evidence of having borne branches. Banks and Davis (1969) remarked about their close resemblance to rhizophore bases such as those in *Selaginella.* Within the axes were terete, exarch protosteles with chiefly scalariform tracheids. Reniform sporangia were grouped on some of the branches in an opposite or subopposite arrangement (Fig. 20-22). Each had a transverse line of dehiscence that opened to produce two unequal portions of the sporangium.

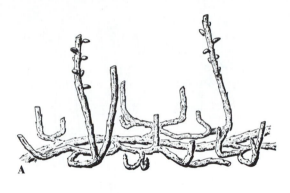

FIGURE 20-21
Kaulangiophyton akantha. **A.** Restoration showing creeping axes with erect branches, two of them fertile. **B.** Restoration of part of fertile axis. **A.** ×½. **B.** ×3. (*After Gensel, Kasper, and Andrews.*)

Serrulacaulis furcatus is a zosterophyllophyte with an unusual habit. On dichotomizing axes were borne sawtoothlike projections, but only in two rows (Fig. 20-23). As in other members of the division, sporangia were situated laterally along the surface of the axes.

The Zosterophyllophyta represent an extremely primitive group of vascular plants that

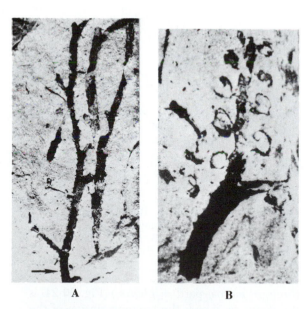

FIGURE 20-22
Crenaticaulis verruculosus. **A.** Branching axes. **B.** Axis with paired sporangia. **A.** ×¾. **B.** ×2.3. (*After Banks and Davis.*)

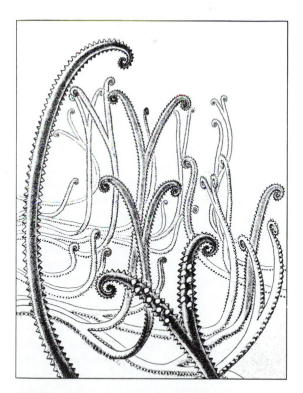

FIGURE 20-23
Reconstruction of *Serrulicaulis furcatus.* (*After Hueber and Banks.*)

coexisted with the Rhyniophyta, also primitive, but each group probably exhibited evolutionary potential in different directions. Most likely the Zosterophyllophyta were the ancestral source of the Microphyllophyta. The former often showed spinelike projections resembling minute leaves, and sporangia were borne laterally. A modification of these spines would result in structures closely resembling leaves of certain *Lycopodium* species. In addition, through time, the sporangia could have come to be intimately associated with the small leaves, as in the microphyllophytes. Anatomically, there is much similarity between the two groups.

The genera discussed within the Zosterophyllophyta may be classified as follows:

Division Zosterophyllophyta
    Order 1. Zosterophyllales
        Family 1. Zosterophyllaceae
            Genera: *Zosterophyllum, Sawdonia, Kaulangiophyton, Crenaticaulis, Serrulacaulis*

DIVISION TRIMEROPHYTOPHYTA

While in many aspects the Trimerophytophyta have much in common with the Rhyniophyta, there seems to be good reason for separating them. As in the Rhyniophyta, members of the trimerophytophytes are typically leafless and rootless, bear terminal sporangia, and have centrarch protosteles. The distinction, however, lies in the branching pattern. Trimerophytophytes tend to show a principal axis from which laterals arose. These laterals may have retained the dichotomous pattern of branching, however. The origin of the Trimerophytophyta is most likely from forms such as those in the Rhyniophyta; indeed, some members of the latter tend to show a departure from a strict dichotomous pattern of branching (e.g., *Cooksonia hemisphaerica*). *Trimerophyton* (Gr. *treis,*

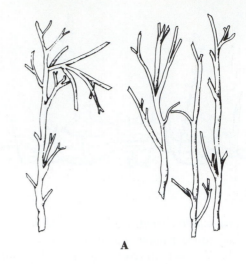

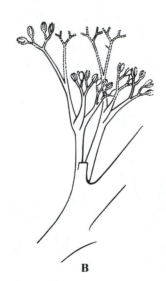

FIGURE 20-24
Trimerophyton robustius. **A.** Branching axes. **B.** Lateral branch with sporangia. **A.** ×¼. **B.** ×6. (**A.** *After Boureau.* **B.** *After Kräusel and Weyland.*)

three, + *meros,* part, + *phyton*) (Fig. 20-24) is the genus on which the circumscription of the division is based. Although it is still imperfectly known, *Trimerophyton* had a fairly robust prin-

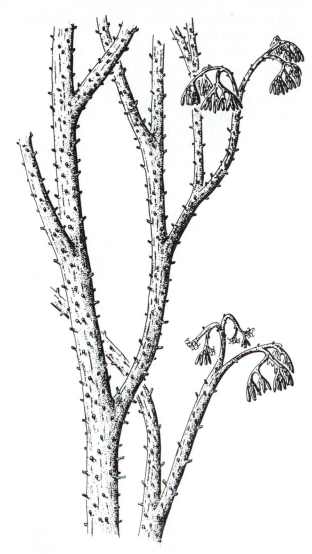

FIGURE 20-25
Psilophyton princeps. Restoration by Hueber (1967). (*After Andrews and Kasper.*)

plant, is *Psilophyton* (Gr. *psilos*, naked, + Gr. *phyton*, plant). *Psilophyton princeps* was first described by Sir J. W. Dawson, a Canadian geologist and university administrator, in 1859 from Lower Devonian rocks of the Gaspé peninsula in Quebec, Canada. That first description depicted a slender, naked, dichotomously branched plant lacking leaves and roots and bearing terminal sporangia. He later (1870, 1871) modified his original diagnosis and described a new variety, *P. princeps* var. *ornatum*, that was said to have borne spines near the basal portions of the plant. Painstaking and critical re-examination of Dawson's specimens and recent new collections of fossil plants from the area have resulted in the recognition of a number of changes in Dawson's original descriptions. For example, *P. princeps* var. *ornatum* has been transferred by Hueber (1971) to the genus *Sawdonia* in the Zosterophyllophyta, with *S. ornata* as the type species. In addition, Hueber and Banks (1967) emended the description of *P. princeps* Dawson, and Hueber (1967) prepared the reconstruction reproduced herein as Fig. 20-25. The new interpretation of *P. princeps* indicates that branching was not strictly dichotomous, but unequal branching resulted in a kind of pseudomonopodial pattern, with laterals less extensively developed than the principal axis. Covering the leafless axes were robust, truncated, spinelike projections. The dichotomizing lateral appendages had pendulous, paired, terminal sporangia. Dehiscence of sporangia was longitudinal. The protostele was centrarch or mesarch. Some species of *Psilophyton*, however, are known to lack spines.

Trimerophyton and *Psilophyton* are classified as follows:

Division Trimerophytophyta
  Order 1. Trimerophytales
    Family 1. Trimerophytaceae
      Genera: *Trimerophyton, Psilophyton*

cipal axis with laterals that tended to trichotomize. The tips of the branches terminated in a cluster of sporangia that dehisced longitudinally.

The best-known genus of the Trimerophytophyta, and the first well-known vascular

DIVISION MICROPHYLLOPHYTA

Although the Rhyniophyta, Zosterophyllophyta, and Trimerophytophyta represent the earliest and most primitive vascular plants, they apparently have no representatives in present-day vegetation. The Microphyllophyta, which also originated in the Devonian period, include forms with features that link them to extant genera such as *Lycopodium* and *Selaginella*. The microphyllophytes, then, represent the oldest of the currently living divisions of vascular plants. Members of the Microphyllophyta have certain characteristics reminiscent of those of the Zosterophyllophyta and most likely were derived from that group. Their closely spaced, generally simple leaves were probably derived from the spiny projections of the zosterophyllophytes. Sporangia are lateral along the axes, and in many they are localized in terminal zones to form strobili. While in the Zosterophyllophyta the sporangia had no special association with the spines, in the Microphyllophyta, both fossil and extant, sporangia are intimately associated with leaves, either attached directly on the adaxial surface or situated in the axil. Members of the Microphyllophyta most frequently have exarch protosteles. A relationship between Zosterophyllophyta and Microphyllophyta was suggested by Banks (1968) and Bierhorst (1971*a*).

Both herbaceous and arborescent fossil Microphyllophyta are known. In the following account, *Asteroxylon*, *Protolepidodendron*, *Baragwanathia*, *Leclercqia*, *Lepidodendron* (and other related genera in the Lepidodendrales), *Lycopodites*, *Selaginella*, *Pleuromeia*, and *Nathorstiana* are among the genera discussed briefly as examples of fossil Microphyllophyta.

Order 1. Asteroxylales

Asteroxylon

Asteroxylon is among the genera found in the famous Rhynie chert beds in Scotland (Fig. 20-26A). It was contemporaneous with *Rhynia*

and *Horneophyton* and was originally included in one order with those two genera. It is distinct enough, however, to warrant considering it among the Zosterophyllophyta or the Microphyllophyta. Its inclusion here within the latter is somewhat arbitrary because it has certain features more typical of those in the former. From smooth, horizontal rhizomes arose erect branches. These upright stems were monopodial, although the laterals tended to dichotomize. Covering the aerial axes were closely spaced, small, *Lycopodium*-like, leaflike structures about 5 mm long. Many botanists hesitate to call them leaves because they lacked vascular tissue. From the stele, traces extended upward and outward toward the leafy structures but stopped near the bases. This situation may be interpreted as one that shows the possible origin of microphylls from enations or spines. Zosterophyllophyte spines, of course, were not vascularized. *Asteroxylon*'s leaflike appendages lacked traces, but there were traces in the stem that were associated with them, and the usual situation in the Microphyllophyta is one in which the trace extends into the leaf.

The stele of *Asteroxylon* is more like that of herbaceous Microphyllophyta than it is like that in the Zosterophyllophyta. In cross section the xylem of the stele appears to be deeply lobed (more or less star-shaped—hence the generic name), with protoxylem near the tips of the ridges (Fig. 20-26B). Stalked sporangia were borne laterally along the aerial shoot, and there seems to be increasing evidence that there is some relationship between a sporangium and a leaflike scale (Fig. 20-26C). The best known species of *Asteroxylon*, *A. mackiei*, probably reached a half meter in height. The organization is somewhat suggestive of the extant *Lycopodium lucidulum* and *L. selago*. (Fig. 20-26A).

Order 2. Protolepidodendrales

The Protolepidodendrales include a single family, Protolepidodendraceae. Principally De-

FIGURE 20-26

Asteroxylon mackiei. **A.** Reconstruction; note horizontal stem system and upright, scale-covered shoot. **B.** Transverse section of stele. ×22. **C.** Reconstruction of portion of axis showing sporangial position. (*A. With permission of the trustees of the British Museum [Natural History]. **B.** Courtesy of the Royal Scottish Museum, Edinburgh.*)

493

vonian in age, members of the Protolepidodendrales were reminiscent of robust *Lycopodium* plants. *Baragwanathia* (named for the paleobotanist, W. Baragwanath) is thought by some (Garratt et al., 1984) to be as old as Late Silurian (certainly no younger than Lower Devonian in age), and although still extremely old, it does not antedate the earliest rhyniophytes. It was a robust plant, with stems 1–6.5 cm in diameter and with crowded, linear leaves up to 4 cm long and 1 mm wide (Fig. 20-27). Reniform sporangia were borne in the axils of some of the leaves in zones. The plant was probably homosporous. *Baragwanathia* was protostelic, with a xylem strand characterized by ridges and furrows and tracheids with annular thickenings.

Protolepidodendron (Gr. *protos*, first, + *lepidos*, scale, + *dendron*, tree) specimens have been recovered from both Lower and Middle Devonian strata. Plants were herbaceous, 20–30 cm tall, with a horizontal, creeping stem covered with closely spaced, helically arranged leaves, forked at the tips (Fig. 20-28). Erect stems, dichotomously branched, were borne on the prostrate part of the plant and they, too, bore small, forked leaves, many of which bore ovoid sporangia on the median adaxial surface. Stems had protosteles that were somewhat crenate in section, with protoxylem produced at the outer surfaces of the ridges.

One interesting feature of *Protolepidodendron* is the manner in which leaves were attached to the stems. The leaf flared at the base, forming an axially elongated swollen portion. At the time of abscission, the entire leaf did not fall off. Rather, the slender portion at the apex of the swollen base became detached, leaving the leaf base behind. Thus, on a twig

FIGURE 20-27
Baragwanathia longifolia. Axis and leaves. Lower Devonian of Australia. ×⅞. (*From Lang and Cookson.*)

FIGURE 20-28
Protolepidodendron scharyanum.
A. Reconstruction of rhizome and aerial branches. **B.** Sporophyll. ×6. (*From Kräusel and Weyland.*)

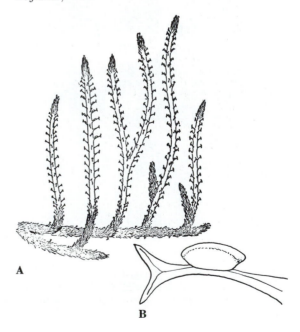

A

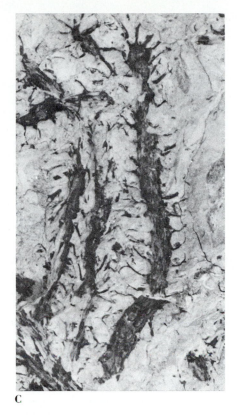

C

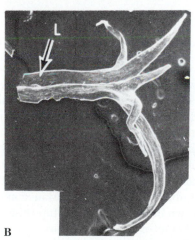

B

from which leaves had fallen, the stem was
covered with the raised, persistent bases. This
feature is even more pronounced in another
group of Microphyllophyta discussed below.

Another well-known member of the Protolepidodendrales is *Leclercqia,* described by Banks,
Bonamo, and Grierson (1972). This herbaceous
plant from the Upper Middle Devonian had
dichotomously branched axes 3.5–7.0 mm in
diameter, with some pieces reaching 46 cm in
length. Leaves were borne in great abundance
in a tight helix (Fig. 20-29A), each divided near
the middle into one elongate, recurved, median
portion and two divided lateral segments (Fig.

495

A

FIGURE 20-30
A. *Lycopodites stackii.* **B.** *Lycopodites pendulus.*
Upper Carboniferous, Yorkshire. **A.** ×0.5. **B.** ×2.
(**A.** *After Bower.* **B.** *Courtesy of Prof. W. G. Chaloner.*)

B

20-29B). A ligule was borne on the upper side of the leaf, making *Leclercqia* the oldest ligulate microphyllophyte. Globose-to-ellipsoidal sporangia were attached to the adaxial surfaces of leaves just proximal to the point of division of the leaf (Fig. 20-29C). The stele is very similar to that of *Protolepidodendron,* with a solid core of primary xylem with 14–18 protoxylem ridges. Tracheids have elongated bordered pits (Grierson, 1976).

Order 3. Lycopodiales

Herbaceous, dichotomously branched plants with closely spaced leaves corresponding closely to the extant genus *Lycopodium* are known from rocks as far back as the Upper Devonian. These fossils, which extend through the Cretaceous, are grouped under the name *Lycopodites* (Fig. 20-30). Some specimens are fertile and indicate that the plants were homosporous.

Order 4. Lepidodendrales

The order Lepidodendrales includes giant, treelike members of the Microphyllophyta that were dominant components of late Paleozoic floras. The order is extinct, with representatives ranging from the Upper Devonian to the Permian. *Lepidodendron* is the best-known genus, with members Carboniferous and Permian in age. Plants exceeded 50 m in height, with a tall trunk, unbranched for some distance and bearing a crown of branches at the top (Figs. 20-31, 20-32). Some of the more massive plants may have had trunks 2 m thick at the base. Leaves were linear and expanded into a swollen base at their points of attachment. After abscission, an elevated leaf cushion was left on the stem surface, and impressions of *Lepidodendron* stems show these closely spaced, persistent leaf

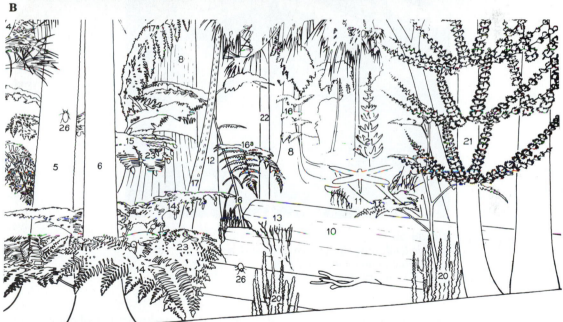

FIGURE 20-31

A. Carboniferous swamp forest as reconstructed at the Field Museum of Natural History. **B.** Key to some of the organisms in **A:** 4, *Lepidodendron obovatum*. 5, *Sigillaria rugosa*. 6, *Sigillaria saulii*. 8, *Sigillaria lacoei*. 10, *Sigillaria*, trunk. 12, *Lepidophloios laricinus*. 13, *Selaginellites* sp. (4–13, lycophytes). 14, *Neuropteris heterophylla*. 15, *Neuropteris decipiens*. 16, *Lyginopteris oldhamia* (seed ferns). 17, *Caulopteris giffordii* (fern); 20, *Sphenophyllum emarginatum*. 21, *Calamites* sp. (20, 21, Arthrophyta). 22, *Cordaites borasifolium* (gymnosperm). 26, *Archeoblattina beecheri*, a roach.

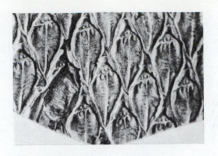

FIGURE 20-32
Lepidodendron sp.
Reconstruction of a mature
plant. (*After Eggert.*)

FIGURE 20-33
Stem surface of *Lepidodendron* showing
characteristic leaf bases.

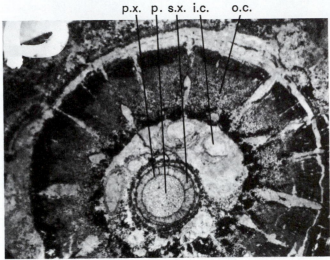

p.x. p. s.x. i.c. o.c.

FIGURE 20-34
Lepidodendron sp. Transverse section of a stem. i.c., inner
cortex; o.c., outer cortex; p., pith;
p.x., primary xylem; s.x., secondary xylem. ×4.

bases (Fig. 20-33). Leaf size varied from species
to species, and even on the same plant, with
larger leaves borne in more proximal regions
and smaller ones distally. Some arborescent
lepidodendralean leaves may have reached a
meter in length. Strictly speaking, these are not
"microphylls" so far as size is concerned, but
traces to leaves do not produce leaf gaps in the

stele of the stem, and most paleobotanists consider these long leaves as having originated
phylogenetically from enations.

From the base toward the top of the unbranched trunk, the stele changed from an
exarch protostele to an ectophloic siphonostele,
with progressively larger pith (Figs. 20-34,
20-35). Steles in branches, however, were

498

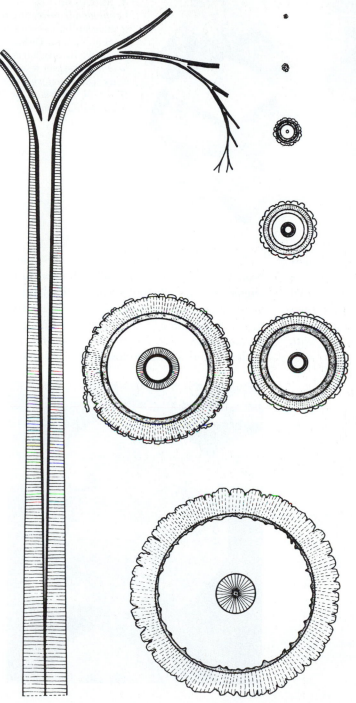

FIGURE 20-35
Diagram showing changes in structure of a *Lepidodendron* stem at different levels. Diagram at left indicates pith (unshaded central region), primary xylem (shaded black), and secondary xylem (striped). Largest transverse section represents a level of the stem near the base; next larger section is near the top of the main trunk; and progressively smaller ones represent sections closer to the tips of the branches. (*After Eggert.*)

A

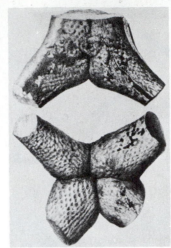

B

FIGURE 20-36
A. Stumps of *Lepidodendron* resting on stigmarian bases; sandstone casts at Victoria Park, Glasgow. B. *Stigmaria ficoides* bases; note root scars. (A. *Courtesy of Prof. Henry N. Andrews, Jr.* B. *After Williamson.*)

smaller than those in regions of the stem below the level of branching, and with each order of branching, the stele diminished in size so that ultimately, in the very tips, only tiny protosteles were present. Secondary xylem is progressively less well developed in more distal regions, with none present at the stem tips.

The massive trunk rested on dichotomously branching bases that are called *Stigmaria* (Gr. *stigma*, mark) (paleobotanical nomenclature involves assigning names to different parts of what might be the same kind of plant; these are called *organ genera*) (Fig. 20-36). On these branching, underground axes are closely spaced, helically

B

A

FIGURE 20-37
A. *Lepidostrobus.* Diagrammatic longitudinal section of a strobilus; note heterospory. B. *L. oldhamius.* Spore. ×2100. (A. *After Smith.* B. *Courtesy of Prof. T. N. Taylor.*)

500

arranged, rootlike appendages. The internal organization of these appendages is quite similar to that of the roots of *Isoetes*.

Terminating some of the branches were elongated strobili, each consisting of an axis with closely spaced, helically arranged sporophylls (Fig. 20-37). The proximal part of each sporophyll bore an elongated sporangium on the adaxial surface, and the distal part was bent abruptly toward the strobilar apex. A ligule often occurred on the sporophyll just distal to the sporangium (Fig. 20-37). Most strobili of *Lepidodendron* (placed in the organ genus *Lepidostrobus*) were heterosporous. In some instances, the number of functional megaspores per sporangium was reduced to one, the spore was not shed, the female gametophyte developed within the spore wall, and lateral margins of the sporophyll enveloped the sporangium, thus forming a seedlike body. This type of cone is placed in the genus *Lepidocarpon* (Fig. 20-38).

Sigillaria (L. *sigillum*, seal) (Figs. 20-31, 20-39) is a genus closely related to *Lepidodendron*, but it differed in being unbranched or branched only once or twice. Its heterosporous strobili were borne along the stem surface inter-

spersed among leaf bases. Leaf bases, although arranged in a close helix, had the appearance of being aligned in vertical rows. *Lepidophloios* resembled *Lepidodendron* even more closely, but leaf bases were tangentially expanded. It bore strobili on short lateral branches (Figs. 20-31, 20-40).

Phytokneme (Gr. *phyton*, plant, + *kneme*, spoke of a wheel) *rhodona* is one of the earliest members of the Lepidodendrales, originating from the Upper Devonian of Kentucky. A sector

FIGURE 20-40
Lepidophloios. Portion of branch system of arborescent lycopod; note strobili on lateral branches at left. *(After Andrews.)*

FIGURE 20-38
Reconstruction of a megasporophyll of *Lepidocarpon* in which lateral edges of the sporophyll overlap and conceal the sporangium in which was produced one functional megaspore. *(After Stewart.)*

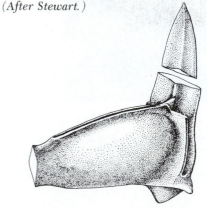

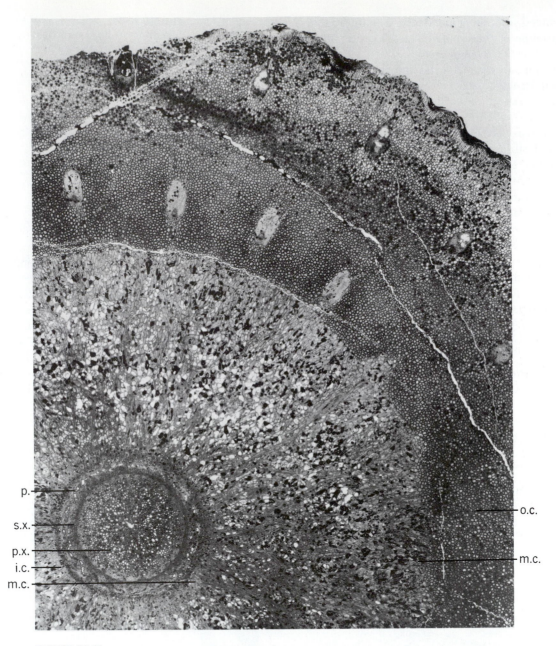

FIGURE 20-41
Phytokneme rhodona. Sector of a transverse section of the axis; i.c., inner cortex; m.c., middle cortex; o.c., outer cortex; p., phloem; p.x., primary xylem; s.x., secondary xylem. ×14. (*After Andrews, Read, and Mamay.*)

of the transverse section of an exceedingly well-preserved specimen is illustrated in Fig. 20-41 and shows a central mass of primary xylem surrounded by a narrow zone of secondary xylem, probably a phloem, a broad "middle cortex," and an "outer cortex."

Order 5. Selaginellales

Selaginella-like plants are known as far back as the Carboniferous. Some of them exhibited anisophylly, a feature in many (but not all) extant species of *Selaginella*. *Selaginellites* is the generic name assigned to most of the fossil forms (Fig. 20-42), but one fossil is so like the extant *Selaginella* it has been included within that genus. Schlanker and Leisman (1969) reported on this Pennsylvanian plant and indicated that it bore a swollen basal root-bearing axis with a microphyll-bearing stem (Fig. 20-43). (Because the root-bearing axis resembled those of *Nathorstiana* and *Isoetes*, some authors prefer not to place the Carboniferous forms within *Selaginella* [Rothwell and Erwin, 1985]).

The stem had an exarch protostele, with the protoxylem borne along the ridges. Strobili were terminal, with microsporophylls and megasporophylls typical of those in *Selaginella*. Megaspores were borne near the base of the strobilus and microspores above (Fig. 20-44). Four megaspores were present in each megasporangium and were 800 μm in diameter. Each of the microspores that were abundant in microsporangia measured 80 μm.

Order 6. Pleuromeiales

Pleuromeia (Gr. *pleuron*, rib, + Gr. *meion*, fewer), the only genus in the order, was a Triassic plant about 2 m tall. The erect, stout stem was unbranched and had a crown of elongated linear leaves (Fig. 20-45). The lower part of the stem was marked by scars left by fallen leaves. The trunk rested on a swollen, stunted

FIGURE 20-42
Selaginellites gutbieri. Carboniferous. (*After Geinitz.*)

base with four upturned lobes, each bearing numerous, helically arranged roots. This basal structure has been considered to represent a very much reduced *Stigmaria* system. A large, club-shaped strobilus terminated the stem. It bore rounded sporophylls, which had deeply

FIGURE 20-43
Selaginella fraipontii. Reconstruction showing habit of growth on fallen log of *Lepidodendron* sp. ×½. (*After Schlanker and Leisman.*)

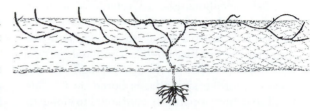

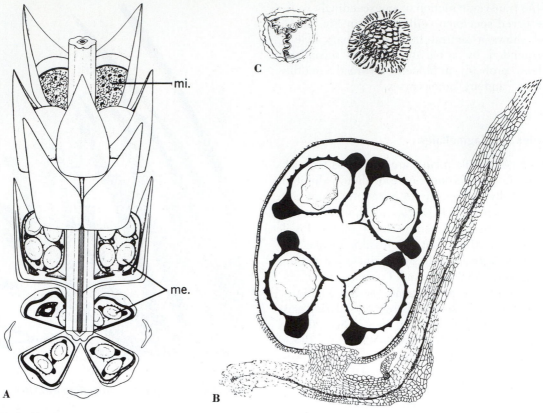

FIGURE 20-44
Selaginella fraipontii. **A.** Reconstruction of the strobilus. me., megasporangium; mi., micro-sporangium. **B.** Diagrammatic longitudinal section of megasporangium; note megaspores, megasporangium, and ligule. **C.** Microspores. (**A.** *After Schlanker and Leisman.* **B, C.** *After Hoskins and Abbott.*)

embedded adaxial sporangia. *Pleuromeia* is regarded by many to be a derivative of the late Paleozoic arborescent Microphyllophyta.

Order 7. Isoetales

Nathorstiana (named for A. G. Nathorst, a Swedish paleobotanist), a Cretaceous member of the order Isoetales, was a small plant, about 20 cm tall, with a squat stem and quill-like terminal leaves (Fig. 20-46). The basal portion was ribbed, with closely spaced, helically arranged, rootlike appendages borne on it. This basal structure apparently continued to elongate

as the plant matured, increasing the root-bearing surface (Karfalt, 1984).

Nathorstiana is considered by some to represent a further reduction in a line of plants that extends back to massive trees in the Paleozoic. In fact, this reduction from arborescent microphyllophytes may have terminated in the extant *Isoetes*, which is also represented in the fossil record of the Cretaceous period. Some of these are included in the fossil genus *Isoetites*.

The history of Microphyllophytan plants extends from the Lower Devonian to the present, although many genera have become extinct, with only *Lycopodium, Selaginella, Phyl-*

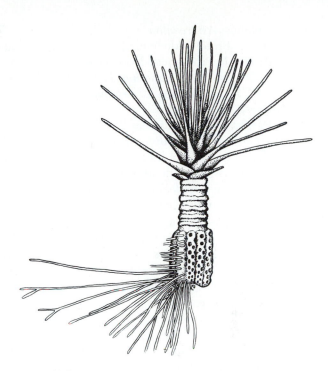

FIGURE 20-45
Pleuromeia sternbergii. Reconstruction. (*After Hirmer.*)

FIGURE 20-46
Nathorstiana arborea. Reconstruction. (*After Mägdefrau, from Delevoryas.*)

loglossum, *Isoetes*, and *Stylites* (if the last is to be regarded as distinct from *Isoetes*) having survived. The first two probably represent persistently primitive types, having survived, with little modification, from the Carboniferous. The Isoetales, on the other hand, very possibly represent the end of a reduction series from arborescent ancestors. Among the extinct genera, both ligulate and eligulate, homosporous and heterosporous, and herbaceous and arborescent types occurred. The last were striking plants, more than 50 m in height, and were abundant in Carboniferous forests. Cambial activity (known only in the Isoetales among extant microphyllophytes) was a noteworthy attribute of the treelike forms, but the bulk of the secondary tissues was periderm rather than secondary xylem. Although some forms had long leaves, all the fossil genera were clearly microphyllous,

with no leaf gaps having formed in the stem steles. Sporangia were always adaxial, probably having been derived from the lateral sporangial attachment in the Zosterophyllophyta.

The Microphyllophyta are an ancient group of plants—probably the most ancient of all vascular plants living on the earth at the present time. Because of some of their unique characteristics, they seem not to have been the progenitors of other groups of vascular plants.

Classification

The genera of Microphyllophyta discussed above may be classified as follows:

Division Microphyllophyta
    Order 1. Asteroxylales
        Family 1. Asteroxylaceae

505

Genus: *Asteroxylon*
Order 2. Protolepidodendrales
  Family 1. Protolepidodendraceae
    Genera: *Baragwanathia, Protolepido-*
        dendron, Leclercqia
Order 3. Lycopodiales
  Family 1. Lycopodiaceae
    Genera: *Lycopodium, Lycopodites*
Order 4. Lepidodendrales
  Family 1. Lepidodendraceae
    Genera: *Lepidodendron, Stigmaria,*
        Lepidostrobus, Lepido-
        carpon, Sigillaria, Lepido-
        phloios, Phytokneme
Order 5. Selaginellales
  Family 1. Selaginellaceae
    Genera: *Selaginella, Selaginellites*
Order 6. Pleuromeiales
  Family 1. Pleuromeiaceae
    Genus: *Pleuromeia*
Order 7. Isoetales
  Family 1. Isoetaceae
    Genera: *Nathorstiana, Isoetes,*
        Isoetites, Stylites

DIVISION ARTHROPHYTA

Although the history of the Arthrophyta parallels that of the Microphyllophyta quite closely, the beginning is somewhat more obscure. It is quite likely that their origin was during the Devonian period, and the division persisted to the present time, represented today only by the genus *Equisetum*. Many fossil Arthrophyta are recognizable by their whorled branching, ribbed stems, and sporangia borne on reduced branches called sporangiophores. The ancestors of the Arthrophyta were probably forms such as those in the Trimerophytophyta, with a prominent principal axis and reduced laterals. Subsequently, the sites of attachment of appendages became telescoped, with a number of them arising at a single level. Sporangium-bearing appendages became recurved.

That Arthrophyta could have had their origins among the Trimerophytophyta is shown by the recently described *Ibyka* (named for the Greek poet Ibykos) *amphikoma* (Skog and Banks, 1973), placed in the order Ibykales (Fig. 20-47). The entire plant is not known, but the distal portions indicate a monopodial habit, with helically arranged appendages. Reduced terminal appendages were dichotomous, with recurved tips. Obovoid sporangia were borne on the tips of some of these ultimate appendages. The Ibykales appear to be another instance where it is impossible to assign a group of plants into existing categories, but they are important in showing important evolutionary trends.

Following are selected orders of Arthrophyta:

FIGURE 20-47
Ibyka amphikoma. Reconstruction of portion of plant. (*After Skog and Banks.*)

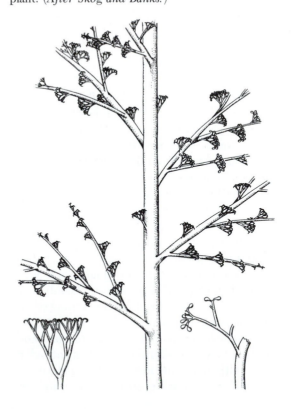

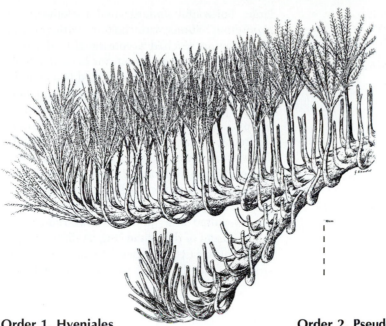

FIGURE 20-48
Hyenia elegans.
Reconstruction of portion of
plant. Scale in figure equals
10 cm. (*After Schweitzer.*)

Order 1. Hyeniales

Schweitzer (1972) is of the opinion that the Middle Devonian *Hyenia*, the type genus for the order, is more probably a fern than an arthrophyte, but because it has certain features like those of arthrophytes, it is included with them and, like *Ibyka*, indicates possible ancestry of the arthrophytes among the trimerophytophytes. The reconstruction of *Hyenia* (Fig. 20-48) shows a massive, horizontal rhizome system from which arise numerous, closely spaced, upright branches that appear digitately branched at higher levels. On these upright shoots numerous, finely dichotomously branched leaflike appendages and, in zones, delicate fertile appendages or sporangiophores were attached. Sporangia were recurved, as they are in many arthrophytes, with delicate dichotomously branched portions of the sporangiophores extending outward. Appendages were not whorled as they are in typical arthrophytes, but the situation in *Hyenia* may easily be interpreted as one that had not yet reached the more highly organized pattern of arrangement of appendages in the Arthrophyta.

Order 2. Pseudoborniales

More typically arthrophytan than members of the Hyeniales is the genus *Pseudobornia* (Gr. *pseudes*, false, + *Bornia*, old generic name for arthrophyte foliage), from Upper Devonian strata. It was an upright tree, 15–20 m tall, with a jointed stem from which arose a small number of branches in whorls (Fig. 20-49). Four leaves were borne at each node on smaller branches. Each leaf dichotomized twice near the point of attachment, with the resulting segments divided into fine, featherlike laminae.

Order 3. Calamitales

Giant, treelike Arthrophyta flourished from the Upper Devonian through the Permian. *Calamites* (Figs. 20-31 [no. 21], 20-50), the giant horsetail, with many species, is abundant in strata ranging from Mississippian to Lower Permian. *Calamites* (Gr. *kalamos*, reed) was treelike in habit, in some instances probably having reached 20–30 m in height, with a trunk that may have been more than 40 cm in diameter. The aerial portion of the plant arose from a

507

FIGURE 20-49

FIGURE 20-49
Pseudobornia ursina.
Suggested reconstruction.
Scale in figure equals 5
m.(*After Schweitzer.*)

large, horizontal underground rhizome, as in *Equisetum*. Stems were hollow, with the pith often having reached a considerable diameter (Fig. 20-51). Primary xylem was in the form of a series of separate strands arranged in a cylinder about the pith; each primary xylem strand had a canal within it representing the disintegrated protoxylem. Secondary xylem of considerable thickness was produced in larger axes by the vascular cambium. During the process of fossilization, pieces of *Calamites* stems may have been covered with sediment, which also washed into the hollow pith cavity. Hardening of the sediment resulted in a "pith cast" that reflects the inner margin of the xylem (Fig. 20-52). Grooves in the pith cast correspond to primary xylem strands, while the ridges represent projection of

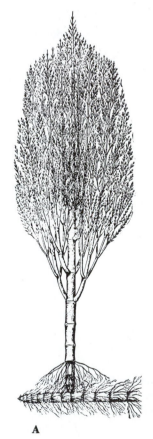

FIGURE 20-50
A. *Calamites carinatus.*
Reconstruction; note rhizome, roots,
and whorled aerial branches.
B. Model reconstruction of a
Calamites trunk (left) and
leaf-bearing branch system (right).
(**A.** *After Hirmer.* **B.** *Courtesy of
the Field Museum of Natural
History.*)

A   B

508

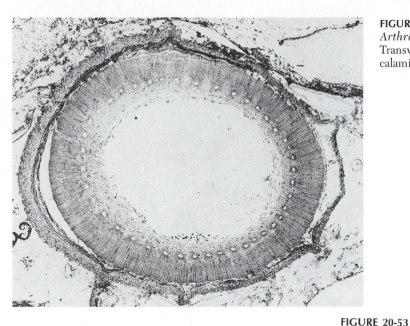

FIGURE 20-51
Arthropitys communis.
Transverse section of a small
calamitean stem. ×8.

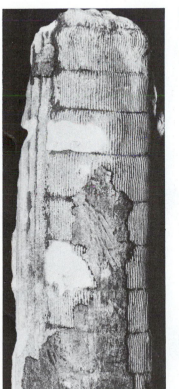

FIGURE 20-52
Calamites sp.
Cast of the
interior of a stem.
×0.33.

FIGURE 20-53
Annularia stellata.
Foliage associated
with *Calamites.*
×0.66.

the sediment into the inner parts of major vascular rays between the primary xylem strands. Horizontal lines were left by remains of pith diaphragms that occurred at nodes. Smaller branches departed from larger axes in whorls, and the ultimate branches, also whorled, were clothed with circles of slender leaves. These leaves have been assigned to the organ genera *Annularia* (L. *annulus*, ring) (Fig. 20-53) and *Asterophyllites* (Gr. *aster*, star, + *phyllon*, leaf). As in arborescent fossil Microphyllophyta, daughter branches were smaller than parent axes, with less potential for growth. They had a smaller pith, fewer primary xylem strands, fewer leaves per whorl, less secondary vascular tissue, and smaller leaves (Fig. 20-54).

Fertile appendages of *Calamites* were borne in compact strobili (some of which were called

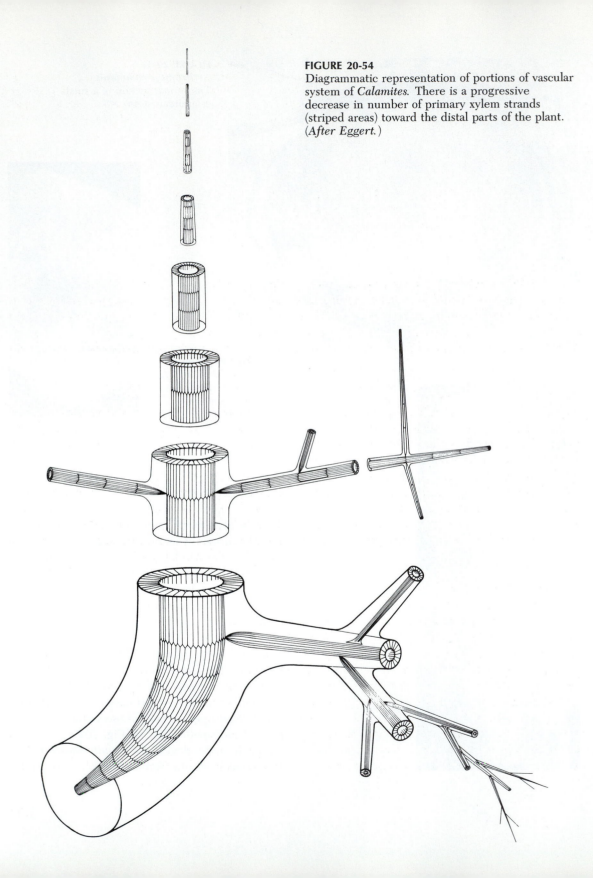

FIGURE 20-54
Diagrammatic representation of portions of vascular system of *Calamites*. There is a progressive decrease in number of primary xylem strands (striped areas) toward the distal parts of the plant. (*After Eggert.*)

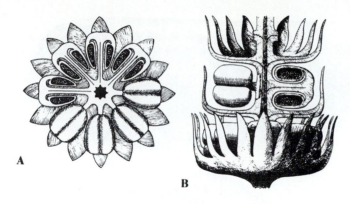

FIGURE 20-55
Calamostachys binneyana. **A.** Transverse section of strobilus; note bracts and sporangiophore whorls from above. **B.** Longitudinal section with alternating bract whorls and sporangiophore whorls. (*After Zimmermann.*)

Calamostachys—Gr. *stachys*, ear of grain), each composed of a central axis that had alternating whorls of bracts and sporangiophores (Fig. 20-55). Sporangiophores were peltate, each with four sporangia directed toward the strobilar axis. Sections of sporangiophores indicate that they were much like those of *Equisetum* and that the sporangial wall was composed of a single layer of cells with annular thickenings. Some species of cones of *Calamites* were heterosporous, and spores have been shown to have borne elaters (Good and Taylor, 1975). One strobilus, in the organ genus *Calamocarpon* (Gr. *karpon*, fruit), demonstrates an extreme manifestation of heterospory in that there is only one functional megaspore per sporangium.

Order 4. Sphenophyllales

Contemporaneous with the arborescent giant horsetails of the Carboniferous and Permian were herbaceous arthrophytes, members of the Sphenophyllales, which actually appeared first in the late Devonian and extended into the Triassic. The slender stems of *Sphenophyllum* (Gr. *sphenos*, wedge, + *phyllon*, leaf) were ribbed when young, but older stems were cylindrical and covered with periderm. Each node bore a whorl of wedge-shaped leaves (Figs. 20-31, 20-56A, B) with dichotomously branched veins. Primary xylem of the stem was triangular

in cross section and was exarch (Fig. 20-56C). Older stems show evidence of secondary vascular tissue. The outline of the xylem in cross section in an older stem is circular. Roots were diarch and also may have had secondary vascular tissue.

Strobili were borne at the tips of certain branches. Each was composed of an axis bearing a whorl of fused bracts that subtended a whorl of sporangiophores (Fig. 20-57). Sporangiophores were variously branched in different species and were partially adnate to the bract whorl, and each branch of the sporangiophore bore a single, recurved sporangium. The most common kind of strobilus attributed to *Sphenophyllum* is the organ genus *Bowmanites* (named after J. E. Bowman, the discoverer). All species appear to have been homosporous.

Order 5. Equisetales

Coexisting with extinct members of the Arthrophyta during late Paleozoic and Mesozoic times were herbaceous arthrophytes, many of which resemble the extant *Equisetum* quite closely (Fig. 20-58). In fact, *Equisetum*-like fossils, often placed in the genus *Equisetites*, are so much like the extant genus that many workers place the fossils in *Equisetum*. *Equisetum* appeared first in the Triassic period and was an abundant component of Mesozoic floras. These

511

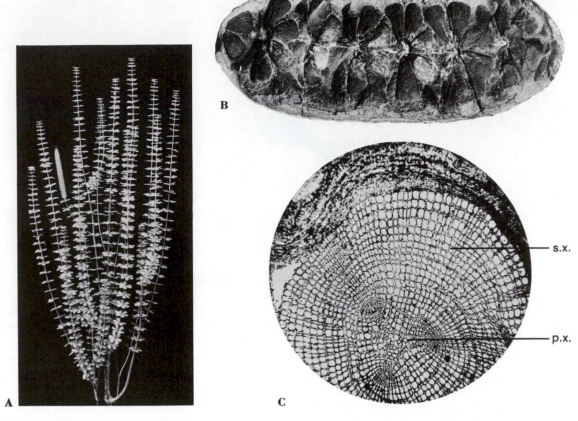

FIGURE 20-56
Sphenophyllum emarginatum. **A.** Model reconstruction; note strobilus. **B.** Portion of axis and whorled leaves. **C.** *S. plurifoliatum.* Transverse section of portion of stem; note triarch protostele surrounded by secondary xylem. p.x., primary xylem; s.x., secondary xylem. **B.** ×1. **C.** ×12. (**A, B.** *Courtesy of the Field Museum of Natural History.*)

Mesozoic plants were considerably more robust than present-day species, but it is quite likely that they had no secondary vascular tissues.

Other herbaceous arthrophytes in the Equisetales include the late Paleozoic–early Mesozoic genus *Phyllotheca* (Gr. *phyllon,* leaf, + *theca,* sheath), which was *Equisetum*-like except that the leaves were more conspicuous, arranged in a whorl that was fused near the point of attachment (Fig. 20-59). It is quite likely that members of the Equisetaceae were derived from arborescent *Calamites*-like ancestors by a progressive reduction in the amount of secondary vascular tissue, a reduction in size of leaves, and a reduction in numbers of branches.

In summary, it may be stated that the fossil record indicates that although currently represented in our flora only by a small number of *Equisetum* species, the Arthrophyta were formerly an important assemblage. Possible primitive arthrophytes are known from the Middle and Upper Devonian. The whorled arrange-

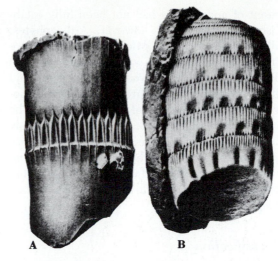

FIGURE 20-58
Equisetites. **A.** *E. platyodon.* **B.** *E. arenaceus.*
(*After Hirmer.*)

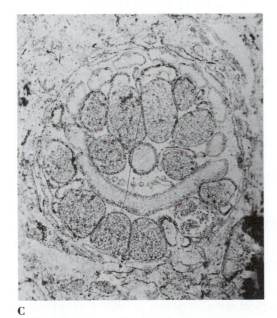

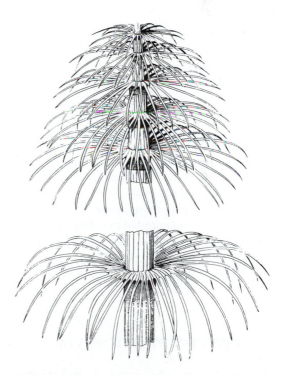

FIGURE 20-57
Bowmanites dawsonii. **A.** Half of one whorl of
bracts with sporangiophores. **B.** Median
longitudinal section of portion of strobilus; note
axis, bracts, and sporangia. **C.** Oblique
transverse section of strobilus. ×4. (**A, B.** *After
Hirmer.*)

FIGURE 20-59
Phyllotheca equisetoides. Reconstruction of portion
of plant. (*After Rasskazova.*)

ment of appendages had not been fully established by that time, however. Arthrophytes, like members of the Microphyllophyta, were important components of Carboniferous and Permian forests. Both woody (arborescent) and herbaceous forms were present, and heterospory occurred in some of the members. The occurrence of leaves with dichotomously branching veins in some fossil genera (*Sphenophyllum*) suggests that the leaves of *Equisetum*, with single, unbranched veins, have arisen from fossil forms with *Annularia*- or *Asterophyllites*-like leaves.

Classification

The members of the Arthrophyta treated in this chapter may be classified as follows:

Division Arthrophyta
    Order 1. Hyeniales
        Family 1. Hyeniaceae
            Genus: *Hyenia*
    Order 2. Pseudoborniales
        Family 1. Pseudoborniaceae
            Genus: *Pseudobornia*
    Order 3. Calamitales
        Family 1. Calamitaceae
            Genera: *Calamites, Calamostachys, Calamocarpon, Annularia, Asterophyllites*
    Order 4. Sphenophyllales
        Family 1. Sphenophyllaceae
            Genera: *Sphenophyllum, Bowmanites*
    Order 5. Equisetales
        Family 1. Equisetaceae
            Genera: *Equisetum, Equisetites, Phyllotheca*

DIVISION PTERIDOPHYTA

In addition to the rootless and leafless Rhyniophyta, Zosterophyllophyta, and Trimerophytophyta, and to the Microphyllophyta and Arthrophyta, ferns and their precursors were present on the earth during early Devonian times. The Pteridophyta, or ferns, have in common the fact that all are megaphyllous vascular cryptogams with foliar sporangia. Most likely the source of the ferns is the Trimerophytophyta, with most botanists believing that a lateral branch system of trimerophytophytes was the precursor of a megaphyll. In fact, recent thought indicates that a megaphyll could have been derived from a branch system bearing microphylls, and that the megaphyll is a kind of compound structure composed of stems and leaves. The fossil record seems to offer some support to this idea.

Ferns and fernlike plants are represented in the fossil record in great abundance, especially in Carboniferous rocks. In fact, many paleobotanists refer to that period of geologic history as the "Age of Ferns." A great many of the fernlike fossils are *form genera*—that is, they have superficially similar appearance but may actually belong to more than one natural group of plants. Unlike most modern ferns, in which the compound leaves are clearly two-dimensional appendages of well-differentiated trunks and/or rhizomes, this distinction between leaves and stems and the two-dimensional organization of the former were absent in the earliest putative pteridophytan precursors.

CLASS 1. CLADOXYLOPSIDA

Order 1. Cladoxylales

Members of the class Cladoxylopsida are not typical fernlike plants in that it is difficult to recognize distinct fronds. Instead, branching of the stem system resulted in formation of smaller, lateral branches that bore minute, often dichotomously branched, appendages. It is difficult to equate these small appendages with branching stem systems. More likely they represent modified enations and were, in a sense,

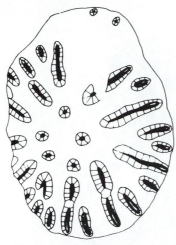

FIGURE 20-60
Cladoxylon taeniatum. Transverse section of axis, diagrammatic. ×7.5. (*After Bertrand, from Andrews.*)

microphylls. The distal branches with the microphylls may be regarded as precursors of large, compound fronds. Although members of the Cladoxylopsida may vary in gross morphological characters, they have in common a stem structure in which, in transverse sections of the axes and larger branches, the xylem consists of radiating, sometimes branching, plates, often with their outer extremities enclosing nests of parenchyma cells (Fig. 20-60).

Representatives of the Cladoxylales are found in strata ranging from Middle Devonian to Mississippian in age. *Cladoxylon scoparium* is a well-known species of the genus. The plant was branched with some of the distal branches bearing small, leaflike appendages and others bearing flattened, dichotomously branched, fan-shaped appendages with terminal, ovoid sporangia (Fig. 20-61). Plants were homosporous.

Pseudosporochnus (Gr. *pseudes*, false, + *Sporochnus*,[6] a genus of Phaeophyta), from the

---

[6]The genus was first described as an alga until it was discovered that the plants contained xylem elements.

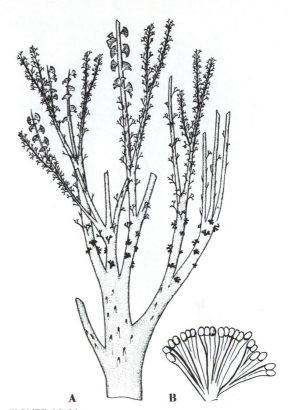

FIGURE 20-61
Cladoxylon scoparium. **A.** Portion of axis with vegetative and fanlike sporangium-bearing axes. **B.** Fertile appendage enlarged; note terminal sporangia. (*After Kräusel and Weyland.*)

Middle Devonian, was until relatively recently assigned to the group of plants that included such simple forms as *Rhynia* and *Cooksonia*. However, discovery of abundant material (in Belgium) of a new species, *P. nodosus*, and intensive study of it have led to the conclusion that *Pseudosporochnus* is actually a member of the Pteridophyta, order Cladoxylales (Leclercq and Banks, 1962). This conclusion is based partly on the complex nature of the "frond" and, more important, on anatomical evidence that demonstrates vascular tissue in the form of radiating plates (Leclercq and Lele, 1968). *Pseudosporochnus nodosus* was a large plant, but the overall height is not known (Fig. 20-62).

515

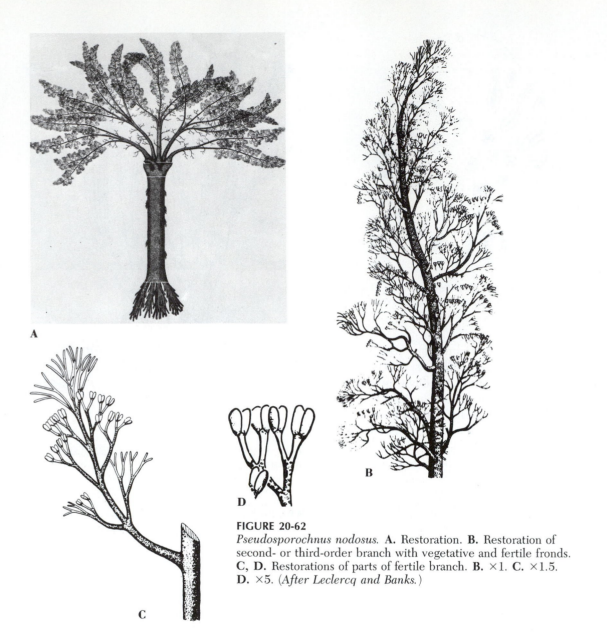

FIGURE 20-62
Pseudosporochnus nodosus. **A.** Restoration. **B.** Restoration of
second- or third-order branch with vegetative and fertile fronds.
C, D. Restorations of parts of fertile branch. **B.** ×1. **C.** ×1.5.
D. ×5. (*After Leclercq and Banks.*)

At the base of the trunklike stem there were a number of principal roots bearing laterals helically. The main stem was unbranched for some distance, but at the apex was a crown of three-dimensional branches of two orders. The second-order branches bore dichotomizing sterile and/or fertile segments arbitrarily called "fronds." Other ultimate branches were fertile, terminating in eight pedicillate, ovoid sporangia. Some fronds were mixed fertile and sterile. *Pseudosporochnus* differs from *Cladoxylon scoparium* in that the former has fertile branches that are not flattened into bladelike structures.

The Middle Devonian *Calamophyton* was

classified earlier with the Arthrophyta on the basis of certain characters it seemed to have in common with *Hyenia*. More recent studies (Leclercq and Schweitzer, 1965; Leclercq, 1969) have resulted in its being assigned to the Cladoxylopsida because it has internal structure more like that of that class than that of the Arthrophyta. Stems of *C. bicephalum* (Fig. 20-63A) had main axes at least 60 cm high. Whether or not they arose from a rhizome has not been ascertained. These axes were digitately, monopodially, and dichotomously branched. The branchlets bore three-dimensional, terete appendages that branched two or three times (leaves). Some of these appendages were somewhat more branched, with most of the tips recurved and bearing elongated sporangia; the distal parts extended outward and were sterile (Fig. 20-63B). Whether *Calamophyton* was homosporous or heterosporous is unknown.

The class Cladoxylopsida most likely represents an artificial group of plants held together principally by the fact that the stele consists of a series of radiating platelike masses of xylem and phloem. Gross morphological features may be quite different among the various genera, and there is no reason to assume that more than one group of plants could not have evolved such an anatomical pattern. Further research based on additional collections may provide insight as to the significance of internal structure and importance of external form in classifying these organisms.

CLASS 2. COENOPTERIDOPSIDA

Order 1. Coenopteridales

Members of this class occur as fossils in strata extending from the Upper Devonian through the Permian, but are predominantly Carboniferous. The single order Coenopteridales may well represent an artificial grouping of diverse organisms. In fact, as more work progresses on plants

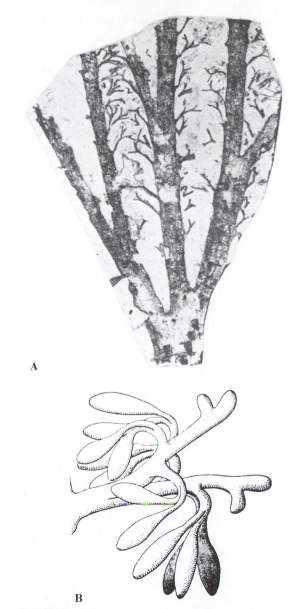

A

B

FIGURE 20-63
Calamophyton. **A.** *C. primaevum.* Portion of branching axis. **B.** *C. bicephalum.* Branched fertile appendage. **A.** ×0.9. **B.** ×5. (*After Leclercq and Andrews.*)

that have been assigned to this order, it is becoming increasingly more apparent that some of them more appropriately may be assigned to

517

other orders of ferns. Our knowledge of these plants is based principally on segments of fronds, petioles, rachises, pinna axes, and fertile regions. In many of the coenopterid ferns it is difficult to distinguish between the rachis of a frond and a stem, and therefore the term **phyllophore** is used for a frondlike structure. The vascular pattern of the phyllophore is most frequently bilaterally symmetrical in either one or two planes and is often delicate and esthetically attractive. Most of the stems were protostelic, and fronds were generally three-dimensional rather than constructed in one plane. The order may be divided, arbitrarily, into four families.

Stauropteris is the single genus in the Carboniferous family Stauropteridaceae and was characterized by a four-lobed protostele (Fig. 20-64A). Appendages were borne in pairs, with the two appendages at a given level arising on the side of the parent axis opposite the position of the two appendages below (Fig. 20-64B). In this genus it is difficult to differentiate between stem and frond. Sporangia were borne terminally. Two species of *Stauropteris* are heterosporous. The megasporangium is small, elongated, and composed of fleshy tissue proximally, with two large megaspores in a distal cavity.

The Botryopteridaceae include only the genus *Botryopteris* (Gr. *botrys*, grape, + L. *pteris*, fern) from the Lower Carboniferous to the Permian. Typically *Botryopteris* has protostelic stems with fronds containing vascular traces, the xylem of which in transverse section is curved, with three arms directed toward the stem (Fig. 20-65A). Typical fern pinnules occurred at the distal parts, and some of the fronds bore massive, globose clusters of tightly packed sporangia (Fig. 20-65B).

In the Anachoropteridaceae, with a range from Carboniferous to Permian, plants were typically protostelic, with leaf traces C-shaped in transverse section. The curvature of the "C" was abaxial, unlike the more common adaxial curvature seen in present-day ferns (e.g., *Osmunda*). *Anachoropteris* is the genus for which the family was named and often shows conspicuously curved petiole traces.

FIGURE 20-64
Stauropteris oldhamia. **A.** Transverse section of rachis. ×15. **B.** Reconstruction. (**B.** *After Delevoryas.*)

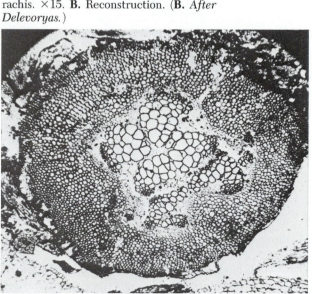

A

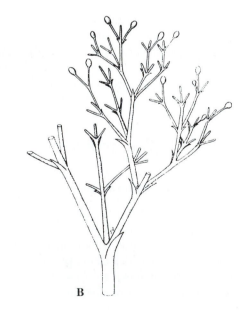

B

518

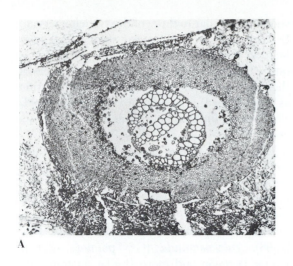

A

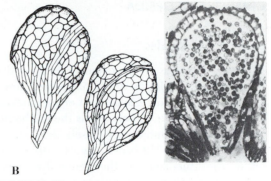

B

FIGURE 20-65
Botryopteris. **A.** *B. forensis.* Transverse section of rachis. **B.** *B. globosa.* Surface and sectional views of sporangia. **A.** ×10. **B.** ×25. (**A.** *After Delevoryas and Morgan.* **B.** *After Murdy and Andrews.*)

Internal structure of petioles in the family Zygopteridaceae (Upper Devonian–Permian) generally involves petiolar xylem strands that were bilaterally symmetrical, with two planes of symmetry (Fig. 20-66A). In transverse section some were elliptical (e.g., *Clepsydropsis* [Gr. *clepsydra*, water clock]), bow-tie-shaped (e.g., *Metaclepsydropsis*), or H-shaped (*Zygopteris*). Pinnae were usually borne in pairs, with the plane of symmetry of a pinna at right angles to that of the rachis, in contrast to the situation in modern ferns in which all pinnae and the rachis are in one plane (Fig. 20-66B). One genus of Zygopteridaceae, *Zygopteris* (Gr. *zygos*, yoke, + L. *pteris*), had secondary xylem, an unusual phenomenon among the Pteridophyta. Another, *Ankyropteris* (Gr. *ankyra*, anchor, + L. *pteris*), showed axillary branching, also not a typical pteridophytan feature.

FIGURE 20-66
Zygopteris. **A.** *Z. illinoiensis.* Transverse section of phyllophore. ×10. **B.** *Z. lacattei.* Portion of frond. (**B.** *After Wettstein.*)

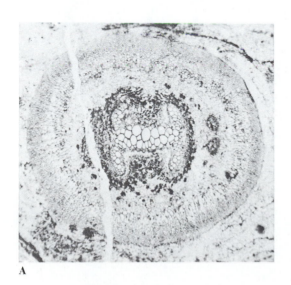

A

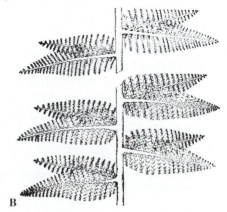

B

CLASS 3. FILICOPSIDA

Order 1. Marattiales

Among orders of extant ferns, the Marattiales, Filicales, and Salviniales are known also as fossils. Many examples of eusporangiate synangia are known as far back as the Carboniferous. These resemble very closely the sporangia of Marattiales. The order was also represented in the Paleozoic by the genus *Psaronius* (NL., a precious stone[7]), known from the Carboniferous and Permian (Fig. 20-67). In general habit *Psaronius* resembled certain extant tree ferns (members of the Filicales). Trunks exceeded 60 cm in diameter near the base and were probably more than 15 m tall. The bulk of the trunk, however, was actually composed of a thick root mantle surrounding a relatively slender stem. These stems were exceedingly complex anatomically, containing a number of concentric dictyosteles (Fig. 20-11). The stem increased in diameter and in complexity from base to apex, changing from protostelic at the very base, to siphonostelic, then dictyostelic at progressively higher levels. The large fronds were pinnately compound with synangia borne on the abaxial surface of some of the pinnules.

The order Marattiales was represented in the Mesozoic era by various kinds of fronds with synangia resembling those of certain extant genera.

Order 2. Filicales

Members of the Filicales are recognized as far back as the Carboniferous. Foliage with sporangia resembling those of the Schizaeaceae is placed in the genus *Senftenbergia*. *Oligocarpia* (Gr. *oligos*, few + *karpon*, fruit) is also Carboniferous and is the name given to foliage with sporangia similar to those of the Gleicheniaceae. In both instances, however, similarity is confined to structure of the sporangia; the foliage is quite different from that of modern members of these two families, and there may be a question as to whether the families are actually that old. There is no doubt about filicalean affinity, however.

The Osmundaceae are probably the oldest family of living ferns, with the first ones appearing in the Permian period. *Thamnopteris* (Gr. *thamnos*, frequent, + L. *pteris*, fern) has a stem with structure like that of *Osmunda* except that there is a protostele, with no leaf gaps (Fig. 20-68). *Palaeosmunda* (Gr. *palaios*, old) is Upper Permian and more similar anatomically to *Osmunda* (Fig. 20-69). Other osmundaceous stems, often included in the genus *Osmun-*

FIGURE 20-67
Psaronius sp. Reconstruction. (*After Morgan.*)

---

[7]From the speckled appearance of the root zone of petrified specimens when polished.

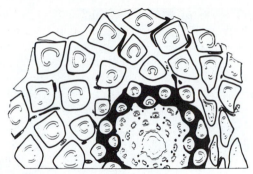

FIGURE 20-68
Thamnopteris schlechtendalii. Transverse section of stem and leaf bases of a Permian osmundaceous fern. (*After Delevoryas.*)

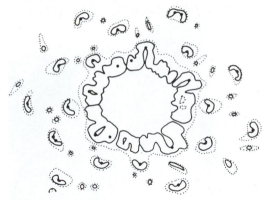

FIGURE 20-69
Palaeosmunda playfordii. Diagrammatic transverse section of stele and inner cortex. (*After Gould.*)

dacaulis, occur throughout the Mesozoic and Tertiary. Foliage, also thought to have been osmundaceous, was common in the Mesozoic era.

The palmately constructed leaf of the Matoniaceae is distinctive, and the family is evident in the Mesozoic, as far back as the Triassic period (Fig. 20-70). *Phlebopteris* (Gr. *phlebos*, vein, + L. *pteris*) has fertile fronds showing sporangial structure and arrangement that are like those in modern members.

Polypodiaceous ferns are known from the Jurassic, but really abundant and diversified remains did not occur until the Cenozoic era (Fig. 20-71). For this reason, the Polypodiaceae are often called "the modern ferns." It is clear that because of their later origin, they could not have been in the ancestral lines that led to the other fern groups.

Order 3. Salviniales

Among the aquatic, heterosporous ferns, the Salviniales extend back into the Cretaceous period. The unusual microspore masses and megaspore apparatuses of both *Azolla* and *Salvinia* are known. Entire plants of these two genera are found in Tertiary rocks.

FIGURE 20-70
Phlebopteris sp. Leaf of a Triassic member of the Matoniaceae. (*Courtesy of Prof. R. C. Hope.*)

521

A

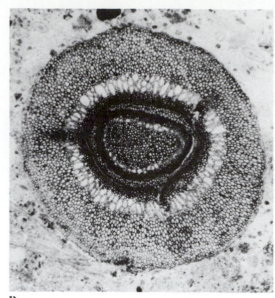

B

FIGURE 20-71

Tertiary fossil ferns. **A.** *Woodwardia* sp., Oligocene of Oregon; portion of frond; note sori. **B.** *Dennstaedtiopsis aerenchymata.* Upper Eocene of Oregon. Transverse section of stem. **A.** ×0.85. **B.** ×10.

Classification

The representatives of the fossil Pteridophyta discussed above may be classified as follows:

Division Pteridophyta
  Class 1. Cladoxylopsida
    Order 1. Cladoxylales
      Family 1. Cladoxylaceae
        Genus: *Cladoxylon*
      Family 2. Pseudosporochnaceae
        Genus: *Pseudosporochnus*
      Family 3. Calamophytaceae
        Genus: *Calamophyton*
  Class 2. Coenopteridopsida
    Order 1. Coenopteridales

      Family 1. Stauropteridaceae
        Genus: *Stauropteris*
      Family 2. Botryopteridaceae
        Genus: *Botryopteris*
      Family 3. Anachoropteridaceae
        Genus: *Anachoropteris*
      Family 4. Zygopteridaceae
        Genera: *Clepsydropsis, Metaclepsydropsis, Zygopteris, Ankyropteris*
  Class 3. Filicopsida
    Order 1. Marattiales
      Family 1. Psaroniaceae
        Genus: *Psaronius*
    Order 2. Filicales

522

Family 1. Schizaeaceae
    Genus: *Senftenbergia*
Family 2. Gleicheniaceae
    Genus: *Oligocarpia*
Family 3. Osmundaceae
    Genera: *Thamnopteris, Osmunda,*
        Palaeosmunda,
        Osmundacaulis
Family 4. Matoniaceae
    Genus: *Phlebopteris*
Family 5. Polypodiaceae
    Genera: *Woodwardia, Dennstaed-*
        tiopsis
Order 3. Salviniales
    Family 1. Salviniaceae
        Genera: *Azolla, Salvinia*

CONCLUSION

From this brief account of representative fossil vascular cryptogams, several conclusions may be drawn. In the first place, vascular land plants first appeared in the late Silurian. During the following Devonian period, which lasted approximately 50 million years, relatively rapid diversification occurred, since representatives of major groups of vascular cryptogams (Rhyniophyta, Zosterophyllophyta, Trimerophytophyta, Microphyllophyta, Arthrophyta, and Pteridophyta) made their appearance (Chaloner, 1970). Of these, representatives of only three (Microphyllophyta, Arthrophyta, and Pteridophyta) have survived. As the fossil record becomes still better known, a greater diversity of these ancient spore-bearing plants will probably be revealed.

DISCUSSION QUESTIONS

1. What conditions favor the preservation of plants as fossils?
2. Are any plants being so preserved at the present time? If so, suggest locations.
3. Summarize the eras of geologic time and their component epochs or periods. In which were the exposed strata in your vicinity deposited?
4. Define or explain petrifaction, cast, and impression. Which is most valuable to the morphologist? Why?
5. Briefly define the peel method of preparing fossil plants for microscopic study.

Chapter 21
INTRODUCTION TO SEED
PLANTS; DIVISION CYCADOPHYTA

INTRODUCTION TO SEED PLANTS

The members of the plant kingdom discussed as representative morphological types in Chapters 3–20 all are seedless plants, the *supposedly obscure* (in the eighteenth century!) reproductive organs and life cycle of which suggested to Linnaeus the name Cryptogamia (Gr. *kryptos*, hidden, + Gr. *gamos*, marriage). The seed plants were later called Phanerogamae (Gr. *phaneros*, visible, + Gr. *gamos*), an allusion to the prominence of their supposed sexual organs, which in fact are spore-bearing organs.

The seed plants are vascular plants usually containing abundant xylem and phloem, except in secondarily reduced aquatics. This attribute they share with the vascular cryptogams, with which they sometimes are grouped in a single division, Tracheophyta (inside back cover). This category has not been adopted in the present text, because in the authors' opinion such a grouping places too much weight on one attribute. The habit of producing seeds formerly was considered to be such an important indication of relationship that all seed-bearing plants were once classified in a single division, Spermatophyta (inside back cover). Paradoxically, however, this division has been replaced in many modern classifications by a still more inclusive taxon, the Pteropsida, under the division Tracheophyta. In spite of its lower rank, the Pteropsida includes a wider assemblage of plants, namely, the megaphyllous ferns as well as the seed plants. This arrangement reflects the views of those who see in the extinct seed ferns (p. 609) a possible bridge between the ferns and the living seed plants. In some systems of classification, the seed plants in which the seeds are not enclosed in a sporophyll are usually grouped in a taxon, Gymnospermae (inside back cover). This is not done in this text because it is considered that this practice also places too much emphasis on a single attribute.

The formal taxonomic designation "Gymnospermae" (Gr. *gymnos*, naked, + Gr. *sperma*, seed) and the informal term "gymnosperm" refer to the lack of enclosure of the seeds of certain plants (Fig. 21-1D, E). By contrast, the terms "Angiospermae" (Gr. *angeion*, a vessel, + Gr. sperma) and "angiosperm" signify that in other seed plants the seeds are enclosed in a structure variously known as the **carpel** or carpels, **pistil,** or **megasporophyll**

(Fig. 21-1B). Angiosperms, the division Anthophyta, comprise the "flowering plants," to be discussed in Chapters 26 and 27. Such topics as the origin and evolution of the seed and the origin and relationships of the seed plants are deferred to Chapter 27, a point by which, it is hoped, the reader will have assimilated adequate basic information to profit from such a discussion.

The classification of the seed plants used in the present text recognizes the important diversities apparent among the plants formerly included in the Gymnospermae. The production of seeds is considered to occur in plants of several *divisions:* the Cycadophyta, Cycadeoidophyta, Pteridospermophyta, Ginkgophyta, Coniferophyta, and Gnetophyta (all gymnospermous) and Anthophyta (angiospermous).

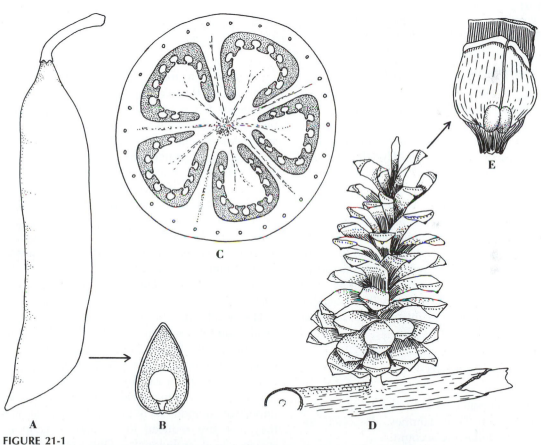

FIGURE 21-1
Location of the seeds of seed plants. **A–C.** Angiospermy. **A.** Fruit (pod) of garden pea. **B.** Transection of the preceding showing enclosed seed. **C.** Transection of the tomato fruit with enclosed seeds. **D, E.** Gymnospermy. **D.** Seed cone or strobilus of pine. **E.** Adaxial view of one of the ovuliferous scales of the preceding showing unenclosed, winged seeds. (*After Bold.*)

A   **B**

FIGURE 21-2
Zamia floridana. **A.** Plant with microstrobili.
B. Plant with megastrobilus. ×0.1.

DIVISION CYCADOPHYTA[1]

The Cycadophyta include both extinct and extant gymnospermous seed plants with fern-like foliage. Three classes—the Pteridospermopsida (seed ferns, p. 609), Cycadeoidopsida (cycadeoids, p. 619), and the Cycadophyta (cycads)—formerly were included in the division Cycadophyta. However, the first and second classes (above), known only as fossils, have herein been elevated to divisional status. The remaining class, Cycadophyta, in addition to fossil members (p. 617), includes a single order (Cycadales) with living genera. Because of their familiarity and availability, the living Cycadales will be discussed first in introducing the seed plants.

CLASS 1. CYCADOPSIDA

The Cycadopsida include the single order Cycadales, with three families, discussed together in the following account.

[1]For discussion of the carcinogenicity and neurotoxicity of cycads, consult Sixth International Cycad Conference, in Federation Proceedings 31: 1465–1546.

General features. The Cycadales, commonly called the cycads, include a small group of nine or ten genera and 100 species of tropical plants that superficially suggest both ferns and palms insofar as the general form of the plant body is concerned (Figs. 21-2, 21-3, 21-4). Formerly included in a single family, the ten genera of cycads are now classified in three families: Cycadaceae, with the single genus *Cycas;* Stangeriaceae, which includes only *Stangeria;* and the Zamiaceae, to which *Zamia, Microcycas, Macrozamia, Lepidozamia, Encephalartos, Bowenia, Ceratozamia,* and *Dioon* belong (Johnson, 1959).

The cycads are sometimes regarded as "living fossils," and it has been predicted that they will be extinct in the future. There is good evidence, indeed, that one of the genera, *Microcycas,* endemic to Cuba, is rapidly approaching extinction at present. Once widely distributed and important components of the earth's vegetation, they now are reduced in number of genera, species, and individuals. Only one genus, *Zamia* (L. *zamiae,* erroneous reading in Pliny for *azaniae,* pine nuts), is represented in the flora of the United States (in Florida), although

A

B

C

FIGURE 21-3
A. *Dioon edule.* **B.** *Ceratozamia mexicana.* **C.** *Dioon edule* with maturing microstrobilus. **A, B.** ×0.015. **C.** ×1/12. (**C.** *Courtesy of the Field Museum of Natural History.*)

the other genera are cultivated in conservatories and tropical gardens.[2]

In addition to *Zamia*, other cycads of Mexico and the West Indies are the genera *Dioon* (Gr. *dis*, two, + Gr. *ōion*, egg) (Fig. 21-3A, C), *Ceratozamia* (Gr. *keras*, horn, + *Zamia*) (Fig. 21-3B), and *Microcycas* (Gr. *mikros*, small, + *Cycas*). In the Southern Hemisphere, cycads occur especially in Australia and Africa. *Stangeria* (after Dr. Stanger[3]) (originally described as and considered to be a fern until it produced strobili!) and *Encephalartos* (Gr. *encephalos*, brain, + Gr. *artos*, bread) are native to South Africa. *Bowenia* (after G. T. Bowen[4]), *Macrozamia* (Gr. *makros*, long, + *Zamia*), and

Cycas occur in Australia. *Cycas* (Gr. *kykos*, a palm) occurs in India and extends to southern Japan; one species grows natively in Madagascar.

Cycas (Fig. 21-4) is widely planted as an ornamental in warmer portions of the United States in the southern portions of states bordering on the Gulf of Mexico, but *Zamia* is emphasized as representative of the cycads in this book because it is usually to be found in university greenhouses or may be obtained readily. Furthermore, the small stature of the plants, usually less than 4 ft tall, makes it feasible to maintain a supply of mature specimens in a limited space.

Two or more species of *Zamia* occur in Florida. Of these, *Z. integrifolia* and *Z. umbrosa* have been studied with respect to reproduction. *Zamia* is widely distributed in the West Indies, Mexico, and northern South America to Chile. Norstog (1980) estimates there are about 30 species.

Vegetative morphology. The plant body of *Zamia* (Figs. 21-2, 21-9) consists of a relatively

---

[2]Such as the Fairchild Tropical Garden at 10901 Old Cutler Road, Miami, Florida, and at the Montgomery Foundation Research Center at 11935 Old Cutler Road. All the genera and approximately 100 species are growing at these two establishments.

[3]Surveyor General of Natal, who sent the plant to England in 1851.

[4]First governor of Queensland.

A

B

FIGURE 21-4
Cycas revoluta. **A.** Plant with crown of new leaves. **B.** Detail of circinately vernate pinnae. **A.** ×0.015. **B.** ×2. (**A.** *Courtesy of Dr. Elsie Quarterman.* **B.** *Courtesy of Dr. Edgar E. Webber.*)

short, vertical, approximately conical stem, which tapers toward the base; the base bears a number of fleshy roots. Remains of leaves of former seasons are visible on the upper portions of the tuberous stem, and the apex supports a crown of spirally arranged, leathery, dark-green leaves. The latter are pinnately compound and decidedly fernlike in appearance. The leaves of *Cycas* (Fig. 21-4B) are even more fernlike because of the circinate vernation of the young pinnae. In *Ceratozamia*, moreover, the entire young fronds are circinately vernate.

As in the Pteridophyta, the leaves of a majority of the cycads are the most striking and dominant organs of the plant. They are arranged spirally on the axis, and under usual greenhouse conditions both *Zamia* and *Cycas* produce a new crown of leaves annually. Leaf crowns of preceding seasons may persist, but they finally become reflexed and disintegrate. The veins run in a direction parallel to the long axes of the pinnae; they arise from basal dichotomies in *Zamia*. The pinnae of *Cycas* have a single midvein and lack laterals. The leaflets of *Zamia* display histological features usually associated with xerophytism (Fig. 21-5). The epidermis of both surfaces is thickened heavily on the outer walls. Stomata are present only on the lower surface of the pinnae and are sunken. A thick-

FIGURE 21-5
Zamia sp. Transection of a pinna. ×60.

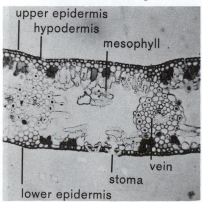

walled **hypodermis,** an additional protective layer, lies beneath the upper epidermis. The mesophyll is differentiated into palisade and spongy zones.

The apex of the cycad stem is occupied by a dome of meristematic cells rather than by the single apical initial that is characteristic of so many vascular cryptogams. The meristematic zone differentiates into permanent tissues during the formation of strobili, and new apical meristems arise successively from subapical meristematic cells lateral to the original 'meristem. A transverse section of the mature stem of *Zamia* (Fig. 21-6) reveals a surprising paucity of xylem. The latter is composed largely of scalariform-pitted tracheids. A large part of the stem is occupied by the central pith and the extensive cortex, the cells of which are full of stored metabolites. The stem surface is protected by an impermeable periderm layer. Mucilage canals are abundant in the cortex and pith. The cortex is traversed by leaf traces, which have an unusual path. Each leaf of *Zamia* receives from seven to nine traces, which depart from different points in the stele and girdle it, rising slightly as they approach the leaf base, which may be 180° from the point of origin of some of the traces.

Although a cambium is present between the xylem and phloem, it is relatively inactive, so that the amount of vascular tissue remains small, and clearly defined growth rings are absent from the xylem of *Zamia*. The roots of cycads are fleshy, although considerable secondary xylem is produced. In some genera, certain root branches are apogeotropic, growing on and above the soil. It is postulated that they serve as vehicles for gaseous interchange. Such roots develop tubercular nodules, which contain the blue-green alga *Anabaena*. The nodular roots can develop in sterile culture (Fig. 21-7). The effects of light on root organization, development, and apogeotropism were investigated by Webb (1981, 1982), D. T. Webb (1981; 1982 *a,b,c*; 1983 *a,b*; 1984), and Webb et al.(1984).

529

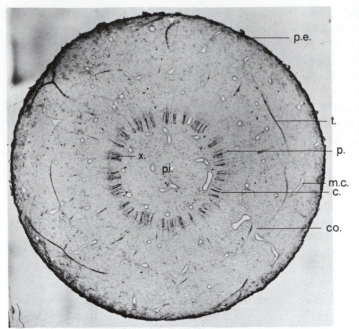

FIGURE 21-6
Zamia sp. Transection of a
stem. c., cambium; co.,
cortex; m.c., mucilage canal;
p., phloem; pe., periderm;
pi., pith; t., leaf trace; x.,
xylem. ×0.75. (*From a
preparation and negative of
Dr. C. J. Chamberlain,
courtesy of the Field Museum
of Natural History.*)

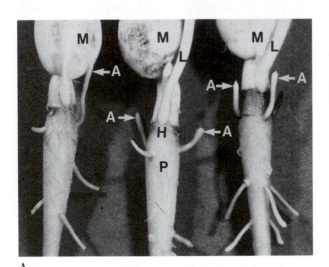

A

B

FIGURE 21-7
Zamia pumila seedlings in axenic culture. **A.**
Three seedlings with apogeotropic (above) and
normally oriented roots. **B.** Apogeotropic root
nodules. A, apogeotropic secondary root; H,
hypocotl; L, leaf; M, female gametophyte; N,
nodule; P, primary root. **A.** ×1. **B.** ×2. (*After
Webb.*)

530

The alga lives in the intercellular spaces between the epidermal cells and forms a distinct layer, which is buried by an extension of the root cap (Wittmann et al, 1965). Obucoivicz et al. (1981), Grilli (1974), Grilli-Caiola (1975), Caiola (1980), and Caiola-de Vecchi (1980) investigated the blue-green algae in the roots of *Macrozamia* and *Encephalartos* and in culture. The alga fixes gaseous nitrogen (Bergersen et al., 1965). Storey (1968) has reported that in seven species of cycads he studied, somatic reduction in chromosome number occurs regularly in the algal zone of the apogeotropic roots.

Reproduction: the sporophyte.   In *Zamia* and other cycads and in all seed plants, two kinds of spores, with respect to function, are produced: One type is genetically conditioned to produce male gametophytes, and the other, female gametophytes. This obligate dioecism differs from that in *Marchantia* (p. 204) and *Polytrichum* (p. 278), however, in that all four spores arising from one sporocyte in the seed plants grow into either male or female gametophytes. Furthermore, in seed plants, the spores that produce male gametophytes arise in sporangia different from those in which female gametophyte-producing spores develop.

It will be recalled that in the vascular cryptogams *Selaginella, Isoetes, Marsilea, Azolla,* and *Salvinia,*[5] a similar dichotomy of spore function prevails, but in those genera it is associated with a dimorphism in the spores themselves, a phenomenon known as heterospory. Now careful measurements of the two kinds of spores in seed plants reveal little significant size difference between the male and female gametophyte-producing spores. Accordingly, it has been suggested that seed plants are homosporous. It is clear, however, that such so-called homospory differs profoundly from that of other homosporous plants, as noted above. In current usage, as in this text, the two kinds of spores of seed plants are considered to be *functionally* heterosporous, in spite of the absence of marked differences in their size. It has been suggested that the permanent retention of the female gametophyte-producing spore within its sporangium in seed plants accounts for diminution of its size. As a matter of fact, this spore in seed plants *ultimately* (during gametogenesis) achieves much greater size than the microspore. The time of its increase in size is later than in heterosporous cryptogams (see p. 356).

Those who are unwilling to accept this broader usage of heterospory have suggested the terms "androspores" and "gynospores" for the spores of seed plants. The assignment to spores of prefixes denoting sex cannot be supported, in the writers' opinion.

After due consideration of these points, the spores that develop into male gametophytes are called **microspores** and those that develop into female gametophytes are designated **megaspores** in the present volume, on the basis of their homology with those of vascular cryptogams and in light of the *ultimate* increase in size of the megaspore. The microsporangia and megasporangia of *Zamia* are segregated in different strobili, which are produced by different individuals (Fig. 21-2). This segregation is known as **dioecism** (Gr. *dis,* two, + Gr. *oikos,* house). The cytological basis for this has been reported for *Cycas revoluta* (Segawa, Kishi, and Tatumo, 1971). Plants that produce megaspores have 20 somatic chromosomes and an XX pair, while microstrobilate plants have 20 somatic chromosomes and an XY (heteromorphic) pair. Marchant (1968), however, was unable to find convincing evidence of chromosomal differences in dioecious cycads.[6]

The microstrobili (Fig. 21-8) are borne among the leaves at the stem apex in *Zamia.* They appear in summer (July) at first as small, conical emergences and gradually enlarge, finally be-

---

[5]And in a number of extinct plants.

[6]Norstog (1980) reported on the chromosome numbers of *Zamia.*

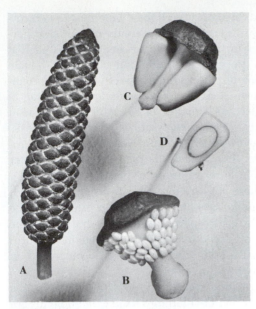

FIGURE 21-8
Zamia floridana. **A.** Microstrobilus at pollen-shedding stage. **B.** Abaxial view of microsporophyll with microsporangia. **C.** Megasporophyll and ovules before fertilization. **D.** Section of an ovule. **A.** ×0.5. **B, C.** ×2. **D.** ×0.75. (*Courtesy of the Field Museum of Natural History.*)

coming as long as 10 cm. They are brown, rather fleshy throughout development, and are composed of a central axis to which the spirally arranged microsporophylls are attached (Fig. 21-10A). There are between 28 and 50 microsporangia attached to the abaxial surface of each microsporophyll. They may show some evidence of being arranged in clusters. Each microsporangium is supported by a short thick stalk and has a several-layered wall. The number of spores is large, and the sporangial wall is composed of several layers. A tapetum functions in nutrition during microsporogenesis, the cells disintegrating to form a tapetal plasmodium. The microsporocytes undergo meiosis, each giving rise to a tetrad of microspores, which finally separate from each other. The walls of mature microspores are composed of several layers.

The megastrobili, which appear among the leaves of plants other than those that bear the microstrobili (Fig. 21-2B), are massive in construction (Figs. 21-9, 21-11A). The megasporophylls are dark brown, peltate structures attached to the axis of the strobilus by stipes (Fig. 21-7C, 21-11B). Each megasporophyll bears two white ovoidal bodies, the **ovules** (Figs. 21-8C, 21-11B), which are attached to the inner surface of the expanded tip of the megasporophyll by extremely short stalks. A median longitudinal section of a young ovule shortly after the appearance of the megastrobilus is shown in Fig. 21-12A. The central cells of the young ovule contain dense protoplasmic contents. These cells constitute the young megasporangium, in which the megasporocyte has not yet differentiated. The megasporangial tissue is surrounded by a multicellular covering, the **integument**, which is incompletely closed, leaving a minute

FIGURE 21-9
Zamia floridana. Portion of plant showing stem, leaf bases, and megastrobilus (at the time of fertilization). ×0.25. (*Courtesy of the Field Museum of Natural History.*)

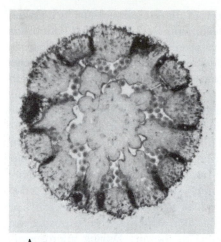

A

FIGURE 21-10

Zamia sp. **A.** Transection of young microstrobilus.
B. Transection of young megastrobilus. ×3.

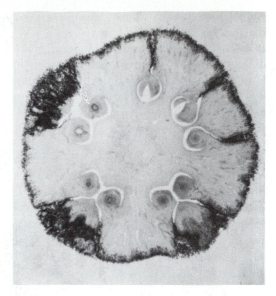

B

A   **B**

FIGURE 21-11

Zamia sp. **A.** Megastrobilus just before
pollination. **B.** Two megasporophylls and ovules
of same. **A.** ×1. **B.** ×3.

passageway, the **micropyle.** The integument is
free from the megasporangial apex in the cycads. This is at the pole opposite the point of
attachment of the ovule. The **ovule** of *Zamia* and
of other seed plants is an indehiscent megaspo-

rangium[7] surrounded by an integumentary layer
or layers. In the cycads, six to nine vascular
bundles arranged in a circle run vertically
through the integument. This may indicate a
phylogenetic union of previously distinct entities (see p. 610). The ovule of *Zamia* is about 10
cm long at the time of pollination. Ovules occur

---

[7]The megasporangium was called the **nucellus** before its
probable homology was appreciated.

FIGURE 21-12

Zamia floridana. **A.** Diagrammatic, median
longitudinal section of ovule showing integument
(i.) and megasporangium (m.), with single
megasporocyte. **B.** Linear tetrad of megaspores;
the lowermost, the functional megaspore. ×620.
(*After F. Grace Smith.*)

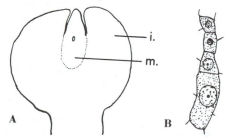

A   **B**

533

only in seed plants. Their possible origin is discussed on page 610.

Somewhat later in its development, one of the deeper cells of the megasporangium enlarges, functions as a megasporocyte, and undergoes two successive nuclear and cell divisions, during which meiosis takes place. The products of these divisions are arranged in linear fashion (Fig. 21-12B). They are interpreted usually as four megaspores arranged as a linear tetrad. The cell that gives rise to the linear tetrad is known as the **megasporocyte**. As noted above, the name megaspore has been assigned to the cells of the linear tetrad on the basis of comparison with the heterosporous cryptogams, in which the megaspores are markedly larger than the microspores.

Soon after the formation of the linear tetrad, three of the four megaspores degenerate. The survivor, usually the one farthest from the micropyle, hence the **chalazal**[8] one, appears to appropriate the products of the degeneration of the nonfunctional megaspores, and it increases in size as a result. This increase in size also depends on the transfer of nutriment from the fleshy megasporangial tissues to the megaspore. The latter never contains large amounts of stored metabolites, as do the megaspores of the heterosporous cryptogams. This surviving megaspore, which gives rise to the female gametophyte, is known as the **functional megaspore.**

Reproduction: the gametophytes. As in many species of *Selaginella*, the microspores begin their development into male gametophytes, and the megaspores their development into female gametophytes, while still enclosed by their respective sporangia. The early stages of the male gametophytes are also intrasporal, and those of the female gametophyte are permanently so in seed plants.

When the microspore (Fig. 21-13A) separates from the tetrad, it is uninucleate.[9] A single nuclear division and cytokinesis then take place within the microspore (Fig. 21-13B). This results in the formation of a small **prothallial cell** and a larger cell. The male gametophyte is considered to consist of a single vegetative cell, the prothallial cell, and an antheridial remnant. A second nuclear cell division in the large cell forms a small **generative cell** and a large **tube cell** (Figs. 21-13C, 21-14). When the male gametophytes have attained this stage of development, the microsporophylls separate slightly, apparently by elongation of the internodes of the axis of the microstrobilus, thus exposing the microsporangia to the air. During microsporogenesis and subsequently, the surface cells of the microsporangium thicken, except for a vertical belt over the summit. When the microsporangia are exposed, drying affects their bivalve-like rupture along a vertical line of dehiscence. The microspore walls containing the immature male gametophyte thus are shed from the microsporangium in the three-celled stage. They are known as **pollen grains** at the time of shedding. In the vicinity of Miami, Florida, this occurs between December and February each year. A **pollen grain** is a microspore containing an immature male gametophyte. The wall of the pollen grain consists of several layers.

Changes now take place in the megastrobili and the ovules within them. A slight elongation of the internodes, beginning at the base of the megastrobilus and extending gradually to the higher nodes, separates the hitherto contiguous sporophylls by 0.3- to 0.6-cm crevices (Fig. 21-15). It is through the latter that the windborne pollen grains sift into the megastrobilus, some apparently insect-pollinated (Faegri and Van der Pijl, 1979; see, however, Niklas and Kyaw Paw U, 1983). During this period, the cells at the apex of the megasporangium disinte-

---

[8]That portion of the ovule of seed plants farthest from the micropyle is called the **chalaza.**

[9]For the details of the development of the male gametophyte of representative gymnosperms, see Terasako (1982).

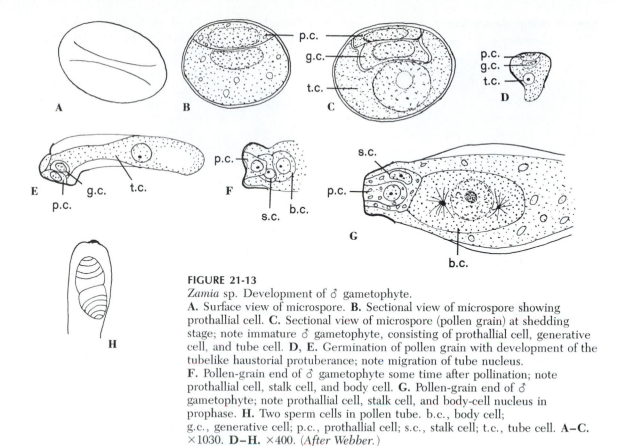

FIGURE 21-13

Zamia sp. Development of ♂ gametophyte.
A. Surface view of microspore. **B.** Sectional view of microspore showing prothallial cell. **C.** Sectional view of microspore (pollen grain) at shedding stage; note immature ♂ gametophyte, consisting of prothallial cell, generative cell, and tube cell. **D, E.** Germination of pollen grain with development of the tubelike haustorial protuberance; note migration of tube nucleus.
F. Pollen-grain end of ♂ gametophyte some time after pollination; note prothallial cell, stalk cell, and body cell. **G.** Pollen-grain end of ♂ gametophyte; note prothallial cell, stalk cell, and body-cell nucleus in prophase. **H.** Two sperm cells in pollen tube. b.c., body cell; g.c., generative cell; p.c., prothallial cell; s.c., stalk cell; t.c., tube cell. **A–C.** ×1030. **D–H.** ×400. (*After Webber.*)

grate, and a droplet of colloidal material known as the **pollination droplet** is secreted. The breakdown of these cells results in the formation of a small depression, the **pollen chamber,** at the apex of the megasporangium. A number of pollen grains come to rest in the pollination droplet when it protrudes through the micropyle, the latter a tubular passageway through the integument. As the pollination droplet dries, it contracts, carrying the pollen grains into the pollen chamber. The actual transfer of the pollen grains from the microsporangia to the micropyle of the ovule is known as **pollination.**

When the pollen grains have come to rest on the walls and base of the pollen chambers, their protoplasts protrude from the pointed pole

through the portion of their wall that lacks an outer layer, and they form tubelike haustoria into which the tube nuclei migrate in each case (Fig. 21-13D, E). These pollen tubes digest their way through the sterile tissue of the megasporangium, at first in a radial direction but then curving toward the apex of the developing female gametophyte. That they accumulate some of the substance digested from the megasporangium tissue is indicated by the appearance of starch grains in the elongating pollen tubes. The tubes are unbranched or sparingly branched. As many as 14 pollen tubes may be present within the apical tissues of a single ovule. The tubes may attain a length of 2–4 mm. Shortly after the pollen tube has been

535

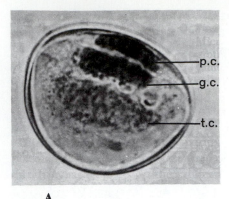

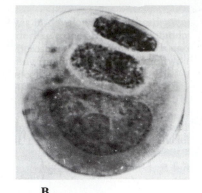

A   **B**

FIGURE 21-14
A. *Zamia* sp. Pollen grain at shedding stage. **B.** *Ceratozamia mexicana.*
Pollen grain at shedding stage. g.c., generative cell; p.c., prothallial cell;
t.c., tube cell. ×1400.

initiated, about a week after pollination, the generative cell divides to form a **stalk cell** and a **body cell** (Fig. 21-13F); the former name is based on the supposed homology with an antheridial stalk, while the latter refers to the antheridium itself. It has recently been suggested that the stalk cell is a sterile spermatogenous cell, while the body cell is, of course, a fertile one. Hence, the stalk cell is sometimes referred to as the sterile cell and the body cell as the spermatogenous cell. The body cell enlarges gradually as the pollen tube lengthens (Fig. 21-18A). The prothallial, stalk, and body cells all remain at the pollen-grain end of the elongating pollen tube (Fig. 21-13F).

Meanwhile, the thin-walled functional megaspore within the megasporangium has begun to enlarge. The tissue immediately surrounding it becomes vacuolated, forming a so-called spongy layer. The latter seems to function like a tapetum, its protoplasmic contents contributing to the enlargement and subsequent development of the functional megaspore. The nucleus of the megaspore now begins a period of repeated free-nuclear divisions. This results, at first, in the formation of a number of nuclei that lie in a peripheral layer of cytoplasm just within the cell wall of the original megaspore; the center of the

latter is occupied by a large vacuole. As free-nuclear divisions continue, however, additional cytoplasm is synthesized, so that the vacuole is obliterated gradually, and the megaspore lumen, which has been increasing constantly by absorption of the surrounding megasporangial

FIGURE 21-15
Zamia floridana. Megastrobili at pollination (January); note fissures between megasporophylls. ×0.75.

536

tissues, is ultimately filled with large numbers of nuclei and watery cytoplasm (Fig. 21-16).

As the female gametophyte matures, it is surrounded by two layers. The innermost, contiguous to the female gemetophyte, is the true **megaspore wall** (Maheshwari and Singh, 1967). On this is deposited a layer of material derived from the degenerated sterile tissue of the megasporangium, sometimes called the "tapetum"; this deposition is called the **tapetal membrane** (Pettit, 1966). Since the megaspore is never discharged from the megasporangium, this augmentation of the megaspore wall by the tapetal membrane is sometimes considered a vestigial attribute persisting from free-spored precursors.

Pettitt (1977*b*) has investigated ultrastructurally the wall of the megaspore of several cycads. He found that the detritus deposited on the wall (to form the tapetal membrane) contains lipids and sporopollenin. He emphasized that there are many similarities in the organization of the megaspore and pollen grain surfaces.

Pettitt (1977*c*) investigated chemically the pollen and ovules of several cycads. He found that enzymatic proteins were present in the walls of pollen grains and suggested that these may be involved in penetration of the megasporangium. He reported further the probability that a lectin was present in the female gametophyte and megaspore wall and suggested that this substance may be involved in discrimination between two sperms at the time of fertilization.

Cell-wall formation is then initiated in the free-nuclear female gametophyte, at first between the peripheral nuclei, and it continues in a centripetal direction until the female gametophyte is entirely cellular (Figs. 21-16, 21-17). The details of the formation of the cellular female gametophyte from its free-nuclear precursor have not been satisfactorily elucidated for any gymnosperm. Four (sometimes fewer) cells of the micropylar pole of the female gametophyte now function as archegonial initials. Each of them, by cell and nuclear division, forms an extremely large archegonium (Figs. 21-17,

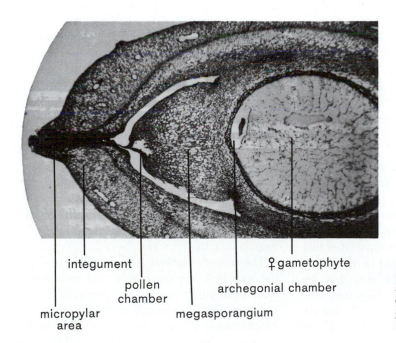

integument

pollen
chamber

micropylar
area

megasporangium

♀ gametophyte

archegonial chamber

FIGURE 21-16
Zamia sp. Longisection of ovule, made just as ♀ gametophyte became cellular. ×60.

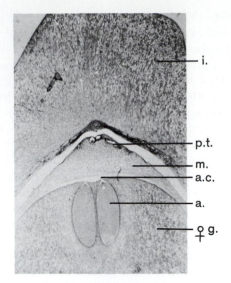

i.

p.t.

m.

a.c.

a.

♀ g.

FIGURE 21-17
Zamia sp. Longisection of ovule apex containing mature ♀ gametophyte. a., archegonium; a.c., archegonial chamber; ♀ g., ♀ gametophyte; i., integument; m., megasporangium; p.t., pollen tube. ×120.

21-18B) consisting of four neck cells and a large central cell (Norstog and Overstreet, 1965). Just before maturation, the nucleus of each central cell undergoes mitosis, forming the egg and ventral canal nuclei, which usually are not separated by cytokinesis.

The vegetative cells of the female gametophyte that surround each archegonium form a **jacket layer,** the cells of which presumably function in the transfer of nutriments to the developing egg and proembryo. The egg cytoplasm projects into these cells through small pitlike apertures. The egg cell may attain dimensions of 3.0 by 1.5 mm and is readily visible to the naked eye. The egg nucleus may attain a diameter of 550 μm.

Megasporogenesis and the maturation of the female gametophyte occupy about 5 months in *Zamia integrifolia.* During this period, all the component tissues of the ovule increase greatly by cell division, as do the megasporophylls and the strobilus axis. It has been demonstrated that the epidermis of the integument of the ovule has stomata, and that stomata also may be present on the surface of the megasporangium. Median sections of ovules containing mature female gametophytes are shown in Figs. 21-16 and 21-18B.

During the development of the female gametophyte, the haustorial pollen tubes progressively digest the tissues of the megasporangium. As the archegonia mature, subapical cells of the female gametophyte increase slightly in size and number, so that the group of archegonia comes to lie in a slight depression, the **archegonial chamber,** sometimes called the **fertilization chamber** (Figs. 21-16, 21-17, 21-18B), which is about 2 mm in diameter and 1 mm deep. Soon after this, the pollen tubes complete digestion of the megasporangium, and the pollen and archegonial chambers (Fig. 21-18A) then become continuous.

Just prior to this, usually late in May, the body cell divides to form two sperm cells (Figs. 21-13G, H; 21-18C; 21-19). These contain very large nuclei invested with a delicate layer of cytoplasm. During the enlargement of the body cell and its nuclear division, two centrosomelike bodies, currently called **blepharoplasts** (Mizukami and Gall, 1966), with astral radiations appear in the body cell, and one becomes associated with each sperm. These blepharoplasts are about 10 μm in diameter. Electron microscopy reveals that each has a single surface layer of about 20,000 procentrioles; these apparently become the centrioles, or basal bodies, of the numerous (70,000 [Norstog, 1982]) flagella (cilia) of the sperm. The flagella are arranged in a spiral pattern (Fig. 21-18C). Each of these spirals ultimately generates a great many short flagella or cilia.

The mature sperms now are liberated from the body cell, become motile, and swim actively within the pollen tube. The sperms of Z. *integrifolia* may attain a length of 300 μm; those of Z. *chigua* reach 400 μm (Norstog, 1977), the largest known in the plant kingdom. The pollen

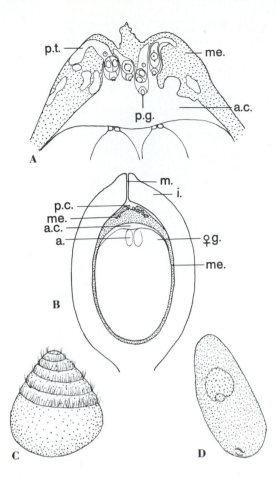

FIGURE 21-18
Zamia sp. **A.** Longitudinal section of apex of megasporangium; note four maturing pollen tubes, elongating at their pollen-grain ends and protruding into archegonial chamber; clear spaces in megasporangium are segments of other pollen tubes. **B.** Median longitudinal section of ovule before fertilization. **C.** Single sperm cell. **D.** Egg cell at karyogamy; remains of flagellar band near apex. a., archegonium; a.c., archegonial chamber; ♀ g., ♀ gametophyte; i., integument; m., micropyle; me., megasporangium; p.c., pollen chamber; p.g., pollen-grain end; p.t., pollen tube. **A.** ×10. **B.** ×2. **C.** ×250. **D.** ×18. (*Modified from Webber.*)

Observations of living material indicate that the sperms swim actively in the pollen tubes before the latter burst and discharge them with some liquid into the archegonial chamber; that the sperms swim actively in the archegonial chamber in the fluid that gathers there from the open pollen tube (Norstog, 1975) (this fluid has been estimated to have a solute concentration of 0.6 M); that the four neck cells separate; and that a portion of the egg cytoplasm protrudes into the archegonial chamber, drawing the sperm or sperms into the egg cytoplasm. At this point the four neck cells collapse and form an opening into the egg cell (Norstog, 1972a). This process is of sufficient violence to sever the flagellar band from the surface of the sperm in some cases (Fig. 21-18D). A sperm nucleus migrates to the vicinity of the egg nucleus, now of tremendous size, and nuclear union occurs (Fig. 21-18D). It should be emphasized at this point that pollination (the transfer of the immature male gametophyte to the micropyle of the ovule) and fertilization (nuclear union and the culmination of sexuality) are very different processes. In *Zamia*, they are separated by an interval of 5 months. Figures 21-17 and 21-18B show median longitudinal sections of an ovule at about the time of fertilization. The archegonia are readily visible by the unaided eye at this stage.

tubes are extremely turgid, probably because of the high osmotic value of the substances that have been digested and absorbed. This is evidenced by the fact that the sperms continue their motility in a 20% solution of cane sugar. The pollen tubes finally burst and discharge the sperm.

Plants with pollen tubes are said to be **siphonogamous;** this means that the male gametes are conveyed by a tube to the egg cells. Strictly speaking, this does not occur in the cycads or *Ginkgo* (p. 556), but it does in *Pinus* (p. 572), *Ephedra* (p. 592), and in *Anthophyta* (p. 651); in the cycads and *Ginkgo* the sperms are discharged into the archegonial chamber from the pollen-grain end of the pollen tube, the function of which is chiefly haustorial.

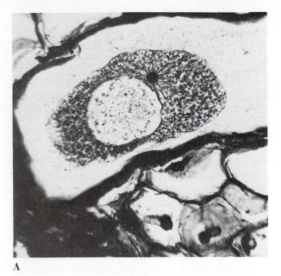

A

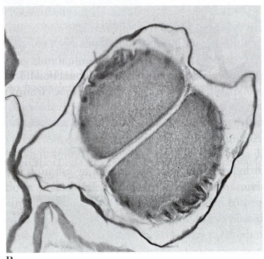

B

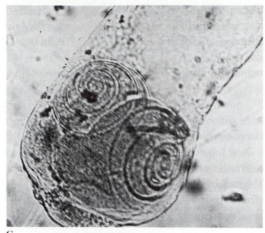

C

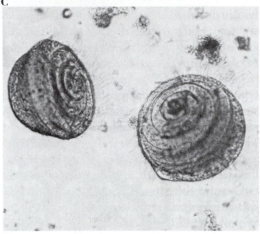

D

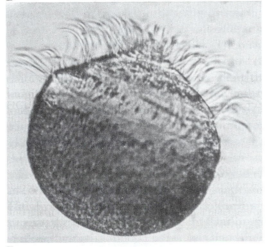

E

FIGURE 21-19
Zamia sp. **A.** Body cell within pollen tube, its nucleus in early prophase. Note blepharoplast and astral rays. **B.** Transection of pollen tube and body cell with two sperms. Note flagella and large nuclei. **C.** Sperms within pollen tube, at pollen-grain end of tube. **D, E.** Living sperms. **A.** ×250. **B.** ×150. **C.** ×170. (**B, C.** *After Norstog;* **D, E.** *Courtesy of Prof. Norstog.*)

540

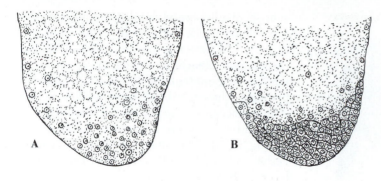

FIGURE 21-20
Zamia umbrosa. Early
embryogeny. **A.** Base of
archegonium with free nuclei
of proembryo. **B.** Cellular
proembryo. **A.** ×50. **B.** ×46.
(*After Bryan.*)

When the female gametophyte is grown in
tissue culture, it can develop roots and leaves,
both of which are haploid (Norstog and Rhamstine, 1967). For experimental morphological
studies of the gymnosperms and other vascular
plants consult Raghavan (1976) and Norstog
(1982*b*).

Reproduction: embryogeny and seed
development.   In the development of the zygote into the embryo in *Zamia* and other cycads,
the zygote nucleus enters a period of free
nuclear mitoses that is variable in extent. In
Zamia, approximately 256 nuclei are formed in
the egg cell (Fig. 21-20). Development of the
embryonic sporophyte has recently been stud-

ied in *Z. umbrosa.* Wall formation here is
initiated among the free nuclei near the base of
each archegonium and extends gradually toward
the neck end. Some free nuclei may remain
unenclosed by walls (Fig. 21-20). The lowermost cells of this intra-archegonial proembryo
function as a meristematic zone, which is covered by a caplike layer. An intermediate layer of
meristematic cells toward the neck region of the
archegonium ceases to divide; its component
cells increase in length and force the basal
meristematic cells and their cap out through the
base of the egg cell into the vegetative tissues of
the female gametophyte (Figs. 21-21, 21-22).
This elongating zone of the **proembryo,** the

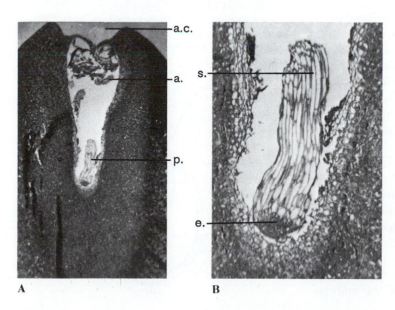

FIGURE 21-21
Zamia sp. **A.** Longisection of
apex of ♀ gametophyte after
fertilization. **B.** Enlarged view
of embryo in **A.** a., region of
archegonia (now disorganized);
a.c., archegonial chamber;
e., embryo-forming cells;
p., proembryo; s., suspensor.
A. ×40. **B.** ×160.

541

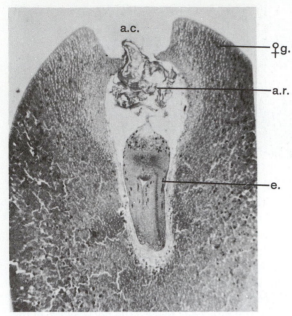

FIGURE 21-22
Zamia sp. Longisection of upper half of ♀
gametophyte with differentiating embryo.
a.c., archegonial chamber; a.r., archegonial
remnants; e., embryo; ♀ g., ♀ gametophyte.
×80.

consists of a **hypocotyl,** which extends from the
cotyledonary node to the **radicle,** the embryonic root. The latter is covered with a special
sheath, the **coleorhiza.** During development of
the female gametophyte, fertilization, and embryogeny, the entire ovule increases in size, and
changes occur in the massive integument, in
which three layers are more clearly differentiated (Fig. 21-23). The outermost of these becomes
fleshy and bright yellow-orange. The middle
layer is hard and stony, and the innermost
remains soft. The morphology of *Cycas* seeds in
relation to their dispersal, evolution and propagation was investigated by Deghan and Yuen
(1983). A median longitudinal section of the
mature ovule, now the **seed** (Fig. 21-23), reveals
that it contains an embryonic sporophyte embedded in vegetative tissue of the female gametophyte, the latter surrounded by the remains of
the megasporangium and an integument. This
statement regarding the seed of *Zamia* will
serve to define seed structure in all gymnosperms.

FIGURE 21-23
Zamia sp. Median longitudinal section of
recently shed seed. col., coleorhiza;
cot., cotyledons; ♀ g., ♀ gametophyte; i.,
integument; i.f., inner fleshy layer; m.,
micropyle; me., megasporangium remains; o.f.,
outer fleshy layer; pl., plumule; ra., radicle or
embryonic root; s., stony layer. ×5/12.

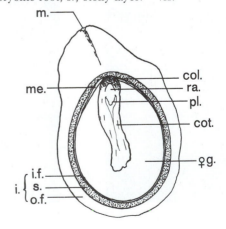

former called the **suspensor,** is augmented from
the meristematic zone below.

Although the zygotes of all the archegonia of
one female gametophyte may initiate embryo
formation (the phenomenon of **polyembryony**),
only one embryo normally is present in the
maturing seed (Fig. 21-22), the others having
aborted. The embryo grows at the expense of
the female gametophytic tissues, which are
gradually digested. The uppermost cells (those
nearest the archegonial neck) of the proembryo
function as a sort of buffer zone that resists the
pressure of the elongating suspensor sufficiently
to cause the latter to become coiled.

The terminal, embryo-forming cells of the
proembryo organize two **cotyledons,** between
which is a minute terminal bud, the **plumule**
(Figs. 21-22, 21-23). The axis of the embryo

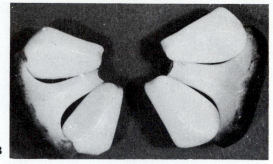

FIGURE 21-24
Zamia sp. **A.** Transection of megastrobilus with maturing seeds. **B.** Two megasporophylls with mature seed. ×1.

As the seeds ripen (Fig. 21-24), the megasporophylls separate and the seeds are abscised. Mature seed-bearing cones are 10–12 cm long. Germination may follow immediately, and no fixed dormant period occurs, except as it may be evoked by the environment. Germination of the seed may be hastened by carefully removing the fleshy and stony layers of the seed coat and then planting the seed on the surface of moist vermiculite (Hooft, 1970).

The first manifestation of germination is the protrusion of the coleorhiza and enclosed root through the micropyle. The coleorhiza soon grows downward and is penetrated by the enclosed primary root. The tips of the cotyledons remain embedded within the seed, where they function in absorbing food stored in the female gametophyte cells. After some weeks the proximal portions of the cotyledons emerge, and ultimately the first foliage leaf appears. Further growth and development of other foliage leaves are very slow in *Zamia*. More than 10 years are required for the young sporophyte to develop strobili.

Webb (1977) investigated the development of excised embryos of *Zamia* in axenic culture in light and darkness. He found that removal of the cotyledons and female gametophyte, as well as light, inhibited the elongation of primary and secondary roots: Light stimulated the formation of root nodules and apogeotropic root growth in two species of *Zamia* and one of *Bowenia* but not in *Dioon*.

Although *Zamia* has been chosen as representative of the cycads, several other genera may be available for observation in tropical regions or in conservatory collections. *Cycas* is of phylogenetic interest because its megasporophylls (Figs. 21-25, 21-26) are woolly, modified, pinnately compound structures, loosely aggregated at the crown of the plant, and not borne in strobili. The microsporophylls of *Cycas* and both types of sporophylls in the remaining genera of cycads are localized in strobili. Rao (1961, 1963, 1964, 1970a,b) has described the reproductive cycle in *C. circinalis*.

Variations of number of ovules and mature seeds in the megasporophylls of the several species of *Cycas* have been interpreted as evidence as to how the markedly nonfoliar megasporophylls of other genera may have evolved (Fig. 21-27). This evidence has, however, been challenged (Meeuse, 1963b).

FIGURE 21-25
Cycas revoluta with crown of megasporophylls.
×¹⁄₁₂. (*Courtesy of Prof. W. G. Moore.*)

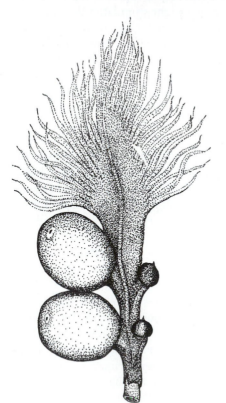

FIGURE 21-26
Cycas revoluta. Pinnate megasporophyll; note
mature seeds and aborted ovules. ×0.5.

FIGURE 21-27
Scheme deriving *Zamia* type (**E**) of
megasporophyll from that of *Cycas*. **A.** *Cycas
revoluta.* **B.** *C. media.* **C.** *Dioon.*
D. *Macrozamia.* (*After Meeuse.*)

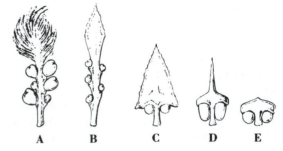

Summary

In general appearance, *Zamia, Cycas, Dioon,* and other cycads are suggestive of *Marattia*-like ferns and/or palms. The tuberous, armored, sparingly branched stems and the pinnately compound leaves and circinate young pinnae or whole fronds (*Ceratozamia*) contribute to this similarity. The stems are characterized by relative paucity of the vascular tissue, in spite of the presence of a cambium, and by their girdling leaf traces. The primary stele is dissected. Beck (1970) is of the opinion that the dissection of the steles in cycads and other gymnosperms is not occasioned by leaf gaps.

All cycads are dioecious, microsporangia and megasporangia being produced on separate strobili (in loosely grouped megasporophylls in *Cycas*) which occur on separate individuals. The megasporangia are covered by a massive protective layer, the integument, which surrounds them completely except for a minute passageway, the micropyle. Such covered megasporangia are known as ovules. The megasporangium wall is not specially differentiated. The megasporocyte gives rise to a linear tetrad of megaspores only one of which is functional. The megaspore does not contain large amounts of stored metabolites, is not markedly larger than the microspore, is relatively thin-walled, and is permanently retained within the megasporangium, in contrast to the megaspores of nonseed plants. As a result, the female gametophyte also develops to maturity within the megasporangium, and fertilization and development of the embryonic sporophyte also take place within the megasporangium and its integument. The thickening of the megaspore surface as the female gametophyte matures by deposition of a tapetal membrane is interpreted as a vestige from free-spored precursors. The permanent retention of these structures within the megasporangium has resulted in the seed habit. It will be recalled that a tendency toward such retention was noted in *Selaginella;* partial retention occurred in certain extinct Microphyllophyta. That the seed habit probably arose independently in more than one group of plants is thus suggested.

The male gametophyte of *Zamia* and other cycads also reveals innovations not present among the vascular cryptogams. Foremost among these is the development of a pollen tube; its primary function seems to be nutritive, since it serves as a haustorial organ to digest the tissues of the megasporangium. Its elongation at the pollen-grain end in cycads, just prior to fertilization, brings the male gametes into close proximity to the archegonia. The sperms of *Zamia* are multiciliate and actively motile. The number of sperms produced by each male gametophyte of *Zamia* is reduced as compared with the vascular cryptogams, two being produced in all the cycads except the genus *Microcycas*, which is said to produce 16–22 sperms. In this respect, the male gametophyte of *Microcycas* would be the most primitive of the seed plants.

In addition to the permanent retention of the female gametophyte within the megasporangium, there are two other adaptive modifications, namely, the pollen chamber and the archegonial or fertilization chamber. Another departure in *Zamia* is the free-nuclear phase in embryogeny, an attribute of gymnosperms not paralleled in the vascular cryptogams. Finally, the abscission of the ovule containing the embryonic sporophyte within the female gametophyte and remains of the megasporangium, all covered by the integument, produces the characteristic structures known as seeds.

DISCUSSION QUESTIONS

1. What characteristics of the plant bodies of cycads suggest *Marattia*-like ferns?
2. In what respect are the leaf traces of *Zamia* anomalous?

3. How do the stem and root apex of *Zamia* differ from those of many vascular cryptogams?

4. What evidence can you cite to indicate that cycad leaves are megaphylls?

5. Define or explain the terms endarch siphonostele, eustele, periderm, dioecism, ovule, seed, micropyle, pollen grain, freenuclear embryogeny, pollination, linear tetrad.

6. What conditions must prevail for seed formation to occur?

7. A possible approach to the seed habit is cited in certain Microphyllophyta that are heterosporous. Would you search for the origin of the seed plants in that group? Explain.

8. Compare the female gametophyte of *Zamia* with the gametophytes of *Thelypteris*, *Selaginella*, and *Isoetes* with reference to both vegetative and reproductive aspects.

9. Compare the male gametophyte of *Zamia* with that of heterosporous cryptogams.

10. What adaptive advantages accrue to the plant that produces seeds?

11. Where does one find cycads in nature?

12. Distinguish between the pollen and archegonial or fertilization chambers. Cite as many differences in these structures as you can.

13. With the aid of labeled drawings, summarize reproduction in *Zamia*.

14. Do all cycads produce strobili? Explain.

Chapter 22
DIVISION GINKGOPHYTA

CLASS 1. GINKGOPSIDA

Order 1. Ginkgoales

Family 1. Ginkgoaceae

Whereas the Cycadophyta, discussed in the preceding chapter, had comparatively small, sparsely branched stems and large, pinnately compound leaves, *Ginkgo* (Chinese,*yin*,silver, + *hing*, apricot), the sole extant genus of the division Ginkgophyta, order Ginkgoales, family Ginkgoaceae, is characterized by large, richly branched stems and smaller simple leaves, attributes that suggest the Coniferophyta, with which *Ginkgo* is sometimes classified. Furthermore, the extensive pith, scanty xylem, and large cortex of cycadean stems are in marked contrast to the small pith, abundant xylem, and narrow cortex in stems of *Ginkgo*.

Ginkgo biloba, the maidenhair tree (Fig. 22-1), has often been called a living fossil (Major, 1967). There has been considerable doubt that it still occurs in habitats where it has not been cultivated, although there are reports of specimens in the wild in eastern China, near the borders of the Anwei, Kiangsu, and Chekiang provinces south of the Yangtze River (Li, 1956). The fossil record indicates that genera related to *Ginkgo* were distributed widely, especially in the northern hemisphere, in earlier geological periods, some records extending back to the Permian (inside front cover).

The name "maidenhair tree" is an allusion to the similarity in appearance between certain leaves of *Ginkgo* trees and the leaflets of the maidenhair fern, *Adiantum*. The tree is widely cultivated in the United States and readily grown from seed. In Washington, D.C., and New York City, a large number of trees have been planted along some of the streets. The leaves are a beautiful yellow-golden color in the autumn. Many Oriental peoples have cultivated *Ginkgo* in their temple grounds. The Japanese once believed that *Ginkgo* exuded water during a fire, probably because these trees are more resistant than others to its disastrous effects.

Vegetative morphology.   Mature specimens of *Ginkgo* (Fig. 22-1) in cultivation may attain a height of more than 30 m. The form of young trees is narrowly conical, with the branches

FIGURE 22-1
Ginkgo biloba. Large trees photographed in China by the botanical explorer, Meyer. (*Courtesy of the Field Museum of Natural History.*)

ascending; but in older specimens, especially ovule-bearing trees, the form is rounded, with the branches somewhat spreading and drooping.

Growth and development of the aerial portions of *Ginkgo*, as in all woody plants of the temperate zone, are marked by seasonal periodicity. *Ginkgo* trees are deciduous, producing entirely new leaves each year; these persist for only one growing season. During the fall and winter, the delicate growing tips of the stems are enclosed by leaf primordia, which will emerge the following spring, and these in turn are covered by the resistant **bud scales** characteristic of woody plants. With the renewal of growth each spring, the bud scales are shed, and the embryonic stem tips undergo rapid cell division and enlargement, thus exposing the rudimentary leaves.

Examination of branches in the leafy condition reveals that there is a dimorphism of branching (Fig. 22-2). The elongate main axes are known as **long shoots.** These produce a series of spirally arranged leaves on widely separated nodes during their first year of growth. Older portions of the long shoots bear a large number of short lateral branches, the **spur shoots.**[1] These develop from the lateral buds of long shoots after the first season. Each spur

---

[1] See J. E. Gunckel and K. V. Thimann, "Studies of Development in Long Shoots and Short Shoots of *Ginkgo biloba* L. III. Auxin Production in Shoot Growth," *Amer. J. Bot.* 36:145–151 (1949).

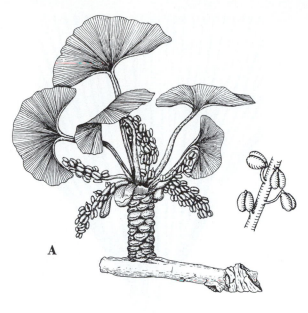

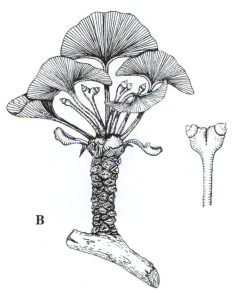

FIGURE 22-2

Ginkgo biloba. **A.** Portion of long shoot and spur shoot in spring, showing numerous, terminal bud-scale scars, bud scales, emerging leaves, and microstrobili; detail of axis and four microsporophylls at the right. **B.** Similar portion of ovulate plant, showing paired pedunculate ovules; detail of latter at the right. ×1.

shoot produces a terminal cluster of as many as 16 leaves every season. Growth in length of the spur shoots is extremely slow. However, on proper stimulation, such as injury to the terminal bud of a long shoot, spur shoots may metamorphose into long shoots.

Development of both spur shoots and long shoots may be traced to the activity of a group of meristematic cells and their derivatives, the **apical meristem.** The young leaves are initiated very close to the stem apex in *Ginkgo.* The vascular system is arranged in the form of a eustele in primary growth. The primary xylem is endarch and two traces serve each leaf.

As in stem development in all woody perennial plants, the completion of primary differentiation is followed early by the initiation of activity by the **cambium;** in fact, these two processes overlap, in part. The latter is a zone of meristematic cells of the procambium, between the primary xylem and phloem, which has remained undifferentiated. Frequently, if not always, cambial activity, with resulting differentiation of secondary xylem and phloem, commences before primary differentiation has been completed. The secondary tissues (Fig. 22-3), which have developed from cambium derivatives, are external to the earlier-produced primary xylem. Cambial activity is seasonal; hence, in older stems the annual zones of secondary xylem are readily distinguishable. Comparison of transverse sections of spur and long shoots of the same age indicates that considerably more secondary xylem (wood) is produced by the cambium of the long shoots. The pith and cortex of the spur shoots are persistent and more extensive than those of the long shoots. With the addition of secondary vascular tissues and expansion of the inner portion of the stem, the outermost cells of the cortex become organized into the **phellogen** or **cork cambium.** This is a meristematic cylinder, derivatives of which augment the cortex and replace the epidermis of the primary stem with **phellem** or **cork cells.** Abundant **lenticels,** which facilitate gaseous inter-

549

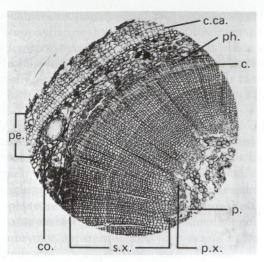

FIGURE 22-3
Ginkgo biloba. Transection of a long shoot early
in its second year. c., cambium; c.ca., cork
cambium; co., cortex; p., pith.; pe., periderm;
ph., secondary phloem; p.x., primary xylem;
s.x., secondary xylem. ×30.

change, develop in the stems after the phellogen becomes active.

A number of meristematic cells also constitute the apical meristem of the root, which is
covered by a protective cap. Each root contains
an exarch protostele which is diarch in arrangement of the protoxylem. The primary tissues of
the root also are supplemented by secondary
tissues developed by cambial activity.

The leaves of *Ginkgo* are perhaps its most
distinctive, readily observable feature. Leaves
of seedlings and those of long shoots are deeply
bilobed. Those of spur shoots are entire or
obscurely lobed (Fig. 22-2). The two vascular
traces of the petiole fork as they enter the blade,
where they undergo repeated dichotomies.
There were occasional connections of four different kinds between the dichotomies (Fig.
22-4) in the veins of 99% of the leaves studied by
Àrnott (1959). The mesophyll cells are differentiated internally (Fig. 22-5) into palisade and
spongy layers in the leaves of long shoots but are
less differentiated in leaves of the spur shoots.

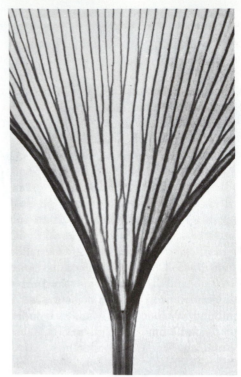

FIGURE 22-4
Ginkgo biloba. Base of leaf blade showing
dichotomous venation with occasional
anastomosis. ×3. (*Courtesy of Prof. H. J.
Arnott.*)

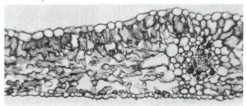

FIGURE 22-5
Ginkgo biloba. Transection of a leaf; note upper
and lower epidermis, the latter with stomata;
palisade and spongy mesophyll; and transection
of vein. ×65.

Stomata occur almost exclusively on the abaxial
surfaces of the leaves. All the organs of *Ginkgo*
are traversed by a series of mucilage canals in
which a sticky substance is secreted.

Reproduction: the sporophyte. The reproductive cycle of *Ginkgo* has been described

550

by Lee (1955) and Favre-Duchartre (1956, 1958). Like the cycads, which are dioecious, the microsporangia and ovules of *Ginkgo* are borne on separate individuals (Figs. 22-2, 22-6). Structural differences have been reported in the chromosomal complements of the microsporangiate and ovulate trees. The latter are said to have four satellited chromosomes and the microsporangiate trees only three. The un-

FIGURE 22-6
Ginkgo biloba. **A.** Spur shoot with microstrobili. **B.** Spur shoot with ovules. **C.** Close-up of ovules. **A, B.** ×0.5. **C.** ×3.

A

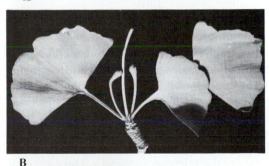

B

C

matched heteromorphic (one satellited, one not so) pair of chromosomes accordingly are interpreted as sex chromosomes, and after meiosis, microspores occur either with two satellited chromosomes or with one such chromosome (Pollock, 1957).

The microsporangia develop in lax strobili (Figs. 22-2A, 22-6A); the ovules occur in pendulous pairs at the tips of short, petiolelike stalks. Both arise among the vegetative leaves of spur shoots and emerge with the latter in the spring. The microstrobili of a given season develop during the summer preceding their emergence from the meristem of the spur shoot and attain considerable size by late autumn. They pass the winter in the microsporocyte state, meiosis and microsporogenesis (Fig. 22-7) occurring the following spring. Wolniak (1976) described distribution of cellular organelles during meiosis of the microsporocytes of *Ginkgo*. Each microstrobilus is composed of an axis that bears spirally arranged microsporophylls, the latter stalked and humped (Fig. 22-2A). Each microsporophyll bears two elongate microsporangia. The wall of the microsporangium is composed of five or six layers of cells, within which there is a

FIGURE 22-7
Ginkgo biloba. Microsporogenesis (aceto-carmine preparations). **A.** Microsporocyte. **B.** End of meiosis, I. **C.** Formation of tetrad of microspores. **D.** Single microspore and remains of common tetrad walls. ×1030.

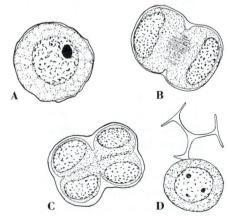

tapetum. Dehiscence of the microsporangia is by a vertical fissure.

The ovules are not borne in strobili but occur in pairs, terminally, at the tips of stalks or **peduncles** that emerge among the leaf bases of the spur shoots in early spring (Figs. 22-2B, 22-6B, C). Two vascular bundles traverse both the stalk of the microsporophyll and the peduncle that bears the ovules. The vascular bundles terminate at the base of the megasporangia. An enlarged, collarlike rim is present at the base of each ovule (Fig. 22-6B). In certain abnormal individuals the collar may be expanded and bladelike. This suggests that the collar may represent the remnant of an expanded sporophyll. The ovule itself consists of a massive integument that rather loosely surrounds the elongate megasporangium except at its tip, where it leaves a micropyle; it is attached only to the base of the megasporangium at this stage of development (Fig. 22-8). As in all seed plants, a single megasporocyte is differentiated, here deep within the megasporangium. Stewart and Gifford (1967) reported that the young megasporocyte has a thick, complex wall. As the megasporocyte approaches the time of meiosis, its organelles become polarized: the plastids and mitochondria move to the chalazal pole while abundant endoplasmic reticulum is present at the micropylar pole. The Golgi bodies do not become polarized. Pettitt (1977*b*) investigated the megaspore wall electron-microscopically. The latter becomes thickened with detritus from the degenerating megasporangium. This material contains lipids and sporopollenin, and the process suggests thickening of the pollen grain surface.

The sterile cells of the megasporangium that immediately surround the megasporocyte form a spongy tapetal tissue that becomes digested and disorganized later. Soon after the ovules have emerged from the buds of the spur shoots, certain of the apical cells of the megasporangia degenerate into mucilaginous masses, forming a deep pollen chamber in each ovule (Figs. 22-10,

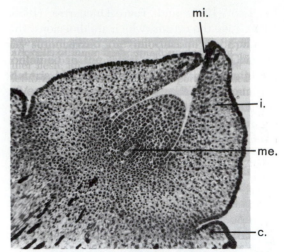

FIGURE 22-8
Ginkgo biloba. Median longisection of young ovule. c., collar; i., integument; me., megasporocyte; mi., micropyle. ×125.

22-11) (De Sloover-Colinet, 1963). Meiosis and megasporogenesis, in which a linear tetrad of megaspores is formed, occur at the time of pollination or soon after.

Reproduction: the gametophytes. Development of the uninucleate microspores (Fig. 22-9A) into male gametophytes is initiated, as in *Zamia,* soon after the completion of microsporogenesis and before the microspores are shed. The mature microspores are slightly protuberant at one pole, which is covered only by a single delicate layer, the **intine** (Fig. 22-9A). The outer layer, the **exine,** is absent from this portion of the microspore surface. Development of the male gametophyte begins with an intrasporal nuclear and cell division which delimits a small prothallial cell from a larger cell. The latter undergoes a second nuclear and cell division to form a second prothallial cell and an antheridial initial (Fig. 22-9B). The first prothallial cell sometimes degenerates promptly, but the second is more persistent. The antheridial cell now divides again, forming a small generative cell adjacent to the second prothallial cell and a tube cell (Fig. 22-9C). The immature male

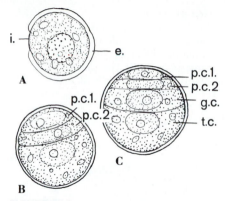

FIGURE 22-9
Ginkgo biloba. Intrasporangial stages of ♂
gametophyte development (I₂−KI preparations).
A. Microspore. **B, C.** Later stages. **C.** At
shedding stage. e., exine; g.c., generative cells;
i., intine; p.c.1, p.c.2, first and second prothallial
cells; t.c., tube cell. ×1033.

gametophyte in the microspore wall, now called
the pollen grain, is shed in this four-celled
condition.

The pollen, which is light and windborne, is
produced in large amounts by the microsporangiate trees. Pollination takes place in late March
or in April each year, although there is variation
in accordance with the temperature and latitude. Some of the pollen grains reach the apex
of each megasporangium and become lodged in
the mucilaginous material at the apex of the
pollen chamber. It has been reported that as
this material dries, it contracts and draws the
pollen grains into the pollen chamber in contact
with the megasporangial cells. A pollen tube,
formed in the region of the pollen grain covered
only by the intine, digests its way into the
tissues of the megasporangium, much as in
Zamia, apparently in *Ginkgo* also serving in an
haustorial capacity. Maturation of the male gametophyte (Gifford and Larson, 1980) is completed as the pollen tube grows through the
tissues of the megasporangium.

It will be recalled that each ovule contains
either a megasporocyte (Fig. 22-8) or a linear
tetrad or triad at the time of pollination. As in
Zamia, usually only one of the megaspores
functions in developing a female gametophyte.
It also will be recalled that the interval between
pollination and fertilization in *Zamia* is about 5
months, and that during this period the functional megaspore develops into the female gametophyte. Similarly, in *Ginkgo,* it takes from
late March until August in the southeastern
United States, or later in higher latitudes, for
the megaspore to produce a mature female
gametophyte.

Development of the female gametophyte involves an extended period of free-nuclear division (Fig. 22-10), during the early stages of
which the nuclei are arranged around the periphery of the enlarging megaspore. The latter
increases in size at the expense of the surrounding megasporangial tissue, the portion immediately surrounding the megaspore being called
the spongy tissue. The period of free-nuclear
division is of about 2 months' duration, during
which all the tissues of the ovule enlarge; approximately 8000 free nuclei are formed. Cell-
wall formation begins at the periphery of the
female gametophyte near the megaspore wall.
The latter may thicken up to 7 μm by the time of
fertilization and consists of two distinct layers.
The formation of cells in the female gametophyte gradually extends centripetally, but the
details of the development of the cellular condition are not well known. Since the innermost
nuclei near the center of the gametophyte are
finally surrounded by complete individual cell
walls, the mature cellular gametophyte (Fig.
22-11) may be split readily into two portions.
Nuclear and cell divisions continue, especially
in the micropylar end of the female gametophyte. Although surrounded by the remains of
the megasporangium as well as by the massive
integument, the vegetative cells of the female
gametophyte are green. Extraction of the pigment and study of its absorption spectra indicate
that chlorophylls *a* and *b* are present (unpub-
lished data of John Ridgway). It is probable that
nutrition of the female gametophyte of *Ginkgo,*

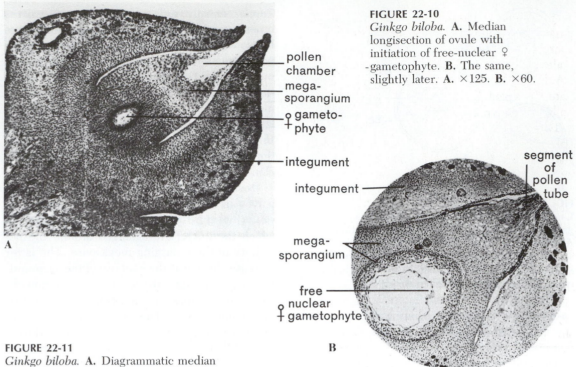

A

FIGURE 22-10
Ginkgo biloba. **A.** Median
longisection of ovule with
initiation of free-nuclear ♀
-gametophyte. **B.** The same,
slightly later. **A.** ×125. **B.** ×60.

pollen
chamber

mega-
sporangium

♀ gameto-
phyte

integument

integument

segment
of
pollen
tube

mega-
sporangium

free
nuclear
♀ gametophyte

B

FIGURE 22-11
Ginkgo biloba. **A.** Diagrammatic median
longitudinal section of entire ovule before
fertilization. a., aborted ovule; a.c., archegonial
chamber; m., micropyle; me., megasporangium;
♂ g., ♂ gametophyte in megasporangium; ♀ g.,
♀ gametophyte in megasporangium; p.c., pollen
chamber. **B.** Pollen-grain end of mature
gametophyte; note prothallial cell nucleus
(second) and two sperm cells. **C.** Blepharoplast
with numerous probasal bodies. **A.** ×300.
B. ×180. **C.** ×18,000. (**B.** *After Hirase.* **C.**
After Gifford and Lin.)

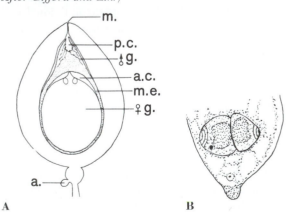

m.

p.c.

♂ g.

a.c.

m.e.

♀ g.

a.

A

B

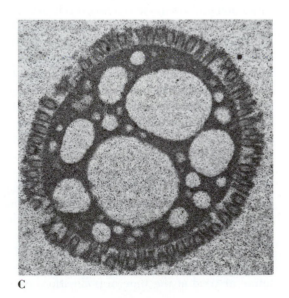

C

as in *Zamia* and other seed plants, is based largely on material derived from the parent sporophyte. Tulecke (1967) has grown the female gametophyte of *Ginkgo* in tissue culture. It remains green and haploid and may differentiate xylem cells. Tissue cultures derived from pollen remained colorless.

Two archegonia develop at the micropylar pole of each female gametophyte, but occasionally there are three or four (Figs. 22-11, 22-12, 22-13). The archegonia are large but not as large as those of *Zamia*. Each has four neck cells and a jacket layer. However, in *Ginkgo* the division of the central cell nucleus into egg and ventral canal nuclei generally is followed by cytokinesis, so that a ventral canal *cell* is formed. The latter disintegrates before fertilization. At that time the archegonia lie in a somewhat circular groove, the archegonial chamber (Figs. 22-11A, 22-13). The vegetative tissue between the archegonia elongates, forming a short column sometimes called a "**tent pole**," the apex of which extends to the megasporangial tissues (Figs. 22-11A, 22-13), and ultimately into the pollen chamber.

FIGURE 22-12
Ginkgo biloba. Apex of nearly mature ♀ gametophyte; central cell nucleus in prophase of division forming egg and ventral canal nuclei. ×80.

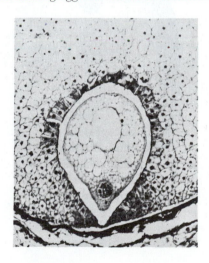

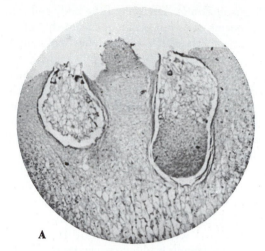

A

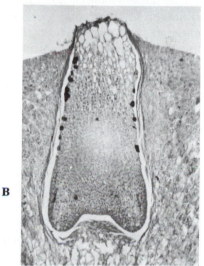

B

FIGURE 22-13
Ginkgo biloba. **A.** Apex of ♀ gametophyte; note "tent pole"-like protuberance and two archegonia, the one at the left, abortive, the one at the right, with a cellular proembryo. **B.** More advanced embryo than in **A;** cotyledons and hypocotyl developed. **A.** ×125. **B.** ×160.

During the development of the female gametophyte, the male gametophyte has been digesting its way through the apical portion of the enlarging megasporangium. Early in this process, the generative cell divides into two cells, the stalk (sterile) and body (spermatogenous) cells, as in *Zamia*. Division of the body cell into two sperms does not occur until shortly before fertilization about 16 weeks after pollination (Gifford and Lin, 1975).

At the poles of the dividing body cell (spermatogenous cell) are present blepharoplasts, as in *Zamia* (p. 535), one of which becomes associated with each of the sperm nuclei. Each blepharoplast has on its surface approximately 1000 probasal bodies, which become arranged as a spiral band in the mature sperm. Each of the probasal bodies becomes a basal body for a short flagellum of the sperm (Gifford and Lin, 1975) (Fig. 22-11B, C). The flagellate sperms of *Ginkgo* and *Cycas* were discovered by Hirase and Ikeno, two Japanese botanists, in 1896 (Ogura, 1967). All the nuclei of the male gametophyte, except sometimes the tube nucleus, remain at the pollen-grain end of the tube until just before fertilization. The function of the pollen tube, therefore, is primarily haustorial.

It has been reported that the archegonial chamber may be filled with fluid at fertilization. Just prior to this, the egg cytoplasm protrudes as a beak-shaped mass, forcing apart the four neck cells, which have previously enlarged to form a hemispherical projection. The egg nucleus may extend into the protuberant cytoplasm. When a sperm has been engulfed by the latter, the egg cytoplasm and nucleus withdraw into the egg cell; the nuclear portion of the sperm is included and unites with the egg nucleus.

The time and place of fertilization in *Ginkgo* are of interest (Eames, 1955; Lee, 1955; Favre-Duchartre, 1958). Fertilization may occur in one and the same tree at various times between August and October (in different seasons) and while the ovules are still attached to the tree or after they have been abscised and have fallen to the ground. In the latter case they are not unlike the vascular cryptogams insofar as the site of fertilization is concerned. The variation in timing of fertilization and embryogeny in *Ginkgo* is probably correlated with variation in the time of pollination.

Reproduction: embryogeny and seed development. Soon after nuclear union, the zygote nucleus enters upon a period of free-nuclear division, the division products being uniformly distributed throughout the cytoplasm of the egg. Eight successive nuclear divisions occur, so that approximately 256 proembryonic nuclei are formed. Cell walls then segregate the nuclei, and the entire embryo becomes cellular (Fig. 22-13). The cells of the embryo in the region of the base of the archegonium divide rapidly; those in the neck region elongate slightly but do not divide. The cells in the intermediate zone then enlarge somewhat, but a highly organized suspensor, such as is found in *Zamia*, does not occur in *Ginkgo*. The actively dividing apical portion of the embryo grows through the base of the archegonium, digesting the vegeta-

FIGURE 22-14
Ginkgo biloba. Longisection of ♀ gametophyte and embryo of seed, some weeks after shedding. c., cotyledons; ♀ g., ♀ gametophyte; p., plumule; r., radicle. ×4.

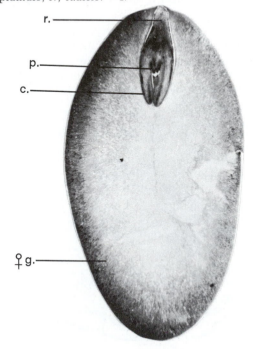

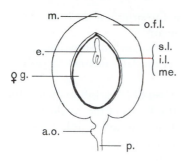

FIGURE 22-15

Ginkgo biloba. Diagram of a median longitudinal section of mature seed. a.o., aborted ovule; e., embryo; ♀ g., ♀ gametophyte; i.l., inner layer; m., micropyle; me., megasporangium remains; o.f.l., outer fleshy layer; s.l., stony layer of integument; p., peduncle.

tive tissues of the female gametophyte (Fig. 22-13) as it develops. The zygotes of both archegonia may initiate embryos, but usually only one is present in the mature seeds. Relatively early in development, the surviving embryo is differentiated into a root, a short hypocotyl, two cotyledons, and a short epicotyl terminated by the primordia of approximately the first five foliage leaves (Figs. 22-14, 22-15).

During the development of the embryo, the integument increases greatly and its three component layers are clearly differentiated. The outermost and most extensive becomes fleshy and a mottled green-purple. The innermost layer is rather dry and papery, and the middle layer is stony, as in the cycads. The mature seeds have the appearance of small plums (Fig. 22-16). A median section of a seed is shown in Figs. 22-14 and 22-15. The fleshy layer has a foul odor and may cause nausea and skin eruptions in certain individuals. The embryo and female gametophyte, however, are edible.

As in the cycads, there apparently is no genetically determined period of dormancy in the embryos of *Ginkgo*, although the environment may impose dormancy secondarily. Embryogeny is continuous, therefore, merging with germination, the latter in this case being not a "renewal of growth" but merely the emergence of the growing embryo from the seed.

In seed germination, *Ginkgo* is **hypogean,** the cotyledons remaining within the seed except for the proximal portions; from it they digest and absorb the remains of the female gametophyte, translocating it in soluble form to other parts of the developing seedling. The primary root emerges early and functions as a taproot. The

FIGURE 22-16

Ginkgo biloba. Branch with mature seeds. ×0.33. (*Courtesy of the Field Museum of Natural History.*)

seedling leaves are deeply bilobed, like those at the tips of long shoots in the mature trees.

Summary

Although the reproductive processes in *Ginkgo* are similar in many respects to those in *Zamia* and other cycads, its vegetative structure is in marked contrast. *Ginkgo* plants are large, richly branched trees, with an active vascular cambium that functions in stems and roots throughout the life of the plant in adding secondary xylem and phloem. The leaves are either almost entire or bilobed, never compound as in the cycads. The spur shoots superficially resemble the cycad stems in their armored surfaces and leaf crowns as well as in their extensive pith and cortex and in the paucity of xylem.

Next to those of the cycads, the sperms of *Ginkgo* are the largest in the plant kingdom, and, with those of the cycads, the only ciliate sperms among the living seed plants. The male gametophyte of *Ginkgo*, with its two prothallial cells, is considered to be more primitive than those of cycads, which have one. The occurrence of cytokinesis to form a ventral canal cell in the archegonium of *Ginkgo* is also a feature more primitive than that seen in other gymnospermous seed plants. An actively functioning suspensor is absent in the developing embryo of *Ginkgo*. The latter also differs from the Cycadophyta and vascular cryptogams in its extremely woody and deciduous habit, in which attributes it resembles the Coniferophyta (Chapter 23).

The fossil history of *Ginkgo*-like plants has been traced back to Permian times (p. 622).

DISCUSSION QUESTIONS

1. Why is *Ginkgo* sometimes called a "living fossil"?

2. What is the origin of the common name for *Ginkgo*?

3. Where does *Ginkgo* grow natively? Where is the tree nearest to your campus?

4. Of what possible significance are the collars at the bases of the ovules?

5. What structures are visible in a median longitudinal section of the ovule made at pollination?

6. Describe the development of the male and female gametophytes and fertilization in *Ginkgo*. Make labeled drawings to illustrate these phenomena.

7. Compare the embryogeny of *Ginkgo* with that of *Zamia*.

8. In what respects are the gametophytes of *Ginkgo* more primitive than those of *Zamia*? On what assumptions do you base your answer?

9. Why is it so easy to split the mature female gametophyte of *Ginkgo* into two portions?

10. What significance do you attach to the chlorophyll of the female gametophyte of *Ginkgo*? Describe procedures to demonstrate the nature of the pigment.

11. Draw and label a mature seed of *Ginkgo* as it would appear in a median longitudinal section.

12. Devise a definition of a seed after you have observed its structure.

13. Define the terms epigean, hypogean, epicotyl, hypocotyl, radicle, cotyledon, coleorhiza.

14. How do the archegonial chambers of *Zamia* and *Ginkgo* differ?

15. Is dormancy a necessary characteristic of seeds? In what plants are the seeds not characterized by dormancy?

16. Of what significance is the fact that roots, leaves, and vascular tissue (xylem) may develop on haploid tissue cultures of the female gametophyte of *Ginkgo*?

Chapter 23
DIVISION CONIFEROPHYTA

Some of the ancient Coniferophyta may well have been contemporaneous with certain Cycadophyta; they differ from the latter in a number of features. These may be summarized comparatively as shown at the top of the following page.

As treated in this text, the division Coniferophyta includes three classes, the Cordaitopsida, the Coniferopsida, and the Taxopsida. The first of these contains no extant members.

CLASS 1. CONIFEROPSIDA

Whereas the cycads are represented by only nine or ten living genera with 100 species and *Ginkgo* by a single living species, some 50 genera and 550 species of the class Coniferopsida have been described. Among these are such familiar trees as *Tsuga* (hemlock), *Abies* (fir), *Picea* (spruce), *Juniperus* (juniper, red cedar), *Sequoiadendron* (Sierra redwood, big tree), and other widely cultivated forms. Although the genera just listed are evergreen in habit, others, like *Larix* (larch, tamarack) and *Taxodium* (cypress), are deciduous. Certain Coniferophyta from the southern hemisphere—among them *Araucaria* and *Podocarpus*—are often cultivated in warm climates and in conservatories and botanical gardens. Members of the Coniferopsida form extensive forests in western North America and in parts of Europe and Asia. Many are large trees, but a few—some of the junipers, for example—are shrublike. In the southern hemisphere, conifers are abundant in temperate South America, New Zealand, and Australia.

Pinus and its relatives may be the dominant type in many forested regions. The great value of these trees as lumber, in the manufacture of paper, and for naval stores and other commercial enterprises has markedly reduced the extent of naturally occurring stands in areas readily accessible to transportation. *Pinus* has been chosen as the representative genus of Coniferophyta because of the great detail in which its structure and reproduction are known and because of its widespread distribution in the northern hemisphere. Representatives of other families are treated after the discussion of *Pinus*.

Cycadophyta	Coniferophyta
1. Leaves, large, complex fern-frondlike, compound	1. Leaves simple, often scalelike or needlelike
2. Xylem loosely arranged with abundant parenchyma, with broad parenchymatous rays, therefore, manoxylic	2. Xylem compact, composed mostly of tracheids, rays narrow, therefore pycnoxylic
3. Pith and cortex extensive, as compared with xylem and phloem	3. Pith and cortex restricted, xylem composing bulk of stem
4. Leaf bases persistent, forming an "armor" on the stem	4. Leaf bases not persistent
5. Stem not differentiated into long and spur shoots	5. Stem often differentiated into long and spur shoots
6. Leaf traces numerous per leaf and complex	6. Leaf traces one to few per leaf, simple
7. Ovules borne singly on leaves, in *simple* strobili, on a megasporophyll (*Cycas*), or on a conical axis	7. Ovules often borne in *compound* strobili or ovules borne singly

Order 1. Coniferales

Family 1. Pinaceae

The Coniferales contain the family Pinaceae, of which *Pinus* is here first discussed, and several other families.

Pinus

A relatively large number of species of *Pinus* occur in North America and in the United States. *Pinus strobus*, the white pine, is a familiar species in the northeastern part of the country and at high altitudes in the Appalachian chain. *Pinus virginiana*, the scrub pine, is abundant in the eastern part of the country; *P. palustris*, the long-leaved pine, is restricted to the coastal plain in the southeast and southwest. *Pinus ponderosa*, the western yellow pine, is one of the largest species of the genus; *P. cembroides* var. *edulis*, the piñon of the western states, produces large edible seeds. The bristlecone pine (formerly *P. aristata*), along with *Sequoiadendron*, is one of the longest-lived organisms known. Specimens from the Inyo National Forest in the White Mountains of California are more than 4600 years old (Schulman, 1958; Ferguson, 1968). The reader

no doubt will discover other species, either native or introduced, in the locality where he or she is living. The following account is based largely on *P. virginiana* and *P. taeda* as they develop in the southern United States.[1]

Vegetative morphology. The habit of *Pinus* is sufficiently familiar to permit dispensing with extensive description. The trees are freely and excurrently branched and evergreen, and, therefore, conspicuous elements of the areas where they occur during the winter months when the surrounding deciduous trees are leafless. Like all woody perennials of the temperate zone, growth is seasonal and periodic. In the winter months, the delicate growing points of the stems and the young leaf primordia are protected by impermeable bud scales. The latter are shed early during renewal of growth in the spring.

Two kinds of branches and two kinds of leaves are produced in *Pinus*. In addition to the familiar needle leaves, less conspicuous leaves, the scale leaves, occur on the main branches and at the bases of the branches that bear the needle

---

[1]The writers are indebted to Ruth B. Thomas for use of her microscopic preparations of *P. virginiana*.

leaves (Fig. 23-1). Only the needle leaves are photosynthetic. They occur singly or in groups that vary from one to eight in number in the several species. The short lateral branches on which the leaves arise are known as spur or dwarf shoots, as in *Ginkgo*. These grow in the axils of scale leaves of the long shoots (Fig. 23-1A), the latter increasing rapidly in length during the growing season. The spur shoots are lateral branches of determinate growth.

The needle leaves of species of *Pinus* have been observed to persist on the trees for periods varying between 2 and 14 years, after which they are abscised with the spur shoots that bear them. They are shed gradually, so that their fall is not as striking as leaf fall is in deciduous plants. Although the needle leaves are small— in some species they have only a single unbranched vein—they are seemingly megaphylls.

The needle leaves of *Pinus* exhibit striking xerophytic attributes (Fig. 23-2). The leaf surface is covered by a heavily cutinized epidermis within which there are one or more thick-walled hypodermal layers. The stomata lie beneath the leaf surface. The mesophyll is compact, with few air spaces; each mesophyll cell has trabecular projections on the walls, on the surfaces of which numerous chloroplasts are arranged. Resin canals are present in the mesophyll. The central portion of the leaf is delimited by a conspicuous endodermis within which, depending on the species, one or two vascular groups are embedded in transfusion tissue. The latter sometimes is interpreted as secondarily simplified, hence reduced, xylem. A detailed investigation of the anatomy of the leaves of *P. resinosa* was made by Gambles and Dengler (1982 *a,b*).

The multicellular apical meristem of the stem

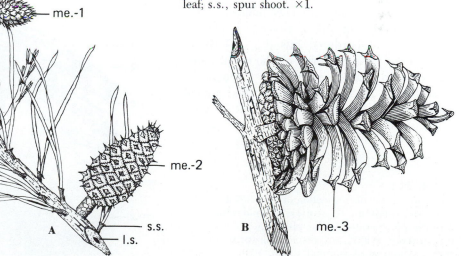

FIGURE 23-1
Pinus virginiana. **A.** Long shoot with megastrobili in spring, just after pollination. **B.** Megastrobilus shedding seed. l.s., long shoot; me.-1, megastrobilus of the season, just after pollination; me.-2, megastrobilus pollinated 1 year earlier; me.-3, megastrobilus pollinated 2 years earlier; s.l., scale leaf; s.s., spur shoot. ×1.

me.-1

s.l.

me.-2

s.s.

l.s.

A

B   me.-3

561

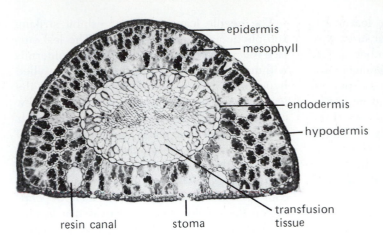

epidermis

mesophyll

endodermis

hypodermis

resin canal

stoma

transfusion tissue

FIGURE 23-2
Pinus nigra. Transection of leaf. ×30.

is active only during the spring and summer months, when the buds are developing the season's branches. The procambium differentiates into separate vascular strands of a eustele (Namboodiri and Beck, 1968), which encloses the pith and is surrounded by pericycle, cortex, and epidermis. Young stems are green and photosynthetic. Neuberger and Evert (1974, 1975, 1976) have investigated the phloem of *Pinus resinosa.* The cambium becomes active even before primary differentiation has been completed (Fig. 23-3), and it adds abundant derivatives that mature internally into secondary xylem and externally into secondary phloem. The xylem of *Pinus* is very homogeneous, being composed mostly of elongate tracheids, the prominent bordered pits (Fig. 23-4B) of which occur in single series on the radial walls. Narrow rays extend centrifugally through the xylem (Figs. 23-3, 23-4). Well-marked annual zones of secondary xylem are present in older stems; sieve elements of the primary and secon-

dary phloem are crushed, except for those most recently added by the cambium (Fig. 23-3).

Since the vascular cylinder is augmented by the cambium, a phellogen or cork cambium arises in the outermost layers of the cortex. Phellem (cork cells) and phelloderm (cork parenchyma or "secondary cortex") are added by this meristematic layer (Fig. 23-3). In older stems, phellogen strips arise progressively deeper in the cortex and finally in the secondary phloem; hence, the bark of older limbs is composed largely of alternating layers of dead secon-

FIGURE 23-3
Pinus sp. Sector of transection of a 3-year-old stem. ca., cambium; co., cortex; e., epidermis; p., pith; pe., periderm; ph., phloem; p.x., primary xylem; r.c., resin canal; spr.x., spring xylem; su.x., summer xylem. ×30.

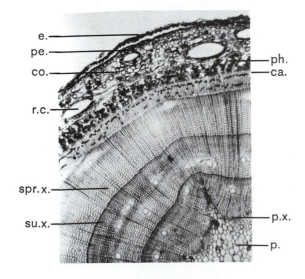

e.

pe.

co.

r.c.

spr.x.

su.x.

ph.

ca.

p.x.

p.

562

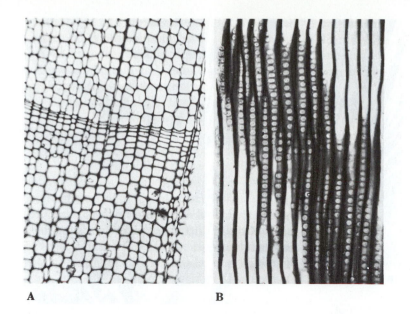

A   B

FIGURE 23-4
Pinus sp. Sections of wood or
secondary xylem. **A.** Transection
at junction of two annual
rings. **B.** Radial longisection;
note tracheids with bordered
pits in their radial walls. ×90.

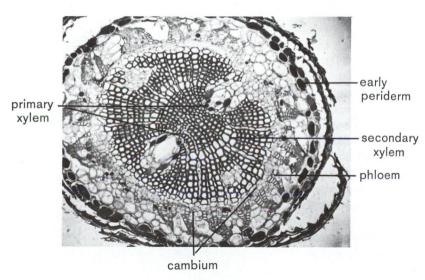

primary
xylem

early
periderm

secondary
xylem

phloem

cambium

FIGURE 23-5
Pinus virginiana. Transection
of a root in early secondary
growth. ×50.

dary phloem and periderm. Resin canals occur in the secondary xylem and in the cortex of the stem, as well as in the root and leaf.

The root of *Pinus* (and other Coniferopsida) is a protostelic taproot and may be diarch, triarch, or tetrarch. The stele is surrounded by a narrow pericycle, a prominent endodermis and exten-

sive cortex, and an epidermis with root hairs. The root-hair zone is very short in *Pinus.* The roots also undergo secondary growth and become extremely woody (Fig. 23-5). The younger portions of the root system frequently are infected with mycorrhizal fungi.

Reproduction. Konar and Oberoi (1969)

563

A

B

C

FIGURE 23-6
Pinus taeda. Successive stages in development of
microstobili from a tree in Bastrop, Texas.
A. January 15. **B.** March 1; meiosis in progress.
C. March 25; pollen being shed; note expansion
of terminal bud in **C. A.** ×0.5. **B.** ×0.05. **C.**
×1.

reviewed reproduction of Coniferophyta. The
microsporangia and megasporangia of *Pinus*
occur in separate strobili, but both are borne on
the same individual. Pine, therefore, is monoecious, in contrast with the cycads and *Ginkgo*.
The average minimum age for the appearance of
ovulate strobili is 5 years, but they may appear
precociously, even in the seedling (Smith and
Konar, 1969).

The microstrobili develop in clusters around
the base of the terminal buds of most branches
on mature individuals and are recognizable all
through the winter preceding the spring in
which they are to emerge. In *P. taeda* they
begin to develop in July prior to the spring in
which they will shed pollen (Greenwood, 1980).
During the dormant season the microstrobili are
covered with brown bud scales (Fig. 23-6A),
which are shed early in the spring as the strobili
enlarge (Fig. 23-6B). The microstrobilus is simple, as in the cycads and *Ginkgo*, being com-

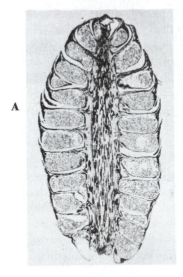

A

B

FIGURE 23-7
Pinus sp. **A.** Median longisection of
microstrobilus containing mature pollen; note
abaxial position of microsporangia. **B.** Abaxial
view of microsporophyll showing both
microsporangia. **A.** ×3. **B.** ×5.

564

posed of an axis bearing spirally arranged micro-sporophylls (Fig. 23-7A). To the abaxial surface of each of these are attached two elongate microsporangia (Fig. 23-7B). The microsporangial wall is about four layers of cells thick, and a prominent tapetum is present. Dehiscence takes place by a longitudinal fissure along a region of thin-walled cells. The microsporocyte stage is attained early in the spring, and microsporogenesis occurs as the strobili enlarge. In *P. virginiana*, the microsporophylls become purple-red as they protrude from the bud scales, and it is during this period that meiosis occurs. In the vicinity of Nashville, Tennessee, the meiotic process (Fig. 23-8) takes place between March 15 and April 1 each year. In *P. taeda* in central Texas, it occurs 4 to 6 weeks earlier. By contrast, in *P. monticola* on Vancouver Island, Canada, microsporogenesis occurs in mid-May (Owens and Molder, 1977a). Microsporogenesis in other species also is vernal, the dates varying with the species, latitude, and temperature. The microsporocytes contain abundant starch, which is digested during sporogenesis. Cytokinesis is accomplished by furrowing at the conclusion of the meiotic nuclear divisions. The individual microspores (Fig. 23-8D) are liberated from the microsporocyte walls through predetermined thin areas. As the microspores enlarge and mature, they develop a two-layered wall composed of an intine and exine. The two layers subsequently separate at two points on the surface of the microspores,

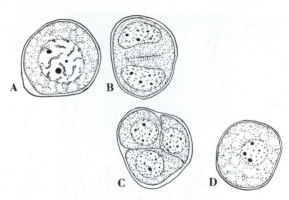

FIGURE 23-8
Pinus virginiana. **A–D.** Sporogenesis. (Late March, near Nashville, Tennessee.)
A. Microsporocyte. **B.** End of meiosis, I. **C.** Tetrad of microspores. **D.** Microspore soon after dissociation of the tetrad. ×1000.

thus forming the characteristic winged cells (Figs. 23-9, 23-14). The haploid number of chromosomes in all species of *Pinus* so far investigated is $n = 12$.

In *Pinus*, the megastrobili, or cones, are borne on short lateral branches near the apices of some of the younger green branches of the current season's growth (Figs. 23-1, 23-10). They are not visible clearly, therefore, until the terminal bud of such a branch has unfolded and elongated. When they first emerge, they are green or purple-red and soft in texture. They begin to harden after pollination.

The ovules are not borne directly on the bractlike appendages that emerge from the stro-

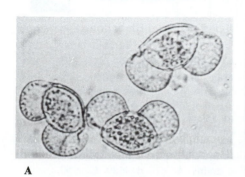

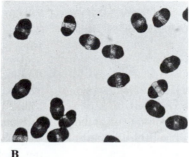

FIGURE 23-9
Pine pollen.
A. *Pinus virginiana.*
B. *P. taeda.*
Note air bladders.
A. ×350. **B.** ×100.

FIGURE 23-10
Pinus taeda. Two megastrobili at pollination.
(March 25, in Bastrop, Texas.) ×0.5.

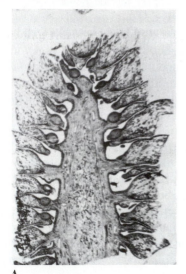

A

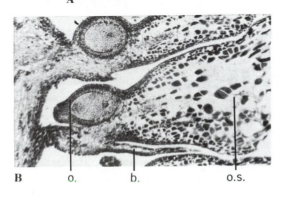

B   o.   b.   o.s.

bilus axis; instead, they develop on **ovuliferous scales** (Fig. 23-11C), which, in turn, are borne on the bracts. This arrangement has suggested, in light of comparison with fossil conifers, that the ovules are borne on reduced fertile spur shoots (= the ovuliferous scale) in the axile of a bract, much as are the vegetative spur shoots. The megastrobili of *Pinus* (and related genera) are, therefore, compound in organization.

Each ovule is composed of a massive integument surrounding a small megasporangium (Figs. 23-11C, 23-12). The integument is attached only to the lower portion of the megasporangium of young ovules. From the apex of the integument, two flaring arms project beyond the micropylar canal (McWilliam, 1958). As in *Zamia* and *Ginkgo*, a single cell of the megasporangium differentiates as a megasporocyte (Figs. 23-11C, 23-12), enlarging somewhat before it

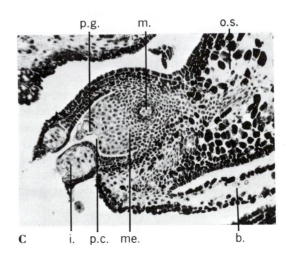

C   i.   p.c. me.   b.

FIGURE 23-11
Pinus virginiana. **A.** Longisection of young ovulate strobilus showing its compound nature. **B.** Two bracts and their ovuliferous scales enlarged. **C.** Median longitudinal section of ovule and associated structures soon after pollination. b., bract; i., integument; m., megasporocyte; me., megasporangium; o., ovule; o.s., ovuliferous scale; p.c., pollen chamber; p.g., pollen grains. **A.** ×8. **B.** ×20. **C.** ×50.

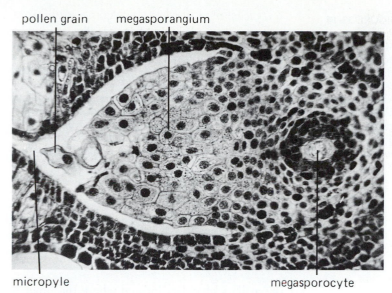

pollen grain   megasporangium

micropyle   megasporocyte

FIGURE 23-12
Pinus virginiana. Enlarged view of section of an ovule soon after pollination (mid-April). ×150.

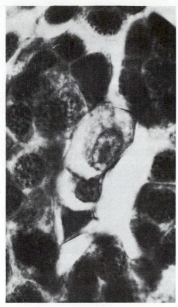

FIGURE 23-13
Pinus virginiana. Linear triad; functional megaspore above. ×1400.

undergoes meiosis to form a linear tetrad of megaspores. In some ovules, one of the daughter nuclei of the first division fails to divide a second time, so that a linear triad results (Fig. 23-13). In either case, however, only one of the megaspores functions, always the one farthest from the micropyle. The remainder degenerate, and their remnants are resorbed. Occasionally, more than one megaspore in a given ovule may function, with the result that two female gametophytes develop.

The microspores begin their development into male gametophytes before they are shed from the microsporangia. As in *Zamia* and *Gink-go,* the first stages in this process involve a series of intrasporal nuclear and cell divisions in which first and second prothallial cells, generative cell, and tube cell are produced, as shown in Fig. 23-14. The prothallial cells disintegrate rapidly, and their remains are incorporated in the wall as it thickens. The immature male

gametophytes, the pollen grains, are shed in this four-celled condition (Fig. 23-14D) by the elongation of the internodes of the microstrobilus and the drying and longitudinal dehiscence of the microsporangia. The pollen grains are produced in enormous numbers and are carried great distances by air currents.

Pollination in *P. virginiana* occurs during April in the vicinity of Nashville, Tennessee, although the date varies within a period of approximately 2 weeks, depending on the temperature. Pollination and other life cycle stages are earlier farther south and later northward in *P. virginiana* and other species. For example, pollination of *P. monticola* on Vancouver Island occurs in mid-June (Owens and Molder, 1977a). At the time of pollination, the distal portions of adjacent ovuliferous scales are slightly separated. The windborne pollen grains thus are carried readily into the fissures (Fig. 23-15). The pollen grains sift between successive scales into

567

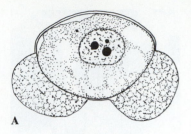

A

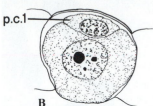

p.c.1

B

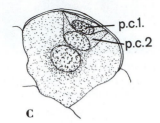

p.c.1.
p.c.2

C

p.c.1
p.c.2
g.n.

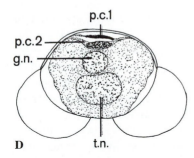

D

t.n.

p.c.
b.c.
t.n.

E
s.c.

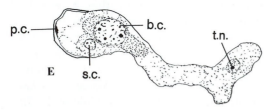

p.c.
s-1 s-2
s.c.
t.c.

F

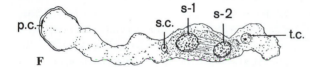

FIGURE 23-14

Pinus virginiana. Stages in development of the ♂ gametophyte. **A.** Microspore with bladderlike exine. **B, C.** Delimitation of first and second prothallial cells. **D.** Microspore (pollen grain) with enclosed ♂ gametophyte at shedding (pollination) stage; note two aborting prothallial cells, small generative cell, and large tube cell. **E.** ♂ gametophyte within megasporangium, 14 months after pollination (compare with Fig. 23-18); note stalk, body and tube cells, and nuclei. **F.** Almost mature ♂ gametophyte; note division of body cell nucleus into two sperm nuclei. b.c., body cell; g.n., generative nucleus; p.c., prothallial cell; s-1, s-2, sperms; s.c., stalk cell; t.c., tube cell: t.n., tube nucleus and cell. **A–D.** ×1030. **E, F.** ×410.

the axillary chamber formed where the bract is attached to the axis of the strobilus. The arms of the integument and its cells lining the entrance to the micropyle are coated with a sticky substance to which the grains adhere at pollination; they are thus similar in function to the stigma of the flowering plants (p. 636). Observation of living ovules at this time has revealed that a prominent pollination droplet develops at the orifice of the micropyle, and that the pollen grains that make contact with the droplet float through it or are drawn up by its evaporation into a chamber between the base of the micropyle and the apex of the megasporangium. This is sometimes called a pollen chamber but differs from the depression formed by the disintegration of apical cells of the megasporangium in the cycads and *Ginkgo*. Only a very small chamber of the latter kind develops in *Pinus* (Fig. 23-12). The megastrobili of *Pinus* are well adapted for wind pollination (Niklas and Paw U, 1983).

After the pollen grains have made contact with megasporangial tissue, the integumentary cells lining the micropyle elongate in a radial direction and decrease the diameter of the micropylar canal, thus "sealing" it. The integumentary arms then wither.

Although it contains sugars, the micropylar

A B

FIGURE 23-15
Pinus taeda. Enlarged views of megastrobilus at pollination (March 25). **A.** Note basal scale leaves and apices of ovuliferous scales. **B.** Note pollen grains windborne to megastrobilus. **A.** ×5. **B.** ×15.

fluid is ephemeral and seems to play only a mechanical, rather than a nutritional, role in transporting pollen grains from the integumentary arms and micropyle to the apex of the megasporangium. Soon after the pollen grain has made contact with the megasporangium, a pollen tube emerges and begins parasitic growth within the former. After pollination, the surface cells on adjacent ovuliferous scales undergo cell division and bridge the fissures between them, thus sealing the megastrobili.

At the time of pollination and germination of the pollen grains, the megasporangium of *P. virginiana* has arrived at the megasporocyte stage. Megasporogenesis follows about a month after pollination (mid-May), but the functional megaspore does not begin development into the female gametophyte for some months, often not until the following October or November. Both the megasporocyte and the functional megaspore apparently secrete substances that diffuse into the tissues of the megasporangium, which forms a nutritive, tapetumlike, spongy tissue (Fig. 23-12), as in *Ginkgo*. Development of the female gametophyte is extremely slow and is effected by a long-continued process of free-nuclear division, which occupies approximately 6 months in *P. virginiana*. As the number of free nuclei increases, additional cytoplasm is produced. The large central vacuole of the early female gametophyte (Fig. 23-16) is in this way replaced by watery cytoplasm. Meanwhile, the entire ovule and megastrobilus increase in size, and the exposed, distal portions of the ovuliferous scales harden (Fig. 23-17).

Early in May of the following year, about 13 months after pollination, the female gametophyte becomes cellular by the formation of delicate walls between the numerous (about 7200) nuclei. The process of wall formation begins at the periphery of the gametophyte and gradually extends in a centripetal direction. A thickened megaspore wall is not demonstrable around the female gametophyte in *P. virginiana*. However, according to Ruth B. Thomas (personal communication, 1952) and Pettitt (1966), an additional layer of material, the **tapetal membrane**, composed of the disorganized cells of the megasporangium, is deposited on the megaspore wall. This thickened layer, frequently visible on the surface of the female gametophyte, is made up, for the most part, of

569

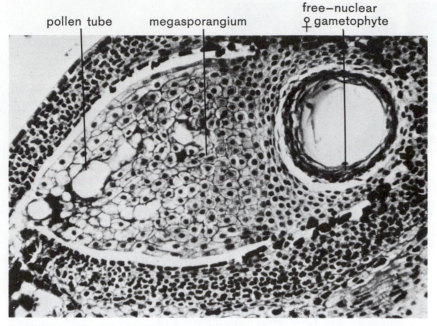

pollen tube   megasporangium   free–nuclear ♀ gametophyte

FIGURE 23-16
Pinus virginiana. Longitudinal section of ovule 1 year after pollination; note free-nuclear ♀ gametophyte surrounded by disintegrating cells of megasporangium; pollen tubes within the latter. ×150.

A   B

FIGURE 23-17
Pinus taeda. **A, B.** Megastrobili 1 year after pollination (March 25); note relation in position to terminal shoot in **A. A.** ×0.5. **B.** ×1.5.

disintegrating cells and nuclei of the megasporangium.

As the gametophyte becomes cellular, several cells at its micropylar end function as archegonial initials (Fig. 23-18A). There are usually two or three archegonia in *P. virginiana.* As the initials

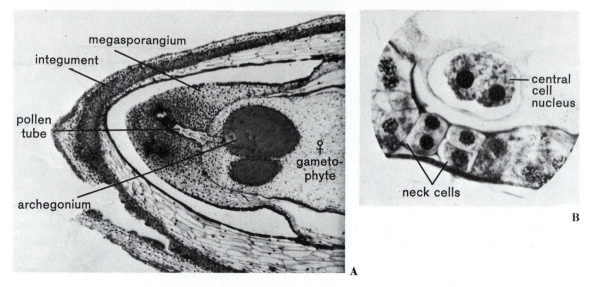

FIGURE 23-18

Pinus virginiana. **A.** Longitudinal section of apex of ovule at fertilization (14 months after pollination); note course of pollen tubes through megasporangium to archegonia. (Section not median at micropyle.) **B.** Apex of archegonium showing central cell nucleus, which will divide to form egg and ventral canal nuclei, and neck cells. **A.** ×30. **B.** ×500.

enlarge, the vegetative cells of the gametophyte immediately surrounding them become modified and organized to form a jacket layer around each archegonium. Prior to extensive enlargement of the archegonial initials, each divides into a neck initial at the surface of the gametophyte and a central cell. The neck initial divides to form a short neck (Fig. 23-18B), which may consist of as many as two tiers of four cells each; but frequently fewer neck cells are formed, the number varying. The central cell and its nucleus enlarge tremendously, as in the cycads and *Ginkgo*. Soon after they have attained their maximum size, early in June, the central cell nucleus migrates to the neck region of the archegonium and there divides, forming a ventral canal nucleus and an egg nucleus. These usually, but not always, are separated by a wall. The cell containing the ventral canal nucleus is very small and promptly disintegrates. The female gametophyte is now mature and ready

for fertilization. A section of an ovule at this stage is shown in Fig. 23-18.

It will be recalled that the pollen grain or immature male gametophyte reached the micropyle and apex of the megasporangium (Fig. 23-12) of the ovule in a four-celled condition (Fig. 23-14D) and that the pollination of the ovule occurs more than a year before the ovule will contain a mature female gametophyte. In the middle of April, 12 months after pollination, the generative nucleus divides; this is followed by cytokinesis, forming a stalk and body cell (Fig. 23-14E). During the intervening year, the pollen tube, initiated soon after pollination, has developed into a branched haustorial organ (Figs. 23-16, 23-18A), which has digested its way through the elongate megasporangium toward the apex of the developing female gametophyte. The pollen tube nucleus lies near the tip of the pollen tube during its growth. Although the pollen tube is

571

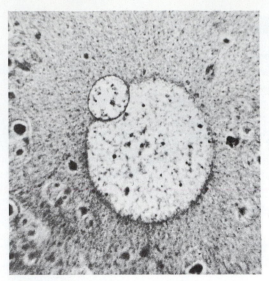

FIGURE 23-19
Pinus virginiana. Fertilization (karyogamy):
union of large egg nucleus and small sperm
nucleus. ×400.

branched and haustorial, as in the cycads and
Ginkgo, it also serves as a conveyor to carry
the male nuclei to the archegonium. This is
in contrast with the pollen tubes of cycads
and *Ginkgo*, which are largely haustorial.
Several days before the pollen tube reaches
the female gametophyte, the nucleus of the
body cell divides to form two sperm nuclei,
which lie close together in the common
cytoplasm of the body cell (Fig. 23-14F);
these are slightly unequal in size. Blepharoplasts do not appear during this division in
Pinus. Sometime before fertilization, the two
sperm nuclei (cells?) and the stalk cell move
nearer the tip of the pollen tube.

As the tube makes contact with the female
gametophyte in the vicinity of an archegonium,
it discharges some of its cytoplasm and generally
all its nuclei into the egg cell. One of the sperm
nuclei migrates toward the egg nucleus, with
which it unites (Fig. 23-19); the mechanism of
this movement is not understood. The remaining nuclei of the male gametophyte disintegrate
in the cytoplasm of the egg.

Doyle (1963) reviewed embryogeny in *Pinus*
and other Coniferophyta. Usually, the egg nuclei of all the archegonia of a single gametophyte
are fertilized by sperms. In such cases, each
zygote initiates embryo development and several embryos may begin to grow, although the
mature seed usually contains only one. The
development of several zygote nuclei into embryos is known as **simple polyembryony;** this
seemingly occurs in the cycads but not in *Ginkgo.* Soon after fertilization, two successive nuclear divisions occur in the zygote to form four
diploid nuclei of the proembryo (Fig. 23-20A,
B). These nuclei, which correspond to the first
four free nuclei in the embryogeny of *Zamia* and
Ginkgo, migrate to the base of the egg cell,
where they arrange themselves in a tier of four;
mitosis and cytokinesis follow to form two tiers
of four. Each tier of cells again undergoes
mitosis and cytokinesis (Fig. 23-20C), so that
four tiers of four cells now constitute the proembryo (Fig. 23-20D, E). The tier of nuclei
nearest the neck cells may or may not remain in
continuity with the cytoplasm of the former egg
cell; these may be designated the upper or open
tier, if they remain open (Doyle, 1963).

In further development, the lowermost tier of
four cells is directly involved in the formation of
the embryo. The next upper tier of four cells
elongates markedly, functioning as the embryonal suspensor, which pushes the cells of the
lowermost tier into the vegetative tissue of the
female gametophyte (Fig. 23-21A). These four
suspensor cells and the lowermost tier, the
embryo-forming cells, often separate along their
longitudinal walls. Since each of the embryo-
forming cells can initiate an embryo, this phenomenon is known as **cleavage polyembryony.**
The lowermost embryo-forming cells cut off
cells, often called **embryonal tubes,** which function as secondary suspensors (Fig. 23-22) between themselves and the primary suspensors.
These, with the primary suspensors, exert pressure on the developing embryos in the direction
of the vegetative cells of the female gameto-

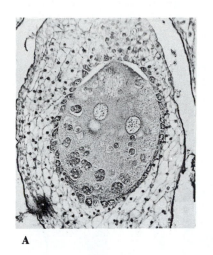

A

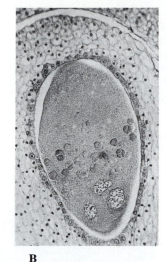

B

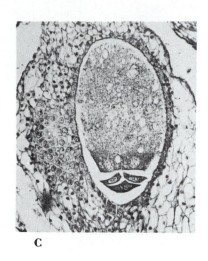

C

FIGURE 23-20
Pinus sp. Early embryogeny. **A.** Zygote nucleus divided into two
proembryonic nuclei (smaller, deep-staining bodies are nutritive
materials). **B.** Embryonic nuclei at base of archegonium (three of four
visible). **C.** Twelve-celled stage of proembryo; upper tier will divide
to form four additional cells. **D.** Sixteen-celled proembryo in median
longisection, semidiagrammatic. a., archegonium; e., embryo-forming
cells; o., open tier; ro., rosette cells; s., suspensor cells. **A, B.** ×90.
C. ×75.

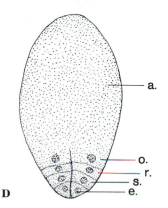

D

a.

o.
r.
s.
e.

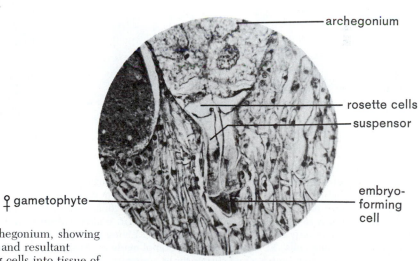

archegonium

rosette cells

suspensor

♀ gametophyte

embryo-
forming
cell

FIGURE 23-21
Pinus virginiana. Base of archegonium, showing
elongation of suspensor cells and resultant
projection of embryo-forming cells into tissue of
the ♀ gametophyte. ×125.

573

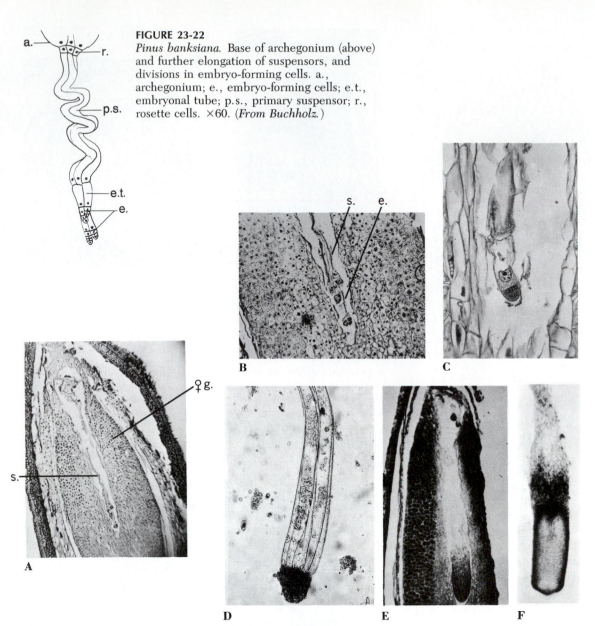

FIGURE 23-22
Pinus banksiana. Base of archegonium (above) and further elongation of suspensors, and divisions in embryo-forming cells. a., archegonium; e., embryo-forming cells; e.t., embryonal tube; p.s., primary suspensor; r., rosette cells. ×60. (*From Buchholz.*)

FIGURE 23-23
Pinus taeda. Later embryogeny. **A.** Longisection of upper part of ovule in late July, 16 months after pollination and about 6 weeks after fertilization; note four young embryos with elongated suspensors in digested center of ♀ gametophyte, disintegrated archegonia, and at least two aborted embryos above. **B.** Embryos in **A** enlarged. **C.** Detail of suspensor, embryonal tube, and embryo-forming cells of one product only of cleavage polyembryo. **D.** Living proembryo showing embryonal tubes and embryo-forming cells. **E.** Longitudinal section of ovule on August 15, one month later than that in **A**. **F.** Note surviving embryo and suspensor at approximately the same stage as in **E.** e., embryo; s., suspensor; ♀ g, ♀ gametophyte. **A.** ×30. **B.** ×90. **C–F.** ×120.

574

phyte, which have become filled with stored metabolites. The gametophyte is digested and liquefied by the advancing embryos, which lie in a cavity into which they have been thrust by the elongating suspensor system. One of the developing embryos outstrips the others, and by rapid nuclear and cell divisions organizes an embryo that occupies a major part of the cavity in the central portion of the female gametophyte (Fig. 23-23).

Berlyn (1962) reported the presence of as many as four embryos of decreasing size in the mature seeds of several species of *Pinus*. All four were viable *in vitro*, but only one usually emerges during natural germination.

The embryo, as it enters the dormant period, has a radicle that lies in the region originally occupied by the archegonia. The remains of the latter and suspensors and embryonal tubes are compressed at the tip of the radicle. The remainder of the axis is the hypocotyl, which in some species of *Pinus* bears as many as 18 or as few as 3 needlelike cotyledons, but 8 is the usual number. Among their bases is the apex of the embryonic stem. The peripheral cells of the female gametophyte continue nuclear and cell divisions during the development of the embryo, so that, as the latter enters dormancy, considerable female gametophyte tissue remains (Fig. 23-24). Although three layers are distinguishable in the integument of the immature seed, the innermost and outermost are disorganized and vestigial by the time the seed is shed. The middle layer becomes hard and stony and actually serves as the seed coat.

In *P. virginiana* and many other species, the seeds are mature and are shed in the autumn of the second year following the appearance[2] of the megastrobili. Although species of *Pinus* in colder climates also require parts of three growing

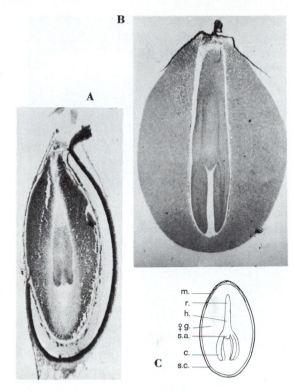

FIGURE 23-24
Pinus sp. **A.** Longisection of almost-mature seed; organization of cotyledons in embryo. **B.** *P. edulis.* Section of megasporangium (black cap), ♀ gametophyte, and embryo of mature seed; note hypocotyl, cotyledons, and plumule. **C.** *P. edulis.* Bisection of living seed. c., cotyledon; ♀ g., ♀ gametophyte; h., hypocotyl; m., remains of megasporangium; r., radicle; s.a., stem apex or plumule; s.c., seed coat. **A, B.** ×30. **C.** ×2.

---

[2]Both the microstrobili and megastrobili are, of course, present as primordia in the late summer and fall of the year before pollination occurs.

seasons for the development of megastrobili to the production of seeds, the active phases are more rapid than in species in the southern United States. Thus *P. taeda* in Texas requires 18 months between pollination and the shedding of seed, while *P. monticola* in Vancouver completes the cycle in 15.5 months (Owens and Molder, 1977*b*). During this period the megastrobilus, which was soft and green or purplered at pollination, increases in size from approx-

A

B

FIGURE 23-25
Pinus taeda. **A.** Megastrobili in August, 17 months after pollination and 2½ months after fertilization; embryos, like those in Fig. 23-24A, are present within them. **B.** Opening of megastrobilus at the time of seed dissemination. ×0.75.

imately 15 mm to a hard, woody cone many times larger (Fig. 23-25A, B). The ovuliferous scales become separated and recurved, making possible the dissemination of the winged seeds (Fig. 23-26).

Germination of the pine seed is epigean. The cotyledons remain within the seed some time after germination, absorbing the metabolites still present in the female gametophyte. Ultimately they emerge, shed the remains of the seed, and function as photosynthetic organs.

The preceding account of the reproductive

process of *Pinus* is based on the studies of Thomas (1951) on *P. virginiana* near Nashville (Ashland City), Tennessee, and on observations of *P. taeda* at Bastrop, Texas, by the authors. The months of the year cited refer to these species at these locations. That these times and intervals vary is evident from the summary table by Lill (1976), which indicates that the interval from pollination and fertilization may be 11½ to 15 months, and from pollination to mature embryos (in seeds), 14 to 20 months.

Other Pinaceae

More than one-half of the Coniferalean genera are members of the family Pinaceae, species of which are well represented in North America, with concentration in the western part of the continent. The Douglas fir, *Pseudotsuga taxifolia*, is one of the largest and most important of the western lumber trees. *Pinus*, with more than 80 species, is restricted to North America with few exceptions. *Larix*, the larch or

A

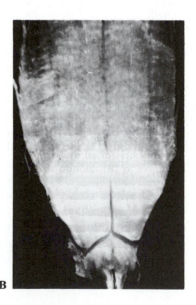

B

FIGURE 23-26
Pinus sp. Winged seeds on adaxial surface of ovuliferous scale. **A.** *P. virginiana.* **B.** *P. taeda.* **A.** ×3. **B.** ×5.

FIGURE 23-27
A. *Tsuga canadensis.* Hemlock; branch with megastrobili shedding seeds. **B.** *Cedrus deodora,* a true cedar. **C.** Close-up of a branchlet of **B. A.** ×0.75. **B.** ×0.5.

male gametophytes and the sperms seem to be naked—that is, without cytoplasmic sheaths—but this requires confirmation. The archegonia are in a ring separated by vegetative cells of the female gametophyte.

Owens and Molder[3] have investigated reproduction in *Abies amabilis* in British Columbia. As in *Pinus,* the microstrobili and megastrobili develop to some extent during the summer and fall of the year before they are to function. The reproductive process from micro- and megasporogenesis through dissemination of mature seed occupies only from the end of April until late September, this being more rapid than in *Pinus,* as it is also in many other conifers. A number of other aspects of reproduction in *Abies amabilis*

tamarack, a northern genus, is distinguished by its deciduous habit. *Abies* (fir), *Picea* (spruce), and *Tsuga* (Fig. 23-27A) (hemlock) are familiar genera of the temperate climates of the world. The stems of *Abies, Picea, Pseudotsuga,* and *Tsuga* are not differentiated into long and spur shoots. The true cedar, *Cedrus* (Fig. 23-27B, C), especially *C. deodara,* native to the Mediterranean and Himalaya regions, is widely cultivated in the United States. In this family the scale leaves, needle leaves, and sporophylls are spirally arranged, and the bract and ovuliferous scales are free from each other in part. Many Pinaceae have winged pollen. The Pinaceae are all monoecious. All have prothallial cells in their

[3]Allen and Owens (1972); Owens (1973); Owens and Molder (1971, 1974, 1975a,b, 1977a,b, 1979a,b,c, 1980a,b); Owens and Simpson (1982); Owens and Singh (1982); Owens et al. (1980, 1981a,b,c); and Singh and Owens (1981a,b, 1982) have contributed greatly to our knowledge of reproduction in various conifers.

577

A B C

FIGURE 23-28
Taxodium distichum. **A.** Pendulous branches with microstrobili. **B.** Megastrobili at pollination. **C.** Maturing megastrobili. **A.** ×⅓. **B.** ×4. **C.** ×0.75.

are different from that in *Pinus*. (1) The pollen grains are 5-celled at shedding because the generative cell has undergone division into a stalk and body cell.[4] (2) No pollination droplet or pollen chamber was observed. (3) After pollination, late in May, the pollen remained dormant on the tip of the integument for a month before it germinated. Both simple and cleavage polyembryony occur in *Abies amabilis*.

Family 2. Taxodiaceae

Of *Taxodium* (United States and Mexico), *Sequoia* and *Sequoiadendron* (the western U.S. redwoods), and *Metasequoia* (native to China), the first three are probably familiar examples. In these, leaves are spirally arranged, but the ovuliferous scales and bracts are closely fused. Winged pollen is absent. *Taxodium* (Fig. 23-28), the bald cypress, is the familiar tree of many swamps in the southern states. Like *Larix* (Pina-

ceae), *Taxodium* is deciduous; its megastrobili form mature seeds during a single growing season. The microstrobili, borne on pendulous axes, appear in the autumn before they release the pollen. *Sequoiadendron gigantea* is one of the giants among trees, up to 340 ft tall with a diameter of 8 to 25 ft and larger than the coastal redwood, *Sequoia sempervirens*. Large specimens of *S. gigantea* are more than 4000 years old; *Metasequoia* (see p. 626) is truly a living fossil. The bracts of the megastrobilus of *Metasequoia* are decussate. All Taxodiaceae are monoecious. Many lack prothallial cells in the male gametophytes. In this family the sperms are cellular, large, and cycadlike, except that they lack cilia. The archegonia are closely contiguous, forming a complex.

Family 3. Cupressaceae

The family Cupressaceae is represented by four genera in North America: *Thuja* (eastern white cedar, including American arborvitae, *T. occidentalis*); *Cupressus* (*C. macrocarpa*, the Monterey cypress); *Chamaecyparis* (*C.*

---

[4]In another species of *Abies* mature sperms have been found in 10% of the pollen before pollination; this occurs also in many angiosperms (see p. 646).

A

B

C

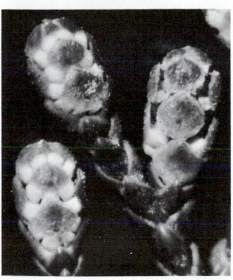

D

E

FIGURE 23-29

Juniperus ashei. **A.** Branchlet with megastrobili at pollination (December–January) in Austin, Texas. **B.** Megastrobili of preceding more highly magnified. **C.** Megastrobili in later March. **D.** Microstrobili, the one at the right shedding pollen. **E.** Megasporangiate branch with maturing seeds. A. ×2. B. ×6. C. ×2. D. ×10. E. ×0.75.

thyoides, the white cedar); and *Juniperus* (juniper), including the red "cedar," *J. virginiana*. The leaves and sporophylls of these are whorled or decussate. The wood of *J. virginiana* is used extensively for chests, closet linings, lead pencils, and fence posts. The leaves are of the scale type, only the juvenile leaves being of the needle type.

Owens and Molder (1974, 1975) reported on reproduction of *Chamaecyparis nootakensis* on

Vancouver Island. The megastrobili are often fleshy at maturity, as in *Juniperus* (L. *Juniperus*, classic Latin name). *Juniperus* is dioecious (Fig. 23-29), but both monoecious and dioecious genera occur in the family. In some species the microspores are shed in uninucleate condition,

579

before any intrasporal divisions to form the male gametophyte. Here, too, the sperms are cellular and cycadlike, except for the absence of cilia. In *Cupressus*, there may be as many as 12 sperms in each male gametophyte. The archegonia, like those of Taxodiaceae, are closely associated to form a complex. There are between four and ten in *Juniperus*.

Family 4. Araucariaceae

Leaves of Araucariaceae are spirally arranged and usually broad (nonneedlelike), and the ovuliferous scale and bract are closely united. The microsporangia (up to 13–15) are borne on a shield-shaped microsporophyll, and the ovules are solitary. The two genera of this family, *Agathis* (Fig. 23-30) and *Araucaria*, are both restricted to the southern hemisphere and may reach 140 ft in height. The trees are used as timber. *Araucaria* species are widely distribut-

FIGURE 23-30
Agathis australis. Model of Kauri pine or Dammara; branchlet with megastrobili. (*Courtesy of the Field Museum of Natural History.*)

ed in cultivation. The monkey-puzzle tree, *A. araucana*, is familiar through its bizarre branching. The Norfolk Island "pine," *A. heterophylla*, is widely cultivated because of the strikingly regular whorled symmetry of its branches. The genus *Araucaria* is of special interest because as many as 40 prothallial cells may develop in the male gametophytes.

Both monoecious and dioecious species are known. In *Araucaria*, 32 free nuclei occur during embryogeny; in *Agathis*, there are 64, the largest number in the Coniferales. Cleavage polyembryony seems not to occur in this family.

Family 5. Podocarpaceae

The genera of this small family are confined to the southern hemisphere except for the genus *Podocarpus* (Gr. *pod-*, *pous*, foot or stalk, + Gr. *karpos*, fruit), which extends northward to the West Indies and Central America. The leaves of *Podocarpus* may be needlelike or broader (Fig. 23-31). The genus is widely cultivated as an ornamental shrub in the southern part of the United States, where it is known in horticulture as "yew"; it is, of course, not at all the same as *Taxus*, the northern yew (see p. 581).

Podocarpus is dioecious (Fig. 23-31A–D). Numerous prothallial cells are present in the male gametophyte. *Podocarpus* is used for timber in Australasia.

CLASS 2. TAXOPSIDA

Order 1. Taxales

The order Taxales is sometimes included in the Coniferales as the family Taxaceae. However, its genera have attributes that seem sufficiently different from those of Coniferales to warrant higher rank; furthermore, the taxads are distinct from the conifers as far back as their fossil record can be followed. Foremost among their distinctive attributes is the absence of a

A

B

C

D

FIGURE 23-31
Podocarpus macrophyllus. **A.** Branch with microstrobili (basal). **B.** Microstrobili. **C.** Branch with ovules. **D.** Ovules with subtending fleshy arils, maturing. **A.** ×½. **B.** ×3. **C.** ×⅔. **D.** ×½.

FIGURE 23-32
Taxus canadensis. Note ovules partially encased in fleshy arils. ×0.3 (*Courtesy of Dr. Raymond Lynn.*)

megastrobilus, the ovules occurring terminally on short lateral branches (Fig. 23-32). Two genera, *Taxus* (L. *Taxus,* yew) and *Torreya* (after John Torrey, 1796–1873, American botanist), occur in the United States.

Genera of Taxales are dioecious. The microsporophylls are shield-shaped, with six to eight microsporangia. The microsporophylls occur in strobili, but the ovules are solitary and have a fleshy, cuplike structure, the **aril,** at the base, which surrounds the seed at maturity (Fig. 23-32).

The microspores are shed in the uninucleate stage, and the male gametophytes lack prothallial cells. The sperms are cellular in *Taxus* but markedly unequal in size. In *Torreya*, only a single archegonium occurs in each female gametophyte, and there are only four free nuclei during embryogeny, as in *Pinus*. Cleavage polyembryony apparently does not occur in this family.

DISCUSSION QUESTIONS

1. With the aid of labeled diagrams, summarize reproduction in *Pinus*, giving careful attention to the time factor.
2. If you were to examine trees of pine as the terminal buds were unfolding, what types of strobili would be present, and where?
3. What functions can you ascribe to the pollen tube in *Zamia, Ginkgo*, and *Pinus*?
4. Why are the needles of pine, hemlock, and spruce not considered to be microphyllous?
5. Why are the megastrobili of pine considered to be compound?
6. How do the sperms of pine differ from those of *Zamia* and *Ginkgo*?
7. Distinguish between simple and cleavage polyembryony.
8. Diagram a median longitudinal section through a pine seed.
9. In what respects do the archegonia of *Pinus* differ from those of *Zamia* and *Ginkgo*?
10. What is the fate of the nuclei, other than the successful sperm, which are discharged into the archegonium by the pollen tube of *Pinus*?
11. Do you consider it appropriate to call microstrobili "male" and megastrobili "female"? Explain.
12. How does embryogeny in pine differ from that in *Zamia* and *Ginkgo*?
13. Describe the process of primary and secondary development in the pine stem, using labeled diagrams of successively older transverse sections.
14. What are resin canals? How are they distributed in the plant? Do corresponding cavities occur in *Zamia* and *Ginkgo*?
15. Ovules containing two female gametophytes occur with considerable frequency in *Pinus virginiana*. How do you explain their presence?
16. The megaspores of *Pinus, Zamia*, and *Ginkgo* are not markedly larger than microspores of the same plants. On what grounds are they called megaspores? Suggest possible reasons for their size relationship.
17. The megaspores of *Selaginella* and *Isoetes* have thick largely impervious walls at maturity. Those of the seed plants lack such walls. Can you suggest an explanation for this?
18. What genera of the Pinaceae and other families of Coniferales and Taxales are familiar to you?

Chapter 24
DIVISION GNETOPHYTA

The final representatives of the gymnospermous seed plants to be considered in this text are *Ephedra* (L. *ephedra*, horsetail) (Fig. 24-1), *Gnetum* (Malay, *gnenom*) (Fig. 24-19), and *Welwitschia* (after F. Welwitsch, its discoverer) (Fig. 24-22). These plants are sometimes considered to constitute a single order, the Gnetales, which includes 71 species. Careful morphological comparisons of these three genera, however, indicate that there are a number of important divergences among them. This is reflected in a modern classification in which each of the genera is placed in a separate order, with a single family in each. *Ephedra* has been chosen for detailed discussion because it usually is more readily available for study in the temperate zones than are the other two genera. Martens (1971) summarized our knowledge of three genera of Gnetales.

CLASS 1. GNETOPSIDA

The Gnetopsida include the orders Ephedrales, Gnetales, and Welwitschiales, each with a single family.

Order 1. Ephedrales

Ephedra, the only representative of the family Ephedraceae, is xerophytic. Approximately 40 species of the genus have been described, all of them shrubby or trailing plants. *Ephedra antisyphilitica* has the habit of a small, shrubby tree, attaining a height of 3–5 m in the Rio Grande valley. Approximately six species occur in the southwestern United States, among them *E. trifurca* and *E. antisyphilitica*. The following account is based on these and on *E. foliata*, an Indian species.

In the arid regions of the southwest and California, *Ephedra*, known as the joint fir, sometimes is important as a range plant and is grazed. The American and Mexican Indians used decoctions of the roots and stems of these plants for genitourinary ailments and as a cooling beverage, and they used the seeds to make a bitter bread. The alkaloid ephedrine is prepared from *E. sinica*, a Chinese species. The Chinese name of the drug is ma huang. The

FIGURE 24-1
Ephedra antisyphilitica. Living plant. ×0.7.

medicinal properties of species of *Ephedra* were known in China as early as 2737 B.C. Hu (1969) has summarized the information on *Ephedra* contained in the latest edition of the Chinese Materia Medica.

Vegetative morphology. The younger, green branches of *Ephedra* plants (Fig. 24-2) superficially resemble species of *Psilotum* and *Equisetum* because of the minuteness and ephemeral photosynthetic activities of the leaves. Many of the lateral branches arise in fasciculate whorls (Fig. 24-2). In *E. foliata* the branching is quite variable. On the main excurrent leaders, it may be either opposite or alternate. In older, woody portions of the axes, the green shoots are fasciculate, and many of them are shed during dry periods by abscission layers that extend across the pith and xylem. The minute leaves are opposite or in whorls of three (Fig. 24-3).

FIGURE 24-3
Ephedra antisyphilitica. Node with three leaves; minute white spots are stomata. ×5.

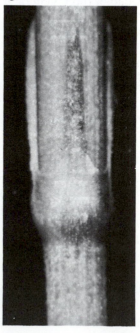

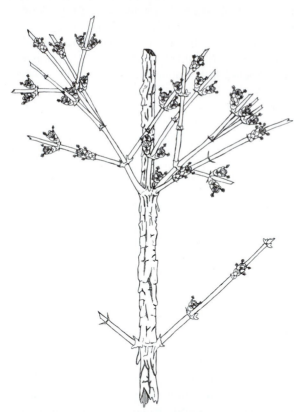

FIGURE 24-2
Ephedra antisyphilitica. Habit of woody branch with photosynthetic shoots bearing microstrobili shedding pollen. ×1.

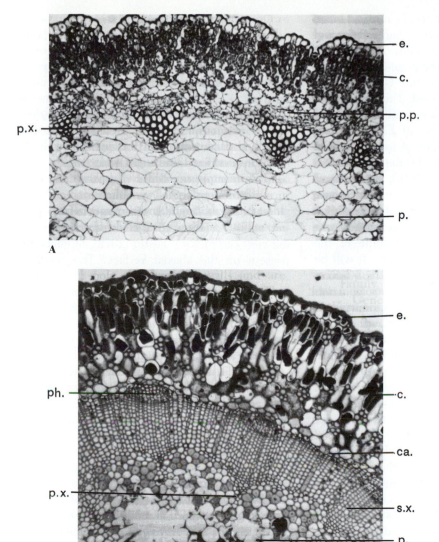

FIGURE 24-4
Ephedra californica. **A.**
Sector of transection of
young branch. **B.** Sector of
transection of stem that has
undergone secondary
growth. c., cortex; ca.,
cambium; e., epidermis;
p., pith; ph., phloem; p.p.,
primary phloem; p.x.,
primary xylem; s.x.,
secondary xylem. ×60.

The younger branches are delicately ribbed and carry on most of the photosynthesis in these plants. The older stems are hard and woody because of secondary growth and are anchored to the substratum by a deep taproot and adventitious roots.

The apices of the young axes are occupied by meristem cells that differentiate into primary meristems, from which the primary, permanent tissues develop. The younger shoots of *E. foliata* contain an extensive parenchymatous pith surrounded by vascular bundles with endarch xylem (Fig. 24-4A). The cortex is composed of patches of photosynthetic parenchyma into which groups of sclerenchymatous cells beneath the ridges of the stem surface intrude as supporting areas. The stomata occur on the slopes of the ridges, as in *Equisetum*, and are sunken and overarched by accessory cells. Inamder and Bhatt (1972) investigated the organization of the stomata in both *Ephedra* and *Gnetum*. The epidermis is heavily thickened. The presence of such an abundance of sclerotic tissue renders even the younger portions of *Ephedra* axes relatively hard and resistant. Annual zones of secondary xylem are added by the activity of a continuous cambium (Fig. 24-4B). The wood is extremely hard and is traversed by multiseriate rays in older axes. The most significant feature in the secondary xylem of *Ephedra* is the perforation of the terminal walls of some of the tracheids (Fig. 24-5) to form vessels. It will be recalled that these are absent from the xylem of most cryptogams, the genera *Selaginella*, *Equisetum*, *Pteridium*, and *Marsilea* being notable exceptions. The wood of the Cycadophyta, Ginkgophyta, and Coniferophyta also lacks vessels, so that *Ephedra*, *Gnetum*, and *Welwitschia* are unique in having them. Vessels are present in the xylem of most flowering plants (Anthophyta) but absent in certain primitive ones (p. 674). The tracheids of *Ephedra* are pitted with bordered pits on both the radial and tangential walls. Although the vessels, for the most part, are larger in diameter than tracheids, they are linked by a series of elements intermediate in size. Many stages transitional between tracheids and vessels may be observed in the xylem of *Ephedra*. The young seedling contains only tracheids at first. It is noteworthy that the perforations in the tracheid end walls (which result in vessels) arise from circular, rather than from scalariform, pits, the latter being the case

in *Selaginella*, *Pteridium*, and flowering plants. Alosi and Alfieri (1972) described details of phloem organization in *Ephedra*. The primary steles of *Ephedra* roots are diarch. They increase by secondary thickening to form woody taproots. Peterson and Vermeer (1980) investigated the root apex of two species of *Ephedra*.

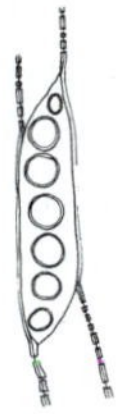

The rudimentary leaves of *Ephedra* are histologically simple; they are composed of a thickened midrib region and thin wings, which soon lose their chlorophyll and turn brown. Two or three (variable in different species and individual leaves) unbranched, convergent veins are present in the leaves of the three species studied very carefully by Foster (1972). Stomata are present on the abaxial surface. The seedling leaves and those of younger branches are less reduced than those in mature specimens.

Reproduction: the sporophyte. Because of a superficial resemblance to angiosperm inflorescences, the reproductive structures of the Gnetophytan sporophyte (especially *Gnetum*) are sometimes referred to as inflorescences or flowers. These terms, however, are inadmissible. Both monoecious and dioecious species are known in *Ephedra*; *E. antisyphilitica* is strictly dioecious (Figs. 24-2, 24-6A, B).

The microsporangiate strobili have rounded apices (Fig. 24-6C, D), while those of the ovulate strobili are acute (Fig. 24-6B, E). The strobili of *E. antisyphilitica* are already visible and well developed in the late autumn in the vicinity of Abilene and elsewhere in Texas, at the nodes among the fasciculate, photosynthetic axes.

The ovulate strobilus (Figs. 24-6E, 24-7A, B) consists of between four and seven pairs of decussate bracts attached to an axis. These correspond to the bracts of the ovulate *Cordaianthus* axis (Fig. 25-35A). The lowermost pairs are sterile, but in the axil of one or occasionally both (Fig. 24-7A) of the terminal pairs of bracts a short-stalked ovule occurs. Sometimes two ovules are present, but one of these may be

FIGURE 24-5
Ephedra antisyphilitica.
Slanting terminal wall of two tracheidlike vessel elements, showing perforations. ×165.

FIGURE 24-6
Ephedra antisyphilitica. **A.** Branch with microstrobili. **B.** Branches with ovulate strobili. **C.** Microstrobili. **D.** Microstrobili with protuberant microsporophylls. **E.** Ovulate strobili, before pollination.

abortive. The ovule (Fig. 24-8) is surrounded by a fleshy cuplike structure, sometimes called an **"outer integument,"** which is attached at the ovule base and free above. It probably represents a united pair of bracteoles. These correspond to the sterile scales of microsporangiate *Cordaianthus* dwarf shoots. A more delicate, **inner** (true) **integument** is prolonged at the time

of pollination into a tubular process that projects beyond the bracts and bracteoles (Fig. 24-8). The inner integument is chlorophyllous at the time of pollination. Comparative study of a number of species of *Ephedra* indicates that the *apparently* terminal ovules are borne on lateral appendages of the strobilus axis. The ovulate strobilus, accordingly, is compound.

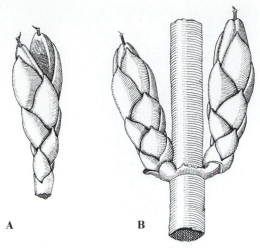

FIGURE 24-7
Ephedra antisyphilitica. **A, B.** Megastrobili at pollination; note protuberant micropylar tubes. ×5.

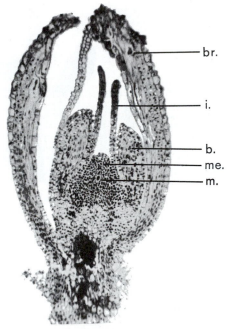

br.

i.

b.
me.
m.

FIGURE 24-8
Ephedra antisyphilitica. Median longisection of apex of very young megastrobilus. b., bracteole; br., bract; i., integument; m., megasporocyte; me., megasporangium. ×45.

The microsporangiate strobilus also is compound. It consists of an axis that bears about seven pairs of decussate bracts (Figs. 24-6C, D, 24-9), most of which subtend a short axis on which microsporangia are borne. Each microsporangiate axis is enclosed early in development by two overlapping, transparent bracteoles (Fig. 24-10); these correspond to sterile appendages of microsporangiate *Cordaianthus* dwarf shoots. The microsporangiate axis of *E. antisyphilitica* is composed of several united microsporophylls, as indicated by anatomical evidence, to form a sterile portion, the **column.** In *E. antisyphilitica*, five two-chambered microsporangia are typically present at the apex of the column. When the pollen is about to be shed, the column elongates, carrying the microsporangia free of the bracteoles and subtending bracts (Fig. 24-10).

The young microsporangia have two-layered walls and a prominent tapetal layer surrounding the sporogenous tissue. The tapetal cells become binucleate during sporogenesis. At maturity, the biloculate microsporangium (Fig. 24-10) is covered only by a single layer of epidermal cells because of the degeneration of the remaining wall cells. Dehiscence of the microsporangium by an apical fissure occurs at maturity. During microsporogenesis, the microsporocytes

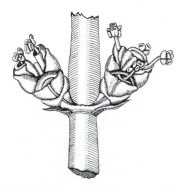

FIGURE 24-9
Ephedra antisyphilitica. Microstrobili with protuberant microsporophylls. ×5.

588

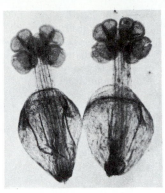

FIGURE 24-10
Ephedra antisyphilitica. Two microsporophylls emerging from their transparent bracteoles. ×12½.

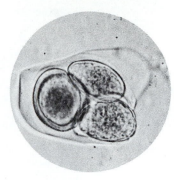

FIGURE 24-11
Ephedra antisyphilitica. Microsporocyte wall containing tetrad of microspores; one being liberated. ×500.

separate from one another, become almost spherical, and undergo the meiotic process, which culminates in the usual tetrad of microspores (Fig. 24-11). Microsporogenesis is not simultaneous in all the microsporangia of a given strobilus, those within the distal bracts being retarded. Microsporogenesis in *E. antisyphilitica* continues over a long period, from December through early February, depending on the temperature. After the microspores have been liberated from the microsporocyte walls, they increase considerably in size and their walls become thickened with ribbed exines (Fig. 24-14C).

A single megasporocyte differentiates within the megasporangium of the ovule (Fig. 24-8). A linear tetrad is produced as a result of meiosis, and the chalazal megaspore alone usually is functional.

Reproduction: the gametophytes. Development of the female gametophyte is freenuclear, as in all seed plants. The number of nuclei so formed varies in the several species of *Ephedra*. In *E. trifurca*, wall formation occurs when approximately 256 free nuclei have formed, in *E. foliata* after 500 nuclei have appeared, and in *E. distachya* after 1000 nuclei

have developed. Two or three archegonia are organized at the micropylar pole of the female gametophyte (Fig. 24-12). The necks of the archegonia are the most massive of any among the gymnosperms; they may consist of 40 or more cells at maturity. The neck cells are arranged in tiers of four or more. The ventral canal nucleus and egg nucleus are rarely separated by cytokinesis in the developing archegonium. The chalazal cells of the female gametophyte are dense and filled with stored metabolites, and those at the micropylar pole are watery and vacuolate. In *E. antisyphilitica*, polynucleate cells are present, particularly at the chalazal portion of the female gametophyte. The walls of the cells of the archegonial jacket are extremely delicate, and the cells may be binucleate. As the female gametophyte develops, the cells at the apex of the megasporangium degenerate to form a deep **pollen chamber** (Fig. 24-12), which, unlike that in other gymnosperms described in earlier chapters, extends to the female gametophyte and archegonia. A prominent **pollination droplet** is visible at the subterminal orifice of the micropyle (Fig. 24-13) at the time of pollination. No thickening of the megaspore wall occurs in *Ephedra*. Haploid

589

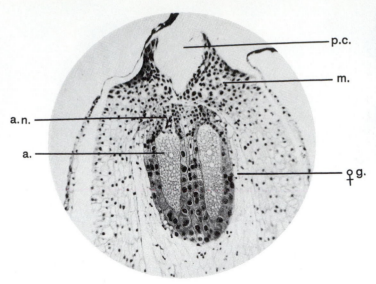

FIGURE 24-12
Ephedra antisyphilitica.
Median longisection of ovule apex just before fertilization. a., archegonium; a.n., archegonial neck; ♀ g., ♀ gametophyte; m., megasporangium; p.c., pollen chamber. ×125.

p.c.

m.

a.n.

a.

♀ g.

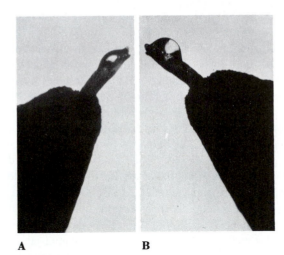

A   B

FIGURE 24-13
Ephedra antisyphilitica. **A.** Micropylar orifice. **B.** The same with pollination droplet. ×125.

plantlets of *E. foliata* were induced to form in culture from the tissue of the female gametophyte (Singh et al., 1984).

As in most gymnosperms, the microspores begin their development into male gameto-phytes before they are shed from the microsporangia. The process may be followed readily in material collected during the winter and forced in laboratory or greenhouse. The microspore nucleus divides first to form a prothallial cell nucleus, which is delimited from a larger cell by cytokinesis. The prothallial cell lies near one of the poles of the microspore (Fig. 24-14A). A small second prothallial nucleus and a larger nucleus then arise by mitosis of the nucleus of the large cell, but usually these are not segregated by a cell wall (Fig. 24-14B). The second prothallial nucleus is more persistent than the first, which gradually shrinks into oblivion. The large nucleus now divides to form the generative and tube nuclei (Fig. 24-14C), which are separated by cytokinesis to form a generative cell, which is ovoidal. The generative nucleus is reported to form a stalk and body nucleus in several species of *Ephedra*, but conclusive evidence of this has not been observed in *E. antisyphilitica.* On the contrary, the generative nucleus is said to give rise to the two sperm nuclei directly (Fig. 24-14D, E), but

590

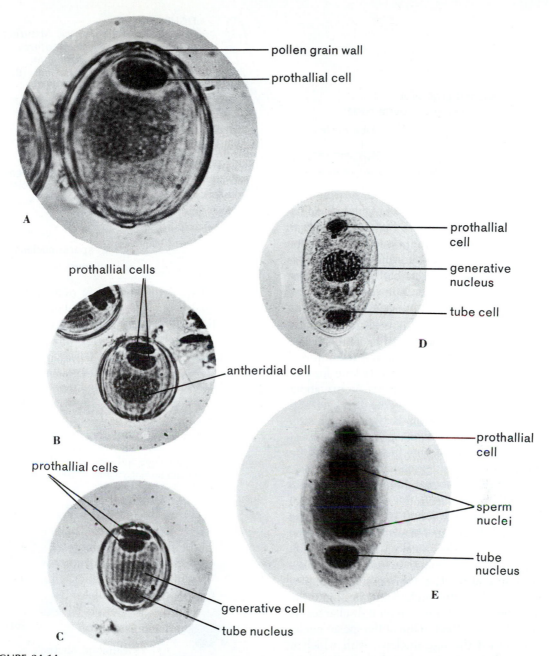

FIGURE 24-14

Ephedra antisyphilitica. Development of the ♂ gametophyte. **A.** Small, first prothallial cell and large cell. **B.** Two prothallial cells and antheridial cell. **C.** Two prothallial cells, generative cell, and tube nucleus (polar). **D.** Prophase of division of generative nucleus; second prothallial cell above, tube cell below. **E.** Telophase of division of generative nucleus; second prothallial cell above, tube nucleus below. **A.** ×1200. **B–E.** ×600.

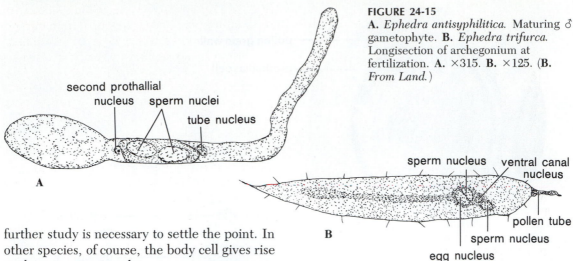

FIGURE 24-15
A. *Ephedra antisyphilitica.* Maturing ♂ gametophyte. **B.** *Ephedra trifurca.* Longisection of archegonium at fertilization. **A.** ×315. **B.** ×125. (**B.** *From Land.*)

further study is necessary to settle the point. In other species, of course, the body cell gives rise to the two sperm nuclei.

Pollination occurs through the agency of wind, and a prominent pollination droplet forms (Fig. 24-13). A number of pollen grains usually are present in the pollen chamber of each ovule. Adequate data on pollination and the interval between it and fertilization are lacking for *Ephedra*. Pollination in *E. trifurca* occurs about the time the archegonia are being organized and are maturing. The interval between pollination and fertilization in that species may be as short as 10 hours, one of the briefest among the gymnosperms.

Within the pollen chamber, the pollen-grain protoplast emerges from the exine and the two male nuclei are formed (Fig. 24-15A). These may be slightly unequal in size. The pollen tube (Figs. 24-15A, 24-16) develops from the region of the male gametophyte to which the tube nucleus migrates. The tube grows down between the cells of the archegonial neck into the apex of the egg cell, into which it discharges its nuclei (Moussel, 1983). One of the sperm nuclei moves toward the egg nucleus, with which it unites, and the other nuclei of the pollen tube remain at the apex of the archegonium in the vicinity of the ventral canal nucleus (Fig. 24-15B). It has been reported (see Martens, 1971, and Moussel, 1978, for a summary) that

the second sperm nucleus may unite with the ventral canal nucleus to effect a "double fertilization," but this requires confirmation.

Reproduction: embryogeny. Development of the embryo of *Ephedra* involves freenuclear divisions, as in other gymnosperms. In *E. trifurca*, the zygote nucleus usually under-

FIGURE 24-16
Ephedra antisyphilitica. Rupture of exine and formation of pollen tube on 10% cane-sugar agar; note distal tube nucleus and second prothallial nucleus. Other nuclei not visible. ×250.

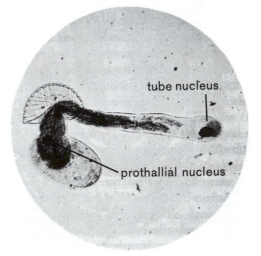

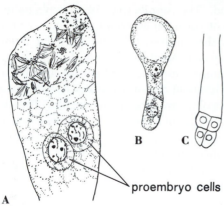

proembryo cells

A

FIGURE 24-17
Ephedra trifurca. Embryogeny. **A.** Two (of eight) cells of proembryo near neck end of egg cell. **B.** Later stage in development of embryo. **C.** Suspensor and early embryo. **A.** ×250. **B, C.** ×125. (*From Land.*)

goes three successive divisions to form eight nuclei, which may be somewhat unequal in size and scattered in the egg cytoplasm. Several of these, usually those near the lower pole of each egg cell, become surrounded by delicate walls

(Fig. 24-17A) formed by free-cell formation with residual cytoplasm; they are called **proembryos.** The small dividing nuclei observed in the neck region of the egg cell (Fig. 24-17A) have been interpreted as dividing nuclei descended from the second sperm nucleus and/or the ventral canal nucleus. The cells of the archegonial jacket break down after fertilization, and their nuclei may mingle with the egg cytoplasm.

All of the proembryonic cells descended from the zygote may begin to develop into embryos, but those at the lower region of the egg cell develop more rapidly. Embryogeny is initiated by a nuclear division and the formation of a tube into which both nuclei migrate. They are separated later by a septum (Fig. 24-17B). The upper cell functions as a **primary suspensor,** and the lower forms the embryo by further nuclear and cell division. The several embryos of each archegonium are thrust into the vegetative tissue of the female gametophyte by elongation of the primary suspensors (Fig. 24-17C). Secondary suspensors are organized by the cells at the micropylar pole of the embryo. The developing embryos (Fig. 24-18A) gradually ap-

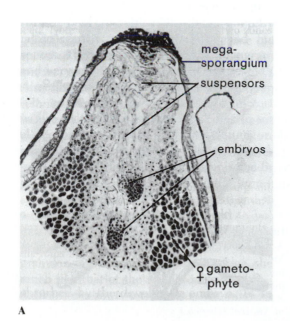

mega-sporangium

suspensors

embryos

♀ gameto-phyte

A

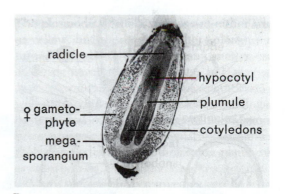

radicle

hypocotyl

♀ gameto-phyte

plumule

mega-sporangium

cotyledons

B

FIGURE 24-18
Ephedra antisyphilitica. **A.** Median longisection of micropylar end of ovule after fertilization. **B.** Median longitudinal section of seed (outer integument removed before sectioning). **A.** ×30. **B.** ×12.5.

propriate the delicate tissue in the center of the female gametophyte and grow at its expense, finally digesting and absorbing all but the most peripheral cells of the female gametophyte. Only one embryo is present in the mature seed (Fig. 24-18B), the others having disintegrated.

Just after fertilization, when the proembryos are developing, certain cells at the micropylar region of the female gametophyte are stimulated to divide; they form a plug, which closes the pollen chamber of the megasporangium. It is interesting to note that although some binucleate cells are present in the vegetative tissue of the female gametophyte of *E. antisyphilitica* at the time of fertilization, the cells of this tissue contain from two to four or five nuclei when the embryo enters the dormant

period. The significance of this phenomenon has not yet been ascertained.

As the embryo becomes dormant (approximately mid-April in western Texas), the inner integument forms a hard seed coat. The seeds require about 5 months to reach maturity after the strobili are first recognizable. The fleshy bracts of the ovulate strobilus become scarlet as the seeds ripen, and the seeds become black. Germination follows immediately, under favorable conditions. It is epigean, and the two cotyledons are long-persistent and photosynthetic.

Orders 2 and 3. Gnetales and Welwitschiales

Brief mention must be made at this point of two other genera of seed plants, *Gnetum* and *Welwitschia*, which formerly were considered to be closely allied to *Ephedra*. *Gnetum* and *Welwitschia* are now classified in separate orders, each in a single family, as is *Ephedra*. On the basis of gradually accumulating knowledge of these genera, it now appears that they represent three independent lines of development. It is regrettable that, because of their geographical distribution and the difficulties involved in cultivating them, these imperfectly known plants are usually unavailable for laboratory study, for what is known of them suggests that in some respects they are similar to the flowering plants.

Gnetum (Fig. 24-19), monographed by Maheshwari and Vasil (1961*b*) and summarized by Martens (1971), is a genus that contains some 30 species of mostly vinelike woody plants that have broad leaves like dicotyledonous angiosperms. *Gnetum* grows natively in northeastern

A

B

FIGURE 24-19
A. *Gnetum gnemon*, young plant. **B.** Model of branches with microsporangiate branches (left) and maturing seeds (right). **A.** ×0.3. **B.** ×¼. (**B.** *Courtesy of the Field Museum of Natural History.*)

South America (Brazil), tropical West Africa, India, and Southeast Asia.

Gnetum gnemon is a tree about 10 m tall. It is cultivated in Java, where the young leaves and reproductive axes are eaten after being cooked in coconut milk. The bark yields strong fibers used for making rope.

The vessels of *Gnetum* (Muhammad and Sattle [1982]) are more highly developed than those of *Ephedra* in that they may have several, or only single, terminal perforations between the segments in the secondary xylem. In the primary xylem there are several perforations, as in *Ephedra*. Behnke and Paliwal (1973) and Paliwal and Behnke (1973) reported that the phloem consists of enucleate sieve cells and phloem parenchyma.

Rodin (1967) has investigated leaf ontogeny in *Gnetum*. Although venation is reticulate, the basic pattern is one of dichotomies in secondary veins that arise from parallel veins of the midrib.

All species of *Gnetum* are dioecious (Fig. 24-19). The microsporophylls occur on elongate axes (Fig. 24-20A) that arise from the axils of bractlike leaves. The fertile axis has prominent nodes with very short internodes. The nodal bracts form a circular collar, or "cupule," above each of which are a number (25–30 in *G. ula*) of microsporangiate units. Each of these consists of a microsporophyll with two to four microsporangia enclosed in a bell-like sheath (Fig. 24-20A). A single ring of abortive ovulate units sometimes is present above each group of microsporangiate axes.

Meiosis of the microsporocytes is followed by liberation of the microspores and their development into male gametophytes. At shedding, the pollen grain contains a single prothallial cell, a generative cell, and a tube cell.

The ovules also occur on axes within nodal collars (Fig. 24-20B). Each ovule consists of a megasporangium enclosed within three structures. The innermost (integument) is prolonged into a micropylar tube, as in *Ephedra*. The

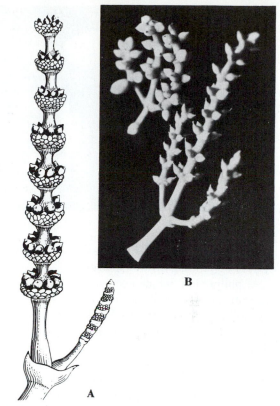

FIGURE 24-20
Gnetum gnemon. **A.** Microsporangiate axis. **B.** Branch with ovules. ×0.5. (**A.** *After Madhulata, from Maheshwari and Vasil.*)

outermost is massive and its epidermis has stomata, as does that of the middle integument. In *G. ula*, it is reported that eight to ten megasporocytes differentiate. All these may undergo meiosis.

Usually only one megasporocyte persists, but on occasion several may do so and give rise to female gametophytes. The four post-meiotic nuclei are not separated by cell walls, and all four (megaspore) nuclei initiate free-nuclear divisions to form the female gametophyte, as in *Lilium*, p. 649. This is the only case known among gymnosperms where four megaspore nuclei are involved in the development of a single female gametophyte. Between 256 and

595

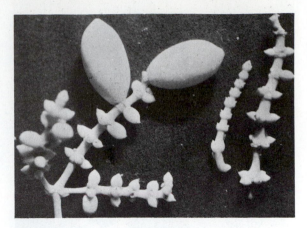

FIGURE 24-21
Gnetum sp. Branch with mature and
abortive seeds (left); younger ovulate
branches at right. ×1.5.

1500 free nuclei develop in the female gametophyte, depending on the species.

At pollination, which occurs during the early
free-nuclear stages of the female gametophyte,
the pollen grains reach the shallow pollen chamber, a depression at the apex of the megasporangium.

The lower portion of the female gametophyte
becomes cellular, but the upper at first remains
free-nucleate except for isolated clusters of
three to eight cells. One or two cells of each
group function as an egg. This report of cellular
eggs in *G. ula* (Vasil, 1959) is at variance with
accounts of other species in which the eggs are
said to consist of free nuclei.

The pollen grains, meanwhile, have germinated in the pollen chamber, their tubes
growing through the megasporangium toward
the apex of the female gametophyte. The generative cells divide to form two sperm cells each
as the pollen tubes penetrate the female gametophyte. Both eggs of a group may be fertilized,
and as many as four zygotes have been found in
a single female gametophyte.

Embryogeny is complex and has not been
completely elucidated. Apparently, depending
on the species, free-nuclear division may or may
not be involved. Polyembryony occurs, and
suspensors function actively. The mature seeds,
which are large and fleshy (Fig. 24-21), normally

contain a single embryo. Development of the
embryo in some species may be completed after
the seeds have been shed, as in *Ginkgo*, so that
germination is often delayed. The primary root
(radicle) persists as a taproot.

Welwitschia (Fig. 24-22) is a truly remarkable
plant. The enlarged, woody-fleshy inverted conical stem, which may attain a diameter of 1.5 m,
is extended below the soil as a long taproot. The
mature plants bear only two enormous leaves,
which persist throughout the life of the plant
through the activity of their basal meristems.

Although *Welwitschia* is often described as a
persistent seedling, Martens (1977) has challenged this view because the plant produces
three decussate leaf whorls (the cotyledons;
persistent, prominent leaves; and two scaly
bodies) as well as the fertile axes, which arise as
axillary buds of the persistent leaves.

The compound microstrobili and megastrobili
of *Welwitschia*, which is dioecious, occur on
branching axes in the axils of the leaves (Fig.
24-23). The mature female gametophyte here,
too, is free-nuclear, and no archegonia are
organized. *Welwitschia* is endemic in the coastal
region of Angola and South West Africa (Rodin,
1953*a*; Kers, 1967; Bornman, 1972), where the
annual rainfall approximates 2.5 cm and where
there may be no precipitation for several years
at a time. In such an environment the plants
probably survive through the water provided by
coastal mists and dews that condense on the
rocky terrain. The plants are slow-growing and
markedly xerophytic in organization (Rodin,
1953*a,b*, 1958). *Welwitschia* may be insect-
pollinated (Faegri and Van der Pijl, 1979).

The female gametophyte arises from one
functional megaspore and is cellular at fertilization, according to Maheshwari and Vasil
(1961*b*), but Martens (1963) stated that all four

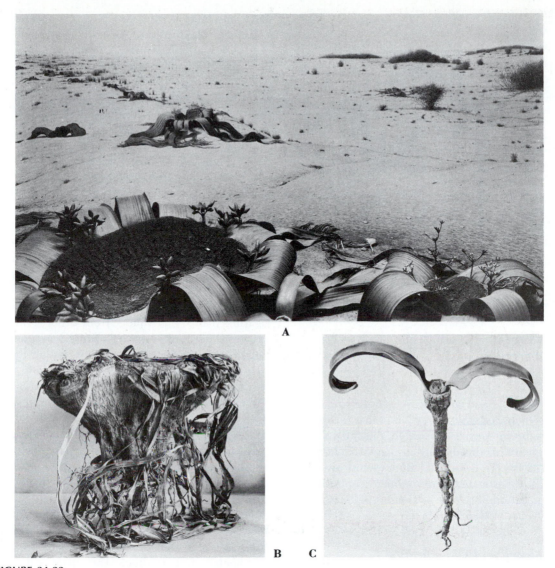

FIGURE 24-22
Welwitschia mirabilis. **A.** Reconstruction of growth habit of plants in Mossandes Desert, Angola; note strobili. **B.** Actual plant; leaves in shreds. **C.** Young plant, showing leaves and part of tap root. *(Courtesy of the Field Museum of Natural History.)*

megaspore nuclei function in forming the female gametophyte. Martens (1959, 1961, 1963, 1964) studied and illustrated many aspects of the organization and reproduction of *Welwitschia*.

Fertilization in *Welwitschia* is unique in that tubular processes from the eggs grow up toward and unite with the pollen tubes. Fertilization occurs within the united tubes.

Butler et al. (1973*a,b,c*) have investigated the

FIGURE 24-23
Welwitschia mirabilis. **A.** Reconstructed microstrobili. **B.** Reconstructed megastrobili. (*Courtesy of the Field Museum of Natural History.*)

morphology and anatomy of developing seedlings of *Welwitschia*. They found that the cotyledons reach maximum size in about 9½ weeks and begin to die distally in one-year-old seedlings. Evert, Bornman, Butler, and Gilliland (1973) investigated electronmicroscopically the organization of the sieve areas of phloem in the leaf veins of *Welwitschia*.

Summary and Classification

Of the three genera of Gnetophyta, it is clear that *Ephedra* is the most primitive in its vegetative and reproductive features. A discussion by Eames (1952) of the relationship of *Ephedra*, *Gnetum*, and *Welwitschia* emphasizes that they are not closely related and recommends that they be classified in separate orders and families. In accordance with this, these plants are classified as follows:

Division Gnetophyta
  Class 1. Gnetopsida
    Order 1. Ephedrales
      Family 1. Ephedraceae
        Genus: *Ephedra*
    Order 2. Gnetales
      Family 1. Gnetaceae
        Genus: *Gnetum*
    Order 3. Welwitschiales
      Family 1. Welwitschiaceae
        Genus: *Welwitschia*

The fossil record of Gnetophyta is scanty. Pollen attributed to plants related to *Ephedra* has been found in Upper Cretaceous strata of Long Island, New York. It had been found previously in Tertiary strata from other parts of the world.

Ephedra is a genus of xerophytic seed-bearing plants that differs from other gymnosperms described in earlier chapters in a number of important attributes. First among these may be cited the occurrence of vessels in the secondary xylem. These arise by perforation of the circular bordered pits of the sloping terminal walls of tracheids. The leaves are reduced to minute, bractlike organs that function only early in development; they are megaphyllous, however. The delicate fasciculate branches are the chief photosynthetic organs of the *Ephedra* plant.

Furthermore, both the microstrobili and the megastrobili of *Ephedra* are compound structures in which the sporogenous organs are appendicular, not cauline. The male gametophytes are primitive in their production of two prothallial cells, but may be advanced, as compared with other gymnosperms, if further study indicates that the generative cells form sperm nuclei directly, without dividing to form a stalk and body cell. Development of the male gametophytes of *Ephedra*, as compared with those of *Zamia*, *Ginkgo*, and *Pinus*, is very rapid. The ovule of *Ephedra* is composed of the megasporangium surrounded by two envelopes. The inner or true integument is delicate, membranous, photosynthetic, and prolonged into a micropylar tube through which pollination is effected. The outer ovular envelope consists of two united bracts. Breakdown of the apical cells of the megasporangium gives rise to a pollen chamber, which, unlike that in the genera described in earlier chapters, is in direct contact with the female gametophyte and archegonia. The latter are primitive in having the largest number of neck cells known among gymnosperms—40 or more. The pollen tube performs no obviously haustorial function. Embryogeny includes a more limited free-nuclear period than that found in other gymnosperms.

Comparative study of the reproductive axes (strobili) of the various species of *Ephedra* reveals that they may correspond more closely to those of the Cordaitales (p. 623) than they do to those of other conifers or to those of *Welwitschia* and *Gnetum*. With respect to vegetative attributes, the same general relationship seems to obtain. The stomata of *Ephedra*, like those of the Coniferophyta, are simple, those of *Gnetum* and *Welwitschia* being complex, but this has been questioned by Maheshwari and Vasil (1961a). Similarities in leaf form, venation, leaf trace number, and wood structure exist between *Ephedra* and the Coniferophyta. *Ephedra*, then, seems to be more closely related to the Coniferophyta than it is to *Gnetum* or to *Welwitschia*.

The literature contains repeated suggestions that the angiosperms evolved from Gnetalean ancestors. Chemical analysis of *Gnetum* wood (Melvin and Stewart, 1969) does reveal significant resemblances to substances in the wood of certain tropical angiosperms.[1]

---

[1]For a summary of extant and extinct gymnosperms, see Chapter 25.

DISCUSSION QUESTIONS

1. What vascular cryptogams are suggested superficially by the vegetative characteristics of *Ephedra?* In what respects are they similar?
2. In what habitats would you seek *Ephedra?*
3. Discuss the medicinal uses of species of *Ephedra*.
4. By what macroscopically visible criteria could you distinguish the microstrobili from the megastrobili of *Ephedra?*
5. Describe the development of the male gametophyte of *Ephedra*.
6. In what genera, other than *Ephedra*, are long micropylar tubes present?
7. How does the pollen chamber of *Ephedra* differ from that in the seed plants discussed in earlier chapters?
8. What functions are ascribable to the pollen tube of *Ephedra?* How does it compare in function with that of *Zamia, Ginkgo*, and *Pinus?*
9. On the basis of supplementary reading, list features that the genera *Ephedra, Gnetum*, and *Welwitschia* have in common.
10. There is evidence that the sperm nuclei of certain gymnosperms differ in size. Can you cite any other examples of dimorphism in reproductive cells in the plant and animal kingdoms?
11. In what respects is the archegonium of *Ephedra* primitive?
12. How do the vessels of Gnetophyta differ from those of other vascular plants?
13. How do the female gametophytes of *Ephedra, Gnetum*, and *Welwitschia* differ?
14. What homologies exist between the sporogenous reproductive structures of *Ephedra* and those of the Cordaitales (see Chapter 25)?

Chapter 25
GYMNOSPERMS: RECAPITULATION AND FOSSIL RECORD

The evolution of the seed habit in vascular plants was a major milestone in the history of the plant kingdom. Along with the development of seeds, there occurred an entirely different mechanism of transfer of the sperm to the egg.

A number of morphological changes had to have taken place in the transition from free-sporing plants to seed-bearing plants. First is the development of heterospory. All seed plants are heterosporous, that is, there are two kinds of spores (megaspores and microspores), each performing a different function. Actually, there may be little or no size difference between microspores and megaspores, but there is definitely a functional difference. In the development of the seed habit, the number of megaspores in the sporangium is reduced typically to only one that is functional. Furthermore, the megaspore is not shed from the megasporangium, and the female gametophyte develops entirely within the megaspore wall. In addition, the megasporangium is surrounded by protective structures, the integuments. Some of these features have been seen already among the vascular cryptogams. For example, *Selaginella* is heterosporous, with a reduced number of megaspores per megasporangium. Although four megaspores are typical, there may be variation, and, conceivably, some megasporangia may have only one functional megaspore. The female gametophyte develops within the confines of the megaspore wall. Although the spores are normally shed, there have been some reported instances of the megaspores, with their contained female gametophytes, having been retained within the sporangium where fertilization takes place.

Another member of the Microphyllophyta, *Lepidocarpon*, is also heterosporous, with the number of megaspores per megasporangium reduced to one functional one. The spore was not shed, and the female gametophyte developed within the megaspore wall. Actually, the spore wall stretched considerably as the female gametophyte enlarged. In *Lepidocarpon*, the lateral edges of the megasporophyll enveloped the megasporangium, leaving a slit at the top. The entire sporophyll and sporangium complex was shed as a unit, and, in effect, functioned like a seed. It must be emphasized, however, that although *Lepidocarpon* closely resembled a seed, seed plants did not originate from the Microphyllophyta.

Along with modifications of the megasporangia, another important change occurred in the evolution of the seed habit. In the vascular cryptogams, the sperm is released from the male gametophyte and somehow makes its way to the egg, usually by swimmimg in water. In seed plants, after the microspores have been produced, one or more cell divisions take place within the microspore wall, representing early stages in the development of the male gametophyte. At this stage, the young male gametophyte, still enclosed within the microspore wall, is transported to the area of the megasporangium. The young male gametophyte is called a **pollen grain**. The pollen grain then continues to develop in that more cells are produced in the male gametophyte, which develops a tube that grows toward the egg. Within the tube are produced two sperm cells that are carried passively to the egg where fertilization occurs. Thus, there is no longer a need for water in the fertilization process.

Although the kinds of changes that were necessary to produce the seed habit were long known, the actual ancestral group or groups involved were a mystery. The discovery of a distinct group of plants placed in the division Progymnospermophyta finally provided insight as to whence seed plants arose.

DIVISION PROGYMNOSPERMOPHYTA

The progymnosperms occupy an interesting position in the plant kingdom because they combine characters of vascular cryptogams (sporangia that shed spores) with those of gymnospermous seed plants (e.g., eusteles without leaf gaps; secondary xylem, often with circular bordered pits; and simple leaves derived from flattened branch systems). Beck (1960*a,b*) first recognized the existence of such a group based on his work with the Upper Devonian *Archaeopteris* (Gr. *archaio*, ancient, + Gr. *pteris*, fern) and the contemporaneous *Callixylon* (Gr. *kalli*, beautiful, + Gr. *xylon*, wood). *Archaeopteris* was a tree, with flattened branch systems bearing wedge-shaped leaves (Fig. 25-1).

These flattened branches had the appearance of fern fronds (hence the name *Archaeopteris*) (Carluccio, Hueber, and Banks, 1966; Beck 1971). Replacing certain leaves were elongated, branched axes bearing slender sporangia. At least two species of *Archaeopteris* were heterosporous (Phillips, Andrews, and Genzel, 1972). *Callixylon* had been thought to represent the trunks of gymnospermous trees because anatomically, with a eustele and compact secondary xylem bearing circular bordered pits, it resembled the stems of conifers quite closely (Fig. 25-1C–E). Beck showed that sections of the axes of *Archaeopteris* that bore leaves showed an internal structure like that of *Callixylon*. The latter genus is readily recognizable in radial section because of the peculiar and characteristic grouping of pits in radial files (Fig. 25-1E).

The Upper Devonian *Tetraxylopteris* is another progymnosperm with stems bearing helically arranged branches (Fig. 25-2). These laterals, in turn, bore decussate branches. Some of these axes were fertile and bore sporangial clusters in a decussate pattern. Primary xylem in the larger axes was cruciform in cross section, with secondary xylem eventually having produced a cylindrical outline to the xylem cylinder.

Rellimia (perhaps better known by its former name, *Protopteridium*), from the Middle Devonian, differed only slightly from *Tetraxylopteris* (Fig. 25-3). It was a shrub or a low arborescent plant with main axes dichotomously or pseudomonopodially branched. Leaf-bearing branch systems were helically arranged and bore leaves that were bifurcated two or three times. Fertile branches were also helically placed and bore pinnately divided sporangium-bearing structures. Earlier work reported that this plant had ultimate sterile portions that were pinnately divided and apparently bladelike. These were interpreted as representing the initiation of fern fronds in a phyletic sense. It has since been shown (Leclercq and Bonamo, 1971) that these structures were actually fertile

A

2 cm

B

C

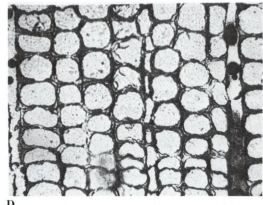

D

FIGURE 25-1
Archaeopteris. **A.** Generalized reconstruction.
B. Detail of reconstruction of leaf-bearing axes.
C. Petrified portion of large trunk (*Callixylon*).
D. Transverse section of a portion of the
secondary xylem. **E.** Radial section of xylem.
Note clusters of circular bordered pits. **B.** ×170.
C. ×350. (**A, B.** *After Beck.*)

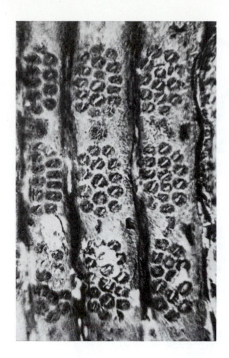

E

organs from which the sporangia had become detached.

Aneurophyton (Gr. *a*, lacking, + Gr. *neuron*, nerve, + *phyton*, plant) (thought to be identical with a plant referred to as *Eospermatopteris* [Gr. *eo*, dawn, + *sperma*, seed,[1] + Gr. *pteris*, a fern]) was more fernlike in habit and was a small tree, between 6 and 12 m tall. The trunks bore a terminal crown of frondlike branches, 1–2 m long (Fig. 25-4). The ultimate divisions of the "fronds" might be interpreted as leaves that were dichotomized and lacked laminae. Some of the appendages terminated in large, ovoid sporangia up to 6 mm long and 3 mm in diameter.

Representatives of the Progymnospermophyta discussed above may be classified as follows:

Division Progymnospermophyta
    Order 1. Aneurophytales
        Family 1. Archaeopteridaceae
            Genera: *Archaeopteris, Callixylon*
        Family 2. Aneurophytaceae
            Genera: *Tetraxylopteris, Rellimia,*
                  Aneurophyton,
                  Eospermatopteris

RECAPITULATION

The plants discussed in Chapters 21–24 all have in common the attribute of producing seeds. As was stated in the opening paragraph of Chapter 21, in earlier classifications the seed habit served as a sufficiently important criterion for uniting a vast assemblage of plants into a single division, the Spermatophyta. This taxon was divided further into two classes; the Gymnospermae (Gr. *gymnos*, naked, + Gr. *sperma*, seed), and the Angiospermae (Gr. *angeion*,

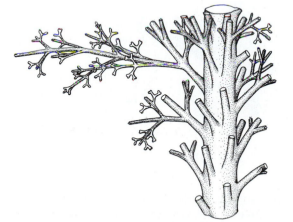

FIGURE 25-2
Tetraxylopteris schmidtii. Reconstruction of a portion of a plant. (*After Beck, from Delevoryas.*)

---

[1]The name is based on an early erroneous interpretation of the plant as a seed fern or pteridosperm.

603

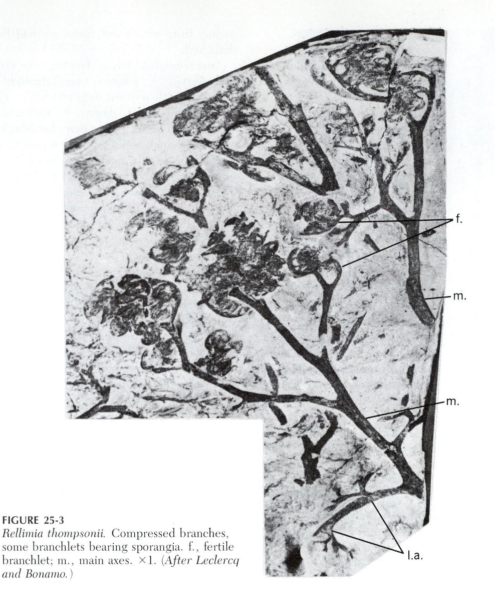

FIGURE 25-3
Rellimia thompsonii. Compressed branches, some branchlets bearing sporangia. f., fertile branchlet; m., main axes. ×1. (*After Leclercq and Bonamo.*)

vessel, + Gr. *sperma*). The Gymnospermae were delimited from the class Angiospermae because the seeds of the former are usually *exposed* on the appendages that bear them, while those of the Angiospermae develop *within*

the structures that bear them. Most modern treatments divide what had been classified as Gymnospermae into a number of independent divisions. In this book, the following divisions of gymnospermous plants are recog-

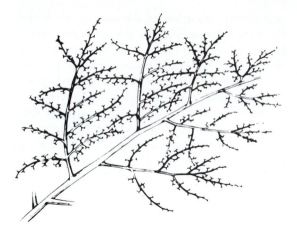

FIGURE 25-4
Aneurophyton germanicum. Part of branch system. (*After Kraüsel and Weyland, from Banks.*)

nized: Pteridospermophyta, Cycadophyta, Cycadeoidophyta, Ginkgophyta, Coniferophyta, and Gnetophyta.

COMPARISON OF VEGETATIVE CHARACTERISTICS

When one considers such extant genera as *Zamia* and *Cycas*, *Ginkgo*, *Pinus* and other conifers, and the genera *Ephedra* and *Gnetum*, the striking divergences among them forcefully crowd their more subtle common attribute of gymnospermy almost into oblivion. The vegetative features of these plants are markedly diverse, although all are megaphyllous. In the Cycadophyta the plant bodies are relatively small, seldom exceeding the status of shrubs or small trees. They are strongly suggestive of the ferns, especially in their pinnately compound leaves. The sparingly branched stems are armored with old leaf bases and contain xylem interspersed with parenchyma tissue, even though they have active cambial layers. The leaflets in *Cycas* and the leaves of *Ceratozamia*, like those of many ferns and pteridosperms (p. 609), exhibit circinate vernation, another fernlike attribute. The Ginkgophyta and Coniferophyta stand in strong contrast to the Cycadophyta with respect to their vegetative organs. All are richly branched, usually large trees in which the leaves are simple, broad-leaved, and deciduous, as in *Ginkgo*, or scalelike, needlelike, or narrow and leathery as in the conifers. In these trees the cambium becomes active in both stem and root during the first season of growth and adds cylinders of secondary xylem with little parenchyma each growing season, so that the stems and roots are woody, not fleshy, as in the cycads. The stems are protected not by the remains of the leaf bases but by periderm layers generated by cork cambiums. Furthermore, the leaves of these trees often are borne in fasciculate fashion on short lateral shoots, the spur or dwarf shoots, which increase very slowly in length.

Finally, in the genera *Ephedra*, *Gnetum*, and *Welwitschia* of the Gnetophyta, the vegetative organization is extremely diverse and in each case is different from that in cycads, *Ginkgo*, and the conifers. *Ephedra*, the representative of the Gnetophyta emphasized in this text, is a shrubby or trailing xerophyte in which the leaves are reduced to scales that function only temporarily in photosynthesis. The ribbed internodes and the absence of well-developed leaves suggest the Arthrophytan genus *Equisetum*, and the presence of vessels in the primary and secondary xylem is unique (except for those in *Gnetum* and *Welwitschia*) among living gymnosperms.

Additional evidence of diversity among the living gymnosperm types is not wanting, of course, but the features cited above should be sufficient to indicate the heterogeneity of the group with respect to vegetative attributes. It seems clear that abandonment of an inclusive taxon Gymnospermae has been justified.

605

REPRODUCTION

A comparative review of the reproductive features of gymnosperms reveals both common and divergent attributes. The sporogenous tissues of some representatives of the four extant gymnospermous divisions are localized in cones or strobili; in others, strobili are absent. For example, the ovules of the genus *Cycas* are borne on pinnately divided sporophylls, while those of *Ginkgo* occur in terminal pairs on a branching peduncle. The microstrobili that produce the microspores are said to be simple, because the spore-bearing appendages are borne directly on the axis of the strobilus, except in the Gnetophyta. In the latter, the microstrobili are compound; the microsporophylls and their enclosing bracteoles are produced in the axils of lateral bracts.

The megastrobili in the cycads are simple, but in *Pinus* and related genera and in *Ephedra* they are compound. The sporangia themselves are in all cases eusporangiate in development, the sporangial walls being relatively massive; the output of microspores in each sporangium is enormous.

The comparative organization of the gymnospermous female gametophyte has been summarized by Maheshwari and Singh (1967) and that of the male gametophyte by Sterling (1963).

In all the representative gymnosperms (and in all seed plants), a single megasporocyte undergoes meiosis to form a linear tetrad or triad of potential megaspores, three of which usually degenerate, although in *Gnetum* all four take part in the formation of the female gametophyte. The heterospory of seed plants is functional, not markedly morphological, as it is in vascular cryptogams. However, the megaspore increases greatly in volume during the development of its enclosed female gametophyte.

The location, nutrition, and course of development of the megaspores into female gametophytes are quite uniform. In all seed plants the permanent retention of the megaspore within the megasporangium and the intimate connection of the megaspore with the surrounding tissue (sterile megasporangium) make possible the seed habit. The megaspore of each genus develops into the female gametophyte after passing through a period of free-nuclear division. In a number of gymnosperms, the megaspore wall is thickened by deposition on its surface during the development of the female gametophyte. This is considered to represent a vestigial attribute retained from free-spored precursors. In a number of cases, however, the thickening has been revealed to be an additional layer composed of material from the digested megasporangium (Pettitt, 1966, 1977*a*). The nutriment for the developing gametophyte diffuses gradually into it by the dissolution of the surrounding sporophytic tissues. This is in marked contrast, timewise, with the storage metabolites in the megaspores of vascular cryptogams. Wall formation at the end of the free-nuclear period (except in *Gnetum* and *Welwitschia*) results in the organization of a completely cellular gametophyte. Archegonia much larger, but less complex,[2] than those of the cryptogams are organized in all genera except *Gnetum* and *Welwitschia*. In *Gnetum*[3] as in the flowering plants (Chapter 26), one or more of the nuclei of the free-nuclear gametophyte function directly as eggs, and archegonia are lacking. In all the representative gymnosperm genera, breakdown of the apical tissues of the megasporangium results in the formation of a chamber for the reception of pollen and is accompanied by the secretion of a pollination droplet, which facilitates the transfer of pollen to the apex of the megasporangium. The pollen chamber in the megasporangium of *Pinus* is extremely rudimentary; that of *Ginkgo* is deep (Fig. 22-11A).

---

[2] In lacking neck canal cells and in most (except *Ephedra*) having short necks.

[3] Except in *G. ula*.

Ephedra, Gnetum, and *Welwitschia* are unique, among the types described, in that their pollen chambers extend to the tissues of the mature female gametophyte.

The development of the microspores into male gametophytes in the gymnosperms is quite uniform except for several variations such as type of motility of the male gametes. In all cases, the process begins before the microspores have been shed from the microsporangia. Flagellate sperms are produced in the cycads and *Ginkgo* but are absent in other living genera. In all divisions except the Gnetophyta, the growth of the pollen tube is slow and prolonged, and the pollen tube is obviously both haustorial and nutritive in function. In the Gnetophyta, however, the proximity of the pollen grain to the archegonia eliminates the circumstance that results in haustorial activities, and the pollen tube accordingly is short in length and ephemeral.

The development of the gymnosperm embryo is unlike that of the vascular cryptogams, on the one hand, and the flowering plants, on the other, because it always involves a period of free-nuclear division of variable duration.[4] In the cycads and *Ginkgo,* approximately 256 nuclei arise in the proembryo before wall formation occurs. In *Pinus* the proembryo consists of only four free nuclei, and the number is often eight in *Ephedra.* In these genera, except *Ginkgo,* the embryo is thrust into the nutritive tissue of the female gametophyte by an active system of suspensors. In *Ginkgo* a suspensor is absent, but the embryo itself grows into the female gametophyte through the base of the archegonium. In gymnosperms, by the time of fertilization, the ovule has achieved essentially the dimensions that will characterize the mature seed. Both simple polyembryony (the development of more than one zygote into an embryo

within a single female gametophyte) and cleavage polyembryony (the division of the cellular progeny of a single zygote into several embryos) occur among the gymnosperms. Cleavage polyembryony, however, is confined to *Pinus* and several other conifers. Normally, the mature seed contains only one functional embryo.

Both hypogean and epigean germination are represented in the gymnosperms. The two cotyledons in *Zamia* and *Ginkgo* remain within the seed at germination and serve as absorptive organs; they emerge later in *Zamia. Pinus,* which is polycotyledonous, and *Ephedra,* which has two or three cotyledons, are both epigean.

Thus, in spite of the diversities in vegetative structure among the gymnospermous plants discussed in Chapters 21–24, there is considerable uniformity in the reproductive process. However, considering both living and fossil genera, it seems that a number of parallel lines of development have long existed among gymnospermous plants.

THE FOSSIL RECORD OF GYMNOSPERMS

Until quite recently, it was agreed that although the vascular cryptogams have been present in the earth's flora since late Silurian times, seed plants (gymnospermous ones) first appeared during the Mississippian (Lower Carboniferous). However, with Pettitt and Beck's (1967) and Chaloner, Hill, and Lacey's (1977) announcement of the discovery of seeds in Upper Devonian strata, the history of seed plants has been extended further back. The seedlike body described by Pettitt and Beck (Fig. 25-5) consisted of cupulelike structures within each of which was a single, large, thick-walled megaspore 4.8 mm long and 2 mm wide. The megaspore was covered with a "tapetal membrane" presumably similar to that deposited on the megaspore membrane of living gymnosperms (p. 569). Pettitt and Beck (1968)

---

[4]Except in *Gnetum.*

607

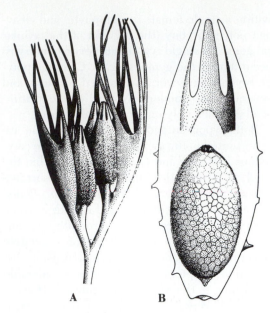

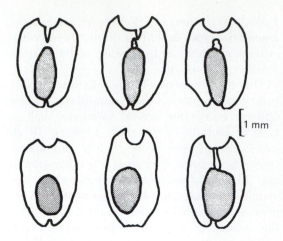

FIGURE 25-6
Spermolithus devonicus. Outline diagrams of six ovules. Note bilaterally symmetrical integument and stippled megaspores. (*After Chaloner et al.*)

FIGURE 25-5
Archaeosperma arnoldii, an Upper Devonian seedlike structure. **A.** Cupulelike appendages surrounding four ovules. **B.** Longitudinal section of a single ovule. Note the large functional megaspore with three degenerated spores and the integumentary lobes. (*After Pettitt and Beck.*)

named their specimen *Archaeosperma* (Gr. *archaio*, ancient, + Gr. *sperma*, seed) *arnoldii*, and their reconstruction of the cupule complex in which the seeds appear to have been borne in pairs is reproduced as Fig. 25-5. The plant that bore these seeds is unknown, but it is possible that one of the progymnosperms may have reached the level of seed-producing plants. *Archaeosperma* occurs as a compression fossil, and it has not been possible to demonstrate a megasporangium within the presumed cupulate integument, but most workers agree with Pettitt and Beck's interpretation of the structure of the seed. *Spermolithus* (Gr. *sperma*, seed, + *lithos*, stone) *devonicus* (Chaloner, Hill, and Lacey, 1977) is more specialized than *Archaeo-*

sperma in that the integument more completely surrounds the megasporangium. Furthermore, the seed is flattened, while *Archaeosperma* has radial symmetry (Fig. 25-6).

Since the report of the discovery of *Archaeosperma* and *Spermolithus*, a still older seedlike body was uncovered from Upper Devonian strata in West Virginia. These seeds were borne in loose clusters surrounded by cupulelike members as were those of *Archaoesperma*, but the protective integumentary apparatus consisted of several separate, fingerlike processes (Figs. 25-7, 25-8)(Scheckler et al., 1981). There is a marked resemblance between these ovules and those of *Genomosperma* (page 610).

Fossil gymnospermous plants of the Cycadophyta, Ginkgophyta, and Coniferophyta, as well as the extinct divisions Pteridospermophyta and Cycadeoidophyta, are discussed in the following account. Fossil Gnetophyta are known from pollen as far back as the Triassic, with megafossil evidence probably having appeared first in the Cretaceous (Chaloner, 1969).

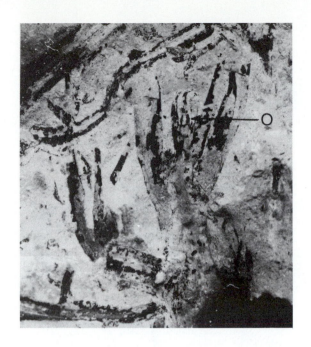

DIVISION PTERIDOSPERMOPHYTA

The pteridosperms, or seed ferns, were plants with fernlike foliage that bore seeds on the leaves (Fig. 25-9). They ranged in time from the late Devonian to the Jurassic. Before the existence of such a group was demonstrated early in the twentieth century, it had been assumed that the fernlike fronds so abundant in the latter part of the Carboniferous all belonged to ferns, and for that reason that geologic period

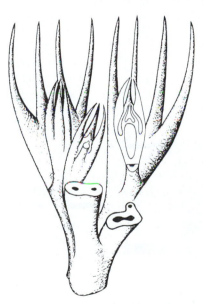

FIGURE 25-8
Suggested reconstruction of Upper Devonian ovules such as that in Fig. 25-7. Parts of three ovules and fingerlike appendages of cupule are evident. (*Courtesy of Prof. G. W. Rothwell.*)

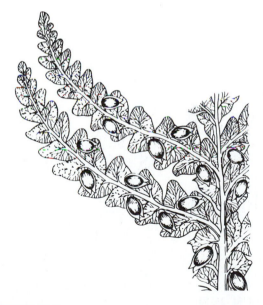

FIGURE 25-9
Emplectopteris triangularis. Fernlike foliage with seeds. (*After Andrews.*)

was designated as the "Age of Ferns." It has since been shown that many of the fernlike leaves lacking sori were actually those of seed ferns, and it is still impossible to determine, in some instances, whether the foliage is pteridophytan or pteridospermophytan.

The existence of seeds in the Devonian and seed ferns in the Carboniferous raises the question of the origin of the seed habit. The ontogenetic precursor of the seed is the ovule, which is defined as a megasporangium covered with one or more integuments. The megasporangium is naked in cryptogams and provided with a covering in phanerogams (cf. Figs. 14-26 and 21-12). The origin of the integument remains obscure. One suggestion is that in groups of megasporangia the peripheral ones became sterile and surrounded a central one as an incipient integument. Another suggestion is that the integument was derived from a number of fingerlike branches that subtended the megasporangium. Fossil evidence seems to lend greater support to the latter interpretation. This integument at first was not fused with the megasporangium except at the base; fusion of the two was a later development. The Lower Carboniferous seed *Genomosperma* (Gr. *gignomai*, to become, +

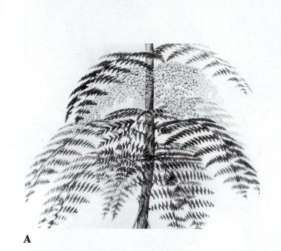

A

B

FIGURE 25-11
Lyginopteris oldhamia. **A.** Habit of growth. **B.** Model reconstruction with details of fronds. (At the extreme left note frond of the seed fern *Neuropteris heterophylla.*) (**A.** *After Hirmer, from Wettstein.* **B.** *Courtesy of the Field Museum of Natural History.*)

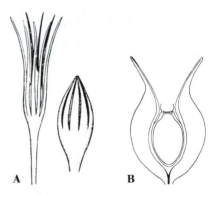

FIGURE 25-10
A. *Genomosperma kidstonii* (left) and *G. latens.* **B.** *Lyrasperma scotica.* Diagrammatic longitudinal section of seed. (**A.** *After Andrews.* **B.** *After Long, from Delevoryas.*)

610

Gr. *sperma*) (Fig. 25-10A) is an example of a seed with a primitive type of integument. Here the megasporangium is surrounded by eight sterile structures, with the tip of the sporangium exposed. In addition to the integument, the presence at the apex of the megasporangium of a special receptive chamber for pollen distinguishes ovules from megasporangia. In *Genomosperma* there was such a chamber. Another Lower Carboniferous seed, *Lyrasperma* (Gr. *lyra*, a stringed instrument, + *sperma*) (Fig. 25-10B), is relevant in this context. Here the integument is more completely united with the megasporangium, and a pollen chamber also is present. Various aspects of ovule evolution are discussed at greater length by Andrews (1961, 1963), Meeuse

(1963c, 1964), Smith (1964), and Long (1966), among others.

Among the Paleozoic Pteridospermophyta, the seed fern *Lyginopteris* (Gr. *lyginos*, twisted or winding, + Gr. *pteris*) is one of the better known. It occurs in Pennsylvanian strata in England and elsewhere in Europe. *Lyginopteris* (Fig. 25-11) is reconstructed as a plant resembling a small tree fern. The stem was slender, a few centimeters thick at the most, and contained a number of conspicuous mesarch primary xylem strands surrounding a pith (Fig. 25-12). A small amount of secondary xylem was developed. In the cortex were conspicuous strands of thick-walled cells that contributed to the support of the slender stem. Leaves were frondlike, with a rachis that dichotomized once, and the

FIGURE 25-12
Lyginopteris oldhamia. Transverse section of stem. i.c., inner cortex; o.c., outer cortex; p., pith; s.x., secondary xylem.

rest of the branching pattern was pinnately constructed. Roots were adventitious, arising from the stem in association with leaf bases. Seeds of *Lyginopteris* (placed in the organ genus *Lagenostoma* [Gr. *lagenos*, flagon, + *stoma*, mouth]) were small (about 5.5 mm long) and were borne within a tulip-shaped cupule (Fig. 25-13A), a characteristic of many Paleozoic seed ferns. The integument was fused to the megasporangium except at the tip. Here the latter was prolonged as a central column within the hollow pollen chamber that was shaped like an inverted funnel (Fig. 25-13B). Pollen grains have been observed within the pollen chamber. The occurrence of characteristic stalked glands (Fig. 25-13A) on the cupule and on leaf fragments and stems helped to prove the co-identity of *Lyginopteris* and *Lagenostoma*, which had not been found actually attached to one another.

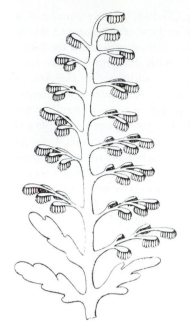

FIGURE 25-14
Crossotheca sp. Restoration. ×1.5. (*After Andrews.*)

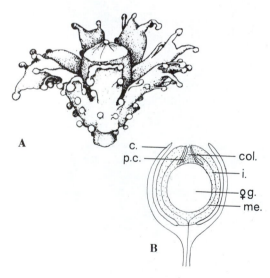

FIGURE 25-13
Lagenostoma lomaxii. **A.** Reconstruction of a seed and the surrounding cupule. **B.** Diagrammatic median longitudinal section of cupule and seed. c., cupule; col., central column of pollen chamber; ♀ g., female gametophyte; i., integument; me., megasporangium; p.c., pollen chamber. (**A.** *After Oliver and Scott, from Andrews.* **B.** *After Oliver, from Arnold.*)

FIGURE 25-15
Callistophyton poroxyloides. Transverse section of young stem axis. Note peeling cortex and epidermis. ×6.5. (*After Delevoryas and Morgan.*)

612

The pollen-bearing phase of *Lyginopteris* is not known with certainty. However, the microsporangiate fructification *Crossotheca* (Gr. *krossos*, fringe, + *theke*, container) (Fig. 25-14) has been found attached to foliage similar to that borne by *Lyginopteris* and could possibly have belonged to it.

The genus *Callistophyton* (Gr. *kallistos*, most beautiful, + *phyton*) (Delevoryas and Morgan, 1954) has not been known as long as has *Lyginopteris* but is becoming the best understood Paleozoic seed fern (Fig. 25-15). An individual of this Pennsylvanian genus is depicted by Rothwell (1975) as a scrambling understory plant with a slender stem and distantly spaced leaves (Fig. 25-16). Adventitious roots arose near nodes. The stem was eustelic, with a number of mesarch primary xylem strands and secondary xylem and phloem. Young stems had strands of thick-walled cells in the outer cortex (although not so pronounced as in *Lyginopteris*), but in older stems the cortex was replaced by periderm after the former had sloughed off. Microsporangiate structures resemble marattiaceous

synangia and were borne on the abaxial surfaces of pinnules. *Idanothekion* (Gr. *idanos*, fair, + *thekion*, diminutive container) (Millay and Eggert, 1970) is the name assigned to these fertile organs (Fig. 25-17), but *Callandrium* (Gr. *kallos*, beautiful, + *andro*, masculine) (Stidd and

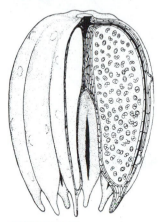

FIGURE 25-17
Idanothekion glandulosum. Diagrammatic reconstruction of a synangium showing some of the internal organization. ×32. (*After Millay and Eggert.*)

FIGURE 25-16
Callistophyton poroxyloides. Restoration of part of a plant. (*After Rothwell.*)

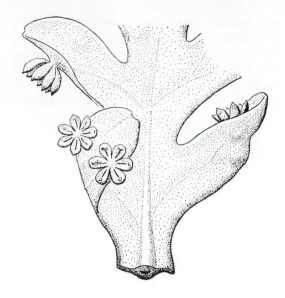

Hall, 1970*b*) (Fig. 25-18) has also been suggested as the pollen-bearing organ of *Callistophyton.* Rothwell (1972*b*) hints that these two entities may be the same. Small, platyspermic seeds placed in the genus *Callospermarion* (Gr. *kallos,* + *spermarion,* diminutive of seed) are believed to be those of *Callistophyton* (Stidd and Hall, 1970*a*) (Fig. 25-19). Rothwell (1972*a*) demonstrated the existence of branching pollen tubes in an ovule of *Callospermarion.* He also (Rothwell, 1977) showed evidence of a pollination-drop mechanism in that pteridosperm.

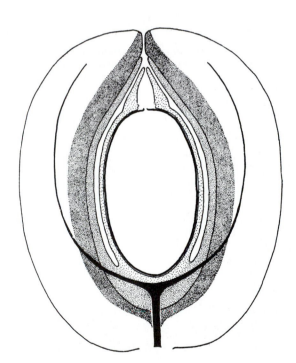

FIGURE 25-19
Callospermarion pusillum. Diagrammatic median longitudinal section through the primary plane (parallel to flat side) of a seed. ×40. (*After Eggert and Delevoryas.*)

FIGURE 25-20
Medullosa noei. Reconstruction depicting habit of growth. (*After Steart and Delevoryas.*)

Medullosa (L., *medullosus*, pithy), a reconstruction of one species of which is illustrated in Fig. 25-20, occurs in strata dating from the Mississippian through the Permian. In habit, *Medullosa* resembled a tree fern with a crown of massive pinnately compound leaves and adventitious roots that originated near the base of the trunk. Stems of *Medullosa* appear to have a number of what appear to be individual steles, ranging from 2 to 25 or more (Fig. 25-21). One interpretation is that the conducting system is a polystele (Stewart and Delevoryas, 1956), but a more recent suggestion (Basinger, Rothwell, and Stewart, 1974) is that each "stele" is simply part of what originally was a protostele that had become dissected. Young stems had a thick, fleshy cortex around the stelar system; this cortex sloughed off as more secondary vascular tissues were produced, and a thick periderm then became the outer protective layer of the stem. Medullosan roots were protostelic, often tetrarch. Leaves of at least two form genera (*Neuropteris* [Gr. *neuron*, nerve, + *pteris*] and *Alethopteris* [Gr. *alethes*, real]) probably belonged to medullosan stems and sometimes bore seeds in place of the terminal pinnule of a pinna (Fig. 25-22). Some medullosan seeds, which are among the largest in the gymnosperms, are members of the organ genus *Pachytesta* (Gr. *pachys*, thick, + *testa*, shell) (Fig. 25-23). These seeds had thick integuments composed of a fleshy outer layer and a hardened inner layer. The megasporangium was attached to the innermost integument only at the base (although there exists some evidence that it may have been in contact with the integument along three major ribs of the integument). A pollen chamber with a small opening was formed at the apex of the megasporangium (Fig. 25-23). Some ovules have been found that contained female gametophytes with archegonia. Medullosan microsporangiate structures were built on the

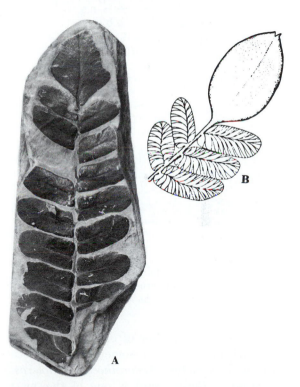

FIGURE 25-22
Neuropteris sp. **A.** Portion of a pinna. **B.** Tip of pinna with a seed replacing terminal pinnule. (**A.** *Courtesy of the Field Museum of Natural History.* **B.** After Stewart and Delevoryas.)

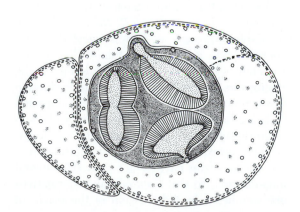

FIGURE 25-21
Medullosa noei. Diagrammatic transverse section of a young stem showing three stelar cylinders and massive cortex. (*After Stewart and Delevoryas.*)

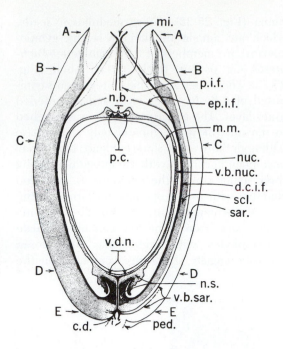

FIGURE 25-23
Pachytesta illinoense. Diagrammatic median longitudinal section. c.d., chalazal dome; d.c.i.f., dark cells of inner flesh; ep.i.f. epidermis of inner flesh; mi., micropyle; m.m., megaspore membrane; n.b., nucellar (megasporangial) beak; n.s., nucellar (megasporangial) stalk; nuc., nucellus (megasporangium); p.c., pollen chamber; ped., pedicel; p.i.f., parenchyma of inner flesh; sar., sarcotesta; scl., sclerotesta; v.b.nuc., vasular bundles of nucellus (megasporangium); v.b.sar., vascular bundles of sarcotesta; v.d.n., vascular disc of nucellus (megasporangium). (*After Stewart.*)

Some seed ferns, quite different from those of the Paleozoic, persisted into the Mesozoic. One of the most interesting of these is the genus *Caytonia* (named for Cayton Bay, Yorkshire) (Fig. 25-25), first described from Middle Jurassic deposits from the Yorkshire coast of England (they have since been found in Triassic and Cretaceous strata). Leaves of *Caytonia* were palmately compound (Fig. 25-25A) and are placed in the form genus *Sagenopteris* (L. *sagena*, fishnet). Pollen organs were borne in clusters (Fig. 25-25E) on branching axes. The ovule-bearing axes (Fig. 25-25B) bore two rows of saclike structures within each of which were a number of ovules (Fig. 25-25C, D). These ovule-bearing sacs at first suggested angiospermy, but it was later found that pollen somehow entered the sacs and ended up in the micropyles of the ovules. Thus reproduction was gymnospermous. The sacs may be regarded as being homologous with the cupules of *Lyginopteris* and certain other Paleozoic seed ferns. In this context, it is of interest that intracarpellary pollen grains have been reported in several angiosperms, so that *Caytonia* and its relatives are significant in a discussion of the origin of angiospermy.

As a group, the pteridosperms are distinguishable from other gymnospermous plants by their fernlike foliage, which bore seeds directly on relatively unmodified pinnules, in place of

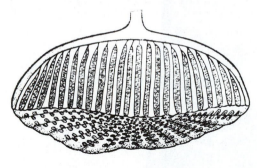

FIGURE 25-24
Dolerotheca formosa. Half of a campanulate synangium showing longitudinal section at the front surface. ×2. (*After Schopf, from Andrews.*)

basic plan of a number of elongated, tubular sporangia fused together, sometimes into a massive, fleshy structure. *Dolerotheca* (Gr. *doleros*, deceitful, + *theke*, case) is one of the most complex (Fig. 25-24), and there is strong indication that it was borne on leaves of the form genus *Alethopteris* (Ramanujam, Rothwell, and Stewart, 1974).

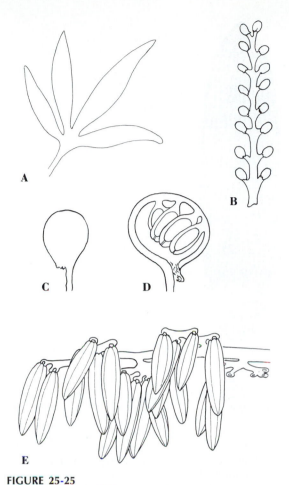

FIGURE 25-25
A. *Sagenopteris phillipsii.* Leaf associated with *Caytonia.* **B, C.** *Caytonia nathorstii.* **B.** Axis with seed-bearing cupules. **C.** Single seed-bearing cupule. **D.** *C. sewardii.* Section of cupule with four seeds. **E.** *Caytonanthus kochii.* Portion of microsporangiate axis. (**A.** *After Andrews.* **B–D**. *After Thomas.* **E.** *After Harris.*)

in the Carboniferous and then declined, with only a relatively few persisting into the Mesozoic.

Because pteridosperms were contemporaneous with many ferns, and because the internal structure of seed ferns is quite different from that of ferns, it is doubtful that the latter could have been seed fern progenitors. It is more probable that the Pteridospermophyta developed from a complex of Progymnospermophyta, with the pteridosperm frond having originated from a flattened branch system of the progymnosperms.

Following is the classification of the pteridosperms discussed above:

Division Pteridospermophyta
    Order 1. Lyginopteridales
        Family 1. Lyginopteridaceae
            Genera: *Lyginopteris, Lagenostoma, Crossotheca*
        Family 2. Callistophytaceae
            Genera: *Callistophyton, Idanothekion, Callandrium, Callospermarion*
    Order 2. Medullosales
        Family 1. Medullosaceae
            Genera: *Medullosa, Neuropteris, Alethopteris, Pachytesta, Dolerotheca*
    Order 3. Caytoniales
        Family 1. Caytoniaceae
            Genera: *Caytonia, Sagenopteris*

DIVISION CYCADOPHYTA

Earliest evidence of true cycads occurs in Lower Permian rocks (Mamay, 1976) in the form of megasporophylls and associated plant parts. The megasporophyll of *Archaeocycas* (Gr. *archaios,* ancient, + *kykas,* name for an African plant) is reconstructed as an entire leaflike structure, with a row of abaxial ovules on either side of the midrib toward the base of the blade (Fig. 25-26). A megasporophyll of *Phasmatocycas* (Gr. *phasma,* apparition, + *kykas*) was also

pinnules, or on modified branches of the leaves. Neither the microspores nor the ovules were produced in strobili as they are among members of the Cycadophyta and Cycadeoidophyta, the groups to which the seed ferns were most closely related. Fertilization may well have occurred after the ovules had been abscised. The pteridosperms reached a degree of prominence

617

FIGURE 25-26
Archaeocycas whitei. Reconstruction of adaxial surface of megasporophyll. (*After Mamay.*)

FIGURE 25-27
Phasmatocycas kansana. Reconstruction of a megasporophyll. (*After Mamay.*)

thought to have been entire, with ovules borne along both sides of the petiole (Fig. 25-27). *Leptocycas* (Gr. *leptos*, slender, + *kykas*) (Delevoryas and Hope, 1971) was an Upper Triassic cycad, the reconstruction of which (Fig. 25-28) is convincing because it is based on a stem segment with an attached leaf and a terminal microstrobilus. Unlike those of modern cycads, the stem of *Leptocycas* was slender and elongate, with an uncrowded crown of pinnately compound leaves at the apex. Unfortunately, the supposed microstrobilus is not well enough preserved to yield pollen grains. This type of

primitive cycad with slender stem and uncrowded leaves is not difficult to relate to the pteridosperms, often considered to be the progenitors of the cycads, unlike the modern genera of cycads with their short, stout stems and densely crowded leaves. *Beania* (named for a Mr. Bean, a fossil collector) is a well-known Middle Jurassic cycad based on megastrobili with peltate sporophylls, each with two seeds directed toward the cone axis. Sporophylls are like those of a cycad such as *Zamia*, differing in wider spacing.

It has been suggested that cycads arose from the pteridosperm complex and that the strobili

618

FIGURE 25-28
Leptocycas gracilis. Suggested reconstruction.
(After Delevoryas and Hope.)

of cycads with their reduced peltate megasporophylls represent a reduction from the frondlike, ovule-bearing structures of pteridosperms. Evidence of reduction of megasporophylls is available even among living cycads in such a series as that illustrated in Fig. 21-27. As was noted earlier (p. 543), this has been challenged, but the argument is not particularly credible.

Fossil members of the Cycadophyta considered in this treatment are classified as follows:

Division Cycadophyta
    Order 1. Cycadales
        Family 1. Cycadaceae
            Genera: *Archaeocycas, Phasmatocycas,*
                Leptocycas
        Family 2. Zamiaceae
            Genus: *Beania*

DIVISION CYCADEOIDOPHYTA

The Cycadeoidophyta include the single order Cycadeoidales (called Bennettitales in some texts). The cycadeoids appeared first in the Triassic and extended into the Cretaceous. The abundance of foliage of fossil cycads and cycadeoids is the reason that the Mesozoic era is sometimes spoken of as the "Age of Cycads."

Cycadeoids resembled cycads superficially in having stems, often with persistent leaf bases, and a terminal crown of generally pinnately compound leaves. In several genera, the trunks were slender and branched (Fig. 25-29). The best known genus, *Cycadeoidea* (Gr. *kykas*, name for an African plant, + *eides*, like) (sometimes referred to as *Bennettites* [named for J. J. Bennett, botanist]) had squat stems that were generally ovoid, globular, or sometimes short columnar, with a massive central pith surrounded by a relatively thin cylinder of secondary vascular tissue (Fig. 25-30). Unlike modern cycads, which have shallowly ascending, girdling leaf traces, those of the cycadeoids passed directly into the leaf bases. Although no mature leaves have ever been found attached to these trunks, some specimens were preserved with young, immature fronds tightly packed at the apex, and these indicate that the leaves were pinnately compound. Leaves of cycads and cycadeoids look alike as fossils, but there are differences in stomatal structure and epidermal features that permit a distinction between the two types.

Some cycadeoids were monoecious, unlike modern cycads, while others were dioecious. In monoecious forms, the microsporangia and ovules were associated on lateral branches that arose among the leaf bases (Fig. 25-30). The central portion of the reproductive axis was in the form of a fleshy, often conical, receptacle on which the ovules were borne. Between them were sterile appendages, often called interseminal scales, with swollen tips that covered the ovules except for the projecting micropyles.

619

FIGURE 25-29
Wielandiella angustifolia.
Model reconstruction.
(*Courtesy of the Field
Museum of Natural
History.*)

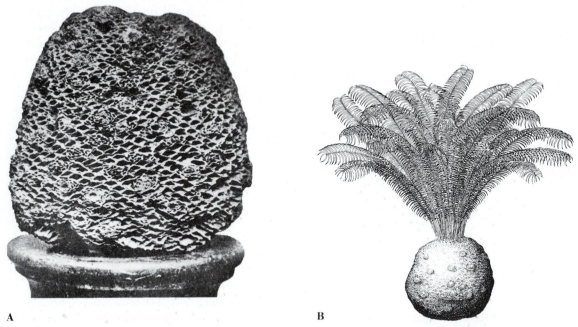

A   B

FIGURE 25-30
A. *Cycadeoidea marylandica*, the first described American cycadeoid, collected between Baltimore and
Washington, D.C. **B.** Restoration of a typical *Cycadeoidea*. (**A.** *After Ward.* **B.** *After Delevoryas.*)

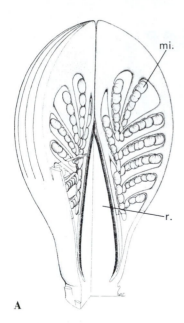

A

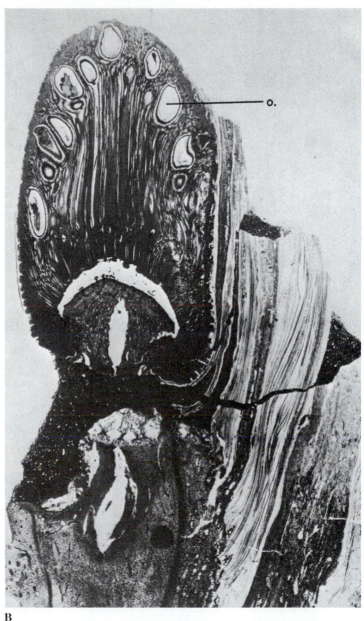

B

FIGURE 25-31
A. *Cycadeoidea* sp. Diagram of partially sectioned cone. Conical ovulate receptacle is surrounded by fleshy microsporangiate complex. mi., microsporangium; r., receptacle. **B.** *C. wielandii.* Longitudinal section of mature, seed-bearing cone. o., seed. (**A.** *After Crepet.* **B.** *After Wieland.*)

Surrounding the ovule-bearing axis were the microsporangiate structures (Fig. 25-31). These were actually fleshy, modified leaves, united at the base but free distally. Pinnalike members bore massive pollen sacs, each of which was a compound synangium, containing a number of elongated, tubular sporangia. Organization of the reproductive structure of *Cycadeoidea* was reported on by Delevoryas (1963, 1965, 1968 *a,b*) and by Crepet (1974). Crepet and Dele-

621

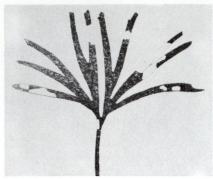

FIGURE 25-32
Baiera gracilis. Fossilized leaf of ginkophyte.
×0.66. (*After Seward.*)

voryas (1972) illustrated cycadeoid ovules that contain linear tetrads in the megasporangia.

The classification of the members of the *Cycadeoidophyta* discussed in this chapter is as follows:

Division Cycadeoidophyta
   Order 1. Cycadeoidales
      Family 1. Williamsoniaceae
         Genus: *Wielandiella*
      Family 2. Cycadeoidaceae
         Genus: *Cycadeoidea*

DIVISION GINKGOPHYTA

Although the extant *Ginkgo* is restricted (except in cultivation) to a localized area of eastern Asia, the fossil record indicates that it was much more widely distributed earlier in geologic time. The leaves of living *Ginkgo* vary in extent of lobing. A number of *Ginkgo*-like leaves found in the fossil record as far back as the Permian seem to be closely related to the extant genus. In fact, some workers prefer to include the fossils in the genus *Ginkgo* rather than *Ginkgoites*, which was established for fossil foliage. Study of these fossil leaves indicates that dissection and lobing of the blades are primitive attributes. The extinct genus *Baiera* (Fig. 25-32),

A

B

FIGURE 25-33
Ginkgoites digitata. Fossilized leaves. ×0.66.
(*After Seward.*)

another member of the Ginkgopsida, was restricted to the Mesozoic. It had finely divided, wedge-shaped leaves that lacked petioles. Approximately 19 extinct genera of ginkgophytan plants, including the fossil members of *Ginkgo* (Fig. 25-33), have been described. Unfortunately, we lack adequate data regarding their reproductive structures.

DIVISION CONIFEROPHYTA

Conifers first appear in the fossil record during the Carboniferous. They probably owe their origin to the progymnosperms, with the leaf of a conifer corresponding to the ultimate leaflike structure of the progymnosperms (e.g., leaf of *Archaeopteris*).

FIGURE 25-34
A, B. *Cordaites borasifolius.* **A.** Model reconstruction of small tree. **B.** Model reconstruction of fertile branch. **C.** *Cordaites* sp. Reconstruction of fertile branch. (**A, B.** *Courtesy of the Field Museum of Natural History.* **C.** *After Grand 'Eury, from Andrews.*)

Order Cordaitales

Among the best-known coniferous fossils are members of the Paleozoic order Cordaitales (Fig. 25-34). *Cordaites* (named for A. J. Corda, German paleobotanist), unlike the pteridosperms, was a tall (reaching 30 m), much-branched tree that was an important constituent of Carboniferous forests. The leaves (Fig. 25-34) of *Cordaites* were straplike and simple, with one report of leaves up to a meter in length; most were shorter, however. Superficially venation appears to have been parallel, but it was actually dichotomous from the base to the apex of the leaves. The latter were restricted to the tips of the youngest branches, the older axes having been leafless.

623

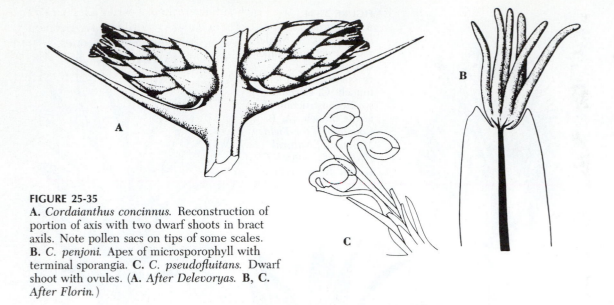

FIGURE 25-35
A. *Cordaianthus concinnus.* Reconstruction of portion of axis with two dwarf shoots in bract axils. Note pollen sacs on tips of some scales. **B.** *C. penjoni.* Apex of microsporophyll with terminal sporangia. **C.** *C. pseudofluitans.* Dwarf shoot with ovules. (**A.** *After Delevoryas.* **B, C.** *After Florin.*)

Internally, stems of *Cordaites* resembled those of conifers quite closely, with a pith, distinct small primary xylem strands, and secondary vascular tissue with compact xylem, the tracheids of which bore circular, bordered pits on the radial walls. The pith was distinctive in that it was septate, with air-filled gaps alternating with plates of parenchyma.

Roots of *Cordaites* (placed in the organ genera *Amyelon* [Gr. *a*, without, + L. *myelos*, pith] and *Premnoxylon* [Gr. *premno*, trunk of a tree, + *xylon*]) contained either exarch protosteles or a number of separate exarch strands that were not quite in contact with the secondary xylem.

Fertile structures of *Cordaites* are included in the organ genus *Cordaianthus* (Corda, + Gr. *anthos*, flower). Specimens of *Cordaianthus* consisted of a slender axis with narrow bracts, in the axils of which were budlike structures (Fig. 25-35A). Each of these buds or short shoots had a reduced axis bearing helically arranged scales. Some of the distal scales, the microsporophylls, bore six elongate microsporangia (Fig. 25-35B).

The ovules were borne in similar axillary buds on separate branches. Instead of microsporan-gia, some of the distal scales of the bud bore one or more terminal ovules or seeds (Fig. 25-35C); these were flattened in one plane and bilaterally symmetrical. As ovules developed, the integuments of certain cordaitean seeds developed a hard, inner seed coat surrounded by a fleshy, outer one (Fig. 25-36). Female gametophytes with archegonia have been seen in some.

Order Voltziales

Several Pennsylvanian and Permian genera of coniferophytes are assigned to the order Voltziales (it extended into the Jurassic period). *Lebachia* (named for Lebach, a Permian plant locality) (Fig. 25-37) is an example of one of these. The ovule-bearing structure of *Lebachia* (Fig. 25-37B) resembles somewhat a telescoped version of the more elongate branch of *Cordaianthus*, but in *Lebachia* the bracts are helically arranged. In the axil of each bifurcate bract there occurred a budlike axis or dwarf shoot, as in *Cordaianthus*. In *Lebachia*, however, the axillary dwarf shoot typically bore only one fertile scale with a terminal ovule (Fig. 25-37B).

624

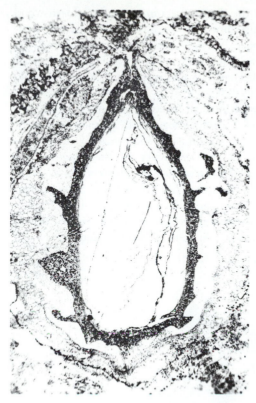

FIGURE 25-36
Cardiocarpus spinatus. Longitudinal section of
seed. Note dark, inner sclerotesta and two zones
of outer sarcotesta. ×9.

The bract and dwarf shoot-bearing axes of *Lebachia* bear appendages in such close arrangement
that they resemble conifer ovulate cones.

Other genera of the Voltziales from the Permian and Triassic show further modification of
the dwarf shoots. These changes involve a dorsiventral flattening, with the ovule-bearing scales
above the sterile part, recurving of the ovules so
that the micropyles are directed toward the
principal axis, and, eventually, a fusing of parts,
so that the sterile structures form a single,
flattened unit, with one or more ovule-bearing
structures fused to the upper surface. These
changes led to formation of a structure such as
the ovuliferous scale of *Pinus* (Fig. 23-11) and
other Pinaceae of the order Coniferales. The
ovulate strobilus of these genera is the counterpart of the more extended compound ovulate
axis of *Cordaianthus*, with its bracts and axillary
dwarf shoots. Both ontogenetic and phylogenetic considerations suggest that the ovulate strobilus of conifers is a compound structure in which
the remains of the dwarf shoot have become
united to its subtending bract.

In contrast to the compound, ovulate strobili
of modern conifers, microsporangiate strobili

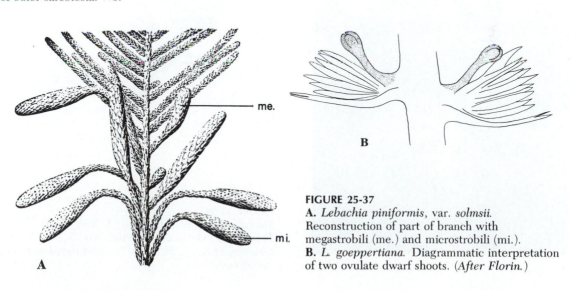

FIGURE 25-37
A. *Lebachia piniformis*, var. *solmsii.*
Reconstruction of part of branch with
megastrobili (me.) and microstrobili (mi.).
B. *L. goeppertiana.* Diagrammatic interpretation
of two ovulate dwarf shoots. (*After Florin.*)

are simple structures (Fig. 23-7A), each corresponding to a single, axillary dwarf shoot of *Cordaianthus*. In spite of the emphasis on the homologies between the conifers and cordaites, the fossil record suggests that the former may not have been derived from the latter, since both groups were contemporaneous for a time. These homologies, however, suggest a close relationship, with possible common origin from earlier precursors.

Order Coniferales

Plants corresponding to families of extant Coniferales have been traced back possibly as far as the Triassic (*Araucarioxylon*, fossil wood bearing some resemblance to wood of the Araucariaceae; members of the Taxaceae) and the Jurassic (Pinaceae, certainly the Araucariaceae, Taxodiaceae, Podocarpaceae, Cupressaceae) (Florin, 1963).

It is of interest to conclude this brief discussion of fossil conifers with a brief statement regarding an unusual sequence of events involved in the discovery of the taxodiaceous conifer *Metasequoia glyptostroboides* (Fig. 25-38). Described first as a fossil *Sequoia*-like genus from the Pliocene in 1941, living specimens were subsequently (1945) discovered to be growing in central China, and seeds and

A

B

FIGURE 25-38
Metasequoia glyptostroboides. **A.** Tree about 15 m tall. **B.** Detail of branchlet. ×0.75. (**A.** *Courtesy of Dr. J. M. Fogg, Jr., and the Morris Arboretum.*)

seedlings have since been distributed all over the world. This seems to have been a unique occurrence in the annals of plant science.

Fossil coniferophytes discussed above may be classified in the following way:

Division Coniferophyta
  Order 1. Cordaitales
    Family 1. Cordaitaceae
      Genera: *Cordaites, Amyelon,*
          Premnoxylon, Cordaianthus,
          Cardiocarpus
  Order 2. Voltziales
    Family 1. Voltziaceae
      Genus: *Lebachia*
  Order 3. Coniferales
    Family 1. Pinaceae
      Genus: *Pinus*
    Family 2. Araucariaceae
      Genus: *Araucarioxylon*
    Family 3. Taxaceae
    Family 4. Taxodiaceae
      Genera: *Metasequoia, Sequoia*
    Family 5. Podocarpaceae
    Family 6. Cupressaceae

The preceding brief account of representative extinct gymnospermous seed plants indicates that while several types have descendants in the earth's flora today, others are no longer present. Thus, the cycads, with precursors in the Paleozoic, *Ginkgo,* and the conifers, with Paleozoic and Mesozoic antecedents, all are members of our living flora. By contrast, the formerly abundant pteridosperms and cycadeoids are no longer represented. The history of seed plants extends back at least to the late Devonian, relatively soon, geologically speaking, after the appearance of vascular cryptograms. The late Paleozoic witnessed a proliferating of the group with maximum development during the Mesozoic. Their decline coincided with the appearance and subsequent radiation of the angiosperms during the Cretaceous.

DISCUSSION QUESTIONS

1. Explain the term "gymnosperm." How do the terms "gymnosperm" and "Gymnospermae" differ?

   (Questions 2–4 refer to *Zamia, Ginkgo, Pinus,* and *Ephedra.*)

2. Compare the vegetative features of these plants.
3. Compare the disposition of sporogenous tissue in these genera.
4. Discuss the ontogeny of the male gametophyte in these genera comparatively. What divergences are apparent in structure, function, and longevity?

5. How does heterospory in seed plants differ from that in vascular cryptogams?
6. How does the embryogeny of gymnosperms differ from that of vascular cryptograms? Are there exceptions?
7. If the formal class "Gymnospermae" is abandoned, may one still refer to a plant as a "gymnosperm"?
8. What attributes does *Ginkgo* have in common with Coniferophyta? With Cycadophyta?
9. In what respects do the Pteridospermophyta, Cycadeoidophyta, and Cycadophyta differ from one another?
10. How do the Cordaitales differ from other gymnosperms?
11. What features of *Caytonia* are of morphological significance?
12. Are the leptosporangiate ferns the precursors of seed ferns?

Chapter 26
DIVISION
ANTHOPHYTA I

The division Anthophyta, the flowering plants, includes the most recent and successful plants that have colonized the earth. It is the largest of the groups of vascular plants in number and in diversity of genera and species, approximately 300,000 species of 12,000 genera having been described (Eames, 1961). The range of these plants in both habit and habitat is extreme. They are represented by such diverse types as trees, shrubs, herbs, vines, floating plants, epiphytes, and even colorless parasites. They have populated a wide variety of xeric, mesic, and hydric habitats, and they form the major portion of the vegetation of many areas. With respect to longevity, they include annual, biennial, and perennial types. They perennate either as woody trees, shrubs, and vines (deciduous or evergreen), or as herbaceous types that survive seasons of dormancy by means of corms, bulbs, rhizomes, or other subterranean organs.

Because of the diversities mentioned above, and many others, it is not feasible to give an adequate description of the morphology of their vegetative organs in a volume of limited size that includes an account of plants other than Anthophyta as well as of the latter. Indeed, in practice, such considerations have come to form much of the subject matter of a separate field of plant science—namely, plant anatomy, while many special courses and texts are devoted to the classification and ecology of Anthophyta. The present chapter, therefore, will summarize only a few of the salient features of vegetative organization. Others of these, it will be recalled, were treated in Chapter 13. The remainder of this chapter will deal largely with the reproductive process in the flowering plants, which occurs in the structure familiarly known as the "flower." The next chapter (Chapter 27) will present a brief consideration of the problem of the origin of the Anthophyta and their fossil record.

VEGETATIVE ORGANIZATION

The plant body of most Anthophyta consists of leaves, stems, and roots, as in all vascular cryptogams except the Psilotophyta and *Salvinia*. These organs are either primary (embryonic) in origin or **adventitious** (arising without connection with the embryo).

The embryonic axis of the seed shows polarity, consisting of an "open" terminal bud or plumule at one pole and a primary root or radicle at the other. The attachment of the leaves (cotyledon or cotyledons) to the axis is at such a level that an intermediate portion, the **hypocotyl,** occurs between stem and root. Branches of the radicle are **secondary roots;** all other roots are adventitious. The probable totipotency of most living cells of angiosperms—demonstrated in the case of carrot phloem (Steward, 1967)—and their widespread powers of regeneration make the occurrence of adventitious organs commonplace.

THE LEAF

Angiosperm leaves, which are megaphyllous, vary in size from the minute organs of certain cacti and spurges to the large ones of palms. A vast majority of leaves, of course, are intermediate in size. They may be arranged on the stems in spiral (hence alternate), opposite, or, more rarely, whorled fashion. Inasmuch as the leaves have buds in their axils, branching is similar in pattern.

The leaf, in many cases, is differentiated into a **blade,** or **lamina,** and a **petiole,** or leaf stalk. The petiole arises from intercalary growth at the base of the lamina late in ontogeny. Other leaves—those of *Iris,* for example—are sessile; that is, they lack petioles. The blade may be simple (undivided), with plain (entire) or variously lobed and dentate margins. Compound leaves (with divided blades) occur in many angiosperms (Fig. 13-17B, C). The leaflets may be arranged pinnately along an extension of the petiole (the **rachis**), or all may originate in digitate (palmate) fashion from the tip of the petiole.

In a number of cases, small, lateral appendages, **stipules,** arise near the leaf base either from the stem or petiole. Ligules occur at the bases of the leaf blades in certain monocotyledonous leaves such as those of grasses and in a few dicotyledons.

The venation of angiosperm leaves is dichotomous in a few species. In most dicotyledons the main veins branch and rebranch frequently to form a network. In most but not all monocotyledons the main veins are prominent and arcuate or longitudinal; these may be connected by very minute, unbranched veins, which are not conspicuous. There are intermediates between these two major patterns of venation.

The leaf blade does not differ markedly, histologically, from that in megaphyllous cryptogams. In angiosperms, the stomata may occur on both leaf surfaces (*Iris*), only on the lower (*Ligustrum*), or only on the upper (water lilies) leaf surface. In most angiosperms the epidermal cells lack chloroplasts. The epidermis may consist of several cell layers.

The mesophyll in most dicotyledons is differentiated into palisade and spongy regions; the palisade layer usually occurs near the adaxial surface of the blade or, more rarely, on both surfaces, in this case with the spongy mesophyll between.

The leaf veins contain both xylem and phloem, the xylem usually nearer the adaxial surface. The delicate vein endings may be composed solely of xylem tracheids. The xylem and phloem of the veins are often surrounded by a sheath of parenchyma, the **bundle sheath.** The latter may be composed of thick-walled cells.

The leaves of many angiosperms are separated from the stem by formation of a special layer, the **abscission layer,** which arises near the base of the petiole. Some leaves wither and are separated by mechanical breakage, without the formation of a special abscission layer.

THE STEM

In angiosperms the stem may be aerial (erect, decumbent, prostrate, or climbing) or subterra-

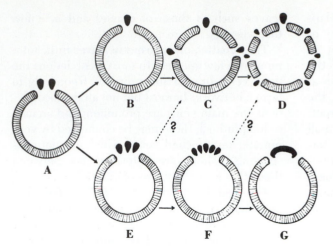

FIGURE 26-1
Nodal anatomy in angiosperms; diagram showing proposed evolutionary relationship. **A, B, E–G.** Unilacunar, with varying number of traces. **C.** With three and **D.** with many traces and gaps. (*After Canright, from Eames.*)

nean (including rhizomes,[1] tubers, bulbs,[2] and corms). They may be extremely woody because of addition of abundant xylem by the cambium, as in trees and shrubs, or they may have little or no secondary growth, as in herbaceous genera. Secondary growth is absent from monocotyledons, except in certain genera such as *Yucca* and *Agave*. Woody angiosperms are considered to be primitive and herbaceous types to be derived from them by decrease in cambial activity.

The steles of dicotyledonous angiosperms are in most cases endarch eusteles, although a few have ectophloic siphonosteles, while in monocotyledons there are eusteles with one or two concentric circles of vascular bundles or atactosteles with scattered bundles.

Of those angiosperms that have been investigated, vessels occur in the xylem of all but about 100. All the genera that lack vessels are members of the putatively primitive order Ranales or a few highly specialized, reduced parasites.

Companion cells are associated with the sieve cells of angiospermous phloem but absent in those dicotyledons that have no vessels in the xylem.

The leaf traces and their associated gaps or lacunae follow several alternative patterns in angiosperms (Fig. 26-1). There may be (1) a single trace associated with a single gap, (2) three traces (a median and two laterals) associated with three gaps, (3) many traces subtending as many gaps, and (4) two or more traces associated with a single gap. Nodes may be described in terms of the number of leaf gaps as **unilacunar, trilacunar,** or **multilacunar.** In stems with eusteles, it is usually not possible to distinguish leaf gaps from the parenchyma between the several vascular bundles.

Although the trilacunar node with three associated leaf traces has often been considered to be the primitive type for angiosperms, the occurrence of single lacunae with two traces in certain primitive living angiosperms and at the cotyledonary node in many others, in *Ginkgo*, and in certain fossil gymnosperms suggests that the two-trace-unilacunar relationship is more primitive.

---

[1] Not all rhizomes are subterranean.

[2] Only a small portion of a bulb is stem; the major portion consists of fleshy leaf bases or entire fleshy leaves.

THE ROOT

A primary root or **radicle** is present in the embryo of angiosperms except in some arborescent monocotyledons. After germination, the emergent radicle may persist as a deep-growing fleshy or woody taproot, especially in dicotyledons. In many monocotyledons the radicle stops growing soon after germination, and a largely adventitious root system functions throughout the life of the plant. Branches of primary roots are called secondary roots. Unlike stem branches, which arise superficially from axillary buds, root branches arise endogenously from the pericycle (Fig. 13-14). Mycorrhizal roots are present in a number of angiosperms. The regular occurrence of nitrogen-fixing bacteria in the root nodules of legumes is of great economic and ecological significance.

The apical meristem of the angiosperm root, as in all roots, is covered by a root cap. Its stele is clearly delimited from the cortex by the modified innermost layer of the latter, the endodermis (Fig. 13-15). In most dicotyledonous roots, the vascular tissue is organized as an exarch protostele, typically diarch or tetrarch (Fig. 13-15). The primary phloem, as in all roots, alternates with the primary xylem ridges. The vascular tissue is surrounded by a more-or-less prominent pericycle, from which branch roots may arise opposite the protoxylem.

In many monocotyledonous, and some dicotyledonous, roots the center of the stele is often parenchymatous and designated as **pith.** The primary xylem is often arranged as many discrete strands alternating with the primary phloem. The exarch protostele of the primary root is connected to the endarch stem stele (eustele or siphonestele) by a **transition zone** in the hypocotyl of seedlings.

Secondary growth occurs in varying degrees in the roots of dicotyledons but is absent from those of monocotyledons. In many roots, addition of abundant secondary xylem and phloem results early in the splitting and destruction of the endodermis, cortex, and epidermis. As this begins, the pericycle undergoes periclinal divisions; one of the outer derivative layers forms a cork cambium (phellogen), which gives rise to an impervious **periderm.**

REPRODUCTION[3]

THE FLOWER: GROSS MORPHOLOGY

In spite of the *apparently* limitless variation of floral structure observable in the numerous genera of flowering plants, a remarkable uniformity of basic organization prevails. This becomes clear on careful analysis, which reveals that the individual flower always is an axis that may have as many as four types of appendages[4] (Fig. 26-2), two of which are fertile. Flowers may occur separately or in groups. The stalk of an individual flower is known as a **peduncle.** When flowers are grouped in **inflorescences,** the stalk of the individual flower is called a **pedicel.** In the center of the flower and at the apex of the axis tip, called the **receptacle** or **torus,** are one or more ovule-bearing structures, the **carpel,** sometimes designated **pistil,** or pistils (L. *pistillum,* pestle) (Fig. 26-2).[5] The fact that the ovules are enclosed during their development is recognized in the term **angiospermy,** as distinct from **gymnospermy.** Surrounding the pistil or pistils, a number of stamens (Fig. 26-2) emerge from the receptacle or from the **hypanthium,** the latter a tubular outgrowth of the receptacle, or they may arise from the corolla

---

[3]For further discussion of many of the topics in this section consult Raghavan (1976) and Johai (1982).

[4]Some botanists consider the floral organs to be modified branch systems and hence cauline and not appendicular (see Esau [1965], pp. 540–542).

[5]See p . 636.

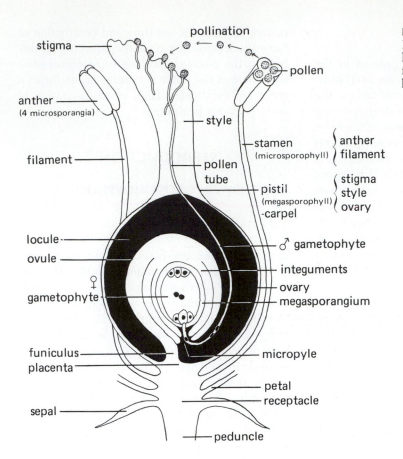

FIGURE 26-2
Diagrammatic median
longisection of a generalized
flower; component structures
labeled.

Labels in figure:
pollination
stigma
pollen
anther (4 microsporangia)
style
stamen (microsporophyll) — anther, filament
filament
pollen tube
pistil (megasporophyll) — stigma, style, ovary
carpel
♂ gametophyte
locule
ovule
integuments
ovary
megasporangium
♀ gametophyte
funiculus
micropyle
placenta
petal
receptacle
sepal
peduncle

tube. The stamen is most often composed of a stalklike portion, the **filament,** and of sporangia that compose the **anther.** The anther may be thought of as a synangium because it is composed of four (usually) microsporangia. The anther is covered by an epidermis, which encloses several layers of the microsporangial walls, within which are the tapetal and microsporogenous tissues. The stamens and pistils are spore-bearing and, therefore, are essential or-

gans of the flower. The great majority of flowers have sterile appendages on the floral axis, the **petals** (Gr. *petalon*, leaf) and **sepals,** collectively known as the **corolla** (L. dim. of *corona*, crown) and the **calyx** (Gr. *kalix*, cup), respectively (Fig. 26-2). The petals and sepals usually, but not always, are distinguishable in both color and place of origin. Petals generally are colored other than green. Sepals arise lower on the floral axis than do petals. The calyx and corolla

FIGURE 26-3
Zea mays, corn. **A.** Tassel or staminate inflorescence. **B.** Portion of preceding, more highly magnified. **C.** Single staminate flower with three stamens; white dots are pollen grains. **D.** Pollen grains. **E.** Pistillate inflorescence at pollination. **F.** Pistillate inflorescence, the surrounding leaves removed; note "silks" or stigma-styles attached to pistillate flowers. **G, H.** Details of young kernels or ovaries showing emergence of stigma-styles. **A.** ×0.2. **B.** ×0.7. **C.** ×3. **D.** ×30. **E.** ×0.3. **F.** ×0.9. **G.** ×9.

A

B

C

D

E

F

G

H

A   B

FIGURE 26-4
Salix niger. Black willow.
Stages in maturation of
inflorescences. **A.** Staminate.
B. Pistillate. ×1.

together constitute the **perianth** of the flower. In a number of flowers, both the petals and sepals are inconspicuous, green, or almost colorless. Flowers that have sepals, petals, stamens, and pistils are said to be **complete;** where one of these is absent, they are **incomplete.**

In a great majority of flowering plants, both stamens and pistils are present in the same flower, which is, therefore, said to be **perfect.** However, in such flowers as those of corn (*Zea mays*) (Fig. 26-3), willow (*Salix* sp.) (Figs. 26-4, 26-5), and oak (*Quercus* sp.) (Fig. 26-6), the stamens and pistils occur in separate flowers, which are then said to be staminate or pistillate and **imperfect.** In both corn and oak, furthermore, staminate and pistillate flowers are borne on the same individual plant; this condition is known as monoecism. In the willow, on the other hand, staminate and pistillate flowers are distributed on different individuals; hence, a state of dioecism prevails. Some flowering plants not only have perfect flowers, but may produce, in addition, separate staminate and carpellate (pistillate) flowers. All of these may be present on one individual or they may be on three separate individuals, depending on the species.

The floral organs are supplied with vascular tissues in the form of traces that depart from the receptacle. Study of the number, path, and arrangement of these traces has contributed much to our understanding of the fundamental

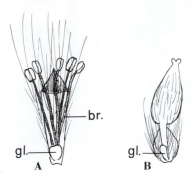

A   B

FIGURE 26-5
Salix sp. **A.** Single staminate flower. **B.** Single pistillate flower; gl., gland; br., bract. **B.** ×8.

patterns and variations of flower structure. It is apparent that a **flower** is an axis of limited growth with shortened internodes, and that it always bears spore-producing appendages and may bear sterile ones in addition. The extreme brevity of the floral axis in many flowers complicates analysis of their floral organization. The sepals are leaflike and usually receive the same number of traces as the foliage leaves of the species, three being a common number. Petals and stamens frequently receive only single traces. Stamens with three traces occur in certain families, such as those of the *Magnolia* group. The pistil (carpel) also receives three traces in many plants that are considered to be primitive. Fewer than three traces, in the past,

634

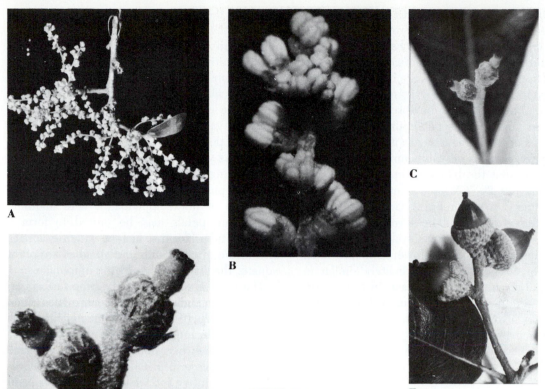

FIGURE 26-6
Quercus virginiana, "live oak." **A.** Staminate inflorescences. **B.** Portion of inflorescence more highly magnified showing staminate flowers. **C.** Two pistillate flowers. **D.** Two pistillate flowers more highly magnified. **E.** Fruits, or acorns. **A.** ×0.5. **B.** ×10. **C.** ×2. **D.** ×6.

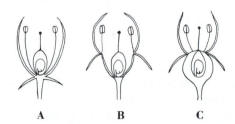

FIGURE 26-7
Median longitudinal section of: **A.** an hypogynous, **B.** a perigynous, **C.** an epigynous flower; diagrammatic.

have been interpreted as evidence of reduction, and more than three as a specialized condition.

The several floral organs exhibit variation in their arrangement on the receptacle and in relation to one another. In many flowers, a median longitudinal section or bisection of the flower (Figs. 26-2, 26-7A, 26-8) reveals that the pistil or pistils are borne at the apex of the receptacle, and that the stamens, petals, and sepals occur on it at lower levels. Such an arrangement exemplifies **hypogyny** (Gr. *hypo*, under, + Gr. *gyne*, female). Epipetalous flowers, in which the stamens arise from a corolla tube, are still considered as hypogynous, since the tube arises below the pistil on the receptacle. In many flowers of the rose alliance, the pistil or pistils occur at the base of a cuplike structure (Fig. 26-7B) on the rim of which the stamens, petals, and sepals are borne. This type of organization is called **perigyny.** In still other flowers, however, it can be observed in longitudinal sections that the pistil, although borne at the apex of the floral axis, has its basal portion, the ovary, partially or wholly united with tissues that represent the bases of sepals, petals, and stamens (Fig. 26-7C). Such a flower is said to be **epigynous** (Gr. *epi*, upon, + Gr. *gyne*).

Further attention will be devoted at this point to the structure and structural variations that occur in the organs of the flower. The individual floral organs may be approximately similar in size and arranged about the axis in radial fashion, a condition called **actinomorphy;** or they may vary in form and insertion so that the flower is **zygomorphic.**

The petals and sepals may be separately in-

FIGURE 26-8
Lilium. Lily. Longitudinal bisection of flower bud; a single stamen to the right. f., filament; m., microsporangia; o., ovary; p., petal; r., receptacle; s., stigma; sta., stamen; sty., style. ×1.

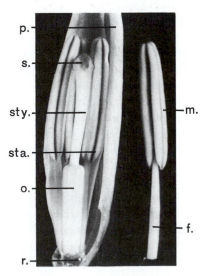

serted on the floral axis (polypetaly and polysepaly), or they may be united at their bases (sympetaly and synsepaly). In flowers considered to be primitive, the stamens and pistils are indefinite in number, spirally inserted, and individually attached to the floral axis. In contrast with this is the cyclic type of flower, in which the several organs appear to arise from the axis in whorls of definite number, members of a whorl alternating with those of succeeding whorls.

Crowding the floral receptacle has also resulted in the phenomenon of **adnation,** the union of one type of floral appendages with another. Sepals and petals may be united to form a perianth tube like that in *Iris*, which is adnate also to the ovary. Petals and stamens are often adnate, the stamens then being epipetalous.

The pistil is enlarged at the base to form an **ovary,** which encloses the ovule or ovules (Figs. 26-2, 26-8, 26-10). From the apex of the ovary arise the **style** and **stigma** (Figs. 26-2, 26-8). The style may be either solid or hollow. The stigma may be simple or branched and variously modified as a receptive surface for pollen. Each pistil may have a single, undivided style and stigma, or both styles and stigmas may be multiple. The multiple condition is often an indication that the pistil is compound rather than simple. A **simple pistil** is one that is composed of a single ovule-bearing unit, known as the **carpel** (Gr. *karpos*, fruit). The carpel is considered by most morphologists to represent a phylogenetically folded megasporophyll that has enclosed one or more ovules (see p. 673). Evidence from external form and internal anatomy indicates that in many flowers a number of carpels have united and function as a unit, the **compound pistil.** Flowers with simple pistils are said to be **apocarpous,** while those with compound pistils are **syncarpous.** A transverse section of the ovary usually[6] indicates whether the pistil is simple

---

[6]But not always—as in violet (*Viola*) and members of the family Asteraceae, for example, in which the compound pistils are unilocular.

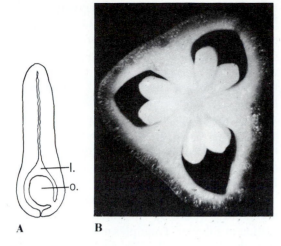

FIGURE 26-9
A. *Pisum sativum*, garden pea. Transection of ovary of simple pistil. **B.** *Lilium* sp. Transection of compound ovary; note three carpels and axile placentation. l., locule; o., ovary. **A.** ×10. **B.** ×4.

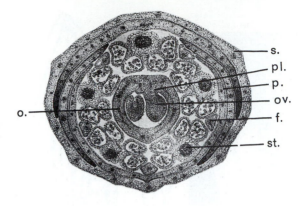

FIGURE 26-10
Pisum sativum, garden pea. Transection of flower bud; note simple ovary and parietal placentation. Five stamens and the filaments of the other five are visible. f., filament; o., ovary; ov., ovule; p., petal; pl., placenta; s., sepal; st., stamen. ×125.

(Figs. 26-9A, 26-10) or compound (Fig. 26-9B) and, in the latter case, the number of carpels involved in the fusion. Compound pistils may have several styles and stigmas or only one of each.

The portion of the ovary to which the ovules are attached, each by **funiculus,** is known as the **placenta** (Figs. 26-2, 26-10). The point of attachment varies in different flowers. In many, the ovules are attached to the ovary wall, a condition known as **parietal placentation** (Figs. 26-2, 26-9A, 26-10). In other flowers, the ovules are borne on the central axis of the compound ovary, a condition known as **axile placentation** (Fig. 25-9B). In still others, the floral axis at maturity is free from the upper portion of the ovary but attached at its base. In this case the ovules may be borne on the distal portion of the axis as well as on its flanks. This last arrangement is described as **free-central placentation.**

The ovules themselves exhibit variations in form and in relation to their stalk, the **funiculus.** Orthotropous, amphitropous, anatropous, and campylotropous types are common (Fig. 26-11). **Orthotropous** (Fig. 26-11A) ovules are those in which the micropyle and funiculus lie on the same longitudinal axis. **Anatropous** ovules (Fig. 26-11B) are those in which greater growth of one surface of the funiculus during development inverts the body of the ovule so that the micropyle and base of the funiculus are adjacent and parallel. In **amphitropous** ovules (Fig. 26-11C) the body of the ovule itself is strongly curved. In **campylotropous** ovules (Fig. 26-11D) the funiculus is attached near the equator of the ovule

FIGURE 26-11
Diagrammatic structure of ovules in median longitudinal section. **A.** Orthotropous. **B.** Anatropous. **C.** Amphitropous. **D.** Campylotropous.

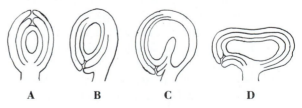

637

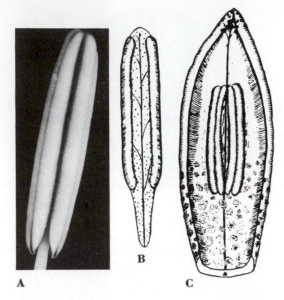

FIGURE 26-12
Stamens of microsporophylls. **A.** *Lilium.* **B.**
Magnolia sp. **C.** *Austrobaileya.* **A.** ×2. **B.** ×2. (**B,**
C. *After Canright.*)

body. Ovules with one or two integuments
occur in most Anthophyta.

Stamens have a stalklike axis, the **filament,** at
the top of which the **anther** is borne (Figs. 26-2,
26-12). The filament may be stalk- or petallike.
The anther is composed of four chambers, the
microsporangia, supported by a zone of sterile
tissue, the **connective,** which contains vascular
tissue extending from the filament (Figs. 26-10,
26-17). Unlike those of other plants, the microsporangia of angiosperms arise in ontogeny from
hypodermal, rather than from superficial, cells.
As the spores mature, the two microsporangia
on either side of the connective usually become
confluent, so that only two **pollen sacs** are
demonstrable in the dehiscent anther. The anthers may dehisce by longitudinal fissures or,
less commonly, by apical pores.

Flowers may occur singly, as they do in
Magnolia (Fig. 26-13), tulip, and buttercups, or

they may be grouped as **inflorescences.** The
latter are of two major types, indeterminate and
determinate. Indeterminate inflorescences keep
developing new flowers over a long period,
often as long as environmental conditions are
favorable. **Spikes** (Fig. 26-14A), with sessile
flowers, and **racemes** (Fig. 26-14B), with
stalked pedicellate flowers on an elongate axis,
are common types, among others, of indeterminate inflorescences. The youngest flowers, of
course, are at the apex of the indeterminate
inflorescence.

In determinate inflorescences, the apical
meristem is "used up" or differentiated in formation of a flower. Examples of these are **cymes**
(Fig. 26-15A), **umbels** (Fig. 26-15B), and **heads**
(Fig. 26-16), or **capitula.** In umbels the floral
pedicels are of equal lengths and arise in whorls
so that the inflorescence is flat-topped. Some
umbels, however, are indeterminate inflorescences, and the outer flowers of these open
before the inner ones (Lawrence, 1951). The
head or **capitulum** (Fig. 26-16) is characteristic
of the large family Compositae (= Asteraceae).
In this type of inflorescence, the axis is shortened and in the form of a convex or flattened
head on which many minute flowers are closely
arranged. These may be of two kinds, the
central disc florets and the marginal ray florets
(Fig. 26-16). The disc florets have bell-like
corollas and are perfect, while the corollas of ray
florets are ligulate and they usually lack stamens. In some genera with capitula, all the
florets may be ligulate or ray type, while in
others they may be all bell-like, as in disc
florets.

Before entering into a study of the details of
reproduction in the flowering plants, one should
familiarize oneself with the gross aspects of
floral morphology summarized in the preceding
paragraphs. Examination of a number of types
of living flowers will be helpful in this connection. It will be shown below that the spores that
arise in the anther become pollen grains that
contain immature male gametophytes. Similar-

638

A

B

C

D

E

F

FIGURE 26-13
Magnolia grandiflora. **A.** Single flower. **B.** Receptacle, enlarged; note stigmas of numerous pistils above, numerous stamens (some abscised, ovaries enlarging) below. **C.** Stamens and pistils, only the stigmas of the latter visible, from a flower bud. **D.** Receptacle with pistils after pollination. **E.** Enlarging fruits (follicles); note their spiral arrangement. **F.** Dehiscent fruits. **A.** ×0.25, **B, D.** ×1. **C, E, F.** ×0.75.

639

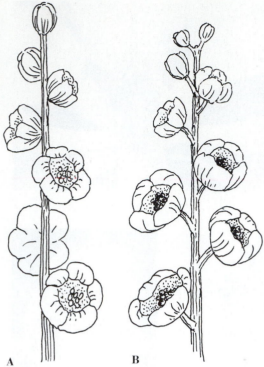

FIGURE 26-14
Inflorescences. **A.** Spike. **B.** Raceme.

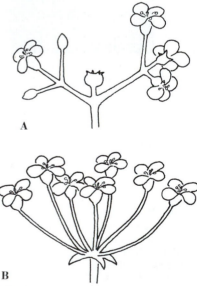

FIGURE 26-15
Inflorescences. **A.** Cyme. **B.** Umbel.

ly, each ovule produces a spore that develops into a female gametophyte. In the light of comparisons with other seed plants, the stamen of flowering plants often is regarded as a microsporophyll that bears four microsporangia that produce microspores, and the pistil or carpel as a megasporophyll enclosing one or more ovules —the latter, megasporangia surrounded by integuments.

In bringing to a close this discussion of the gross morphology of the flower, we are brought finally to the problem of defining it. It is clear that a **flower** is essentially an axis with short internodes on which spore-bearing appendages are borne. This statement, of course, also defined the strobilus, and it is practically impossible to distinguish flowers and strobili by definition. If it is argued that flowers have colored petals and sepals, unlike strobili, numerous examples of incomplete flowers lacking these may be cited. The flower then, *of itself*, does not distinguish Anthophyta from other seed plants. More subtle characters are available, however, which make this distinction reliable (see p. 663).

THE REPRODUCTIVE PROCESS

The process of reproduction in the flowering plants may be discussed conveniently under the following topical headings: (1) microsporogenesis, (2) megasporogenesis, (3) male gametogenesis, (4) female gametogenesis, (5) pollination and fertilization, and (6) development of the embryo, seed, fruit, and seedlings. Many aspects of the sexual process in the flowering plants, not included here, have been discussed by Heslop-Harrison (1972,1979,1983).

MICROSPOROGENESIS

Early stages in the development of the stamen (microsporophyll) may be found by dissect-

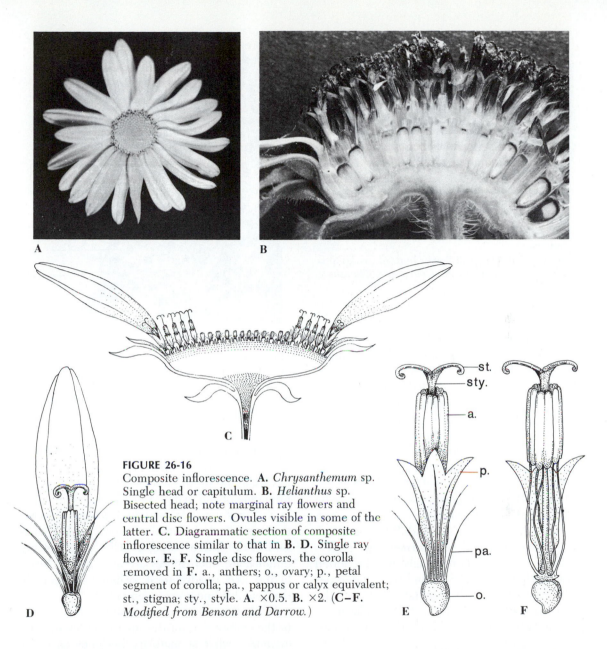

FIGURE 26-16
Composite inflorescence. **A.** *Chrysanthemum* sp. Single head or capitulum. **B.** *Helianthus* sp. Bisected head; note marginal ray flowers and central disc flowers. Ovules visible in some of the latter. **C.** Diagrammatic section of composite inflorescence similar to that in **B. D.** Single ray flower. **E, F.** Single disc flowers, the corolla removed in **F.** a., anthers; o., ovary; p., petal segment of corolla; pa., pappus or calyx equivalent; st., stigma; sty., style. **A.** ×0.5. **B.** ×2. (**C–F.** *Modified from Benson and Darrow.*)

ing or sectioning very young flower buds. In these, transverse sections of the anthers exhibit a rudimentary lobing into four parts, but the anther tissue is homogeneous except for slightly differentiated epidermal cells. One or more cells in each lobe function in generating sporogenous tissue. In older anthers (Figs. 26-17, 26-18), the wall is composed of several layers of cells, as in typical eusporangiate sporangia, the outermost of which differentiates as an epidermis. The layer of cells immediately within the epidermis is known as the **endothecium.** As the

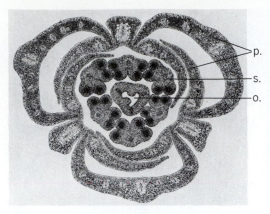

FIGURE 26-17
Lilium sp. Transection of flower bud. o., ovary; p., perianth; s., stamen. ×12.5 (*Courtesy of Triarch, Inc.*)

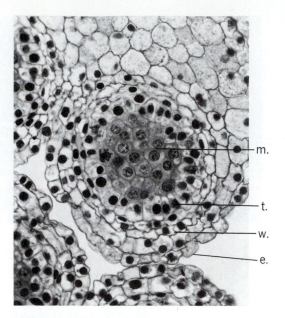

FIGURE 26-18
Lilium sp. Transection of one immature anther lobe (microsporangium). e., epidermis; m., microsporocytes; t., tapetum; w., microsporangial wall. ×120.

anther matures, the endothecial cells may develop fibrous bands on their walls. The innermost layer of the microsporangium wall functions as a well-differentiated **tapetum** (Fig. 26-18). The nuclei of tapetal cells frequently undergo either normal or modified mitoses, not followed by cytokinesis; hence, binucleate or multinucleate cells result.

In some cases the tapetal cells may be polyploid. Echlin (1971) emphasized that all metabolites for the pollen must pass through it or be metabolized by the tapetum. There are two types of tapetum: (1) The glandular or secretory and (2) the amoeboid. The former type undergoes autolysis, and its materials thus are added to the anther fluid. The second type undergoes early dissolution of the tapetal walls, and the liberated protoplasts form a tapetal plasmodium; the latter occurs also in *Psilotum*, for example.

As development proceeds, the sporogenous cells become differentiated as microsporocytes, which ultimately separate from one another and become spherical and suspended in the locular fluid of the anther. Each microsporocyte becomes surrounded by a special wall of callose.

The meiotic process follows, and tetrads of microspores are organized (Fig. 26-19). Cytokinesis may be by successive bipartition with the formation of cell plates after each division of the nuclei, or it may be centripetal cleavage furrows that are initiated after the second nuclear division in the microsporocyte. The tetrads of the uninucleate microspores are arranged in tetrahedral and quadrilateral fashion; occasionally they exhibit linear arrangement.

While the microspores are still enclosed within the callose wall of the microsporocyte, each organizes what at maturity becomes an outer wall layer, the exine. This exine is composed of sporopollenin possibly derived from the exudation of carotenoids. The callose wall to some degree acts as a barrier between the microspores within it and the anther fluid external to it. It has been hypothesized that the micro-

642

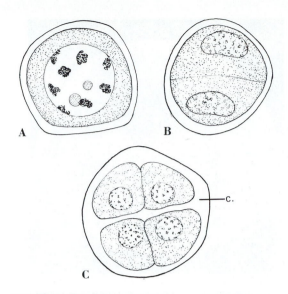

FIGURE 26-19
Lilium sp. Microsporogenesis. **A.** Microsporocyte. **B.** End of meiosis I. **C.** Tetrad of microspores. c., callose wall. ×770.

spores themselves presumably control the formation of the exine (Heslop-Harrison, 1972). The microspores are finally released from the enclosing callose walls by an enzyme, β, 1,3-glucanase, and their exines begin to thicken by apposition of substances probably derived from the tapetum. Soon after the microspores have been released from the callose walls, another wall layer, the **intine,** pectocellulosic in composition, is deposited within the **exine.** The intine is exclusively the product of the haploid microspore (Heslop-Harrison, 1976*b*,1979).

As noted above, the pollen-grain wall consists of two major layers, the exine and intine, the details of whose structure, as revealed by the electron microscope, vary with the species. The intine is a continuous wall layer that contains proteins from the microspore and male gametophyte. It later expands to form the wall of the pollen tube. These are derived in part from excluded fragments of the cell surface. Many enzymes seem to be localized in the exine (Heslop-Harrison, 1975*b*; Knox et al., 1975)

which may be variously ornamented and chambered (Muller, 1979). The minute chambers are filled with proteins, especially with glycoproteins, which are derived from the parent sporophyte via the tapetum. Their role in further development is discussed below.

Each microspore contains a single haploid nucleus. The microspores enlarge as they separate from the tetrad and become sculptured by deposition of more or less highly ornamented surface layers (Fig. 26-20). In a few plants— *Rhododendron* and cattail (*Typha*), for example —the microspores remain permanently in the tetrad condition like the spores of some species of the liverworts, *Sphaerocarpos,* and *Riccia.* In certain orchids and members of the milkweed family, the microspores adhere in a waxy mass, the **pollinium.**

It is of interest that when young anthers of a few species have been cultivated in basal inorganic medium supplemented with 2% sucrose and coconut milk, small *haploid* plantlets (sporophytes) develop in the anther sacs. This has occurred in tobacco and rice and is a further evidence that the differences between gametophytes and sporophytes are not dependent on chromosome number alone.

MEGASPOROGENESIS

The ovule, a megasporangium surrounded with integuments, develops by cell division from the superficial cells of the placenta (Fig. 26-9B, 26-21). The number of ovules in each ovary varies with the genus. There may be one, as in corn (*Zea mays*), buckwheat (*Fagopyrum esculentum*), and sunflower (*Helianthus* sp.); or a small number as in pea (*Pisum sativum*) and sweet pea (*Lathyrus odoratus*) (Fig. 26-36C); or they may be very numerous and minute as in the orchids, begonias, and snapdragons (*Antirrhinum majus*). Each ovule is attached to its placenta by the funiculus, as noted above. The megasporangium tissue is covered with either one or two closely adpressed (to each other and

643

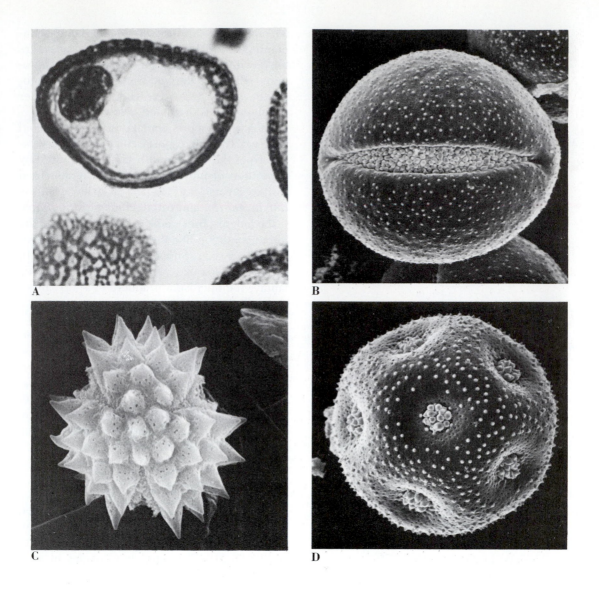

A

B

C

D

to the megasporangium) integumentary layers except for a minute passageway, the **micropyle** (Figs. 26-2, 26-21).

Herr (1971) described a rapid method for studying megasporogenesis and the development of the female gametophyte. Early in the development of each ovule, a subepidermal cell in the micropylar region differentiates into a primary archesporial cell. This may form several

sporogenous cells, or it may function directly as a megasporocyte (Figs. 26-21, 26-22, 26-26). The latter is readily recognizable because of its large size as compared with the sterile cells of the megasporangium. In a great majority of the flowering plants, the megasporocyte undergoes two successive nuclear and cell divisions, during which meiosis occurs. The products of these divisions are the four megaspores arranged in

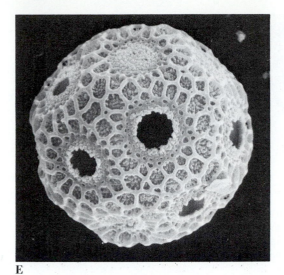

E

FIGURE 26-20
A. *Lilium* sp. Microspores, sectional and surface
view. **B–E.** Scanning electron micrographs of
various pollen grains. **B.** *Cometes surattensis.* **C.**
Pelucha trifida. **D.** *Cerastium alpinum.* **E.** *Opuntia
lindheimeri.* **A.** ×900. **B.** ×2100. **C.** ×2000. **D.**
×1000. **E.** ×1600. (**B–E.** *Courtesy of Prof. John
Skvarla.*)

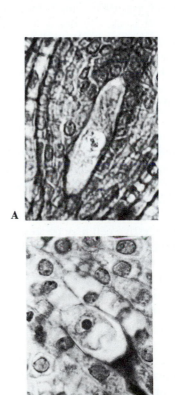

A

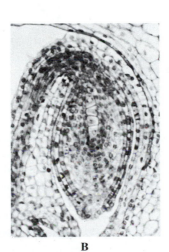

B

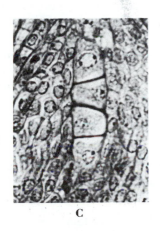

C

D

FIGURE 26-21
Oenothera sp. Megasporogenesis. **A.** Megasporocyte. **B.**
Longisection of ovule with linear tetrad. **C.** Linear tetrad
of megaspores. **D.** Functional and degenerating megaspores.
A. ×800. **B.** ×250. **C, D.** ×800.

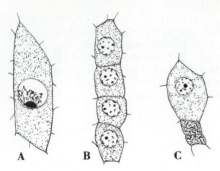

FIGURE 26-22
Oenothera sp. Megasporogenesis. **A.**
Megasporocyte. **B.** Linear tetrad. **C.** Functional
megaspore and three abortive. ×770.

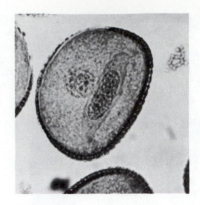

FIGURE 26-23
Lilium sp. Intrasporal development of male
gametophyte; note tube nucleus and cell and
elongate generative nucleus and cell. ×900.

linear fashion (Figs. 26-21B, C; 26-22B). Several
variations[7] in this process are known. One of the
daughter cells of the megasporocyte may fail to
undergo division, so that three instead of four
cells result. Furthermore, the megaspores are
not always arranged in a strictly linear fashion.
Normally, as in the gymnosperms, the three
megaspores nearest the micropyle degenerate,
and the chalazal megaspore persists and functions (Figs. 26-21D, 26-22B, C). This is permanently retained within the megasporangium,
where it continues its development into the
female gametophyte.

MALE GAMETOGENESIS

As in other seed plants, the uninucleate microspore (Fig. 26-20) begins its development
into the male gametophyte before it is shed
from the microsporangium, by undergoing nuclear division and cytokinesis. The two daughter
nuclei and cells differ in size and often in form as
well; the larger represents the **tube cell** (or
vegetative cell, Jensen, 1973) (Fig. 26-23). The
generative cell may be attached to the intine as

---

[7]See also p. 648.

in rye (Karas and Cass, 1976). The cytoplasm of
the generative cell is usually dense and lacks
plastids. In many flowering plants the generative cell may be ellipsoid, lenticular, or somewhat elongate (Fig. 26-23). It should be noted
that a prothallial cell is not formed in the male
gametophyte of angiosperms. The generative
nucleus may divide to form two sperm nuclei
before the pollen is shed, or this division may
occur in the pollen tube. In either case, the
sperm nuclei usually are surrounded by specially differentiated cytoplasm so that they are, in
fact, *cells* (Figs. 26-24, 26-25) (Karas and Cass,
1976).

The walled generative cell may be free in the
microspore cytoplasm or attached to the pollen-
grain wall. Its cytoplasm seemingly lacks plastids, as does that of the sperm cells, at least in
rye (Karas and Cass, 1976).

Dehiscence of the anther results in dissemination of the immature male gametophytes or
pollen grains. Depending on the species, some
of these reach the receptive stigmatic surface of
the pistil by means of gravity, wind, insects,
rain, or water currents (Fig. 26-2).

The ornamentation of the surface of the pollen grain is a feature of taxonomic value. The
pollen grains of angiosperms vary in size, shape,

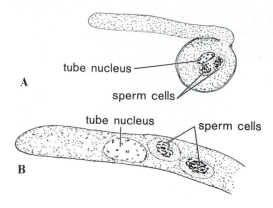

tube nucleus

sperm cells

tube nucleus

sperm cells

FIGURE 26-24
Polygonatum sp. **A.** Mature ♂ gametophyte, tube nucleus, and two sperm cells in microspore. **B.** Tip of pollen tube showing detail of tube nucleus and two sperm cells. **A.** ×515. **B.** ×1030.

and ornamentation of the outer wall, or exine. The pollen grains of most living angiosperms seem to range from 25 to 100 µm in greatest diameter; grains as small as 5 µm are known as well as those over 200 µm. Pollen may be radially or bilaterally symmetrical.

Pollen grains may have thin areas in their walls, either in the form of germinal pores or germinal furrows, or both. In pollen without furrows, the germinal pores are usually numerous (up to 30 in some cases). Pollen may be classified into two major types, the **monocolpate** with one germinal furrow, and the **tricolpate**

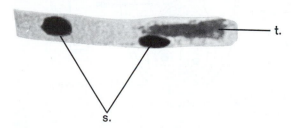

FIGURE 26-25
Hippeastrum belladona. Segment of pollen tube near its tip showing two sperm nuclei (s.) and tube nucleus (t.). ×960 (*From a preparation by Dr. James Pipkin.*)

with three. The monocolpate type is thought to be primitive and occurs in monocotyledons and primitive woody families of dicotyledons (and in cycads, cycadeoids, and pteridosperms). The tricolpate type of pollen characterizes most dicotyledons.

The study of pollen grains and spores of other plants, known as **palynology,** has become an increasingly important phase of plant science. Fossil pollen has been used by geologists as an aid in locating ancient shorelines, since deposits of pollen are more abundant near the shores of seas. Fossil pollen is used to identify coal beds through the relative abundance in the deposits of certain types of grains. Study of fossil pollen, of course, gives insight into the nature and diversity of extinct floras and past changes in climate.

The identification of pollen of living plants is, of course, of paramount importance in diagnostic procedure and treatment of certain allergic conditions. Finally, the study of both extant and extinct pollen has provided helpful data regarding phylogenetic relationships among taxa. A great deal of additional information about pollen is available in depth in the publications of Heslop-Harrison (1971, 1979, 1983), Stanley and Linskens (1974), and Ferguson and Muller (1976).

FEMALE GAMETOGENESIS

The functional megaspore normally gives rise to a single female gametophyte, as in gymnosperms. This process is usually accompanied by a series of three consecutive free-nuclear divisions within the functional megaspore, which enlarges during this period (Fig. 26-26 [1-8]); there is no evidence in angiosperms of thickening of the megaspore wall during gametogenesis. At the conclusion of these divisions, the developing female gametophyte contains a quartet of nuclei at each pole. Rearrangement, cell wall formation, and differentiation of the resulting cells effect the maturation of the fe-

647

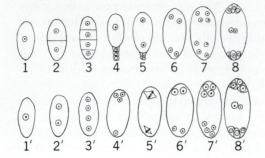

FIGURE 26-26
Megasporogenesis and ontogeny of the ♀
gametophyte; 1–8, the "normal type"; 1'–8',
Lilium (Fritillaria) type; diagrammatic.

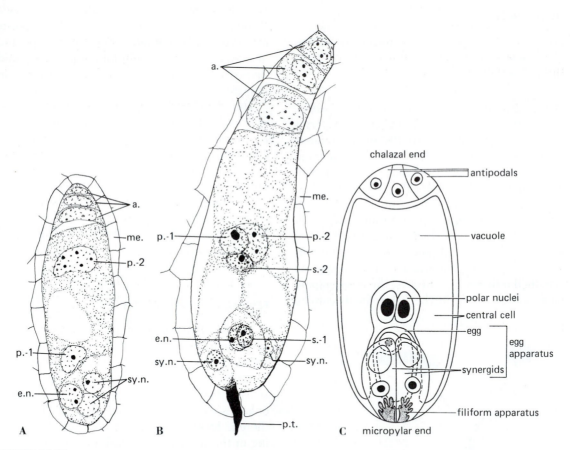

FIGURE 26-27
Lilium sp. **A.** Immature eight-nucleate ♀ gametophyte. **B.** Mature ♀ gametophyte and double
fertilization. **C.** Diagram of normal type of ♀ gametophyte in angiosperms. a., antipodal cells; e.n.,
egg nucleus; me., megasporangium; p.-1, haploid polar nucleus; p.-2, triploid polar nucleus; p.t.,
pollen tube; s.-1, s.-2, sperm nuclei; sy.n. synergid nuclei. **B.** ×315. (**C.** *After Jensen.*)

male gametophyte[8] (Fig. 26-26), the size, duration, and organization of which are obviously much reduced as compared with those of gymnospermous female gametophytes. Differentiated archegonia are entirely lacking (as in *Gnetum* and *Welwitschia*).

Three of the four nuclei at the micropylar pole differentiate as the **"egg apparatus,"** consisting of an egg cell in contact with and two cellular **synergids** (Fig. 26-27B). The synergids are characterized by a **filiform apparatus** (Fig. 26-27C) at their micropylar poles. The filiform apparatus is composed of fingerlike projections of the synergid wall (Jensen, 1972, 1973); the projections are composed of pectic substances and hemicellulose. Three of the four nuclei at the chalazal pole develop cell walls and are known as **antipodal cells** (Fig. 26-27B). The remaining two nuclei, one from each pole, migrate toward each other away from the poles. They are therefore known as **polar nuclei** (Fig. 26-27B). These may unite before or at fertilization to form the **secondary nucleus.** The mature female gametophyte of most angiosperms thus is a seven-celled structure consisting of three uninucleate cells at each pole with a larger binucleate cell (the **central cell**) between them. As far as is known, the synergids are always sister cells.

The process of female gametophyte development just described is essentially that which occurs in more than 70% of the flowering plants. It should be noted, however, that a great many regularly recurring variations characteristic of specific genera have been described. The genus *Lilium*, usually used to illustrate reproduction in angiosperms because of the large size of its nuclei and female gametophyte, exhibits one of these "deviations." Although the development of the female gametophyte in lily is often spoken

of as a "deviation" from the "normal" angiosperm type, it is, of course, quite "normal" in lily. Since this plant is still used in most laboratories, and because the process of female gametophyte development can be readily demonstrated, the divergent ontogeny will be described at this point (Figs. 26-26 [1'–8'], 26-28).

The first point at which the lily differs from the normal type of female gametophyte development is during the division and meiosis of the megasporocyte. Two successive nuclear divisions occur as usual, but these are not followed by cytokinesis (Figs. 26-26 [1'–3'], 26-28). As a result, the four megaspore nuclei are embedded in common cytoplasm of the megasporocyte. A second deviation is that in *Lilium:* Three of the four megaspore nuclei do not degenerate, and, furthermore, all four are involved in the formation of the female gametophyte. The latter, therefore, is tetrasporic in origin. The linear arrangement of the megaspore nuclei is followed by a change in position in which one remains near the micropylar pole of the developing female gametophyte and three migrate to the other (chalazal) end (Figs. 26-26 [4'], 26-28D). As the three chalazal nuclei enter mitosis, either in late prophase or in metaphase, they join together, making a single large division figure (Figs. 26-26 [5'], 26-28E). Since each of the haploid nuclei had undergone chromosome reduplication in preparation for this mitosis, the single large spindle includes three haploid sets of dual chromosomes. The result is that two large triploid nuclei are organized at telophase (Figs. 26-26 [6'], 26-28F). The megaspore nucleus at the micropylar pole, meanwhile, has divided to form two haploid nuclei (Figs. 26-26 [6'], 26-28F). This "second four-nucleate stage" is readily recognizable by the presence of two large (triploid) nuclei at the chalazal pole and two smaller (haploid) nuclei at the micropylar pole of the female gametophyte. Following this, a fourth mitosis takes place, resulting in the

---

[8]Before its homologies were appreciated, called the "embryo sac." The latter has been variously interpreted (Favre-Duchartre, 1977).

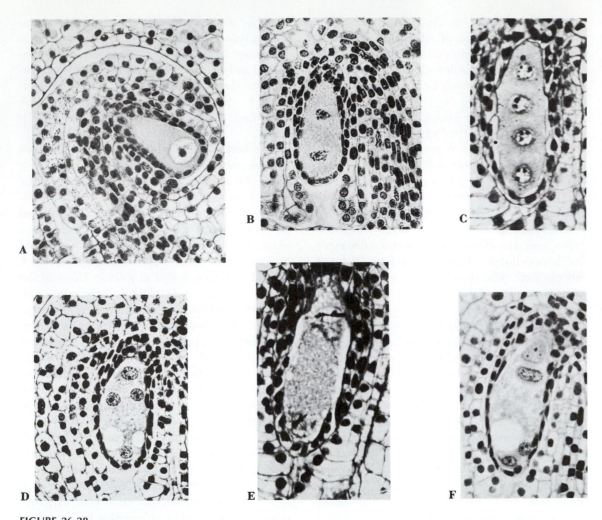

FIGURE 26-28

Lilium sp. Development of the ♀ gametophyte. **A.** Median longitudinal section of ovule showing integuments, micropyle, megasporangium, and megasporocytes. **B.** End of meiosis. I. **C.** Four haploid megaspore nuclei ("first four-nucleate stage"). **D.** Migration of three haploid nuclei to chalazal pole; prophases of division. **E.** Metaphase of triploid division, above; metaphase of haploid division, nearer micropyle. **F.** Two triploid nuclei, above; two haploid nuclei nearer micropyle ("second four-nucleate stage"). **A.** ×125. **B–F.** ×250.

formation of four triploid nuclei and four haploid nuclei (Fig. 26-26 [7']). Further development is normal, one triploid nucleus and one haploid nucleus functioning as polar nuclei (Figs. 26-26 [8'], 26-27). In the mature female gametophyte of *Lilium*, the antipodal cells are triploid, and the egg and the synergids are haploid. One polar nucleus is triploid and one is haploid.

POLLINATION AND FERTILIZATION

In many angiosperms, the maturation of the female gametophyte occurs just as the flower opens or just prior to its opening. Each ovule in the ovary contains a single mature gametophyte. The pollen grains, in reality immature male gametophytes, are transferred to the receptive surface of the stigma after the flower has opened, in the process of **pollination** (Fig. 26-2). Pollination here differs from that in gymnosperms in that in the latter the pollen is transferred directly to the vicinity of the micropyle of the ovule. A few flowering plants—many species of *Viola*, for example—produce cleistogamous flowers,[9] which do not open. These are regularly self-pollinated; in *V. odorata* it has been shown that the pollen grains germinate within the anther and grow through its walls to the stigma. In a few angiosperms with hollow stigmas and styles, intracarpellary pollen grains have been observed; this suggests the pollination process in gymnosperms.

The agents of pollination may be gravity, wind, water (in certain aquatics), snails, insects, birds, or bats. In self-pollinated flowers, anthers and stigma may be in contact, or the pollen may fall on the stigma if the anthers are above it. Insect pollination is considered by some to be the primitive condition in angiosperms, while others consider wind pollination to be so. Pollination by beetles, which occurs in certain ranalian and magnolialian genera, among others, is considered to be the most primitive type of insect pollination.

A number of specialized relationships between pollinating insects and certain species of angiosperms are well known. Among these are the wasp, *Blastophaga*, and the edible fig, and the moth *Tegaticula yuccasella*, on which *Yucca* depends for pollination.

The pollen of wind-pollinated species is small and dry, while insect-pollinated plants have sticky, adherent pollen with a waxy or oily surface. The ornamentation and secretions of the walls of some pollen grains are thought to be adaptations to attract insects.

That moist pollen releases substances that are allergenic is well known and has been for more than 100 years, while its liberation of enzymes has been recognized for more than 80 years. The role of these released substances has not been elucidated until more recently.

Successful pollination involves more than a mere mechanical contact between pollen grains and stigma;[10] in addition compatibility must prevail between tissues and secretions of the stigma and the substances (lipids, proteins, and glycoproteins) in the pollen-grain wall. These substances include a broad spectrum of enzymes. Those of the exine are involved in compatibility relations, while those of the intine have to do with penetration, maturation, and nutrition of the pollen tube within the stigma (Heslop-Harrison, 1975a,b, 1976a,b,; Heslop-Harrison et al., 1976; Linskens, 1975; Knox, 1984; and Knox et al., 1974).

Pollen grains germinate on the stigmatic surface (Figs. 26-2, 26-29). In most cases, germination occurs through a predetermined germ pore or thin place in the wall of the grain. A tubular process emerges from the pollen grain and grows into the stigma and through the style into the ovary (Fig. 26-2).

Germination of the pollen grains in most angiosperms is relatively rapid, occurring soon after the pollen has reached the stigma. In some cases, pollen grains germinate readily in sugar solutions, the optimum concentration varying with the species. Normally, only one tube emerges from each pollen grain. However, in some species as many as 14 tubes have been observed developing from a single pollen grain. The style through which the pollen tube grows may be hollow, as in *Viola*, or solid, as in cotton (*Gossypium*). In the latter case, the pollen tube

---

[9]These species of *Viola* have open or chasmogamous flowers as well; these are produced earlier in the season.

[10]Consult Shivanna (1982) and Knox (1984).

651

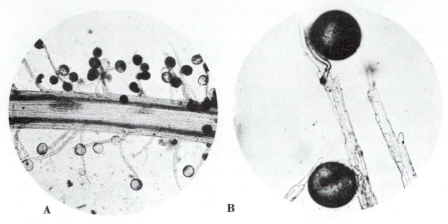

FIGURE 26-29
Zea mays, corn. **A.** "Silk" (stigma-style) with hairs and attached pollen.
B. Hairs with germinating pollen grains. **A.** ×30. **B.** ×250.

forces its way through the intercellular spaces of the stylar tissue.

Rosen (1975) reviewed investigations on pollen-tube relationships. He pointed out that most angiosperm pollen does not contain enough materials to build the long pollen tube necessary to reach the ovule (for example, a distance of about 10 cm in lily and sometimes longer in corn). It was demonstrated (Labarca and Loewus, 1973) in these investigations that labeled materials secreted by the cells lining the stylar canal become incorporated in the pollen tube. It also seems that some of the secretion on the stigma originates from the stylar canal.

Jensen and Fisher (1969) made a detailed investigation of the path of the pollen tube in cotton, while Cass and Peteya (1979) reported on the same phenomenon in barley. In cotton, the style is covered by an epidermis (with stomata and guard cells), within which, in centripetal order, are a cortex, vascular bundles, and a special central region of thick-walled cells, the **transmitting tissue.** The pollen tubes never enter any of the cells of the stigma or style but grow between the stigmatic cells and between the cells of the transmitting tissue. The secretions of the stigma and/or style must react

favorably with the growing apex of the pollen tube if fertilization is to occur (Shivanna, 1982; Heslop-Harrison, 1983).

As the pollen tubes are growing, one of the synergids shows signs of degeneration. The pollen tube penetrates through the filiform apparatus of that synergid and into its cytoplasm and then discharges its two sperms, the tube's nucleus, and some cytoplasm through a subapical pore into the synergid. From the synergid one sperm enters the egg, and its nucleus unites with that of the latter, while the second sperm enters the central cell and unites with the two polar nuclei (Fig. 26-27B, 26-30). Since both sperms are involved in nuclear unions, the phenomenon is spoken of as **double fertilization.** The details of the process provided above are based largely on the investigations of Cass and Jensen (1970) on barley and Jensen (1972, 1973) on cotton. It may be that the process of double fertilization may differ in detail in other flowering plants. In plants in which the polar nuclei have united to form the so-called secondary nucleus prior to fertilization, the second sperm unites with it. Double fertilization seems to be limited to angiosperms, in which it was discovered in 1898. It is clear, in the female

652

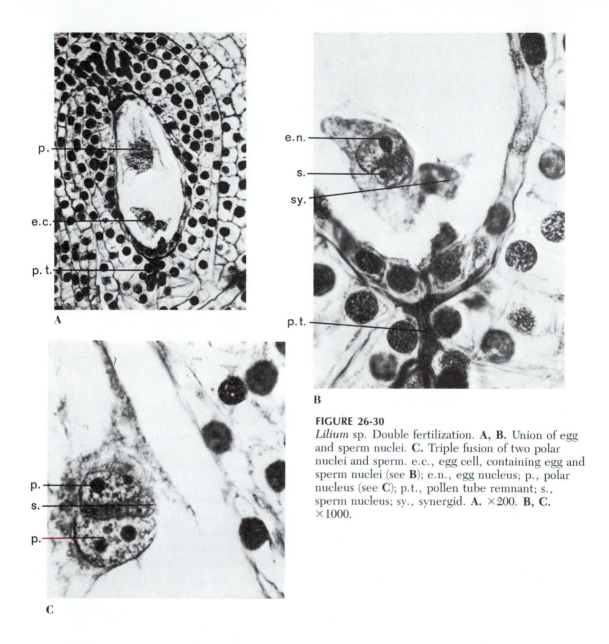

A

B

C

FIGURE 26-30
Lilium sp. Double fertilization. **A, B.** Union of egg
and sperm nuclei. **C.** Triple fusion of two polar
nuclei and sperm. e.c., egg cell, containing egg and
sperm nuclei (see **B**); e.n., egg nucleus; p., polar
nucleus (see **C**); p.t., pollen tube remnant; s.,
sperm nucleus; sy., synergid. **A.** ×200. **B, C.**
×1000.

gametophytes that have developed in the "normal" fashion from one haploid megaspore, that
union of the two haploid polar nuclei with one
sperm nucleus or of a secondary nucleus with a
sperm nucleus results in the formation of a
triploid nucleus. The latter is called the **primary**

endosperm nucleus because of its subsequent
activity. In a plant such as *Lilium*, however, it
will be recalled that one of the polar nuclei is
triploid and one haploid. Union of these with
the sperm nucleus produces a pentaploid primary endosperm nucleus.

653

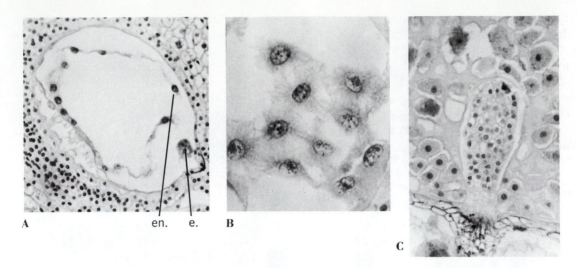

A   en.   e.   B

C

FIGURE 26-31
Lilium sp. Embryogeny. **A.** Section of early post-fertilization ovule; note embryo at right, near micropyle, and ring of free endosperm nuclei. **B.** Surface view of free-nuclear endosperm. **C.** Micropylar region of dormant seed; note embryo (small cells) embedded in endosperm. e., embryo; en., endosperm. **A, C.** ×60. **B.** ×125.

DEVELOPMENT OF THE EMBRYO, SEED, FRUIT, AND SEEDLINGS

In most cases, the zygote, enclosed by a delicate cell wall, remains undivided for an interval after fertilization. The primary endosperm nucleus, in contrast, soon enters upon a period of rapid division, which may be free-nuclear (Fig. 26-31A, B), or in some species the mitoses may be followed by successive cytokineses after each nuclear division. This forms at once a cellular **endosperm.**[11] If free-nuclear division has occurred, walls are usually, but not always, developed between the free nuclei, and the tissue that contains them is called endosperm. The endosperm of *Lilium* passes through a free-nuclear period before cell walls are laid down (Fig. 26-31A, B). When the first and subsequent divisions of the primary endosperm nucleus are followed by cytokineses, the endosperm is cellular from its inception. This occurs in many plants, among them *Adoxa*, *Lobelia*, and *Nemophila*. The descendants of the primary endosperm nucleus give rise to more or less extensive endosperm within the enlarging lumen of the female gametophyte, now perhaps appropriately called the **embryo sac.** The endosperm cells are often filled with starch grains or other stored metabolites that have diffused into them from the parent sporophyte. The endosperm is a storage tissue that provides readily available nutriment to the developing embryo. It is triploid tissue in a great majority of angiosperms, but a number of exceptions are known, such as the pentaploid type in *Lilium*. The origin and development of the endosperm and phenotypic attributes residing in it are involved in the phenomenon of **xenia.** This term denotes the direct influence of the male parent, as expressed in endosperm characteristics, an influence readily observable in certain maize hy-

---

[11]A few angiosperms, e.g., orchids, lack endosperm (Srivistava, 1982).

bridizations. For example, when varieties of white or light-yellow corn are pollinated by pollen from dark varieties, the kernels of the former are noticeably darker as they mature.

Sometime after the initiation of endosperm formation, the zygote begins the nuclear and cell divisions that result in the formation of the embryonic sporophyte (Fig. 26-31C). The cytokinesis that follows the first nuclear division forms a two-celled embryo. Free-nuclear stages are absent from the embryogeny of angiosperms. The division product that lies more deeply within the female gametophyte is known as the terminal cell, and the other is called the basal cell. Considerable variation exists in the subsequent stages of development. The following account is based largely on the process as it occurs in the "shepherd's purse," *Capsella bursa-pastoris*, a dicotyledonous weed with amphitropous ovules (Fig. 26-32).

This species is widespread in distribution, and microscopic preparations of the stages of embryogeny are readily obtainable. After the first division of the zygote, the basal cell divides in a transverse plane, and the terminal cell undergoes vertical division at right angles to the plane of division of the basal cell. A second vertical division in a plane perpendicular to the first results in the formation of a quartet of cells (quadrant stage) from the original terminal cell. This is followed by transverse divisions in all four cells to form an octant stage. The four cells of the octant farthest from the basal cell give rise to the cotyledons and stem apex of the embryo. The other four initiate an axis or hypocotyl. Periclinal divisions in each of the cells of the spherical eight-celled embryo segregate the eight superficial protodermal from eight inner cells.

While these changes have been taking place in the descendants of the terminal (embryoforming) cell, the products of the division of the basal cell have divided to form a short chain that functions as a suspensor, the basal cell of which is attached to the wall of the embryo sac (Fig.

26-32D). This cell enlarges markedly and possibly functions in absorption. By regular sequential divisions, the lowermost (nearest the embryo-forming cells) of the suspensor cells give rise to the root and root cap of the embryo. In subsequent development, the original octant region becomes organized into an elongate hypocotyl bearing two large cotyledons (Fig. 26-32E). Between them lies the apical meristem of the stem. The embryo is curved in later development in conformity with the amphitropy of the *Capsella* ovule.

The developing embryo digests the bulk of the endosperm and assimilates it. As the latter enters a period of dormancy, histological changes in the cells of the integuments result in the formation of seed coats, the inner known as the **tegmen** and the outer as the **testa**. It should be noted that in *Capsella* the antipodal cells, having undergone several divisions, persist at the antipodal pole of the embryo sac (Fig. 26-32E). The embryos of *Capsella* are dicotyledonous.

Monocotyledonous embryos differ in their organization, having a single, large cotyledon and the axis or stem lateral (Fig. 26-33). In a number of flowering plants—lily (Fig. 26-31C) and many orchids, for example—the embryo remains minute and undifferentiated. In such seeds, further development is delayed until after the seeds have been shed.

Peonies (*Paeonia*) are probably unique among angiosperms in that their embryogeny involves a period of free-nuclear division (Cave, Arnott, and Cook, 1961). After the free-nuclear proembryo has become cellular, several embryos form at its surface. Only one of these usually survives in the mature seed. The free-nuclear embryogeny of *Paeonia* is considered to be an example of parallel evolution in angiosperms and gymnosperms.

Although endosperm is produced in the development of almost all angiosperm seeds, the orchids being a notable exception, whether or not endosperm is present in the mature seed depends on the degree to which it has been

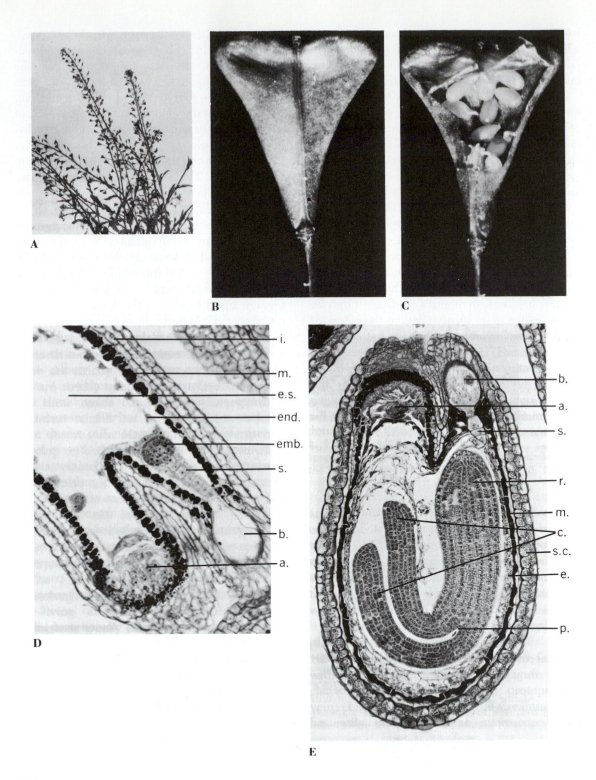

A

B

C

i.
m.
e.s.
end.
emb.
s.
b.
a.

D

b.
a.
s.
r.
m.
c.
s.c.
e.
p.

E

FIGURE 26-32

Capsella bursa-pastoris (shepherd's purse). **A.** Plants in flower and fruit. **B.** Immature fruit containing ovules with embryos in stage shown in **E. C.** Fruit wall partially removed to show maturing seeds. **D.** Longisection of ovule with early embryo. **E.** Longisection of ovule with well-differentiated embryo. a., antipodal tissue; b., basal cell; c., cotyledons; emb., embryo; end., endosperm; e.s., embryo sac or ♂ gametophyte; i., integument; m., megasporangium; p., plumule; r., radicle; s., suspensor; s.c., seed coat. **A.** ×0.1. **B, C.** ×4. **D.** ×150. **E.** ×90.

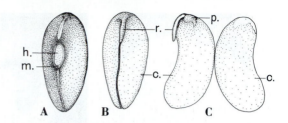

FIGURE 26-34

Phaseolus vulgaris, garden bean. Seed morphology. **A.** External appearance. **B.** With seed coats removed. **C.** Embryo with cotyledons separated. c., cotyledon; h., hilum or funiculus scar; m., micropyle; p., plumule; r., radicle.

embryo is massive and its cotyledons are rich in stored metabolites. In seeds with endosperm, the cotyledons are more like foliage leaves, as in the castor bean (*Ricinus*) (Fig. 26-35), basswood (*Tilia*), and *Magnolia*. The **seed** of the angiosperms, therefore, may be defined as an embry-

FIGURE 26-35

Ricinus communis. Castor-oil plant. Seed structure. **A.** External view. **B.** Median sagittal section. **C.** Longisection parallel to flat surface. ×1.5. (*After Holman and Robbins.*)

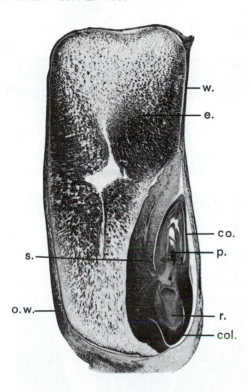

FIGURE 26-33

Zea mays, corn. Longisection of immature grain. co., coleoptile; col., coleorhiza; e., endosperm; o.w., ovary wall; p., plumule; r., radicle; s., scutellum or cotyledon; w., united integuments and ovary wall. ×25. (*Courtesy of Dr. J. E. Sass and the Iowa State University Press.*)

absorbed by the developing embryo. In seeds that lack endosperm at maturity (Fig. 26-34), such as bean, peanut, and garden pea, the

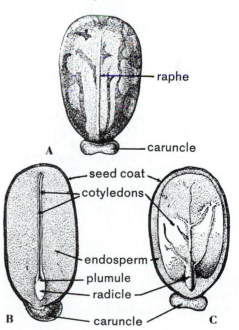

657

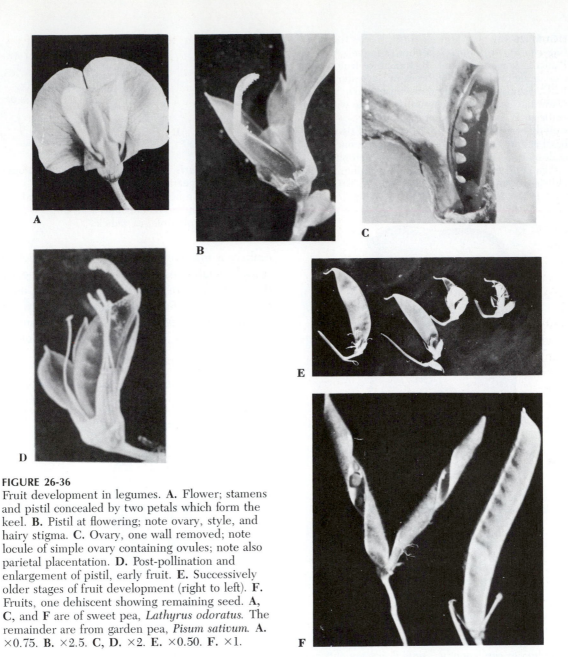

FIGURE 26-36

Fruit development in legumes. **A.** Flower; stamens and pistil concealed by two petals which form the keel. **B.** Pistil at flowering; note ovary, style, and hairy stigma. **C.** Ovary, one wall removed; note locule of simple ovary containing ovules; note also parietal placentation. **D.** Post-pollination and enlargement of pistil, early fruit. **E.** Successively older stages of fruit development (right to left). **F.** Fruits, one dehiscent showing remaining seed. **A, C,** and **F** are of sweet pea, *Lathyrus odoratus.* The remainder are from garden pea, *Pisum sativum.* **A.** ×0.75. **B.** ×2.5. **C, D.** ×2. **E.** ×0.50. **F.** ×1.

onic sporophyte often, but not always, in a dormant condition, either surrounded by endosperm or gorged with food (Fig. 26-34, 26-35); furthermore, the embryo is enclosed also by the remains of the megasporangium and by the matured integuments, the seed coats. The angiosperm seed differs fundamentally from that of gymnosperms in the origin of the tissue that

658

nourishes the developing embryo. In the gymnosperms, the haploid vegetative tissue of the female gametophyte serves this purpose, whereas in the angiosperms, the food tissue or endosperm is a postfertilization development initiated and stimulated by the union of the sperm nucleus and the polar nuclei or secondary nucleus.

Fertilization (and sometimes pollination alone) stimulates some of the cells of the ovary wall to multiply and to develop into the fruit wall, or **pericarp.** Soon after pollination and fertilization, the stamens and petals of most flowers wither and may be abscised from the receptacle. The ovary of the pistil, which contains the ovule or ovules, the latter with developing embryos, enlarges rapidly after pollination and fertilization (Fig. 26-36). This is in contrast to gymnosperms, in which the ovule at fertilization is essentially the same size as the mature seed. The pistil (carpel or carpels), and in some cases the receptacle and other floral organs as well, matures into the structure known as the fruit. The structure of the mature fruit varies in the families of angiosperms with respect to form, texture, and dehiscence, if any. These variations serve, in part, as criteria for the delimitation of various taxonomic categories. The **fruit,** then, is the matured pistil or pistils of the angiosperm flower along with, in some types, certain associated accessory structures.

Fruits are often classified in three categories on the basis of origin—namely, simple, aggregate, and multiple. Some examples of these are presented in the following abbreviated treatment.

Simple Fruits

These arise from the pistil of a single flower; the pistil may be simple or compound. Simple fruits may be subdivided into those that are dry when their contained seeds mature and those that are moist and fleshy.

Dry Fruits

(1) Dehiscent types: these open and shed their seed at maturity. Here are included, among others, **follicles** (Fig. 26-37A) (opening along one side), like those of willow and colum-

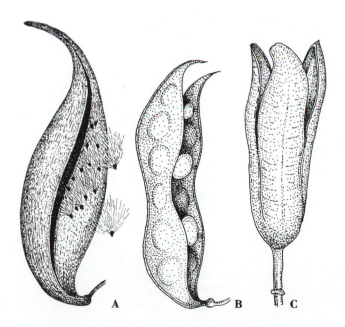

FIGURE 26-37
Simple dehiscent fruits. **A.** *Asclepias tuberosa.* Follicle of milkweed. **B.** *Robinia pseudoacacia.* Pod or legume of locust. **C.** *Iris* sp. Capsule.

659

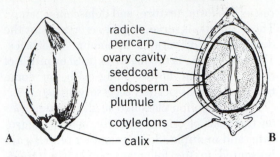

radicle
pericarp
ovary cavity
seedcoat
endosperm
plumule
cotyledons
calix

A   B

FIGURE 26-38
Achene of *Fagopyrum esculentum*, buckwheat. Fruit
(achene). **A.** External features. **B.** Median
longisection. ×7. (*After Robbins, from Holman and
Robbins.*)

bine; **pods** or **legumes** (Fig. 26-37B) (opening
along two sides), as in peas, beans, and many
other legumes; and **capsules** (like those of *Iris*),
with various loci of dehiscence (Fig. 26-37C).
Both follicles and pods arise from simple pistils,
while capsules are matured compound ones.

(2) Indehiscent types: here the seeds are
permanently retained within the fruit. The **caryopsis,** or **grain** (Fig. 26-33), and **achene** (Fig.
26-38) are well-known examples of this type.
The seed of caryopsis is nowhere free from the
fruit wall; in the achene, the seed is attached to
the fruit wall at only one point.

Fleshy Fruits

In fleshy fruits, the wall of the ovary increased markedly in thickness after pollination
and fertilization. It is called the **pericarp** and
often has three component layers, termed exocarp, mesocarp, and endocarp. The berry,
drupe, and pome are familiar examples of simple fleshy fruits.

In the **berry** (grape and tomato, for example)
the exocarp is skinlike, the mesocarp fleshy, and
the endocarp succulent or slimy. (The berry is
sometimes defined less restrictively as a pulpy
fruit with several seeds.) Berries may arise from
either simple or compound pistils (Fig. 26-39A).

Drupes (Fig. 26-39B) are much like berries,

FIGURE 26-39
Simple fleshy fruits, bisected. **A.** *Vitis* sp., grape, a berry. **B.** *Prunus armeniaca*, apricot, a drupe. **C.** *Pyrus
malus*, apple, a pome. c., fleshy floral tube; en., endocarp; ex., exocarp; m., mesocarp; p., periderm; s.,
seed; v.b., vascular bundle of receptacle.

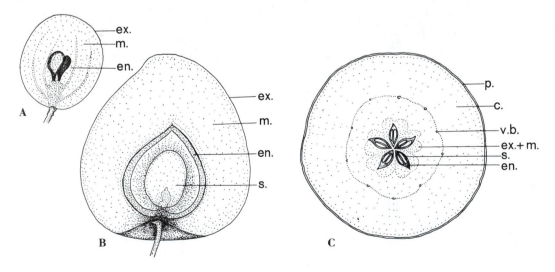

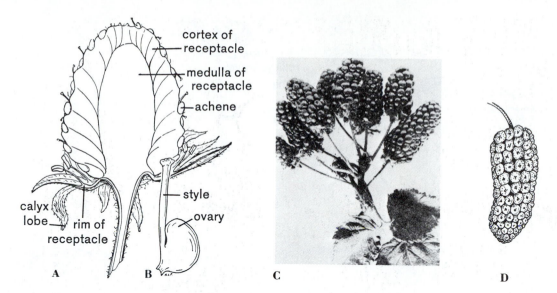

FIGURE 26-40

A–C. Aggregate fruits. **A, B.** *Fragaria* sp. Strawberry; note enlarged receptacle bearing the true fruits, which are achenes. **C.** *Rubus* sp., blackberries. The receptacle bears a number of minute, fleshy drupes. **D.** *Morus alba.* Multiple fruit of mulberry. **A.** ×1.5. **B.** ×5. **C.** ×0.5. **D.** ×2. (*After Holman and Robbins.*)

except that here the endocarp is stony and single-seeded.

Pomes (Fig. 26-39C) arise from a compound pistil embedded in the very fleshy hypanthium or floral tube of epigynous flowers. Here, the edible portion is not part of the fruit (*sensu stricto*) or core.

Aggregate Fruits

Aggregate fruits arise from several simple pistils of a single flower and its receptacle. Strawberries, raspberries, and blackberries are familiar examples. Their differences are described in Fig. 26-40A and its legend.

Multiple Fruits

Multiple fruits arise from an inflorescence axis and the pistils of a number of associated flowers. An "ear" of corn, a pineapple, and a mulberry all exemplify this type of fruit (Fig. 26-40C).

Seed Germination

Seed germination in angiosperms varies both morphologically and physiologically over a wide range. To some extent, rapidity of germination is correlated with the degree of development of the embryo at the time the seed is freed from the parent plant. Seed dormancy may be morphological, occasioned by incomplete differentiation of the embryo, as in the coconut and a number of other angiosperms, both mono- and dicotyledons, and in the cycads, *Gnetum*, and *Ginkgo*. Completion of embryo development after the seed has been shed is known as **afterripening.** In general, seeds with fully developed embryos germinate more rapidly than those with rudimentary embryos, although such factors as seed coat texture, temperature, humidity, and light may have a bearing on the speed of the germination process. In some angiosperms, including a number of tropical forms, no dormant period occurs, so that embryogeny is continuous into the seedling stages.

661

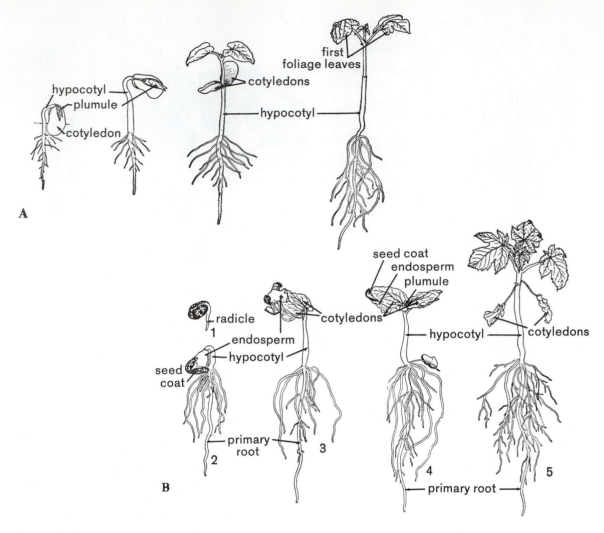

FIGURE 26-41

Seed germination and seedlings. **A.** *Phaseolus vulgaris*, garden bean. **B.** *Ricinus communis*, castor bean. ×0.5. (*After Holman and Robbins.*)

As in gymnosperms, both epigean and hypogean germination occur in angiosperms (Figs. 26-41, 26-42). A majority of dicotyledons are epigean, while most monocotyledons are hypogean. Epigean germination is considered to be more primitive than hypogean. Germination and seedling stages of four common angiosperms are illustrated in Figs. 26-41 and 26-42.

In bringing to a close this brief account of

reproduction in Anthophyta, it should be emphasized, as would be expected in such a large and diversified assemblage of plants, that there are many variations in detail from the general account presented above. Furthermore, as was indicated previously both in the Bryophyta and Pteridophyta, deviations from normal sexual reproduction also occur in angiosperms, both spontaneously and as a result of artificial stimu-

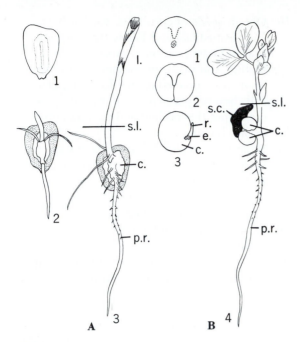

FIGURE 26-42
Seed germination and seedlings. **A.** *Zea mays*,
corn. 1. Ungerminated grain; embryo with large
scutellum visible. 2. Early germination.
3. Six-day-seedlings. **B.** *Pisum sativum*, garden pea.
1. Seed covered with seed coats; embryonic root
barely visible. 2. Seed coat removed showing two
fleshy cotyledons and radicle. 3. Hemisected seed.
4. Eight-day-old seedling. c., cotyledon;
e., epicotyl; l., leaf; p.r., primary root; r., radicle;
s.c., seed coat; s.l., soil line. (*After Bold.*)

lation. These deviations in angiosperms also
may be grouped under the phenomena of apogamy, or **apomixis,** and apospory. Among them
may be mentioned the development of haploid
embryos from unfertilized egg cells, development of haploid embryos from cells of the
female gametophyte other than the egg, development of diploid embryos because of parthenogenetic development of diploid eggs in apo-
sporously produced female gametophytes, and
development of diploid embryos from the megasporangium or integument. Kimber and Riley
(1963) listed 71 species of haploid angiosperms.

Here again is evidence that chromosome number *per se* does not determine whether a sporophyte or a gametophyte develops. Details of
these deviations are beyond the scope of this
account, but they are mentioned as examples of
similar phenomena seen in other plant groups.

Summary

Because of the tremendous range in their
diversity, organization of the vegetative organs
of the angiosperms has been given only cursory
consideration in the present chapter. Instead,
this chapter contains a brief discussion of the
gross morphology of the flower and the details
of the reproductive process. The latter may be
considered to consist of microsporogenesis and
megasporogenesis, maturation of the male and
female gametophytes, double fertilization, embryogeny, and development of the seed and
fruit. Brief reference to variations from the
usual reproductive cycle has been made.
Maheshwari and Rangaswamey (1965) have reviewed a number of aspects of angiosperm
reproduction that supplement the brief account
presented in this chapter. The flowering plants
differ from other seed plants in having the seeds
enclosed in the megasporophylls (angiospermy),
in their pollination, in their double fertilization,
and in their postfertilization endosperm. It has
been suggested (Meeuse, 1963*a*) that double
fertilization arose when the pollen tube first
entered an ancestral archegonium, as it now
enters the female gametophyte of angiosperms.
Unlike those of gymnosperms, angiospermous
ovules enlarge markedly after fertilization until
the seed matures. Except for peony (*Paeonia*),
angiosperms lack the free-nuclear phases in
embryogeny characteristic of gymnosperms
other than *Gnetum*. Embryogeny does not involve free-nuclear stages, which characterize
the process in gymnosperms. The widespread
occurrence of vessels in their xylem is a characteristic they share only with the Gnetophyta
among seed plants.

1. What reasons can you suggest to explain the fact that the flowering plants exceed any other group of plants in number of species?

2. In what respects are flowers and strobili similar? In what respects do they differ?

3. What attributes do angiosperms and gymnosperms have in common?

4. In what respects do they differ?

5. How do the seeds of gymnosperms and angiosperms differ?

6. Botanists used to group all seed plants in a single division, Spermatophyta. Give reasons for and against such a disposition of these plants.

7. On what basis are the spores of angiosperms called microspores and megaspores? Express your opinion regarding the propriety of this practice.

8. What evidence can you cite that sepals and petals differ in origin?

9. What evidence can you cite that a carpel (simple pistil) is foliar in origin? Are there alternate possibilities?

10. Summarize the terms used to describe the gross morphology of the flower and explain each in a single complete sentence.

11. Explain the use of the terms perfect and imperfect; monoecious and dioecious.

12. Distinguish between hypogynous, perigynous, and epigynous flowers, citing familiar examples of each.

13. What is meant by the term fruit?

14. What effect does epigyny have on the fruit?

15. Explain the term seed as it applies to angiosperms. Describe types of seed structure and variations in seed germination.

16. Obtain seeds of corn, pea, garden bean, and castor bean and follow their germination in moist peat moss or sand.

17. Outline the major phenomena in the reproductive process of the flowering plant and then prepare an account of the process, using this outline and making appropriate drawings.

18. What is an ovule? What variations in ovular form occur? Give an example of a plant with each type.

19. Are the Anthophyta unique, among seed plants, in lacking archegonia? Explain.

20. How are the gametophytes of flowering plants nourished?

21. What interval may elapse between pollination and fertilization in Anthophyta? How does this compare with gymnosperms?

22. Distinguish between pollination and fertilization in both angiosperms and gymnosperms.

23. How does the reproductive cycle of *Lilium* deviate from the "normal" type? Why has the word "normal" been enclosed in quotation marks?

24. Compare the mature male and female gametophytes of the flowering plants with those of the several genera of gymnosperms described in an earlier chapter.

25. What is meant by cleistogamy? Give examples. Of what genetic significance is this phenomenon?

26. Comment on the possible benefits that accrue to plants through cross-pollination.

27. What is meant by xenia? Give an example.

28. With the aid of labeled diagrams, describe the embryogeny of *Capsella*.

Chapter 27
DIVISION
ANTHOPHYTA II

From the preceding chapter it should be evident that the flower, of itself, is an inadequate criterion by which to differentiate between Anthophyta and other seed plants. The differentiating attributes of Anthophyta are, in fact, their angiospermy, site of pollination, double fertilization, postfertilization endosperm, and lack of free-nuclear embryogeny.[1] This chapter will discuss some of the important problems concerning the origin, possible precursors, and evolution of the flowering plants.

THE ORIGIN AND FOSSIL RECORD OF ANGIOSPERMS

More than a century ago, Charles Darwin wrote of the mystery surrounding what appeared to be the sudden appearance in abundance of the flowering plants in relatively recent rock strata (Cretaceous). Progress in unraveling this mystery has been slow, but there have been noticeable gains in recent years.

Critical appraisal of the fossil record emphasizes that at present there is little convincing paleobotanical evidence of the existence of pre-Cretaceous flowering plants (Scott, Barghoorn, and Leopold, 1960). Angiosperm remains in Cretaceous and later rocks are in the form of pollen, leaves, fruits, and even flowers (Fig. 27-1). Until recently it had been customary to assign angiosperm fossils to genera and families with extant members with the implication that they differ little from modern flowering plants. This is especially true with leaf fossils. However, recent work based on leaves involves a much more critical analysis than was used by earlier investigators (Dilcher, 1974), and it is quite likely that many Cretaceous angiosperms belonged to orders and families not represented in modern floras (Doyle and Hickey, 1976). Furthermore, pollen from the Cretaceous is in many cases unassignable to genera of extant angiosperms.

The apparent sudden abundance of a diverse angiosperm flora in the Cretaceous has

---

[1]Except in *Paeonia*, insofar as known.

A

B

C

FIGURE 27-1
Cretaceous angiosperm fossils.
A. Part of a compound leaf
resembling those of *Rhus*.
Scale is in millimeters. **B.**
Elongate inflorescence with
closely spaced staminate
flowers. **C.** *Archaeanthus
linnenbergeri*. Cluster of
follicles on an elongated
receptacle. **B.** ×1.25. **C.**
×0.93. (*Courtesy of Prof. D.
L. Dilcher.*)

suggested to some workers that the flowering plants must have had a long period of evolution beginning earlier in the Mesozoic or even the Paleozoic, and that for one or more reasons they do not appear as fossils until the Cretaceous. To account for the absence of pre-Cretaceous fossils, there have been several hypotheses. One popular idea postulates that the first angiosperms (either a monophyletic or polyphyletic series) originally evolved in upland habitats and that therefore they were not preserved as fossils because they were too far removed from depositional basins. But even if upland angiosperms had been preserved in localized basins, these sites would have been eroded away and the fossilized remains would have been lost. Many evolutionists feel that the more pronounced environmental extremes and the diversified habitats of upland regions favor appearance of new adaptive types and that it would be conceivable that the primitive angiosperms arose in such upland habitats. Time involved in migration downslope would explain what appears to be a sudden burst of angiosperms in the floras.

Another school of thought postulates that there were no pre-Cretaceous angiosperms and that the appearance of flowering plants in the fossil record reflects the time of their appearance on the earth. In criticizing the "upland" hypothesis, it has been pointed out that certain plant remains (pollen, fruits, seeds, wood) have been transported great distances from the point of origin to the site of deposition and fossilization. The fossil record indicates quite clearly the absence of indisputable angiosperm pollen, readily transportable, from pre-Cretaceous deposits. Furthermore, the present-day lowland tropical regions of the world have the most richly diversified floras, suggesting that these lowlands could have served as centers of diversification and subsequent radiation.

In summary, a quotation from a discussion (Scott et al., 1960) of the origin and fossil record of angiosperms seems singularly appropriate:

FIGURE 27-2
Archaeanthus linnenbergeri. Suggested reconstruction of a leafy twig and terminal flower. (*After Dilcher and Crane.*)

Despite its vagaries and imperfections, the fossil record remains our best index to relationship involving geologic time; and it does not bear out speculations on the origin of the angiosperms in the Paleozoic era.

For a quite different viewpoint, see Meeuse (1965).

Remains of clearly angiospermous plants are present in lower Cretaceous deposits and by middle Cretaceous time there had already been considerable diversity. Some of the forms known as fossils have flowers and fruits that are considered to be primitive (Figs. 27-2, 27-3)(Dilcher and Crane, 1984).

It appears that the earliest angiosperms are not assignable to extant taxa. Even though they represent the most primitive types of flowering plants, they seem to have attained such a level

FIGURE 27-3
Archaeanthus linnenbergeri. Suggested reconstruction of a leafy twig with a cluster of follicles on an elongated receptacle. (*After Dilcher and Crane.*)

has been postulated by some to have been the fertile structures of the cycadeoids. The latter have been viewed with favor as a possible ancestral line of the angiosperms by those who consider the dicotyledonous ranalian and magnoliaceous flowers (Figs. 26-13, 27-4, 27-5) to be the most primitive type of flowering plants. The aggregation of ovule-bearing appendages on a central axis surrounded by microspore-bearing appendages in cycadeoids (Fig. 25-31) has suggested the organization of a primitive flower (Figs. 27-4, 27-5), in spite of the absence of true angiospermy and the massive, compound structure of the microspore-bearing organ of cycadeoids.

In the complete absence of a fossil record in support, the Gnetophyta have been suggested as angiosperm precursors for a number of reasons, especially by those who consider the inconspicuous, imperfect type of flower (Figs. 26-3, 26-4) to represent the primitive condition in angiosperms. The presence of vessels in the secondary xylem and the absence of archegonia and the free-nuclear female gametophyte of two of the genera (*Gnetum* and *Welwitschia*) are often cited in support of this suggestion.

of evolution as to provide few clues regarding their progenitors. Most of the remains are of vegetative organs, such as leaves and stems, with fruits less abundant and flowers even more rare. Pollen has been well preserved under certain circumstances. However, botanists have not failed to speculate on the problem of angiosperm origin, and almost every one of the gymnospermous series (and even some vascular cryptogams) has been suggested as a possible point of origin for the flowering plants. The cycadeoids, the *Gnetum-Ephedra* alliance, the pteridosperms, the cycads, and, among others, a hypothetical group, the "Hemiangiospermae," have featured prominently in such discussions.

The origin of the "perfect" condition (having both stamens and pistils, or microsporophylls and megasporophylls) of the majority of flowers

THE ORIGIN OF ANGIOSPERMY

The problem of recognizing angiosperm precursors is complex. In the search for them it is important to realize that angiosperms differ from other seed plants in a number of attributes

FIGURE 27-4
Ranunculus macranthus. Floral structure. **A.** Single flower; note numerous petals, stamens, and pistils. At right, receptacle with numerous one-ovuled simple pistils (carpels) enlarging after pollination with fertilization. **B.** Bisection of flower; note massive, elongate receptacle to which are attached (in order of base to apex): sepals, petals, stamens, and pistils. **C.** Single pistil with ovary bisected longitudinally; note parietal placentation of single ovule.

C   A

B

A   B   C

D   E   F

FIGURE 27-5

Ranunculus macranthus. Structure of the flower and fruit. **A.** Flowers *in situ.* **B.** Single flower: note numerous petals, stamens, and pistils (sepals not visible). **C.** Stamens and stigmas and styles of pistils, more highly magnified. **D.** Stigmas and anthers at pollination; the black stigmas have been pollinated; note the dehiscent anthers and pollen grains, some of the latter on the stigmas. **E.** Portion of the preceding more highly magnified. **F.** Pistils after pollination; all but one stamen and the sepals and petals have been shed from the receptacle. **G, H.** Pistils enlarging as fruits (achenes); note single ovules in the ovaries of **H.** **A.** ×.33. **B.** ×67. **C.** ×3. **D.** ×5. **E.** ×12. **F.** ×2.5. **G, H.** ×3.

in addition to their angiospermy. These include the flower (however difficult of definition); its typical production of both microsporangia and megasporangia; a method of pollination in which pollen does not reach the ovule; double fertilization; postfertilization endosperm formation; and absence of free-nuclear stages during embryogeny. Furthermore, there is some difference of opinion regarding which extant angiosperms represent the primitive type (see p. 674).

With reference to the origin of angiospermy, there are several prominent hypotheses, as fol-

H

G

lows: (1) On the basis of paleobotanical evidence, it has been suggested that the origin of angiospermy lies in the Pteridospermophyta and their cupules (Andrews, 1963); that is, the cupule is looked upon as the precursor of the angiosperm carpel or pistil (Fig. 27-6). (2) The Cycadeoidophyta have also been considered to be angiosperm ancestors by Meeuse (1964, 1965) and others, in part because both microsporangia and ovules occur on the same receptacle as they do in most flowers. Meeuse strongly proposes a polyphyletic origin of the angio-

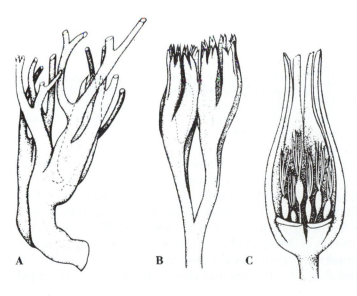

FIGURE 27-6
A. *Eurystoma angulare.*
B. *Stamnostoma huttonense.*
C. *Calathospermum scoticum.*
Seed-bearing structures; position of seeds indicated (in part) by dotted lines. (**A, B.** *After Long.* **C.** *After Walton. All from Andrews.*)

671

sperms. (3) It has also been suggested that the angiosperm carpel has arisen by longitudinal folding (conduplication) of an ovule-bearing leaf (see p. 607). In this connection a brief report by Meyer (1970) regarding facultative gymnospermy in cotton hybrids is of interest. Some of the hybrids develop ovules on the margins of petals. None of these, however, has produced seeds.

Very rudimentary, pteridosperm, cuplike enclosures around seeds are known from the Mississippian (Long, 1966). These have been discussed comprehensively by Andrews (1963). *Eurystoma angulare* and *Stamnostoma huttonense* are examples of these. In the former (Fig. 27-6A, B) the cupule is formed by dichotomously branching cylindrical axes. In the latter, the enveloping axes are more cuplike in arrangement. In some pteridosperms, such as *Calathospermum scoticum* (Fig. 27-6C), the cupule enclosed more than one seed. The saclike enclosures of the ovules in *Caytonia* (Fig. 25-25C, D) have also been suggested as possible near-precursors to the angiospermous state, because their phylogenetic closure is not difficult to envisage.

Recently, Long (1977) described ovule-bearing cupules from the Lower Carboniferous and suggested a possible means of deriving carpels from two-lobed cupules. It must be remembered, however, that although, through imagination, one could derive a carpel by picturing a closing of a cupule, there are many other features possessed by angiosperms that could not be derived from seed ferns.

Because of the lack of strong evidence from paleobotany, clues to the origin of the angiosperms have been sought in the ontogeny and morphology of their flowers, especially those of supposed primitive types in the order Ranales. In this order, long considered to be primitive on the basis of the gross structure of the flower (Figs. 27-4, 27-5) (elongate floral axis, spirally arranged appendages, and indefinite number and separate attachment of the latter, and also

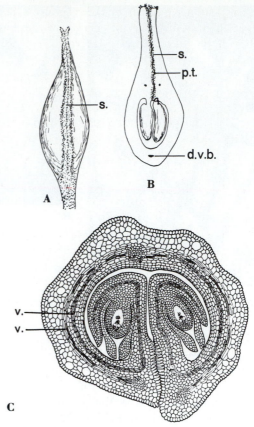

FIGURE 27-7
Primitive angiosperm carpel, diagrammatic. **A.** Ventral view showing paired stigmatic areas (s), running the full length of the carpel. **B.** Transverse section showing dorsal (d.v.b.) and two lateral vascular bundles; stigmatic area traversed by a pollen tube (p.t.). **C.** *Cananga* sp. Transection of conduplicate carpel; note closure by interlocking of epidermal cells. v., vascular bundle. (**A, B.** *From Bailey and Swamy.* **C.** *After Periasamy and Swamy, from Eames.*)

because of the presence of some of its members in early rock strata), there occur a number of genera (about 100 species) with vesselless xylem, a gymnosperm attribute, and others with a primitive type of carpel (Fig. 27-7) and stamen (Fig. 26-12). In some species, both primitive attributes are present.

To some morphologists, the primitive carpel

672

is clearly a longitudinally folded (conduplicate), three-veined leaf blade that is not firmly sealed along the line of union of the margins until the initiation of fruit formation; the margins may then be merely appressed. These margins bear interlocking glandular hairs that function as stigmatic surfaces for the reception of pollen (Fig. 27-7A, B). There is no style or localized stigma, but the entire edge of the conduplicate carpel is stigmatic. A progressive series in modification of such primitive carpels has been traced in certain of these genera. This involves localization of the stigmatic surface to form a stigma, formation of a style, and firm closure of the carpels with formation of a locule.

The ovules in these primitive Ranalian carpels are laminar (i.e., not marginal) in origin (Fig. 27-7B). In modifications of the primitive conduplicate carpel, loss of the sterile portions of the lamina lateral to the ovules and sealing of the carpel give the false appearance that the ovules arise at the margins of the blade, but they are truly laminar in origin. In spite of such evidence from comparative morphology regarding the origin of carpels from leaves, there is no clear evidence of the derivation of this type of angiospermy from gymnospermous types.

It is possible that angiospermy also arose by the enclosure of ovules (hence seeds) by the overgrowth of pteridospermous cupules. Were such the case, angiosperms would be at least biphyletic, and their carpels would not be strictly homologous.

THE NATURE OF PRIMITIVE FLOWERS

As noted earlier, it is difficult if not impossible to formulate a definition of the flower that would delimit it from the cone or strobilus. Both are, essentially, determinate, or growth-limited, axes with appendicular organs bearing sporangia. Sepals and petals, present in most (but not all) flowers but absent in strobili, are clearly modified leaves that seem to have several possible origins. It has been suggested that the perianth represents a modified continuation of the sterile bracts associated with certain strobili. Another hypothesis holds that sepals and petals represent secondarily sterilized sporophylls. Evidence for this is available in petaloid stamens of water lilies and certain "double" flowers. Still another suggestion is that the perianth was a new development that appeared with the angiosperms. According to this, primitive flowers lacked a perianth, and the latter evolved from protective scales and bracts. Only later did some of the perianth segments (petals) develop color. Some of these suggested steps in development seem, indeed, to be present in the order Ranales, in such genera as *Liriodendron* (tulip tree) and *Magnolia*.

Inasmuch as the fossil record has not yet provided us with clear and unequivocal evidence regarding the nature of primitive angiosperms and their precursors, data from the comparative morphology of living angiosperms have been marshaled from time to time in attempts to decide which of the extant angiosperms show primitive attributes. That this practice involves difficulties is clear from lack of complete agreement regarding which living angiosperms are, indeed, primitive. An eminent student of the group (Lawrence, 1951) wrote as follows, in this connection:

The primitive characters of contemporary, and presumedly phylogenetic, systems of angiosperm classification are not often based on paleobotanical evidence illustrative of ancestral conditions, and for the most part their primitiveness may be a matter of personal opinion or of judgments based on circumstantial evidence. Many of the so-called primitive characters in published lists . . . are alleged to be primitive because they occur in members of primitive taxa, and the taxa are primitive because they have primitive characters. It is difficult to find, by objective methods, devices to break this cyclic reasoning, and, among the angiosperms at least, there is inadequate paleobotanical evidence to support one view and reject the other.

There are two (among others) quite different

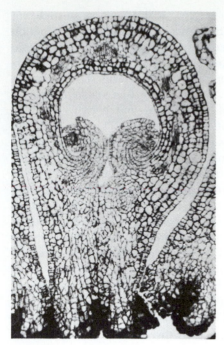

FIGURE 27-8

Drimys winteri var. *chilensis.* Transection of carpel; note line of appression of carpel margins. Dark cells below are stigmatic area. ×115. (*After Tucker.*)

major hypotheses regarding which extant angiosperms are most primitive. According to one, imperfect flowers that lack petals and that have cyclic arrangement of the floral members are considered to be primitive. This would be supported by the occurrence of microsporangia and megasporangia on different axes or in different strobili in all gymnosperms except the Cycadeoidophyta. Examples of such allegedly primitive flowers are those of the Salicales (willow order), Fagales (beech order), and Juglandales (walnut order), among others. The minute apetalous flowers of willow (*Salix*) (Figs. 26-4, 26-5), oak (*Quercus*) (Fig. 27-8), and pecan (*Carya illinoensis*) are exemplified by familiar genera in these orders. In *Salix*, both the staminate and pistillate flowers are very simple, consisting of a few stamens or a pistil subtended by a bract. The flowers of *Quercus* and *Carya* are equally

simple. The staminate flowers (and the pistillate ones in *Salix*) are borne in pendulous inflorescences known as **aments**.[2] The simple pistillate flowers of *Quercus* and *Carya* are borne in terminal clusters of a few flowers. The conclusion that flowers of this type are primitive has been challenged, however, on the basis of anatomical evidence that seems to indicate that their apetalous condition represents secondary simplification or reduction from petaliferous precursors. Furthermore, the xylem of these ament-bearing angiosperms is relatively highly specialized, not primitive.

A more widely accepted (at present) view is that the flowers of many members of the Ranales and Magnoliales exhibit primitive attributes and that certain attributes of the vegetative plant body are correspondingly primitive. The floral attributes considered to be primitive include (1) an elongate receptacle to which (2) are attached separately in spiral arrangement an indefinite (large) number of stamens and pistils (Figs. 26-2, 26-13), (3) the presence of unfused perianth segments, often not differentiated as sepals and petals, (4) radially symmetrical flowers, (5) the occurrence of leaflike stamens with microsporangia embedded in the blade, (6) the presence of simple pistils (carpels), in some cases not completely closed or closing only relatively late in ontogeny (Figs. 27-8, 27-9), (7) seeds with two integuments, and (8) seeds with vascularized integuments.

Among supposedly primitive attributes of the vegetative plant body may be cited: (1) the woody habit, (2) absence of vessels from the xylem, (3) spirally (alternately) arranged leaves, (4) glandular and stipulate leaves.

The fossil record supports this second (Ranalian and Magnolialian) hypothesis with respect to the primitive nature of spiral and separate attachment of appendages, as contrasted with the whorled or cyclic one. According to this

---

[2]Hence the group name, Amentiferae, for these orders.

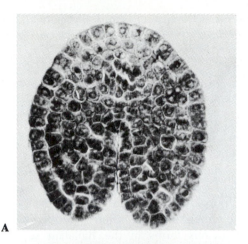

A

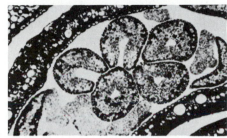

B

FIGURE 27-9
A. *Aquilegia formosa* var. *truncata*. Transection of carpel primordium; note closure due to activity of marginal cells. **B.** Transection of (five) slightly older carpels; the margins of some are appressed to form a locule, but others are still open. (*After Tepfer.*) **A.** ×500. **B.** ×75.

same hypothesis, shortening of the receptacle has resulted in crowding and a change to a cyclic condition of the floral organs. Other results have been the reduction in number of the floral organs and, in some cases, their union to form compound structures.

With respect to stamens, certain Magnoliacean flowers bear leaflike microsporophylls (Fig. 26-12B, C) in which the two pairs of microsporangia are embedded near the midvein of the microsporophyll. Reduction in the vegetative tissue of the microsporophyll has resulted in the type of stamen with a slender filament and distal anther (Fig. 26-12A).

Certain Ranalian and Magnoliaceous carpels are considered to be primitive because of their seemingly incipient closure and their lack of a localized stigma (p. 672). The genus *Aquilegia* and many other genera provide ontogenetic evidence that carpels are infolded megasporophylls (Fig. 27-9).

THE ORIGIN OF OTHER ANGIOSPERMOUS CHARACTERISTICS

The origin of double fertilization, a unique common attribute of angiosperms, has been under discussion (Gerassimova-Navashina, 1961; Meeuse, 1963a). The fossil record here is of little help, since the delicate reproductive tissues are not preserved to provide sufficient details for cytological study. It has been suggested that the beginnings of double fertilization must be sought in the seed plants, in which both sperms of a given pollen tube were discharged into or engulfed by the cytoplasm of one archegonium. This seems to occur with frequency in the conifers and Gnetophyta. In *Gnetum*, it will be recalled, the female gametophyte is free-nuclear at fertilization in at least some species, and two sperms usually are discharged into it. One unites with one of the potential egg nuclei, and the other sperm may, on occasion, also unite with another free nucleus of the female gametophyte. However, only one zygote may represent a precursor to the fusion that leads to the primary endosperm nucleus in angiosperms.

This brings us now to a consideration of the origin of the endosperm of angiosperms. Here again, the fossil record is of little value. It is perhaps significant to note that although double fertilization probably occurs in all angiosperms, the formation of a primary endosperm nucleus does not always result in development of endosperm. It has been suggested that lack of endosperm is a primitive condition in such organisms.

Finally, the absence of free-nuclear stages in angiosperm embryogeny seems to have no counterpart in a majority of extant gymnosperms and is, clearly, an attribute that became fixed early in angiosperm evolution. It is noteworthy, however, that comparative embryology of the Coniferales reveals a series with decreasing numbers of free embryonic nuclei, culminating in *Pinus*, which has only four, and *Sequoia sempervirens*, which has none. Furthermore, it has been reported that in the embryogeny of certain species of *Paeonia*, a free-nuclear phase does, indeed, occur. These exceptions indicate the need for caution in considering distinguishing characteristics as universal and absolute.

Not only does the origin of angiosperms remain in the realm of speculation, but the true relationship of the numerous orders and families of angiosperms to each other is to some extent conjectural. However, Sporne (1949, 1956, 1969, 1972) has been studying statistically a number of primitive characteristics of angiosperms, generally conceded to be primitive largely on the basis of paleobotanical evidence. He has demonstrated significant correlations among groups of 22 such characteristics, both vegetative and reproductive. For example, the occurrence in angiospermous families of such a primitive characteristic as ovules with a massive megasporangium (nucellus) is correlated with such other primitive characteristics as radially symmetric flowers, free (not united) petals, and so on. On the basis of presence or absence of various combinations of primitive attributes, Sporne (1969) has arranged (his Table 2) families of dicotyledons in order from primitive to more advanced. From such data he concludes that

perhaps seven or eight of the largest orders of dicotyledons diverged very early in the evolution of the group, that each has retained some primitive characters, but that a remarkable amount of parallel evolution has occurred subsequently. This has involved such trends as the following: among woody plants, from primitive to advanced types of wood, but in any case from woody to herbaceous; from alternate leaf arrangement to opposite; loss of stipules; from actinomorphic to zygomorphic flowers; from polypetaly to gamopetaly; reduction in the number of stamens and carpels; towards fusion of carpels and parietal placentation, etc.

There are those who raise the question of whether the angiosperms are truly monophyletic (Meeuse, 1962). Whether monophyletic or polyphyletic in origin, the relation of the monocotyledons and dicotyledons to each other is a further perplexing problem. The monocotyledons are sometimes interpreted as derived from dicotyledonous forms in the Ranalian series. That woody genera in each family are primitive and that herbaceous ones have arisen from woody forms by curtailment of cambial activity has been proposed on the basis of comparative anatomy and other considerations. This point of view has found wide acceptance.

A most lucid and succinct illustrated summary of the several suggestions regarding the origin of the angiosperms has been prepared by Sporne (1971); the reader is referred to this publication for more comprehensive discussion of the subject.

If this brief discussion of the possible origin of the angiosperms seems somewhat nebulous and unsubstantial, it will correspond quite appropriately to the state of our knowledge about the problem. Since there are such a large number of orders and families of flowering plants, an eloquent manifestation of the great range of variation and complexity, no summary of their classification will be presented in the present text. This aspect of the flowering plants usually is treated in texts and courses in plant taxonomy.

In the preceding pages an attempt has been made to present a glimpse of plants of the past and their significance in relation to our extant flora. The data in this chapter will be drawn upon as the basis for certain conclusions presented in Chapter 37.

1. On what basis have cycadeoids been suggested as angiosperm precursors?
2. What explanation has been suggested for the absence of angiosperms from fossiliferous strata earlier than Cretaceous?
3. What attributes have been cited as those of primitive extant flowers? On what basis?

Chapter 28[1]
INTRODUCTION
TO THE
KINGDOM MYCETEAE (FUNGI)

The fungi are eukaryotic, achlorophyllous, typically filamentous, spore-bearing organisms with the cell walls of most species consisting of chitin combined with other complex carbohydrates, sometimes including cellulose. As such, they are unable to manufacture their own food and consequently lead a saprobic or parasitic existence. The saprobes (saprophytes) live on nonliving organic matter, causing it to decay; the parasites infect mostly plants or sometimes animals, including human beings, causing disease. Fungi require already elaborated organic food in order to survive. This they digest outside their bodies (thalli) by secreting enzymes and making the substratum soluble so that it may be absorbed into their cells through their walls and plasma membranes.

Although the fungi are plantlike in the cellular structure of their thalli and certainly in their reproduction by spores, **mycologists** (biologists who study fungi) now believe that most fungi have originated from some ancestral protozoanlike flagellate and are therefore not related to plants. Fungi are consequently segregated into a kingdom of their own: Kingdom Myceteae (Fungi). Nevertheless, botanists still claim them, study them, place them into two or three divisions of the plant kingdom, and classify and name them according to the rules of the International Code of Botanical Nomenclature.

Exactly what organisms should be included in the fungi is a matter of controversy. Many biologists, for example, believe the slime molds to have greater affinities with the Protozoa than with the Fungi, but the fact remains that they are studied almost entirely by mycologists. For this reason, we shall include them here with the Fungi without claiming relationships. Among the more comprehensive accounts of the fungi that have been written in English in the last half of the twentieth century are those by Bessey (1950), Gäumann (English translation by Wynd) (1952), Alexopoulos and Mims (1979), Burnett (1968), Webster (1970), Müller and Loeffler (English translation by Kendrick and Bärlocher) (1976), the very detailed five-volume treatise by a number of authors, edited by Ainsworth and Sussman (1965, 1966,

---

[1]Because of prolonged, serious illness, C. J. Alexopoulos was unable to revise the chapters on fungi. He is grateful to Professor James W. Kimbrough, mycologist, Department of Botany, University of Florida, Gainesville, for revising the material on fungi.

1968, 1973), and Ross (1979). *The Dictionary of the Fungi*, by Ainsworth and Bisby, now in its sixth edition (1971), is also a remarkably useful reference work.

IMPORTANCE

About 80,000 species of fungi have been described, and many more probably await discovery in the tropics and in the oceans. Fungi are ubiquitous over the earth, being more abundant in areas of high moisture but also having been isolated from desert soils. Fungi are of great importance in human affairs. They have been decaying plant and animal bodies for the last 2 billion years at least, thus liberating various elements such as nitrogen, phosphorus, potassium, sulfur, iron, calcium, magnesium, and zinc, which would forever be locked up in these bodies without the activity of fungi and bacteria, and they liberate CO_2 into the atmosphere to be used again in photosynthesis by green plants. The role of fungi in the decay of plant bodies, especially, is very great indeed, because of the ability of fungi to utilize cellulose. In addition, the yeasts, which are true fungi, are the basis of the baking and brewing industries and have played an important part in the nutrition of the human race ever since man discovered fermentation and baked his first loaf of leavened bread. Fungi are also used industrially in the manufacture of citric acid, the basis of the enormous soft drink industry; in the manufacture of certain types of cheese (Roquefort and Camembert types); in the production of a large number of useful antibiotics, notably penicillin and griseofulvin; and in the manufacture of certain vitamins, such as riboflavin, and of various important drugs such as ergotamine and cortisone.

Fungi may be destructive too. The majority of plant diseases are caused by parasitic fungi. Those human and animal skin diseases commonly called ringworms, including athlete's foot, and some serious pulmonary infections, such as histoplasmosis and coccidioidomycosis, are of fungal origin. Fungi are also instrumental in the destruction of lumber and timber; they cause clothing apparel to mildew and etch the lenses of cameras, binoculars, telescopes, and other optical instruments in humid climates. In addition, they destroy foods in the home and the supermarket, causing millions of dollars' worth of spoilage.

We must not leave the topic of importance of fungi without mentioning the mushrooms, which are the **sporophores** (spore-bearing organs) of the most complex fungi and which have played a great role in human affairs. For centuries some mushrooms have been used in religious ceremonies of many ancient peoples and primitive tribes, who knew, collected, and used the kinds that contain hallucinogenic chemicals. *Mushrooms, Russia, and History*, by Wasson and Wasson (1957), now a collector's item, and *Soma, Divine Mushroom of Immortality*, by Wasson (1968), are two fascinating books on this subject. The more recent *Maria Sabina and Her Mazatec Mushroom Velada*, by Wasson and coauthors (1974), which describes in detail, with photographs and sound, the use and effect of hallucinogenic mushrooms in the mystic rituals of a Mexican Indian tribe, appears to be a classic, judging from the review in *Mycologia* 68:953–959. Lowy (1971, 1974, 1977) has also written extensively on the legends about mushrooms and mushroom stones (Fig. 28-1) of some Indian tribes of Central America.

And let us not forget the use of mushrooms as food. The mushroom-growing industry of Europe, the United States, and the Orient (China and Japan) is "mushrooming." It is based on the cultivation of but a few species. Now that the French have discovered a method of establishing "truffle farms," the industry will receive a new impetus, for truffles are considered by epicureans to be the most exquisite of all fungi in flavor. Many wild mushrooms are edible and of excellent flavor—much better than that of

679

FIGURE 28-1
A mushroom stone from Mixco Viejo, Guatemala. Cap diameter, 15.5 cm; stipe, 11 cm. (*Courtesy Bernard Lowy. Private collection of Bernard Lowy, #963.*)

most cultivated varieties—but the novice is cautioned to be absolutely certain of what he collects for the table, for he may not have the chance of making a second mistake! Some mushrooms are deadly poisonous when consumed, even in small quantities. Poisonous mushrooms are commonly referred to as "toadstools," but there is no way to distinguish between the two except by eating them, a dangerous practice! For a delightful evening's reading of some of the fascinating aspects of the fungi, read Brodie's (1978) *Fungi—Delight of Curiosity.*

Summary and classification

The fungi are plantlike organisms without chlorophyll, which lead a saprobic or parasitic existence and reproduce by spores. As such,

they are of great significance in nature and in human affairs. Together with the bacteria they are responsible for the disintegration of all bodies of organisms, particularly those of plants, that have lived and died on earth in the last 2 billion years and for the recycling of the chemical elements of which those bodies consisted. Fungi are the causes of most plant diseases and of some serious diseases of human beings and animals. They are also of great importance industrially in the manufacture of organic acids, various vitamin preparations and other drugs, as also in the fermentation of beer and wine, the baking of bread, and the manufacture of certain types of cheese. Many wild mushrooms are edible, but some are poisonous, a few deadly; a few are hallucinogenic.

Whether the slime molds should be included in the fungi is a matter of controversy; we have decided to include them in the Kingdom Myceteae. A general outline for the classification of fungi, including the slime molds, through the level of subclass (based on Alexopoulos and Mims [1979]) follows:

Kingdom Myceteae
    Division 1. Gymnomycota
        Subdivision 1. Acrasiogymnomycotina
            Class 1. Acrasiomycetes
        Subdivision 2. Plasmodiogymnomycotina
            Class 1. Protosteliomycetes
            Class 2. Myxomycetes
    Division 2. Mastigomycota
        Subdivision 1. Haplomastigomycotina
            Class 1. Chytridiomycetes
            Class 2. Hyphochytridiomycetes
            Class 3. Plasmodiophoromycetes
        Subdivision 2. Diplomastigomycotina
            Class 1. Oömycetes
    Division 3. Amastigomycota
        Subdivision 1. Zygomycotina
            Class 1. Zygomycetes
            Class 2. Trichomycetes
        Subdivision 2. Ascomycotina
            Class 1. Ascomycetes
                Subclass 1. Hemiascomycetidae

DISCUSSION QUESTIONS

1. Cite some plantlike characteristics of fungi.
2. Why are fungi assigned to a separate kingdom?
3. Cite both harmful and beneficial activities of fungi.
4. What types of nutrition occur in fungi?
5. Discuss the habitats of fungi.
6. How do fungi differ from plants?
7. What rules are used in the classification of the fungi?
8. How does one distinguish between a mushroom and a toadstool?

Chapter 29
DIVISION GYMNOMYCOTA: THE SLIME MOLDS

The slime molds, also known as Mycetozoa (fungus animals), have been comprehensively summarized by Olive (1975). They differ from the true fungi in their phagotrophic nutrition and in that their somatic parts lack cell walls, consisting only of protoplasts bounded by plasma membranes. Their spores, however, are each enveloped by a rigid cell wall as are those of the true fungi. Three groups of slime molds are known: the cellular slime molds (Acrasiomycetes), the Protostelids (Protosteliomycetes), and the true, or plasmodial, slime molds (Myxomycetes). Olive also includes the endoparasitic slime molds (Plasmodiophoromycetes) and the net slime molds (Labryinthulales) in his discussion, but most mycologists believe these two groups to be more properly assigned to another division of fungi.

SUBDIVISION ACRASIOGYMNOMYCOTINA

CLASS ACRASIOMYCETES

The Acrasiomycetes (cellular slime molds) have the following characteristics:

1. They produce no flagellated cells.[1]
2. Their **myxamoebae** (special types of amoebae formed by slime molds) aggregate to form a pseudoplasmodium in which they do not fuse but retain their individuality.
3. The stalks of all but a few species consist of cells.
4. The spore walls contain cellulose.

Cellular slime molds have been discovered in all continents, occurring in cultivated as well as in native soils, but are most abundant in the upper layers of humus in well-established deciduous forests (Cavender and Raper, 1965, 1968). They are important in human affairs only in that they provide excellent biological systems for the study of morphogenesis and

---

[1]Olive has reported flagellated cells in one species, *Pocheina rosea.*

molecular biology. The best-known system is *Dictyostelium* (Gr. *dictyon*, net + Gr. *stele*, a post) *discoideum* discovered by Dr. K. B. Raper in 1935.

The **sporocarps** (a special type of sporophore) of the Acrasiomycetes are delicate and ephemeral. As such, they are seldom encountered in nature and are known almost entirely from laboratory cultures. These organisms are easy to isolate. If a suspension of finely divided surface humus is streaked onto a weak nutrient agar medium, such as glucose-peptone or hay-infusion agar, and cross-streaked with a suspension of the bacterium *Escherichia coli*, sporocarps often appear and may then be isolated and identified. They can be recognized by their delicate, cellular stalks,[2] which may be simple or branched and which may arise from a disc or a cramponlike base or just be plain without a differentiated basal structure. The stalk apex of each branch holds a droplet of mucus, in which the spores are held. There is no common envelope around all the spores as there is around the spores of the true slime molds (Fig. 29-1D).

Life cycle. The spores of the cellular slime molds are generally ovoid, and each is enveloped by a cellulose wall. As the stalk that supports them bends over and falls, the droplet of spores at its apex is released and the spores eventually germinate, each producing a single, uninucleate, haploid myxamoeba (Fig. 29-1A). The myxamoebae feed by engulfing bacteria and multiply by simple cell division until a large amoebal population results. Under unfavorable conditions, the myxamoebae encyst, forming microcysts. A **microcyst** is an amoeba that has become rounded and has secreted a delicate but rigid cellulose wall around itself. Upon return of favorable conditions the microcysts germinate, liberating myxamoebae that resume their normal activities. When the food supply is exhaust-

ed, an aggregation of myxamoebae begins. What triggers this stage is as yet unknown. It is believed that a single cell in the population begins to secrete the attractant **acrasin,** recently identified as cyclic AMP in some species, toward which other myxamoebae migrate. These, in turn, develop the ability to secrete acrasin and so on down the stream, thus setting up a gradient of the attractant. As the myxamoebae reach the center of each aggregation (Fig. 29-1B), they come in intimate contact one with another and form a gelatinous unit commonly called the **slug** or **grex** but more technically known as the **pseudoplasmodium** (Fig. 29-1C). This is a sausage-shaped mass of intimately associated amoebal cells, which, however, are not fused but retain their individuality and may be separated by mechanical means. Nevertheless, the amoebae in a pseudoplasmodium become differentiated. Those in the front end of the slug become potential stalk cells, whereas those in the rear end will give rise to the spores of the sorocarp.

In some species, such as *Dictyostelium discoideum*, the pseudoplasmodium migrates over the culture dish before it comes to rest in preparation for sporulation. When it finally rests, it becomes flattened below but assumes a hemispherical shape above, with a papilla protruding from its center. As the mass of myxamoebae elongates vertically, certain of the cells at the apex migrate downward through the whole mass. A cellulose cylinder is formed at this time, within which the cellular stalk of the sporocarp develops (Fig. 29-1E) by the sacrifice of some of the amoebae, while the cells at the bottom migrate up the stalk to form the spores at the tip. This completes the life cycle.

Sexual reproduction. For many years the matter of sexuality in the Acrasiomycetes was one of great controversy. In 1972, however, Erdos, Nickerson, and Raper found that the amoebae of *Polysphondylium* (Gr. *poly,* many, + *sphondylos,* vertebra) *violaceum* aggregate into large clumps, in the center of which a large

---

[2]In the genus *Acytostelium* (Gr. *a,* not, + *kytos* = hollow vessel, i.e., cell, + *stele,* column), the stalks are not composed of cells.

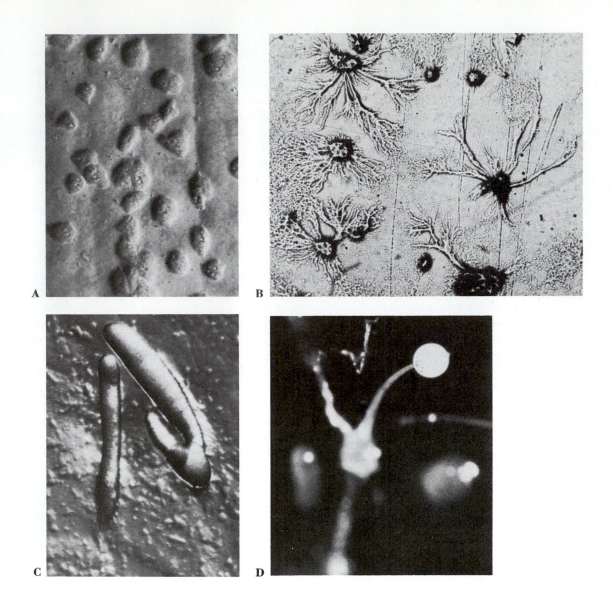

FIGURE 29-1

Dictyostelium discoideum. **A.** Myxamoebae.
B. Convergence of myxamoebae at centers of
aggregation. **C.** Migrating aggregation or slug.
D. Sporocarp. **E.** Portion of a sporocarp stalk
enlarged; note component myxamoebae with cell
walls. **F.** Crushed sporocarp showing spores. **A.**
×400. **B.** ×15. **C.** ×45. **D.** ×30. **E.** ×525. **F.**
×230. (*Courtesy of Dr. A. C. Lonert and the
General Biological Supply House.*)

cell originates, which engulfs the surrounding
myxamoebae and secretes a thick wall around
itself, thus becoming a **macrocyst.** Macrocysts
have also been found in *Dictyostelium mucoroides* (Fig. 29-2) and in other species. At first
binucleate, the macrocyst becomes uninucleate
probably through karyogamy. The evidence for
karyogamy's taking place is derived from the

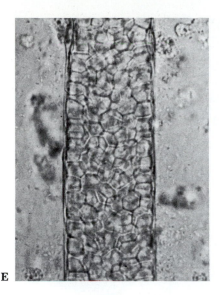

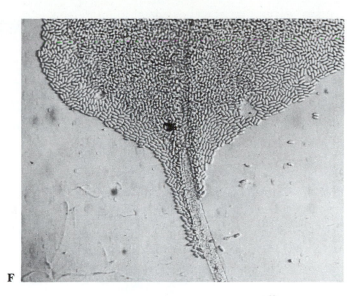

E   F

discovery of structures resembling synaptonemal complexes at a somewhat later stage of development, which is followed by a multinucleate condition. This is a strong indication that meiosis and several mitotic divisions have occurred. Erdos, Raper, and Vogen (1973) discovered that two mating types were necessary for macrocyst formation in *Dictyostelium discoideum* and later (1975) in *Dictyostelium giganteum*. Sexuality and the existence of heterothallism in the Acrasiomycetes are thus well established, but the sexual cycle appears not to be necessary to the organism. The fate of the macrocysts has been followed by Erdos, Nickerson, and Raper (1973), who found that they germinate, releasing myxamoebae, which then repeat the life cycle.

The literature of the Acrasiomycetes consists of hundreds of research articles. It has been ably summarized by Bonner (1967), Raper (1973), and Olive (1975). The best known of the Acrasiales are the Dictyostelidae which have been beautifully monographed by Professor Kenneth B. Raper of the University of Wisconsin (Raper, 1984).

Classification

For an up-to-date classification of the Acrasiomycetes, see Raper (1973) and Olive (1975).

SUBDIVISION PLASMODIOGYMNOMYCOTINA

In contrast to the Acrasiogymnomycotina, which form pseudoplasmodia by aggregation, many of the Plasmodiogymnomycotina form true plasmodia. There is no aggregation stage in this subdivision. The subdivision includes two classes, Protosteliomycetes and Myxomycetes.

CLASS PROTOSTELIOMYCETES

The Protosteliomycetes are cosmopolitan, microscopic organisms invisible to the unaided human eye. They appear on dried leaves, flowers, fruits, and bark when these are wet and placed in moist chambers and may be observed under a stereomicroscope. Their sporophores produce from one to several spores, each of

685

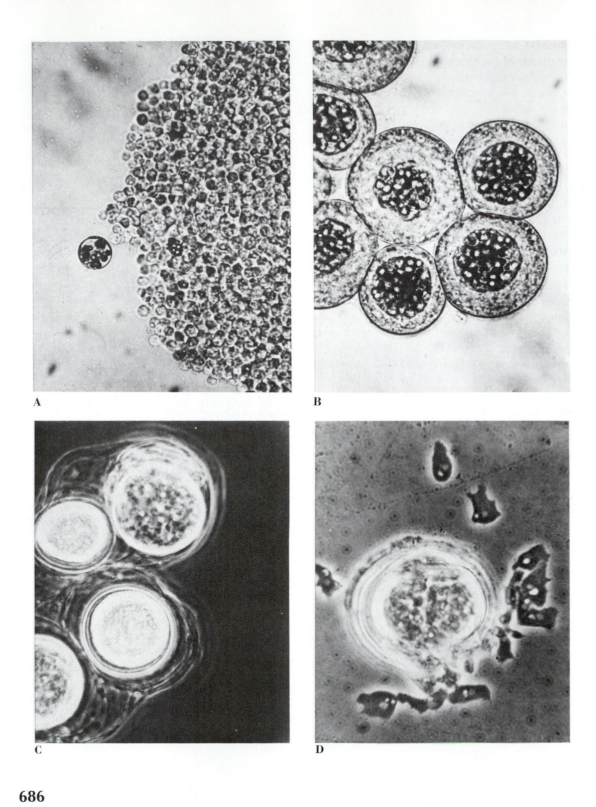

A

B

C

D

FIGURE 29-2

A. *Dictyostelium mucoroides.* Giant cell expressed from the center of a premacrocyst-forming mass of amoebae. **B.** *D. discoideum.* Macrocyst formation; the well-defined central area represents the giant cell; the granules are the engulfed amoebae. **C.** *D. mucoroides.* Macrocyst at top near germination, dormant mature cysts below. **D.** *D. mucoroides.* Germinating macrocyst. **A.** ×175. **B–D.** ×400. (*From Bold, 1977, by permission of Prentice-Hall, Inc.*)

which, upon germination, gives rise to a single myxamoeba. In the simple genus *Protostelium* (Gr. *protos*, first, + *stele*, post), the sporophores produce only one or two spores each. The spore germinates to release a single myxamoeba, which feeds on bacteria and yeasts and eventually, after nuclear and cell division, forms a **sporocarp** (designation of the protostelid sporophore). *Nematostelium* (Gr. *nema*, thread, + *stele*, post) (Fig. 29-3) is also a very simple organism.

The more complex protostelids, such as

Ceratiomyxella (diminutive of *Ceratiomyxa*)[3] (Fig. 29-4), form flagellated cells that become converted into myxamoebae by withdrawing their flagella and by nuclear divisions develop into minute **plasmodia** (sing. **plasmodium**) in which protoplasmic streaming is unidirectional, as contrasted to the "shuttle streaming" of the myxomycete plasmodium (see p. 689). A plasmodium is a multinucleate, creeping mass of protoplasm without cell walls, moving and feeding like a giant amoeba. In *Ceratiomyxella* the plasmodium eventually cleaves into prespore cells, each of which then develops a minute sporocarp with a single spore at its tip. Such protostelids are believed by some to represent extant forms of types thought to be ancestral to the Myxomycetes, but this, of course, is only

[3]*Ceratiomyxa* (Gr. *keraton*, horn, + Gr. *myxa*, slime) is a myxomycete whose life cycle somewhat resembles that of *Ceratiomyxella.*

FIGURE 29-3

Nematostelium ovatum sporogenesis. **A.** Prespore cell. **B.** Hat-shaped stage. **C.** Appearance of steliogen. **D.** Beginning of stalk formation. **E.** Extension of steliogen into stalk tube. **F.** Mature sporocarp. (*From Olive and Stoianovitch [1966], Journal of Protozoology, vol. 13, p. 168, by permission of Society of Protozoologists. Courtesy of Dr. L. S. Olive.*)

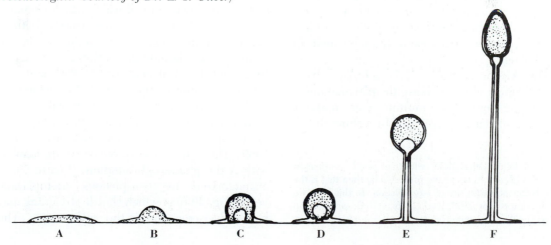

A   B   C   D   E   F

conjecture. No sexual reproduction has been found in any protostelid.[4]

Classification

For a comprehensive classification of the protostelids, see Olive (1975).

CLASS MYXOMYCETES

The true, or plasmodial, slime molds, class Myxomycetes, have been treated in four comprehensive summaries in English: that of Gray and Alexopoulos (1968), a summary of their biology, and three taxonomic treatises, those of Martin and Alexopoulos (1969), Alexopoulos (1973), and Farr (1976).

There are about 450 known species of Myxomycetes. Most are widely distributed from the far north and far south regions to the tropics, but some are strictly tropical or subtropical and a few are strictly temperate-zone forms. Plasmodial slime molds occur mostly in regions of abundant rainfall but have also been isolated from desert soils. Many of the smaller species develop on bark from living trees, or on plant debris, placed in moist chambers. In nature, slime molds occur on various types of organic matter such as moist logs, dead leaves on the forest floor, city lawns, or sometimes living plants. A number of mountain species sporulate below the snow and appear as the snow melts in the spring.

The Myxomycetes are characterized by a phagotrophic somatic phase, the **plasmodium**, a free-living, creeping, multinucleate mass of protoplasm usually enveloped by a slime sheath

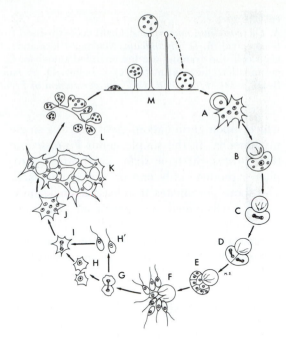

FIGURE 29-4
Life cycle of *Ceratiomyxella tahitiensis*.
A. Germinating spore. **B–F.** Stages in zoocyst development. **G, H, H.'** Proliferation of amoeboid and flagellate cells. **I, J.** Early stages in plasmodial development. **K.** Reticulate plasmodium.
L. Prespore cell formation. **M.** Stages in sporogenesis. (*From Olive [1970], by permission of the Botanical Gazette. Courtesy of Dr. L. S. Olive.*)

and totally devoid of cell walls, which becomes converted into one or more fruiting bodies, which bear the spores. The protoplasm in the veins of a plasmodium exhibits reversible streaming. This is in contrast to the streaming in the plasmodium of the protostelids, which, as we have emphasized, is unidirectional.

Morphologically, there are three types of plasmodia recognized (Alexopoulos, 1960, 1969). The type that most biologists are familiar with is the **phaneroplasmodium** (Figure 29-5), characterized by conspicuous protoplasmic veins (Fig. 29-6) in which shuttle streaming may be easily seen under the microscope. The

---

[4]If we follow Olive (1970), who transferred *Ceratiomyxa* from the Myxomycetes to the protostelids, then that is the only protostelid with sexual reproduction. In this book we follow Alexopoulos (1973), Farr (1976), and Alexopoulos and Mims (1979), who retain *Ceratiomyxa* in the Myxomycetes.

688

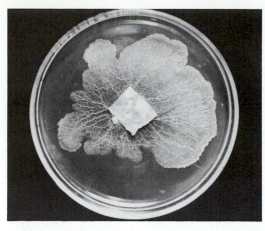

FIGURE 29-5
Phaneroplasmodium of *Physarum polycephalum* on agar in Petri dish. ×½.

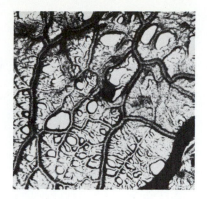

FIGURE 29-6
Physarum polycephalum. Photomicrograph of portion of a plasmodium showing main veins. ×12.5.

phaneroplasmodium is granular and is quite conspicuous even when small. As it grows, it forms large, gelatinous, reticulate protoplasmic sheets that may cover an area as large as several feet square. The phaneroplasmodium is enveloped by a slime sheath, which it leaves behind as a trace as the plasmodium creeps over the surface of the substratum feeding on various microorganisms such as bacteria and fungal spores. The **aphanoplasmodium** is nongranular and very transparent. Its veins are devoid of a slime sheath. As such, it is, as its name indicates (Gr. *aphanes*, invisible), difficult to detect in nature. Both these types usually produce several sporophores from each plasmodium under conditions favorable for sporulation. The third type of plasmodium is the **protoplasmodium,** characteristic of the genus *Echinostelium* (Gr. *echinos*, hedgehog, + Gr. *stele*, post) and other minute Myxomycetes. It is believed to be the most primitive type. A protoplasmodium never grows larger than 1 mm in diameter. It is granular in structure and has a slime sheath much like that of the phaneroplasmodium, but unlike the other two types, it is not differentiat-

ed into veins and its protoplasm streams irregularly. Also, a protoplasmodium produces but a single sporophore when it fruits.

The most commonly encountered plasmodia in nature are phaneroplasmodia. These are of various sizes and colors, ranging from minute to large gelatinous sheets and from white to black through various hues such as cream, yellow, greenish, orange, red, brown, violet, and blue. The plasmodia of most species are either white or yellow, but plasmodia of other colors are not rare. Plasmodia creep over organic matter, such as dead leaves or rotting logs, but may also appear, sometimes in massive quantities, on well-watered lawns and gardens, feeding on bacteria and other microorganisms. Myxomycetes are not parasitic on plants, but sometimes their plasmodia may smother some plants such as strawberries.

Many species of Myxomycetes can be grown easily in laboratory culture on agar, with *Escherichia coli, Enterobacter aerogenes,* or other bacteria that they use for food. Large plasmodia are often developed in such cultures and can be easily propagated by cutting off small portions

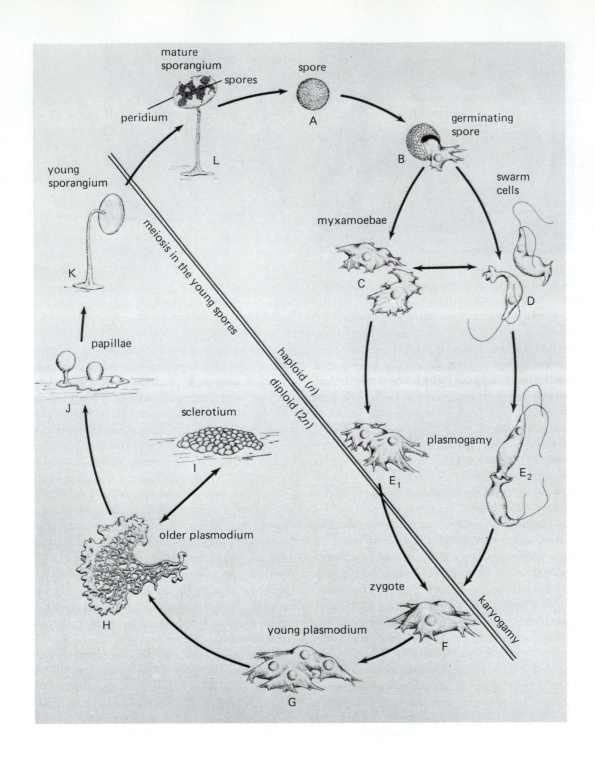

mature
sporangium
spores
spore
peridium
A
germinating
spore
B
young
sporangium
swarm
cells
L
meiosis in the young spores
myxamoebae
K
C
D
haploid (n)
papillae
diploid (2n)
plasmogamy
J
sclerotium
E_1
E_2
I
older plasmodium
zygote
karyogamy
H
young plasmodium
F
G

FIGURE 29-7
Life cycle of a typical, homothallic myxomycete. **A.** Mature spore. **B.** Spore germinating by the split method. **C.** Two myxamoebae derived from the division of the one that emerged from the spore. **D.** Swarm cells derived from the myxamoebae (the two stages—myxamoebal and flagellate—are interconvertible). E_1 and E_2. Plasmogamy by fusion of two myxamoebae (E_1); of two swarm cells (E_2). **F.** Karyogamy and formation of zygote. **G.** Young plasmodium derived from zygote by karyokinesis without cytokinesis. **H.** Older plasmodium. **I.** Sclerotium. **J.** Plasmodium forming papillae in the first stages of sporulation. **K.** Young sporangium before sporal meiosis has occurred. **L.** Postmeiotic sporangium with dehisced peridium. Figures not to scale. (*Drawn from living material by R. W. Scheetz. From Alexopoulos and Mims [1979], by permission of John Wiley & Sons.*)

and transferring them to fresh media with a suitable bacterium, as shown in Fig. 29-5. The rhythmical shuttle streaming in the veins of such plasmodia is an unforgettable demonstration of protoplasmic streaming. Kamiya (1950), investigating the rate of flow in the plasmodium of *Physarum* (Gr. *physa,* bubble), *polycephalum,* the species commonly used in the laboratory, found that the maximum rate of flow in the plasmodial veins reached 1.45 mm per second, which, he stated, is the greatest velocity of protoplasmic flow recorded in any living organism. The motive force in plasmodium is generated by the interaction of ATP with two contractile proteins, actin and plasmodium-myosin A, in the plasmodial protoplasm. Fibrils, which occur in the plasmodium, probably consist of these proteins, and it appears virtually certain that they are instrumental in generating the motive force required for protoplasmic streaming (Wohlfarth-Botterman, 1964). It has also been suggested that there is a primitive neuromotor system in the plasmodium that controls streaming.

All pigmented and some white plasmodia require light in order to sporulate. Under the most favorable environmental conditions, the entire plasmodium is converted into a mass of sporophores, which may be stalked, sessile, or a mixture of both, depending on the species. Under unfavorable conditions of moisture and nutrition, the plasmodium, instead of sporulating, changes into a hard, dormant body, the **sclerotium** (pl. **sclerotia**; Gr. *skleros,* hard) (Fig.

29-7I), composed of uninucleate or multinucleate round **spherules,** each enveloped by a cell wall. Upon the return of a favorable environment the sclerotium again becomes plasmodial (Fig. 29-7H).

When the plasmodium of *Physarum polycephalum* and other similar species reaches a certain stage of maturity, and when the food supply is nearly exhausted, under environmental conditions favorable for fruiting, it becomes concentrated at various points, forming small, papillalike mounds, which soon develop into sporophores (Fig. 29-7J, K), often objects of great and delicate beauty. This has been appreciated by artistically inclined biologists, who have published beautiful paintings or color photographs of slime molds. Among the best of these are the paintings in Crowder's article in the *National Geographic Magazine* (April 1926), Lister (1925), Hattori (1935, 1964), Martin and Alexopoulos (1969), Emoto (1977), and the color photographs by Alexopoulos (1973).

Sporophores and spores. There are four main types of myxomycete sporophores: **sporangia, plasmodiocarps, aethalia,** and **pseudoaethalia. Sporangia** are simple fruiting bodies, which may be stalked (Fig. 29-8) or sessile (Fig. 29-9). They vary in size from minute (less than 0.5 mm in diameter when sessile, as in *Licea tenera*) to as much as 25 mm tall, as in *Stemonitis splendens*. Their stalks may be hollow or stuffed. Many are covered with lime ($CaCO_3$), which may be powdery or scaly or which may form a hard crust over the **peridium** (sporo-

FIGURE 29-8
Stipitate sporangia of *Physarum pusillum*. Note heavy coat of lime over the entire sporangium. ×22.

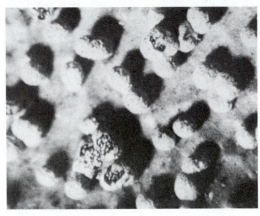

FIGURE 29-9
Sessile sporangia of *Physarum cinereum*. The peridium of several sporangia is broken, revealing the lime nodes of the capillitium. ×22.

FIGURE 29-10
Sporangium of *Dictydium cancellatum*, with spores removed to show the basketlike peridial net. The order Liceales, to which *Dictydium* belongs, has no capillitium, but the peridial net of this species could easily be confused with capillitial threads. ×22. (*Scanning electron micrograph by R. W. Scheetz.*)

phore wall around the spores). In a few species, the peridium is composed of a number of ribs over which a fugacious membrane is spread. When the membrane disappears, as in *Dictydium cancellatum*, the ribs remain and can easily be mistaken for capillitium (see below) by the inexperienced observer (Fig. 29-10). *Physarum polycephalum* is an important slime mold, extensively used for research in molecular biology. This organism, as its name indicates (Gr. *poly*, many, + *kephale*, head), produces its spores in many heads on a single stalk. Its sporangium is, therefore, said to be lobed (Fig. 29-11). **Plasmodiocarps** are typically sessile (Fig. 29-12). They are irregular in shape, varying from ovoid to elongated and even branched or netlike, as in *Hemitrichia serpula* (Gr. *hemi*, half + *Trichia*). They are formed by the secretion of a peridium around the veins or parts of the veins of a plasmodium just after it stops streaming and prepares to sporulate. **Aethalia** (sing. **aethalium**) are relatively massive sporophores (Fig. 29-13) thought to have been

692

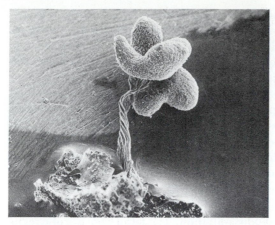

FIGURE 29-11
Physarum polycephalum (Physaraceae, Physarales). A sporophore with a single stalk carrying several sporangial lobes. ×22. (*Scanning electron micrograph by R. W. Scheetz.*)

FIGURE 29-13
Aethalium of *Lycogala conicum.* ×14. (*Scanning electron micrograph by R. W. Scheetz.*)

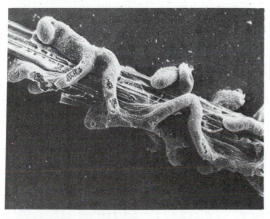

FIGURE 29-12
Plasmodiocarps of *Physarum bivalve.* The sporophores are covered with a crust of lime (CaCO₃). ×14. (*Scanning electron micrograph by R. W. Scheetz.*)

FIGURE 29-14
Pseudoaethalium of *Tubifera microsperma.* Note single, stalklike, spongy hypothallus. ×5. (*Scanning electron micrograph by R. W. Scheetz.*)

formed by the complete fusion of large numbers of sporangia during their evolutionary development. If this is true, fusion is, indeed, complete, for no traces of individual sporangial walls are found in most aethalia, as those of the genus *Lycogala* (Gr. *lycos*, wolf, + Gr. *gala*, milk), for example. **Pseudoaethalia** (sing. **pseudoaethalium**) (false aethalia), on the other hand, are tight aggregations of sporangia, often on a single stalk or spongy hypothallus,[5] which resemble single spore-bearing units. The individual sporangia in a pseudoaethalium are, however, easily discernible (Fig. 29-14).

When the sporophore of a myxomycete has

---

[5]A hypothallus is a horny, spongy, membranous, or calcareous base on which the sporophores are usually seated.

693

tured and colored, depending on the species, which aid in the dissemination of the spores by absorbing water from a moist atmosphere or releasing it to a dry atmosphere. This causes the capillitial threads to expand or contract and agitate the spores that cling to them. In some species, the capillitium is very elastic and expands greatly, bringing the spores with it as the peridium breaks (Fig. 29-15). The appearance of the capillitium varies greatly in different species, and capillitial structure and ornamentation are important taxonomic characters. Figure 29-16 illustrates several types of capillitium and **pseudocapillitium.** The latter consists of irregular threads or of plates. At the time of capillitial

attained its maximum size, its multinucleate protoplast begins, in most species, to form a network of vacuoles, in which various substances are deposited to form the **capillitium.** This is a mass of nonliving threads, variously sculp-

FIGURE 29-16
Several types of capillitium. **A–F.** True capillitium. **A.** Limeless surface net (*Stemonitis,* Stemonitales). **B.** Lime nodes connected by slender, tubular filaments (*Physarum,* Physarales). **C.** Spinulose network of tubular threads (*Arcyria,* Trichiales). **D.** Calcareous tubules (*Badhamia,* Physarales). **E.** Single, spiny elater (*Trichia,* Trichiales). **F.** Slender, branching, and anastomosing capillitial threads (*Didymium,* Physarales). **G–H.** Pseudocapillitium. **G.** Threadlike (*Lycogala,* Liceales). **H.** Platelike, perforated (*Reticularia,* Liceales). (*Drawings by R. W. Scheetz.*)

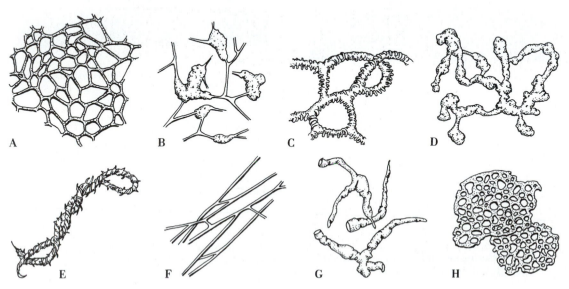

formation, the nuclei of the sporophore protoplasm undergo a single mitotic division, and the protoplast, by means of furrowing, cleaves progressively into uninucleate diploid segments, which become enveloped by cell walls and separate into spores. Myxomycete spores are typically globose but may be ovoid in a few species. They vary in size from 4 to 20 μm or more in diameter. Their walls are variously ornamented, ranging from smooth through spiny, verrucose (warty), and reticulate (bearing a network of ridges) (Fig. 29-17). In color they

FIGURE 29-17
Spores of four species of Myxomycetes all at the same magnification, showing different types of wall ornamentation. **A.** Verrucose (warty) with large and small warts, the large ones clustered. **B.** Uniformly verrucose. **C.** Spinulose. **D.** Reticulate. ×4500. (*Scanning electron micrographs by R. W. Scheetz.*)

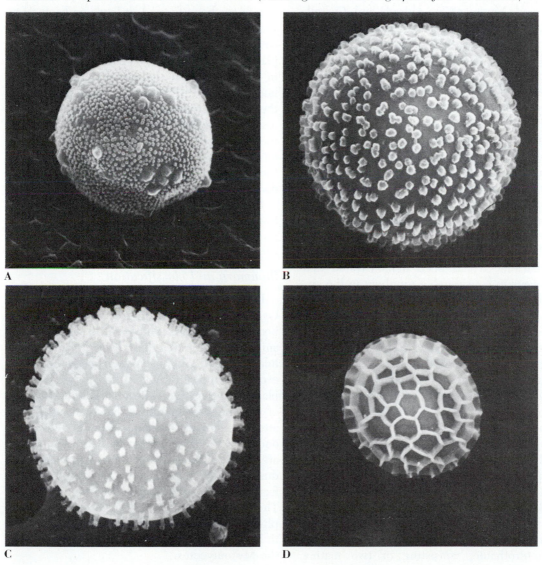

A   B

C   D

range from nearly colorless to yellow, reddish, violet, purple, brown, gray, or nearly black. The chemical composition of the spore wall has not been determined, except in *Physarum polycephalum*, in which it is composed of 81% galactosamine polymer, 1.4% phosphate, 2.1% amino acids, and 15.44% melanin (McCormick, Blomquist, and Rusch, 1970).

Soon after spore formation, meiosis takes place in the young spores (Aldrich, 1967; Aldrich and Mims, 1970; Aldrich and Carroll, 1971). There is evidence that three of the four resultant nuclei disintegrate (Aldrich, 1967), so that the mature spore is usually uninucleate and haploid.

Life-cycle pattern. When the spores are mature, they are released from the sporophores and are wind-borne. They eventually settle, and in the presence of water under favorable temperatures (around 20–25°C), they germinate (Figure 29-7B), releasing usually one but up to four or more myxamoebae (Fig. 29-7C) or flagellate, comma-shaped, swarm cells (Fig. 29-7D). Myxomycete spores are very resistant to desiccation and remain alive for long periods. Some have been germinated after 76 years' storage in a herbarium. It has been demonstrated that they occur in considerable numbers in the air, and those of *Fuligo* (L. *fuligo*, soot) *septica* at least, cause allergic reactions in susceptible individuals.

Myxamoebae feed on bacteria and yeasts, undergo mitosis, and divide (Fig. 29-7C). Thus a large population of myxamoebae results. When a critical mass has been reached, compatible myxamoebae fuse in pairs (Fig. 29-7E$_1$), producing zygotes (Fig. 29-7F). Swarmers, equipped with usually two anterior, whiplash flagella, may also behave as gametes, fusing by their posterior ends (Fig. 29-7E$_2$) to form flagellate zygotes, which eventually withdraw their flagella. Sexual reproduction in the Myxomycetes takes place, therefore, by the fusion of two myxamoebae or two swarm cells. Some species are heterothallic, consisting of two mating types, $(+)$ and $(-)$, and only gametes of opposite mating type are able to fuse. In other species, all gametes appear to be compatible. There is also evidence that in some strains the life cycle is completed without sexual fusions. Thus, in *Didymium* (Gr. *didymos*, double, twin) *iridis*, some strains are sexual, exhibiting mating types, some are sexual without mating types (homothallic), and some are asexual (apogamic) (Therrien, Bell, and Collins, 1977). After the zygote has been formed, it feeds on bacteria or yeasts, creeps, and grows as successive, synchronous mitoses occur, thus developing into a plasmodium (Fig. 29-7G). The nuclear divisions in the zygote and the plasmodium are intranuclear (closed), the spindle developing within the nuclear envelope, which breaks down only near the conclusion of nuclear division to form the envelopes of the daughter nuclei. Such intranuclear divisions are commonly found in the other divisions of fungi. In contrast, nuclear divisions in the myxamoebae are centric (open), as they are in the majority of plants and animals. Swarm cells do not divide as such; they change into myxamoebae before they multiply.

Summary and classification

In conclusion it may be stated that the Gymnomycota are of interest to the biologist because of the combination of fungus- and animallike attributes they display. The noncellular plasmodia of the Myxomycetes afford an excellent opportunity for the study of protoplasm, whereas the delicate beauty and design of the minute sporangia have attracted the attention of many students of morphology, taxonomy, and differentiation. The Acrasiomycetes are of special interest for their morphogenesis, during which independent cells aggregate to form a multicellular spore-bearing structure. The Protosteliomycetes are of particular interest phylogenetically for they have been postulated to be ancestral both to the Acrasiomycetes and the Myxomycetes.

As noted above, the slime molds are classified by many biologists in the kingdom Protista or the phylum Protozoa of the animal kingdom. Because mycologists study them almost exclusively, we have included them in the kingdom Myceteae but without claiming relationships to the fungi. The organisms mentioned or illustrated in this chapter may be classified as follows:

Division 1. Gymnomycota
  Subdivision 1. Acrasiogymnomycotina
    Class 1. Acrasiomycetes
      Order 1. Acrasiales
        Family 1. Guttulinaceae
          Genus: *Pocheina*
      Order 2. Dictyosteliales
        Family 1. Dictyosteliaceae
          Genera: *Dictyostelium, Polysphon-dylium*
        Family 2. Acytosteliaceae
          Genus: *Acytostelium*
  Subdivision 2. Plasmodiogymnomycotina
    Class 1. Protosteliomycetes
      Order 1. Protosteliales
        Family 1. Protosteliaceae
          Genera: *Protostelium, Nemato-stelium*
        Family 2. Cavosteliaceae
          Genus: *Ceratiomyxella*
    Class 2. Myxomycetes
      Subclass 1. Ceratiomyxomycetidae
      Order 1. Ceratiomyxales
        Family 1. Ceratiomyxaceae
          Genus: *Ceratiomyxa*
      Subclass 2. Myxogastromycetidae
      Order 1. Liceales
        Family 1. Liceaceae
          Genus: *Licea*
        Family 2. Reticulariaceae
          Genera: *Tubifera, Lycogala*
        Family 3. Cribrariaceae
          Genus: *Dictydium*
      Order 2. Trichiales
        Family 1. Trichiaceae
          Genera: *Arcyria, Hemitrichia*

      Order 3. Echinosteliales
        Family 1. Echinosteliaceae
          Genus: *Echinostelium*
      Order 4. Physarales
        Family 1. Physaraceae
          Genera: *Physarum, Fuligo*
        Family 2. Didymiaceae
          Genus: *Didymium*
      Subclass 3. Stemonitomycetidae
      Order 1. Stemonitales
        Family 1. Stemonitaceae
          Genus: *Stemonitis*

For modern, detailed classifications of Myxomycetes, see Martin and Alexopoulos (1969), Alexopoulos (1973), and Farr (1976).

DISCUSSION QUESTIONS

1. In what respects do Gymnomycota differ from other fungi?
2. Name the classes of organisms that constitute the Gymnomycota.
3. Distinguish, by listing the characteristics of each, among the three classes of Gymnomycota.
4. Where would you seek representatives of the three classes of Gymnomycota in nature?
5. How do the plasmodium and sporocarps of Myxomycetes differ in their tactic responses?
6. How does the nutrition of Gymnomycota differ from that of bacteria and fungi?
7. With the aid of drawings summarize the life cycle of *Dictyostelium discoideum* and *Polysphondylium violaceum.*
8. Define or explain: sporocarp, microcyst, acrasin, slug or grex, pseudoplasmodium, plasmodium, phaneroplasmodium, aphanoplasmodium, protoplasmodium, sclerotium, spherule, myxamoeba, plasmodiocarp, aethalium, pseudoaethalium, peridium, capillitium, pseudocapillitium, closed and open nuclear division.

9. With the aid of drawings, describe the life cycle of a typical Myxomycete.

10. When and where does meiosis presumably occur in Acrasiomycetes? in Myxomycetes?

11. If you were cultivating Myxomycetes in the laboratory, in what forms could you supply the essential elements?

12. If you were observing a plasmodium of one of the protostelids and one of the Myxomycetes under the microscope, how would you determine which is which?

13. About how many species of Myxomycetes have been discovered?

14. Name a species of Myxomycetes that has been used extensively for research in molecular biology.

15. What is the function, if any, of the capillitium? Explain.

16. How do homothallic Myxomycetes differ from heterothallic ones?

17. Does mitosis in Myxomycetes differ in any significant way from that in plants? If so, how?

Chapter 30
FUNGI WITH ABSORPTIVE NUTRITION

In contrast to the Gymnomycota, other members of the Kingdom Myceteae, except for a few species, have definite cell walls. This is not to say that many species do not produce unwalled cells such as zoöspores or **planogametes** (flagellate gametes), but the somatic structures even of these have cell walls.

SOMATIC STRUCTURES

Some fungi[1] with absorptive nutrition are unicellular, but the majority have a differentiated thallus consisting of threadlike, tubular filaments, the **hyphae** (sing. **hypha**). The network of hyphae constituting the body (thallus, soma) of a fungus is called the **mycelium.**

In the simpler fungi, the hyphae are **coenocytic**—i.e., they are long, tubular filaments filled or lined with cytoplasm in which many nuclei are embedded and are not separated into cells or compartments (Fig. 30-1A). It was once believed that such fungi had originated from algal ancestors, and they were therefore given the name **Phycomycetes** (Gr. *phykos,* seaweed, i.e., alga, + *myketes,* fungi). Although this theory of their origin is no longer accepted, the term Phycomycetes is still used in a general way for them but without taxonomic significance.

In the more complex groups, the hyphae are **septate**—i.e., divided into compartments or cells by cross walls we call **septa** (sing. **septum**) (Fig. 30-1B). In some of these, the septa are solid, whereas in others they are perforated and, in fact, may be very complex in structure. The cells or compartments may be uninucleate, binucleate, or multinucleate, depending on the species. Perforated septa permit cytoplasmic strands to pass through them so that there is an organic connection of all parts of the mycelium. Such septa also permit nuclei to travel through the mycelium from one cell to the other. Moving pictures have been taken of migrating nuclei, and Dowding (1958) has calculated that in *Gelasinospora* (Gr. *gelasinos,* dimple) *tetrasperma,* one of the Ascomycetes, streaming nuclei reach speeds of 40 mm per hour. In the more complex fungi it is believed that septa may have the power to regulate the

---

[1]From here on, the term "fungi" will be understood to refer to all but the slime molds.

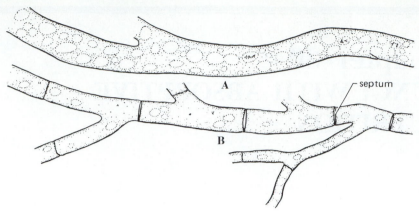

FIGURE 30-1
Somatic hyphae. **A.** Portion of a coenocytic (aseptate) hypha. **B.** Portion of a septate hypha. ×500.
(*From Alexopoulos [1962], by permission of John Wiley & Sons.*)

passage of nuclei or other organelles through them, but this has not been proved. The presence of plasmodesmata through the septa of fungi has been adequately demonstrated.

The cell walls of most fungi contain chitin and other complex carbohydrates but little or no cellulose except in one group, the Oömycetes, which appear to have had a different origin from all the other fungi.

NUCLEAR DIVISION

Nuclear division in the fungi has been studied for a very long time, and the type of division that occurs in this group of organisms was the subject of great controversy until the advent of the electron microscope. Now it appears that mitosis in most fungi is typically intranuclear—i.e., the nuclear envelope remains intact through mitosis until late anaphase or telophase, when it breaks down and forms the nuclear envelopes of the daughter nuclei. Associated with the mitotic process in fungi are **centrioles** in the aquatic fungi with motile cells (Mastigomycota) or small electron-opaque structures called **"spindle pole bodies"** (SPBs) in most other fungi. Otherwise, it appears that

mitosis in most fungi is rather typical (Lu, 1974). There are variations in different groups, to be sure. Two fine reviews of the whole subject of fungal nuclear division are by Heath (1974) and Fuller (1976). Meiotic divisions are also quite typical but also intranuclear. The position of meiosis in the life cycle of the fungi has been pinpointed through the search for synaptonemal complexes and by measuring the amount of DNA in various stages of the life cycle by spectrophotometric analysis, inasmuch as chromosomes are too small to count accurately with the light microscope and do not show up clearly under the electron microscope.

REPRODUCTION

The chief method of fungal propagation is by means of spores, which may be motile (flagellate) or nonmotile and which are produced in various ways. Spores that are formed asexually are sometimes designated as **mitospores.** If they are formed as a result of karyogamy and meiosis, they are **meiospores.** Most fungi reproduce both asexually and sexually. Asexual reproduction may take the form of spores, fission, budding, or fragmentation. Fission and budding are

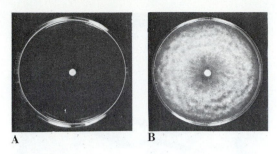

FIGURE 30-2
Mycelial fragmentation as a method of propagating fungi in the laboratory. **A.** Agar disc containing mycelium. **B.** Same culture after 7 days. × ⅓.

usually employed by unicellular fungi, such as the yeasts (see Chapter 33). Fragmentation occurs in mycelial fungi when the hyphae break into fragments accidentally in nature or are fragmented purposely in the laboratory, each fragment resuming growth and establishing a new colony. This is a common method of propagating fungi in the laboratory (Fig. 30-2). Sexual reproduction takes many forms in the fungi and it is better to discuss this topic in connection with various fungal groups. Suffice it to say here that fungi may be homothallic or heterothallic. In the former all gametes are compatible; in the latter, gametes of different mating types are produced, which must meet before plasmogamy or karyogamy can occur. In heterothallic fungi two thalli must be present for sexual reproduction.

PHYSIOLOGY

Fungi obtain their food by secreting various enzymes outside their thalli and digesting the substratum on or in which they live so that nutriments in solution may pass through their cell walls and plasma membranes into their cells. As we have already pointed out, fungi must have elaborated carbohydrate molecules inasmuch as they are unable to photosynthesize them. The various mineral elements, such as nitrogen, phosphorus, potassium, sulfur, iron, magnesium, manganese, and boron, are also essential to fungus nutrition. These, too, are obtained from the substratum in the form of salts in solution. Various fungi differ in their ability to utilize certain sugars and salts. Thus, most species thrive on a substratum that contains glucose as a sugar, ammonium salts to provide nitrogen, and KH_2PO_4 for potassium and phosphorus. Others, however, may be unable to utilize ammonium salts, preferring some other source of nitrogen, such as amino acids or perhaps nitrates.

Most fungi seem to grow optimally in the laboratory at 20–25°C but can withstand considerably higher and very much lower temperatures. Many fungi will grow—slowly, to be sure—at 6°C, as evidenced by the growth of molds in household refrigerators. The fact that fungal cultures are often stored under liquid nitrogen at temperatures of −196°C shows how resistant some of them are to cold. Of course at temperatures below freezing little or no growth takes place, the fungal structures remaining dormant. At the other end of the scale, there are some thermophilic fungi that will not grow when the temperature falls below 20°C and actually prefer temperatures in the upper 30s or in the 40s. The role that these fungi play in composting and their economic importance in causing spontaneous combustion—which, measured in terms of U.S. dollars, exceeds $20 million a year in the United States alone—is discussed by Cooney and Emerson (1964), as is their importance in industrial processes, such as the fermentation of cacao, the sweating of tobacco, the composting of mushroom beds, and the disposal of refuse and sewage in large cities.

Fungi prefer an acid medium in which to grow, most of them thriving best at pH 6.

Light is not required for the growth of fungi, but a few, such as *Blastocladiella* (dimin. of *Blastocladia*) *emersonii* (Chytridiomycetes, Blastocladiales), actually fix CO_2 in the presence of light. Many species, too, require light for sporulation, but the stimulus does not appear to be transmitted from one portion of the mycelium to another (Koehn, 1971). Many fungal sporophores are phototropic, bending and dispersing their spores toward the light. For a detailed discussion of the physiology of fungi, see Hawker (1950, 1957, 1966) and Cochrane (1958) and the references therein.

Growth of fungal hyphae is terminal, being confined at the very tip of each hyphal branch. Fungi, however, may be propagated by transferring a small section of the mycelium to new media. The cut areas regenerate new hyphal tips, which grow normally thereafter.

FOSSIL FUNGI

Although many fossil fungi have been found and more are being constantly discovered, we still do not know when the fungi originated or how they evolved in succeeding millennia. Fungal hyphae have been reported in Precambrian strata (along with Cyanochloronta and green algae), 900 million years old (Schopf, 1970) (Fig. 30-3A), but the supposed fossil fungi of the Precambrian cannot be unquestionably identified as such (Schopf, Ford, and Breed, 1973).

By the end of the Paleozoic, all the major groups of fungi are represented in the fossil record. In the stems and associated remains in an ancient bog at Rhynie, Scotland, which is of Lower Devonian age (inside front cover), very well-preserved fossil fungi (Fig. 30-3B) are present. Mycorrhizal fungi were functioning in the superficial root cells of certain Carboniferous gymnosperms (Fig. 30-4). Andrews and Lenz (1943) illustrate a fungus (Fig. 30-5) within the cells of a Carboniferous (inside front cover) rhizome. More recent fossil fungi from the Eocene appear to be closely related to modern genera, according to Dilcher (1965). Much work is being done, at present, on fossil fungal spores (Elsik, 1976). The following references summa-

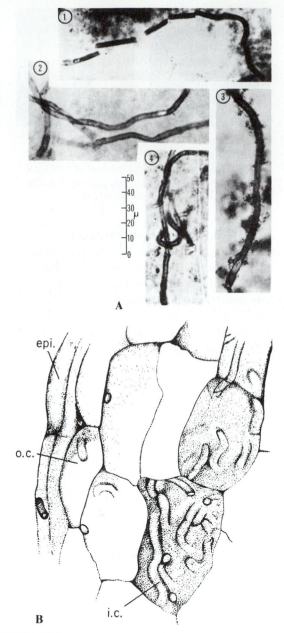

FIGURE 30-3
A. Precambrian fungi, about 1 billion years old; *Eomycetopsis filiformis* (1) and *E. robusta* (2–4). **B.** Fungus in rootlet of *Amyelon radicans*, Upper Carboniferous. epi., epidermis; i.c., inner cortex; o.c., outer cortex. ×200. (**A.** *After Schopf.* **B.** *After Osborn.*)

702

FIGURE 30-4
Fossilized fungus in cell of a rhizome of a
Carboniferous plant. (*After Andrews and Lentz.*)

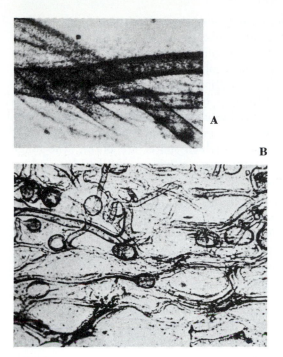

FIGURE 30-5
Devonian fungi. **A.** Mycelium in decayed stem of
Asteroxylon. **B.** *Palaeomyces asteroxyli.* Hyphae
and vesicles in the cortex of *Asteroxylon.* **A.** ×250.
B. ×125. (*After Kidston and Lang.*)

rize our present knowledge of fossil fungi admir-
ably: Dilcher, 1965; Tiffney and Barghoorn,
1974; Elsik, 1976, 1977*a,b*; Pirozynski, 1976.

Summary and classification

In the previous chapter we discussed a group
of organisms with phagotrophic nutrition and
without cell walls in their somatic structures. As
we have already said, the slime molds are
regarded by many biologists as having their
affinities with the Protozoa rather than with the
fungi, but they have been studied almost exclusively by mycologists.

In this chapter we introduced the study of
what might be called the true fungi, all of which
have absorptive nutrition and almost all of
which have cell walls. Whether all these organisms, however, constitute a monophyletic series
is highly controversial, as we shall see in the
following chapters. In general, they are all

considered to be fungi by most biologists, although there is considerable disagreement
among phylogenists as to their origin and proper
classification.

The points emphasized in this chapter are the
presence of cell walls in and the hyphal structure of the soma; nutrition by absorption
through the cell wall of simple organic molecules derived by the extracellular digestion of
more complex carbohydrates and proteins in
living or nonliving substrata; and propagation by
spores. The presence of both asexual and sexual
reproduction in most fungi has also been
brought out.

As stated in the discussion of fossil algae, the
fossil record of fungi, in part because of its
fragmentary nature, has shed little light on
fungal evolution. Most ancient fungi seem to

703

have been very similar to extant forms, so that the fungi, like the algae, are a long-lived race and have been relatively little modified since they first evolved.

Classification of the fungi with absorptive nutrition is based on the structure of their hyphae, the chemical composition of their cell walls, the type of sexual reproduction, and the manner in which spores are borne. The fungi with absorptive nutrition may be classified as follows:

Division 1. Mastigomycota
Division 2. Amastigomycota
    Subdivision 1. Zygomycotina
    Subdivision 2. Ascomycotina
    Subdivision 3. Basidiomycotina
    Subdivision 4. Deuteromycotina

In the chapters that follow, we shall discuss each of these groups briefly. For a more complete discussion, read Alexopoulos and Mims (1979).

DISCUSSION QUESTIONS

1. Define or explain: planogamete, hypha, mycelium, coenocytic, septum (septa), mitospore, meiospore.

2. How does the nutrition of other fungi differ from that of the Gymnomycota?

3. What conditions of nutrition and environment favor the growth of most fungi in the laboratory?

4. Briefly summarize what is known about fossil fungi.

5. What is meant by "absorptive nutrition"?

6. Why were the so-called "lower fungi" called Phycomycetes in former years, and why is the term no longer used in a taxonomic sense?

7. What is the distinguishing chemical component of the cell walls of most fungi? How does the fungal cell wall differ in chemical composition from that of plants?

8. How does mitosis in most fungi differ from that in plants?

9. Since fungal chromosomes are very minute and difficult to count accurately, what other criterion is used to determine the occurrence of meiosis in fungi?

10. How are fungi commonly propagated in the laboratory?

11. Briefly discuss the role of light in the growth and reproduction of fungi.

12. Where does growth take place in a hypha?

Chapter 31
DIVISION MASTIGOMYCOTA: THE FLAGELLATE FUNGI

The term Mastigomycota is derived from the Greek word *mastix (mastigos)*,[1] which means "whip," combined with the Greek word for fungus, *mykes (myketos)*. The group, therefore, contains all the fungi (exclusive of the slime molds) that produce flagellate cells in their life cycles, be these zoospores or planogametes. There are two types of flagella produced by fungi: the whiplash and the tinsel (Fig. 4-1, p. 43). The whiplash flagellum is a long, relatively rigid filament with a flexible whip at the tip, which might be of considerable length also. The tinsel flagellum is long, bearing many filamentous extensions, the **mastigonemes** (Gr. *mastix*, *mastigos*, whip + *nema*, thread, skein). The presence of one or the other or both of these flagellar types and their position on the motile cells (zoospores or planogametes) is the basis for the classification of the Mastigomycota. One other important characteristic of this division is that **centrioles** are functional during cell division and in the formation of the flagella, for which they serve as basal bodies. The Mastigomycota produce their spores in **sporangia** (sing. **sporangium**). These are saclike structures, the entire protoplast of which is cleaved into spores termed **sporangiospores**. Sporangiospores in this division are almost always motile— i.e., they are zoospores. Sporangia release their spores through apical papillae or opercula, by bursting, or by deliquescing. Fungi that produce sporangia and have a coenocytic mycelium are often grouped under the general term Phycomycetes. Because they are regarded as primitive by most mycologists, they are also termed the "lower fungi." These organisms are discussed in interesting detail in a recent book edited by M. L. Fuller (1978), *The Lower Fungi in the Laboratory*.

The division Mastigomycota includes two subdivisions: the Haplomastigomycotina and the Diplomastigomycotina.

SUBDIVISION HAPLOMASTIGOMYCOTINA

In the subdivision Haplomastigomycotina meiosis is either zygotic or meiosporangial (sporic)—i.e., it takes place either during the germination of the zygote or in special types of

---

[1]According to the rules of Greek grammar, combining forms are derived from the genitive of a noun. When the genitive form differs sufficiently from the nominative to warrant it, the genitive form is placed in parentheses after the nominative.

diploid sporangia, which, as a result, produce meiospores (haploid spores). There are two types of life cycle in this subdivision, monobiontic-haploid (M, h), or dibiontic (D, h + d). The subdivision contains three classes of fungi—Chytridiomycetes, Hyphochytridiomycetes, and Plasmodiophoromycetes.

CLASS CHYTRIDIOMYCETES

The motile cells of these fungi bear a single whiplash flagellum inserted posteriorly on the motile cells; the cell walls of those that have been examined in this regard contain chitin and glucans but no cellulose. Sexual reproduction, when it has been found, takes place usually by the fusion of two planogametes or by the fusion of hyphalike, delicate filaments, the rhizoids (some other methods of plasmogamy have also been reported in a few special species), and results in the formation of a thick-walled resting body. Meiosis, where it has been discovered, is either zygotic, or, in the case of a dibiontic life cycle, takes place in special types of sporangia (meiosporangia), which release meiospores to start the haploid generation.

Classification

The class Chytridiomycetes is usually divided into five orders:

Class Chytridiomycetes
  Order 1. Chytridiales
  Order 2. Blastocladiales
  Order 3. Monoblepharidales
  Order 4. Harpochytridiales
  Order 5. Spizellomycetales
We shall discuss only two of these.

Order Chytridiales

The Chytridiales includes aquatic—freshwater or marine—fungi, commonly referred to as "chytrids." These may be **holocarpic,** converting their entire thallus into a reproductive organ, or **eucarpic,** with a portion of the thallus remaining trophic and supporting one or more reproductive organs.

Few chytrids are of great direct economic importance, but *Synchytrium endoboticum* does cause a serious disease of potatoes (potato wart) in all potato-growing regions of the world, and a disease of watercress, an important crop in Great Britain and other regions. Members of the genera *Olpidium* and *Physoderma* also cause diseases of economic plants.

Many chytrids are indirectly injurious to humans by parasitizing and destroying algae that form an important link in the food chain of aquatic animals. The variety of fungi included in this order is so great that it is impossible to select any one organism to typify the order. We shall, therefore, discuss several to illustrate the variation that exists.

Chytriomyces

The life cycle of *Chytriomyces* (Gr. *chytra,* jug) *hyalinus* has been thoroughly studied by its discoverer, Karling (1945), and subsequently by Miller and his coworkers (Miller 1967, 1977; Moore and Miller, 1973). The posteriorly uniflagellate zoospore of this organism swims about for some time and eventually settles on some dead plant material or on exuviae of water insects. There it germinates, producing a system of rhizoids. Rhizoids are not true hyphae but very delicate hyphalike filaments with protoplasm but without nuclei. Nuclei, however, travel through rhizoids to different parts of the thallus. As the thallus develops, the main body of the zoospore becomes converted into a prosporangium in which nuclear divisions and protoplasmic synthesis occur, converting it into a sporangium. When a large number of nuclei have been formed, the multinucleate protoplasmic mass cleaves into posteriorly uniflagellate zoospores and is released into the sporangium through a hinged, caplike structure, the operculum. The zoospores now separate and swim away, completing the asexual cycle of the organism.

Some of the zoospores, however, develop into sexual thalli. They germinate, producing a rhizoidal system, but they themselves enlarge into a thin-walled cell in which nuclear divisions take place. When the rhizoids of two compatible thalli come in contact, plasmogamy takes place, the protoplasm of each contributing thallus streaming toward the point of fusion, where a resting, thick-walled body is formed, (Fig. 31-1) receiving the protoplasts of the two contributing thalli, leaving them empty. Karyogamy now occurs, converting the resting body into a zygote. Eventually, the resting body functions as a sporangium. Several nuclear divisions take place, two of which are presumed, but have not been proved, to be meiotic; zoospores are eventually delimited and the resting sporangium releases them into the water. Note that the two contributing (sexual) thalli can in no way be morphologically distinguished as male and female and that neither receives the protoplasm of the other. Instead, they both contribute to a new structure, which becomes the resting sporangium. In the next example to be discussed, *Polyphagus*, the beginnings of sexual differentiation in the chytrids may be noted.

Polyphagus

Polyphagus (Gr. *poly*, many, + Gr. *phagein*, to eat) is a chytrid genus with 10 or 11 species (Karling, 1977). *Polyphagus euglenae*, which attacks the green alga *Euglena*, is the best known of these through the work of Wager (1913). Its life cycle is illustrated in Fig. 31-2. The zoospore swims in the water together with the *Euglena* cells and eventually comes to rest, becomes rounded, and produces a number of delicate rhizoidal filaments (Fig. 31-2A–C). These become attached to a *Euglena* cell and penetrate it. The zoospore outside the cell develops into a sporangium, which becomes multinucleate by successive mitoses of its nucle-

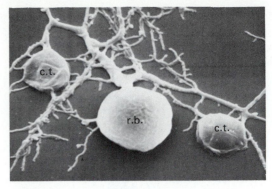

FIGURE 31-1
Development of a resting body (r.b.,) of *Chytriomyces hyalinus*. Two empty contributing thalli (c.t.) are shown attached to the rhizoidal system. ×5400. (*Courtesy of Dr. C. E. Miller.*)

us. Its protoplasm now cleaves into zoospores, which are liberated in water (Fig. 31-2D–F).

In sexual reproduction, some zoospores develop into small (male) thalli and others into larger (female) thalli (Fig. 31-2G–H). The small thallus produces a long, rhizoidlike copulation tube, which becomes attached to one of the larger thalli. Just below the point of attachment a swelling develops, into which a nucleus from the male thallus enters, traveling through the copulation tube (Fig. 31-2I–J). The nucleus from the female thallus also enters the swelling, which eventually secretes a thick wall around itself and is termed a "zygote" even though no karyogamy has occurred as yet (Figure 31-2K). The "zygote" eventually produces a prosporangium, into which the two nuclei pass and there fuse (Fig. 31-2L–M). The resulting diploid nucleus undergoes several divisions, the first two of which are presumed, but have not been proved, to be meiotic, and the prosporangium then develops into a sporangium in which the zoospores are delimited and from which they are released (Fig. 31-2N–P). Here, then, seems to be the beginning of sexual differentiation, although it is represented only as a difference in size between the two contributing thalli and by

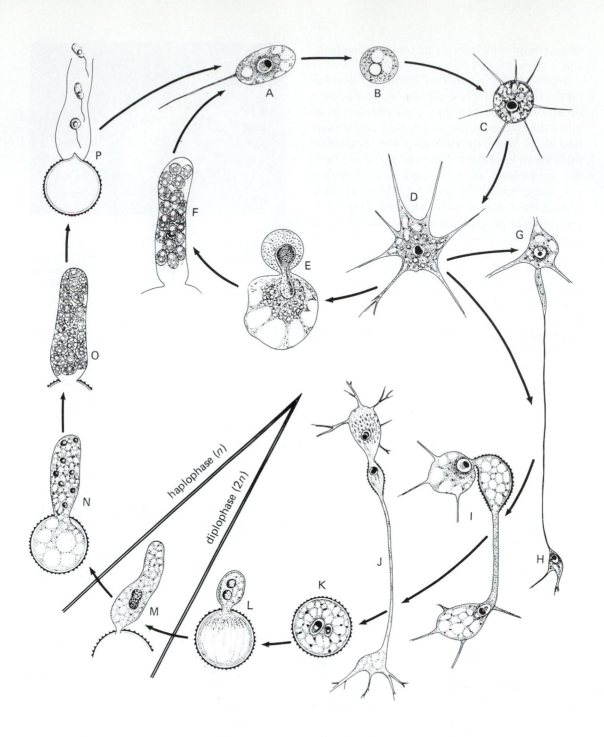

haplophase (*n*)

diplophase (2*n*)

FIGURE 31-2

Life cycle diagram of *Polyphagus euglenae*. **A.** Zoospore. **B.** Encysted zoospore. **C, D.** Germinating zoospores. **E.** Formation of prosporangium. **F.** Sporangium. **G.** Large (female) thallus. **H.** Small (male) thallus with copulation tube already in contact with female thallus. **I.** Plasmogamy. **J.** Male nucleus in incipient zygote. **K.** "Zygote" with the two nuclei before fusion. **L.** Germination of "zygote." **M.** Karyogamy in prosporangium. **N.** Multinucleate, developing young sporangium. **O.** Mature sporangium with zoospores. **P.** Release of zoospores. Figures not to scale. (*Redrawn from Karling [1977] and arranged as a life-cycle diagram by R. W. Scheetz, by permission of Verlag von Cramer. From Alexopoulos and Mims [1979], by permission of John Wiley & Sons, New York.*)

the growth of a copulation tube from one (male) to the other (female). The zoospores are eventually liberated, thus completing the life cycle.

Whether a hormonal mechanism operates to attract the copulation tube from the small sexual thallus to the larger one is not known. Secretion of sex attractants will be mentioned in the next order to be discussed.

Order Blastocladiales

The characteristics of the fungi assigned to this order are (1) the presence of a **nuclear cap** in the zoospores, (2) sexual reproduction by fusion of anisogamous planogametes (motile gametes of different size), and (3) the production of characteristically pitted, thick-walled, resting sporangia (RS). Our example of this order will be *Allomyces macrogynus*.

Unlike the chytrids, which are either strictly unicellular or which have only rhizoidal processes, *Allomyces* (Gr. *allo*, other, + Gr. *mykes*) consists of a well-developed branching mycelium (Figs. 31-3A, 31-4A), anchored by rhizoidal absorptive branches that penetrate the substratum. The organism, which occurs in moist soil and aquatic habitats, mostly in warmer climates, grows readily in laboratory cultures on split hemp seeds or other organic substrata. Branching in *Allomyces* is typically dichotomous, and growth is apical. Superficially, the mycelium appears to be septate, but careful scrutiny of the septations indicates that they are incomplete and that the protoplasm is continuous throughout the fungus (Fig. 31-3B). The mycelium,

with its many nuclei, is coenocytic. The cell walls contain chitin and glucan.

After a period of vegetative growth, the mycelium enters the reproductive phase. Cultural and cytological studies have demonstrated the occurrence of isomorphic alternation of a diploid, asexual, zoospore-producing phase (**sporothallus**) with a haploid, sexual gamete-producing phase (**gametothallus**) in *A. macrogynous*. Its life cycle may be summarized as in Chart 31-1. The dibiontic life cycle here is similar to that of *Cladophora* and *Ectocarpus* type D, h + d (p. 85). In *A. macrogynus*, the terminal portions of the mycelium of the sporothallus become delimited as zoosporangia (mitosporangia) (Figs. 31-3C, 31-4C). The portion of the hypha just below the sporangial septum may form a new branch; hence, the originally terminal sporangial initial becomes secondarily lateral in position. As development proceeds, it becomes apparent that two types of sporangia may be produced. The first of these are thin-walled, ephemeral, and colorless **mitosporangia** (Figs. 31-3C, 31-4C) and are produced early in development. The second type, the **meiosporangia,** are thick-walled, pitted, persistent, and brown and occur later (Figs. 31-3C, 31-4B, E). Both types of sporangia contain a number of nuclei at the time of their formation; this number is increased by division as the sporangia mature.

During development, the thin-walled mitosporangia undergo progressive cleavage to form a number of posteriorly uniflagellate zoospores. These are liberated at maturity through a pore in the sporangial wall (Fig. 31-4D). The zoo-

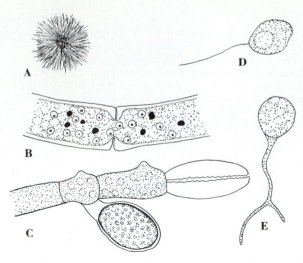

FIGURE 31-3
Allomyces macrogynus. **A.** Habit of growth on submerged hemp seed. **B.** Median longitudinal section of hypha; note multinucleate condition and partial septum. **C.** Three seriate, thin-walled mitosporangia (the apical one empty) and a thick-walled, lateral meiosporangium. **D.** Zoospore. **E.** Zoospore germination. **A.** ×0.75. **B, D, E.** ×700. **C.** ×300.

spores are attracted chemotactically by amino acids, a nutritional adaptation. The zoospores serve as agents for increasing the number of thalli (Fig. 31-3D, E), and under suitable conditions a large number of asexual generations is produced in this manner. Only 30–48 hours are required for a mature thallus to develop from a zoospore.

The thick-walled, resistant meiosporangia (Figs. 31-3C, 31-4E) can withstand long periods of desiccation and temperatures up to 100°C for short periods and still retain their viability. These meiosporangia also are multinucleate. It has been demonstrated that their nuclei persist for long periods in the prophase stages of meiosis, even when the sporangia have been dried. Transfer to water breaks their dormancy and stimulates further development. This consists of the completion of the nuclear divisions, which are meiotic, and the formation of approximately 48 uniflagellate zoospores. The haploid chromosome number in *A. macrogynus* is 14, 28, or 56 (depending on the race); the haploid number in *A. arbuscula* is 8 or 16. The zoospores from resistant sporangia settle on available substrata and develop into gametothalli that form only gametangia, not zoosporangia.

The gametangia, as they first appear on the somatic branches of a clonal culture, occur in pairs (Figs. 31-4F, G, 31-5A), but those produced subsequently may be borne in chainlike series. In their development the male gametangia produce an orange-red, carotenoid pigment, which is dissolved in droplets of oil in the gametes. The female gametangia remain colorless throughout their development. Both types of gametangia undergo progressive cleavage to form uninucleate gametes. The male gametangia contain more nuclei than the female, and as a result the male gametes are considerably smaller than the female (Fig. 31-5B). Both are liberated through pores and unite in pairs under suitable environmental conditions (Fig. 31-5B).

Mycelia of *A. macrogynus* derived from single zoospores are anisogamous and monoecious. The female gametes secrete a hormone called **sirenin** (Gr. *seirin*, siren), which attracts the male gametes chemotactically. Sirenin, which is now available in pure form, was shown by its discoverer, Dr. Leonard Machlis (1958), and his coworkers to attract male gametes in concentrations as low as 10^{-10}M. The chemical structure of sirenin (Fig. 31-6) was elucidated in 1968 by Nutting, Rapoport, and Machlis. Machlis (1972) has summarized the topic of plant hormones in algae and fungi. According to him, sirenin is an example of an **erotactin,** because it attracts male gametes.

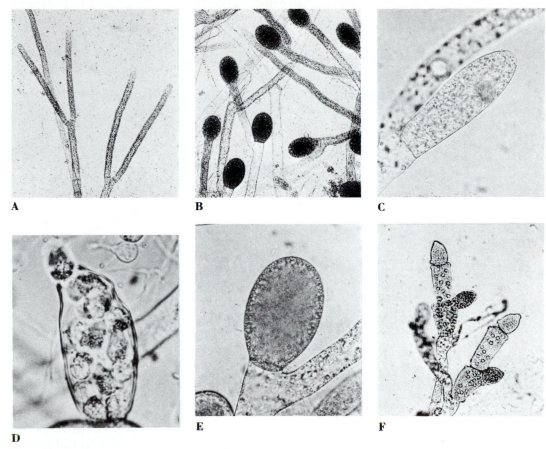

A   B   C

D   E   F

FIGURE 31-4

Allomyces sp. **A.** Dichotomously branching vegetative hyphae.
B. Meiosporangia. **C.** Mitosporangium. **D.** Release of mitospores.
E. Meiosporangium. **F.** Sexual phase: ♂ gametangia terminal, the
♀ just below each. **G.** Male and ♀ gametangia; note exit papillae.
A. ×70. **B.** ×125. **C.** ×300. **D.** ×500. **E.** ×700. **F.** ×125.
G. ×500.

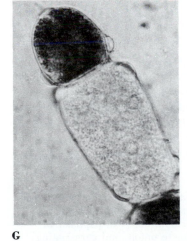

G

The zygote settles on an available, nutriment-
rich substratum and grows (Fig. 31-5B) into a
mycelium that develops only zoosporangia, not
gametangia. Cultural and cytological studies
have demonstrated the occurrence of isomorphic alternation of a diploid, asexual, zoospore-
producing phase with a haploid, sexual, gamete-
producing phase (**sporothallus**) with a haploid,
sexual, gamete producing phase (**gametothallus**)

711

CHART 31-1
Life cycle of *Allomyces macrogynus*.

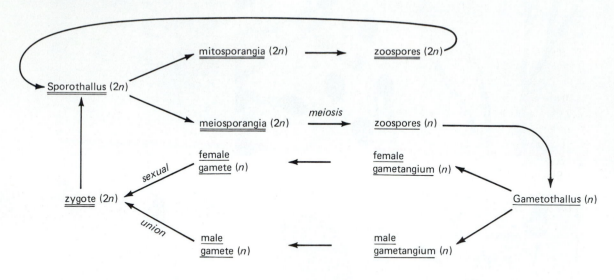

in *A. macrogynus*. Its life cycle may be summarized as in Chart 31-1. The dibiontic life cycle here is similar to that of *Cladophora* and *Ectocarpus* type D, h + d (p. 85). Deviations from this life cycle may occur in certain strains.

Another interesting member of the Blastocladiales is *Coelomomyces* (Gr. *koeloma*, cavity, + Gr. *mykes*, fungus), a genus of fungi that parasitize mosquito larvae. Of interest is the fact that the hyphae of *Coelomomyces* are unwalled. They do, however, develop septa to cut off the sporangia of the fungus. These sporangia are brown, thick-walled, and pitted, like those of *Allomyces*, and are one of the reasons for classifying *Coelomomyces* in the Blastocladiales.

The life cycle of *Coelomomyces psorophorae* is of particular interest because it requires two entirely different species of hosts to come to completion. This situation, known as **heteroe-**

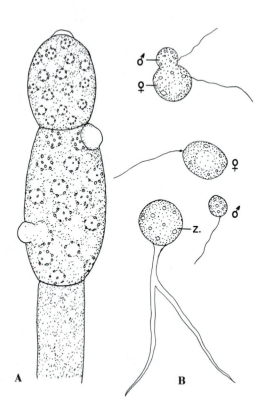

FIGURE 31-5
Allomyces macrogynus. **A.** Male (upper) and female gametangia; clusters of granules (mitochondria?) indicate nuclear position; note exit papillae. **B.** Heterogametes, gamete union, and germinating zygote (z). **A.** ×700. **B.** ×1500.

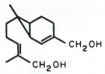

FIGURE 31-6
Chemical structure of sirenin. (*After Nutting et al.*)

cism, was hitherto known to exist only in the rust fungi (Uredinales, Basidiomycetes). *Coelomomyces psorophorae* produces its sporothalli in the mosquito larvae and its gametothalli in the copepod *Cyclops vernalis* (Whisler, Zebold, and Shemanchuk, 1975). Whether other species of *Coelomomyces* are also **heteroecious** remains to be discovered.

CLASS HYPHOCHYTRIDIOMYCETES

This is a small group of predominantly marine fungi with a single tinsel flagellum inserted anteriorly on the zoospores. Of particular interest is the fact that both cellulose and chitin occur in the cell walls of the few species that have been investigated in this regard. The group is very thoroughly discussed by Sparrow (1960, 1973) and many fungi in this class have been meticulously illustrated by Karling (1977). The best known species is *Rhizidiomyces* (Gr. *rhiza*, root, + Gr. *mykes*, fungus) *appophysatus* through the researches of Karling (1944), Fuller (1960, 1962), who grew the organism in pure culture, and Fuller and Reichle (1965). One order and three families belong here. Their life cycles parallel those of the Chytridiomycetes.

CLASS PLASMODIOPHOROMYCETES

The Plasmodiophoromycetes, commonly known as the endoparasitic slime molds, differ from fungi other than slime molds in that their thalli are plasmodial. For this reason they have been classified by Ainsworth, Sparrow, and Sussman (*The Fungi*, vol. 4B, 1973, p. 5) with the Myxomycota (our Gymnomycota), but neither Sparrow himself (1960) nor Waterhouse (1973) nor Alexopoulos and Mims (1979) subscribe to that view. Two of these organisms are serious plant pathogens. *Plasmodiophora* (Gr. *plasmodium* + Gr. *phoreus*, bearer) *brassicae* causes club root of cabbage, and *Spongospora* (Gr. *spongia*, sponge, + Gr. *spora*, spore) *subterranea* is the cause of powdery scab of potatoes. The zoospores of the Plasmodiophoromycetes are equipped with two anteriorly inserted whiplash flagella of unequal size. This character, together with their plasmodial thallus, links them to the Myxomycetes, but the fact that they produce spores in funguslike sporangia places them in the Mastigomycota. An interesting character of this group is the so-called cruciform nuclear division in their sporangiogenous plasmodia.[2] During nuclear division, the nucleolus elongates and assumes the shape of a dumbbell, around which the chromosomes become arranged in a ring formation at metaphase, so that the whole configuration, viewed from the side, resembles a cross—hence the name cruciform division. The life cycle of these organisms is not well known; it appears that it may be diobiontic.

SUBDIVISION DIPLOMASTIGOMYCOTINA

In the Diplomastigomycotina, meiosis, wherever it has been investigated by modern methods, has been found to occur in the gametangia. The life cycle of these organisms is, therefore, monobiontic-diploid (M, d), with the gametes as the only haploid cells. There is but a single class, the Oomycetes.

---

[2]There are two types of plasmodia in the life cycle of the Plasmodiophoromycetes: sporangiogenous, which produce zoosporangia, and cystogenous, which produce thick-walled cysts (resting bodies).

CLASS OOMYCETES

This is a large class in which the organisms included differ so markedly from all other fungi that some mycologists (Shaffer, 1975) do not include them in the fungi at all. The characteristics that set them apart are (1) the general absence of chitin and the presence of cellulose in their cell walls;[3] (2) biflagellate zoospores with two different flagella, one whiplash and one tinsel, anteriorly or laterally inserted on the zoospores; (3) formation of two types of zoospores in some species, one pear-shaped (primary), the other reniform (secondary); (4) oogamous sexual reproduction by the passage of gamete nuclei from a well-defined antheridium to a well-defined oogonium containing oospheres (eggs), thus resulting in thick-walled oospores;[4] and (5) gametangial instead of zygotic meiosis, rendering the thallus diploid. The first four of these characteristics have been well known for almost a century. The fifth has been a matter of controversy ever since Stevens (1899), working with *Albugo*, one of the Peronosporales, found two nuclear divisions occurring in the gametangia and suggested that they represented meiosis. This did not find general acceptance, however, and for the first sixty years of this century mycologists believed that the Oomycetes were haploid, like other fungi, and that meiosis occurred during the germination of the zygote. There were, however, only a few researchers who claimed to have proof positive of zygotic meiosis. Since 1961, when Sansome published a paper claiming that gametangial meiosis occurs in *Pythium debaryanum*, a well-known member of the order Peronosporales of the Oomycetes, so many researchers

have found gametangial meiosis in various Oomycetes that the evidence is now overwhelmingly convincing as summarized by Sansome (1966), Flanagan (1970), Dick and Win-Tin (1973), and Win-Tin and Dick (1975), and as referred to by Ellzey and Huizar (1977). To be sure, there are some who still hold out for zygotic meiosis (Timmer et al., 1970), but as Dick (1973) stated, if future research should show this to be the case in any family of the Oomycetes, its relationships will have to be reevaluated.

Classification

The class Oomycetes is usually divided into five orders:

Class Oomycetes
    Order 1. Lagenidiales
    Order 2. Thraustochytriales
    Order 3. Saprolegniales
    Order 4. Leptomitales
    Order 5. Peronosporales

We shall discuss only two of these.

Order Saprolegniales

There are several families in the Saprolegniales (Dick, 1973; Sparrow, 1976), of which the Saprolegniaceae is the most typical and, by far, the best known. We shall confine our discussion to that family alone.

The Saprolegniaceae are the classical "water molds," although that designation has also been given to the Blastocladiales. The Saprolegniaceae are widespread, occurring in aquatic, mostly freshwater, habitats and also in the top layers of moist soils. A few have been found in brackish and salt waters. When the mycology of the oceans and seas is better known than it is at present, more Saprolegniaceous species may be found in that habitat. These fungi are very easy to obtain for laboratory study by baiting pond water or aqueous soil suspensions with split

---

[3]Both chitin and cellulose, however, have been found in the genus *Apodachlya* by Lin, Sicher, and Aronson (1976), and in *Leptomitus* by Aronson and Lin (1978).

[4]The zygotes formed in oogamous sexual reproduction are often called oospores.

hemp seed or sesame seed, around which they form fluffy colonies (Fig. 31-7).

Mostly saprobic, the Saprolegniaceae are of little economic importance. A few, however, such as *Saprolegnia parasitica*, cause disease of fish and destroy fish roe in commercial hatcheries. A few others, such as *Aphanomyces euteiches*, are parasitic on plants, causing root rot of beets and other crops.

The coenocytic somatic hyphae of the Saprolegniaceae are large in diameter and easy to observe under the microscope. They grow rapidly and branch profusely in a favorable nutrient medium. After a period of somatic development, the reproductive phases, asexual at first, are initiated by the development of hyphalike cylindrical zoosporangia (Figs. 31-8A), the contents of which become cleaved and differentiate into a considerable number of zoospores. How

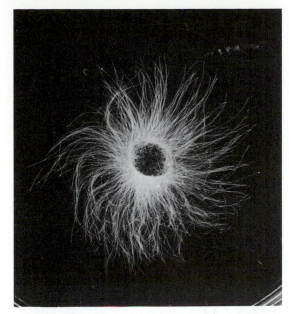

FIGURE 31-7
Achlya ambisexualis growing on hemp seed. × 1. (*After Barksdale.*)

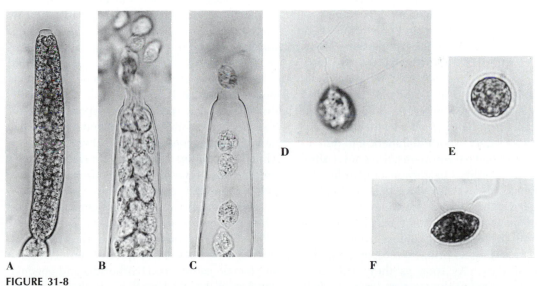

FIGURE 31-8
Saprolegnia sp. **A.** Zoosporangium filled with zoospores; note exit papilla at apex. **B, C.** Release of zoospores. **D.** Primary zoospore, more highly magnified. **E.** Encysted primary zoospore. **F.** Secondary zoospore emerged from a cyst. **A.** ×150. **B, C.** ×300. **D.** ×800. **F.** ×900. (*After Holland, courtesy of the Carolina Biological Supply Company.*)

the zoospores are liberated and how they behave, varies in different genera. The two most common genera of the Saprolegniaceae are *Saprolegnia* (Gr. *sapros*, rotten, + Gr. *legnon*, border), monographed by Seymour (1970), and *Achlya* (Gr. *achlys*, mist), monographed by Johnson (1956). A still useful though outdated monograph of the family Saprolegniaceae is the one by Coker (1923). A modern, updated treatment of this family is in preparation by Seymour and Johnson.

Two types of zoospores occur in the Saprolegniaceae: pyriform (primary) zoospores with two flagella, one whiplash, one tinsel, attached anteriorly; and reniform (secondary) zoospores with their two flagella, again one whiplash and one tinsel, attached at the concave side. Species that produce only one type of zoospore are **monomorphic;** those forming both types are **dimorphic.**

Life cycle.   In the life cycle of *Saprolegnia* there are typically two swarming periods of the zoospores separated by an encystment stage. The occurrence of two motile periods is called **diplanetism.** The zoosporangia release their primary zoospores from their tips, one by one (Fig. 31-8B–D). After a period of motility these zoospores withdraw their flagella, come to rest, become spherical, and encyst by secreting walls around themselves (Fig. 31-8E). After a period of rest, the cysts germinate to form secondary zoospores (Fig. 31-8F), one from each cyst. Thus, *Saprolegnia* is both **diplanetic** and dimorphic. The secondary zoospores also encyst after a period of motility. In germination to form a new thallus, the cysts produce delicate hyphae called **germ tubes,** which grow into the mycelium. *Achlya* (note the spelling of this genus; it gives trouble to students) differs from *Saprolegnia* in releasing its primary zoospores rapidly and all at once. As soon as they strike the outside environment they encyst, forming a ball of spores that clings to the tip of the sporangium (Fig. 31-9). There, the cysts germinate and each

FIGURE 31-9
Sporangium of *Achlya* sp. showing encysted zoospores at the apical opening. (*Drawn by Stephen Chase.*)

liberates a secondary (reniform) zoospore, which continues the life cycle.

In *Dictyuchus* (Gr. *dictyon*, net), another genus of the Saprolegniaceae, no primary, motile zoospores are formed. Instead, the sporangium develops a network of walls, which are thought to be the walls of encysted primary zoospores. It is believed that such zoospores may actually be formed but encyst within the sporangium, their cysts being represented by the network of walls that divides the sporangium into a large number of small compartments (Fig. 31-10). An ultrastructural study is needed here to elucidate the development of these stages. Be that as it may, the wall of each "cell" dissolves and a secondary zoospore emerges. These spores swim and encyst repeatedly, each new emerging zoospore being always reniform. Thus, *Dictyuchus* has many swarming periods but only one type of zoospore. It is, therefore, **polyplanetic** but monomorphic.

Following a series of asexual generations and when the food in the medium has been almost exhausted, the mycelium of the Saprolegniaceae initiates sexual reproduction. The male and female gametangia (antheridia and oogonia) develop as lateral branches on the mycelium (Fig. 31-11). Some clones are bisexual and others are unisexual. Careful study of the sex

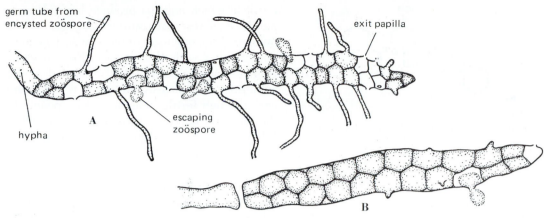

FIGURE 31-10
Dictyuchus sp. **A.** Sporangium attached to somatic hypha. Some zoospores have escaped through individual exit papillae, leaving empty cells; two are in the process of escaping; several have encysted within the sporangium and have produced germ tubes. **B.** Detached sporangium. ×300. (*From Alexopoulos, 1962. By permission of John Wiley & Sons.*)

organs of *Achlya* has revealed a process of unsuspected complexity, involving the secretion of a number of complex chemicals that influence the course of development. It has

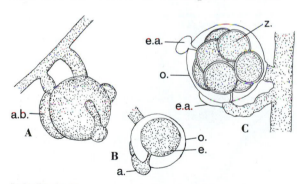

FIGURE 31-11
Achlya sp. Sexual reproduction, bisexual species. **A.** Immature antheridial and oogonial branches. **B.** Oogonium with single egg, antheridium, and fertilization tube. **C.** Oogonium containing dormant zygotes (oospores). a., antheridium; a.b., antheridial branch; e., egg; e.a., empty antheridium; o., oogonium; z., zygote. ×300.

been demonstrated that the formation and maturation of the sex organs in *Achlya* are controlled by a hormonal system that operates when potentially male hyphae and female hyphae of *A. ambisexualis* are in close proximity in an aquatic environment. The hormonal system is successively responsible for the proliferation of male somatic hyphae to form antheridial branches; the stimulation of the potentially female hyphae to form oogonial branches; the attraction of the antheridial branches toward the oogonial branches; the delimitation of the antheridia from their subtending branches; and finally, the septation of the oogonia from their stalks and the differentiation of the eggs. This research of Dr. John R. Raper was summarized by him in 1957.

One of these sequential stages in the sexual reproduction of *A. ambisexualis* and *A. bisexualis*, the formation of antheridial branches on the male mycelium, has been studied by Barksdale (see her summary, 1969). As noted above, a substance secreted by the female filaments evoked the production of antheridial branches on the male mycelium. McMorris and Barksdale (1967) prepared a crystalline substance from the cultures of female thalli of *A. bisexualis* that could induce antheridium pro-

717

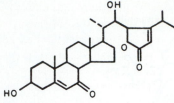

FIGURE 31-12
Chemical structure of antheridiol. (*After Edwards et al.*)

duction and called it **antheridiol.** The chemical structure (Fig. 31-12) of this compound was later ascertained (Arsenault, Biemann, Barksdale, and McMorris, 1968), and antheridiol was later synthesized (Edwards, Mills, Jundeen, and Fried, 1969).

As the antheridia become appressed to the oogonia, meiosis occurs within the gametangia (Steffens, 1976; Ellzey and Huizar, 1977), as a result of which oogonia produce haploid, nonmotile gametes, the oospheres (eggs), their number, from one to 20 or more in each oogonium, varying with the species (Fig. 31-13A). Haploid gamete nuclei are also formed in the antheridia after meiosis, and these travel to the eggs through fertilization tubes that penetrate the oogonia. Karyogamy then occurs, reestablishing the diploid condition in the zygote. After fertilization, the zygotes rapidly develop thick walls, which obscure their contents (Fig. 31-13B). The dormant zygotes are known as **oospores.** The oogonia eventually break down, releasing the thick-walled oospores into the water. After a period of rest the oospore germinates, giving rise to a diploid hypha, which establishes a new thallus. Accordingly, the life cycle of the Saprolegniaceae is monobiontic, with the mycelium diploid and only the gamete nuclei haploid (M, d).

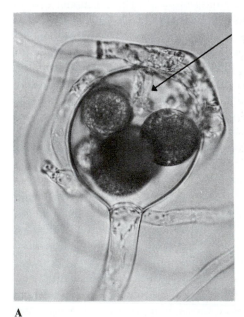
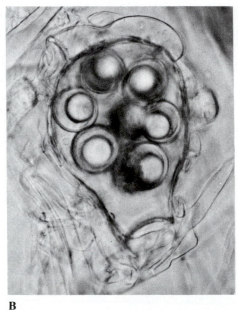

A   B

FIGURE 31-13
Achlya ambisexualis. **A.** Oogonium with eggs surrounded by antheridial branches; note fertilization tubes (arrow) from antheridia making contact with the eggs. **B.** Oogonium with spent antheridia and zygotes (oospores). **A.** ×500. **B.** ×550. (*After Barksdale.*)

718

Order Peronosporales

In contrast to the Saprolegniales, the Peronosporales are of great economic importance, at least two of them having had a great impact on human affairs. Not long after the discovery of America by Columbus, the potato was introduced into Europe and eventually was established as an important food crop. In some countries, such as Ireland, the potato was grown almost to the exclusion of other crops, and the Irish lived on the potato alone. In 1845, the fungus *Phytophthora* (Gr. *phyton*, plant, + Gr. *phthora*, destruction) *infestans*, one of the Peronosporales, invaded the island and spread like wildfire in the potato fields. The destruction was enormous. Resulting in the complete failure of the potato crop, the fungus was directly responsible for the Irish famine of 1845 and 1846, during which over a million people died of starvation and more than that were forced to emigrate to foreign shores, notably those of the United States. Because of the famine it caused, the fungus was also indirectly responsible for the repeal of the infamous corn laws of Great Britain which imposed heavy duty on imported grain to protect home farmers. In 1846, they were repealed so as to admit foreign grain into the country and so make it available to the starving population of Ireland (Large, 1940).

Sometime during the latter part of the nineteenth century, *Phylloxera,* an aphid that attacks the roots of the grape, was accidentally introduced into southern France from the New World, where it is native. The European grapes (*Vitis vinifera*) proved to be extremely susceptible to it, and the whole French grape and wine industry was threatened when the vines began to die. A French scientific mission, after studying *Phylloxera* in the United States, decided that the only solution to the problem was the introduction into France of resistant American stocks, on which the French wine grapes could be grafted. Large quantities of American grapevines were thus shipped to France, but while the *Phylloxera* problem was being solved, another equally severe one appeared. It seems that some of the imported vines carried the spores of *Plasmopara* (N.L. *plasma*, mold, + L. *parere*, to bring), *viticola* another of the Peronosporales, and this destructive fungus, native to America, the cause of downy mildew of grapes, spread over the very susceptible European varieties. The French wine industry appeared to be doomed. Fortunately, through a chance observation, Alexis Millardet, Professor of Botany at the University of Bordeaux, developed in 1885 Bordeaux mixture, the first fungicide to be discovered, and saved the grapes of France. This introduced the era of plant disease control and the eventual development of a large number of different fungicides now used to combat plant diseases, of which Bordeaux mixture is still one of the most effective. These stories are related in detail by Large (1940) in his fascinating book *The Advance of the Fungi*.

The Peronosporales differ from the Saprolegniales in a number of important ways. Most, but not all, are terrestrial rather than aquatic fungi, although the genus *Pythium* (Gr. *pythein*, to cause to rot) does contain many aquatic species. The terrestrial species produce their sporangia on specialized, differentiated hyphae, the **sporangiophores,** rather than directly on the somatic hyphae as do the Saprolegniales; the Peronosporales are monoplanetic and produce only reniform (secondary) zoospores; the most complex Peronosporales (*Plasmopara, Peronospora* [N.L. *perono*, fibula, + Gr. *spora*, spore], *Albugo*, etc.) are obligate parasites, developing their entire life cycle on the living host, and cannot be grown in laboratory culture from spore to spore; with very few exceptions, only a single oosphere (egg) is formed within the oogonium, and this is surrounded by leftover cytoplasm (**periplasm**) rather than being free within the gametangium as in the Saprolegniaceae.

The genus *Albugo* will serve to illustrate the life cycle of the Peronosporales.

719

B

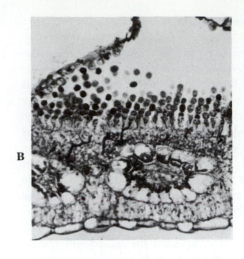

FIGURE 31-14
Albugo bliti. **A.** Portion of infected *Amaranthus* leaf.
B. Transection of infected leaf showing ruptured
epidermis, sporangiophores, and sporangia. ×125.

Albugo

Except for *Polyphagus* (Fig. 31-2) and a few
species of the Saprolegniaceae, the fungi so far
described are all similar in that they are saprobic. *Albugo* (L. *albus*, white), on the contrary,
is a parasitic genus that occurs on a number of
hosts. *Albugo candida* grows on certain mus-
tards (Cruciferae). *Albugo impomeaepandur-
anae* is widespread on sweet potatoes and certain morning glories, and *A. blitii* occurs on
some species of *Amaranthus*. Leaves of infected
plants become covered with conspicuous mealy
white spots or patches (Fig. 31-14A), which are
caused by eruptions of large numbers of sporelike sporangia below the epidermis (Fig.
31-14B). Because of their lesions, infected
plants are said to have white rust. The formation
of the spores is preceded by the development of
a vegetative mycelium, which spreads through
the host tissues from the site of the primary
infection; the mycelium is entirely intercellular.
The fungus obtains its metabolites by forming
small, protuberant, papillate branches, **haustoria** (sing. **haustorium**; L. *haustor*, one who
drinks), which penetrate some of the host cells.

In sporangial formation, a number of multinucleate hyphal tips push out between the mesophyll cells, forming a bedlike mass, and enlarge
terminally (Figs. 31-14B, 31-15A). After several
nuclei have migrated into the enlarged tip, the

latter is delimited by an annular centripetal
ingrowth of the wall. The tip of the hypha below
the delimited sporangium now enlarges and is
ultimately cut off like the first. The sporangia
thus are produced in chains in basipetal succession, the oldest being farthest from the sporangiophore. Continued sporangial production
brings about a localized uplifting of the epidermis (Fig. 31-14B), which is finally ruptured, and
the mature sporangia escape freely. The sporan-

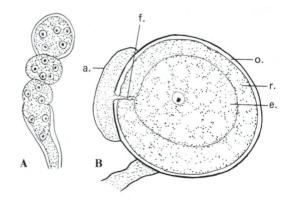

FIGURE 31-15
Albugo bliti. **A.** Sporangiophore and seriate
sporangia. **B.** Mature sex organs at fertilization. a.,
antheridium; e., egg; f., fertilization tube; o.,
oogonium; r., residual cytoplasm. ×700.

720

gia may function as zoosporangia, undergoing cleavage to form as many reniform zoospores as there are nuclei present.

The sporangia are disseminated by wind, rain, or other agents, and if they settle on moist surfaces of leaves of the host species, they germinate either by germ tubes or by producing zoospores. The biflagellate zoospores become spherical after a period of motility, encyst, and then develop delicate hyphal tubes, which usually enter the host through a stoma. The infection is spread in this way.

Sexual reproduction may follow asexual later in the growing season. The sex organs are produced from the tips of the hyphae among the mesophyll cells of the leaf and are suggestive of those of the water molds. After the hyphal tips have enlarged considerably, they are segregated from the remainder of the hyphae by cell walls. The antheridia are smaller than the oogonia, but both are multinucleate. As the oogonium matures, the protoplasm becomes rather densely aggregated in the center, leaving a more watery, vacuolate periplasm at the periphery. After the antheridium comes in contact with the oogonium, meiosis occurs in both gametangia.

By the time of fertilization, all but one of the nuclei of the oogonium disintegrate in certain species, notably *A. candida*; the remaining nucleus functions as the egg nucleus. The egg is delimited from the peripheral cytoplasm by a delicate membrane. This type of cytokinesis, in which a cell is delimited within another, leaving residual cytoplasm (**periplasm**), is known as **free cell formation.** The antheridium, which is appressed to the oogonium, now produces a small hyphal protuberance (fertilization tube) that penetrates the oogonial wall and grows through the periplasm into the central dense cytoplasm containing the egg nucleus (Fig. 31-15B). After nuclear union, a multilayered wall is secreted by the zygote, the nucleus of which divides soon after fertilization, until about 32 nuclei are formed. In the spring, further nuclear division

and, finally, cleavage occur. Upon germination, usually in the spring after it is formed, the zygote forms numerous biflagellate, reniform, diploid zoospores, which renew the infection.

The sexual organs of *A. candida* are superficially similar to those of *Saprolegnia* and *Achlya.* The oogonia differ, however, in the production of a single egg, which is delimited from the periplasm by free cell formation. In *Saprolegnia* and *Achlya*, on the other hand, even in those cases when only one egg is developed, there is no residual cytoplasm.

Summary and classification

The former divisional name Phycomycota or class name Phycomycetes still informally used by mycologists, without taxonomic significance, for the fungi discussed in this chapter, reflected the views of those who speculated that these fungi were derived from algal progenitors that have lost their chlorophyll and have entered upon either a saprobic or parasitic mode of life. This concept was based largely on the similarity of their nonseptate mycelium to the tubular, nonseptate filaments of the siphonous Chlorophyceae and Xanthophyceae such as *Dichotomosiphon, Vaucheria*, and *Botrydium.*

With the exception of some species of *Allomyces* and *Coelomomyces,* the Chytridiomycetes are monobiontic in their life cycle, with meiosis occurring during the germination of the zygote (often a resting body) or in the meiosporangia on the sporothallus of the dibiontic species. Many types of sexuality occur, such as the fusion of rhizomycelia in *Chytriomyces*, the conjugation of large with small thalli in *Polyphagus,* or the union of motile, isogamous gametes in *Allomyces* and *Coelomomyces.*

With the acquisition of newer knowledge, particularly on cell wall composition, many mycologists look at the Chytridiomycetes as a group with protozoan ancestry representing the most primitive of the fungi.

The Hyphochytridiomycetes differ from the

721

Chytridiomycetes in their anteriorly flagellate zoospores, which bear a tinsel flagellum, and in the presence of cellulose along with chitin in the cell walls of at least some species. In common with those of the Chytridiomycetes the zoospores of these fungi have a second, functionless basal body, possibly indicating the descent of both groups from a biflagellate ancestor.

The place of the Plasmodiophoromycetes in the classification system is even more uncertain and will remain so until their life cycles are elucidated. Their somatic phases are plasmodial, which links them to the slime molds, as does the fact that they have anteriorly biflagellate motile cells, but their endoparasitic mode of life, their probably dibiontic life cycle, their characteristic cruciform division, the presence of· zoosporangia, and the formation of resting spores in masses rather than in distinct sporophores tend to exclude them from the slime molds.

The similarity of the oogamous reproduction of the Oomycetes to that of oogamous algae such as *Vaucheria*, and the presence of cellulose coupled with the absence of chitin in all but a few species, have served to make the hypothesis of their algal origin an attractive one. The Oomycetes also differ from all other fungi in their monobiontic diploid life cycle and in their uniquely, among the fungi, gametangial meiosis. Many mycologists take the view that the Oomycetes have had a separate origin—possibly algal—from that of all other fungi. Some exclude them from the "true fungi" altogether.

The organisms mentioned in this chapter may be classified as follows in the system we have adopted in this book:

Division Mastigomycota
  Subdivision 1. Haplomastigomycotina
    Class 1. Chytridiomycetes
      Order 1. Chytridiales
        Family 1. Chytridiaceae
          Genus: *Chytriomyces*
        Family 2. Rhizidiaceae
          Genus: *Polyphagus*
      Order 2. Blastocladiales
        Family 1. Blastocladiaceae
          Genus: *Allomyces*
        Family 2. Coelomomycetaceae
          Genus: *Coelomomyces*
    Class 2. Hyphochytridiomycetes
      Order 1. Hyphochytriales
        Family 1. Rhizidiomycetaceae
          Genus: *Rhizidiomyces*
    Class 3. Plasmodiophoromycetes
      Order 1. Plasmodiophorales
        Family 1. Plasmodiophoraceae
          Genera: *Plasmodiophora, Spongospora*
  Subdivision 2. Diplomastigomycotina
    Class 1. Oomycetes
      Order 1. Saprolegniales
        Family 1. Saprolegniaceae
          Genera: *Saprolegnia, Achlya, Aphanomyces*
      Order 2. Peronosporales
        Family 1. Pythiaceae
          Genera: *Pythium, Phytophthora*
        Family 2. Peronosporaceae
          Genera: *Plasmopara, Peronospora*
        Family 3. Albuginaceae
          Genus: *Albugo*

DISCUSSION QUESTIONS

1. Explain the etymology of "Mastigomycota."
2. Distinguish between the whiplash and tinsel types of flagellum.
3. Define or explain: mastigoneme, sporangium, sporangiospore, zoospore, holocarpic, eucarpic, nuclear cap, sporothallus, gametothallus, mitosporangium, meiosporangium, sirenin, erotactin, heteroericism, monomorphic, dimorphic, diplanetism, polyplanetic, germ tube,

haustorium, periplasm, free cell formation.

4. How do the Haplomastigomycotina differ from the Diplomastigomycotina?

5. Distinguish among the Chytridiomycetes, Hyphochytridiomycetes, and Plasmodiophoromycetes.

6. Describe the structure and reproduction of *Chytriomyces hyalinus, Polyphagus euglenae,* and *Allomyces macrogynus.*

7. What characteristics distinguish the Oomycetes from other Mastigomycota?

8. Name the orders of Oomycetes and distinguish among them.

9. Describe the structure and reproduction of *Saprolegnia.* How do *Achlya* and *Dictyuchus* differ from it?

10. Describe the hormonal control mechanisms in *Achlya.* What is antheridiol?

11. How are members of the Peronosporales important to mankind?

12. Describe the structure and life cycle of *Albugo.*

Chapter 32
DIVISION AMASTIGOMYCOTA I: SUBDIVISION ZYGOMYCOTINA

In contrast to the fungi we classify in the Mastigomycota, the Amastigomycota produce no flagellate cells whatsoever even in the aquatic species, and all but a few, therefore, have spindle-pole bodies (SPBs) (p. 700) instead of centrioles functioning in nuclear division. Because these fungi are considerably more complex than those we have already discussed, they are often referred to as "higher fungi" to distinguish them from the so-called "lower fungi"—designations we have studiously avoided—by those who believe them to have been evolved from the latter, and, therefore, more recently in the course of evolution. Although such origin seems probable, it is merely conjecture. The division Amastigomycota contains four subdivisions: Zygomycotina, Ascomycotina, Basidiomycotina, and Deuteromycotina. In this chapter we shall deal with the first of these subdivisions.

SUBDIVISION ZYGOMYCOTINA: BREAD MOLDS, FLY FUNGI, AND ARTHROPOD COMMENSALS

The Zygomycotina reproduce sexually by the copulation of two usually equal gametangia, resulting in the formation of a thick-walled **zygosporangium** which contains a **zygospore**.

Two classes constitute this subdivision: the Zygomycetes and the Trichomycetes. The former has been studied intensively for over 100 years. The latter is still not well known. We shall, therefore, devote most of our discussion to the Zygomycetes, which include the fungi commonly known as the bread molds and the fly fungi.

CLASS ZYGOMYCETES

It will be recalled that in *Albugo* and the other Mastigomycota we have studied in the previous chapter, the asexual sporangia either produce zoospores upon germination or themselves germinate directly by germ tube. In the Zygomycotina, sporangia are still formed, but contain nonmotile spores called **aplanospores**.[1] These are released from the sporangia and germinate by germ tubes, which grow into mycelia. No flagellated cells are ever formed by any of these fungi, even though some live in aquatic or semiaquatic environments. About 600 species of Zygomycetes are known, which have been classified in as many as 7 orders (Benjamin, 1979). We will discuss only the Mucorales, Entomophthorales, and Endogonales.

Order Mucorales

Fungi in this order are mostly saprobes but may be weak parasites of flowers and fruits. Some are pathogenic to humans, causing a group of diseases known as **mucormycoses**. Perhaps the most familiar of all the Zygomycetes is *Rhizopus* (Gr. *rhiza*, root, + Gr. *pous*, foot), one of several genera in the family Mucoraceae. *Rhizopus stolonifer*, commonly known as bread mold because of its occurrence on that substratum, frequently appears in damp, warm weather. It is also the chief cause of "leak," a serious disease of strawberries while they are in transit to the market.

Its mycelium is nonseptate (coenocytic). *Rhizopus* and related genera may be present on all sorts of organic matter, including dung, fruit, and fleshy fungi, when there is sufficient moisture to support growth. Its spores are almost always present in the atmosphere, as evidenced by the frequency with which it can be obtained when moistened bread that does not contain preservatives and other organic substances is exposed to air currents and then maintained in a humid atmosphere.

The mycelium of the Mucorales is a cottony white mass during the somatic phase but presents a sooty appearance at the time of sporulation. This is caused by the presence of large numbers of black-walled spores. Although the mycelium is coenocytic, it exhibits considerable differentiation. Certain branches creep over the substratum much like stolons of strawberries, for example. Also like stolons, the portions of the horizontal hyphae that make contact with the substratum produce rhizoidal branches (Fig. 32-1B), which serve as absorptive organs and secrete digestive enzymes.

Asexual reproduction. After a short period of somatic development in *Rhizopus*, groups of unbranched, elongate hyphae arise from the absorptive branches opposite a group of **rhizoids**, forming erect hyphae, the tips of which become enlarged with nuclei and cytoplasm as increase in length ceases. These sporangiophores bear single sporangia (Fig. 32-1A).

The sporangia are formed in the following manner: As the enlarging tips of sporangiophores attain their characteristic size (Fig. 32-1C), the peripheral cytoplasm becomes dense and the central portion remains vacuolate. The two regions are segregated from each other by the coalescence of a series of vacuoles present in a domelike arrangement. A wall finally is secreted between the two portions of the protoplasm (Fig. 32-2A). The central sterile portion is called the **columella** (L. *columella*, dim. of column); the peripheral portion is fertile and sporogenous. As the sporangium matures, the sporogenous protoplasm undergoes progressive cleavage, with the ultimate production of large numbers of minute spores (Fig. 32-2A), each of which contains several nuclei and develops a black wall. The outer sporangial wall is extremely delicate and readily torn. When this occurs, the exposed spores are quickly carried away by air currents; the naked columella remains (Fig. 32-1D). The spores germinate readily on moist substrates (Fig. 32-2B).

Sexual reproduction. *Rhizopus stolonifer* is heterothallic, consisting of two mating types (+ and −). Clonal cultures are, therefore, self-sterile. They are also considered to be isoga-

---

[1]For a different definition of "aplanospore," see the Glossary.

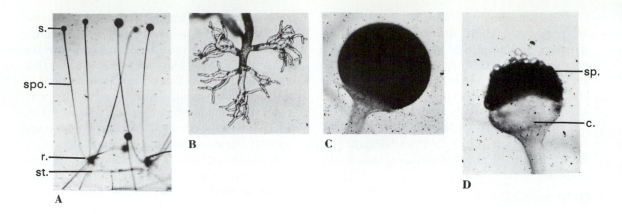

FIGURE 32-1
Rhizopus stolonifer. **A.** Habit of growth. **B.** Rhizoidal hyphae. **C.** Mature sporangium.
D. Sporangium after wall rupture, showing columella and spores. c., columella; r., rhizoidal hyphae;
s., sporangium; sp., spore; spo., sporangiophore; st., stolon. **A.** ×12.5. **B.** ×125. **C, D.** ×135.

mous in sexual reproduction. When spores of opposite mating types are planted in reasonably close proximity in agar cultures, sexuality soon becomes manifest. Hyphae of the two strains that approach each other increase in size at their tips (Fig. 32-3A, B) and rise above the substrate.

These are **progametangia**. Transverse septa are soon laid down, so that the multinucleate tip of each branch is delimited from the remainder of the hypha (Figs. 32-3B, 32-4). The delimited portions, called **gametangia,** are considered to be multinucleate gametes, and the subtending

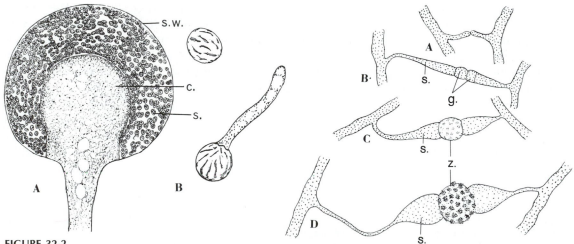

FIGURE 32-2
Rhizopus stolonifer. **A.** Median longitudinal section of mature sporangium. **B.** Spore and its germination. c., columella; s., spores; s.w., sporangial wall. **A.** ×260. **B.** ×380.

FIGURE 32-3
Rhizopus stolonifer. **A–D.** Stages in sexual reproduction. g., multinucleate gametes; s., suspensor; z. zygosporangium. ×65.

726

A

B

FIGURE 32-4
Rhizopus stolonifer. Photomicrographs of sexual reproduction. **A.** Delimitation of gametes.
B. Zygosporangium and suspensors. ×125.

hyphae are known as **suspensors**. After the gametangia have made contact, the walls between the tips of contiguous gametangia dissolve, with the result that the cytoplasm and nuclei then lie free within a single lumen (Fig. 32-3C). During this period the nuclei in uniting gametangia increase in number; subsequently, many nuclei unite in pairs, but some supernumerary nuclei remain. A thick wall is secreted by the zygote (Figs. 32-3D, 32-4B). The supernumerary nuclei have been reported to disintegrate, so that the dormant zygote contains only diploid nuclei. It has been recently shown that the thick-walled zygote is actually a sporangium, now called a zygosporangium, in which a zygospore is enclosed.

Germination of zygospore when the zygosporangium splits open has been observed infrequently (Gauger, 1961), but there is some cytological evidence that the nuclear divisions that occur during germination are meiotic. This evidence is also supported by genetic studies of genera related to *Rhizopus.* The spores from such a sporangium are either all (+), or all (−), or both (+) and (−) spores have been found. The cytological basis for this has not been elucidated with complete clarity, but it is thought that when all spores in a germ sporangium are of one mating type, three of the four nuclei resulting from meiosis disintegrate, and all the spores, therefore, contain nuclei that are the progeny of the one surviving nucleus.

It was proved more than 80 years ago (for the first time in fungi) that the sexual process in *Rhizopus*-like molds (e.g., *Mucor mucedo* and *Blakeslea trispora*) is under chemical control. Van den Ende and Stegwee (1971) have summarized the literature on this subject. It has been shown that progametangia do not develop if the maturing types are grown separately. It has been postulated that a "progamone" (not yet demonstrated to exist) "informs" the compatible mycelia of one another's proximity. This causes the mycelium to secrete trisporic acids (Fig. 32-5), which evoke the differentiation of the progametangia in both the compatible strains. Other, as yet chemically undefined, substances effect contact, gametangial union, and zygote formation.

A number of genera similar to *Rhizopus* are widespread on organic substrata. *Mucor* (L. *muceo*, be moldy) is similar to *Rhizopus*, except that its sporangiophores arise from the main branches, rhizoids being absent at their bases, and its sporangiospores adhere closely one to another, forming a tight mass, and are therefore not windblown. They are disseminated by splashing raindrops.

Phycomyces (Gr. *phykos*, alga, + Gr. *mykes*, fungus) (Fig. 32-6) produces sporangiophores that may attain a length of 10 cm. Its zygosporangia are made conspicuous by the development of dark, branching projections on arched suspensors (Fig. 32-6B, C).

FIGURE 32-5
Trisporic acid C, chemical structure.

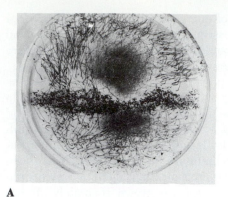

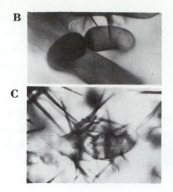

FIGURE 32-6
Phycomyces blakesleeanus.
Sexual reproduction. **A.** Petri
dish inoculated with
compatible mating types; note
zone of dark zygosporangia.
B. Early approach of sexually
compatible hyphae.
C. Delimitation of gametes
and initiation of antlerlike
suspensor appendages.
A. ×½. **B, C.** ×125.

A

B

C

The genus *Pilobolus* is an interesting dung-inhabiting mold. Its resistant spores pass unharmed through the digestive tract of animals. Horse dung, if stored in a moist chamber, soon becomes covered with the positively phototropic sporangiophores of this organism (Fig. 32-7A). Each of them bears a terminal black sporangium (Figs. 32-7B, 32-8A, B). Unlike that of *Rhizopus*, the sporangial wall of *Pilobolus* is firm, and the sporangium is abscised as a unit (Page, 1964). The sporangiophores begin development early from the mycelium just below the surface of the substratum. In early evening their tips enlarge to form sporangia. Shortly after midnight a subsporangial swelling appears (Fig. 32-8B), which explodes as a propulsive jet late the following morning because of excess turgor pressure. The sporangia and their spores (Fig. 32-8C) thus are forcibly ejected for distances as great as 6 ft and at a rate of 16 m/sec.[2] *Pilobolus* has an interesting mechanism that detects light and causes the sporangiophore to turn toward it. When the sticky sporangia are ejected, they strike and adhere to the stem or leaf of a nearby plant and are then eaten by herbivorous animals. They are eventually voided, unharmed, in the excreta, where the sporangiospores germinate and repeat the life cycle. *Pilobolus*, the hat thrower (Gr. *pilos*, hat, + *bole*, a throw), is often referred to as the fungus gun. Brodie (1978) in his book has an interesting short chapter on this fungus, entitled "Gunnery in the Fungus World."

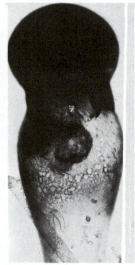

A

B

FIGURE 32-7
Pilobolus sp. **A.** Habit of growth on horse dung;
unilateral light was from the right during
development. **B.** Enlarged view of maturing
sporangium. **A.** ×1.5. **B.** ×315.

---

[2]For a more detailed discussion of this and spore dispersal in general, see Ingold (1971).

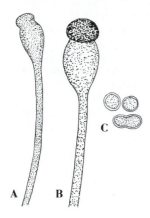

Order Entomophthorales

These are the so-called fly fungi which attack flies on long-unwashed windows of garages and university classrooms. The Entomophthorales shoot their one-spored sporangia off the tips of the sporangiophores. These propagative units are much like conidia[3] in that in most instances they germinate directly by germ tube. A dead fly stuck on a windowpane will usually be surrounded by a white, halolike zone on the glass. This white zone consists of thousands of "conidia" of *Entomophthora* (Gr. *entomon*, insect, + *phthora*, destruction), which have been forcibly expelled by the "conidiophores"[4] issuing in large numbers from the infected body of the fly the fungus has killed and in which it is growing.

An interesting fungus in this order is *Basidiobolus ranarum*, abundant in soils, which grows on the dung of lizards and frogs and shoots off its sporangia. These are eaten by beetles, which in turn are eaten by frogs or lizards. The beetles are digested in the amphibian's stomach and the sporangia of the fungus are freed. There they divide internally to produce spores which multiply by budding in the body of the amphibian and are excreted with the dung. *Basidiobolus haptosporus* is pathogenic to animals and human beings, causing a disease generally known as **Entomophthoromycosis** (Rippon, 1974), or **Zygomycosis** (Dworzack et al., 1978).

Order Endogonales

These fungi encompass what we currently recognize as the vesicular-arbuscular mycorrhizae (VAM). With the exception of saprobic species of *Endogone*, all other genera develop a special symbiotic relationship on the roots of numerous plants; since 95% of all seed plants have V-A mycorrhizae. VAM fungi are characterized by the production of large **zygospores**, **azygospores**, or **chlamydospores** on sparse mycelium in the root hair regions of most plants. Aside from *Endogone*, which has been shown to have true zygospores as a method of sexual reproduction, little is known about the reproduction of other genera (Fig. 32-9).

Close to a dozen genera have been recognized in the single family Endogonaceae. *Gigaspora*, *Glomus*, *Sclerocystis*, and *Acaulospora* have species commonly found on the roots of agricultural crop plants. Research has shown that they are vitally important to a number of crops, especially those growing in low-nutrient soils, becuase of the innate ability of these fungi to absorb phosphorus beyond the capabilities of their host plants (Harley and Smith, 1983). The VAM have evolved interesting features that enable them to transfer nutrients into the root and to store reserve food. Most species form

---

[3]Conidia are asexually produced spores formed usually externally on the apexes or sides of specialized reproductive hyphae (conidiophores).

[4]Conidiophores are specialized reproductive hyphae that bear conidiogenous cells, which form conidia.

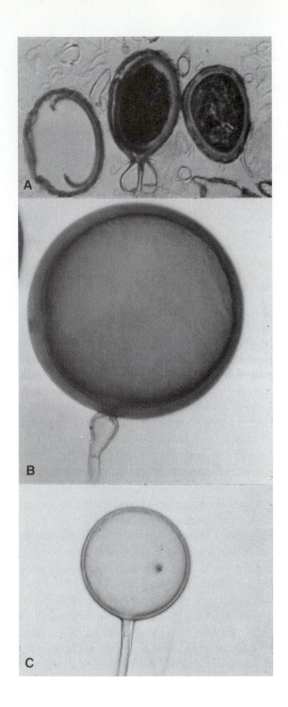

FIGURE 32-9
Spores of the Endogonaceae. **A.** Zygospores of
Endogone pisiformis. **B.** An azygospore of
Gigaspora albida. **C.** A chlamydospore of *Glomus
aggregatus.* **A.** × 600. **B.** × 250. **C.** × 400.
(*Courtesy of Dr. Jack L. Gibson*)

enlarged, thick-walled structures within the
cortical cells of the root referred to as **vesicles.**
The fungus penetrates the cortical cells and
becomes highly branched forming feathery
lobes called **arbuscules,** hence the name
vesicular-arbuscular. Colonized cortical cells
differ from adjacent uncolonized cells only in
their inability to store starch.

CLASS TRICHOMYCETES

The Trichomycetes are always found in association with arthropods, often attached to the gut
of centipedes or millipedes. They do not seem
to be parasitic, however, and are regarded as
commensals.[5]

Trichomycetes do not usually produce an
extensive mycelium. The somatic stage of some
species is confined to a foot cell or holdfast, from
which a sporangium grows. In others, however,
a limited filamentous branched or unbranched
mycelium is produced. The hyphae may be
coenocytic or septate.

Asexual reproduction takes place by amoeboid cells, **arthrospores,** or sporangiospores. All
are aplanospores. Stuctures referred to as zygospores have been reported in almost all genera
of Trichomycetes (Moss and Lichtwardt, 1977).
Although they have not been confirmed cytologically to be zygospores, they typically form
after conjugation between different thalli. Some
display sclariform conjugation similar to that
observed in *Spirogyra.* Such zygospores are
believed to be formed in the same manner as

---

[5]A commensal is an organism that lives with or on
another, not parasitizing it but partaking of the same food
(Webster's International Dictionary, 2d ed., 1933).

those of the Zygomycetes. Lichtwardt (1973*a*) recognized 4 orders: Eccrinales, Harpellales, Amoebidiales, and Asellariales. The Harpellales and Asellariales, because of their peculiar septal pore plugging and pseudophialidlike trichospores, are believed to be related to the Kickxellales of the Zygomycetes. It is chiefly for these reasons that the Trichomycetes are grouped with the Zygomycetes in the subdivision Zygomycotina.

Alexopoulos and Mims (1979) discuss these organisms more fully. For a detailed modern discussion of the Trichomycetes, see Lichtwardt (1973*a,b*, 1976).

Summary and classification

The group we have called Amastigomycota is designated by some mycologists as the Eumycota (true fungi), implying, at least, that the groups we have discussed in the previous chapter are not related to them. One of the great distinctions between the Mastigomycota and the Amastigomycota is in the presence of flagella and, consequently, of centrioles in the former and the absence of both in the latter, with the substitution of spindle-pole bodies (SPBs) for centrioles in most Amastigomycota.

The Zygomycotina included here were formerly placed in the Phycomycetes because their mycelium is typically coenocytic and because they reproduce asexually by sporangiospores, as do the Mastigomycota, which were also in the Phycomycetes. The absence of flagella and centrioles, however, links the Zygomycotina to the Ascomycotina and Basidiomycotina rather than to the Mastigomycota, in accordance with modern phylogenetic theory. Their sexual reproduction is isogamous, and meiosis is zygotic.

The majority of the Zygomycotina (Mucorales) are saprobes or weak parasites of plants. Some, however, cause human diseases (Mucormycoses). Another group (Entomophthorales) causes diseases of insects, and those in still another group (Trichomycetes) are associated with arthropods as commensals. A few (Zoopagales), which we have not discussed, are small animal trappers.

As with most fungal groups, the origin of the Zygomycotina is controversial. The most widely held theory is that they have originated from chytridiaceous ancestors, reproducing sexually through the fusion of isogametangia. The old theory of their oomycetous ancestry has been discarded.

The representative organisms mentioned in this chapter may be classified as follows:

Division Amastigomycota
   Subdivision 1. Zygomycotina
     Class 1. Zygomycetes
       Order 1. Mucorales
         Family 1. Mucoraceae
           Genera: *Mucor, Rhizopus, Phycomyces*
         Family 2. Pilobolaceae
           Genus: *Pilobolus*
       Order 2. Entomophthorales
         Family 1. Entomophthoraceae
           Genus: *Entomophthora*
         Family 2. Basidiobolaceae
           Genus: *Basidiobolus*
       Order 3. Endogonales
         Family 1. Endogonaceae
           Genera: *Endogone, Gigaspora, Glomus, Sclerocystis,* and *Acaulospora.*

DISCUSSION QUESTIONS

1. In what respects do the Amastigomycota differ from the Mastigomycota?
2. Define or explain: zygosporangium, zygospore, aplanospore, mucormycosis, rhizoid, sporangiophore, columella, progametangium, gametangium, suspensor, progamone, arthrospores.

3. Describe the structure and life cycle of *Rhizopus stolonifer*.
4. How do *Rhizopus*, *Phycomyces*, and *Pilobolus* differ from one another?
5. How do the Entomophthorales differ from the Mucorales?
6. What types of nutrition occur among the Zygomycotina?
7. How do the Zygomycotina differ from other Amastigomycota?

Chapter 33
DIVISION AMASTIGOMYCOTA II: SUBDIVISION ASCOMYCOTINA

The three final subdivisions of the fungi—the Ascomycotina, the Basidiomycotina, and the Deuteromycotina—are structurally more complex than the Mastigomycota and the Zygomycotina and are thought to have originated more recently in geologic time. These subdivisions share an important character with the Zygomycotina—that is, the complete absence of flagellate cells from their life cycles, and this is why they are grouped with them in the division Amastigomycota. Even the aquatic and marine species among them lack such cells. Denison and Carroll (1965), among others, believe that the Ascomycotina originated in the sea from nonflagellate ancestors closely related to the red algae, but others favor a zygomycetous ancestry for them. Many mycologists believe that the Basidiomycotina arose from ascomycetous ancestors for reasons that will become evident when we discuss the Basidiomycotina in Chapter 34.

In both Ascomycotina and Basidiomycotina meiosis is zygotic, but in both subdivisions, as we shall see, a dikaryotic phase is interspersed between the haploid portions of the thallus and the zygote. In a recent classification system these two groups are put together in a division Dikaryomycota to emphasize this point (Shaffer, 1975).

CLASS ASCOMYCETES

The subdivision Ascomycotina consists of the single class Ascomycetes, the chief characters of which are summarized below:

1. Spores resulting from karyogamy and meiosis are enclosed in an **ascus** (pl. **asci**), which is a saclike cell containing usually a definite number of ascospores (typically eight), developed by free cell formation. This is a process in which one or more cells are

delimited within a cell in such a way as to leave residual cytoplasm (**epiplasm**)[1] around the spores. It must be emphasized that if a fungus has asci, it is an ascomycete, regardless of other characters it may or may not have; if it does not, it cannot be placed in this class, again regardless of its other characteristics.

2. Absence of flagellate cells.

3. A septate mycelium (in mycelial forms) with centrally perforated septa that divide the hyphae into uninucleate, binucleate, or multinucleate segments and through which the protoplasm of adjacent cells is continuous. Nuclei and other organelles can pass through the septal perforations and travel through the mycelium.

4. Presence of **Woronin bodies** in the hyphal cells. These are ultrastructural elements of a crystalline nature but of unknown chemical composition.

5. Hyphal walls with a large proportion of chitin and, with very few exceptions, devoid of cellulose.

6. Asexual reproduction typically by means of **conidia** (sing. **conidium**), which are spores borne on specialized reproductive hyphae (**conidiophores**).

7. Formation by most species, as a result of sexual reproduction, of sporulating bodies (**ascocarps**) containing the asci.

There are a large number of fungi that closely resemble the Ascomycetes except that they do not reproduce sexually and, therefore, form no asci. Such fungi, classified in the form-class Deuteromycetes, are believed to be Ascomycetes, for the most part, which have lost their sexual (ascus) stages or whose asci are rarely formed and have not been discovered, or which, perhaps, never had a sexual stage. If and when, as often happens, a sexual stage is discovered,

such a fungus is transferred to the proper genus of Ascomycetes.[2]

Importance. Ascomycetes are enormously important in human affairs. A large number are parasitic on plants, causing the serious diseases apple scab, apple bitter rot, brown rot of stone fruits, stem rot of strawberries, and a large number of others. *Endothia* (Gr. *endothen*, from within) *parasitica*, an ascomycete imported to the United States from Asia, which causes chestnut blight, has reduced the American chestnut (*Castanea dentata*) to a minor understory shrub. The Dutch elm disease, caused by another ascomycete, was imported into the United States from Europe and is similarly threatening the American elm (*Ulmus americana*), which was once the dominant shade tree on "Main Street, U.S.A." Oak wilt, a very serious disease of various species of oak (*Quercus* spp.), is caused by still another destructive ascomycete. In addition, most of the fungi that cause human diseases are Ascomycetes or related Deuteromycetes.

On the other side of the coin, there are some very useful Ascomycetes. Yeasts are the basis of the baking and brewing industries. Digesting sugar, they produce alcohol, important in the manufacture of beers and wines, and liberate CO_2, which causes the dough to rise in the baking of bread. Morels and truffles are eagerly sought by "mushroom hunters" for their excellent flavors. Unfortunately, the former have not been grown to the ascocarp stage commercially. The French, however, appear to have succeeded in cultivating truffles by inoculating the roots of seedling, symbiotic trees with the mycelium of the truffles and planting "truffle orchards," so to speak. This requires some explanation. Many soil fungi, not only Ascomycetes, become asso-

---

[1]Epiplasm in the ascus is comparable to periplasm in the oogonium of *Albugo* (Peronosporales). See Chapter 31.

[2]A case in point is the discovery, in the last decade, of ascus stages of many fungi that cause human diseases. These fungi were, until recently, classified as Deuteromycetes. Some, whose sexual stages, if they exist, have not yet been found, are still in that form-class.

ciated with the roots of green plants, forming **mycorrhizae** (sing. **mycorrhiza**) (Gr. *mykes*, fungus, + *rhiza*, root) and living symbiotically with the plants, the fungus obtaining carbohydrates from the plant roots and the plant absorbing water with dissolved salts from the mantles of fungal hyphae that surround them and extend into the soil. The truffle has long been known to be a mycorrhizal fungus living in association with oak and beech trees and producing its ascocarps (the truffles of commerce) below the ground. For an interesting discussion of how truffles are harvested in France and Italy, read Christensen's (1965) *The Molds and Man.*

Some Ascomycetes are both destructive and beneficial. *Claviceps* (N.L. *clavi*, club, + L. *ceps*, headed) *purpurea*, the ergot fungus, causes a destructive diesease of rye and other grasses by infecting their flower parts and replacing the grain with the sclerotia of the fungus. **Sclerotia** (sing. **sclerotium**) are stony-hard bodies consisting of tightly cemented hyphae. When these bodies (known as **ergot**) are consumed by cattle grazing in infected fields, the alkaloids in the sclerotia cause abortion in cows and gangrene of the hooves and tails of the animals. These same alkaloids are deadly to human beings, who contract a disease known as St. Anthony's Fire when they consume bread made with ergotized flour (flour made from grain that has not been thoroughly freed from ergot). Yet some of these ergot alkaloids are eagerly sought by pharmaceutical companies for manufacturing drugs (which cannot as yet be economically synthesized in the laboratory) useful in inducing labor and preventing *postpartum* hemmorhage. It is also of interest that ergot contains lysergic acid, from which LSD is easily synthesized. Although the ergot fungus can be grown in the laboratory on artificial media, no one has succeeded in inducing it to form sclerotia, in which the useful alkaloids are concentrated, apart from its natural hosts. Pharmaceutical companies are, therefore, forced to purchase ergot from farmers who are willing to

collect it before harvesting the grain. In some areas, rye and wheat fields are artificially inoculated with conidia of the fungus in order to increase ergot production.[3]

Somatic structures. Most species of Ascomycetes are mycelial, producing septate hyphae with perforated septa, with the perforations sometimes plugged and with Woronin bodies nearby. The hyphae originate by the germination of ascospores or conidia and grow and branch rapidly.

Asexual reproduction. In the classical life-cycle pattern, the mycelium begins to produce conidiophores soon after it has attained some growth. Conidiophores are specialized, reproductive hyphae, usually with determinate growth, which produce conidia in various ways. Some species produce erect conidiophores from almost any hyphal cell without any seeming organization. These form conidia at their tips or sides, or the cells of the conidiophore may themselves become differentiated and be cut off as a chain of conidia. In other Ascomycetes the conidiophores are organized in **pycnidia** (sing. **pycnidium**), which are generally flask-shaped bodies, their bases lined inside with short conidiophores, producing conidia (**pycnidiospores**) at their tips (Fig. 33-1A). A pycnidium is usually provided with a pore at the top, the ostiole, through which conidia are exuded in a mucoid, tendril- or ribbonlike mass (**cirrhus**), much as toothpaste issues from a tube. In other species, especially in the fungi that cause a disease of plants called anthracnose, the short conidiophores of the parasite form a bedlike mass (**acervulus**; pl. **acervuli**) under the epidermis or cuticle of the infected part (stem, leaf, or fruit), which breaks through to the surface (Fig.

---

[3]In recent years some saprobic strains of *Claviceps purpurea* have been discovered that produce sufficient quantities of alkaloids in their hyphae while growing in submerged culture to make it feasible to extract them from the mycelium. Most ergot alkaloids are, however, still extracted commercially from naturally occurring sclerotia.

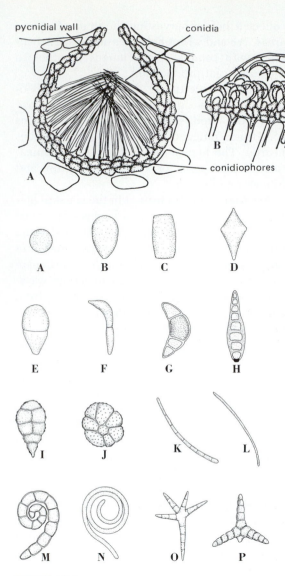

FIGURE 33-1
Asexual reproduction in
Ascomycetes. **A.** Pycnidium
B. Acervulus. Note the
needle-shaped conidia in the
pycnidium of this species.
*(From Alexopoulos [1962], by
permission of John Wiley &
Sons, Inc.)*

FIGURE 33-2
Various types of conidia. **A–D.** Nonseptate. **E–H,
K, M, P.** Transversely septate. **I, J.** Muriform. **K,
L.** Needle-shaped. **M, N.** Helical. **O, P.** Irregular
(Drawn by R. W. Scheetz.)

33-1B) much like the asexual stage of *Albugo*.
This produces enormous numbers of conidia,
which are either splashed by raindrops or carried by insects or wind to other nearby susceptible plants. There are other ways, too, in which

conidiophores are produced, but the three described above are probably the most common.
Conidia vary greatly as to size, shape, number
of cells, and wall ornamentation (Fig. 33-2). In
general we recognize conidia as being hyaline
(colorless) (Fig. 33-2A–F, K–O) or brown (Fig.
33-2G–J, P). However, some are yellow, pink,
or black. In shape, conidia vary from globose
(Fig. 33-2A) to needle-shaped straight or curved
(Fig. 33-2K–L). Some are cross- or star-shaped
(33-2O–P) and some even resemble a tree or a
pair of trousers. In size conidia vary from smaller than 2 μm to as much as 40 μm in length; in
number of cells, from one to many. Some
conidia are septate transversely, whereas others
have both vertical and transverse septa, being
muriform[4] (Fig. 33-2I,J).

Sexual reproduction. Plasmogamy in the
Ascomycetes is accomplished by a variety of
methods. In a typical species, the mycelium
forms coiled **ascogonia** (female sex organs), each
usually bearing a long, hairlike hypha, the **trichogyne**. It also forms either club-shaped **antheridia** (sing. **antheridium**) or large numbers of
very tiny sporelike **spermatia** (sing. **spermatium**) in various ways. Antheridia and spermatia
are male sex organs.

As is the case with the Zygomycetes, the
Ascomycetes, too, may be homothallic or he-

---

[4]The cells of muriform spores are arranged somewhat
like bricks in a wall (L. *murus*, wall).

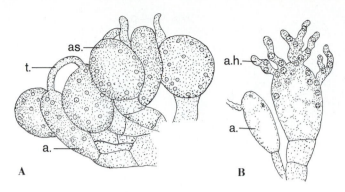

FIGURE 33-3
Pyronema omphalodes. Sexual
reproduction. **A.** Sex organs at
fertilization. **B.** Postfertilization. a.,
antheridium; a.h., ascogenous
hyphae; as., ascogonium; t.,
trichogyne. ×400. (*After
Gwynne-Vaughan and Williamson.*)

terothallic. In homothallic species there are no
mating types and any male sex organ is compatible with any female sex organ of the same
species, whether these are produced on the
same or on different individual thalli. Heterothallic species consist of two mating types,
designated (A) and (a) or A_1 and A_2. Both must
be present so that the sex organs of opposite
mating types may copulate. This implies, of
course, that ascogonia and antheridia (or sper-
matizing cells) produced by the same mycelium
are incompatible and that two individual thalli
of opposite mating types are necessary for sexual
reproduction to take place.

When a trichogyne contacts a compatible
antheridium or spermatium, depending on the
species, the walls dissolve and the contents of
the male gametangium enter the ascogonium
(Fig. 33-3). Nuclei of the male and female
gametangia are thus brought together in the
same cell by plasmogamy in preparation for
eventual fusion. The nuclei, however, do not
fuse at this point. Instead, they appear to form
pairs distributed under the ascogonial wall (Fig.
33-3A), where the oogonium forms buds that
grow into ascogenous hyphae in which the
nuclear pairs migrate (Fig. 33-3B). As the ascogenous hypha grows, the nuclei divide repeatedly by mitosis and become distributed in the
ascogenous hypha, which produces a septum
between the two daughter nuclei of each mitosis. The ascogenous hypha thus becomes sep-

tate, with each cell containing two nuclei, descendants one of the original antheridial and the
other of the original ascogonial nucleus. This is
the dikaryotic phase of the *Ascomycetes*. In the
majority of species, each ascogenous hypha produces a branch from one or more of its dikaryotic cells, the tip of which bends to form a hook
(**crozier**), containing two compatible nuclei,
which now proceed to divide so that the spindles are oriented perpendicularly. With septa
formed between the two daughter nuclei of
each division, the crozier is divided into three
cells (Fig. 33-4A). The penultimate (hook cell) is
dikaryotic, containing one nucleus of each
"sex," whereas the tip and basal cells of the
crozier contain one nucleus each, of different
origin.

Karyogamy takes place in the binucleate hook
cell of the crozier, which is now termed the
ascus-mother-cell (Fig. 33-4B), because as it
elongates it will develop into the ascus. In fact,
elongation often takes place first and karyogamy
occurs in the young ascus (Fig. 33-4C). As the
ascus grows, the diploid nucleus undergoes
meiosis, producing four haploid nuclei (Fig.
33-4D–F), which, in most species, divide once
more by mitosis (Fig. 31-4G), resulting in an
ascus with eight nuclei. A system of membranes
now results in an ascus vesicle, which surrounds
all eight nuclei and which invaginates around
the nuclei and, by the addition of new membrane, enfolds them and cuts them off from the

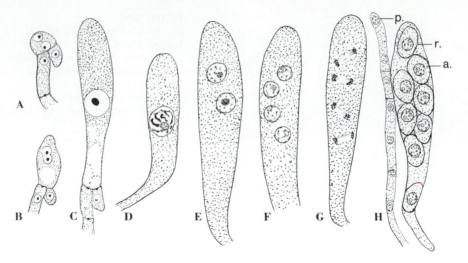

FIGURE 33-4

Pyronema omphalodes. Ascus and ascospore development. **A–C.** Origin of the ascus by crozier formation. **D–G.** Successive nuclear divisions in the ascus. **H.** Paraphysis and ascus. a., ascospore; p., paraphysis; r., residual cytoplasm (epiplasm).

surrounding cytoplasm. A thicker wall then forms around each uninucleate unit, which then develops into an ascospore (Fig. 33-4H). The surrounding cytoplasm in which the spores are embedded is the **epiplasm**, and the whole process is called free cell formation. In the meantime, the basal cell of the crozier fuses with the tip cell and then grows out into a new ascogenous hypha, wherein the history is repeated. Thus, ascogenous hyphae proliferate so that many more asci are formed than there were ascogenous hyphae originally.

In fungi such as *Neurospora* (page 751), in which the asci are cylindrical and narrow, the arrangement of the spores (in pairs) represents the tetrad, and by dissecting and growing the ascospores in the order in which they were arranged in the ascus, it is possible to perform tetrad analysis and determine the way characters are inherited and segregated during meiosis. This is one reason why *Neurospora crassa* has become such an important tool in the study of inheritance and has provided data on which the one-gene–one-enzyme theory is based.

While this procedure is developing asci and ascospores, the stalk of the ascogonium and the hyphal cells below it are stimulated to proliferate and form a wall, which soon surrounds the sexual apparatus and forms an ascocarp.

The ascocarp. There are many kinds of ascocarps produced by the Ascomycotina but in general they may be organized into four main types: The **cleistothecium**, which is globose and has no opening, the **perithecium**, a flask-shaped body with an opening (ostiole) at the top, the **apothecium**, which assumes various forms but is always open with asci exposed, and the **ascostroma**, which is a cushion of somatic tissue (**stroma**, pl. **stromata**) in which one or more cavities are formed by the developing asci. There are also some Ascomycetes that produce their asci naked, without forming ascocarps. Some methods in which Ascomycetes bear their asci are shown in Figs. 33-5 and 33-33.

The ascospores. As is true of conidia, ascospores, too, may be of an infinite variety of sizes, shapes, number of cells, colors, and wall ornamentations (Fig. 33-6). As you might guess,

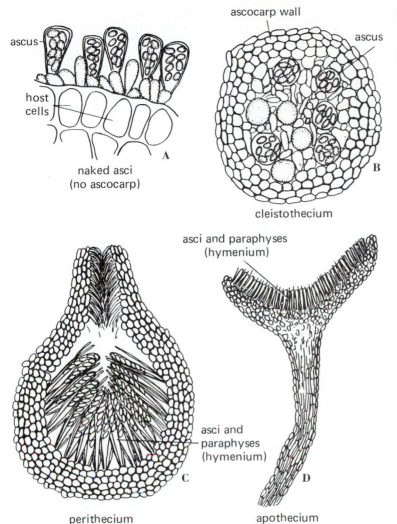

ascus

host cells

naked asci
(no ascocarp)

A

ascocarp wall

ascus

cleistothecium

B

FIGURE 33-5
Four ways in which
Ascomycetes bear their asci.
(*From Alexopoulos, [1962]. By
Wiley & Sons, Inc.*)

asci and paraphyses
(hymenium)

asci and
paraphyses
(hymenium)

C

perithecium

D

apothecium

these characters serve to distinguish genera and species within the group. The release of the ascospores from the asci is apparently due to osmotic pressure that develops in the asci when the complex polysaccharides in the epiplasm are converted to osmotically active simpler sugars at the time the ascospores are mature and ready to be expelled.

Life cycle. In nature many Ascomycetes produce several generations of conidia before sexual reproduction takes place. The asexual cycle is thus chiefly responsible for the propagation and dissemination of the fungus. Sexual reproduction takes place in many species in late summer or early fall, with the ascospores maturing at the end of the season or in very early spring. A typical life cycle diagram of the Ascomycetes is represented in Chart 33-1.

In spite of what we have said above, there is a group of Ascomycetes that form neither asco-

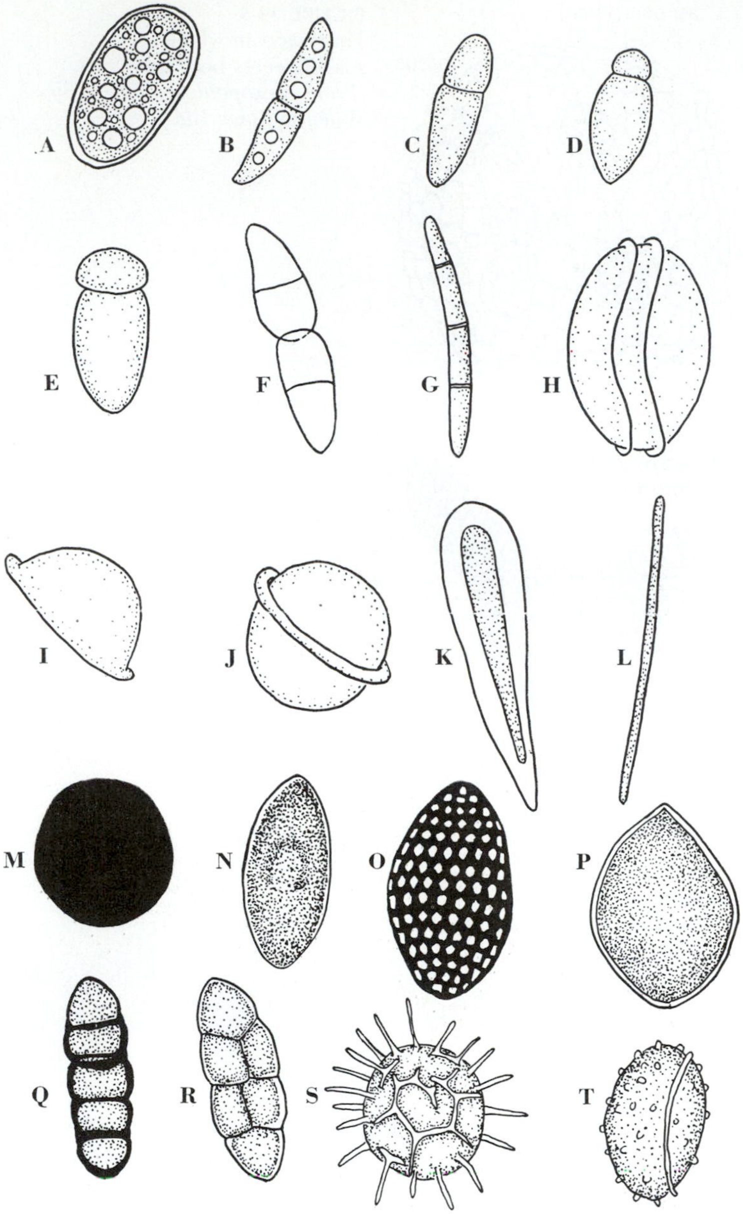

FIGURE 33-6
Various types of ascospores.
(*From Alexopoulos [1962]. By permission of John Wiley & Sons, Inc.*)

carps nor ascogenous hyphae. These are the yeasts and their relatives, which we shall discuss presently.

Classification.   The class Ascomycetes is divided into five subclasses on the basis of ascus and ascocarp characters. These are Hemiascomycetidae, Plectomycetidae, Hymenoascomycetidae, Laboulbeniomycetidae, and Loculoascomycetidae (see classification scheme, p. 763).

740

CHART 33-1
Life-cycle pattern of typical ascocarpic Ascomycete.

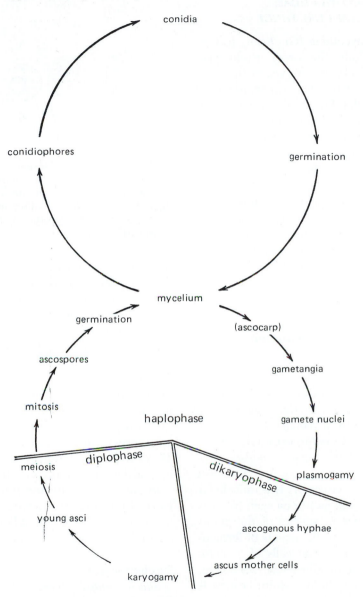

conidia

germination

conidiophores

mycelium

germination

(ascocarp)

ascospores

gametangia

mitosis

haplophase

gamete nuclei

diplophase

dikaryophase

meiosis

plasmogamy

young asci

ascogenous hyphae

ascus mother cells

karyogamy

THE NONASCOCARPIC ASCOMYCETES

SUBCLASS HEMIASCOMYCETIDAE: THE YEASTS AND LEAF-CURL FUNGI

The **Hemiascomycetidae** (Gr. *hemi*, half, + Ascomycetes) include the order Endomycetales, in which we classify the yeasts that produce ascospores (true or ascosporogenous yeasts), and the order Taphrinales (the leaf-curl fungi), among others of less interest to the general student. These fungi are morphologically simple; they produce no ascocarps or ascogenous hyphae, hence the name Hemiascomycetidae.

The true yeasts are unicellular organisms that reproduce asexually either by binary fission or by budding. Sexual reproduction occurs under conditions of reduced food supply and is accomplished by the fusion of two compatible cells to form a zygote, which develops directly into an ascus.

Order Endomycetales: The Ascosporogenous Yeasts

The ascosporogenous yeasts are classified in the family Saccharomycetaceae of the order Endomycetales but are sometimes placed in a separate order Saccharomycetales.

Schizosaccharomyces octosporus (Gr. *schizo*, to cleave, + *saccharon*, sugar, + *mykes*, fungus; *okto*, eight, + *spora*, spore) is the best known of the fission yeasts. It is a unicellular organism that occurs in nature on such fruits as grapes and figs. Certain species of the genus *Schizosaccharomyces* are the agents of fermentation in tropical beers. The cells of *S. octosporus* are spherical to ellipsoidal, vacuolate, and uninucleate (Fig. 33-7A). Multiplication is effected by simple cell division, which follows nuclear division (Fig. 33-7B). Recently divided cells may remain adherent, or they may separate promptly. After several days' growth in laboratory cultures, sexuality occurs. In this process (Fig. 33-7C–G) two adjacent cells pro-

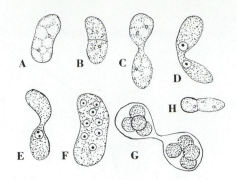

FIGURE 33-7
Schizosaccharomyces octosporus. **A.** Somatic cell. **B.** Cell division. **C.** Cell union (plasmogamy). **D.** Cell union stained; note nuclei. **E.** Zygote with diploid fusion nucleus. **F.** Free nuclei of ascus. **G.** Ascus with eight ascospores. **H.** Ascospore germination. ×1400.

duce short protuberances that meet; their tips dissolve, and plasmogamy and karyogamy follow. The zygote nucleus soon undergoes three successive nuclear divisions, resulting in the formation of eight nuclei (Fig. 33-7F). Two of these divisions constitute meiosis. An ascospore is delimited around each of the eight nuclei thus formed (Fig. 33-7G), leaving epiplasm around them. The zygote, thus, is transformed directly into a single ascus, which ultimately liberates the ascospores when its wall deliquesces. The ascospores contain abundant food reserves. They produce new generations of cells asexually by nuclear and cell division (Fig. 32-7H).

In the related *S. pombe*, it has been shown by Egel (1971) that the nitrogen of the culture medium must be depleted before sexual reproduction and ascus formation can occur.

Saccharomyces

Saccharomyces cerevisiae, a brewer's yeast (Figs. 33-8 to 33-11), is representative of budding yeasts that occur in nature on various fruits. The ovoidal cells of *Saccharomyces* (Gr. *saccharon*, sugar, + *mykes*, fungus) contain a rather large vacuole and an excentric nucleus (Fig. 33-8).

742

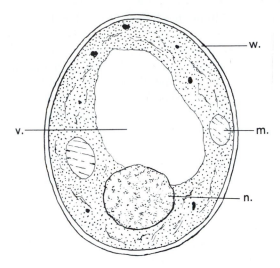

FIGURE 33-8
Saccharomyces cerevisiae. Cellular organization based on electron microscopy. m., mitochondrion; n., nucleus; v., vacuole; w., wall. ×8500.

Multiplication occurs by budding (Figs. 33-9A–C, 33-10, 33-11), during which nuclear division takes place (Robinow and Marak, 1966; Santandreu and Northcote, 1969). One of the daughter nuclei migrates into the bud, which subsequently enlarges and becomes segregated from the parent cell. Rapid budding may result in the formation of short chains of cells. Budding takes place in predetermined spots on the cell wall (Belin, 1972; Hartwell, 1974), which remain as bud scars after the buds have separated from the mother cell.

Saccharomyces cerevisiae is heterothallic, consisting of two mating types, (A) and (a).

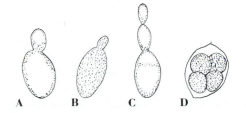

FIGURE 33-9
Saccharomyces cerevisiae. **A–C.** Budding somatic cells. **D.** Ascus with ascospores. ×1700.

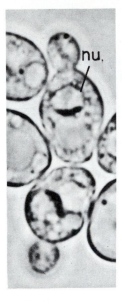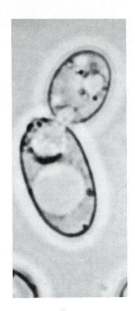

FIGURE 33-10
Saccharomyces cerevisiae. Note large vacuole, stages in budding, and nuclei, nu. ×2700. (*After Robinow and Marak.*)

Under certain environmental conditions sexual reproduction results in the formation of asci, each of which usually produces four ascospores (Fig. 33-9D), two of each mating type. The ascospores from a single ascus, if isolated into

FIGURE 33-11
Saccharomyces cerevisiae. Freeze-etched electron micrograph. Note nucleus preparing for division, partially in parent cell and bud; note 1-μ line on micrograph. (*Courtesy of Balzers Artengeselschaft.*)

743

individual culture vessels, will germinate to form spherical somatic cells, which will continue to reproduce by budding as long as the four cultures remain separated. However, when cells of the two mating types are brought together into one culture, union of the haploid cells in compatible pairs forms diploid zygotes that multiply by budding and establish a diploid population. In this process, the cells of one mating type elongate toward the cells of the other mating type when they are in proximity (Leir, 1956).

A diffusible sex substance (hormone) has been reported to be formed by cells of one mating type of *S. cerevisiae* by Duntze, MacKay, and Manney (1971). The diffusible substance stimulates the cells of the opposite mating type to elongate as they do in preparation for sexual union. Ascospores of different mating type may unite immediately (before undergoing asexual budding) to initiate the diploid phase.

Ascosporogenesis in yeasts differs somewhat from that in other Ascomycetes in that a typical ascus vesicle is not formed. When the diploid nucleus divides, a nuclear plaque, functioning in the same manner as a centrosome in flagellate organisms, divides, and the daughter plaques migrate to opposite sides of the nucleus. Microtubules that are formed become a part of the spindle, and the first meiotic division occurs, but without the nuclear envelope breaking up. Immediately after meiosis I, the two plaques divide again and migrate at an angle to the first spindle, so that the two parallel spindles are formed, while the original nucleus has now formed four lobes. During meiosis II, a double membrane forms partially around each nuclear lobe, and at the completion of meiosis II the membranes close around the nuclear lobes, thus cutting off four ascospores from the surrounding protoplasm (epiplasm). The details of ascosporogenesis in *S. cerevisiae* are beautifully summarized by Tingle et al. (1973). The ascus wall of the yeasts is thin, and when the asci are mature it deliquesces and releases the asco-spores. Although differing in detail, ascosporogenesis in the yeasts is the same in principle as that in other Ascomycetes.

With an alternation of haploid and diploid generations, the first through the budding of ascospores, the second through the budding of zygotes, the life cycle of *S. cerevisiae* is dibiontic. This is reminiscent of the life cycle of *Allomyces* in the Chytridiomycetes.

The importance of various types of yeast to man, because of their biochemical activities, can scarcely be summarized adequately in a short account. Yeasts are used as agents of alcoholic fermentation in brewing and as the leavening agent in baking, and for these reasons have been the subject of intensive cytological, genetic, and physiological investigations.

Order Taphrinales: The Leaf-Curl Fungi

The order Taphrinales includes simple Ascomycetes with a binucleate mycelium parasitic on flowering plants, causing puckering of the leaves, a disease known as leaf curl, and proliferation of small branches that characterize the disease called witches' broom. The most important and best known species is *Taphrina* (*Gr. taphre*, trench or ditch) *deformans*, which causes peach and almond leaf curl. *Taphrina cerasi* causes witches' broom of cherry.

The ascospores of *T. deformans* are released from the asci, on the surface of the peach leaves (Fig. 33-12), and bud like yeasts. Because of this behavior, they are called blastospores. This budding results in a population of free cells, which then germinate, producing germ tubes that penetrate into the peach leaf and form an intercellular mycelium. At the time of germination, the nucleus divides, and the resultant pair of nuclei migrates into the germ tube. As the germ tube develops into a hypha, conjugate divisions of the nuclei perpetuate the binucleate condition of the hyphal cells. The binucleate hyphae continue to grow and branch and eventually become massed just below the leaf cuti-

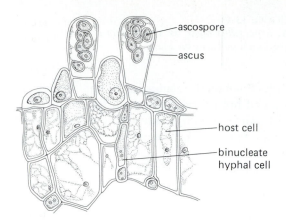

ascospore

ascus

host cell

binucleate
hyphal cell

FIGURE 33-12
Taphrina deformans. Asci with ascospores on the
surface of a host leaf. *(From Martin [1940], Amer.
J. Bot. 27: 743–751. Redrawn from Alexopoulos
[1962] by R. W. Scheetz. By permission of John
Wiley & Sons.)*

cle; here they break up into their component
binucleate cells. Karyogamy, followed by a single mitotic division and septum formation between the daughter nuclei, results in the formation of a diploid basal stalk cell and upper
ascogenous cell. The ascogenous cells continue
to elongate and soon break through the leaf
cuticle to the surface of the leaf. In the meantime, meiosis followed by a single mitosis occurs
in the ascogenous cells. The ascus plasma membrane invaginates to cut off eight ascospores.
These are released through the ascus tip, which
bursts open. Peach and almond leaf curl is
worldwide in distribution and a serious disease
in all peach- and almond-growing regions.

THE ASCOCARPIC ASCOMYCETES

SUBCLASS PLECTOMYCETIDAE:
BLUE MOLDS, BLACK MOLDS, AND HUMAN
PATHOGENS

The Plectomycetidae produce their asci at
various levels in a cleistothecium or more rarely
a perithecium. The asci are thin and evanes-
cent, releasing the ascospores within the ascocarp, from which they escape when the wall of

the cleistothecium breaks or decays. The asci
are usually formed from croziers on ascogenous
hyphae.

Eurotiaceae

Most of these fungi are placed in the order
Eurotiales, the two most important families of
which are the Eurotiaceae and the Gymnoascaceae. In the Eurotiaceae, the genera *Eurotium* (Gr. *euros*, a mold) and *Emericella* have
conidia that are classified in the form-genus
Aspergillus. *Talaromyces* and *Eupenicillium*
have conidia that belong to the form-genus
Penicillium. This requires some explanation.

There are many fungi that reproduce solely
by conidia and never—so far as is known—
sexually. Such fungi are classified in the subdivision Deuteromycotina, form-class Deuteromycetes (*Fungi Imperfecti*). At the same time,
many common Ascomycetes are encountered in
nature mostly in their conidial stages because
they form their asci but once a year. The
conidial stages of such Ascomycetes have the
same characteristics as some Deuteromycetes
and are conveniently classified in the same
form-genera. Thus, the form-genera *Aspergillus*
(L. *aspergillum*, a special type of brush) and
Penicillium (L. *penicillus*, a brush) contain over
100 form-species each, only a few of which also
reproduce sexually and form cleistothecia with
asci and ascospores. For example, *Aspergillus
niger* and *Penicillium notatum* (Figs. 33-13,
33-14) are Deuteromycetes, no sexual stage of
these fungi ever having been found; *Aspergillus
chevalieri*, on the other hand, in addition to its
Aspergillus conidial stage, also produces cleistothecia sexually (Fig. 33-15C, D), which we
place in the ascomycete genus *Eurotium*. Such
fungi actually have two names, one for the
ascocarp genus and one for the conidial genus—
Eurotium chevalieri (= *Aspergillus chevalieri*).[5]

---

[5]Names of Ascomycetes are often written followed by the
name of their conidial stages in parentheses. Thus, here,
Eurotium chevalieri is the name for the ascomycete which
has *Aspergillus chevalieri* as its conidial stage.

A

B

FIGURE 33-13
A. *Aspergillus niger.* B. *Penicillium notatum.*
Petri-dish cultures on agar. × ½.

This situation is confusing to the beginning student but is very convenient for the mycologist, who can simply say that the ascomycete genus *Eurotium* produces *Aspergillus*-type conidia.

Returning to the Eurotiaceae, fungi in the genera *Eurotium*, and *Emericella*, as we have said, form conidiophores and conidia characteristic of the form-genus *Aspergillus* (Figs. 33-14A, B, 33-15A). Some species of *Aspergillus* form dark-brown almost black conidia, whereas in others the conidia are brightly colored in yellow or green hues. The same is true of the genus *Emericella*, which differs from the other two in that its cleistothecia are covered with characteristic "hülle" cells, the function of which is not known. Hülle cells are able to germinate and form mycelium (Ellis, Reynolds, and Alexopoulos, 1973), but it is doubtful whether they are instrumental in propagating the fungus in nature. The ascospores of *Emericella* are of interest in that they are brilliant red in color and have pointed ridges, which makes them appear starlike in side view.

A similar relationship exists between the as-

A   B   C

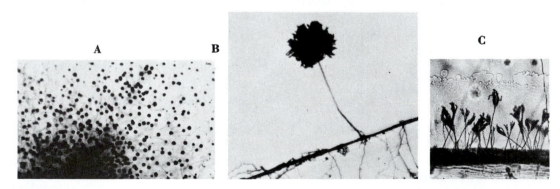

FIGURE 33-14
Photomicrographs of conidiophores and conidia. **A, B.** *Aspergillus niger.* **C.** *Penicillium notatum.*
A. ×12. **B, C.** ×135.

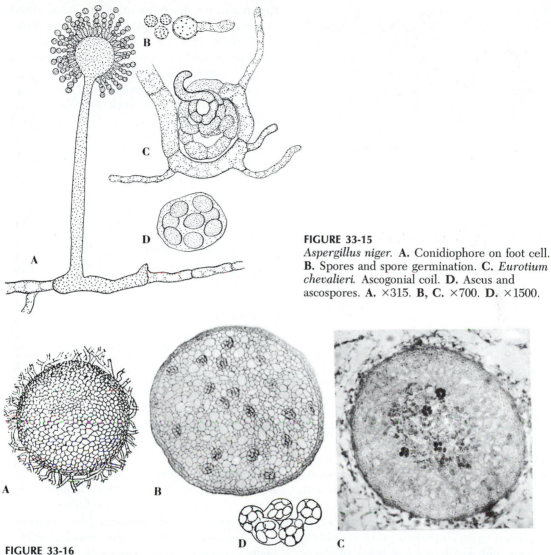

FIGURE 33-15

Aspergillus niger. **A.** Conidiophore on foot cell.
B. Spores and spore germination. **C.** *Eurotium chevalieri.* Ascogonial coil. **D.** Ascus and ascospores. **A.** ×315. **B, C.** ×700. **D.** ×1500.

FIGURE 33-16

A. *Eupenicillium crustaceum.* Cleistothecium surface view. **B.** *Emericella nidulans.* Cleistothecium in section; note asci and ascospores. **C.** *Eupenicillium* sp. Photomicrograph of section of cleistothecium; note asci and ascospores. **D.** *Eupenicillium crustaceum,* asci and ascospores.
A. ×450. **B.** ×600. **C.** ×450. **D.** ×1750. (**A, D.** *After Brefeld from Wolf and Wolf.* **B.** *After Eidam from Raper and Fennell.*)

comycete genera *Talaromyces* and *Eupenicillium* (Gr. *eu*, well or good + *Penicillium*) (Fig. 33-16) and the deuteromycete form-genus *Penicillium.* All known species of *Talaromyces* and *Eupenicillium* form conidiophores and conidia characteristic of the form-genus *Penicillium* (Figs. 33-16C, 33-17).

Of the four ascomycete genera listed above—*Eurotium, Emericella, Talaromyces,* and *Eupenicillium*—only a few species are of importance to

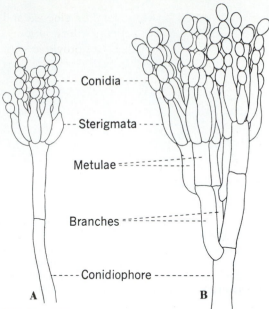

Morphology of the conidiophore (penicillus) of *Penicillium*. **A.** Single-branched penicillus as in *P. frequentans*. **B.** Asymmetric, branching type as in *P. expansum*. ×1100. (*After Raper and Thom.*)

man. *Eurotium* is the perfect stage of *Aspergillus flavus* and related species responsible for the production of **aflatoxins** in stored grains. A number of farm animals, especially turkeys and other fowl, are highly susceptible. Death results from internal hemorrhaging, which is particularly acute in the liver. There has been a direct correlation of liver cancer with the intake of contaminated grains or grain products. Rigid inspections and certification of grain products are maintained to prevent the marketing of contaminated foods. Many other species of *Aspergillus* and *Penicillium* are economically important, but none of these has a perfect stage of any kind. They will therefore be discussed in Chapter 35 with the Deuteromycetes.

Gymnoascaceae

Relatively few of the Eurotiaceae as such are of importance in human affairs. On the contrary, the Gymnoascaceae—the other family of the Eurotiales we have mentioned—is of direct importance because it contains the ascus stages of all fungi that cause human skin diseases (ringworms) and of some that cause diseases of internal organs in humans.

The asci of the Gymnoascaceae are usually enveloped by a loose network of hyphae, sometimes ornamented with various types of appendages. This is the cleistothecium, which is sometimes nothing more than a loose cottony hyphal mass. The ringworm fungi—usually known as the dermatophytes—never form cleistothecia on the infected skin and seldom in nature, where they occur on such animal substrata as hair, feathers, and dung. Some are also found in the soil or even on plant debris. The majority of the dermatophytes whose perfect stages have been discovered are heterothallic. It appears, however, that the ascospores are of little importance biologically and that these fungi depend for their multiplication on their conidia, most of which belong to the form-genera *Microsporum* (Gr. *mikron*, small) and *Trichophyton* (Gr. *thrix, trichos*, hair, + *phyton*, plant) of the Deuteromycetes. The deuteromycete form-genus *Epidermophyton* (Gr. *epidermis*, skin, + *phyton*, plant) is also important in causing athlete's foot. No ascus stage has been discovered for this form-genus.

There are two genera of Gymnoascaceae that are implicated in the causation of human skin diseases: *Arthroderma* (Gr. *arthron*, joint, + Gr. *derma*, skin) and *Nannizzia* (after Nannizzi, an Italian mycologist). *Arthroderma* has *Trichophyton* conidia, and *Nannizzia* has *Microsporum* conidia. The various skin diseases caused by fungi are medically known as *Tineas*. Thus we have *Tinea barbae* (barber's itch), *Tinea corporis* (body ringworm), *Tinea pedis* (foot ringworm, or athlete's foot), *Tinea cruris* (ringworm of the groin), and others. Up to recent years it was very difficult to treat these diseases, but with the discovery of griseofulvin, an antifungal antibiotic, treatment is now assured, even if long. More about griseofulvin when we discuss the Deuteromycetes.

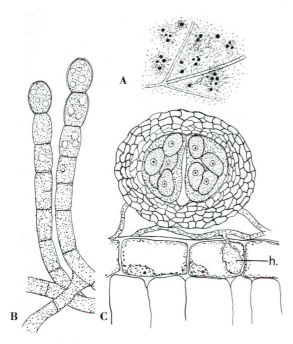

FIGURE 33-18
Microsphaera alni. **A.** Mycelium and cleistothecia on lower surface of lilac leaf. **B.** Conidiophore with seriate conidia. **C.** Stained section of cleistothecium on upper surface of lilac leaf. h., haustorium in epidermal cell. **A.** ×3. **B.** ×275. **C.** ×400.

In addition to the dermatophytes, the Gymnoascaceae include the perfect stages of at least two fungi that cause deep mycoses. Deep mycoses are fungal diseases that affect human internal organs such as lungs. *Ajellomyces dermatitidis* (after Ajello, an American mycologist) is the perfect stage of the fungus that causes blastomycosis, a serious disease of the skin, lungs, and bones, and *Ajellomyces capsulatus* causes histoplasmosis, a widespread, serious respiratory disease.

SUBCLASS HYMENOASCOMYCETIDAE

The subclass Hymenoascomycetidae includes the powdery mildews, the perithecial fungi, and the cup fungi. Their chief characteristic is the formation of asci in a basal layer, the hymenium, inside an ascocarp, which may be globose and completely closed as in the powdery mildews (Erysiphaceae), flask-shaped and provided with an opening (ostiole) at its apex, as in *Neurospora* and *Sordaria*, or open from the beginning and variously shaped.

This is a very large subclass with several orders and many hundreds of species. We shall mention only a few forms as examples.

Order Erysiphales: Powdery Mildews

The powdery mildews are so called because they form a mealy, powdery white stratum on the surfaces of leaves in a number of plants. Their mycelium is obligately parasitic on a specific host, because all attempts to grow them for prolonged periods in artificial culture have thus far failed. Examples of powdery mildews that infect well-known plants are the following: *Microsphaera alni* (Gr. *mikro*, small, + *sphaera*, sphere) on lilacs and oaks (Fig. 33-18), *Erysiphe cichoracearum* on garden plantain, *Sphaerotheca pannosa* (Gr. *sphaera*, sphere + *theke*, box, sheath) on roses and peaches, *Erysiphe* (Gr. *erys*, red, + Gr. *siphon*, tube) *graminis* on cereal grains, and *Uncinula* (L. *uncinus*, a hook) *necator* on grapes. Ascocarps typical of the three genera are shown in Fig. 33-19. The mycelium spreads over the leaf from the original point of infection (Fig. 33-18A) and obtains nourishment by means of haustoria that penetrate into the epidermal cells (Fig. 33-18C). The hyphal cells of most species are uninucleate. After a period of vegetative growth, certain hyphae in most species produce erect branches, which form conidia in chains (Fig. 33-18B). These are blown about by air currents and germinate, initiating new infections. Unlike those of most other fungi, the conidia of many powdery mildews not only do not require free water for germination but are able to germinate in very dry air, even at 0% relative humidity. What it is that enables them to do this is not known (Brodie, 1945). The asexual cycle, frequently repeated, rapidly spreads the fungus.

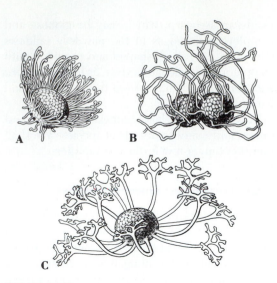

FIGURE 33-19
Cleistothecia of powdery mildews. **A.** *Uncinula salicis.* **B.** *Sphaerotheca humuli.* **C.** *Microsphaera berberidis.* (After Bessey.)

Sexual reproduction and ascocarp formation occur later in the growing season. About mid-summer, in the temperate zones, the mycelium begins to form cleistothecia. These first appear as minute white knots on the mycelium but soon grow and change color to cream, yellow, orange, and finally dark red to brown, nearly black. The sex organs, which precede the cleistothecium, are not highly differentiated but consist of short hyphae that curve around each other. One has been identified as an antheridium and the other as the ascogonium. The walls between these dissolve at one point of contact, and the antheridial nucleus is reported to migrate into the ascogonium. Nuclear union is probably delayed, as in other Ascomycota. Descendants of the sexual nuclei are distributed among the cells of the ascogenous hyphae. The latter, depending on the genus, give rise to one or more asci. Soon after the sex organs have developed, somatic hyphae at their base form a sterile protective layer, which becomes the wall

of the cleistothecium. Three successive nuclear divisions occur in each ascus, so eight potential ascospore nuclei are developed. In some species each of these is delimited to form an ascospore by free cell formation. In others, fewer ascospores are produced, and the supernumerary nuclei disintegrate in the epiplasm. The cleistothecia (Figs. 33-18A, C, 33-19) usually remain on the leaves when the latter are shed, and dissemination of the ascospores does not occur until the following growing season.

In most localities in North America the ascospores do not mature until late fall or early winter, the fungi overwintering in the cleistothecial stage. Ascospore discharge has been studied in *Podosphaera* (Gr. *pous, podos,* foot, + Gr. *sphaera,* sphere), which produces a single large globose ascus with eight spores in each cleistothecium. As the cleistothecium absorbs water in the spring, the ascus expands and the cleistothecium bursts, expelling the entire ascus forcibly. In flight, the ascus explodes and also releases its spores explosively. These, in turn, germinate on the surface of a susceptible host and start the life cycle over again. Powdery mildews as a rule form no ascocarps in the tropics, where they reproduce entirely by conidia.

Cleistothecia of the powdery mildews bear appendages of different types, which are useful in anchoring the cleistothecium to the leaf surface. The type of appendage combined with the number of asci—one or many—within the cleistothecium are the characters on which the genera of the Erysiphaceae are separated. Thus, *Erysiphe* and *Sphaerotheca* have mycelioid indefinite appendages (Fig. 33-19B). In *Microsphaera* and *Podosphaera* the tips of the appendages are dichotomously branched (Figs. 33-19C, 33-20E, F); *Phyllactinia* (Gr. *phyllon,* leaf, + Gr. *aktis,* ray) has appendages with bulbous bases and pointed tips (Fig. 33-20C, D), whereas *Uncinula* and *Pleochaeta* (Gr. *pleio,* more, + Gr. *chaete,* mane) have appendages with uncinoid (curled) tips (Figs. 33-19A, 33-20A, B). Of these genera—all that occur in North America—*Sphaerotheca* and *Podosphaera* produce only a single large ascus in each cleistothecium, whereas each cleistothecium of all the other genera mentioned contains many asci.

Perithecial Ascomycetes (Pyrenomycetes)

The genera *Sordaria* (N.L. *sordes*, dirt or filth) and *Neurospora* (Gr. *neuron*, nerve, + Gr. *spora*, seed, spore), both easily grown in the laboratory, are excellent for classroom demonstration of perithecial fungi. *Neurospora*, you will remember, is the fungus that has been used so successfully by geneticists to elucidate the laws of heredity. It has become the *Drosophila* of the fungus world.

In *Neurospora* and *Sordaria* the ascocarps are perithecia. All species of these genera are easily grown on laboratory media. In addition, *Neurospora sitophila* and certain isolates of *Sordaria fimicola* are indispensable for demonstrating heterothallism and the segregation of characters in the ascus, as we shall see.

Neurospora

The conidiophores of *Neurospora* are not markedly differentiated from the multinucleate vegetative hyphae (Fig. 33-21A). Some species of this genus produce minute, uninucleate microconidia in addition to those of ordinary size; the latter are multinucleate and are called macroconidia. The conidial walls are responsible for the pink color of the fungus. In *N. sitophila* the young ascogonium, a curved septate hypha with several nuclei in each cell, becomes covered with several layers of interwoven sterile hyphae. This structure has been called a **protoperithecium**. Certain cells of the ascogonium produce long, tenuous, trichogynelike branches, which penetrate the ster-

ile hyphal layers surrounding the ascogonium. It has been demonstrated that not only the microconidia but also the macroconidia and even vegetative hyphae and trichogynes of one strain may unite with the trichogynes and vegetative hyphae of another compatible strain. Fusion of compatible vegetative hyphae is called **somatogamy.** Inasmuch as *Neurospora sitophila* is heterothallic, two different mating types are required for sexual reproduction to take place. Even though a mycelium derived from a single ascospore or conidium produces both ascogonia and microconidia (i.e., is monoecious), these sex organs are incompatible and cannot function among themselves to produce ascospores because they are of the same mating type.

In various ways cited above nuclei of the two mating types are brought together into the same mycelium and ultimately into the ascogonial cell. The latter now gives rise to ascogenous hyphae, the tips of which enlarge to form elongate asci. Meanwhile, the sterile layer of the protoperithecium has increased in extent and organized itself into a perithecium (Fig. 33-21B), at the apex of which a small aperture, the ostiole, develops.

The young asci of *Neurospora* are binucleate. The two nuclei of each ascus represent descendants of nuclei of the two mating types originally brought together in trichogynal or other types of plasmogamy. In further development, nuclear fusion takes place in each young ascus. This is soon followed by three successive nuclear divisions, during which meiosis is accomplished. The asci at this stage contain eight linearly arranged nuclei. These, with a portion of their surrounding cytoplasm, are finally segregated from the residual cytoplasm of the ascus by free cell formation (Fig. 33-22). The mature ascospores become binucleate as a result of mitosis within each spore. At maturity they are discharged from the perithecium through its ostiole. The mature spore walls are ribbed, an attribute that suggested the generic name.

FIGURE 33-20
Cleistothecia of three genera of powdery mildews, showing characteristic appendages. **A, B.** *Uncinula.*
A. Cleistothecium with appendages. **B.** Appendage tips. **C, D.** *Phyllactinia.* **C.** Cleistothecium with appendages. **D.** Close-up of one appendage and base of a second (right). **E, F.** *Podosphaera.*
E. Cleistothecium with appendages. **F.** Close-up of appendage tips. **A.** ×220. **B.** ×690. **C.** ×140.
D. ×690. **E.** ×230. **F.** ×800. (*Scanning electron micrographs by R. W. Scheetz.*)

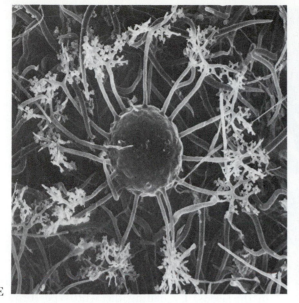

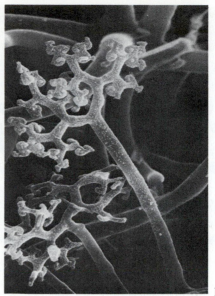

E

F

The ascospores germinate in laboratory culture after suitable treatment (with heat or furfural), giving rise to a mycelium that produces only protoperithecia and conidia, unless contact is made with a mycelium or conidia of a compatible strain. It has been shown experimentally that four of the eight ascospores of each ascus of *N. sitophila* give rise to one mating type and that the other four are of the opposite mating type.

Various species and races of *Neurospora* have provided the basis for important genetic and

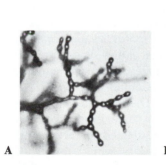

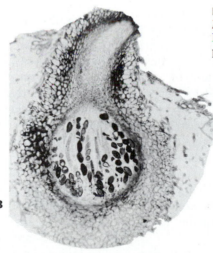

A   B

FIGURE 33-21
Neurospora sitophila. **A.** Conidia.
B. Median longitudinal section of
perithecium. **A.** ×175. **B.** ×135.

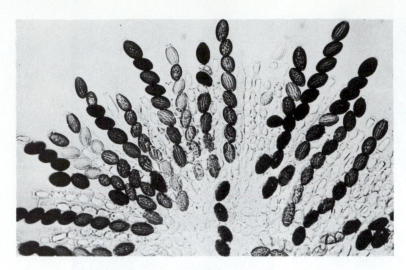

FIGURE 33-22
Neurospora sitophila.
Asci and ascospores.
×1000. (*After B. O. Dodge.*)

biochemical studies. It should be noted that the necessity for fusion between two strains of *N. sitophila* for maturation of the perithecia is analogous to the self-incompatibility found in certain types of flowers. As in certain flowers, both types of reproductive organs are present but fail to function; the controlling factor here is apparently physiological.

Sordaria, Gelasinospora

Sordaria, an organism similar to *Neurospora,* differs from the latter, among other respects, in producing smooth-walled ascospores. *Sordaria fimicola* (Fig. 33-23) does not produce conidia or microconidia but reproduces solely by ascospores. The latter are surrounded by a gelatinous sheath. The perithecial necks are positively phototropic, as they are in many ascocarpic Ascomycetes. As asci mature within, one of them at a time enlarges and protrudes through the ostiole. Its ascospores are violently discharged, the ascus collapses, and then, in turn, another protrudes to liberate its ascospores; such proliferation of asci is common in the ascocarpic Ascomycetes.

By subjecting cultures of *Sordaria fimicola* to ultraviolet irradiation, Olive (1956) obtained ascospore color mutants. This is extremely useful for demonstrating segregation of characters in the ascus. By mating the wild type (brown ascospores) with the mutant (colorless or gray ascospores) and examining the hybrid asci produced, the student has a visual demonstration of how the color character segregates in the asci.

Gelasinospora, with pitted ascospores (Fig. 33-24), is another genus closely resembling *Sordaria* and *Neurospora.*

The Apothecial Ascomycetes (Discomycetes)

The Discomycetes are Ascomycetes that produce their asci in apothecia. An apothecium is an open ascocarp with its hymenium exposed from the time the ascospores are mature and in many species at an even earlier stage. Typically, paraphyses (sterile threads) are interspersed with the asci in the hymenium (Fig. 33-27A, B). The apothecia of most Discomycetes are cup- or saucer-shaped, but in some groups they are club-shaped or tongue-shaped, or even mushroom-shaped. The morels, which also belong here, are spongelike, and the apothecia of the false morels are saddle-shaped or convoluted like brains perched on a stalk. The bell-morels have bell-like apothecia. In some species the apothecia are funnel-shaped.

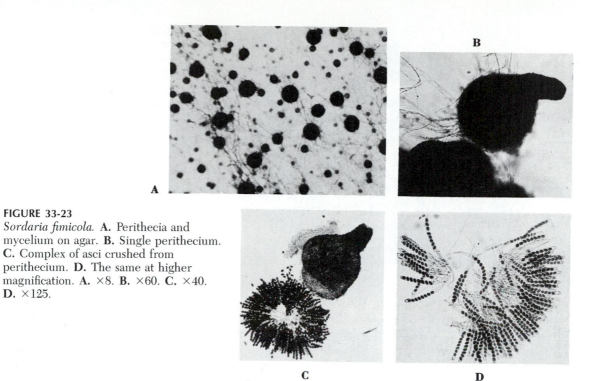

A

B

C

D

FIGURE 33-23
Sordaria fimicola. **A.** Perithecia and mycelium on agar. **B.** Single perithecium. **C.** Complex of asci crushed from perithecium. **D.** The same at higher magnification. **A.** ×8. **B.** ×60. **C.** ×40. **D.** ×125.

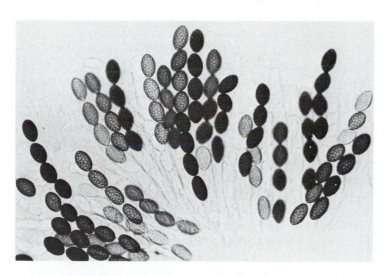

FIGURE 33-24
Gelasinospora autosteira. Asci and ascospores. ×1000.

The majority of Discomycetes are saprobic, but a few are parasitic, causing serious diseases of economic plants. Such is *Monilinia* (L. *monile*, necklace) *fructicola*, the cause of brown rot of stone fruits, probably the most serious disease of peaches. The life history of this fungus is as follows: Ascospores are released from the apothecia in early spring when the peach trees

FIGURE 33-25
Humaria axillaris, a cup fungus (among plants of the moss *Funaria*). ×6.

are in bloom. Those that fall on the flowers germinate, and the mycelium invades the flowers and causes them to blight. Conidia, produced in long chains on the flowers from conidiophores, are blown by the wind and spread the disease to other blossoms and eventually to the young peach leaves, causing leaf and twig blight. The young fruit is not susceptible to the fungus, but when the fruit matures, the conidia germinate on the fruit surface and the germ tubes penetrate through insect punctures or other small wounds such as broken hairs. The enzymes secreted by the hyphae dissolve the middle lamellae of the fruit cells and cause a softening of the invaded tissue, which turns brown. Eventually the fruit rots and shrivels, becoming a **mummy**. Some mummies remain on the trees after the ripe peaches are harvested, but many fall to the ground and are partially buried. Before winter sets in, the mycelium in the mummies produces ascogonia and microconidia. The details of sexual reproduction have not been completely determined, but it appears that plasmogamy takes place between trichogynes and microconidia, which results in the formation of apothecial primordia. In early spring, these develop into funnel-shaped apothecia on the surface of the mummies that have

overwintered on the orchard floor. Weather conditions that bring about blossoming of peach trees also favor the maturing and release of the ascospores, which shoot up from the ground and are deposited by air currents on the peach blossoms, starting the cycle over once more. It is of interest that mummies that hang on the trees through the winter do not produce apothecia but form new crops of conidia from the overwintered mycelium. The probable explanation for this is that conditions on the ground are favorable for spermatization but not so on the mummies hanging from the trees, which probably do not remain wet long enough for spermatia and ascogonia to be formed and spermatization to take place. This is only conjecture, however.

Many saprobic Discomycetes often produce brightly colored conspicuous apothecia (Fig. 33-25). An example is the scarlet cup fungus (*Sarcoscypha* [Gr. *sar, surkos,* flesh + Gr. *kyphos,* crooked]) *coccinea* (Fig. 33-26A), a widely distributed fungus you can find in the spring growing on dead wood in the forests. Several other genera, among them *Pyronema* (Gr. *pyr,* fire, + *nema,* thread, skein, i.e., hypha), are inhabitants of burned-over or sterilized soil (Fig. 32-26B, C). The apothecia of Disco-

FIGURE 33-26

A. *Sarcoscypha coccinea*, apothecium. **B, C.** *Pyronema domesticum.* **B.** Apothecia on soil.
C. Section of apothecium; note hymenial layer with asci. **A.** ×2. **B.** ×8. **C.** ×60. (**A.** *After Alexopoulos, 1962. By permission of John Wiley & Sons.*)

mycetes vary from a millimeter in diameter up to the size of small teacups. In some genera the apothecia are stalked.

Although the apothecium is somewhat ephemeral, its formation is in all cases preceded by an extended period of vegetative activity on the part of the mycelium, which ramifies in the substratum, absorbing nutriment. In a few genera the vegetative mycelium reproduces itself asexually by conidia, but these are entirely absent in most others. There is good reason to believe that the apothecium typically arises as a result of sexuality, but the latter has been clearly demonstrated only in a few species, and

757

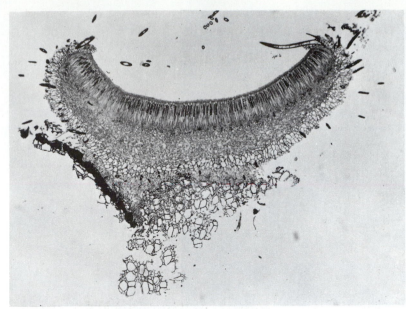

A

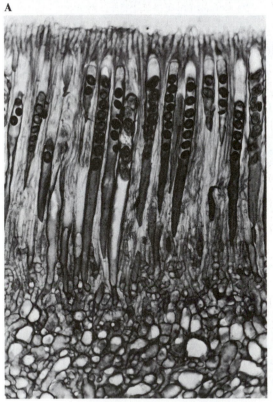

B

even in these there is a difference of opinion
regarding the cytological details of the process.
Pyronema has been investigated frequently regarding its cytological and sexual features. In
this genus well-differentiated antheridia and
ascogonia are developed (Fig. 33-3). The explanation of ascus formation through croziers on p.
737 and in Fig. 33-4 is based mostly on investigations of *Pyronema*.

The inner surface of the apothecium in cup
fungi, the **hymenium**, is composed of intermingled columnar asci and sterile paraphyses (Figs.
33-26C, 33-27). The remainder of the apothecium is made up of sterile interwoven hyphae that
form pseudoparenchyma, as viewed in section.
Spore discharge is explosive in many species,
and large numbers of spores may be disseminated simultaneously in visible puffs.

The apothecia of *Pyronema* are relatively
small and inconspicuous, and plane or almost
convex. The apothecia of *Peziza* (Gr. *pezis*,
puffball), *Humaria* (L. *humus*, earth) (Fig.
33-25), *Urnula* (L. *urnus*, pitcher jar), and *Sarc-*

FIGURE 33-28
Morchella angustipes. (Courtesy of Dr. Stanley Flegler.)

France morels have been grown by crumpling ascocarps over soil with which apple mash from nearby cider factories had been mixed, but apparently this method cannot be depended upon for commercial purposes because there is no morel-growing industry anywhere in the world. Nevertheless, it is said that growing morels out of doors as a horticultural crop presents no problem (Heim, 1936; Singer, 1961). Such morels are harvested in the spring at the same time that wild morels appear. No one has succeeded in growing them commercially like mushrooms, which are harvested at least twice a year.

The saddle fungi produce large apothecia that resemble saddles on a central stalk. There are a

oscypha (Fig. 33-26A) are widely distributed, larger, and more conspicuous than those of *Pyronema*; they are flat or concave. Unfortunately, however, their life cycles have not been worked out as completely as that of *Pyronema*. *Chorioactis* (Gr. *chori*, apart, distinct), with the single species *C. geaster*, is of interest because the apothecium, closed at first, splits open like an earthstar (p. 778) and because of its peculiar distribution: Texas and Japan!

Morels, False Morels, and Saddle Fungi

A word should be added about morels, saddle fungi, and false morels. Morels belong to the genus *Morchella* (Gr. *morchel*, morel), which includes several species. All morels have a stalked, spongelike apothecium (Fig. 33-28, 33-29). All are edible and delicious. Unfortunately, the morel season is short, not exceeding one month in the spring. They can be grown in culture from spores to the mycelial stage, but no one has succeeded in producing ascocarps on laboratory media. It has been reported that in

FIGURE 33-29
Morchella esculenta. × 1. *(Courtesy of Dr. Stanley Flegler.)*

759

FIGURE 33-30
Helvella crispa. × 1. *(Courtesy of Dr. Stanley Flegler.)*

FIGURE 33-31
Discina gigas. × ½

FIGURE 33-32
Gyromitra esculenta. × 1. *(Courtesy of Dr. Stanley Flegler.)*

number of genera, among which are *Helvella* (L. *helvella*, a small pot herb) (Fig. 33-30) and *Discina* (Gr. *diskos*, disc, + L. *ina*, like) (Fig. 33-31). Some of these are edible, but some are poisonous. The false morels are in the genus *Gyromitra* (Gr. *gyros*, rounded + Gr. *mitra*, girdle) (Fig. 33-32). The heads (pilei) of the apothecia of some species are highly convoluted, resembling brains on top of thick, ridged stalks. Again, these may be eaten safely by some people but are poisonous to others and should therefore be avoided by mycophagists.[6] The

poison referred to in earlier books as **gyromitrin** or **helvellic acid** turns out to be **monomethylhydrazine** (MMH). The aerospace industry

---

[6]A mycophagist is one who collects mushrooms for the table, a mushroom eater.

uses large quanities of MMH and other hydrazins as propellants for rockets. Illness of workers exposed to MMK was identical to that from *Gyromitra* poisoning.

Truffles

Truffles are mycorrhizal fungi that produce their ascocarps below the ground. Their spores are generally dispersed by animals digging up the truffles for food. Truffle ascocarps emit a strong odor, which enables animals to find them. These fungi have been found in many parts of the world, but only the kinds native to southern France, northern Italy, and the shores of the Mediterranean generally are edible and choice. Those in other areas, although not poisonous, are not worth hunting for the table. The edible truffles belong chiefly to the genera *Tuber* (L. *tuber*, truffle) and *Terfezia* (Arabic *terfez*, truffle). The latter has been known since ancient times, when it was believed truffles were formed by thunderbolts striking the earth. The Arabs of North Africa prize them as food. Most of the truffles known from the United States were described from California and Oregon, chiefly because Dr. Helen Gilkey, who was interested in studying them and became an authority on this group, studied in California and later worked at Oregon State University. It is axiomatic that the known distribution of living organisms coincides with the distribution and the activities of the biologists who hunt them!

Truffle mycelium can be grown in artificial culture in the laboratory but does not produce ascocarps. In recent years, Italian and French scientists have succeeded in artificially synthesizing truffle mycorrhizae by inoculating susceptible tree seedlings with the mycelium or spores of the edible truffles. This research was published in Italian with English summaries (Palenzona, 1969; Palenzona, Chevalier, and Fontana, 1972). The last comprehensive report on truffles in English is Singer's (1961) *Mushrooms and Truffles*.

SUBCLASS LABOULBENIOMYCETIDAE

This is a very large group of minute, very specialized fungi mostly on insects and some mites. A few are also known from marine red algae in the Southern Hemisphere. They seem to do no damage to their hosts and are of little interest to the general student although they are quite abundant.

SUBCLASS LOCULOASCOMYCETIDAE

The Loculoascomycetidae produce their asci in cavities (locules) in a stroma. A stroma is a cushion- or mattress-shaped structure that bears reproductive organs. Often, only a single locule is produced in a stroma and is hard to distinguish from a perithecium. Such a stroma is called a **pseudothecium** (Fig. 33-33). There are a large number of Loculoascomycetidae that are of great economic importance. The most common perhaps is *Venturia inaequalis*, which causes the widespread apple scab, found everywhere apples are grown. In *Venturia* (after A. Venturi, nineteenth-century Italian botanist), sterile threads, the **pseudoparaphyses,** originate at the apex of the locule and grow down, anastomosing among the asci, which grow up from the base among them. Pseudoparaphyses are produced by a number of other Loculoascomycetidae. Among the most economically important Loculoascomycetidae that form no sterile threads in their pseudothecia is *Guignardia bidwellii* (in honor of Guignard, a French mycologist). In addition to asci in pseudothecia, this fungus produces conidia in pycnidia (Fig. 35-1).

Summary and classification

The second large group of Amastigomycota comprises the so-called sac fungi or Ascomycetes, all placed in the subdivision Ascomycotina. These, together with the Basidiomycotina, are sometimes grouped under the term

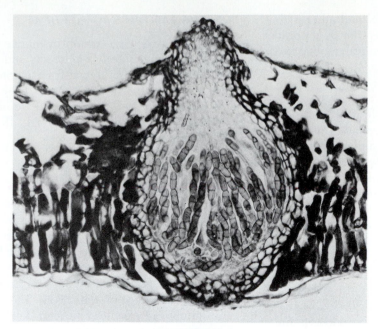

FIGURE 33-33
Venturia inaequalis. Median
longitudinal section of a
pseudothecium showing asci
with two-celled ascospores.
(*Courtesy of Triarch, Inc.*)

Dikaryota because of a characteristic dikaryotic phase, which is typically interspersed between plasmogamy and karyogamy.

Of the representative genera of Ascomycetes mentioned in this chapter, some, such as *Monilinia, Guignardia,* and *Venturia,* are parasitic on plants but can also survive saprobically; others, such as the powdery mildew, are obligate parasites on plants, and some are parasitic on human beings and animals, causing skin ailments or deep mycoses. Ascomycetes in such genera as *Eurotium, Emericella, Talaromyces,* and *Eupenicillium,* as well as the majority of the Discomycetes (cup fungi), are saprobic.

Except for the unicellular yeasts, the somatic phase regularly consists of a branched, septate mycelium, which may be composed of unicellular (in the powdery mildews) or multicellular compartments delimited by perforate septa of simple construction and with Woronin bodies. Some yeasts may develop short myceliumlike stages under certain conditions. For this reason they often have been interpreted as organisms reduced from "higher," strictly mycelial forms.

The life cycle of most Ascomycetes consists of an asexual phase, in which conidia are produced and a sexual phase, in which the zygote, either directly as in the yeasts or indirectly as in other genera, produces asci. Indirectly ascus formation involves the production of dikaryotic ascogenous hyphae and generally of croziers.

The diploid somatic cells of *Saccharomyces* and other yeasts become transformed directly into single asci. In most other Ascomycetes there is a dikaryotic phase between plasmogamy and karyogamy; the fusion cell, in which no karyogamy occurs, produces dikaryotic ascogenous hyphae, which give rise to asci in which karyogamy and meiosis occur. In the yeast *Saccharomyces,* as a result of one sexual union and budding of the diploid cells, many agents of propagation (in this case, ascospores) are produced. Further examples of the same phenomenon among the algae and other groups of plants will occur to the reader.

Both the conidial and the ascogenous (ascusforming) stages may be well developed, as in the powdery mildews, in *Neurospora,* in *Eurotium,* and in *Eupenicillium,* or one or the other phase may be absent from the life cycle. Thus, some genera of Ascomycetes have *Aspergillus* and *Penicillium* imperfect (conidial) stages, but for most species of these form-genera no ascogenous stages have been discovered, and such species must, therefore, be classified with the Deuteromycetes or imperfect fungi. In many cup fungi and in some species of *Sordaria*, among others, on the other hand, no conidial stages have been observed.

The Ascomycetes exhibit considerable range of variation with reference to the production of differentiated sex organs. Among the yeasts, the haploid somatic cells, or even the ascospores, function directly as gametes, as in the alga *Chlamydomonas* (Figs. 4-2 to 4-5). In other genera, markedly differentiated sex organs may be present, as in *Pyronema*. Among the powdery mildews the so-called antheridia and ascogonia are scarcely distinguishable from each other or from vegetative hyphal branches. Ascogonia have been observed in a number of species of *Eurotium* and *Eupenicillium*, but differentiated male organs are rarely present. In *Neurospora* the number of alternative mechanisms by which approximation of sexually compatible nuclei can be effected suggests that the ancestral male sex organs may have been lost.

As to the origin of the Ascomycetes, several hypotheses have been suggested. According to one, they have been derived from the Rhodophyta. Evidence listed in support of this view is the absence of flagellate cells in both groups, the similarities between the ascocarp (especially cleistothecia and perithecia) and the cystocarp of Rhodophyta, the occurrence in some Ascomycetes of nonmotile spermatia and trichogynes, and the resemblance between diploid gonimoblasts of certain Rhodophyta and the ascogenous hyphae of the Ascomycetes. At first glance, the marked physiological differences between the Rhodophyta and the Ascomycetes would seem to present an insurmountable barrier to relationship. It has been pointed out, however, that several species of extant Rho-

dophyta have lost their pigments and are parasitic on other Rhodophyta.

The organisms mentioned in this chapter may be classified as follows:

Division Amastigomycota
 Subdivision 1. Ascomycotina
  Class 1. Ascomycetes
   Subclass 1. Hemiascomycetidae
    Order 1. Endomycetales
     Family 1. Saccharomycetaceae
      Genera: *Schizosaccharomyces*, *Saccharomyces*
    Order 2. Taphrinales
     Family 1. Taphrinaceae
      Genus: *Taphrina*
   Subclass 2. Plectomycetidae
    Order 1. Eurotiales
     Family 1. Gymnoascaceae
      Genera: *Arthroderma, Nannizia, Ajellomyces*
     Family 2. Eurotiaceae
      Genera: *Eurotium, Emericella, Talaromyces, Eupenicillium*
   Subclass 3. Hymenoascomycetidae
    Order 1. Erysiphales
     Family 1. Erysiphaceae
      Genera: *Erysiphe, Sphaerotheca, Uncinula, Pleochaeta, Microsphaera, Podosphaera, Phyllactinia*
    Order 2. Xylariales
     Family 1. Sordariaceae
      Genera: *Neurospora, Sordaria, Gelasinospora*
    Order 3. Diaporthales
     Family 1. Diaporthaceae
      Genus: *Endothia*
    Order 4. Clavicipitales
     Family 1. Clavicipitaceae
      Genus: *Claviceps*
    Order 5. Helotiales
     Family 1. Sclerotiniaceae

Genus: *Monilinia*
Order 6. Pezizales
 Family 1: Sarcoscyphaceae
  Genera: *Sarcoscypha, Urnula,*
   Chorioactis
 Family 2. Pyronemataceae
  Genus: *Pyronema*
 Family 3. Pezizaceae
  Genus: *Peziza*
 Family 4. Humariaceae
  Genus: *Humaria*
 Family 5. Morchellaceae
  Genus: *Morchella*
 Family 6. Helvellaceae
  Genera: *Helvella, Gyromitra,*
   Discina
Order 7. Tuberales
 Family 1. Tuberaceae
  Genus: *Tuber*
 Family 2. Terfeziaceae
  Genus: *Terfezia*
Subclass 4. Loculoascomycetidae
 Order 1. Dothideales
  Family 1. Dothideaceae
   Genus: *Guignardia*
 Order 2. Pleosporales
  Family 1. Venturiaceae
   Genus: *Venturia*

For the classification of the conidial form-genera that were mentioned in this chapter, see Chapter 35 (subdivision Deuteromycotina).

DISCUSSION QUESTIONS

1. In what respects do Ascomycotina (Class Ascomycetes) differ from other Amastigomycota?
2. What evolutionary origins have been postulated for the Ascomycotina?
3. Define or explain ascus, free cell formation, Woronin bodies, conidium, conidiophore, ascocarp, pycnidium, pycnidiospore, cirrhus, acervulus, ascogonium, trichogyne, antheridium, spermatium, heterothallic, homothallic, dikaryotic, ascogenous hypha, crozier, ascus-mother-cell, ascospore, epiplasm, cleistothecium, apothecium, ascostroma, perithecium, ascus vesicle, protoperithecium, pseudothecium, pseudoparaphysis.
4. In what respects are Ascomycetes important in human affairs?
5. What types of nutrition occur among Ascomycetes; cite illustrative genera.
6. Describe the generalized life cycle of the Ascomycetes.
7. Distinguish among the Hemiascomycetidae, Plectomycetidae, Hymenoascomycetidae, Laboulbeniomycetidae, and Loculuascomycetidae.
8. Describe the structure and life cycles of *Schizosaccararomyces octosporus* and *Saccharomyces cerevisiae.*
9. In what respects are yeasts of importance in human affairs?
10. Describe the structure and life cycle of *Taphrina deformans*. What disease does it cause?
11. Discuss the relationship between *Eurotium* and *Aspergillus* and *Eupenicillium* and *Penicillium*. Are all species of *Aspergillus* and *Penicillium* related to these genera?
12. Name some Ascomycetes that cause human dermatitis.
13. Describe the structure and life cycle of powdery mildews.
14. Describe the structure and life cycle of *Neurospora sitophila.*
15. Of what importance is *Neurospora* to the science of genetics?
16. In what respects do *Sordaria* and *Gelasinospora* differ from one another and from *Neurospora*?
17. Name a parasitic discomycete and describe its structure and life cycle.
18. Briefly discuss morels and truffles.
19. Describe the structure of *Venturia inaequalis.*

Chapter 34
DIVISION AMASTIGOMYCOTA III: SUBDIVISION BASIDIOMYCOTINA

The Basidiomycotina differ from the Ascomycotina chiefly in that they produce their meiospores externally on basidia instead of internally in asci. Other important characteristics of the Basidiomycotina, some of which they share with the Ascomycotina, are the following:

1. They produce no flagellate cells.
2. A dikaryotic phase is interspersed between plasmogamy and karyogamy. In the Basidiomycotina the dikaryotic phase is of long duration, being represented by the extensive dikaryotic mycelium, which produces the sporophores (**basidiocarps**).
3. The septa of most Basidiomycotina that have been investigated are swollen, more or less barrel-shaped, and centrally perforated. The pore is covered by a membranous pore cap. This is the **dolipore septum** (see below).
4. The dikaryotic mycelium of many, but not all, Basidiomycotina has **clamp connections** (see below).
5. No differentiated gametangia are produced except in the rusts (Uredinales), the sexual function having been relegated to less specialized structures, usually the somatic hyphae.

Of the above listed characteristics, the presence of the basidium is the *sine qua non* of the Basidiomycotina. It must be emphasized here that there are several types of basidia known, all of which are the organs in which karyogamy and meiosis take place.

The dolipore septum.   This septum has been found in all groups of the Basidiomycotina except the subclass Teliomycetidae (the rusts and the smuts). It is a complex apparatus consisting of a central, perforated swelling over which there is a **pore cap** (Fig. 34-1), sometimes referred to as the **parenthesome**. This type of septum has also been found in some of the mycelial members of the subclass Hemiascomycetidae but is generally regarded as a typical basidiomycetous structure. We know that nuclei and other cytoplasmic organelles can and do pass through the dolipore, and it has been suggested that this septum is able to regulate the passage of organelles from one cell to another, by opening and closing at the appropriate moment.

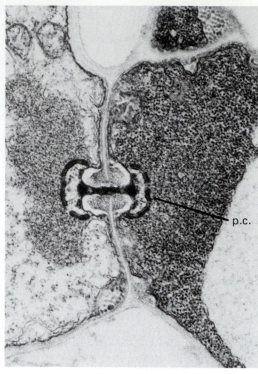

FIGURE 34-1
Dolipore septum of a basidiomycete. PC, pore cap.
×10,000. (*Transmission electron micrograph
courtesy of Dr. Charles Mims.*)

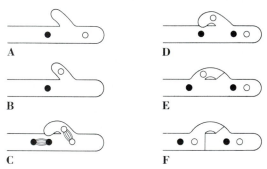

FIGURE 34-2
Stages in cell division and clamp connection
formation in the mycelium of the Basidiomycetes,
schematized; black nuclei represent one mating
type and white nuclei the other. Successive stages
show mechanism by which the dikaryotic condition
is maintained. (*Modified from Alexopoulos [1962].*)

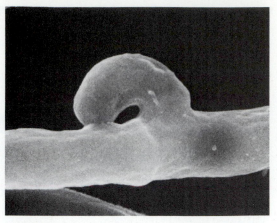

FIGURE 34-3
Scanning electron micrograph of a clamp
connection in the mycelium of *Panus rudis.* × 600.
(*Courtesy of Dr. Stanley Flegler.*)

The clamp connections. Clamp connections (Fig. 34-2F) have also been found in all groups of Basidiomycotina except the subclass Teliomycetidae. In many species they are restricted to the dikaryotic mycelium and are often used in breeding experiments to indicate whether dikaryotization has actually occurred, because they are much easier to see than the nuclei in the hyphal cells. A clamp connection begins as a short branch from an end cell of a growing hypha (Fig. 34-2A) and forms a bent hook, the tip of which connects with the cell wall near the septum (Fig. 34-3). As the walls between the clamp and the cell dissolve, the nuclei divide (Fig. 34-2C), one division so oriented that one of its daughter nuclei is formed in the clamp and the other in the mother cell (Fig. 34-2D). The other division is oriented horizontally in the mother cell. The daughter nucleus in the clamp then migrates into the mother cell and the two septa are formed (Fig. 34-2E, F), one cutting off the clamp from its mother cell and the other vertically just below the point where the clamp fused with the cell wall of its mother cell. Thus the mother cell has divided into two cells while the dikaryotic con-

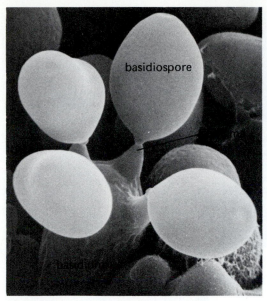

FIGURE 34-4
Upper portion of the basidium of a mushroom, bearing four basidiospores on sterigmata. ×5300. (*Scanning electron micrograph courtesy of Dr. Stanley Flegler.*)

dition is perpetuated, with nuclei of different mating types always present in the newly formed cells as the hypha grows.

The basidium. The basidium is the organ in which karyogamy and meiosis take place in the Basidiomycotina, as a result of which generally four meiospores, termed basidiospores in this subdivision, are produced on the surface of the basidium (Fig. 34-4). As we have already said, the basidium is the one characteristic structure of the Basidiomycotina that supersedes all others in importance. If a fungus produces its meiospores on basidia, it is placed in the Basidiomycotina regardless of its other characteristics. If it does not, it cannot be classified in this subdivision, again regardless of other characteristics.

There are three main types of basidia: the **holobasidium**, a one-celled structure of varying shape (Fig. 34-5A, B); the **phragmobasidium**, a septate basidium with one or more transverse (Fig. 34-5C) or vertical septa; and the **teliobasidium**, which consists of a generally thick-

FIGURE 34-5
Four types of basidia. **A.** Club-shaped holobasidium. **B.** Divided (tuning-fork) holobasidium. **C.** Phragmobasidium with transverse septa. **D.** Teliobasidium with promycelium bearing spores, emerging from thick-walled resting spore. b., basidiospores; s., sterigma. (*Drawings by Stephen Chase.*)

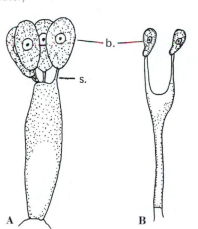

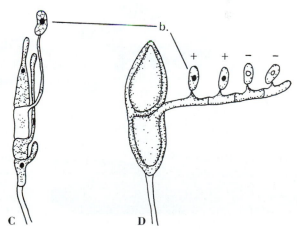

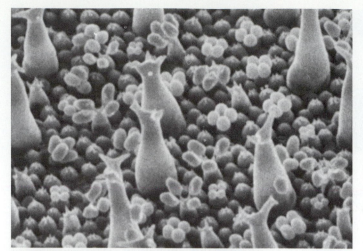

FIGURE 34-6
Hymenium of the
hymenobasidiomycete *Pluteus
cervinus*. Note the numerous
basidia, each with four spores.
The large bottle-shaped
structures are cystidia, which
serve to keep the lamellae
apart. ×600 (*Scanning
electron micrograph courtesy
of Dr. Stanley Flegler.*)

A

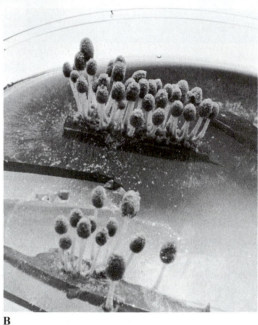

B

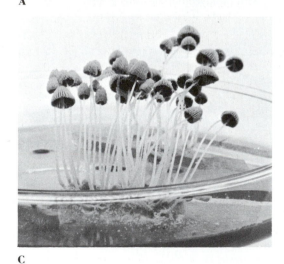

C

FIGURE 34-7
A. *Chlorophyllum molybdites*, a wild, poisonous
mushroom. **B, C.** *Coprinus stercorarius* in culture
on agar. **B.** Young basidiocarps; note mycelium.
C. Mature basidiocarps. **A.** ×0.25. **B, C.** ×1 (i.e.,
natural size). (**A.** *Courtesy of Prof. L. R. Hesler.*
B, C. *Courtesy of Dr. Milton Rogers.*)

walled resting spore and a finite germ tube (promycelium), which issues from it and which bears the basidiospores (Fig. 34-5D).

In species with clamp connections, a clamp is always formed at the base of a basidium. This has been likened to the crozier of the Ascomycetes and together with the dikaryotic phase in the two groups has been used as an indication of the origin of the Basidiomycetes from an ancestral ascomycete.

Classification. The subdivision Basidiomycotina consists of the single class Basidiomycetes, which we subdivide into three subclasses: Holobasidiomycetidae, Phragmobasidiomycetidae, and Teliomycetidae. We shall discuss each of these briefly.

SUBCLASS 1. HOLOBASIDIOMYCETIDAE

The Holobasidiomycetidae have one-celled, usually club-shaped basidia. This subclass consists of two large groups of fungi, Hymenobasidiomycetes and Gasteromycetes. These are convenient groupings without taxonomic rank.

Hymenobasidiomycetes

The Hymenobasidiomycetes produce their basidia in hymenial layers (Fig. 34-6), reminiscent of those of the Hymenoascomycetidae (Chapter 33, p. 749). Their sporophores (basidiocarps) are either open from the beginning or they open, exposing their basidia, before the spores mature. Familiar examples of Hymenobasidiomycetes are the mushrooms or toadstools (Fig. 34-7), the boletes (Figs. 34-15B, 34-21), and the shelf (bracket) (Figs. 34-15A, 34-16), coral, and toothfungi (Fig. 34-20). The student must remember that these familiar structures are but the sporophores (basidiocarps) of the fungi that produce them, the most extensive portion of which is the mycelium from which the sporophores originate and which obtains nourishment from the substratum.

The mycelium of some mushrooms and some other large basidiocarps, such as the puffballs (in the Gasteromycetes), when it grows on the ground, particularly on lawns, golf courses, or other grassy places, forms a large, circular colony that continues growing year after year if undisturbed, producing a crop of basidiocarps at the periphery of the colony, with the grass much greener within the circle than outside. Such mushroom circles are called **fairy rings** (Figure 34-8), because in the Middle Ages mycelium was unknown, and it was believed by superstitious people, of whom there were even more than there are today, that the circle of mushrooms represented the path of dancing fairies. As the fairy ring grows, the older hyphae in the center die and disintegrate by bacterial action, which breaks down the fungal proteins and releases available nitrogen to the grass roots, causing the grass to have a brighter green color. In some places it was considered bad luck to step inside a fairy ring; in other areas it was good luck to do so. No one ever reported seeing a dancing mushroom fairy except in Walt Disney's *Fantasia*.

Mushrooms are basidiocarps of fungi in the order Agaricales, which is subdivided into some 15 or more families. A mushroom (Fig. 34-9B) consists of a stem (**stipe**), supporting a cap (**pileus**) and sometimes rising from a basal cup (**volva**). Gills (**lamellae**), platelike structures radiating from the edge inward toward the stem, under the pileus, are covered with basidia on both their surfaces (Fig. 34-10), each basidium generally producing four spores, after karyogamy and meiosis have occurred. In some mushrooms there is a ring (**annulus**) around the stem below the edge of the pileus (Fig. 34-9B). This represents the remnants of the **velum**, which covers the gills before the spores mature. The pileus may also be covered with scalelike membranes, the remnants of the universal veil, a

FIGURE 34-8
A fairy ring of the mushroom *Marasmius oreades*. (*Courtesy of Dr. James Lampky.*)

membrane that covered the entire sporophore in the young "button" stage, before its final expansion (Fig. 34-9A). In some mushrooms, also, a weblike membrane (**cortina**) hangs down from the edge of the pileus but soon wears off. The rows of basidia on the membranelike lamellae represent the hymenium that gives this group its name. This is comparable to the hymenium of the Hymenoascomycetidae. We cannot overemphasize the fact, which many students find hard to believe, that a mushroom consists of tightly woven hyphae in its entirety. Mounting a small piece of tissue from any part of the mushroom and examining it under the microscope will convince even the most skeptical. It cannot be otherwise, as the following short account of the mushroom life cycle indicates (Fig. 34-11).

Life cycle. Basidiospores are forcibly ejected from their basidia on the lamellae, to a distance of a millimeter or two, just far enough to free them from the basidia and propel them midway between two adjacent lamellae. By force of gravity they then fall on the substratum below, or, more likely, are carried by air currents to new sites, where they germinate and form a uninucleate mycelium (Fig. 34-11E). This stage is ephemeral because of the process of dikaryotization, which takes place soon. Most species of mushrooms are heterothallic and **tetrapolar**—that is, they consist of four mating types because there are four genes (A_1, A_2; B_1, B_2) on two chromosomes, controlling mating type. There are, therefore, four types of basidiospores on each basidium (A_1B_1), (A_1B_2), (A_2B_1), (A_2B_2). In order for sexual reproduction to take

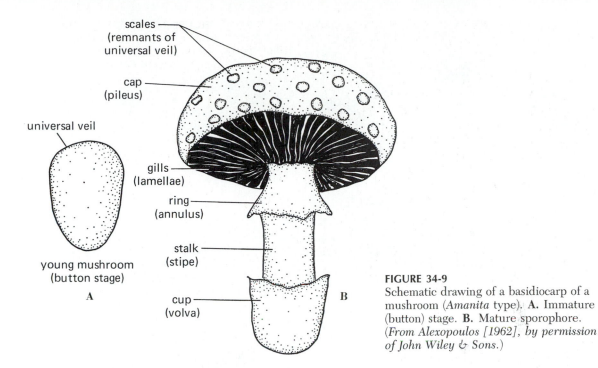

scales
(remnants of
universal veil)

cap
(pileus)

universal veil

gills
(lamellae)

ring
(annulus)

young mushroom
(button stage)

A

stalk
(stipe)

cup
(volva)

B

FIGURE 34-9
Schematic drawing of a basidiocarp of a
mushroom (*Amanita* type). **A.** Immature
(button) stage. **B.** Mature sporophore.
(*From Alexopoulos [1962], by permission
of John Wiley & Sons.*)

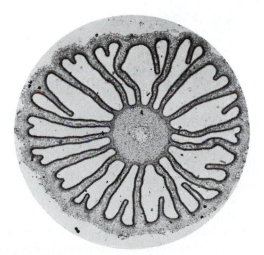

FIGURE 34-10
Coprinus sp. Transection of pileus and gills; the
dark margins of the latter constitute the hymenium.

place, it is necessary for two compatible mating
types to meet (Fig. 34-11F). Such a union must
result in an $A_1B_1A_2B_2$ genotype for a sporulating
mycelium to form. When the uninucleate myce-

lia from two compatible basidiospores meet, the
hyphal tips fuse (plasmogamy) and nuclei are
exchanged, so that both mycelia behave both as
donors and receptors of nuclei. The invading
nuclei then multiply in the invaded mycelium in
such a way so as to result in dikaryotization. This
means that all the cells of each mycelium will
contain one nucleus of each mating type that
engaged in plasmogamy. The details of exactly
how this is done do not concern us here. They
usually involve the presence of clamp connections. For a detailed explanation of this process
read Alexopoulos and Mims (1979) or any other
general mycology book. Much work has been
done on the genetics of the Basidiomycetes,
particularly with regard to the incompatibility
system, using *Schizophyllum commune* as the
experimental organism. Some of this work is
summarized by Raper and Flexer (1971).

The dikaryotic hypha thus formed grows into
a binucleate mycelium (Fig. 34-11G), the two

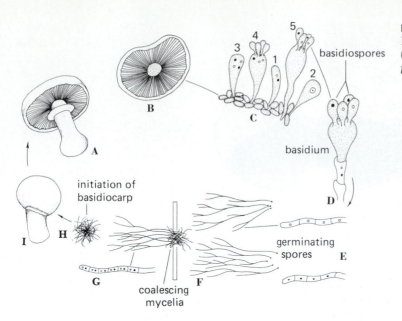

FIGURE 34-11
Life cycle of a mushroom.
(*From Bold [1974], by
permission of Prentice-Hall, Inc.*)

nuclei in each cell dividing conjugately each time, forming new dikaryotic cells.

When the mycelium has stored sufficient food, small hyphal knots, the basidiocarp primordia (Fig. 34-11H), form on the periphery of the colony and develop into tiny sporophores (buttons) (Fig. 34-11I), the tissues of which, in the presence of adequate moisture, expand rapidly into the familiar mushrooms (Fig. 34-11A).

As the binucleate basidia develop from hyphal tips on the lamellae, the two nuclei of opposite mating type in each basidium fuse (karyogamy), and the zygote nucleus soon undergoes meiosis, during which the mating type genes, as well as those governing other characters, segregate into the resulting four daughter nuclei (Fig. 34-11C). In the meantime, the tetranucleate basidium produces four stalklike, tubular extensions (**sterigmata**; sing. **sterigma**), the tips of which expand into spores (Fig. 34-11D). The haploid nuclei now migrate through the sterigmata into the basidiospores, thus completing the life cycle.

Identification of mushrooms.   Mushroom spores vary in color, shape, and markings in

FIGURE 34-12
Spore prints of black-spored mushrooms.

different species. Spore color is very important in identifying mushrooms. However, looking at individual spores under the microscope is sometimes deceiving because the pigment is often very dilute and the spore color difficult to determine. For this reason, a spore print is the first thing one needs in attempting to identify a mushroom. This is made by cutting the stem of the mushroom just below the cap, even with the

FIGURE 34-13
Amanita verna. (Courtesy of Dr. Stanley Flegler.)

cautioned that mushroom identification is not a simple matter, and if the mushrooms are to be consumed, it is best to have the opinion of a mushroom specialist. There are many hundreds of mushroom species known. The amateur collector cannot hope to learn all of them, nor is it necessary. He can learn a few edible, tasty species occurring in his region and leave the rest alone. It is of interest, however, to learn to recognize the genus *Amanita*, in which the most deadly of mushrooms belong. The best advice we can give you here is: Don't touch any mushroom that has a volva (cup) at the base of the stipe. Sometimes the volva is just below the surface of the ground and one must dig out the entire sporophore before he can be sure there is no volva. If there is any doubt, do not take a chance. *Amanita phalloides* (the death cup) is present in the United States, particularly in the eastern states, but is not very common. *Amanita verna* (Fig. 34-13) and *Amanita virosa* (the destroying angel) are two very similar, very beautiful, and very deadly mushrooms. They are almost pure white. *Amanita muscaria* (the fly agaric) (Fig. 34-14), the red form of which is spectacular, is hallucinogenic when eaten in small quantities but poisonous in larger doses. Wasson (1968) believes this mushroom to be the ancient Indian god Soma.

Polypores

In addition to mushrooms, other sporophores of the Hymenobasidiomycetes are the so-called bracket, or shelf, fungi (Polyporaceae) (Figs. 34-15A, 34-17). These are also called **polypores** (Gr. *poly*, much, many, + *poros*, pore), because basidia line the inner surface of pores or tubes beneath the cap (Fig. 34-16). These basidiocarps may or may not have a stem, which, if present, may be centrally or laterally inserted. *Polyporus sulphureus*, the sulfur mushroom (Fig. 34-17), growing on oaks and other trees, is easy to recognize and is said to be of excellent flavor when young, soft, and spongy. It becomes woody, however, with age.

edge of the gills, placing the cap—gills down—on a white piece of paper, and covering it with a bell jar or other convenient cover for a few hours. The spores, forcibly expelled from the basidia, fall on the paper under the influence of gravity and pile up to form a spore print (Fig. 34-12) the same color as the pigment in the spores. Starting with spore color, one then identifies the mushroom by using any one of a number of available mushroom manuals, listed at the end of this chapter.[1] The student must be

---

[1] Because many students are interested in identifying mushrooms rather than other fungi, we are including a list of mushroom manuals at the end of this chapter. For identification of other fungi see vols. 4A, 4B of Ainsworth, Sparrow, and Sussman, eds., *The Fungi: An Advanced Treatise* (Academic Press, New York, 1973).

FIGURE 34-14
*Amanita muscaria. (Courtesy
of Dr. Stanley Flegler.)*

A

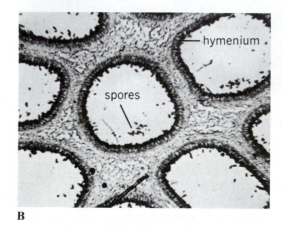

B

FIGURE 34-15
A. *Polyporus* sp.; basidiocarps on wood. **B.** Transection of pores of *Boletus* sp.; note that pores are lined
with basidiospore-producing hymenial layer. **A.** ×⅓. **B.** ×60.

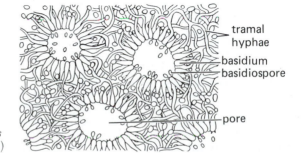

FIGURE 34-16
Hymenium of a polypore. (*From Alexopoulos
[1962], by permission of John Wiley & Sons.*)

FIGURE 34-18
Cantharellus cibarius. (*Courtesy of Dr. R. Pomerleau.*)

FIGURE 34-17
Polyporus sulphureus basidiocarps growing on a tree. (*Courtesy of Dr. Stanley Flegler.*)

Coral and Cantharelloid Fungi

Some of the coral fungi (Clavariaceae) and the famous funnel-shaped chanterelle (*Cantharellus cibarius*), which has shallow or deep ridges running from the underside of the funnel down onto the stem, is eminently edible (Fig. 34-18). Many other coral fungi are beautiful, with their many upright branches (Fig. 34-19) and their colors ranging from white, through yellow, orange, and amethyst. Most are not poisonous.

Tooth Fungi and Boletes

The tooth fungi (Hydnaceae), too, contain a number of edible species. *Hericium coralloides*, with many-branched pendant teeth, resembling

FIGURE 34-19
Clavicorona pyxidata, a common coral fungus. (*Courtesy of Dr. Stanley Flegler.*)

FIGURE 34-20
Hericium coralloides, a common tooth fungus. (*Courtesy of Dr. Stanley Flegler.*)

a large white coral, is one of these (Fig. 34-20). It can attain a considerable size. We must also mention the boletes (Boletaceae), which look like mushrooms (Fig. 34-21) but produce their basidia on the underside of the fleshy cap in pores or tubes instead of on gills (Fig. 34-15B). *Boletus edulis*, as the name indicates, is edible and highly prized. It is often dried and pulverized and is offered for sale, particularly in central Europe (Switzerland and elsewhere), to flavor soups. Most species of boletes are not poisonous, but some have a disagreeable, bitter taste.

Cultivated Mushrooms

There are at least two species of mushrooms that form the basis of the everexpanding mushroom-growing industry throughout the world. *Agaricus* (Gr. *agarikon*, mushroom) *brunescens*, called until recently *Agaricus bisporus*, is extensively cultivated in Europe and the United States. *Cortinellus berkelianus*, the

FIGURE 34-21
Boletus rubellus, one of the common boletes. (*Courtesy of Dr. Stanley Flegler.*)

A

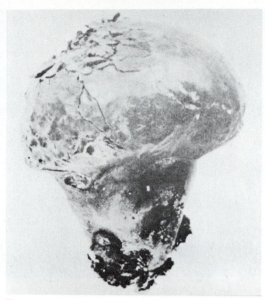

B

FIGURE 34-22
Two types of puffballs. **A.** *Scleroderma cepa*, a common hard-skinned puffball with dark-colored spores. **B.** *Calvatia* sp., a thin-skinned puffball with light-colored spores. **A, B.** ×½.

shiitake mushroom, is widely cultivated in the Orient—Japan and Taiwan chiefly. Needless to say, cultivated mushrooms offered for sale in reputable stores are safe to eat, if not as flavorful as some of the wild species. Wild mushrooms, offered for sale in farmer's and village markets in the United States, Mexico, and Central America, are not always reliably safe. In Latin American countries, hallucinogenic mushrooms can be bought on the streets and in some markets.

Hallucinogenic Mushrooms

Much interest has been aroused in recent years in hallucinogenic mushrooms, and much has been written about them ever since the distinguished French mycologist Roger Heim and his American sponsor, the ethnomycologist R. G. Wasson, visited Mexico and took part in the sacred mushroom rites of an Indian tribe conducted by the now famous curandera Maria Sabina. Heim described the mushroom used in those rites as *Psilocybe mexicana*, and the Swiss chemist Hoffman isolated the hallucinogenic principles, **psilocybin** and **psylocin**. The writ-

ings by Heim and Wasson (1957), Wasson and Wasson (1957), and Wasson alone (1959, 1961, 1968) very well summarize our knowledge of hallucinogenic mushrooms. The more recent book by Wasson and coauthors (1974), entitled *Maria Sabina and Her Mazatec Mushroom Velada*, appears to be as complete an account of the use and effects of *Psilocybe mexicana* as anyone could wish for.[2]

Gasteromycetes

These are the puffballs (Figs. 34-22, 34-23), the earthstars (Fig. 34-24), the stinkhorns (Fig. 34-25), and the bird's nest fungi (Fig. 34-26) and include some of the largest fungal sporophores known. The largest puffball of *Calvatia* (L. *calvus*, bald) *gigantea* on record was 1.6 m in length, 1.35 m in width and 24 cm in height (Bessey, 1950).

---

[2]A review of this book was published in *Mycologia* 68:953–954, 1976.

FIGURE 34-23
Lycoperdon perlatum, a common puffball. (*Courtesy of Dr. Stanley Flegler.*)

Puffballs and Earthstars

The Gasteromycetes produce their spores in closed sporophores. In some groups, however, the sporophores do open and expose the spores but only after the latter are completely mature and ready for dispersal. Otherwise, the spores escape when the basidiocarp breaks or begins to disintegrate. In these fungi, the basidia are not arranged in hymenia when the spores are ma-

ture. Hymenia may be present in the early stages of sporophore formation but become disrupted at a later stage. Puffballs (Figs. 34-22, 34-23) are so called because when the wall (peridium) of the mature basidiocarp breaks, the spores are puffed out by raindrops or when some object such as a twig falls on the sporophore or some animal touches it in an attempt to eat it. Earthstars are puffballs with two or more

FIGURE 34-24
Geastrum saccatum, a common earthstar. (*Courtesy of Dr. Stanley Flegler.*)

peridia, the outer of which split open and turn over, resembling a star with a ball in the center. (Fig. 34-24).

Stinkhorns

Stinkhorns (Phallales) are egglike when immature. On maturity, the egg hatches and a long stalk usually expands, bearing a gelatinous gleba at the top, in which the spores are embedded (Fig. 34-25). As their common name indicates, most stinkhorns emit an incredibly disagreeable odor, which, however, attracts flies and some other insects. The insects fly to them and alight on the gleba to sip the sweet jelly and, in so doing, pick up the spores on their feet and mouthparts and so distribute them when they fly away.

FIGURE 34-25
Phallus ravenellii, a common stinkhorn. (*Courtesy of Dr. Stanley Flegler.*)

Bird's Nest Fungi

The bird's nest fungi (Nidulariales) have developed an ingenious method for spore distribution, elegantly described by Brodie (1975, 1978) in his two books *The Bird's Nest Fungi* and *Fungi—Delight of Curiosity*. The sporophores of the bird's nest fungi are cups that contain small, waxy, hard peridioles, thus resembling birds' nests with eggs (Fig. 34-26). It is inside these peridioles that the basidia and basidiospores are formed. The cups are so shaped that when, during a driving rain, raindrops hit the peridioles with great force, those spore-containing bodies slide out of the cups, attaining a great velocity and traveling considerable distances. In most species, peridioles are attached to their cups, each by a long cord neatly wound up under each peridiole, which becomes gelatinous and sticky when wet. In flight, the cord trails behind and, striking a twig or other plant part nearby, adheres to it, causing the peridiole to wind itself around the object and hang down a few centimeters. There it is eventually devoured by some herbivorous animal and voided with excreta, where it releases the spores, which germinate to form mycelium. Consequently, bird's nest fungi are usually found on horse or cow dung or in gardens fertilized with manure. The largest genus of the Nidulariales is *Cyathus*. Other genera are *Crucibulum*, *Nidula*, *Nidularia*, and *Mycocalia*. The fungus artillery *Sphaerobolus* also belongs here (see Brodie, 1978).

Jelly Fungi with Tuning-Fork Basidia

Mention must be made of those jelly fungi with holobasidia. These are the Dacrymycetales. Their basidia are deeply divided, resembling a tuning fork (Fig. 34-5B). Their basidiocarps are gelatinous and usually brightly colored, mostly yellow or orange. They can be spotted easily on a rainy winter day in deciduous forests, where they contrast vividly with the dark tree branches devoid of leaves. *Dacrymyces* (Gr. *dakry*, tear, + *mykes*, fungus) forms

FIGURE 34-26
Cyathus stercorius, a bird's nest fungus. ×4.5. *(Courtesy of Dr. Harold J. Brodie.)*

cushion-shaped, gelatinous, orange masses several millimeters in diameter. The very common *Calocera* (Gr. *kalos*, beautiful, + *keras*, horn) *cornea* grows on dead logs. It resembles tiny gelatinous horns 1–2 mm tall arising from the wet wood (Fig. 34-27).

SUBCLASS 2. PHRAGMOBASIDIOMYCETIDAE

The Phragmobasidiomycetidae are also jelly fungi. Their chief distinguishing feature is that their basidia are septate, either transversely, as in *Auricularia* (L. *auricula*, ear) (Fig. 34-5C), or

FIGURE 34-27
Calocera cornea, a jelly fungus with tuning-fork basidia. ×2. *(Courtesy of Dr. Bernard Lowy.)*

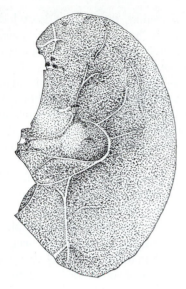

FIGURE 34-28
Auricularia sp.; an ear jelly fungus. (*Drawn by Stephen Chase from a Kodachrome slide courtesy of Dr. Bryce Kendrick.*)

longitudinally, as in *Exidia*. These fungi are saprobes on decaying logs in damp situations. Their basidiocarps are most conspicuous after periods of prolonged rains, when they have absorbed water and expanded. When dry they form inconspicuous crusts. *Auricularia* is called the ear fungus (Figs. 34-28, 34-29) because its basidiocarp somewhat resembles a human ear in shape. There are several species, many tropical but some also in the temperate zones. *Auricularia auricula* is cultivated in the Orient and is almost indispensable in Chinese cuisine. It is canned or dried and offered for sale in western markets, particularly in big cities with a large Chinese population. *Phlogiotis* (Gr. *phlox, phlogos,* flame) *helveloides,* a pink or rose funnel-shaped jelly fungus, is not uncommon in the United States. The white *Tremella* (Gr. *tremo,* tremble) *reticulata* is a showy, rather large jelly fungus. The two largest and most important orders are the Tremellales and Auriculariales.

FIGURE 34-29
Auricularia polytricha. Natural size. (*Courtesy of Dr. Bernard Lowy.*)

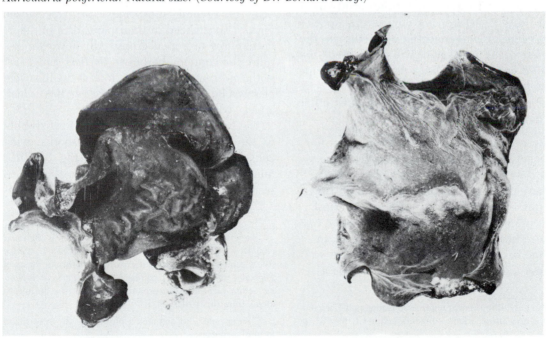

These are the rusts (order Uredinales) and the smuts (order Ustilaginales), the most economically important Basidiomycetes because of the serious diseases they cause on a large number of crops. The chief characteristic of these fungi is the production of thick-walled resting spores (**teliospores**), in which karyogamy takes place. The teliospore is, therefore, a part of the basidial apparatus, which constitutes what we have called a **teliobasidium**.

Order Uredinales

It is impossible to do justice to this large and important group of fungi in the space we can devote to them in this book. We shall, therefore, do nothing more than to try to summarize some of the important facts concerning one or two species, to give you a general idea about the group.

Up to a few years ago the Uredinales were thought to be obligately parasitic on plants. We still believe that they do not grow outside their living hosts in nature, but now several species have been grown on artificial media in the laboratory and many more will undoubtedly be cultured in the future. For a discussion of axenic culture of rusts, see Coffey (1975).

The rusts, in general, differ from most other Basidiomycetes in a number of ways:

1. The mycelium bears no dolipore septa, nor does it produce clamp connections. Haustoria develop within host cells.
2. Many rusts are heteroecious—that is, they require two different species of plant host to complete their life cycle.[3] Incredibly, these hosts are usually very dissimilar, in some cases one an angiosperm and the other a gymnosperm or a fern, or, one a monocotyledon, the other a dicotyledon. Many rusts, however, are autoecious, completing their entire life cycle on a single species of host.
3. All rusts are heterothallic, as are many other Basidiomycetes, but in the rusts there are only two mating types for each species, (+) and (−). In this respect the rusts resemble the Ascomycetes and the Zygomycetes.
4. Sexual reproduction is accomplished by spermatization (the union of spermatia with receptive hyphae, similar to ascomycetous trichogynes), but somatogamy (hyphal fusion as in the mushrooms) also occurs.
5. The life cycle of most rusts is very complex, several types of spores being produced during their development, and both microcyclic and macrocyclic species occurring.

Life cycle.   We have the space to consider only one life cycle, that of *Puccinia* (after T. Puccini, Italian anatomist) *graminis*, the cause of black stem rust of cereals. This species consists of several subspecies, or "specialized forms," specialized in their parasitism of different groups of cereal crops. All, however, require the common barberry (*Berberis vulgaris*)[4] as the alternate host. Let us look at the life cycle of *Puccinia graminis* form spec.[5] *tritici*, which attacks wheat (Fig. 34-30).

The fungus overwinters on the wheat stubble in the form of thick-walled teliospores, each consisting of two binucleate cells (Fig. 34-30A). Sometime during the winter, karyogamy takes place, rendering each cell uninucleate and dip-

---

[3]The only other known heteroecious fungus is *Coelomomyces psorophorae*, which we mentioned in Chapter 31, p. 712.

[4]Several other species of *Berberis*, such as *Berberis canadensis* (the Canadian barberry) and *Berberis cretica* (the Cretan barberry), may also serve as alternate hosts of *Puccinia graminis*. The Japanese barberry (*Berberis japonica*), often planted as an ornamental, is not susceptible to attack from this rust.

[5]"Form. spec." designates a physiologically specialized form of a species.

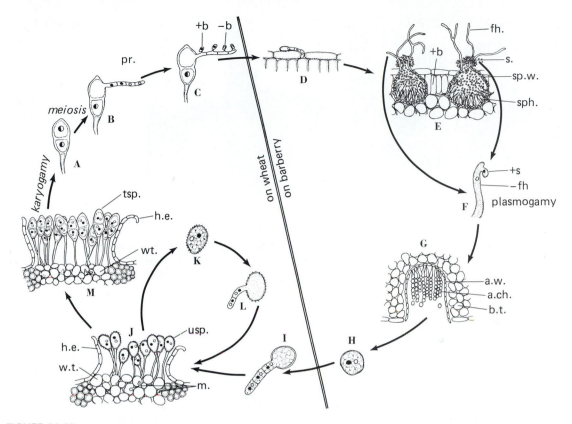

FIGURE 34-30

Life cycle of *Puccinia graminis tritici* (not to scale). **A.** Mature teliospore. **B.** Formation of promycelium. **C.** Formation of basidiospores of the two mating types, (+) and (−). **D.** Basidiospore germination and infection of barberry leaf. **E.** Spermogonia and flexuous (receptive) hyphae. **F.** Spermatization. **G.** Aecium with aeciospore chains. **H.** Mature, binucleate aeciospore. **I.** Germinating aeciospore. **J.** Uredinium with binucleate urediniospores. **K.** Mature urediniospore. **L.** Germination of urediniospore to form binucleate mycelium. **M.** Telium with binucleate, two-celled teliospores. pr., promycelium; +b., −b., basidiospores of two mating types; fl., flexuous (receptive) hypha; s., spermatia; sp.w., spermogonium wall; sph., spermatiophore; +s., −fh., spermatium and flexuous hypha of opposite mating types; a.w., aecium wall; a.ch., aeciospore chain; b.t., barberry tissue; m., mycelium; w.t., wheat tissue; h.e., host epidermis; usp., urediniospore; tsp., teliospore. (*Constructed by Dr. R. W. Scheetz.*)

loid. In the spring the spore germinates, each cell producing a long, finite germ tube the **promycelium** (also called **metabasidium** or **epibasidium**), in which the diploid nucleus migrates and undergoes meiosis. This makes the teliospore and the promycelium both parts of the basidial apparatus (teliobasidium) (Fig. 34-30B). A septum is formed after each division

of meiosis, separating the four nuclei into four cells, and a short sterigma develops from each cell of the promycelium, on the tip of which a small basidiospore is formed. It is important to note that the genes controlling mating type segregate during meiosis so that two of the basidiospores carry the (+) factor and two carry the (−) factor (Fig. 34-30C). The basidiospores

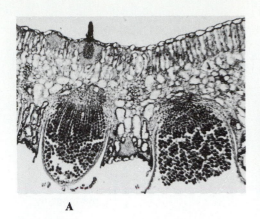 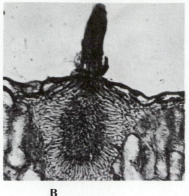

A   B

FIGURE 34-31
Puccinia graminis. **A.** Transection of infected barberry leaf; spermogonium above, two aecial cups with aeciospores below. **B.** Spermogonium in medium longitudinal section; note projecting periphyses and wall lined with columnar hyphae (spermatiophores) bearing spermatia. (See Fig. 34-30B.) **A.** ×75. **B.** ×250.

are forcibly ejected from the sterigmata and are distributed by the wind. Those that happen to fall on a barberry bush continue the life cycle; the others perish.

The importance of the barberry in the development of black stem rust of wheat had been noticed in the seventeenth century by French farmers, who, in 1660, persuaded the City Council of Rouen to enact a law ordering the eradication of all barberry bushes growing in the vicinity of wheat fields. The reason for this was unknown and remained so until about two hundred years later, when Anton deBary, the great German mycologist, elucidated the complicated life cycle of *Puccinia graminis.*

On the barberry, the basidiospores germinate and the germ tubes issuing from them invade the leaf (Fig. 34-30D), growing through the epidermal cells and then in between the mesophyll cells to the lower side of the leaf. The uninucleate mycelium developing from the germination of the basidiospores now produces some important structures. Near the upper side of the barberry leaf, it forms the first of these, flask-shaped **spermogonia,** resembling pycnidia of Ascomycetes and Deuteromycetes, which are

filled with long, fertile hyphae (Fig. 34-30E). These are the **spermatiophores** (Figs. 34-31B, 34-32B), which cut off chains of minute uninucleate, conidiumlike spermatia mixed with a mucous jelly in the central cavity of the spermogonium. Eventually, they come out in fragrant droplets (nectar) from the necks of the spermogonia, which by now protrude from the barberry leaf in pustules (Fig. 34-31). Other long hyphae, issuing from the spermogonia, and also pushing out from the leaf beside them, are the female receptive hyphae (also known as flexuous hyphae), on which the male spermatia *from the opposite mating type* are brought by flies or other insects that visit the fragrant pustules to sip the sweet nectar full of spermatia. Thus spermatization is accomplished (Figs. 34-30F, 34-33). It appears, however, that (+) and (−) mycelia from mixed infections also meet in the barberry leaf and exchange nuclei much as mushroom mycelia do. It is of interest to note that repeated attempts to germinate the conidiumlike spermatia have failed. They appear to be specialized male reproductive cells. Other important structures formed by the uninucleate mycelium within the barberry leaf are knots of

784

A

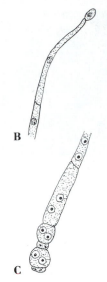

B

C

FIGURE 34-32
Puccinia graminis. **A.** Aecia (cluster cups) on lower surface of barberry leaf. **B.** Spermatiophore bearing spermatium. **C.** Chain of binucleate aeciospores. **A.** ×1½. **B, C.** ×600. (**A.** *Courtesy of Canada Department of Agriculture.*)

uninucleate cells, the aecial primordia. These become dikaryotized soon after spermatization of receptive hyphae has occurred (Figs. 34-30F, 34-33). It appears that the spermatial nuclei shed into the receptive hyphae, travel down the

FIGURE 34-33
Puccinia graminis tritici. Flexuous hypha with attached spermatium. ×1850. (*After Craigie and Green.*)

mycelial network in the barberry leaf, passing through the septal perforations, and enter the cells of the aecial primordia, which now become dikaryotic, each with one (+) and one (−) nucleus. In a relatively short time, the aecial primordia develop into **aecia** (cluster cups) (sing. **aecium**), which contain chains of binucleate **aeciospores** when they break through the lower epidermis of the barberry leaf (Figs. 34-30G, 34-31A, 34-32A, C, 34-34). When these spores are shed, they are distributed by the wind, and those that fall on wheat plants continue the life cycle of the fungus; all others perish.

Aeciospores germinating on wheat plants produce binucleate mycelium, which grows between the cells of the wheat and saps the strength out of the plant, severely reducing the yield of grain. In early summer,[6] the mycelium in the wheat stalks and leaves produces lesions (**uredinia**) that produce relatively large binucleate, straw-colored spiny **urediniospores** (Figs. 34-30J, 34-35, 34-37A), which, in mass, appear red. This is the so-called red, or summer, stage of the rust, which gives it its name,

---

[6]Earlier in mild climates.

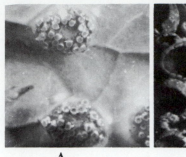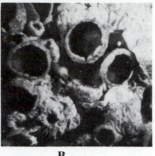

A   **B**

FIGURE 34-34
Puccinia graminis. Photograph of aecial cups of successively larger magnifications. **A.** ×1.5. **B.** ×15.

because the wheat fields appear rusty. So great are the numbers of these spores in a heavily infected field that the author of these lines, walking across such a field in Illinois one summer, wearing a light-blue suit, came out the other side of the field in a red suit!

The urediniospores are windblown and have the ability to reinfect wheat plants. This is the repeating stage of the rust, which is responsible for spreading the disease from wheat plant to wheat plant, from wheat field to wheat field, from Texas to North Dakota, gradually but steadily. When the wheat is nearing maturity, the pustules (uredinia) that produced the urediniospores now begin to produce teliospores,

the black stage of the rust. By the time wheat is harvested, the stalks are covered with long, black streaks (telia) composed of dark, reddish-brown teliospores (Figs. 34-30M, 34-36, 34-37B, C), which appear black in mass and which give the disease its name, "black stem rust of wheat." These spores overwinter on the straw and start the life cycle over again in the spring.

Eradication of the barberry is the best and most economical way to control this disease, as the farmers of Rouen had discovered in the seventeenth century. Yet, this has not eliminated the disease from the United States and Canada, where barberry eradication has been practiced for over half a century. The reason for this is the ability of the urediniospores to overwinter in the South. Surviving until spring, they

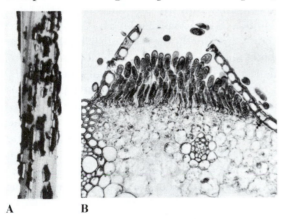

A   **B**

FIGURE 34-35
Puccinia graminis. **A.** Uredinial lesions on wheat stem. **B.** Section of uredinium on wheat leaf; note ruptured leaf epidermis and urediniospores. **A.** ×½. **B.** ×135 (**A.** *Courtesy of Canada Department of Agriculture.*)

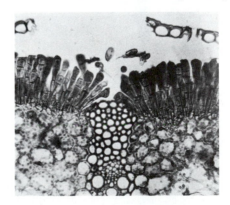

FIGURE 34-36
Puccinia graminis. Section of telium on wheat leaf. ×135.

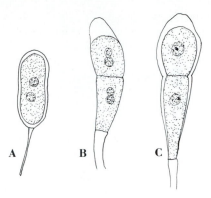

FIGURE 34-37
Puccinia graminis. **A.** Single urediniospore. **B.** Immature teliospore before karyogamy. **C.** Mature two-celled teliospore, each cell with a diploid nucleus.

perpetuate the disease in the wheat fields the following year without needing the barberry. Nevertheless, barberry eradication has proved to be very important in reducing crop losses. When this is coupled with early-maturing varieties in the northern wheat states, the onslaught from the southern urediniospores is greatly lessened.

Other Genera of Rust Fungi

Two other important genera of rusts are *Gymnosporangium* and *Cronartium. Gymnosporangium* does not produce urediniospores and thus lacks a repeating stage. *Gymnosporangium juniperi-virginianae*, the cause of apple and juniper rust, also infects the crab apple (*Crataegus*), as does *Gymnosporangium clavipes* (Fig. 34-38). Both produce their spermogonia and aecia on the angiosperm host (*Malus* or *Crataegus*). The dikaryotic mycelium in the junipers causes the formation of fairly large cells, galls, commonly known as "cedar apples" because the junipers are commonly but erroneously called "cedars." In early spring after some good rains, the cedar apples, full of dikaryotic mycelium, produce long, gelatinous, orange-colored, fingerlike projections full of two-celled teliospores with long stalks. These

germinate, producing promycelia and basidiospores (Fig. 34-38B–F), which are blown over to nearby apple or crab-apple trees and infect them. Obviously if one grows apples, one should see that there are no junipers nearby.

Cronartium ribicola is a very serious parasite of the white pine (*Pinus strobus*); it causes blister rust. The alternate hosts are species of *Ribes*—currants and gooseberries. Unlike *Gymnosporangium, Cronartium ribicola* has a complete life cycle. The spermogonia and aecia are produced in blisters on large galls on the pine, and the aeciospores infect *Ribes*, where the urediniospores and teliospores are formed. The teliospores in *Cronartium* are sessile (stalkless) but are fused, forming fingerlike columns that protrude from the underside of *Ribes* leaves (Fig. 34-39). A campaign to eradicate wild *Ribes* has been in progress in the United States for many decades in an effort to save the white pine, which is an important forest tree. Fortunately, the repeating stage of this rust is produced on the less important host.

Another genus of rusts, *Phragmidium*, has many-celled teliospores on wide stalks. *Phragmidium disciflorum*, among other species, causes rust of roses.

Order Ustilaginales

These are the smut fungi. They resemble the rusts in producing teliospores that germinate to form promycelia and basidiospores. They differ from the rusts in that their life cycles are much simpler and in that none is heteroecious. The mycelium of some smuts has clamp connections.

Covered smut of wheat caused by *Tilletia caries* is a world problem. Maize (corn) smut, caused by *Ustilago zeae*, is very prevalent in the maize-growing regions of the world (see below).

In addition to causing plant disease, smut spores, produced in enormous numbers, form explosive dusts in thrashing machines and storage bins and are also common allergens to man.

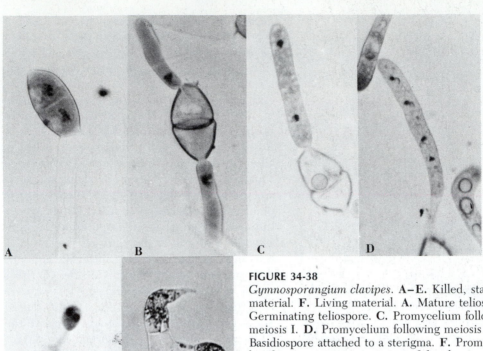

FIGURE 34-38
Gymnosporangium clavipes. **A–E.** Killed, stained material. **F.** Living material. **A.** Mature teliospore. **B.** Germinating teliospore. **C.** Promycelium following meiosis I. **D.** Promycelium following meiosis II. **E.** Basidiospore attached to a sterigma. **F.** Promycelium and basidiospores in various stages of development. All approximately ×850. (*Courtesy of Dr. Charles Mims.*)

Ustilago

Ustilago zeae, the corn smut, occurs in most spectacular fashion in the ears and tassels of the plant, where it causes immense, enlarged growths (Fig. 34-40B). These tumorlike galls of *Ustilago* (L. *ustus,* burned) are black at maturity. The color results from the transformation of the mycelium in the swollen host tissue into a mass of countless dark-walled, globose teliospores (Fig. 34-41), which are binucleate. The two nuclei in the young teliospore unite as the spore wall thickens.

The epidermis of the host, which at first covers the growing gall-like enlargement, is finally ruptured. The interior of the mass is composed of large numbers of teliospores, intermingled with the remains of sterile hyphae and host cells. The spores are readily disseminated by air currents and can germinate immediately, or they may undergo dormancy until the next growing season of the host. Upon germination, the thick spore wall is ruptured by the protrusion of a delicate promycelium, which becomes divided into four linearly arranged cells (Fig.

FIGURE 34-39
Cronartium ribicola. Telia. (*From Alexopoulos [1962], by permission of John Wiley & Sons.*)

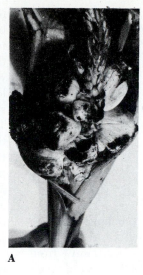

A

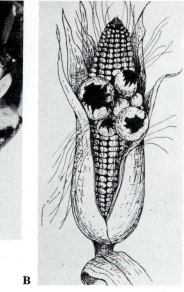

B

FIGURE 34-40
Ustilago zeae. **A.** Infection on emergent staminate inflorescence of corn. **B.** Infected ear of Golden Bantam corn. **A.** ×½. **B.** ×⅓.

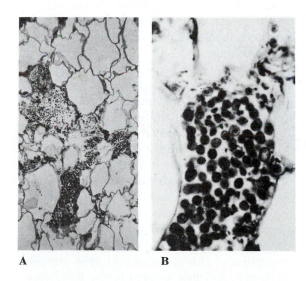

A   B

FIGURE 34-41
Ustilago zeae. Hypertrophied corn cells containing smut spores. **A.** ×60. **B.** ×770.

34-42), each containing a single haploid nucleus. The nucleus in each cell of the basidium divides mitotically to form two nuclei, one of which migrates into a thin-walled basidiospore (sporidium), which is budded off each basidial segment. Each of the latter may continue to produce additional basidiospores. It has been demonstrated, by the technique of single-spore isolation and culture, that meiosis occurs during the division of the primary basidial nucleus—that is, the fusion nucleus of the teliospore. Therefore, the basidiospores are usually of two kinds, (+) and (−), in their sexual potentialities. Host plants innoculated with a single basidiospore fail to develop typical smut galls.

The basidiospores that chance to fall on the meristematic epidermis of young host tissues form delicate germination tubes, which penetrate it and develop an intercellular mycelium, nourished by intracellular haustorial branches.

789

The cells of this mycelium are uninucleate. The binucleate mycelium that produces the gall-like growth is initiated by somatogamy from basidiosporal infections of different mating type that chance to be in close proximity within the host. The infected region of the plant undergoes cell enlargement as a result of the presence of the fungus, and a new gall is ultimately produced. Other species of *Ustilago* cause smut diseases in different cereal grains. *Ustilago tritici* causes "loose smut" of wheat, and *U. avenae* causes a similar disease of oats. In addition to the origin of the binucleate (dikaryotic) mycelium by somatogamy described above, other mechanisms occur among smuts to effect the same result. For example, the dikaryotic mycelium is initiated in some smuts by conjugation of compatible basidiospores, often called sporidia. Once initiated, the binucleate condition of the mycelium of many smuts and other Basidiomycota is maintained by the formation of clamp connections, a phenomenon illustrated in Fig. 34-2 and described in its legend. For a comprehensive discussion of the smut fungi, read Fischer and Holton (1957) and Duran (1973).

Summary and classification

The representative genera of Basidiomycetes described in the preceding sections of this chapter constitute a rather heterogeneous assemblage of fungi, which, however, are bound together by at least two attributes: the production of meiospores on the outside of a structure we call a basidium, and the presence of a rather lengthy dikaryotic phase in the life cycle, interspersed between plasmogamy and karyogamy. The basidium itself is not uniform in structure throughout the class or in time of occurrence in the life cycle. In the mushrooms, pore fungi, puffballs, stinkhorns, bird's nest fungi, and a few jelly fungi, the basidia are one-celled and always appear on the basidiocarp as elongated hyphal tips. These are the holobasidia. In most jelly fungi, the basidia are transversely or longitudinally septate but are still formed on basidiocarps as enlarged hyphal tips. These are the phragmobasidia. In the rusts and smuts, on the other hand, the basidium consists of a thick-walled spore, the teliospore, in which karyogamy takes place, and a finite germ tube, the promycelium (metabasidium or epibasidium), in which meiosis occurs and which bears the basidiospores. The promycelium may or may not be septate. This type of basidium, consisting of a teliospore and a promycelium, we have designated as a **teliobasidium**. This is a new term we are introducing for the first time.

Obvious manifestations of sexuality, such as specially differentiated gametes and gametangia, are usually absent in the Basidiomycetes (the rusts excepted), and somatogamy prevails. For this reason, sexuality is said to be *reduced* in this assemblage of fungi. On the other hand, in many other organisms considered to be highly developed—as in isogamous species of *Chlamydomonas*, *Spirogyra*, and *Rhizopus*, for example—highly differentiated gametangia and gametes are also lacking. The primary manifestations of sexuality, the union of cells and nuclei, and the association of chromosomal and gene complements are obviously present in the Basidiomycetes, in spite of the absence of such secondary criteria as specialized gametes and gametangia.

The compatibility mechanism, in fact, is much more highly developed in the Basidiomycetes than in other fungi. It is quite true that plasmogamy and karyogamy may be separated in time for an exceptionally long interval occupied by a dikaryotic phase (as in *Puccinia graminis*), but descendants of the original pairs of nuclei of different mating types brought together by plasmogamy are maintained by conjugate nuclear division, and one or more pairs of these descendant nuclei ultimately unite in the basidium or teliospore. Clamp connections are associated with the maintenance of the dikaryotic condition in many Basidiomycetes.

On the basis of basidial structure we subdivide the Basidiomycetes into three subclasses: Holobasidiomycetidae, Phragmobasidiomycetidae, and Teliomycetidae. The illustrative genera mentioned in this chapter may be classified as follows:

Division Amastigomycota
  Class Basidiomycetes
    Subclass 1. Holobasidiomycetidae
      Order 1. Aphyllophorales
        Family 1. Schizophyllaceae
          Genus: *Schizophyllum*
        Family 2. Clavariaceae
          Genus: *Clavaria*
        Family 3. Cantharellaceae
          Genus: *Cantharellus*
        Family 4. Hydnaceae
          Genus: *Hericium*
        Family 5. Polyporaceae
          Genus: *Polyporus*
      Order 2. Agaricales
        Family 1. Amanitaceae
          Genus: *Amanita*
        Family 2. Lepiotaceae
          Genus: *Chlorophyllum*
        Family 3. Hypholomataceae
          Genera: *Cortinarius, Psilocybe*
        Family 4. Volvariaceae
          Genus: *Panus*

        Family 5. Agariceaceae
          Genus: *Agaricus*
        Family 6. Coprinaceae
          Genus: *Coprinus*
        Family 7. Boletaceae
          Genus: *Boletus*
      Order 3. Lycoperdales
        Family 1. Lycoperdaceae
          Genera: *Calvatia, Lycoperdon*
        Family 2. Geastraceae
          Genus: *Geastrum*
      Order 4. Sclerodermatales
        Family 1. Sclerodermataceae
          Genus: *Scleroderma*
      Order 5. Phallales
        Family 1. Phallaceae
          Genus: *Phallus*
      Order 6. Nidulariales
        Family 1. Nidulariaceae
          Genera: *Cyathus, Crucibulum, Nidularia, Nidula, Mycocalia*
        Family 2. Sphaerobolaceae
          Genus: *Sphaerobolus*
      Order 7. Dacrymycetales
        Family 1. Dacrymycetaceae
          Genera: *Calocera, Dacrymyces*
    Subclass 2. Phragmobasidiomycetidae
      Order 1. Tremellales
        Family 1. Tremellaceae
          Genera: *Tremella, Exidia, Phlogiotis*
      Order 2. Auriculariales
        Family 1. Auriculariaceae
          Genus: *Auricularia*
    Subclass 3. Teliomycetidae
      Order 1. Uredinales
        Family 1. Pucciniaceae
          Genera: *Puccinia, Gymnosporangium, Phragmidium*
        Family 2. Melampsoraceae
          Genus: *Cronartium*
      Order 2. Ustilaginales
        Family 1. Ustilaginaceae
          Genus: *Ustilago*

BOOKS FOR MUSHROOM IDENTIFICATION

An asterisk (*) before each listing indicates an exceptionally useful book. "C" after the asterisk indicates that color illustrations are included.

* **Christensen, C. L.** 1965. *Common Fleshy Fungi.* Burgess Publishing Co., Minneapolis, Minn. Wire-spiral bound.

Coker, W. C., and Alma H. Beers. 1943. *The Boletaceae of North Carolina.* University of North Carolina Press, Chapel Hill.

Coker, W. C., and J. N. Couch. 1928. *The Gasteromycetes of the Eastern United States and Canada.* University of North Carolina Press, Chapel Hill. Puffballs, stinkhorns, bird's-nest fungi.

*C **Groves, Walter.** 1966. *Edible and Poisonous Mushrooms of Canada.* The Queen's Printer. Publications Division, Ottawa, Canada.

Hard, M. E. 1908. *The Mushroom, Edible and Otherwise.* The Ohio Library Co., Columbus.

* **Hesler, L. R.** 1960. *Mushrooms of the Great Smokies.* University of Tennessee Press, Knoxville.

Kauffman, C. H. 1918. *The Agaricaceae of Michigan.* 2 vols. Vol. 1: text. Vol. 2: black/white photographs. I Series: Michigan Geological Survey. Publ. 26.

Krieger, L. C. C. 1936. *The Mushroom Handbook.* Macmillan, New York.

Lange, M., and F. B. Hora. 1963. *A Guide to Mushrooms and Toadstools.* E. P. Dutton, New York.

C **McIlvaine, C., and R. K. Macadam.** 1973. *One Thousand American Fungi.* Dover Publications, New York. Paperback.

*C **Miller, O. K., Jr.** 1972. *Mushrooms of North America.* E. P. Dutton, New York. This is probably the best mushroom manual now available. Also in paperback.

* **Pomerleau, R.** 1951. *Mushrooms of the Eastern United States and Canada.* Les Editions Chantecler, Montreal, Canada.

Smith, A. H. 1938. *Edible and Poisonous Mushrooms of Southeastern Michigan.* Cranbrook Institute of Science Bulletin 14, Cranbrook, Michigan.

*C ————. 1949. *Mushrooms in Their Natural Habitats.* Sawyer's, Portland, Ore. Illustrated with Viewmaster stereochromes. Photocopied reprint available; also complete sets of stereochromes available.

————. 1951. *Puffballs and Their Allies in Michigan.* University of Michigan Press, Ann Arbor.

*C ————. 1963. *The Mushroom Hunters Field Guide.* University of Michigan Press, Ann Arbor. Very useful. Good color pictures. Relatively inexpensive for what you get.

* **Smith, H. V., and A. H. Smith.** 1973. *How to Know the Non-gilled Fleshy Fungi.* William C. Brown, Dubuque, Iowa. Wire-spiral bound. Particularly good for boletes.

*C **Snell, W. H., and E. A. Dick.** 1970. *The Boleti of Northeastern North America.* C. Cramer Verlag, Lehre, Germany. xii + 115 pp., 87 pls. (71 col.). Indispensable for boleti identification.

DISCUSSION QUESTIONS

1. List the characteristics that distinguish Basidiomycotina from other fungi.
2. Define or explain: basidium, basidiospore, basidiocarp, dolipore septum, clamp connection, holobasidium, phragmobasidium, teliobasidium, fairy ring, stipe, pileus, volva, annulus, velum, cortina, dikaryotic, sterigma, teliospore, heteroecious, autoe-

cious, somatogamy, promycelium, epibasidium, basidiospore, spermatiophore, spermogonium, aecium, aeciospore, uredinium, urediniospore.

3. Distinguish among the Holobasidiomycetidae, Phragmobasidiomycetidae, and Teliomycetidae.

4. Summarize the structure and life cycles of mushrooms.

5. What is meant by "tetrapolar" with reference to mushroom reproduction?

6. Describe the structure of the sporophores of polyporous fungi and tooth fungi.

7. What species of mushrooms are grown commercially?

8. What is meant by "tuning-fork" basidia? Cite a member of the Basidiomycetes that produces these.

9. Name examples of jelly fungi.

10. Summarize, with the aid of drawings, the life cycle and nuclear cycle of *Puccinia graminis*.

11. What are "cedar apples"? Explain.

12. Name two heteroecious rusts, other than *Puccinia graminis*, and name their hosts.

13. Describe the life cycle of corn smut, including reference to the nuclear cycle.

14. Why has the eradication of barberry bushes not caused the disappearance of wheat rust from North America?

15. Sexual reproduction in rusts has been compared with that in the Rhodophyta. What evidences of parallelism can you cite? What differences?

16. In view of the tremendous number of spores produced by such organs as the mushroom basidiocarp and the aecium of the rusts, why is the world not overrun with mushrooms and rusts?

17. How would you determine whether or not a given rust fungus is self-incompatible or not? A mushroom? A smut?

18. How do rusts and smuts reduce the yield of cereal grains?

19. How are the peridioles of the bird's nest fungi disseminated?

Chapter 35
DIVISION AMASTIGOMYCOTA IV: SUBDIVISION DEUTEROMYCOTINA

The Deuteromycotina, all placed in the form-class Deuteromycetes, have no sexual reproduction and are consequently called the *Fungi Imperfecti*, because their life cycles are "imperfect." Most reproduce by means of **conidia**. The imperfect yeasts (those that form no ascospores) reproduce by budding. A few mycelial species of Deuteromycotina form no spores of any kind. They reproduce by other asexual methods such as sclerotia or chlamydospores.

We have already described conidial production in the Ascomycetes. Most Deuteromycetes are so similar to the conidial stages of well-known Ascomycetes that they are believed to be Ascomycetes that have lost their ability to form asci through evolution, or whose ascus stages are seldom formed, under unknown conditions, and have not been discovered as yet. When cultivated under controlled environmental conditions properly manipulated, a number of fungi once included in this alliance have been induced to undergo sexual reproduction and, thus, to complete their life cycles. The spores produced as a result of sexual reproduction have been ascospores in a majority of species, but basidia and basidiospores have developed in a few. When the ascosporic or basidiosporic phase is known, the fungus can also be classified in the appropriate genus of Ascomycetes or Basidiomycetes.

It is probable that many Deuteromycetes never produce sexual stages, ascospores, or basidiospores. They are classified in what is called an artificial system, which emphasizes the location and manner of formation of the conidiogenous cells and the color and structure of the conidia.

Although a detailed consideration of this very large group of fungi is beyond the scope of the present volume, a short general discussion of their classification is certainly warranted. Because of the artificial[1] nature of this classification, the designation of the taxa is often preceded by the prefix "form" as in form-subclass, form-order, form-family, form-genus, and form-species. Many mycologists, however, do not use those designations.

---

[1] An artificial classification is one of convenience and does not attempt to reflect anyone's phylogenetic views, as opposed to a "natural" system, which does reflect current phylogenetic theory.

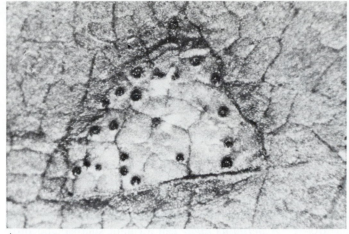

A

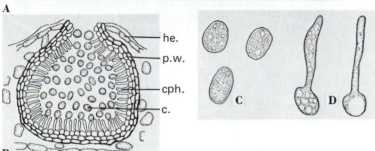

B

C

D

FIGURE 35-1
Guignardia bidwellii (Phylosticta labruscae). **A.** Pycnidia on grape leaf. **B.** Longitudinal median section through a pycnidium. **C.** Conidia (pycnidiospores). **D.** Germinating conidia. **A.** ×5. **B.** ×50. **C, D.** ×400. (**B–D.** *from Alexopoulos [1962], by permission of John Wiley & Sons, Inc.*)

he.

p.w.

cph.

c.

We place all the Deuteromycetes in three form-subclasses: Coelomycetidae, Hyphomycetidae, and Agonomycetidae.

COELOMYCETIDAE

The Coelomycetidae produce their conidiophores in **pycnidia** (sing. **pycnidium**) or in **acervuli** (sing. **acervulus**). Pycnidia are generally flask-shaped structures with an ostiole at the top through which the conidia escape (see p. 736). Pycnidia of *Phyllosticta* (Gr. *phyllon*, leaf, + Gr. *stiktos*, spotted) *labruscae*, the imperfect stage of *Guignardia* (after Guignard, a French plant pathologist) *bidwellii* (one of the Loculoascomycetidae), are illustrated in Fig. 35-1. *Guignardia bidwellii* (= *Phyllosticta labruscae*) is the cause of the serious black rot of grapes.

An acervulus is a mass of hyphae (stroma) formed subepidermally or subcuticularly in an infected plant, which produces a bed of short conidiophores with one conidium at the tip of each (Fig. 35-2). The acervuli of some species produce **setae** among the conidiophores, as shown in Fig. 35-2B. These are sterile, dark, bristle-shaped structures apparently induced by certain environmental conditions. Plant diseases caused by fungi with acervuli are called **anthracnoses**.

HYPHOMYCETIDAE

The Hyphomycetidae (commonly referred to as Hyphomycetes) is an enormous group of fungi that produce their conidia on conidio-

795

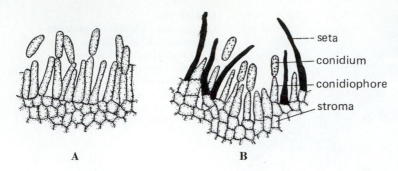

FIGURE 35-2
Two types of acervuli. **A.** Without setae. **B.** With setae. ×200. (*From Alexopoulos [1962], by permission of John Wiley & Sons, Inc.*)

phores originating directly from the somatic hyphae and are more or less unorganized.[2]

In our discussion of the Plectomycetidae (Ascomycetes), we mentioned the ascus stages of some species of *Aspergillus* (L. *Aspergillum*, brush) and *Penicillium* (L. *penicillus*) and pointed out that these are form-genera of Deuteromycetes. Both, of great importance in human affairs, are classified as Hyphomycetidae. The conidiophores of *Aspergillus* originate from special mycelial cells called **foot cells**. They stand erect above the somatic hyphae and form globose swellings (**vesicles**) at their tips. These are covered by one or two rows of bottle-shaped **phialides** (sterigmata), from which small, globose conidia issue forth adhering one to the other in long chains (see p. 747 and Fig. 33-15A). Thus the conidiophore of *Aspergillus* somewhat resembles the aspergillum used in the Roman Catholic Church and is named after it. As we have said before, the form-genus *Aspergillus* has over 100 species. It has been beautifully monographed by Raper and Fennell (1965). Many species are of economic importance but perhaps none more than *Aspergillus niger*, the very common black mold that is used for the fermentative manufacture of citric acid, the basis of the soft drink industry. Citric acid

used to be extracted from lemon juice, a rather expensive process, until it was discovered that *Aspergillus niger* can ferment sucrose and produce citric acid much more cheaply. Almost all the commercial citric acid is now produced by this process. *Aspergillus niger* and some other species have been implicated in a human lung infection called **aspergillosis**, which resembles tuberculosis symptomatologically and which is difficult to diagnose. A number of species of *Aspergillus* and other genera of hyphomycetes contaminate stored grains and forage and form **mycotoxins**. The most important of these are the **aflatoxins** produced by *A. flavus* (see p. 748).

The form-genus *Penicillium* produces its conidia in long chains on repeatedly branched, broomlike conidiophores (Figure 33-17) called **penicilli** (sing. **penicillus**). Among the more than 100 form-species (Raper and Thom, 1949) are several of economic importance. *Penicillium notatum* (Fig. 35-3) is historically famous for being responsible for the discovery of penicillin when in 1928 it contaminated a *Staphylococcus* culture, with which Alexander Flemming was working in a London hospital, and inhibited the growth of the *Staphylococcus*. Since then it has been discovered that *Penicillium chrysogenum* is much more productive and is now used exclusively for penicillin production.

Penicillium griseofulvum is another important species, used to produce the antifungal antibiotic griseofulvin, the only effective treatment for ringworm diseases, including athlete's foot, the

---

[2]There are a number of Hyphomycetidae in which the conidiophores are assembled on superficial stromata (sporodochia) or are closely adherent, forming synnemata. Such fungi are not discussed in this chapter.

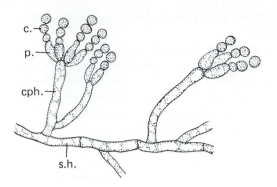

FIGURE 35-3
Penicillium notatum. Somatic hyphae bearing conidiophores (penicilli) with phialides and chains of conidia. ×770.

latter caused mostly by another hyphomycete *Epidermophyton* (Gr. *epidermis*, skin + *phyton*, plant) *floccosum.*

Penicillium roqueforti is used in the manufacture of various types of blue cheese such as Roquefort, Danish blue, and Gorgonzola, and *Penicillium camemberti* is the fungus used to

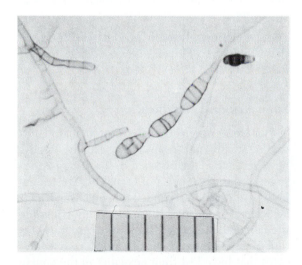

FIGURE 35-4
Alternaria sp. Somatic hyphae and conidial chain. Note vertical septum in penultimate cell of spore at end of spore chain. The scale at the bottom is divided into 10 μm sections. (*Courtesy of Dr. Emory Simmons.*)

impart its special flavor to Camembert cheese. It is quite probable that all these species are truly *Fungi Imperfecti.* They have been studied and selected for so long that someone would have probably found their perfect stages if they existed. However, the form-genera *Microsporum* and *Trichophyton*, all known species of which cause ringworm of human beings and animals, were, as we have seen, also thought to be strictly deuteromycetous until recent years, when the ascus stages of all were discovered in relatively rapid succession.

A very large number of Hyphomycetidae cause serious plant diseases, but space does not permit us to discuss them here. We must, however, mention the ubiquitous *Alternaria* (L. *alternis*, interchangeable), a very large form-genus that produces easily recognized conidia (Fig. 35-4). These are brown and many-septate, with some of the septa vertical. Some species form their conidia singly, but many form them in long chains. *Alternaria* spores are common in house dust and are one of the primary fungal causes of allergy. Some species of *Alternaria* also cause plant disease.

It is among the Hyphomycetidae that we classify the asporogenous (imperfect) yeasts. They are of great significance to us because they are an important source of protein. As the world population increases and food supply becomes even more critical than it is today, food yeast may play an important role in feeding the human race.

AGONOMYCETIDAE

The form-subclass Agonomycetidae is a small group of fungi that produce no spores. They reproduce by fragmentation of the mycelium or by sclerotia. Some of these have proved to be Basidiomycetes when their sexual stages were discovered.

One more point should be made. Many Deuteromycetes such as *Aspergillus fumigatus* exhibit what is known as **parasexuality**. This is a process in which a heterokaryotic condition is established in a mycelium through hyphal anastomosis. Occasionally, two different nuclei fuse somewhere in that mycelium and perpetuate themselves through mitosis, spreading through the hyphae. Again occasionally, mitotic crossing over takes place and a mitotic reduction division occurs, resulting in daughter nuclei with new genetic combinations. These nuclei propagate themselves by entering conidia and establishing new genetic strains. Although anastomosis, nuclear fusion, and haplodization do occur, they do not take place at a specified place or time in the life cycle, which thus differs from a regularly occurring sexual cycle but provides some of the advantages of sexuality.

PREDACIOUS FUNGI

No account of the fungi, however necessarily abbreviated, should fail to present at least a brief discussion of a remarkable and miscellaneous group of organisms, nutrition of which, at least facultatively, is based on capture and digestion of minute animals. The latter include amoebae, rotifers, nematodes, and springtails. The fungi involved are chiefly Zygomycetes and Deuteromycetes.

Of especial interest are some nematode-destroying organisms. Nematodes are minute (0.1–1.0 mm long), wormlike animals, many of which live in soil and are important enemies to horticultural and crop plants. More than 50 species of nematode-destroying fungi are known. The mechanisms used by the fungi to trap the nematode are various and remarkable. In some species undifferentiated hyphae adhere to the animals, but in others special networks of adherent branches (Fig. 35-5A, B), stalked adhesive knobs (Fig. 35-5C, D), and noncon-

stricting and constricting rings occur (Fig. 35-5E).

The last are especially interesting, consisting of three curved cells in the form of a closed ring (Fig. 35-5F) at the end of a short stalk. As the nematode enters the ring, the ring cells rapidly increase to three times their original size, thus constricting and holding the organism; this may occur in 0.1 sec. After capture, hyphae of the fungus penetrate the animal and digest and absorb its substance. It is possible that the fungus produces a toxin.

In writing of these organisms, Duddington[3] has remarked succinctly:

It must be remembered that nematodes are, for their size, powerful and enormously active; they move from place to place by means of a rapid thrashing of their bodies, so that a vigorous specimen will cross the vision field of a microscope with the ferocious speed of a conger eel on the deck of a trawler. To capture such an animal is no mean task for a fungus that is itself composed of threads so delicate that the finest gossamer would by comparison be as a steel hawser to a piece of string, and the means by which this is accomplished by the predaceous fungi are as extraordinary as they are efficient.

For a modern, beautifully illustrated book on these fungi, see Barron's *Nematode Destroying Fungi* (1977).

Summary and classification

The Deuteromycotina are fungi that, so far as we know, do not reproduce sexually. Many, however, have a parasexual cycle that provides some of the advantages of sexuality. Most fungi in this subdivision reproduce by means of conidia, which, in most species, so closely resemble the conidial stages of well-known Ascomycetes that it is presumed that such fungi are Ascomycetes that have lost their sexuality in the course of evolution. Inasmuch as Ascomycetes are classified almost entirely on the basis of the charac-

[3]C. L. Duddington, *The Friendly Fungi*, Faber, London, 1957, p. 51.

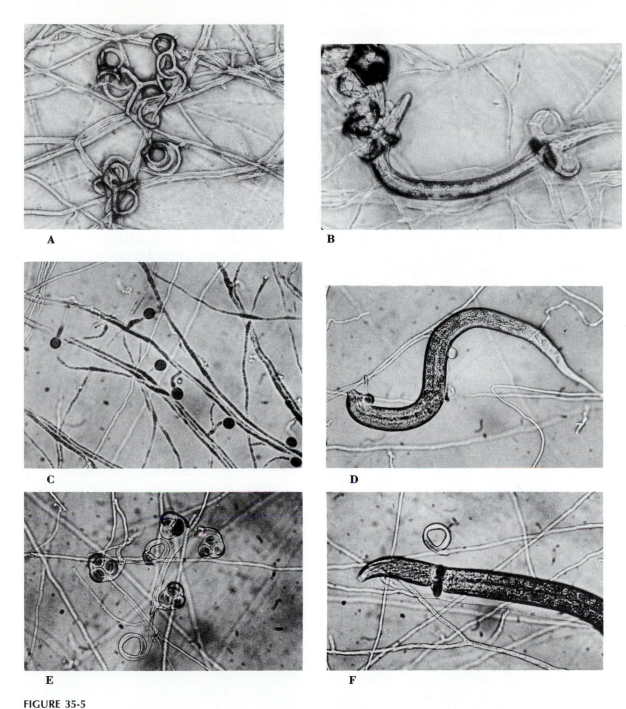

FIGURE 35-5
Nematode-trapping fungi. **A.** *Arthrobotrys connoides*; note hyphal loops, the latter coated with a mucilaginous layer. **B.** Capture of a nematode. **C.** *Dactylella dreschsleri;* note short-stalked, adhesive knobs which capture nematodes upon contact, as in **D. E.** *Arthrobotrys dactyloides;* note constricting rings and their role in capturing a nematode, in **F.** ×300. (*Courtesy of Dr. David Pramer.*)

799

ters of the ascus stages, the conidial fungi cannot be classified in the class Ascomycetes. On the contrary, the conidial (imperfect) stages of Ascomycetes are conveniently classified in the deuteromycete "form-genera" in which they fit.

The Deuteromycotina are of very great importance as plant, animal, and human pathogens or allergens; as contaminants of human food and animal feed, sometimes producing dangerous mycotoxins; in industry for the manufacture of organic acids, certain types of cheese, and some important antibiotics (penicillin and griseofulvin); and in many other ways.

Among the Deuteromycotina we also find some common nematode-destroying fungi, which produce ingenious traps for capturing their prey.

Fungi forming conidia in pycnidia or acervuli are placed in the form-subclass Coelomycetidae; those producing conidia in some other way are lumped together in the form-subclass Hyphomycetidae. The few fungi that produce no spores of any kind are all placed in the form-subclass Agonomycetidae. Imperfect yeasts that reproduce asexually by budding and never form ascospores are sometimes placed in the form-subclass Blastomycetidae, not recognized in this book.

The form-genera mentioned in this chapter may be classified as follows:

Form-subdivision Deuteromycotina
    Form-class Deuteromycetes

Form-subclass 1. Coelomycetidae
    Form-order 1. Sphaeropsidales
        Form-family 1. Sphaeropsidaceae
            Form-genus: *Phyllosticta*
Form-subclass 2. Hyphomycetidae
    Form-order 1. Moniliales
        Form-family 1. Moniliaceae
            Form-genera: *Aspergillus, Penicillium, Epidermophyton, Microsporum, Trichophyton, Arthrobotrys, Dactylella*
        Form-family 2. Dematiaceae
            Form-genus: *Alternaria*

DISCUSSION QUESTIONS

1. Why are the Deuteromycotina also called *Fungi Imperfecti*?
2. How does the genus *Aspergillus* impinge on human affairs?
3. How does the genus *Penicillium* relate to human affairs?
4. What types of nutrition occur among the Deuteromycetes?
5. Define or explain "parasexuality."
6. What is the significance of predacious fungi?
7. What is the relationship, if any, between the Deuteromycetes and the Ascomycetes? Basidiomycetes?

Chapter 36
THE LICHENS

The associations known as lichens,[1] of which there are approximately 18,000–20,000 species, might be classified as a separate division of the plant kingdom were it not for the marked artificiality of such a grouping. Such a hypothetical division might be named the Mycophycophyta (Gr. *mykes*, fungus, + Gr. *phykos*, alga, + Gr. *phyton*, plant) or Phycomycophyta, names that emphasize that these organisms are dual in nature, consisting of a **phycobiont**[2] (an alga) and a **mycobiont** (a fungus) growing together to form a plant body of consistently recognizable structure and appearance. The most frequently encountered phycobiont are the unicellular green algae *Trebouxia* and *Pseudotrebouxia* (Figs. 36-6B and 36-7), which reproduce by zoospores and aplanospores when growing in liquid culture medium. *Nostoc* and *Scytonema* are the most frequent blue-green-algal partner. The lichen thallus is composed mostly of interwoven fungal hyphae, while the algae are usually limited to a subsurface layer and represent only about 7% of the total thallus volume (Collins and Farrar, 1981).

Because the component organisms are members of other divisions—the Cyanophyta, the Chlorophyta, and the Eumycota (Ascomycetes, Basidiomycetes, and Deuteromycetes)—the lichens usually are not classified as a separate division but are grouped with the fungi. In most lichens, fungal morphology determines the morphology of the lichen as a whole. Single algal species may reside in several morphologically distinct lichen thalli. In some cases there is direct evidence, however, that the phycobiont plays a morphogenetic role in the determination of the lichen growth form (James and Henssen, 1976; Ahmadjian, 1982). This is further supported by the fact that when the fungus is grown free from the alga in laboratory culture (Fig. 36-5D), its growth pattern is usually different from that assumed when it is associated with the lichen alga. In some gelatinous lichens the phycobiont may determine thallus morphology, and the fungus is loosely associated with the algal colony.

Lichens are ubiquitous plants that occur in a variety of habitats. One finds these organisms

---

[1]The authors are greatly indebted to Professor Vernon Ahmadjian and to Dr. Robert D. Slocum for critically reading this chapter on lichens and for providing excellent suggestions for its improvement.

[2]By those who consider blue-green algae to be bacteria called by the inclusive term "photobionts."

on the bare surfaces of exposed desert rocks or on the frozen substrata of polar regions. Some endolithic lichens grow inside the sandstone formations that occur in the dry valleys of Antarctica. This harsh environment is similar to the planet Mars and suggests that endolithic lichens may also be growing there. Ahmadjian (1970a) reported that five species of lichens can survive a temperature of −198°C and that lichens can photosynthesize actively at −18.5°C in the Antarctic. Lichens commonly colonize rocks (saxicolous-species) and undisturbed soil (terricolous species), and tree bark supports an extensive lichen flora (corticolous species). There are also reports of lichens growing on a number of unusual substrates, ranging from iron grave markers to the backs of large forest weevils (Gressitt et al., 1965).

A few species of lichens are submerged aquatics, both freshwater and marine, that are periodically exposed to the air, while others survive long periods of desiccation. Lichens can absorb water in both liquid and vapor form, and the absorption of water seems to be an entirely physical process. In the desiccated state, they can survive with a water content of 2–15% of their dry weight; during a rain, they may contain 100–300% of their dry weight as water (Smith, 1962).

Rock-inhabiting lichens have been implicated as agents initiating soil formation. The underlying substrate is apparently chemically weathered by lichen compounds (Ascaso and Galvan, 1976) as well as subjected to mechanical destruction by the growing thallus (Frey, 1927). Partially weathered rock is then readily broken down by ice and other physical agents to form a primitive type of soil. As organic remains from lichen vegetation become incorporated among the rock particles, other forms of vegetation become established in this primitive soil.

Lichens are extremely slow-growing organisms, many increasing in radial diameter no more than 1 mm to 1 cm per year. Accordingly, large plants are probably quite ancient. The age of some arctic lichens has been calculated to be 4500 years, and lichen growth rates are sometimes used to set approximate dates of glacial retreat in arctic and alpine regions. Their slow rate of growth is probably correlated with their lack of adaptations to conserve water. In the dry state their vital activities continue only at an extremely reduced rate. Hydrated lichens, however, have the capacity to absorb from their environment large quantities of solutes—for example, phosphate (Smith, 1973b).

Lichens produce a variety of organic compounds, some of which occur as crystalline encrustations on the thallus surface. These are known as "lichen acids," although not all are acids (Smith, 1962; Culberson, 1969, 1970), and the majority of these compounds are synthesized only by the lichen symbionts. These compounds have become increasingly useful in the identification and classification of lichen taxa. Litmus was prepared from lichen compounds until it was replaced by a synthetic substitute. Usnic acid, one of these substances, has been shown to be antibiotic to Gram-positive bacteria and to inhibit the growth of *Trebouxia*, a common algal member of lichens (Kinraide and Ahmadjian, 1970). Culberson (1969) and Huneck (1973) have summarized our knowledge of these lichen compounds.

A number of electron microscopic studies of lichens (e.g., Jacobs and Ahmadjian, 1969; Gallun, et al., 1970) have attempted to elucidate the relationships between the alga and fungus constituting various lichen thalli, the exact nature of which is only now beginning to be understood (Ahmadjian and Jacobs, 1982; Galun and Burbrick, 1984).

Some view the lichen association as a fungus parasitizing an alga. Recent studies have tried to determine if a precise recognition process occurs between the symbionts of a lichen and whether the recognition is mediated by lectins (Hersoug, 1983). Support for this view is afforded by lichens in which the fungal hyphae characteristically penetrate the algal protoplasts via specialized absorptive organs called **haustoria** (Fig. 36-7A,B). Fungal haustoria penetrate the algal cell walls, but not their plasmalemmas, presumably by both mechanical and chemical (enzymatic) means. However, this process occurs only to a limited extent, not to the point of complete elimination of the algal host. Webber and Webber (1970) suggested that this process represented a controlled "harvesting" of the algal cells by the fungus.

The fungal component of a number of lichens has been grown successfully in artificial culture media (Fig. 36-5D) (Ahmadjian, 1963), so the supposed parasitism is not obligate or highly specialized. Furthermore, in spite of the inferred parasitism, the algal cells grow and multiply for long periods within the lichen thallus without apparent injurious effects from the fungus to most of the algae.

Other investigators interpret lichens as manifestations of a form of mutualism. Symbiosis, as originally defined by De Bary, meant the phenomenon of dissimilar organisms living together in a constant and intimate association. This was a broad definition that included all types of association, including parasitism, as well as associations in which the associated organisms benefit. Smith (1973a,b) has suggested that such different types of association be designated, respectively, **parasitic symbiosis** and **mutualistic symbiosis**. Those proponents of the latter definition of the lichen association argue that the alga is surrounded and mechanically protected by the meshwork of fungal hyphae, which absorb and adsorb water, mineral salts, and organic materials from the substrate, thus extending the habitat range of the alga. The fungus, presumably by means of **appressoria**[3] or haustoria, diffusion from the algal cells, or autolysis of the latter, is supplied with a source of carbohydrate, organic nitrogen compounds, and vitamins. Our knowledge of the physiological relations between fungi and algae in lichens has in the past been colored to some extent by teleological considerations and speculation.

It has been shown that in *Peltigera* (N.L. *pelta*, shield, + L. *genere*, to bear) *aphthosa*, which has two phycobionts—a green alga (*Coccomyxa*) and the cyanophyte *Nostoc*—the latter fixes nitrogen actively. Almost all of the fixed nitrogen is secreted into the lichen plant body, where it is used by the mycobiont and the green phycobiont (Rai et al.,1981). In unlichenized cynanophytes, a small aliquot of the nitrogen compounds is released from the algal cells, further suggesting that the mycobiont controls the rate of secretion from the algal cells. The fungus inhibits enzymes of the alga that would normally assimilate fixed ammonia (NH_3) (Stewart, 1978).

Richardson et al. (1968) demonstrated that a number of phycobionts excrete various sugar alcohols (ribitol, sorbitol, and erythritol) or glucose, which are used by the mycobiont. In the absence of the latter, this excretion ceases or is markedly reduced. Cultured phycobionts also develop a pectic sheath, which may be part of a presymbiotic condition that helps build the symbionts together (Ahmadjian and Jacobs 1982). Drew and Smith (1967), after exposing *P. polydactyla* to $^{14}CO_2$, detected ^{14}C-labeled compounds first in the algal cortex and later in the fungal medulla of that lichen. Jacobs and Ahmadjian (1971) detected ^{14}C-labeled compounds in the chloroplast and pyrenoid of the green phycobiont in *Cladonia* (Gr. *klados*, branch or sprout) *cristatella* (Fig. 36-3) and also in the cell walls of the mycobiont.

In summary, it is difficult to generalize with assurance concerning the physiological relations between the component organisms of lichens. Further studies of these questions are now being carried on by a number of investigators. These are based on pure culture studies of the organisms grown separately and together, as well as on growth of lichens in field and in the laboratory (Galun and Burbrick, 1984).

The great majority of lichens have mycobionts that are ascomycetous, probably related to the cup fungi or to perithecium-forming

---

[3]An appressorium is a specialized hyphal cell of a fungus adpressed to a cell surface; by contrast, a haustorium penetrates the cell.

genera. In a few lichens the fungus is one of the Basidiomycetes. Their phycobionts are alternately cyanochlorontan organisms such as *Gloeocapsa, Nostoc,* and *Stigonema,* or Chlorophyceae such as *Trebouxia,* and *Pseudotrebouxia* (*Chlorococcum*-like genera) (Fig. 36-7); in a few lichens, Cyanophyta, in addition to the green algae, are present secondarily in the fungal fruiting body.

The lichen thallus may be leaflike in organization, or **foliose**; crustlike, or **crustose**; or branching and cylindrical, or **fruticose**. Examples of foliose organization are *Peltigera* (Fig. 36-1) and *Parmelia* (L. *parma,* small shield) (Fig. 36-2B). *Cladonia* (Fig. 36-3) and *Usnea* (Ar. *ushnea,* moss) (Fig. 36-2A) are two fruticose genera. A crustose lichen thallus may be seen in Fig. 36-4A. In transverse sections of the typical stratified lichen thallus (Fig. 36-4B), the dense layer of surface hyphae, or **upper cortex,** may assume an epidermislike configuration and may have a colloidal, water-retaining surface layer. The fungal hyphae of many lichens are densely interwoven, so that in section they seem to form a parenchyma.[4] Lining the upper cortex are the alyal layer and the **medulla,** or central portion of the thallus, below which a dense **lower cortex** may be present. The medulla is composed of a loose network of hyphae in which the alga is usually present in a distinct zone (Fig. 36-4). The medullary hyphae are longitudinal in orientation in *Peltigera* and function in conduction. Special attachment hyphae, the **rhizines,** enter the substratum from the lower surface of the thallus. Rhizines (Fig. 36-5A) are complex bundles of hyphae that have anastomosed by fusion of young hyphal tips. They may conduct water rapidly in the capillary spaces among the com-

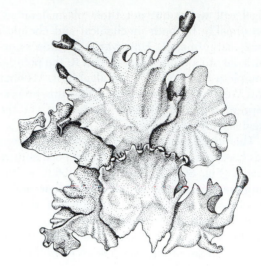

FIGURE 36-1
Peltigera rufescens. Portion of plant with fertile (ascospore-bearing) lobes. ×⅔.

ponent hyphae, but their primary role is to anchor the lichen to the substrate.

Propagation of the usually slowly growing lichen thallus is generally effected by fragmentation as the older portions of the plant body die and leave the growing regions isolated. A number of lichens also produce specialized asexual propagules called **soredia** (Gr. *soros,* heap) (Fig. 36-6) or **isidia** (origin uncertain) (Fig. 36-5B). Isidia are small fragments of the thallus, consisting of algal cells surrounded by fungal hyphae. Insida are protuberances of the upper cortex, while soredia do not have a cortical covering. These propagules are readily detached from the thallus and, when borne to suitable environments, by air currents or other dispersive agents, may develop into new lichen thalli. In *Cladonia chlorophaea* the soredia are borne in gobletlike structures. It is probable that fragments containing both algal and fungal components are the primary agents of lichen dispersal and colonization.

In addition to multiplication of the lichen as a unit, the component organisms also reproduce independently. Under conditions of abundant moisture in the laboratory, as well as under

---

[4]Parenchyma cells are living plants, closely contiguous, with thin walls and large central vacuoles. They are genetically related through a series of cell divisions. The fungal hyphae by interweaving form a pseudoparenchyma, because in sectional view, although the cells are contiguous, they are not derived from one another by cell division but are contiguous because of interweaving.

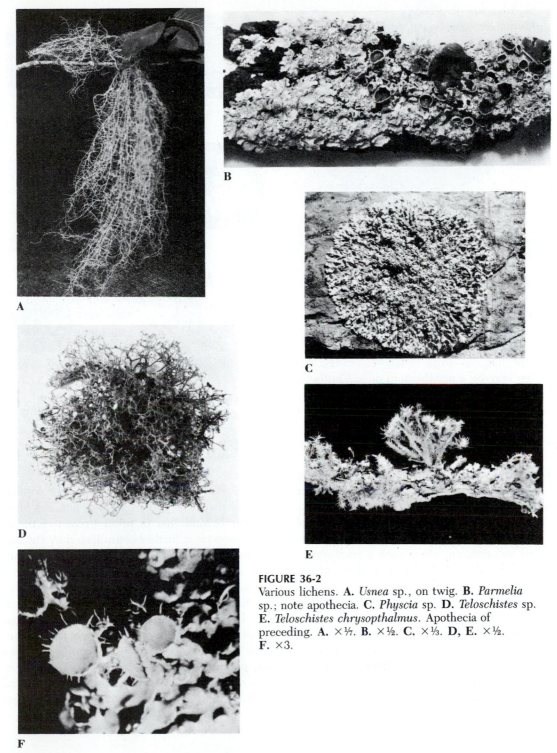

FIGURE 36-2
Various lichens. **A.** *Usnea* sp., on twig. **B.** *Parmelia* sp.; note apothecia. **C.** *Physcia* sp. **D.** *Teloschistes* sp. **E.** *Teloschistes chrysopthalmus*. Apothecia of preceding. **A.** ×⅐. **B.** ×½. **C.** ×⅓. **D, E.** ×½. **F.** ×3.

FIGURE 36-3
Cladonia cristatella, "British Soldiers." ×2.5. (*After Ahmadjian.*)

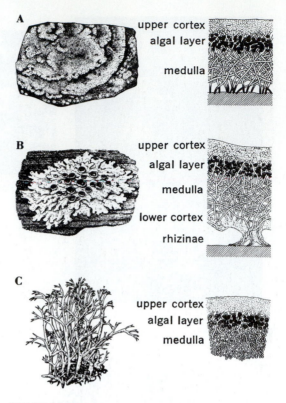

FIGURE 36-4
Growth habits of lichens, diagrammatic and transections thereof to the right. **A.** Crustose. **B.** Foliose, note apothecia. **C.** Fruticose. (*After Ahmadjian.*)

natural conditions, the algal member may undergo rapid growth and cell division, thus outstripping the enveloping fungus; groups of algal cells are thus set free into the substratum. In many lichens the fungus regularly produces ascocarps and asci (Figs. 36-2B–E, 36-4B); the ascocarp may be a perithecium or an apothecium (Figs. 36-2B, 36-4B, 36-8), or some modified form thereof (Fig. 36-5). These fruiting bodies discharge numerous spores throughout the year, peak periods of discharge being in the spring and autumn in the temperate zone.

The picturesque red tips of certain ascending branches of *Cladonia cristatella* (Fig. 36-3), for example, are apothecia. It has been demonstrated in a number of genera that the formation of the ascocarp is preceded by the development of a coiled ascogonium, often with a trichogyne, which is probably fertilized by a spermatium. Cytological details are unknown in most cases.

Lichenized fungi produce **pycnospores** in reproductive structures called **pycnidia** (Fig. 36-9C). It is believed that these spores may act as spermatia in the sexual reproductive process in some lichens (Ahmadjian, 1966). In other lichens, they may function simply as conidia, as some pycnospores are capable of germination in culture (Vobis, 1977). In general, our knowledge of sexual reproduction in lichenized fungi lags far behind that in other groups of fungi. This situation is further complicated by the fact that isolated mycobionts in axenic culture exhibit an apparent loss of sexuality.

The mature ascocarp may be elevated above the thallus on a stalk called a **podetium**, or it may be sessile or even sunken. The ascocarps are perennial. The hymenial layer of the apothecium is composed of densely intermingled asci and paraphyses (Figs. 36-8, 36-9). The wall of the ascocarp may be composed only of fungal hyphae, or it may be surrounded by an alga-

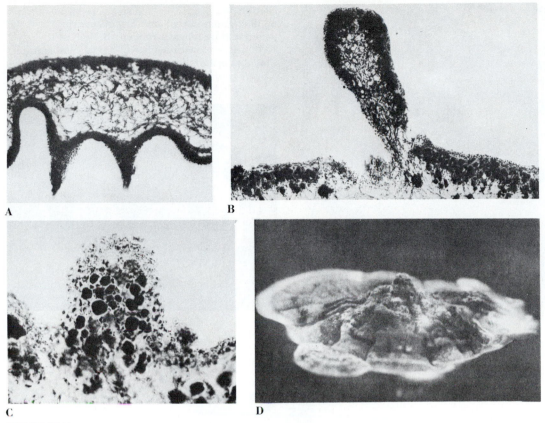

FIGURE 36-5
A. *Parmelia caperata.* Transection showing rhizines. **B, C.** *Parmelia rudecta.* Sections of mature isidia. **D.** Mycobiont of *Cladonia cristatella*: 6-month-old culture on agar. **A, B.** ×100. **C.** ×240. **D.** ×2. (**A–C.** *Courtesy of Mr. Robert D. Slocum.* **D.** *After Ahmadjian.*)

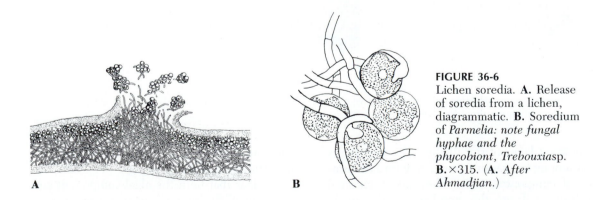

FIGURE 36-6
Lichen soredia. **A.** Release of soredia from a lichen, diagrammatic. **B.** Soredium of *Parmelia*: note fungal *hyphae and the phycobiont, Trebouxia*sp. **B.**×315. (**A.** *After Ahmadjian.*)

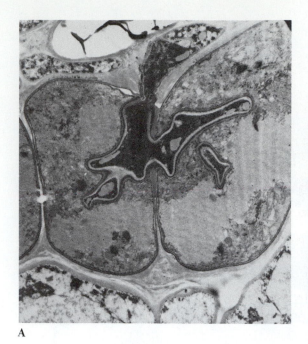

A

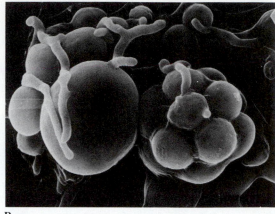

FIGURE 36-7
A. Blue-green phycobiont (*Scytonema* sp.)
penetrated by fungal symbiont. **B.** cells of
Trebouxia erici with hyphae of fungal symbiont (in
Cladonia cristatella). **A.** × 27050. **B.** × 1400.
(*Courtesy of Professor Vernon Ahmadjian.*)

B

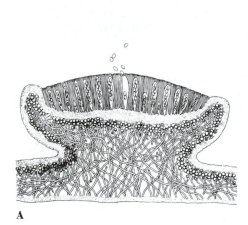

A

B

FIGURE 36-8
A. Section of the apothecium of a mycobiont of a lichen, diagrammatic. **B.** *Physcia*
sp., section of an apothecium. ×60. (**A.** *After Ahmadjian.*)

containing layer of the thallus. The ascospores
vary among the various genera in form and number of component cells. They are discharged
explosively from the apothecia and usually
germinate readily if they fall on a favorable
substratum.

It is clear that both the algal component and
the spores of the fungus, once separated from

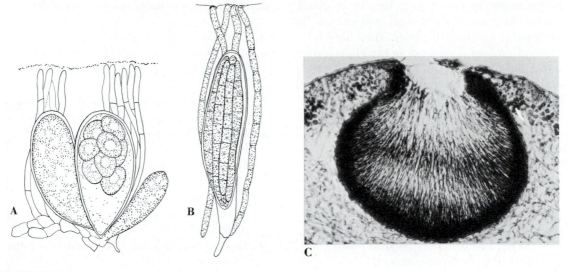

FIGURE 36-9
A. *Parmelia* sp. Asci and paraphyses. **B.** *Peltigera* sp. Ascus with septate ascospores and paraphyses.
C. *Parmotrema perforatum*. Section of pycnidium immersed in lichen thallus. **A, B.** ×600. **C.** ×230.
(**C.** *Courtesy of Dr. Robert D. Slocum.*)

the parent thallus, can initiate an independent existence. The synthesis of a new lichen thallus would then depend upon the fortuitous proximity of a germinating fungal spore and algal cells of the species with which the fungus had been associated in the parent thallus.

For this reason it has been suggested that ascospores do not play an important role in the multiplication of lichens as such. It is probable that fragmentation and asexual propagule production are more efficient in this connection.

The researches of Ahmadjian have demonstrated clearly that assumption of the lichen relationship by an alga and a fungus depends on the existence of minimal nutrition for both organisms. Attempts to induce or initiate a lichen synthesis in culture media rich in nutrients are consistently failures. However, in low-nutrient, minimal culture media, lichen synthesis has been successfully evoked. A number of lichens can now be synthesized in axenic cultures in the laboratory (Ahmadjian et al., 1980; Ahmadjian and Jacobs, 1981, 1982; and Ahmadjian, 1982).

Until recently, lichens have been neglected by most botanists, except for a few who have occupied themsleves almost exclusively with the taxonomy of these interesting composite organisms. Careful physiological studies of the components, grown separately and together in laboratory culture under controlled conditions, are augmenting our understanding of these organisms. It seems clear that the lichen association is in a delicate state of equilibrium in nature, a condition that is perhaps reflected by the sensitivity of these organisms to sulfur dioxide and other gaseous pollutants. Lichens are becoming increasingly important as indicators of air pollution, being adversely affected or even eliminated following prolonged exposure to relatively low levels of pollution in urban areas.

It is difficult to assign lichens a natural position in any plan of classification of the plant kingdom. If they were classified with the group of ascomycetous or basidiomycetous fungi to which their fungus component is related, as is usually done, the importance of the algal component would be minimized. The opposite ob-

jection might be raised were they to be classified on the basis of their algal components. In
the present volume, therefore, they have been
assigned no formal position in the classification
of the plant kingdom but are treated separately
in the present chapter. Most michologists and
lichenologists classify lichens with the fungi.

However, the current prevailing consensus
seems to be that these organisms are lichenized
fungi. After all, the same phycobiont occurs in a
great variety of different lichens that are seemingly not closely related (Poelt, 1973).

Finally, the reader is referred to the publications of Smith (1962), Ahmadjian (1963, 1970*b*),
Ahmadjian and Hale (1973), Seaward (1977),
Hale (1983), Lawrey (1984), and Kershaw (1984)
for further information on lichens.

DISCUSSION QUESTIONS

1. What types of algae occur in lichens?

2. What types of fungi are present in lichens?
3. How do lichens reproduce?
4. How would you go about isolating and
   growing the components of a given lichen?
5. How would you go about resynthesizing
   the lichen?
6. Why are lichens sometimes called "soil
   builders"?
7. Define or explain: phycobiont, photo-
   biont, mycobiont, rhizine, soredium, isidium, parenchyma, symbiosis, appressorium, pycnidium, pycnospore,
   podetium.
8. Discuss adaptations for survival in lichens.
9. Describe the physiological relationship between the symbionts of lichens.
10. List the types of thallus formed by li-
    chens.

Chapter 37
GENERAL SUMMARY

The preceding chapter brought to a conclusion the discussion of representative types, both extant and extinct, of the various divisions of plants and fungi. Although the treatment of these has not been exhaustive in scope, thoughtful reading of it, along with the study of the relevant laboratory materials and collateral reading, will have provided a substantial basis on which to formulate some generalizations and conclusions. The purpose of this final chapter, therefore, is to summarize some of the more important principles and phenomena to which we have alluded in earlier pages. These are discussed under four categories: (1) vegetative organization, (2) reproduction, (3) the fossil record, (4) classification. The discussion that follows presupposes that the reader is familiar with the more important characteristics of the representative genera of plants described up to this point.

ORGANIZATION

FORM AND GROWTH

With respect to form of the organism, plants and fungi include a rather wide range of organisms. The simplest are those in which the organism exists as a single cell, the great multiplicity of vital processes taking place within an extremely minute unit of protoplasm. This concept of the unicellular organism, as exemplified by such an assemblage as certain bacteria, and algae such as *Chlamydomonas*, *Chlorella*, and *Pinnularia*, among others, is here restricted in scope. It does not include such siphonous and coenocytic organisms as *Bryopsis*, the Myxomycetes, and *Rhizopus*, for example, which may, perhaps, be interpreted more properly as acellular organisms in which partition into cells has not accompanied nuclear multiplication.

Increasing complexity of body form probably first arose when single-celled organisms remained together after division, either in the form of colonies in which the cells were remote from one another and separated by polysaccharide matrices (*Volvox*), or in the form of aggregates of cells—some of them filaments or membranes. That some colonies and aggregates are relatively loosely organized assemblages of cells is attested by the ease with which fragments survive and regenerate. Other types of colonies with contiguous cells are illustrated by *Pediastrum* and *Hydrodictyon*, among others.

It is doubtful, however, that the colonial type of organism was the precursor from which the more complex bodies of the so-called higher plants have developed. However, *Tetracystis* (p. 63) and similar organisms, with their capacity to form incipient tissues, suggest that multicellularity evolved early. Such organisms and/or simple filaments may well have been the precursors of more histologically complex plants. Ontogeny furnishes us with the clue that unbranched filaments preceded branched types as well as membranous, parenchymatous types of plant bodies. In the development of the germlings of such membranous plants as *Ulva* and *Laminaria*, the reproductive cells pass through an unbranched, filamentous stage before forming a parenchymatous plant body. The same phenomenon occurs also in developing gametophytes of liverworts and ferns, in the protonemata of mosses, in the sporophyte of *Sphagnum*, and even in the development of the embryonic sporophyte of *Capsella* and other Anthophyta.

The fungi, of course, and many siphonous algae have remained permanently filamentous, although they often consist of complex pseudo-parenchymatous plant bodies (as in *Codium*, for example) that result from interweaving of their filaments. This type of construction, however, like the colonial one, probably was not involved in the ancestry of parenchymatous organisms.

By continuous cell division in more than two directions, the unbranched filament may have given rise to another type of plant body, the solid cylindrical axis, exemplified by a few green algae and many genera of brown algae. A parenchymatous cylindrical type of construction occurs widely in the plant kingdom, as, for example, in the axes of mosses and vascular plants.

Another aspect of filamentous algae must be emphasized in discussing the origin of plant body types—namely, the phenomenon of **heterotrichy,** in which a much-branched or even disclike prostrate system supports an erect filamentous one (*Stigeoclonium*, Fig. 4-30). There are some algae in which the erect system has been almost entirely suppressed, so that the prostrate system expands as a membranous parenchymatous layer. It is possible that some of the early hepatophytan land plants may have arisen from such prostrate algae, and the cylindrical axes of other land plants may have been derived from the erect portions of algal ancestors (see, however, pp. 308 and 310). Of considerable significance in any discussion of form and growth must be the patterns and mechanisms of cell division in the building of plant bodies. Four patterns (Fig. 37-1) occur among Chlorophyta and Charophyta, one or the other, or both, of these groups, the putative progenitors of the so-called "higher plants." In some green algae, cytokinesis is accomplished by simple furrowing, without (Fig. 37-1A) or with (Fig. 37-1C) the association at the furrow of microtubules. Alternately, cytokinesis may be by cell plate formation with associated and parallel microtubules (**phycoplast**) (Fig. 37-1B); or finally, cytokinesis may involve a cell plate with associated perpendicular microtubules (phragmoplast) (Fig. 37-1D). The last pattern occurs in a few green algae such as the Charophyta (Chapter 5) and *Coleochaete* (p. 82) and is apparently universally present in liverworts, hornworts, mosses and vascular plants. This is considered to be of special phylogenetic significance (McBride, 1970; Floyd et al., 1971, 1975; and Graham, 1982). Whatever its origin, the plant body increases in size throughout its individual existence by the process known as **growth**. The latter has been defined as increase in mass usually accompanied by differentiation. Increase in mass without differentiation occurs in the plasmodial phase of the Myxomycetes and in algae with generalized growth. Increase in mass without differentiation occurs in the plasmodial phase of the Myxomycetes and in algae with

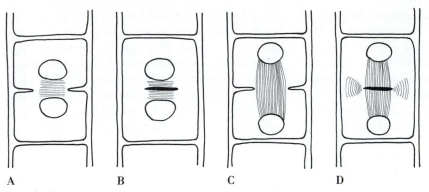

FIGURE 37-1
Some patterns of cell division in green plants (diagrammatic). **A., C.** Furrowing with microtubles present. **B.** Cell plate with associated horizontal microtubles (a phycoplast). **D.** Cell plate with persistent associated perpendicular microtubles (a phragmoplast). (*Based on the investigations of, and discussions with, Pickett-Heaps, Floyd, Mattox, and Stewart*).

generalized growth. Increase in mass of multi-cellular organisms is expressed by increase in cell number and cell size, and these, in turn, are made possible by the synthesis of additional protoplasm. Increase in cell number and size may be entirely separate processes, as illustrated by the rapid enlargement of the fruiting bodies of fleshy fungi after a rain or the rapid increase in length of the *Pellia* seta, by increase in the size (only) of their component cells, and the final phase of growth in all organs of vascular plants.

In the plant kingdom, growth may be generalized—as in free-floating filamentous algae and noncoenobic colonies—or localized. The growing region may be apical, basal, inter-calary, or peripheral, the last type exemplified by the cambium of the vascular plants. Representative plants displaying one or another type of localization of growth have been referred to repeatedly in the preceding pages. Growth also may be classified as determinate or indeterminate and as continuous or periodic. **Determinate growth** is illustrated by the acrogynous liverworts, the differentiation of branches of algae and fungi into sporangia or gametangia, and by strobili and flowers in which the axis usually differentiates completely, thus "using up" the meristem. **Indeterminate growth**, that is, continuing increase in size, is manifested in vegetative apices of many plants in all divisions of the plant kingdom. Periodic, as contrasted with continuous, growth involves periods of inactivity. Although this may occur in all plants, it is most strikingly manifested in woody perennial plants of the temperate zone with their closed or dormant buds. Differentiation, which usually accompanies growth, will be discussed in the next paragraphs.

DIFFERENTIATION

In unicellular organisms, differentiation is entirely intracellular, except for the cell wall. Special portions of the protoplasm and cell wall perform all the functions that support the individual and maintain the species. Similarly, in many colonial algae the individual cells of the aggregate differ little from one another in structure or function. In other colonies, however, the cells are differentiated into so-called vegetative (somatic) and reproductive (germ) cells. The implication of these terms in these and similar cases often is misunderstood. Vegetative activities are those that are involved in the maintenance of the individual; reproduction results in the maintenance of the species or race. The difference between vegetative and reproductive cells probably arose when cells that originally performed both types of functions lost, during subsequent evolutionary specialization, the capacity for reproduction. All cells are vegetative (somatic) in the sense that they carry on the functions essential to the life of the individual cell. Although this segregation of function seems quite absolute in such a plant as *Volvox* and annual vascular plants, because the vegetative cells survive for only one generation, segregation of vegetative from reproductive activities in other plants seems to be less complete. This is evidenced by the widespread occurrence of vegetative reproduction and regeneration from specialized tissues, a phenomenon that will be discussed below.

In addition to this primary dichotomy into vegetative and reproductive phases, further differentiation and specialization are manifest in plant bodies. Differentiation of a variety of vegetative tissues and functions in multicellular organisms with resulting complexity is a measure of degree of advancement in the plant kingdom. This has resulted in the concepts of "lower" and "higher" plants, the adjectives by implication indicating different levels in the evolutionary hierarchy. The terms are unfortunate in view of the almost universally similar ultraorganization of all plants (except bacteria and blue-green algae, which are by some not considered to be members of the plant kingdom [Stanier et al., 1971]) and because they are relative terms with indefinable limits.

As a group, the fungi offer few indications that differentiation of the vegetative system, their mycelium, has progressed to any marked degree. The algae, on the other hand, include an instructive series in this connection. The colonial forms have already been cited. In the filamentous and membranous green algae, the plant bodies are histologically uniform, except for the lower holdfast or rhizoidal cells. In many Rhodophyta and Phaeophyta, however, the plant bodies are differentiated into a medullary region composed of cells with few or no plastids, presumably a storage region, and more superficial chlorophyllous cells, which form the phytosynthetic cortex. Histological simplicity, of course, characterizes algae, the plant bodies of which are immersed in a solution containing the elements essential for metabolism. No part of the plant body is far removed from the source of water and the inorganic salts dissolved in it. But in the larger, longer-lived kelps, among the Phaeophyta, there has developed a specialized conducting system (phloem) that transports organic materials from the site of their manufacture to more remote portions of the plant. It is of interest, but hardly surprising, that no water-conducting system has been discovered in such algae. That extensive development of water-conducting tissues does not usually occur in plants with aquatic habitats is indicated also by the paucity of xylem in submerged aquatic representatives of the vascular plants.

The differentiation of a water-conducting system and other specializations have made possible plant life on land. It is true that a number of terrestrial plants are scarcely more differentiated than membranous green algae. Such liverworts as *Sphaerocarpos* and *Pellia*, for example, are not markedly more complex internally than *Ulva*. It is noteworthy, however, that such simple terrestrial plants as the liverworts, as a group, are restricted mostly to moist habitats, although there are such conspicuous exceptions as the corticolous *Frullania* and *Porella*. The evolutionary change from a habitat in which the entire plant surface is bathed in a solution containing its required raw materials to one in which at least one surface is exposed to the atmosphere was probably an important force in selecting differentiations and specializations of survival value. Special cells and tissues for absorption, cuticular mechanisms and stomata,

814

some types of epidermal hairs, and silicification are all manifestations of adaptation to a terrestrial habitat. The development of supporting tissues and of those for the efficient conduction of water and solutes and organic substances throughout the plant body resulted in the possibility of increase in size and extension of plant habitats to less hydric surroundings. Such conducting tissue occurs in the form of xylem and phloem in vascular plants, but phloemlike cells occur also in the kelps of the brown algae. Furthermore, water-conducting cells (hydroids) and cells (leptoids) that conduct organic substances are present in the axes of a number of mosses. Elongate, pitted conducting cells occur in the central strands of certain Hepatophyta.

Differentiation in the plant kingdom occurred not only at the intracellular level but also in the grouping and arrangement of the cells and tissues in plant bodies as specialized organs—namely, roots, stems, and leaves. True roots, stems, and leaves are those that have vascular tissues, xylem, and phloem. However, rootlike organs, leaves, and stems, all lacking vascular tissues, are present in certain algae, in the liverworts, in the mosses, and in fungi. Although these vascularized and nonvascularized structures differ, there is every indication of parallelism in function. The root hairs of the vascular plants, the multicellular rhizoids of mosses, vascularized roots, and the unicellular rhizoids of *Botrydium* among the algae, of the liverworts, and of the fern gametophyte all perform the functions of anchorage and absorption. Similarly, the bladelike branches of *Sargassum* and the kelps, those of the leafy liverworts and mosses, the microphyllous leaves of *Lycopodium*, *Selaginella*, and *Isoetes*, and the megaphyllous leaves of other plants all represent expanded surfaces that function primarily in photosynthesis. The axes of *Chara*, of the kelps, and of leafy liverworts and mosses, and the stems of vascular plants, all are organs supporting photosynthetic appendages and performing translocation and are also themselves

actively photosynthetic. The several series just cited indicate that differentiation may follow diverse paths in relation to a given function.

NUTRITION AND HABITAT

Growth and differentiation of living organisms involve syntheses of additional protoplasm and its nonliving adjuncts from basic materials of their environment. The preceding chapters have referred to organisms that vary in the materials they require and the pathways by which syntheses are accomplished, and these will be reviewed briefly at this point.

Autotrophic organisms are those that synthesize their protoplasm from entirely inorganic sources. Their enzyme systems, therefore, are the most extensive and complex in the world of living things. As noted in Chapter 2, autotrophism has been interpreted both as a primitive and as a derived attribute. Autotrophic organisms obtain their primary energy either from light, as in photoautotrophism, or from chemical processes, as in chemoautotrophism. Although most organisms that have chlorophyll are assumed to be photoautotrophic, there are relatively few in which this has actually been demonstrated by cultivation in a controlled environment. On the other hand, in such organisms as *Euglena* and probably many other flagellates, in spite of the presence of chlorophyll, one or more organic substances—such as vitamins and acetate—may be required or enhance growth. It seems probable, too, that although organisms can grow in a completely autotrophic environment in the laboratory, such conditions rarely prevail in natural habitats, and it is probable that syntheses from inorganic sources may be augmented by the organic materials of the surrounding medium.

Organisms that are not autotrophic are heterotrophic. Of the organisms considered in this text, the fungi and bacteria that lack chlorophyll are heterotrophic, with the exception of certain chemosynthetic bacteria. Parasitism, saprophytism, and phagotrophism are exemplified in the various groups of these colorless organisms. Heterotrophic nutrition is not confined exclusively to fungi and bacteria, however. Parasitism and lack of chlorophyll are found in the angiosperms in such a plant as dodder (*Cuscuta*) and Indian pipe (*Monotropa*). In recent years, however, it has been shown that *Monotropa* has a mycorrhizal fungus that is also mycorrhizal on neighboring photosynthetic plants. This triangle of partners has proven to be one of the most interesting forms of symbiosis in higher plants.

Finally, it should be noted that both photoautotrophic and heterotrophic nutrition may prevail in different phases of the same organisms. The nutrition of the developing carpospore-producing filaments of the red algae and the embryos of the vascular cryptogams and seed plants is certainly heterotrophic during early stages of development, although the parent plants are photoautotrophic. Similarly, the gametophytes of heterosporous plants are either saprophytic, in those in which the spores are shed, or parasitic, in those in which the spores are retained, although the mature sporophytes are presumably photoautotrophic. Whatever the methods and pathways of syntheses, they culminate in the production of additional protoplasm and its subsequent differentiation.

The representative genera described in this text are found in a wide range of natural habitats. As a group, the algae are predominantly aquatics, although a number occur both in the soil and in moist aerial habitats. A majority of fungi must be classified as hydrophytes, because they cannot grow actively except in the presence of abundant moisture. Most liverworts and mosses also are plants with high moisture requirements, although many are able to withstand desiccation for long periods, during which they remain dormant. A few grow exclusively under xeric situations. The vascular plants, probably primarily because of the efficiency of their absorbing, conducting, and epidermal systems, are less restricted and occur in a wide range of habitats, insofar as available water is involved. There is every indication that morphological adaptations for absorbing, conducting, and conserving water are correlated closely with the habitats in which the various plant types can survive. For example, many xerophytes become quite dehydrated at times. It is the abundance of their supporting tissues that holds them in position and masks their wilting. It is likely that this role of the supporting tissues in maintaining the organisms' shape and in preventing mechanical damage to flaccid tissues makes it possible for the organisms to survive in xeric habitats.

REPRODUCTION

Once generated, the plant body ultimately reaches a degree of maturity, at which the phenomenon of reproduction begins. Reproduction is simply the reduplication of the individual, and it results in increase in numbers, directly or indirectly. The various reproductive process cited in earlier chapters have been treated as examples of two basic types, sexual and asexual reproduction. The former usually involves the union of cells and nuclei, the association of chromosomes and genes in a zygote, from which a new individual sooner or later arises; meiosis is inevitably involved. In contrast to the unions involved in sexual reproduction, asexual reproduction includes methods of reduplication of individuals in which no such cell and nuclear union takes place.

Perhaps the simplest type of asexual reproduction is that of unicellular organisms, in which the division of the cell duplicates the individual. In colonial aggregates and multicellular organisms, however, cell division results only in the reduplication of the component cells and in growth of the plant body. In these, a variety of reproductive phenomena have been described. One of the most widespread may be designated as fragmentation; it occurs throughout the whole plant kingdom. It involves the separation of the original plant body into segments, often specialized parts, and their distribution and subsequent development into new individuals. It is illustrated, for example, in the breaking up of colonial algae, in hormogonium formation in Cyanophyta, in the separation of branches by posterior decay in liverworts and rhizomatous plants, and by the natural and artificial propagation of segments of leaves, stems, and roots. In a number of cases, special portions of the plant body are set aside as reproductive fragments. This is exemplified by the gemmae of certain fungi, liverworts, and mosses; by the soredia and isidia of lichens and the bulbils of *Lycopodium* and certain ferns; and by the formation of bulbs, rhizomes, and stolons in many vascular plants.

Asexual reproduction is accomplished in the cryptogams also by the formation of special bodies usually known as spores. These may be unicellular or few-celled; in the latter case they are not very different from certain types of gemmae. Such structures as zoospores, aplanospores, autospores, and many types of fungal spores all may be classified as asexual, because they continue their development without prior union with other reproductive cells. According to this concept, the spermatia of rusts and the uniting basidiospores of certain smuts would not be interpreted as asexual spores. Although many spores are asexual in the sense that they germinate without union, they are in many cases more or less closely related to the sexual process. Examples of this are seen in the auxospores of diatoms, the dormant zygotes and the carpospores of algae, the ascospores and basidiospores of fungi, and especially the spores of the land plants.

The origin of sexual reproduction remains obscure in spite of numerous investigations of the simple algae—for example, *Chlamydomonas* and *Chlorococcum*—where sexuality seems to be incipient, or sporadic and not obligate. Sexuality in such organisms as *Chlamydomonas* is thought to be incipient both because the individual vegetative cells themselves may function directly as gametes without special morphological differentiation, and because the gametes may develop into new individuals whether or not they undergo sexual union. There is good evidence that even in such organisms, in the absence of morphological differentiations, the process of sexual union is regulated by a complex series of hormones, and that disturbances in their functioning prevent the culmination of sexuality. Regarding the actual origin of sexuality, however, nothing is known with certainty. Machlis and Rawitscher-Kunkel (1967) and Wiese (1984) reviewed our knowledge of the mechanisms of gametic approach in plants.

Sexual reproduction[1] occurs in the kingdoms Monera, Phyta, and Myceteae. It often involves the union of cells and their nuclei, with the resultant association of DNA in the same nucleus. This occurs in most species of plants and fungi, although there are, of course, exceptions. In the bacteria, the phenomena of conjugation, transduction, and transformation produce merozygotes—that is, "partial zygotes," "partial" in the sense that two *complete* sets of genes are not involved in their formation. Furthermore, sexuality in bacteria, and probably in the Cyanophyta, differs in the absence of meiosis, the latter a universal component of sexuality in the Eukaryonta. Among the Phyta (Plantae) and Myceteae (Fungi), although sexual reproduction is widespread, there are notable exceptions, such as the members of the divisions Euglenophyta and the subdivision Deuteromycotina.

In some organisms, although sexuality is present potentially, the race is maintained through many generations by asexual means, and sexuality may be considered a purely genetic phenomenon, a sporadic opportunity for gene combina-

---

[1]That is, some form of genetic interchange.

tion and segregation. The plant kingdom includes organisms that exhibit a high degree of differentiation in the sexual process. Examples of a supposedly primitive type of sexual reproduction are plants such as *Chlamydomonas eugametos* and *Spirogyra*, in which there is no *morphological* differentiation of vegetative cells from gametes. In other plants, special reproductive cells are differentiated when the individual attains maturity. These may be scattered in the plant body, as in certain filamentous green algae (*Ulothrix*, *Stigeoclonium*), or they may be restricted to certain regions where they are borne in special structures called gametangia (*Codium*, *Ectocarpus*, *Chara*, *Allomyces*, etc.).

The gametes may be morphologically similar to each other, as in isogamy, or differentiated, as in anisogamy (both gametes flagellate but differing in size) and oogamy (large, nonflagellate egg and minute sperm). When gametes produced by one clone unite sexually, the clone is said to be homothallic. If by contrast the gametes of a clone are self-incompatible and require other compatible gametes with which to mate, the condition is called heterothallism. Individual organisms that produce both male and female gametes are said to be monoecious; when the male and female gametes are produced on different individuals, dioecism prevails. The terms monoecious and dioecious, although originally used to describe the location of microspores and megaspores on sporophytes and still so employed, later were applied to the distribution of male and female gametes on gametophytes.

The segregation of plants into Cryptogams and Phanerogams was based originally on the premise that stamens and pistils are sex organs. Allen (1932) called attention succinctly to this problem in the following sentences:

Any genetic analysis of sex in Angiosperms must deal almost exclusively with characters of the so-called asexual generation, since those of the much reduced haploid "sexual" generation have yet afforded little material for genetic study. To speak of sexual characters in an asexual generation is paradoxical; but the paradox inheres in the terminology, not in the facts. The diploid sporophyte helps through various devices to effect union of gametes produced by the filial gametophytes, and to provide for shelter and nutrition of the embryonic grand-filial sporophyte; and such devices are sexual characters under any usable definition of the term.

This last may be true, but in such plants as *Fucus* and angiosperms, in which the gametophytic phases are reduced almost to oblivion, it is almost true that the sporophyte has become sexual.

The advent of the meiotic process must have been coincident with the evolution of sexuality. As pointed out about 90 years ago by the German botanist Strasburger, for every act of sexuality in which two chromosome complements (and genotypes) are brought together, there is a preceding or following meiosis in which the chromosomes and the genes they bear are again segregated. The relation between these two phenomena, in time, differs in various plants (and animals). Three fundamental types of life cycles based on these differences have been described. In the first (M + h), the vegetative phase is haploid and potentially sexual and is designated as the gametophyte. Gametic union produces a zygote, the only diploid cell in the life cycle. This zygote is interpreted by some to represent the precursor of and to be the homologue of the sporophyte of other types of life cycles. Zygotic meiosis restores the haploid gametophytic phase; hence, alternation of generations is absent in a morphological sense, although cytological alternation is present.

At the other extreme lies a second type of life cycle (M, d), characteristic of most members of the animal kingdom but occurring also in certain algae. In this type, the sexual individual is diploid, and meiosis is gametic. The gametes alone in the life cycle are haploid, and here again morphological alternation is absent.

Many algae, certain fungi, and all the land plants are characterized by a third type of life cycle (D, h + d), which involves morphological as well as cytological alternation of generations or phases. In these, the gametophytic and sporophytic phases both exceed a single cell generation (gamete or zygote) in duration and complexity. The diploid zygote gives rise to the sporophyte, in which meiosis occurs at sporogenesis. The products of meiosis (meiospores) develop into haploid gametophytes. The two alternates may be free-living (dibiontic cycle), or one may be physically associated with the others. Furthermore, the alternates may be morphologically equivalent (isomorphic alternation) or divergent (heteromorphic alternation). Both generations may be propagated indefinitely either by naturally occurring reproductive cells or by artificial manipulation. That the chromosome number of itself is not responsible for the divergence of the alternates in ontogeny is clear from the phenomena of apogamy and apospory discussed in Chapters 11, 12, and 18. The causal factors in heteromorphic alternation of generations remain to be elucidated.

In bringing to a conclusion this summary of reproductive processes, brief consideration must again be given to the phenomena that culminate in the production of seeds. The initiation of seed formation has certainly changed the face of the earth, since the seed plants, especially modern flowering plants, constitute its dominant vegetation, and seed-eating animal species (including man) are correspondingly abundant.

A number of coordinated morphological phenomena are associated in the formation of the seed. A seed consists of an embryonic sporophyte embedded within a female gametophyte (in gymnosperms), which is surrounded by the delicate remains of the megasporangium tissue and covered by an integumentary layer. In angiospermous seeds, an additional nutritive tissue, the postfertilization endosperm, which replaces the nutritive female gametophyte in function, may persist. Among these coordinated phenomena may be cited the permanent retention of a single functional spore that produces the female gametophyte within the tissues of its sporangium and surrounding integument and the transportation of the male gametophyte to the proximity of the female gametophyte. The latter is known as **pollination**. In relation to pollination and proximity of the gametophytes, several further correlated phenomena may occur. In gymnosperms, these include the formation of a pollen chamber at the apex of the megasporangium for the reception of pollen grains and the formation of a pollen tube. The latter may be long or short, branched or unbranched. It is chiefly haustorial in the cycads and *Ginkgo*, but it is also a mechanism for sperm transfer in other gymnosperms and the angiosperms. Pollen chambers are absent in the latter, the receptive function residing in the stigma of the megasporophyll.

Although chlorophyll is present in the female gametophyte of *Ginkgo*, the gametophytes of other seed plants are colorless, and their nutrition is based ultimately on the materials synthesized by the parent sporophyte. The nutrition of the male gametophyte of such genera as *Zamia*, *Ginkgo*, and *Pinus*, which involves breakdown of megasporangium tissue through enzymatic activity, probably should be classified as parasitism. In *Ephedra* and many angiosperms, however, the pollen tube is so ephemeral that there is little probability of its absorbing much nutriment, unless it be from the mucilaginous secretions present in the micropyle- or style. In other angiosperms, however, where the pollen tube grows slowly or over a great distance, parasitism may prevail.

The origin and steps in the development of these coordinated processes associated in seed formation remain in the realm of speculation, since the most trustworthy area of evidence, the fossil record, is not very helpful. As far as is known, these phenomena characterize both extinct and extant seeds and their development. A great majority of botanists cite still another morphological attribute, namely, **heterospory**, as the inevitable concomitant of the seed habit.

It will be recalled that this term is applied to such plants as *Selaginella*, *Isoetes*, and *Marsilea*, and to certain fossil genera, in which the spores that produce male gametophytes are markedly smaller than those that develop into female gametophytes. This is the reason for the terms "microspores" and "megaspores." Pettitt (1970) has recently discussed heterospory and the origin of seed production.

Careful measurements of the microspores and megaspores of gymnosperms and angiosperms, however, supply no evidence that so-called megaspores are consistently larger than microspores. In fact, the contrary is true in a number of genera. This is in striking contrast to such heterosporous cryptogams as *Selaginella scandens*, in which the ratio, by volume, of megaspore to microspore is in the order of 30,000 to 1. It is probable that permanent retention of the megaspore within the sporangium has been responsible for the initial reduction of its size. As evidence of its pristine free existence, some botanists emphasize the thickness of the megaspore membrane that surrounds the female gametophyte in such genera as *Ginkgo* and its augmentation by the tapetal membrane. However, the essential factor that distinguishes the heterospory of seed plants from that of cryptogams seems to be the timing of possible nutritional inflow: In cryptogams, food for the anticipated development of the female gametophyte is very early stored in the megaspore, with resulting increase in the size of the latter. In seed plants, by contrast, provision of the nutriment is delayed until the initiation of the gametophyte development. Here the megaspore does, in fact, enlarge, but only later.

THE FOSSIL RECORD

The somewhat detailed discussion of living or extant representatives of the 19 divisions in which the plant kingdom is classified has been supplemented in the preceding pages of this text with brief treatments of certain extinct organisms presumably related to living plants. Following is an attempt to summarize briefly our knowledge regarding the course of evolution of the plant kingdom as it is revealed in the fossil record.

The table on the inside front cover summarizes the various divisions of geologic time together with the intervals during which the various plant groups have been represented in the earth's flora.

A number of significant facts emerge from examination of the table and from the discussion of fossil organisms in earlier chapters. Some of these may be summarized as follows:

1. There was a long prebiotic period in the earth's history, the so-called period of chemical evolution, before the first living organisms[2] appeared.
2. What are possibly the earliest known organisms with cellular organization have been found in Lower Precambrian rocks (Swartkoppie Formation) of South Africa that are almost 3½ billion years old. If these were, indeed, biogenic, it is not known to what group they might have belonged. In structure and possible occurrence of binary division, these structures resemble certain unicellular Cyanophyta (Knoll and Barghoorn, 1977).
3. There is good evidence, reviewed earlier in the text, that both unicellular and filamentous blue-green algae were present in the Precambrian between 1.6 and 2.0 billion years ago and that some of their descendants appeared to have persisted with little modification to the present (Schopf, 1970, 1975).
4. Late Precambrian rocks, such as those of the Bitter Springs strata of Australia, about 900 million years old, contain dozens of

---

[2]Not *life*, it should be noted.

species of blue-green and green algae (Schopf, 1968; Schopf and Blacic, 1971).

5. Land plants seemed to have appeared first in the late Silurian and did not become significant in the earth's flora until the Devonian (408–360 million years ago), more than 3 billion years after the first appearance of living organisms. Most Devonian vascular plants were free-sporing and lacked seeds.

6. The earliest vascular plants appear to have been of two principal types. One was represented by naked, dichotomizing axes and terminal sporangia. The other was also typically dichotomized, with sporangia borne laterally along the axes. All other groups of vascular plants could have been derived from such body types.

7. The vascular cryptogams did not have hepatophytan or bryophytan ancestors, while, in fact, some morphologists consider that the liverworts and mosses represent a reduction series from certain vascular cryptogams.

8. The progymnosperms (Progymnospermophyta) of the Devonian had developed vegetative features (secondary xylem, etc.) characteristic of gymnosperms, but their reproduction was at the level of vascular cryptogams.

9. The discovery of seeds such as *Archaeosperma* and *Spermolithus* in the Upper Devonian indicates that seed plants have had a longer existence on the earth than had been suspected previously.

10. Unequivocal gymnospermous seed plants were present and diverse during Mississippian and Pennsylvanian times.

11. The angiosperms, however, are the most recent (Lower Cretaceous), yet most successful, plants to appear on the earth; their immediate ancestors have yet to be discovered. The possession of the seed habit together with a wide range of adaptive vegetative features may explain the successful colonization of the earth by angiosperms.

CLASSIFICATION

In concluding this volume on the morphology of representative plants and fungi, brief reference must be made again to the subject of classifying the groups of which the types discussed are representatives. The problem of classification was referred to in Chapter 1, when it was stated that fruitful discussion of systems of classification should be deferred until one has become familiar with the plants to be classified. It is assumed that the reader is now in this position, at least to some degree.

In initiating such a discussion, a word of caution from an eminent morphologist (Arnold, 1948) seems particularly appropriate. He wrote:

Once a system of classification becomes widely adopted, it takes on many of the attributes of a creed. Not only does it constitute the framework about which the botanist does his thinking, but it rapidly becomes a substitute for it. It comes to be looked upon as having emanated from some authoritative and inspired source. It is accepted as final, and anyone who would be so bold as to suggest that it be altered to conform with modern knowledge is promptly squelched with a smooth flow of well-rehearsed oratory.

Happily, a less dogmatic atmosphere prevails at present.

The old saying "A little knowledge is a dangerous thing" is an appropriate and chastening thought for anyone who attempts to erect a phylogenetic system of classification of plants and animals. A phylogenetic classification, it will be recalled, is one in which the organisms are arranged in categories or taxa in an order that describes their supposed relationship based on evolutionary development. As expressed by a distinguished taxonomist: "A phylogenetic system classifies organisms according to their evolutionary sequence, it reflects genetic relationships, and it enables one to determine at a glance the ancestors or derivatives (when present) of any taxon" (Lawrence, 1951). Attempts at phylogenetic classification multiplied after Darwin's *Origin of Species* was published in 1859. In spite of the paucity of evidence from paleobotany and comparative morphology of living plants, faith that plants could be arranged in phylogenetic systems apparently was unbounded, and in some cases enthusiasm outstripped critical judgment. One might well assume, therefore, that as rich as they have been in scientific inquiry, the intervening decades, more than a century, would have provided firm and convincing data on which to erect the one definitive, phylogenetic scheme. Actually, many such schemes have been presented, and although incontrovertible evidence of evolution and kinship has been educed at the species level and below it, increasing knowledge has counseled caution regarding the degree of assurance with which we should postulate relationship and the *course* of evolution among the higher categories. This caution is reflected in a marked increase in the number and degree of polyphyletic systems of the plant kingdom that have been suggested in the recent literature.

The table on the inside back cover presents in comparative fashion a summary of the widely used system of Eichler,[3] somewhat modified by others, the system of Tippo (1942), and the system used in the first and with slight modification in the second, third, fourth, and present editions of this text. The legend of the table explains the mechanics of comparison. It is clear at once that the number of divisions of plants has increased from Eichler's 4, through Tippo's 12, to the 19 in the present volume. The table shows the progressive subdivision of earlier categories. Classification of the old Eichlerian class Algae into seven or eight separate taxa, raised to divisional rank, was the first manifesta-

tion of a more polyphyletic viewpoint and first appeared in an American textbook in 1933. The partition of the Eichlerian Pteridophyta, which included the ferns and their "allies" (inside back cover), came later. Abandonment of the class name Gymnospermae was proposed in 1948 (Arnold), and is accepted in the present volume. The classification of plants here presented differs mainly in that plants other than algae also have been classified polyphyletically. These innovations will now be considered.

The fungi have been transferred from the Plant Kingdom to a Kingdom of their own. The liverworts, hornworts, and mosses were treated here as separate divisions, the Hepatophyta, anthocerotophyta, and Bryophyta, as explained at the end of Chapter 12. Doubt that liverworts and mosses are closely related was expressed many years ago by Goebel. In the opinion of the writer, no convincing evidence to contradict Goebel's view has yet been educed. Furthermore, it seems clear that the hornworts differ significantly enough from liverworts to be classified in a division of their own, as formalized recently by Stotler and Crandall-Stotler (1977).

The most striking departures of the authors' system are apparent in their disposition of the vascular plants, often widely conceived as constituting a single division, Tracheophyta (inside back cover), although several recent texts have abandoned this concept. As noted in Chapter 13, uniting such a large and diverse assemblage of morphological types in a single division on the basis of a single common attribute—namely, the presence of xylem and phloem—is open to question. "The result of the widespread practice of classifying plants on single sets of characters has been to encourage overemphasis on certain morphological phenomena, to the neglect or exclusion of others of equal significance, and to try to construct phylogenetic lines on them alone" (Arnold, 1948).

The division Tracheophyta is subject to this very criticism in the writers' opinion. Accordingly, the vascular plants have been classified

---

[3]G. M. Smith, *Cryptogamic Botany* (McGraw-Hill, New York, 1955, Vol. I) presented evidence that this system was used, at least in part, before Eichler.

here in a number of separate divisions. This has occasioned raising the subdivisions Psilotopsida, Lycopsida, and Sphenopsida to divisional rank and has required changing the suffixes of the group names. The quite meaningless name Lycopsida ("having the appearance of a wolf") has been replaced by the designation Microphyllophyta, an allusion to an important attribute of the group. Arthrophyta, an older name, here is used as the divisional name to replace Sphenopsida, because the latter is based on a fossil genus not well known to many students. Boureau (1964–1975) and his collaborators, in their *Traité de Paléobotanique*, have classified the vascular plants approximately this way.

The classification of megaphyllous vascular plants in this text (inside back cover) also deviates from the practice of some other botanists, who include both the ferns and seed plants in a "subphylum" Pteropsida. This practice groups in the same category cryptogams, which have free spores and free-living gametophytes, with the seed plants, which have permanently retained megaspores and parasitic male and female gametophytes; this is unacceptable to the authors. Accordingly, in this book the "subphylum" Pterophyta has been abandoned and has been replaced by six divisions of living plants—the Pteridophyta (ferns), Cycadophyta, Ginkgophyta, Coniferophyta, Gnetophyta, and Anthophyta. The rationale for this arrangement has been presented in earlier chapters and need not be repeated.

Although the number of divisions of living plants has been increased by these changes to 19, the authors are strongly of the opinion that in light of the available evidence a tentatively "polyphyletic" view really is the properly conservative one. This opinion in this connection and the system of classification used in this volume should not be construed as representing the final solution to the question of plant relationship. The authors themselves at present look upon the suggested classification as, at best, a *tentative approximation*.

Does the classification presented in the present volume then imply or deny that there is evidence available for the relationship through common lineage of living plants to one another and to extinct plants? Quite the contrary, as stated in the introductory chapter to this volume. Such characteristics as the cellular organization of protoplasm (at both the light-microscopic and ultrastructural levels), the basic similarities of the motile cells (Paolillo, 1981) methods of nuclear and cell replication, biochemical patterns in metabolism, and the mechanisms of genetic interchange, recombination, and segregation, among others, are so universal that they seemingly cannot be explained satisfactorily on any basis other than that of a common origin of all plants and animals.

It is, however, when we come to consider the actual *course* or *lineage* in the subsequent diversification of organisms, both extinct and extant, as this can be expressed in a system of classification, that we meet with disappointment and frustration if we rigorously distinguish between *evidence* and *speculation*. This becomes increasingly true as we consider taxa or groups successively higher than species. The writers, after carefully weighing the currently available evidence of comparative morphology, cytology, biochemistry, and fossil record, are *at present* unwilling to amalgamate any two or more of the 19 divisions in which they have tentatively classified the organisms of the plant kingdom. When and if additional relevant data become available, such amalgamations will undoubtedly be required, but at this time there are no known living or fossil forms that unequivocally link any two of the proposed divisions.

Divisions 1–7 of the Kingdom Phyta, while at the algal level of morphological organization, are cytologically (with respect to wall composition and flagellar insertion and organization) and biochemically (with respect to accessory pigmentation and stored metabolites) diversified. By contrast, a second series, exemplified by some members of division 1 and all members of divisions 8–19, while retaining essentially the same cytological organization (including possession of stacked or grouped chloroplast lamellae, either granoids or grana), also have conservatively retained the same chlorophyll pigments (*a* and *b*), storage of starch, and often cellulosic wall composition. The evolutionary diversification here was in the direction of increasingly complex gross morphology, as is apparent in such subseries as the liverworts, mosses, vascular cryptogams, gymnosperms, and angiosperms. The Chlorophyta, of all the algae, are here considered to have been the most likely progenitors of the remaining divisions of green plants. Certainly all available evidence supports that conclusion.

As mentioned earlier (p. 98), ultrastructural and biochemical evidence has suggested to some investigators (Stewart and Mattox, 1975, 1978; Syrrett and Al-Houty, 1984; and Mattox and Stewart, 1984) that one special group of Chlorophyta, the Charophyceae, were the progenitors of the so-called "higher plants."

Certain criticisms of the classification, as presented in the several editions of this book, have been expressed by several colleagues. The strongest of these held that it is a negative, retrogressive system that overlooks evidences of relationship and separates subdivisions of plants grouped together in other classifications.[4] This criticism has been directed especially against the treatment of the vascular plants and the author's abandonment of the "phylum" Tracheophyta. Similar criticisms were voiced when the Eichlerian division Thallophyta, and also its component classes, the Algae and Fungi, were dismembered a number of years ago. We still, however, meaningfully speak of "thallophytes," "algae," and "fungi," and one can continue to refer to "tracheophytes" or "vascular plants," "seed plants," "gymnosperms," and "angiosperms" in informal designation of groups of taxa, even if one is reluctant to use these names as proper nouns in the system of classification, thereby formalizing relationship.

In final defense of the *tentative* classification used in this book, the writers can only reiterate that the appraisal and interpretation of the evidence of evolutionary development and phylogeny vary with the appraiser and interpreter. As Leedale (1974) pointed out, in spite of all our best efforts "the system is *still* artificial, of course. Living organisms do not fit neatly into man-made categories; diversity is always continuous in some aspects." The translation of the interpreter's conclusions into a system of classification will, therefore, vary accordingly. Variations in and prolonged consideration of diverse systems of classification should serve as the inspiration for discussion and should not detract from the reader's understanding of the plants themselves, as they have been presented in earlier chapters.

After all, nature mocks at human categories, and our present set of classifications represents only our current, subjective appraisals of the significance of available data and their bearing on phylogeny. "No single system (of classification) can be accepted as final as long as a single fact concerning any kind of plant remains unknown" (Arnold, 1948).

---

[4]"You have not *reclassified* the plant kingdom but *declassified* it," wrote the outraged colleague in 1956; in the interim, he, himself, has modified his views.

GLOSSARY

Abaxial   Away from the axis.

Abscission layer   A zone of cells at the base of an appendage (petiole, fruit stalk, etc.) that results in the separation of that appendage from another organ.

Acervulus   A mass of hyphae formed subepidermally or subcuticularly on a plant host, which produces a bed of short conidiophores, each with a conidium at its apex.

Achene   A dry, indehiscent, one-seeded fruit in which the seed is free from the pericarp except at the point of attachment of the seed.

Acrasin   An attractant—sometimes cyclic AMP—secreted by myxamoebae of Acrasiomycetes; effective in evoking aggregation.

Acrocarpous   Usually erect, sparsely branched mosses with sporophyte at the apex of a main axis.

Acrogynous   Having the apical cell of an axis "used up" or differentiated in archegonium formation, as in *Porella* and others.

Actinomorphy   Radial symmetry.

Actinostele   A protostele with ridged xylem.

Adaxial   Toward the axis.

Adnation   Fusion of one organ to another, as in epipetalous stamens.

Advanced   Derived or descended with modification from the primitive.

Adventitious root   One neither primary nor secondary nor arising therefrom—for example, roots on cuttings.

Aecial primordium   The precursors of an aecium composed of uninucleate cells.

Aeciospore   The spore produced in the aecial stage of a rust.

Aecium   A cuplike structure in which aeciospores are produced.

Aethalium   A rather massive sporophore of some Myxomycetes believed to have evolved by the complete fusion of several sporangia.

Aflatoxin   Toxin produced by *Aspergillus flavus* and related species of fungi in stored grains. May be both hemorrhagic and carcinogenic to animals, including man.

After-ripening   Period of dormancy in some seeds before germination.

Agar   A gel-forming polysaccharide, a polymer of galactose, derived from certain red algae and used as a solidifying agent in culture media.

Aggregate fruit   A receptacle bearing a number of matured pistils (fruits) of a single flower.

Akaryota   See PROKARYONTA.

Akinete   A vegetative cell transformed by wall thickening into a nonmotile spore; corresponds to chlamydospore of fungi.

Alar cells   Cells at the basal angles of moss leaves.

Alga   A photosynthetic eukaryotic organism lacking sex organs; or with unicellular ones; or, if the gametangium is multicellular, having every cell fertile.

Allophycocyanin   A blue biliprotein pigment of the blue-green algae.

Alternation of generations or phases   A reproductive cycle in which a haploid organism (or tissue) gives rise sexually to a diploid organism (or tissue); the meiotic products of the latter grow into a haploid organism (or tissue).

Ament   A spike of apetalous flowers, either staminate or pistillate, and often deciduous, as in willow.

Amphigastrium   The ventral leaf (often reduced in size) in leafy liverworts.

Amphiphloic   Having phloem on both sides of the xylem.

Amphithecium   An external or superficial cell

layer or layers differentiated in embryonic sporophytes from an internal endothecium.

Amphitropous ovule One in which the ovular body itself is bent in the form of a "U."

Anacrogynous Not having the apical cell of an axis involved in archegonium formation.

Anatomy In botany, area that deals with organization of organs of vascular plants.

Anatropous ovule One in which the ovule is inverted with respect to its funiculus; the micropyle, accordingly, at the same level as the base of the funiculus.

Androspore A motile zoospore of Oedogoniales that generates a dwarf-male plant.

Angiosperm A seed plant with ovules and seeds enclosed in a carpel.

Anisogamy Male and female gametes differing in size, both flagellate.

Anisophyllous Having more than one form of leaf.

Anisospory The formation of different-sized spores in mosses.

Annulus (1) A specialized or differentiated layer of cells in the moss capsule; or (2) in the sporangial wall of vascular cryptogams involved in sporangial dehiscence; or (3) ring around the stipe below the pileus of a mushroom.

Antheridial cell The larger product of internal division (along with the smaller prothallial cell) of the microspore.

Antheridiogen A substance secreted by certain precocious fern gametophytes that stimulates the production of antheridia in younger gametophytes.

Antheridiol A substance secreted by female filaments that evokes the formation of antheridia.

Antheridium A multicellular sperm-producing organ consisting of spermatogenous tissue and a sterile jacket in plants other than algae and fungi; in the latter, a male gametangium, often unicellular.

Antherozoid A sperm.

Anthesis Opening of the flower bud.

Anticlinal Perpendicular to the surface.

Antipodal cell One or more cells of the mature female gametophyte of angiosperms, located at the opposite end from the micropyle.

Aphanoplasmodium A transparent, nongranular plasmodium, usually devoid of a slime sheath, produced by Myxomycetes in the order Stemonitales; difficult to observe in nature until it is ready to sporulate.

Apical cell A prominent apical meristem cell, derivatives of which organize the plant body or an organ.

Apical growth Growth localized in an apical meristem.

Aplanospore A nonmotile spore in algae (with the potentiality of motility) and in fungi.

Apocarpous Having simple pistils or carpels, these not united.

Apogamy Formation of a sporophyte without gametic union.

Apogeotropic Negatively geotropic.

Apomixis A collective term for absence of sexual union and failure of meiosis.

Apophysis Basal region of moss capsule, enlarged in some.

Apospory Formation of the gametophyte in a manner other than from spores and not involving meiosis.

Apothecium An open ascocarp, often cup- or saucerlike.

Appressorium A flattened hypha from which a minute process may enter a host cell.

Archegonial chamber A cavity or groove at the apex of the female gametophyte in certain gymnosperms; the archegonial neck cells are on its surface.

Archegoniophore A branch bearing archegonia.

Archegonium A multicellular, egg-producing gametangium with a jacket of sterile cells.

Aril A cuplike structure partially surrounding an ovule.

Ascocarp A fruiting body in which asci are formed.

Ascogenous hyphae Hyphae that give rise to asci.

826

Ascogonium   The female sex organ of the Ascomycetes.

Ascospore   A spore produced by free cell formation following meiosis in an ascus.

Ascostroma   A stroma containing asci.

Ascus   A saclike cell of the Ascomycetes in which karyogamy is followed immediately by meiosis and in which ascospores of definite number arise by free cell formation.

Ascus mother cell   The penultimate, binucleate, hook cell of the crozier in the Ascomycetes.

Asexual reproduction   Reproduction not involving plasmogamy and karyogamy.

Aspergillosis   A disease of the lungs caused by species of *Aspergillus.*

Atactostele   A scattered arrangement of xylem and phloem groups (vascular bundles).

Autocolony   A miniature daughter colony formed by coenobic algae.

Autoecious   Completing the life cycle on a single host species, as in certain rusts (Uredinales).

Autospore   A nonmotile, asexual daughter cell that lacks the ontogenetic potentiality of motility.

Autotrophic   Capable of synthesizing protoplasm from entirely inorganic substances.

Auxiliary cell   That cell to which the zygote nucleus migrates in certain red algae.

Auxospore   The zygote of diatoms.

Axenic culture   One that contains a population of only one species.

Axil   The adaxial angle between a leaf and a stem.

Axile placentation   Attachment of ovules to the central axis of a compound ovary.

Axillary bud   One borne in the axil of the leaf.

Axis   The stem and root (if present) of a plant.

Azygospore   A spore that is morphologically similar to zygospores but has no proven sexual origin.

Bacillus   A rod-shaped bacterium.

Basidiocarp   The sporophore of the Basidiomycetes.

Basidiospores   A spore borne outside of a basidium and arising after meiosis.

Basidium   A nonseptate or septate hypha bearing (usually) four basidiospores exogenously following karyogamy and meiosis.

Berry   A pulpy fruit with immersed seeds.

Biochemistry   The chemistry of living matter.

Blastospore   A spore formed by budding.

Body cell   That cell of the male gametophyte of gymnosperms that divides to form the sperm cells.

Bracteole   The inner of two appendages surrounding the microsporophylls in *Ephedra;* small bracts in general.

Bryology   The study of liverworts, hornworts, and mosses.

Bud   (1) A minute stem with short internodes bearing the primordia of vegetative leaves or sporophylls; (2) a cellular protuberance from a single cell or hypha.

Bud scale   A modified basal leaf that encloses the more delicate leaf or sporophyll primordia.

Budding   Abstriction of a cellular protuberance resulting in multiplication, as in certain yeasts.

Bulb   A short, vertical subterranean stem covered by fleshy bases or scales, as in lily and onion.

Bulbil   A gemma or miniature plantlet produced in certain species of *Lycopodium* and other plants.

Bundle sheath   Differentiated tissue, often thick-walled, surrounding a leaf vein.

Callus   Undifferentiated tissue that heals a wound or that may be grown in tissue culture.

Calyptra   The enlarged and modified archegonium that for a while encloses the embryonic sporophyte of liverworts, mosses, and vascular cryptogams.

Calyx   The collective term for the sepals of a flower.

Cambium vascular   A zone of meristematic cells, between primary xylem and phloem, the division products of which differentiate as secondary xylem and phloem.

Campylotropous ovule   An ovule in which the funiculus is perpendicular to the long axis of the ovular body.

Capillitium   Threadlike structures among the spores of certain Myxomycetes and puffballs.

Capitulum   (1) The aggregations of branches of limited growth at the apices of the main axes in *Sphagnum;* or (2) the cell or cells from which the antheridial filaments of charophytes grow; or (3) the headlike inflorescence of composite plants, as in daisy.

Capsule   (1) The sporangium of liverworts and mosses; or the colloidal sheath in algae and bacteria; (2) a dry, dehiscent fruit arising from a compound ovary.

Carinal canal   A cavity within the arthrophytan stem lying on the same radius as a ridge, as in *Equisetum.*

Carotene   A red and orange pigment (oxygen-free hydrocarbon).

Carotenoid   Collective term that includes both carotenes and xanthophylls.

Carpel   Structure surrounding angiospermous ovules often considered to represent a folded megasporophyll.

Carpogonium   The female gametangium of red algae.

Carposporangium   A carpospore-containing cell in red algae.

Carpospore   The single protoplast contained within a carposporangium.

Carposporophyte   A group of carposporangia and carpospores in red algae that arise directly or indirectly from the zygote.

Carrageenan   Polysaccharide extracted by hot water from certain red algae, composed of D-galactose units and sulfate groups.

Caryopsis   A dry, indehiscent, one-seeded fruit in which the seed coat is completely fused to the pericarp, as in the grass family.

Cast   A type of fossil that represents a three-dimensional replica of the original plant part but which has no original plant material remaining.

Cauline   Pertaining to the stem.

Caulonema   That portion of the protonema of *Funaria* (and other mosses?) with brown cell walls, oblique terminal walls, and spindle-shaped chloroplasts.

Cell plate   The first partition between two dividing cells.

Central canal   The central cavity in the internodes of arthrophytes.

Central cell   The precursor of the egg and ventral canal cells that arise by its division.

Central nodule   A complex, thickened region in the center of the valves of motile, pennate diatoms.

Centriole   A cell organelle functioning in cell division of flagellate plants and fungi, which also serves as a basal body of a flagellum.

Chalaza   The base of an ovule, often conspicuous as the region below the point of union of integuments with the megasporangium.

Chasmogamous flower   A flower that opens for pollination and reproduction.

Chemosynthetic   Autotrophic and using chemical energy for synthesis.

Chemotaxis   Movement of cells to or away from a chemical stimulus.

Chert   Fine-grained rock formed by precipitation of silicon dioxide.

Chlamydospore   A hyphal cell that becomes thick-walled and segregated from the parent mycelium and functions as a spore (probably equivalent to an algal akinete).

Chlorenchyma   Parenchyma containing chloroplasts.

Chloronema   That portion of the protonema of *Funaria* (and other mosses?) with hyaline cell walls, perpendicular terminal walls, and lens-shaped chloroplasts.

Chlorophyll   The green pigment complex in plants; several types are known; tetrapyrrolic compounds containing a magnesium atom.

Chlorophyllose   Containing chlorophyll.

Chloroplast   A membrane-bounded area of cytoplasm containing photosynthetic lamellae.

Chromoplast   A plastid low in chlorophyll content and high in carotenoid content.

Chrysolaminaran   A polysaccharide composed of glucose units in β-1 : 3-linkage. (Synonym: leucosin.)

Cilia   (1) Short flagella; or (2) filiform structures associated with the peristome of certain mosses.

Circinate vernation   Curled arrangement of leaves and leaflets in the bud occasioned by their more rapid growth on one surface.

Cirrhus   A ribbon- or tendril-like exudate of spores.

Clamp connection   A lateral connection between adjacent cells of the mycelium of certain Basidiomycetes.

Cleavage polyembryony   The formation of as many as four proembryos from a single zygote by separation of the cleavage products.

Cleistocarpous   Fruits or spore-bearing bodies (such as moss capsules or puffballs) not opening.

Cleistogamous flower   One that does not open for pollination but is pollinated and fertilized by germination of pollen within the closed system.

Cleistothecium   An ascocarp that does not open.

Clonal culture   A population descended asexually from a single individual.

Coccus   A spherical bacterium.

Coenobium   A colony with the cell number fixed at origin and not subsequently augmented.

Coenocyst   A multinucleate cyst or dormant spore.

Coenocyte   A multinucleate mass of protoplasm lacking internal septa.

Coenocytic   See COENOCYTE.

Coleorhiza   A sheath of tissue surrounding the radicle in some seed plants.

Collenchyma   Living cells whose walls are thickened at multiple cell junctions by ridges of cellulose.

Columella   The central sterile tissue or structure surrounded or covered by sporogenous tissue or spores (in Anthocerotes, *Sphagnum*, certain Zygomycotina—e.g., *Rhizopus*—and in many Myxomycetes).

Column   (1) The anterior protuberance of the female gametophyte in *Ginkgo*; (2) the stalk that bears the microsporangia in *Ephedra*.

Commensal   Pertaining to two organisms living together without either harming the other but with at least one deriving some benefit from the association.

Companion cell   A small nucleated cell associated with some sieve-tube elements of phloem.

Complete flower   One having sepals, petals, stamens, and pistil(s).

Complicate-bilobed   Leaves having dorsal and ventral lobes adpressed to each other.

Compound leaf   One with a divided blade.

Compound pistil   One composed of two to several united simple pistils or carpels.

Compression   A type of fossil produced by extreme flattening of the plant organ(s) by pressure of overlying sediment.

Conceptacle   A chamber in a receptacle of the rockweeds in which gametangia are borne.

Conidiophore   A hypha that produces conidia.

Conidium   An asexually produced spore in Ascomycetes, Deuteromycetes, and some Basidiomycetes.

Conjugate nuclear division   Synchronous mitosis of dikaryotic cells.

Conjugation   Union of gametes and/or gametangia in certain algae and fungi.

Connective   The tissue between the pairs of microsporangia in an anther.

Contractile vacuole   One that excretes liquid after previous dilation by the same.

Cork cambium   The phellogen, a layer of meristematic cells that function in secondary growth, forming cork cells (phellem) to the periphery and cork parenchyma (phelloderm) internally.

Corm   A short, vertical, fleshy, subterranean stem.

Corolla   Collective term for petals.

Corona   The crown cells on the oogonia of Charophyta.

Corpus   The portion of the apical meristem internal to the tunica.

Cortina   A weblike membrane hanging down from the edge of the pileus of a mushroom.

Costa   A riblike region on a diatom valve.

Cotyledon   A primary embryonic leaf.

Cover cell   One or more sterile cells at the apex of antheridia in cryptogams.

Crozier   (1) The configuration of the ultimate, penultimate, and antepenultimate cells during ascus formation; (2) the "fiddlehead" of ferns.

Cruciform division   A type of mitotic division in which the nucleolus elongates into a dumbbell shape and is surrounded by a ring of chromosomes at metaphase, the whole appearing like a cross in side view.

Cryoflora   Plants of ice and snow.

Cryptoblast   A sterile invagination of the surface in rockweeds.

Cryptogam   A seedless plant.

Culture   A laboratory population of organisms and/or tissues.

Cuticle   A waxy layer on the surface of plants.

Cyanophycean starch   Glycogenlike storage product of Cyanochlorontan cells.

Cyclosis   Protoplasmic streaming.

Cyme   An inflorescence in which the apex differentiates, with the result that lateral branches differentiate later to produce younger flowers.

Cyst   A dormant, walled cell in the life cycle.

Cystidium   An enlarged sterile cell in the hymenium of Basidiomycetes.

Cystocarp   The carposporangia and associated sterile covering cells in red algae.

Cytological alternation of generations or phases   Term applied to that type of life cycle (H + h or H + d) in which one genera-

tion is represented only by reproductive cells of different chromosome number.

Cytology   The study of cellular organization.

Decussate   Opposite but with successive pairs emerging 90° from the pair above and below.

Dehiscence   Opening of a structure during maturation.

Derived   Descended or modified from an older, more primitive precursor.

Dermatophyte   A fungus that causes a disease of the skin (ringworm).

Desmid   One of a group of green algae in which the cells are composed of mirror-image halves; their reproduction is by union of amoeboid gametes.

Diarch   Having two sites of protoxylem differentiation.

Diatomaceous earth   Remains of the frustules (cell walls) of fossil diatoms.

Dibiontic   Having two free-living alternates (organisms) in the life cycle.

Dichotomous   Branching into two equal parts.

Dictyostele   A solenostele dissected by leaf gaps.

Dikaryotic   Binucleate; especially when the two nuclei originate in male and female gametangia or come from different mating types.

Dimorphic   Having structural duality, as in, for example, Oomycetes, such as *Saprolegnia*, producing two kinds of zoospores (primary and secondary) in the same life cycle.

Dioecious   (1) Having microspores and megaspores produced on different individual sporophytes; or (2) having male and female gametes produced on different individuals; or (3) having staminate and pistillate flowers produced on different individuals.

Diplanetism   The phenomenon of zoospores having two motile periods.

Diploid   Having two complements of haploid chromosomes.

Disc floret   One of the central, usually radially symmetrical flowers of the capitulum of composites.

DNA Deoxyribonucleic acid.

Dolipore septum A complex, consisting of a central, perforated swelling over which there is a pore cap, or parenthesome.

Double fertilization In angiosperms, union of one sperm with the egg and the other with two polar nuclei or a secondary nucleus.

Drupe A simple, fleshy, indehiscent, single-seeded fruit with stony endocarp, fleshy mesocarp, and skinlike exocarp.

Ecology The study of interrelationships between organisms and the environment.

Ectocarpene A volatile sex attractant produced by the female gametes of *Ectocarpus*.

Ectophloic Having phloem external to the xylem.

Egg A large, nonflagellate female gamete.

Egg apparatus The egg cell and synergids in angiosperms.

Elater (1) A sterile, hygroscopic cell in the capsule of certain liverworts; (2) the appendages formed from the outer spore wall in *Equisetum*.

Elaterophore Region of attachment of elaters in certain liverworts.

Embryo sac A name sometimes used for the female gametophyte of angiosperms.

Embryonal tube Secondary suspensorlike cells in the proembryos of certain gymnosperms.

Endarch Differentiation of primary xylem centrifugally.

Endocarp The innermost layer of the ovary wall at the fruit stage.

Endodermis The differentiated tissue between the pericycle and the cortex in roots, rhizomes, and certain cryptogamous stems.

Endophyte An organism living within another organism.

Endoplasmic reticulum Lamellar or tubular system in the colorless cytoplasm.

Endosperm Nutritive tissue for the embryo in angiosperms.

Endospore (1) An internally formed, thick-walled spore of bacteria and certain blue-green algae; or (2) the innermost layer of the spore wall in most land plants. See also INTINE.

Endothecium (1) The internal tissue of an embryonic sporophyte surrounded by amphithecium; (2) the inner portion of an anther wall.

Epibasidium That part of a phragmobasidium or teliobasidium in which meiosis occurs.

Epicotyl The young stem above the cotyledons.

Epidermis The surface tissue of plant organs, composed of living cells.

Epigean (1) Having the cotyledons emerging from the seed and raised above the soil during germination; (2) fungi producing their sporophores above the ground.

Epigyny The overgrowth of the ovary by the receptacle and/or bases of other floral organs.

Epipetalous Borne on the petals.

Epiphragm A membranous structure to which the peristome is attached in Polytrichalean mosses.

Epiplasm The residual cytoplasm resulting from free cell formation in an ascus in which ascospores are embedded.

Epitheca The larger of the two valves of the diatom frustule.

Ergot The sclerotium of the ascomycete genus *Claviceps*.

Ergotism A disease of cattle and humans caused by the consumption of ergotized grain or bread made from ergotized flour.

Erotactin A hormone that attracts gametes of the opposite sex.

Essential organs Stamens and pistils of a flower.

Eucarpic Only part of the thallus converted into a reproductive organ, as in most fungi.

Eukaryotic Organisms having membrane-bounded nuclei, plastids, Golgi apparatus, and mitochondria.

Eusporangiate Having the sporangial wall develop from superficial cells and the sporogenous tissue from internal cells of the sporophyll or sporangiophore.

Eustele Cylindrical arrangement of primary xylem and phloem in discrete strands separated by parenchymatous tissue.

Evolution Descent with modification.

Exarch Differentiation of primary xylem centripetally.

Exine Outer layer of the wall of a spore or pollen grain.

Exocarp The outermost layer of the ovary wall in the fruit.

Exospore (1) Apically abstricted spores in certain blue-green algae; or (2) the outer layer of the spore wall in most land plants. See also EXINE.

Eye spot Stigma; the hypothetical site of light perception.

Fairy ring A circle or arc of mushrooms on a usually grassy place, with the grass greener inside the circle.

False branching Branching originating by rupture of an unbranched trichome from its sheath.

False indusium The inrolled leaf margin when it covers sporangia in certain ferns.

Fertile spike The spore-bearing axis in the Ophioglossales.

Fertilization Term often used for oogamy in the sense that union of a sperm with an egg makes the latter "fertile."

Fertilization chamber Archegonial chamber.

Fertilization tube A protuberance from the antheridium that penetrates each egg in the Oomycetes.

Fiber An elongate, pointed, thick-walled, usually nonliving, supporting sclerenchyma cell.

Filament (1) A chain of cells; or (2) the stalk of the angiosperm stamen.

Filiform apparatus Fingerlike projections of the synergid wall, composed of pectic substance and hemicellulose.

Flagellum A threadlike extension of the protoplasm, the beating of which propels the cell.

Floridean starch Extraplastid polysaccharide of red algae similar to the branched amylopectin fraction of other plant starches.

Flower A determinate axis with spore-bearing appendages (and usually sterile appendages) and short internodes occurring in the angiosperms.

Follicle A simple, dry, dehiscent fruit originating from a simple pistil that dehisces along one line.

Foot The absorbing organ of the embryonic sporophyte in liverworts, hornworts, mosses, and vascular cryptogams.

Foot cell A hyphal cell that gives rise to a conidiophore of *Aspergillus*.

Free cell formation Cytokinesis within a cell to form one or more cells, leaving residual cytoplasm (epiplasm or periplasm).

Free central placentation Attachment of ovules to a central axis, the latter free from the margins of the carpels.

Free-nuclear division Mitoses without cytokineses.

Frond The fern leaf; a large compound leaf, also in palms.

Fruit The matured ovary or ovaries of one or more flowers and sometimes associated structures.

Frustule The siliceous cell wall of diatoms.

Fuco-serratene A volatile sperm attractant produced by the eggs of *Fucus serratus*.

Fucoxanthin A xanthophyll produced in brown algae and Chrysophyta.

Functional megaspore The surviving megaspore of the tetrad in seed plants that produces the female gametophyte.

Fungi In a broad sense, all achlorophyllous nonvascular cryptogams, except bacteria.

Funiculus The stalk of the ovule that attaches it to the placenta.

Gametangium A structure containing gametes.

Gamete A sex cell that unites with another to form a zygote.

Gametic meiosis Meiosis occurring during the production of gametes.

Gametophore A branch bearing gametes.

Gametophyte A plant that produces gametes.

Gametothallus The gamete-producing thallus of certain dibiontic fungi—for instance, *Allomyces*.

Gamone A chemical substance involved in effecting sexual union.

Gap A parenchymatous interruption of a siphonostele occasioned by the departure of a vascular trace.

Gas vacuoles Minute gas-filled cylindrical cavities in many planktonic Cyanochloronta.

Gemma A bud or fragment of an organism that functions in asexual reproduction.

Generalized growth Increase in size, not localized.

Generative cell (1) In many gymnosperms, the cell that gives rise to stalk and body cells; (2) in angiosperms, the cell that divides to form two sperms.

Genetics The science of heredity and variation.

Germ tube A hypha that protrudes from a germinating fungal spore.

Germinal pore A thin place in the wall of a pollen grain through which the tube emerges.

Gills The radiating lamellae in certain basidiocarps.

Girdle band In diatoms, the structure to which hypotheca and epitheca are attached.

Globule Name sometimes assigned to the antheridium of Charophyta.

Golgi apparatus Cellular organelles consisting of stacks of sacs or cisternae, probably secretory in function.

Gonidium An asexual reproductive cell.

Gonimoblast A filament that arises after fertilization in red algae, giving rise to carposporangia.

Gradate An arrangement of sporangia on the fern receptable, the oldest at the apex and younger ones lateral.

Grex A more common designation for PSEUDOPLASMODIUM.

Ground meristem The primary meristematic tissue other than protoderm and procambium.

Guard cells Specialized cells flanking stomata.

Gymnosperm A seed plant with seeds not enclosed by a pistil or carpel.

Gynandrosporous In *Oedogonium* and related algae, having the androspore produced by the female filaments.

Haploid Having a single chromosome complement.

Haplostele A cylindrical protostele with a smooth margin in transection.

Haustorium An absorptive hyphal branch that penetrates a host cell.

Heterocyst A transparent, thick-walled Cyanochlorontan cell.

Heteroecious A parasitic fungus that requires two species of host to complete its life cycle, as in many rust fungi (Uredinales).

Heteroecism The requirement of two species of host for the completion of life cycle of a parasitic fungus—for example, *Coelomomyces psorophorae* and many of the Uredinales (rusts).

Heteromorphic alternation Having alternates that differ morphologically.

Heterospory Production of microspores that grow into male gametophytes and megaspores that develop into female gametophytes; the two kinds of spores may or may not differ in size (see p. 531).

Heterothallism Self-incompatibility, thus requiring two compatible thalli for sexual reproduction.

Heterotrichous Consisting of prostrate and erect filaments in algae.

Heterotrophic Nutrition based on organic compounds, in contrast to autotrophic nutrition.

Histology The study of tissues.

Holdfast   An attaching organ or cell in certain algae.

Holobasidium   One-celled (nonseptate) basidium of varying shape.

Holocarpic   Converting the entire thallus into a reproductive organ, as in the Gymnomycota and certain Chytridiomycetes.

Holozoic   Phagotrophic—that is, ingesting solid food particles.

Homospory   Condition of producing monomorphic spores.

Homothallism   Sexual reproduction occurring in a self-compatible individual or strain.

Hormogonium   A segment (usually motile) of a trichome of a blue-green alga that can grow into a new trichome.

Hülle cells   Ellipsoidal or globose cells with very heavy walls that almost obliterate the cell lumen; occurring in the genus *Emericella* of the Plectomycetidae (Ascomycetes).

Hydroid   A water-conducting cell of mosses.

Hydrophyte   A plant living in water or in a very moist environment.

Hymenium   A fertile layer of asci or basidia.

Hypanthium   A floral or receptacular tubelike structure surrounding the carpel or carpels of a flower.

Hypha   One branch of mycelium in a fungus.

Hypocotyl   The portion of the axis between the cotyledonary node and primary root.

Hypodermis   A layer or layers of thick-walled cells below the epidermis.

Hypogean   (1) Germination in which the cotyledons are not elevated but remain in the seed within the soil; (2) fungi that produce their sporophores below the ground.

Hypogyny   Floral organization in which the pistil is at the apex of the receptacle, the other floral organs originating below it.

Hypophysis   The enlarged neck or sterile basal portion of the capsule in mosses.

Hypothallus   A thin deposit under the fruiting body in certain Myxomycetes.

Hypotheca   The smaller valve of a diatom frustule.

Idioandrosporous   In *Oedogonium* and related algae, having the androspore produced in a special androsporangial filament.

Imperfect   Referring to (1) flowers that lack either stamens or pistils; or (2) fungi without a known sexual stage—for example, the Deuteromycotina.

Imperfect flower   One lacking either stamens or pistils.

Imperfect fungus   One that is not known to reproduce sexually.

Imperfect stage   The conidial stage of an ascomycete.

Impression   A type of fossil produced by extreme flattening of the plant organ(s) by pressure of overlying sediment with subsequent loss of all organic material, leaving an imprint in the rock matrix.

Incomplete flower   One lacking any one or more of the four types of floral organ.

Incubous   Leaf arrangement in which the distal margin of a given leaf overlaps the proximal portion of the next younger leaf.

Indusium   The thin covering layer of a group of sporangia in ferns.

Inflorescence   An axis or axes bearing flowers.

Integument   The tissues covering the megasporangium in ovules.

Intercalary growth   Growth localized (sometimes at intervals) between base and apex of an organism.

Internode   The region of the stem between two nodes.

Intine   The inner layer of the spore or pollen-grain wall.

Involucre   (1) Tissue that encloses a group of archegonia and hence a sporophyte in liverworts; (2) the bracts below a capitulum.

Isidium   A gemmalike outgrowth of a lichen thallus that becomes detached and may propagate the thallus.

Isogamy   The type of sexual reproduction in which the gametes are morphologically indistinguishable.

Isomorphic alternation of generations   A life

cycle in which the two alternates are morphologically similar.

Isophyllous   Producing one type of leaf.

Isthmus   The region of the protoplast connecting the semicells of incised desmids.

Jacket layer   (1) The wall of an antheridium in cryptogams; (2) the cells surrounding the egg in gymnosperms.

Karyogamy   Nuclear union.

Lacuna   A chamber or air space.

Lamellae   The gills of the mushrooms.

Laminaran   A polymer of glucose and mannitol with 1 : 6 glucosidic linkages; a storage product in brown algae.

Lateral conjugation   Union of two algal cells of the same filament by formation of lateral protuberances near common transverse septa.

Leaf gap   A parenchymatous interruption in a siphonostele associated with departure of a leaf trace.

Leaf primordium   A miniature leaf in the bud.

Leaf trace   Xylem and phloem connecting that of the petiole with that of the stele.

Legume   (1) A member of the legume family; or (2) a simple, dry fruit that dehisces along two sutures at maturity (fruits in some genera of the family indehiscent).

Lenticel   A region in the bark of woody stems where gaseous interchange occurs.

Leptoid   A cell of mosses that seemingly functions like the sieve elements of vascular plants.

Leptosporangiate   The development of an entire sporangium from the periclinal division of a superficial cell.

Leucosin   See CHRYSOLAMINARAN.

Ligule   A minute, membranous appendage at the base of grass leaves and those of certain Microphyllophyta (and certain Asteraceae).

Linear tetrad   Arrangement of four spores in a single series.

Lip cells   Thin-walled cells that interrupt the annulus in certain fern sporangia.

Lobule   In complicate-bilobed leaves, the ventral, usually smaller lobe.

Localized growth   Growth occurring in one or more definite regions of an organism.

Locule   (1) A cavity, as in a stroma of the Loculoascomycetidae; (2) the lumen of a carpel.

Long shoot   In certain woody plants, branches that increase rapidly in length during the first season in which they emerge from the bud (in contrast to spur shoots).

Macrandrous   In *Oedogonium* and related algae, having male filaments about as long as the female.

Macroconidium   A relatively large conidium of certain ascomycetous fungi, as distinct from a microconidium.

Macrocyclic   Long-cycled; a rust that produces one or more types of binucleate spores in addition to teliospores.

Macrocyst   The probable dormant zygote of the Acrasiomycetes.

Macrophyll   See MEGAPHYLLOUS.

Mannitol   An alcoholic storage product of brown algae.

Manoxylic   Xylem with broad parenchymatous rays between groups of conducting elements.

Manubrium   A prismatic cell that bears capitula and antheridial filaments in Charophyta.

Marginal meristem   The mersitem that gives rise to the leaf blade in ontogeny.

Mastigoneme   A delicate fibrillar appendage of tinsel flagella.

Medulla   Pith or central region.

Medullation   Occurrence of pith in the center of a stele.

Megaphyll   Leaves thought to be derived phylogenetically from a stem and its branches.

Megaphyllous   Having leaves with branching veins, the traces of which leaves are associated with gaps in the stem stele; same as macrophyll.

Megasporangium   The sporangium in which megaspores are produced.

Megaspore A spore arising by meiosis from a megasporocyte; potentially developing a female gametophyte; often, but not always, larger than microspores.

Megasporocyte The megaspore mother cell that forms megaspores after meiosis.

Megasporophyll A leaf bearing one or more megasporangia or ovules.

Meiosis Two successive nuclear divisions in which the chromosome number is halved and genetic segregation and recombination occur.

Meiosporangium A sporangium in which meiosis occurs.

Meiospore A spore produced immediately following meiosis.

Meristem A tissue composed of embryonic unspecialized cells.

Merophyte A segment of a plant.

Merozygote A bacterial cell that has conjugated with another with resulting change in its genetic components.

Mesarch A pattern of primary xylem differentiation in which the metaxylem develops both centrifugally and centripetally with reference to the protoxylem.

Mesocarp The layer of a fruit between exocarp and endocarp.

Mesome A segment of the axis proximal to a telome.

Mesophyll Tissue other than veins between lower and upper epidermis of a leaf.

Mesophyte A plant intermediate in moisture requirements and habitats between a hydrophyte and xerophyte.

Metabolism The chemical processes occurring in living organisms.

Metaboly Term sometimes used to describe the changing form of certain Euglenid organisms.

Metaphloem Primary phloem that differentiates from procambium later than the protophloem and after elongation of the organ has ceased.

Metaxylem The primary xylem that differenti-

ates from procambium after elongation of the organ has ceased and after the protoxylem has matured.

Microconidium A small conidium of ascomycetous fungi that acts like a spermatium.

Microcyclic Short-cycled; a rust in which the teliospore is the only binucleate spore.

Microcyst Encysted myxamoeba of Gymnomycota.

Micrometer (μm). Current term for micron.

Micron (μ). Equivalent of 0.001 mm, or $1/_{25,000}$ inch.

Microphyll A leaf derived phylogenetically from an enation of a stem or a product of the latter's reduction.

Micropyle A passageway between the apex of the integument or integuments.

Microsporangium A sporangium producing microspores.

Microspore A product of a microsporocyte, often, but not always, smaller than the megaspore of the species, and producing a male gametophyte.

Microsporocyte The microspore mother cell that forms four microspores after meiosis.

Microsporophyll A leaf bearing one or more microsporangia.

Midrib The differentiated midaxis of a thallus or leaf.

Mitochondrion A double membrane-bounded cytoplasmic organelle, site of energy release in cellular respiration.

Mitosis Nuclear division involving chromosomes that are replicated and distributed equally between the nuclear progeny.

Mitosporangium A sporangium in which the spores arise by mitotic, rather than meiotic, nuclear divisions.

Mitospore A spore that arises by mitotic division.

Mixed A fern sorus in which the sporangia are in various stages of development at a given time.

Mold (1) A type of fossil produced by hardening of sediment around a three-dimensional plant part, forming a negative of the plant part; or (2) a fungus.

Monarch Having one protoxylem group.

Monobiontic Having one free-living organism in the life cycle.

Monocolpate Pollen grains having one germ furrow.

Monoecious (1) Producing both microspores and megaspores on one and the same individual sporophyte; or (2) producing male and female gametes on the same individual; or (3) staminate and pistillate flowers produced on the same individual.

Monomethylhydrazine A type of hydrazine gas similar to rocket propellants that is formed naturally in *Gyromitra* and related cup-fungi. It is a common cause of mushroom poisoning in inexperienced mycophagists.

Monomorphic Oomycetes producing only one type of zoospore.

Monophyletic Representing a single or direct line of evolutionary descent.

Monosporangium A sporangium producing a single spore (monospore).

Monospore A single spore produced within the sporangium in red algae.

Monostromatic Composed of a single layer of cells.

Morphological alternation of generations Having two morphologically recognizable phases (plants or tissues) in the life cycle.

Morphology The study of structure or organization and development, both ontogenetic and phylogenetic.

Mucormycosis A disease of human beings caused by a member of the Mucorales.

Multiple fruit One developing from the maturing ovaries of more than one flower.

Mummy A fruit (usually one of the stone fruits) that has shriveled as a result of infection with *Monilinia fructicola* and has overwintered on the ground, producing apothecia of the fungus in the spring.

Mutation A sudden change in an organism that is transmitted to offspring.

Mutualistic symbiosis The phenomenon of two organisms living together with mutual benefit (commensalism).

Mycelium The collective term for the hyphae of a fungus; the somatic thallus of a fungus.

Mycobiont The fungal member of a lichen.

Mycologist A scientist who studies fungi.

Mycology The study of fungi.

Mycorrhiza An association of root and/or rhizome with a fungus that may be superficial or internal.

Mycosis A disease caused by fungi.

Myxamoeba The amoebal stage of the Gymnomycota.

Nacreous Pearly, referring to the walls of certain sieve elements.

Neck The slender portion of an archegonium.

Neck canal cells Cells that fill the center of an immature archegonial neck.

Necridium A dead cell between hormogonia in filamentous blue-green prokaryotes.

Nitrogen fixation Conversion of gaseous nitrogen to a combined form useful in metabolism.

Node Point of attachment of a leaf; also, point of branch emergence.

Nucellus The tissue surrounding the megasporocyte of seed plants; the megasporangium.

Nuclear cap A mass of ribosomes enclosed in a membrane, located above the nucleus in certain Chytridiomycetes.

Nucule Name sometimes assigned to the oogonium of Charophyta.

Oidium A thin-walled sporelike hyphal cell.

Ontogeny Development of an individual.

Oogamy Sexual union of a large nonmotile egg and a small motile sperm.

Oogonium A unicellular gametangium that contains an egg.

Oosphere Name applied to the egg in certain fungi—for example, Oomycetes.

Oospore The thick-walled, dormant zygote of oogamous reproduction in the Oomycetes.

Open-dichotomous venation Dichotomous venation in leaves with free vein endings.

Operculum (1) The coverlike apex of a moss capsule freed by rupture at the annulus, or (2) the coverlike apex of certain fungal sporangia or asci.

Orthotropous ovule An ovule in which the micropyle is 180° from the point of attachment of the funiculus to the ovule.

Ostiole Opening of a conceptacle, perithecium, or pycnidium.

Ovary The ovule-bearing region of a pistil.

Overtopping Dominance of one fork of a dichotomy.

Ovule A megasporangium covered by an integument; the precursor of the seed.

Ovuliferous scale The appendage to which the ovule is attached in certain conifers.

Paleae Chaffy scales on fern petioles; also occur in grasses.

Paleobotany The study of fossil plants.

Palmella stage A nonmotile stage of motile algae embedded in a colloidal sheath.

Palynology The study of spores and pollen grains.

Paramylon A β-1 : 3-linked glucan, the reserve carbohydrate in Euglenid cells.

Paraphysis(es) A sterile structure among reproductive cells or organs.

Parasexuality Reproduction involving heterokaryosis, nuclear fusion, somatic crossing over, somatic recombination, and haploidization but not in a specified place or sequence in the life cycle of an individual; frequently occurring in the Deuteromycotina.

Parasite An organism that lives on and at the expense of another.

Parasitic symbiosis The phenomenon of two organisms living together, one of which derives benefit and may injure the other (the host).

Parenchyma Thin-walled living cells often with large vacuoles; photosynthetic or storage cells.

Parenthesome The pore cap in a dolipore septum.

Parietal placentation Attachment of ovules to the ovary wall.

Parthenogenesis Development of an individual without gametic union, usually from a female gamete.

Pedicel The stalk that attaches a flower to the axis of the inflorescence.

Peduncle The stalk of a flower and fruit when these are not members of an inflorescence.

Pellicle A specialized layer just within the plasma membrane of Euglenid cells.

Penicillus The special conidiophore of the fungi in the genus *Penicillium* and other similar genera of Deuteromycetes.

Perfect flower One that has both stamens and pistils.

Perfect stage The sexual stage in Ascomycetes.

Perianth (1) The collective term for sepals and/or petals; (2) in liverworts, leaves or other tissue surrounding a group of archegonia.

Pericarp (1) The cells forming a wall around a group of carpospores; or (2) the name given to the ovary wall in the fruit.

Perichaetial Pertaining to the more or less modified leaves surrounding the archegonia in mosses.

Periclinal Parallel to the surface.

Pericycle A thin zone of living cells just within the endodermis.

Periderm A composite layer consisting of cork cells, cork cambium, and sometimes parenchymatous cells that replace the epidermis in organs undergoing secondary growth.

Peridiole Hard, egglike structure of bird's-nest fungi that contains basidia and basidiospores.

Peridium The wall of a sporophore in fungi and most slime molds.

Perigonial Pertaining to the more or less modified leaves surrounding the antheridia of mosses.

Perigynium A fleshy covering, derived from the stem, that surrounds the archegonia of certain liverworts.

Perigyny A condition in certain flowers in which the sepals, petals, and stamens are borne on the rim of a cuplike receptacle, the carpel being at its base.

Periphyses Sterile filaments internally lining and sometimes protruding from an ostiole.

Periplasm Residual cytoplasm surrounding the egg in Peronosporales.

Periplast A specialized layer just within the plasma membrane in Euglenids.

Perispore Wall material deposited on the surface of spores upon the exospore or exine.

Peristome Cellular or acellular structures at the mouth of the capsule in many mosses.

Perithecium An ascocarp with a terminal ostiole.

Petal A colored (usually) sterile appendage of the angiospermous flower.

Petiole The stalk that attaches the leaf blade to the stem.

Petrifaction A kind of fossil formed by infiltration of mineral material in solution throughout the tissues of a plant part with subsequent precipitation of the mineral, thereby embedding the plant material within a rock matrix.

Phagotrophic Ingesting solid food particles.

Phanerogam A seed plant.

Phaneroplasmodium A granular, often massive, easily visible plasmodium of the Myxomycetes in the order Physarales.

Phellem Cork tissue.

Phellogen Cork cambium.

Phialide A conidiogenous cell, usually bottle-shaped, which produces conidia internally—for example, in *Aspergillus* and *Penicillium*.

Phloem Living, thin-walled cells, typified by sieve areas in the walls of some of the cells, food-conducting in function.

Photoautotrophic Using light energy in synthesizing protoplasm from inorganic compounds.

Photosynthesis The synthesis in light from carbon dioxide and water of carbohydrates with the liberation of oxygen in chlorophyllous plants.

Phototaxis Movement as stimulated positively or negatively by light.

Phragmobasidium A septate basidium with one or more transverse or vertical septa.

Phragmoplast A zone of fibers, with a cell plate, between two telophasic nuclei; a mechanism of cytokinesis.

Phycobilin A protein-linked algal pigment of blue-green and red algae.

Phycobilisome A minute granule of a phycobilin pigment present on the surface of thylakoids.

Phycobiont The algal member of a lichen.

Phycocyanin A blue phycobilin pigment of blue-green and red algae.

Phycoerythrin A red phycobilin pigment of blue-green and red algae.

Phycology The study of algae.

Phyllophore The rachis of a leaflike organ having features of both stems and leaves.

Phylogeny The real relationships of organisms through evolutionary descent.

Physiology The study of the functioning of organisms.

Phytogeography The study of plant distribution.

Pileus The cap of a mushroom basidiocarp and that of certain ascocarps.

Pilus A tubular projection of a cell that makes contact with another cell in genetic interchange of bacteria.

Pinna The primary division of the blade of a pinnately compound leaf.

Pinnule The ultimate unit or division of a bipinnately compound fern leaf.

Pistil A megasporophyll or carpel; or, if compound, a group of united megasporophylls in angiosperms.

Pistillate flower One having only pistils, not stamens.

Pit connection A modification in the common wall between contiguous cells in red algae.

Pith The central, parenchymatous (usually) tissue within steles other than protosteles.

Placenta The region to which ovules are attached.

Placental cell A cell in red algae that gives rise to carposporangia.

Placentation The pattern of ovular attachment.

Planation Flattening of branches.

Plankton Suspended, free-floating, aquatic microorganisms.

Planogamete A flagellate gamete.

Plasma membrane The membrane bounding the protoplast.

Plasmodesma A delicate strand of protoplasm connecting protoplasts of adjacent cells through their walls.

Plasmodiocarp A sessile sporophore of Myxomycetes developed from portions of the plasmodial veins and retaining their shape.

Plasmodium An unwalled, amoeboid, multinucleate mass of protoplasm.

Plasmogamy Union of sex cells or gametes.

Plectostele A protostele in which xylem and phloem are intermingled in zones.

Pleurocarpous Usually prostrate, much branched mosses, the sporophytes borne on short lateral branches.

Plumule The terminal bud of an embryo.

Plurilocular A multicellular structure, each cell of which produces a reproductive cell.

Pneumatocyst The gas-filled, bulbous chamber in certain kelps.

Pod See LEGUME.

Podetium An elevated stiped lichen ascocarp.

Polar nodule The polar wall thickening in heterocysts and certain diatom frustules.

Polar nuclei The nuclei of the angiosperm female gametophyte that migrate to its center from the poles.

Pollen chamber A modification of the apex of the megasporangium; it receives pollen.

Pollen grain A microspore wall containing a mature or immature male gametophyte in seed plants.

Pollen sac A chamber within the anther that contains mature pollen.

Pollen tube The tubular protuberance of maturing male gametophytes in seed plants.

Pollination The transfer of pollen from the microsporangium to the micropyle of the ovule (in gymnosperms) or to the stigma of the pistil (in angiosperms).

Pollination droplet A droplet of fluid at the micropyle at pollination.

Pollinium A cohesive mass of pollen grains.

Polyeder A tetrahedral cell in *Hydrodictyon* and *Pediastrum* that forms zoospores.

Polypetaly Having separate unfused petals.

Polyphyletic Descended from more than one phyletic line.

Polyplanetic Having more than two periods of motility as in some Saprolegniales.

Polysepaly Having separate, unfused sepals.

Polystelic Containing more than one or two steles.

Pome A simple, fleshy, indehiscent fruit formed from an epigynous flower, of which the true fruit is surrounded by enlarged floral tube and/or fleshy receptacle.

Pore-cap (parenthesome) A perforated membrane over the opening of a dolipore septum.

Primary archesporial cell The earliest recognizable differentiated sporogenous cell.

Primary endosperm nucleus The nucleus formed by the union of two polar nuclei (or a secondary nucleus) with a sperm in angiosperms.

Primary germination Nuclear and cell divisions within the spore wall; precocious germination.

Primary meristems The three meristematic derivatives (protoderm, procambium, and ground meristem) of the apical meristem.

Primary mycelium The haploid hyphae arising from germinating basidiospores.

Primary phloem The phloem derived from procambium.

Primary root The emergent or emerged embryonic root or radicle.

Primary suspensor The tier of cells above the embryo-forming cells in certain conifers.

Primary tissues Those differentiated from a primary meristem.

Primary xylem The xylem derived from procambium.

Primary zoospore Anteriorly flagellated, pyriform zoospores of Oomycetes.

Primitive Ancient; not changed or only slightly modified with time or in evolution.

Primordium Precursor of an organ.

Procambium The primary meristem that differentiates into vascular tissue and cambium (if present).

Proembryo A mass of cells produced by division of the zygote before organization of the organs of the embryo.

Progametangium An enlarged hyphal tip of Zygomycetous fungi from which a gametangium is delimited.

Progamone The precursor of a sexual substance.

Progressive cleavage Gradual cytokinesis of multinucleate protoplasm ultimately into uninucleate or few-nucleate segments.

Prokaryonta Organisms lacking membrane-bounded nuclei, plastids, Golgi apparatus, and mitochondria.

Proliferation Repeated development of spores from the same sporangium.

Promeristem The apical meristematic region of an axis.

Promycelium The germ tube produced by a teliospore of a rust or smut fungus (also called epibasidium).

Propagule An agent of propagation or reproduction.

Prophylls Scalelike appendages lacking vascular tissue, as in *Psilotum*.

Prosporangium Precursor of the zoosporangium in certain Chytridiomycetes.

Protandrous (1) Earlier production of male than of female gametes in bisexual individuals; or (2) maturation of stamens before carpels in flowers.

Prothallial cell The small, often lenticular, cell or cells produced in primary germination of microspores, thought to represent vestigial vegetative tissue of the male gametophyte.

Prothallium or prothallus The gametophyte of vascular cryptogams.

Protoderm The primary meristem that is the precursor of epidermis.

Protogynous (1) Earlier production of female than of male gametes in bisexual individuals; or (2) maturation of carpels before stamens in flowers.

Protonema The product of spore germination in mosses and certain liverworts; the precursor of the leafy gametophores.

Protoperithecium The precursor of the perithecium in certain ascomycetous molds.

Protophloem The first-differentiated primary phloem.

Protoplasmodium A usually microscopic, granular plasmodium with irregular streaming, giving rise to a single sporangium; typical of the Myxomycetes of the order Echinosteliales but also found in some of the Liceales.

Protoplast The unit of protoplasm within the cell wall.

Protostele A solid core of xylem with peripheral phloem.

Protoxylem The first-differentiated primary xylem, usually in a region where the organ is increasing in length.

Pseudoaethalium A tight aggregation of sporangia in Myxomycetes.

Pseudoelater The sterile cells among the

spores of Anthocerotopsida.

Pseudoparaphyses Sterile vertical growths among the asci in a pseudothecium, attached at both ends.

Pseudoparenchymatous Giving the appearance of parenchyma, when sectioned, through interweaving of filaments.

Pseudoperianth An envelope surrounding a single archegonium.

Pseudophialide A small, vial-shaped, asexually formed spore in members of the Kickxellales that resembles phialides in the Deuteromycete fungi.

Pseudoplasmodium An aggregate of amoebae in cellular slime molds.

Pseudopodium The elongating portion of the gametophore axis that elevates the sporophyte in *Sphagnum* and *Andreaea*.

Pseudothecium A unilocular ascostroma resembling a perithecium.

Psilocin A hallucinogenic chemical present in the hallucinogenic species of the genus *Psilocybe*.

Psilocybin A hallucinogenic chemical present in the hallucinogenic species of the mushroom genus *Psilocybe*.

Pure culture See AXENIC CULTURE.

Pycnidiospores Conidia produced in a pycnidium.

Pycnidium A usually flask-shaped structure in fungi that produces conidia (pycnidiospores).

Pycnoxylic Xylem with narrow rays, composed mostly of conducting elements.

Pyrenoid (1) A differentiated region of the plastid in certain algae; in green algae, a region of condensation of glucose into starch; (2) presumably analogous organelles occurring in the chloroplasts of Anthocerotophyta.

Raceme An elongate inflorescence in which the short-pedicellate flowers are arranged with the youngest at the apex.

Rachis The axis of a compound leaf on which the leaflets are borne.

Radicle The primary or embryonic root.

Raphe A longitudinal fissure in the frustules of motile pennate diatoms.

Ray flower A ligulate flower at the margin of a capitulum.

Receptacle A fertile area on which reproductive organs are borne.

Receptive hypha A protuberant hypha in rust fungi that is receptive to compatible spermatia; also called flexuous hypha.

Reduced Secondarily simplified in evolution.

Reproduction Replication or multiplication of individuals.

Resin canal A secretory-cell-bounded tube in conifers into which resin is secreted.

Reticulate venation Branching, rebranching, and anastomosis of veins.

Rhizine Ropelike strands of absorptive hyphae in lichens.

Rhizoid (1) A unicellular or multicellular absorptive filament lacking vascular tissue and a root cap; (2) a short, thin, hyphalike, enucleate extension of the thallus of some fungi.

Rhizome A fleshy, elongate nonerect stem often, but not always, subterranean.

Rhizomorph Ropelike, twisted hyphal strands.

Rhizophore A root-bearing organ or region.

Root cap A covering of parenchymatous cells over the apical meristem of the root.

Root hair An absorptive unicellular protuberance of the epidermal cells of the root.

Sagittal section A section cut parallel to the long axis of a dorsiventral structure and perpendicular to its surface.

Saprobe A fungus that uses nonliving organic matter in its nutrition.

Saprophyte A plant that uses nonliving organic matter in its nutrition.

Scalariform conjugation Formation of conjugation tubes laterally, producing a ladderlike configuration in certain algae.

Sclerenchyma Lignified supporting tissue.

Sclerotium (1) A hard fungal structure consisting of tightly cemented hyphae, resistant to unfavorable environmental conditions; or (2) the hardened, dormant stage of Myxomycetes.

Secondary growth   That derived from a secondary meristem.

Secondary meristem   A lateral meristem such as cambium and phellogen.

Secondary nucleus   The nucleus formed by precocious union of the polar nuclei.

Secondary phloem   That derived from the vascular cambium.

Secondary root   A branch of the primary root.

Secondary suspensor   Elongating cells formed by division of the primary suspensors.

Secondary xylem   That derived from the vascular cambium.

Seed   An embryonic sporophyte embedded in the female gametophyte (in gymnosperms) or in the endosperm, or gorged with digested products of the latter (in angiosperms), within the remains of the megasporangium and covered with one or more integuments.

Sepal   The lowermost sterile appendages, usually green, on a floral receptacle.

Septate   Divided into compartments (cells) by cross walls (septa).

Septum (septa)   A transverse wall.

Serology   The study of immunological phenomena.

Seta   (1) The stalk of the sporophyte in liverworts and mosses; (2) a bristlelike structure of certain fungi.

Sexual reproduction   Reproduction involving nuclear union and meiosis and often plasmogamy.

Sheath   A colloidal capsule about a cell or trichome.

Shoot   The leaf-bearing portion of the axis.

Sieve element   A single cell with sieve plates or areas.

Sieve plate   An aggregate of sieve areas in the wall of sieve cells.

Sieve tube   A series of sieve tube elements.

Simple fruit   One derived from a single pistil (simple or compound) of a single flower.

Simple leaf   One with an undivided blade.

Simple pistil   One derived from one megasporophyll or composed of one carpel.

Simple polyembryony   Development of more than one embryo through formation of several zygotes.

Siphonogamous   Having pollen tubes.

Siphonostele   A cylinder of xylem and phloem surrounding the pith.

Sirenin   A sexual substance attracting male gametes in *Allomyces*.

Solenostele   An amphiphloic siphonostele.

Somatogamy   Union of vegetative hyphae.

Soredia   Clusters of hyphae and algal cells of lichens.

Sorocarp   The fruiting structure of cellular slime molds.

Sorophore   The gelatinous structure that emerges from germinating sporocarp of Marsileales.

Sorus   A group or cluster—for example, of sporangia in ferns.

Specialized   Derived; modified.

Sperm   The motile male gamete (or nucleus).

Spermatangium   The cell that produces a spermatium in red algae.

Spermatium   A minute nonflagellate male gamete.

Spermatogenous cells   Those that will give rise to sperms.

Spermogonium   The structure producing spermatia in rust fungi.

Spherule   A segment of a Myxomycetous sclerotium.

Spike   An elongate inflorescence with sessile flowers, the youngest at the apex.

Spirillum   A spirally twisted bacterium.

Spongy layer   Specialized nutritive tissue of the megasporangium surrounding the developing female gametophyte in gymnosperms.

Sporangiophore   (1) In fungi, a hypha bearing sporangia; (2) in vascular cryptogams, a branch bearing sporangia.

Sporangium   A spore sac.

Sporocarp A group of fern sporangia enclosed in an indusium (Salviniales) or hardened pinna (Marsileales); also the sporulating bodies of slime molds and fungi.

Sporophyll A leaf bearing one or more sporangia.

Sporophyte The spore-producing alternate of plants with alternating generations or phases.

Sporothallus Diploid, spore-forming phase of certain fungi; similar to the sporophyte of green plants.

Spur shoot A lateral dwarf shoot in certain woody plants.

Stalk cell One of the products of division of the generative cell in gymnosperms, said to be the homologue of the antheridial stalk.

Stamen The pollen-bearing organ or microsporophyll of flowering plants.

Staminate flower One containing only stamens of the essential organs and usually a perianth.

Stele The primary vascular tissue of axes, in some types including the pericycle, endodermis, and pith.

Sterile jacket The cells covering an antheridium.

Stigma (stigmata) (1) The "red eyespot" of algae; or (2) in angiosperms, the receptive region of the pistil.

Stipe (1) The stalk of a basidiocarp; or (2) the portion of a kelp between blade and base.

Stolon An elongate, nonfleshy, procumbent stem that roots at the nodes.

Stoma (stomata) A minute, intercellular fissure in the epidermis flanked by guard cells.

Stomium The lip-cell region of the sporangium in ferns.

Striate venation Having a series of arcuate or parallel main veins, mostly not interconnected.

Strobilus A stem with short internodes and spore-bearing appendages.

Style The portion of a pistil between stigma and ovary.

Supporting cell That cell that bears the carpogonial branch in red algae.

Suspensor (1) A sterile cell adjacent to the gametangium in zygomycetous molds; (2) a cell or cells that by elongation project the embryo into nutritive tissue.

Sympetaly Condition of having petals united.

Synangium United sporangia.

Synergid Cell associated with the angiosperm egg.

Synsepaly Union of sepals.

Tapetal membrane Deposit on the megaspore wall on gymnosperms composed of detritus from the digested megasporangium.

Tapetal plasmodium A multinucleate mass of protoplasm arising by the breakdown of the tapetal walls.

Tapetum A nutritive layer within sporangia.

Taxis A movement toward or away from a stimulus.

Taxon (taxa) A category in classification.

Taxonomy The science of classification.

Tegmen The inner of two seed coats in angiosperms.

Teliobasidium A basidium consisting of a thick-walled resting spore and a finite germ tube.

Teliospore The thick-walled, resting spore of the rust and smut fungi (Teliomycetidae), the basal portion of the teliobasidium.

Telium A sorus of teliospores.

Telome A hypothetical unit of organization consisting of the ultimate segment of a dichotomously branched axis, either fertile or sterile.

Telome truss A combination of a fertile and sterile telome.

Terminal bud That one at the apex of an axis.

Terminal conjugation Conjugation by tubular connection of two adjacent cells in the same filament in *Spirogyra*.

Testa The outer of two seed coats in angiosperms; the seed coat in gymnosperms.

Tetrad   A group of four. Often used to designate a spore tetrad resulting from meiosis.

Tetrapolar   A type of sexual reproduction in many Basidiomycetes characterized by four mating-type genes on two chromosomes.

Tetrasporangium   A sporangium in red algae that gives rise to four spores after meiosis.

Tetraspore   The meiotic product of a tetrasporangium.

Tetrasporophyte   In the red algae, the diploid plant that produces tetraspores.

Thallose   Lacking roots, stems, and leaves or their analogues.

Thylakoid   A flattened sack in whose membranes chlorophyll may be present.

Tinea   A skin disease (ringworm) caused by fungi.

Tinsel flagellum   One with filiform appendages (mastigonemes).

Tooth   A division of the peristome or outer peristome, when the latter is double.

Trabecula   (1) A beamlike structure in certain large sporangia (*Isoetes*); (2) bridging lacunae in siphonalean algae and their stems and those of *Selaginella*.

Tracheid   A single-celled, lignified, nonliving water-conducting element of xylem with tapered end walls.

Trama   Interwoven fungal hyphae composing the pileus, or the lamellae on which basidia are borne.

Transduction   Genetic modification (in bacteria) effected by the DNA of viruses (bacteriophages).

Transformation   Genetic modification in bacteria effected by their incorporation of DNA from another strain.

Transfusion tissue   Tissue surrounding a leaf vein within the endodermis in gymnosperm leaves, some of its cells thickened like xylem cells.

Transition zone   That portion of the axis between root and stem in which the xylem-phloem arrangement changes from radial to collateral.

Translocation   Movement of substances in plants.

Transpiration   The loss of water in vapor form from the aerial parts of a plant.

Triarch   Having three protoxylem groups.

Trichoblast   A sterile, branching, colorless filament in red algae.

Trichogyne   A receptive protuberance for spermatia in red algae and certain Ascomycetes and Basidiomycetes.

Trichome   (1) A chain of cells (in Cyanochloronta); or (2) an epidermal hair (in vascular plants).

Trichothallic   Intercalary growth at the base of a hair.

Tricolpate   Having three germ furrows, as in pollen grains.

Triradiate ridge   A triradiate thickening on the walls of meiospores formed while they are associated in the tetrad.

Tropism   A growth response of bending toward or away from a stimulus.

True indusium   An epidermal outgrowth covering receptacle and sporangia in certain ferns.

Trumpet hyphae   Elongate, hyphalike cells inflated near the transverse septa.

Tube cell   The precursor of the pollen tube in male gametophytes.

Tuberculate rhizoid   One with peglike thickenings.

Tunica   Superficial layer or layers of meristematic cells covering the corpus.

Umbel   A flat-topped, cymose inflorescence.

Unilacunar   Having a single leaf gap in the stele.

Unilocular   Having a single cavity.

Universal veil   A membranous covering of an entire mushroom sporophore when immature.

Urediniospore   A binucleate spore of the rust fungi (Uredinales) capable of reinfecting the host on which it was produced.

Uredinium   A mass (sorus) of binucleate hyphae forming urediniospores.

Vallecular canal   A canal opposite the valleys between the ridges of an arthrophytan stem.

Valve   (1) One-half of a diatom frustule; (2) a segment of the capsule wall of liverworts and hornworts at dehiscence.

Vascular plant   One containing xylem and phloem.

Vegetative   Somatic; usually not reproductive.

Vein   Xylem and phloem strand in a leaf and/or parts of a flower.

Velum   A membrane (inner veil) that covers the gills of a mushroom before it opens to expose them. The velum gives rise to the annulus of a mature mushroom.

Venation   Pattern of vein arrangement in leaves and/or parts of a flower.

Venter   The enlarged basal portion of an oogonium, which contains an egg.

Ventral canal cell   The sister cell of the egg, derived with it by division of the central cell.

Ventral scale   A multicellular, monostromatic appendage in certain liverworts.

Vernation   Arrangement of leaves in the bud.

Vesicle   Enlarged hyphal tip upon which conidia are borne in chains from phialides—for example, in *Aspergillus*.

Vessel   A series of lignified conducting cells of xylem with perforate end walls.

Vessel element   One cellular component of a vessel.

Volva   A cup at the base of the stipe in certain mushrooms—for example, *Amanita*.

Water bloom   A dense population of planktonic algae.

Webbing   Hypothetical filling of areas between branches with parenchyma tissue.

Whiplash flagellum   One that is sheathed except for a free tip; lacking fibrillar appendages.

Woronin bodies   Ultrastructural organelles of a crystalline nature found in mycelium of Ascomycetes, some Deuteromycetes, and in the ascomycetous lichens.

Xanthophyll   A carotenoid pigment, differing from carotine in containing oxygen.

Xenia   Direct influence of pollen on seed or fruit.

Xerophyte   A plant of dry habitats.

Xylem   Lignified water-conducting tissue.

Zoosporangium   A sporangium that produces zoospores.

Zoospore   A flagellate, asexual reproductive cell formed by a nonmotile organism.

Zygomorphy   The condition of being bilaterally symmetrical.

Zygosporangium   The sporangium enclosing a zygospore.

Zygospore   A thick-walled, dormant zygote of the Zygomycetes.

Zygote   The cell produced by the union of the two gametes and/or gametic nuclei.

Zygotic meiosis   Meiosis at zygote germination.

BIBLIOGRAPHY

The names of authors of books and articles of general interest are preceded by an asterisk.

Aaronson, S., B. De Angelis, O. Frank, and H. Baker. 1971. Secretion of vitamins and amino acids into the environment by *Ochromonas danica.* J. Phycol. 7:215–218.

Aaronson, S., S. W. Dhawale, N. J. Patni, B. De Angelis, O. Frank, and H. Baker. 1977. The cell content and secretion of water-soluble vitamins by several freshwater algae. Arch. Microbiol. 112:57–59.

Ahluwalia, A. S., and H. D. Kumar. 1982. Pattern of akinete differentiation in the blue-green alga *Nostoc* sp. Beitr. z. Biol. der Pflanzen. 57:459–467.

*Ahmadjian, V. 1963. The fungi of lichens. Sci. Amer. 208:122–131.

*———. 1965. Lichens. Annual Rev. Microbiol. 19:1–20.

———. 1966a. Artificial re-establishment of the lichen *Cladonia cristatella.* Science 151:199–201.

*———. 1966b. Lichens. *In* S. M. Henry, ed. Symbiosis, vol. 1. Academic Press, New York.

———. 1970a. Adaptations of Antarctic terrestrial plants. Antarctic Ecol. 2:801–811.

*———. 1970b. The lichen symbiosis: its origin and evolution. *In* T. Dobzhansky, M. H. Hecht, and W. C. Steere, eds., Evolutionary biology, vol. 4, chap. 6. Prentice-Hall, Englewood Cliffs, N.J.

———. 1982. Algal/fungal symbioses. *In* F. E. Round and D. Chapman, eds., Progress in phycological research, chap. 5. Elsevier Biomedical Press B.V.

*Ahmadjian, V. and M. E. Hale, Jr., eds. 1973. The lichens. Academic Press, New York.

Ahmadjian, V., and J. B. Jacobs. 1981. Relationship between fungus and alga in the lichen *Cladonia cristatella* Tuck. Nature 289:169–172.

———. 1982. Artificial reestablishment of lichens. III. Synthetic development of *Usnea strigosa* J. Hattori Bot. Lab. 52:393–399.

Ahmadjian, V., L. A. Russell, and K. C. Hildreth. 1980. Artificial reestablishment of lichens. I. Morphological interactions between the phycobiont of different lichens and the mycobionts of *Cladonia cristatella* and *Lecanora chrysoleuca.* Mycologia 72:73–89.

*Ainsworth, G. C. 1971. Ainsworth and Bisby's dictionary of the fungi. 6th ed. Commonwealth Mycological Institute, Kew, Surrey.

*Ainsworth, G. C., F. K. Sparrow, and A. S. Sussman, eds. 1973. The fungi, vols. IVA, Taxonomic reviews with keys, and IVB. Academic Press, New York.

*Ainsworth, G. C., and A. S. Sussman, eds. 1965–1968. The fungi. I. 1965. The fungal cell. II. 1966. The fungal organisms. III. 1968. The Fungal population. Academic Press, New York.

Aldrich, H. C. 1967. The ultrastructure of meiosis in three species of *Physarum.* Mycologia 59:127–148.

Aldrich, H. C., and G. W. Carroll. 1971. Synaptonemal complexes and meiosis in *Didymium iridis:* a reinvestigation. Mycologia 63:308–310.

Aldrich, H. C., and C. W. Mims. 1970. Synatonemal complexes and meiosis in Myxomycetes. Amer. J. Bot. 57:935–941.

Alexopoulos, C. J. 1960. Gross morphology of the plasmodium and its possible significance in the relationships among the Myxomycetes. Mycologia 52:1–20.

———. 1969. The experimental approach to the taxonomy of the Myxomycetes. Mycologia 61:219–239.

*———. 1973. The Myxomycetes. The fungi, vol. IVB, chap. 3: pp. 39–60. Academic Press, New York.

*Alexopoulos, C. J., and C. W. Mims. 1979. Introductory mycology. 3d ed. Wiley, New York.

Allen, C. E. 1932. Sex inheritance and sex determination. Amer. Natur. 66:97–107.

Allen, G. S., and J. N. Owens. 1972. The life history of Douglas fir. 139 pp. Information Canada.

Allen, M. M. 1968. Ultrastructure of the cell wall and cell division of unicellular blue-green algae. J. Bacteriol. 96:842–852.

Allsopp, A. 1969. Phylogenetic relationships of the Procaryota and the origin of the eucaryotic cell. New Phytol. 68:591–612.

Alosi, M. C., and F. J. Alfieri. 1972. Ontogeny and structure of the secondary phloem in *Ephedra.* Amer. J. Bot. 59:818–827.

Ananiev, A. R., and S. A. Stepanov. 1969. The first finding of the Psilophyton flora in Lower Devonian Salairsky Ridge (West Siberia). Treatise of the Tomsk Order of the Worker's Red Banner of V. V. Kujbyshev State University, vol. 203:13–28.

*Anderson, L. E. 1963. Modern species concepts: mosses. Bryologist 66:107–119.

———. 1980. Cytology and reproductive biology of mosses. A.A.A.S. Symposium on Mosses, Seattle, June 1978. *In* R. J. Taylor and A. E. Leviton, eds., Mosses of North America. Pacific Division. A.A.A.S. San Francisco.

Andrews, H. N. 1960. Notes on Belgian specimens of *Sporogonites.* Palaeobotanist 7:85–89.

*———. 1961. Studies in paleobotany. Wiley, New York.

———. 1963. Early seed plants. Science 142:925–931.

Andrews, H. N., and H. Lenz. 1943. A mycorrhizome from the Carboniferous of Illinois. Bull. Torrey Bot. Club 70:120–125.

Andrews, H. N., C. B. Read, and S. H. Mamay. 1971. A Devonian lycopod stem with well-preserved cortical tissues. Palaeontology 14:1–9.

Antia, N. J. 1977. A critical appraisal of Lewin's Prochlorophyta. Brit. Phycol. J. 12:271–276.

Apostolakos, P., B. Galatis, and K. Mitrakos. 1982. Studies on the development of the air pores and air chambers of *Marchantia paleacea*. Ann. Bot. 49:377–396.

Arnold, C. A. 1948. Classification of the gymnosperms from the viewpoint of paleobotany. Bot. Gaz. 110:2–12.

———. 1953. Origin and relationship of the cycads. Phytomorphology 3:51–65.

———. 1968. Current trends in paleobotany. Earth-Sci. Rev. 4:283–309.

Arnott, H. J. 1959. Anastomoses in the venation of *Ginkgo biloba*. Amer. J. Bot. 46:405–411.

Arnott, H. J., and P. L. Walne. 1967. Observations in the fine structure of the pellicle pores of *Euglena granulata*. Protoplasma 64:330–344.

Aronson, J. M., and C. C. Lin. 1978. Hyphal wall chemistry of *Leptomitus lacteus*. Mycologia 70:363–369.

Arsenault, G. P., R. Biemann, A. W. Barksdale, and T. C. McMorris. 1968. The structure of antheridiol, a sex hormone in *Achlya bisexualis*. J. Amer. Chem. Soc. 90:5635–5636.

Ascaso, C., and J. Galvan. 1976. Studies on the pedogenic action of lichen acids. Pedobiologia 16:321–331.

Ashton, P., and R. Walmsley. 1976. The aquatic fern *Azolla* and its *Anabaena* symbiont. Endeavor 35:39–43.

Atkinson, L. R. 1973. The (fern) gametophyte and family relationships. *In* A. C. Jermy, J. A. Crabbe, and B. A. Thomas, eds. The phylogeny and classification of the ferns. Bot. J. Linn. Soc. 67, Supplement 1:73–90.

Balakrishnan, M. S., and B. B. Chaugule. 1980. Cytology and life history of *Batrachospermum mahafaleshuarensis*. Cryptogamie: Algologie 1:83–97.

Banks, H. P. 1968. The early history of land plants. *In* E. T. Drake, ed., Evolution and environment, pp. 73–107. Yale University Press, New Haven and London.

———. 1970a. Major evolutionary events and the geological record of plants. Biol. Rev. 45:451–454.

———. 1970b. Evolution and plants of the past. Wadsworth, Belmont, Calif.

Banks, H. P., C. B. Beck, W. G. Chaloner, J. Muller, J. Pettitt, and J. W. Schopf. 1970. Major evolutionary events and the geological history of plants. Biol. Rev. (Cambridge) 45:317–454.

Banks, H. P., P. M. Bonamo, and J. D. Grierson. 1972. *Leclercqia complexa* gen. et sp. nov., a new lycopod from the late Middle Devonian of eastern New York. Rev. Palaeobot. Palyn. 14:19–40.

Banks, H. P., and M. R. Davis. 1969. *Crenaticaulis*, a new genus of Devonian plants allied to *Zosterophyllum*, and its bearing on the classification of early land plants. Amer. J. Bot. 56:436–449.

*Barghoorn, E. S. 1971. The oldest fossils. Sci. Amer. 225:230–242.

Barksdale, A. W. 1969. Sexual hormones in *Achlya* and other fungi. Science 166:831–837.

*Barron, G. L. 1977. The nematode-destroying fungi. Topics in Mycology 1. Canadian Biological Publications, Guelph, Ontario.

Basile, D. V. 1967. The influences of hydroxy-L-proline on ontogeny and morphogenesis of the liverwort, *Scapania nemorosa*. Amer. J. Bot. 54:977–983.

———. 1969. Toward an experimental approach to the systematics and phylogeny of leafy liverworts, pp. 120–133. *In* J. E. Gunckel, Current topics in plant science. Academic Press, New York.

———. 1970. Hydroxy-L-proline and 2,2'-dipyridyl-induced phenovariations in the liverwort *Nowellia curvifolia*. Science 170:1218–1220.

Basile, D. V., and M. R. Basile. 1983. Desuppression of leaf primordia of *Plagiochila arctica* (Hepaticae) by ethylene antagonist. Science 220:1051–1053.

———. 1984. Probing the evolutionary history of bryophytes experimentally. J. Hattori Bot. Lab. 56:173–185.

Basinger, J. F., G. W. Rothwell, and W. N. Stewart. 1974. Cauline vasculature and leaf trace production in medullosan pteridosperms. Amer. J. Bot. 61:1002–1015.

Battaglia, E. 1982. Embryological questions: 4. Gynogonium vs. archegoniums and the generalization of the prefixes andro- and gyno- in plant reproduction. Ann. di Botanica 40:1–45.

Bauer, L. 1963. On the stabilization of the male sexual tendency in Musci. J. Linn. Soc. (Bot.) 58:337–342.

———. 1966. Über die Induktion apogamer Sporogonbildung bei Laubmoosen durch Chloralhydrat. Beitr. Biol. Pflanzen 42:113–125.

Bazin, M. J. 1968. Sexuality in a blue-green alga: genetic recombination in *Anacystis nidulans*. Nature 28:282–283.

Beck, C. B. 1960a. Connection between *Archaeopteris* and *Callixylon*. Science 131:1524–1525.

———. 1960b. The identity of *Archaeopteris* and *Callixylon*. Brittonia 12:351–388.

———. 1962. Reconstruction of *Archaeopteris* and further consideration of its phylogenetic position. Amer. J. Bot. 49:373–382.

———. 1970. The appearance of gymnospermous structure. Biol. Rev. 45:329–400.

———. 1971. On the anatomy and morphology of the lateral branch system of *Archaeopteris*. Amer. J. Bot. 58:758–784.

Behnke, H.-D., and G. S. Paliwal. 1973. Ultrastructure of phloem and its development in *Gnetum gnemon*, with some observations on *Ephedra campylopoda*. Protoplasma 78:305–319.

Belin, J. M. 1972. A study of budding of *Saccharomyces uvarum* Beijerinck with the scanning electron microscope. Anton. v. Lewwenh. J. Microbiol. Serol. 38:341–349.

Bell, P. R. 1959. The experimental investigation of the pteridophyte life cycle. J. Linn. Soc. Lond. (Bot.) 56:188–203.

———. 1970. The archegoniate revolution. Sci. Prog. Oxford 58:27–45.

———. 1975. Physical interactions of nucleus and cytoplasm in plant cells. Endeavour 34:19–22.

Bell, P. R., and J. G. Duckett. 1976. Gametogenesis and fertilization in *Pteridium*. Bot. J. Linn. Soc. 73:47–78.

*Bell, P. R., and C. L. F. Woodcock. 1968. The diversity of green plants. Addison-Wesley, Reading, Mass.

Benjamin, C. R. 1955. Ascocarps of *Aspergillus* and *Penicillium*. Mycologia 47:669–687.

Benjamin, R. K. 1979. Zygomycetes and their spores, pp. 573–621. In B. Kendrick, ed., The Whole Fungus, II. Proc. II Int. Mycol. Conf., Kananaskis, Nat. Mus. Nat. Sci. and Nat. Mus. Canada. Ottawa, Canada.

*Benson, L. 1957. Plant classification. Heath, Lexington, Mass.

Bergersen, F. J., G. S. Kennedy, and W. Wittmann. 1965. Nitrogen fixation in the coralloid roots of *Macrozamia communis* L. Johnson. Austral. J. Biol. Sci. 18:1135–1142.

Berlyn, G. P. 1962. Developmental patterns in pine polyembryony. Amer. J. Bot. 49:327–333.

Bernal, J. D., and A. Synge. 1972. The origin of life. In J. J. Head, ed., Oxford Biology Reader 13.

Bernstein, I. L., and R. S. Safferman. 1970. Viable algae in home dust. Nature 227:851–852.

Berrie, G. K. 1963. Cytology and phylogeny of liverworts. Evolution 17:347–357.

Berrie, G. K., and J. M. O. Eze. 1975. The relationship between an epiphyllous liverwort and host leaves. Ann. Bot. 39:955–963.

*Bessey, E. A. 1950. Morphology and taxonomy of fungi. Harper & Row (Blakiston), New York.

Bhardwaja, T. N., and S. Abdullah. 1971. Some observations on parthenogenetic sporelings of the water fern *Marsilea*. Nova Hedwigia 21:521–528.

Bhardwaja, T. N., and T. Takker. 1979. Tracheary elements in *Pilularia*. Phytomorph. 29:388–389.

Bierhorst, D. W. 1958. Vessels in *Equisetum*. Amer. J. Bot. 45:534–537.

———. 1968. On the Stromatopteridaceae (fam. nov.) and on the Psilotaceae. Phytomorphology 18:232–268.

———. 1969. On *Stromatopteris* and its ill-defined organs. Amer. J. Bot. 56:160–174.

*———. 1971a. Morphology of vascular plants. Macmillan, New York.

———. 1971b. Seral and pinnule ontogeny in *Gleichenia linearis*. Amer. J. Bot. 58:417–423.

———. 1973. Non-appendicular fronds in the Filicales. In A. C. Jermy, J. A. Crabbe, and B. A. Thomas, eds., The phylogeny and classification of ferns. Bot. J. Linn. Soc. Lond. 67, Supplement 1.

———. 1974. Variable expression of the appendicular status of the megaphyll in extant ferns with particular reference to the Hymenophyllaceae. Ann. Mo. Bot. Garden 61:408–426.

———. 1977. On the stem apex, leaf initiation and early leaf ontogeny in filicalean ferns. Amer. J. Bot. 64:125–152.

Bilderback, D. E. 1978a. The development of the sporocarp of *Marsilea vestita*. Amer. J. Bot. 65:629–637.

———. 1978b. The ultrastructure of the developing sporophore of *Marsilea vestita*. Amer. J. Bot. 65:638–645.

Bilderback, Dianne E., D. E. Bilderback, T. L. John, and J. R. Fonseca. 1973. The release mechanism and locomotor behavior of *Equisetum* sperm. Amer. J. Bot. 60:796–801.

Billmire, E., and S. Aaronson. 1976. The secretion of lipids by the freshwater phytoflagellate *Ochromonas danica*. Limnol. and Oceanog. 21:138–140.

Bischler, H., and S. Jovet-Ast. 1981. The biological significance of morphological characters in Marchantiales (Hepaticae). Bryologist 84:208–215.

Blank, R., B. Grobe, and C. Arnold. 1978. Time-sequence of nuclear and chloroplast fusions in the zygote of *Chlamydomonas reinhardii*. Planta 138:63–64.

Bloom, W. W. 1955. Comparative viability of sporocarps of *Marsilea quadrifolia* L. in relation to age. Trans. Ill. Acad. Sci. 47:72–76.

Bogorad, L. 1975. Evolution of organelles and eukaryotic genomes. Science 188:891–898.

Boland, W., F. Marner, D. G. Müller, and E. Fölster. 1983. Comparative receptor study in gamete chemotaxis of the seaweeds *Ectocarpus siliculosus* and *Cutleria multifida*. Eur. J. Biochem. 134:97–103.

Bold, H. C. 1970. Some aspects of the taxonomy of soil algae. Ann. N.Y. Acad. Sci. 175:601–616.

Bold, H. C., A. Cronquist, C. Jeffrey, L. A. S. Johnson, L. Margulis, H. Merxmüller, P. H. Raven, and A. L. Takhtajan. 1978. Proposal (10) to substitute the term "phylum" for "division" for groups treated as plants. Taxon 27:121–122.

*Bold, H. C., and M. J. Wynne. 1985. Introduction to the algae: Structure and reproduction. Prentice-Hall, Englewood Cliffs, N.J.

*Boney, A. D. 1965. Aspects of the biology of the seaweeds of economic importance. Advanc. Marine Biol. 3:105–253.

*———. 1966. A biology of marine algae. Hutchinson Educational, London, 1966.

*Bonner, J. T. 1967. The cellular slime molds. 2d ed. Princeton University Press.

Bonnet, A. L. M. 1955. Contribution a l' étude des Hydropterideces. I. Rechèrches sur *Pilularia globulifera* L. et *P. minuta* Dur. La Cellule 57:137–239.

———. 1955. II. Rechèrches sur *Salvinia auriculata* Aubl. Ann. der Sc. Nat., Bota., 11e:529–600.

———. 1957a. III. Rechèrches sur *Azolla filiculoides* Lamk. Rev. Cytol. Biol. Veg. 18:1–86.

———. 1957b. Commentaries et conclusions generales.

Rec. Trav. Lab. Bot. Geol. Zool., Fac. Sciences. Montpellier, Serie Bot. 8:37–104.

Bopp, M. 1963. Development of the protonema and bud formation in mosses. J. Linn. Soc. Lond. (Bot.). 58:305–309.

———. 1976. External and internal regulation of the differentiation of the moss protonema. J. Hattori Bot. Lab. 41:167–177.

Bopp, M., U. Ericksen, M. Nessel, and B. Knoop. 1977. Connection between the synthesis of differentiation-specific protein and the capacity of cells to respond to cytokinins in the moss *Funaria*. Physiol. Plant 42:73–78.

Borden, Carol A., and J. R. Stein. 1969. Reproduction and early development in *Codium fragile* (Suringar) Hariot. Phycologia 8:91–99.

Bornman, C. H. 1972. *Welwitschia mirabilis:* a paradox of the Nanub desert. Endeavour 31:95–99.

Bothe, H. 1982. Nitrogen fixation. *In* B. O. Carr and B. A. Whitton, eds., chap. 4. The Biology of Cyanobacteria, University of California Press, Berkeley and Los Angeles.

*Boureau, E. 1967–1975. Traitè de paleobotanique. II. 1967. Bryophyta, Psilophyta, Lycophyta. III. 1970. Sphenophyta, Noeggerathiophyta, 1964. IV. 1970. Fasc. 1. Filicophyta. IV. 1975. Fasc. 2, part 1. Pteridophylla. Masson, Paris.

*Bourrelly, P. 1966–1972. Les algues d'eau douce. I. 1966, 1972. Algues vertes. II. 1968. Les algues jaunes et brunes. III. 1970. Algues bleues et rouges. N. Boubee, Paris.

*Bower, F. O. 1908. Origin of a land flora. Macmillan, London.

*———. 1923, 1926, 1928. The ferns. Cambridge University Press.

*———. 1935. Primitive land plants. Macmillan, London.

Brachet, J. L. A. 1965. *Acetabularia.* Endeavour 93:155–161.

Bradley, S., and N. G. Carr. 1977. Heterocyst development in *Anabaena cylindrica:* the necessity for light as an initial trigger and sequential stages of commitment. J. Gen. Micro. 101:291–297.

Brandes, H. 1973. Gametophyte development in ferns and bryophytes. Ann. Rev. Plant Physiol. 24:115.

Brandes, H., and H. Kende, 1968. Studies in cytokinin-controlled bud formation in moss protonemata. Plant Physiol. 43:827–837.

*Breed, R. S., et al. 1957. Bergey's manual of determinative bacteriology. 7th ed. Williams & Wilkins, Baltimore.

Brian, P. W. 1967. Obligate parasitism in fungi. Proc. Roy. Soc. Lond. 168B:101–118.

Bristow, J. M. 1962. The controlled *in vitro* differentiation of callus derived from a fern, *Pteris cretica* L., into gametophyte or sporophyte tissues. Develop. Biol. 4:361–375.

*Brock, T. D. 1970. Biology of microorganisms. Prentice-Hall, Englewood Cliffs, N.J.

Brodie, H. J. 1945. Further observations on the mechanism of germination of the conidia of various species of powdery mildews at low humidity. Canad. J. Res. C. 23:198–210.

*———. 1975. The bird's nest fungi. University of Toronto Press.

*———. 1978. Fungi—delight of curiosity. University of Toronto Press.

Brotherus, V. F. 1924–25. Musci (Laubmoose). *In* A. Engler and K. Prantl, Die natürlichen Pflanzenfamilien, vol. 10–111. 2d ed. Engelmann, Leipzig.

Brown, R. C., and B. E. Lemmon. 1984. Spore wall development in *Andreaea* (Musci: Andreaeopsida). Am. J. Bot. 71:412–420.

Brown, R. C., B. E. Lemmon, and Z. R. Carothers. 1982. Spore wall development in *Sphagnum lescurii.* Can. J. Bot. 60:2394–2409.

Brown, R. M., Jr., Sr. Clement Johnson, O. P., and H. C. Bold. 1968. Electron and phase-contrast microscopy of sexual reproduction in *Chlamydomonas moewusii.* J. Phycol. 4:100–120.

Browning, A. J., and B. E. S. Gunning. 1979a. Structure and function of transfer cells in the sporophyte haustorium of *Funaria hygrometrica* Hedw. I. The development and ultrastructure of the haustorium. J. Exp. Bot. 30:1233–1246.

———. 1979b. Structure and function of transfer cells in the sporophyte haustorium of *Funaria hygrometrica* Hedw. II. Kinetics of uptake of labelled sugars and localization of absorbed products by freeze substitution and autoradiography. J. Exp. Bot. 30:1247–1264.

———. 1979c. Structure and function of transfer cells in the sporophyte haustorium of *Funaria hygrometrica* Hedw. III. Translocation of assimilate into the attached sporophyte and along the sets of attached and excised sporophytes. J. Exp. Bot. 30:1265–1273.

Bruce, J. G. 1972. Observations on the occurrence of the prothallia of *Lycopodium inundatum.* Amer. Fern J. 62:82–87.

———. 1976a. Development and distribution of mucilage canals in *Lycopodium.* Amer. J. Bot. 63:481–491.

———. 1976b. Gametophytes and subgeneric concepts in *Lycopodium.* Am. J. Bot. 63:919–924.

———. 1976c. Comparative studies in the biology of *Lycopodium carolanum.* Amer. Fern J. 66:125–137.

———. 1979a. Gamephyte of *Lycopodium digitatum.* Amer. J. Bot. 66:1138–1150.

———. 1979b Gametophyte and young sporophyte of *Lycopodium carolinianum.* Amer. J. Bot. 66:1156-1163.

Bruce, J. G., and J. M. Beitel. 1979. A community of *Lycopodium* gametophytes in Michigan. Amer. Fern J. 69:33–41.

Bruchmann, H. 1898. Über die Prothallien und die Keimplanzen mehrerer Europäischer Lycopodien. Gotha.

Buck, W. R., and T. W. Lucansky. 1976. An anatomical and morphological comparison of *Selaginella apoda* and *Selaginella ludoviciana.* Bull. Torrey Bot. Club 103:9–16.

*Buetow, D. E., ed. 1968. The biology of Euglena. I. General biology and ultrastructure. II. Biochemistry. Academic Press, New York.

*Buller, A. H. 1909–1950. Researches on the fungi. Longmans, Harlow, Essex, England, vols. 1–6; University Press, Toronto, vol. 7.

*Burnett, J. H. 1968. Fundamentals of mycology. St. Martin's Press, New York.

Burr, F. A. 1970. Phylogenetic transitions in the chloroplasts of the Anthocerotales. I. The number and ultrastructure of mature plastids. Amer. J. Bot. 57:97–110.

Burr, F. A., and R. F. Evert. 1973. Some aspects of sieve-element structure and development in *Selaginella kraussiana*. Protoplasma 78:81–97.

Burr, R. J., B. G. Butterfield, and C. Hébant. 1974. A correlated scanning and transmission electron microscope study of the water-conducting elements in the gametophytes of *Haplomitrium gibbsiae* and *Hymenophyton flabellatum*. Bryologist 77:612–617.

Butler, V., C. H. Bornman, and R. F. Evert. 1973a. *Welwitschia mirabilis:* morphology of the seedling. Bot. Gaz. 134:52–59.

———. 1973b. *Welwitschia mirabilis:* vascularization of a four-week-old seedling. Bot. Gaz. 134:59–63.

———. 1973c. *Welwitschia mirabilis:* vascularization of a one-year-old seedling. Bot. Gaz. 134:63–73.

Caiola, M. G., and L. deVecchi. 1980. Akinete ultrastructure of *Nostoc* species isolated from cycad coralloid roots. Can. J. Bot. 58:2513–2519.

*Cairns-Smith, A. G. 1982. Genetic takeover and the mineral origins of life. Cambridge University Press. 477 pp. Cambridge.

Caldicolt, A. B., and G. Eglinton. 1976. Cutin acids from Bryophytes: an −1 hydroxy-alkanoic acid in two liverwort species. Phytochem. 15:1139–1143.

*Campbell, D. H. 1928. The structure and development of mosses and ferns. Macmillan, New York.

*———. 1940. The evolution of the land plants (Embryophyta). Stanford University Press.

Carlile, M. J. 1970. The photoresponses of fungi. *In* Per Halldal, ed., Photobiology of microorganisms, chap. 11. Wiley, New York.

Carluccio, L. M., F. M. Hueber, and H. P. Banks. 1966. *Archaeopteris macilenta*, anatomy and morphology of its frond. Amer. J. Bot. 53:719–730.

Carothers, Z. B. 1975. Comparative studies on spermatogenesis in Bryophytes. *In* J. G. Duckett and P. A. Racey, eds., Biology of the male gamete, pp. 71–84. Academic Press, New York.

Carothers, Z. B., and J. G. Duckett. 1978. A comparative study of the multilayered structure in developing bryophyte spermatozoids. Bryophytarum Biblio. 13:95–112.

———. 1979. Spermatogenesis in the systematics and phylogeny of the Hepaticae and Anthocerotae. *In* G. C. S. Clarke and J. G. Duckett, eds., chap. 18, 1979.

———. 1980. The Bryophyte spermatozoid: A source of new phylogenetic information. Bull. Torrey Bot. Club 107:281–297.

Carothers, Z. B., and G. L. Kreitner. 1967. Studies of spermatogenesis in Hepaticae. I. Ultrastructure of the Vierergruppe in *Marchantia*. J. Cell Biol. 33:43–51.

———. 1968. Studies of spermatogenesis in Hepaticae. II. Blepharoplast structure in the spermatid of *Marchantia*. J. Cell Biol. 36:603–616.

Carothers, Z. B., R. C. Brown, and J. G. Duckett. 1983. Comparative spermatogenesis in the Sphaerocarpales. I. Blepharoplast structure in *Sphaerocarpus* and *Riella*. Bryol. 86:97–105.

Carothers, Z. B., J. W. Moser, and J. G. Duckett. 1977. Ultrastructural studies of spermatogenesis in the Anthocerotales. II. The blepharoplast and anterior mitochondrion in *Phaeoceros laeirs:* later development. Amer. J. Bot. 64:1107–1116.

Carothers, Z. B., R. R. Robbins, and D. L. Haas. 1975. Some ultrastructural aspects of spermatogenesis in *Lycopodium complanatum*. Protoplasma 86:334–350.

Carr, N. G., and B. A. Whitton. eds. 1982. The biology of Cyanobacteria. Botanical Monograph 9, University of California Press, Berkeley. xi + 688 pp.

Cass, D. D., and W. A. Jensen. 1970. Fertilization in barley. Amer. J. Bot. 57:62–70.

Cass, D. D., and D. J. Peteya. 1979. Growth of barley pollen tubes *in vivo*. I. Ultrastructural aspects of early tube growth in the stigmatic hair. Canad. J. Bot. 57:386–396.

Castenholz, R. W. 1969. Thermophilic blue-green algae and the thermal environment. Bacteriol. Rev. 33:476–504.

Caussin, C., P. Fleurat-Lessard. and J. L. Bonnemain. 1983. Absorption of some amino acids by sporophytes isolated from *Polytrichum formosuus* and ultrastructural characteristics of the haustorium transfer cells. Ann. Bot. 51:167–173.

Cavalier-Smith, T. 1975. The origin of nuclei and of eukaryotic cells. Nature 256:463–468.

Cave, M. S., H. J. Arnott, and S. A. Cook. 1961. Embryogeny in the California peonies with special reference to their taxonomic position. Amer. J. Bot. 48:397–404.

Cavender, J. C., and K. B. Raper. 1965. The Acrasieae in nature. I. Isolation. II. Forest soil as a primary habitat. III. Occurrence and distribution in forests of eastern North America. Amer. J. Bot. 52:294–308.

———. 1968. The occurrence and distribution of Acrasiae in forests of tropical and subtropical America. Amer. J. Bot. 55:504–513.

Chaloner, W. G. 1969. Triassic spores and pollen. *In* R. H. Tschudy and R. A. Scott, Aspects of palynology, pp. 291–309. Wiley, New York.

———. 1970. The rise of the first land plants. Biol. Rev. 45:353–377.

Chaloner, W. G., A. J. Hill, and W. S. Lacey. 1977. First Devonian platyspermic seed and its implications in gymnosperm evolution. Nature 265:233–235.

*Chamberlain, C. J. 1919. The living cycads. University of Chicago Press.

*———. 1937. Gymnosperms: structure and evolution. University of Chicago Press.

*Chao, L., and C. C. Bowen. 1971. Purification and properties of glycogen isolated from a blue-green alga, *Nostoc muscorum*. J. Bacteriol. 105:331–338.

*Chapman, V. J. 1962. The algae. Macmillan, London.

*———. 1970. Seaweeds and their uses. Methuen, London.

Chardard, R. 1977. La secretion de mucilage chêz quelques Desmidiales. I. Les Pores. Protoplasma 13:241–251.

*Chase, F. M. 1941. Useful algae. Smithsonian Publication 3667, Smithsonian Institution, Washington, D.C.

Chen, C., and J. Lewin. 1969. Silicon as a nutrient element for *Equisetum arvense*. Canad. J. Bot. 47:125–131.

Chen, L. D. M., T. Edelstein, E. Ogata, and J. McLachlan. 1970. The life history of *Porphyra miniata*. Canad. J. Bot. 48:385–389.

Chevallier, D., F. Nurit, and H. Pesey. 1977. Orthophosphate absorption by the sporophyte of *Funaria hygrometrica* during maturation. Ann. Bot. 41:527–531.

Chopra, R. N., and S. C. Bhatla. 1981. Effect of physical factors on gametangial induction, fertilization and sporophyte development in the moss *Bryum argenteum* grown *in vitro*. New Phytol. 89:430–447.

———. 1983. Regulation of gametangial formation in bryophytes. Bot. Rev. 49:29–63.

Chopra, R. N., and A. Rashid. 1967. Apogamy in *Funaria hygrometrica* Hedw. Bryologist 70:206–208.

Chopra, R. S. 1968. Relationship between liverworts and mosses. Phytomorphology 17:70–78.

*Christensen, C. M. 1965. The molds and man. 3d ed. University of Minnesota Press.

Chu, M. C-Y. 1974. A comparative study of foliar anatomy of *Lycopodium* species. Amer. J. Bot. 61:681–692.

*Clarke, G. C. S., and J. G. Duckett, eds. 1979 (1980). Bryophyte systematics. 582 pp. Academic Press. London.

*Cochrane, V. W. 1958. Physiology of fungi. Wiley, New York.

*Coffey, M. D. 1975. Obligate parasitism and the rust fungi. Sympos. Soc. Exp. Biol. 29. Symbiosis. Cambridge University Press.

*Coker, W. C. 1923. The Saprolegniaceae with notes on other water molds. University of North Carolina Press.

Cole, G. T., and M. J. Wynne. 1974. Endomytosis of *Microcystis aeruginosa* by *Ochromonas danica*. J. Phycol. 10:397–410.

Cole, K., and E. Conway. 1980. Studies in the Bangiaceae: reproductive modes. Bot. Marina 23:545–553.

Coleman, A. W. 1979. Sexuality in colonial green flagellates. *In* S. H. Hutner and M. Levandowsky, eds., Biochemistry and Physiology of Protozoa vol. 1, chap. 10. Academic Press, New York.

Collins, C. R., and J. F. Farrar. 1978. Structural resistance to mass transfer in the lichen *Xanthoria parietina*. New Phytol. 81:71–83.

Collins, N. J., and W. C. Oechel. 1974. The pattern of growth and translocation of photosynthate in a tundra moss, *Polytrichum alpinum*. Canad. J. Bot. 52:255–263.

*Conrad, H. S. 1956. How to know the mosses and liverworts. Brown, Dubuque, Iowa.

*Conger, P. S. 1936. Significance of shell structure in diatoms, pp. 325–344. Smithsonian Report, Smithsonian Institution, Washington, D.C.

Cooney, D. G., and R. Emerson. 1964. Thermophilic fungi. Freeman, San Francisco.

Cooper, K. M. 1977. Endomycorrhizas affect growth of *Dryopteris felix-mas*. Trans. Brit. Mycol. Soc. 69:161–169.

Cooper-Driver, G. 1977. Chemical evidence for separating the Psilotaceae from the Filicales. Science 198:1260–1262.

Copeland, E. B. 1942. Edible ferns. Amer. Fern J. 32:121–126.

Courtoy, R. 1966a. Contribution a l'étude du rôle de la lumière dans la séxualisation du gametophyte de *Marchantia polymorpha* L. Phytochem. Photobiol. 5:441–447.

———. 1966b. De la séxualisation du gametophyte de *Marchantia polymorpha* L., en milieu conditionné. Les Congrés et Coloques de L'Université de Liege 38:233–236.

Craigie, J. H., and G. C. Green. 1962. Nuclear behavior leading to conjugal association in haploid infections of *Puccinia graminis*. Canad. J. Bot. 40:163–178.

Crandall, B. J. 1969. Morphology and development of branches in the leafy Hepaticae. Beih. z. Nova Hedwigia 30:1–261.

Crandall-Stotler, B. 1980. Morphogenetic designs and a theory of bryophyte origins and divergence. Bio Science 30:580–585.

———. 1981. Morphology/anatomy of hepatics and anthocerotes. *In* W. Schultze-Motel, ed., *Advances in Bryology*, vol. 1, pp. 315–398.

———. 1984. Musci, Hepatics, an essay on analogues. *In* R. M. Schuster, ed., chap. 17, pp. 1093–1129. 1984.

Crayton, M. A. 1982. A comparative cytochemical study of volvocacean matrix polysaccharides. J. Phycol. 18:314–336.

Crepet, W. L. 1974. Investigations of North American cycadeoids: the reproductive biology of *Cycadeoidea*. Palaeontographica 148B:144–169.

Crepet, W. L., and T. Delevoryas. 1972. Investigations of North American cycadeoids: early ovule ontogeny. Amer. J. Bot. 59:209–215.

*Cronquist, A. 1968. The evolution and classification of flowering plants. Houghton Mifflin, Boston.

*Crowder, W. 1926. Marvels of Mycetozoa. Nat. Geogr. Mag. 49:421–443.

Croxdale, J. G. 1978. *Salvinia* leaves. I. Origin and early differentiation of floating and submerged leaves. Can. J. Bot. 56:1982–1991.

———. 1979. *Salvinia* leaves. II. Morphogenesis of the floating leaves. Can. J. Bot. 57:1951–1959.

———. 1981. *Salvinia* leaves. III. Morphogenesis of the submerged leaf. Can. J. Bot. 59:2065–2072.

Crum, H. 1976. Mosses of the Great Lakes Forest. Rev. ed. University Herbarium, University of Michigan, Ann Arbor.

Crum, H. A., and L. E. Anderson. 1981. Mosses of eastern North America. Vols. 1 and 2. 1328 pp. Columbia University Press. New York.

Crundwell, A. C. 1979. Rhizoids and taxonomy. *In* G. C. S. Clarke and J. G. Duckett, eds., chap. 15. 1979.

Culberson, C. F. 1969. Chemical and botanical guide to lichen products. University of North Carolina Press.

———. 1970. Supplement of "Chemical and botanical guide to lichen products." Bryologist 73:177–377.

*Cutter, E. G. 1969. Plant anatomy: experiment and interpretation. I. Cells and tissues. Addison-Wesley, Reading, Mass.

———. 1971. Plant anatomy: experiment and interpretation. II. Organs. Addison-Wesley, Reading, Mass.

Darden, W. H., Jr. 1966. Sexual differentiation in *Volvox aureus.* J. Protozool. 13:329–355.

———. 1970. Hormonal control of sexuality in the genus *Volvox.* Ann. N.Y. Acad. Sci. 175:757–763.

*Darrah, W. C. 1960. Principles of paleobotany. Ronald, New York.

Davidonis, G. H., and M. Ruddat. 1973. Allelopathic compounds. Thelypterin A and B in the fern *Thelypteris normalis.* Planta 111:23–32.

———. 1974. Growth inhibition in gametophytes and oat coleoptiles by Thelypterin A and B released from roots of the fern *Thelypteris normalis.* Amer. J. Bot. 61:925–930.

*Davis, B. D., R. Dulbecco, H. N. Eisen, H. S. Ginsberg, and W. B. Wood, Jr. 1967. Microbiology. Harper & Row, New York.

*Davis, P. H., and V. H. Heywood. 1963. Principles of angiosperm taxonomy. Van Nostrand Reinhold, New York.

*Dawson, E. Y. 1956. How to know the seaweeds. Brown, Dubuque, Iowa.

*———. 1966. Marine botany, an introduction. Holt, Rinehart and Winston, New York.

Dawson, J. W. 1859. On fossil plants from the Devonian rocks of Canada. Quart. J. Geol. Soc. Lond. 15:477–488.

———. 1870. The primitive vegetation of the earth. Nature 2:85–88.

———. 1871. The fossil plants of the Devonian and Upper Silurian formations of Canada. Geol. Surv. Canad. 1–92.

Dayanandan, P., and P. B. Kaufman. 1973. Stomata in *Equisetum.* Canad. J. Bot. 51:1555–1564.

Decker-Eisel, C., and W. Hagemann. 1978. Untersuchungen über den Ursprung der fertilen Fieder von *Ophioglossum pedunculatum.* Pl. Syst. Evol. 130:143–155.

Dehgan, B., and C.K.K.H. Yuen. 1983. Seed morphology in relation to dispersal, evolution, and propagation of *Cycas* L. Bot. Gaz. 144:412–418.

*Delevoryas, T. 1962. Morphology and evolution of fossil plants. Holt, Rinehart and Winston, New York.

———. 1963. Investigations of North America cycadeoids: cones of *Cycadeoidea.* Amer. J. Bot. 50:45–52.

*———, ed. 1964. Origin and evolution of ferns. Mem. Torrey Bot. Club 21:1–95.

———. 1965. Investigations of North American cycadeoids: microsporangiate structures and phylogenetic implications. Palaeobotanist 14:89–93.

———. 1968a. Some aspects of cycadeoid evolution. J. Linn. Soc. Lond. (Bot.) 61:137–146.

———. 1968b. Investigations of North American cycadeoids: structure, ontogeny and phylogenetic considerations of cones of *Cycadeoidea.* Palaeontographica 121B:122–133.

———. 1970. Plant life in the Triassic of North Carolina. Discovery 6:15–22.

Delevoryas, T., and J. Morgan. 1954. A new pteridosperm from Upper Pennsylvanian deposits of North America. Palaeontographica 96B:12–23.

Delevoyars, T., and R. C. Hope. 1971. A new Triassic cycad and its phyletic implications. Postilla 150, Peabody Museum, Yale University, New Haven, Conn.

De Maggio, A. E. 1963. Morphogenetic factors influencing the development of fern embryos. J. Linn. Soc. Lond. (Bot.) 58:361–376.

———. 1968. Meiosis *in vitro:* sporogenesis in cultured fern plants. Amer. J. Bot. 55:916–922.

———. 1977. Cytological aspects of reproduction in ferns. Bot. Rev. 43:427–448.

———. 1982. Experimental embryology of pteridophytes. *In* B. M. Johri, ed., chap. 2, 1982.

De Maggio, A. E., and D. A. Stetler. 1977. Protonemal organization and growth in the moss *Dawsonia superba:* ultrastructural characteristics. Amer. J. Bot. 64:449–454.

De Maggio, A. E., and R. H. Wetmore. 1961. Morphogenetic studies of the fern *Todea barbara.* II. Experimental embryology. Amer. J. Bot. 48:551–565.

De Maggio, A. E., R. H. Wetmore, J. E. Hannaford, D. A. Stetler, and V. Raghavan. 1971. Ferns as a model system for studying polyploidy and gene dosage effects. BioScience 21:313–316.

De Maggio, A. E., R. H. Wetmore, and G. Morel. 1963. Morphogenesis in the fern life cycle. Amer. J. Bot. 50:620–621.

Dengler, N. G. 1980. The histological basis of leaf dimorphism in *Selaginella martensii.* Can. J. Bot. 58:1225–1234.

―――. 1983*a,b*. The developmental basis of anisophylly in *Selaginella martensii*. I. Initiation and morphology of growth. II. Histogenesis. Amer. J. Bot. 70:181–206.

*Denison, N. C. and G. C. Carroll. 1965. The primitive ascomycete: a new look at an old problem. Mycologia 58:249–269.

Dennis, W. M., and D. H. Webb. 1981. The distribution of *Pilularia americana* A. Br. (Marsileaceae) in North America north of Mexico. Sida 9:19–24.

De Sloover-Colinet, A. 1963. Chambre polinique et gametophyte male chez *Ginkgo biloba*. La Cellule 64:129–145.

*Dick, M. W. 1973. Saprolegniales. *In* G. C. Ainsworth, F. K. Sparrow, and A. S. Sussman, eds., The fungi; vol. IVB, pp. 113–144. Academic Press, New York.

*Dick, M. W. and Win-Tin. 1973. The development of cytological theory in the Oomycetes. Biol. Rev. 48:133–158.

Diehn, B. 1969. Action spectra of the photactic responses in *Euglena*. Biochem. Biophys. Acta 177:136–143.

Diers, L. 1967. Der Feinbau des Spermatozoids von *Sphaerocarpos donnellii* Aust. (Hepaticae). Planta 72:119–145.

Dietert, M. F., 1980. The effect of temperature and photoperiod on the development of geographically isolated populations of *Funaria hygrometrica* and *Weisseu controversa*. Amer. J. Bot. 67:369–380.

*Dilcher, D. L. 1965. Epiphyllous fungi from Eocene deposits in western Tennessee, U.S.A. Palaeontographica 116B:1–54.

―――. 1974. Approaches to the identification of angiosperm leaf remains. Bot. Rev. 40:1–157.

Dilcher, D. L., and P. R. Crane. 1984. *Archaeanthus*; an early angiosperm from the Western interior of North America. Ann. Mo. Bot. Gard.71:380–388.

Dixon, P. S. 1963. The Rhodophyta: Some aspects of their biology. Oceanogr. Marine Biol. Annual Rev. 1:177–196.

―――. 1970. The Rhodophyta: some aspects of their biology. II. Oceanogr. Marine Biol. Annual Rev. 8:307–352.

―――. 1973. Biology of the Rhodophyta. Macmillan (Hafner Press), New York.

Dodge, J. O. 1971*a*. Fine structure of Pyrrophyta. Bot. Rev. 37:481–508.

―――. 1971*b*. A dinoflagellate with both a mesocaryotic and eucaryotic nucleus. I. Fine structure of the nuclei. Protoplasma 73:145–157.

Dodge, J. O., and R. M. Crawford. 1970. A survey of thecal fine structure in the Dinophyceae. J. Linn. Soc. Lond. (Bot.) 63:53–57.

*Dodson, E. O. 1971. The kingdoms of organisms. Syst. Zool. 20:265–281.

Döpp, W. 1962. Weitere Untersuchungen über die Physiologie der Antheridienbildung bei *Pteridium aquilinum*. Planta 58:483–508.

Doyle, J. 1963. Proembryogeny in *Pinus* in relation to

that in other conifers—a survey. Proc. Roy. Irish Acad. 62B:181–216.

Doyle, J. A., and L. J. Hickey. 1976. Pollen and leaves from the mid-Cretaceous Potamic Group and their bearing on early angiosperm evolution. *In* C.B. Beck, ed., Origin and early evolution of angiosperms, 139–206. Columbia University Press, New York and London.

*Doyle, W. T. 1970. The biology of the higher cryptogams. Macmillan, New York.

Dowding, E. S. 1958. Nuclear streaming in *Gelasinospora*. Can. J. Microbiol. 4:295–301.

Drebes, G. 1966. On the life history of the marine plankton diatom *Stephanopyxis palmeriana*. Helgoländer Wiss. Meeresunters. 13:101–114.

Drew, E. A., and D. C. Smith. 1967. Studies in the physiology of lichens. VIII. Movement of glucose from alga to fungus during photosynthesis in the thallus of *Peltigera polydactyla*. New Phytol. 66:389–400.

*Drouet, F. 1968. Revision of the classification of the Oscillatoriaceae. Monograph 15, Academy Natural Sciences, Philadelphia.

―――. 1973. Revision of the Nostocaceae with cylindrical trichomes. Macmillan (Hafner Press), New York.

―――. 1978. Revision of the Nostacaceae with constricted trichomes. Beih. Nova. Hedw. 57:1–258.

―――. 1981. Revision of the Stigonemataceae with a summary of the classification of blue green-algae. Beih. Nova. Hedw. 66:221 pp.

Drouet, F., and W. A. Daily. 1956. Revision of the coccoid Myxophyceae. Butler University Botanical Studies 12, Indianapolis.

Drouet, F., and W. A. Daily. 1957. Revision of the coccoid Myxophyceae: additions and corrections. Trans. Ann. Microsc. Soc. 76:219–222

Drum, R. W., and J. T. Hopkins. 1966. Diatom locomotion: an explanation. Protoplasma 62:1–33.

Duckett, J. G. 1970*a*. Spore size in the genus *Equisetum*. New Phytol. 69:333–346.

―――. 1970*b*. Sexual behavior in the genus *Equisetum*, subgenus *Equisetum*. J. Linn. Soc. Lond. (Bot.) 63:327–352.

―――. 1972. Sexual behavior of the genus *Equisetum*, subgenus *Hippochaete*. Bot. J. Linn. Soc. Lond. 65:87–108.

―――. 1973. Comparative morphology of the gametophytes of *Equisetum*, subgenus *Equisetum*. Bot. J. Linn. Soc. Lond. 66:1–22.

―――. 1974*a*. An ultrastructural study of the differentiation of the spermatozoid of *Equisetum*. J. Cell Sci. 12:95–129.

―――. 1974*b*. An ultrastructural study of spermatogenesis in *Anthoceros laevis* L. with particular reference to the multilayered structure. Bull. Soc. Bot. France 121(suppl):81–91.

―――. 1975. Spermatogenesis in pteridophytes. *In* J. G. Duckett and P. A. Racey, eds., The biology of the male gamete, pp. 97–127. Biol. J. Linn. Soc. 7, supplement 1.

———. 1977. Toward an understanding of sex determination in *Equisetum:* an analysis of regeneration in gametophytes of the subgenus *Equisetum.* Bot. J. Linn. Soc. 74:215–242.

———. 1979*a.* An experimental study of the reproductive biology and hybridization in the European and North American species of *Equisetum.* Bot. J. Linn. Soc. 79:205–229

———. 1979*b.* Comparative morphology of the gametophytes of *Equisetum* subgenus *Hippochaete* and the sexual behaviour of *E. ramosissimum* subsp. *debile* (Roxb.) Hauke, *E. hyemale* var. *affine* (Engelm.) A. A., and *E. laevigatum* A. Br. Bot. J. Linn. Soc. 79:179–203.

Duckett, J. G., and **P. R. Bell.** 1972*a.* Studies on fertilization in archegoniate plants. I. Changes in the structure of spermatozoids of *Pteridium aquilinum* (L.) Kuhn during entry into the archegonium. Cytobiol. 4:421–436.

———. 1972*b.* Studies on fertilization in archegoniate plants. V. Egg penetration in *Pteridium aquilinum* (L.) Kuhn. Cytobiol. 6:35–50.

———. 1977. An ultrastructural study of the mature spermatozoid of *Equisetum* Phil. Trans. Roy. Soc. Lond. B277:131–158.

Duckett, J. G., and **Z. B. Carothers,** 1979. Spermatogenesis in the systematics and phylogeny of the Musci. *In* G. C. S. Clarke and J. G. Duckett, eds., chap. 17. 1979.

———. 1982. The assembly of flagella during spermatogenesis in land plants. *In* W. B. Amos and J. G. Duckett, eds., Prokaryotic and Eukaryotic Flagella. Soc. Exper. Biol. Symp. 35, pp.533–561. Cambridge University Press.

———. 1984. Gametogenesis. *In* R. M. Schuster, chap. 5, pp. 232–275.

Duckett, J. G., **Z. B. Carothers,** and **C. C. J. Miller.** 1982. Comparative spermatology and bryophyte phylogeny. J. Hattori Bot. Lab. 53:107–125.

Duckett, J. G., **Z. B. Carothers,** and **J. W. Moser.** 1980. Ultrastructural studies of spermatogenesis in the Anthocerotales. III. Gamete morphogenesis: from spermatogenous cell through midstage spermatid. Gamete Res. 3:149–167.

Duckett, J. G., and **A. R. Duckett.** 1980. Reproductive biology and population dynamics of wild gametophytes of *Equisetum.* Bot. J. Linn. Soc. 80:1–40.

Duckett, J. G., and **W. C. Pang.** 1984. The origins of heterospory: a comparative study of sexual behavior in the fern *Platyzoma microphyllum* R. Br. and the horsetail *Equisetum giganteum* L. Bot. J. Linn. Soc. 88:11–34.

Duckett, J. G., **W. C. Pang,** and **Z. B. Carothers.** 1981. *Pellia neesiana:* the biggest spermatozoid in the bryophytes. Bull. Br. Bryol Soc. 37:9–10.

Duckett, J. G., **A. K. S. K. Prasad, D. A. Davies,** and

S. Walker. 1977. A cytological analyses of the *Nostoc*–Bryophyte relationship. New Phytol. 79:349–362.

*Duddington, C. L. 1957. The friendly fungi. Faber, London.

Duntze, W., **V. MacKay,** and **T. R. Manney.** 1971. *Saccharomyces cerevisiae:* a diffusible sex factor. Science 168:1472–1473.

*Durán, R. 1973. Ustilaginales. *In* G. C. Ainsworth, F. K. Sparrow, and A. S. Sussman, eds., The fungi. vol. IVB, pp. 281–300. Academic Press, New York.

Dute, R. R., and **R. F. Evert.** 1977*a.* Sieve-element ontogeny in the root of *Equisetum hyemale.* Amer. J. Bot. 64:421–438.

———. 1977*b.* Primitive-like metaphloem sieve elements in the aerial stem of *Equisetum hyemale.* Protoplasma 91:257–266.

———. 1977*c.* Sieve-element ontogeny in the root of *Equisetum hyemale.* Amer. J. Bot. 64:421–438.

———. 1978. Sieve-element ontogeny in the aerial shoot of *Equisetum hyemale* L. Ann. Bot. 42:23–32.

Dworzack, D. L., **A. S. Pollock, G. R. Hodges, W. G. Barnes, L. Ajello,** and **A. Padhye.** 1978. Zygomycosis of the maxillary sinus and palate caused by *Basidiobolus haptosporus.* Arch. Intern. Med. 138:1274–1276.

Dyer, A. F., and **D. G. Cran.** 1976. The formation and ultrastructure of rhizoids on protonemata of *Dryopteris borreri* Neum. Ann. Bot. 40:757–765.

*Eames, A. J. 1936. Morphology of vascular plants. McGraw-Hill, New York.

———. 1952. Relationships of the Ephedrales. Phytomorphology 2:79–100.

———. 1955. The seed and *Ginkgo.* J. Arnold Arboretum 36:165–170.

———. 1961. Morphology of the angiosperms. McGraw-Hill, New York.

*Eames, A. J., and **L. H. MacDaniels.** 1947. An introduction to plant anatomy. 2d ed. McGraw-Hill, New York.

Echlin, P. 1971. The role of the tapetum during microsporogenesis of Angiosperms. *In* J. Heslop-Harrison, ed., Pollen: development and physiology. Prentice-Hall, Englewood Cliffs, N.J.

Eddy, A. 1979. Taxonomy and evolution of *Sphagnum. In* G. C. S. Clarke and J. G. Duckett, eds., chap. 5. 1979.

Edwards, D. 1970. Fertile Rhyniophytina from the Lower Devonian of Britain. Palaeontology 13:451–461.

Edwards, D. S. 1980. Evidence for the sporophytic status of the lower Dermian plant *Rhynia gwynne-vaughanii* Kidston and Lang. Rev. Paleobot. Palynol. 29:177–188.

Edwards, J. A., **J. S. Mills, J. Jundeen,** and **J. H. Fried.** 1969. The synthesis of the fungal sex hormone antheridiol. J. Amer. Chem. Soc. 91:1248–1249.

Edwards, P. 1968. The life history of *Polysiphonia*

denudata (Dillwyn) Kütz. in culture. J. Phycol. 4:35–37.

————. 1969. The life history of *Callithamnion byssoides* in culture. J. Phycol. 5:266–268.

————. 1976. A classification of plants into higher taxa baxed on cytological and biochemical criteria. Taxon 25:529–542.

Edwards, S. 1984. Homologies and interrelationships of moss peristomes. *In* R. M. Schuster, ed., chap. 12, pp. 658–695.

Egel, R. 1971. Physiological aspects of conjugation in fission yeast. Planta 98:89–96.

Eggert, D. A. 1961. The ontogeny of Carboniferous aborescent Lycopsida. Palaeontographica 108B:43–92.

————. 1974. The sporangium of *Horneophyton lignieri* (Rhyniophytina). Amer. J. Bot. 61:405–413.

Ehresmann, D. W., E. F. Deig, M. T. Hatch, L. H. DiSalvo, and N. A. Vedros. 1977. Antiviral substances from California Marine algae. J. Phycol. 13:37–40.

Ellis, T. T., D. R. Reynolds, and C. J. Alexopoulos. 1973. Hülle cell development in *Emericella nidulans*. Mycologia 65:1028–1035.

Elmore, H. W., and D. P. Whittier. 1975a. Ethylene production and ethylene-induced apogamous and formation in nine gametophytic strains of *Pteridium aquilinum* (L.) Kuhn. Am. Bot. 39:965–971.

————. 1975b. The involvement of ethylene and sucrose in the inductive and developmental phases of apogamous bud formation in *Pteridium* gametophytes. Canad. J. Bot. 53:375–381.

El-Saadawy, W., and W. S. Lacey. 1979. The sporangia of *Horneophyton lignieri* (Kidston and Lang) Barghoorn and Darrah. Rev. Palaeobot. Palynol. 28:137–144.

Elsik, W. C. 1976. Fossil fungal spores. *In* D. J. Weber and W. M. Hess, eds., The fungal spore: form and function. Wiley, New York.

————. 1977a. Classification and geologic history of dispersed microthyriaceous fungi. Second Int. Mycol. Congress (Tampa) Abstracts, p. 168.

————. 1977b. Morphologic phylogeny of dispersed fossil fungal spores-intimation. Second Int. Mycol. Congress (Tampa) Abstracts, p. 169.

Ellzey, J. T., and E. Huizar. 1977. Synaptonemal complexes in antheridia of *Achlya ambisexualis* E. 87. Arch. Microbiol. 112:311–313.

Emigh, V. D., and D. R. Farrar. 1977. Gemmae: a role in sexual reproduction with fern genus *Vittaria*. Science 198:297–298.

*Emoto, Y. 1977. The Myxomycetes of Japan. Sangyo Tosho, Tokyo.

Enderlin, C. S., and J. C. Meeks. 1983. Pure culture and reconstitution of the *Anthoceros-Nostoc* symbiotic association. Planta 358:157–165.

Endo, M., K. Nakanishi, U. Näf, W. McKeon, and R. Walker. 1972. Isolation of the antheridiogen of *Anemia phyllitides*. Physiol. Plant. 26:183–185.

Erdos, G. W., A. W. Nickerson, and K. B. Raper.

1972. The fine structure of macrocysts in *Polysphondylium violaceum*. Cytobiol. 6:351–356.

————. 1973. The fine structure of macrocyst germination in *Dictyostelium mucoroides*. Develop. Biol. 32:321–330.

*Erdos, G. W., K. B. Raper, and L. K. Vogen. 1973. Mating types and macrocyst formation in *Dictyostelium discoideum*. Proc. Nat. Acad. Sci. 70:1828–1830.

————. 1975. Sexuality in the cellular slime mold *Dictyostelium giganteum*. Proc. Nat. Acad. Sci. 72:970–973.

*Esau, K. 1965. Plant Anatomy. 2d ed. Wiley, New York.

Eschrich, W., and M. Steiner. 1967. Autoradiographische. Untersuchungen zum Stofftransport bei *Polytrichum commune*. Planta 74:330–349.

————. 1968. Die Struktur des Leitgewebessystems von *Polytrichum commune*. Planta 83:33–49.

Evans, A. M. 1964. Ameiotic alternation of generations: a new life cycle in the ferns. Science 143:261–263.

Evans, A. W., and H. D. Hooker, Jr. 1913. Development of the peristome in *Ceratodon purpureus*. Bull. Torrey Bot. Club. 40:97–109.

Evans, L. S., and D. M. Bazonne. 1977. Effect of buffered solutions and sulfate on vegetative and sexual development in gametophytes of *Pteridium aquilinum*. Amer. J. Bot. 64:897–902.

Evans, L. V. 1965. Cytological studies in the Laminariales. Ann. Bot. N.S. 29:541–562.

Evert, R. F. 1976. Some aspects of sieve-element structure and development in *Botrychium virginianum*. Israel J. Bot. 25:101–126.

————. 1977. Phloem structure and histochemistry. Annual Rev. Plant Physiol. 28:199–222.

————. 1984. Comparative structure of phloem. *In* R. A. White and W. C. Dickison, eds., Contemporary problems in plant anatomy, pp.145–234. Academic Press, Orlando, Florida.

Evert, R. F., and S. E. Eichorn. 1974. Sieve element ultrastructure in *Platycerium bifuricatum* and some other polypodiaceous ferns: the refractive spherules. Planta 119:319–334.

————. 1976. Sieve-element ultrastructure in *Platycerium bifurcatum* and some other polypodeaceous ferns: the nacreous wall thickening and maturation of the protoplast. Amer. J. Bot. 63:30–48.

Evert, R. F., C. H. Bronman, V. Butler, and M. G. Guilland. 1973. Structure and development of sieve areas in leaf veins of *Welwitschia*. Protoplasma 76:23–34.

Faegri, K., and L. van der Pijl. 1979. The principles of pollination ecology. 3rd ed. 244 pp. Pergamon Press, Oxford, New York, Toronto, Sydney, Paris, Frankfurt.

Fagerlind, F. 1961. The initiation and early development of the sporangium in vascular plants. Svensk. Bot. Tidskr. 55:299–312.

*Fahn, A. 1982. Plant anatomy. 3rd ed. Pergamon Press, Oxford.

*Farr, M. L. 1976. Myxomycetes. Flora neutropica. Monograph 16, N.Y. Botanical Garden, New York.

Farrar, D. R. 1967. Gametophytes of four tropical fern genera reproducing independently of their sporophytes in the southern Appalachians. Science 155:1266–1267.

Farrar, D. R. 1974. Gemmiferous fern gametophytes—Vittariaceae. Amer. J. Bot. 61:146–155.

———. 1976. Spore retention and release from overwintering fern fronds. Amer. Fern J. 66:49–52.

———. 1978. Problems in the identity and origin of the Appalachian *Vittaria* gametophyte, a sporophyteless fern of eastern United States. Amer. J. Bot. 64:1–12.

Farrar, D. R., and R. D. Gooth. 1975. Fern reproduction at Woodman Hollow, Central Iowa: preliminary observations and a consideration of the feasibility of studying fern reproductive biology in nature. Proc. Iowa Acad. Sci. 82:119–122.

Farrar, D. R., and W. H. Wagner, Jr. 1968. The gametophyte of *Trichomanes holopterum* Kunze. Bot. Gaz. 129:210–219.

Favali, M. A., and M. Bassi. 1974. Seta ultrastructure in *Polytrichum commune* L. Nova Hedwigia 25:451–462.

Favre-Duchartre, M. 1956. Contribution à l'étude de la reproduction chez le *Ginkgo biloba*. Rev. Cytol. Biol. Veg. 17:1–218.

———. 1958. *Ginkgo*, an oviparous plant. Phytomorphology 8:377–390.

———. 1977. Eight interpretations of embryo sac. Phytomorph. 27:407–418.

Fay, P., and N. J. Lang. 1971a. The heterocysts of blue-green algae. I. Ultrastructural integrity after isolation. Proc. Roy. Soc. Lond. 178B:185–192.

———. 1971b. The heterocysts of blue-green algae. II. Details of ultrastructure. Proc. Roy. Soc. Lond. 178B:193–203.

Ferguson, C. W. 1968. Bristlecone pine: science and esthetics. Science 159:839–846.

Ferguson, I. K., and J. Muller, eds. 1976. The evolutionary significance of the exine. Linnaean Society Symposium Series 1, Academic Press, London.

*Fink, B. 1949. The lichen flora of the United States. University of Michigan Press.

*Fischer, G. W., and C. S. Holton. 1957. Biology and control of the smut fungi. Ronald Press, New York.

Flanagan, P. W. 1970. Meiosis and mitosis in Saprolegniaceae. Canad. J. Bot. 48:2069–2076.

Florin, R. 1963. The distribution of conifer and taxad genera in time and space. Acta Horti Berg. 20:121–312.

Flowers, S. 1974. Mosses: Utah and the West. Brigham Young University Press, Provo, Utah.

Floyd, G. L., K. D. Stewart, and K. R. Mattox. 1971. Cytokinesis and plasmodesmata in *Ulothrix*. J. Phycol. 7:306–309.

Flügel, G., ed. 1977. Fossil algae: recent results and development. Springer, Berlin, Heidelberg, and New York.

Fogg, G. E., W. D. P. Stewart, P. Fay, and A. E.

Walsby. 1973. The blue-green algae. Academic Press, New York.

Foster, A. S. 1972. Venation patterns in leaves of *Ephedra*. J. Arnold Arboretum 53:364–378.

*Foster, A. S., and E. M. Gifford, Jr. 1974. Comparative morphology of vascular plants. Freeman, San Francisco.

*Foster, J. W. 1949. Chemical activities of fungi. Academic Press, New York.

*Fott, B. 1971. Algenkunde. 2d ed. Gustav Fischer, Jena.

Fowke, L. C., and J. D. Pickett-Heaps. 1978. Electron microscope study of vegetative cell division in two species of *Marchantia*. Canad. J. Bot. 56:467–475.

Freeberg, J. A. 1957. The apogamous development of sporelings of *Lycopodium cernuum* L., *L. complanatum* var. *flabelliforme* Fernald and *L. selago* L. *in vitro*. Phytomorphology 7:217–229.

Freeberg, J. A., and R. H. Wetmore. 1957. Gametophytes of *Lycopodium* as grown *in vitro*. Phytomorphology 7:204–217.

———. 1967. The Lycopsida—a study in development. Phytomorphology 17:78–91.

Freeland, R. O. 1957. Plastid pigments of gametophytes and sporophytes of Musci. Plant Physiol. 32:64–66.

French, J. C. 1972. Dimensional correlations in developing *Selaginella* sporangia. Amer. J. Bot. 59:224–227.

French, J. C., and D. J. Paolillo, Jr. 1975a. Intercalary meristem activity in the sporophyte of *Funaria* (Musci). Amer. J. Bot. 62:86–96.

———. 1975b. The effect of the calyptra on the plane of guard cell mother cell division in *Funaria* and *Physcomitrium* capsules. Ann. Bot. 39:233–236.

———. 1975c. Effect of exogenously supplied growth regulators on intercalary meristematic activity and capsule expansion in *Funaria*. Bryologist 78:431–437.

———. 1975d. On the role of the calyptra in permitting expansion of capsules in the moss *Funaria*. Bryologist 78:438–446.

———. 1976a. Effect of the calyptra on intercalary meristematic activity in the sporophyte of *Funaria* (Musci). Amer. J. Bot. 63:492–498.

———. 1976b. Effect of light and other factors on capsule expansion in *Funaria hygrometrica*. Bryologist 79:457–465.

Frey, E. 1927. The mechanical action of crustaceous lichens on subtrata of shale, gneiss, limestone and obsidian. Ann. Bot. 41:437–460.

Friedmann, E. I., A. L. Colwin, and L. H. Colwin. 1968. Fine structural aspects of fertilization in *Chlamydomonas reinhardi*. J. Cell Sci. 3:115–128.

Fries, L. 1967. The sporophyte of *Nemalion multifidum* (Weber et Mohr). J. Ag. Svensk. Bot. Tidskr. 61:452–462.

*Fritsch, F. E. 1935, 1945. Structure and reproduction of the algae. Vols. 1 and 2. Cambridge University Press, London.

*Frobisher, M. 1968. Fundamentals of microbiology. Saunders, Philadelphia.

*Frye, T. C., and L. Clark. 1937–1947. Hepaticae of North America. University of Washington Publications in Biology 6, pp. 1–1018.

Fulford, M. 1964. Contemporary thought in plant morphology: Hepaticae and Anthocerotae. Phytomorphology 14:103–119.

———. 1965. Evolutionary trends and convergence in the Hepaticae. Bryologist 68:1–31.

Fuller, M. S. 1960. Biochemical and microchemical study of the cell walls of *Rhizidiomyces* sp. Amer. J. Bot. 47:838–842.

———. 1962. Growth and development of the water mold *Rhizidiomyces* in pure culture. Amer. J. Bot. 49:64–71.

*———. 1976. Mitosis in fungi. Int. Rev. Cytol. 45:113–153.

*———, ed. 1978. The lower fungi in the laboratory. Department of Botany, University of Georgia, Athens.

Fuller, M. S., and R. Reichle. 1965. The zoospore and early development of *Rhizidiomyces apophysatus*. Mycologia 57:946–961.

Galun, M., and P. Burbrick. 1984. Physiological interactions between the partners of the lichen symbiosis. *In* H. F. Linskens and J. Helsop-Harrison, eds., Encyclopedia of Plant Physiology, chap. 18, pp.362–401. New Series, vol. 17, 743 pp. Springer-Verlag, Berlin, Heidelberg, New York, and Tokyo.

Galun, M., N. Paran, and Y. Ben-Shaul. 1970. The fungus-alga association in the Lecanoraceae: an ultrastructural study. New Phytol. 69:599–603.

Gambardella, R., R. Ligrone, and R. Castaldo. 1981. Ultrastructure of the sporophyte foot in *Phaeoceros*. Cryptogamie, Bryologie Lichenologie 2:23–45.

Gambles, R. L., and R. E. Dengler. 1982a. The anatomy of the leaf of red pine, *Pinus resinosa*. I. Nonvascular tissues. Can. J. Bot. 60:2788–2803.

———. 1982b. The anatomy of the leaf of the red pine, *Pinus resinosa*. II. Vascular tissues. Can. J. Bot. 60:2804–2824.

Gantt, E., and S. F. Conti. 1965. The ultrastructure of *Porphyridium cruentum*. J. Cell Biol. 26:365–381.

———. 1966. Granules associated with the chloroplast lamellae of *Porphyridium cruentum*. J. Cell Biol. 29:423–434.

———. 1967. Phycobiliprotein localization in algae. Brookhaven Symp. Biol. 19:393–405.

———. 1969. Ultrastructure of blue-green algae. J. Bacteriol. 97:1486–1493.

Garner, D., and D. J. Paolillo, Jr. 1973a. A time-course of sporophyte development in *Funaria hygrometrica* Hedw. Bryologist 76:356–360.

Garner, D. L. B., and D. J. Paolillo, Jr. 1973b. On the functioning of stomates in *Funaria*. Bryologist 76:423–427.

Garratt, M. J., J. D. Tims, R. B. Rickards, T. C. Chambers, and J. G. Douglas. 1984. The appearance of *Baragwanathia* (Lycophytina) in the Silurian. Bot. Jour. Linn. Soc. 89:355–358.

Gastony, G. J. 1977. Chromosomes of the independently reproducing Appalachian gametophyte: a new source of taxonomic evidence. Syst. Bot. 2:43–48.

Gates, J., R. Fisher, and R. Candler. 1980. The occurrence of coryniform bacteria in the leaf cavity of *Azolla*. Arch. Microbiol. 127:163–165

Gaudet, J. J. 1964. Morphology of *Marsilea vestita*. II. Morphology of the adult land and submerged leaves. Amer. J. Bot. 51:591–597.

———. 1973. Growth of a floating aquatic weed, *Salvinia*, under standard conditions. Hydrobiologia 41:77–106.

Gauger, W. 1961. The germination of zygospores of *Rhizopus stolonifer*. Amer. J. Bot. 48:427–429.

*Gäumann, E., 1952. The fungi. Transl. by F. L. Wynd. Macmillan (Hafner Press), New York.

———. 1964. Die Pilze. Birkhäuser Verlag, Basel and Stuttgart.

Geiger, H. 1934. Untersuchungen an *Selaginella*. (Makroprothallien, Befruchtung und Apomixis). Flora 129:140–157.

Gemmrich, A. R. 1976. Keimungsinduktien durch Salzionen, Licht und Gibberellinsäure bei sporen von *Marchantia polymorpha* L. Flora 165:479–480.

Gensel, P. G., and H. N. Andrews. 1984. Plant life in the Devonian. Praeger Publishers, New York.

Gensel, P., A. Kasper, and H. N. Andrews. 1969. *Kaulangiophyton*, a new genus of plants from the Devonian of Maine. Bull. Torrey Bot. Club 96:261–276.

Gerassimova-Navashina, H. 1961. Fertilisation and events leading up to fertilisation and their bearing on the origin of the angiosperms. Phytomorphology 11:139–146.

Gibbs, S. P. 1970. The comparative ultrastructure of the algal chloroplast. Ann. N.Y. Acad. Sci. 175:454–473.

———. 1981. The chloroplasts of some algal groups may have evolved from symbiotic green algae. Can. J. Bot. 56:2883–2889.

Gibor, A. 1966. *Acetabularia:* a useful giant cell. Sci. Amer. 215:118–124.

Giddy, C. 1974. Cycads of South Africa. Purnell, Cape Town, New York, Johannesburg, and London.

Gifford, E. M., 1983. Concept of apical cells in bryophytes and pteridophytes. Ann. Rev. Pl. Phys. 34:419–440.

Gifford, E. M., and D. D. Brandon. 1978. Gametophytes of *Botrychium multifidum* as grown in axenic culture. Amer. Fern. J. 68:71–75.

Gifford, E. M., and E. Kurth. 1982. Quantitative studies of the root apical meristem of *Equisetum scirpoides*. Amer. J. Bot. 69:464–473.

———. 1983. Quantitative studies off vegetative shoot apex of *Equisetum scirpoides*. Amer. J. Bot. 70:74–79.

Gifford, E. M., and S. Larson. 1980. Developmental features of spermatogenous cell in *Ginkgo biloba*. Amer. J. Bot. 67:119–124.

858

Gifford, E. M., and J. Lin. 1975. Light microscope and ultrastructural studies of the male gametophyte in *Ginkgo biloba:* the spermatogenous cell. Amer. J. Bot. 62:974–981.

Gifford, E. M., and V. S. Polito. 1981*a.* Growth of *Azolla filiculoides.* Bio. Science 31:526–527.

———. 1981*b.* Mitotic activity at the shoot apex of *Azolla filiculoides.* Amer. J. Bot. 68:1050–1055.

Gifford, E. M., V. S. Polito, and S. Nitayangkura. 1979. The apical cell in shoots and roots of certain ferns: a re-evaluation of its functional role in histogenesis. Pl. Sci. Lett. 15:305–311.

Giraud, A., and F. Magne. 1968. La place de la meiose dans le cycle devéloppement de *Porphyra umbilicalis.* Compt. Rend. Sci. Paris 267:586–588.

Godward, M. B. E. 1961. Meiosis in *Spirogyra crassa.* Heredity 16:53–62.

*Goebel, K. 1905. Organography of plants, especially of the Archegoniatae and Spermatophyta, part 2. English ed., transl. I. B. Balfour. Oxford University Press, London.

Goldstein, M., and S. Morrall. 1970. Gametogenesis and fertilization in *Caulerpa.* Ann. N.Y. Acad. Sci. 175:660–672.

Good, C. W., and T. N. Taylor. 1975. The morphology and systematic position of calamitean elater-bearing spores. Geoscience and Man 11:133–139.

*Gorham, P. R. 1964. Toxic algae. *In* D. F. Jackson, ed., Algae and man. Plenum, New York.

Gorham, P. R., and W. W. Carmichael. 1979. Phycotoxins from blue-green algae. Pure Appl. Chem. 52:165–174.

Graham, L. E. 1980. Ultrastructure of spermatogenesis in *Colochaete.* J. Phycol. 16:15.

———. 1982. The occurrence, evolution, and phylogenetic significance of parenchyma in *Coleochaeta* Bréb. (Chlorophyta). Amer. J. Bot. 69:447–454.

———. 1984. *Coleochaete* and the origin of land plants. Amer. J. Bot. 71:603–608.

———. 1985. The origin of the life cycle of land plants. Amer. Scientist 73:178–186.

Graham, L. E., and G. E. McBride. 1979. The occurrence and phylogenetic significance of a multilayered structure in *Coleochaete* spermatozoids. Amer. J. Bot. 66:887–894.

Graham, L. E., and L. W. Wilcox. 1983. The occurrence and phylogenetic significance of putative placental transfer cells in the green algae *Coleochaete.* Amer. J. Bot. 70:113–120.

Granetti, B. 1968. Studio comparativo della struttura e del ciclo biologico de due diatomes di aqua dolce: *Navicula minima* Grun. e *Navicula seminulum* Grun. I. Struttura e ciclo biologico di *Navicula minima* Grun. cultivata *in vitro.* Giornale Bot. Ital. 102:133–158.

Grant, V. 1977. Organismic evolution. Freeman, San Francisco.

———. 1985. The evolutionary process: a critical review of evolutionary theory. Columbia University Press, New York.

Gray, J., and A. J. Boucot. 1971. Early Silurian spore tetrads from New York: earliest new world evidence for vascular plants? Science 173:918–921.

Gray, J., S. Laufeld, and A. J. Boucot. 1974. Silurian trilete spores and spore tetrads from Gotland: their implications for land plant evolution. Science 185:260–263.

*Gray, W. D. 1959. The relation of fungi to human affairs. Holt, Rinehart and Winston, New York.

*Gray, W. D., and C. J. Alexopoulos. 1968. Biology of the Myxomycetes. Ronald Press, New York.

Green, J. W. 1980. A revised terminology for the spore-containing parts of anthers. New Phytol. 84:401–406.

Green, K. J., and D. L. Kirk. 1982. A revision of the cell lineages recently reported for *Volvox carteri* embryos. J. Cell Biol. 94:741–742.

Greenwood, M. S. 1980. Reproductive development in loblolly pine. I. The early development of male and female strobili in relation to the long shoot growth behavior. Amer. J. Bot. 67:1414–1422.

Greguss, P. 1968. Die Isoetales sind-keine Lycopsida. Ber. Deutsch. Bot. Ges. 81:187–195.

Grenville, D. J., and R. L. Peterson. 1981. Structure of the aerial and subterranean roots of *Selaginella kraussiana* A. Br. Bot. Gaz. 142:73–81.

Gressitt, J. L., J. Sedlacek, and J. J. H. Szent-Ivany. 1965. Flora and fauna on backs of large Papaun moss-forest weevils. Science 150:1833–1835.

Gretz, M. R., J. M. Arsonson, J. M. Sommerfeld, and M. R. Sommerfeld. 1984. Taxonomic significance of cellulosic cell walls in the Bangiales (Rhodophyta). Phytochem. 23:2513–2514.

Grierson, J. D. 1976. *Leclercqia complexa* (Lycopsida, Middle Devonian): its anatomy, and the interpretation of pyrite petrifactions. Amer. J. Bot. 63:1184–1202.

Grilli, C. M. 1974. A light and electron microscope study of blue-green algae living either in the coralloid-roots of *Macrozamia communis* or isolated in culture. Giornale Bot. Ital. 108:161–173.

Grilli-Caiola, M. 1975. A light and electron microscope study of blue-green algae growing in the coralloid-roots of *Encephalartos altensteinii* and in culture. Phycologia 14:25–33.

———. 1980. On the phycobionts of the cyad corolloid roots. New Phytol. 85:537–544.

Grote, M. 1977. On sexual reproduction of *Spirogyra majuscula* under defined cultural conditions. Z. Pflanzenphysiol. 83:95–108.

*Grout, A. J. 1928–1941. Moss flora of North America north of Mexico., vols. 1–3. Published by the author, Newfane, Vt.

*———. 1903. Mosses with a hand-lens and microscope. Published by the author.

*———. 1924. Mosses with a hand-lens. Published by the author.

Grubbs, P. J. 1970. Observations on the structure and biology of *Haplomitrium* and *Takakia,* hepatics with roots. New Phytol. 69:303–326.

Guedes, M., and P. Dupuy. 1970. Further remarks on the "leaflet theory" of the ovule. New Phytol. 69:1081–1092.

*Gunslaus, I. C., and R. Y. Stanier. 1960–1964. The bacteria: a treatise on structure and function, vols. 1–4. Academic Press, New York.

Hagemann, W. 1980. Über den Verzwiergungsvorgand bei Psilotum und Selaginella mit Anmerkungen zum Begriff der Dichotomie. Pl. Syst. Evol. 133:181–197.

*Hale, M. 1961. Lichen handbook. Smithsonian Institution, Washington, D.C.

*———. 1983. The biology of lichens. 3d ed. Edward Arnold, London. 190 pp.

Halfen, L. N., and R. W. Castenholz. 1971. Gliding in a blue-green alga: a possible mechanism. Nature 225:1163–1164.

Hancock, J. A., and G. R. Brassard. 1974. Element content of moss sporophytes: Buxbaumia aphylla. Canad. J. Bot. 52:1861–1865.

Harley, J. L. 1969. The biology of mycorrhiza. 2d ed. Leonard Hill, London.

Harley, J. L., and S. E. Smith. 1983. Mycorrhizal symbiosis. Academic Press, New York.

Harris, T. M. 1961. The Yorkshire Jurassic flora. I. Thallophyta-Pteridophyta. British Museum of Natural History, London.

*Hartwell, L. H. 1974. Saccharomyces cerevisiae cell cycle. Bacteriol. Rev. 38:164–198.

Haselkorn, R. 1978. Heterocysts. Ann. Rev. Pl. Phys. 29:319–344.

*Hattori, H. 1935, 1964. Myxomycetes of Nasu District. 1st and 2d eds. Text in Japanese. Many colored paintings.

*Hattori, S. A., J. Sharp, M. Mizutani, and Z. Iwatsuki. 1968. Takakia ceratophylla and T. lepidozioides of Pacific North America and a short history of the genus. Misc. Bryol. Lichenol. 4:137–149.

Hauke, R. L. 1957. The stomatal apparatus of Equisetum. Bull. Torrey Bot. Club 84:178–181.

———. 1963. A taxonomic monograph of the genus Equisetum subgenus Hippochaete. Beih. z. Nova Hedwigia. 8:1–123.

———. 1967a. Sexuality in a wild population of Equisetum arvense gametophytes. Amer. Fern J. 57:59–66.

———. 1967b. Stalking the giant horsetail. Ward's Bulletin 6(44):1–2.

———. 1968. Gametangia of Equisetum bogotense. Bull. Torrey Bot. Club 95:341–345.

———. 1969. Gametophyte development in Latin American horsetails. Bull. Torrey Bot. Club. 96:568–577.

———. 1970. Terminology of sporangial structures of Equisetum. Amer. Fern J. 60:162–163.

———. 1971. The effect of light quality and intensity on sexual expression in Equisetum gametophytes. Amer. J. Bot. 58:373–377.

———.1980. Gametophytes of Equisetum diffusum. Amer. Fern J. 70:39–44.

*Haupt, A. W. 1953. Plant morphology. McGraw-Hill, New York.

Hausmann, M. K., and D. J. Paolillo, Jr. 1977. On the development and maturation of antheridia in Polytrichum. Bryologist 80:143–148.

———. 1978a. The tip cells of antheridia of Polytrichum juniperinum. Canad. J. Bot. 56:1394–1399.

———. 1978b. The ultrastructure of the stalk and base of the antheridium of Polytrichum. Amer. J. Bot. 65:646–653.

Hawker, L. E. 1950. Physiology of fungi. University of London Press.

———. 1957. The physiology of reproduction in fungi. Cambridge University Press, London and New York.

———. 1966. Environmental influences on reproduction. In G. C. Ainsworth, F. K. Sparrow, and A. S. Sussman, eds., The fungi, 2:435–469. Academic Press, New York.

Hawkes, M. W. 1978. Sexual reproduction in Porphyra gardneri (Smith and Hollenberg) Hawkes (Bangiales, Rhodophyta). Phycologia 17:326–350.

Hawkins, A. F., and G. F. Leedale. 1971. Zoospore structure and colony formation Pediastrum spp. and Hydrodictyon reticulatum (L.) Lagerheim. Ann. Bot. 35:201–211.

Hayes, W. 1967. The mechanism of bacterial sexuality. Endeavour 26:33–38.

Heaney, S. I., and G. H. M. Jaworski. 1977. A simple separation technique for purifying micro-algae. Brit. Phycol. J. 12:171–174.

Heath, I. B., ed. 1978. Nuclear division in the fungi. Academic Press, New York.

Hébant, C. 1971. A new look at the conducting tissues of mosses (Bryopsida): their structure, distribution and significance. Phytomorphology 20:390–410.

———. 1975. On the occurrence of lysozomal acid phosphatase activity in the differentiating water-conducting strand of Takakia and its evolutionary significance. Phytomorphology 25:279–282.

———. 1976. Evidence for the presence of sieve elements in the vascularised gametophytes of Psilotum from Holloway's Collections. New Zealand J. Bot. 14:187–191.

———. 1977. The conducting tissues of bryophytes. J. Cramer, Weinheim.

———.1979. Conducting tissues in bryophyte systematics. In G. C. S. Clarke and J. G. Duckett, eds., chap. 16.

Hébant, C., R. Guiraud, J. Barthounet, and A. Ba. 1978. Le phloème de Lycopodium clavatum: organisation, ultrastructure et histochimie. Can. J. Bot. 56:2973–2980.

Hébant, C., R. Guiraud, and M. G. Martin. 1980. Le phloème de Selaginella willdenowii: histophysiologie comparée. Pl. Syst. Evol. 135:159–169.

Hébant, C., R. Hébant-Mauri, and J. Barthonet. 1978.

Evidence for division and polarity in apical cells of bryophytes and pteridophytes. Planta 138:49–52.

Hébant, C., and R. P. C. Johnson. 1976. Ultrastructural features of freeze-etched water-conducting cells in *Polytrichum* (Polytrichales, Musci). Cytobiol. 13:354–363.

Heim, R. 1936. La culture des morilles. Rev. Mycol. Paris Supplement 1, 10–11, 19–25.

*Heim, R., and R. G. Wasson. 1957. Les champignons hallucinogènes du Mexique. Musée d'Histoire Naturelle, Paris.

Hepler, P. K. 1976. The blepharoplast of *Marsilea:* its *de novo* formation and spindle association. J. Cell. Sci. 21:361–390.

Herr, J. M. 1971. A new clearing-squash technique for the study of ovule development in angiosperms. Amer. J. Bot. 58:785–790.

Hersong, L. G. 1983. Lichen protein affinity towards walls of cultured and freshly isolated phycobionts and its relationship to cell wall cytochemistry. FEMS Microbiology Letters 20:417–420.

Heslop-Harrison, J., ed. 1971. The pollen wall: structure and development. *In* J. Heslop-Harrison, ed., Pollen: development and physiology. Prentice-Hall, Englewood Cliffs, N.J.

———, ed. 1971. Pollen development and physiology. Butterworth, London.

———. 1972a. Sexuality of angiosperms. *In* F. C. Steward, ed., Plant physiology, vol. VIC, pp. 134–289. Academic Press, New York.

———. 1972b. Sexuality of angiosperms. *In* F. C. Steward, ed., Plant physiology, vol. VIC, pp. 133–289. Academic Press, New York.

———. 1975a. Incompatibility and the pollen-stigma interaction. Annual Rev. Plant Physiol. 26:403–425.

———. 1975b. The physiology of the pollen grain surface. Proc. Roy. Soc. Lond. B190:275–299.

———. 1976a. A new look at pollination. Rep. E. Malling Res. Stn. for 1975:141–157.

———. 1976b. The adaptive significance of the exine. *In* J. K. Ferguson and J. Muller, eds., The evolutionary significance of the exine, pp. 27–38. Linnaean Society Symposium Series 1, 1975.

———. 1979. Pollen walls as adaptive systems. Ann. Missouri Bot. Gard. 66:813–829.

———. 1983. Self-incompatibility: phenomenology and physiology. Proc. Roy. Soc. Lond. B. 218:371–395.

Heslop-Harrison, J., R. B. Knox, Y. Heslop-Harrison, and O. Mattsson. 1976. Pollen-wall proteins: emission and role in incompatibility responses. *In* J. G. Duckett and P. A. Racey, eds., The biology of the male gamete. Supplement 1, Biol. J. Lin. Soc. 7: 189–202.

Hibberd, D. J., and G. F. Leedale. 1970. Eustigmatophyceae—a new algal class with unique organization of the motile cell. Nature 225:758–760.

Hill, D. J. 1977. The role of *Anabaena* in the *Azolla-Anabaena* symbiosis. New Phytol. 78:611–616.

Hill, G. J. C., and L. Machlis. 1968. An ultrastructural study of vegetative cell division in *Oedogonium borisianum.* J. Phycol. 4:261–271.

Hirsch, A. M. 1976. The development of aposporous gametophytes and regenerated sporophytes from epidermal cells of excised fern leaves: an anatomical study. Amer. J. Bot. 63:263–271.

Hoare, D. S., L. O. Ingram, E. L. Thurston, and R. Walkup. 1971. Dark heterotrophic growth of an endophytic blue-green alga. Arch. Mikrobiol. 78:310–321.

Hoffman, F. M., and C. J. Hillson. 1979. Effects of silicon on the life cycle of *Equisetum hyemale* L. Bot. Gaz. 140:127–132.

Hoffman, L. R. 1960. Chemotaxis of *Oedogonium* sperms. Southwestern Natur. 3:111–116.

Holdsworth, R. H. 1971. The isolation and partial characterization of the pyrenoid protein of *Eremosphaera viridis.* J. Cell Biol. 51:499–513.

Holm-Hansen, O., R. Prasad, and R. A. Lewin. 1965. Occurrence of α, ε-diaminopimelic acid in algae and flexibacteria. Phycologia 5:1–14.

Holst, R. W. 1977. Anthocyanins in *Azolla.* Amer. Fern J. 67:99–100.

Holst, R., and J. Yopp. 1979. Studies of the *Azolla-Anabaena* symbiosis using *Azolla mexicana.* I. Growth in nature and laboratory. Amer. Fern J. 69:17–25.

Honegger, R. 1980. Zytologie der Blaualgen-Hornmoos-Symbiose bei *Anthoceros laevis* ans Island. Flora 170:290–302.

Hooft, J. 1970. *Zamia* from seed. Carolina Tips 33:21–22.

Hopkins, F. G. 1931. Problems of specificity in biochemical catalysis. Thirty-third Bolye lecture. Oxford University Press.

Hopkins, J. T., and R. W. Drum. 1966. Diatom motility: an explanation and a problem. Brit. Phycol. Bull. 3:63–67.

Horner, H. T., Jr., and H. J. Arnott. 1963. Sporangial arrangement in North American species of *Selaginella.* Bot. Gaz. 124:371–383.

Horner, H. T., Jr., and C. K. Beltz. 1970. Cellular differentiation of heterospory in *Selaginella.* Protoplasma 71:335–361.

———. 1970. Cellular differentiation of heterospory in *Selaginella.* Protoplasma 71:335–341.

Horner, H. T., Jr., C. K. Beltz, R. Jagels, and R. E. Boudreau. 1975. Ligule development and fine structure in two heterophyllous species of *Selaginella.* Canad. J. Bot. 53:127–143.

Hsiao, S. I. C., and L. D. Druehl. 1971. Environmental control of gametogenesis in *Laminaria saccharina.* I. The effects of light and culture media. Canad. J. Bot. 49:1503–1508.

Hu, S. 1969. *Ephedra* (Ma-Huang) in the new Chinese Materia Medica. Econ. Bot. 23:346–351.

Hueber, F. M. 1961. *Hepaticites devonicus,* a new fossil

liverwort from the Devonian of New York. Ann. Mo. Bot. Garden 48:125–132.

————. 1971. *Sawdonia ornata:* a new name for *Psilophyton princeps* var. *ornatum.* Taxon 20:641–642.

Hueber, F. M., and H. P. Banks. 1967. *Psilophyton princeps:* the search for organic connection. Taxon 16:81–85.

Huneck, S. 1973. Nature of lichen substances. *In* V. Ahmadjian and M. E. Hale, Jr., eds., The lichens, pp. 495–522. Academic Press, New York.

Hustede, H. 1964. Entwicklungsphysiologische Untersuchungen über den Generationswechsel zwischen *Derbesia neglecta* Berth. und *Bryopsis halymeniae* Berth. Biol. Mar. 6:134–142.

Inamdar, J. A., and D. C. Bhatt. 1972. Epidermal structure and ontogeny of stomata in vegetative and reproductive organs of *Ephedra* and *Gnetum.* Ann. Bot. 36:1041–1046.

*Ingold, C. T. 1971. Spore liberation in cryptogams. J. J. Head, ed., Oxford Biology Readers. Oxford University Press.

Ishiura, M. 1976. Gametogenesis of *Chlamydomonas* in the dark. Plant and Cell Physiol. 17:1141–1150.

*Jackson, D. F., ed. 1964. Algae and man. Plenum, New York.

*Jacobs, F., and E. Wollman. 1961. Sexuality and genetics of bacteria. Academic Press, New York.

Jacobs, J. B., and V. Ahmadjian. 1969. The ultrastructure of lichens. I. A general survey. J. Phycol. 5:227–240.

————. 1971. The ultrastructure of lichens. IV. Movement of carbon products from alga to fungus as demonstrated by high resolution radioautography. New Phytol. 70:47–50.

Jaenicke, L. 1977. Sex hormones of brown algae. Naturwiss. 64:69–75.

Jaenicke, L., and S. Waffenschmidt. 1981. Liberation of reproductive units in *Volvox* and *Chlamydomonas:* proteolytic processes. Ber. Deutsch. Bot. Ges. 94:375–386.

Jagels, R. 1970. Photosynthetic apparatus in *Selaginella.* I. Morphology photosynthesis under different light and temperature regimes. II. Changes in plastid ultrastructure and pigment content under different light and temperature regimes. Canad. J. Bot. 48:1843–1860.

Jagels, R., and J. Garner. 1979. Variation in callose deposition in the ligules of seven species of *Selaginella.* Amer. J. Bot. 66:963–969.

Jain, R. K. 1971. Pre-Tertiary records of Salviniaceae. Amer. J. Bot. 58:487–496.

James, P. W., and A. Henssen. 1976. The morphological and taxonomic significance of cephalodia. *In* D. H. Brown, D. L. Hawksworth, and R. H. Bailey, eds.,

Lichenology: progress and problems, pp. 27–77. Academic Press, New York.

*Jeffrey, E. C. 1930. The anatomy of woody plants. University of Chicago Press.

Jensen, W. A. 1972. The embryo sac and fertilization in angiosperms. Harold L. Lyon Arboretum Lecture 3, University of Hawaii, Honolulu.

————. 1973. Fertilization in flowering plants. BioScience 23:21–27.

Jensen, W. A., and D. B. Fisher. 1969. Cotton embryogenesis: the tissues of the stigma and style and their relation to the pollen tube. Planta 84:97–121.

Jermy, A. C., J. A. Crabbe, and B. A. Thomas, eds. 1973. The phylogeny and classification of the ferns. Bot. J. Linn. Soc. 67. Supplement. 1.

*Johansen, D. A. 1950. Plant embryology. Chronica Botanica, Waltham, Mass.

Johnson, L. A. S. 1959. The families of cycads and the Zamiaceae of Australia. Proc. Linn. Soc. N.S.W. 84:64–177.

*Johnson, T. W., Jr. 1956. The genus *Achlya:* morphology and taxonomy. University of Michigan Press.

*Johri, B. M., ed. 1982. Experimental embryology of vascular plants. 273 pp. Springer-Verlag, Berlin, Heidelberg, and New York.

Kamitsubo, E. 1980. Cytoplasmic streaming in characean cells: role of subcortical fibrils. Can. J. Bot. 58:760–765.

Kamiya, N. 1950. The rate of protoplasmic flow in the myxomycete plasmodium. I. Cytologia 15:183–193. II. 194–204.

Kaplan, D. R. 1977. Morphological status of the shoot systems of Psilotaceae. Brittonia 29:30–53.

Kara, I., and D. D. Cass. 1976. Ultrastructural aspects of sperm cell formation in rye: evidence for cell plate involvement in generative cell division. Phytomorph. 26:36–45.

Karling, J. S. 1944. Brazilian anisochytrids. Amer. J. Bot. 31:391–397.

————. 1945. Brazilian chytrids. VI. *Rhopalophlyctis* and *Chytriomyces* two new chitinophylic operculate genera. Amer. J. Bot. 32:362–369.

*————. 1977. Chytridiomycetarum iconographia. J. Cramer, Monticello, N.Y.

Karrfalt, E. E. 1977. Substrate penetration by the corm of *Isoetes.* Amer. Fern J. 67:1–4.

————. 1981. The comparative and developmental morphology of the root system of *Selaginella selaginoides* (L.) Link. Amer. J. Bot. 68:244–253.

————. 1982. Secondary development in the cortex of *Isoetes.* Bot. Gaz. 143:439–445.

————. 1984. Further observations on *Nathorstiana* (Isoetaceae) Amer. Jour. Bot. 71:1023–1030.

Karrfalt, E. E., and D. A. Eggert. 1977. The comparative morphology and development of *Isoetes* L. I. Lobe and furrow development in *I. tuckermanii.* Bot. Gaz. 138:236–247. II. Branching of the base of the corm in *I. tuckermannii* A. Br. and *I. nuttallii* A. Br. Bot. Gaz. 138:357–368.

Karrfalt, E. E., and **D. M. Hunter.** 1980. Notes on the natural history of *Stylites gemmifera*. Amer. Fern J. 70:69–72.

Kasai, F., and **T. Ichimura.** 1983. Zygospore germination and meiosis in *Closterium ehrenbergii* Meneghini (Conjugatophyceae). Phycologia 22:267–275.

Kashiwagi, M., and **T. R. Norton.** 1977. Antileukemia activity in the Oscillatoriaceae: isolation of Debromoaplysiatoxin from *Lyngbya*. Science 196:538–540.

Kass, L. B., and **D. J. Paolillo, Jr.** 1974*a*. On the light requirement for replication of plastids in *Polytrichum*. Plant Science Letters 3:81–85.

———. 1974*b*. The effect of darkness and inhibitors of protein synthesis in the replication of chloroplasts in the moss, *Polytrichum*. Z. Pflanzenphysiol. 73:198–207.

Kaufman, P. B., W. C. Bigelow, R. Schmid, and **N. S. Ghosheh.** 1971. Electron microprobe analysis in epidermal cells of *Equisetum*. Amer. J. Bot. 58:309–316.

Keever, C. 1957. Establishment of *Grimmia laevigata* on bare granite. Ecol. 38:422–429.

Kelley, C. 1969. Wall projections in the sporophyte and gametophyte of *Sphaerocarpus*. J. Cell Biol. 41:910–914.

Kelley, C. B., and **W. T. Doyle.** 1975. Differentiation of intracapsular cells in the sporophyte of *Sphaerocarpos donnellii*. Amer. J. Bot. 62:547–559.

Kenyon, C. N., and **R. Y. Stanier.** 1970. Possible evolutionary significance of polyunsaturated fatty acids in blue-green algae. Nature 227:1164–1166.

Kers, L. E. 1967. The distribution of *Welwitschia mirabilis* Hook F. Svensk. Bot. Tidskr. 61:97–125.

*****Kershaw, K. A.** 1984. Physiological ecology of lichens. Cambridge University Press, Cambridge, London, New York, New Rochelle, Melbourne, Sydney.

Kershaw, K. A., and **J. W. Milbank.** 1970. Nitrogen metabolism in lichens. II. The partition of cephalodial-fixed nitrogen between the mycobiont and phycobiont of *Peltigera aphthosa*. New Phytol. 69:75–79.

Khoja, T., and **B. A. Whitton.** 1971. Heterotrophic growth of blue-green algae. Arch. Mikrobiol. 79:280–282.

Kimber, G., and **R. Riley.** 1963. Haploid angiosperms. Bot. Rev. 29:480–531.

Kingsbury, J. M. 1964. Poisonous plants of the United States and Canada. Prentice-Hall, Englewood Cliffs, N.J.

Kinraide, W. T. B., and **V. Ahmadjian.** 1970. The effects of usnic acid on the physiology of two cultured species of the lichen alga *Trebouxia* Puym. Lichenologist 4:234–247.

Kito, H., E. Ogata, and **J. McLachlan.** 1971. Cytological observations on three species of *Porphyra* from the Atlantic. Bot. Mag. Tokyo 54:141–148.

Klekowski, E. J., Jr. 1967. Observations of pteridophyte

life cycles: relative lengths under cultural conditions. Amer. Fern. J. 57:49–51.

———. 1968–1970. Reproductive biology of the Pteridophyta. I. 1968. General considerations and a study of *Onoclea sensibilis* L. J. Linn. Soc. Lond. (Bot.) 60:315–324. II. 1969*a*. Theoretical considerations. J. Linn. Soc. Lond. (Bot.) 62:347–354. III. 1969*b*. A study of the Blechnaceae, J. Linn. Soc. Lond. (Bot.) 62:361–377. 1970*a*. IV. An experimental study of mating systems in *Ceratopteris thalictroides* (CL.) Brongn., J. Linn. Soc. Lond. (Bot.) 63:153–169.

———. 1970*b*. Evidence against self-incompatibility for genetic lethals in the fern *Stenochlaena tenuifolia* (Desv.) Moore. J. Linn. Soc. Lond. (Bot.) 63:171–176.

———. 1970*c*. Populational and genetic studies of a homosporous fern—*Osmunda regalis*. Amer. J. Bot. 57:1122–1138.

———. 1971. Ferns and genetics. BioScience 21:317–322.

———. 1972. Evidence against self-incompatibility in the homosporous fern *Pteridium aquilinum*. Evolution 26:66–73.

Klekowski, E. J., Jr., and **H. G. Baker.** 1966. Evolutionary significance of polyploidy in the Pteridophyta. Science 153:305–307.

Klekowski, E. J., Jr., and **R. M. Lloyd.** 1968. Reproductive biology of the Pteridophyta. I. General conditions and a study of *Onoclea sensibilis* L. J. Linn. Soc. Lond. (Bot.) 60:315–324.

Klepper, Betty. 1963. Water relations in *Dicranum scoparium*. Bryologist 66:41–54.

Knaggs, F. W. 1969. A review of florideophycidean life histories and of the culture techniques employed in their investigation. Nova Hedwigia 18:293–330.

Knoll, A. H., and **E. S. Barghoorn.** 1975. Precambrian eukaryotic organisms: a reassessment of the evidence. Science 190:52–54.

———. 1977. Archean microfossils showing cell division from the Swaziland System of South Africa. Science 198:396–398.

Knoll, A. H., E. S. Barghoorn, and **S. Golubic.** 1975. *Paleopleurocapsa wopfnerii* gen. et sp. nov.: A late Precambrian alga and its modern counterpart. Proc. Nat. Acad. Sci. 72:2488–2492.

Knox, R. B. 1984. Pollen-pistil interactions. *In* H. F. Linskens and J. Heslop-Harrison, Encyclopedia of plant physiology, chap. 24, pp. 508–608. New series. vol. 17, 743 pp. Springer-Verlag, Berlin, Heidelberg, New York, Tokyo.

Knox, R. B., J. Heslop-Harrison, and **Y. Heslop-Harrison.** 1974. Pollen-wall proteins: localization and characterization of gametophytic and sporophytic fractions. *In* J. W. Duckett and P. A. Racey, eds., The biology of the male gamete. Supplement 1, Biol. J. Linn. Soc. 7:177–187.

Kochert, G. 1968. Differentiation of reproductive cells in *Volvox carteri*. J. Protozool. 15:438–435.

———. 1975. Developmental mechanisms in *Volvox* reproduction. *In* C. L. Markert and J. Papuconstantinou,

The developmental biology of reproduction, pp. 55–90. Academic Press, New York.

———. 1982. Sexual processes in the Volvocales. *In* progress in Phycological Research., chap. 6., vol. 1. Round and Chapman, eds., Elsevier Biomedical Press B. V.

Koehn, R. D. 1971. Laboratory culture and ascocarp development of *Podosordaria leporina.* Mycologia 63:441–458.

Kollenbach, H. W., and **T. Geier.** 1970. Untersuchungen an *Selaginella kraussiana* (Kunze) A. Br. zur Funktion der Ligula. Beitr. Biol. Pflanzen 47:141–153.

Konar, R. N., and **Y. P. Oberoi.** 1969. Recent work on reproductive structures of living conifers and taxads—a review. Bot. Rev. 35:89–116.

Kondratyeva, N. V. 1982. On difference of opinions of phycologists and bacteriologists concerning the nomenclature of Cyanophyta. Arch. f. Protistenk 126:247–259.

Koop, H.-U. 1975*a*. Germination of cysts in *Acetabularia mediterranea.* Protoplasma 84:137–146.

———. 1975*b*. Über den Ort der Meiosis bei *Acetabularia mediterranea.* Protoplasma 85:109–114.

Kott, L. S., and **D. M. Britton.** 1980. Chromosome numbers for *Isoetes* in northeastern North America. Can. J. Bot. 58:980–984.

Krasselov, V.A. 1977. The origin of angiosperms. Bot. Rev. 43:143–176.

*Krauss, R. W.** 1962. Mass culture of algae for food and other organic compounds. Amer. J. Bot. 49:425–435.

*———, ed. 1978. The marine biomass of the Pacific Northwest Coast. Oregon State University Press.

*Krieger, L. C. C.** 1920. Common mushrooms of the United States. Nat. Geogr. Mag. 37:387–439.

*———. 1936. The mushroom handbook. Macmillan, New York.

Krietner, G. L. 1977*a*. Transformation of the nucleus in *Marchantia* spermatids: morphogenesis. Amer. J. Bot. 64:464–475.

———. 1977*b*. Influence of the multilayered structure on the morphogenesis of *Marchantia* spermatids. Amer. J. Bot. 64:57–64.

Krisko, M. E. P., and **D. J. Paolillo, Jr.** 1972. Capsule expansion in the hairy-cap moss, *Polytrichum.* Bryologist 75:509–515.

Kronestedt, E. 1981. Anatomy of *Ricciocarpus natans* (L.) Corda, studied by scanning electron microscopy. Ann. Bot. 47:817–827.

Kruatrachue, M., and **R. F. Evert.** 1974. Structure and development of sieve elements in the leaf of *Isoetes muricata.* Amer. J. Bot. 61:253–266.

———. 1977. The lateral meristem and its derivatives in the corm of *Isoetes muricata.* Amer. J. Bot. 64:310–325.

———. 1978. Structure and development of sieve elements in the root of *Isoetes muricata* Dur. Ann. Bot. 42:15–21.

Krupa, J. 1969. Photosynthetic activity and productivity

of the sporophyte of *Funaria hygrometrica* during ontogenesis. Acta Soc. Bot. Poloniae. 38:207–215.

Kubitzki, K., and **R. Borchert.** 1964. Morphologische Studien an *Isoetes triquetra* A. Braun und Bermerküngen über das Verhältnis der Gattung *Stylites* E. Amstutz zur Gattung *Isoetes* L. Ber. Deutsch Bot. Ges. 77:227–233.

Kumra, P. K., and **R. N. Chapra.** 1980. Occurrence of apogamy and apospory from the capsules of *Funaria hygrometrica* Hedw. Cryptogamme Bryologie Lichenologie 1:197–206.

Kurth, E. 1981. Mitotic activity in the root apex of the water fern *Marsilea vestita* Hook. and Grev. Amer. J. Bot. 68:881–896.

Kurz, W. G. W., and **T. A. La Rue.** 1971. Nitrogenase in *Anabaena flosaquae* filaments lacking heterocysts. Naturwiss. 58:417.

Labarca, C., and **F. Loewus.** 1973. The nutritional role of pistil exudate in pollen tube wall formation in *Lilium longiflorum.* II. Production and utilization of exudate from stigma and stylar canal. Plant Physiol. 52:87–92.

Ladha, J. K., and **H. D. Kumar.** 1978. Genetics of blue-green algae. Biol. Rev. 53:355–386.

Lal, M. 1961. *In vitro* production of apogamous sporangia in *Physcomirium coorgense* Broth. Phytomorphologia 11:263–269.

Lal, M., and **P. R. Bell.** 1975. Spermatogenesis in mosses. *In* J. G. Duckett and P. A. Racey, eds., Biology of the male gamete, pp. 85–93. Academic Press, New York.

Lal, M., and **E. Chauhan.** 1981. Transfer cells in the sporophyte gametophyte junction of *Physcomitrium cyathicarpum* Nutt. Protoplasma 107:79–83.

Lal, M., G. Kaur, and **E. Chauhan.** 1982. Ultrastructural studies on archegonial development in the moss *Physcomitrium cyathicarpum.* New Phytol. 92:441–452.

*Lamb, I. M.** 1959. Lichens. Sci. Amer. 201:144–156.

Lamont, B. B., and **R. A. Ryan.** 1977. Formation of corolloid roots by cycads under sterile conditions. Phytomorph. 27:426–429.

Lamont, H. C. 1969. Sacrificial cell death and trichome breakage in an oscillatoriacean blue-green alga: the role of murein. Arch. Mikrobiol. 68:257–259.

Lang, N. J. 1968. The fine structure of blue-green algae. Annual Rev. Microbiol. 22:15–46.

Lang, N. J., and **P. Fay.** 1971. The heterocysts of blue-green algae. II. Details of ulstrastructure. Proc. Roy. Soc. Lond. B178:193–203.

*Large, E. C.** 1940. The advance of the fungi. Holt, Rinehart and Winston, New York.

*Lawrence, G. H. M.** 1951. Taxonomy of vascular plants. Macmillan, New York.

*Lawrey, J. D.** 1984. Biology of lichenized fungi. Praeger Scientific. 407 pp.

Lazaroff, N. 1966. Photoinduction and photoreversal of the nostacacean development cycle. J. Phycol. 2:7–17.

Lazaroff, N., and **J. E. Scheff.** 1962. Action spectrum for development photoinduction of the blue-green alga *Nostoc muscorum.* Science 137:603–604.

Lazaroff, N., and W. Vishniac. 1961. The effect of light on the developmental cycle of *Nostoc muscorum,* a filamentous blue-green alga. J. Gen. Microbiol. 26:365–374.

———. 1964. The relationship of cellular differentiation to colonial morphogenesis of the blue-green alga *Nostoc muscorum.* J. Gen. Microbiol. 35:447–457.

Leclercq, S. 1969. *Calamophyton primaevum:* The complex morphology of its fertile appendage. Amer. J. Bot. 56:773–781.

Leclercq, S., and H.N. Andrews. 1960. *Calamophyton bicephalum,* a new species from the Middle Devonian of Belgium. Ann. Missouri Bot. Gard. 47:1–23.

Leclercq, S., and H. P. Banks. 1962. *Pseudosporochnus nodosus* sp. nov., a Middle Devonian plant with cladoxylalean affinities. Palaeontographica 110B:1–34.

Leclercq, S., and P. M. Bonamo. 1971. A study of the fructification of *Milleria (Protopteridium) thomsonii* Lang from the Middle-Devonian of Belgium. Palaeontographica 136B:83–114.

Leclercq, S., and K. M. Lele. 1968. Further investigations of the vascular system of *Pseudosporochnus nodosus* Leclercq et Banks. Palaeontographica 123B:97–110.

Leclercq, S., and H. J. Schweitzer. 1965. *Calamophyton* is not a sphenopsid. Acad. Roy. Belg. Ser. 5. t. 51:1395–1403.

Lee, C. L. 1955. Fertilization in *Ginkgo.* Bot. Gaz. 117:79–100.

Lee, R. E., and S. A. Fultz. 1970. Ultrastructure of the *Conchocelis* stage of the marine red alga *Porphyra leucosticta.* J. Phycol. 6:22–28.

*Leedale, G. F. 1967. Euglenoid flagellates. Prentice-Hall, Englewood Cliffs, N.J.

———. 1974. How many are the kingdoms of organisms? Taxon 23:261–270.

Leir, J. D. 1956. Mating reaction in yeast. Nature 177:753–754.

Lemoigne, Y. 1970. Nouvelles diagnoses du genre *Rhynia* et de l'espèce *Rhynia gwynne-vaughanii.* Bull. Soc. Bot. France 117:307–320.

*Levring, T., H. A. Hoppe, and O. J. Schmid. 1969. Marine algae. A survey of research and utilization. De Gruyter, Berlin and New York.

*Lewin, R. A., ed. 1962. Physiology and biochemistry of algae. Academic Press, New York.

———. 1975. A marine *Synechocystis* (Cyanophyta, Chroococcales) epizoic on ascidians. Phycologia 14:153–160.

———, ed. 1976. The genetics of algae. Bot. Monogr. 12. Blackwell Science Publications, Oxford.

———. 1977. *Prochloron,* type genus of the Prochlorophyta. Phycologia 16:217.

———. 1984. *Prochloron*—a status report. Phycologia 23:203–208.

*Lewis, J. R. 1964. The ecology of rocky shores. English Universities Press, London.

*Lewis, K. R. 1961. The genetics of Bryophytes. Trans. Brit. Bryol. Soc. 4:111–130.

Lewis, M. 1981. Human uses of bryophytes. I. Use of mosses for chinking log structures in Alaska. Bryologist 84:571–572.

Li, H. 1956. A horticultural and botanical history of *Ginkgo.* Bull. Morris Arboretum 7:3–12.

Lichtwardt, R. W. 1973a. Trichomycetes. *In* G. C. Ainsworth, F. K. Sparrow, and A. S. Sussman, eds., The fungi, vol. IVB, pp. 237–243. Academic Press, New York.

———. 1973b. The Trichomycetes: what are their relationships? Mycologia 65:1–20.

———. 1976. Trichomycetes. *In* E. B. G. Jones, ed., Recent advances in aquatic mycology, pp. 651–671. Elek Science, London.

Lill, B. S. 1976. Ovule and seed development in *Pinus radiata:* postmeiotic development, fertilization, and embryogeny. Canad. J. Bot. 54:2141–2154.

Lin, C. C., R. C. Sicher, and J. M. Aronson. 1976. Chitin and cellulose in the cell walls of the oomycete *Apodachlya* sp. Arch. Mikrobiol. 72:111–114.

Lin, C. K., and J. L. Blum. 1977. Recent invasion of a red alga (*Bangia atropurpurea*) in Lake Michigan. J. Fisheries Res. Board of Canada 34:2413–2416.

*Lindegren, C. C. 1949. The yeast cell, its genetics and cytology. Educational, St. Louis, Mo.

Linskens, H. F. 1975. The physiological basis of incompatibility in angiosperms. *In* J. G. Duckett and P. A. Racey, eds., Biology of the male gamete, pp. 143–152. Academic Press, New York.

Linskins, H. F., and J. Heslop-Harrison, eds. 1984. Cellular interactions. Encyclopedia of Plant Physiology. New Series. vol. 17, 743 pp. Springer-Verlag, Berlin, Heidelberg, New York and Tokyo.

*Lister, A. (Revised by G. Lister.) 1925. Mycotozoa. 3d ed. British Museum (Natural History), London.

Lloyd, R. M. 1974. Reproductive biology and evolution in the Pteridophyta. Ann. Mo. Bot. Garden 61:318–331.

Lloyd, R. M., and E. J. Klekowski, Jr. 1970. Spore germination and viability in Pteridophyta: evolutionary significance of chlorophyllous spores. Biotropica 2:129–137.

*Lobban, S., and M. J. Wynne, eds. 1981. The biology of seaweeds. Bot. Monograph 17. Blackwell Scientific Publications, Oxford, London, Edinburgh, Boston, Melbourne. 786 pp.

Loeblich, A. R., III. 1966. Aspects of the physiology and biochemistry of Pyrrhophyta. Phykos, Iyengar Memorial Volume 5:216–255.

Loiseaux, S., and J. A. West. 1970. Brown algal mastigonemes: comparative ultrastructure. Trans. Amer. Microbiol. Soc. 89:524–532.

Long, A. G. 1966. Some Lower Carboniferous fructifications from Berwickshire, together with a theoretical account of the evolution of ovules, cupules and carpels. Trans. Roy. Soc. Edinburgh 66:345–375.

———. 1977. Lower Carboniferous pteridosperm

cupules and the origin of angiosperms. Trans. Roy. Soc. Edinburgh 70:13–35.

Lorenzen, H., U. Kaiser, and **M. Foerster.** 1981. Intensives Wachstum von *Ricciocarpus natans* (Lebermoos) in Durchluftungs-Kultur. Ber. Deutsch. Bot. Gesell. 94:719–725.

Loveland, H. 1956. Sexual dimorphism in the moss genus *Dicranum* Hedw. Ph.D. dissertation, University of Michigan.

Lowy, B. 1971. New records of mushroom stones from Guatemala. Mycologia 63:983–993.

———. 1974. *Amanita muscaria* and the thunderbolt legend in Guatemala and Mexico. Mycologia 66:188–190.

———. 1977. Hallucinogenic mushrooms in Guatemala. J. Psychedelic Drugs 9:123–125.

Lu, B. C. 1974. Meiosis in *Coprinus*. V. The role of light in basidiocarp initiation, mitosis, and hymenium differentiation in *Coprinus lagopus*. Can. J. Bot. 52:299–305.

Lucas, R. C., and **J. G. Duckett.** 1980. A cytological study of the male and female sporocarps of the heterosporous fern *Azolla filiculoides* Lam. New Phytol. 85:409–418.

Lyon, F. M. 1901. A study of the sporangia and gametophytes of *Selaginella apus* and *Selaginella rupestris*. Bot. Gaz. 37:124–141; 170–194.

———. 1904. The evolution of sex organs in plants. Bot. Gaz. 37:280–293.

McAlpin, B., and **D. R. Farrar.** 1978. *Trichomanes* gametophytes in Massachusetts. Amer. Fern J. 68:97–98.

McCormick, J. J., J. C. Blomquist, and **H. R. Rusch.** 1970. Isolation and characterization of a galactos-amine wall from spores and spherules of *Physarum polycephalum*. J. Bacteriol. 104:1119–1125.

McCracken, M. D., and **R. C. Starr.** 1970. Induction and development of reproductive cells in the K-32 strains of *Volvox rousseletii*. Arch. Protistenk. 112:262–282.

McCully, M. E. 1966–1968. Histological studies on the genus *Fucus*. I. 1966. Light microscopy of the mature vegetative plant. Protoplasma 62:287–305. II. 1968. Histology of the reproductive tissues. Protoplasma 66:205–230.

*McElhenney, T. R., H. C. Bold, R. M. Brown, Jr.,** and **J. P. McGovern.** 1962. Algae: a cause of inhalant allergy in children. Ann. Allergy 20:739–743.

Machlis, L. 1958. Evidence for a sexual hormone in *Allomyces*. Physiol. Plantarum 11:185–192.

———. 1972. The coming of age of sex hormones in plants. Mycologia 64:235–247.

Machlis, L., and **E. Rawitscher-Kunkel.** 1967a. The hydrated megaspore of *Marsilea vestita*. Amer. J. Bot. 54:689–694.

*———. 1967b. Mechanisms of gametic approach in plants. *In* C. B. Metz and A. Monroy, eds.,

Fertilization, vol. 1, chap. 4. Academic Press, New York.

McLachlan, J. L., C. M. Chen, and **T. Edelstein.** 1971. The culture of four species of *Fucus* under laboratory conditions. Canad. J. Bot. 49:1463–1469.

McMorris, T. C., and **A. W. Barksdale.** 1967. Isolation of a sex hormone from the water mold *Achlya bisexualis*. Nature 215:320–321.

McWilliam, J. P. 1958. The role of the micropyle in the pollination of *Pinus*. Bot. Gaz. 120:109–117.

Magne, F. 1961. Sur le cycle cytologique de *Nemalion helminthoides* (Velley) Batters. Comp. Rend. Acad. Sci. Paris 252:157–159.

Maguire, M. P. 1976. Mitotic and meiotic behavior of the chromosomes of the octet strain of *Chlamydomonas reinhardtii*. Genetica 46:479–502.

Maheshwari, P., and **N. S. Rangaswamy.** 1965. Embryology in relation to physiology and genetics. Advanc. Bot. Res. 2:219–321.

Maheshwari, P., and **H. Singh.** 1967. The female gametophyte. Biol. Rev. 42:88–130.

Maheshwari, P., and **V. Vasil.** 1961a. The stomata of *Gnetum*. Ann. Bot. 25:313–319.

———. 1961b. *Gnetum*. Botanical Monograph 1, Council of Scientific and Industrial Research, New Delhi.

Maier, K. 1967. Dehiscence of the moss capsule. I. Die Ablösung des Annulus von Kapsel; dargestellt an *Funaria hygrometrica* L. Österr. Bot. Zeit. 114:51–65.

———. 1973a. II. The annulus: analysis of its functional apparatus. Österr. Bot. Zeit. 122:75–98.

———. 1973b. Annulus function and the lid stability: a study with light and scanning electron microscope. Österr. Bot. Zeit. 122:99–114.

———. 1973c. IV. Prinzipien der Annulusfunktion: Teil I. Österr. Bot. Zeit. 122:237–257.

Major, R. T. 1967. The *Ginkgo*, the most ancient living tree. Science 157:1270–1273.

Mamay, S. H. 1969. Cycads: fossil evidence of late Paleozoic origin. Science 164:295–296.

———. 1976. Paleozoic origin of the cycads. U.S. Geol. Survey Prof. Paper 934:1–48.

*Manton, I.** 1950. Problems of cytology and evolution in the Pteridophyta. Cambridge University Press, London.

Marchant, C. J. 1968. Chromosome patterns and nuclear phenomena in the cycad families Stangeriaceae and Zamiaceae. Chromosoma 24:100–134.

Marchant, H. J., and **J. D. Pickett-Heaps.** 1970–1972. Ultrastructure and differentiation of *Hydrodictyon reticulatum*. I. 1970. Mitosis in the coenobium. Austral. J. Biol. Sci. 23:1173–1186. II. 1971. Formation of zoids within the coenobium. Austral. J. Biol. Sci. 24:471–486. III. 1972a. Formation of the vegetative daughter net. Austral. J. Biol. Sci. 25:265–278. IV. 1972b. Conjugation of gametes and the development of zygospores and azygospores. Austral. J. Biol. Sci. 25:279–291. V. 1972c. Development of polyhedra. Austral. J. Biol. Sci. 25:1187–1197. VI. 1972d.

866

Formation of the germ net. Austral. J. Biol. Sci. 25:1199–1213.

———. 1973. Mitosis and cytokinesis in *Coleochaete scutata*. J. Phycol. 9:461–471.

Margulis, L. 1968. Evolutionary criteria in the Thallophytes: a radical alternative. Science 161:1020–1022.

———. 1969. New phylogenies of the lower organisms: possible relation to organic deposits in precambrian sediment. J. Geol. 77:606–617.

*———. 1970. Origin of eukaryotic cells. Yale University Press, New Haven, Conn.

———. 1975. Symbiotic theory of the origin of eukaryotic cells: criteria for proof. Sympos. Soc. Exp. Biol. 29:21–38.

*———. 1980. Evolution of cells. Harvard University Press, Cambridge, Mass. In press.

Margulis, L., L. To, and D. Chase. 1978. Microtubules in prokaryotes. Science 200:1118–1124.

Martens, P. 1959. Études sur les Gnétales III. Structure et ontogenèse du cone et de la fleur femelles de *Welwitschia mirabilis*. La Cellule 50:169–286.

———. 1961. Études sur les Gnétales V. Structure et ontogenèse du cone et de la fleur mâles de *Welwitschia mirabilis*. La Cellule 62:6–91.

———. 1963. Études sur les Gnétales VI. Recherches sur *Welwitschia mirabilis*. III. L'ovule et le sac embryonnaire. La Cellule 63:308–329.

———. 1964. Études sur les Gnétales VII. Recherches sur *Welwitschia mirabilis*. IV. Germination et plantules. La Cellule 55:6–68.

*———. 1971. Les Gnétophytes. Borntraeger, Berlin.

———. 1977. *Welwitschia mirabilis* and neoteny. Amer. J. Bot. 64:916–920.

*Martin, G. W., and C. J. Alexopoulos. 1969. The Myxomycetes. University of Iowa Press.

———, C. J. Alexopoulos, and M. L. Farr. The genera of Myxomycetes. University of Iowa Press, Iowa City.

Mascarenhas, J. P., and L. Machlis. 1962a. The hormonal control of the directional growth of pollen tubes. Vitamins and Hormones 20:347–372.

———. 1962b. The pollen-tube chemotropic factor from *Antirrhinum majus*: bioassay, extraction and partial purification. Amer. J. Bot. 49:482–489.

Massalski, A., and G. F. Leedale. 1969. Cytology and ultrastructure of the Xanthophyceae. I. Comparative morphology of the zoospores of *Bumilleria sicula* Borzi and *Tribonema vulgare* Pascher. Brit. Phycol. J. 4:159–180.

Masuda, M., and I. Umezaki. 1977. On the life history of *Nemalion vermiculare* Suringar (Rhodophyta) in culture. Bull. Jap. Soc. Phycol. 25. Supplement (Mem. Iss. Yamada):129–136.

Mattox, K. R., and K. D. Stewart. 1984. The classification of the green algae: a concept based on comparative cytology. *In* D. E. G. Ironie and D. M. John, eds., Systematics of the green algae. Academic Press, London.

Matzke, E. B., and L. Raudzens. 1968. Aposporous diploid gametophytes from sporophytes of the liverwort *Blasia pusilla* L. Proc. Nat. Acad. Sci. 59:752–755.

Meeuse, A. D. J. 1962. The multiple origin of the angiosperms. Advanc. Frontiers Plant Sci. (New Delhi) 1:105–127.

———. 1963a. Some phylogenetic aspects of the process of double fertilization. Phytomorphology 13:237–244.

———. 1963b. The so-called "megasporophyll" of *Cycas*-a morphological misconception. Its bearing on the phylogeny and the classification of the Cycadophyta. Acta Bot. Neerland. 12:119–128.

———. 1963c. From ovule to ovary: a contribution to the phylogeny of the megasporangium. Acta Biotheoretica 16:127–182.

———. 1964. The bitegmic spermatophytic ovule and the cupule: a re-appraisal of the so-called pseudomomerous ovary. Acta Bot. Neerland. 13:97–112.

———. 1965. Angiosperms—past and present. Advanc. Frontiers Plant Sci. (New Delhi) 11:1–228.

Mehra, P. N. 1968. Phyletic evolution in the Hepaticae. Phytomorphology 17:47–58.

———. 1969. Evolutionary trends in Hepaticae with particular reference to the Marchantiales. Phytomorphology 19:203–218.

Mehra, P. N., and D. S. Sulkyan. 1969. *In vitro* studies on apogamy, apospory and controlled differentiation of rhizome segments of the fern *Ampelopteris prolifera* (Retz.) Copel. J. Linn. Soc. Lond. (Bot.) 62:431–443.

Melvin, J. F., and C. M. Stewart. 1969. The chemical composition of the wood of *Gnetum gnemon*. Holzforshung 23:51–56.

Menon, M. K. C., and M. Lal. 1974. Morphogenetic role of kinetin and abscisic acid in the moss *Physcomitrum*. Planta 115:319–328.

———. 1977. Regulation of a subsexual life cycle in a moss: evidence for the occurrence of a factor for apogamy in *Physcomitrium*. Amer. Bot. 41:1179–1190.

Mesland, D. A. M. 1976. Mating in *Chlamydomonas eugametos*: a scanning electron microscopical study. Arch. Microbiol. 109:31–35.

Mesler, M. R. 1973. Sexual reproduction in *Ophioglossum crotalophoroides*. Amer. Fern J. 63:28–33.

———. 1974. The natural history of *Ophioglossum palmatum* in South Florida. Amer. Fern J. 64:33–40.

———. 1975. The gametophytes of *Ophioglossum palmatum*. Amer. J. Bot. 62:982–992.

———. 1976. Gametophytes and young sporophytes of *Ophioglossum crotalophoroides*. Amer. J. Bot. 63:443–448.

Mesler, M. R., and K. L. Lu. 1977. Large gametophytes of *Equisetum hyemale* in northern California. Amer. Fern J. 67:97–98.

Metting, B. 1981. The systematics and ecology of soil algae. Bot. Rev. 47:195–312.

Meyer, V. G. 1970. A facultative gymnosperm from an interspecific cotton hybrid. Science 169:886–888.

Michanek, G. 1975. Seaweed resources of the ocean. FAO Fisheries Tech. Paper, No. 138, pp. v and 127. UN Food and Agricultural Organization, Rome.

Mickel, J. T. 1974. Phyletic lines in the modern ferns. Ann. Mo. Bot. Garden 61:474–482.

Millay, M. A., and D. A. Eggert. 1970. *Idanothekion* gen. n.: a synangiate pollen organ with saccate pollen from the Middle Pennsylvanian of Illinois. Amer. J. Bot. 57:50–61.

Millbank, J. W. 1974. Associations with blue-green algae. *In* A. Quispel, ed., The biology of nitrogen fixation. pp. 238–264. American Elsevier, New York.

Miller, C. E. 1967. Isolation and pure culture of aquatic Phycomycetes by membrane filtration. Mycologia 59:524–527.

———. 1977. A developmental study with the SEM of sexual reproduction in *Chytriomyces hyalinus.* Bull. Soc. Bot. France 124:281–289.

Miller, C. J., and J. G. Duckett. 1979. A study of stelar ultrastructure in the heterosporous water fern *Marsilea quadrifolia* L. Ann. Bot. 44:231–238.

Miller, H. A. 1974. Rhyniophytina, alternation of generations and the evolution of Bryophytes. J. Hattori Bot. Lab. 38:161–168.

———. 1979. The phylogeny and distribution of Musci. *In* G. C. S. Clarke and J. G. Duckett, eds., chap. 2.

———. 1982. Bryophyte evolution and geography. Biol. J. Linn. Soc. 18:145–196.

Miller, J. H. 1968. Fern gametophytes as experimental material. Bot. Rev. 34:361–440.

Miller, J. H., and P. M. Miller. 1961. The effect of different light conditions and sucrose on the growth and development of the gametophyte of the fern *Onoclea sensibilis.* Amer. J. Bot. 48:154–159.

Miller, N. G., and L. J. H. Ambrose. 1976. Growth in culture of wind-blown bryophyte gametophyte fragments from Arctic Canada. Bryologist 79:55–63.

*Milliger, L. E., K. W. Stewart, and J. K. G. Silvey. 1971. The passive dispersal of viable algae, protozoans, and fungi by aquatic and terrestrial Coleoptera. Ann. Entomol. Soc. Amer. 64:36–45.

Mitchell, J. C., W. B. Schofield, B. Singh, and G. H. N. Towers. 1969. Allergy to *Frullania.* Arch. Dermatol. 100:46–49.

Mizukami, I., and J. Gall. 1966. Centriole replication. II. Sperm formation in the fern *Marsilea* and the cycad *Zamia.* J. Cell Biol. 29:97–111.

Moestrop, O. 1970a. The fine structure of mature spermatids of *Chara corallina,* with special reference to microtubules and scales. Planta 93:295–308.

———. 1970b. On the fine structure of the spermatozoids of *Vaucheria sesculpicaria* and in the later stages of spermatogenesis. J. Marine Biol. Assoc. U.K. 50:513–523.

Mohr, H. 1963. The influence of visible radiation in the germination of archegoniate spores and the growth of fern protonema. J. Linn. Soc. Lond. (Bot.) 58:287–296.

Moikeha, S. N., and G. W. Chu. 1971. Dermatitis-producing alga *Lyngbya majuscula* Gomont in Hawaii. II. Biological properties of the toxic factor. J. Phycol. 7:8–13.

Moikeha, S. N., G. W. Chu, and L. R. Berger. 1971. Dermatitis-producing alga *Lyngbya majuscula* Gomont in Hawaii. I. Isolation and chemical characterization of the toxic factor. J. Phycol. 7:4–8.

Monroe, J. H. 1965. Some factors evoking formation of sex organs in *Funaria.* Bryologist 68:337–339.

———. 1968. Light and electron microscopic observations on spore germination in *Funaria hygrometrica.* Bot. Gaz. 129:247–258.

Moore, E. D., and C. E. Miller. 1973. Resting body formation by rhizoidal fusion in *Chytriomyces hyalinus.* Mycologia 65:145–154.

Moore, R. E. 1977. Toxins from blue-green algae. BioScience 27:197–802.

Morel, G. 1963. Leaf regeneration in *Adiantum pedatum.* J. Linn. Soc. Lond. (Bot.) 58:381–383.

Mortimer, M. G., and W. G. Chaloner. 1972. The palynology of the concealed Devonian rocks of southern England. Bull. Geol. Surv. Great Britain 39:1–56.

Moser, J. W., J. G. Duckett, and G. Carothers. 1977. Ultrastructural studies of spermatogenesis in the Anthocerotales. I. The blepharoplast and anterior mitochondrion in *Phaeoceros laevis:* early development. Amer. J. Bot. 64:1097–1106. II. The blepharoplast and anterior mitochondrion in *Phaeoceros laevis:* later development. Amer. J. Bot. 64:1107–1116.

Moss, S. T., and R. W. Lichtwardt. 1977. Zygospores of the Harpellales: an ultrastructural study. Can. J. Bot. 55:3099–3110.

Moussel, B. 1978. Double fertilization in the genus *Ephedra.* Phytomorph. 28:336–345.

———. 1983. Données temporelles concernant les diverses phases du cycle de reproduction sexuée l' *Ephedra distachya* L. Rev. de Cytol. et de Biol. Vey. Le Botaniste 6:103–109.

Mueller, D. J. 1972. Observations on the ultrastructure of *Buxbaumia* protonema. Plasmodesmata in the cross walls. Bryologist 75:63–68.

———. 1973. The peristome of *Fissidens limbatus.* Sullivan University of California Publications in Botany 63.

———. 1974. Spore wall formation and chloroplast development during sporogenesis in the moss *Fissidens limbatus.* Amer. J. Bot. 61:525–534.

Mueller, R. J. 1982a. Shoot morphology of the climbing fern *Lygodium* (Schizeaaceae): general organography, leaf initiation, and branching. Bot. Gaz. 143:319–330.

———. 1982b. Shoot ontogeny and the comparative root development of the heteroblastic leaf series in *Lygodium japonicum* (Thunb.) Sw. Bot. Gaz. 143:424–438.

———. 1983a. Indeterminate growth and ramification of

the climbing leaves of *Lygodium japonicum* (Schizaeaceae). Amer. J. Bot. 70:682–690.

———. 1983*b*. Indeterminate growth of the climbing leaves of a fern. Bio. Sci. 33:586–587.

Mues, R., and **H. D. Zinsmeiste.** 1978. Studies on phenolic constituents of Jungermanniales in relation to their taxonomy. Congress Int. de Bryologie, Bordeaux. Bryophytarum Biblio. 13:399–409.

Muhammad, A. F., and **R. Sattler.** 1982. Vessel structure of *Gnetum* and the origin of angiosperms. Amer. J. Bot. 69:1004–1021.

Müller, D. G. 1966. Untersuchungen zur Entwicklungsgeschichte der Braunalge *Ectocarpus siliculosus* aus Neapel. Planta 68:57–68.

———. 1967. Generationswechsel, Kernphasenwechsel und Sexualität der Braunalge *Ectocarpus siliculosus* in Kulturversuch. Planta 75:39–54.

———. 1977. Sexual reproduction in British *Ectocarpus siliculosus* (Phaeophyta). Brit. Phycol. J. 12:131–136.

Müller, D. G., L. Jaenicke, N. Donike, and **J. Akintobi.** 1971. Sex attractant in a brown alga: chemical structure. Science 171:815–817.

Müller, D. G., and **K. Seferiadis.** 1977. Specificity of sexual chemotaxis in *Fucus serratus* and *Fucus vesiculosus* (Phaeophyceae). Z. Pflanzenphysiol. 84:85–94.

*Müller, E.,** and **W. Loeffler.** 1976. Mycology. Trans. B. Kendrick and F. Bärlocher. Georg Theme, Stuttgart.

Muller, J. 1979. Form and function in angiosperm pollen. Ann. Missouri Bot. Gard. 66:593–632.

*Muller, K.** 1940. Die Lebermoose. Rabenhorst's Kryptogamenflora VI, Leipzig.

Muniyamma, M. 1977. Triploid embryos from endosperm *in vivo*. Ann. Bot. 41:1077–1079.

Munroe, M. H., and **I. M. Sussex.** 1969. Gametophyte formation in bracken fern root cultures. Canad. J. Bot. 47:617–621.

Munther, W. E., and **D. E. Fairbrothers.** 1980. Allelopathy and autotoxicity in three eastern North American ferns. Amer. Fern J. 70:124–135.

Muscatine, L. 1971. Experiments on green algae coexistent with zooxanthellae in sea anemones. Pacific Sci. 25:13–21.

Muscatine, L., and **J. W. Porter.** 1977. Reef corals: mutualistic symbioses adapted to nutrient-poor environments. BioScience 27:454–460.

Myles, D. G. 1975. Structural changes in the sperm of *Marsilea vestita* before and after fertilization. *In* J. G. Duckett and P. A. Racey, eds., Biology of the male gamete, pp. 129–134. Academic Press, New York.

Myles, D. G., and **Hepler, P. K.** 1977. Spermiogenesis in the fern *Marsilea*: microtubules, nuclear shaping, and cytomorphogenesis. J. Cell Sci. 23:57–83.

Myles, D. G., D. Southworth, and **P. K. Hepler.** 1978. Cell surface topography during *Marsilea* spermiogenesis: flagellar reorientation and membrane particle arrays. Protoplasma 93:405–417.

Mynderse, J. S., R. E. Moore, M. Kashiwagi, and **T. R. Morton.** 1977. Antileukemia in the Oscillatoreaceae: isolation of Debromoaplysiatoxin from *Lyngbya*. Science 196:538–540.

Näf, U. 1959. Control of antheridium formation in the fern species *Anemia phyllitidis*. Nature 184:798–800.

———. 1960. On the control of antheridium formation in the fern species *Lygodium japonicum*. Proc. Soc. Exp. Biol. Med. 105:82–86.

———. 1962. Developmental physiology of the lower archegoniates. Annual Rev. Plant Physiol. 13:507–532.

———. 1968. On the separation and identity of fern antheridiogens. Plant and Cell Physiol. 9:27–33.

———. 1969. On the control of antheridium formation in ferns. *In* J. E. Gunckel, ed., Current topics in plant science, pp. 97–116. Academic Press, New York.

Näf, U., K. Nakanishi, and **M. Endo.** 1975. On the physiology and chemistry of fern antheridiogen. Bot. Rev. 41:315–359.

Näf, U., and **V. Raghavan.** 1976. Cytolinin induced bud formation in *Microdus miquelianus*. Bryologist 79:495–499.

Näf, U., J. Sullivan, and **M. Cummins.** 1969. New antheridiogen from the fern *Onoclea sensibilis*. Science 163:1357–1358.

Nakanishi, K., M. Endo, U. Näf, and **L. F. Johnson.** 1971. Structure of the antheridium-inducing factor of the fern *Anemia phyllitides*. J. Amer. Chem. Soc. 93:5579–5581.

Namboodiri, K. K., and **C. B. Beck.** 1968. A comparative study of the primary vascular system of conifers. Amer. J. Bot. 55:447–457.

Nayar, B. K., and **S. Kaur.** 1971. The gametophytes of homosporous ferns. Bot. Rev. 37:295–396.

Nebel, B. J., and **A. W. Naylor.** 1968*a*. Initiation and development of short buds from protonemata in the moss *Physcomitrium*. Amer. J. Bot. 55:33–37.

———. 1968*b*. Light, temperature and carbohydrate requirements for short bud initiation from protonemata in the moss *Physcomitrium turbinatum*. Amer. J. Bot. 55:38–44.

Neckelmann, N., and **L. Muscatine.** 1983. Regulatory mechanisms maintaining the *Hydra-Chlorella* symbiosis. Proc. Roy. Soc. B 219:193–210.

Nehlsen, W. 1979. A new method for examining induction of moss buds by cytokinin. Amer. J. Bot. 66:601–603.

Neilson, A., R. Rippka, and **R. Kunisawa.** 1971. Heterocyst formation and nitrogenase synthesis in *Anabaena* sp. Arch. Mikrobiol. 76:139–150.

Neuberger, D., and **R. F. Evert.** 1974. Structure and development of the sieve-element protoplast in the hypocotyl of *Pinus resinosa*. Amer. J. Bot. 61:360–374.

Neuberger, D. S., and **R. F. Evert.** 1975. Structure and development of sieve areas in the hypocotyl of *Pinus resinosa*. Protoplasma 84:109–125.

———. 1976. Structure and development of sieve cells in the primary phloem of *Pinus resinosa*. Protoplasma 87:27–37.

Neuburg, M. F. 1960. Beblätterte Moose aus den Perm

des Angara Beckens. (Russian). Trud. Geol. Inst. Akad. Nauk. U.S.S.R. 19:1–104.

Newton, M. E. 1972. An investigation of photoperiod and temperature in relation to the life cycles of *Mnium hornum* Hedw. and *M. undulatum.* Sw. (Musci) with reference to their histology. Bot. J. Linn. Soc. Lond. 65:189–209.

Nicholson, N. L., and W. R. Briggs. 1972. Translocation of photosynthate in the brown alga *Nereocystis.* Amer. J. Bot. 50:97–106.

Niklas, K. J. 1979. Simulations of apical developmental sequences in bryophytes. Ann. Bot. 44:339–352.

——, and Kyaw Tha Paw U. 1983. Conifer ovulate cone morphology: implications on pollen impaction patterns. Amer. J. Bot. 70:568–577.

Niklas, K. J., and T. L. Phillips. 1976. Morphology of *Protosalvinia* from the Upper Devonian of Ohio and Kentucky. Amer. J. Bot. 63:9–29.

Nishida, Y. 1978. Studies on the sporeling types in mosses. J. Hattori Bot. Lab. 44:371–454.

Nitayankura, S., E. M. Gifford, Jr., and T. L. Rost. 1980. Mitotic activity in the root apical meristem of *Azolla filiculoides* Lam. with special reference to the apical cell. Amer. J. Bot. 67:1484–1492.

Norstog, K. 1972. Role of archegonial neck cells of *Zamia* and other cycads. Phytomorphology 22:125–130.

——. 1974. Fine structure of the spermatozoid of *Zamia:* the Vierergruppe. Amer. J. Bot. 61:449–456.

——. 1975. The motility of cycad spermatozoids in relation to structure and function. *In* J. G. Duckett and P. A. Racey, eds., Biology of the male gamete, pp. 135–142. Academic Press, New York.

——. 1977. The spermatozoid of *Zamia chigua* Sum. Bot. Gaz. 138:409–412.

——. 1980. Chromosome numbers in *Zamia* (Cycadales). Caryologica 33:419–428.

——. 1982a. Some aspects of spermatogenesis in *Zamia.* Phyta, Studies on living and fossil plants, Pant Comm. Vol. 1982:199–206.

——. 1982b. Experimental embryology of gymnosperms. *In* B. M. Johri, ed., 1982.

Norstog, K., and R. Overstreet. 1965. Some observations on the gametophytes of *Zamia integrifolia.* Phytomorphology 15:46–49.

Norstog, K., and E. Rhamstine. 1967. Isolation and cultivation of haploid and diploid cycad tissues. Phytomorphology 117:374–381.

*North, W. J., ed. 1971. The biology of giant kelp beds (*Macrocystis*) in California. Beih. z. Nova Hedwigia 32:123–168.

Nozu, Y. 1961. The gametophyte of *Helimithostachys zeylonica* and *Ophioglossum vulgatum.* Phytomorphology 11:199–206.

Nutting, W. H., H. Rapoport, and L. Machlis. 1968. The structure of sirenin. J. Amer. Chem. Soc. 90:6434–6438.

Nygren, A. 1967. Apomixis in angiosperms. Handbuch der Pflanzenphys. 18:551–96.

Obukowicz, M., M. Schaller, and G. S. Kennedy. 1981. Ultrastructure and phenolic histochemistry of the *Cycas revoluta-Anabaena* symbiosis. New Phytol. 87:751–759.

Oehler, D. Z. 1976. Transmission electron microscopy of organic microfossils from the late Precambrian Bitter Springs Formation of Australia: techniques and survey of preserved ultrastructure. J. Paleontol. 50:90–106.

Oehler, J. H., D. Z. Oehler, and M. D. Muir. 1976. On the significance of tetrahedral tetrads of Precambrian algal cells. Origins of Life 7:259–267.

Ogura, Y. 1967. History of discovery of spermatozoids in *Ginkgo biloba* and *Cycas revoluta.* Phytomorphology 17:109–114.

*Olive, L. S. 1956. Genetics of *Sordaria fimicola.* I. Ascospore color mutants. Amer. J. Bot. 43:97–106.

*——. 1970. The Mycetozoa: a revised classification. Bot. Rev. 36:59–87.

*——. 1975. The Mycetozoans. Academic Press, New York.

*Oparin, A. I. 1953. The origin of life. 2d ed. Macmillan, New York.

Orkwizewski, K. G., and A. R. Kaney. 1974. Genetic transformation of the blue-green bacterium *Anacystis nidulans.* Arch. Mikrobiol. 98:31–37.

Owens, J. N. 1973. Reproductive patterns in conifers. Proc. of all Union Symposium of sexual reproduction of conifers. Acad. Sci. USSR, Siberian branch, Novosibirsk 1:46–49.

Owens, J. N., and M. Molder. 1971. Meiosis in conifers: prolonged pachytene and difuse diplotene stages. Can. J. Bot. 49:2061–2064.

——. 1974. Cone initiation and development before dormancy in yellow cedar (*Chamaecyparis nootkatensis*). Can. J. Bot. 52:2075–2084.

——. 1975a. Pollination, female gametophyte, and embryo and seed development in yellow cedar (*Chamaecyparis nootkatensis*). Can. J. Bot. 53:186–199.

——. 1975b. Sexual reproduction in mountain hemlock (*Tsuga mertensiana*). Can. J. Bot. 53:1811–1826.

——. 1977a. Seed cone differentiation and sexual reproduction in the western white pine (*Pinus monticola*). Can. J. Bot. 55:2574–2590.

——. 1977b. Sexual reproduction in *Abies amabilis.* Can. J. Bot. 55:2653–2667.

——. 1979a. Sexual reproduction in white spruce (*Picea glauca*). Can. J. Bot. 57:152–169.

——. 1979b. Bud development in *Larix occidentalis.* II. Cone differentiation and early development. Can. J. Bot. 57:1557–1572.

——. 1979c. Sexual reproduction of *Larix occidentalis.* Can. J. Bot. 57:2673–2690.

——. 1980a. Sexual reproduction of Sitka spruce (*Picea sitchensis*). Can. J. Bot. 58:886–901.

——. 1980b. Sexual reproduction in western red cedar (*Thuja plicata*). Can. J. Bot. 58:1376–1393.

Owens, J. N., and A. J. Simpson. 1982. Further observations on the pollination mechanism and seed production of Douglas fir. Can. J. Forest Res. 12:431–434.

Owens, J. N., A. J. Simpson, and M. Molder. 1980. The pollination mechanism in yellow cypress (*Chamaecyparis nootkatensis*). Can. J. Forest Res. 10:564–572.

———. 1981a. Sexual reproduction of *Pinus contorta*. I. Pollen development, the pollination mechanism and early ovule development. Can. J. Bot. 59:1828–1843.

———. 1981b. Postdormancy ovule, embryo and seed development. Can. J. Bot. 60:2071–2083.

———. 1981c. The pollen mechanism and the optimal time of pollination in Douglas fir (*Pseudotsuga menziesii*). Can. J. Forest Res. 11:36–50.

Owens, J. N., and H. Singh. 1981. Vegetative bud development and the time and method of cone initiation in subalpine fir (*Abies lasiocarpa*). Can. J. Bot. 60:2249–2262.

Page, C. N. 1972. An assessment of interspecific relationships in *Equisetum* subgenus *Equisetum*. New Phytol. 71:355–369.

———. 1973. An interpretation of the morphology and evolution of the cone and shoot of *Equisetum*. Bot. J. Linn. Soc. Lond. 65:359–377.

Page, R. M. 1964. Sporangium discharge in *Pilobolus*. Science 146:925–927.

Palenzona, M. 1969. Sintesi micorrizica tra "*Tuber aestivum*" Vitt., "*Tuber brumale*" Vitt., "*Tuber melanosporum*" Vitt. e semenzali di "*Corylus avellana*" L. Allionia 15:121–131.

Palenzona, M., G. Chevalier, and A. Fontana. 1972. Sintesi micorrhizica tra i miceli in coltura di "Tuber brumale," "T. melanosporum," "T. rufum" e semenzali di conifere e latifoglie. Allionia 18:41–52.

Paliwal, G. S., and H. D. Behnke. 1973. Light microscopic study of the organization of phloem in the stem of *Gnetum gnemon*. Phytomorphologia 23:183–193.

*Palmer, C. M. 1959. Algae in water supplies. U.S. Public Health Service Publication No. 657, Washington, D.C.

Pant, D. D. 1962. The gametophyte of the Psilophytales. Proc. Summer School Bot. Darjeeling June 2–15, 1960, pp. 276–301.

Paolillo, D. 1963. Developmental anatomy of *Isoetes*. University of Illinois Press.

———. 1968. The effect of the calyptra on capsule symmetry in *Polytrichum juniperinum* Hedw. Bryologist 71:327–334.

Paolillo, D. J., Jr. 1964. The plastids of *Polytrichum commune*. I. The capsule at meiosis. Protoplasma 58:667–680.

———. 1967. On the structure of the axoneme in flagella of *Polytrichum juniperinum*. Trans. Amer. Microbiol. Soc. 86:428–433.

———. 1969. The plastids of *Polytrichum*. II. The sporogenous cells. Cytologia 34:133–144.

———. 1975. The release of sperms from the antheridia of *Polytrichum juniperinum*. Hedw. New Phytol. 74:287–293.

———. 1977a. On the release of sperms in *Atrichum*. Amer. J. Bot. 64:81–85.

———. 1977b. Release of sperms in *Funaria hygrometrica*. Bryologist 80:619–629.

———. 1981 The swimming sperms of land plants. Bio. Sci. 31:367–373.

Paolillo, D. J., Jr., and F. A. Bazzaz. 1968. Photosynthesis in sporophytes of *Polytrichum* and *Funaria*. Bryologist 71:335–343.

Paolillo, D. J., Jr., and M. Cukierski. 1976. Wall developments and coordinated cytoplasmic changes in spermatogenous cells of *Polytrichum* (Musci). Bryologist 79:466–479.

Paolillo, D. J., Jr., and R. H. Jagels. 1969. Photosynthesis and respiration in germinating spores of *Polytrichum*. Bryologist 72:444–451.

Paolillo, D. J., Jr., and L. B. Kass. 1973. The germinability of immature spores in *Polytrichum*. Bryologist 76:163–168.

———. 1977. The relationship between cell size and chloroplast number in the spores of a moss, *Polytrichum*. J. Exper. Bot. 28:457–467.

Paolillo, D. J., Jr., G. L. Kreitner, and J. A. Reighard. 1968. Spermatogenesis in *Polytrichum juniperinum*. I and II. Planta 78:226–261.

Paolillo, D. J., Jr., and M. E. Payne. 1970. Carbon dioxide exchange during spore germination in *Polytrichum*. Bryologist 73:654–661.

Papenfuss, G. F. 1946. Proposed names for the phyla of algae. Bull. Torrey Bot. Club 73:217–218.

*———. 1955. Classification of the algae. Century Progress Nat. Sci. 1853–1953 (pp. 115–224). California Academy of Science, San Francisco.

*Parihar, N. S. 1965. An introduction to embryophyta: I. Bryophyta. 5th rev. ed. Central Book Depot, Allahabad.

Parker, B. C. 1970a. Significance of cell wall chemistry to phylogeny in the algae. Ann. N.Y. Acad. Sci. 175:417–428.

———. 1970b. Life in the sky. Natur. Hist. 79:54–59.

Parker, B. C., and E. Y. Dawson. 1965. Noncalcareous marine algae from California Miocene deposits. Nova Hedwigia 10:273–295.

Paton, J., and J. R. Pearce. 1957. The occurrence, structure and functions of the stomata in British Bryophytes. Trans. Brit. Bryol. Soc. 3:228–259.

Patrick, R., and C. W. Reimer. 1966. The diatoms of the United States exclusive of Alaska and Hawaii, vol. 1. Monograph 13, Academy of Natural Sciences, Philadelphia.

———. 1975. The diatoms of the United States exclusive of Alaska and Hawaii, vol. 2, part 1. Entomoneidaceae, Cymbellaceae, Gomphonemaceae, Epithemiacene. Monograph 13, Academy of Natural Sciences, Philadelphia.

Patterson, G. M. L., and **D. O. Harris.** 1983. The effect of *Pandorina morum* (Chlorophyta) toxin on the growth of selected algae, bacteria and higher plants. Br. Phyc. J. 18:259–266.

Perry, J. W., and **R. F. Evert.** 1975. Structure and development of the sieve elements in *Psilotum nudum.* Amer. J. Bot. 62:1038–1052.

Peters, G. A. 1978. Blue-green algae and algal associations. BioScience 28:580–585.

Peters, G. A., O. Ito, V. V. S. Tyagi, and **D. Kaplan.** 1981. Physiological studies on N$_2$-fixing *Azolla. In* J. M. Lyons, R. C. Valentine, D. A. Phillips, D.W. Rains, and R. C. Huffaker, eds., Genetic engineering of symbiotic nitrogen fixation and conservation of fixed nitrogen. pp. 343–362.

Peters, G. A., R. E. Toia, J., D. Raveed, and **N. J. Levine.** 1978. The *Azolla-Anabaena azollae* relationship. VI. Morphological aspects of the association. New Phytol. 80:583–593.

Petersen, R. L., and **J. B. Brisson.** 1977. Root cap structure in the fern *Ophioglossum petiolatum:* light and electron microscopy. Canad. J. Bot. 55:1861–1878.

Petersen, R. L., and **D. E. Fairbrothers.** 1973. Allelopathy: chemoantagonism between *Dryopteris* and *Osmunda* or supporting the "weakest link" hypothesis. Amer. J. Bot. 60 (Supplement 4):32.

Petersen, R. L., and **D. L. Fairbrothers.** 1980. Reciprocal allelopathy between the gametophytes of *Osmunda cinnamomea* and *Dryopteris intermedia.* Amer. Fern J. 70:73–78.

Petersen, R. L., M. L. Howarth, and **D. P. Whittier.** 1981. Interactions between a fungal endophyte and gametophyte cells in *Psilotum nudum.* Can. J. Bot. 59:711–720.

Petersen, R. L., and **J. Vermeer.** 1980. Root apex structure of *Ephedra monosperma* and *Ephedra chilensis* (Ephedraceae). Amer. J. Bot. 67:815–823.

Pettitt, J. M. 1966. A new interpretation of the structure of the megaspore membrane in some gymnospermous ovules. J. Linn. Soc. Lond. (Bot.) 59:253–263.

———. 1970. Heterospory and the origin of the seed habit. Biol. Rev. 45:401–415.

———. 1974. Developmental mechanisms in heterospory. II. Evidence for pinocytosis in the microspores of *Selaginella.* Bot. J. Linn. Soc. Lond. 69:79–87.

———. 1976a. A route for passage of substances through the developing pteridophyte exine. Protoplasma 88:117–131.

———. 1976b. Developmental mechanisms in heterospory. III. The plastid cycle during megasporogenesis in *Isoetes.* J. Cell Sci. 20:671–685.

———. 1977a. Developmental mechanisms in heterospory: features of post-meiotic regression in *Selaginella.* Ann. Bot. 41:117–125.

———. 1977b. The megaspore wall in gymnosperms: ultrastructure in some zooidogamous forms. Proc. Roy. Soc. Lond. B195:497–515.

———. 1977c. Detection in primitive gymnosperms of proteins and glycoproteins of possible significance in reproduction. Nature 266:530–532.

Pettitt, J. M., and **C. B. Beck.** 1967. Seed from the Upper Devonian. Science 156:1727–1729.

———. 1968. *Archaeosperma arnoldii*—a cupulate seed from the Upper Devonian of North America. Contr. Mus. Paleontol., Univ. Mich. 22:139–154.

Pfiester, L. A. 1975. Sexual reproduction of *Peridinium cinctum* f. *ovoplanum* (Dinophyceae). J. Phycol. 11:258–265.

———. 1976. Sexual reproduction of *Peridinium willei* (Dinophyceae). J. Phycol. 12:234–248.

Phillips, T. L., H. N. Andrews, and **P. G. Genzel.** 1972. Two heterosporous species of *Archaeopteris* from the Upper Devonian of West Virginia. Palaeontographica 139B:47–71.

Pickett-Heaps, J. D. 1967–1968. Ultrastructure and differentiation in *Chara* sp. I. 1967. Vegetative cells. Austral. J. Biol. Sci. 20:539–551. II. 1968a. Formation of the antheridia. Austral. J. Biol. Sci. 21:255–274 IV. 1968b. Spermatogenesis. Austral. J. Biol. Sci. 21:655–690.

———. 1971. Reproduction by zoospores in *Oedogonium.* I. Zoosporogenesis. Protoplasma 72:275–314.

———. 1972. Reproduction by zoospores in Oedogonium. II. Emergence of the zoospore and the motile phase. Protoplasma 74:149–169; III. Differentiation of the germling. Protoplasma 74:169–193. IV. Cell Division in the germling and evidence concerning the possible evolution of the wall rings. Protoplasma 74:195–212.

———. 1975. Green algae. Sinauer Associates, Sunderland, Mass.

Pickett-Heaps, J. D., and **L. C. Fowke.** 1969. Cell division in *Oedogonium.* I. Mitosis, cytokinesis, and cell elongation. Austral. J. Biol. Sci. 22:857–894.

Pieterse, A. H., L. Delange, and **J. P. Van Vliet.** 1977. A comparative study of *Azolla* in the Netherlands. Acta Bot. Neerland. 26:433–449.

***Pirozynski, K. A.** 1976. Fossil fungi. Annual Rev. Phytopath. 14:237–246.

Poelt, J. 1973. Systematic evaluation of morphological characters. *In* V. Ahmadjian and M. E. Hale, Jr., eds. The lichens, pp. 91–115. Academic Press, New York.

Pollock, E. G. 1957. The sex chromosomes of the maidenhair tree. J. Hered. 48:290–294.

———. 1970. Fertilization in *Fucus.* Planta 92:85–99.

***Prescott, G. W.** 1954. How to know the fresh-water algae. Brown, Dubuque, Iowa.

*———. 1962. Algae of the Western Great Lakes area. Brown, Dubuque, Iowa.

Price, J. A. 1972. Zygote development in *Caulerpa* (Chlorophyta, Caulerpales). Phycologia 11:217–218.

***Pringsheim, E. G.** 1946. Pure cultures of algae, their preparation, and maintenance. Cambridge University Press, London.

Proctor, M. C. F. 1979. Structure and eco-physiological

adaptation in bryophytes. *In* G. C. S. Clarke and J. D. Duckett, eds., chap. 20. 1979.

Proskauer, J. 1960. Studies on Anthocerotales. VI. On spiral thickening in the columella and its bearing on phylogeny. Phytomorphology 10:1–19.

Puiseux-Dao, S. 1970. *Acetabularia* and cell biology. Loycs Press, London.

*Ragan, M. A., and D. J. Chapman. 1978. A biochemical phylogeny of Protists. Academic Press, New York.

*Raghavan, V. 1976. Experimental embryogenesis in vascular plants. 603 pp. Academic Press, London, New York, and San Francisco.

Rai, A. N., P. Rowell, and W. D. P. Stewart. 1981. 15N₂ incorporation and metabolism in the lichen *Peltigera apthosa* Willd. Planta 152:544–552.

Ramanujam, C. G. K., G. W. Rothwell, and W. N. Stewart. 1974. Probable attachment of the *Dolerotheca campanulum* to a *Myeloxylon-Alethopteris* type frond. Amer. J. Bot. 61:1057–1066.

Ramsay, H. P. 1966. Sex chromosomes in *Macromitrium*. Bryologist 69:293–311.

———. 1979. Anisopory and sexual dimorphism in the Musci. *In* G. C. S. Clarke and J. G. Duckett, chap. 13. 1979.

Ramsay, H. P., and G. K. Berrie. 1982. Sex determination in bryophytes. J. Hattori Bot. Lab. 52:255–274.

*Ramsbottom, J. 1953. Mushrooms and toadstools. Collins, New York.

Ramus, J. 1971. *Codium:* the invader. Discovery 6:59–68.

Rao, L. N. Life history of *Cycas circinalis* L. 1961–1970a. I. 1961. Microsporogenesis, male and female gametophytes and spermatogenesis. J. Indian Bot. Soc. 60:601–619. II. 1963. Fertilization, embryogeny and germination of the seed. J. Indian Bot. Soc. 62:319–332. III. 1964. Polyembryony in *Cycas circinalis.* Current Sci. 33:375–376. IV. 1970a. (No title.) Proc. Indian Acad. Sci. 72:179–186.

———. 1970b. Embryonal haustorium of *Cycas circinalis* L. Current Sci. 39:239–240.

Rao, V. M. R., and T. V. Desikachary. 1970. Macdonald-Pfitzer hypothesis and cell size in diatoms. Nova Hedwigia, Beihefte 31:485–493.

*Raper, J. R. 1957. Hormones and sexuality in lower plants. Sympos. Soc. Exp. Biol. 11:143–165.

Raper, J. R., and A. S. Flexer. 1971. Mating systems and evolution of the Basidiomycetes. *In* R. S. Petersen, ed., Evolution in the Higher Basidiomycetes, pp. 149–176. University of Tennessee Press.

*Raper, K. B. 1973. Acrasiomycetes. G. C. Ainsworth, F. K. Sparrow, and A. S. Sussman, eds., The fungi, vol. IVB, pp. 9–36. Academic Press, New York.

———. 1984. The Dictyostelids. Princeton University Press.

*Raper, K. B., and D. I. Fennell. 1965. The genus *Aspergillus.* Williams & Wilkins, Baltimore.

*Raper, K. B., and C. Thom. 1949. Manual of the Penicillia. Williams & Wilkins, Baltimore.

Rashid, A., and R. N. Chopra. 1969. The apogamous sporophytes of *Funaria hygrometrica* and their cultural behavior. Phytomorphology 19:170–178.

Rastorfer, J. R. 1962. Photosynthesis and respiration in moss sporophytes and gametophytes. Phyton 19:169–177.

Rau, W., and H. Falk. 1959. *Stylites* E. Amstutz, eine neue Isoëtaceae ausden Hochanden Perus. I and II. Sitzber. Heidelberger Akad. Wiss. Springer, Berlin-Göttingen-Heidelberg.

*Raven, P. H. 1970. A multiple origin of plastids and mitochondria. Science 169:641–646.

Rawitscher-Kunkel, E., and L. Machlis. 1962. The hormonal integration of sexual reproduction in *Oedogonium.* Amer. J. Bot. 49:177–183.

*Reid, R. D., and M. J. Pelczar, Jr. 1972. Microbiology. McGraw-Hill, New York.

Reiman, B. E. F., J. Lewin, and B. C. Volcani. 1966. Studies in the biochemistry and fine structure of silica shell formation in diatoms. II. The structure of the cell wall of *Navicula pelliculosa* (Bréb.) Hilse. J. Phycol. 2:74–84.

Reinhart, D. A., and R. J. Thomas. 1981. Sucrose uptake and transport in conducting cells of *Polytrichum commune.* Bryol. 84:59–64.

Renzaglia, K. S. 1978. A comparative morphology and developmental anatomy of the Anthocerotophyta. J. Hattori Bot. Lab. 44:31–90.

Rice, E. L. 1979. Allelopathy: an update. Bot. Rev. 45:15–109.

*———. 1984. Allelopathy. Academic Press. 422 pp.

*Richardson, D. H. S. 1981. The biology of mosses. 220 pp. Wiley, New York, Toronto.

Richardson, D. N. S., D. J. Hill, and D. C. Smith. 1968. Lichen physiology XI. The role of the alga in determining the patterns of carbohydrate movement between lichen symbionts. New Phytol. 67:469–486.

Richardson, N. 1970. Studies on the photobiology of *Bangia fuscopurpurea.* J. Phycol. 6:215–219.

Ridgway, J. E. 1967a. Factors initiating antheridial formation in six Anthocerotales. Bryologist 70:203–205.

———. 1967b. The biotic relationship of *Anthoceros* and *Phaeoceros* to certain Cyanophyta. Ann. Mo. Bot. Garden 54:95–102.

Rietema, H. 1969. A new type of life history in *Bryopsis* (Chlorophyceae, Caulerpales). Acta Bot. Neerland. 18:615–619.

———. 1970. Life histories of *Bryopsis plumosa* (Chlorophyceae, Caulerpales) from European waters. Acta Bot. Neerland. 19:859–866.

———. 1971. Life-history studies in the genus *Bryopsis* (Chlorophyceae). V. Life histories in *Bryopsis hypnoides* Lamx. from different points along the European coasts. Acta Bot. Neerland. 20:291–298.

Rippka, R., A. Neilson, R. Kunisawa, and G. Cohen-Bazire. 1971. Nitrogen fixation by unicellular blue-green algae. Arch. Mikrobiol. 76:341–348.

*Rippon, J. W. 1974. Medical Mycology. Saunders, Philadelphia.

Robbins, R. R., and Z. B. Carothers. 1975. The occurrence and structure of centrosomes in *Lycopodium*. Protoplasma 86:279–284.

———. 1978. Spermatogenesis in *Lycopodium:* the mature spermatozoid. Amer. J. Bot. 65:440–443.

Robert, D. 1971a. Le gamétophyte femelle de *Selaginella kraussiana* (Kunze) A. Br. I. Organisation générale de la megaspore. Le diaphragme et l'endospore les réserves. Rev. Cytol. Biol. Veg. 34:93–164.

———. 1971b. Le gamétophyte femelle de *Selaginella kraussiana* (Kunze) A. Br. II. Organisation histologique du tissu reproducteur et principaux aspects à la dédifférenciation cellulair préparatoire a l'oogénèse. Rev. Cytol. Biol. Veg. 34:189–232.

———. 1974. Étude ultrastructurale de la spermiogenése, notamment de la différenciation de l'appareil nucleaire chez *Selaginella kraussiana* (Kunze) A. Br. Ann. des Sci. Nat., Bot., Paris 15:65–118.

———. 1977. The nucleus of the spermatozoid of *Selaginella kraussiana*. J. Ultrastr. Res. 58:178–195.

Roberts, K., G. J. Hills, and R. J. Shaw. 1982. The structure of algal cell walls. *In* J. Harris, ed., Electron microscopy of proteins, vol. 3:1–40. Academic Press, London.

Robinow, C. F., and J. Marak. 1966. Fiber apparatus in the yeast cell. J. Cell Biol. 29:129–151.

Robinson, B., and J. H. Miller. 1970. Photomorphogenesis in the blue-green alga *Nostoc commune* 584. Physiol. Plant 23:461–472.

Rodgers, G. A., and W. D. P. Stewart. 1977. The cyanophyte-hepatic symbiosis. I. Morphology and physiology. New Phytol. 78:441–458.

Rodin, R. J. 1953a. Distribution of *Welwitschia mirabilis*. Amer. J. Bot. 40:280–285.

———. 1953b. Seedling morphology of *Welwitschia*. Amer. J. Bot. 40:371–378.

———. 1958. Leaf anatomy of *Welwitschia*. I. Early development of the leaf. Amer. J. Bot. 45:90–95. II. A study of mature leaves. Amer. J. Bot. 45:96–103.

———. 1967. Ontogeny of the foliage leaves in *Gnetum*. Phytomorphologia 117:118–128.

Rosen, W. G. 1961. Studies on pollen tube chemotropism. Amer. J. Bot. 48:889–895.

———. 1968. Ultrastructure and physiology of pollen. Annual Rev. Plant Physiol. 19:435–462.

———. 1975. Pollen pistil interactions. *In* J. G. Duckett and P. A. Racey, eds., Biology of the male gamete, pp. 153–164. Academic Press, New York.

Rothwell, G. W. 1972a. Evidence of pollen tubes in Paleozoic pteridosperms. Science 175:772–774.

———. 1972b. Pollen organs of the Pennsylvanian Callistophytaceae (Pteridospermopsida). Amer. J. Bot. 59:993–999.

———. 1975. The Callistophytaceae (Pteridospermopsida). I. Vegetative structures. Palaeontographica 151B:171–196.

———. 1977. Evidence for a pollination-drop mechanism in Paleozoic pteridosperms. Science 198:1251–1252.

Rouffa, A. S. 1971. An appendageless *Psilotum*. Introduction to aerial shoot morphology. Amer. Fern J. 61:75–86.

———. 1978. On phenotypic expression, morphogenetic pattern and synangium evolution in *Psilotum*. Amer. J. Bot. 65:692–713.

Rubin, G., and D. J. Paolillo, Jr. 1983. Sexual development of *Onoclea sensibilis* on agar and soil media without the addition of antheridiogen. Amer. J. Bot. 70:811–815.

Ruf, E. W., Jr. 1975. The morphology and reproduction of certain Selaginellas. Ph.D. dissertation, University of Texas at Austin.

Rury, P. M. 1978. A new and unique mat-forming Merlin's-grass (*Isoëtes*) from Georgia. Amer. Fern J. 68:99–108.

Rusch, H. P. 1968. Some biochemical events in the life cycle of *Physarum polycephalum*. *In* D. M. Prescott, ed., Advances in cell biology, vol. 1. Prentice-Hall, Englewood Cliffs, N.J.

Sack, F. D. and D. J. Paolillo, Jr. 1983a. Structure and development of walls in *Funaria* stomata. Amer. J. Bot. 70:1019–1030.

———. 1983b. Stomatal pore and cuticle formation in *Funaria*. Protoplasma 116:1–13.

———. 1983c. Protoplasmic changes during stomatal development in *Funaria*. Can. J. Bot. 61:2515–2526.

*Sagan, L. 1967. On the origin of mitosing cells. J. Theoretical Biol. 14:225–275.

Sager, R., and S. Granick. 1954. Nutritional control of sexuality in *Chlamydomonas reinhardi*. J. Gen. Physiol. 37:729–742.

Sam, S. J. 1982. A germination method for *Isoetes*. Amer. Fern J. 72:61.

Sansome, E. 1961. Meiosis in the oogonium and antheridium of *Pythium debaryanum* Hesse. Nature (Lond.) 191:827–828.

———. 1966. Meiosis in the sex organs of the Oomycetes. *In* C. D. Darlington and K. R. Lewis, eds., Chromosomes today, pp. 77–83. Plenum, New York.

Santandreu, R., and D. Northcote. 1969. The formation of buds in yeast. J. Gen. Microbiol. 55:393–398.

Satterthwait, S. F., and J. W. Schopf. 1972. Structurally preserved phloem zone tissue in *Rhynia*. Amer. J. Bot. 59:373–376.

Saville, D. B. O. 1971. Coevolution of the rust fungi and their hosts. Quart. Rev. Biol. 146:211–218.

*Scagel, R. F., R. J. Bandoni, G. E. Rouse, W. B. Schofield, J. R. Stein, and T. M. C. Taylor. 1965. An evolutionary survey of the plant kingdom. Wadsworth, Belmont, Calif.

Schantz, R., E. Blee, A. Chammai, and B. Wolff. 1981. The triple-layered organization of the *Euglena* chloroplast envelope (significance and functions). Ber. Deutsch Bot. Gesellsch. 94:463–476.

Schedlbauer, M. D. 1976. Fern gametophyte development: controls of dimorphism in *Ceratopteris thalictroides*. Amer. J. Bot. 63:1080–1087.

Schedlbauer, M. D., and E. J. Klekowski, Jr. 1972. Antheridogen activity in the fern *Ceratopteris thalictroides* (L.) Brongn. Bot. J. Linn. Soc. Lond. 65:399–413.

Scheirer, D. C. 1977. The thickened leptoid (sieve element) wall of *Dendrotigotrichum* (Bryophyta): Cytochemistry and fine structure. Am. J. Bot. 64:369–376.

———.1980. Differentiation of bryophyte conducting tissues: structure and histochemistry. Bull. Torrey Bot. Club 107:298–307.

———. 1983. Leaf parenchyma with transfer cell-like characteristics in the moss, *Polytrichum commune* Hedw. Amer. J. Bot. 70:987–992.

Scheirer, D. C., and H. M. Brasell. 1984. Epifluorescence microscopy for the study of nitrogen fixing blue-green algae associated with *Funaria hygrometrica* (Byrophyta). Amer. J. Bot. 71:461–465.

Scheirer, D. C., and I. J. Goldklang. 1977. Pathway of water movement in hydroids of *Polytrichum commune* Hedw. (Bryopsida). Amer. J. Bot. 64:1046–1047.

*Schellbach, L., and J. Lacke. 1955. Grand Canyon: Nature's story of creation. Nat. Geogr. Mag. 107:589–629.

Schertler, M. M. 1977. Morphology and developmental anatomy in the leafy hepatic *Nowellia curvifolia* (Dicks.) Nutt. J. Hattori Bot. Lab. 42:241–271.

Schieder, O. 1968. Untersuchungen über das chemotaktisch wirksame Gamon von *Sphaerocarpos donnellii* Aust. Z. Pflanzenphysiol. 39:258–273.

———. 1973. Untersuchungen an Nicotinsäure-auxotrophen Stämmen von *Sphaerocarpos donnellii* Aust. Z. Pflanzenphysiol. 70:185–189.

———. 1974. Selektion einer somatischen Hybride nach Fusion von Protoplasten auxotropher Mutanten von *Sphaerocarpos donnellii* Aust. Z. Pflanzenphysiol. 74:357–365.

———. 1976. The spectrum of auxotrophic mutants from the liverwort *Sphaerocarpos donnellii* Aust. Molec. Genet. 144:63–66.

Schlanker, C. M., and G. A. Leisman. 1969. The herbaceous carboniferous lycopod *Selaginella fraiponti* comb. nov. Bot. Gaz. 130:35–41.

Schlösser, U. G. 1981. Release of reproduction cells by action of cell wall autolytic factors in *Chlamydomonas* and *Geminella*. Ber. Deutsch. Bot. Gesellsch. 94:373–374.

Schmid, R. 1982. The terminology and classification of steles: historical perspective and the outlines of a system. Bot. Rev. 48:817–931.

Schmiedel, G., and E. Schnepf. 1979a. Side branch formation and orientation in the caulonema of the moss, *Funaria hygrometrica*: normal development and fine structure. Protoplasma 100:367–383.

———. 1979b. Side branch formation and orientation in the caulonema of the moss *Funaria hygrometrica*: Experiments with inhibitors and with centrifugation. Protoplasma 101:47–59.

Schmitz, K. 1981. Translocation. *In* Lobban and Wynne, chap. 15. 1981.

Schneider, M. J., and A. J. Sharp. 1962. Observations on the reproduction and development of the gametophyte of *Tetraphis pellucida* in culture. Bryologist 65:154–166.

*Schnepf, E., and R. M. Brown, Jr. 1971. On relationships between endosymbiosis and the origin of plastids and mitochondria. *In* J. Reinert and H. Ursprung, eds., Results and problems in cell differentiation. Vol. 2: Origin and continuity of cell organelles. Springer, Berlin.

Schofield, W. B. 1981. Ecological significance of morphological characteristics in the moss gametophyte. Bryologist 84:149–165.

Schofield, W. B., and C. Hébant. 1984. The morphology and anatomy of the moss gametophore. *In* R. M. Schuster, ed., chap. 11, pp. 627–657. 1984.

Schonberr, J., and H. Zeigler. 1975. Hydrophobic cuticular ledges prevent water entering the air pores of liverwort thalli. Planta 124:51–60.

Schopf, J. W. 1968. Microflora of the Bitter Springs Formation, late Precambrian, central Australia. J. Paleontol. 42:651–688.

*———. 1970. Precambrian microorganisms and evolutionary events prior to the origin of vascular plants. Biol. Rev. 45:319–352.

———. 1975. Precambrian paleobiology: problems and perspectives. Annual Rev. Earth Planet. Sci. 3:213–249.

Schopf, J. W., and J. M. Blacic. 1971. New microorganisms from the Bitter Springs Formation (Late Precambrian) of the North-Central Amadeus Basin, Australia. J. Paleontol. 45:925–959.

Schopf, J. W., T. D. Ford, and W. J. Breed. 1973. Microorganisms from the late precambrian of the Grand Canyon, Arizona. Science 179:1319–1321.

Schopf, J. W., and D. Z. Oehler. 1976. How old are the eukaryotes? Science 193:47–49.

Schopf, J. W., D. Z. Oehler, R. J. Horedyeki, and K. A. Kvenvolden. 1971. Biogenicity and significance of the oldest known stromatolites. J. Paleontol. 45:477–485.

Schrandholf, H., and U. Richter. 1978. Elektronmikroscopische Analyse der Musterbildung in Antheridium der Polypodiaceae. Pl. Syst. Evol. 129:291–297.

Schulman, E. 1958. Bristlecone pine, oldest known living thing. Nat. Geogr. Mag. 113:355–372.

Schulz, D., and W. Schmidt. 1974. Entwicklung des Peristoms von *Funaria hygrometrica*. Flora 163:451–465.

*Schuster, R. M. 1966–1980. The Hepaticae and Anthocerotae of North America. Vol. 1, 1966; Vol. 2,

1969; Vol. 3, 1974; Vol. 4, 1980. Columbia University Press, New York.

———. 1979. The phylogeny of the Hepaticae. *In* G. C. S. Clarke and J. G. Duckett, eds., chap. 3. 1979.

*———. ed. 1983–1984*a*. New manual of bryology. Vol. 1, 2—1295 pp. Hattori Bot. Laboratory.

———. 1984*b*. Comparative anatomy and morphology of the Hepaticae. *In* R. M. Schuster, ed., chap. 14, pp. 760–891.

———. 1984*c*. Evolution, phylogeny, and classification of the Hepaticae. *In* R. M. Schuster, ed., chaps. 1–5, pp. 893–1070. 1984.

———. 1984*d*. Morphology, phylogeny, and classification of the Anthocerotae. *In* R. M. Schuster, ed., chap. 16, pp. 1071–1092. 1984.

Schweitzer, H.-J. 1967. Die Oberdevon-Flora der Bäreninsel 1. *Pseudobornia ursina* Nathorst. Palaeontographic 120B:116–137.

———. 1972. Die Mitteldevon-Flora von Lindlar (Rheinland) 3. Filicinae—*Hyenia elegans* Kräusel and Weyland. Palaeontographica 137B:154–175.

*Schwimmer, D., and M. Schwimmer. 1964. Algae and medicine. *In* D. F. Jackson, ed., Algae and man. Plenum, New York.

Scott, R. A., E. S. Barghoorn, and E. B. Leopold. 1960. How old are the angiosperms? Amer. J. Sci. 258A:284–299.

Searles, R. B. 1980. The strategy of the red algal life history. Amer. Nat. 115:113–120.

*Seaward, M. R. D. 1977. Lichen ecology. 550 pp. Academic Press.

Segawa, M., S. Kishi, and S. Tatumo. 1971. Sex chromosomes of *Cycas revoluta*. Jap. J. Genet. 46:33–41.

Seymour, R. L. 1970. The genus *Saprolegnia*. Nova Hedwigia 19:1–124.

Seymour, R. L., and T. W. Johnson, Jr. The Saprolegniaceae. In press.

Shaffer, R. L. 1975. The major groups of Basidiomycetes. Mycologia 67:1–18.

Shah, A. K., and G. S. Vaidya. 1977. Detection of vitamin B_{12} and pantothenic acid in cell exudates of blue-green algae. Biol. Plantarum (Praha) 19:426–429.

Sharma, B. D., and R. Singh. 1984. The ligule of *Isoetes*. Am. Fern. J. 74:22–28.

Sharp, A. J. 1974*a*. Bryopsida. *In* Encyclopaedia Britannica 3:351–354.

———. 1974*b*. Hepatopsida. *In* Encyclopaedia Britannica 8:779–781.

Shen, E. Y. F. 1967*a*. The amount of nuclear DNA in *Chara zeylanica* measured by microspectrophotometry. Taiwainia B:111–114.

———. 1967*b*. Amitosis in *Chara*. Cytologia 32:481–488.

Shivanna, K. R. 1982. Pollen-pistil interaction and control of fertilization. *In* B. M. Johri, ed., chap. 7. 1982.

Shyam, R. 1980. On the life cycle, cytology and taxonomy of *Cladophora callicoma* from India. Amer. J. Bot. 67:619–624.

Sideman, E. J., and D. C. Scheirer. 1977. Some fine structural observations on developing and mature sieve elements in the brown alga *Laminaria saccharina*. Amer. J. Bot. 64:649–657.

Sievers, A., and B. Buchen. 1970. Uber den Feinbau der wachsenden Megaspore von *Selaginella*. Protoplasma 71:267–279.

Sigee, D. C. 1974. Structure and function in the ligule of *Selaginella kraussiana*. I. Fine structure and development. Protoplasma 79:359–375.

———. 1975. II. The cytoplasmic distribution of complex carbohydrates. Protoplasma 85:133–145.

———. 1976. III. The uptake of tritiated glucose. Protoplasma 90:333–341.

Sigle, D. C. 1974–1976. Structure and function in the ligule of *Selaginella kraussiana*. I. 1974. Fine structure and development. Protoplasma 79:359–375. II. 1975. The cytoplasmic distribution of complex carbohydrates. Protoplasma 85:133–145. III. 1976. The uptake of tritiated glucose. Protoplasma 90:333–341.

*Singer, R. 1961. Mushrooms and truffles. Leonard Hill, London. Wiley, New York.

Singh, H., and J. N. Owens. 1981*a*. Sexual reproduction of Engelmann spruce (*Picea engelmannii*). Can. J. Bot. 59:743–810.

———. 1981*b*. Sexual reproduction in subalpine fir (*Abies lasiocarpus*). Can. J. Bot. 59:2650–2666.

———. 1982. Sexual reproduction in grand fir (*Abies grandis*). Can. J. Bot. 60:2197–2214.

Singh, M. N., R. N. Konar, and S. P. Bhatnagar. 1981. Haploid plantlet formation from female gametophytes of *Ephedra foliata* Boiss. *in vitro*. Ann. Bot. 48:215–220.

Singh, P. K. 1977. Growth and nitrogen fixation of the unicellular blue-green alga *Aphanotheca castagnei*. Biol. Plantarum (Praha) 19:156–177.

Singh, R. N., and D. Tiwari. 1970. Frequent heterocyst germination in the blue-green alga *Gloeotrichia ghosei*. J. Phycol. 6:172–176.

Skog, J. E., and H. P. Banks. 1973. *Ibyka amphikoma*, gen. et sp. n., a new protoarticulate precursor for the late Middle Devonian of New York state. Amer. J. Bot. 60:366–380.

Slade, J. S. 1965. Physical factors influencing seta elongation in *Pellia epiphylla*. Bryologist 68:440–445.

Sluiman, H. J. 1983. The flagellar apparatus of the zoospore of the filamentous green alga *Coleochaete pulvinata*: Absolute configuration and phylogenetic significance. Protoplasma 115:160–175.

*Smith, A. J. E., ed. 1982. Bryophyte Ecology. 511 pp. Chapman and Hall, London, New York.

Smith, D. C. 1962. The biology of lichen thalli. Biol. Rev. 37:537–570.

———. 1973*a*. Symbiosis of algae with invertebrates. Oxford Biology Readers 43, Oxford University Press, London.

———. 1973*b*. The Lichen Symbiosis. Oxford Biology Readers 42, Oxford University Press.

876

————. 1981. The role of nutrient exchange in recognition between symbionts. Ber. Deutsch. Bot. Ges. 94:517–528.

Smith, D. C., L. Muscatine, and D. H. Lewis. 1969. Carbohydrate movement from autotrophs to heterotrophs in parasitic and mutualistic symbiosis. Biol. Rev. Biol. Proc. Cambridge Phil. Soc. 44:17–90.

Smith, D. L. 1964. The evolution of the ovule. Biol. Rev. 39:137–159.

Smith, G. L. 1971. A conspectus of the genera of Polytrichaceae. Mem. N.Y. Bot. Garden 21:1–83.

*Smith, G. M. 1950. Freshwater algae of the United States. McGraw-Hill, New York.

*————. 1955. Cryptogamic botany. 2d ed. Vol. II: Bryophytes and Pteridophytes. McGraw-Hill, New York.

*————. 1969. Marine algae of the Monterey Peninsula. Stanford University Press.

*Smith, G. M., et al., eds. 1951. Manual of phycology. Chronica Botanica, Waltham, Mass.

Smith, J. L. 1966. The liverworts *Pallavicinia* and *Symphyogyna* and their conducting system. University of California Publications in Botany 39:1–46.

Smith, R. V., and M. C. W. Evans. 1970. Soluble nitrogenase from vegetative cells of the blue-green alga *Anabaena cylindrica*. Nature 225:1253–1254.

Smith, W. H., and R. N. Konar. 1969. Initiation of ovulate strobili in cotyledon-stage seedings of *Pinus elliotti*. Canad. J. Bot. 47:624–626.

Snider, J. A. 1975a. Sporophyte development in the genus *Archidium* (Musci). J. Hattori Bot. Lab. 39:85–104.

————. 1975b. A revision of the genus *Archidium* (Musci). J. Hattori Bot. Lab. 39:105–201.

Sommerfeld, M. R., and H. W. Nichols. 1970a. Comparative studies in the genus *Porphyridium* Naeg. J. Phycol. 6:67–68.

————. 1970b. Developmental and cytological studies of *Bangia fuscopurpurea* in culture. Amer. J. Bot. 57:640–648.

Sparrow, F. K. 1958. Interrelationships and phylogeny of the aquatic phycomycetes. Mycologia 50:797–873.

————. 1960. Aquatic phycomycetes. University of Michigan Press.

————. 1973. Hyphochytridiomycetes. *In* G. C. Ainsworth, F. K. Sparrow, and A. S. Sussman, eds., The fungi, vol. IVB, p. 71. Academic Press, New York.

————. 1976. The present state of classification of biflagellate fungi. *In* E. B. Gareth Jones, ed., Recent advances in aquatic mycology, pp. 213–222. Wiley, New York.

Spiess, L. D., B. B. Lippincott, and J. A. Lippincott. 1971. Development and gametophore initiation in the moss *Pylaisiella selwynii* as influenced by *Agrobacterium tumefaciens*. Amer. J. Bot. 58:726–731.

————. 1977. Comparative response of *Pylaisiella selwynii* to *Agrobacterium* and *Rhizobium* sp. Bot. Gaz. 138:35–40.

————. 1982. Bacteria-moss interaction in the regulation of protonemal growth and development. J. Hattori Bot. Lab. 53:215–220.

Sporne, K. R. 1949. A new approach to the primitive flower. New Phytol. 48:259–276.

————. 1956. The phylogenetic classification of angiosperms. Biol. Rev. 31:1–29.

————. 1964. Self-fertility in a thallus of *Equisetum telmateia* Ehrh. Nature 201:1345–1346.

————. 1969. The ovule as an indicator of evolutionary status in angiosperms. New Phytol. 68:555–566.

————. 1969. The morphology of gymnosperms. Hutchinson University Library, London.

————. 1970. The morphology of pteridophytes. Hutchinson University Library, London.

*————. 1971. The mysterious origin of flowering plants. Oxford Biology Readers 3. Oxford University Press.

————. 1972. Some observations on the evolution of pollen types in dicotyledons. New Phytol. 71:181–185.

————. 1975. The morphology of angiosperms. St. Martin's Press, New York.

Srinivasan, J. C., and P. B. Kaufman. 1978. Sex expression in *Equisetum scirpoides in vitro*. Phytomorph. 28:331–335.

Srivastava, P. S. 1982. Endosperm culture. *In* B. M. Johri, chap. 8. 1982.

*Stafleu, F. A., et al., eds. 1978. International Code of Botanical Nomenclature. Bohn, Schelteman, and Hoekema. Utrecht.

Stange, L. 1977. Meristem differentiation in *Riella helicophylla* (Bory et Mont.) Mont. under the influence of auxin or antiauxin. Planta 135:289–295.

Stanier, R. Y., and G. Cohen-Bazire. 1977. Phototrophic prokaryotes: the cyanobacteria. Annual Rev. Microbiol. 31:225–274.

*Stanier, R. Y., M. Doudoroff, and E. A. Adelburg. 1970. The microbial world. Prentice-Hall, Englewood Cliffs, N.J.

Stanier, R. Y., R. Kunisawa, M. Mandel, and G. Cohen-Bazire. 1971. Purification and properties of unicellular blue-green algae (Order Chroococcales). Bacteriol. Rev. 35:171–205.

Stanier, R. Y., N. Pfennig, and H. G. Trüper. 1981. Introduction to the phototrophic prokaryotes. *In* M. P. Starr, H. Stolp, H. G. Trüper, A. Balows, and H. G. Schlegel, eds., The Prokaryotes, chap. 7, pp. 197–211. Vol. I, 1022 pp. plus 156 pp. Index. Springer-Verlag, Berlin.

Stanley, R. G., and H. F. Linskens. 1974. Pollen: biology, biochemistry, management. Springer, New York.

Star, A., and T. Weber. 1978. Sporophyte exudate inhibition of gametophyte development in *Pityogramma calemolanos*. Bot. Soc. Amer. Misc. Publ. 136:16.

Starr, R. C. 1968. Cellular differentiation in *Volvox*. Proc. Nat. Acad. Sci. 59:1082–1088.

————. 1969. Structure, reproduction and differentiation in *Volvox carteri* f. *nagariensis* Iyengar, Strains H, K9 and 10. Arch. Protistenk. 111:204–222.

―――. 1970. *Volvox pocockiae*, a new species with dwarf males. J. Phycol. 6:234–239.

―――. 1971. Control of differentiation in *Volvox*. Twenty-ninth Symposium of the Society for Developmental Biology. Develop. Biol. Supplement 4:59–100.

―――. 1984. Colony formation in algae. Chap. 13, pp. 261–290. *In* H. F. Linskens and J. Heslop-Harrison, eds., chap. 13 pp., 261–290. 1984.

Starr, R. C. and **L. Jaenicke.** 1974. Purification and characterization of the hormone initiating sexual morphogenesis in *Volvox carteri* f. *nagariensis* Iyengar. Proc. Nat. Acad. Sci. 71:1050–1054.

Steere, W. C. 1958. Evolution and speciation in mosses. Amer. Nat. 92:5–20.

―――. 1969. A new look at evolution and phylogeny of Bryophytes. *In* J. E. Gunckel, ed., Current topics in plant science, pp. 134–143. Academic Press, New York.

Steeves, T. A. 1963. Morphogenetic studies of fern leaves. J. Linn. Soc. Lond. (Bot.) 58:401–415.

*__**Steeves, T. A.,**__ and **I. M. Sussex.** 1972. Patterns in plant development. Prentice-Hall, Englewood Cliffs, N.J.

Steffens, W. L. 1976. A developmental study of *Achlya recurva* utilizing light and electron microscopy. Master's thesis, University of Texas at El Paso.

Steidinger, K. A. 1973. Phytoplankton ecology: a conceptual review based on eastern Gulf of Mexico research. CRC Critical Rev. Microbiol. 3:49–68.

―――. 1975. Basic factors influencing red tides. *In* V. R. Lo Cicero, ed., Proceedings of the First International Congress on Toxic Dinoflagellate Blooms, November 1974, pp. 153–162. Boston, Mass.

*__**Steidinger, K. A.,**__ and **J. Williams.** 1970. Dinoflagellates, vol. 2. Florida Department of Natural Resources, St. Petersburg.

Steinkamp, M. P., and **W. T. Doyle.** 1979. Spore wall ultrastructure in four species of the liverwort *Riccia*. Amer. J. Bot. 66:546–556.

Sterling, C. 1963. Structure of the male gametophyte in gymnosperms. Biol. Rev. 38:167–203.

Stevens, F. L. 1899. The compound oosphere in *Albugo bliti*. Gametogenesis and fertilization in *Albugo*. Bot. Gaz. 28:149–176; 225–245.

Stevenson, D. W. 1974. Ultrastructure of the nacreous leptoids (sieve elements) in the polytrichaceous moss *Atrichum undulatum*. Amer. J. Bot. 61:414–421.

―――. 1976. Observations on phyllotaxis, stelar morphology, the shoot apex gemmae of *Lycopodium lucidulum* Michaux (Lycopodiaceae). Bot. J. Linn. Soc. Lond. 72:81–100.

―――. 1980. Ontogeny of the vascular system of *Botrychium multifidum* (S.G. Ginelin) Rupr. (*Ophioglossaceae*) and the bearing on stelar theories. Bot. J. Linn. Soc. 80:41–52.

Steward, F. C. 1967. Totipotency of angiosperm cells: its significance for morphology and embryology. Phytomorphology 17:499–507.

Stewart, K. D., and **E. M. Gifford, Jr.** 1967.
Ultrastructure of the developing megaspore mother cell of *Ginkgo biloba*. Amer. J. Bot. 54:375–383.

Stewart, K. D., and **K. R. Mattox.** 1975. Comparative cytology, evolution and classification of the green algae with some consideration of the origin of other organisms with chlorophylls *a* and *b*. Bot. Rev. 41:105–131.

―――. 1978. Structural evolution in the flagellated cells of green algae and land plants. Biosystems 10:145–152.

Stewart, W. D. P. 1970. Nitrogen fixation by blue-green algae in Yellowstone thermal areas. Phycologia 9:261–268.

―――, ed. 1974. Algal physiology and biochemistry. University of California Press.

―――. 1977. A botanical ramble among the blue-green algae. Proc. Phyc. J. 12:89–165.

―――. 1978. Nitrogen-fixing Cyanobacteria and their associations with Eukaryotic plants. Endeavour N.S. 2:170–179.

―――. 1980. Some aspects of structure and function of N_2-fixing Cyanobacteria. Ann. Rev. Microbiol. 34:497–536.

Stewart, W. D. P., A. Haystead, and **H. W. Pearson.** 1969. Nitrogenase activity in heterocysts of filamentous blue-green algae. Nature 224:226–228.

Stewart, W. D. P., and **M. Lex.** 1970. Nitrogenase activity in the blue green alga *Plectonema boryanum* strain 594. Arch. Mikrobiol. 73:250–260.

Stewart, W. D. P., and **H. W. Pearson.** 1970. Effects of aerobic and anaerobic conditions on growth and metabolism of blue-green algae. Proc. Roy. Soc. Lond. B175:293–311.

Stewart, W. D. P., and **G. A. Rogers.** 1977. The cyanophyte-hepatic symbiosis. II. Nitrogen fixation and the interchange of nitrogen and carbon. New Phytol. 78:459–471.

*__**Stewart, W. H.**__ 1983. Paleobotany and the Evolution of Plants. Cambridge University Press, Cambridge.

*__**Stewart, W. N.**__ 1964. An upward outlook in plant morphology. Phytomorphology 14:120–134.

―――. 1983. Paleobotany and the evolution of plants. Cambridge University Press.

Stewart, W. N., and **T. Delevoryas.** 1956. The medullosan pteridosperms. Bot. Rev. 22:45–80.

Stidd, B. M. 1974. Evolutionary trends in the Marattiales. Ann. Mo. Bot. Garden 61:388–407.

Stidd, B. M., and **J. W. Hall.** 1970*a*. The natural affinity of the Carboniferous seed, *Callospermarion*. Amer. J. Bot. 57:827–836.

―――. 1970*b*. *Callandrium callistophytoides*, gen. et sp. nov., the probable pollen-bearing organ of the seed fern *Callistophyton*. Amer. J. Bot. 57:394–403.

Storey, W. B. 1968. Somatic reduction in cycads. Science 159:648–650.

Stosch, H. A. von. 1964. Zum Problem der sexuellen Fortpflanzung in der Perideengattung *Ceratium*. Helgoländer Wiss. Meeresunters. 10:140–152.

————. 1965a. Sexualität bei *Ceratium cornuti* (Dinophyta). Naturwiss. 5:111–113.

————. 1965b. Manipulierung der Zellgrosse von Diatomeen in Experiment. Phycologia 5:21–44.

Stosch, H. A. von, and **G. Drebes**. 1964. Entwicklungsgeschichtliche Untersuchungen an zentrischen Diatomeen. 4. Die Plankton diatomee *Stephanopyxis turris*, ihre Behandlung und Entwicklungsgeschichte. Helgoländer Wiss. Meeresunters. 11:209–257.

Stosch, H. A. von, and **G. Theil**. 1979. A new mode of life history in the freshwater algae genus *Batrachospermum*. Amer. J. Bot. 66:105–107.

Stotler, R., and **B. Crandall-Stotler**. 1977. A checklist of the liverworts and hornworts of North America. Bryologist 80:405–428.

Suire, C. 1970. Recherches cytologiques sur deux Hepatiques. *Pellia* et *Radula complanata*. Ergastome, sporogenese et spermatogenese. Le Botaniste 53:125–392.

Swamy, B. G. L. 1974. On the presumed ancestry of the angiosperm embryo sac. Phytomorphology 24:102–106.

Syrett, P. J., and **F. A. A. Al-Houty**. 1984. The phylogenetic significance of the occurrence of urease/urea amidolayse and gylcollate oxidase/glycollate dehydrogenase in green algae. Br. Phyc. J. 19:11–23.

Talley, S., and **D. Rains**. 1980. *Azolla filiculoides* Lam. as a fallow-season green manure for rice in a temperate climate. J. Agron. 22:11–18.

Tatuno, S., and **I. Liyama**. 1971. Cytological studies in *Spirogyra*. I. Cytologia 36:86–92.

Taylor, D. L. 1969. Some aspects of the regulation and maintenance of algal numbers in zooxanthellae-coelenterate symbioses, with a note on the nutritional relationships of *Anemonia sulcata*. J. Marine Biol. Assoc. U.K. 49:1057–1065.

————. 1970. Chloroplasts as symbiotic organelles. Int. Rev. Cytol. 27:29–64.

Taylor, T. N. 1969. Cycads: evidence from the Upper Pennsylvanian. Science 164:294–295.

*————. 1981. Paleobotany. An introduction to fossil plant biology. McGraw-Hill, New York.

Taylor, T. N., and **J. T. Mickel**. 1974. Evolution of systematic characters in the ferns. Ann. Mo. Bot. Garden 61:307–474.

*Taylor, W. R. 1957. Marine algae of the northeastern coast of North America. University of Michigan Press.

*————. 1960. Marine algae of the eastern tropical and subtropical coasts of the Americas. University of Michigan Press.

Terasako, O. 1982. Nuclear differentiation of male gametophytes in gymnosperms. Cytologia 47:27–46.

Terui, K. 1981. Growth and gemma-cup formation in relation to archegoniophore protrusion in *Marchantia polymorpha* L. Annual Report of Fac. of Educ. Iwate Univ. 40:19–28.

Thaithong, O. 1982. Fine structure of spore wall in fourteen species of *Riccia*. J. Hattori Bot. Lab. 53:133–146.

Therrien, C. D. 1966. Microspectrophotometric measurement of nuclear deoxyribonucleic acid content in two Myxomycetes. Canad. J. Bot. 44:1667–1675.

Therrien, C. D., **W. R. Bell**, and **O. R. Collins**. 1977. Nuclear DNA content of myxamoebae and plasmodia in the six nonheterothallic isolates of a myxomycete *Didymium iridis*. Amer. J. Bot. 61:286–291.

Thomas, J. 1970. Absence of the pigments of Photosystem II of photosynthesis in heterocysts of a blue-green alga. Nature 228:181–183.

Thomas, R. B. 1951. Reproduction in *Pinus virginiana* Miller. Ph.D. dissertation, Vanderbilt University.

Thomas, R. J. 1975. Lipid composition of maturing and elongate liverwort sporophytes. Phytochem. 14:623–626.

————. 1976. Correlation between protein content and cell elongation in setae of *Lophocolea heterophylla* sporophytes. J. Hattori Bot. Lab. 40:87–90.

————. 1977a. Wall analyses of *Lophocolea* setae cells (Bryophyta) before and after elongation. Plant Physiol. 59:337–340.

————. 1977b. Water relations of elongating seta cells in sporophytes of the liverwort *Lophocolea heterophylla*. Bryologist 80:345–347.

————. 1980. Cell elongation in hepatics: the seta system. Bull. Torrey Bot. Club 107:339–345.

Thomas, R. J., and **W. T. Doyle**. 1976. Changes in the carbohydrate constituents of elongating *Lophocolea heterophylla* setae. (Hepaticae). Amer. J. Bot. 63:1054–1059.

Thomas, R. J., **D. S. Stanton**, and **M. A. Grusak**. 1979. Radioactive tracer study of sporophyte nutrition in hepatics. Amer. J. Bot. 66:398–403.

Thomas, R. J., **D. S. Stanton**, **D. H. Longendorfer**, and **M. E. Farr**. 1978. Physiological evaluation of the nutritional autonomy of a hornwort sporophyte. Bot. Gaz. 139:306–311.

Trachtengberg, S., and **E. Zaniski**. 1979. The apoplastic conduction of water in *Polytrichum juniperinum* Willd. gametophytes. New Phytol. 83:49–52.

Thrower, S. L. 1964. An investigation of translocation in *Dawsonia intermedia* C. M. Trans. Brit. Bryol. Soc. 4:664–667.

*Tiffany, L. H. 1958. Algae, the grass of many waters. Thomas, Springfield, Ill.

*Tiffney, B. H., and **E. S. Barghoorn**. 1974. The fossil record. *In* R. C. Rollins and K. Roby, eds., Occasional papers of the Farlow Herbarium of Cryptogamic Botany 7. June 1974, Harvard University.

*Tilden, J. E. 1935. The algae and their life relations. University of Minnesota Press.

Timmer, L. W., **J. Castro**, **D. C. Erwin**, **W. L. Belzer**, and **G. A. Zentmeyer**. 1970. Genetic evidence to zygotic meiosis in *Phytophthora caprici*. Amer. J. Bot. 57:1211–1218.

Tingle, M., **Singh Klar**, **S. A. Henry**, and

H. O. Halvorson. 1973. Ascospore formation in yeast. *In* J. M. Ashworth and J. E. Smith, eds., Microbial differentiation. University Press, Cambridge.

Tippo, O. 1942. A modern classification of the plant kingdom. Chron. Bot. 7:203–206.

Torrey, J. G., D. E. Fosket, and P. K. Hepler. 1971. Xylem formation: a paradigm of cytodifferentiation in higher plants. Amer. Scientist 59:338–352.

Trainor, F. R., and C. A. Burg. 1965. *Scenedesmus obliquus* sexuality. Science 148:1094–1095.

Trench, R. K. 1971. The physiology and biochemistry of zooxanthellae symbiotic with marine coelenterates. I. The assimilation of photosynthetic products of zooxanthellae by two marine coelenterates, Proc. Roy. Soc. Lond. B177:225–235. II. Liberation of fixed ^{14}C by zooxanthellae in vitro, Proc. Roy. Soc. Lond. B177:237–250. III. The effect of homogenates of host tissue on the excretion of photosynthetic products in vitro by zooxanthellae from two marine coelenterates. Proc. Roy. Soc. Lond. B177:251–264.

Trench, R. K., M. E. Trench, and L. Muscatine. 1972. Symbiotic chloroplasts: their photosynthetic products and contribution to mucus synthesis in two marine slugs. Biol. Bull. 142:335–349.

Triemer, R. E., and R. M. Brown, Jr. 1977. Ultrastructure of meiosis in *Chlamydomonas reinhardtii*. Brit. Phycol. J. 12:23–44.

*Tryon, A. F. 1964. *Platyzoma*—a Queensland fern with incipient heterospory. Amer. J. Bot. 51:939–942.

*Tryon, R. M., and A. F. Tryon. 1982. Ferns and allied plants with special reference to tropical America. 857 pp. Springer-Verlag, New York, Heidelberg, Berlin.

Tryon, R. M., and G. Vitale. 1977. Evidence for antheridogen production and its mediation of a mating system in natural populations of fern gametophytes. Bot. J. Linn. Soc. Lond. 74:343–349.

Tulecke, W. 1967. Studies in tissue cultures derived from *Ginkgo biloba* L. Phytomorphology 17:381–386.

Tung, H. F., and T. C. Shen. 1981. Studies of the *Azolla pinnata–Anabaena azollae* symbiosis: growth and nitrogen fixation. New Phytol. 87:743–749.

Turner, F. R. 1968. An ultrastructural study of plant spermatogenesis. Spermatogenesis in *Nitella*. J. Cell Biol. 37:370–393.

Umezaki, L. 1967. The tetrasporophyte of *Nemalion vermiculare* Suringar. Rev. Algol. 9:19–24.

Van Baalen, C., and R. M. Brown, Jr. 1969. The ultrastructure of the marine blue-green alga, *Trichodesmium erythraeum*, with special reference to the cell wall, gas vacuoles, and cylindrical bodies. Arch. Mikrobiol. 69:79–91.

Van Baalen, C., D. S. Hoare, and E. Brandt. 1971. Heterotrophic growth of blue-green algae in dim light. J. Bacteriol. 105:685–689.

Van Cotthem, W. R. J. 1970a. Comparative morphological study of the stomata in the Filicopsida. Bull. Jard. Bot. Nat. Belg. 40:81–239.

——. 1970b. A classification of stomatal types. J. Linn. Soc. Lond. (Bot.) 63:235–246.

——. 1971. Vergleichende morphologische Studien über Stomata und eine neue Klassifikation ihrer Typen. Ber. Deutsch. Bot. Ges. 84:141–168.

Van den Ende, H., and D. Stegwee. 1971. Physiology of sex in Mucorales. Bot. Rev. 37:22–36.

Vasil, V. 1959. Morphology and embryology of *Gnetum ula* Brongm. Phytomorphology 9:167–215.

*Verdoorn, F. 1932. Manual of bryology. Martinus Nijhoff, The Hague.

*——. 1938. Manual of pteridology. Martinus Nijhoff, The Hague.

Vitt, D. H. 1968. Sex determination in mosses. Mich. Bot. 7:195–203.

——. 1981. Adaptive modes of the moss sporophyte. Bryologist 84:166–186.

——. 1984. Classification of the Bryopsida. Chap. 13, pp. 696–759. *In* R. M. Schuster, ed., chap. 13, pp. 696–759. 1984.

Vobis, G. 1977. Studies on the germination of lichen conidia. Lichenologist 9:131–136.

Voeller, B. 1971. Developmental physiology of fern gametophytes: relevance for biology. BioScience 21:266–275.

Voeller, B. R., and E. S. Weinberg. 1969. Evolutionary and physiological aspects of antheridium induction in ferns. *In* J. E. Gunckel, ed., Current topics in plant science, pp. 77–93. Academic Press, New York.

Waaland, J. R. 1981. Commercial utilization (of algae). *In* C. S. Lobban and M. J. Wynne, eds., 1981.

Waaland, J. R., S. D. Waaland, and D. Branton. 1971. Gas vacuoles; light shielding in blue-green algae. J. Cell Biol. 48:212–215.

Wager, H. 1913. Life-history and cytology of *Polyphagus euglenae*. Ann. Bot. 27:173–202.

Wagner, W. H. 1964. The evolutionary patterns of living ferns. Mem. Torrey Bot. Club 21:86–95.

Wagner, W. H., Jr. 1973. Some future challenges of fern systematics and phylogeny. *In* A. C. Jermy, J. A. Crabbe, and B. A. Thomas, eds., The phylogeny and classification of ferns, pp. 245–256. Bot. J. Linn. Soc. 67. Supplement 1.

——. 1974. Structure of spores in relation to fern phylogeny. Ann. Mo. Bot. Garden 61:332–353.

——. 1977. Systematic implications of the Psilotaceae. Brittonia 29:54–63.

Wagner, W. H., Jr., J. M. Beitel, and F. S. Wagner. 1982. Complex venation patterns in the leaves of *Selaginella*: megaphyll-like leaves in Lycophytes. Science 218:793–794.

Wagner, W. H., Jr., and W. E. Hammitt. 1970. Natural proliferation of floating stems of scouring-rush, *Equisetum hyemale*. Mich. Bot. 9:166–174.

*Wahl, H. A. 1965. Alternation of generations—again. Turtox News 43:206–209, 248–251.

*Waksman, S. A., and H. A. Lechavalier. 1953.

880

Actinomycetes and their antibiotics. Williams & Wilkins, Baltimore.

Wallace, J. W., and K. R. Markham. 1978a. Flavonoids of the Psilotaceae, Stromopteridaceae and other primitive ferns and their phylogenetic significance. Bot. Soc. Amer. Misc. Publ. 156:29.

———. 1978b. Flavonoids of the primitive ferns: *Stromatopteris, Schizaea, Gleichenia, Hymenophyllum*, and *Cardiomanes*. Amer. J. Bot. 65:965–969.

———. 1978c. Apigenin and Amentoflavone glycosides in the Psilotaceae and their phylogenetic significance. Phytochem. 17:1313–1317.

Walne, P. L., and H. J. Arnott. 1967. The comparative ultrastructure and possible function of eyespots: *Euglena granulata* and *Chlamydomonas eugametos*. Planta 77:325–353.

Walsby, A. E. 1972. Structure and function of gas vacuoles. Bacterid. Rev. 36:1–32.

———. 1978. The properties and buoyancy-providing role of gas vacuoles in *Trichodesmium* Ehrenberg. Brit. Phycol. J. 13:103–116.

*Walton, J. 1953. An introduction to the study of fossil plants. A. & C. Black, London.

Ward, M. 1963. Developmental patterns of adventitious sporophytes in *Phlebodium aureum* J. Sm. J. Linn. Soc. Lond. (Bot.) 58:377–380.

Ward, M., and R. H. Wetmore. 1954. Experimental control of development in the embryo of the fern *Phlebodium aureum* J. Sm. Amer. J. Bot. 41:428–434.

*Wardlaw, C. W. 1952. Phylogeny and morphogenesis. Macmillan, London.

*———. 1955. Embryogenesis in plants. Wiley, New York.

Warmbrodt, R. D. 1980. Characteristics of structure and differentiation in the sieve elements of lower vascular plants. Ber. Deutsch. Bot. Gesellsch. 93:13–28.

Warmbrodt, R. D., and R. F. Evert. 1974. Structure and development of the sieve element in the stem of *Lycopodium lucidulum*. Amer. J. Bot. 61:267–277.

———. 1978. Comparative leaf structure of six species of heterosporous ferns. Bot. Gaz. 139:393–429.

———. 1979. Comparative leaf structure of six species of eusporangiate and protoleptosporangiate ferns. Bot. Gaz. 140:153–167.

Wasson, R. G. 1959. The hallucinogenic mushrooms of Mexico: An adventure in ethnomycological exploration. Trans. N.Y. Acad. Sci. 21:325–339.

———. 1961. The hallucinogenic fungi of Mexico. Harvard Botanical Museum Leaflet 19:137–162.

*———. 1968. Soma—divine mushroom of immortality. Harcourt Brace Jovanovich, New York.

*Wasson, R. G., G. Cowan, F. Cowan, and W. Rhodes. 1974. Maria Sabina and her Mazatec Mushroom Velada. Reviewed in Mycologia 68:953–959.

*Wasson, V. P., and R. G. Wasson. 1957. Mushrooms, Russia, and history. 2 vol. Pantheon Books, New York.

Watanabe, T., C. Espanas, N. Berju, and B. Alimagno. 1977. Utilization of the *Azolla-Anabaena* complex as a nitrogen fertilization for rice. IRRI Research Paper Series 11:1–15.

Waterhouse, G. M. 1973. Plasmodiophoromycetes. *In* G. C. Ainsworth, F. K. Sparrow, and A. S. Sussman, eds., The fungi, vol. IVB, pp. 75–82. Academic Press, New York.

*Watson, E. V. 1964. The structure and life of Bryophytes. Hutchinson University Library, London.

Webb, D. 1981. Effects of light on root nodulation and elongation of seedlings in sterile culture of *Bowenia serrulata*. Phytomorph. 31:121–123.

———. 1982. Effects of light on root growth, nodulation, and apogeotropism of *Zamia pumila* L. seedlings in sterile culture. Amer. J. Bot. 69:298–305.

Webb, D. T. 1977. Root photomorphogenesis and nodulation in *Zamia floridana*. Ph.D. dissertation, University of Montana.

———. 1981. Effects of light quality on root elongation and nodulation of *Zamia floridana* DC. seedlings in sterile culture. Z. Pflanzenphysiol. 106:253–258.

———. 1982a. Structure and ultrastructure of plastids of light and dark-grown *Zamia floridana* DC. seedling roots *in vitro*. New Phytol. 91:721–725.

———. 1982b. Light-induced callus formation and root growth inhibition of *Dioon edule* Lindl. seedlings in sterile culture. Zeit. f. Pflanzenphys. 106:223–228.

———. 1982c. Importance of the megagametophyte and cotyledons for root growth of *Zamia floridana* DC. embryos *in vitro*. Zeit. f. Pflanzenphys. 106:37–42.

———. 1983a. Developmental anatomy of light-induced root nodulation in *Zamia pumila* L. seedlings in sterile culture. Am. J. Bot. 70:1109–1117.

———. 1983b. Nodulation in light and dark-grown *Macrozamia communis* L. Johnson seedlings in sterile culture. Ann. Bot. 52:543–547.

———. 1984. Developmental anatomy and histochemistry of light-induced callus formation by *Dioon edule* (Zamiaceae) seedling roots *in vitro*. Amer. J. Bot. 71:65–68.

Webb, D. T., M. Nevarez, and S. de Jesus. 1984. Further *in vitro* studies on light-induced root nodulation in the Cycadales. Exper. Bot. 24:37–44.

Webb, E. 1981. Stem anatomy, phyllotaxy and stem protoxylem tracheids in several species of *Ophioglossum*. I. *O. petiolatum* and *O. crotalophoroides*. Bot. Gaz. 142:597–608.

Webber, M. M., and P. J. Webber. 1970. Ultrastructure of lichen haustoria: symbiosis in *Parmelia sulcata*. Canad. J. Bot. 48:1521–1524.

*Webster, J. 1970. Introduction to fungi. Cambridge University Press.

Webster, T. R., and R. Jagels. 1977. Morphology and development of aerial roots of *Selaginella martensis* grown in moist containers. Canad. J. Bot. 55:2149–2158.

Webster, T. R., and T. A. Steeves. 1963. Morphology and development of the root of *Selaginella densa* Rydb. Phytomorphology 13:367–376.

———. 1964. Developmental morphology of the root of

Selaginella kraussiana A. Br. and *Selaginella wallacei* Hieron. Canad. J. Bot. 42:1665–1676.

———. 1967. Developmental morphology of the root of *Selaginella martensii* Spring. Canad. J. Bot. 45:395–404.

Weincke, C., and D. Schulz. 1975. Sporophyten-entwicklung von *Funaria hygrometrica* Sebth. I. Strukturelle Grundlagen der Wasser—und Näbrstoffaufuahme in Haustorium. Protoplasma 86:107–117.

Weiss, D. S. 1977. Synchronous development of symbiotic Chlorellae within *Paramecium bursaria*. Trans. Amer. Microbiol. Soc. 96:82–86.

Went, F. W., and N. Stark. 1968. Mycorrhiza. BioScience 18:1035–1039.

Werner, D., ed. 1977. The biology of diatoms. Bot. Monogr. 13. Blackwell Scientific Publications, Oxford.

Wetmore, R. H., A. E. De Maggio, and G. Morel. 1963. Morphogenetic look at the alternation of generations. J. Indian Bot. Soc. (Maheshwari Comm. Vol.) 42A:306–320.

Wetmore, R. H., and J. P. Rier. 1963. Experimental induction of vascular tissues in callus of angiosperms. Amer. J. Bot. 50:418–430.

Wharton, R. A., Jr., and W. C. Vinyard. 1983. Distribution of snow and ice algae in western North America. Madrõno 30:201–209.

Whatley, J. M. 1977. The fine structure of *Prochloron*. New Phytol. 79:309–313.

———. 1982. Ultrastructure of plastid inheritance: green algae to Angiosperms. Biol. Rev. Cambridge Phil. Soc. 57:527–569.

Whatley, J. M., and F. R. Whatley. 1981. Chloroplast evolution. New Phytol. 87:233–247.

———. 1984. Evolutionary aspects of the eukaryotic cell and its organelles. *In* H. F. Linskens and J. Heslop-Harrison, eds., chap 3. 1984.

Whisler, H. C., S. L. Zebold, and J. S. Shemanchuk. 1975. Life-history of *Coelomomyces psorophorae*. Proc. Nat. Acad. Sci. 72:693–696.

White, R. A. 1971a. Experimental and developmental studies of the fern sporophyte. Bot. Rev. 37:509–540.

———. 1971b. Experimental studies of the sporophytes of ferns. BioScience 21:271–275.

———. 1974. Comparative anatomical studies of the ferns. Ann. Mo. Bot. Garden 61:379–387.

White, R. A., D. W. Bierhorst, P. G. Gensel, D. R. Kaplan, and W. H. Wagner. 1977. Taxonomic and morphological relationships of the Psilotaceae. Brittonia 29:1–68.

Whittaker, R. H., and L. Margulis. 1978. Protist classification and the kingdoms of organisms. Bio Systems. 10:3–18.

Whittier, D. P. 1964. The influence of cultural conditions on the induction of apogamy in *Pteridium* gametophytes. Amer. J. Bot. 51:730–736.

———. 1972. Gametophytes of *Botrychium dissectum* as grown in sterile culture. Bot. Gaz. 133:336–339.

———. 1973a. The effect of light and other factors on spore germination in *Botrychium dissectum*. Canad. J. Bot. 51:1791–1794.

———. 1973b. Germination of *Psilotum* spores in axenic culture. Canad. J. Bot. 51:2000–2001.

———. 1975. The origin of the apical cell in *Psilotum* gametophytes. Amer. Fern J. 65:83–86.

———. 1976. Tracheids, apogamous leaves, and sporophytes of *Botrychium dissectum*. Bot. Gaz. 137:237–241.

———. 1977. Gametophytes of *Lycopodium obscurum* as grown in axenic culture. Can. J. Bot. 55:563–567.

———. 1978. Gametophytes of *Ophioglossum engelmanni* from axenic culture. Bot. Soc. Amer. Misc. Publ. 156:28.

Whittier, D. P., 1981a. Gametophytes of *Lycopodium digitatum* (formerly *L. complanatum* var. *flabelliforme*) as grown in axenic culture. Bot. Gaz. 142:519–524.

———. 1981b. Spore germination and young gametophyte development of *Botrychium* and *Ophioglossum* in axenic culture. Amer. Fern. J. 71:13–19.

———. 1983. Gametophytes of *Ophioglossum engelmannii*. Can. J. Bot. 61:2369–2373.

Whittier, D. P., and L. H. Pratt. 1971. The effect of light quality on the induction of apogamy in prothalli of *Pteridium aquilinum*. Planta 99:174–178.

Whittier, D. P., and R. L. Peterson. 1980. Archegonial opening in *Psilotum*. Can. J. Bot. 58:1905–1907.

———. 1984. The surface coating on the gametophytes of *Ophioglossum engelmannii*. Can. J. Bot. 62:360–365.

Whittier, D. P., and T. A. Steeves. 1962. Further studies on induced apogamy in ferns. Canad. J. Bot. 40:1525–1531.

Whitton, B. A., N. G. Carr, and I. W. Craig. 1971. A comparison of the fine structure and nucleic acid biochemistry of chloroplasts and blue-green algae. Protoplasma 72:325–357.

Wiencke, C., and D. Schulz. 1983. The fine structural basis of symplasmic and apoplasmic transport in the "nerve" of the *Funaria* leaflet. Zeit. Pflanzenphysiol. 112:337–350.

Wiese, L. 1984. Mating system in the unicellular algae. Chap. 12. *In* H. F. Linskens and J. Heslop-Harrison, eds., Encyclopedia of plant physiology, chap. 12. 1984.

Wilce, J. H. 1972. Lycopod spores. I. General spore patterns and the generic segregates of *Lycopodium*. Amer. Fern J. 62:65–79.

Williams, M. F. 1981. On the evolution of complex life cycles in plants: A review and ecological perspective. Ann. Missouri Bot. Gard. 68:275–300.

Williams, P. G. 1967. Sporulation and pathogenicity of *Puccinia graminis* f. sp. *tritici* grown on an artificial medium. Phytopathology 57:326–327.

Williams, P. G., K. J. Scott, and J. L. Kuhl. 1966. Vegetative growth of *Puccinia graminis* f. sp. *tritici in vitro*. Phytopathology 56:1418–1419.

Wilson, C. L. 1953. The telome theory. Bot. Rev. 19:417–437.

882

Win-Tin and **M. W. Dick.** 1975. Cytology of the Oomycetes: evidence for meiosis and multiple chromosome association in Saprolegniaceae and Pythiaceae, with an introduction to the cytotaxonomy of *Achlya* and *Pythium.* Arch. Mikrobiol. 105:283–293.

Wittmann, W., F. J. Bergersen, and **G. S. Kennedy.** 1965. The coralloid roots of *Macrozamia communis* L. Johnson. Austral. J. Biol. Sci. 18:1129–1134.

Wochok, Z. S., and **I. M. Sussex.** 1976. Redetermination of cultured root tips to leafy shoots in *Selaginella wildenovii.* Plant Sci. Letters 6:185–192.

Wohlfartt-Bottermann, K. E. 1964. Differentiations of the ground cytoplasm and their significance for the motive force of amoebic movements. *In* R. D. Allen and N. Kamiya, eds., Primitive motile systems in cell biology, pp. 79–109. Academic Press, New York.

Wolk, C. P. 1965. Heterocyst germination under defined conditions. Nature 205:201–202.

———. 1966. Evidence of a role of heterocysts in the sporulation of a blue-green alga. Amer. J. Bot. 53:260–262.

———. 1967. Physiological basis of the pattern of vegetative growth of a blue-green alga. Proc. Nat. Acad. Sci. 57:1246–1251.

———. 1973. Physiology and cytological chemistry of blue-green algae. Bacteriol. Rev. 37:32–101.

Wolniak, S. M. 1976. Organelle distribution and apportionment during meiosis in the microsporocyte of *Ginkgo biloba* L. Amer. J. Bot. 63:251–258.

Woodfin, C. M. 1972. An investigation of the family Ricciaceae. Ph.D. dissertation, University of Texas at Austin.

———. 1976. Physiological studies on selected species of the liverwort family Ricciaceae. J. Hattori Bot. Lab. 41:179–183.

Wyatt, J. T., and **J. K. G. Silvey.** 1969. Nitrogen fixation by *Gloeocapsa.* Science 165:908–909.

Wyatt, R. 1977. Spatial pattern and gamete dispersal distances in *Atrichum angustatum,* a dioecious moss. Bryologist 80:284–291.

Yabu, H. 1969. Observations on chromosomes in some species of *Porphyra.* Bull. Fac. Fish. Hokaido Univ. 19:239–243.

Yadow, B. L., and **T. N. Bhardwaja.** 1980. Morphology and histochemistry of stomata of *Ophioglossum.* Phytomorph. 30:403–406.

Yeh, P., and **A. Gibor.** 1970. Growth patterns and motility of *Spirogyra* sp. and *Closterium acerosum.* J. Phycol. 6:44–48.

Zajic, J. E. 1970. Properties and products of algae. Proceedings of the symposium in the culture of algae sponsored by the Division of Microbiol. Chemistry and Technology of the American Chemical Society, New York, Sept. 7–12, 1969.

Zimmerman, H.-P. 1973. Elektronmikroskopische Untersuchungen zur Spermiogenese von *Sphaerocarpos donnellii* Aust. (Hepaticae). I. Mitochondrien und Plastide. Cytobiol. 7:42–54.

*****Zimmerman, W.** 1952. The main results of the telome theory. Palaeobotanist 1:456–470.

———. 1965. Die Telomtheorie. Fortschritte der Evolutionsforschung I. Gustav Fischer, Stuttgart.

Zingmark, R. G. 1970a. Ultrastructural studies on two kinds of mesokaryotic nuclei. Amer. J. Bot. 57:586–592.

———. 1970b. Sexual reproduction in the dinoflagellate *Noctiluca miliaris* Suriray. J. Phycol. 6:122–126.

Zinsmeister, D. D., and **Z. B. Carothers.** 1974. The fine structure of oogenesis in *Marchantia polymorpha.* Amer. J. Bot. 61:499–512.

ACKNOWLEDGMENTS

Where not acknowledged here or in the captions, the illustrations have been provided by the authors. Acknowledgment is gratefully extended for permission to reproduce the following figures:

Fig. 2-1A: From E. S. Barghoorn and J. W. Schopf, "Microorganisms Three Billion Years Old from the Precambrian of South Africa," *Science* 152:758–763 (1966), Figs. 1, 2, 5. Copyright ©1966 by the American Association for the Advancement of Science.

Fig. 2-1B: From J. W. Schopf, E. S. Barghoorn, M. D. Maser, and R. O. Gordon, "Electron Microscopy of Fossil Bacteria Two Billion Years Old," *Science* 149:1365–1366 (1966). Copyright ©1965 by the American Association for the Advancement of Science.

Fig. 2-5: From E. E. Clifton, *Introduction to the Bacteria*, Fig 3-1. Copyright ©1950 by McGraw-Hill Book company, Inc., New York. Reprinted by permission of the publishers.

Fig. 2-9A From *J. Bacteriol.* 53:527–537, Fig. 5. Copyright ©American Society for Microbiology.

Fig. 2-11A: From D. A. Hopwood, "Phase-Contrast Observations on *Streptomyces coelicolor*," *J. Gen. Microbiol.* 22:295–302 (1960), Fig. 18.

Fig. 2-12: From J. Hunnicutt, "Oil-Eating Bacteria," *Texas Eng. Sci. Mag.* 8:2–3 (1971), upper figure.

Fig. 2-14A: From D. S. Hoare, L. O. Ingram, E. L. Thurston, and R. Walkup, "Dark Heterothrophic Growth of an Endophytic Blue-Green Alga," *Arch. Mikrobiol.* 78:310–321 (1971), Fig. 4.

Fig. 2-32B, D: From R. N. Singh and D. N. Trivari, "Frequent Heterocyst Germination in the Blue-Green Alga *Gloeotrichia ghosei* Singh," *J. Phycol.* 6:172–176 (1970), Figs. 3, 6.

Fig. 4-5C E: From D. A. M. Mesland, "Mating in *Chlamydomonas eugametos*, a scanning electron microscopical study," *Arch. Microbiol.* 109:31–35 (1976), Figs. 2, 3, 4.

Fig. 4-5G: From R. A. Lewin and J. O. Meinhart, "Studies on the Flagella of Algae, III." Reproduced by permission of the National Research Council of Canada from *Can. J. Bot.* 31:711–717 (1953).

Fig. 4-25C, D: From G. L. Floyd, K. D. Stewart, and K. R. Mattox, "Cytokinesis and Plasmodesmata in *Ulothrix*," *J. Phycol.* 7:306–309 (1971), Fig. 1.

Fig. 4-31: From R. N. Singh, "On Some Phases in the Life History of the Terrestrial Alga *Fritschiella tuberosa* Iyeng. and Its Autecology," *New Phytol.* 60:170–182 (1941), Fig. 1.

Fig. 4-38E: From P. W. Cook, "Variation in Vegetative

and Sexual Morphology Among the Small Curved Species of *Closterium*," *Phycol.* 3:1–18 (1963), Fig. 20.

Fig. 4-43: From G. M. Smith, *Cryptogamic Botany*, 2nd ed., Vol. 1, Fig. 53A, F, G. Copyright ©1955 by McGraw-Hill Book Company, Inc., New York. Reprinted by permission of the publishers.

Fig. 4-45A: From J. Ramus, "*Codium:* The Invader," *Discovery*, 6:59–68 (1971), Fig. 1.

Fig. 4-47A, B, C: From G. M. Smith, *Cryptogamic Botany*, 2nd ed., Vol. 1, Fig 65. Copyright ©1955 by McGraw-Hill Book Company, Inc., New York. Reprinted by permission of the publishers.

Fig. 4-47D: From G. M. Smith, *Cryptogamic Botany*, 2nd ed., Vol. 1, Fig. 64B. Copyright ©1955 by McGraw-Hill Book Company, Inc., New York. Reprinted by permission of the publishers.

Fig. 5-4: From J. D. Pickett-Heaps, "Ultrastructure and Differentiation in *Chara*. I. Vegetative cells," *Austral. J. Biol. Sci.* 20:539–551 (1967), Figs. 1, 2.

Fig. 5-9: From F. R. Turner, "An Ultrastructural Study of Plant Spermatogenesis," *J. Cell Biol.* 37:370–393 (1968), Fig. 6.

Fig. 7-6D: From K. Schmitz and L. M. Srivastava, "Fine Structure and Development of Sieve Tubes in Laminaria groenlandica," *Rosen. Cytobiologie* 10:66–87.

Fig. 7-9: From T. Kanda, "On the Gametophytes of Some Japanese Species of Laminariales," *Scientific Papers*, Institute Algological Research, Hokkaido Imp. Univ., 1936, Vol. 1, No. 2, pp. 221–260.

Fig. 7-10A, B: From S. I. C. Hsiao and L. D. Druehl, "Environmental Control of Gametogenesis in *Laminaria saccharina*. I. The Effects of Light and Culture Media." Reproduced by permission of the National Research Council of Canada from the *Canad. J. Bot.* 49:1503–1508 (1971), Fig. 6.

Fig. 7-11: Reprinted from G. M. Smith, *Marine Algae of the Monterey Peninsula*, by permission of the publishers, Stanford University Press, Stanford, Calif. Copyright ©1944, 1966, 1969 by the Board of Trustees of the Leland Stanford Junior University. Plate 26.

Fig. 7-14B: Reprinted from G. M. Smith, *Marine Algae of the Monterey Peninsula*, by permission of the publishers. Stanford University Press, Stanford, Calif. Copyright ©1944, 1966, 1969 by the Board of Trustees of Leland Stanford Junior University. Fig. on p. 465.

Fig. 7-15B: From W. R. Taylor, *Marine Algae of the Eastern Tropical and Subtropical Coasts of the Americas*, University of Michigan Press, Ann Arbor (1960), Fig. 1, Plate 34.

Fig. 7-24: From E. G. Pollock, "Fertilization in *Fucus*," *Planta* 92:85–99 (1970), Fig. 1-7.

Fig. 8-3: From A. Massalski and G. F. Leedale, "Cytology and Ultrastructure of the Xanthophyceae. I. Comparative Morphology of the Zoospores of *Bumilleria sicula* Borzi and *Tribonema vulgare* Pascher," *Brit. Phycol. J.* 4:159–180 (1969), Fig. 11.

Fig. 8-7C: From W. J. Koch, "A Study of the Motile Cells of *Vaucheria,*" *J. Elisha Mitchell Sci. Soc.* 67:123–131 (1951), Fig. 18, Plate 5.

Fig. 8-7D, E: From Ø. Moestrup, "On the Fine Structures of the Spermatozoids of *Vaucheria sescuplicaria* and on the Later Stages of Spermatogenesis," *J. Marine Biol. Ass. U.K.* 50:513–523 (1970), Figs, C, D, Plate I.

Fig. 8-8A: From G. M. Smith, *Cryptogamic Botany*, 2nd ed., Vol. 1, Fig. 101A, B. Copyright ©1955 by McGraw-Hill Book Company, Inc., New York. Reprinted by permission of the publishers.

Fig. 8-8B: From J. N. Couch, "Gametogenesis in *Vaucheria,*" *Bot. Gaz.* 94:272–296 (1932), Fig. 8.

Fig. 8-11A: From G. M. Smith, *Freshwater Algae of the United States*, Fig. 336B. Copyright ©1950 by McGraw-Hill Book Company, Inc., New York. Reprinted by permission of the publishers.

Fig. 8-15: From R. Subrahmanyan, "On Somatic Division, Reduction Divison, Auxospore Formation and Sex-Differentiation in *Navicula halophila* (Grun.)," Cl. *Indian Bot. Soc.*, The M.O.P. Iyengar Commemoration Volume (1945), pp. 238–266, Figs. 16, 17, 27, 28, and 39.

Fig. 8-19: From H. A. Von Stosch and G. Drebes, "Entwicklungsgeschichtliche Untersuchungen an zentrischen Diatomeen IV. Die Planktondiatomee *Stephanopyxis turris*—ihre Behandlung und Entwicklungsgeschichte, *Helgoland. Wiss. Meeresunters.* 11:209–257 (1964), Fig. 18.

Fig. 8-20: From G. Drebes, "On the Life History of the Marine Plankton Diatom *Stephanopyxis palmeriana,*" *Helgoland. Wiss. Meeresunters.* 13:101–114 (1966), Figs. 2, 4–7.

Fig. 8-21: From A. W. Haupt, *Plant Morphology*, McGraw-Hill Book Company, New York, 1953, Fig. 247D. Used with permission of McGraw-Hill Book Company.

Fig. 9-2: From E. Gantt and S. F. Conti, "The Ultrastructure of *Porphyridium cruentum.*" Reprinted by permission of the Rockefeller University Press, New York, from *J. Cell Biol.* 26:365–381 (1965), Fig. 1.

Fig. 9-3B–D: From M. R. Sommerfeld and H. W. Nichols, "Developmental and Cytological Studies of *Bangia purpurea* in culture," *Amer. J. Bot.* 57:640–648 (1970), Figs. 2, 6, 9.

Fig. 9-11A–C: From L. Umezaki, "The Tetrasporophyte of *Nemalion vermiculare* Suringar," *Rev. Algol.* 9:19–24 (1967), Figs. 9–11.

Fig. 9-11D, E: From M. Masuda and L. Umezaki, "On the Life History of *Nemalion vermiculare* Surnigar," (Rhodophyta) in Culture," *Bull. Jap. Soc. Phycol.* 25. Suppl. (Mem. Iss. Yamada):129–136 (1977).

Fig. 9-14: From P. Bourrelly, *Les Algues d'eau douce.* III.

Les Algues bleues et rouges, Editions N. Boubée et Cie, Paris (1970), Figs. 2, 6, Plate 54.

Fig. 9-15: From H. C. Bold, *The Plant Kingdom*, 3rd ed., Prentice-Hall, Englewood Cliffs, N.J. (1970), Fig. 2–23. Courtesy of P. Edwards. Reprinted by permission of Prentice-Hall, Inc.

Fig. 9-17: From P. Edwards, "The Life History of *Polysiphonia denudata* (Dillwyn) Kützing in Culture," *J. Phycol.* 4:35–37 (1968), Fig. 1–6.

Fig. 10-1: From A. H. Knoll and E. Barghoorn, Precambrian Eukaryote organisms: a reassessment of the evidence, *Science* 190:52–54. Copyright ©the American Association for the Advancement of Science.

Fig. 10-2: From J. W. Schopf and E. S. Barghoorn, "Alga-like Fossils from the Early Precambrian of South Africa," *Science* 156:508–512 (1967), Fig. 1-3. Copyright ©1967 by the American Association for the Advancement of Science.

Fig. 10-3A, B, D: From J. W. Schopf and J. M. Blacic, "New Microorganisms from the Bitter Springs Formation (Late Precambrian) of the North Central Amadeus Basin, Australia," *J. Paleontol.* 45:925–959 (1971), Figs. 1, 4.

Fig. 10-3C: From J. W. Schopf, "Precambrian Microorganisms and Evolutionary Events Prior to the Origin of Vascular Plants," *Biol. Rev.* 45:319–352 (1970), Fig. 5, Plate I; Fig. 46, Plate IV.

Fig. 10-5A: From J. Pia, "Die Siphoneae Verticillatae von Karbon bis zur Kreide," *Abh. Zool.-bot. Ges.* Bd. 11(2) Wien (1920), Fig. 21, p. 112.

Fig. 10-5B: From J. Pia, "Die Diploren der deutschen Trias und ide Frage der Gleichsetzung der deutschen und Alpinen Triasstufen," *Z. Deut. Geol. Ges.* 78, Monatsch, 192.

Fig. 10-5C: From R. A. Baschnagel, "New Fossil Algae from the Middle Devonian of New York," *Trans. Amer. Microbiol. Soc.* 85:297–302 (1966), Fig. 3.

Fig. 10-6: From R. E. Peck and J. A. Eyer, "Pennsylvanian, Permian and Triassic Charophyta of North America." *J. Paleontol,,* 37:835–844 (1963), Fig. 3-8.

Fig. 10-7: From A. Mann, "The Economic Importance of the Diatoms," from the *Annual Report of the Smithsonian Institution*, 1916. Reprinted with permission of the Smithsonian Institution, Washington, D.C.

Figs. 11-36B–E, 11-37A, B: From G. M. Smith, *Cryptogamic Botany*, 2nd ed., Vol. 2, Figs. 9F, G; 7G, H; 10A, F. Copyright ©1955 by McGraw-Hill Book Company, Inc., New York. Reprinted by permission of the publishers.

Fig. 11-54D, E: From R. M. Schuster, *Hepaticae and Anthroceroteae of North America*, Columbia University Press, New York (1966-1974), Vol. 1 (1966), Fig. 13.

Fig. 11-59A: From E. V. Watson, *British Mosses and Liverworts*, Cambridge University Press, London (1968), Fig. 199.

Fig. 11-59B: From R. M. Schuster, *Hepaticae and Anthoceroteae of North America*, Columbia University

Press, New York (1966, 1969), Vol. 1 (1966), Fig. 80, plate 1.

Fig. 11-59C, D: From R. M. Schuster, *Hepaticae and Anthoceroteae of North America*, Columbia University Press, New York, 1966–1974, Vol. 2 (1969), Plate 89, Figs. 5, 7; Plate 90, Fig. 11.

Fig. 11-60A–F: From R. M. Schuster, *Hepaticae and Anthoceroteae of North America*, Columbia University Press, New York, 1966–1974, Vol. 2 (1969), Plate 113, Fig. 3; Plate 257, Fig. 5; Plate 260, Figs. 1 and 2; Plate 114, Fig. 4. Vol. 3 (1974), Plate 424, Figs. 1–3.

Fig. 11-61B–D: From R. M. Schuster, *Hepaticae and Anthoceroteae of North America*, Columbia University Press, New York, 1966–1974, Vol. 3 (1974), Plate 447; Plate 465, Figs. 1, 2.

Fig. 11-62A: From R. M. Schuster, *Hepaticae and Anthoceroteae of North America*, Columbia University Press, New York, 1966–1974, Vol. 3 (1974), Plate 467, Figs. 3, 5.

Fig. 11-63D: From E. V. Watson, *British Mosses and Liverworts*, Cambridge University Press, London, 1968, Fig. 242.

Fig. 11-64C, D: From R. M. Schuster, *Hepaticae and Anthoceroteae of North America*, Columbia University Press, New York, 1966–1974, Vol. 1 (1966), Plate 68, Figs. 1, 2.

Figs. 11-65C, D: From R. M. Schuster, *Hepaticae and Anthoceroteae of North America*, Columbia University Press, New York, 1966–1974, Vol. 1 (1966), Plate 18, Fig. 1.

Fig. 11-67: From W. D. P. Steward and G. A. Rogers, "The Cyanophyta-Hepatic Symbiosis. II. Nitrogen Fixation and the Interchange of Nitrogen and Carbon," *New Phytologist* 78:459–471 (1977).

Fig. 11-80: From D. H. Campbell, "Studies on Javanese Anthocerotaceae, I, II. *Ann. Bot.* 21:467–468 (1907), Fig. 1; *Ann. Bot.* 22:91–102 (1908), Fig. 1.

Fig. 11-81: From J. Walton, "Carboniferous Bryophyta; I. Hepaticae," *Ann. Bot.* 39 (1925), Plate 13, Fig. 1.

Fig. 11-83: From T. M. Harris, *The British Rhaetic Flora*, British Museum (Natural History), London, 1938, Figs. 15A, 18D. Used by permission of the Trustees of the British Museum (Natural History).

Fig. 11-84: From H. N. Andrews, Jr., "Notes on Belgian Specimens of *Sporogonites*," *Palaeobotanist* 7:85–89 (1960), text Fig. 1.

Fig. 12-9: From G. M. Smith, *Cryptogamic Botany*, 2nd ed., Vol. 2, Fig. 64. Copyright ©1955 by McGraw-Hill Book Company, Inc., New York. Reprinted by permission of the publishers.

Fig. 12-10B: From G. M. Smith, *Cryptogamic Botany*, 2nd ed., Vol. 1, Fig. 65B. Copyright ©1955 by McGraw-Hill Book Company, Inc., New York. Reprinted by permission of the publishers.

Fig. 12-10D: From S. Flowers, *Mosses: Utah and the West*, Brigham Young University Press, Provo, Utah, 1973, Plate 3, Figs. 194, 195.

Fig. 12-11A–C: From Engler and Prantl, "Die Natürlichen Pflanzenfamilien," Figs. 10, 11.

Fig. 12-15A, 12-16A, 12-17: From N. S. Parihar, *An Introduction to Embryophyta*, Vol. 1, *Bryophyta*, Central Book Depot, Allahabad (1957), Figs. 64, 79.

Fig. 12-24: From D. J. Paolillo, Jr., G. L. Kreitner, and J. A. Reighard, "Spermatogenesis in *Polytrichum juniperinum.* I and II. The Mature Sperm," *Planta* 78:248–261 (1968), Fig. 15.

Fig. 12-25A: From D. H. Campbell, *Mosses and Ferns*, Macmillan, New York (1928), Fig. 106A.

Fig. 12-26I: From J. C. French and D. J. Paolillo, Jr., "Intercalary Meristem Activity in the Sporophyte of *Funaria* (Musci)," *Amer. J. Bot.* 62:286–296 (1975), Fig. 2.

Fig. 12-27A–C; 12-29B: From J. C. French and D. J. Paolillo, Jr., "Intercalary Meristem Activity in the Sporophyte of *Funaria* (Musci)," *Amer. J. Bot.* 62:286–296 (1975), Figs. 3, 7, 10.

Fig. 12-30E: From D. Schulz and W. Schmidt, "Entwicklungs des Peristoms von *Funaria hygrometrica*," *Flora* 163:451–465.

Figs. 12-33F–H, J: From S. Flowers, *Mosses: Utah and the West*, Brigham Young University Press, Provo, Utah, 1973, Figs. 190, 192, 193, 170.

Figs. 12-35A, B: From J. A. Snider, "A Revision of the Genus *Archidium* (Musci)," *J. Hattori, Bot. Lab.* 39:105–201 (1975), Figs. 335, 338.

Figs. 12-35C, D: From J. A. Snider, "Sporophyte Development in the Genus *Archidium* (Musci)," *J. Hattori, Bot. Lab.* 39:85–104 (1975), Figs. 22, 25.

Figs. 12-36, 12-37, 12-38, 12-41, 12-43, 12-44, 12-45A, C: From S. Flowers, *Mosses: Utah and the West*, Brigham Young University Press, Provo, Utah, 1973, Plate 3, Figs. 2, 7, 8; Plate 8, Fig. 5; Plate 64, Fig. 2; Plate 88, Fig. 1; Plate 140, Figs. 1, 9; Plate 133, Figs. 1, 3, 10.

Fig. 12-45B: From K. Denning, "Untersuchungen über sexuellen Dimorphismus der Gametophyten heterothallischen Laubmoosen," *Flora* N. F. 30:57–86, Fig. 2.

Fig. 12-46: From S. Flowers, *Mosses: Utah and the West*, Brigham Young University Press, Provo, Utah, 1973, Figs, 185, 188.

Fig. 12-48A: From W. C. Steere, "Cenozoic and Mesozoic Byophytes of North America," *Amer. Midland Natur.* 36(2):298–324 (1946), Plate 2, lower right-hand figure.

Fig. 13-2: From H. C. Bold, *The Plant Kingdom* 3rd ed., Prentice-Hall, Englewood Cliffs, N.J. (1970), Fig. 5-6. Reprinted by permission of Prentice-Hall, Inc.

Fig. 13-3: From O. Schüepp, *Meristeme*, Vol. IV, Handb. Pflanzenanatonie 1 abt., 2 teil, Histologie, Fig. 16, 1926.

Fig. 13-11: From H. C. Bold, *The Plant Kingdom* 3rd ed., Prentice-Hall, Englewood Cliffs, N.J. (1970), Fig. 5-7. Reprinted by permission of Prentice-Hall, Inc.

Fig. 13-12: From A. J. Eames and L. H. MacDaniels, *An Introduction to Plant Anatomy*, Fig. 121. Copyright ©1947 by McGraw-Hill Book Company, Inc., New York. Reprinted by permission of the publishers.

Fig. 13-14: From M. R. Darnley Gibbs, *Botany*, McGraw-Hill (Blakiston), New York, 1950, Fig. 154.

Fig. 13-16: From G. S. Avery, "Structure and Development of the Tobacco Leaf," *Amer. J. Bot.* 20:565–592 (1933), Fig. 6.

Fig. 13-19A, B: From H. C. Bold, *The Plant Kingdom* 3rd ed., Prentice-Hall, Englewood Cliffs, N.J. (1970), Fig. 5-21. Reprinted by permission of Prentice-Hall, Inc.

Figs. 14-11A, 14-12B, 14-13A, 14-14: From H. Bruchmann, "Die Keimung der Sporen und die Entwicklung der Prothallien von *Lycopodium clavatum*, *L. annotinum* und *L. selago*," *Flora* 101:262, 263 (1910).

Figs. 14-16A B, 14-20A, B: From D. S. Correll, *Ferns and Fern Allies of Texas*, Texas Research Foundation, Renner (1956), Fig. 1, Plate 4; Fig. 5, Plate 5.

Fig. 14-22: From T. R. Webster and T. A. Steeves, "Developmental Morphology of the Root of *Selaginella kraussiana* A.Br. and *Selaginella wallacei* Hieron." Reproduced by permission of the National Research Council of Canada from the *Canad. J. Bot.* 42:1665 (1964).

Fig. 14-30A, B: From D. Robert, "Le gamétophyte femelle de *Selaginella kraussiana* (Kunze) A.Br. I. Organisation générale de la megaspore, etc." *Rev. Cytol. Biol. Végétales* 34:189–232 (1971), Fig. 1, Plate 2; text Fig. 2.

Figs. 14-32B, 14-33: From M. Bruchmann, "Die Keimung der Sporen und die Entwicklung der Prothallien von *Lycopodium clavatum*, L. *annotinum* und L. *selago*, *Flora* 101:263 (1910).

Fig. 14-34: From E. Steiner, A. S. Sussman, and W. H. Wagner, Jr., *Botany Laboratory Manual*, Holt, Rinehart and Winston, New York (1965), Fig. 22-2.

Fig. 14-37B: From D. J. Paolillo, "The Development and Anatomy of Isoetes," *Illinois Biol. Monograph* 31 (1963), Fig. 1.

Fig. 14-40A: From J. Liebig, "Erganzungen zur Entwicklungsgeschichte von *Isoetes lacustris* L.," *Flora* 125:343 (1931).

Fig. 14-40B, C: From C. LaMotte, "Morphology of the Megagametophyte and the Embryo Sporophyte of *Isoetes lithophila*," *Amer. J. Bot.* 20:217–233 (1933), text Fig. 16.

Figs. 15-4C, 15-5C: From P. B. Kaufman, W. C. Bigelow, R. Schmid, and N. S. Ghosheh, "Electron Probe Analysis in Epidermal Cells of *Equisetum*," *Amer. J. Bot.* 58:309–316 (1971), Fig. 20.

Fig. 15-11: From J. G. Duckett, "Toward an Understanding of Sex Determination in *Equisetum*: An Analysis of Regeneration in Gametophytes of the Subgenus *Equisetum*," *Bot. J. Linn. Soc.* 74:215–242 (1977), Plate 2, Figs. A, C.

Fig. 15-14B: From G. M. Smith, *Cryptogamic Botany*, 2nd ed., Vol. 2, Fig. 160G. Copyright © 1955 by McGraw-Hill Book Company, Inc., New York. Reprinted by permission of the publishers.

Fig. 15-15D: From M. R. Mesler and K. L. Lu, "Large

Gametophytes of *Equisetum hyemale* in Northern California," *Amer. Fern Journal* 67:97–98, Fig. 1.

Fig. 16-9B: From G. M. Smith, *Cryptogamic Botany*, 2nd ed., Vol. 2, Fig. 185E. Copyright © 1955 by McGraw-Hill Book Company, Inc., New York. Reprinted by permission of the publishers.

Fig. 16-10A, C-F, H: From M. R. Mesler, "The gametophytes of *Ophioglossum palmatum*. *Amer. J. Bot.* 62:982–992 (1975), Figs. 1, 2, 6, 28, 34, 44.

Fig. 16-10B, G: From M. R. Mesler, "Gametophytes and Young Sporophytes of *Ophioglossum crotalophoroides*, *Amer. J. Bot.* 63:443–448.

Fig. 16-11A: From G. M. Smith, *Cryptogamic Botany*, 2nd ed., Vol. 2, Fig. 186. Copyright © 1955 by McGraw-Hill Book Company, Inc., New York. Reprinted by permission of the publishers.

Fig. 16-14: From A. G. Stokey, "Gametophytes of *Marrattia sambricina* and *Macroglossum Smithii*," *Bot. Gaz.* 103, University of Chicago Press, Chicago (1942), Fig. 3.

Fig. 17-15: From A. F. Tryon, "*Platyzoma*—A Queensland Fern with Incipient Heterospory," *Amer. J. Bot.* 51:939–942 (1964), Figs. 7, 8.

Fig. 17-20A, B: From U. Näf, "Control of Antheridium Formation in the Fern Species *Anemia phyllitides*," *Nature* 184:798–800 (1959), Figs. la, b.

Fig. 17-21A–D: From A. W. Haupt, *Plant Morphology*, Fig. 247D. Copyright © 1953 by McGraw-Hill Book Company, Inc., New York. Reprinted by permission of the publishers.

Fig. 17-23: From H. A. Gleason, *New Britton and Brown Illustrated Flora*, Vol. 1 (1952), parts of figures on *Osmunda*, p. 27. used by permission of the New York Botanical Garden.

Figs. 17-26, 17-33: From H. N. Andrews, Jr., *Studies in Paleobotany* (1961), Figs. 4-6A, B, D.

Fig. 17-32A: From D. S. Correll, *Ferns and Fern Allies of Texas*, Texas Research Foundation, Renner (1956), Plate 10, Fig. 1.

Fig. 17-32C: From C. A. Brown and D. S. Correll, *Ferns and Fern Allies of Louisiana*, Louisiana State University Press, Baton Rouge (1942), Fig. 37.

Fig. 17-34: From G. M. Smith, *Cryptogamic Botany*, 2nd ed., Vol. 2, Fig. 197A, B. Copyright © 1955 by McGraw-Hill Book Company, Inc., New York. Reprinted by permission of the publishers.

Fig. 17-35B, C: From D. W. Bierhorst, "The Fleshy, Cylindrical, Subterranean Gametophyte of *Schizaea melanesica*," *Amer. J. Bot.* 53:123–133 (1966), Figs. 1, 2.

Fig. 17-35D–G: From D. W. Bierhorst, "On the Stromopteridaceae (fam. nov.) and on the Psilotacae," *Phytomorphology* 18:233–268 (1968), Figs. 1, 52, 82, 109.

Fig. 17-36A: From C. A. Brown and D. S. Correll, *Ferns and Fern Allies of Louisiana*, Louisiana State University Press (1942), Fig. 5.

Fig. 17-36B: From G. M. Smith, *Cryptogamic Botany*, 2nd ed., Vol. 2, Fig. 214. Copyright © 1955 by

McGraw-Hill Book Company, Inc., New York. Reprinted by permission of the publishers.

Fig. 17-37: From G.M. Smith, Cryptogamic Botany, 2nd ed., Vol. 2, Figs. 202, 204F & G, and 205A. Copyright © 1955 by McGraw-Hill Book Company, Inc., New York. Reprinted by permission of the publishers.

Figs. 18-1A, 18-3, 18-7: From A. J. Eames, *Morphology of Vascular Plants*, Figs. 126, 133B. Copyright © 1936 by McGraw-Hill Book Company, Inc., New York. Reprinted by permission of the publishers.

Fig. 18-9: From A. W. Haupt, *Plant Morphology*, Fig. 251, Copyright © 1953 by McGraw-Hill Book Company, Inc., New York. Reprinted by permission of the publishers.

Fig. 18-12A: From L. W. Sharp, "Spermatogenesis in *Marsilia*," *Bot. Gaz.* 58, University of Chicago Press, Chicago (1914), Plate 33, Figs. 7, 8.

Fig. 18-14E: Reprinted by permission of the Rockefeller University Press from *J. Cell Biol.* 29:97–111 (1966).

Fig. 18-15A: From D. H. Campbell, *Mosses and Ferns*, The Macmillan Company, New York (1928), Fig. 250A.

Fig. 18-18D: From B. P. Kremer, "Mikroalgen als Symbiosepartner," *Mikrokosmos* 67:206–209, Fig. 2.

Fig. 19-4: From A. S. Rouffa, "An Appendageless *Psilotum:* Introduction to Aerial Shoot Morphology," *Amer. Fern J.* 61:75–86 (1971), Figs. 3, 4.

Fig. 19-11: From D. W. Bierhorst, "Structure and Development of the Gametophyte of *Psilotum nudum*," *Amer. J. Bot.* 40 (1953), Figs, 5, 6, 7.

Figs. 20-1, 20-2, 20-4 through 20-9: From W. N. Stewart, "An Upward Look in Plant Morphology," *Phytomorphology* 14:120–124 (1964), parts of Figs. 1–16.

Fig. 20-14A: From R. Kidston and W. H. Lang, "On Old Red Sandstone Plants Showing Structure from the Rhynie Chert Bed, Aberdeenshire," *Trans. Roy. Soc. Edinburgh* 51, 52 (1917–1921), Part I: Plate IX, Fig. 62.

Fig. 20-15: From Y. Lemoigne, "Nouvelles diagnoses du genre *Rhynia* et de l'espèce *gwynnevaughanii*," *Bull. Soc. Bot. France* 117:307–320 (1970), Fig. on p. 310.

Fig. 20-17A: From D. Edwards, "Fertile Rhyniophytina from the Lower Devonian of Britain," *Paleontology* 13:451–461 (1970), Fig. 4.

Fig. 20-17B: From A. R. Ananjev and S. A. Stepanov, "Traces of Spore-Bearing Organs of *Psilophyton princeps* Dawson Emend. Halle in the Lower Devonian in the Southern Minusina Basin (Western Siberia)," *Records of Tomsk State University* 203 (1969), Fig. 3a.

Fig. 20-19: From J. Walton, "On the Morphology of *Zosterophyllum* and Some Early Devonian Plants," *Phytomorphology* 14:155–160 (1964), Fig. 1.

Fig. 20-20: From H. N. Andrews and A. E. Kasper, "Plant Fossils of the Trout Valley Formation," *Maine State Geol. Surv. Bull.* 23:3–16 (1970), Fig. 5.

Fig. 20-21: From P. Genzel, A. Kasper, and H. N. Andrews, "*Kaulangiophyton*, A New Genus of Plants from the Devonian of Maine," *Bull. Torrey Bot. Club* 96:265–276 (1969).

Fig. 20-22: From H. P. Banks and M. R. Davis, "*Crenaticaulis*, a New Genus of Devonian Plants Allied to *Zosterophyllum*, and Its Bearing on the Classification of the Early Land Plants," *Amer. J. Bot.* 56:436–449 (1969), Figs. 2, 3.

Fig. 20-24A: From E. Boureau, *Traité de Paléobotanique*, Vol. II, Masson et Cie, Paris (1967), Fig. 305.

Fig. 20-24B: From R. Kräusel and H. Weyland, "Uber *Psilophyton robustius* Dawson," *Palaeontographica* 108B:11–21 (1961), Fig. 1.

Fig. 20-26A: With permission of the Trustees of the British Museum (Natural History).

Fig. 20-26C: Courtesy Royal Scottish Museum.

Fig. 20-27: From W. H. Lang and I. Cookson, "On a Flora Including Vascular Land Plants, Associated with *Monograptus*, in Rocks of Silurian Age from Victoria, Australia," *Phil. Trans. Roy. Soc. London* 2.24B:421–449 (1935), Plate 29, Fig. 1.

Fig. 20-28: From R. Kräusel and H. Weyland, "Pflanzenreste aus den Devon," *Senckenbergiana* (1932), Fig. 14.

Fig. 20-32: From D. A. Eggert, "The Ontogeny of Carboniferous Arborescent Lycopsida," *Palaeontographica* 108B:43–92 (1961), Fig. 75.

Fig. 20-37A: From G. M. Smith, *Cryptogamic Botany*, 2nd ed., Vol. 2, Fig. 130A. Copyright © 1955 by McGraw-Hill Book Company, Inc., New York. Reprinted by permission of the publishers.

Fig. 20-39: From M. Hirmer, *Handbuch der Palaobotanik*, Vol. 1, Oldenbourg, München, and Berlin (1927), Fig. 285.

Fig. 20-40: From H. N. Andrews, Jr., *Studies in Paleobotany*, Wiley, New York (1961), Fig. 8-14.

Fig. 20-41: From H. N. Andrews, C. B. Read, and S. H. Mamay, "A Devonian Lycopod Stem with Well-Preserved Cortical Tissues," *Palaeontology* 14:1–9 (1971), Fig. 1, Plate I.

Fig. 20-43, 20-44A: From C. M. Schlanker and G. A. Leisman, "The Herbaceous Carboniferous Lycopod *Selaginella fraiponti* comb. nov," *Bot. Gaz.* 130:35–41 (1969), Figs. 11, 13.

Fig. 20-44B, C: From J. H. Hoskins and M. L. Abbott, "*Selaginellites crassicinctus*, a New Species from the Desmoinesian Series of Kansas," *Amer. J. Bot.* 43:36–46 (1956), Figs. 5, 7, 8.

Fig. 20-45: From M. Hirmer, "Paläophytologische Notizen: I. Reckonstruction von *Pleuromeia sternbergi* Corda, nebst. Bemerkungen zur Morphologie der Lycopodiales," *Palaeontographica* 78:48 (1933).

Fig. 20-50A: From M. Hirmer, *Handbuch der Paläobotanik*, Vol. 1, Oldenbourg, München, and Berlin (1927), Fig. 537.

Fig. 20-55: From W. Zimmerman, *Die Phylogenie der Pflanzen*, Springer Verlag, Jena (1959), Fig. 107.

Fig. 20-57A, B: From M. Hirmer, *Handbuch der Paläobotanik*, Vol. 1, Oldenbourg, München, and Berlin (1927), Fig. 417.

Fig. 20-58: From M. Hirmer, *Handbuch der Palaobotanik*, Vol. 1, Oldenbourg, München, and Berlin (1927), Figs. 553, 554.

Fig. 20-60: From H. N. Andrews, Jr., *Studies in Paleobotany*, Fig. 3-12. Copyright © 1961 by John Wiley & Sons, Inc., New York. Reprinted by permission of the publishers.

Fig. 20-61: From R. Kräusel and H. Weyland, "Beitrag zur Kenntnis der Devon-flora II," *Abh. Senckenbergische Natur. Ges.* 40:113–155 (1926), Fig. 24.

Fig. 20-62: From S. Leclercq and H. P. Banks, "*Pseudosporochnus nodosus* sp. nov., a Middle-Devonian Plant with Cladoxylalean Affinities," *Palaeontographica* 110B:1–34 (1962), Figs. 2, 3, 7.

Fig. 20-63B: From S. Leclercq and H. N. Andrews, Jr., "*Calamophyton bicephalum*, a New Species from the Middle Devonian of Belgium" *Ann. Missouri Bot. Garden* 47:1–23 (1960), Fig. 8.

Fig. 20-64B, C: From T. Delevoryas, *Morphology and Evolution of Fossil Plants*, Holt, Rinehart and Winston, New York (1962), Fig. 7-4B.

Fig. 20-65B: From W. H. Murdy and H. N. Andrews, Jr., "A Study of *Bryopteris globosa* Darrah," *Bull. Torrey Bot. Club* 84:252–267 (1957), Fig. 8.

Fig. 20-66B: From R. Wettstein, *Handbuch der Systematischen Botanik*, Franz Deuticke, and Asher and Co., Leipzig (1962), Fig. 262-261.

Fig. 20-67: From J. Morgan, "The Morphology and Anatomy of American Species of the Genus *Psaronius*," *Illinois Biol. Monogr.* No. 27, pp. 1–108 (1959), frontispiece.

Fig. 20-68: From T. Delevoryas, *Morphology and Evolution of Fossil Plants*, Holt, Rinehart and Winston, Inc., New York (1962), Fig. 7-11A.

Fig. 21-1: From H. C. Bold, *The Plant Kingdom*, 3rd ed., Prentice-Hall, Englewood Cliffs, N.J. (1970), Fig. 8-1. Reprinted by permission of Prentice-Hall, Inc.

Fig. 21-11: From F. Grace Smith, "Development of the Ovulate Strobilus and Young Ovule of *Zamia floridana*," *Bot. Gaz.* 50, University of Chicago Press, Chicago (1910), Figs. 9, 105.

Fig. 21-18: From K. Norstog, "The Spermatozoids of *Zamia chigua*," *Biol. J. Linn. Soc.* (1977) Suppl. 1 to Vol. 7, Plate I, Figs. A, B.

Fig. 21-19: From G. S. Bryan, "The Cellular Proembryo of *Zamia* and Its Cap Cells," *Amer. J. Bot.* 39 (1952), Fig. 5.

Fig. 21-26: From A. D. J. Meeuse, "The So-Called Megasprorophyll of *Cycas*, etc.," *Acta Bot. Neerland.* 12:119–128 (1963), Fig. 1.

Fig. 22-11C: From E. M. Gifford and J. Liu, "Light Microscope and Ultrastructural Studies of the Male Gametophyte in *Ginkgo biloba*: the Spermatogenous Cell, *Amer. J. Bot.* 62:974–981 (1975), Fig. 10.

Fig. 23-22: From J. T. Buchholz, "Suspensor and Early Embryo of *Pinus*," *Bot. Gaz.* 66, University of Chicago Press, Chicago (1918), Plate VIII, Fig. 40.

Fig. 24-15B: From W. J. G. Land, "Fertilization and Embryogeny in *Ephedra trifurca*," *Bot. Gaz.* 44, University of Chicago Press, Chicago (1907), Fig. 17.

Fig. 24-17: From W. J. G. Land, "Fertilization and Embryogeny in *Ephedra trifurca*," *Bot. Gaz.* 44, University of Chicago Press, Chicago (1907), Figs. 4, 16, 21.

Fig. 24-20A: From P. Maheshwari and V. Vasil, *Gnetum*, Botanical Monograph No. 4, CSIR, new Delhi (1961).

Fig. 25-1A: From C. B. Beck, "Reconstructions of *Archaeopteris* and Further Consideration of Its Phylogenetic Position," *Amer. J. Bot.* 49:373382.

Fig. 25-3: From S. Leclercq and P. M. Bonamo, "A Study of the Fructification of *Milleria (Protopteridium) thomsonii* Lang from the Middle Devonian of Belgium," *Palaeontographica* 122B:83–114 (1971), Fig. 1a.

Fig. 25-4: From H. P. Banks, *Evolution and Plants of the Past*, Fig. 5-4a. Wadsworth Publishing Co., Belmont, Calif. (1970).

Fig. 25-5: From J. M. Pettitt and C. B. Beck, "*Archaeosperma arnoldii*—A Cupulate Seed from the Upper Devonian of North America," *Contrib. Mus. Paleontol. Univ. Mich.* 10:139–154 (1968), text Figs. 1, 2.

Fig. 25-9: From H. N. Andrews, Jr., *Ancient Plants and the World They Lived In*, Comstock Publishing Co., Ithaca, New York (1947), Fig. 92.

Fig. 25-10A: From H. N. Andrews, "Early Seed Plants," *Science* 142:925 (1963). Copyright © 1963 by the American Association for the Advancement of Science.

Fig. 25-10B: From A. G. Long, "On the Structure of *Samaropsis scotica* Calder (emended) and *Eurystoma angulare* gen, et sp. nov., Petrified Seeds from the Calciferous Sandstone Series of Berwickshire," *Trans. Roy. Soc. Edinburgh* 64:261–280 (1960), Fig. 1B, Fig. 6.

Fig. 25-13B: From C. A. Arnold, *An Introduction to Paleobotany*, Fig. 101A. Copyright © 1947 by McGraw-Hill Book Company, Inc., New York. Reprinted by permission of the publishers.

Fig. 25-14: From H. N. Andrews, Jr., *Studies in Paleobotany*, Fig. 5-4. Copyright © 1961 by John Wiley & Sons, Inc., New York. Reprinted by permission of the publishers.

Fig. 25-20, 25-21, 25-22B: From W. N. Stewart and T. Delevoryas, "The Medullosan Pteridosperms," *Bot. Rev.* 22:45–80 (1956), Figs. 9, 8.

Fig. 25-23: From W. N. Stewart, "The Structure and Affinities of *Pachytesta illinoense* comb nov.," *Amer. J. Bot.* 41:500–508 (1954), Fig. 1.

Figs. 25-24, 25-25A: From H. N. Andrews, Jr., *Studies in Paleobotany*, Figs. 5-11B, 6-3C. Copyright © 1961 by John Wiley & Sons, Inc., New York. Reprinted by permission of the publishers.

Fig. 25-28: From T. Delevoryas and R. C. Hope, "A New Triassic Cycad and Its Phyletic Implications," *Postilla*, Peabody Mus. Nat. Hist., Yale Univ., New Haven, Conn. (1971)

Fig. 25-31B: From T. Delevoryas, "Investigations of

North American Cycadeoids: Cones of *Cycadeoidea*," *Amer. J. Bot.* 50:45–52 (1963), Fig. 13A.

Figs. 25-32, 25-33: From A. C. Steward, *Fossil Plants,* No. IV, Cambridge University Press, Cambridge (1919), Figs. 639, 650.

Fig. 25-34C: From H. N. Andrews, *Studies in Paleobotany,* Fig. 11-1. Copyright © 1961 by John Wiley & Sons, Inc., New York. Reprinted by permission of the publishers.

Fig. 25-35A: From T. Delevoryas, "A New Male Cordaitean Fructification from the Kansas Carboniferous," *Amer. J. Bot.* 40:144–150 (1953), Fig. 5.

Fig. 25-35B, C; 25-37: From R. Florin, "Die Koniferen des Oberkarbons und des Unteren Perms," *Palaeontographica* 85:365–654 (1944), Figs. 58B, 45C (in part), Fig. 28, and text Fig. 33a.

Fig. 26-1: From J. E. Canright, "The Comparative Morphology and Relationships of the Magnoliaceae: IV. Wood and Nodal Anatomy," *J. Arnold Arboretum* 36:119–140 (1955), Fig. 1.

Fig. 26-12B, C: From J. E. Canright, "The Comparative Morphology and Relationships of the Magnoliaceae: I. Trends of Specialization in the Stamen," *Amer. J. Bot.* 39:484–497 (1951), Fig. 13 (in part).

Fig. 26-16: From L. Benson and R. A. Darrow, "The Trees and Shrubs of Southwestern Desserts," University of Arizona Press, Tucson (1954), Fig. 74.

Fig. 26-27C: From W. A. Jensen, "The Embryo Sac and Fertilization in Angiosperms," Harold L. Lyon Arboretum Lecture No. 3, May 10, 1972, Fig. 1. Used by permission of the Harold L. Lyon Arboretum, University of Hawaii at Manoa.

Fig. 26-33: Reproduced by permission from J. E. Sass, *Botanical Microtechnique* 3rd ed. Copyright © 1958 by the Iowa State University Press, Ames.

Figs. 26-35, 26-38, 26-40, 26-41: From R. M. Holman and W. W. Robbins, *Textbook of General Botany* 4th ed. Wiley, New York, 1934, Figs. 216, 224, 242, 243, 283. Used by permission of Barbara Robbins.

Fig. 27-6A: From A. G. Long, "On the Structure of *Samaropsis scotia* Calder (emended) and *Eurystoma angulare* gen. et sp. nov., Petrified Seeds from the Calciferous Sandstone Series of Berwickshire, *Trans. Roy. Soc. Edinburgh* 64:261–280 (1960), Figs. 1B, 3.

Fig. 27-6B: From A. G. Long, "*Stamnostoma Huttonense*" gen. et sp. nov.—a Pteridosperm Seed and Cupule from the Calciferous Sandstone Series of Berwick-shire," *Trans. Roy. Soc. Edinburgh* 64:201–215 (1960), Fig. 4.

Fig. 27-6C: From J. Walton, *An Introduction to the Study of Fossil Plants,* Black, London (1940), Fig. 110B.

Fig. 27-7A, B: From I. W. Bailey and B. G. L. Swamy, "The Conduplicate Carpel of Dicotyledons and Its Initial Trends of Specialization," *Amer. J. Bot.* 38:373–379 (1950), Figs. 2, 3.

Fig. 27-7C: From K. Periasany and B. G. L. Swamy, "The Conduplicate Carpel of *Cananga odorata*," *J. Arnold Arboretum* 37:366–372 (1956).

Fig. 27-8: From S. Tucker, "Ontogeny of the Inflorescence and the Flower in Drimys winteri var. chilensis," *Univ. Calif. (Berkeley) Publ. Bot.* 30(4) (1959), Plate 32.

Fig. 27-9: S. Tepfer, "Floral Anatomy and Ontogeny in Aquilegia formosa var. truncata and Ranunculus repens," *Univ. Calif. (Berkeley) Publ. Bot.* 25(7) (1953), Plate 59B; 60A.

Fig. 29-1: From A. C. Lonert, "Dictyostelium," *Turtox News* 43(2):50–53 (1965). Reprinted by permission of General Biological Supply House, Inc., and A. C. Lonert.

Fig. 30-3A: From J. W. Schopf, "Microflora of the Bitter Springs Formation, Late Precambrian, Central Australia," *J. Paleontol.* 42:651–687 (1968), Plate 82, Fig. 2; Plate 83, Fig. 1-3.

Fig. 30-4: From H. N. Andrews and H. Lentz, "A Myccorhizome from the Carboniferous of Illinois," *Bull. Torrey Bot. Club* 70:120–125 (1943), Fig. 4.

Fig. 30-5: From R. Kidston and W. H. Lang, "On Old Red Sandstone Plants Showing Structure from the Rhynie Chert Bed, Aberdeenshire," *Trans. Roy. Soc. Edinburgh* 51, 52 (1917–1921), Plate I, Fig. 4; Plate III, Fig. 34.

Figs. 31-7: From A. W. Barksdale, "Sexual Hormones of *Achlya* and Other Fungi," *Science* 166:831–837 (1969), Fig. 2, 3. Copyright © 1969 by the American Association for the Advancement of Science.

Fig. 31-8: From M. M. Holland, *Saprolegnia* Life Cycle," *Carolina Tips* 34:1–3 (1971), Fig. 3-9.

Fig. 31-13: From A. W. Barksdale, "Sexual Hormones of *Achlya* and Other fungi," *Science* 166:831–837 (1969), Fig. 2, 3. Copyright © 1969 by the American Association for the Advancement of Science.

Fig. 33-3: From H. C. I. Gwynne-Vaughan and H. S. Williamson, "Contributions to the Study of *Pyronema confluens*," *Ann. Bot.* 45:355–371.

Fig. 33-11: From E. F. Robinow and J. Marak, "A Fiber Apparatus in the Nucleus of the Yeast Cell," *J. Cell Biol.* 29:129–152 (1966), Figs. 1, 2.

Fig. 33-16A, D: From F. A. Wolf and F. T. Wolf, *The Fungi,* Vol. 1, Fig. 54F, G. (1947).

Fig. 33-17: From K. B. Raper and C. Thom, *A Manual of the Penicillia,* Williams & Wilkins, Baltimore (1949), Fig. 7.

Fig. 33-19: From E. A. Bessey, *Morphology and Taxonomy of Fungi,* Fig. 101. Copyright © 1950 by McGraw-Hill Book Company, Inc., New York. Reprinted by permission of the publishers.

Fig. 33-22: From B. O. Dodge, "Production of Fertile Hybrids in the Ascomycete Neurospora," *J. Agri. Res.* 36:1–14 (1928).

Fig. 33-26A: From C. J. Alexopoulos, *Introductory Mycology,* Copyright © 1952 by John Wiley & Sons, Inc., New York. Reprinted by permission of the publishers.

Fig. 34-32: From J. H. Craigie and G. C. Green, "Nuclear Behavior Leading to Conjugate Association of Haploid Infections of *Puccinia graminis*," *Canad. J.*

Bot. 40:163–178 (1962). Reproduced by permission of the National Research Council of Canada.

Fig. 35-5: From D. Pramer, "Nematode-Trapping Fungi," *Science* 144:382–388 (1964). Copyright © 1964 by the American Association for the Advancement of Science.

Fig. 36-3: From *Evolutionary Biology*, Vol. 4, 1970, Fig. opposite p. 174. Used by permission of Plenum Publishing Corporation.

Figs. 36-4, 36-5D, 36-6, 36-7: From V. Ahmadjian, "The Fungi of Lichens," *Sci. Amer.* 208:122–131 (1963), Fig. on p. 124. Copyright © 1963 by Scientific American, Inc. All rights reserved.

INDEX

Powdery mildews, 749–751
Predacious fungi, 798–800
Premnoxylon, 624, 627
Primary archesporial cell, **409**
Primary capitulum, 103
Primary embryonic leaf, 424
Primary endosperm nucleus, 653, 654
Primary germination, 357
Primary leaf, 417
Primary mature tissues, 317
Primary meristem, 317, 459
Primary root, 359, **395**, 474, 543, 557, 629, 631
Primary suspensor, 593
Primary wall, 112
Primary xylem, 325, 508, 511, 595
Primary zoospore, **715**, 716
Primicorallina, **177**
Primitive, 63
Primitive flowers, 673–674
Primordium, 329
Procambium, 317, 321, 323, **328**, 376, 404, 459, **460**, 562
Procentriole, 538
Prochloron, 45
Prochlorophyta, 8
Proembryo, **541**, **555**, 572, 593, 607
Progametangia, 726
Progymnospermophyta, 314, 601–603
Prokaryonta, 8–36, 37
Prokaryotes, defined, 6
Promeristem, **363**, **460**
Promycelium, 769, 783, **788**
Prophylls, **347**, 459
Prosporangium, 707
Protandrous, 192, 246, 413, 419, 424
Proteus, **15**
Prothallia, 413
Prothallial cell, **358**, 367, 446, 454, 534, **536**, 552, **554**, 558, 567, 578, 580, 590
Protista, 6
Protoderm, 317, **328**, 459, **460**
Protogynous, 192, 381, 419
Protolepidodendron, 335, 492, **494**, 496, 506
Protonema(ta), 233, 265, **266**, 268, **269**, 270, **271**, **273**, 274, **303**, 306

Protoperithecium, 751
Protophloem, 323
Protoplasmic connections, **61**
Protoplasmodium, 689
Protoplast, 21, 107, 112
Protopteridium, 601
Protosiphon, 65–**67**, **66**, 135
Protosiphon life cycle, **67**
Protosphagnum, 306
Protostele, 320, **321**, **338**, 339, 364, 376, 388, 422, 430, 433, 442, 451, 460, 476, 492, 494, 503, 550
Protostelic, 404
Protosteliomycetes, 680, 685–687, 697
Protostelium, 687, 697
Prototaxites, 178, **179**
Protoxylem, **324**, 376, 404, 492
Protozoa, 51
Prunus, **328**
Psaronius, **481**, 520, 522
Pseudoaethalium(a), 691, **693**
Pseudobornia, 507, **508**, 514
Pseudocapillitium, 694
Pseudoelater, 249, **250**, 253
Pseudoparaphyses, 761
Pseudoparenchyma, 82
Pseudoperianth, **207**, 210, **224**, **225**
Pseudoplasmodium, 682, 683
Pseudopodium, 265, 266, **268**
Pseudosporochnus, 515, **516**, 522
Pseudothecium, 761, **762**
Pseudotrebouxia, 801, 804
Pseudotsuga, 576, 577
Psilocybe, 777, 791
Psilocybin, 777
Psilophyta, 327, 332
Psilophyton, 457, **491**
Psilotophyta, 457–468, 472, 474, 476, 478, 628
Psilotopsida, 823
Psilotum, 362, 427, 437, 457–468, **458**, **459**, **460**, **461**, **462**, **463**, **464**, **465**, **473**, 474, 475, **477**, 485, 486, 584, 642
  embryogeny, 464–465
  reproduction, 461–467
  vegetative morphology, 459–461
Psylocin, 777
Pteretis, 406

Pteridium, 401, 402, 404, 413, 414, 416, 419, 435, 436, 472, 586
  P. aquilinum, 401
Pteridologist, 3
Pteridology, 3
Pteridophyta, 39, 40, 314, 321, 472, 475, 476, 478, 482, 514–523, 529, 662, 823
Pteridospermophyta, 314, 601–617, 605, 608, 671
Pteridospermopsida, 526
Pteridosperms, 609–617
Pteris, 402, 406, **408**, 412
Pteropsida, 388, 524
Ptilidium, **234**, 235, 256
Puccinia, 782, **783**, **784**, **785**, **786**, **787**, 791
  life cycle, **783**
Puffballs, **777**, 778–779
Pycnidiospore, 735
Pycnidium(a), 735, **736**, **795**, 806
Pycnospore, 806
Pylaisiella, 274, 305
Pyrenoid, **48**, **50**, 51, 57, 67, 70, 72, 82, **89**, 91, 94, **109**, 155, **156**, 161, **163**
Pyrenoid bodies, 245
Pyrenomycetes, 751–754
Pyronema, **737**, **738**, 756, **757**, 758, 759, 763, 764
Pyrrhophyta, 132, 150–153, 180
Pythium, 714, 719, 722

Quaking bogs, 260
Quercus, 674
Quercus flowers, **635**
Quillworts, 363

Raceme, 638, **640**
Rachis, 405, 629
Radicle, 359, **362**, **395**, 404, 474, 542, **575**, 629, 631, **657**
Radula, 229, **239**, 256
Ranales, 674
Ranunculus, **329**, 668, **669**, **670**, 671
Raphe, 143
Ray flower, **641**
Reboulia, **210**, 211, 212, 255
Receptacle, **125**, 126, **130**, 406, 442, 443, 631, **636**, **639**, 668
Receptive hyphae, 784
Recurvation, **473**